INTRODUCTION TO

INSTRUMENTAL ANALYSIS

Second Edition

INTRODUCTION TO

INSTRUMENTAL ANALYSIS

Second Edition

Robert D. Braun

Lafayette, Louisiana, USA

PharmaMed Press

An imprint of Pharma Book Syndicate

A unit of BSP Books Pvt. Ltd.

4-4-316, Giriraj Lane,

Sultan Bazar, Hyderabad - 500 095.

Introduction to Instrumental Analysis
Second edition

Published by

PharmaMed Press
An imprint of Pharma Book Syndicate

A unit of BSP Books Pvt. Ltd.
4-4-309/316, Giriraj Lane, Sultan Bazar, Hyderabad - 500 095.
Phone: 040-23445605, 23445688; Fax: 91+40-23445611
E-mail: info@pharmamedpress.com

ISBN : 978-93-85433-20-7 (HB)

Content

Preface to Second Edition

Since "Introduction to Instrumental Analysis" was originally published, a multitude of changes have occurred in laboratory analytical instrumentation. The rapid development of computers and computer interfacing has revolutionized instrumental control and data processing. Nearly all instruments have versions that either include microprocessors as part of the instrumentation package, or have circuits that can be easily interfaced to laboratory computers. Since many laboratory instruments consequently have fewer available manual adjustments, laboratory instruments often are treated as "black boxes" by users owing to the difficulty in understanding the manner in which they function. Nevertheless, the non-computer portion of most analytical instruments, such as spectrophotometers, chromatographs, spectrometers and electroanalytical instruments, remain basically unchanged. Individual components of some instruments have been altered, but the basic design of most instruments remains.

The author remains convinced that, in spite of instrument and electronic changes, it is important to the accuracy and efficiency of the analytical process for the analyst to have a basic understanding of the equipment. Consequently, much of the original instrumental design description from the first edition of the text has been kept in the second edition. All of the typographical and other errors in the first edition that could be identified have been removed, and many of the explanations have been modified so as to make the descriptions more easily understandable. The text, of course, also has been updated to reflect current knowledge and practice. It is the author's hope that the second edition will provide a solid, modern underpinning upon which chemical analysts and analytical chemists can base their careers.

Robert D. Braun

Preface to First Edition

Writing a text for an instrumental analysis course is a formidable task. The number of types of instrumental methods of chemical analysis is staggering. Because space and time are limited, it is impossible to describe all of the instrumental methods in a single text. Decisions that sometimes appear to be arbitrary must be made as to the topics that are excluded from the text. The problem is further complicated because, in some branches of instrumental analysis, progress is being made so rapidly that nearly anything that is written is outdated before it can be put into print.

In the present text the topics that were chosen for inclusion in the text are either well-established methods of chemical analysis with which nearly all analytical chemists are familiar or relatively new methods that show such promise that, in the author's opinion, they soon will become well established in many analytical laboratories. Examples of the former category include atomic absorption spectrophotometry (Chapter 6), ultraviolet-visible spectroscopy (Chapter 9), potentiometry (Chapter 22), and chromatography (Chapter 24 through 26). Instrumental methods that fall into the latter category include laser-enhanced ionization (Chapter 8), photoacoustic spectrometry (Chapter 12), and the use of laboratory robots (Chapter 28). It was the original intention to include chapters on distillation and extraction. Those techniques were not included because they are not truly instrumental and because they often are adequately described in organic chemistry courses. Because nearly all analytical instruments are electrically operated and because a knowledge of the basics of electricity and electronics permits the analyst to utilize those instruments most efficiently, Chapters 2 through 4 of the text are devoted to a description of electricity, electronics, and logic devices.

It did not take long to realize the difficulties inherent in describing specific instrumental designs. Sometimes it seems as through the instrument manufacturers change designs as rapidly as the automobile industry changes car models. Because the designs change rapidly and because a considerable period was required to write the text (about 4½ years) and to put it into print (another year), most instruments are described in general terms. Occasionally a design of a specific instrument is shown in a figure when, in the author's opinion, the instrument is likely to remain in production throughout the useful life of the text. The author appreciates the cooperation of the instrument manufacturers that have contributed diagrams and photographs of their instruments.

Perhaps the most important requirements when writing a text are to make it accurate and understandable. The comments of the reviewers have made both of those requirements more attainable. The author appreciates the helpful comments of all of the reviewers and friends who have read all or part of the manuscript. Revised versions of the manuscript were significantly improved owing to those comments. Hopefully students as well as instructors will find the text to be useful. Among the author's colleagues who particularly should be acknowledged for their assistance are Dr. James L. Reed and Atlanta University, Dr. K. Rick Lung of Duke University, Dr. Duane Bartak of the University of North Dakota, Dr. Eugene F. Barry of the University of Lowell, and Drs. August A. Gallo. Frederick H. Walters, and Dean F. Keeley of University of South-western Louisiana. The helpful comments of the students on whom portions of the manuscript were class-tested also proved useful and are appreciated by the author.

Robert D. Braun

CHAPTER

ONE

INTRODUCTION TO CHEMICAL INSTRUMENTAL ANALYSIS

The purposes of Chapter 1 are to review the definitions of some of the terms that are used in the study of analytical chemistry and to present a brief overview of the instrumental methods of chemical analysis. It is hoped that the introduction can provide insight into the organization of both the remainder of the text and the field of analytical instrumentation. No attempt is made in Chapter 1 to provide a detailed description of the analytical methods.

Some analysts distinguish between a chemical analysis and an assay. Those analysts define a chemical analysis as the entire process that leads to determining the identity or amount of a substance in a sample. The chemical analysis consists of collecting a sample, possibly treating the sample either physically or chemically, performing a laboratory or nonlaboratory measurement on the sample, mathematically manipulating the data as required to obtain a meaningful result, and reporting the result. The chemical assay consists only of the laboratory or nonlaboratory measurement. Other analysts use analysis and assay interchangeably. Most analysts define assay as the laboratory or nonlaboratory measurement, and analysis as either the entire process described previously or as the measurement. The latter definitions are used in the text.

Chemical analysis is concerned with determining either the identity of the chemical substances or the amount of a particular substance in a sample. The former type of analysis is a qualitative chemical analysis. The latter type is a quantitative chemical analysis.

Sometimes chemical analysis is divided into classical and instrumental analysis. Although the division probably is not as important as it once was, many analysts continue to distinguish between the two categories. Classical or non-instrumental analysis is the group of analytical methods that only requires the use of chemicals, a balance, calibrated glassware, and other commonplace laboratory apparatus, such as funnels, burners or hot plates, flasks, and beakers. Instrumental analysis requires the use of an analytical instrument in addition to the apparatus that is used for classical analyses. Classical and instrumental methods can be used for qualitative and quantitative analysis. An analytical instrument is a physical, often electrically operated, device that is used to determine the identity or amount of one or more components in the analyzed substance (the analyte).

Regardless of whether classical or instrumental analysis is used, many quantitative analyses can be classified as being gravimetric or volumetric. A gravimetric analysis relies upon a critical mass measurement of the product of a chemical reaction, or a measurement of a mass change during a chemical reaction to determine the amount of a chemical reactant in the sample. The mass measurement is made with an accurate balance. A classical gravimetric analysis usually consists of a precipitation of a salt of the assayed substance. The precipitate is collected by filtration, dried, and weighed. Instrumental gravimetric analysis normally consists of heating the sample on a balance pan in an oven while observing the mass change. The temperature of the sample is increased during the heating and the readout from the device is a plot of mass as a function of sample temperature. That technique is thermogravimetric analysis (Chapter 27).

A volumetric analysis relies upon a critical measurement of the volume of a chemical reactant to determine the concentration of the sample. Volumetric analyses are titrations in which a solution of one of the chemical reactants in a buret is added to a solution of a second chemical reactant. The solution in the buret is the titrant, and the solution in the reaction vessel is the titrand. The sample can be either the titrant or the titrand. The volume of titrant added at the endpoint of the titration is measured and used to calculate the concentration of the sample. A classical volumetric analysis uses a chemical indicator to locate the endpoint of those titrations in which no natural color change is observed. An instrumental volumetric analysis uses a laboratory instrument to determine the endpoint.

INSTRUMENTAL ANALYSIS

Essentially all analytical instruments are electrically operated. An understanding of the operation of the electrical components of an instrument can aid in locating a malfunctioning portion of the instrument and can make it possible for the analyst to obtain maximal use and information from the instrument. In addition, some research analytical chemists design and develop new instruments that can be used for chemical analysis. Consequently, Chapters 2 through 4 are an introductory description of electrical circuits.

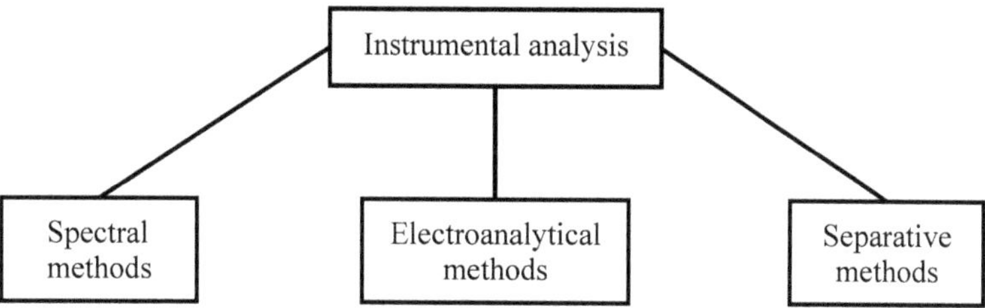

Fig. 1-1 The three major categories of instrumental methods of chemical analysis.

Analytical instruments are devices that measure a physical or chemical property of the assayed substance or that measure some factor that enables determination of a property of the substance. Traditionally, instrumental analyses are divided into three categories (Fig. 1-1) according to the type of property of the assayed substance that is measured or used during the assay. The *spectral methods* use or measure some form of radiation during the assay. The *electroanalytical methods* apply an electrical signal to the sample and/or monitor an electrical property of the sample. The *separative methods* rely upon separation of the components of a sample prior to measuring a property of the components. In the following sections the more important instrumental techniques are mentioned. Because numerous techniques exist, no attempt is made to be comprehensive.

SPECTRAL METHODS

The spectral methods of analysis use an instrument to measure the amount of radiation that is absorbed, emitted, or scattered by the sample. If the amount of absorbed radiation is measured, the technique is absorptiometry or absorption spectrophotometry. Except for naturally occurring radioactive materials, radiation can be emitted from a sample only after the sample has absorbed energy from an outside source. If the absorbed energy is electromagnetic radiation in the x-ray, ultraviolet, or visible region of the spectrum, the subsequently emitted electromagnetic radiation is a form of luminescence termed either fluorescence or phosphorescence depending upon the manner in which the deexcitation takes place. A description of the difference between fluorescence and phosphorescence is included in Chapter 5. Absorption of ultraviolet and visible radiation by atoms and polyatomic species is described in Chapters 6 and 9 respectively. Fluorescence from atoms is described in Chapter 8, and fluorescence and phosphorescence from molecules are described in Chapter 11. A description of the absorption and fluorescence of x-rays is given in Chapter 18. Generally a laboratory instrument that is designed to provide the energy required to excite the sample (if necessary) and to monitor the emitted radiation or particles is a spectrometer.

It is not necessary for the energy that is absorbed by a sample prior to emission to be in the form of electromagnetic radiation. During assays in which chemiluminescence and electrochemiluminescence (Chapter 10) are measured, energy emitted from a chemical reaction or electrical energy is absorbed. Sometimes thermal energy from a flame or electrical energy from an electrical discharge (Chapter 7) can be used to initiate emission. Similarly, the energy required to cause emission of radiation or particles from the sample can come from collisions between the sample and electrons (Chapter 19), or ions (Chapter 18).

During assays using the radiochemical methods (Chapter 20), the radioactive products that have decayed from a sample are measured. If the decay follows energy absorption from neutrons that have bombarded the sample, the technique is neutron activation analysis. A radiometric analysis uses a radioactive reagent to chemically react with the assayed substance. The radioactivity of the product of the reaction is either directly measured and related to concentration or the endpoint of a titration with the radioactive reagent is determined by measuring the radioactivity of the titrand during the titration.

Radiation that is scattered from sample particles can be used for analyses. Nephelometry, turbidimetry, and Raman scattering (Chapter 14) are examples of analytical techniques that rely upon scattered radiation. The wavelength of the scattered radiation is identical to that of the incident radiation in nephelometry and turbidimetry. The wavelengths are not identical during Raman scattering.

The ratio of the speed of electromagnetic radiation in a vacuum to the speed of radiation of the same wavelength in a sample is the refractive index of the sample. The refractive index is usually determined by measuring the extent to which the direction of travel of the radiation is altered as it enters the sample. Because the refractive index is a characteristic of a substance, refractometry (Chapter 15) can be used for analysis.

During analyses that use photoacoustic spectroscopy (Chapter 13) chopped incident radiation in the infrared, visible, or ultraviolet region is absorbed by a sample in an enclosed space. A portion of the absorbed radiation is converted to heat that warms the gas adjacent to the sample. The resulting pressure waves in the gas are monitored with a microphone or other device. The incident radiation is chopped at a frequency that is characteristic of the measured sound waves.

The remaining spectral methods of analysis are divided according to the energy of the radiation that is used for the assay. Assays can be performed using radiation in the ultraviolet-visible (Chapters 6 to 11 and 13 to 15), the infrared (Chapter 12), the radiofrequency (Chapter 16), the microwave (Chapter 17), and the thermal (Chapter 27) regions. Absorbance measurements can be made in each of the regions. Fluorescent measurements are usually restricted to excitation in the x-ray and ultraviolet-visible regions. Phosphorescent measurements normally are used to assay polyatomic species after excitation in the ultraviolet-visible region. Nephelometry and turbidimetry generally involve measurements in the visible region. Raman scattering occurs in the ultraviolet-visible region.

Electron spin resonance spectroscopy uses electromagnetic radiation that is in the microwave region of the spectrum. Nuclear magnetic resonance spectroscopy uses electromagnetic radiation in the radiofrequency region. Any of several methods using thermal radiation can be utilized. The energy, wavelength, and frequency ranges for each type of electromagnetic radiation are listed in Chapter 5. The spectral methods of analysis and the chapter in which each method is described are listed in Table 1-1.

Table 1-1 A list of the major spectral methods of chemical analysis and the chapter in which each method is described

Method	Chapter
Atomic absorption n	6
Atomic fluorescence and ionization	7
Flame and atomic emission	8
Ultraviolet-visible absorption by polyatomic species	9
Chemiluminescence and electrochemiluminescence	10
Fluorescence and phosphorescence	11
Infrared absorption	12
Photoacoustic spectrometry	13
Scattering	14
Refractometry	15
Nuclear magnetic resonance	16
Electron spin resonance	17
X-ray methods	18
Electron spectroscopy	19
Radiochemical methods	20
Thermal analysis	27

ELECTROANALYTICAL METHODS

Those instrumental methods of chemical analysis in which either an electrical signal is applied to one of the electrodes dipping into the sample solution or an electrical property of the solution is measured arc the electroanalytical methods. Most electroanalytical techniques apply an electrical signal while a different electrical parameter of the solution is monitored. The electroanalytical methods are divided into categories, as shown in Fig. 1-2, according to the type of electrical measurement that is made and type of applied electrical signal.

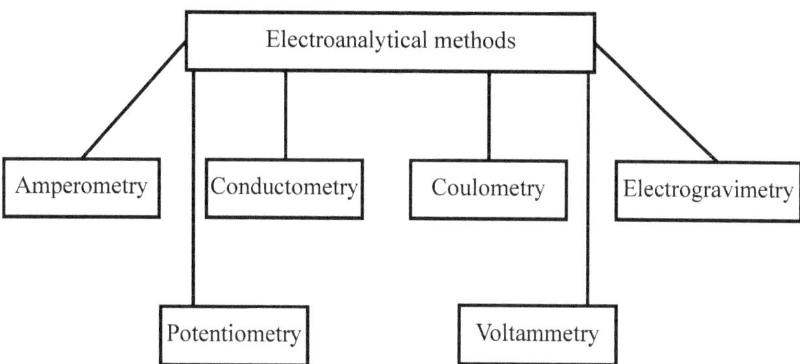

Figure 1-2 The major categories of electroanalytical methods.

During analysis with most of the electroanalytical methods, electrical contact with the sample is completed by dipping at least two electrodes into the solution. Amperometry is the method in which the potential between the two electrodes is controlled and the current is measured. When the current between the electrodes is controlled (usually at nearly zero) and the potential is measured, the technique is potentiometry.

During electroanalytical studies using coulometry and electrogravimetry, either a potential or a current can be applied to the electrodes in the solution. When the quantity Q of electricity that is consumed during an electrochemical reaction is measured, the technique is coulometry. When the mass of a reaction product (often a metal) after an exhaustive electrolysis is measured, the method is electrogravimetry. Because a separation occurs when the assayed metal plates onto one of the electrodes, electrogravimetry also can be classified as a separative method of chemical analysis.

Conductometry is the electroanalytical method in which the electrical conductance G of the sample solution is measured. Conductance is defined as the inverse of the electrical resistance. Normally an alternating electrical potential is applied to the electrodes during the measurement.

Voltammetry is the technique in which a potential is applied to one of the electrodes by a potentiostat while the current flowing through the electrode is measured. During the current measurement the potential is varied in some predetermined manner. The many types of voltammetry differ in the way in which the applied potential varies. All of the electroanalytical methods with the exception of potentiometry are described in Chapter 23. Potentiometry is described in Chapter 22.

SEPARATIVE METHODS

The separative methods take advantage of physical or chemical properties of the components of a mixture to separate the components. After the separation, the components can be individually assayed either qualitatively or quantitatively.

Sometimes the separating instrument simultaneously performs the separation and the assay. In other cases the separation is done prior to an assay by another method.

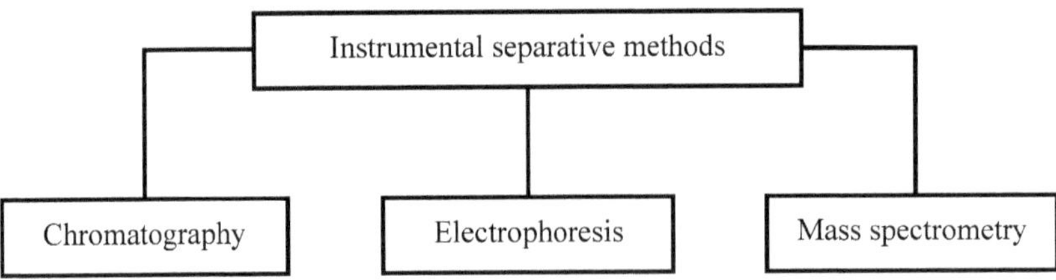

Figure 1-3 The major categories of separative methods of chemical analysis.

Among the noninstrumental separative methods are distillation, extraction, precipitation, filtration, osmosis, and reverse osmosis. Because descriptions of those methods normally are included in other courses, they are not described in the text. The instrumental separative techniques are divided into categories, as shown in Fig. 1-3, according to the method used to effect the separation.

Chromatography is the method by which a mixture is separated into its components as a result of the relative ability of each component to be flushed along or through a stationary phase by a mobile phase. The sample is placed on the edge of the stationary phase (a solid or liquid) and a mobile phase (a liquid or gas) is allowed to flow over the stationary phase and to sweep the sample along the length of the stationary phase. Components which more strongly adhere to the stationary phase are swept less rapidly along the length of the stationary phase than are those components that less strongly adhere to the stationary phase. The result is a separation of sample components in space at a fixed time after the start of the separation, or a separation in time at a fixed distance along the length of the stationary phase. Chromatography is divided into liquid chromatography and gas chromatography, depending upon the state or nature of the mobile phase. Further subdivisions of liquid and gas chromatography are commonly used and are described in Chapters 24 through 26.

Electrophoresis is the separative method that takes advantage of the relative mobility of ions toward an electrode of opposite charge (to the ion) and away from an electrode of similar charge. The buffered solution through which the ions normally travel is either supported by porous paper or is in a gel. Because highly mobile ions move farther toward the electrode of opposite charge than less mobile ions, a separation occurs. Electrophoresis is described in Chapter 25. Because electrophoresis requires application of an electrical signal, it is sometimes classified as an electroanalytical method.

Separations using mass spectrometry are based upon the relative motion in an electrical or magnetic field of the components of a gaseous mixture of sample ions. The electrical or magnetic field causes the ions to move in different paths or at different velocities. The relative motion of each ion is dependent upon the mass-charge ratio of the ion. Mass spectrometry can be used either alone or in combination with some other analytical technique, such as gas or liquid chromatography. Mass spectrometry is described in Chapter 21.

COMPUTERIZATION AND AUTOMATION

Computers that are interfaced to or a component of analytical instruments are commonly used for data treatment and to control operation of the instrument. The use of computers in the analytical laboratory has significantly reduced the time that must be spent on each analysis by the analyst. An introduction to the use of computers for chemical analysis is included in Chapter 28.

In many instances routine analyses can be performed more accurately and more rapidly by automating the apparatus that are used for the analyses. The results from automated analyses of an industrial process can be used to automatically control the process. Process-control apparatus usually operate in one of two fashions. Either the apparatus automatically samples a product of the industrial process and adjusts one or more of the variables in the process, so as to keep the product within specifications, or it samples one variable of the process while controlling a second variable. Process-control apparatus is especially useful when judgments must be made rapidly or when the sampling or control process is unsafe.

In addition to automating analytical instruments, laboratory robots can be used to perform many of the tasks that have traditionally been performed by human analysts. Regardless of the method of automation, modern automated laboratories are often controlled by computers. Laboratory automation and process control are described in Chapter 28.

CHOOSING AN ANALYTICAL METHOD

A wide array of analytical methods are available to the analyst. In most cases more than one instrumental method can be used for a particular assay. Among the factors that should be considered when choosing an analytical method are:

1. The type of readily available instrument
2. The experience of the analyst in using a particular instrument
3. The expected concentration range within which the sample falls
4. The required precision and accuracy of the assays
5. Potential interferences in the assays
6. The number of samples that must be assayed with the instrument
7. The rate at which the results must be obtained
8. The expense of each assay and the entire analysis

While reading the text it is useful to keep in mind the preceding eight points.

CHAPTER

TWO

SIMPLE DC AND AC ELECTRIC CIRCUITS

Most analytical instruments are electrically operated. The design philosophy used in many analytical instruments is illustrated in Fig. 2-1. Typically the measured property during an assay is either an electric signal, that might require conversion to a different electric parameter, or a nonelectric signal that is converted to an electric signal before measurement. The product of the conversion is measured, electrically manipulated, if necessary, and forwarded to a readout device. Conversion of a measured property to an electric signal is done with a *transducer*. A transducer is a device which converts one form of energy into another. Examples of transducers include phototubes and photomultiplier tubes which convert electromagnetic radiation into electric currents or potentials.

Although it is rarely necessary for an analytical chemist to design an instrument it is important for the chemist to have an understanding of the operation of the instrument. Such an understanding enables the chemist to make optimal use of the analytical instrument and can prevent introduction of a determinate error into the analysis by improper use of the instrument. Furthermore, a basic knowledge of electric circuits sometimes makes it possible to repair a malfunctioning instrument rather than relying on the services of a repair technician. For those reasons, a description of electricity and several analytically important circuits is included in the text. Owing to space restrictions in the text and time restrictions in analytical courses, Chapters 2 through 4 contain relatively brief descriptions of simple, electric circuits. Further information is available in the Bibliography and in numerous texts.

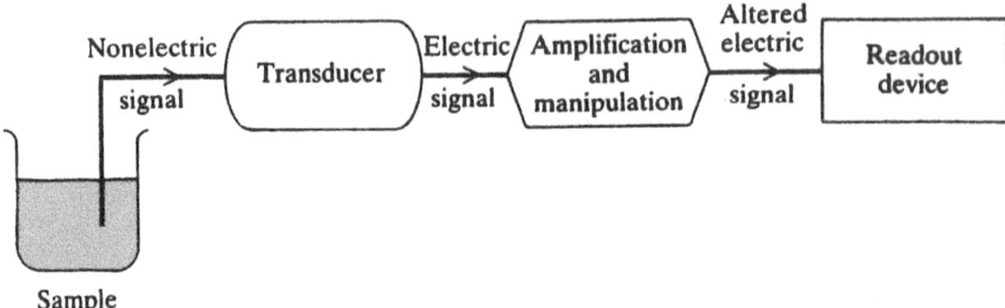

Figure 2-1 A flow diagram showing the design of many analytical instruments.

DC CIRCUITS

Electricity consists of a flow of charged particles. A *direct current* (dc) circuit is one in which the current flow is unidirectional. It always flows in the same direction at a particular location in the circuit, although it does not necessarily flow at the same rate. In metals, such as wires, electricity consists of a flow of electrons. In solution electricity is a flow of ions. The *quantity* Q of charged particles is measured in units of *coulombs* C. A coulomb is defined by chemists as the quantity of electric charge that is required to convert a silver nitrate solution to 0.0011180 g of elemental silver:

$$Ag^+ + e^- = Ag(s)\ (0.0011180\text{ g}) \tag{2-1}$$

The charge that corresponds to Avogadro's number (6.022×10^{23}) of electrons is a *faraday* F. A faraday corresponds to 96, 485 C. The rate at which an electric charge flows through a conducting medium is the electric *current* I:

$$I = \frac{dQ}{dt} \tag{2-2}$$

Electric current is measured in units of *amperes* A. An ampere is a current flow of 1 C/s. Current is regarded as being composed of positive particles even though the electric current in a wire is actually a flow of negatively charged electrons. The practice of regarding electric current as a flow of positive particles is conventional and was established prior to knowledge of the actual composition of the electric current.

Sample problem 2-1 A constant current was allowed to flow between two platinum electrodes in a silver nitrate solution for 3 min 45 s. Before the electrolysis one of the electrodes had a mass of 10.337 g. During the electrolysis, silver plated on one of the electrodes (the cathode) and the electrode mass increased to 10.462 g. Calculate the number of coulombs and the number of faradays that were used during the electrolysis. Determine the constant current that was used.

SOLUTION The mass of the plated silver was 10.462 – 10.337 = 0.125 g. The quantity of electricity that was used can be calculated from the mass of the plated silver:

$$0.125 \text{ g Ag} \times \frac{1 \text{ C}}{0.001\,118 \text{ g Ag}} = 112 \text{ C}$$

The number of faradays is calculated from the number of coulombs:

$$112 \text{ C} \times \frac{1 \text{ F}}{96{,}485 \text{ C}} = 1.16 \times 10^{-3} \text{ F}$$

The current used during the electrolysis is obtained by rearrangement and integration of Eq. (2-2):

$$\int_0^Q dQ = \int_0^t I \; dt$$

Because I is a constant, the result is $Q = It$. The measured values of Q and t can be substituted into the equation and the equation solved for I. From the definition in Eq. (2-2), it is apparent that t must be measured in units of seconds if I is measured in amperes and Q is in coulombs:

$$t = 3 \text{ min } 45 \text{ s} = 225 \text{ s}$$
$$112 \text{ C} = I\,(225 \text{ s})$$
$$I = 0.498 \text{ C/s} = 0.498 \text{ A}$$

An electric current only flows when it is energetically feasible, i.e., when the charged particles possess potential energy. The *electromotive force* (*emf*) or *potential difference* is the difference in potential energy between the ends of the conducting medium through which the charge flows. The potential difference between two points is the work that is necessary to transport a unit of positive charge between the two points. Potential difference is often shortened to *potential* E.

In electroanalytical studies potential is defined as the potential difference between a test electrode and a reference electrode. Hopefully no confusion will develop from using the term in two ways. The unit of potential is the *volt* V. A volt is the work W (joules) required to move a charge (coulombs) between two points:

$$\text{E (V)} = \text{W (J)} / \text{Q (C)}$$
$$1 \text{ volt} = 1 \text{ joule/coulomb} \tag{2-3}$$

The potential difference between two points in a circuit is directly proportional to the current flowing between the points:

$$E = RI \tag{2-4}$$

The proportionality constant in Eq. (2-4) is the *resistance R* of the circuit. Resistance is a measure of the opposition to current flow in the circuit. For a fixed potential difference, the current flow decreases with increasing resistance.

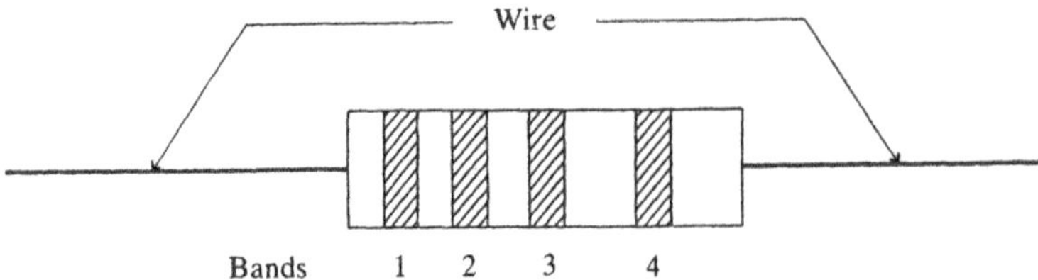

Figure 2-2 A side-view sketch of a resistor. The four bands identify the resistance to electric current flow through the device.

Equation (2-4) is a statement of *Ohm's law*. The unit of resistance is the *ohm* Ω. An ohm is the resistance in a circuit in which a current of 1 A flows between two points that have a potential difference of 1 V:

$$1 \text{ ohm} = 1 \text{ volt/ampere} \tag{2-5}$$

A *resistor* is an electric component that offers resistance to current flow. Resistance can purposely be introduced into an electric circuit with a resistor. A sketch of a resistor is shown in Fig. 2-2. Generally, a resistor is a solid cylindrical device with a wire emerging from each end. The color-coded bands on the resistor identify the resistance through the device. The first three bands specify the amount of resistance and the fourth band specifies the relative accuracy with which the resistance is known. A list of the color coding used on resistors is given in Table 2-1. As an example of the use of the coding, a resistor that has bands that respectively are yellow, violet, red, and silver for bands 1 through 4 has a resistance of 4700 Ω and a relative accuracy of 10 percent, that is, ± 470 Ω.

Table 2-1 A list of the color coding used on resistors

Color	Value: Bands 1 and 2	Value: Band 3	Value: Band 4
Black	0	$\times 1$	
Brown	1	$\times 10$	
Red	2	$\times 10^2$	
Orange	3	$\times 10^3$	
Yellow	4	$\times 10^4$	
Green	5	$\times 10^5$	
Blue	6	$\times 10^6$	
Violet	7		
Gray	8		
White	9		
None			± 20%
Silver		$\times 10^{-2}$	± 10%
Gold		$\times 0.1$	± 5%

Precision resistors, in which the resistance is known with more accuracy than 5 percent, either have the resistance and accuracy stamped on each resistor or are identified with a fifth band. A resistor is an example of a *linear device.* Current flow through a linear device increases linearly with increasing potential across the device.

Conductance G is the inverse of resistance:

$$G = \frac{1}{R} \tag{2-6}$$

With a fixed potential difference, conductance is directly proportional to the current flow as indicated by rearranging Eq. 2-4 and substituting G for 1/R. The unit of conductance is the *siemens* S. A siemens is the conductance that corresponds to a resistance of 1Ω:

$$1 \text{ siemens} = 1 \text{ ohm}^{-1} \tag{2-7}$$

Power P is the rate at which work is done:

$$P = \frac{\text{work}}{\text{time}} \tag{2-8}$$

The unit of power is the *watt* W which corresponds to 1 J/s. In an electric circuit, power is consumed when current flows through a resistance. The consumed power is converted by a resistor to heat which dissipates to the environment of the resistor. The power (watts) that is dissipated in a resistance is given by

$$P = EI \tag{2-9}$$

where E is measured in volts and I is in amperes.

If the potential difference in the circuit is unknown but the resistance is known, Eq. (2-9) can be replaced with

$$P = I^2 R \tag{2-10}$$

Equation (2-10) is obtained from Eq. (2 9) by substitution for E from Ohm's law [Eq. (2-4)]. By similar substitution into Eq. (2-9) for current, Eq. (2-11) is obtained:

$$P = \frac{E^2}{R} \tag{2-11}$$

Equation (2-11) can be used to calculate power when the potential difference and resistance are known.

Sample problem 2-2 A constant current of 0.400 A flowed across a resistor after application of a potential of 50.0 V. Calculate the resistance and the conductance of the resistor. Determine the amount of power that is dissipated in the resistor.

SOLUTION Ohm's law is used to calculate the resistance of the resistor:

$$E = IR$$
$$50.0\ V = (0.400\ A)\ R$$
$$R = 125\ \Omega$$

The conductance can be determined with the aid of Eq. (2-6):

$$G = \frac{1}{R} = \frac{1}{125\ \Omega} = 8.00 \times 10^{-3}\ \mathrm{S}$$

The dissipated power can be calculated with Eqs. (2-9), (2-10), or (2-11):

$$P = EI = (50.0\ \mathrm{V})\ (0.400\ \mathrm{A}) = 20.0\ \mathrm{W}$$

Circuit Diagrams

An *electric circuit* consists of one or more electric components attached to each other by conductors (usually by wires or the foil on a circuit board). An electric circuit can be symbolically represented by a *circuit diagram*. In a circuit diagram each electric component is represented by a symbol. A resistor is represented by -^^^-, a variable resistor, i.e., a resistor whose resistance can be changed, by -^^^- or -^^^-, a single *voltaic cell* by -|⊢, and a combination of voltaic cells, i.e., a *battery*, by -||||⊢. A voltaic cell is a device which causes an electric current to flow through the closed circuit as a result of the potential difference that occurs between two physically separated oxidation-reduction half reactions. It is regarded as a constant-potential source. In the representations for a voltaic cell and a battery, the end which has the smaller line is negative and the end with the larger line is positive. The number of cells in a battery can be represented by the number of pairs of lines in the symbol for the battery, for example, -||⊢ could be used to indicate that two voltaic cells make up the battery. Generally the practice is only followed when it is important to know the number of cells in a battery. Wires which connect electric components are represented by lines. A *switch* is generally represented by a break in the line as shown in Fig. 2-3. The symbols for other electric components will be introduced later.

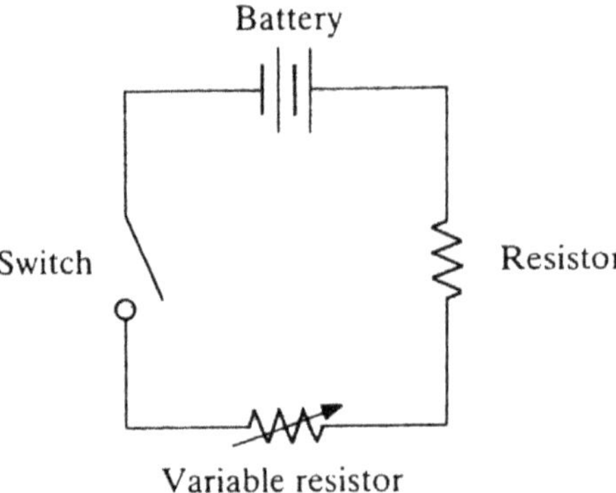

Fig.2-3 A Simple electric circuit.

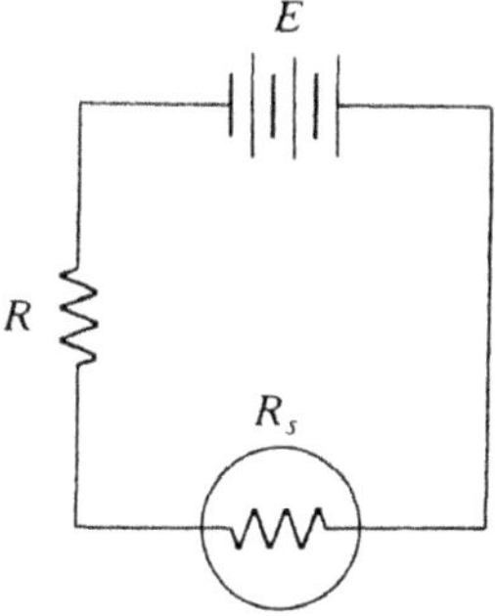

Figure 2-4 A constant-current source. E is a battery; R is a resistor; and R_s is the resistance in the device that is attached to the source. R is much greater than R_s.

An example of a simple dc circuit that can be used as a constant-current source during potentiometric measurements using two indicator electrodes (Chapter 22) is shown in Fig. 2-4. Current regulation can be accomplished in several ways. The use of operational amplifiers for current regulation is described in Chapter 4. A straightforward method of current regulation is to pass current from a constant-potential source, such as a battery or voltage regulator, through a resistor. The current that flows depends upon the size of the resistor (Ohm's law). If a variation in resistance throughout the remainder of the circuit (excluding the resistor) is possible, the current that flows through the circuit can vary. Consequently, the resistor is chosen so that its resistance is large in comparison with any possible resistance in the remainder of the circuit. The constant-potential source is chosen to provide the required constant current which is calculated by using Ohm's law (see Prob. 2-14).

Complex DC Circuits

The application of Ohm's law to a simple circuit that contains a single source of potential and a single resistor is obvious. For more complicated circuits the application is not as obvious. If several resistors are connected together as shown in Fig. 2-5, the current flow through each resistor is identical, because only a single path for the current exists. The resistors in the circuit are said to be connected in *series* (one after the other) and the circuit is a *series circuit*. From Ohm's law it is apparent that the potential across each resistor is a function of

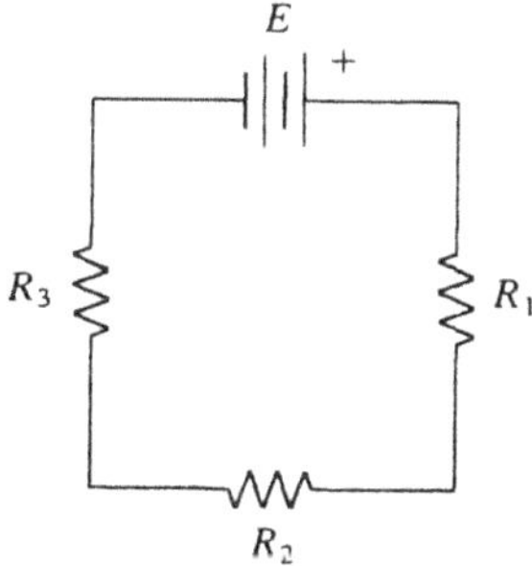

Figure 2-5 A simple series circuit.

both the current I flowing through the resistors and of the resistances of the resistors. For the particular case that is shown in Fig. 2-5, $E_1 = IR_1$, $E_2 = IR_2$, and $E_3 = IR_3$. Because energy is neither created nor destroyed in the circuit, the sum of the potentials across each resistor must equal the potential across the battery:

$$E = E_1 + E_2 + E_3 = IR_1 + IR_2 + IR_3 \tag{2-12}$$

Rearrangement of Eq. (2-12) yields

$$E = I(R_1 + R_2 + R_3) = IR_{eq} \tag{2-13}$$

It is apparent from Eq. (2-13) that the equivalent resistance R_{eq} of series resistors is the sum of the resistances of the resistors:

$$R_{eq} = R_1 + R_2 + R_3 + \ldots + R_n \tag{2-14}$$

A circuit that is equivalent to that shown in Fig. 2-5 could be prepared by substituting a single resistor with a resistance of R_{eq} for the three resistors of resistances R_1, R_2, and R_3.

An important series circuit that is often found in analytical equipment is the *voltage* (or *potential) divider* (Fig. 2-6). A voltage divider is used to decrease the potential from a battery or other power supply to a more useful value. An example of the application of a voltage divider during an assay is found in amperometry with two indicator electrodes (Chapter 23). A small potential is applied between two electrodes using a battery as the potential source. The current flow through the circuit is measured.

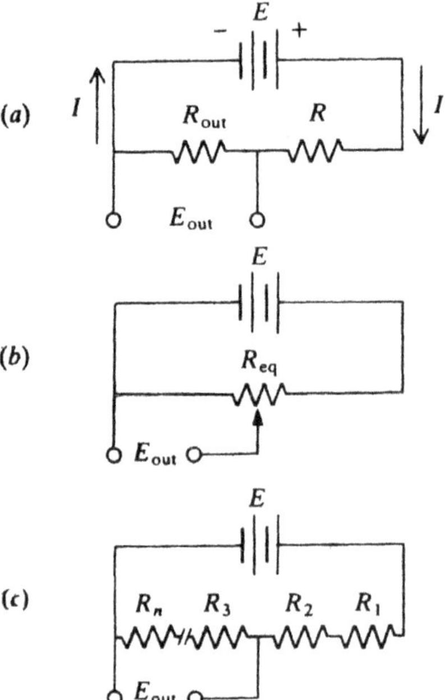

Figure 2-6 Typical voltage dividers. (*a*) A circuit containing two resistors; (*b*) A circuit containing a variable resistor; (*c*) A circuit containing numerous resistors.

For the voltage divider that is shown in Fig. 2-6*a*, Eq. (2-13) can be applied to yield

$$E = I\,(R_{\text{out}} + R) = IR_{\text{eq}}$$

The output potential E_{out} from the voltage divider is given by $E_{\text{out}} = IR_{\text{out}}$ and the ratio of the output potential to the battery potential is given by

$$\frac{E_{\text{out}}}{E} = \frac{IR_{\text{out}}}{IR_{\text{eq}}} = \frac{R_{\text{out}}}{R_{\text{eq}}} \tag{2-15}$$

Rearrangement yields

$$E_{\text{out}} = E\frac{R_{\text{out}}}{R_{\text{eq}}} \tag{2-16}$$

While applying Eq. (2-16) it is important to realize that the assumption was made that negligible current flows through the external circuit that is attached to the output of the device. Often that means that the resistance in the external circuit is large compared to that in the voltage-divider circuit.

Sample problem 2-3 Calculate the output potential of the voltage divider that is shown in Fig. 2-6*b*, if E is 50.0 V, R_{eq} (the total resistance) is 1250 Ω, and R (the portion of R_{eq} between the contact points to the circuit for E_{out}) is 50.0Ω.

SOLUTION The output potential is calculated with the aid of Eq. (2-16):

$$E_{\text{out}} = E\frac{R_{\text{out}}}{R_{\text{eq}}} = 50.0\text{V}\frac{50.0\ \Omega}{1250\ \Omega} = 2.00\ \text{V}$$

Resistors connected as shown in Fig. 2-7 are connected in *parallel.* The current that flows through the parallel circuit has at least two optional paths.

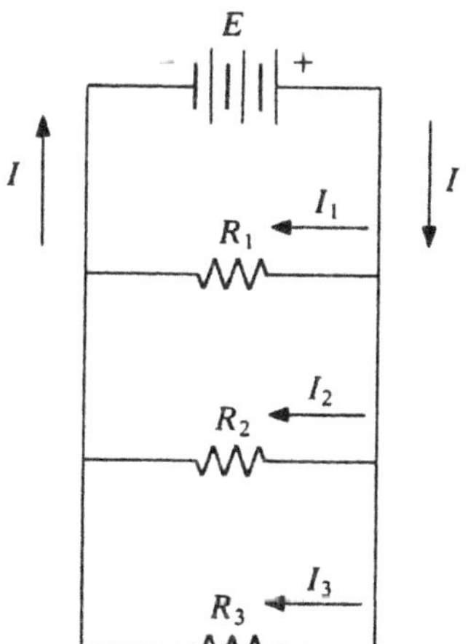

Figure 2-7 A simple parallel circuit. The arrows indicate the direction of current flow through the several alternate paths.

The amount of current flow through each path depends upon the resistance in each path. As the resistance in each path decreases, the current that flows through the path increases. The several paths through which the current can flow and the direction of current flow are shown by the arrows in Fig. 2-7.

Because the potential source in a simple parallel circuit, such as that shown in Fig. 2-7, is attached directly across each resistor, the potential across each resistor is identical to the potential of the battery. For the circuit that is shown in Fig. 2-7, the current flow through the battery is equal to the sum of the currents that flow through each of the alternate paths:

$$I = I_1 + I_2 + I_3 \tag{2-17}$$

The situation is analogous to the flow of water in several small rivers which all combine to form a larger river. The total current flow in the larger river at a location downstream from the small rivers is equal to the sum of the current flow in each of the small rivers.

Individual values of I_1, I_2, and I_3 can be calculated by application of Ohm's law to each parallel branch of the circuit followed by substitution into Eq. (2-17):

$$I = \frac{E}{R_1} + \frac{E}{R_2} + \frac{E}{R_3} = E\left(\frac{1}{R_1} + \frac{1}{R_2} + \frac{1}{R_3}\right) \tag{2-18}$$

Substitution into Eq. (2-18) of E/R_{eq} for I and rearrangement of the resulting equation enables calculation of a value of an equivalent resistance for the circuit:

$$\frac{E}{R_{eq}} = \frac{E}{R_1} + \frac{E}{R_2} + \frac{E}{R_3}$$

$$\frac{1}{R_{eq}} = \frac{1}{R_1} + \frac{1}{R_2} + \frac{1}{R_3}$$

Similarly, for the general case in which n resistors are connected in parallel, either of the following equations is obeyed:

$$\frac{1}{R_{eq}} = \frac{1}{R_1} + \frac{1}{R_2} + \cdots + \frac{1}{R_n} \tag{2-19}$$

$$G_{eq} = G_1 + G_2 + \cdots + G_n \tag{2-20}$$

In series circuits an equivalent resistance can be determined by addition of the individual resistances [Eq. (2-14)] whereas in parallel circuits an *equivalent conductance* G_{eq} can be determined by addition of the individual conductances. In circuits which contain some resistors that are connected in series and other resistors that are connected in parallel, it is sometimes possible to take advantage of equivalent resistances to simplify calculations involving the circuit.

Sample problem 2-4 Determine the current that flows through the 50.0 Ω resistor in Fig. 2-8.

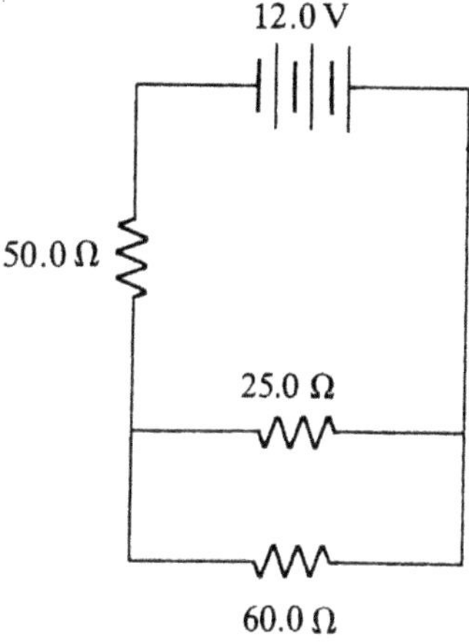

Figure 2-8 The combination series-parallel circuit that is used in Sample problem 2-4.

SOLUTION The equivalent resistance for the 25.0 and 60.0 Ω resistors can be calculated using Eq. (2-19):

$$\frac{1}{R_{\text{eq}}} = \frac{1}{25.0} + \frac{1}{60.0} = 0.0567$$

$$R_{\text{eq}} = 17.6\ \Omega$$

The equivalent resistance is used in the circuit in place of the 25.0 and 60.0 Ω resistors and the new series circuit is used to calculate the current flow through the 50.0 Ω resistor. Equation (2-13) is used to calculate I:

$$E = I\ (R_1 + R_2)$$
$$12.0 - I\,(50.0 + 17.6) = 67.6I$$
$$I = 0.178\ \text{A}$$

For complex dc circuits it often is easier to use Kirchoff's *laws* for calculations. Kirchoff's laws are based upon the laws of conservation of mass and energy. *Kirchoff's first law* (also known as Kirchoff's current law) states that the sum of the currents around a point in a circuit at which three or more conductors are joined is zero:

$$\sum I = 0 \tag{2-21}$$

It is an electrical equivalent to the law of conservation of mass. The point is called a *branch point.* As an example, in Fig. 2-9 the sum of the currents around branch point F is $I_3 + I_4 - I_2 = 0$. In obtaining the sum, the sign of the current must be used. It is conventional to assign a current a positive sign if it is flowing toward the branch point and a negative sign if it is flowing away from the point. Keep in mind that in circuit diagrams, current is thought of as a flow of positive charges. If a negative current is obtained after solving the equation, the assumed direction of current flow was incorrect.

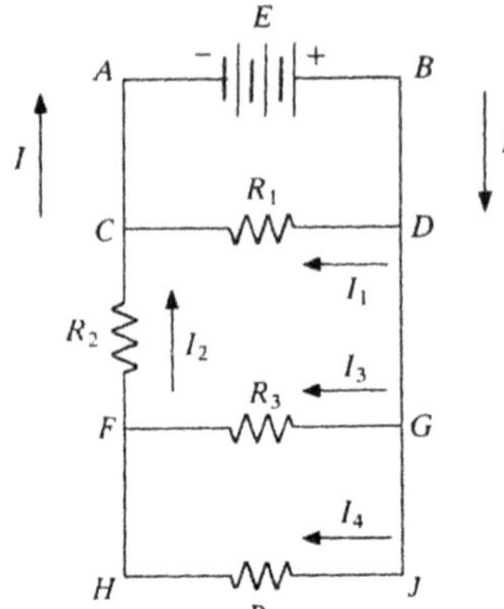

Figure 2-9 A complex dc circuit.

Kirchoff's second *law* (also known as Kirchoff's voltage law) states that the sum of the potentials around any loop in a circuit is zero:

$$\sum E = 0 \tag{2-22}$$

It is the electrical equivalent to the law of conservation of energy. Application of Kirchoff's second law to the circuit in Fig. 2-9 reveals that for the loop *ABDC* (the letters correspond to the labeled points in the diagram), $E - I_1R_1 = 0$. While using Kirchoff's second law, the potential across a resistor is calculated from Ohm's law. Application of Eq. (2-22) to loop *CDGF* yields $I_1R_1 - I_3R_3 - I_2R_2 = 0$. Application of Kirchoff's second law to a circuit that contains two potential sources is illustrated in sample problem 2-5. In order to determine the sign of each potential in a loop, draw a circular arrow within the loop. In this case the arrow tip (loop *CDGF*) is pointing in a clockwise direction. Potentials across resistors which have currents flowing in the direction of the arrow tip are considered negative whereas potentials corresponding to currents flowing in the direction opposite to that of the arrow head are positive.

Sample problem 2-5 Determine the current that flows through the 12– Ω resistor in the circuit of Fig. 2-10.

SOLUTION Application of Kirchoff's second law to loop *ABFD* results in

$$10.0 - I_1\,(50.0) = 0$$

$$I_1 = 0.200 \text{ A}$$

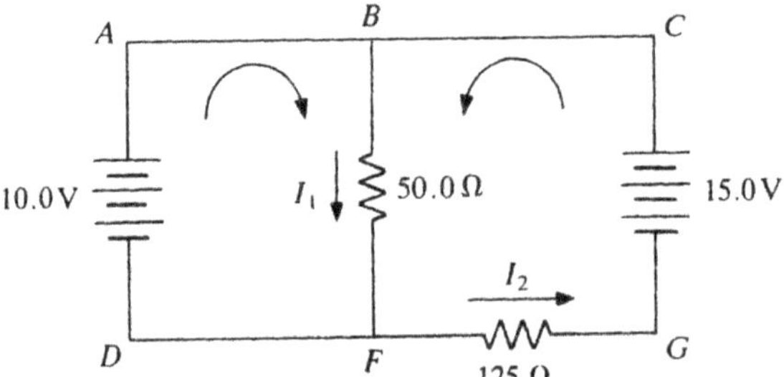

Figure 2-10 The circuit diagram for Sample problem 2-5.

Similarly, application of Kirchoff's second law to loop *CB FG* yields

$$15.0 - I_2\,(125) - I_1\,(50.0) = 0$$

Substitution of the calculated value for I_1 into the equation and rearrangement gives the desired current:

$$15.0 - I_2\,(125) - (0.200)\,(50.0) = 0$$

$$I_2 = 0.040 \text{ A}$$

CURRENT – AND POTENTIAL – MEASURING DEVICES

Several methods are available for measuring current. Regardless of the method that is used, it is important to understand that the insertion of a current-measuring device into a circuit introduces a resistance that could affect current flow in the circuit. A device which measures current is an *ammeter*. A relatively rugged and inexpensive ammeter uses a *moving coil meter*. A coil of wire is wound on an aluminum frame around an iron core and placed between the poles of a magnet. A current flow through the coil creates a magnetic field that interacts with the magnetic field of the permanent magnet causing rotation of the aluminum frame and the coil. In the D'Arsonval meter a pointer which is attached to the aluminum frame indicates the extent of the rotation. The extent

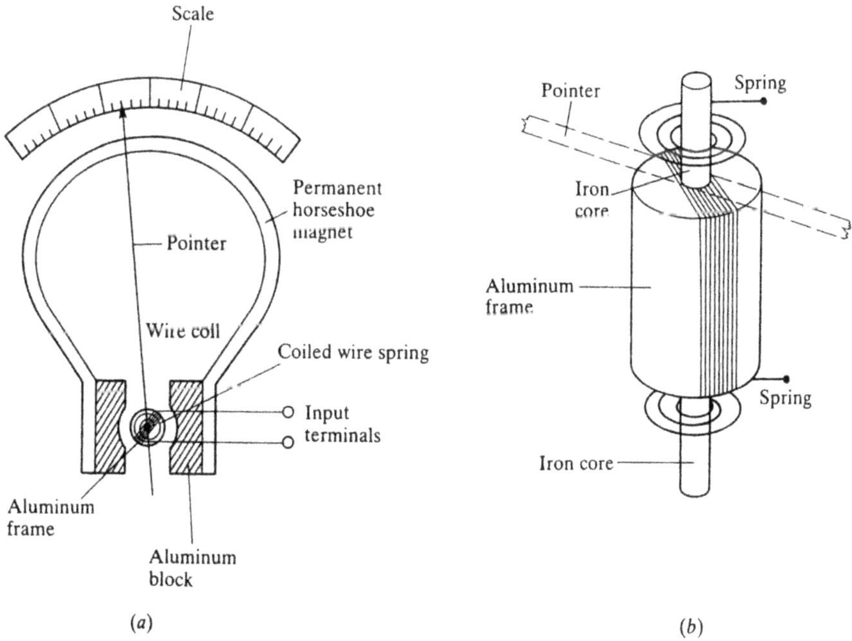

Figure 2-11 (a) A sketch of a D' Arsonval meter. Wire springs are located above and below the aluminum frame and are electrically connected to the terminals and to a coil of wire that is wound around the frame. (b) A sketch of the aluminum-frame assembly.

of the rotation is proportional to the strength of the created magnetic field and consequently to the current in the coil. The wire coil is attached to the external circuit through metallic springs which cause the pointer to return to the zero position on the calibrated scale when the current ceases. Figure 2-11 is a sketch of a moving coil meter.

A *galvanometer* is a moving coil meter in which the pointer of the D'Arsonval meter is replaced by light which is reflected from a mirror that is attached to the pivotal point of the aluminum frame. As the mirror rotates, the spot of reflected light moves on a translucent scale. Galvanometers generally are more sensitive than D'Arsonval meters. Often moving coil meters are damped in order to minimize oscillations of the pointer or light spot.

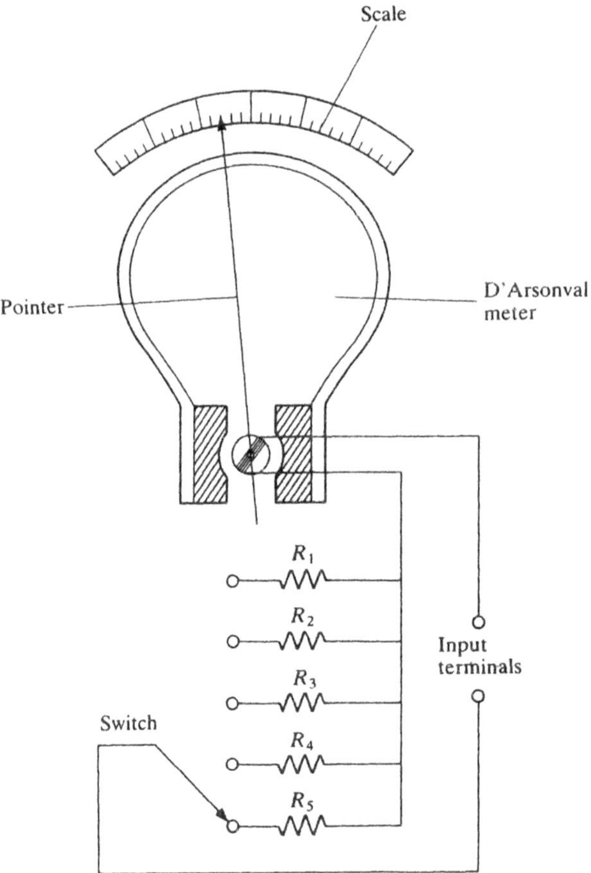

Figure 2-12 A sketch of a voltmeter. Resistors R_1 through R_5 can be placed in the circuit by changing the position of the switch.

The sensitivity of moving coil meters can be changed with the aid of a low-resistance conductor that is connected in parallel with the meter terminals. Since the resistance through the meter is greater than that through the low-resistance conductor in the parallel circuit, the low-resistance conductor carries most of the current and allows moving coil meters to be used to measure larger currents. The low-resistance conductor is a *shunt.* By varying the shunt resistance (by turning a knob on the meter), the sensitivity of the apparatus is altered.

A device which is used to measure potential differences is a *voltmeter.* A voltmeter through which a significant current flows during a measurement can alter the potential in the circuit. An often-used voltmeter uses a moving coil meter connected in series with a resistor of known resistance. The current that is measured by the meter is related to the potential by Ohm's law. The scale on the meter is calibrated in units of potential rather than of current. By changing the resistor the range of the meter is altered.

A *potentiometer* is another device which has been used to measure potential. It is described in Sample problem 2-6. Frequently an ammeter and a voltmeter are combined into a single instrument that is termed a *volt-ohmmeter* (VOM). A digital version of a voltmeter is a *digital voltmeter* (DVM) and a digital version of a multipurpose meter is a *digital multimeter* (DMM). Figure 2-12 is a sketch of a voltmeter. Typically DVMs and DMMs use operational amplifier circuits (Chapter 4) in order to decrease the current passing through the devices to nearly zero during potential measurements.

Sample problem 2-6 Figure 2-13 is a typical circuit diagram for a potentiometer. Potentiometers are used to measure potentials. Although

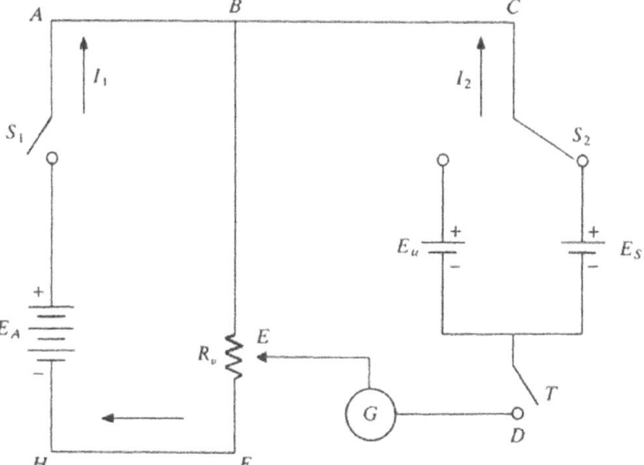

Figure 2-13 A circuit diagram of one form of potentiometer, where E_A is a battery; E_s, a standard cell of known potential; E_u, the device whose potential is measured; G, a galvanometer or other current-measuring device; R_v, a variable resistor; S_1, S_2, switches; T, a tapping key.

easier methods for measuring potential are now available, potentiometers are still used in some laboratories. Switch S_1 was closed to place the battery E_A in the circuit and switch S_2 was adjusted to place the standard cell E_s in the circuit. While depressing the tapping key T, the variable resistor was adjusted to yield zero current through galvanometer G. The adjusted value of the resistor, i.e., the resistance between B and E, was 32.3 Ω and the resistance between B and F was 100.0 Ω. The standard cell was a Weston cell with a potential of 1.018 3 V (at 25°). After the initial adjustment, switch S_2 was thrown to place E_u in the circuit; the tapping key was depressed; and R_v was again adjusted to yield zero current at G. After adjustment, the resistance between B and E was 48.5 Ω. Calculate the potential E_u.

SOLUTION The potential at the output (between B and E) to the voltage divider in the loop $ABFH$ can be calculated using Eq. (2-16):

$$E_{BE} = \frac{R_{BE}}{R_{BF}} E_A$$

When Es is in the circuit,

$$E_{BE(s)} = \frac{R_{BE(s)}}{R_{BF}} E_A$$

and when E_u is in the circuit,

$$E_{BE(u)} = \frac{R_{BE(u)}}{R_{BF}} E_A$$

When the current at G is zero, Kirchoff's second law applied to loop $BCDE$ yields

$$E_{BE\,(S)} - E_S = 0;\ E_{BE(S)} = E_s$$

when E_S is in the circuit, and

$$E_{BE\,(u)} - E_u = 0;\ E_{BE\,(u)} = E_u$$

when E_u is in the circuit. Substitution of the values into the voltage-divider equations yields

$$E_s = \frac{R_{BE(S)}}{R_{BF}} E_A$$

and
$$E_u = \frac{R_{BE(u)}}{R_{BF}} E_A$$

Division of the second equation by the first equation gives

$$\frac{E_u}{E_s} = \frac{\dfrac{R_{BE(u)}E_A}{R_{BF}}}{\dfrac{R_{BE(s)}E_A}{R_{BF}}}$$

which upon rearrangement gives

$$E_u = E_s \frac{R_{BE(u)}}{R_{BE(s)}} \tag{2-23}$$

Equation (2 -23) can be used with potentiometers to calculate an unknown potential. Substitution of the appropriate values into Eq. (2-23) allows calculation of E_u:

$$E_u = 1.0183 \frac{48.5}{32.3} = 1.53 \text{ V}$$

A measurement that relies upon adjustment of some parameter, e.g., a potential or a current, to zero is a *null measurement.* A potentiometer (see Sample problem 2-6 and Fig. 2-13) is an example of a device that uses a null measurement. The galvanometer in the potentiometer is an example of a *null-point detector.*

AC CIRCUITS

An *alternating current* (*ac*) is a current that periodically changes direction of flow. The change in flow direction is caused by a periodically changing potential. The current can change with time in several ways. The two most-often-encountered forms of alternating currents are sine wave currents and square wave currents (Fig. 2-14). In circuit diagrams an ac power supply (potential source) is represented by a sketch of one repetitive unit of the wave inside a circle, e.g., a sine wave potential source is represented by -⊚-.

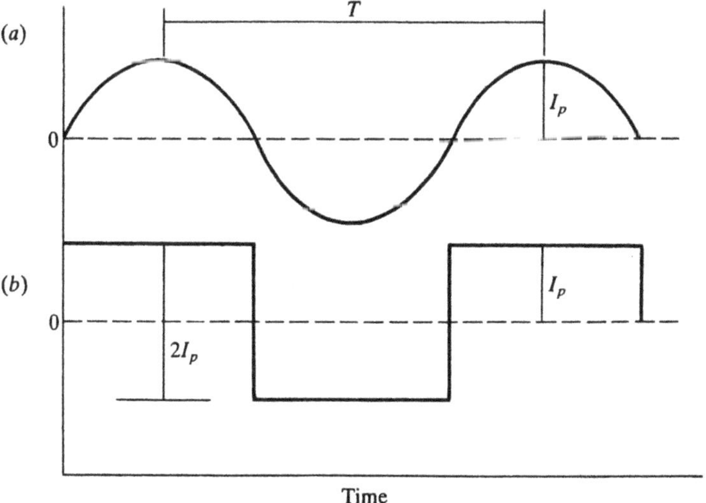

Figure 2-14 A sketch of the variation of current (or potential) with time for two common types of alternating current. (*a*) A sine wave current and (*b*) a square wave current; where I_p is the wave amplitude; 2 I_p, peak-to-peak current; *T*, wave period.

Repetitive waveforms such as those encountered with alternating currents are characterized by several parameters. The *amplitude* of the wave is the height of the wave at its maximum. The *current amplitude* I_p of the wave is the maximum alternating current that flows during each cycle. Similarly, the *potential amplitude* E_p is the maximum alternating potential that causes the alternating current.

The difference between the largest current (or potential) during a wave and the smallest current (or potential) is the *peak-to-peak current* (or *peak-to-peak potential*). The peak-to-peak current $I_{P\text{-}P}$ is $2I_P$. The *root mean square* (rms) current (or potential) is given by

$$I_{\text{rms}} = \sqrt{\frac{I_p^2}{2}} = \frac{I_p}{\sqrt{2}} = 0.707 I_p \qquad \text{(2-24}a\text{)}$$

$$E_{\text{rms}} = \sqrt{\frac{E_p^2}{2}} = \frac{E_p}{\sqrt{2}} = 0.707 E_p \qquad \text{(2-24}b\text{)}$$

An alternating current with an amplitude of I_P produces the same heating in a resistor as a direct current of I_{rms}.

Alternating current voltmeters and ammeters are generally calibrated to measure rms values of potential and current. When applying Kirchoff's laws to ac circuits, rms currents and potentials are generally used.

The *period T* of the wave is the time that is required to complete a single cycle of the wave. The *frequency f* is the number of cycles that pass a fixed point in a fixed period of time. Frequency is related to period by

$$f = \frac{1}{T} \qquad \text{(2-25)}$$

The most-often-used unit of period is seconds s and the corresponding frequency unit is hertz Hz, that is, 1 Hz is 1 s^{-1}.

The following equations describe the variation in current and potential with time for sinusoidal waves:

$$i = I_p \sin 2\pi ft \qquad \text{(2-26)}$$

$$e = E_p \sin 2\pi ft \qquad \text{(2-27)}$$

In ac circuits the potential and the current vary with time. Lowercase letters [*i* and *e* in Eqs. (2-26) and (2-27)] are used to represent instantaneous values of time-varying parameters. Capital letters are used to represent time-independent parameters such as wave amplitudes or dc parameters.

Sample problem 2-7 A sinusoidal, alternating current has a peak current of 12.3 mA and a period of 5.75×10^{-4} s. Calculate the frequency, peak-to-peak current, and rms current.

SOLUTION The frequency can be calculated with Eq. (2-25):

$$f = \frac{1}{T} = \frac{1}{5.75 \times 10^{-4}} = 1.74 \times 10^3 \text{ Hz}$$

The peak-to-peak current is 2 I_P:

$$I_{\text{p-p}} = 2(12.3) = 24.6 \text{ mA}$$

The rms current is calculated with the aid of Eq. (2.24*a*):

$$I_{\text{rms}} = \sqrt{\frac{I_p^2}{2}} = \sqrt{\frac{12.3^2}{2}} = 8.70 \text{ mA}$$

Sample problem 2-8 Determine R_x in the circuit diagram shown in Fig. 2-15 if the rms current measured at H is zero when R_1 is 15.0 Ω R_2 is 32.0, and R_v is adjusted to 53.0 Ω.

SOLUTION Because the current measured at H is zero, the potential between B and D is zero. Consequently, the current flowing through R_1 equals that flowing through R_x and the current flowing through R_v equals that flowing through R_2. Kirchoff's second law applied to loop ABD yields

$$I_1 R_x - I_2 R_v = 0$$

Rearrangement of the equation results in

$$R_x = \frac{I_2 R_v}{I_1}$$

It should be noted that Kirchoff's laws can be applied to ac circuits in the same manner in which they are applied to dc circuits. The rms currents and

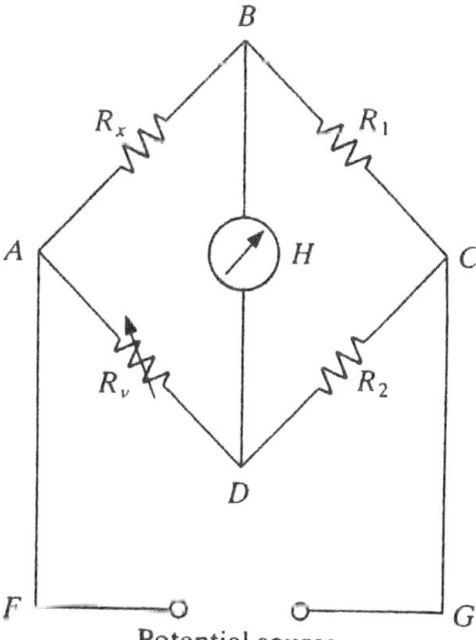

Figure 2-15 A circuit diagram of a Wheatstone bridge. Refer to Sample problem 2-8.

potentials generally are used in the equations. Kirchoff's second law applied to loop *CBD* gives

$$-I_1R_1 + I_2 R_2 = 0$$

which rearranges to yield

$$\frac{I_2}{I_1} = \frac{R_1}{R_2}$$

Substitution of the ratio into the equation for R_x yields an expression which is independent of the magnitude of the applied potential:

$$R_x = \frac{R_1R_v}{R_2} \tag{2-28}$$

Substitution of the appropriate values into Eq. (2-28) enables calculation of R_x:

$$R_x = \frac{(15.0\ \Omega)(53.0\ \Omega)}{32.0\Omega} = 24.8\ \Omega$$

The circuit shown in Fig. 2-15 is a *Wheatstone bridge*. The Wheatstone bridge or an adaptation is commonly used in various instruments to measure resistance. It is used in both dc and ac circuits. The resistance to be determined, R_x, is attached between terminals *A* and *B* of the bridge and the variable resistor is adjusted to yield zero potential (and zero current) between terminals *B* and *D*. Equation (2-28) can be used to calculate the resistance or the variable resistor can be calibrated to directly yield the resistance of the sample.

Reactance and Capacitance

In ac circuits the current is continuously changing. Whenever a current is changed (either increased or decreased), energy is required to create the magnetic and electric fields that accompany the change. Briefly stated, electric circuits resist change. The resistance to the changing current in ac circuits is *reactance X*. *Capacitance* and *inductance* cause two forms of reactance, X_c and X_L, that occur in ac circuits.

Capacitance can be purposely added to an ac circuit by adding a *capacitor* to the circuit. A capacitor in a circuit, diagram is symbolized by two parallel lines ⊣⊢ or by ⊣(⊢ . A variable capacitor is represented by ⊣⊢ with an arrow.

The most common type of capacitor consists of two, parallel, conducting plates that are separated from each other either by a narrow air gap or by some other insulator. If one terminal of a battery is attached to one of the plates and the other terminal is attached to the second plate, a current momentarily flows through the circuit as electrons from the negative terminal flow into one plate and electrons from the second plate flow to the positive terminal of the battery. The result is a charge differential between the two plates. When the charge differential becomes sufficiently large to yield a potential that is equal to (but opposing) that from the battery, the current ceases. When the current stops, the capacitor is *charged.*

If the battery possessed sufficient potential energy, the accumulated charges on the capacitor plates would become sufficiently large to cause electrons to jump across the gap between the plates. The capacitor can be *discharged* by shorting together the two plates. That can be accomplished by simultaneously connecting a single conductor to both plates. During the discharging process the stored charge on one plate flows to the other plate. Capacitors are often utilized to store electric charges for later use.

The ability of a capacitor to hold an electric charge in a dc circuit is dependent upon the physical dimensions of the plates, their separation from each other, and the applied potential. For a particular capacitor the dimensions and plate separation are constant and the amount of charge held by a capacitor is directly proportional to the applied potential:

$$Q = CE \tag{2-29}$$

The proportionality constant C in Eq. (2-29) is the *capacitance* of the capacitor. If Q is measured in units of coulombs and E is measured in volts, C has units of coulombs per volt, which is *farads* F. The capacitance is directly proportional to the plate area and inversely proportional to the width of the gap.

An *electrolytic capacitor* consists of an aluminum- or tantalum-foil plate separated by a thin oxide layer from a conducting paste or solution. The curved plate in the symbol for electrolytic capacitors corresponds to the negative terminal of the capacitor. In order to prevent destruction of the oxide layer, electrolytic capacitors must always be placed in the circuit so that the metal foil never acquires a negative charge when compared to the conducting paste.

Nonelectrolytic capacitors usually consist of two metal foils separated by mica, paper, ceramic, or plastic. *Air-dielectric* capacitors consist of two metal plates separated by an air gap. Variable air-dielectric capacitors are often used for tuning radiofrequency circuits.

For ac circuits Eq. (2-29) can be rewritten to show the change in charge and potential with time:

$$\frac{dq}{dt} = C\frac{de}{dt} = i \tag{2-30}$$

According to Eq. (2-2), *dq/dt* is *i,* that is, in an ac circuit a current flows without the necessity of an electric charge jumping the gap between the plates of the capacitor. Charge flows into a plate during one half of the ac cycle and flows out during the second half. Because the presence of a capacitor in a circuit permits the flow of an alternating current while preventing the flow of direct current, capacitors can be used to remove direct currents from a total current that has both direct and alternating components.

The equation for the variation of potential with time in a sinusoidal ac circuit that contains a capacitor is given by Eq. (2-27). The variation of current with time in the same capacitor can be determined by substituting the value of *e* from Eq. (2-27) into Eq. (2-30):

$$i = C\frac{de}{dt} = C\frac{d(E_p \sin 2\pi ft)}{dt} = 2\pi fCE_p \cos 2\pi ft$$

The cosine function is 90° (π/2 rad) ahead of the sine function. After making that substitution,

$$i = 2\pi fCE_p \sin\left(2\pi ft + \frac{\pi}{2}\right) \qquad (2\text{-}31)$$

Comparison of the equation with Eq. (2-27) reveals that the current in a sinusoidally charged capacitor is out-of-phase and leads the potential by π/2 rad (90°).

When a sine wave is compared to another sine wave, the equation for the first wave relative to the second wave is given by Eq. (2-32) for currents and Eq. (2-33) for potentials:

$$i = I_p \sin(2\pi ft + \phi) \qquad (2\text{-}32)$$
$$e = E_p \sin(2\pi ft + \phi) \qquad (2\text{-}33)$$

The angle ϕ in the equations is the *phase angle*. When the phase angle is zero, i.e., when the two waves simultaneously rise and fall, the two waves are *in-phase*. If the phase angle is not zero, the two waves are *out-of-phase*. From Eq. (2-31) it is apparent that the phase angle for the current in a capacitor when compared to the potential is π/2 rad (90°). The phase angle in ac circuits is advantageously used in ac polarography for chemical analysis (Chapter 23). In Eqs. (2-26), (2-27), (2-29), (2-31), (2-32) and (2-33), $2\pi f$ is often replaced by the *angular frequency* ω.

Capacitive reactance X_c is defined by

$$X_C = \frac{1}{2\pi fC} \qquad (2\text{-}34)$$

The units of reactance are ohm. The *impedance* Z of a circuit is a measure of the combined effect of reactance and resistance. For a series *RC* circuit the impedance is given by

$$Z = \sqrt{X_C^2 + R^2} \qquad (2\text{-}35)$$

The capacitance and resistance cannot be directly added to give the impedance because they are 90° out-of-phase. The result of the vector addition is Eq. (2-35). Impedance has units of ohms and is the ac equivalent of resistance in dc circuits. The impedance of a lone capacitor is the reactance of the capacitor. The ac version of Ohm's law is

$$E = IZ \qquad (2\text{-}36)$$

It can be used to calculate the potential drop across an impedance.

RC Circuits

Series resistance-capacitance (*RC*) circuits are often used as *high-pass filters*. A high-pass filter allows high-frequency potentials to pass through it while filtering out most low-frequency potentials. High-pass filters are often used to remove the 60-Hz line-frequency potential when high-frequency information is carried (superimposed) on the 60-Hz signal. A circuit diagram of an *RC* high-pass filter is shown in Fig. 2-16.The output potential E_o is obtained across the resistor.

The rms output potential of the *RC* high-pass filter is given by

$$E_o = \frac{E_i}{\sqrt{1 + (1/2\pi fRC)^2}} \qquad (2\text{-}37)$$

(see Reference 1, pp. 47-50). As the frequency f of the potential approaches infinity, the output potential approaches the input potential. As the frequency approaches zero, the output potential approaches zero; i.e., low-frequency signals are diminished. The frequency at which $2\pi fRC$ is unity is the *half-power frequency* $f_{1/2}$. At the half-power frequency E_o is $0.707E_i$. In ac circuits that contain a capacitor *RC* is the *time constant*. The time constant has units of ohms-coulombs per volt, which is equivalent to seconds. As shown in Eq. (2-37), both *R* and *C* have an effect upon the output potential.

Sample problem 2-9 Determine the half-power frequency, the ac frequency at which the output signal is 1.0 percent of the input signal, and the frequency at which the output signal is 99.0 percent of the input signal for an *RC* circuit (Fig. 2-16) in which *R* is $5.0 \times 10^5\ \Omega$ and *C* is 0.25 μF.

SOLUTION The half-power frequency $f_{1/2}$ is the frequency at which $2\pi f_{1/2} RC$ is unity. Substitution of the appropriate values of *R* and *C* into the equation and solving for $f_{1/2}$ yields

$$2\pi f_{1/2}(5.0 \times 10^5)(0.25 \times 10^{-6}) = 1$$

$$f_{1/2} = 1.3 \text{ Hz}$$

At 1.3 Hz about 71 percent of the input potential passes through the filter.

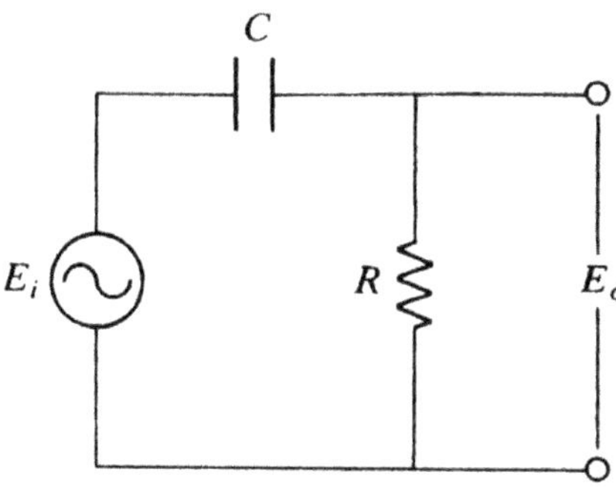

Figure 2-16 A series *RC* circuit which is used as a high-pass filter. *C*, a capacitor; E_i, input rms potential; E_o, output rms potential; *R*, a resistor.

Equation (2-37) can be used to calculate f when 1.0 and 99.0 percent of the input signal are passed through the filter. When 1.0 percent of the signal is passed, $E_o = 0.010\,E_i$ and

$$0.010E_i = \frac{E_i}{\sqrt{1+\left[1/2\pi f\left(5.0\times10^5\right)\left(0.25\times10^{-6}\right)\right]^2}}$$

$$f = 0.013 \text{ Hz}$$

Similarly, when 99.0 percent of the signal is passed, $E_o = 0.990\,E_i$, and

$$0.990E_i = \frac{E_i}{\sqrt{1+\left[1/2\pi f\left(5.0\times10^5\right)\left(0.25\times10^{-6}\right)\right]^2}}$$

$$f = 9.0 \text{ Hz}$$

At frequencies above 9.0 Hz, more than 99.0 percent of the signal is passed through the filter

Series RC circuits can be used as *low-pass filters* as well as for high-pass filters. A low-pass filter preferentially allows low-frequency signals to pass. An RC low-pass filter is similar to an RC high-pass filter except that the output potential is obtained across the capacitor rather than across the resistor. Low-pass filters are often used to remove ac interferences prior to making dc measurements. Fig. 2-17 is a circuit diagram of an RC low-pass filter. The rms output potential E_o for the low-pass filter of Fig. 2-17 is given by

$$E_o = \frac{E_i}{\sqrt{1+\left(2\pi fRC\right)^2}} \qquad (2\text{-}38)$$

As with the high-pass filter, the half-power frequency is the frequency at which $2\pi fRC$ is equal to unity.

Inductance

Inductance L is the second cause of reactance in ac circuits. If an electric current is allowed to flow through a coil of wire, a magnetic field is formed. In an ac circuit the current that flows through the coil fluctuates with time, as previously

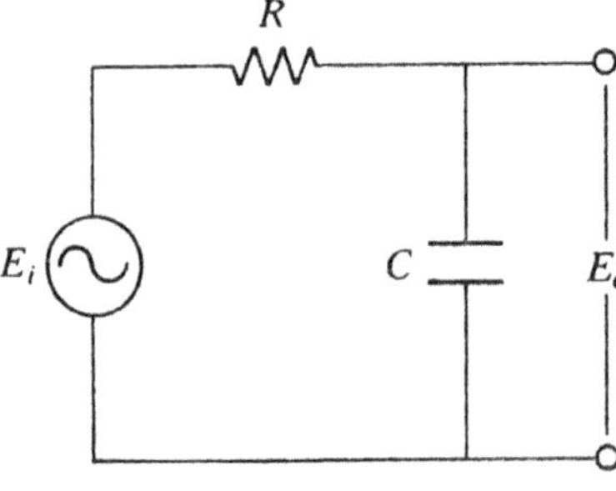

Figure 2-17 A series RC circuit that is used as a low-pass filter. C, a capacitor; E_i, input rms potential; E_o, output rms potential; R a resistor.

described. The fluctuating magnetic field results in an induced potential that opposes the potential in the electric circuit and consequently opposes electron flow. The device in which the magnetic field is generated is an *inductor*. In electric circuit diagrams, inductors are given a symbol ⌇⌇⌇ that resembles a coil of wire.

For many inductors a linear relationship exists between the magnitude of the potential that is formed as a result of the magnetic field and the current that flows through the inductor:

$$e = -L\frac{di}{dt} \tag{2-39}$$

The negative sign in Eq. (2-39) indicates that the potential that is induced in the inductor opposes the potential in the circuit. The proportionality constant L in the equation is the inductance of the inductor. When e, i, and t respectively are measured in volts, amperes, and seconds, L is in units of volts per ampere per second, which is defined as *henrys* H. Although inductance exists in all ac circuits, the inductance generally is small and can be ignored, except at extremely high frequencies. An inductor is a device that is used to purposely introduce inductance into a circuit.

Inductors consist of multiple adjacent turns of wire on the same support. The magnitude of the inductance is dependent upon the number of coils in the inductor. By using inductors of multiple coils, the magnetic field produced by the inductor can be made relatively large. The inductances that are used in most electric circuits are between 1 μH and 1 mH. An iron core is used in some inductors. Iron-core inductors are given the symbol ⌇⌇⌇ in circuit diagrams.

It is informative to compare the functions that are performed by capacitors with those that are performed by inductors. The comparison can be accomplished by comparing Eq. (2-30) with Eq. (2-39). From Eq. (2-30) it is apparent that a changing potential in a capacitor results in a current. Equation (2-39) reveals that a changing current in an inductor results in a potential. Consequently, capacitors and inductors serve opposite purposes in an ac circuit, i.e., their functions complement each other.

Transformers

A *transformer* is a device that consists of two magnetically coupled inductors (Fig. 2-18). When an alternating current (or potential) is passed through one of the inductors (the *primary coil*), the resulting magnetic flux generates an alternating current in the second inductor (the *secondary coil*). The physical support upon which the inductors are wound is the *core* of the inductor. Although air-core transformers are sometimes used, the iron-core transformer is more common.

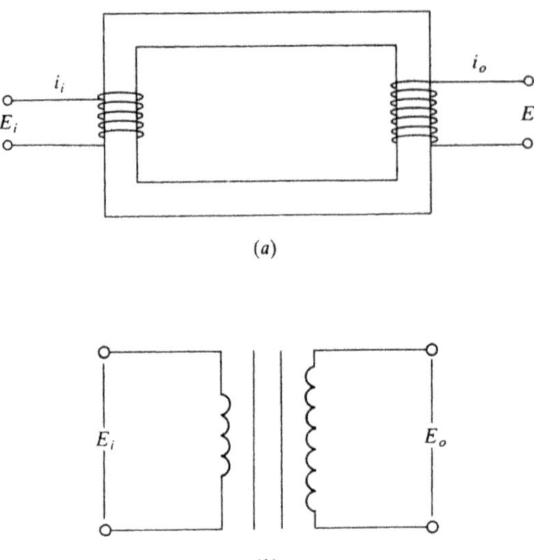

Figure 2-18 (*a*) A step-up transformer and (*b*) the corresponding symbolic representation in circuit diagrams. E_i, input potential; E_o, output potential; i_i, input current; i_o, output current.

Recall that the inductance of an inductor is a function of the number of coils in the inductor. In an iron-core transformer, the magnetic flux in the core links the two coils. In an ideal iron-core transformer, the ratio of the input potential to the output potential is given by

$$\frac{E_i}{E_o} = \frac{n_1}{n_2} \qquad \text{(2-40)}$$

where n_1 and n_2 are the number of turns of wire in the primary (input) and secondary (output) coils respectively. The ratio of the input to the output current of the device is given by

$$\frac{i_i}{i_o} = \frac{n_2}{n_1} \qquad \text{(2-41)}$$

Derivations of Eqs. (2-40) and (2-41) can be found in the references listed in the Bibliography.

A *step-up transformer* has fewer turns of wire in the primary coil than in the secondary coil. Consequently the output potential of a step-up transformer is greater than the input potential. Similarly, a *step-down transformer* has more turns of wire in the primary coil than in the secondary coil and a lower output potential than input potential. Many transformers are tapped in more than one location so that it is possible to obtain several different output potentials from a transformer with a single input.

Sample problem 2-10 Determine the output potential for a transformer in which the primary coil is connected to a 110-V ac source if the primary coil has 150 turns of wire and a the secondary coil has 25 turns of wire.

SOLUTION Equation (2-40) is used to solve the problem:

$$\frac{E_i}{E_o} = \frac{n_1}{n_2}; \quad \frac{110}{E_o} = \frac{150}{25}; \quad E_o = 18 \text{ V}$$

RL Circuits

A series ac circuit that contains a resistor and an inductor can be used as a low-pass filter (Fig. 2-19). The rms output for the filter that is shown in Fig. 2-19 is given by

$$E_o = \frac{E_i}{\sqrt{1 + (2\pi fL / R)^2}} \tag{2-42}$$

(see Reference 1, pp. 45-47). As f approaches infinity, E_o approaches zero. Consequently, high-frequency signals are attenuated while low-frequency signals are allowed to pass through the filter with little change. The frequency at which $2\pi fL/R$ is one is the half-power frequency. At the half-power frequency, E_o is 0.707 E_i.

For practical purposes the half-power frequency is regarded as the cutoff point for the filter, i.e., the filter is used to remove or *choke off* frequencies that are greater than the half-power frequency. The time constant in *RL* circuits is *L/R*. The time constant has units of seconds when *L* is in henrys and *R* is in ohms. For series *RL* circuits the impedance of the circuit is given by

$$\text{Z} = \sqrt{\text{X}_\text{L}^2 + \text{R}^2} \tag{2-43}$$

[cf. Eq. (2-35)] and *inductive reactance* is defined by

$$X_L = 2\ \pi fL \tag{2-44}$$

The impedance of a lone inductor is the reactance of the inductor.

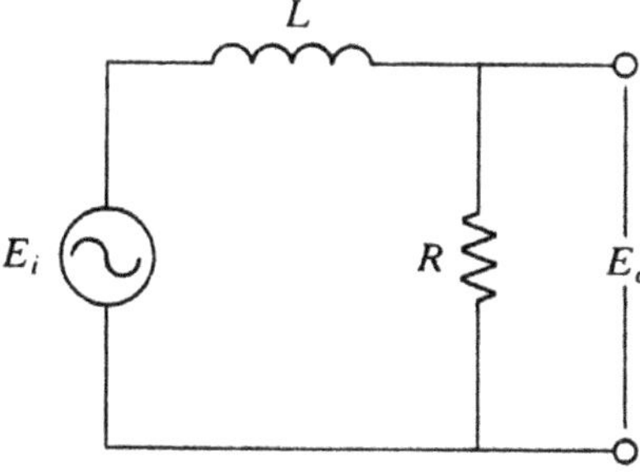

Figure 2-19 A series *RL* circuit that is used as a low-pass filter. E_i, input potential; E_o, output potential; *L*, inductor; *R*, resistor.

RLC Circuits

Alternating current circuits, that contain resistors, inductors, and capacitors are *RLC* circuits. Calculations that involve *RLC* circuits are generally most easily done with the aid of Kirchoff's laws and Eq. (2-36). The impedance of the circuit is substituted into Eq. (2-36) and the equation is solved for the desired quantity.

The equivalent impedance Z_{eq} across several capacitors, inductors, and resistors *in series* can be derived from Eq. (2-36) in a manner that is analogous to the way in which the equivalent resistance of several series resistors was determined [Eq. (2-14)]. The result is

$$Z_{eq} = Z_1 + Z_2 + \ldots + Z_n \tag{2-45}$$

Admittance Y is defined as the reciprocal of impedance;

$$Y = \frac{1}{Z} \tag{2-46}$$

Admittance in ac circuits that contain capacitance and inductance is analogous to conductance in dc circuits that contain only resistance. The equivalent admittance Y_{eq} for a circuit that contains several components that are connected in parallel is given by

$$Y_{eq} = Y_1 + Y_2 + \ldots + Y_n \tag{2-47}$$

The addition signs in Eqs. (2-45) and (2-47) are intended to indicate *vector addition*.

Several useful equations which can be used to calculate the equivalent impedance of several common circuits are listed in Eqs. (2-48) through (2-53). The equations can be derived from Eqs. (2-45) and (2-47) by substituting for each term and performing the vector additions.

$$Z_{eq} = X_L - X_C \quad \text{(series inductance, capacitance)} \tag{2-48}$$

$$Z_{eq} = \frac{RX_C}{\sqrt{R^2 + X_C^2}} \quad \text{(parallel resistance, capacitance)} \tag{2-49}$$

$$Z_{eq} = \frac{RX_L}{\sqrt{R^2 + X_L^2}} \quad \text{(parallel resistance, inductance)} \tag{2-50}$$

$$Z_{eq} = \frac{X_L}{1 - (2\pi f)^2 LC} \quad \text{(parallel inductance, capacitance)} \tag{2-51}$$

$$Z_{eq} = \sqrt{R^2 + (X_L - X_C)^2} \quad \text{(series resistance, capacitance, inductance)} \tag{2-52}$$

$$Z_{eq} = \frac{RX_L X_C}{\sqrt{R^2 (X_L - X_C)^2 + X_L^2 X_C^2}}$$

(parallel resistance, capacitance, inductance) (2-53)

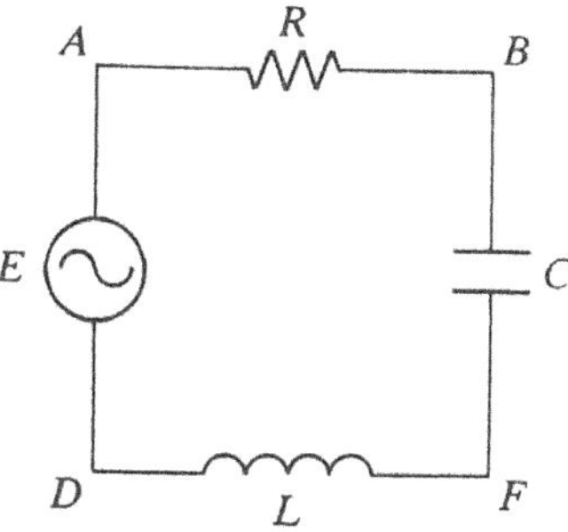

Figure 2-20 A series *RLC* circuit.

As an example of the derivations, Eq (2-52) will be derived for a series resistance, capacitance, and inductance (Fig 2-20). From Eq. (2-45) it is apparent that, for the case when the components are in series, the reactances of the components must be added vectorially. The vector diagram for the addition is shown in Fig. 2-21.

As is apparent from Fig. 2-21, the output from an inductive reactance precedes the output from a resistance by 90° and that from a capacitive reactance follows the output from a resistance by 90°. Because X_L and X_C are 180° out-of-phase, the vector sum of the two reactances is $X_L - X_C$. The resulting vector is added to the vector that represents the resistance and the result is the equivalent impedance of the circuit. The addition is shown in Fig. 2.21*b*. From the pythagorean theorem it is apparent that the *magnitude* of the equivalent impedance is given by

$$Z_{\text{eq}}^2 = (X_L - X_C)^2 + R^2$$

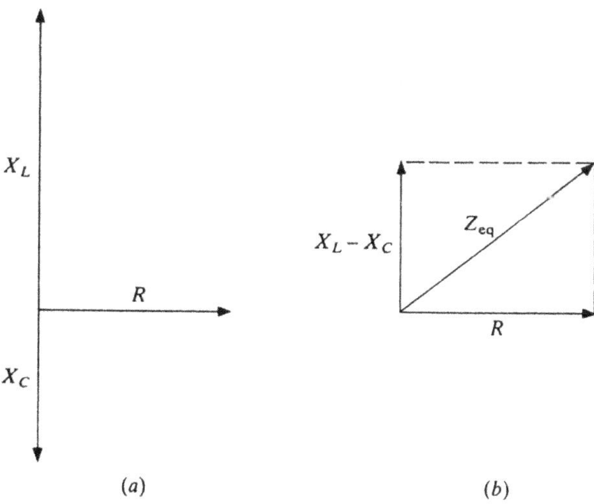

Figure 2-21 A vector diagram for the addition of the reactances that are caused by capacitance, resistance, and inductance in a series circuit. (*a*) A diagram showing the vectors of the components; (*b*) a diagram showing the vector addition.

By taking the square root of both sides of the equation, Eq. (2-52) is obtained. The equations for parallel circuits are similarly derived when the inverse of the reactances are vectorially added [Eq. (2-47)].

Equation (2-48) results from the potential drop across a capacitor and across an inductor being 180° out-of-phase. If X_L equals X_C, no potential drop occurs across the couple. When X_L equals X_C, the energy stored in the capacitor during one half-cycle of the alternating current is just sufficient to force current through the inductor during the second half-cycle. Similarly, the energy stored in the magnetic field of the inductor is just sufficient to charge the capacitor. That condition is termed *resonance* or *series resonance*. The frequency f_o at which resonance occurs can be calculated by setting the inductive impedance equal to and the capacitive impedance:

$$X_L = X_C$$

$$\frac{1}{2\pi f_o C} = 2\pi f_o L$$

$$f_o = \frac{1}{2\pi\sqrt{LC}} \tag{2-54}$$

If the inductor and the capacitor are connected in parallel, resonance or *parallel resonance* occurs when X_L and X_C are equal and Eq. (2-54) applies. For the series circuit the impedance at resonance is zero [Eq. (2-48)]. That leads to a maximum current flow through the circuit. For the parallel circuit the impedance at resonance is infinite [Eq. (2-51)] and the current that flows through the impedance is at a minimum. Resonant circuits can be used as filters.

Sample problem 2-11 Determine the resonant frequency for the circuit shown in Fig. 2-20. Calculate the equivalent impedance of the circuit and the rms current that flows through the circuit at the resonant frequency. Assume that the rms potential of E is 15 V, the resistance of R is 50.0 Ω, the capacitance of C is 0.15 μF, and the inductance of L is 0.75 mH.

SOLUTION The resonant frequency can be calculated with the aid of Eq.(2-54):

$$f_o = \frac{1}{2\pi\sqrt{LC}} = \frac{1}{2\pi\sqrt{(0.75 \times 10^{-3})(0.15 \times 10^{-6})}}$$

$$f_o = 1.5 \times 10^4 \text{ Hz}$$

The equivalent impedance of the circuit is calculated using Eq. (2-52):

$$Z_{eq} = \sqrt{R^2 + (X_L - X_C)^2}$$

At the resonant frequency $X_L = X_C$ and

$$Z_{eq} = \sqrt{R^2 + 0} = R = 50.0\ \Omega$$

The current that flows through the circuit is calculated using Eq. (2-36):

$$E = IZ; \qquad 15 = I\,(50.0); \qquad I = 0.30 \text{ A}$$

Sample problem 2-12 Determine the potentials between *A* and *B*, *B* and *F*, and *F* and *D* at the resonant frequency for the circuit described in Sample problem 2-11 and shown in Fig. 2-20.

SOLUTION The potential across the resistor is

$$E_R = IR = (0.30)(50.0) = 15 \text{ V}$$

The potentials across the inductor and the capacitor are identical at the resonant frequency because $X_L = X_C$, but they are 180° out-of-phase:

$$X_L = 2\pi fL = 2\pi\left(1.5\times10^4\right)\left(0.75\times10^{-3}\right) = 71\ \Omega$$

$$X_C = \frac{1}{2\pi fC} = \frac{1}{2\pi\left(1.5\times10^4\right)\left(0.15\times10^{-6}\right)} = 71\ \Omega$$

The potentials can be obtained from Ohm's law for the particular component:

$$E_L = I X_L \qquad \text{and} \qquad E_C = I X_C$$

$$E_L = E_C = (0.30)\,(71) = 21 \text{ V}$$

The potential across the inductor and across the capacitor at the resonant frequency is greater than the potential from the ac power supply.

Sample problem 2-13 Determine the equivalent impedance of the circuit shown in Fig. 2-22 at the resonant frequency and at 1.5×10^2 Hz. If the rms potential of *E* is 11 V, determine the current that flows through *E* at the two frequencies. The resistance of *R* is 45 Ω, the inductance of *L* is 1.0 mH, and the capacitance of *C* is 0.85 mF.

SOLUTION At resonance X_L equals X_C and Eq. (2-54) can be used to determine the resonant frequency:

$$f_o = \frac{1}{2\pi\sqrt{LC}} = \frac{1}{2\pi\sqrt{\left(1.0\times10^{-3}\right)\left(0.85\times10^{-3}\right)}} = 1.7\times10^2 \text{ Hz}$$

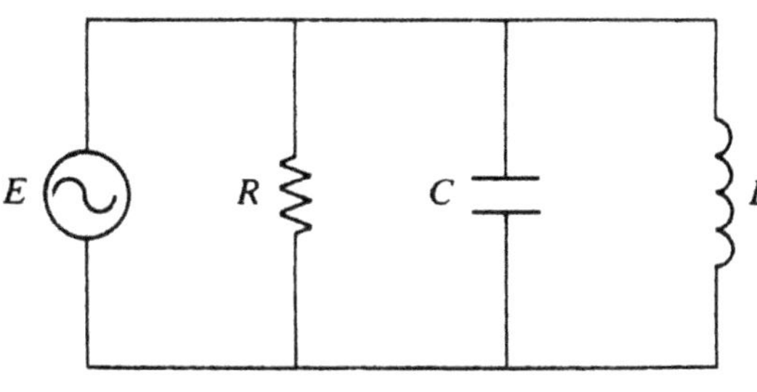

Figure 2-22 The *RLC* circuit that is described in Sample problem 2-13.

The capacitive and inductive reactances are calculated with the aid of Eqs. (2-34) and (2-44) (of course, $X_L = X_C$ at resonance):

$$X_C = \frac{1}{2\pi fC} = \frac{1}{2\pi\left(1.7 \times 10^2\right)\left(0.85 \times 10^{-3}\right)} = 1.1\ \Omega$$

$$X_L = 2\pi fL = 2\pi\left(1.7 \times 10^2\right)\left(1.0 \times 10^{-3}\right) = 1.1\ \Omega$$

The calculated values are used with Eq. (2-53) to calculate the equivalent impedance of the circuit:

$$Z_{eq} = \frac{RX_LX_C}{\sqrt{R^2\left(X_L - X_C\right)^2 + X_L^2X_C^2}}$$

$$= \frac{(45)(1.1)(1.1)}{\sqrt{(45)^2(1.1-1.1)^2 + (1.1)^2(1.1)^2}}$$

$$= 45\ \Omega$$

As expected, at resonance X_L and X_C, balance each other and the total impedance of the circuit is attributable to the resistance in the circuit. The current that flows through E is calculated using Eq. (2-36):

$$E = IZ; \qquad 11 = I(45); \qquad I = 0.24\ \text{A}$$

X_c and X_L are similarly determined at 1.5×10^2 Hz:

$$X_C = \frac{1}{2\pi fC} = \frac{1}{2\pi\left(1.5 \times 10^2\right)\left(0.85 \times 10^{-3}\right)} = 1.25\ \Omega$$

$$X_L = 2\pi fL = 2\pi\left(1.5 \times 10^2\right)\left(1.0 \times 10^{-3}\right) = 0.94\ \Omega$$

Substitution of the values into Eq. (2-53) yields

$$Z_{eq} = \frac{(45)(0.94)(1.25)}{\sqrt{(45)^2(0.94-1.25)^2 + (0.94)^2\ (1.2)^2}}$$

$$= 3.8\ \Omega$$

Now the current that flows through E is

$$E = IZ; \qquad 11 = I(3.8); \qquad I = 2.9\ \text{A}$$

The current that flows through the parallel RLC circuit is a minimum at the resonant frequency and dramatically increases at frequencies even slightly removed from the resonant frequency.

IMPORTANT TERMS

Admittance
Alternating current
Amplitude
Battery
Capacitance
Capacitive reactance

Conductance
D'Arsonval meter
Direct current
Faraday
Filters
Frequency
Galvanometer
Half-power frequency
High-pass filter
Impedance
Inductance
Inductive reactance
Inductor
In-phase
Kirchoff's laws
Linear device
Low-pass filter
Moving coil meter
Null measurement
Null-point detector
Ohm's law
Out-of-phase
Parallel circuit
Parallel resonance
Peak-to-peak current
Peak-to-peak potential
Period
Phase angle
Potential
Potentiometer
Power
Reactance
Resistor
Resonance
Root mean square current
Root mean square potential
Series circuit
Series resonance
Shunt
Transducer
Transformer
Voltage divider
Voltmeter
Wheatstone bridge

PROBLEMS

DC Circuits

2-1 Determine the number of coulombs and the number of faradays that are required to plate 5.34 g of silver on an electrode.

2-2 How long must a 0.240A current flow in order to plate 0.525 g of copper from a copper (1I) nitrate solution?

2-3 Determine the current if 5.42×10^{18} electrons/s flow through a wire.

2-4 Calculate the work (joules) required to move a positive charge of 0.115 C from a negative electrode to a positive electrode if the potential drop between the electrodes is 1.37 V.

2-5 Calculate the potential across a 50.0– Ω resistor if a 0.675A current flows through the resistor.

2-6 What current would flow through a 125-Ω resistor that is connected across the terminals of a 1.5-V battery?

2-7 What is the conductance of an electric component through which a 3.47×10^{-3} A current flows when a potential difference of 10.0 V is applied across the component?

2-8 Determine the power that is dissipated in a 1.20×10^{3} Ω resistor when a potential of 2.49 V is applied across the resistor.

2-9 What current must flow through a 4.50-M Ω resistor in order to dissipate 25.0 W in the resistor?

2-10 If an applied potential of 15.0 V results in a current flow of 42.8 mA, determine the power dissipated in the circuit and the resistance of the circuit.

Circuit Diagrams

2-11 Prepare a circuit diagram that shows a 15.0 V battery in series with a 150 Ω resistor and a variable resistor.

2-12 Prepare a circuit diagram for a circuit that has a 22.5-V battery in series with a 500- Ω resistor and a switch-selectable resistor that could be 10, 20, or 50 Ω.

2-13 Prepare a circuit diagram for a circuit that contains 10-, 25-, 50-, and 100 Ω resistors in parallel with a 1.5 V battery.

2-14 A 1.00-mA constant-current source is to be prepared from a dc constant-potential source and a load resistor. If the anticipated resistance of the circuit in which the constant-current source is to be used can vary from 1 to 175 Ω, determine the minimum resistance of the load resistor and the potential of the constant-potential source if a maximum error of 1.0 percent can be tolerated in the output current.

Complex DC Circuits

2-15 Calculate the potential drop across the 50.0-Ω resistor in the circuit in Fig. P2-15.

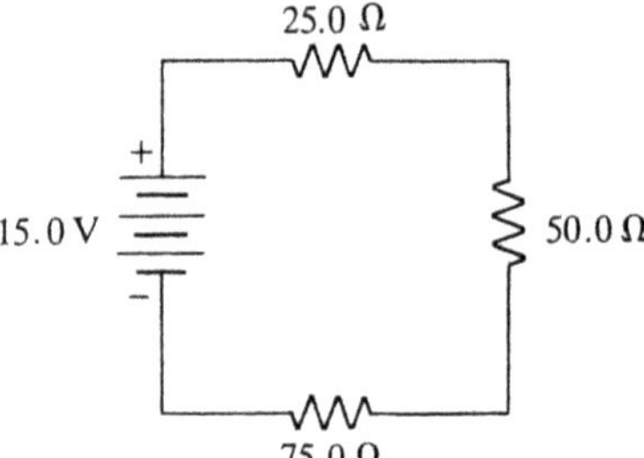

Figure P2-15

2-16 The variable resistor R_v in the circuit in Fig. P2-16 was adjusted until the current flowing through meter *I* was 50.0 mA. Determine the resistance of the adjusted resistor.

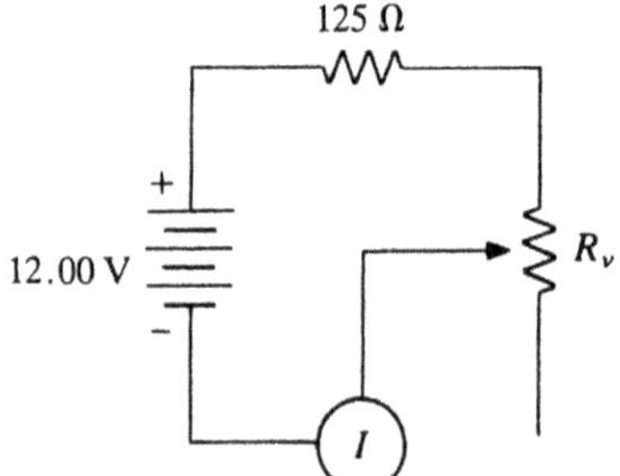

Figure P2-16

2-17 Determine the output potential for a voltage divider that is adjusted as shown in the circuit diagram of Fig. P2-17 if each d the five resistors is 20.0 Ω.

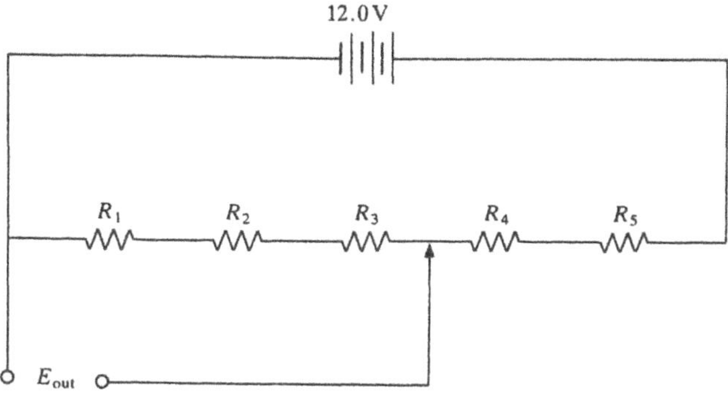

Figure P2-17

2-18 Determine the output potential for the voltage divider shown in Fig. P2-17 if R_1 is 5.0 Ω, R_2 is 10.0 Ω, R_3 is 25.0 Ω R_4 is 50.0 Ω, and R_5 is 100.0 Ω.

2-19 The variable resistor in the voltage divider illustrated in the circuit diagram in Fig. P2-19 was adjusted until the output potential was 15.0 V. If the maximum (full-scale) resistance of the resistor is 495 Ω, determine the output resistance R_{out} for the resistor.

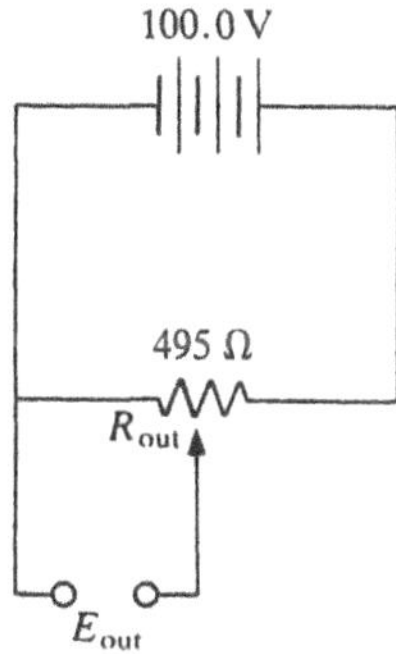

Figure P2-19

2-20 Determine I, I_1, I_2, and I_3 in the circuit in Fig. P2-20 if E is 25.0 V, R_1 is 25.0 Ω, R_2 is 42.3 Ω ,and R_3 is 62.5 Ω.

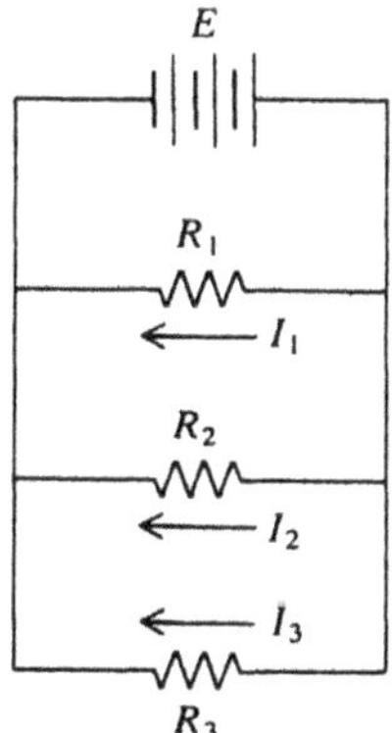

Figure P2-20

2-21 What potential must the battery supply in order to cause a total current of 0.477 A to flow through the circuit that is shown in the circuit diagram in Fig. P2-21 ?

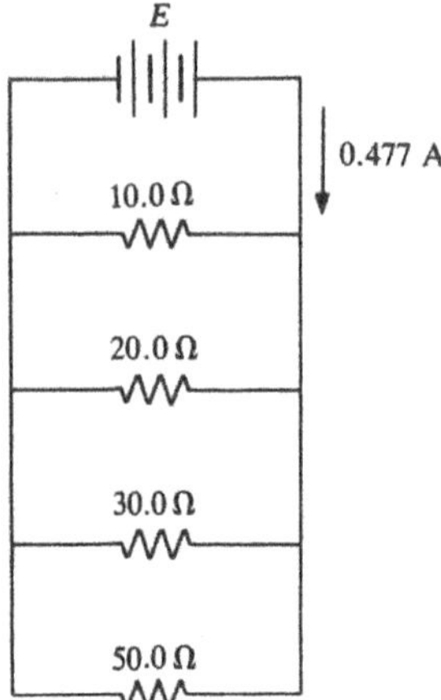

Figure P2-21

2-22 Determine the current that flows through each of the four resistors in the circuit in Fig. P2-22.

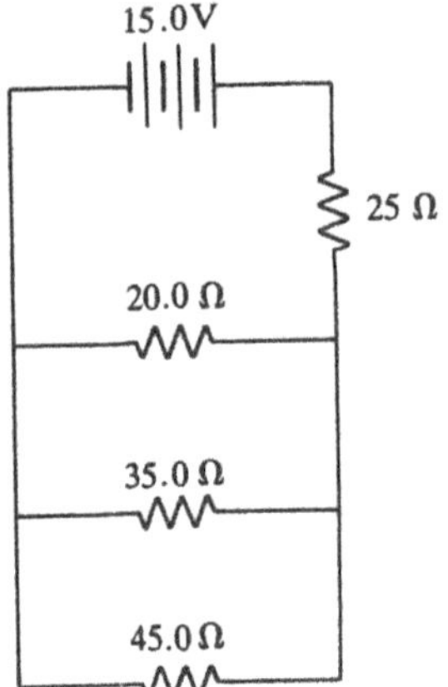

Figure P2-22

2-23 Determine the value of resistor R in the circuit shown in the circuit diagram in Fig. P2-23.

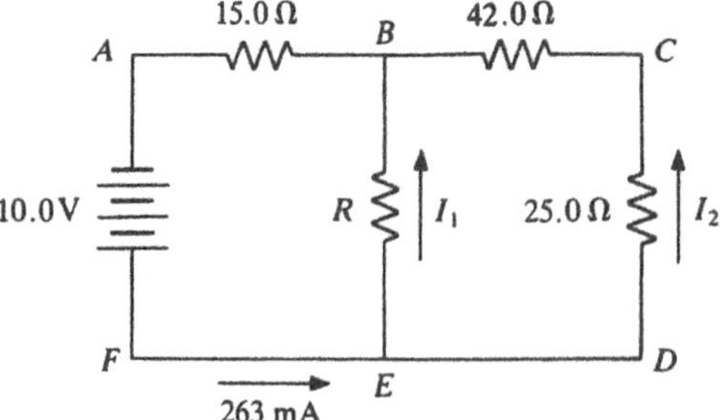

Figure P2-23

2-24 Determine the current I_2 that flows through the 20.0 Ω resistor in the circuit in Fig. P2-24.

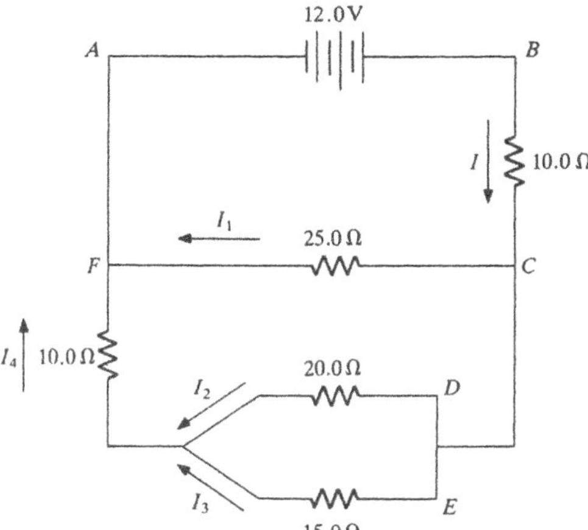

Figure P2-24

Current- and Potential-Measuring Devices

2-25 The meter resistance of a particular moving coil meter was 53.8 Ω and a full-scale deflection on the meter corresponded to a current flow through the meter of 1.20 mA. Draw a circuit diagram that shows the meter, the shunt resistor, and the input terminals. Determine the shunt resistance that would be necessary to make a full-scale deflection correspond to 20.0 mA.

2-26 The voltmeter shown in the circuit diagram in Fig. P2-26 was designed to have full-scale deflections of 0.100, 0.200, 0.500, and 1.000 V when the switch was set to place resistors R_1, R_2, R_3, and R_4 in the circuit. A full-scale deflection on the moving coil meter corresponded to 1.1 mA and the internal resistance R_M of the meter was 75.0 Ω. Determine the resistances of the four resistors.

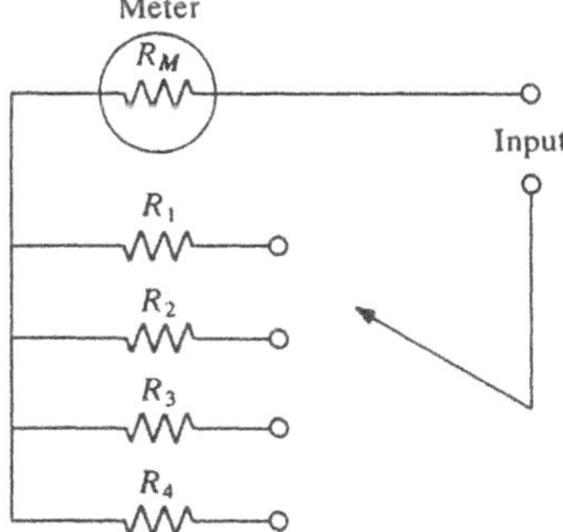

Figure P2-26

2-27 The potentiometer shown in Fig. 2-13 and described in Sample problem 2-6 was used to measure an unknown potential E_u. With the Weston cell in the circuit, no current flowed when the variable resistor was adjusted to 35.7 Ω. When the unknown potential was placed in the circuit the variable resistor had to be adjusted to 68.3 Ω in order to achieve zero current flow. Determine the value of the unknown potential.

AC Circuits

2-28 A sinusoidal alternating current has a current amplitude of 5.85 mA and a period of 2.50×10^{-3} s. Calculate the frequency, peak-to-peak current, and rms current.

2-29 The rms peak potential of a sinusoidal alternating potential is 50.0 mV and the frequency is 60.0 Hz. Determine the potential amplitude, the peak-to-peak potential, and the period of the wave.

2-30 The Wheatstone bridge that is shown in Fig. 2-15 was used to measure a resistance. Determine the resistance R_x when a null point is achieved on the meter. R_1 is 100.0 Ω, R_2 is 200.0 Ω, and R_v is73.6 Ω.

2-31 If the Wheatstone bridge that is shown in Fig. 2-15 is to be used to measure resistances within the range of 10.0 and 1250 Ω, what must be the minimal range of the variable resistor where R_1 is 500 Ω and R_2 is 750 Ω ?

Capacitance

2-32 Determine the charge held by a capacitor which has a capacitance of 0.500 mF when a potential difference of 15.0 V is applied across the capacitor.

2-33 What is the capacitance of a capacitor that holds a charge of 823 mC when a 12.0-V potential is applied across its leads?

RC Circuits

2-34 Calculate the ratio of the rms output potential E_o to the rms input potential E_i for the high-pass filter that is shown in Fig. 2-16 when the frequency is 0.1, 1.0, 5.0, 10, 20, 50, 100, 200, 500, 1000, and 5000 Hz . The capacitance of C is 1.0 μF and the resistance of R is 2.2×10^3 Ω. Prepare a plot of E_o / E_i as a function of the logarithm of the frequency.

2-35 Determine the half-power frequency and the time constant for the high-pass filter that was described in Prob. 2-34.

2-36 Determine the half-power frequency, the frequency at which the output potential is 5.0 percent of the input signal, and the frequency at which the output potential is 95.0 percent of the input potential for the circuit that is shown in Fig. 2-16 if C is 2.0 μF and R is 1.5×10^4 Ω.

2-37 Calculate the ratio of the rms output potential E_o to the rms input potential E_i for the low-pass filter that is shown in Fig. 2-17 if C and R have the same values that were used in Prob. 2-34. Use the same frequencies that were used in Prob. 2-34. Plot the ratio of E_o to E_i as a function of the logarithm of the frequency.

2-38 Calculate the capacitive reactance and the impedance of a circuit that contains a 50.0-μF capacitor in series with a 175- Ω resistor when the circuit is operating at 60.0 Hz.

2-39 Determine the impedance of the circuit in Fig. P2-39.

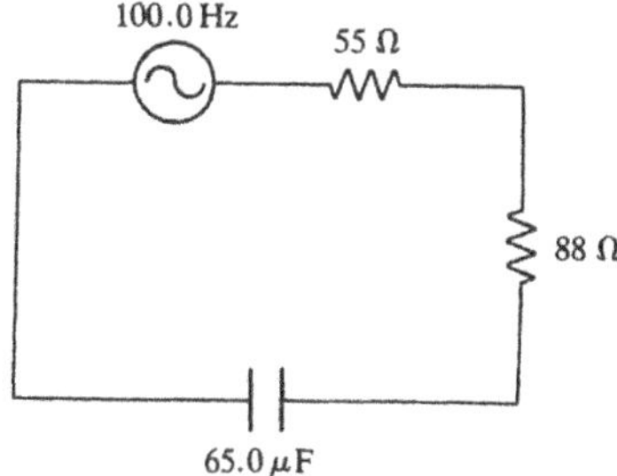

Figure P2-39

2-40 If the ac potential source in the circuit that was described in Prob. 2-39 operates at an rms potential of 20.0 V, calculate the rms current that flows through the circuit.

Transformers

2-41 The primary coil of an iron-core transformer is connected to a 120-V ac source. Determine the ratio of the number of turns of wire in the secondary coil to the number of turns in the primary coil if an output potential of 15 V is desired.

2-42 The primary coil of an iron-core transformer contains 1250 turns of wire and is connected to a 115V ac source. Output potentials of 5.0, 10, 15, and 20 V are desired from the transformer. Determine the number of turns of wire in the secondary coil that is required to yield the output potential of 20 V. Determine the number of turns that are required in the secondary coil to yield the other required potentials.

RL Circuits

2-43 Use frequencies of 50, 100, 200, 500, 1000, 2000, 5000, 10,000, and 100,000 Hz to calculate the ratio of the output potential to the input potential for the low-pass filter that is shown in Fig. 2-19. Assume that L is 0.500 mH and R is 2.50 Ω. Prepare a plot of the ratio as a function of the logarithm of the frequency.

2-44 Determine the half-power frequency and the time constant for the filter that is described in Prob. 2-43.

2-45 Determine the inductive reactance of a 0.475 mH inductor when the inductor is in a circuit that is operating at 60.0 Hz.

2-46 Determine the impedance of the circuit shown in Fig. P2-46 when the frequency of the alternating current is 100 and 10,000 Hz. Take note of the increase in inductive reactance with increased frequency and of the increased contribution of the inductive reactance to the impedance at the higher frequency.

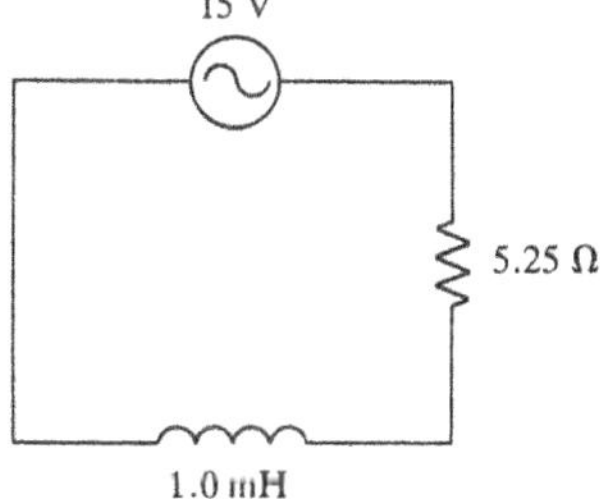

Figure P2-46

RLC Circuits

2-47 Determine the equivalent impedance of the circuit in Fig. P2-47 which is operated at a frequency of 1000 Hz.

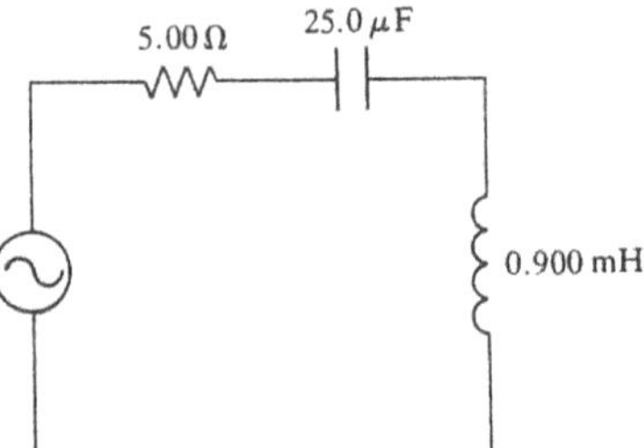

Figure P2-47

2-48 Determine the rms current that flows through the circuit that is shown in Prob. 2-47 if the rms potential is 15.0 V.

2-49 Determine I, I_1, I_2, and I_3 in the circuit shown in Fig. P2-49 when E is a potential source that delivers 10.0 V at 250 Hz. R is 50.0 Ω, L is 1.00 mH, and C is 35.0 μF.

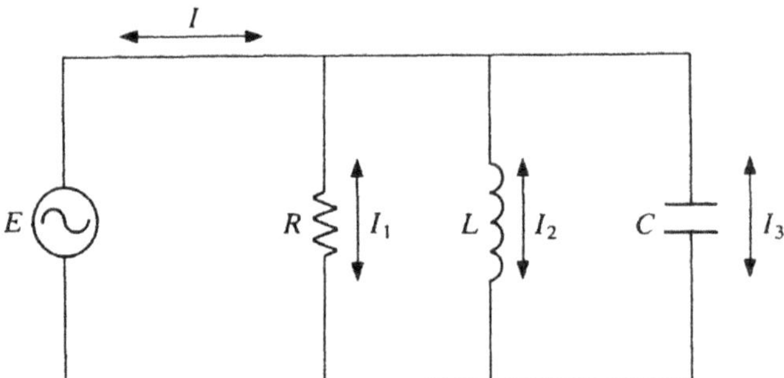

Figure P2-49

2-50 Derive the equation [Eq.(2-49)] for the equivalent impedance of a parallel resistance and capacitance.

2-51 Determine the resonant frequency for the series circuit that is shown in Fig. 2-20 when L is 0.500 mH, C is 25 μF, and R is 75 Ω. Determine the current that flows through the circuit at the resonant frequency when the rms potential of E is 10.0 V.

2-52 Calculate the potential drop at the resonant frequency across the inductor in the circuit that is described in Prob. 2-51.

2-53 Calculate the resonant frequency for the circuit that is shown in Fig. 2-22 when R is 95 Ω, C is 0.450 mF, and L is 0.75 mH. Determine the equivalent impedance of the circuit at the resonant frequency at 2.5×10^2 Hz and at 2.9×10^2 Hz.

2-54 For the circuit that is described in Prob. 2-53, calculate the current that flows through the resistor, capacitor, and inductor at the resonant frequency when the rms applied potential from the ac power supply is 78 V.

REFERENCE

1. Brophy, J. J.: *Basic Electronics for Scientists,* 3d ed., McGraw-Hill, New York, 1977.

BIBLIOGRAPHY

Benedict, R. R.: *Electronics for Scientists and Engineeers,* 2d ed., Prentice-Hall, Englewood Cliffs, New Jersey, 1976.

Malmstadt, H. V., C. G. Enke, and S. R. Crouch: *Electronics and Instrumentation for Scientists*, Benjamin/Cummings, Menlo Park, California, 1981.

Vassos, B. H., and G. W. Ewing: *Analog and Digital Electronics for Scientists,* 2d ed., Wiley, New York, 1980.

CHAPTER

THREE

ELECTRONIC CIRCUITS

In Chapter 2 the use of resistors, capacitors, and inductors in electric circuits was described. The current flow through those devices linearly increases with increasing potential, i.e., the current is directly proportional to the potential. Devices which behave in that manner are linear devices or components. Devices in which the current is not directly proportional to the applied potential are *nonlinear* devices or components. *Electronic circuits* are those electric circuits which contain one or more nonlinear devices.

Kirchoff's laws can be applied to nonlinear circuits; however, the mathematical calculations are difficult. Generally, graphical techniques are used to solve problems relating to circuits with nonlinear devices. In Chapter 3 nonlinear devices and their applications are described. The descriptions primarily are qualitative. The intent of the chapter is to provide an understanding of the way in which various constructed circuits function. Because the purpose is not to describe ways to design circuits, a description of the various specific components which can be obtained from each of the manufacturers of nonlinear devices is excluded.

SEMICONDUCTORS

Metals are good conductors of electricity while nonmetals are insulators, i.e., poor conductors. The difference in resistance between the best metallic conductor and the best nonmetallic insulator is a factor of about 10^{30}. As learned in introductory chemistry, the metals are those elements which are found on the left of the periodic table. The nonmetals are found on the right of the periodic table. The elements which are found adjacent to the line that separates metals from nonmetals are metalloids. Metalloids have a resistance which is in the approximate range of 10^4 to 10^7 times that of the metallic conductors.

The metalloids include B, Si, Ge, As, Sb, Te, Po, and At. Materials that have an ability to conduct electricity that is intermediate between that of the conductors and the insulators are *semiconductors*.

It is not necessary for a semiconductor to be a pure metalloid although the metalloids are semiconductors. Semiconductors are advantageously used in electric circuits. Of the semiconductors that are used in electric circuits, silicon, germanium, a few intermetallic compounds such as silicon carbide, and several organic compounds are used most often. Semiconductors which are composed of a pure element or compound are *intrinsic semiconductors* whereas semiconductors to which a fixed amount of an impurity is added are *extrinsic semiconductors*.

The manner in which semiconductors conduct electricity is exemplified by the conductive mechanism of silicon and germanium. Conduction in other semiconductors occurs in a similar manner. Both silicon and germanium exist in the crystalline diamond structure. Each atom is tetrahedrally surrounded by four other atoms. A single pair of electrons is covalently shared between adjacent atoms. At low temperatures the electrons in the covalent bonds are not free to move and the crystal is an electric insulator. At higher temperatures (room temperature is sufficient), however, the increased thermal energy is sufficient to displace a few of the electrons from their bonds. If an electric potential is applied across the crystal, the freed electrons can travel toward the positive side of the crystal giving rise to a current

The place within the crystal from which the electron was removed is termed a *hole*. When an electron from a nearby bond jumps into and fills the hole, it leaves another hole. In that manner holes can migrate through the crystal. In the presence of an electric field, electrons migrate toward the positive end of the crystal while holes migrate toward the negative end. The process by which holes migrate through the crystal is a second method by which an electric current is carried through a semiconductor. The holes act like and can be treated like positive particles with a charge of the same magnitude as that of an electron. When a free electron combines with (or falls into) a hole, the process is *recombination* and the hole is eliminated. In semiconductors which consist of pure materials (intrinsic semiconductors), the number of free electrons is identical to the number of positive holes.

The conductivity of a pure silicon or germanium semiconductor can be increased considerably by adding a small known amount of a selected impurity. Usually about one atom of impurity for each million silicon or germanium atoms is adequate. The process by which the impurity is added is *doping* and the added substance is the *dopant*. The result is an extrinsic semiconductor. Both silicon and germanium have four valence electrons. The dopant is an element which can replace one of the atoms within the crystalline structure. The dopant is chosen so that it can fit into a hole that would have been occupied by a silicon or germanium atom.

If the dopant is an element, such as Al, B, Ga, or In, which has three valence electrons, then around each dopant atom each of the three valence-shell electrons combines with an electron from an adjacent silicon or germanium atom to form a covalent bond. The fourth silicon or germanium atom around the dopant has an electron but no available dopant electron with which to pair. The lack of the extra electron is essentially identical to the positive hole that is formed in an intrinsic semiconductor. The increased conductivity of the crystal in comparison to that of the comparable intrinsic semiconductor is due to migration of positive holes. The number of holes within the crystalline structure is determined by the number of dopant atoms that are added. A semiconductor which is formed by adding a dopant that has fewer valence electrons than those found in atoms of the major component of the semiconductor is a *p-type semiconductor*, because positive holes are formed within the crystal.

If a dopant, such as arsenic or antimony, which has five valence electrons is added to germanium or silicon, four of the electrons combine with electrons from adjacent silicon or germanium atoms. The fifth electron remains unpaired and is free to move throughout the crystal, resulting in increased conductivity. Because the result is similar to adding negatively charged free electrons to an intrinsic semiconductor, the device is an *n-type semiconductor*. Of course, the conductivity of both *p*-type and *n*-type semiconductors is also partially due to the creation of free electrons and positive holes by thermal agitation as occurs in intrinsic semiconductors.

DIODES

A diode is usually a cylindrical device, similar to a resistor in shape, which has a wire emerging from each end. Diodes preferentially allow current to flow through them in a single direction. Most diodes have a band around the cathodic end; i.e., around the negative end when the diode is forward-biased. Typically the resistance as measured in the forward direction through a diode is about 10 Ω, while in the reverse direction the resistance is about 100 MΩ.

A potentially faulty diode in a circuit can be easily checked. First, a meter is used to measure the resistance through the diode in the allowed direction of current flow. The meter's leads are reversed and the resistance in the reversed direction is measured. The ratio of the resistance in the reversed direction to that in the forward direction should be at least 25. If the ratio is less than 25, the diode probably should be replaced. Sometimes it is necessary to remove the diode from the circuit prior to making the measurement since the meter will measure the lowest resistance through the circuit, which might not be the resistance through the diode.

In circuit diagrams a diode is represented as shown in Fig. 3-1, where the arrowhead points in the allowed direction of flow of positive current. Electrons preferentially flow in the direction opposite to the direction in which the arrowhead points. When a diode is connected in a circuit so that current can easily flow through the device, it is *forward-biased*, and when the leads to the diode are reversed, it is *reverse-biased* (Fig. 3-1).

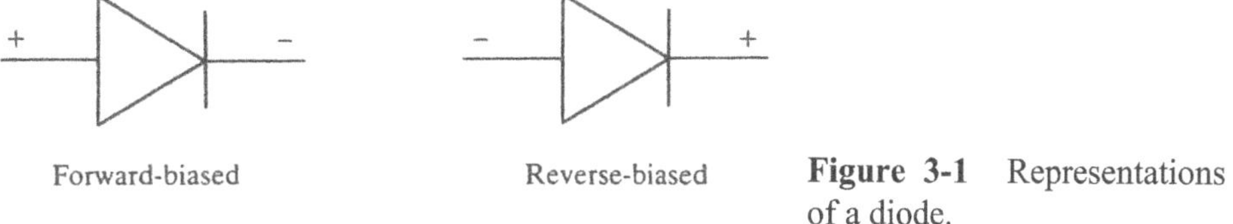

Figure 3-1 Representations of a diode.

A crystalline diode consists of a *p*-type semiconductor in contact with an *n*-type semiconductor (Fig. 3-2). Usually a crystalline diode is made from a single crystal (called a *wafer* or *chip*) of a semiconductor (often silicon or germanium). One end of the crystal is doped with an element that causes it to become a *p*-type semiconductor and the other end is doped with an element that causes it to become an *n*-type semiconductor.

When the diode is forward-biased (Fig. 3-3), free electrons from the *n*-type region and positive holes from the *p*-type region flow toward each other. When a

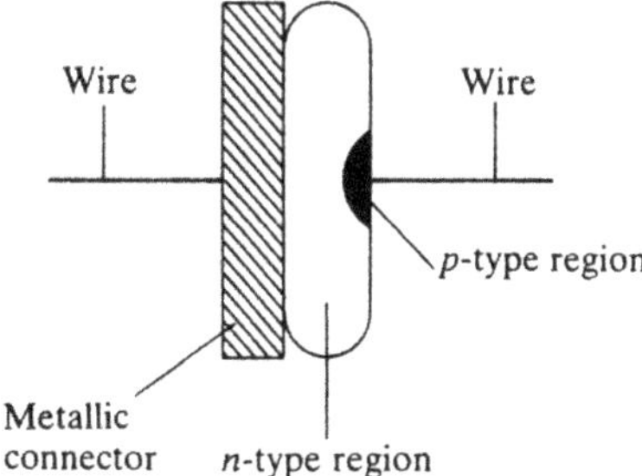

Figure 3-2 A sketch of a single-crystal diode.

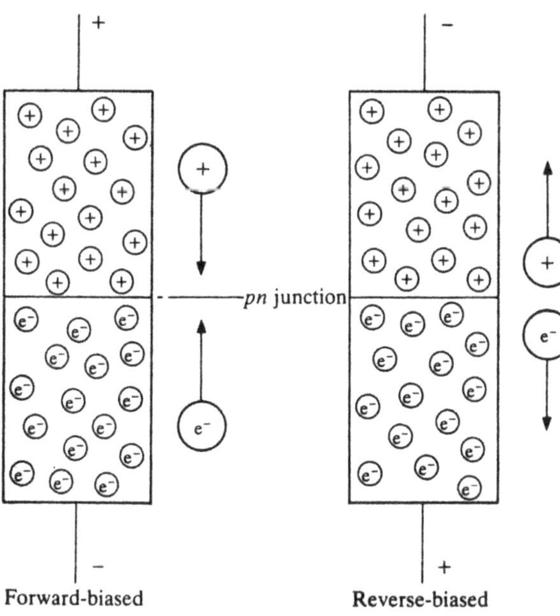

Figure 3-3 A representation of the flow of charge through forward-biased and reverse-biased diodes. The arrows to the right of each diode represent the direction of flow. Annihilation of the free electrons and the positive holes occurs at the *pn* junction in the forward-biased diode.

free electron contacts a positive hole at the junction (called a *pn junction*) between the *p*-type region and the *n*-type region, the two combine to annihilate each other. The movement of charge through the diode is an electric current.

When the diode is reverse-biased, the holes and electrons flow away from the *pn* junction and away from each other. The separation in charge within the diode rapidly causes a potential difference across the diode which is equal to but opposite in sign to the applied potential. The result is nearly zero current flow and a high resistance through the device. Because diodes only allow current to flow in one direction, they are often used as *current rectifiers*. A rectifier is a device that changes an alternating current or potential to a direct current. The diode rectifier removes the half of the ac current which tries to flow backward through the diode. It removes that portion of each ac cycle during which the diode is reverse-biased. Rectifiers use one or more diodes to perform the rectification. The diagrams of two circuits that can be used for rectification are shown in Fig. 3-4.

The single diode in the half-wave rectifier (Fig. 3-4*a*) allows only that half of the ac signal to pass through it which is flowing in the direction that corresponds to forward-biasing in the diode. Consequently, the output dc

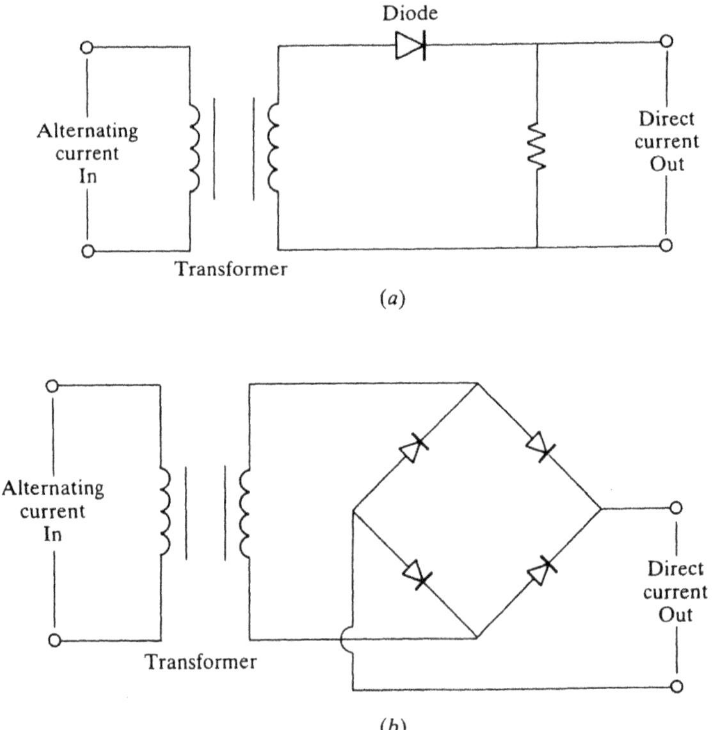

Figure 3-4 Circuit diagrams for (*a*) a half-wave rectifier and (*b*) a full-wave rectifier. The diodes perform the rectification.

potential corresponds to only the top (or bottom) half of the ac sine wave potential from the transformer. The full-wave rectifier (Fig. 3-4*b*) provides two alternate paths for current flow. When the alternating current from the transformer flows in one direction, the first of the two paths is forward-biased and provides rectification as in the half-wave rectifier. When the alternating current flows in the opposite direction, the second path is forward-biased, thereby rectifying the second half of the alternating current as well as the first. The output from a full-wave rectifier consists of the top of the ac sinusoidal wave from the transformer during half of each cycle and the inverse of the bottom of the ac sinusoidal wave during the alternate half-cycle, i.e., the rectified potential (or current) has a ⌒⌒⌒ shape.

Zener Diodes

If the potential drop across a reverse-biased diode is gradually increased, eventually sufficient potential is applied to break some covalent bonds within the crystal and to force a current to flow backward through the diode. The potential at which the current starts to flow is the *breakdown potential* or *breakdown voltage*. A generalized sketch of the current that flows through a silicon diode is shown as a function of the applied potential in Fig. 3-5. The breakdown potential of a germanium diode is not as well defined as that for a silicon diode, i.e., the current does not fall off as rapidly as shown in Fig. 3-5. Because the current in the forward-biased diode increases logarithmically with applied potential, diodes are nonlinear devices. In the reverse-biased diode the small current that flows

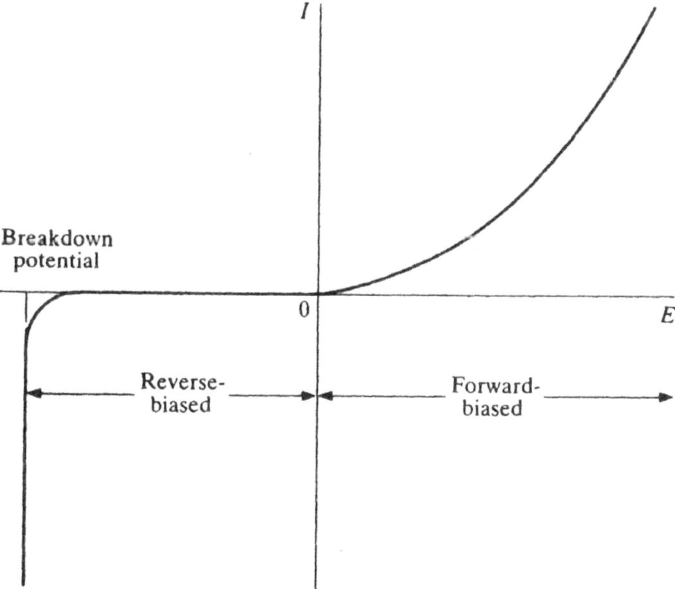

Figure 3.5 A sketch of the variation in current as a function of potential for a silicon diode.

prior to the breakdown potential is typically in the microampere region. The logarithmic variation in current with applied potential in the forward-biased diode is advantageously used in some circuits to logarithmically multiply and divide electric quantities.

Advantage can be taken of the breakdown potential (also called the *zener breakdown voltage*) of a reverse-biased diode for use as a low-voltage reference source or for voltage regulation. A diode which is designed to be used in the reverse-biased mode is a *zener diode*. Zener diodes are given the symbol ⏄ in circuit diagrams. At low potentials nearly zero current passes through a zener diode. When the breakdown potential is achieved, current flows through the diode. Because a large current can flow at the breakdown potential (Fig. 3-5), the potential drop across the diode becomes constant.

The breakdown potential of a zener diode and the controlled potential in a circuit that contains a zener diode can be regulated by control of the thickness and number of impurities in the junction layer between the *p* and *n* portions of the diode. In some ac circuits a pair of zener diodes which are connected in a head-to-head arrangement is used to protect the remainder of the circuit from unduly large alternating potentials. During one half-cycle of the alternating potential one diode is forward-biased and allows current to pass through it, but

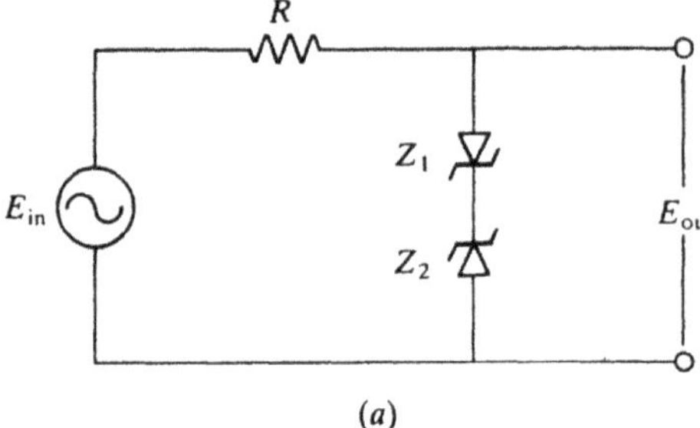

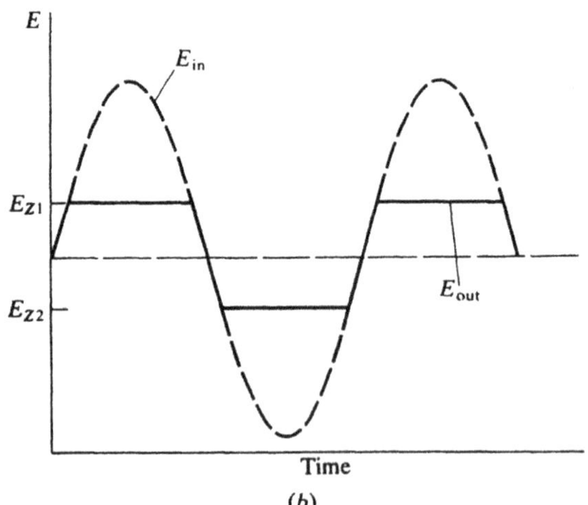

Figure 3-6 (*a*) A circuit diagram of a clipping circuit that contains two head-to-head zener diodes. (*b*) A plot of the input potential E_{in} and output potential E_{out} for the circuit shown in (*a*). E_{Z1} and E_{Z2} are the respective breakdown potentials of zener diodes Z_1 and Z_2.

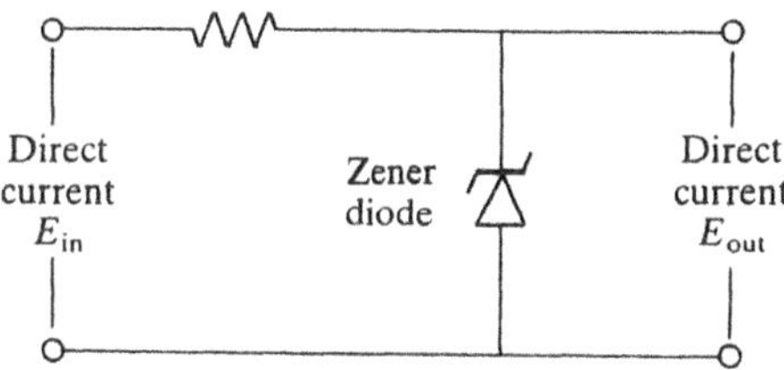

Figure 3-7 A circuit diagram of a voltage regulator that relies upon a zener diode.

the other diode is reverse-biased and regulates the potential at the breakdown potential. During the second half-cycle, the two diodes exchange functions. A circuit of that type is a *clipping circuit* (Fig. 3-6).

Potential regulation is accomplished with a *voltage regulator*. Voltage regulators usually either rely upon zener diodes or upon operational amplifiers to perform the regulation. Voltage regulators that contain zener diodes regulate the potential at the breakdown potential of the diode. A circuit diagram of a voltage regulator that contains a zener diode is shown in Fig. 3-7. Voltage regulators that rely upon operational amplifiers are described in the next chapter.

Photodiodes and LEDs

In some cases diodes can be used either as detectors for light or to emit light upon application of a proper potential. In *photodiodes* radiation which strikes the diode and penetrates into the depletion layer in a reverse-biased diode provides the energy necessary to change an electron in a bond to a free electron. The free electron contributes to the current through the diode. Essentially, the diode is the active element in a transducer which changes electromagnetic radiation into an electric current.

A *light-emitting diode* (LED) operates in the opposite manner. In LEDs, application of the proper, forward biased potential (often about 1.7 V) to the device causes a free electron to fall into a hole while simultaneously losing energy in the form of emitted light. Among other things, such devices can be used as panel lights on instruments, in numerical readout devices, and as radiative sources. Only certain semiconductor materials, e.g., gallium arsenide which emits red radiation at 660 nm, can be used in LEDs. A more thorough description is included in the section on Decimal Displays in Chapter 4.

TRANSISTORS

Bipolar Transistors

Transistors are constructed from extrinsic semiconductors. Each transistor has three regions of alternating *p*- and *n*-type semiconductor. Transistors can be of either the *pnp* type or the *npn* type. The order of the letters represents the order of the type of semiconductor within the transistor. Transistors are generally constructed from single crystals (chips) of silicon which are doped in different regions to provide the proper electric properties. Transistors also have been

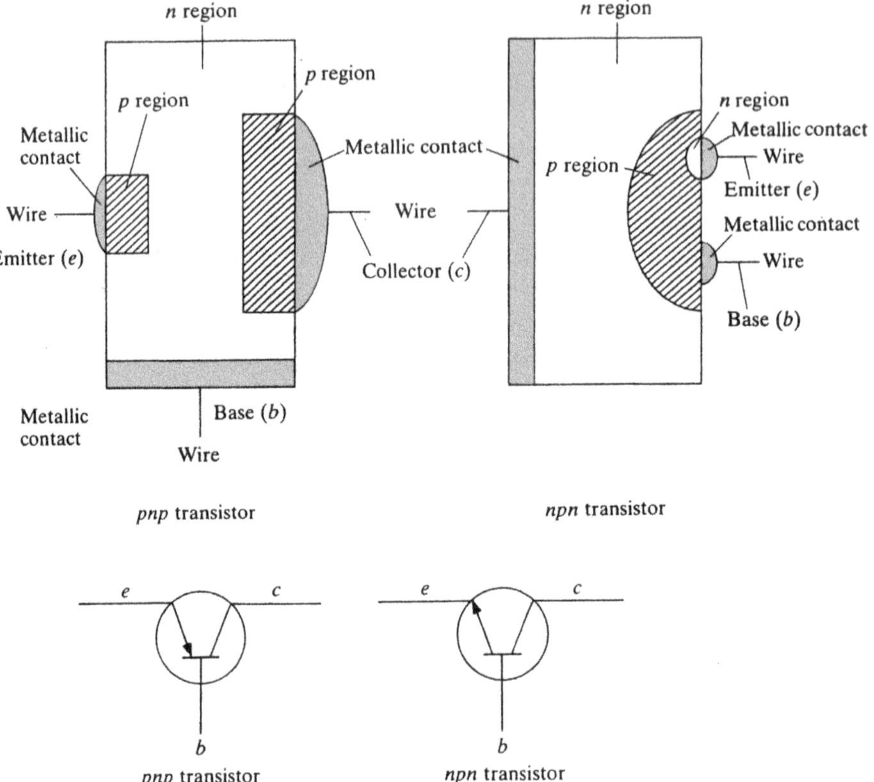

Figure 3-8 Sketches and circuit diagram symbols of two types of transistors.

constructed from germanium, gallium arsenide, and several other compounds. A sketch of two forms of *bipolar* transistors and their symbols in circuit diagrams is shown in Fig. 3-8. The arrowhead indicates the direction of positive current flow. A separate wire makes contact with each of the three regions of a bipolar transistor as shown in the figure. Transistors can take any of several physical shapes, but always have three emerging wires.

The two outer regions of a transistor consist of the same type (p or n) of semiconductor. The smaller of the two outer regions is the *emitter e* and the large region is the *collector c*. The central region of opposite type (n or p) to the outer regions is the *base b*. A forward bias is maintained between the emitter and the base, while a reverse bias is maintained between the collector and the base. With a few important exceptions, a transistor can be thought of as back-to-back diodes.

Transistors are *active devices*. They accept energy in the form of a potential or current, and amplify that energy. Consequently, a major use of transistors is in circuits which are designed to amplify electric signals. A description of the operation of an *npn* transistor will serve as an example of the manner in which

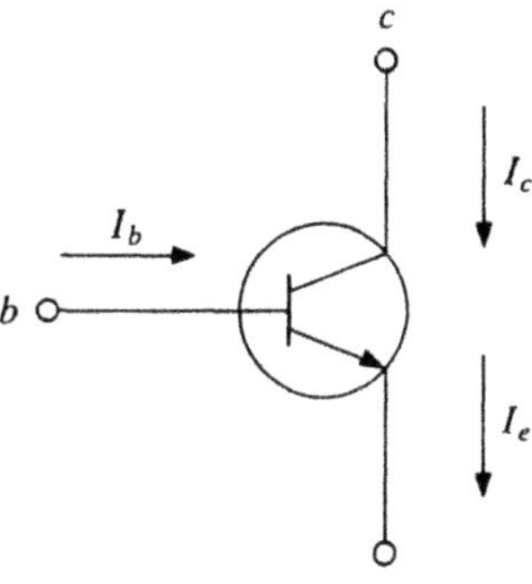

Figure 3-9 The current flow through an *npn* transistor. I_b, base current; I_c, collector current; I_e, emitter current. I_c is much larger than I_b and $I_e = I_b + I_c$.

amplification takes place in transistors. A *pnp* transistor behaves in an identical manner except that the potentials are reversed on the leads and the carriers (free electrons or holes) of the majority of the current through the device are opposite to those used in the *npn* transistor.

Because the junction between the emitter and the base is forward-biased, electrons easily flow across the junction from the emitter to the base. Upon reaching the base, the electrons can be annihilated by falling into positive holes, can flow into the collector and out of the transistor through the collector lead, or can flow out of the transistor through the base lead. In order to result in amplification, transistors are designed so that a relatively small amount of annihilation takes place and the majority of the electrons flow into the collector (Fig. 3-9).

Annihilation is kept low in a transistor by minimizing the extent to which the base is doped. In a diode the two regions are approximately equally doped and the current through the diode is nearly equally controlled by the flow of free electrons in one direction and of holes in the opposite direction. In a transistor the emitter is more highly doped than the base and the current is primarily carried by the type of carrier (electrons or holes) that has been doped into the emitter.

The electrons or holes which carry the major portion of the current are the *majority carriers*. In *npn* transistors electrons are the majority carriers. In *pnp* transistors positive holes are the majority carriers.

The portion of the electrons that flow from the base to the collector is maximized by making the base layer as thin as possible. Because the base is thin, the electrons that enter the base from the emitter can easily pass into the collector and out of the transistor through the collector lead. A relatively small, but usually constant, proportion (typically 1 percent) of the electrons which enter the base region depart the base through the base lead. The remainder of the electrons (or current) leave the base through the collector and the collector lead. The collector current I_c is usually directly proportional to, or nearly directly proportional to, the base current I_b:

$$I_c = \beta I_b \tag{3-1}$$

The signal which is to be amplified is attached to the base terminal and defines I_b in Eq. (3-1). When I_b increases, I_c increases proportionately. Consequently, I_c corresponds to an amplification of the base signal I_b. The proportionality between the collector current and the base current is the *transistor effect*. The proportionality constant β in Eq. (3-1) is the *current gain* or the *beta*. *Typically* the current gain is between 50 and 200. Input signals are usually in the microampere range with corresponding output signals in the milliampere range.

Minority carriers in the base region that are annihilated by combination with majority carriers or which exit (or leak from) the transistor through the electric contact must be replaced from the input signal that is attached to the base. It is important to keep loss of the minority carriers to a minimum in order to keep the input impedance to the device large, i.e., in order to prevent the necessity of drawing a significant current from the input signal.

Three leads come from a transistor. In nearly all transistor circuits, two of the leads are used as input for a potential signal and two are used as output. Consequently one of the leads is shared between the input and the output. The shared lead is the *common* lead. Because three leads are available, three possibilities exist for the common lead (Fig. 3-10). A transistor which has a common base lead is in the *common-base* or *grounded-base* mode. In most cases a *ground* in an electric circuit is a point to which several leads are connected and with respect to which potential measurements are made. That is the meaning of ground in this instance.

Input and output potential measurements are made with respect to the common lead. In each case one of the two input leads is connected to the base. In the common-base mode, the input signal is applied to the emitter and the output appears at the collector. The common-base mode generally is used in high-frequency ac applications. The ratio of current flowing from the output of the transistor to current flowing into the transistor is the gain. The gain for transistors that are in the common-base mode is nearly equal to but less than one.

In the common-collector configuration the input potential is applied to the base and the output is at the emitter. In that orientation the ratio of the current

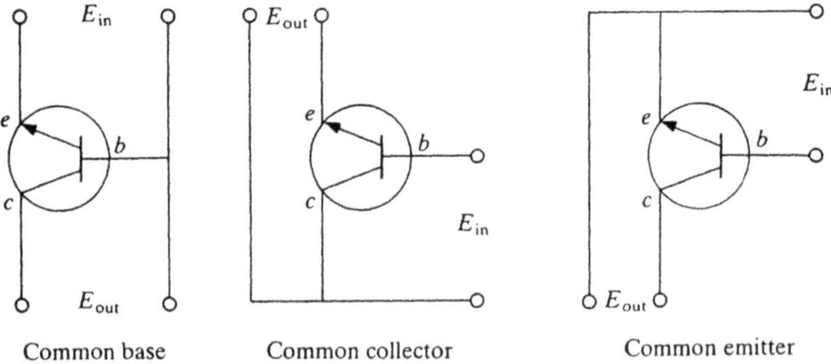

Figure 3-10 The three possible ways in which a transistor can be connected in a circuit. A *pnp* transistor has the same three possibilities that are shown for an *npn* transistor.

flow from the output to the current flow into the input is relatively large and is nearly equal to beta. A slight potential loss occurs in the common-collector configuration. It is primarily used to deliver large currents.

The common-emitter mode is used to control the current between the emitter and the collector. The input signal is applied to the base and the output appears at the collector. The output current is a factor of beta larger than the input current [Eq. (3-1)]. In many cases integrated circuits which contain transistors as well as other components are more convenient to use and consequently are used in place of individual transistors in electric circuits. Integrated circuits are described in Chapter 4.

Field-Effect Transistors

Field-effect transistors (FETs) are used to control current through a circuit. A field-effect transistor has three electric connections to it as in a bipolar transistor, but the operating principle of an FET differs from that of the bipolar transistor. A *junction field-effect transistor* (JFET) consists of a bar of doped silicon which has an electric contact at each end and a region of oppositely doped material in the center of the bar (Fig. 3-11).

The doped silicon at one end of the FET is the *source*. The opposite end is the *drain* and the central oppositely doped region is the *gate*. The source and drain can consist of either *p*- or *n*-doped regions. The gate is oppositely doped. The *channel* is the region of the FET near the gate that connects the source and the drain. A *p*-channel FET has a *p*-doped source, drain, and channel. An *n*-channel FET has an *n*-doped source, drain, and channel. The circuit diagram symbols for the two forms of JFETs are shown in Fig. 3-11.

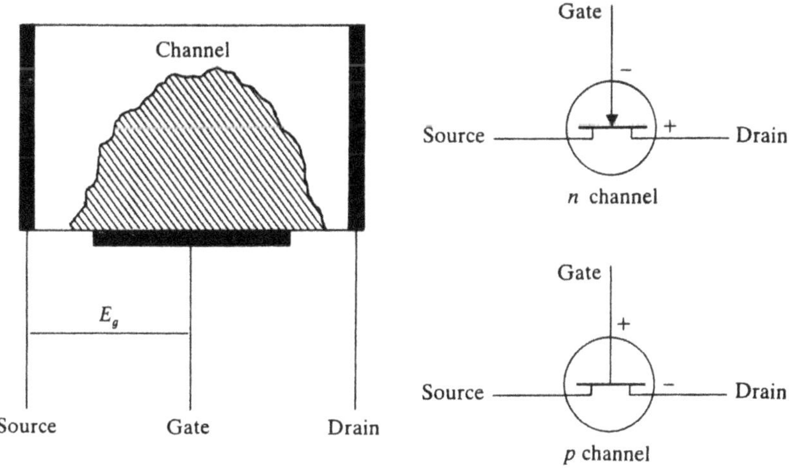

Figure 3-11 A sketch of a junction field-effect transistor and the circuit diagram symbols.

The operating principles of an *n*-channel FET will be described. A *p*-channel FET operates similarly, although all of the applied potentials are of opposite sign. When no potential is applied between the source and the gate, the resistance between the source and the drain is small. Consequently, application of a potential between the source and the drain leads to a relatively large current flow. In an *n*-channel FET the drain is made positive with respect to the source and electrons travel from the source through the channel to the drain. While some FETs are constructed so that the connections to the source and to the drain can be interchanged, other FETs can only be connected to the circuit so that the current flows through the device in one direction.

Application of a negative potential (with respect to the source) to the gate of the *n*-channel FET results in repulsion of the electrons from the gate as they pass through the channel. Essentially the result is to narrow the channel, thereby increasing electric resistance through the FET and decreasing current flow. The electric field caused by the applied potential controls the current through the device. Often the gate potential in FETs is sufficiently large to prevent flow of any current between the source and the drain.

Current through an FET varies with the potential difference between the gate and the source. As the gate potential in an *n*-channel FET becomes more negative, the current decreases, as shown in Fig. 3-12. A plot of current through an FET as a function of gate potential is the *transfer characteristic* of the FET. Gate potentials that are more negative (*n*-channel FET) or more positive (*p*-channel FET) than the *pinchoff potential* (the potential at which no current flows) have no effect upon the current through the FET until the breakdown potential across the source-gate junction is reached. Because transfer characteristics for FETs (and for bipolar transistors) are temperature-dependent, some method of temperature control of circuits that contain FETs is often used.

Field-effect transistors can be used in several ways in electric circuits. FETs are often used in circuits of amplifiers. The input potential is applied to the gate while the output potential is obtained at either the source or the drain, depending

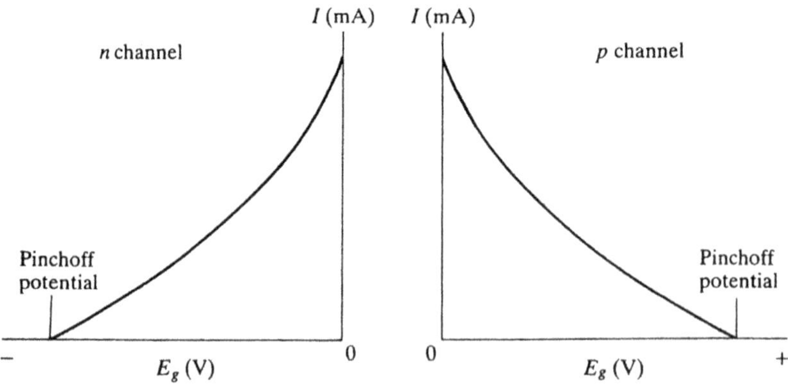

Figure 3-12 Sketches of typical transfer characteristics for field-effect transistors.

upon the circuit. As is apparent from the preceding description, FETs behave as variable resistors which are controlled by the gate potential. That can be used to advantage in circuits, such as Wheatstone bridges, which require variable resistors and in circuits which require voltage-controlled switches. With no potential applied to the gate, the "switch" is open and current readily flows from the source to the drain. With the pinchoff potential (or more) applied to the gate, the switch is closed. A *multiplexer* is a series of several FET switches in a single integrated circuit. Recently FETs and insulated gate field effect transistors (next section) have found use as electroanalytical probes.

Insulated-Gate Field-Effect Transistors

Insulated-gate field-effect transistors (IGFETs) are similar to FETs. Unlike the gate in an FET, the gate in an IGFET is insulated from the remainder of the transistor. The most common insulation is a metal-oxide layer. In silicon devices the insulation consists of a thin layer of silicon dioxide. An IGFET in which the layer of insulation consists of a metal oxide is a *metal-oxide semiconductor field-effect transistor* (MOSFET). Diagrams of two forms of *n*-channel MOSFETs and the corresponding symbols in circuit diagrams are shown in Fig. 3-13. MOSFETs are termed either *n* or *p* channel depending upon the nature of the channel that connects the source and the drain. The *p*-channel MOSFETs have the same symbols as the *n*-channel MOSFETs except that the direction of the arrow and the sign of the potentials are reversed. In addition to

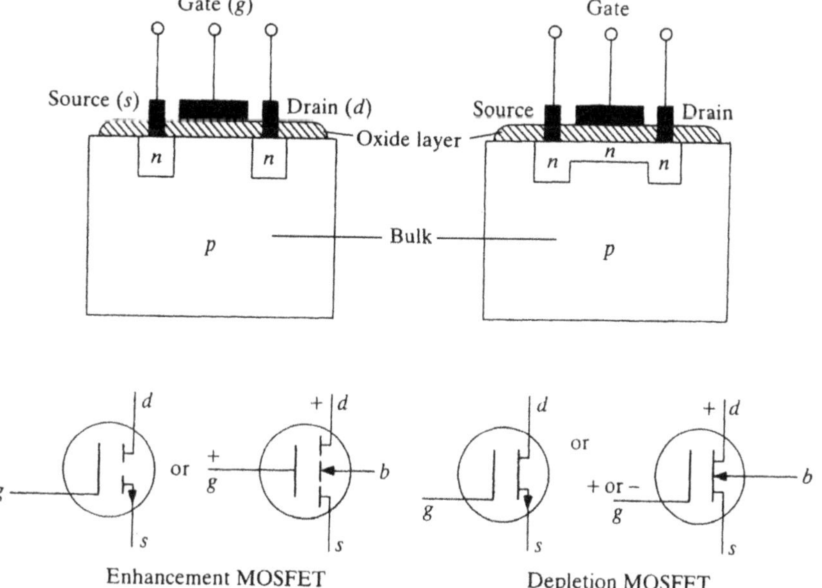

Figure 3-13 Sketches and circuit diagram symbols of two forms of n-channel MOSFETs.

electric leads to the source, gate, and drain, MOSFETs have a fourth lead which is labeled *b* in Fig. 3-13 and is connected to the bulk, slightly doped material in which the transistor is embedded. Generally the bulk lead is connected either to ground or to the source by an internal connection so that only three leads are visible.

As was done with FETs, *n*-channel MOSFETs will be described. The design and operation of *p*-channel MOSFETs is completely analogous to that of *n*-channel MOSFETs when the applied potentials are of opposite polarity. An *enhancement n-channel MOSFET* contains an *n*-doped source and an *n*-doped drain that are formed within a lightly *p*-doped substrate which is labeled, *b* (for bulk) in the symbol for the MOSFET. The gate is insulated from the remainder of the MOSFET by a thin layer of a metal oxide (usually silicon dioxide). The drain potential is made positive with respect to the source. With no potential applied to the gate, the source and drain are completely separated, and the resistance between the source and the drain is large. Essentially zero current flows through the MOSFET when the gate potential with respect to the source is zero.

Application of a positive potential to the gate results in an electrostatic field that repels the majority *p* carriers in the lightly doped bulk region, thereby creating an *n*-region channel that can connect the two *n* regions of the source and the drain. As the positive gate potential is increased, the size of the channel increases; the resistance between the source and drain decreases; and current flow through the MOSFET increases. The gate acts as a valve that controls current flow through the MOSFET. Enhancement MOSFETs are often used in amplificatory circuits. In simple circuits the input potential is applied between the gate and drain and the output potential is obtained across a resistor that is in series with a power supply that is connected between the source and the drain.

A depletion *n-channel MOSFET* (Fig. 3-13) contains connected *n*-doped source, drain, and channel regions embedded in lightly *p*-doped bulk. As in the enhancement MOSFET, the gate is insulated from the channel and the remainder of the MOSFET with a thin layer of silicon dioxide. Application of a negative potential (with respect to the source) to the gate results in an increased resistance and a decreased current between the source and the positively charged drain as in FETs. The pinchoff potential corresponds to the gate potential at which current ceases to flow between the source and the drain.

In *n*-channel FETs the gate potential could not be increased above 0 V with respect to the source without changing the bias at the *np* junction between the source and the gate. Because the gate in an *n*-channel MOSFET is insulated from the remainder of the device, the gate potential can be increased above 0 V without fear of having the source and gate behaving as a forward-biased diode. Increasing the gate potential above 0 V results in a further increase in current flow through the MOSFET, as illustrated in the transfer coefficients that are shown in Fig. 3-14. If the gate of a MOSFET is operated at both positive and negative potentials, the MOSFET is termed a *depletion-enhancement MOSFET* because it acts as a depletion MOSFET when the gate potential is of one polarity and as an enhancement MOSFET when the potential is of the opposite polarity.

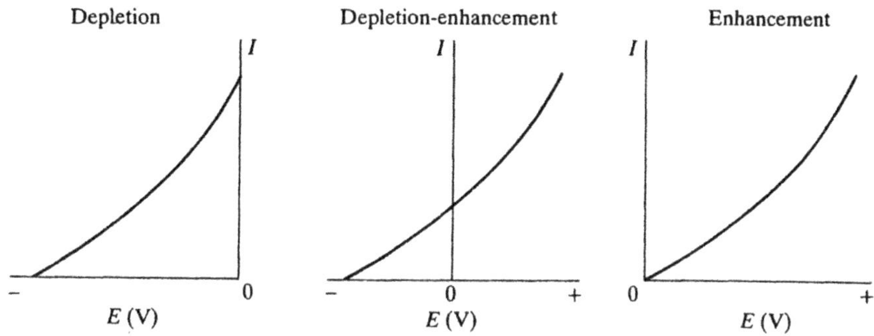

Figure 3-14 Sketches of the transfer characteristics of n-channel MOSFETs.

Dual-gate MOSFETs contain two gates along the channel. They are sometimes used in circuits that are operated at high frequencies because their performance in those circuits is generally better than that of single-gate MOSFETs.

PRACTICAL CIRCUITS

The electric components that have been described in Chapters 2 and 3 are found in many circuits that analytical chemists use daily. A few of the circuits have already been described. The circuits that are described in the following sections are usually simple versions of popular circuits. It should be understood that other more-complicated versions of the circuits are in common use. The circuits that are described were chosen to illustrate the operation of various electric components. These or similar circuits are used in most analytical instruments.

Oscillators

Oscillators are used to convert a direct current or potential to an alternating current or potential. Generally the term oscillator is used to describe a device which changes direct current to a sinusoidal alternating current. If some ac waveshape other than sinusoidal is generated, the device is usually termed a *signal generator* rather than an oscillator.

Oscillators are required in many circuits. Low-potential direct current can be converted to high-potential direct current with the aid of an oscillator. The oscillator is used to convert the low-potential direct current to alternating current which subsequently is increased in potential with a *transformer* and converted to high-potential direct current with a rectifier. Oscillators are also used as frequency or time standards and as signal sources.

Oscillators use a transistor, FET, or other amplifying device in combination with an *LC* or *RC* circuit. Essentially an oscillator is an amplifier in which a portion of the output from the amplifier is fed back into the amplifier. A device in which a portion of the output is used as input to the device uses *positive feedback*. In oscillator circuits the output from the amplifier owing to the feedback is in-phase with the output from the amplifier owing to the original input signal. The amplifier connections are arranged so that an increase in the feedback signal from the amplifier causes a decrease in output from the amplifier which in turn causes an increase in the output, etc. The result is an ac output. The *LC* or *RC* network that is associated with the amplifier in an oscillator is used to maintain positive feedback.

Many types of oscillators are available. Only a single type is used to illustrate the manner in which oscillators operate. If a parallel *LC* circuit is used in the feedback path of the amplifier, the oscillator is a *Hartley oscillator*. The *LC* circuit is called a *tank circuit*. A circuit diagram for one form of a Hartley circuit which uses an *n*-channel FET in the common-source mode as the amplifying device is shown in Fig. 3-15.

The output from the drain of the FET is fed through the tank circuit which determines the frequency of the output potential. As was described in Chapter 2, the resonant frequency f_o of the *LC* circuit is given by

$$f_o = \frac{1}{2\pi\sqrt{LC}} \tag{3-2}$$

In a parallel *LC* circuit the impedance is a maximum at resonance. The variable capacitor C_v in the tank circuit is used to select the resonant frequency of the circuit. The output potential from the tank circuit is fed to the gate of the FET and is used to control the channel size. When the output from the tank circuit is high, the size of the channel is decreased and the output at the drain of the FET becomes low. Similarly, when the output from the tank circuit is low,

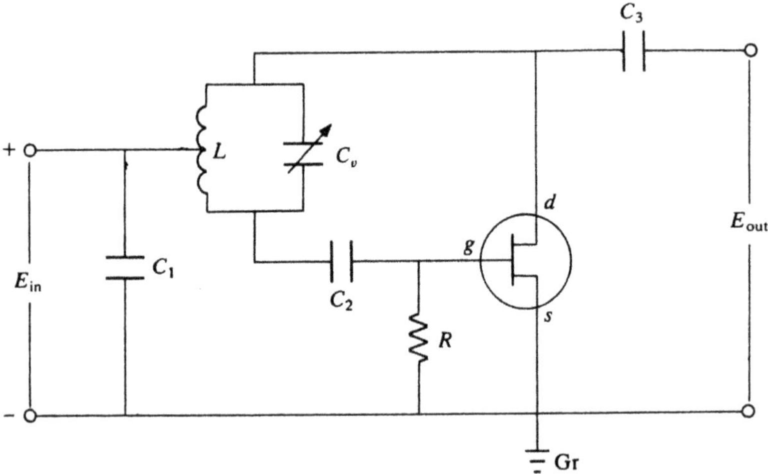

Figure 3-15 A Hartley oscillator circuit. C_1, C_2, C_3, capacitors; C_v, a variable capacitor; *d*, drain of FET; E_{in}, a dc supply; E_{out}, an ac output; *g*, gate of FET; Gr, ground; *L*, inductor; *R*, resistor; *s*, source of FET.

the output from the FET becomes high. The output from the oscillator varies at the resonant frequency of the *LC* circuit. Because positive feedback is maintained within the circuit, the amplitude of the oscillator output increases at the resonant frequency (in-phase with the resonant frequency) and decreases at all other frequencies until the resonant frequency is the only output frequency of the oscillator.

As shown in Fig. 3-15, the inductor is tapped in such a way that the upper portion of the inductor is part of the drain load and the lower portion is connected to the gate. The capacitors C_1 and C_3 are used to prevent passage of a direct current. Bias to the FET is supplied by capacitor C_2 and resistor R. Under the operating conditions of the oscillator, the potential across C_2 is nearly constant and equals the dc gate bias. Because the gate is held at ground potential and because R is between the gate-source diode, the charge from C_2 can drain. That makes it possible for the circuit to automatically adjust to amplitude changes in the gate input potential

During most of each cycle the potential at the gate is sufficient to pinch off the drain. A circuit operated under that condition is termed *class C operation.* While the drain is closed, the output potential is maintained by the resonating potential of the tank circuit. The circuit is excited by periodic pulses from the drain of the FET.

Sample problem 3-1 If C_v in Fig. 3-15 is adjusted to 25 μF and L is 0.32 mH, determine the output frequency of the oscillator.

SOLUTION The output frequency is the resonant frequency of the tank circuit. Equation (3-2) is used for the calculation:

$$f_o = \frac{1}{2\pi\sqrt{LC}} = \frac{1}{2\pi\sqrt{(0.32 \times 10^{-3})(25 \times 10^{-6})}}$$

$$= 1.8 \times 10^3 \text{ Hz}$$

Amplifiers

An amplifier circuit generally contains at least one bipolar transistor, FET, or MOSFET which performs the amplification. Most amplifiers use bipolar transistors. FETs are used less often than bipolar transistors but more often than MOSFETs. The remainder of the circuit is used to provide input and output connections and to properly bias the transistor. As was described earlier, transistors can be used in any of several configurations.

The common-emitter (or grounded emitter) mode is most often encountered in bipolar-transistor amplifiers. A typical amplificatory circuit that uses a transistor in the common-emitter mode is shown in Fig. 3-16. The power supply is connected between ground and E_{cc}. In circuit diagrams for transistor circuits, the connection that is common to the input and output signals is the ground. In this case ground is the connection at the bottom of Fig. 3-16. Although the ground

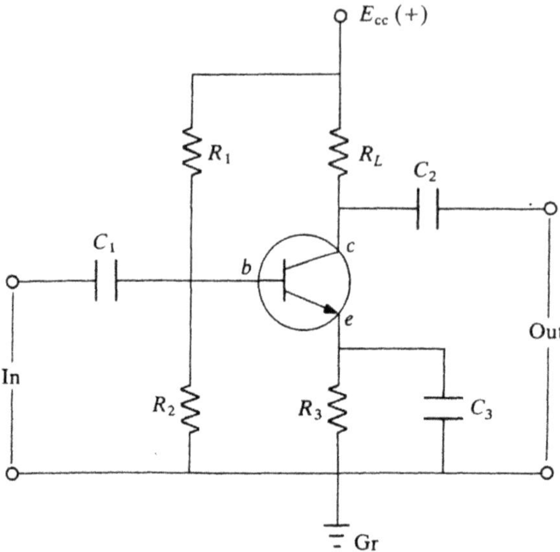

Figure 3-16 A typical transistor amplifier in the common-emitter mode. E_{cc} is the collector potential as applied with a power supply.

connection in the circuit diagram that is shown in Fig. 3-16 is indicated by the ground symbol ⏚, the symbol is often omitted. The resistor R_L that is connected to the collector *c* is the load resistor. The proper potential is applied to the emitter with the aid of resistor R_3. The two resistors R_1 and R_2 form a voltage divider that provides the proper potential (bias) to the base. The circuit is designed so that any change in current flow though the transistor alters the potential drop at R_3 and consequently changes the potential that is applied to the base in such a way as to partially restore the original current through the device.

The input and output capacitors C_1 and C_2 allow an alternating current to pass while preventing passage of a direct current. Capacitor C_3 provides an alternate path for an alternating current to enter the transistor by bypassing R_3. The output potential and current from the common-emitter amplifier is greater than the input potential and is 180° out-of-phase with the input potential and current.

A typical common-base transistor amplifier is shown in Fig. 3-17. The potentials to the emitter, base, and collector are applied from the connection to the power supply E_{cc} with the aid of resistors R_1, R_2, and R_3 as in the common-emitter amplifier. The capacitors C_1, C_2, and C_3, serve the same purposes as described for the common-emitter amplifier. The input and output signals are in-phase. The input resistance (between the "In" terminals) is low (typically 30 to 40 Ω) and *the* output resistance (between the "Out" terminals) is high (typically 2 MΩ). The potential gain can be large with the amplifier if the load

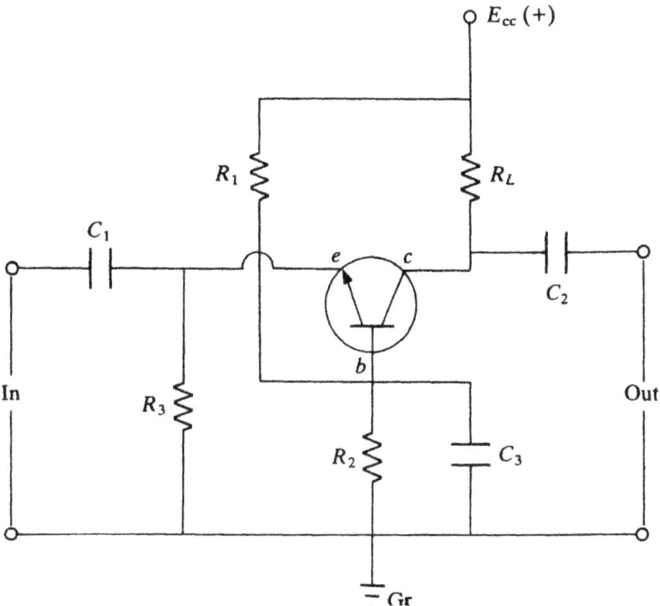

Figure 3-17 A transistor amplifier in the common-base mode. E_{cc} is the potential that is applied to the collector from a power supply.

resistor R_L is large. Typically the potential gain for the common-base amplifier is approximately equal to that for the common-emitter amplifier. A typical value is between 250 and 300.

The common-collector transistor amplifier is also called the *emitter-follower amplifier*. As shown in Fig. 3-18, the required potentials are again supplied from a single source E_{cc} with the aid of the appropriate resistors. The input capacitor also performs the same function as described earlier. The emitter-follower amplifier has a potential gain of nearly unity (no amplification) but a current gain that is approximately the same as the common-emitter amplifier. The output signal is in-phase with the input signal. The input resistance is large (typically 0.5 MΩ) while the output resistance is low (typically 20 to 30 Ω). That is essentially the opposite of the situation with the common-base amplifier. The amplifier is termed an emitter-follower amplifier because the output potential at the emitter follows (is the same as) the input potential at the base.

Amplifiers that are used to deliver relatively large amounts of power generally use a transformer in place of the load resistor in the collector circuit. The circuit diagram for a typical *power amplifier* that uses transformer coupling to the output load is shown in Fig. 3-19. Use of transformer coupling reduces the dc power loss which takes place in the load resistance (the resistance of the device that is attached to the output of the transformer).

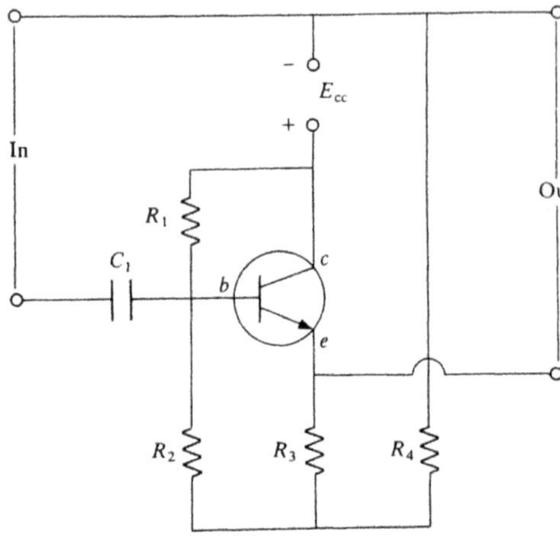

Figure 3-18 A transistor amplifier in the common-collector mode. E_{cc} represents the potential applied to the collector from a power supply.

FETs or IGFETs (primarily MOSFETs) can also be used in amplificatory circuits as the amplifying device. The devices are connected in essentially the same manner that was used with bipolar transistors. The most common FET amplifier uses the common-source mode. The common-drain amplifier is also called the *source-follower amplifier.* It is analogous to the emitter-follower amplifier. Depletion and enhancement IGFETs are used in much the same way that FETs are used in amplifiers. Further information is available in the Bibliography.

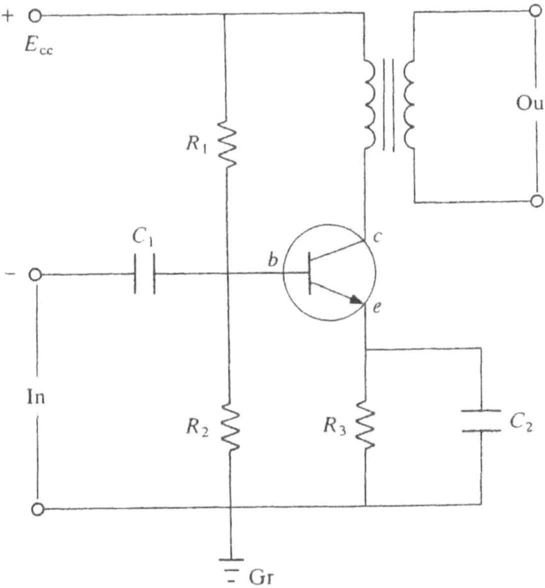

Figure 3-19 A circuit diagram for a power amplifier.

Multitransistor Amplifiers

Some amplificatory circuits contain more than one bipolar transistor, FET, or MOSFET. Several examples of commonly encountered multitransistor amplifiers are described in this section.

A particularly useful two-transistor amplifier is the *emitter-coupled pair*. It is also sometimes called a long-tailed pair. One version of an emitter-coupled pair circuit is shown in Fig. 3-20. The two transistors have connected (coupled) emitters. E_{cc} and E_e are the supply voltages to the collectors and emitters respectively. Transistor T_1 functions as an emitter-follower amplifier. The input resistance is large and no potential amplification occurs as a result of T_1. The output from the emitter of T_1 is applied to the emitter of the transistor T_2 which functions as a common-base amplifier. The amplified output potential from T_2 occurs at the collector. The capacitor C grounds the base of T_2 for alternating current. Resistors R_1, R_2, R_3, and R_4 and E_{cc} and E_e are used to properly bias the transistors. The emitter-coupled amplifier has the advantage of providing high input resistance with an amplified potential output. A high input resistance is particularly important in sensitive measuring instruments because a significant current flow through the instrument can affect the measured parameter.

A useful version of the emitter-coupled amplifier which is often used in laboratory instruments is shown in Fig. 3-21. The amplifier differs in several ways from the amplifier that was shown in Fig. 3-20. The two transistors T_1 and T_2 are as identical as possible; both transistors have load resistors R_L connected

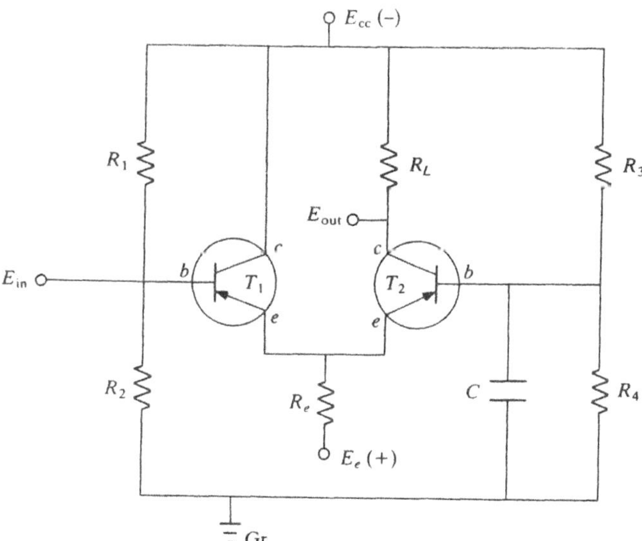

Figure 3-20 A two-transistor emitter-coupled amplifier. The input E_{in} and output E_{out} potentials are relative to ground.

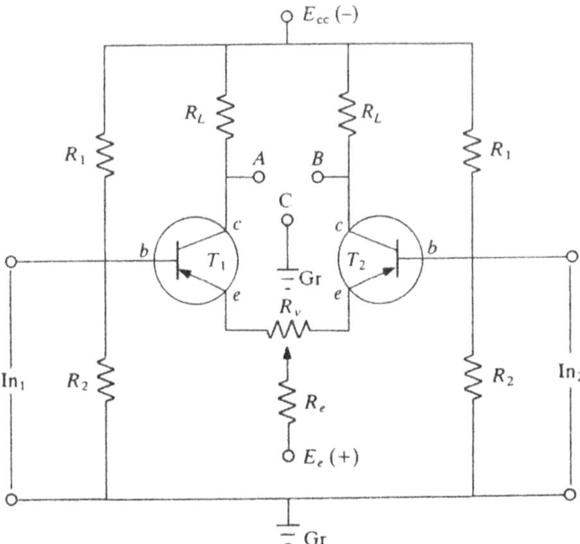

Figure 3-21 A symmetrical two-transistor emitter-coupled amplifier.

to the collectors; and the amplifier is symmetrical, i.e., the resistors in the biasing circuits of the two transistors are identical. The combination of the applied potential across R_e and the resistor R_e is used to cause a constant *total* current to flow through the two emitters. The constant current can be supplied by a third transistor network. The variable resistor is used to compensate for small differences between the two transistors. If the two transistors behave identically, R_v is not required. Although the simple amplifier that is described here is often replaced with more complex, integrated circuit amplifiers, the operating principles for the amplifiers are identical.

The circuit that is shown in Fig. 3-21 can be used in several ways. When zero or identical input signals are applied to the two bases (In_1 and In_2), the two transistors equally conduct current. If a negative input potential is applied at In_1 while In_2 is zero (grounded), the emitter current through T_1 is forced to increase. Because a constant, total, emitter current flows through the two transistors, the emitter current that flows through T_2 decreases. As a consequence the collector current of T_1 increases and the collector current of T_2 decreases. Of course, the increased current in T_1 must be exactly equal to the decreased current in T_2.

The output signal can be taken between the two collectors (*AB*) or between either collector and ground (*AC* or *BC*). The output signal between *A* and *B* is twice the output signal between either *A* and *C* or *B* and *C*. Because the current at one collector increases while the current at the second collector decreases, the output signals between *A* and *C* and between *B* and *C* are 180° out-of-phase. If two output signals are simultaneously taken between *A* and *C* and between *B* and *C*, the amplifier is a *paraphrase* or *phase-splitting amplifier*.

If two different input signals are applied at the two bases, the output, as measured between *A* and *B*, is directly proportional to the *difference* between the two input signals. While used in that mode, the circuit is a *differential amplifier*. With identical inputs the output should be zero. The variable resistor R_v can be used to adjust the output signal to zero when identical input signals are used. An advantage of a differential amplifier is that noise in both input signals does not appear in the amplified output signal.

Another often-encountered two-transistor amplifier consists of the *Darlington pair* or *Darlington connection* (Fig. 3-22). The Darlington pair is a set of two transistors in which the emitter of one transistor is connected to the base of the second transistor. The two collectors are usually connected. The proper bias potentials are applied to the pair with a potential source and biasing network of resistors such as that used in the common-emitter amplifier (Fig. 3-16). A circuit diagram of an amplifier that uses the Darlington pair is shown in Fig. 3-23. The three connections (*A*, *B*, and *C*) to the Darlington pair are made where the base, collector, and emitter connections were made in the common-emitter amplifier. The input signal is applied at *A* and the output signal is obtained at *B*, *C*, or both *B* and *C* (Fig. 3-23).

The first transistor T_1 functions as an emitter-follower (common-collector) amplifier. It gives the pair a high input resistance (typically 1 MΩ), current amplification entering the base of the second transistor, and a potential gain of nearly unity. The second transistor T_2 functions either as a common-emitter amplifier when the output is taken at *B* or as an emitter-follower amplifier when the output signal is taken at *C*. If the signal is alternately taken at *B* and *C*, the device can be used as an inverter that supplies alternating current. In either case the output current is further amplified. The current amplification [Eq. (3-1)] of the amplifier is the product of the amplification β_1 from transistor T_1 and the amplification β_2 from transistor T_2, that is, $\beta_1 \beta_2$. Typically $\beta_1 \beta_2$ varies from about 1000 to 25,000 or more.

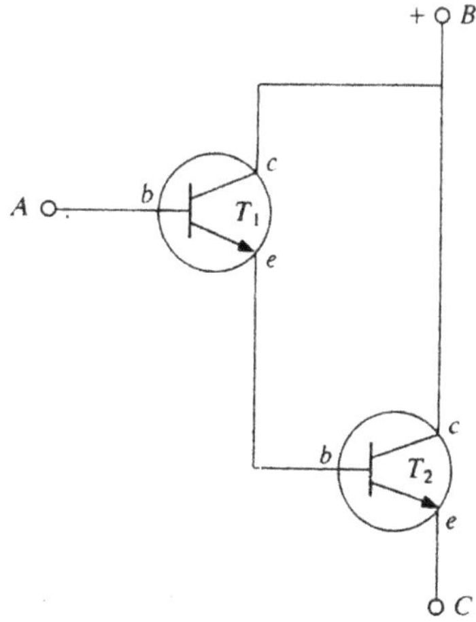

Figure 3-22 The Darlington pair.

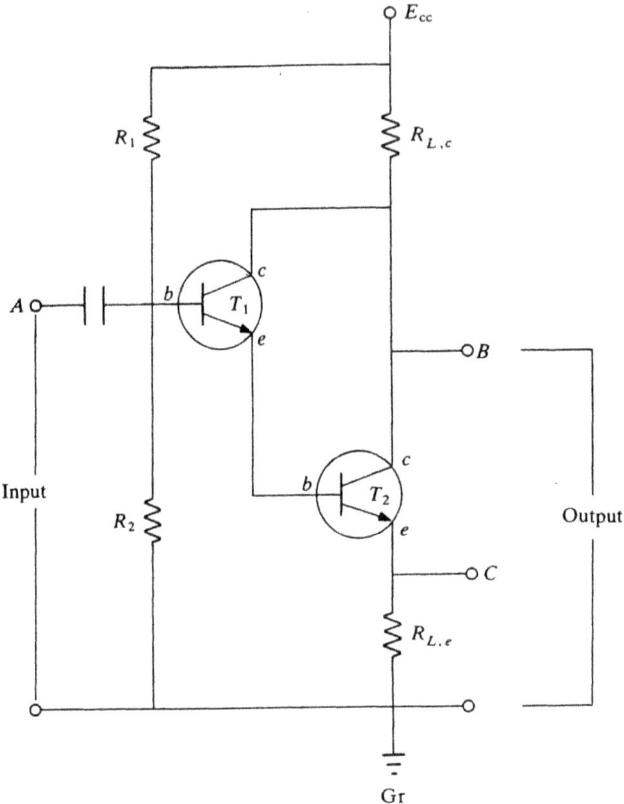

Figure 3-23 A circuit diagram for an amplifier that uses a Darlington pair. The output signal is obtained between either *B* or *C* and ground.

A *cascode amplifier* is sometimes used in the input stage of laboratory counters and in oscillator circuits. It functions well at high frequencies. The amplifier consists of two transistors connected in series so that the output from the collector of the first transistor is connected to the emitter of the second transistor (Fig. 3-24). The first transistor functions as a common-emitter amplifier and the second transistor functions as a common-base amplifier.

If greater amplification is required than can be obtained from any of the amplifier circuits that have been described, sometimes it is possible to increase the amplification by *cascading* several amplifiers. Cascading is the process by which the output from one amplifier is made the input to a second amplifier. The output from the second amplifier similarly can be made the input to a third amplifier, etc. Each transistor is termed a *stage* in the cascaded amplifier.

A *push-pull amplifier* uses two transistors with input and output transformers (Fig. 3-25). Because the input transformer is tapped in the center, one transistor is driven 180° out-of-phase with respect to the second transistor. The name of the device is derived from its ability to drive the two transistors

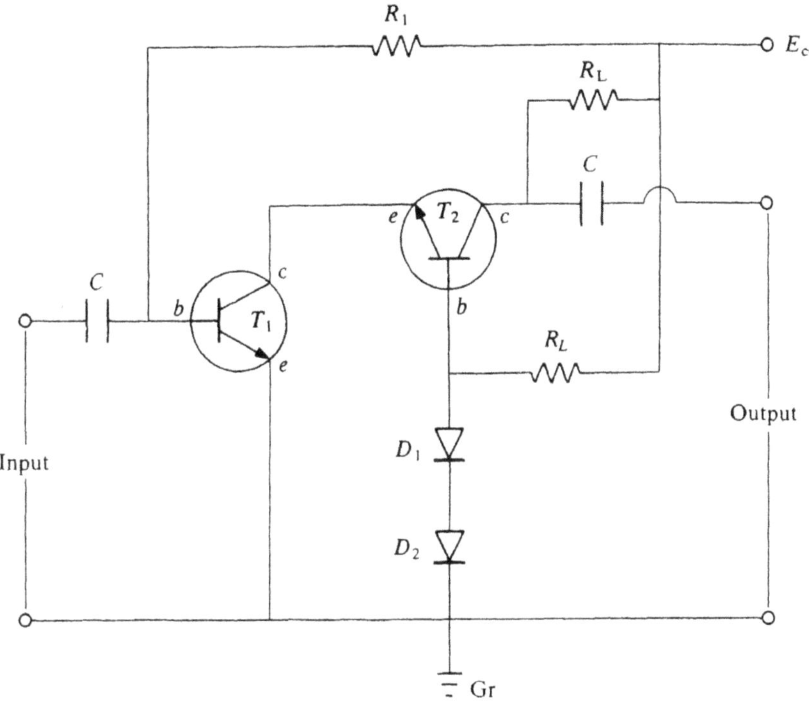

Figure 3-24 A diagram of a cascode amplifier.

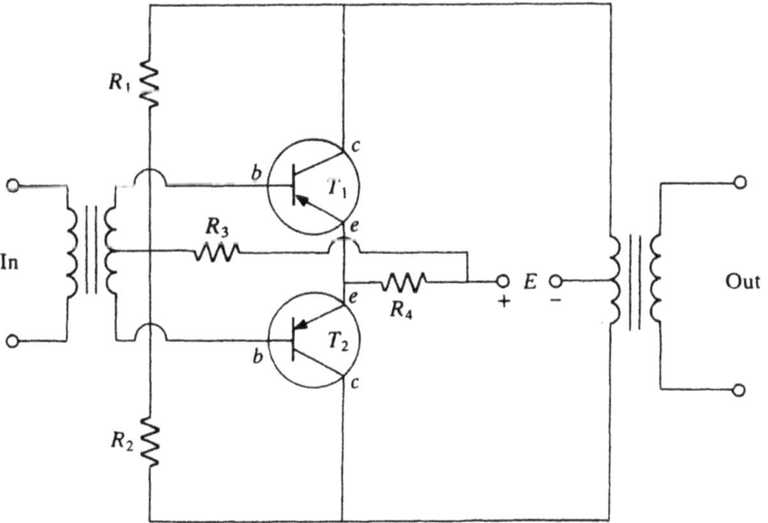

Figure 3-25 A circuit diagram of a push-pull amplifier. E is a dc source potential.

at 180° with respect to each other. During one half-cycle, current is "pushed" through one transistor and "pulled" through the other transistor. The outputs from the two transistors are combined in the center-tapped output transformer. The transistors are biased with the potential source E and resistors R_1 through R_4. The push-pull amplifier has a greater output power than single-transistor circuits. Generally the push-pull amplifier has higher efficiency and less distortion than single-transistor amplifiers.

With the exception of the amplifier that uses a Darlington pair, the amplifiers that have been described up to this point are primarily used for amplification of ac signals. An amplifier that is often used in laboratory instruments to amplify dc signals is the *chopper amplifier.* A chopper amplifier converts the dc signal to an ac signal with a mechanical or electronic device that is known as a *chopper.* The resulting ac signal is amplified with an ac amplifier and converted back to a dc signal with a rectifier.

Mechanical choppers in electric circuits consist of electromagnetically driven vibrating switches (or "reeds"). Sometimes a chopper is used to convert the steady-state nonelectric input signal to the detector in an analytical instrument into an alternating signal prior to allowing the signal to strike the transducer. The transducer (detector) changes the alternating signal to an alternating electric current. An ac amplifier is used to increase the output from the transducer, if necessary. Mechanical choppers are encountered in some atomic absorption, infrared, and ultraviolet-visible spectrophotometers. The electromagnetic radiation from the source is converted to a pulsed signal with a rotating blade that is placed in the path of the radiation between the radiative source and the detector.

Mechanical chopping in electric circuits usually occurs at line frequency (60 Hz). High-frequency chopping generally is accomplished with electronic devices such as diodes, bipolar transistors, FETs, or MOSFETs. The transistors are driven by an alternating square wave potential which alternately allows and prohibits the electric signal to flow through the circuit.

Simple chopper amplifiers convert dc signals of either positive or negative potential to an ac signal. With simple chopper amplifiers, it is not possible to tell, from the output signal, the polarity of the input signal. When it is necessary to be able to distinguish between positive and negative input signals, a *synchronous chopper amplifier* is used. A synchronous chopper amplifier uses a vibrating switch that has two contact points rather than a single contact point as is found in a simple chopper amplifier. By using two switches it is possible to use a circuit in which the phase of the ac signal is related to the sign of the dc input. Consequently, the output from a synchronous chopper amplifier is a direct current with the same sign as the input direct current. A sketch of the major parts of a synchronous chopper amplifier is shown in Fig. 3-26.

A *lock-in* or *phase-sensitive amplifier* is an amplifier that uses a reference ac signal in a circuit to selectively amplify an input signal that occurs at the same frequency and at a particular phase with respect to the reference signal. The reference signal is usually the result of an electronic chopper. Lock-in amplifiers

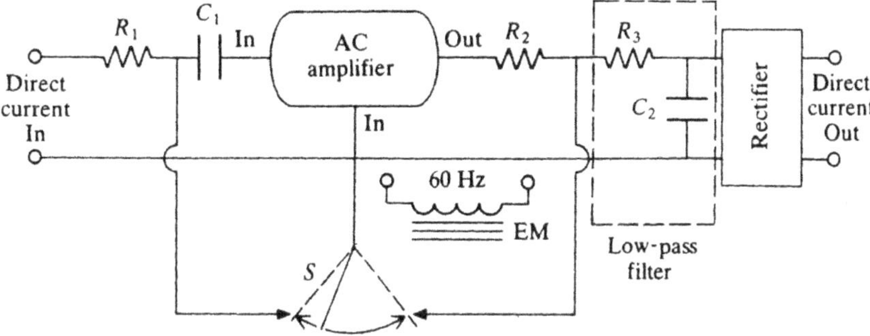

Figure 3-26 A diagram of a synchronous chopper amplifier. EM is an electromagnet that is operated at 60 Hz and used to control the vibrating switch *S*.

are used to amplify weak ac signals of a particular frequency and phase in the presence of significant background noise. Usually lock-in amplifiers consist of an amplifier (or preamplifier) which increases the amplitude of the signal and the noise, a bandpass filter which eliminates any noise at a frequency that is well removed from the signal frequency, the reference signal, a device which uses the reference signal to control amplification at a particular phase with respect to the reference signal, and a rectifier which converts the signal to a dc output.

POWER SUPPLIES

The type of power supply that is most likely to be encountered is the ac-to-dc power supply. The input to the device is an alternating current, usually from a wall socket that is connected to an ac power line. The output is a direct current at some useful potential. The power supply accomplishes several functions. The alternating current is rectified to yield a varying direct current (one half of each ac cycle with half-wave rectifiers or both halves converted to the same polarity with full-wave rectifiers). The varying direct current is filtered to form a

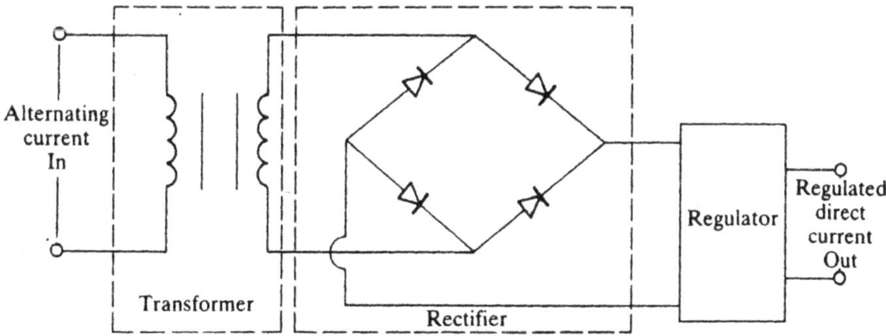

Figure 3-27 A diagram of the major components of a series-regulated linear power supply.

constant-potential direct current. Circuitry generally is required to maintain the output potential at a constant level regardless of the load placed upon the power supply, i.e., regardless of the circuit that is attached to the power supply.

The *series-regulated linear supply* is the power supply that is used most often. It consists of the three components that are shown in the block diagram in Fig. 3-27. The transformer is usually the heaviest portion of the power supply.

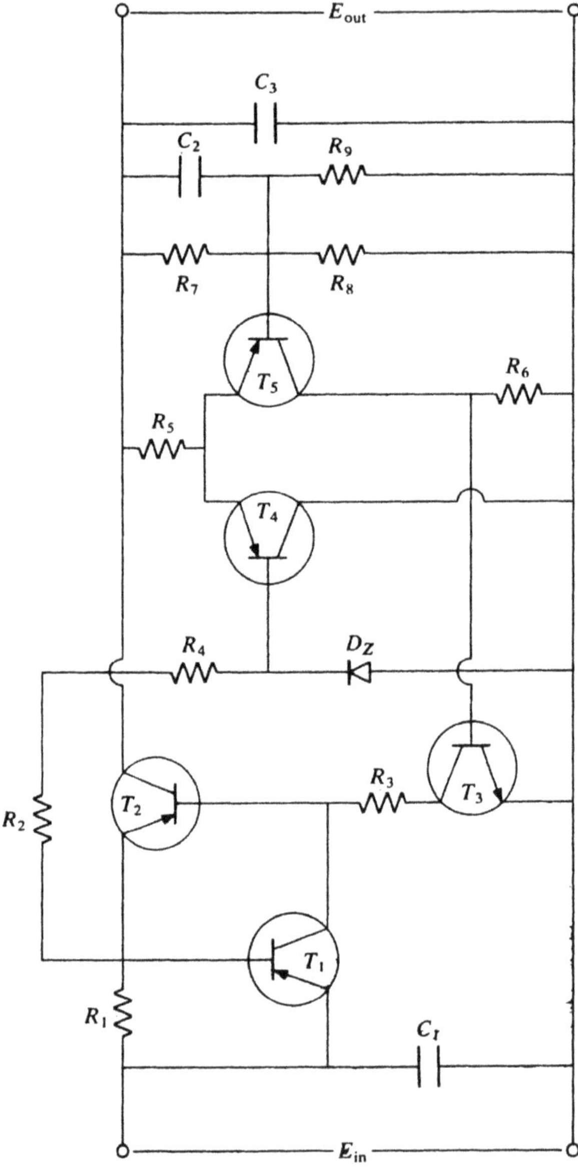

Figure 3-28 A circuit diagram of a series potential regulator.

Unfortunately the size of the transformer is inversely proportional to the frequency of the ac input. For the relatively low frequency (60 Hz) that is most likely to be encountered in power lines, the transformer must be relatively large. Transformers and rectifiers were previously described. The rectifier is usually a bridge such as the one shown in Fig. 3-4*b*.

Regulation is often accomplished with a circuit that contains one or more transistors that are termed *pass* transistors. A pass transistor is placed in series with the load and functions as a variable resistor while regulating current flow. A typical series potential regulator is shown in Fig. 3-28.

The varying dc potential from the rectifier enters the device at E_{in}. The potential charges capacitor C_3 and is applied to the differential amplifier that is composed of transistors T_4 and T_5 and the accompanying resistors (refer to the earlier description of differential amplifiers). The differential amplifier determines the conductivity of transistor T_3 (called the *driver transistor*) which provides the required power to the pass transistor T_2. The output from the differential amplifier (which is dependent upon the input potential from the rectifier) controls transistor T_3 which provides power to the pass transistor, consequently regulating the output potential E_{out}. Transistor T_1 and the associated resistor R_1 provide overload protection to the circuit. Further information on power supplies can be found in Reference 1.

IMPORTANT TERMS

Active device
Amplifier
Base
Beta
Breakdown potential
Cascode amplifier
Channel
Chip
Chopper
Chopper amplifier
Clipping circuit
Collector
Common-emitter amplifier
Core
Darlington connection
Depletion MOSFET
Differential amplifier
Diode
Dopant
Drain
Electronic circuit
Emitter
Emitter-follower amplifier
Enhancement MOSFET
Extrinsic semiconductor
Field-effect transistor
Forward-biased
Gate
Ground
Hartley oscillator circuit
IGFET
Intrinsic semiconductor
Junction field-effect transistor
Light-emitting diode
Linear device
Lock-in amplifier
Majority carrier
MOSFET
Multiplexer
Nonlinear device
N-type semiconductor
Pass transistor

Phase-splitting amplifier
Photodiode
Pinchoff potential
Positive feedback
Power amplifier
Primary coil
P-type semiconductor
Push-pull amplifier
Recombination
Rectifier
Reverse-biased
Secondary coil
Semiconductor
Series-regulated linear power supply
Signal generator
Source
Source-follower amplifier
Synchronous chopper amplifier
Tank circuit
Transfer characteristic
Transistor
Transistor effect
Voltage regulator
Zener diode

PROBLEM

Semiconductors

3-1 If a *p*-type semiconductor is to be constructed from a silicon chip, would it be better to dope the material with Al or Ga? Why?

3-2 What element would be used to dope a Ge chip to prepare an *n*-type semiconductor?

Diode

3-3 Sketch an ac sine wave. Sketch the dc output potentials if the ac wave were input to the half-wave and the full-wave rectifiers that are shown in Fig. 3-4.

3-4 Why must a photodiode be reverse-biased?

Transistors

3-5 What is the major advantage of a MOSFET as compared to a JFET?

3-6 What is the sign of the potential that must be applied to the gate of a *p*-channel FET to prevent current flow between the source and the drain?

3-7 Define the pinchoff potential.

3-8 Sketch a dual-gate MOSFET.

Practical Circuits

3-9 What is a tank circuit?

3-10 Determine the output frequency of a Hartley oscillator if the variable capacitor and the inductor in the tank circuit have values of 35 μF and 0.45 mH respectively.

3-11 If the desired output frequency of the Hartley oscillator that was described in Prob. 3-10 is 1 kHz, determine the required capacitance of the variable capacitor.

3-12 What is the advantage of using a transformer in place of the load resistor in the output circuit of an amplifier?

3-13 A Darlington pair was used in the circuit shown in Fig. 3-23. Assuming β_1 for transistor T_1 is 115 and β_2 for transistor T_2 is 150, calculate the current amplification through the amplifier.

REFERENCES

1. Hnatek, E. R.: *Design of Solid-State Power Supplies,* 2d ed, Van Nostrand-Reinhold, New York, 1981.

BIBLIOGRAPHY

Benedict, R. R.: *Electronics for Scientists and Engineers.* 2d *ed.,* Prentice-Hall, Englewood Cliffs, New Jersey, 1976.

Brophy, J. J.: *Basic Electronics for Scientists,* 3d ed., McGraw-Hill, New York, 1977.

Malmstadt, *H.* V., C. G. Enke, and S. R. Crouch: *Electronics and Instrumentation for Scientists,* Benjamin/Cummings, Menlo Park, California, 1981.

Price, *L. W.*: *Electronic Laboratory Techniques,* 3d ed., McGraw-Hill, New York, 1977.

Pullen, Jr., K. A.: *Design of Transistor Circuits with Experiments,* Howard *W.* Sams, Indianapolis, 1981.

Vassos, B. H, and *G.* W. Ewing: *Analog and Digital Electronics for Scientists,* 2d ed., Wiley, New York, 1980.

CHAPTER

FOUR

OPERATIONAL AMPLIFIERS, LOGIC DEVICES, AND COMPUTERS

Chapters 2 and 3 contained a description of simple ac and dc circuits and a few practical applications of those circuits. In addition to the circuits previously described, most modern analytical instruments contain more complex circuits. Among the components of the more useful circuits are *operational amplifiers*. Many laboratory instruments are controlled by a computer or a microprocessor that relies upon logic devices for storage of data and instructions. Chapter 4 includes an introduction to the use of operational amplifiers, logic devices, and computers.

OPERATIONAL AMPLIFIERS

An operational amplifier (op amp) is an amplifier that has a high input impedance and a high gain. Operational amplifiers can be used for dc and ac signals over a wide frequency range. The performance of an operational amplifier is controlled by feedback from the output of the device to the input. The name of the amplifier is derived from its ability to perform mathematical operations such as addition, subtraction, multiplication, division, integration, and differentiation.

Operational amplifiers consist of several interconnected bipolar or field-effect transistor amplifiers that are similar to those described in Chapter 3. Different circuits are used within different operational amplifiers. Generally, it is not necessary to be concerned with the specific circuit of a particular operational amplifier because the entire device can be treated as a single component. Operational amplifiers are available in prepackaged units with only the input and output connections to the circuit visible.

Most modern operational amplifiers are available in *integrated circuits*. An integrated circuit is formed using masking and deposition methods on a single semiconductor chip. The circuit can consist of transistors, diodes, resistors, and capacitors. Integrated circuits can be highly complex while occupying a small space. The circuit is enclosed in a case and only the connections to the circuit are visible.

A *monolithic* operational amplifier is one in which the entire circuit is formed on a single chip. The components of a monolithic operational amplifier can be categorized according to function. Each category occupies a different section of the amplifier. Generally, the first section consists of a carefully designed input differential amplifier similar to that described in Chapter 3. The input amplifier provides high input impedance, determines the gain stability of the operational amplifier, and can perform other functions. A second differential amplifier is sometimes used to increase the gain of the operational amplifier. The input impedance of the second amplifier is usually not as large as that of the first.

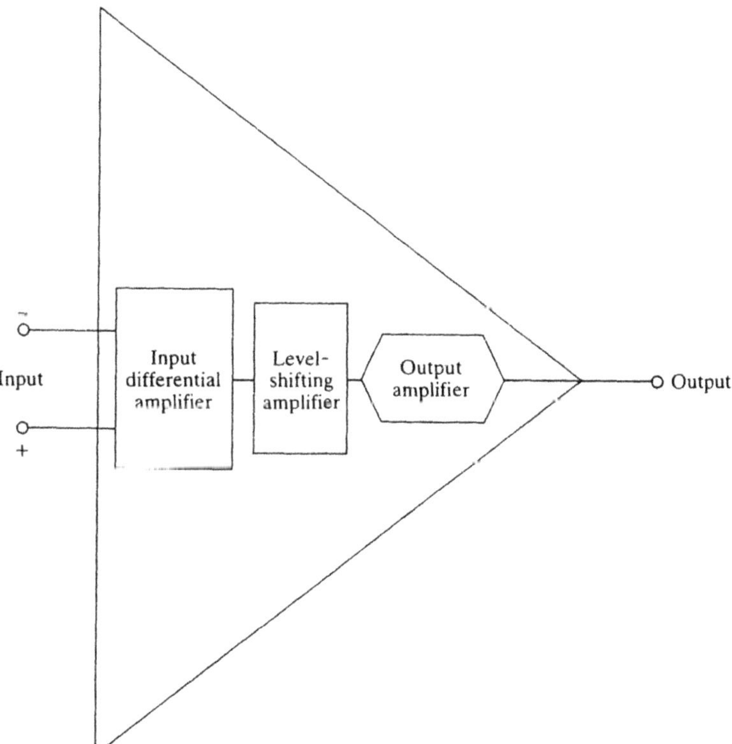

Figure 4-1 A block diagram of the several sections that are found in many operational amplifiers.

The second section contains a level-shifting amplifier. It is used to provide a dc output-level shift so that the output potential from the operational amplifier is zero when the input signal is zero. That is necessary because the output from most differential amplifiers is offset from 0 V by a fixed potential when no input signal is applied. Often a *pnp* common-emitter amplifier is used as the second stage.

The final section is an output amplifier which provides the power that is required to operate devices that are outside the operational amplifier. An emitter-follower amplifier which provides high current gain and a low output impedance can be used. The output potential of most operational amplifiers is limited to 10 V. A block diagram illustrating the sections of many operational amplifiers is shown in Fig. 4-1, and a schematic diagram of a commercially available operational amplifier is shown in Fig. 4-2.

The symbol for an operational amplifier in a circuit diagram is a triangle (Fig. 4-3). By convention, the grounded (common) output connection to the amplifier is not shown in the circuit diagram. In most cases the grounded input lead to the amplifier is shown. The output potential E_o from the operational amplifier is directly proportional to the difference between the applied, input potentials:

$$E_o = A\,(E_2 - E_1) \tag{4-1}$$

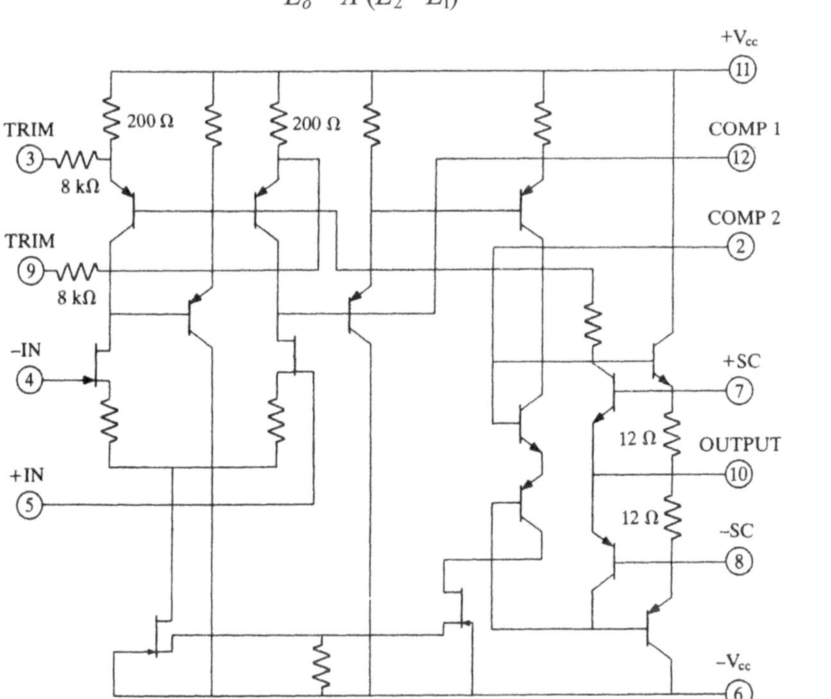

Figure 4-2 A schematic diagram of an OPA 605 operational amplifier. (*Courtesy of Burr-Brown Corp., Tucson, Arizona.*)

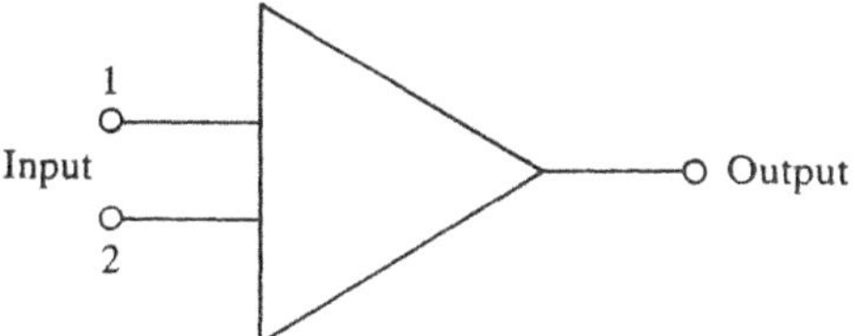

Figure 4-3 The symbolic representation of an operational amplifier.

The proportionality constant A in Eq. (4-1) is the *open-loop potential gain* of the amplifier. Gain is sometimes measured in units of *decibels* dB. A decibel is one-tenth of a *bel* which was named after Alexander Graham Bell. It is defined as

$$\text{dB} = 20 \log \frac{E_o}{E_{\text{in}}} \tag{4-2}$$

If either of the input leads is grounded, the output has the sign that is predicted using Eq. (4-1). In most applications, input terminal 2 (Fig. 4-3) is grounded and the output potential is of opposite sign to the input potential that is applied at terminal 1 because $E_2 = 0$ in Eq. (4-1). For that reason, terminal 1 is labeled as the negative input terminal and is called the *inverting input.*

If a signal is applied to both terminals 1 and 2, the operational amplifier is a *comparator.* The output from a comparator is used to determine the relative sizes of the two input signals. If the output potential is negative, the input potential at terminal 1 is greater than that at terminal 2. If the output potential is positive, the potential at terminal 2 is greater than that at terminal 1. The output potential is zero when the two input potentials are equal. A comparator is an example of a *double-input* operational amplifier. A *single-input* operational amplifier has one of the input terminals connected to ground.

Feedback in Operational Amplifiers

Most of the advantages of operational amplifiers derive from the use of feedback between the output and the input of the amplifier. Feedback enables the device to operate linearly rather than nonlinearly as with the comparator. The feedback circuit connects the output from the amplifier with the inverting input. Feedback of the type used in operational amplifiers is either *operational feedback* because it allows the amplifiers to perform mathematical manipulations or *negative*

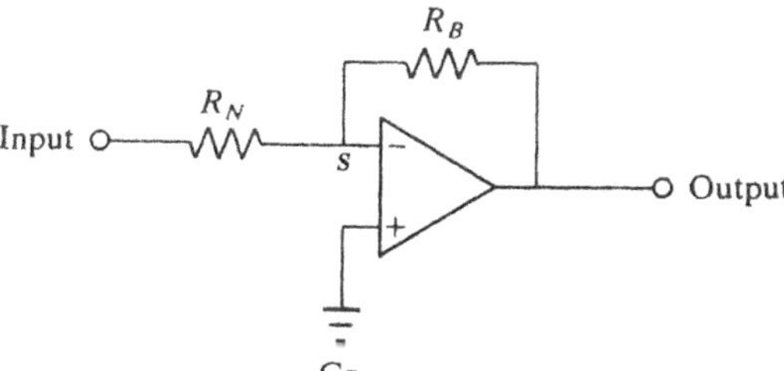

Figure 4-4 An operational amplifier with a feedback circuit.

feedback because the potential at the output of the amplifier is of opposite sign to that at the input. An operational amplifier with feedback is shown in Fig. 4-4. The inverting input in Fig. 4-4 is labeled with a negative sign and the non-inverting input is shown attached to ground. Some people prefer not to show the grounded input terminal.

The point S at which the feedback circuit attaches to the input circuit of an operational amplifier is the *summing junction.* From Ohm's law the current i_N that flows into the summing junction from the input to the amplifier is given by

$$i_N = \frac{\Delta E}{R_N} = \frac{E_N - E_S}{R_N} \tag{4-3}$$

where E_N is the applied potential at the input terminal and E_S is the potential at the summing junction. Similarly, the *feedback current* i_B that flows into the summing junction from the feedback circuit is given by

$$i_B = \frac{\Delta E}{R_B} = \frac{E_o - E_S}{R_B} \tag{4-4}$$

where E_o is the output potential of the amplifier.

Because operational amplifiers have high input impedances, the current that flows from the summing junction through the amplifier is negligible. It is the high input impedance that permits operational amplifiers to perform mathematical functions. Because essentially zero current flows from the summing junction into the amplifier, Kirchoff's first law as applied to the summing junction is

$$i_N + i_B = \frac{E_N - E_S}{R_N} + \frac{E_o - E_S}{R_B} = 0 \tag{4-5}$$

Rearrangement of Eq. (4-5) yields

$$\frac{E_N - E_S}{R_N} = \frac{E_S - E_o}{R_B} \tag{4-6}$$

From Eq. (4-1) it is apparent that $E_o = - AE_S$ because E_2 is zero (grounded) and $E_1 = E_S$. Substituting the value of E_S into Eq. (4-6) and rearranging results in

$$\frac{E_o}{E_N} = \frac{-AR_B}{R_B + (1 + A)\,R_N} \tag{4-7}$$

Equation (4-7) only assumes a high input impedance to the operational amplifier. Because in nearly every practical circuit the gain A of the amplifier is large, R_B is negligible when compared to $(1 + A)R_N$ and 1 is negligible when compared to A. Consequently, AR_N can be substituted for the term in the denominator of Eq. (4-7). The result is

$$\frac{E_o}{E_N} = \frac{-AR_B}{AR_N} = \frac{-R_B}{R_N} \tag{4-8}$$

Equation (4-8) can be used to calculate the resistances of the resistors that are required to obtain the desired gain for an amplifier. Gain is usually defined as the positive ratio of the output signal to the input signal, i.e., as $-E_o/E_N$ in this case.

Unlike Eq. (4-7) which contains the open-loop gain A, Eq. (4-8) contains no term which is dependent upon the circuitry of the amplifier. As long as the assumptions that were used in deriving Eq. (4-8) are correct (i.e., the input impedance and A are large), the gain of an operational amplifier is dependent only upon the external circuit. If precision resistors are used in the external circuit, the accuracy obtained while using Eq. (4-8) is excellent. The circuit of Fig. 4-4 can be used to multiply the input potential by a constant. In order to change the value of the constant it is only necessary to change the ratio of R_B to R_N.

If the input and feedback resistors (R_N and R_B) are identical, the amplifier is an *inverter*. The output potential is identical to the input potential but of opposite sign. Besides changing the sign of the input potential, inverters can be used in the input circuit of an analytical instrument to ensure a high input impedance and a resulting negligible effect of the instrument on the measured parameter.

Sample problem 4-1 An operational amplifier has an open-loop gain of 1.0×10^4. Use Eqs. (4-7) and (4-8) to calculate the output potential of the amplifier when the feedback resistance is $1.2 \times 10^5\,\Omega$, the input resistance is $1.0 \times 10^3\,\Omega$, and the input potential is 1.0 mV.

SOLUTION Substitution into Eq. (4-7) yields

$$\frac{E_o}{1.0 \times 10^{-3}} = \frac{-(1.0 \times 10^4)(1.2 \times 10^5)}{1.2 \times 10^5 + (1 + 1.0 \times 10^4)(1.0 \times 10^3)}$$

$$= -1.2 \times 10^2$$

$$E_o = -0.12\ \text{V}$$

Substitution of the parameters into Eq. (4-8) yields

$$\frac{E_o}{1.0 \times 10^{-3}} = \frac{-1.2 \times 10^5}{1.0 \times 10^3} = -1.2 \times 10^2$$

$$E_o = -0.12\ \text{V}$$

The effective impedance Z between the summing junction in Fig. 4-4 and ground is the ratio of the potential E_s ($\Delta E = E_S - E_g = E_S$) at the summing junction to the input current i_N (Ohm's law):

$$Z = \frac{E_S}{i_N} \tag{4-9}$$

Because the input current of an operational amplifier is equal to the feedback current, but of opposite sign, $-i_B$ can be substituted for i_N. The result is

$$Z = \frac{E_S}{-i_B} \tag{4-10}$$

Substitution of the value of i_B from Eq. (4-4) into Eq. (4-10) yields

$$Z = \frac{E_S\, R_B}{E_S - E_o} = \frac{R_B}{1 - E_o/E_S}$$

$$= \frac{R_B}{1 + A} \tag{4-11}$$

The impedance between the summing junction and ground is dependent upon the resistance in the feedback circuit and upon the open-loop gain of the amplifier. Because the open-loop gain of the amplifier is large, the impedance is small. As an example, for the amplifier that was described in Sample problem 4-1, the effective impedance between the summing junction and ground is $1.2 \times 10^5/(1 + 1.0 \times 10^4) = 12\ \Omega$. Because the effective impedance between the summing junction and ground is small, the summing junction is termed a *virtual ground.*

Properties of Operational Amplifiers

The behavior of all operational amplifiers should be identical according to Eq. (4-8). Actually the internal circuitry of operational amplifiers does have an effect upon their operation. Several characteristics of operational amplifiers can be used to compare their behavior.

The open-loop gain of an operational amplifier is the gain of the amplifier in the absence of a feedback circuit. A large value is desirable because Eq. (4-8) is only obeyed when A is large relative to R_B and 1. Typical open-loop gains vary from about 3×10^2 to 3×10^7.

The *bandwidth* of an operational amplifier is the frequency range of the input signal which results in a large gain. Often the open-loop bandwidth between zero frequency (direct current) and the frequency at which the gain falls to unity is used for comparison. Although the gain varies with frequency, it does not affect the results of the amplification as long as it is sufficiently large to allow Eq. (4-8) to be obeyed.

The input impedance ranges from about 10^4 to $10^{12}\ \Omega$ for operational amplifiers. A high input impedance is important because zero current flow into the operational amplifier from the summing junction was assumed while deriving Eq. (4-8).

The output signal of many operational amplifiers is a finite value in the absence of an input signal. The amount by which the output signal differs from zero volts is the *offset voltage.* A low offset voltage is desirable. The offset voltage rarely exceeds 5 mV. With many operational amplifiers a variable resistor in the circuit (similar to that in the circuit of the differential amplifier shown in Fig. 3-21) allows adjustment of the offset voltage to zero.

The *drift* or change in the offset voltage with temperature is another property of operational amplifiers. A low drift is desirable to prevent errors from the amplifier as the temperature changes. Typical values are within the range from 0.2 to 30 μV/degree.

Mathematical Functions with Operational Amplifiers

Several mathematical functions can be performed with operational amplifiers. In computer-controlled analytical instruments, mathematical manipulations are normally done with the computer. In those instruments that are not interfaced to a computer or microprocessor, operational amplifiers can be used to rapidly perform mathematical manipulations that otherwise would have been performed by the analyst. The following sections describe the use of operational amplifiers for mathematical manipulations. In analytical instruments the manipulations are often performed on data as it is obtained from the transducer of the analytical instrument.

Addition, subtraction, and multiplication Single-input operational amplifiers are those in which the noninverting terminal of the amplifier is grounded and the input to the amplifier is through the inverting terminal. Addition can be accomplished by use of the circuit that is shown in Fig. 4-5. The potentials (E_1, E_2, E_3, etc.) that are added enter the circuit through parallel resistors (R_1, R_2, R_3, etc.), and combine at the summing junction. Because the summing junction is a virtual ground, no danger exists that current from one of the parallel branches will flow into any other parallel branch of the circuit; i.e., the resistance to current flow into a branch is greater than that to flow into the junction. As long as the input impedance to the amplifier is large, Kirchoff's first law as applied to the summing junction is

$$I_1 + I_2 + I_3 + I_B = 0 \tag{4-12}$$

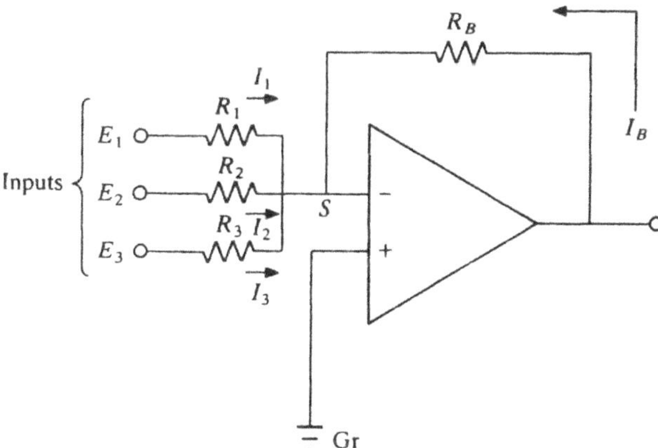

Figure 4-5 A summing operational amplifier.

Keeping in mind the fact that the summing junction is a virtual ground (nearly zero potential), Ohm's law can be used to rewrite Eq. (4-12) in terms of potentials:

$$\frac{E_1}{R_1} + \frac{E_2}{R_2} + \frac{E_3}{R_3} + \frac{E_o}{R_B} = 0 \tag{4-13}$$

Rearrangement of Eq. (4-13) yields

$$E_o = -R_B\left(\frac{E_1}{R_1} + \frac{E_2}{R_2} + \frac{E_3}{R_3}\right) \tag{4-14}$$

From Eq. (4-14) it is apparent that the output potential of the operational amplifier is the weighted sum of the several input potentials. If the input resistors are all identical, the weighting for each input potential is identical, and the circuit is a *summer* or *summer circuit*. If all resistors, including the feedback resistor, are identical the output potential is the negative sum of the input potentials. If the resistors are not identical, the circuit is a *weighted summer circuit*. If currents rather than potentials are to be added, the parallel resistors can be eliminated from the circuit. If subtraction rather than addition is required, the sign of one or more of the input potentials is reversed.

Sample problem 4-2 Calculate the output potential for the weighted summer circuit shown in Fig. 4-5 if R_1 is 575 Ω, R_2 is 425 Ω, R_3 is 1.50×10^3 Ω, and R_B is 1.50×10^4 Ω when E_1, E_2, and E_3 are all 0.100 V.

SOLUTION Substitution into Eq. (4-14) yields

$$E_o = -1.50 \times 10^4\left(\frac{0.100}{575} + \frac{0.100}{425} + \frac{0.100}{1.50 \times 10^3}\right) = -7.14 \text{ V}$$

Integration Integration with respect to time can be accomplished by replacing the feedback resistor in Fig. 4-4 with a capacitor (Fig. 4-6). The integrator operates with both ac and dc inputs. When the switch SW is open and an input potential is applied, the feedback current charges the capacitor C. The feedback current i_B through the capacitor varies with time, as described earlier [Eq. (2-30)]:

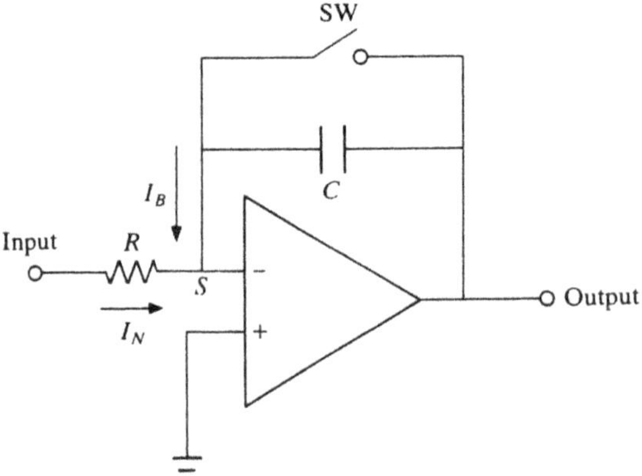

Figure 4-6 An integrating operational amplifier.

$$i_B = C\frac{de_o}{dt} \tag{4-15}$$

At the summing junction, application of Kirchoff's first law results in

$$i_B + i_N = C\frac{de_o}{dt} + \frac{e_N}{R} = 0 \tag{4-16}$$

Rearrangement and integration of Eq. (4-16) between the limits of 0 and t yield

$$e_{o,\,t} = -\frac{1}{RC}\int_0^t e_N\, dt + e_{o,\,0} \tag{4-17}$$

where RC is the time constant for the circuit, $e_{o,\,t}$ is the output potential at time t after the start of the integrations, and $e_{o,0}$ is the output potential before the start of the integration.

After each integration the integrator can be reset by closing the switch and discharging the capacitor. Operational amplifier integrators are used to integrate peaks obtained while using several analytical instruments including those usually older NMR spectrometers that are not computer-controlled. Because the errors involved in performing integrations with operational amplifiers are additive, it generally is not feasible to use operational amplifier integrators for integrations which require more than about an hour to complete.

Linear-signal generator If the input to the operational amplifier integrator is a constant potential, Eq. (4-17) can be solved to yield

$$e_{o,\,t} = \frac{-E_N(t-0)}{RC} + E_{o,\,0} = E_{o,\,0} - \frac{E_N t}{RC} \quad \text{V} \tag{4-18}$$

Because E_N, R, and C are all constants, Eq. (4-18) can be rewritten as

$$e_{o,\,t} = E_{o,\,0} - kt \text{ V} \tag{4-19}$$

It is apparent from Eq. (4-19) that the output potential linearly varies with time. The variation can be used as a voltage ramp in several instruments including polarographs. In linear-sweep voltammetry, operational amplifier linear-signal generators are often used to apply a linearly varying potential to the test electrode.

Differentiation By interchanging the positions of the capacitor and the resistor in the integrator of Fig. 4-6, operational amplifiers can be used to obtain time derivatives (Fig. 4-7). The capacitor is charged by the input current. Application of Kirchoff's first law to the summing junction (a virtual ground) of the circuit shown in Fig. 4-7 yields

$$i_N + i_B = C\frac{de_N}{dt} + \frac{e_o}{R} = 0 \tag{4-20}$$

Solution of Eq. (4-20) for the output potential e_o yields

$$e_o = -RC\frac{de_N}{dt} \tag{4-21}$$

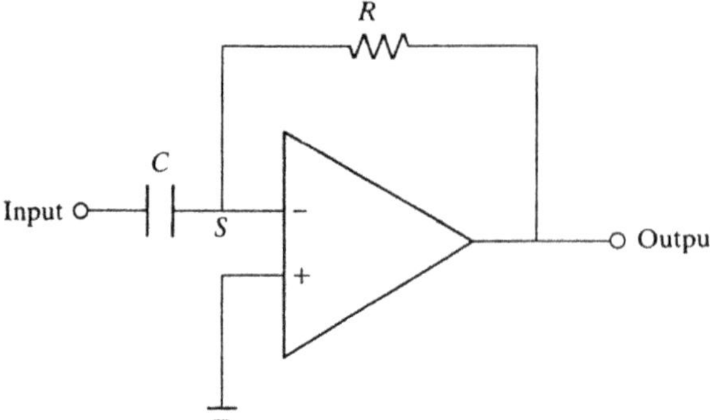

Figure 4-7 An operational amplifier differentiator.

The output potential is a function of the first derivative of the input potential.

Unfortunately, the gain for that differentiator increases as the frequency of the input potential increases and at high frequencies the circuit oscillates. A variation of the differentiator which functions better over a wide frequency range is shown in Fig. 4-8. The circuit contains one capacitor-resistor couple arranged as in the integrator of Fig. 4-6 and a second couple arranged as in the differentiator of Fig. 4-7. The time constant $R_N C_B$ for the integrator is short, and the time constant for the differentiator $R_B C_N$ is typically 10^4 times larger than that for the integrator. More details of the circuit can be found in any of the texts that are listed in the Bibliography and References which deal with operational amplifiers.

Logarithmic amplifiers Nonlinear outputs from operational amplifiers are obtained by using nonlinear components in the feedback circuit. A circuit that has an output potential that is proportional to the logarithm of the input signal is shown in Fig. 4-9. Logarithmic amplifiers can also be constructed with diodes rather than transistors in the feedback circuit. The collector current I_c of the transistor is given by

$$I_C = \alpha I_S (e^{eEo/kT} - 1) \tag{4-22}$$

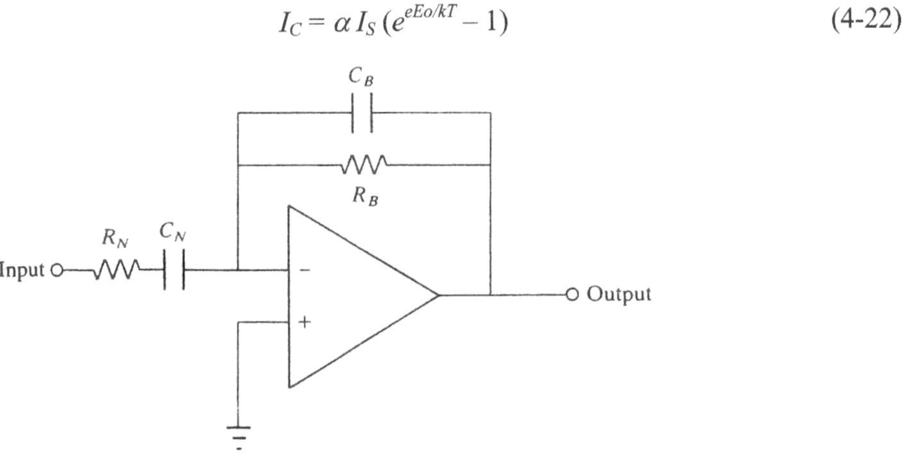

Figure 4-8 A practical differentiator.

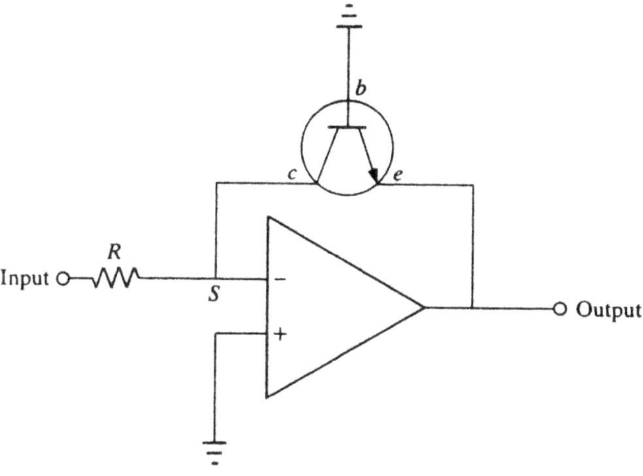

Figure 4-9 A logarithmic operational amplifier.

where I_s is the saturation current for the transistor, α is the current-gain factor (the ratio of the change in collector current to the change in emitter current at constant-collector potential), and k is Boltzmann's constant.

Because 1 is negligible in comparison to the exponential term, Eq. (4-22) can be rewritten as

$$I_C = \alpha I_S e^{eEo/kT} \tag{4-23}$$

At the summing junction the feedback current (the collector current) and the input current must be equal (Kirchoff's first law):

$$I_N = I_C \tag{4-24}$$

Substitution of I_N for I_C in Eq. (4-23) and application of Ohm's law (assuming S is a virtual ground) results in

$$\frac{E_N}{R} = \alpha I_S e^{eEo/kT} \tag{4-25}$$

Rearrangement of Eq. (4-25) yields

$$e^{eEo/kT} = \frac{E_N}{\alpha I_S R} \tag{4-26}$$

Upon taking the logarithm of Eq. (4-26) and rearranging, the following equation is obtained:

$$E_o = \frac{2.303kT}{e} \log E_N - \frac{2.303kT}{e} \log \alpha I_S R \tag{4-27}$$

The conversion from natural logarithms to base 10 logarithms is accomplished by multiplying by 2.303 in Eq. (4-27). Combining all the constant terms in Eq. (4-27) results in

$$E_o = K_1 \log E_N - K_2 \tag{4-28}$$

It is apparent from Eq. (4-28) that the output potential of the amplifier is directly proportional to the logarithm of the input potential. Logarithmic amplifiers can be used to multiply two or more terms by summing the outputs from two or more logarithmic amplifiers and passing the result through an antilogarithmic amplifier. Division can be accomplished in a similar manner by subtracting the output of one logarithmic amplifier from the output of a second logarithmic amplifier and passing the result through an antilogarithmic amplifier. Antilogarithms of an input potential are obtained by interchanging the positions of the transistor (or diode) and resistor in Fig. 4-9 (see Prob. 4-9).

Analog computers An *analog computer* is an electric device that contains operational amplifiers and is used to perform various mathematical functions. Usually an analog computer is used to simulate a physical or chemical system which can be described mathematically. The devices are rarely used in modern laboratories owing to the increased use of digital computers. An electric circuit is constructed that obeys the same mathematical laws as those obeyed by the physical system. By studying the change in the output signal of the electric system as a function of the input signal, information can be obtained relating to the behavior of the chemical or physical system. Generally, the electric circuits consist of the inverter (Fig. 4-4), the summer (Fig. 4-5), the integrator (Fig. 4-6), and sometimes logarithmic and antilogarithmic amplifiers.

Commercially available analog computers consist of a group of operational amplifiers which are enclosed in the same cabinet and have their input and output connections readily available for external connections. By appropriately attaching wires between the operational amplifiers of the analog computer, the apparatus can be made to simulate the desired physical system. Typically, commercial analog computers contain a minimum of 24 operational amplifiers.

After the electric circuit that simulates the chemical or physical system is prepared, the extent of the chemical or physical property is assigned a potential. In that way the input and the output potentials from the circuit can be related to physical or chemical properties. For example, if the rate of a chemical reaction is to be studied as a function of the concentration of one of the reactants, the concentration of the reactant must be converted to an input potential. The process of conversion from units that are used in the simulated system to potential for use in the computer is *scaling*. In this particular case, the molarity of the reactant would be assigned a potential, e.g., a 1 *M* solution might be assigned a potential of 0.1 V. When deciding how to do the scaling, it is important to remember that the output potential cannot exceed the saturation level of the electric circuit. Usually the output of an operational amplifier cannot exceed 10 V.

Sample problem 4-3 Design a circuit that simulates the pressure of an ideal gas when the volume and temperature of the gas are simultaneously changed.

SOLUTION The ideal gas law is $PV = nRT$. When the number of moles n of the gas is constant, the equation can be simplified to $PV = kT$ or to

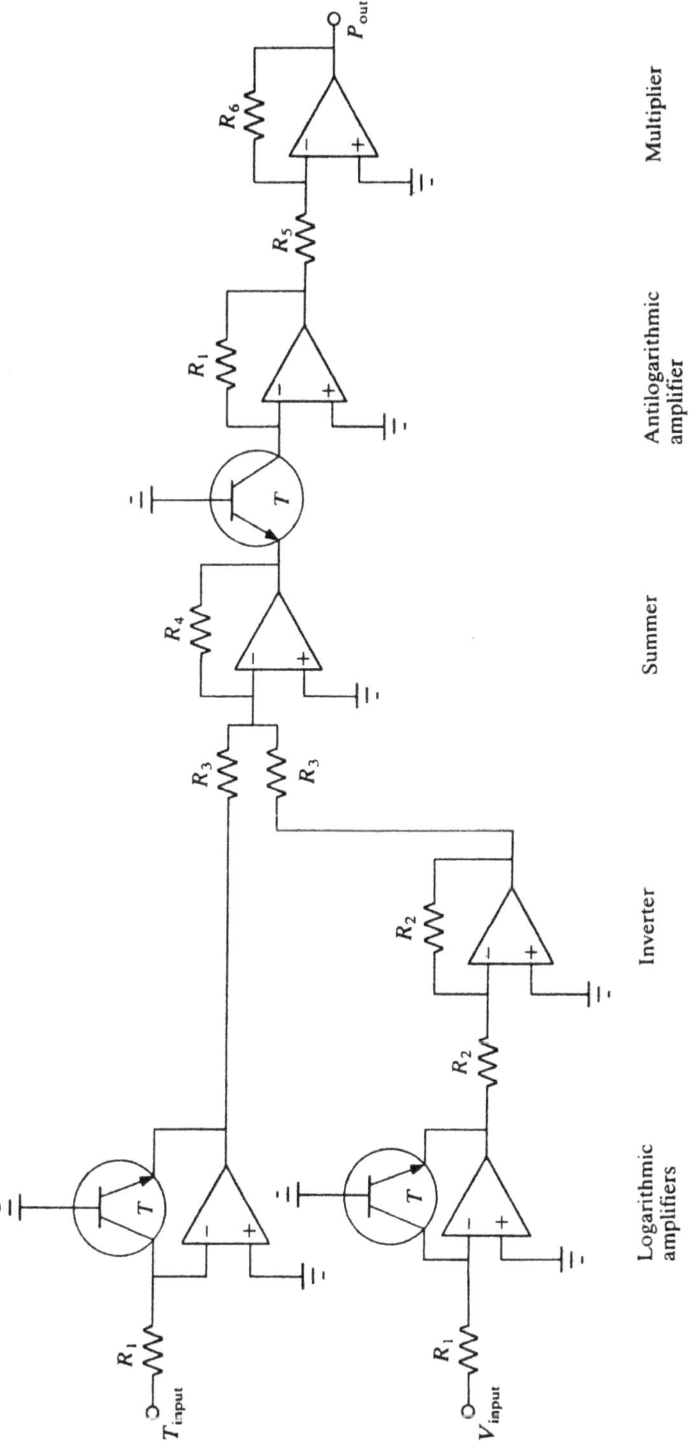

Figure 4-10 A solution to Sample problem 4-3.

$P = kT/V$. A circuit is needed that accepts two inputs (for T and V), that divides the temperature input by the volume input, and that multiplies the result by a constant. A solution to the problem is to use logarithmic amplifiers to obtain logarithms of T and V; use a summer to subtract one from the other; and use an antilogarithmic circuit to obtain the antilogarithm of the result:

$$\log P = \log k + \log T - \log V$$

A circuit which should perform the proper functions is shown in Fig. 4-10. Simpler circuits can probably be designed to perform the same function, but are not as illustrative. The two logarithmic amplifiers obtain the logarithms of the two input potentials (one each for T and V). The inverter is used to change the sign of the log V signal so that the summer can be used to combine the log T signal with the log V signal. Because the two resistors in the inverter circuit are identical, the magnitude of the logarithmic term is not altered. The antilogarithm of the result is obtained by the antilogarithmic amplifier. Multiplication by the constant (if needed) is accomplished with the final amplifier. The values of the various resistors depend upon the scaling for P, V, and T (i.e., upon the desired amplification, etc.). Either the proper values can be calculated with the aid of the equation for the simulated system and the corresponding equations for the operational amplifiers [Eqs. (4-8), (4-14), and (4-27)] or the scaling can be done experimentally after the circuit is functioning by inputting known signals and measuring the output.

Noninverting operational amplifiers If the input signal is applied to the positive (noninverting) terminal of the operational amplifier, the output signal is of the same sign as the input signal, i.e., the output is in-phase with the input. The magnitude of the output signal is, of course, dependent upon the external circuitry of the amplifier. The feedback circuit connects the output from the amplifier with the inverting input, as with inverting operational amplifiers. The circuit for a noninverting amplifier is shown in Fig. 4-11.

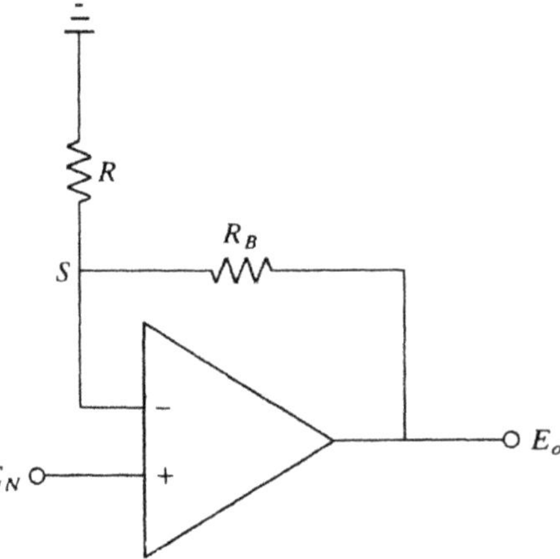

Figure 4-11 A noninverting operational amplifier.

The equation which relates the input potential E_N of the noninverting amplifier to the output potential E_o is derived in a manner that is similar to the derivation of Eq. (4-9). The output potential is given by substitution into Eq. (4-1):

$$E_o = A\,(E_N - E_S) \tag{4-29}$$

At the summing junction the sum of the currents is zero (Kirchoff's first law). Assuming negligible current flow from the summing junction into the inverting terminal of the operational amplifier leads to

$$\frac{E_o - E_S}{R_B} + \frac{0 - E_S}{R} = 0 \tag{4-30}$$

The zero in Eq. (4-30) results from the fact that one end of resistor R is grounded. Rearrangement of Eq. (4-30) yields

$$E_s = E_o \frac{R}{R + R_B} \tag{4-31}$$

Substitution of the value of E_s from Eq. (4-31) into Eq. (4-29) and rearrangement yields the result that is given by

$$E_o \left(1 + A \frac{R}{R + R_B}\right) = AE_N \tag{4-32}$$

When the gain of the amplifier is large, 1 is negligible in comparison with $A[R/(R + R_{B},)]$, and Eq. (4-32) simplifies to

$$E_o = \frac{R + R_B}{R} E_N \tag{4-33}$$

It is apparent from Eq. (4-33) that the output potential of the amplifier is dependent upon the values of the two resistors and has the same sign as the input potential.

Double-input operational amplifiers Occasionally it is advantageous to use both input terminals of an operational amplifier. An example of an operational amplifier in which both input terminals are used is the comparator that was described earlier. Generally, both input terminals are used whenever the difference between two signals or functions is required.

A typical circuit that is used for subtraction of two input potentials is shown in Fig. 4-12. Application at the summing junction of Kirchoff's first law and substitution from Ohm's law results in

$$\frac{E_S - E_1}{R_N} + \frac{E_S - E_o}{R_B} = 0 \tag{4-34}$$

When both input terminals to an operational amplifier are used, the potentials at the two terminals are virtually equal. The virtual-ground assumption at the

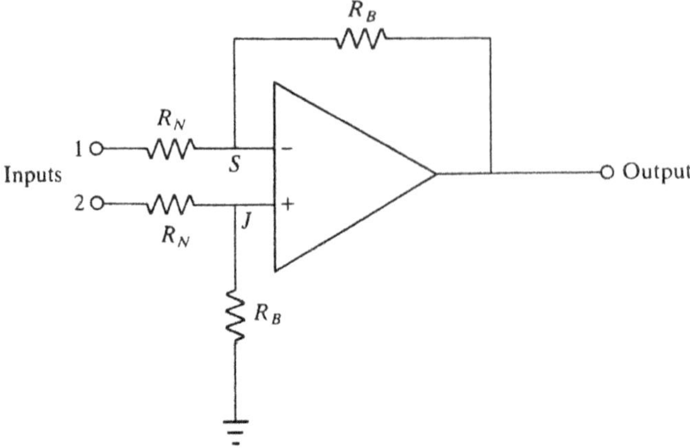

Figure 4-12 A double-input operational amplifier that is used for subtraction.

summing junction, which was made when the noninverting terminal was grounded, is a special case of that assumption. Substitution of E_J for E_S in Eq. (4-34) results in

$$\frac{E_J - E_1}{R_N} + \frac{E_J - E_o}{R_B} = 0 \tag{4-35}$$

In the circuit of input 2, R_N and R_B form a voltage divider. The change in potential between terminal 2 and the ground is E_2. The potential E_J at J can be calculated with the aid of Eq. (2-16) for voltage dividers:

$$E_J = \frac{R_B\, E_2}{R_B + R_N} \tag{4-36}$$

Substitution of the value of E_J from Eq. (4-36) into Eq. (4-35) and rearrangement yields

$$E_o = R_B \frac{E_2 - E_1}{R_N} \tag{4-37}$$

From Eq. (4-37) it is apparent that the output potential is proportional to the difference between the two input potentials. Double-input operational amplifiers also can be used to perform difference integration as well as other differential functions. Descriptions of other double-input operational amplifier circuits are given in References 1 through 4.

Nonmathematical-Uses of Operational Amplifiers

Operational amplifiers have many uses in addition to the performance of mathematical operations. A few of the applications that are of particular interest to analytical chemists are described in this section.

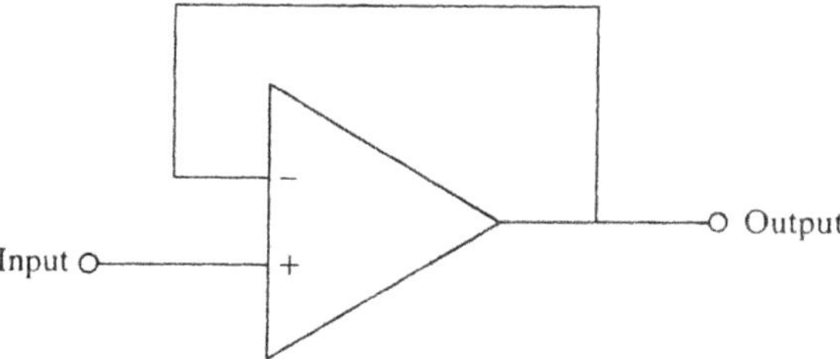

Figure 4-13 A voltage follower.

The *voltage* (or *potential*) *follower* (Fig. 4-13) is often encountered in various circuits. The feedback circuit consists of an electric connection between the output and the inverting input terminal of the operational amplifier. The input signal is introduced at the noninverting terminal. The output potential is equal to the input potential in both sign and magnitude [Eq. (4-33) where R_B is equal to zero]. The voltage follower has a high input impedance and a low output impedance.

The importance of a high input impedance in many analytical instruments cannot be overemphasized. A high input impedance prevents a significant current from flowing into the instrument from the transducer, thereby minimizing the effect of the measurement on the measured parameter. As an example, pH meters must have a high input impedance because a significant current flow through the high resistance membrane of the pH electrode significantly alters the potential of the electrode resulting in an apparent pH change. A potential follower is often placed in the input circuit of an analytical instrument to give the instrument a high input impedance.

Current-to-voltage conversion can be accomplished by eliminating the input resistor in the circuit of Fig. 4-4 or by eliminating the input resistors in the summer circuit of Fig. 4-5. The input signal is a current rather than a potential and the output of the current-to-voltage converter is a potential.

Rectification can be accomplished by inserting one or more diodes in the external circuit of an operational amplifier. *Precision rectifiers* rectify signals that appear at the inputs of the operational amplifiers. They are advantageous when compared to the diode rectifiers that were described earlier because they rectify signals at much lower (millivolt region) potential levels than can be rectified with diode rectifiers. A simple, half-wave precision rectifier is shown in Fig. 4-14 and a full-wave precision rectifier is shown in Fig. 4-15.

Operational amplifier circuits can be used as *constant-current amplifiers*. Constant-current amplifiers must be capable of delivering a constant current at the output while the load resistance (the resistance of the device to which the output is attached) changes. Consequently, they must be able to function over a wide output-potential range and they must have a large output impedance. A simple constant-current amplifier is shown in Fig. 4-16. The input signal (potential) is applied to the noninverting terminal of the amplifier. The current i_o

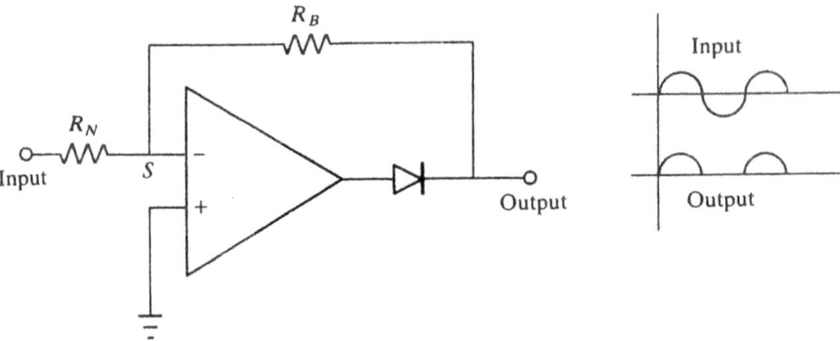

Figure 4-14 A half-wave precision rectifier with inverted output.

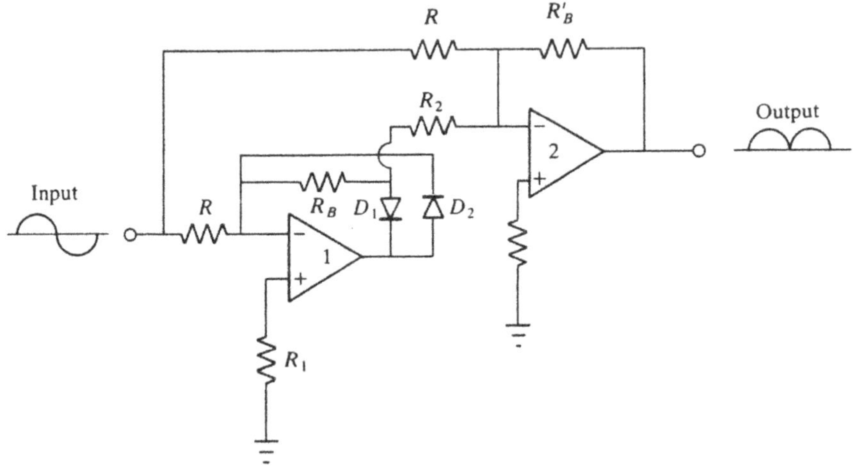

Figure 4-15 A full-wave precision rectifier.

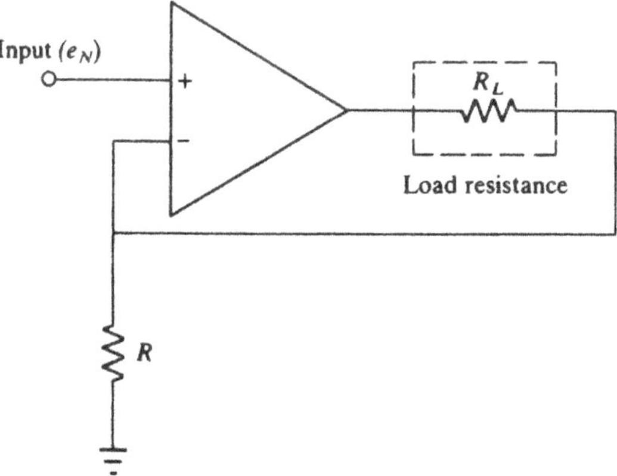

Figure 4-16 A constant-current amplifier.

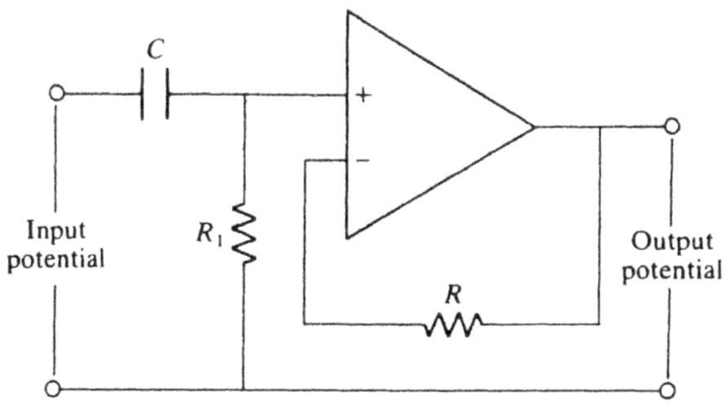

Figure 4-17 An active high-pass filter.

that flows through the load resistance can be shown to be given by

$$i_o = \frac{e_N}{R} \frac{(1 + A)R}{R(1 + A)R + R_L + r_o} \tag{4-38}$$

where r_o is the output impedance of the amplifier.

Passive filters were described in Chapter 2. *Active filters* have a filter circuit that is similar to those used for passive filters followed by a voltage follower which provides a low output impedance. Providing a low output impedance is another common use for a voltage follower. A low output impedance is important if significant current must be withdrawn from the circuit. An active high-pass filter is shown in Fig. 4-17; an active low-pass filter is shown in Fig. 4-18; and an active bandpass filter is shown in Fig. 4-19. It is instructive to compare these circuits with those described in Chapter 2 (Figs. 2-16 and 2-17).

Operational amplifiers are used in several types of oscillators. The oscillator with the least components is the *Wien-bridge oscillator* which is shown in Fig. 4-20. It generates a sine wave. The output frequency f_o of the oscillator is given by

$$f_o = \frac{1}{2\pi\sqrt{R_1 R_2 C_1 C_2}} \tag{4-39}$$

It is informative to compare that oscillator with the Hartley oscillator (Fig. 3-15).

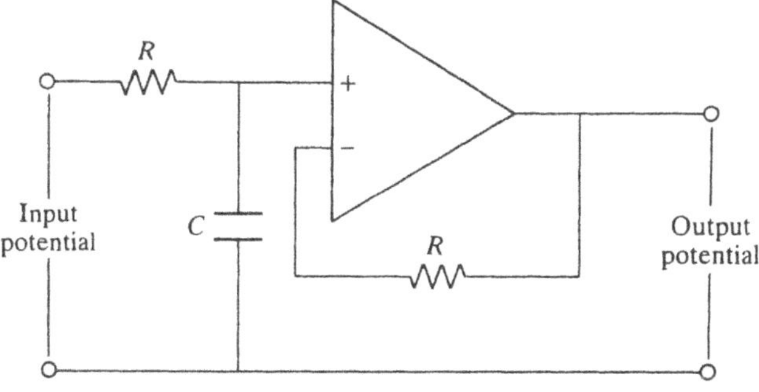

Figure 4-18 An active low-pass filter.

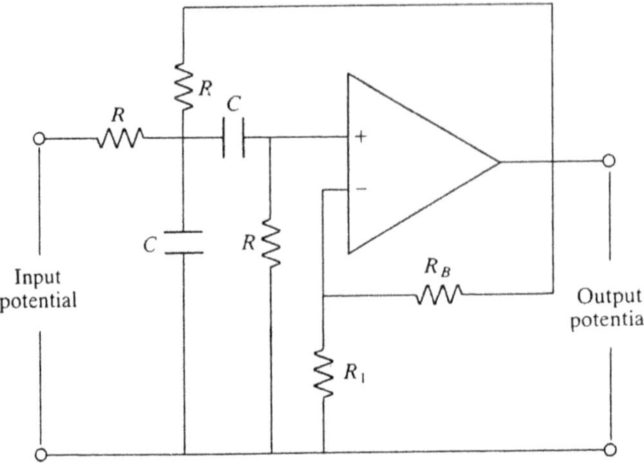

Figure 4-19 An active bandpass filter.

The voltage follower (operational amplifier 2) in the oscillator prevents significant current loss from the circuit while providing a low output impedance. The feedback resistors R_B are variable in order to compensate for distortion of the output waveshape. Distortion occurs if the gain of amplifier 1 is not carefully selected. The variable resistor R_2 is used to change the output frequency [Eq. (4-39)].

Operational amplifiers are often used in instruments which measure an electric parameter. They are commonly used in circuits that are used for the measurement of potential, current, resistance, and conductance. They are also used in instruments that measure a null point, i.e., that compare two potentials and adjust some parameter of the circuit so that the potentials are identical. The electric output from transducers that are used in many spectrophotometric and

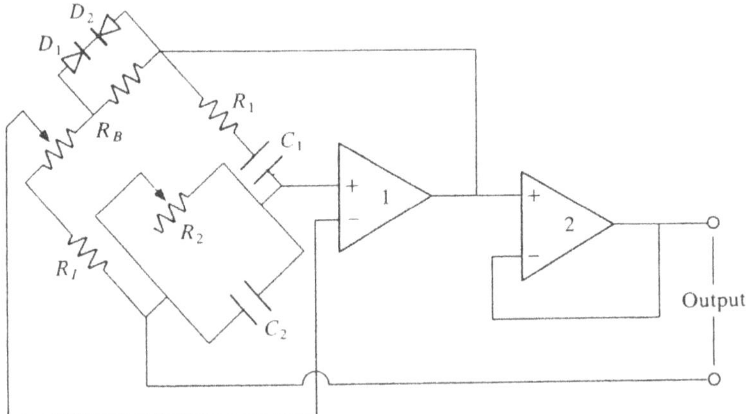

Figure 4-20 A Wien-bridge oscillator.

other analytical instruments are connected to the input terminal of an operational amplifier. Operational amplifiers in analytical instruments are used to amplify the signal from the transducer, to perform mathematical manipulations on the signal, to increase the input impedance of the instrument, to decrease the output impedance, and to perform other, functions.

Digital Uses of Operational Amplifiers

Digital computers can be interfaced to analytical instruments for the purpose of automatically acquiring data and controlling the instrument. The computers can be either built into the laboratory instrument or a separate laboratory computer that might be attached to multiple laboratory devices. The transducers that are used in many analytical instruments have continuous (analog) output signals which must be converted to discrete (digital) signals prior to use by a computer. Consequently, the need arises for a device that can convert the analog signal from an analytical instrument to the digital signal that can be used by a digital computer. The device which performs the conversion is an *analog-to-digital converter* (ADC, or A/D converter). An ADC accepts a dc potential from the instrument, converts the potential to a binary number which is proportional to the magnitude of the potential, and stores the number in a register that can be accessed by the computer.

When a computer is used to control an analytical instrument, it is often necessary to convert the digital command from the computer to an analog signal which is used to control the instrument. The device which performs the conversion from a digital to an analog signal is a *digital-to-analog converter* (DAC, or D/A converter). Operational amplifiers are used in the circuits of both ADCs and DACs. The DAC accepts the binary number from the computer and stores the number in a counter. The outputs from the counter are attached to a variable-gain operational amplifier which has weighted gain resistors. The output from the operational amplifier is a dc potential that is proportional to the number from the computer. An ADC and sometimes a DAC are built into some analytical instruments to facilitate interfacing to a computer.

Several designs of DACs are available. A single design is described here. Further information relating to DACs and ADCs is given in Reference 2, pp. 207-232. In the *inverting summer DAC,* the input resistors of a summer operational amplifier are connected to the outputs from the counter in which the binary number is stored (Fig. 4-21). Although that DAC is not encountered as often as some of the other forms, it serves well as an example of the manner in which DACs function.

The counter in which the binary number is stored is often a transistor transistor logic (TTL; see later description) register in which each digit of the stored number is followed by an electronic (transistor) switch. The output from the register is a series of high and low voltages. The high voltage corresponds to a 1 in a binary number and the low voltage to a 0. Consequently, the binary number 100110 (38 is the decimal equivalent) requires a register with six outputs and six switches (one for each digit). The output voltages from the switches would be (left to right) high (1), low (0), low, high, high, and low.

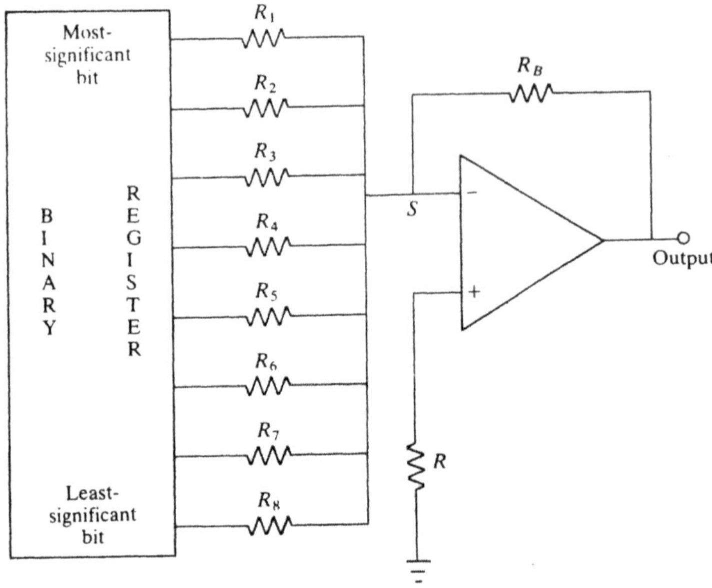

Figure 4-21 An eight-bit inverting summer digital-to-analog converter.

Each digit of a binary number is a *bit*. The least-significant bit is the bit which represents the number of smallest magnitude, i.e., the bit furthest to the right. The most-significant bit is the bit that is furthest to the left. Because each successive bit between the least significant and the most significant represents an increase in the magnitude of the number by a factor of two (for binary numbers), the resistors in the input circuits of the summer are weighted so that the most-significant bit of the register is attached to the smallest resistor [Eq. (4-14)].

Each successively less-significant bit of the register is attached to an input resistor that is twice the value of the next more-significant bit; i.e., as the significance of the bit decreases, the resistance increases. The weighting of the resistors is identical to the magnitude associated with each of the bits of the register. For example, if R_1 in Fig. 4-21 (attached to the most-significant bit of the register) is 100 Ω (100×2^0), R_2 would be 200 Ω (100×2^1), R_3 would be 400 Ω (100×2^2),..., and R_8 would be 12.8 kΩ (100×2^7). Precision resistors (± 0.1 percent) must be used for the eight-bit DAC that is shown in Fig. 4-21. As the number of bits increases, the precision of the resistors must increase.

Sample problem 4-4 Convert the binary number 011100101 into its decimal equivalent.

SOLUTION As in base 10 numbers, the least-significant digit corresponds to the number in the digit multiplied by the base to the zero order. Likewise, the next more-significant digit corresponds to the number in the digit multiplied by the base to the first order, etc. The decimal equivalent is the sum of the equivalents of each digit.

Least-significant bit: $1 \times 2^0 = 1$
$0 \times 2^1 = 0$
$1 \times 2^2 = 4$
$0 \times 2^3 = 0$
$0 \times 2^4 = 0$
$1 \times 2^5 = 32$
$1 \times 2^6 = 64$
$1 \times 2^7 = 128$
Most-significant bit: $0 \times 2^8 = 0$
Sum = decimal equivalent = 229

Sample problem 4-5 Determine the number of bits of a DAC that would be required to express 1055 to the nearest whole number. Determine the binary equivalent of the decimal number.

SOLUTION The problem is most easily solved by converting the decimal number to its binary equivalent. Each digit of the binary number requires a bit in the DAC. Table 4-1 lists the decimal equivalent of a 1 in each of the first 32 bits of binary numbers. The easiest way to perform the conversion is to find the most-significant bit number in Table 4-1 that has a decimal equivalent that is equal to or less than the decimal number. Because 1024 (bit11) is less

Table 4-1 The decimal equivalent (dec. eq.) of a 1 (high) in the first 32 bits of a binary number

Bit no.	Dec. eq.	Bit no.	Dec. eq.
1	1	17	65,536
2	2	18	131,072
3	4	19	262,144
4	8	20	524,288
5	16	21	1,048,576
6	32	22	2,097,152
7	64	23	4,194,304
8	128	24	8,388,608
9	256	25	16,777,216
10	512	26	33,554,432
11	1024	27	67,108,864
12	2048	28	134,217,728
13	4096	29	268,435,456
14	8192	30	536,870,912
15	16,384	31	1,073,741,824
16	32,768	32	2,147,483,648

than 1055, a 1 must belong in the eleventh bit. Consequently, 1055 requires 11 bits.

To convert 1055 to a binary number, determine the bits which must contain a 1, starting with the most-significant bit. The eleventh bit was already determined to contain a.1. Subtract the decimal equivalent of the 1 in the eleventh bit from the original number in order to determine the remainder that must be converted to a binary number:

$$1055 - 1024 = 31$$

Again consult the table to find the most-significant bit number which has a decimal equivalent that is equal to or less than 31. Because 16 is less than 31, the next bit which contains a 1 is bit 5. Repeat the procedure until the entire number has been converted:

$$31 - 16 = 15 \qquad \text{(bit 4 is 1)}$$
$$15 - 8 = 7 \qquad \text{(bit 3 is 1)}$$
$$7 - 4 = 3 \qquad \text{(bit 2 is 1)}$$
$$3 - 2 = 1 \qquad \text{(bit 1 is 1)}$$

The binary number is 10000011111.

As was the case with DACs, several forms of ADCs are available. A single example is described in order to illustrate the operation of ADCs. The *successive-approximation* ADC (Fig. 4-22) performs the conversion rapidly and is one of the more useful forms of ADCs. Essentially the device consists of a DAC similar to that in Fig. 4-21 and a comparator. The input analog signal is compared (with the comparator) to the analog signal from the DAC as the input to the DAC is altered by the digital processor until the two signals are nearly identical.

The ADC can operate in either of two ways. The first method is to initially set all of the outputs from the digital processor to 0 (low). Starting with the most-significant bit, the output from the digital processor is switched to 1 (high) and the output from the comparator is examined. If the output from the comparator changes sign, the bit is returned to 0; otherwise it remains at 1. The next lower bit is then changed to 1, and the process is repeated. The comparisons continue until all of the bits have been changed. After the process is complete, the digital reading is obtained from the output of the digital processor at the terminals between the processor and the resistors. The digital output is between zero and one step less than the analog input potential.

The second method is similar to the first except that all of the bits are initially set at 1 rather than 0. The output from the comparator is examined by the digital processor as each of the bits, starting with the most significant, is changed to 0. When a change in sign from the comparator occurs, the bit is changed back to 1. After the process is completed, the digital output from the processor has a value that is between zero and one step greater than that corresponding to the magnitude of the analog input.

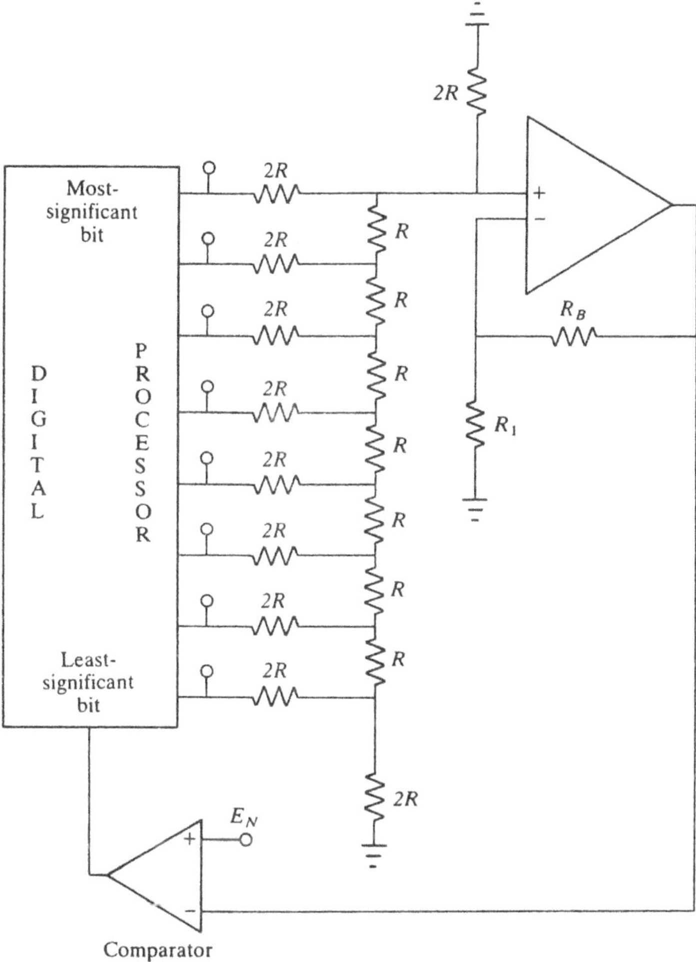

Figure 4-22 An eight-bit successive-approximation analog-to-digital converter.

Integrated Circuits

An integrated circuit (IC) is an electric circuit that consists of more than one electric component (capacitors, diodes, resistors, and/or transistors) on a single chip. Integrated circuits are classified according to their degree of complexity. *Small-scale integration* (SSI) is the least complex form. An SSI circuit usually contains the electric components that are required to perform a single simple function. A single gate or flip-flop (refer to later description) on an IC chip is an example of SSI. Typically an SSI circuit might consist of four or five transistors and the associated diodes, resistors, and capacitors.

Medium-scale integration (MSI) is more complex than SSI. Typically it contains the circuitry that is the equivalent of 10 to 50 SSI circuits. MSI circuits perform more complicated functions than SSI circuits. Devices of the complexity of counters are classified as MSI circuits.

Large-scale integration (LSI) combines several circuits which are the equivalent of MSI circuits onto a single chip. Devices of the complexity of ADCs are usually regarded as LSI circuits. *Very large-scale integration* (VLSI) consists of highly complex circuits on a single chip. An example of a VLSI chip is the Intel 8087 chip which contains an entire numeric data processor. The circuit contains over 75,000 transistors and the associated resistors, etc., on a single silicon chip. It is capable of performing floating-point arithmetic and of calculating trigonometric functions, logarithmic functions, powers, and square roots. It has an accuracy corresponding to 80 bits.

Some modern computers and other devices utilize even more components on a single chip. Those chips utilize *ultra large scale integration* (ULSI). It is possible to place more components on a single chip because the size of each component is smaller and the distance between the components has decreased. It is now possible to manufacture chips which have individual components with dimensions in the micrometer and nanometer range. It is advantageous to use multiple components on a single chip because generally the device is less expensive than the larger equivalent circuit, and because the electric connections between individual components are shorter. Shorter connections result in decreased communication time between the components and in less heat generation owing to electric resistance in the connections. Because of the low cost of ICs, individual electric components are currently not being used as much as they have been used in the past.

The *fan-in* of an integrated circuit is the number of inputs to the circuit. The *fan-out* is the number of outputs, i.e., the number of external devices which can be driven by the IC. The *propagation delay* T_d is the time that is required for a signal to pass through the circuit. The *power dissipation* P_d is the power that is required to operate the circuit.

Logic Devices

Logic devices are electronic devices that can be used to perform logical functions in computers and other equipment. They are the devices that allow computers to function as they do. The result of a logical function is dependent upon the type of function performed and the input signal to the logic device. For example, a logic device might be used to output a high potential whenever both input 1 *and* input 2 to the device are high potentials. If either input is a low potential or if both inputs are low potentials, the device outputs a low potential. The device performs a logical AND on the inputs.

Logic devices have inputs and outputs that are either high or low. A high input corresponds to a high input potential and a low input to a low potential at the input terminal. Logic devices are manufactured from the electric components that have been described earlier.

If the device is constructed from resistors, capacitors, and transistors, the device is said to use *resistor-capacitor-transistor logic* (RCTL). Similarly, a device that is composed of resistors and transistors uses *resistor-transistor logic* (RTL). Other devices use *diode-transistor logic* (DTL), *transistor-transistor logic* (TTL), *emitter-coupled* (transistors) *logic* (ECL), and *complementary metal-oxide semiconductor* (CMOS) *logic*. CMOS logic uses both *p*-channel and *n*-channel FETs.

Logic devices that are designed in such a way that the more positive of the two possible output potentials is regarded as 1 (high) utilize *positive* logic. If the more positive output potential is regarded as 0, the device uses *negative* logic. The two categories of components that are found in logic devices are gates and flip-flops.

Gates and Boolean Algebra

A gate is a device in which the output is a logical function of the inputs to the device. The devices can be designed to have any desired number of inputs. The most common gates are classified as *AND*, *OR*, *NOR*, *EXCLUSIVE-OR*, or *EXCLUSIVE-NOR*, depending upon the logical function that is performed on the input signals.

Boolean algebra can be used to predict the output from a gate. Boolean algebra was developed by George Boole in the nineteenth century. It is used for single-digit binary mathematical manipulations. Boolean algebra recognizes only two mathematical operations (addition and multiplication). Addition is symbolized by " + " and multiplication by "·". The logical OR is equivalent to addition in boolean algebra and the logical AND is equivalent to multiplication. Several of the more useful theorems of boolean algebra are listed in Table 4-2. A bar over a letter represents negation. Because boolean algebra deals with binary numbers, a negated letter has a specific meaning. As an example, if A is 1, then $\overline{A}$ must be 0. Likewise, if A is 0, then $\overline{A}$ is 1.

An AND gate gives a high output only when all of the input signals are high. It can be thought of as a series circuit that contains a switch for each input. A current can only be conducted through the circuit when all of the switches are closed (corresponding to a high input), i.e., when switch 1 *and* switch 2 *and* switch 3... are all closed. A simple AND gate can be constructed from diodes, as shown in Fig. 4-23.

With no signal applied to the inputs, the positive applied potential E biases the diodes in the forward direction resulting in a negligible output potential (low). If a positive signal that is greater than E is simultaneously applied to all of the input terminals, all of the diodes becomes reverse-biased and the output potential becomes E (high). If even one of the diodes is forward-biased, the output remains low.

The symbol for an AND gate is shown in Fig. 4-24, and the *truth table* for the gate is listed in Table 4-3. A truth table is a listing of the output signal of a

Table 4-2 A list of several theorems of Boolean algebra

	Number	Theorem
Logical AND		
	1	$A \cdot A = A$
	2	$A \cdot 1 = A$
	3	$A \cdot 0 = 0$
	4	$A \cdot \overline{A} = 0$
Logical OR		
	5	$A + A = A$
	6	$A + 1 = 1$
	7	$A + 0 = A$
	8	$A + \overline{A} = 1$

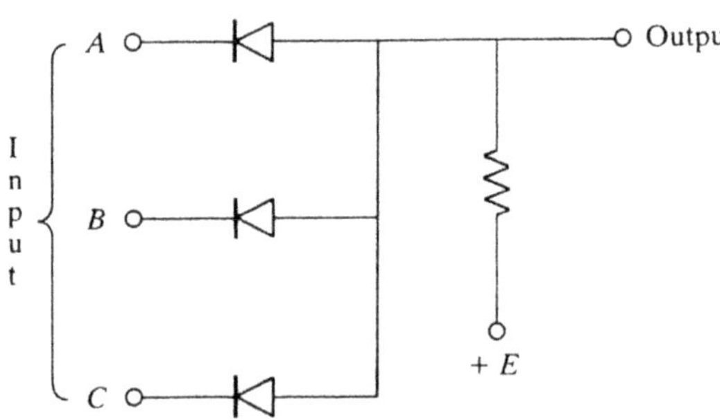

Figure 4-23 A three-input AND gate.

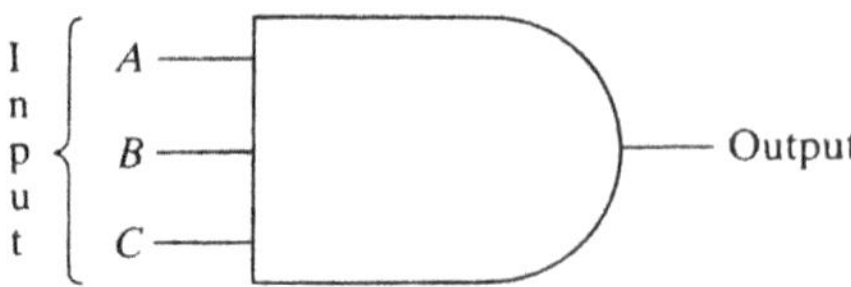

Figure 4-24 The symbol for a three-input AND gate.

device for all of the possible input signals. The output of an AND gate can be predicted using theorems 1 through 4 of Table 4-2.

A NAND (negated AND) gate functions in the opposite manner to an AND gate. It gives a high (1) output for all possible combinations of inputs except when all the inputs are high. The symbol for a three-input NAND gate is shown in Fig. 4-25 and the corresponding truth table is in Table 4-3. In the symbols for gates, a small circle at an input or output represents negation (inversion) of the signal.

An OR gate yields a high output when at least one of the inputs is high, i.e., when input *A or* input *B or* input *C or* any combination of inputs is high.

Table 4-3 The truth tables for the three-input AND gate shown in Fig. 4-24 and for the three-input NAND gate shown in Fig. 4-25

Inputs			Output	
A	*B*	*C*	AND	NAND
0	0	0	0	1
1	0	0	0	1
0	1	0	0	1
0	0	1	0	1
1	1	0	0	1
1	0	1	0	1
0	1	1	0	1
1	1	1	1	0

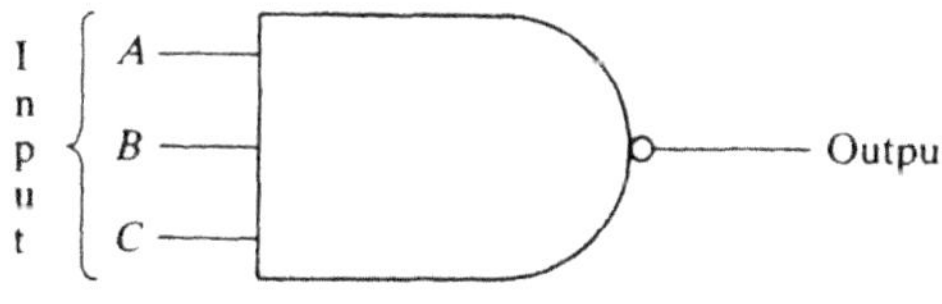

Figure 4-25 The symbol for a three-input NAND gate.

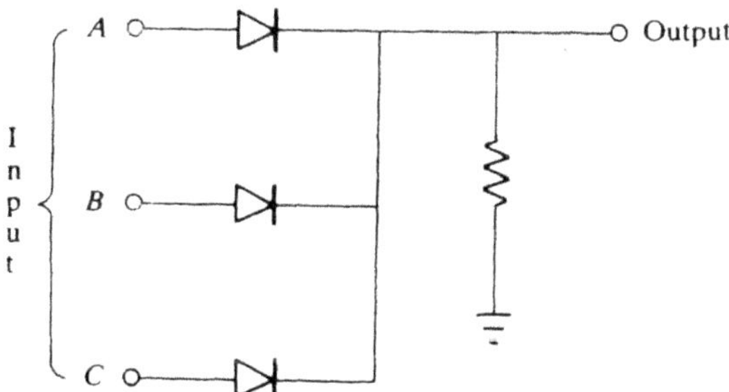

Figure 4-26 A three-input OR gate.

Figure 4-27 The symbol for a three-input OR gate.

A simple OR gate can be constructed from diodes as shown in Fig. 4-26. With no signal (low) applied to all of the inputs, the output is grounded (low). If a positive potential is applied to at least one of the inputs, the corresponding diode becomes forward-biased and a positive potential appears at the output. The symbol for a three-input OR gate is shown in Fig. 4-27 and the truth table for the device is listed in Table 4-4.

Table 4-4 Truth tables for the three-input OR and NOR gates shown in Figs. 4-27 and 4-28

Input			Output	
A	*B*	*C*	OR	NOR
0	0	0	0	1
1	0	0	1	0
0	1	0	1	0
0	0	1	1	0
1	1	0	1	0
1	0	1	1	0
0	1	1	1	0
1	1	1	1	0

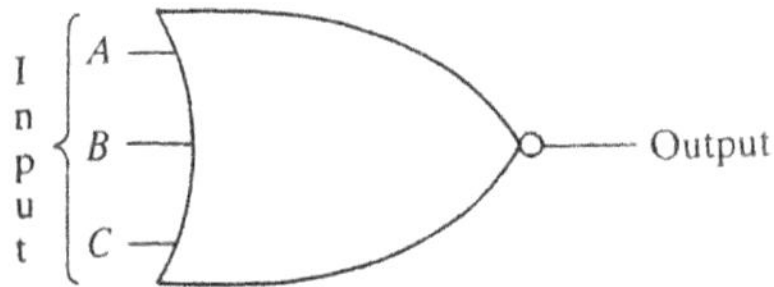

Figure 4-28 The symbol for a three-input NOR gate.

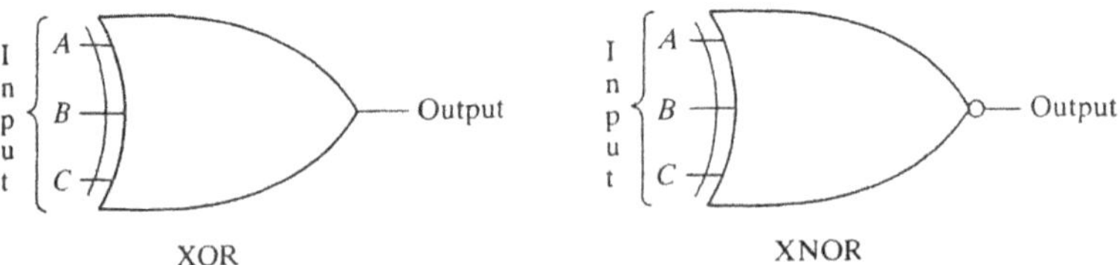

Figure 4-29 The symbols for a three-input EXCLUSIVE-OR gate and for a three-input EXCLUSIVE-NOR gate.

A NOR (negated OR) gate yields a high output only when all of the inputs are low. With all other possible combinations it yields a low output. The symbol for a three-input NOR gate is shown in Fig. 4-28. The corresponding truth table is listed in Table 4-4.

An EXCLUSIVE-OR gate yields a high output whenever one of the inputs is high, but not when more than one input is high. Similarly, an EXCLUSIVENOR gate yields a low output only when one of the inputs is high. The symbols for EXCLUSIVE-OR (XOR) and EXCLUSIVE-NOR (XNOR) gates are shown in Fig. 4-29. The corresponding truth tables for three-input gates are given in Table 4-5.

Table 4-5 Truth tables for three-input EXCLUSIVE-OR and three-input EXCLUSIVE-NOR gates

Input			Output	
A	*B*	*C*	XOR	XNOR
0	0	0	0	1
1	0	0	1	0
0	1	0	1	0
0	0	1	1	0
1	1	0	0	1
1	0	1	0	1
0	1	1	0	1
1	1	1	0	1

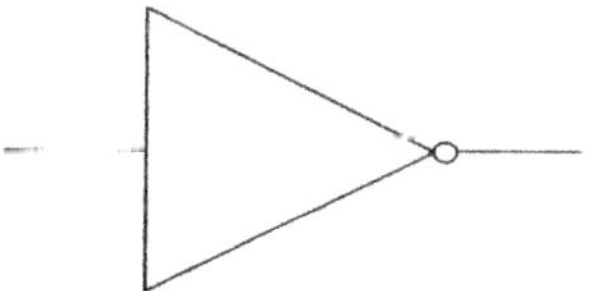

Figure 4-30 The symbol for an inverter.

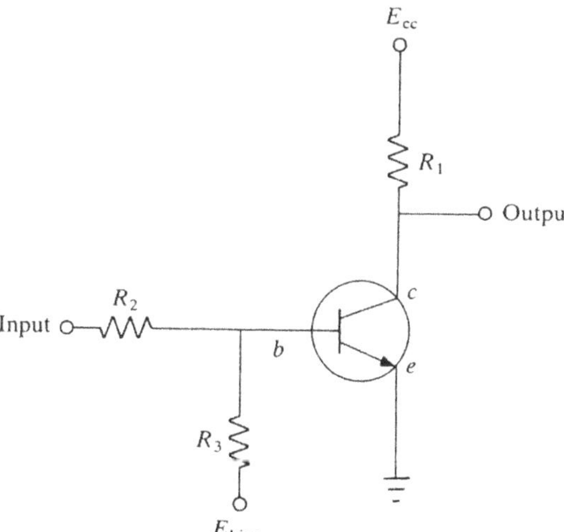

Figure 4-31 A transistor inverter.

All of the negated gates (NAND, NOR, and XNOR) can be created from the corresponding nonnegated gates (AND, OR, and XOR) with an *inverter*. An inverter converts a high signal (1) into a low signal (0) and converts a low signal into a high signal. Because inverters are widely used and perform a valuable function, they are given a symbol. The symbol for an inverter is shown in Fig. 4-30. Recall that an inverter can be an operational amplifier.

The circuit diagram for a simple inverter is shown in Fig. 4-31. With no input (low), the *npn* transistor is prevented from conducting a current by application of the proper bias potential to the base. The output potential is positive (high) as determined by E_{cc} and R_1. When a proper positive potential (high) is applied to the input terminal, the transistor is driven to saturation and the output potential decreases to a low value that is dependent upon the saturation potential between the collector and the emitter of the transistor.

Sample problem 4-6 Prepare a truth table for the circuit shown in Fig. 4-32.

SOLUTION The output from the AND gate is 1 (at *E*) only when inputs *A, B,* and *C* are all 1. The output from the inverter (at *F*) is the opposite of the input. The output from the OR gate is 1 whenever the signal at *E* or *F*, or *E* and *F* is 1. The truth table for the circuit is listed in Table 4-6.

Table 4-6 The truth table for the circuit shown in Fig. 4-32

Inputs				
A	*B*	*C*	*D*	Output
0	0	0	0	1
1	0	0	0	1
0	1	0	0	1
0	0	1	0	1
1	1	0	0	1
1	0	1	0	1
0	1	1	0	1
1	1	1	0	1
0	0	0	1	0
1	0	0	1	0
0	1	0	1	0
0	0	1	1	0
1	1	0	1	0
1	0	1	1	0
0	1	1	1	0
1	1	1	1	1

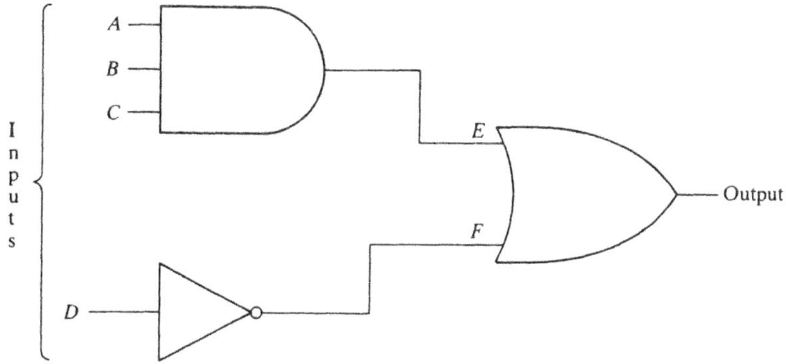

Figure 4-32 The circuit that is described in Sample problem 4-6.

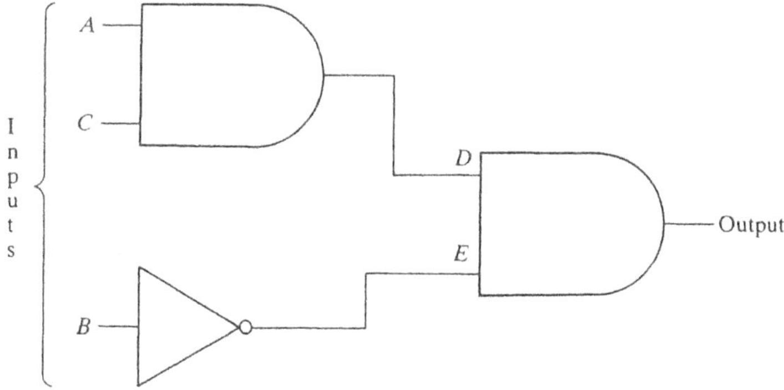

Figure 4-33 A circuit that performs the function that is described is Sample problem 4-7.

Sample problem 4-7 Draw the circuit diagram for a circuit with three inputs (*A*, *B*, and *C*) which yields a high output only when inputs *A* and *C* are high and input *B* is low.

SOLUTION The desired function can be accomplished with two AND gates and an inverter (Fig. 4-33). Other circuits can be designed that accomplish the same objective. The output is high only when the signals at *D* and *E* are high. The signal at *D* is high only when inputs *A* and *C* are high. The signal at *E* is high only when the input *B* to the inverter is low.

Flip-Flops

A *flip-flop* is a device that alternates between two output signals (0 and 1) depending upon the input signal. The major uses for flip-flops are as counters and as memory units, i.e., as devices that are used to store an input signal. Originally flip flops were developed for use in analog circuits. The major use of flip-flops at present is in digital circuits.

The simplest flip-flop can be constructed from two inverters connected as shown in Fig. 4-34. When the input terminal is momentarily contacted with either logic signal (0 or 1), the output from the first inverter becomes the opposite of the input and the output of the second inverter becomes identical to the input to the first inverter. Because the output from the second inverter is fed back to the input of the first inverter, the output to the device remains at the initially set value even when the external contact is absent.

A circuit that holds an applied signal in the manner described above is a *latch*. A latch stores 1 bit of information. The only way to change the stored bit in the flip-flop that is shown in Fig. 4-34 is by momentarily attaching the input or output terminal to the desired signal.

More convenient flip-flops, which have permanently attached input lines, can be constructed from logic gates. A flip-flop which is constructed from NAND gates is shown in Fig. 4-35. The two inputs to the device are labeled S (for set) and R (for reset). The output can be taken either at Q or $\overline{Q}$. Except when a bit is to be stored, both R and S are maintained at logical 1. If S is momentarily changed to 0 (grounded), the output from gate 1 at Q becomes 1 and $\overline{Q}$ becomes 0. The two outputs are maintained at those values by feedback from the output of gate 2 to the lower input of gate 1.

If the R input is momentarily changed to *0* (S is at 1), $\overline{Q}$ becomes 1 and Q becomes 0. Grounding R resets Q to 0 while grounding S sets Q to 1. Once S has been grounded, Q remains at 1 and $\overline{Q}$ remains at 0 regardless of further changes at S. The values are reset ($Q = 0$, $\overline{Q} = 1$) by grounding R.

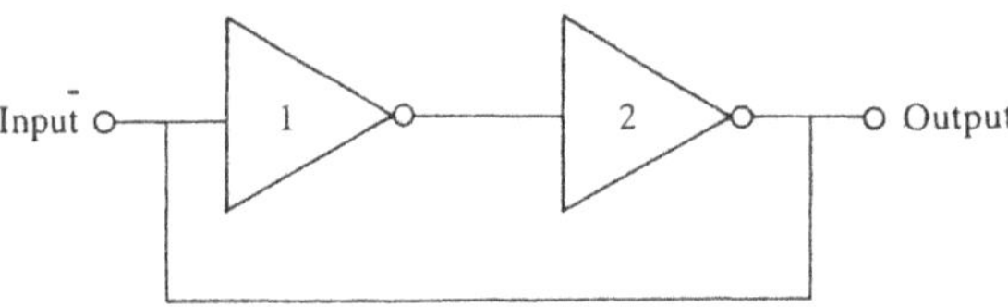

Figure 4-34 A flip-flop.

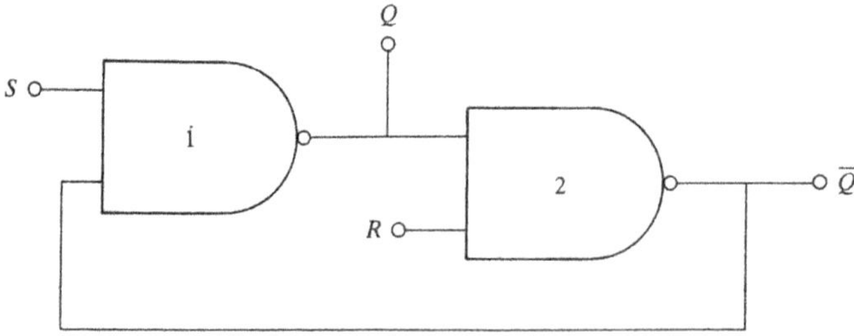

Figure 4-35 An *RS* flip-flop that is constructed from two-input NAND gates.

Simultaneously, momentarily grounding *S* and *R* leads to unpredictable outputs at *Q* and $\overline{Q}$. The flip-flop is termed an *RS flip-flop*. *RS* flip-flops also can be constructed from NOR gates.

An *RS* flip-flop which accepts an input signal only when prompted to do so by a second signal is a *clocked* (or *gated*) *RS flip-flop*. A clocked *RS* flip-flop can *be* constructed by inserting a two-input NAND gate in each of the *S* and *R* input circuits of the *RS* flip-flop. An input to the new NAND gate, which has an output attached to the set *S*' terminal of the original *RS* flip-flop, takes on the set *S* function of the clocked *RS* flip-flop. One of the input terminals to the second new NAND gate takes on the reset *R* function. The other two leads of the new NAND gates are connected together and to the clock (or gate) terminal *C* (Fig. 4-36). Of course, the flip-flop can have more than one *R* and *S* terminal.

When the clocked input is 1, the new pair of NAND gates function as inverters and the output signals of the clocked *RS* flip-flop are the opposites of those in the *RS* flip-flop. When *C* is 1, *R* is 0, and *S* is 1, the signals at *S*' and *R*' are 0 and 1 respectively and *Q* and $\overline{Q}$ become 1 and 0. If *C* is 1, *R* is 1, and *S* is 0, *S*' and *R*' become 1 and 0, and *Q* and $\overline{Q}$ respectively become 0 and 1. When *C* is 0, S' and *R*' are both 1 regardless of the values of *S* and *R*, and the outputs are held at the values that were determined when *C* was 1. The clocked *RS* flip-flop only accepts input signals when a pulse (1) is applied at *C*. All three inputs (*S*, C, and *R*) should never be 1 simultaneously because that results in unpredictable outputs. The logic symbols for flip-flops are rectangles with labeled terminals. The type of flip-flop usually can be determined from the type and number of terminals. The logic symbol for a clocked *RS* flip-flop is shown in Fig. 4-37.

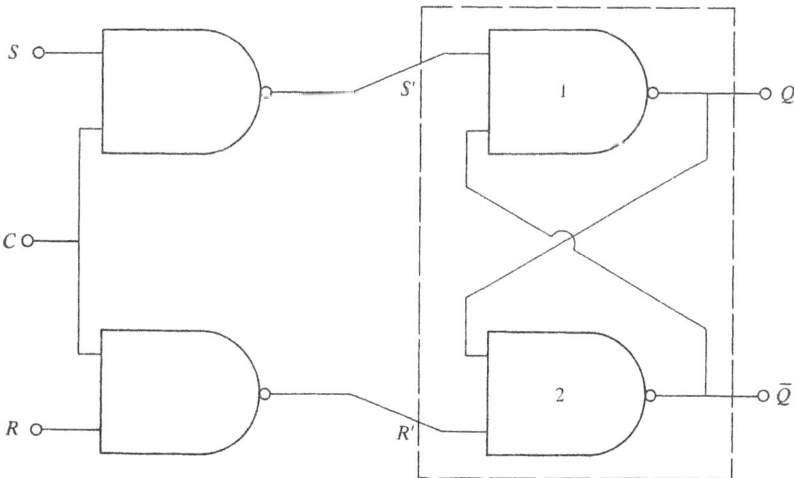

Figure 4-36 A clocked RS flip-flop. The enclosed portion is the RS flip-flop that is shown in Fig. 4-35.

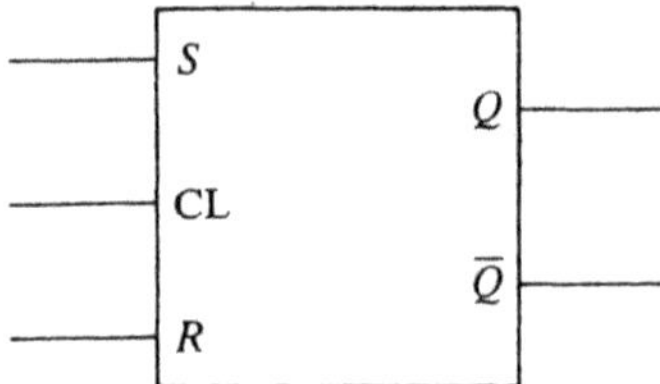

Figure 4-37 The logic symbol for a clocked RS flip-flop.

Another flip-flop which is often encountered and which is more versatile than *RS* flip-flops is the *JK master-slave flip-flop*. It consists of two clocked *RS* flip-flops that are connected in such a way that one (the master) controls the other (the slave). A *JK* master-slave flip-flop is shown in Fig. 4-38 and its logic symbol is shown in Fig. 4-39.

The two gated flip-flops in the *JK* flip-flop are controlled by the same clock pulse. The slave flip-flop is controlled by an inverted clock pulse which is slightly advanced (as compared to the pulse at the master flip-flop) by the resistor *R*. As with the clocked *RS* flip-flop, the master and slave flip-flops can only accept a signal when the clock input is 1. The inverter prevents the clock pulse at the two flip-flops from simultaneously being 1. When the clock pulse is 1, a bit of data is transferred to the master flip-flop. When the clock pulse goes to 0, further signals cannot be accepted at the master flip-flop, but the clock pulse at the slave becomes 1 and the slave flip-flop accepts the bit from the master. When the clock pulse changes back to 1, the pulse at the slave changes to 0 slightly before the 1 appears at the master, and the slave can no longer accept input from the master. A new bit is accepted by the master. The slave

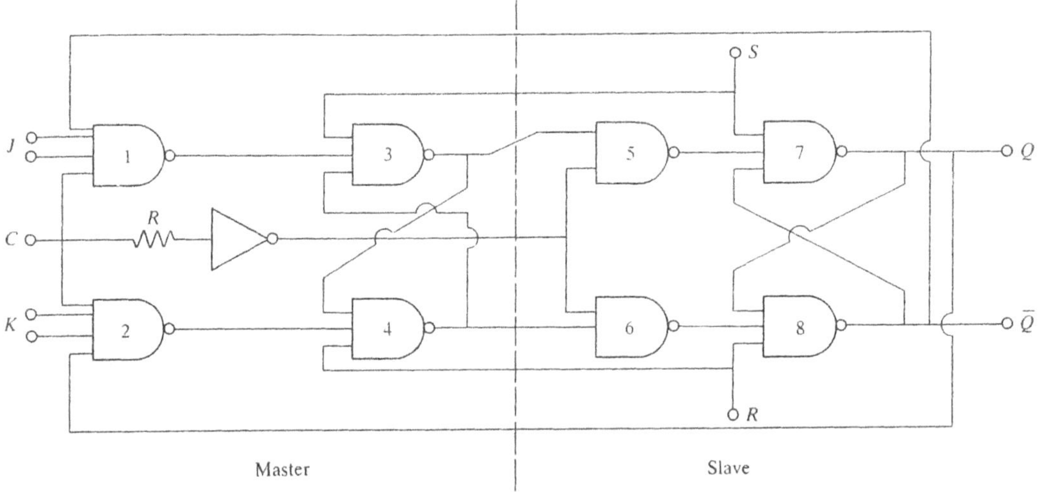

Figure 4-38 A *JK* flip-flop which is connected to toggle.

contains the old bit and the master contains the new bit. The flip-flop is made to *toggle* by connecting $\overline{Q}$ to J and Q to K. Toggling means that the flip-flop changes state (0 or 1) at Q and $\overline{Q}$ whenever a new clock pulse of 0 is applied. In the JK flip-flop, toggling occurs when R , S, J, and K are all 1. It can be prevented by changing one of the inputs.

Applications of flip-flops Flip-flops are primarily used for counting and for information storage. The flip-flops that have been described are *bistable*. They can remain in either the 0 or 1 state until changed by an external signal. *Monostable* flip-flops are stable in only one state. An external signal can cause the flip-flop to momentarily change states, but the device automatically returns to its stable state. An *astable* flip-flop has circuitry that causes the flip-flop to continuously change state. The output from an astable flip-flop is a square wave.

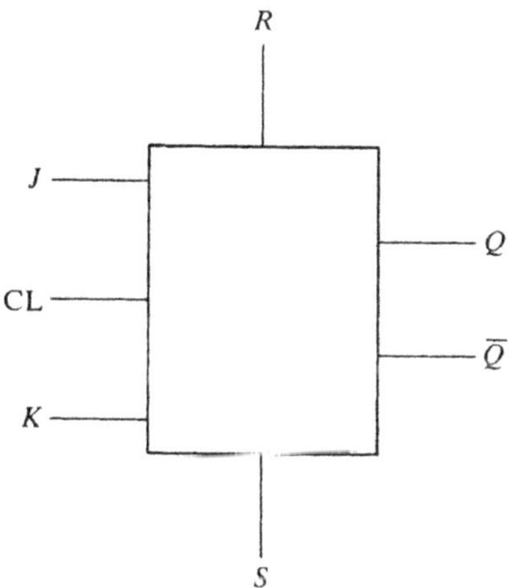

Figure 4-39 The logic symbol for a *JK* flip-flop.

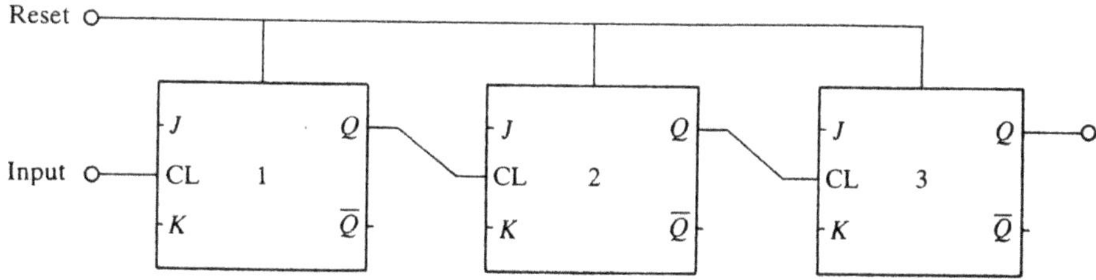

Figure 4-40 A binary counter that is constructed from *JK* flip-flops.

A series of cascaded flip-flops can be used as a binary counter. As an example, a series of *JK* flip-flops that are arranged to toggle can be connected as shown in Fig. 4-40. Initially all of the flip-flops are reset to state 0. Application of a clock pulse at the input causes flip-flop 1 to become 1 while the other flip-flops remain at 0. A second pulse at the input changes flip-flop 1 back to 0 and flip-flop 2 to 1. Flip-flop 2 changes to 1 because it is triggered by the output from the slave of *JK* flip-flop 1. Pulse 3 changes flip-flop 1 to 1 but does not alter flip-flop 2 because the flip-flops only toggle upon application of a pulse (1) at the clock input to the slave (a 0 at the master). Pulse 4 changes flip-flop 1 to 0, flip-flop 2 to 0, and flip-flop 3 to 1. The state of each flip-flop after each input pulse is shown in Table 4-7. Flip-flop 1 stores the least-significant bit of the binary number while flip-flop 3 stores the most-significant bit.

From Table 4-7 it is apparent that a counter that is composed of three flip-flops can store eight numbers (2^3; 0 through 7) before it returns to its initial setting. Likewise, a counter containing four flip-flops can store $2^4 = 16$ numbers (0 through 15). Any number of flip-flops can be combined to store as many numbers as required. If the state of each flip-flop can be examined, e.g., if a lamp attached to each flip-flop illuminates when the flip-flop is in state 1, the number of input pulses can be directly read in binary form.

A *register* is a series of flip-flops that is used to store logic signals (not necessarily numbers). A *shift register* is designed to move (shift) each bit of a logic signal to an adjacent flip-flop upon application of a clock pulse. The bits can either shift left or shift right through the device. The logic symbol for a shift register is shown in Fig. 4-41.

Table 4-7 The response of the counter shown in Fig. 4-40 to input pulses at the clock terminal of flip-flop 1

	Flip-flop state		
Pulse	**1**	**2**	**3**
0	0	0	0
1	1	0	0
2	0	1	0
3	1	1	0
4	0	0	1
5	1	0	1
6	0	1	1
7	1	1	1

Among other uses, shift registers can be used to change a *parallel* signal to a *serial* signal or to change a serial signal to a parallel signal. A parallel input signal is one in which each flip-flop of the register is simultaneously loaded with a bit of information through a separate connection; i.e., each bit of the logic signal is simultaneously carried by its own conductor. After parallel information is loaded into the shift register shown in Fig. 4-41, clock pulses cause each bit of the information to be shifted from left to right. Each successive bit of the information sequentially reaches the serial output of the register. Information which is sequentially transmitted through a single conductor is serial information. The parallel information which was loaded into the register is converted to serial information at the serial output of the register. *Synchronous counters* are designed to accept parallel data and *asynchronous counters* accept serial data.

The shift register of Fig. 4-41 can be used to convert serial information to parallel information by loading one bit of the serial information into the leftmost flip-flop at each clock pulse. At the next clock pulse the bit is shifted to the right and a new bit is added. When the register is full, the parallel information is read from the parallel output of the register.

If the serial output of a shift register is connected to the serial input, the information in the register is not lost at the output, but rather is cycled back through the register. A circulating register of that type is a *ring counter*. Ring counters can be used in automated laboratory equipment to sequentially control a series of repetitive laboratory steps such as inserting a sample, obtaining an instrumental reading, removing the sample, and rinsing the apparatus. After the last step, the ring counter is back to its original position and ready to start again.

Information-Storage Conventions

Several widely used methods are available that make it possible to code information in binary form. Numbers are usually written in octal, hexadecimal, or decimal form. Octal and hexadecimal numbers are particularly well suited to storage in binary systems such as flip-flop registers. An octal digit (0 to 7) can be represented by three bits ($2^3 = 8$) and a hexadecimal digit (0 to 15) by four

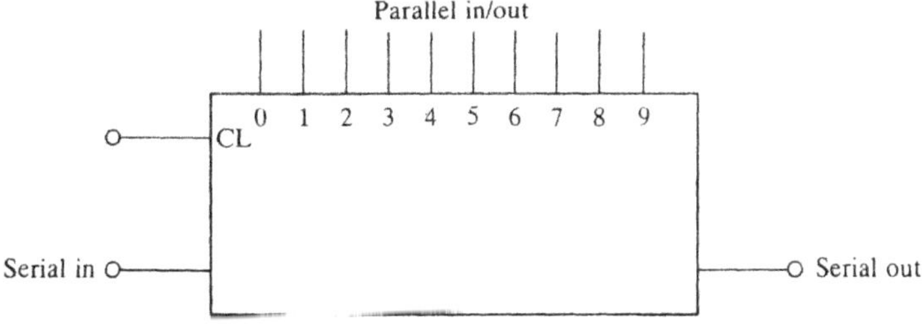

Figure 4-41 A 10-bit shift register.

bits ($2^4 = 16$). In order to represent a decimal digit (0 to 9), four bits are required, but not all of the possible combinations of 1's and 0's in the four bits are used. The three numbering systems and their binary equivalents are listed in Table 4-8. Several methods are used to represent a digit greater than 9 in hexadecimal notation. In Table 4-8 letters are used.

It is often desirable to express a four-bit binary number in decimal form. Many laboratory instruments perform measurements with binary logic systems but have decimal digital readouts. Several standard methods are available for storing binary data in such a way that it can be easily converted to decimal data. The most straightforward way is to use the binary equivalent of the decimal digits (0 to 9) that are listed in Table 4-8 to store each decimal digit in four bits. If a five-digit decimal number must be stored, 20 bits (5 decimal digits × 4 binary digits for each decimal digit) are required. Decimal numbers stored in that manner are stored in *binary-coded decimal* (BCD).

When information in addition to numbers must be digitally stored, the storage system requires more bits. The *American Standard Code for Information Interchange* (ASCII) uses seven-bit words to represent decimal numbers, the alphabet (upper and lower case), various symbols, and punctuation. Many of the ASCII characters are abbreviations of terms that

Table 4-8 A list of the possible numbers in each digit of three numbering systems and the binary equivalents

Octal	Decimal	Hexadecimal	Binary equivalent
0	0	0	0
1	1	1	1
2	2	2	10
3	3	3	11
4	4	4	100
5	5	5	101
6	6	6	110
7	7	7	111
	8	8	1000
	9	9	1001
		A	1010
		B	1011
		C	1100
		D	1101
		E	1110
		F	1111

are used with digital computers or printers; e.g., CR is an abbreviation for carriage return and LF for line feed. ASCII representations are used with many digital computers. Because seven bits are used, ASCII can be used to represent $2^7 = 128$ characters. The ASCII characters and the corresponding octal, decimal, hexadecimal, and binary values are listed in Appendix B. An extended set of ASCII characters use 8-bit words to increase the number of available characters to 256. The extended set was originally developed for use in computers by IBM.

Decimal Displays

Older binary displays often consisted of a series of lamps that are either on (1) or off (0). Decimal displays are more convenient. Many modern laboratory instruments use *light-emitting diodes* (LEDs), see the description on p. 57, or *liquid crystalline displays* (LCDs) to display measured decimal data. LEDs are constructed from diodes which emit light when they are forward-biased. The color of the display is dependent upon the material of which the diode is constructed. The emitted colors of available LEDs include red (gallium arsenide), orange and yellow (gallium arsenide phosphide), green (gallium phosphide), and blue and violet (indium gallium nitride). Other LEDs also are available.

The most common type of visual display is the seven-segment device (Fig. 4-42). Each character in the display consists of seven LEDs or LCDs (*a* through *g*). By properly activating combinations of the LEDs, all of the decimal digits and many letters can be displayed. The activated segments for each of the 10 decimal characters are listed in Table 4-9.

Liquid crystals are composed of long narrow molecules with a polar functional group at one end of the molecule. To prepare an LCD, a layer of twisted, nematic, liquid crystalline molecules is sandwiched between two polarizers. The upper polarizer is rotated 90° relative to the lower polarizer. Light is plane-polarized as it passes through the upper polarizer. With no applied potential the helical liquid crystal rotates the plane-polarized radiation by 90° so that it can pass through the lower polarizer. When a potential is

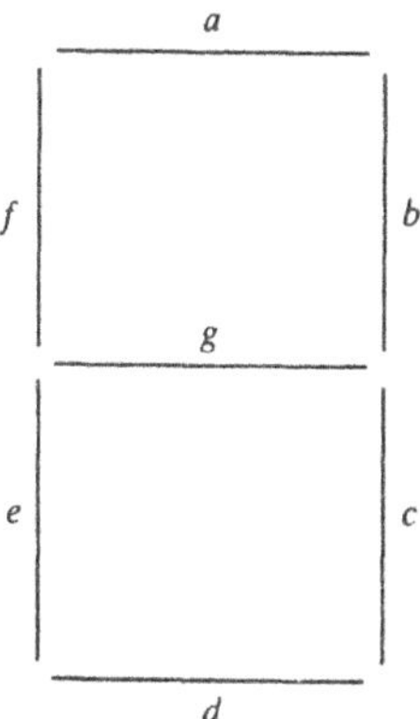

Figure 4-42 A seven-segment display.

Table 4-9 The portions of a seven-segment display that must be activated in order to display the decimal characters

Character	Activated segments
1	*e*, *f*
2	*a*, *b*, *d*, *e*, *g*
3	*a*, *b*, *c*, *d*, *g*
4	*b*, *c*, *f*, *g*
5	*a*, *c*, *d*, *f*, *g*
6	*a*, *c*, *d*, *e*, *f*, *g*
7	*a*, *b*, *c*
8	*a*, *b*, *c*, *d*, *e*, *f*, *g*
9	*a*, *b*, *c*, *d*, *f*, *g*
0	*a*, *b*, *c*, *d*, *e*, *f*

applied to the liquid crystals they align themselves with the upper polarizer causing the plane-polarized light to pass through the crystal without rotation. Because the lower polarizer is at 90° relative to the upper polarizer, the plane-polarized light is absorbed by the lower polarizer causing a darkened image to appear. Simultaneously applying potentials at several locations within the liquid crystal allows decimal digits to be displayed in the seven-segment form described for LEDs. A more thorough description of liquid crystals is given in Chapter 25.

The electronic device that accepts the four-bit (BCD) or seven-bit (ASCII) signal and activates the proper segments of the seven-segment LED or LCD display is a *decoder driver*. Seven-segment displays also can be constructed from incandescent filaments. Incandescent filaments require more current than LCDs or LEDs, however, and consequently are used less often.

A second type of device which occasionally is used in decimal displays is the *NIXIE tube* (NIXIE is a registered trademark of Burroughs). A NIXIE tube is a neon glow lamp which contains 10 cathodes in the shapes of the decimal figures (0 to 9) and a single anode. Whenever electric contact is made to a particular cathode, the corresponding figure glows. A disadvantage of the use of NIXIE tubes is the relatively high potential (about 55 V) that is required for their operation.

DIGITAL COMPUTERS

Digital computers are devices that use digital electronics to store and process data. A digital computer consists of a *memory,* a *central processor unit* (CPU), and one or more *input-output* (I / O) devices. The memory is used to store data and instructions. The data and instructions are stored in groups of sequential bits called *bytes*. In some cases more than one byte is required to store a particular instruction or datum point. A *word* is the combination of one or more required bytes. In most laboratory computers, a byte consists of 16 or 32 bits and a word is one or two bytes.

The CPU processes the data. The I/O devices transfer the results of the processed data to the computer operator or a device. A sketch of the major parts of a typical laboratory computer system is shown in Fig. 4-43.

The memory of a computer can be thought of as a series of flip-flops that are used to store binary data and instructions. A series of instructions to the CPU is a *program*. Generally, the memory of a computer is divided between *read-only memory* (ROM) and *random access memory* (RAM).

Read-only memory is memory which contains data or instructions that cannot easily be altered; i.e., the data is essentially permanent. ROM cannot be used to store newly obtained data. The data or instructions in ROM, however, can be *read* (copied) into the CPU and used to perform a desired function. ROM is hardwired in such a way that it permanently contains the desired information. Generally ROM is used to store information that frequently is used. The *bootstrap* loading program containing the instructions that permit the computer to load more complicated programs is often stored in ROM. Usually the bootstrap loader is automatically read from ROM into the CPU when the computer is activated. In some computers a programming language (see later description) such as BASIC, which requires relatively little space, is stored in ROM.

Programmable read-only memory (PROM) can be programmed by the user to contain any desired information. Once the information is in place, a mechanism is used to permanently store the information. After programming, PROM becomes ROM. *Erasable programmable read-only memory* (EPROM) allows the user to change information which has been entered into it.

Various types of devices can be used to store binary information. Sometimes shift registers are used. A disadvantage of the use of shift registers is the potentially long time that is required to access the information. The information only becomes available when it is shifted to the output of the register. Consequently, the time that is required to access a particular piece of information is dependent upon the distance between the information and the register output.

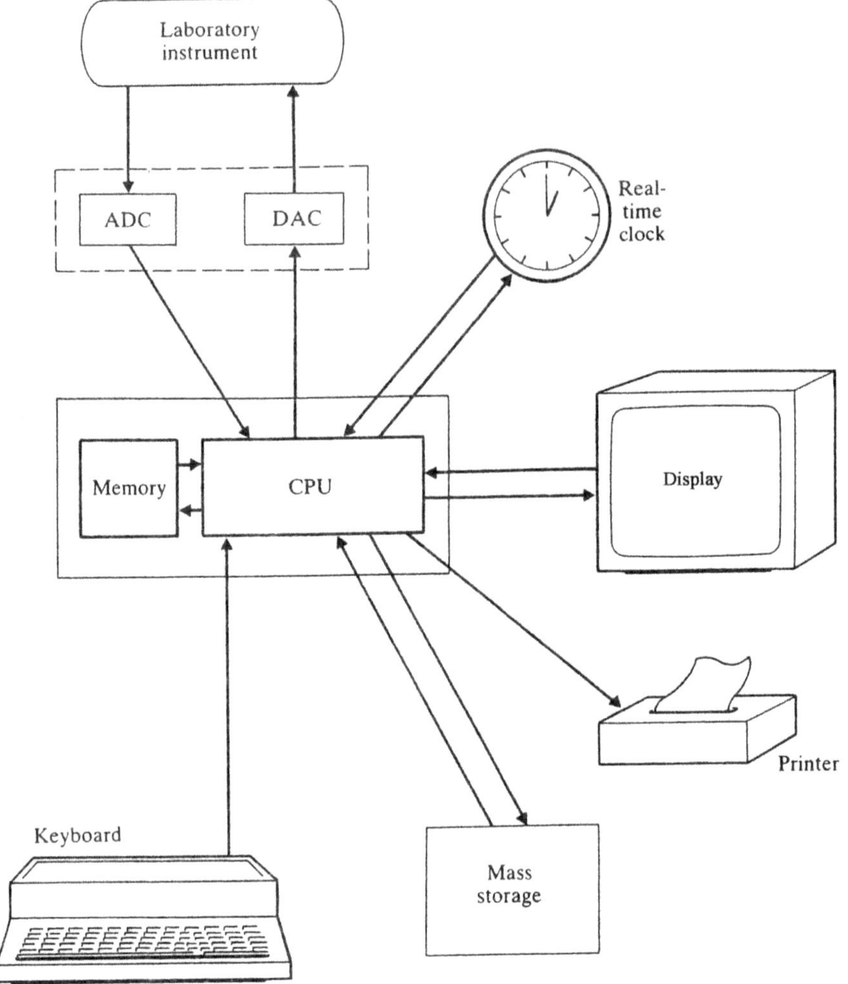

Figure 4-43 A diagram of the major portions of a laboratory computer system. The portions within the dashed line are not required by all laboratory instruments.

That disadvantage is overcome with the use of random access memory. The time that is required to access any piece of information that is stored in RAM is independent of the location of the information. All information is equally available. The access time depends upon the manner in which the memory was manufactured. Typically, MOS dynamic RAM has an access time of 300 ns or less.

Static memory retains information that is stored in it as long as power is supplied to the memory. Because the potentials in *dynamic* memory slowly decay with time, the information that is stored in dynamic memory must periodically be restored. Dynamic memory requires less space than static memory and consequently is frequently used in computers. Each location within the memory of a computer is given a unique *address*.

If the memory elements are composed of semiconductor devices, the information that is stored in the memory is lost when the power is turned off. Permanent storage is possible with memory which is composed of iron-core magnets. The magnetic field around each element in *core memory* remains after the power is discontinued. Several other types of memory also have been used.

Central Processor Unit

The central processor unit consists of an *arithmetic-logic unit* (ALU), several registers and counters, and the accompanying control circuits. It is contained in a single IC chip. The CPU controls the remainder of the computer. It can read or *fetch* instructions or data from memory, write to memory, interpret instructions in a program, and process or execute the instructions. The CPU can also respond to signals from outside the computer.

The operations of the CPU are controlled by an internal clock that has a period that varies with the computer. Older computers can have periods as long as 2 μs while more modern computers can have periods around 0.3 ns. A machine cycle consists of several (typically 1 to 6) clock cycles. During one machine cycle (also called a *state*) the CPU reads the address of an instruction or a datum in memory; during the next cycle it reads the instruction from the memory location; during the third cycle it executes the instruction or writes to memory. The series of machine cycles that is required to perform a task is an *instruction cycle*. Long instructions that require more than one machine cycle to read are loaded from consecutive memory locations into a register of the CPU during successive fetch phases. The instructions are not executed until after the entire instruction is in the CPU.

A datum which, as directed by an instruction in a program, is to be operated upon by the CPU is usually stored in a register that is called an *accumulator*. The instruction is performed and the result is placed either in an accumulator or in a memory location. The result can be placed back into the same accumulator that contained the original datum. As an example, the program might instruct the CPU to load an accumulator with the datum in a specific memory location, to add 7 to the datum, and to put the answer back into the accumulator.

The program is a logically related set of instructions. Except when a *jump* instruction is given, the instructions are written and stored in memory in the order in which they are to be executed. The instruction in memory location 155 is executed before the instruction in location 156, which is executed before the instruction in location 157, etc. A jump instruction tells the CPU to fetch and execute the instruction from a particular nosequential location in memory.

The memory address of the next sequential instruction is contained in a register that is termed the *program counter* (PC). After each instruction is read from memory, the PC is incremented by 1, thereby giving the address of the next instruction. The CPU uses the address in the PC to *point* to the next instruction; i.e., it fetches the instruction from the memory location that is listed in the PC.

Sometimes a particular portion of a program, known as a *subroutine,* is to be repeated several times during execution of a program. An example is a subroutine that is used to calculate the logarithm of a number. Rather than recopying the subroutine each time it is needed, it is more efficient to store the subroutine in a single set of memory locations and to instruct the CPU to take instructions from the memory locations when needed. That process is accomplished by loading the address of the next instruction to be executed, after processing the subroutine, into a separate counter that is called a *stack*. The last instruction in the subroutine tells the CPU to *return* to the main program. When a subroutine is called, the location of the subroutine is loaded into the PC and the next sequential location of the main program is loaded into the stack. The instructions in the subroutine are sequentially executed. When the return instruction is read, the PC is loaded with the last address that was loaded into the stack and the main program continues.

The stack can either be a reserved area in memory or a counter within the CPU. If the stack is a reserved area in memory, the CPU contains a *pointer register* which holds the address of the last entry in the stack. The stack is also used to hold the next memory location of a continuing program that has been interrupted by a signal from outside the computer, such as from a laboratory instrument.

The arithmetic-logic unit is the portion of the CPU in which all arithmetic and logical operations are performed on the data. One portion of the ALU is the adder, which adds the binary contents of two registers. The adder is usually constructed from AND and XOR gates. The rules followed for addition of binary numbers are the same as those used for decimal numbers.

Sample problem 4-8 Write the binary equivalents of the decimal numbers 18 and 23 and add the binary numbers. Convert the answer to a decimal number.

SOLUTION The binary equivalent of 18 is 10010 and the binary equivalent of 23 is 10111 (see Sample problem 4-5). The addition is performed as with decimal addition ($1 + 1 = 10$; $0 + 1 = 1$; $0 + 0 = 0$):

$$\begin{array}{r} 10010 \\ +10111 \\ \hline 101001 \end{array}$$

The answer can be converted to a decimal number as follows:

$$1 \times 2^0 + 1 \times 2^3 + 1 \times 2^5 = 1 + 8 + 32 = 41$$

As expected, the answer to $23 + 18$ is 41.

Subtraction is often performed in the ALU adder with the aid of *twos-complement arithmetic*. The leftmost bit of a binary number that is to be used for twos-complement arithmetic represents the sign of the number. A 0 indicates positive and a 1 indicates negative. As an example, if eight bits are used to represent a number, the bit furthest to the left is used to indicate the sign of the number and the remaining seven bits represent the magnitude of the number. Consequently, 00010111 is + 23 (decimal).

Subtraction can be performed by adding the number from which the subtraction is to be performed to the twos complement of the number which is to be subtracted. The number which is to be subtracted is written in twos-complement form and added to the other number. When the twos complement of the number is obtained, the sign bit is included in the operation and any overflow (after addition) beyond the number of bits in the word is ignored. The twos complement of an n-digit number X is given by

$$\text{Twos complement} = 2^n - X \tag{4-40}$$

Calculation of the twos complement can be simplified by first calculating the *ones complement* of the number. The ones complement of a binary number is obtained by replacing each 1 in the number with a 0 and each 0 in the number with a 1:

$$\text{Ones complement} = (2^n - 1) - X \tag{4-41}$$

The relationship between the twos complement and the ones complement can be obtained by rearranging Eq. (4-41) and substituting for the twos complement:

$$\begin{aligned}\text{Ones complement} &= (2^n - 1) - X = (2^n - X) - 1\\ &= \text{twos complement} - 1\\ \text{Twos complement} &= \text{ones complement} + 1\end{aligned} \tag{4-42}$$

The twos complement of a number is easily obtained by adding 1 to the ones complement of the number. If the answer to the subtraction is a negative number (1 in the most-significant bit), the number is expressed in twos-complement form. To convert a negative answer to a decimal number, the above procedure is reversed (try Prob. 4-24).

Sample problem 4-9 Subtract 15 from 42 using twos-complement arithmetic. Assume that an eight-bit accumulator is used for the subtraction. Convert the answer to its decimal equivalent.

SOLUTION The binary equivalent of 15 is 00001111 and the binary equivalent of 42 is 00101010. The subtraction can be performed by obtaining the twos complement of 00001111 (the subtracted number) and adding it to 00101010. The ones complement of 00001111 is 11110000. The twos complement is the ones complement + 1 [Eq. (4-42)]:

$$\text{Twos complement} = 11110000 + 1 = 11110001$$

The subtraction is accomplished by adding 00101010 to 11110001:

$$\begin{array}{r} 00101010 \\ +11110001 \\ \hline 100011011 \end{array}$$

Because the accumulator only has room for eight bits, the bit furthest to the left is discarded, and the answer is 00011011. The leftmost bit corresponds to the sign (0 is positive) and the remainder of the bits correspond to the magnitude of the answer. The decimal equivalent of the answer is + 27. Take note that since the answer is positive (0 in the leftmost bit), the answer was expressed in normal binary form rather than in twos complement form.

Multiplication can be performed either by adding the number the required number of times (28 × 3 is 28 + 28 + 28) or in the same manner in which multiplication is performed by hand. Each individual step in the process is stored in a register and added together in the accumulator to obtain the answer. Division can similarly be done by properly manipulating the adder. All mathematical functions can, in theory, be performed with only the adder. Most modern computers have a built-in hardwired capacity to perform subtraction, logic operations, shift capabilities, and sometimes other mathematical functions. The use of those capabilities usually shortens the time required to obtain an answer. Mathematical functions are often performed with the aid of an *algorithm*. An algorithm is a particular method for solving a particular type of mathematical problem. The program for the algorithm is stored in the computer as a subroutine.

In-Out Devices

Communication between a computer and other devices or between a computer and a human can be accomplished in either a serial or a parallel fashion. Serial devices require a single line for the transmission (plus other lines for other purposes such as indicating the direction of the communication and the start of a word) while parallel devices require a separate line for each bit of the transmitted datum or instruction.

Older computers communicated through separate serial and/or parallel ports. Some modern computers also have serial and/or parallel ports primarily for communication with older peripheral devices. Nearly all modern computers now communicate through a *universal serial bus* (USB). Laboratory and personal computers have multiple USB ports available. USBs were invented by Ajay Bhatt during his tenure at Intel Corporation. USB ports can supply electric connections to peripheral devices and have nearly completely replaced other manners of connecting computers to external devices. USB ports have the advantages of offering more rapid one and two way communications and providing a power source to run some external devices, such as USB flash drives which are used as data storage devices.

Information transmission from a computer to a device that cannot accept the information as rapidly as it can be transmitted by the computer requires the use of a *buffer*. The buffer stores some of the information until it can be accepted by the device and prohibits transmission of further information until the accepting device is ready. The rate at which information is transmitted is measured in *bauds* Bd. A baud is 1 bit/s. Serial transmission can be accomplished while using either potentials or currents to indicate the two logic states. A *current-loop* device uses a 20-mA current to indicate 1 and 0 mA to indicate 0. The currents are changed to the desired potential prior to use in the receiving device.

Several devices which can communicate with a computer are in common use. A keyboard is often used to enter instructions into the CPU. A *teletypewriter* (TTY) with a keyboard can be used for two-way communications with a computer but now is rarely used. Most computers communicate with humans through a *video display terminal* (VDT) which operates similarly to the screen of a television. Modern terminals use LCD or LED displays. Data and programs also can be transmitted between the computer and any of several mass-storage devices or between the devices and the computer. The most-popular mass-storage devices in the analytical laboratory are internal and external hard drives, *compact discs* (CDs) and USB flash drives.

When an unusually high rate of data transmission from a device to the computer is required, some computers are capable of permitting the data to be directly entered into reserved locations in memory while bypassing the usual I/O control equipment. That process is *direct memory access* (DMA).

Communication with or between computers over long distances is often accomplished over telephone lines or through wireless cell phone devices. A phone *modem* converts the binary signal from the transmitting device to a binary audio signal which is transmitted through the telephone network to the receiving device. A second modem at the receiving end converts the audio signal back to an electric signal which can be accepted by the receiving device. Older telephone modems usually operated at 300 or 1200 Bd although significantly faster transmission rates are now possible.

Many phone modems use two sets of two frequencies each. One pair of frequencies is used to transmit binary information from a particular modem and the second pair of frequencies is used for reception at the same modem. Typically 2225 (logical 1) and 2025 (logical 0) Hz are used as one pair of frequencies and 1270 (logical 1) and 1070 (logical 0) Hz as the second pair.

Communication with Laboratory Instruments

Laboratory instruments are attached to a serial, parallel or USB port of a computer. Usually communication to the instrument from the computer is through a DAC. Data communication from the instrument to the computer is through an ADC. The DAC or ADC is built into some laboratory instruments.

Those instruments that have digital outputs can be directly connected to the input port of some computers.

Timing of the communications is performed with a clock inside the computer. A typical procedure will be described. If the computer is to accept datum from the computer every 5 s, the required number of clock cycles that add up to 5 s is entered into a counter in the computer. After each cycle the number in the counter is decreased by 1. When the counter gets to 0, the computer uses a communications line to transmit a ready signal to the instrument (the ADC) and the original number is reentered into the counter. The instrument responds by transmitting the datum to the computer. Alternatively, the data can continuously be sent to the computer, but it is only read when the number in the counter is 0. Programming the computer to accept data requires proper determination of the number of clock cycles between readings, indicating the port into which the data is received and a series of memory addresses where the data is to be stored after it has been received.

Programming

The physical equipment that makes up a computer and its accessories is *computer hardware*. The instruction sets or programs which the computer stores and processes are *computer software*. Computer programs can be written in several forms. The most direct approach is to write the instructions in the binary form which can be directly understood by the CPU. Programs which are written in binary form are written in *machine language*.

Writing programs in machine language is tedious and prone to error because mistakes cannot be easily recognized by looking at the written program. It is easier to write the program in a *language* which uses mnemonics in place of the machine language instructions. Mnemonics are designed to make it easier to associate the instruction to the CPU with the process that the instruction requires the CPU to perform. As an example, ADD A,n means add the value *n* to the term that is in the accumulator and LDA, (BC) means load the accumulator with the data in memory location *BC*. It is easier for a human to remember a mnemonic than to remember the corresponding instruction in machine language. Different types of CPUs have different mnemonics. Those used as examples above are for a 280 processor. A program which is written with a set of mnemonics is written in *assembly language*.

Languages which are generally easier to use by humans than assembly language are termed *higher level languages*. Higher level languages use simple words to represent a series of relatively complex instructions to the CPU. Because of the direct correspondence between human language and the instructions in higher level languages, the languages are easy to use. Use of higher level languages do not require the human computer operator to understand the manner in which the computer performs the instructions. Some of the popular higher level languages are *Beginner's All-purpose Symbolic*

Instruction Codes (BASIC), *Common Business-Oriented Language* (COBOL), *FORmula TRANslation* (FORTRAN), *Pascal,* and A *Programming Language* (APL). If a higher level language is used, a program called a *compiler* is used to convert the written higher level program to a machine language program which can be executed by the CPU. Assembly language programs are converted to computer-executable programs with a program which is termed an *assembler.*

IMPORTANT TERMS

A/D converter
Accumulator
Active filter
Algorithm
ALU
Analog signal
AND gate
ASCII
Assembly language
Asynchronous counter
Bandwidth
Baud
Binary number
Binary-coded decimal
Bistable flip-flop
Boolean algebra
Byte
Clocked flip-flop
CMOS logic
Comparator
Constant-current amplifier
Core memory
CPU
Current-loop device
Current-to-voltage converter
D/A converter
Decibel
Decoder driver
Digital signal
Diode-transistor logic
DMA
Drift
Dynamic memory
ECL
EPROM
Fan-in
Fan-out
Flip-flop
Gain
Gate
Hardware
Integrated circuit
Inverter
JK flip-flop
Latch
Logic device
LSI
Machine language
Modem
Monolithic device
Monostable flip-flop
MSI
NAND gate
Negative logic
NIXIE tube
NOR gate
Offset voltage
Open-loop gain
Operational amplifier
OR gate
Parallel signal
Positive logic
Power dissipation
Program
Program counter
PROM
Propagation delay
RAM
RCTL
Register
Ring counter
ROM
RS flip-flop
RTL
Scaling
Serial signal
Shift register
Software

SSI
Stack
Static memory
Summer circuit
Summing junction
Synchronous counter
Time constant
Toggling
Truth table
TTL
Twos complement
ULSI
Virtual ground
VLSI
Voltage follower
Wien-bridge oscillator
Word
XNOR gate
XOR gate

PROBLEMS

Operational Amplifiers

4-1 Substitute $E_S = -E_o/A$ into Eq. (4-6) and rearrange the equation to yield Eq. (4-7).

4-2 Determine the gain of an operational amplifier if the output potential is 2.45 V when the input potential at the inverting terminal of the amplifier is – 3.8 mV.

4-3 If the feedback resistor is $1 \times 10^4\ \Omega$ for the circuit of Fig. 4-4, determine the value of the input resistor which will result in a gain of 8×10^2.

4-4 Determine the gain of the operational amplifier shown in Fig. 4-4 if the input resistor is 15 Ω and the feedback resistor is 10.2 kΩ.

4-5 Calculate the output potential for the summer circuit shown in Fig. 4-5 if all four resistors are 550 Ω and the three input potentials are 0.175, 0.248, and 0.124 V respectively.

4-6 Draw the circuit diagram for an operational amplifier circuit which will perform the following function:

$$E_o = -1000\,(E_1 + 2E_2 + 3E_3)$$

4-7 Starting with Eq. (4-16), derive Eq. (4-17).

4-8 Starting with Eq. (4-26), derive Eq. (4-28).

4-9 Show that the circuit in Fig. P4-9 is an antilogarithmic amplifier [refer to Eqs. (4-23) and (4-24)].

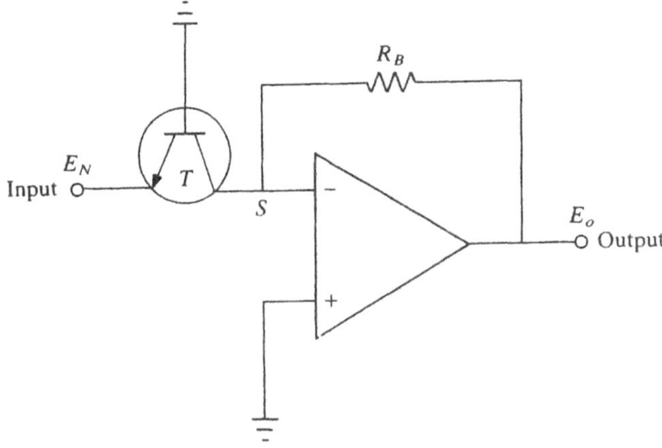

Figure P4-9

4-10 Design a circuit which simulates the first-order decay of a chemical reactant.

4-11 Substitute Eq. (4-36) into Eq. (4-35) and rearrange the result to yield Eq. (4-37).

Binary Numbers

4-12 Convert the binary number 10111010 into its decimal equivalent.

413 What is the decimal equivalent of the binary number 1110001011?

4-14 What is the largest decimal number that can be stored in 8, 10, and 12 bits? If the most-significant bit of the number is used to indicate the sign of the number, what is the range of decimal numbers which can be stored in 8, 10, and 12 bits?

4-15 Convert decimal 157 to a binary number.

4-16 What is the binary equivalent of 7497?

Gates and Flip-Flops

4-17 Write the truth table for the circuit in Fig. P4-17.

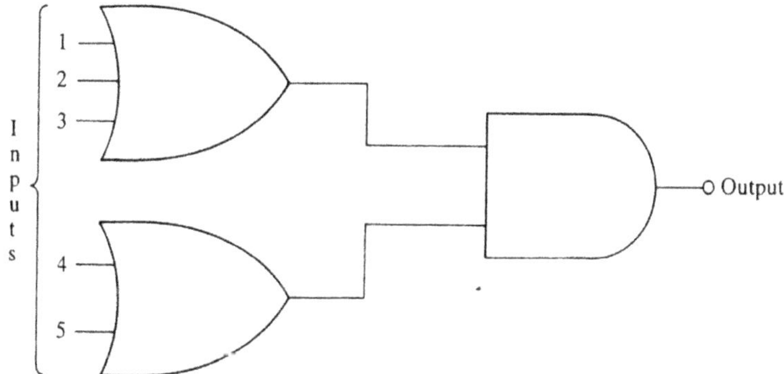

Figure P4-17

4-18 Write the truth table for the circuit in Fig. P4-18.

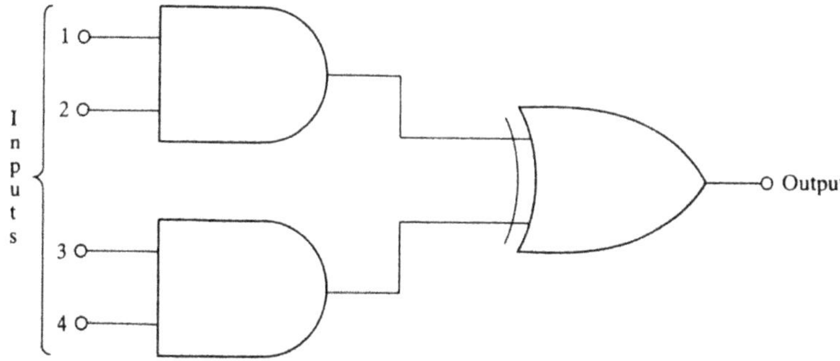

Figure P4-18

4-19 Design a four-input logic circuit which will yield a high output only when inputs 1 and 2 are high and inputs 3 and 4 are low.

4-20 Design a four-input logic circuit which will yield a high output when input 1 or 2 (but not both) is simultaneously high with both inputs 3 and 4.

Binary Arithmetic

4-21 Write the ones complement and the twos complement of the following binary numbers:

(*a*) 101101 (*b*) 01110001
(*c*) 11000111 (*d*) 00111111
(*e*) 10101010 (*f*) 01010100

4-22 Write the decimal equivalent of the binary numbers in Prob. 4-21. Assume the most-significant bit of the numbers is used to indicate the sign of the number (0 is positive and 1 is negative).

4-23 Add the following binary numbers:

(*a*) 110111 + 000111 (*b*) 010101 + 110000
(*c*) 100001 + 001111 (*d*) 010100 + 001001
(*e*) 000111 + 111111

4-24 Use twos-complement arithmetic to perform the following subtractions. Convert the answers to the decimal equivalents.

(*a*) 01001011 – 00110010 (*b*) 00111010 – 00111111
(*c*) 00001111 – 00010000 (*d*) 01111100 – 00000111
(*e*) 00101101 – 00000100

4-25 Convert the following decimal numbers to eight-bit, signed, binary numbers and perform the binary multiplications:

(*a*) 8 × 5 (*b*) 7 × (– 5)
(*c*) 9 × 12 (*d*) (– 5) × (– 17)

REFERENCES

1. Graeme, J. G, G. E. Tobey, and L. P. Huelsmand (eds.): *Operational Amplifiers, Design and Applications*, McGraw-Hill, New York, 1971.
2. Irvine, R. G.: *Operational Amplifier Characteristics and Applications*, Prentice-Hall, Englewood Cliffs, New Jersey, 1981:
3. Roberge, J. K.: *Operational Amplifiers, Theory and Practice*, Wiley, New York, 1975.
4. Stout, D. F., and M. Kaufman (eds.): *Handbook of Operational Amplifier Circuit Design*, McGraw-Hill, New York, 1976.

BIBLIOGRAPHY

Benedict, R.: *Electronics for Scientists and Engineers,* 2d ed., Prentice-Hall, Englewood Cliffs, New Jersey, 1976.

Brophy, J. J.: *Basic Electronics for Scientists,* 3d ed., McGraw-Hill, New York, 1977.

Diefenderfer, A. J.: *Principles of Electronic Instrumentation*, 2d ed., Saunders, Philadelphia, 1979.

Enke, C. G., S. R. Crouch, F. J. Heller, H. V. Malmstadt, and J. P. Avery: *Anal. Chem.*, 54: 367A (1982).

Floyd, Thomas L: *Digital Fundamentals*, Prentice-Hall, Upper Saddle River, New Jersey, 2000.

Jung, W. G.: *IC Op-amp Cookbook*, 2d ed., Howard W. Sams, Indianapolis, 1981.

Lenk, J. D.: *Manual for Operational Amplifier Users*, Reston, Reston, Virginia, 1976.

Mims III, F. M.: *103 Projects for Electronics Experimenters*, TAB Books, Blue Ridge Summit, Pennsylvania, 1981.

Vassos, B. H., and G. W. Ewing: *Analog and Digital Electronics for Scientists*, 2d ed., Wiley, New York, 1980.

Wait, J. V., L. P. Huelsman, and G. A. Korn: *Introduction to Operational Amplifier Theory and Applications*, McGraw-Hill, New York, 1975.

CHAPTER

FIVE

INTRODUCTION TO SPECTRAL METHODS OF ANALYSIS

The purposes of Chapter 5 are to introduce the basic principles, topics, and terminology that are common to all or most of the spectral methods. It is not intended to provide a detailed description of any particular method or of the apparatus that is used exclusively for a particular method. Detailed descriptions of individual analytical methods are provided in later chapters.

The spectral methods of analysis use measurements of the amount of *electromagnetic radiation* (EMR) that is absorbed, emitted, or scattered by a sample to perform an assay. Electromagnetic radiation is a form of energy. EMR possesses properties of both discrete particles which are termed photons and of waves. In some instances it is more convenient to think of EMR as photons while in other cases the wave model of EMR is easier to use. The two descriptions are related by the Planck equation

$$E = h\nu \tag{5-1}$$

in which E is the energy of a single photon of the EMR, ν is the frequency of the wave, and h is Planck's constant (6.625×10^{-34} J · s).

The frequency of a wave is the number of repetitive units of the wave which pass a fixed point in space every second. The unit of frequency is reciprocal seconds s^{-1} or *hertz* Hz. One hertz is one reciprocal second. The *period* T of a wave is the time that is required for a single wavelength to pass a fixed point in space. The period of a wave is the inverse of the frequency of the wave. The wavelength λ is the distance that corresponds to a single repetitive portion of the wave. It is the distance through which a wave moves before the wave starts to repeat itself. Frequency and wavelength are related by

$$\lambda \nu = \upsilon \tag{5-2}$$

where υ is the velocity of the EMR in the medium through which the radiation is traveling.

The velocity of EMR varies with the medium through which it travels. The velocity of EMR in a particular medium is conveniently expressed in terms of the *refractive index n* of the medium. The refractive index of a medium is defined as the ratio of the velocity c of EMR in a vacuum (2.9979×10^8 m/s) to the velocity υ in the medium

$$n = \frac{c}{\upsilon} \tag{5-3}$$

Consequently, the refractive index of a vacuum is exactly unity. Because the refractive index of a medium can vary with wavelength and with temperature, it is necessary to specify the wavelength and temperature at which the measurement of the refractive index is made. Often refractive index measurements are made at or near room temperature and at the emission wavelength (589.3 nm) of the D line from a sodium lamp. It is relatively easy to make measurements of n with an accuracy of seven or eight significant figures. Because n often does not significantly vary (within an accuracy of three or four significant figures) with wavelength in the visible region of the spectrum and with temperature (within 15-20°C), calculations obtained with Eqs. (5-3) - (5-5) can utilize values of n measured at wavelengths and temperatures that are slightly different from the desired values.

Solving Eq. (5-3) for ν and substituting the result into Eq. (5-2) yields

$$\lambda \nu = \frac{c}{n} \tag{5-4}$$

which relates the frequency and wavelength of EMR to the refractive index of the medium through which the radiation is traveling.

Solving Eq. (5-4) for frequency followed by substituting the result into Eq. (5-1) gives

$$E = \frac{hc}{\lambda n} \tag{5-5}$$

which relates the energy of a photon to the wavelength of the EMR and to the refractive index of the medium through which it is passing.

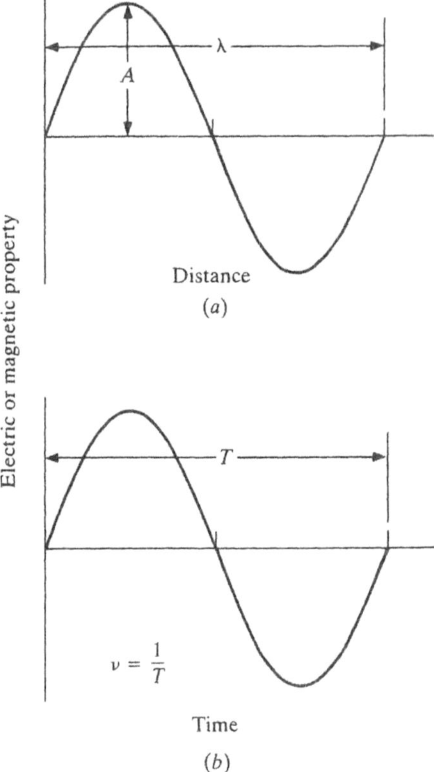

Figure 5-1 A sketch of a wave as viewed (*a*) instantaneously through a distance and (*b*) at a single location through time. *A*, amplitude; *T*, period; *λ*, wavelength; *ν*, frequency.

It is apparent from Eq. (5-1) that the energy of a photon is directly proportional to the frequency of the radiation. Equation (5-5) reveals that the energy of a photon is inversely proportional to the wavelength of the radiation. High-energy EMR has a high frequency and a short wavelength. Low-energy EMR has a low frequency and a long wavelength. Electromagnetic radiation can be described using the energy, wavelength, frequency, or period of the radiation. Because any one of the terms can be calculated from any of the other terms, it is not important which term is used to characterize the EMR. Wavelength and frequency are used more often than the other terms.

Amplitude A is the height of a wave. Sketches that illustrate the relationships between wavelength, period, frequency, and amplitude are shown in Fig. 5-1.

Sample problem 5-1 An assay was performed using electromagnetic radiation that has a wavelength of 443 nm in methanol (*n* is 1.329). Calculate the velocity of the EMR in methanol, the frequency of the radiation, the energy that is possessed by each photon of the radiation, and the period of the radiation.

SOLUTION The velocity of EMR in methanol can be calculated using Eq. (5-3):

$$\upsilon = \frac{c}{n} = \frac{2.9979 \times 10^8 \text{ m/s}}{1.329} = 2.256 \times 10^8 \text{ m/s}$$

The frequency can be calculated using either Eq. (5-2) or Eq. (5-4):

$$\lambda v = \frac{c}{n}$$

$$\left(443 \times 10^{-9}\text{m}\right)v = \frac{2.9979 \times 10^8 \text{ m/s}}{1.329}$$

$$v = 5.09 \times 10^{14} \text{ Hz}$$

The energy possessed by each photon is determined using Eq. (5-1):

$$E = hv = \left(6.625 \times 10^{-34}\text{J}\cdot\text{s}\right)\left(5.09 \times 10^{14} \text{ s}^{-1}\right)$$

$$= 3.37 \times 10^{-19} \text{ J}$$

The period of the radiation is the inverse of the frequency:

$$T = \frac{1}{v} = \frac{1}{5.09 \times 10^{14} \text{ Hz}} = 1.96 \times 10^{-15} \text{ s}$$

In Sample problem 5-1 the wavelength of the radiation in methanol was provided. Analytical instruments are usually calibrated to measure wavelengths in air rather than in a specific solvent. Consequently, the wavelengths on spectra that are recorded with those instruments do not correspond to the wavelengths that actually exist in a particular solvent. Throughout the remainder of the text, the listed wavelengths are those measured in air. Frequencies do not vary with medium because the energy of each photon [Eq. (5-1)] is constant as it passes between media.

Sample problem 5-2 Determine the wavelength in air *(n* is 1.0003) of the radiation that was described in Sample problem 5-1.

SOLUTION The frequency of the radiation was 5.09×10^{14} Hz in all media. Equation (5-4) can be used to calculate the wavelength in air:

$$\lambda\left(5.09 \times 10^{14}\right) = \frac{2.9979 \times 10^8}{1.0003}$$

$$n = 5.89 \times 10^{-7} \text{ m} = 589 \text{ nm}$$

Compare this wavelength to the wavelength in methanol which was 443 nm (Sample Problem 5-1).

The wavelength or frequency of EMR can be used to specify the energy that is possessed by a single photon of radiation. Those parameters are useful for qualitative analysis. In a beam of radiation the *power* P or *intensity* I is often measured because those parameters can be used for quantitative analysis. The power of a beam of EMR is defined as the rate at which energy is transported by the beam. It is the sum of the energy that is associated with each of the photons in the beam divided by the time during which the energies are summed. If the energy is measured in units of joules and the time in units of seconds, the corresponding unit of power is the *watt* W (1 W = 1 J/s). The average power of an EMR wave is directly proportional to the square of the frequency of the wave.

The intensity of a beam of electromagnetic radiation, such as that used in many spectrophotometers, is the power transmitted through a unit area that is perpendicular to the direction of movement of the radiation. Intensity is the power of the radiation divided by the area through which it is passing. The SI unit of intensity is watts per square meter W/m^2.

SPECTRAL REGIONS

Electromagnetic radiation is divided into regions according to energy. The several spectral regions that are useful for chemical analysis and their approximate energy, wavelength, and frequency boundaries are listed in Table 5-1. The boundaries between adjacent regions are not strictly defined. Radiation that is near a boundary between two regions can be defined as being in either region. Often the region in which EMR is classified is determined by the type of instrument that is used to perform the assay.

Table 5-1 A list of the spectral regions that are particularly useful for chemical analysis

	Approximate boundaries		
Region name	**Energy, J**	**Wavelength**	**Frequency, Hz**
X-ray	$2 \times 10^{-14} - 2 \times 10^{-17}$	$10^{-2} - 10$ nm	$3 \times 10^{19} - 3 \times 10^{16}$
Vacuum ultraviolet	$2 \times 10^{-17} - 9.9 \times 10^{-19}$	10-200 nm	$3 \times 10^{16} - 1.5 \times 10^{15}$
Near ultraviolet	$9.9 \times 10^{-19} - 5 \times 10^{-19}$	200-400 nm	$1.5 \times 10^{15} - 7.5 \times 10^{14}$
Visible	$5 \times 10^{-19} - 2.5 \times 10^{-19}$	400-800 nm	$7.5 \times 10^{14} - 3.8 \times 10^{14}$
Near infrared	$2.5 \times 10^{-19} - 6.6 \times 10^{-20}$	0.8-2.5 μm	$3.8 \times 10^{14} - 1 \times 10^{14}$
Fundamental infrared	$6.6 \times 10^{-20} - 4 \times 10^{-21}$	2.5-50 μm	$1 \times 10^{14} - 6 \times 10^{12}$
Far infrared	$4 \times 10^{-21} - 6.6 \times 10^{-22}$	50-300 μm	$6 \times 10^{12} - 1 \times 10^{12}$
Microwave	$6.6 \times 10^{-22} - 4 \times 10^{-25}$	0.3 mm-0.5 m	$1 \times 10^{12} - 6 \times 10^{8}$
Radiowave	$4 \times 10^{-25} - 6.6 \times 10^{-28}$	0.5-300 m	$6 \times 10^{8} - 1 \times 10^{6}$

TYPES OF ANALYSIS WITH EMR

Electromagnetic radiation is generally used for chemical analysis in one of three manners. If EMR is absorbed by the sample, the wavelength or frequency at which absorption occurs can be used for qualitative analysis. The extent to which the absorption occurs can be used for quantitative analysis.

If the sample either naturally emits radiation or can be forced to emit radiation by application of energy, the wavelength at which emission occurs can be used for qualitative analysis and the intensity of the emission can be used for quantitative analysis. If the emission follows absorption of energy from incident EMR or from a chemical reaction, the emitted radiation is *luminescence*. EMR that is *scattered* from a sample can sometimes be used for qualitative and quantitative analysis.

Atoms, molecules, and ions can exist in many energetic states or levels. During absorption energy from an incident photon can be absorbed by the chemical species with a resulting increase in the energetic state of the species. Luminescence occurs when a species in a higher energetic state relaxes to a lower state while simultaneously emitting EMR.

Different energetic transitions within a chemical species occur in different regions of the EMR spectrum. High-energy x-ray tansitions are associated with movement of an electron between an inner-shell atomic orbital and a higher atomic orbital. Absorption in the ultraviolet and visible regions is associated with electron transitions between the highest occupied electron orbital and a higher unoccupied orbital. EMR emission corresponds to the opposite process. An outer-shell electron that is in the lowest available electron level is in the *ground electron state*.

Emission and absorption in the infrared region generally correspond to energetic transitions between vibrational or rotational states of polyatomic species. Because discrete atoms are not attached to other atoms, they cannot possess vibrational or rotational energy levels, and cannot absorb or emit in the infrared region. Changes in rotational levels are observed in the microwave region. Changes between vibrational levels are of higher energy than those between rotational levels and usually occur in the infrared region. In a polyatomic species the total energetic change corresponds to the sum of the electron, vibrational, and rotational energetic changes. Radiowave radiation is used in nuclear magnetic resonance spectrophotometry as a method of orienting spinning nuclei in an externally applied magnetic field. Absorption corresponds to reorientation to a less-stable (higher energy) orientation within the magnetic field.

Absorption

Measurements of absorption are often made as illustrated in Fig. 5-2. EMR that is emitted from a *source* passes through a wavelength-limiting device, and

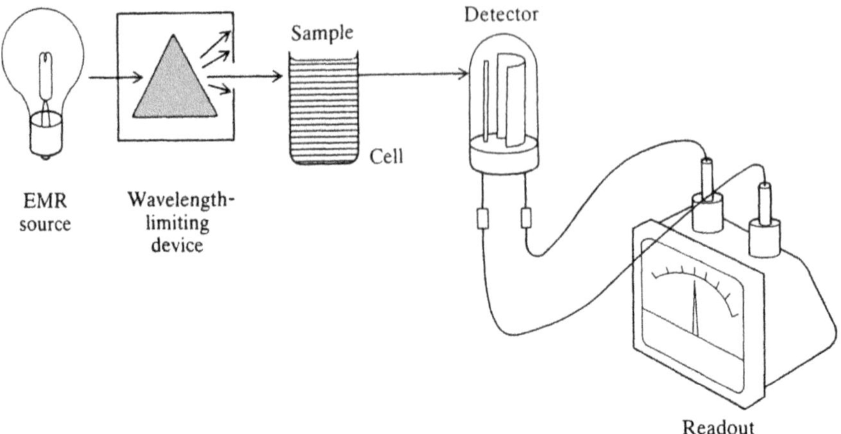

Figure 5-2 A diagram of the apparatus that is used for absorptive measurements.

impinges upon the sample that is held in a *cell* or *cuvet*. EMR that is not absorbed by the sample passes through the cell and strikes a *detector*.

The type of source that is used is dependent upon the spectral region in which radiation is required. The wavelength-limiting device restricts the EMR that reaches the cell to a relatively narrow band. Usually the device is either a monochromator or a filter. Both devices are described later in the chapter.

A cell is a vessel that contains the sample during the measurement. In most cases the cell has rigid walls. In some forms of atomic absorption and atomic emission spectroscopy the cell consists of a flame, plasma or other energy source. The walls of the cell are transparent to the EMR that is used for the assay.

The detector is a transducer that converts the intensity of the impinging radiation to an electrical signal. The electric output from the detector is amplified or manipulated as required and directed to a readout device. Typically the readout device is a recorder, analog meter, digital meter, or computer-controlled device.

BEER'S LAW

Several things can happen to the EMR as it passes through a cell (Fig. 5-3). A portion of the incident radiation can be reflected from its original path by the interior or exterior cell walls. The radiation can be scattered by the cell walls or a suspended particle within the cell. Some of the radiation can be absorbed by the cell walls and by the sample within the cell.

If a measurement of the amount of absorption by the sample is be used to perform the assay, it is desirable to minimize all mechanisms other than sample

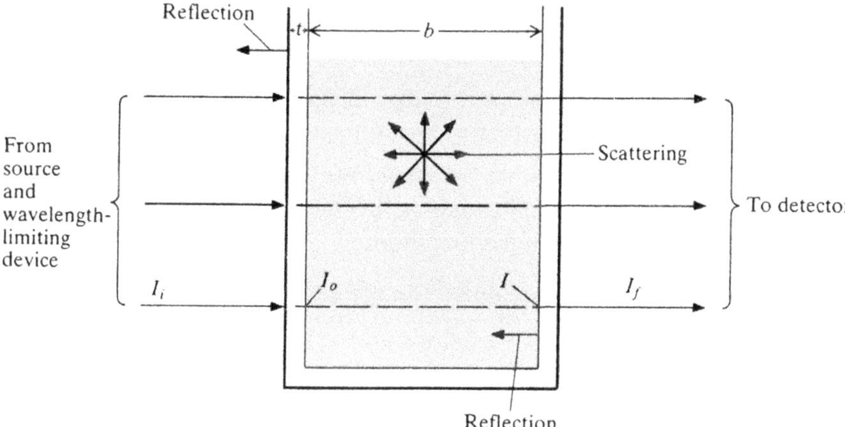

Figure 5-3 A sketch of several potential paths of EMR as it passes through a cell. I_i, incident intensity; I_o, intensity at inner entrance wall; I, intensity at interior exit wall; I_f, intensity after exiting cell; b, cell pathlength; t, thickness of the cell wall.

absorption by which the intensity I_f of the radiation that exits the cell is lessened. Absorption and scattering by the cell walls is minimized by careful choice of the materials from which the walls are constructed. As an example, glass can be used for cell walls when studies are performed in the visible region because it does not appreciably absorb or scatter visible radiation. Scattering by suspended particles within the sample solution can be minimized or eliminated by careful solution preparation.

If the material from which the cell walls are constructed is carefully chosen, the intensity of the incident radiation I_i that strikes the cell is nearly identical to the intensity I_o of the radiation as it contacts the sample. Likewise, the intensity I of the radiation after it has passed through the sample is identical to the intensity I_f of the radiation after it has passed through the sample and exited from the cell (passed through the cell wall).

As radiation passes through an absorptive sample in the cell, some of the photons are absorbed by the sample and the intensity of the radiative beam is decreased. At a single wavelength, all of the photons in a beam of EMR possess identical energy [Eq. (5-5)]. The need to use a single wavelength is the reason for the wavelength-limiting device in Figure 5-2. The total energy in the beam is the product of the energy possessed by each photon and the number of photons in the beam. In that case the intensity I of the beam can be defined as the number of photons in the beam divided by the time in seconds of operation of the beam and its cross-sectional area A. EMR that consists of a beam of photons of identical or nearly identical wavelength is *monochromatic* radiation.

Polychromatic EMR consists of photons that do not have identical wavelengths. The following derivation assumes the use of monochromatic radiation.

Absorption can occur when a photon of the proper energy strikes an atom, molecule, or ion of the absorbing species. The number of absorbed photons from the radiative beam is proportional to the number of photons of proper energy in the beam and to the number n of species in the sample that can potentially absorb the radiation:

$$\text{Absorbed photons} = kn\ (\text{photons in beam}) \tag{5-6}$$

The number of particles (atoms, molecules, or ions) of the substance that can absorb the radiation in a thickness db of the cell can be calculated from the concentration C (M) of the absorbing species, the cross-sectional area of the radiative beam, and Avogadro's number. If the volume V of solution within the cell which has a width db and cross-sectional area A is known, the number of particles of the absorbing species within the volume is given by

$$n = C(\text{moles/liter}) \times 1\ \text{liter}/1000\ \text{cm}^3 \times V\,(\text{cm}^3) \times 6.023 \times 10^{23}\ (\text{particles/mole}) \tag{5-7}$$

The volume of the sample that has a cross-sectional area that is equivalent to that of the radiative beam which passes through the cell and has a thickness db is Adb. Substitution of Adb for V in Eq. (5-7) and rearrangement yields

$$n = 6.023 \times 10^{23}\ ACdb/1000 = k'\,Cdb \tag{5-8}$$

The constant k' in Eq. (5-8) is equivalent to $6.023 \times 10^{20}\ A$.

The number of photons in the radiative beam is directly proportional to the intensity of the beam:

$$\text{Photons in beam} = k''\,I \tag{5-9}$$

The number of absorbed photons is proportional to $-\,dI$ (the negative sign indicates the decreased intensity):

$$\text{Absorbed photons} = -\,k''\,dI \tag{5-10}$$

The constant k'' in Eqs. (5-9) and (5-10) is the proportionality constant. Substitution from Eqs. (5-8), (5-9), and (5-10) into Eq. (5-6) gives

$$-\,k''\,dI = k(k'Cdb)(k''I) \tag{5-11}$$

Rearrangement and combination of constants yields

$$\frac{dI}{I} = -\,k'''\,Cdb \tag{5-12}$$

where k''' is kk'. Integration of Eq. (5-12) between the limits of intensity (I_o to I) and throughout the pathlength b of the cell gives

$$\log \frac{I}{I_o} = -\frac{k'''}{2.303} Cb \tag{5-13}$$

The ratio of I to I_o is the *transmittance* T:

$$T = \frac{I}{I_o} \tag{5-14}$$

Percent transmittance $\% T$ is $100T$:

$$\% T = 100\ T \tag{5-15}$$

Substitution from Eq. (5-14) into Eq. (5-13) and combination of the constants yields

$$\log T = -\varepsilon bC \tag{5-16}$$

where ε is the *molar absorptivity* of the absorbing species. The *absorbance* A is defined as $-\log T$:

$$A = -\log T \tag{5-17}$$

Substitution of A into Eq. (5-16) gives the Bouger-Beer law, the Lambert-Beer law, or Beer's law:

$$A = \varepsilon bC \tag{5-18}$$

While deriving Beer's law the unit of concentration of the absorbing species was molarity and the cell pathlength was measured in units of centimeters. If the concentration is measured in units other than molarity and/or the cell pathlength is measured in units other than centimeters, Beer's law is

$$A = abC \tag{5-19}$$

where the molar absorptivity is replaced by the *absorptivity* a. Equation (5-18) is used only when the concentration is measured in units of molarity and the cell pathlength is measured in units of centimeters. The value of a in Eq. (5-19) is dependent upon the units of b and C.

From Beer's law it is apparent that the absorbance of a solution is directly proportional to the concentration of the absorbing species. Advantage is taken of that relationship for quantitative analysis. If the cell pathlength and the absorptivity or molar absorptivity of the absorbing species is known, Beer's law can be used directly to calculate the concentration of the absorbing species.

Sample problem 5-3 The molar absorptivity of the iron(II)-2,2′,2″-terpyridyl complex is 1.11×10^4 at 522 nm. Calculate the concentration of the complex in a solution which has a percent transmittance of 38.5 at 522 nm in a cell with a pathlength of 1.00 cm.

SOLUTION Equation (5-15) is used to convert % T to T and Eq. (5-17) is used to convert T to A:

$$100T = 38.5; \quad T = 0.385$$

$$A = -\log T = -\log 0.385 = 0.415$$

Beer's law is used to calculate concentration:

$$A = \varepsilon bC = 1.11 \times 10^4 \times 1.00 \times C$$

$$0.415 = 1.11 \times 10^4\, C$$

$$C = 3.74 \times 10^{-5}\, M$$

Deviations from Beer's Law

If Beer's law is obeyed, a plot of absorbance as a function of concentration is linear with a slope of εb (or ab) and the plot goes through the origin (A is 0 when C is 0). If Beer's law is not obeyed, the measured absorbance can be either larger or smaller than that predicted using Beer's law. If the actual absorbance is greater than that predicted using Beer's law, the solution exhibits a *positive deviation* from Beer's law. Similarly, an absorbance that is less than predicted has a *negative deviation* from Beer's law. Typical plots of absorbance as a function of concentration for a solution which obeys Beer's law and for solutions which deviate from Beer's law are shown in Fig. 5-4. Deviations from Beer's law can be caused either by problems related to the instrumentation that is used to perform the analysis or by chemical reactions that are taking place within the solution. The major causes for deviations from Beer's law are described in the following sections.

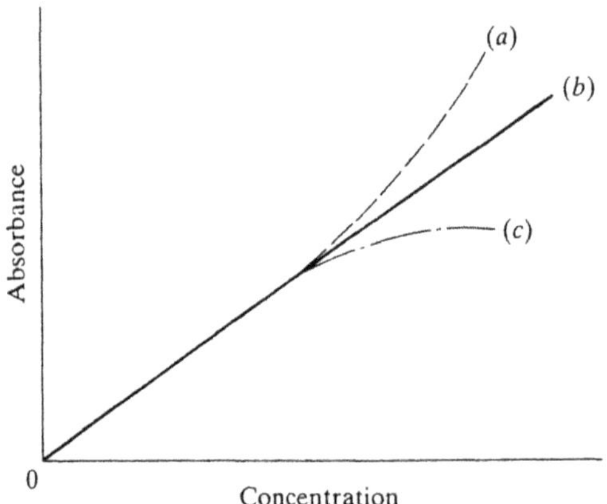

Figure 5-4 Plots (working curves) of absorbance as a function of concentration for (*a*) a solution which exhibits a positive deviation from Beer's law, (*b*) a solution which obeys Beer's law, and (*c*) a solution which exhibits a negative deviation from Beer's law.

Instrumental causes for deviations from Beer's Law Fluctuations in the electric power can cause a deviation from Beer's law in single-beam instruments by changing the intensity of the radiation that is emitted by the EMR source and by changing the response of the detector to the EMR which strikes it. If I and I_o [Eq. (5-14)] are not measured at the same power level because the level is fluctuating, the transmittance and absorbance [Eq. (5-17)] can be in error.

Stray radiation that can strike the detector without passing through the sample can lead to a negative deviation from Beer's law. Stray radiation can result when EMR is reflected from the walls of the cell and the walls of the cavity within the instrument into which the cell is placed. The effect of stray radiation is usually greatest at wavelengths that are near the edge of the available spectral range for the particular instrument. As an example, stray radiation is often observed at wavelengths below about 220 nm when an instrument is used that measures absorbance in the ultraviolet-visible region. Stray radiation is often minimized or eliminated by painting the cell compartment with flat black paint and by carefully sealing the door to the compartment so that radiation from outside the compartment is prevented from entering.

While deriving Beer's law it was necessary to assume the use of monochromatic radiation. In an actual assay the radiation has a finite wavelength bandwidth. If the absorptivity of the sample varies significantly throughout the wavelength band that is used for the study, a deviation from Beer's law can be expected. The deviation can be either positive or negative depending upon the situation. Cheng (see Reference 1, p. 332) provides a more detailed description of the manner in which polychromatic radiation affects measurements of absorbance. Deviations from Beer's law owing to the use of polychromatic radiation can be minimized by using narrow bands of radiation and by choosing a wavelength for the assay in a region in which the absorptivities at neighboring wavelengths do not significantly differ from that at the chosen wavelength. Practically, that generally means choosing a wavelength on a spectral peak or in a spectral valley.

Error resulting from an inaccurate response of the detector is most likely to be encountered when the intensity of the EMR that strikes the detector is either very large or very small. When the concentration of the absorbing species is large, the intensity of radiation that strikes the detector is small. Because detectors cannot accurately measure small intensities, measurement of absorbance at high sample concentrations is inaccurate. When the concentration of the absorbing species is small, the intensity I of the EMR that strikes the detector is nearly identical to the intensity I_o in the absence of an absorbing species. Because absorbance is log (I_o/I), a small error in measuring I when I_o is nearly equal to I can lead to a relatively large error in absorbance. This is particularly a problem with single-beam spectrophotometers. Error owing to inaccurate detector response at high and low concentrations is random. It can be minimized by restricting absorbance measurements to the concentration range in which the absorbance varies from about 0.2 to 1.

Instrumental noise In some cases accurate instrumental measurements are limited by the amount of instrumental noise. The noise that is associated with photometers and spectrophotometers can be classified as *Johnson noise*, *signal shot noise*, or *flicker noise*. Johnson noise is caused by thermally induced motion of electrons in the current-measuring circuit of the detector. The resulting increase or decrease in current flow through the circuit causes a random variation in the measured current.

Signal shot noise results from the occurrence of discrete events such as the emission of an electron from the photocathode of a photomultiplier tube or phototube (detectors). The number of emitted electrons for each impinging photon is not constant, but occurs in a normal (Gaussian) distribution. The variation in current flow as a result of different numbers of emitted electrons is shot noise. Shot noise increases with the temperature of the detector. It is necessary to cool some detectors in order to minimize shot noise.

Flicker noise in spectral instruments is the result of the altering intensity of emission from the radiative source. It is particularly important during low-absorbance measurements. It logarithmically decreases in importance as the absorbance increases.

Chemical causes for deviations from Beer's law Any chemical reaction that can alter the concentration of an absorbing species can result in a deviation from Beer's law. If the concentration is decreased because of the chemical reaction and the product does not absorb radiation at the wavelength at which the measurement is made, a negative deviation occurs. If a product of the chemical reaction absorbs more strongly than the assayed substance, a positive deviation occurs. Among the types of chemical reactions which can lead to a deviation from Beer's law are association-dissociation reactions, acid-base reactions, polymerization reactions, complexation reactions, and reactions with the solvent.

Many examples of association-dissociation reactions that can alter the concentration of an absorbing species exist. The formation of an absorbing complex by the addition of a ligand to a metal ion is a common example. Copper(II) in the solvent dimethylformamide forms a complex with chloride which absorbs radiation at 438 nm:

$$Cu^{2+} + 4Cl^{-} = CuCl_4^{2-} \tag{5-20}$$

The copper ion does not absorb radiation at that wavelength. If the equilibrium position of the reaction is shifted by changes in the reactant concentrations, such as occur upon dilution or after addition of another reactant, then the concentration of the absorbing species changes (Le Chatelier's principle) with an accompanying deviation from Beer's law. As an example, if the concentration of chloride increased from 0.01 to 1 *M*, the reaction shown in Eq. (5-20) would be expected to shift to the right, causing an increased amount of the absorbing complex to be formed.

The chemical reactions of acid-base indicators are examples of acid-base reactions that alter the concentration of an absorbing species. The acid-base reaction of *p*-nitrophenol is as follows:

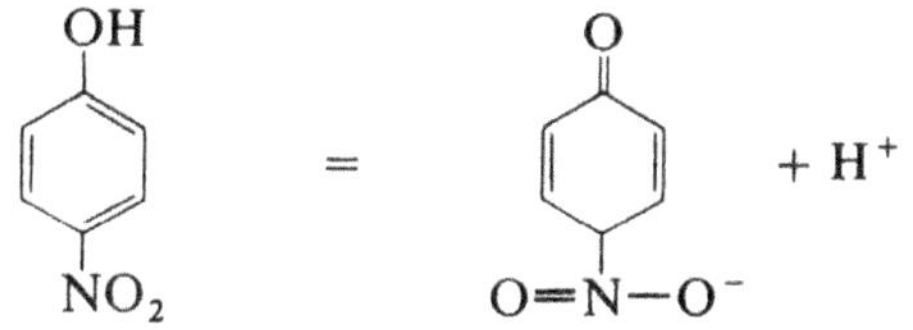

$$HO\text{-}C_6H_4\text{-}NO_2 = O{=}C_6H_4{=}NO_2^- + H^+ \qquad (5\text{-}21)$$

p-Nitrophenol (colorless)   *p*-Nitrophenolate (yellow)

In an unbuffered solution changes in concentration result in a shift in the equilibrium position and a consequent deviation from Beer's law. Of course, if the solution had been buffered, the pH of the solution would have been fixed and the ratio of the concentrations of *p*-nitrophenolate to the *p*-nitrophenol would have been constant with changing concentration. The result would have been adherence to Beer's law. It is apparent that in any system containing an absorbing weak acid-base couple the system should be buffered to ensure that Beer's law is obeyed.

Polymerization reactions (dimerizations, trimerizations, etc.) are common with certain types of compounds such as basic dyes. Because the monomeric and polymeric forms of the absorbing species normally do not have identical spectra, the absorbance of a solution that contains any of the species varies as a function of the concentration of each polymeric form in solution. An example of a compound that polymerizes in that manner is the indicator methylene blue (abbreviated as MeBl):

$$\text{MeBl} = (\text{MeBl})_2 = (\text{MeBl})_3 \qquad (5\text{-}22)$$

At concentrations up to about 1×10^{-5} *M*, methylene blue exists nearly completely in the monomeric form, which has an absorptive maximum at a wavelength of about 660 nm. At concentrations approximately between 1×10^{-5} and 5×10^{-3} *M*, a mixture of the monomer and the dimer (absorptive maximum is about 610 nm) is present in solution. At concentrations between about 5×10^{-3} and about 0.1 *M* a mixture of the dimer and the trimer (absorptive maximum is about 580 nm) is in solution. Above about 0.1 *M* methylene blue exists nearly completely as the trimer. The variations in the absorptivities of the various forms of methylene blue at a fixed wavelength cause a deviation from Beer's law. For methylene blue Beer's law is obeyed at 660 nm only at concentrations below about 1×10^{-5} *M*, where the monomer is the only polymeric form in solution.

Errors are also likely to occur at concentrations above about 0.01 *M* because at high concentrations the proximity of the absorbing species in solution prevents them from acting independently of each other. The derivation of Beer's law assumed that the absorptivity of the analyte did not vary with concentration; i.e., the species behaved independently of each other.

When more than one chemical species in an equilibrium mixture absorbs radiation, it is sometimes possible to avoid deviations from Beer's law by choosing a wavelength for the analysis at which all of the possible absorbing species have the same absorptivity. In that case it makes no difference which chemical species is present in solution. Beer's law is obeyed regardless of the relative amounts of reactants and products. A wavelength at which multiple absorbing species have identical absorptivities is an *isobestic point*. The presence of an isobestic point in a mixture of two absorbing species is sometimes considered evidence for a reversible reaction in which one of the absorbing species is the reactant and the other is the product.

Temperature changes can cause shifts in chemical equilibria which can alter the absorption at a specific wavelength. Deviations from Beer's law that result from temperature changes can be avoided by maintaining the cell at constant temperature. It is usually not necessary to control the temperature to closer than two or three degrees.

Other reasons for deviations from Beer's law include the presence of a luminescing substance in solution and a solvent change. A substance that luminesces at the wavelength of the absorbance measurement emits radiation that can strike the detector, causing a negative deviation from Beer's law. The solvent can affect both the location of an absorbance maximum and the absorptivity of the compound.

SPECTRAL QUANTITATIVE ANALYSIS

The direct use of Beer's law for quantitative analysis, as illustrated in Sample problem 5-3, has several drawbacks. Of course, Beer's law cannot be used when a deviation from the law occurs. Unfortunately deviations from Beer's law are relatively common. Any assay based upon a single measurement is prone to error because an undetected determinate error can easily occur. Furthermore, the presence of an interference during a single assay is not usually apparent. For those reasons quantitative analysis using spectrophotometry or some other instrumental technique is rarely performed using a single measurement. The two methods that are used most often for spectrophotometric quantitative analysis and that eliminate some of the errors associated with a single measurement are the *working- curve method* and the *standard-addition technique.*

Working-Curve Method

A working curve can be prepared by plotting the absorbance of several standard solutions of the analyzed substance as a function of the concentration of the solutions. The absorbance of the sample is measured and the concentration is determined from the working curve. If Beer's law is obeyed, absorbance is directly proportional to concentration. A plot of absorbance as a function of concentration is linear (with a slope of εb or ab) and goes through the origin (Fig. 5-4).

It is not necessary to have a working curve that is linear, but it is important that the standard solutions that were used to prepare the curve have a sufficiently broad range of concentrations to encompass the measured absorbance of all sample solutions. A major advantage of the use of a working curve when compared to direct application of Beer's law is that use of a working curve does not assume a linear variation of absorbance with concentration. Beer's law does not have to be obeyed in order to use the working-curve method. Generally, it is not good practice to extrapolate a linear working curve to absorbances and concentrations beyond those used during preparation of the curve because a linear response in regions beyond the range of the curve has not been demonstrated. Working curves can be prepared from the responses of many analytical instruments in addition to those from instruments that measure absorbance.

Sample problem 5-4 The tabulated data were obtained during the analysis of calcium by atomic absorption spectrophotometry at 422.7 nm. Prepare a working curve of the data and determine the concentration of calcium in the sample.

Calcium concentration, μg/mL	Absorbance
2.00	0.155
4.00	0.395
6.01	0.615
8.01	0.805
10.01	1.000
Sample	0.425

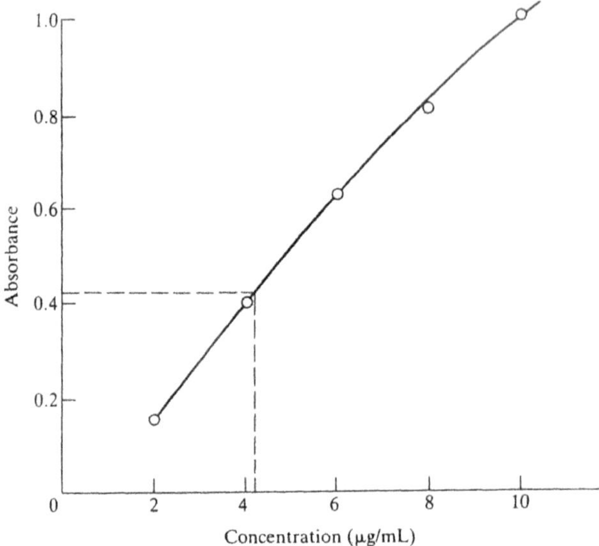

Figure 5-5 The working curve for the data in Sample problem 5-4.

SOLUTION The working curve for the data is shown in Fig. 5-5. The curve is not linear. The point on the curve that corresponds to a sample absorption of 0.425 represents a concentration of 4.2 μg/mL.

Standard-Addition Technique

If Beer's law is known to be obeyed for a particular sample solution, the *standard-addition technique* can be used to determine the concentration of the sample solution. The standard-addition technique can be used with any type of assay in which the measured parameter (usually an instrumental reading) varies linearly with concentration and is zero at zero concentration.

Varying but known amounts (standard additions) of the assayed substance are added to several equal portions of the sample solution. The absorbances of each solution to which a standard addition has been made and of the sample solution to which no addition has been made are measured and plotted as a function of the concentration of the added standard in the analyte. If the solution was diluted during the additions, the dilution factor must be considered when calculating the concentrations.

The resulting linear plot is extrapolated to the concentration axis. The distance on the concentration axis between the intercept of the plot and the point that corresponds to the sample solution is equal to the concentration of the sample solution. The standard-addition technique assumes that the absorbance is zero at zero concentration and that the plot is linear. The standard-addition technique is particularly useful when the sample solution potentially contains an interference which does not itself absorb at the wavelength that is used for the analysis, but which can affect the extent of absorption of the assayed substance. A chemical substance that can react with the assayed substance (e.g., a complexing agent) in a reversible reaction to form a product that does not absorb or that has a different absorptivity from that of the assayed substance is such an interference. Because the concentration of the interference is identical in the sample solution and the solutions to which standard additions have been made, the effect of the interference is automatically taken into consideration during the assay.

Sample problem 5-5 Five 25.0-mL volumetric flasks were labeled S, 1, 2, 3, and 4. A 20.0-mL aliquot of a sample solution was added to each flask with a pipet, and pipets were used to add 1.00 mL of a 5.00×10^{-3} *M* solution of the analyzed substance to flask 1, 2.00 mL to flask 2, 3.00 mL to flask 3, and 4.00 mL to flask 4. The solution in each flask was diluted to the mark with solvent and stirred. The absorbance of each solution in a 1.00-cm (pathlength) cell at 528 nm was measured and tabulated. Determine the concentration of the sample solution.

Flask	Absorbance
S	0.311
1	0.470
2	0.630
3	0.788
4	0.949

SOLUTION The concentration of the added standard in each flask can be calculated from the volume of the standard that was added to each flask and the total volume of each solution.

Flask	Concentration
1	5.00×10^{-3} mmol/mL std × 1.00 mL std/25.0 mL = 2.00×10^{-4} *M*
2	$5.00 \times 10^{-3} \times 2.00/25.0 = 4.00 \times 10^{-4}$ *M*
3	$5.00 \times 10^{-3} \times 3.00/25.0 = 6.00 \times 10^{-4}$ *M*
4	$5.00 \times 10^{-3} \times 4.00/25.0 = 8.00 \times 10^{-4}$ *M*

The absorbance of each solution is plotted as a function of the concentration of the added standard in each solution. The concentration of the solution in flask S is obtained from the extrapolated intercept of the linear plot. The plot is shown in Fig. 5-6. The triangle represents the solution in flask S and the circles represent the other solutions. The x intercept (3.90×10^{-4} *M*) corresponds to the concentration of the absorbing solution in flask S. The concentration of the original sample solution is obtained from the concentration in flask S, the volume of sample added to the flask, and the total volume in the flask:

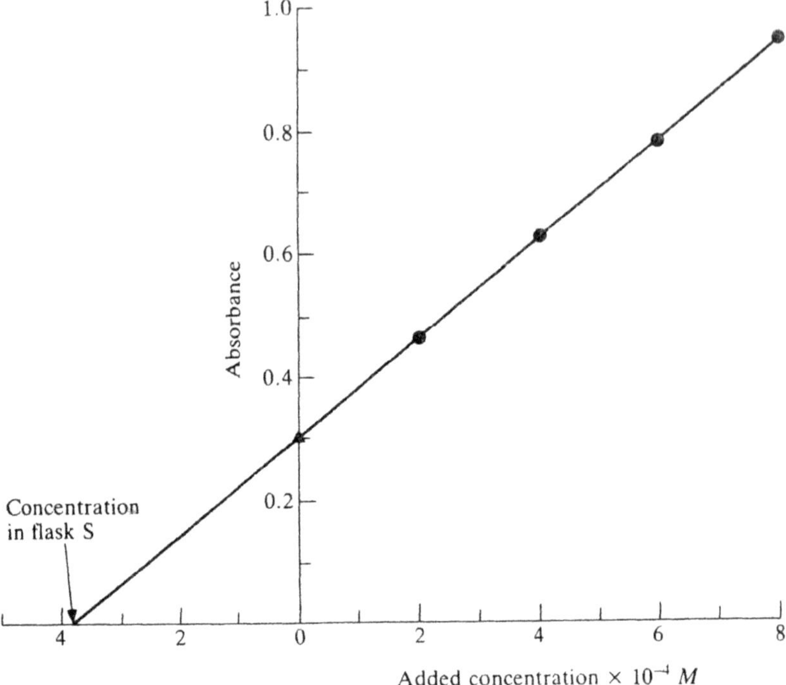

Figure 5-6 Use of the standard-addition technique to solve Sample problem 5-5. The circles represent the solutions to which standard additions have been made. The triangle represents the diluted sample solution that is in flask S.

$$\text{Sample concentration} = 3.90 \times 10^{-4}\ \text{mmol/mL} \times 25.0\ \text{mL/20.0 mL}$$

$$\text{Sample} = 4.88 \times 10^{-4}\ M$$

If the assumptions that were made when using the standard-addition technique are correct, it also is possible to determine the x-axis (concentration axis) intercept of the plot from the equation for the best straight line through the data points in Figure 5-6. The least-squares fit for the data can be obtained from an appropriate computer program or scientific calculator. Inputs to the program or calculator are the standard-addition concentrations and corresponding absorbances.

Luminescence

Luminescence occurs when an electron that is in an excited electron level falls into a lower energy electron level with the simultaneous emission of radiation. The emitted radiation has a wavelength that is characteristic of the energetic difference between the two electron levels. The energetic difference between electron levels corresponds to radiation in the ultraviolet-visible or x-ray region.

The most stable electron configuration for an atom or a molecule occurs when the electrons in the atom or molecule occupy the lowest available electron levels. A molecule or atom with that electron configuration is in the ground state. The ground electron state corresponds to the energetic level of the outermost electron in a molecule or atom that is in the ground state. In order for luminescence to occur, one or more electrons in a molecule or atom that is in the ground state must be excited to a higher electron level. The excitation can be the result of energy absorption from any of several sources. If the source of the excitative energy is a chemical reaction, the resulting luminescence is *chemiluminescence.* If the excitative energy is the result of an electrochemical reaction that occurs at one of two or more electrodes that are placed in the solution, the luminescence is *electrochemiluminescence.* Chemiluminescence and electrochemiluminescence are rarely used for chemical analysis. Luminescence that occurs after the absorption of electromagnetic radiation is used most often for chemical analysis.

A molecule or an atom that has all of its electrons paired is in a *singlet state.* A molecule in a singlet state has a total electron spin of zero because it contains equal numbers of electrons of opposite spin. A molecule that has a single unpaired electron (a free radical) is in a *doublet state* and a molecule that has two unpaired electrons with identical spin is in a *triplet state.* Most organic molecules have singlet ground states. The energy-level transitions that occur during absorption and luminescence in the ultraviolet-visible region within a molecule are shown in Fig. 5-7. Similar transitions are made during absorption and luminescence within atoms; however atoms do not have vibrational and rotational energy levels. In the x-ray region absorption raises an electron in an inner shell to an outer shell or completely removes the electron from the atom.

After a molecule has absorbed radiation in the uv-visible region, the electron in the excited electron energy level possesses potential energy which can be released in several ways. If the molecule loses some of its energy by colliding with another molecule (*collisional deactivation*), if energy is lost by emission of heat, or if the energy is lost by some other means in which uv visible radiation is not emitted, the process is called a *radiationless loss of energy.* In Fig. 5-7 radiationless transitions are indicated by dashed lines.

In solution radiationless loss of energy most often occurs because of collisions with other molecules. Typically each collision results in loss of some vibrational energy (*vibrational relaxation*) within the molecule until the lowest vibrational level within an electron level is reached. Because the highest vibrational levels of the next lower electron level are often about the same energetic level as the lowest vibrational level of the excited electron level, the molecule can continue to lose energy in small steps by changing to a high vibrational level of the lower electron level. That process is *internal conversion* (Fig. 5-7). Eventually the molecule can return to its lowest energy state (the ground state) without emitting any uv-visible radiation.

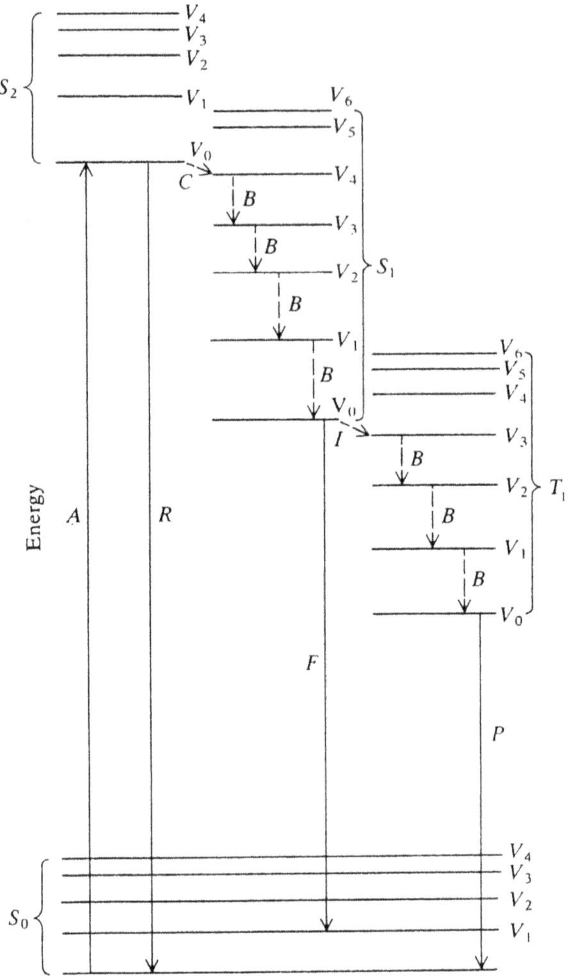

Figure 5-7 Molecular energetic transitions that are associated with absorption A, resonant fluorescence R, normal fluorescence F, phosphorescence P, internal conversion C, intersystem crossing I, and vibrational relaxation B. S_0, ground singlet electron state; S_1, S_2, excited singlet electron states; T_1, lowest energy triplet electron state; V_0, V_1, V_2, V_3, V_4, V_5, V_6, V_7, vibrational states. Rotational states are not shown and the sketch is not to scale.

After absorption one of the two paired electrons $\left(\underline{\uparrow\ \downarrow}\right)$ that originally was in the ground electron level is excited to a higher singlet electron level $\left(\frac{\downarrow}{\underline{\uparrow}}\right)$. The horizontal lines in the brackets represent the ground and excited energy levels. The arrows represent the electrons and their spin orientations. If the higher energy electron in an excited singlet state reverses its spin so that it now spins in the same direction as the remaining unpaired electron in the ground state $\begin{pmatrix}\uparrow \\ \overline{\uparrow}\end{pmatrix}$ the resulting molecule is in an excited triplet state. Because the electron which reversed its spin in order to form the triplet level gave up some energy in the process, the energy of the triplet state is slightly less than that of the corresponding singlet state. The process by which a molecule changes from a singlet to a triplet state or from a triplet to a singlet state is *intersystem crossing*. Intersystem crossing rarely occurs because considerable opposition to electron spin reversal exists.

Sometime prior to the loss of all the potential energy in the molecule, a molecule in an excited electron level can return directly to the ground electron state. When that occurs, luminescence of an energy equivalent to the difference between energetic levels traversed during the electron transition is emitted by the molecule. Luminescence that occurs after excitation during absorption can be classified as either *fluorescence* or *phosphorescence*. Fluorescence is luminescence that occurs during an electron transition from an excited singlet state to a lower energy singlet state. X-ray fluorescence occurs when one of the electrons in an outer shell falls into the vacancy in the inner shell. Phosphorescence corresponds to a transition from an excited triplet state to a lower energy singlet state and requires electron spin reversal. Fluorescence and phosphorescence generally are the result of a transition from the lowest excited singlet or triplet state to the ground electron state.

If the wavelength of the emitted radiation is identical to the wavelength of the exciting radiation, the fluorescence is *resonant fluorescence*. Resonant fluorescence is never observed with molecules in solution, but it sometimes occurs in solids where no collisional deactivation is possible. In atoms, resonant fluorescence usually predominates. In solution, the fluorescent radiation is of lower energy than the absorbed radiation; i.e., the fluorescent radiation occurs at a wavelength that is longer than the wavelength of the absorbed radiation. That type of fluorescence is *normal fluorescence*.

Fluorescence occurs nearly instantaneously after absorption. The time an electron spends in the excited state prior to falling to the ground state is about to 10^{-6} to 10^{-10} s. Although the process is rapid, it is relatively slow when compared to the approximately 10^{-15} s that is required for absorption to occur.

Phosphorescence occurs throughout a longer period than does fluorescence. Intersystem crossing from the excited singlet state to the excited triplet state requires 10^{-7} to 10^{-8} s to occur. Because it requires spin reversal to which there is considerable resistance, the ratio of singlet-singlet transitions to singlet-triplet transitions during relaxation typically is between 10^5 and 10^6; i.e., the quantum yield for fluorescence is about 10^5 or 10^6 times that for phosphorescence. After intersystem crossing has occurred, considerable resistance to the transition from the excited triplet state to the ground singlet state exists. Consequently, many molecules spend a relatively long time in the excited triplet state prior to falling to the ground singlet state. Typically, phosphorescence can occur 10^{-6} to 10 s or more after the exciting radiation has been absorbed.

Phosphorescence is most likely to be observed when the compound is rigid. Rigidity is more easily attained in a glass or solid state rather than in the liquid state. In glass and solid states collisional deactivation is minimized and the time in the excited singlet states is increased. The result is an increase in the probability of intersystem crossing and consequently in phosphorescence. Rigidity and the consequent phosphorescence can also be attained by adsorption of the molecule on a surface and by using a micelle to stabilize the molecule.

Because many vibrational and rotational energetic levels are associated with each electron level of a particular molecule, absorption and luminescence can occur at many frequencies within the ultraviolet-visible region of the EMR spectrum. Because the vibrational and rotational energy levels are closely spaced, absorptive and luminescent spectra for molecules appear as broad bands of radiation. Atoms do not have vibrational and rotational levels and consequently have narrow absorptive and fluorescent bands which are called spectral *lines*.

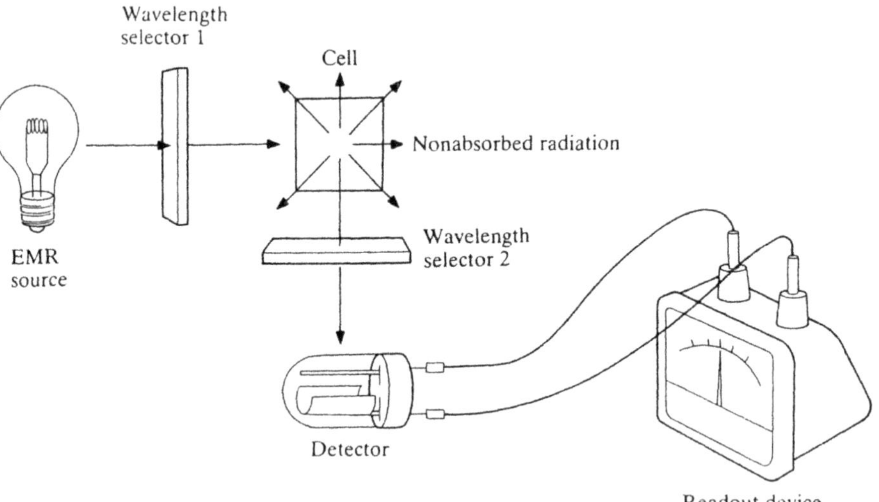

Figure 5-8 A diagram of an instrument that is used to measure luminescence.

Apparatus for Luminescence

A diagram of the major components of the apparatus that is used to perform luminescent measurements is shown in Fig. 5-8. EMR from an EMR source passes through a device that limits the width and center wavelength of the band of radiation. The resulting monochromatic radiation enters the cell and is partially absorbed by the sample. Radiation that is not absorbed passes through the cell and is lost.

Luminescence from the cell is emitted in all directions. Consequently, it should be possible to measure the intensity of the emitted radiation at any angle relative to the incident radiation. In practice the intensity of the luminescence is nearly always measured at 90° to the incident radiation. Other angles may be used in specific apparatus. Generally it is not advisable to make the measurement at 180° (in-line) to the incident radiation because the incident radiation that is not absorbed in the cell might interfere with the measurement. The popularity of measurement at 90° is primarily the result of the ease with which rectangular cells are constructed. A measurement at 90° in a rectangular cell permits the incident radiation and the measured luminescence to pass perpendicularly through the cell walls. This minimizes reflection of the incident radiation and the luminescence by the walls.

After the luminescence exits from the cell, it passes through a second wavelength-limiting device and into the detector. The signal from the detector is amplified, if necessary, and sent to a readout device. It is necessary to have two wavelength-limiting devices because the incident radiation and the luminescence usually have different wavelengths and because more than one band of luminescence can result from excitation at a single wavelength.

If the apparatus is to be used to measure phosphorescence, an additional device is required which enables the measurement of phosphorescence in the presence of fluorescence. Most phosphorescent substances also fluoresce. Usually the device permits examination of the luminescence while the exciting radiation is not striking the sample. In older instruments this was often accomplished with a rotating opaque cylinder that is placed around the cell. The cylinder contains two or more slits that permit radiation to enter or exit from the cell.

As the cylinder rotates, a slit becomes aligned with the exciting radiation and the cell and allows incident radiation to enter the cell. Because a pair of slits is never simultaneously aligned with the path of the exciting radiation to the cell and the path from the cell to the detector, any fluorescence that occurs in the cell in the direction of the detector strikes the inside of the cylinder and is lost. Because the lifetime of fluorescence is short, no fluorescence occurs by the time a slit is aligned with the cell and the detector. The long lifetime of phosphorescence permits the intensity of the emission to be measured when a slit aligns with the path of emitted radiation between the cell and detector. As an alternative, most modern instruments use pulsed incident radiation and electronically delay the measurement for a defined period after the pulse of incident radiation has been turned off.

Because phosphorescence sometimes requires samples that are in the glass or solid state, some *phosphorimeters* have provision for cooling the sample. Generally the cooling is done by submerging the cell in a Dewar flask that contains liquid nitrogen.

Quantitative Analysis Using Luminescence

The measured intensity of luminescence I_L, which is initiated by absorption of incident radiation, is directly proportional to the intensity of the incident radiation I_o and the concentration C of the luminescent species:

$$I_L = KI_o C \tag{5-23}$$

The equation is derived in Chapter 11. From the equation it is apparent that the *detection limit* (the lowest concentration that can be detected) when using fluorescence or phosphorescence decreases as the intensity of the incident radiation increases. By using sufficiently intense incident radiation, it is possible to use luminescence to assay luminescent compounds at much lower concentrations than those which can be measured using absorbance.

In most analyses the intensity of the incident radiation is held constant. In that case the intensity of the luminescence should be directly proportional to the concentration of the luminescent sample. Equation (5-23) is rarely used directly in a manner that is analogous to the direct use of Beer's law for absorbance measurements (as in Sample problem 5-3). Usually analyses are performed by use of the working-curve method or the standard-addition technique.

SCATTERED RADIATION

Radiative scattering is *the* change in direction of motion of an incident photon as it strikes a particle of the sample. Usually the change in direction is random or nearly random. Radiation scattering can occur as the result of several mechanisms. The three types of scattering that are of most concern in analytical chemistry are *Tyndall scattering, Rayleigh scattering,* and *Raman scattering.*

Tyndall Scattering

Tyndall scattering occurs when radiation is passed through a colloidal or turbid solution. The diameters of the scattering particles in the solution are approximately as large or larger than the wavelength of the scattered radiation. Visible light is scattered when the diameter of the particles is between about 10 and 1000 nm. The mechanism by which Tyndall scattering occurs includes a complex combination of reflection, internal reflection from within the particles, diffraction, and refraction. The wavelength of the scattered radiation is identical to that of the incident radiation. Tyndall scattering is primarily used for quantitative analysis. The intensity of the scattered radiation increases with increasing concentration of the scatterer.

Rayleigh and Raman Scattering

Rayleigh and Raman scattering occur when the dimensions of the particles that cause the scattering are small in comparison to the wavelength of the incident radiation. Dissolved particles can result in Rayleigh and Raman scattering Rayleigh-scattered radiation occurs at the same wavelength as that of the incident radiation. Raman scattering is considerably less intense than Rayleigh scattering and occurs at wavelengths that are greater than and less than that of the incident radiation. The change in wavelength between that of the incident and the scattered radiation corresponds to the difference between the vibrational and rotational levels within the scatterer and can be used for qualitative analysis of the scatterer in a manner similar to that in which infrared absorptive spectra are used.

APPARATUS NOMENCLATURE

Not all chemists agree on the names to be given to various pieces of equipment that are used for making optical measurements. The definitions in this section appear to be agreed upon by a majority of analytical chemists and chemical spectroscopists. A *spectrometer* is an instrument that is used to measure the intensity of radiation emitted by the sample. The wavelength can be either fixed or scanned during the measurement. A spectrometer is not usually regarded as an instrument that is used to measure absorption. A *spectrograph* is an instrument that uses a photographic plate as a detector to measure the intensity of radiation that is emitted by the sample.

The names of the instruments that are used to measure absorption by the sample are determined by the type of device which is used to select the wavelength for the study. A *photometer* is an instrument in which a filter provides radiation of the proper wavelength range. The wavelength at which the measurement is made cannot be scanned automatically in a photometer. A photometer that is used solely with radiation in the visible spectral region and which relies on the human eye as the detector is a *colorimeter*. A *spectrophotometer* relies on a monochromator to select the wavelength of the radiation to be used. The wavelength of the radiation can be either held constant during the study or scanned.

An instrument that measures fluorescence is *a fluorometer* or *fluorimeter,* and an instrument that measures phosphorescence is a *phosphorimeter*. If the fluorometer or phosphorimeter contains monochromators, the instruments are a spectrofluorometer and a spectrophosphorimeter. An instrument that is used to measure turbidance exclusively is a *turbidimeter* and an instrument that is used to make nephelometric measurements is a *nephelometer*. Turbidance is an indirect method to measure Tyndall scattered radiation. It measures the decrease in intensity of incident radiation owing to the scattering. It is analogous to absorbance. Nephelometry is the direct measurement of the intensity of Tyndall scattered radiation. It is analogous to fluorescence. Raman spectra are obtained with a *Raman spectrometer*. Devices that measure fluorescence, phosphorescence, or scattering are forms of spectrometers. Details of specific instruments are included in later chapters with the detailed descriptions of the specific methods of spectral analysis.

Wavelength Selectors

Devices that enable selection of a relatively narrow band of wavelengths from a broader band are used in nearly all spectral instruments. In most instruments the wavelength band is narrowed with either one or more filters or a monochromator.

Filters

Filters prevent the passage of radiation at all wavelengths except in a particular-wavelength region. Filters are classified as either *absorptive filters* (also called absorption filters) or *interference filters*. Absorptive filters absorb radiation at wavelengths beyond a certain value. They can be constructed from any material that has the desired absorptive properties. Typically, absorptive filters are constructed from tinted glass, plastic; or an absorptive liquid. Absorptive filters are most often used in the visible region. *Sharp-cutoff short-wavelength filters* (also called *sharp-cut filters*) transmit light of wavelengths greater than a fixed value and absorb light of shorter wavelengths. Maximum transmission through a sharp-cut filter corresponds to between 80 and 90 per cent of the intensity of the incident radiation. The transition from complete absorption to maximum transmission typically occurs over a wavelength range of about 40 nm. The cutoff wavelength for a particular filter, as specified by the manufacturer of the filter, corresponds to the wavelength (± 10 nm) at which 37 per cent of the incident radiation is transmitted. A *sharp-cutoff long-wavelength filter* absorbs radiation at wavelengths greater than the cutoff wavelength and passes radiation at shorter wavelengths.

Bandpass filters transmit radiation within a range that is typically 20 to 70 nm wide. Bandpass filters can be constructed by combining spectrally overlapping sharp-cutoff short-wavelength filters and sharp-cutoff long-wavelength filters. Radiation is transmitted through the combination only in the spectral range which can be transmitted through both filters. The intensity of maximum transmission through a bandpass filter is typically about 25 per cent or less of the intensity of incident radiation in the same wavelength region.

Interference filters use destructive interference from out-of-phase radiation to eliminate all but relatively narrow bands of radiation. A sketch that illustrates the operation of an interference filter is shown in Fig. 5-9. The filter consists of two transparent parallel plates that are separated by a transparent film. The two inner walls of the plates are partially coated with a reflecting material which usually is a metal. Often the inner walls are half-silvered.

A portion of the incident polychromatic radiation passes through the first wall and the transparent film and exits through the second wall. A second portion of the radiation is reflected back through the filter after it strikes the inner surface of the second wall. Upon striking the inner surface of the first wall, some of the radiation is again reflected and leaves the filter after passing through wall 2. The radiation that directly passed through the filter is coincident with the radiation that was reflected through the filter. If the reflected radiation is out-of-phase with the nonreflected radiation, destructive interference occurs and the radiation is decreased in intensity or eliminated. If the reflected radiation is in-phase with the nonreflected radiation, constructive interference occurs and the intensity of the radiation is the sum of that from the two sources. The filter transmits that radiation in which the two components are in-phase and eliminates radiation of other wavelengths.

In order for the reflected and nonreflected radiation to be in-phase, it is necessary for the distance through which the reflected radiation travels within the filter to be a whole-number multiple of the wavelength of the radiation. Because the total distance within the filter through which the reflected radiation travels is twice the width of the filter ($2d$), it is necessary for $2d$ to be a whole multiple of the wavelength. Because the wavelength of radiation varies from a medium of one refractive index to that of a second refractive index, it is necessary to consider the refractive index of the transparent film during the calculation. The results are summarized as follows:

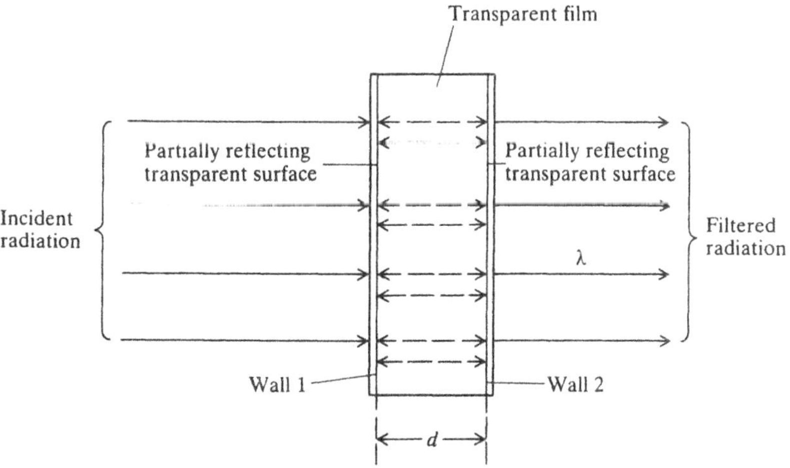

Figure 5-9 A sketch of an interference filter. The distance d between the partially reflective inner surfaces of the filter is usually $\lambda/2$, λ, or $3\lambda/2$.

$$2dn = m\lambda \tag{5–24}$$

where n is the refractive index of the film, λ is the central transmitted wavelength, and m is the whole-number multiple of the transmitted wavelength. The equation assumes that the radiation passes through air, which has a refractive index that is nearly 1, before and after entering the filter.

The central wavelength of the transmitted band of radiation is controlled by careful construction of the filter so that it has the proper width d. The filters are usually constructed so that d is equivalent to $\frac{1}{2}$, 1, or $\frac{3}{2}$ of the wavelength of the transmitted radiation, i.e., m in Eq. (5-24) is 1, 2, or 3. The corresponding filters are first-, second-, and third-order filters. It is apparent from Eq. (5-24) that the radiation that is transmitted through an interference filter actually consists of the several possible harmonic bands (m = 1, 2, 3, etc.) of radiation. Typically, the width *(bandpass)* of the transmitted band is about 15 nm or less and the intensity of the transmitted band is equal to or less than about 40 per cent of the intensity of the incident radiation within the same bandwidth.

In some filters d is continuously varied from one end of the filter to the other. Such a *wedge filter* can be used to transmit any wavelength between that corresponding to the smallest and the largest value of d depending upon the location on the filter at which the incident radiation is focused. A *dichroic filter* is an interference filter that acts as a sharp-cutoff short-wavelength filter, i.e., it transmits radiation of wavelengths that are longer than that of a cutoff wavelength.

Monochromators

In all instruments in which the wavelength is automatically scanned and in many instruments that are not automatically scanned, a monochromator is used as the device which converts polychromatic radiation from the EMR source to monochromatic radiation. A monochromator usually consists of entrance and exit slits, a prism, diffraction grating, reflective diffraction grating, or, less frequently, a wedge filter, and lenses or mirrors that are used to collimate (make the beams parallel) or focus the radiation. A sketch of a simple monochromator is shown in Fig. 5-10. Monochromators of many designs are commercially available.

The prism or diffraction grating separates (*disperses*) the polychromatic radiation in space into bands of "monochromatic" radiation. In most instruments reflective diffraction gratings are used. The wavelength band that is focused on the exit slit of the monochromator exits the device and is used in the remainder of the spectral instrument. Mirrors are used most often within monochromators to focus and collimate the radiation.

The operator usually has control of the width of the entrance and/or exit slits. Generally, a single control simultaneously alters the width of both the entrance

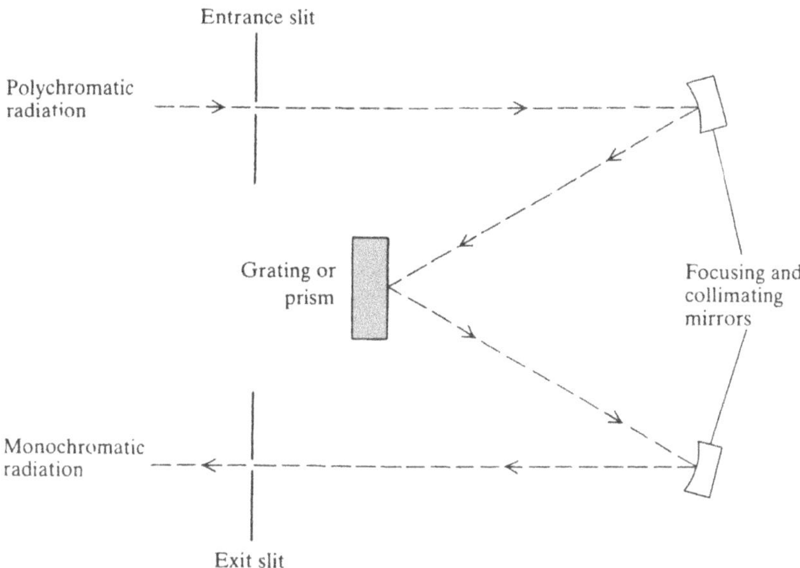

Figure 5-10 A diagram of a monochromator. The path of the radiation is indicated by the dashed line.

and exit slits. As the slit width is increased, more radiation passes through the monochromator. Increasing the slit width also increases the wavelength bandwidth of the radiation that exits the monochromator. The resulting decrease in monochromaticity of the exiting radiation results in decreased resolution between adjacent spectral peaks. The slit width is normally adjusted to the minimal value that allows sufficient radiation to pass through the monochromator to enable accurate measurements by the detector of the spectral instrument. A description of the effect of slit width on spectrophotometric measurements can be found in Reference 2.

The wavelength of the radiation to be used in the study can be selected by rotating the prism or diffraction grating. The rotation can be done either manually by adjusting a knob on the instrument or automatically. In either case the rotation causes the proper wavelength band to be focused on the exit slit.

Instruments that contain monochromators are usually more convenient to use than instruments that contain filters, but they are more expensive. Some instruments have monochromators that can continuously scan through a wavelength region with the aid of a motor that automatically rotates the prism or diffraction grating. Depending upon the design, the wavelength band that is passed through the monochromator is usually narrower than that passed through a filter and can be as narrow as 0.1 nm.

Prisms

In some monochromators a triangular prism is used to separate polychromatic radiation into a broad band of radiation of continuously varying wavelength. Because prisms normally are more expensive and require more space than diffraction gratings and reflective diffraction gratings, they are rarely used in modern instruments. Prisms take advantage of the change with wavelength in refractive index of the substance from which the prism is constructed. For the substances from which prisms are constructed, the refractive index decreases as the wavelength increases. Prisms are primarily used to disperse EMR in the ultraviolet, visible, and infrared spectral regions. Diagrams of a full prism and a Littrow-mounted half-prism are shown in Fig. 5-11.

If a wavefront of EMR strikes the surface of a prism at any angle other than perpendicularly, it is impossible for the entire wavefront to simultaneously enter

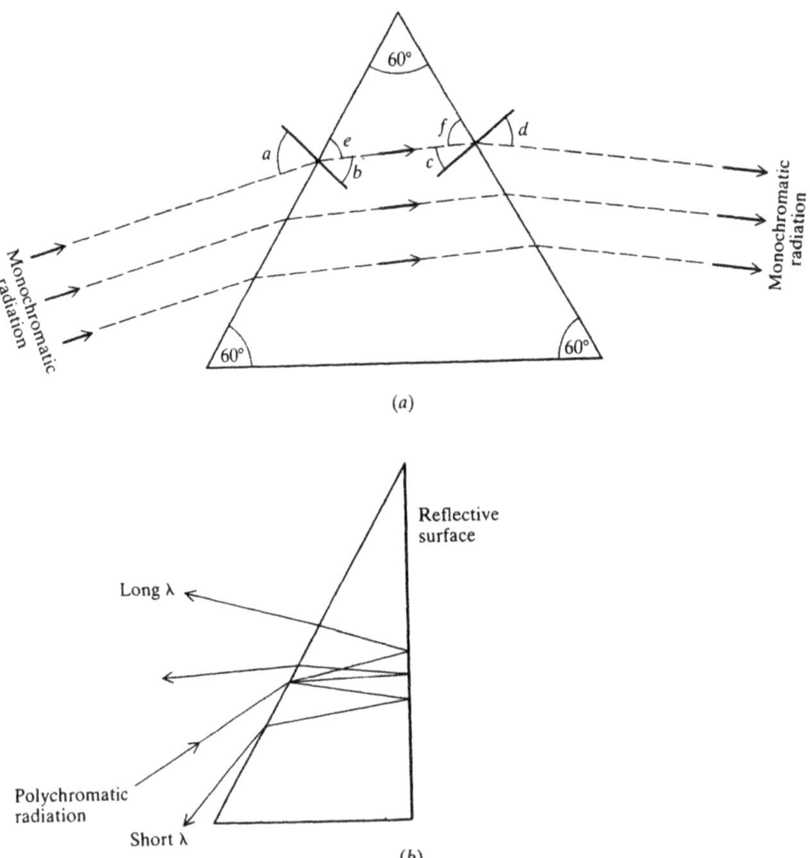

Figure 5-11 Side-view sketches of two forms of triangular prisms: (*a*) a full prism; (*b*) a Littrow-mounted half-prism. The arrows indicate the direction of motion of radiation. In (*a*) the incident radiation is monochromatic. In (*b*) the incident radiation is polychromatic.

the prism. Because the refractive indices of the relatively dense materials from which prisms are constructed are larger than the refractive index of air, the first portion of the wavefront to enter the prism is slowed with respect to the portion that is still in the air. As a result the wavefront changes direction (is refracted) as it enters the prism (Fig. 5-11). The amount of the change depends upon the relative velocity of EMR in the prism as compared to the velocity in air. The relative velocity is inversely proportional to the refractive indices in the two media. It can be shown geometrically that

$$\frac{\sin a}{\sin b} = \frac{n_2}{n_1} \tag{5-25}$$

is obeyed as radiation of a fixed wavelength enters a medium of refractive index n_2 from a medium of refractive index n_1. The angles, *a* and *b,* are measured with respect to a line perpendicular to the surface between the two media, as shown in Fig. 5-11*a*.

When the radiation exits from the prism, it is refracted again. If the entrance and exit walls of the prism were parallel, the radiation would have exited traveling in the same direction in which it entered the prism. Because the walls are not parallel, the angle at which the radiation strikes the wall as it leaves the prism causes the radiation to exit the prism at an angle other than that at which it entered the prism.

When the incident radiation strikes the surface of the prism, some or all of the radiation can be reflected from the surface rather than pass into the prism. As the angle (measured from the perpendicular to the surface) at which the wave-front strikes the prism increases, a greater portion of the incident radiation is reflected. The *critical angle* θ_c is the smallest angle of incidence at which total reflection occurs. All of the incident radiation is reflected from the surface if the impinging radiation strikes the surface at an angle greater than the critical angle. The critical angle is dependent upon the refractive index n_1 of the medium in which the radiation initially is traveling and the refractive index n_2 of the medium upon which the radiation impinges:

$$\sin \theta_c = \frac{n_2}{n_1} \tag{5-26}$$

Sample problem 5-6 Monochromatic radiation enters the prism that is shown in Fig. 5-11*a* at an incident angle of 62.8° as measured from the perpendicular to the surface. The refractive index of the radiation in air is 1.00 and in the prism is 1.60. Calculate the angle at which the radiation exits from the prism.

SOLUTION Equation (5-25) is used to calculate the angle of refraction (angle *b*) as the radiation enters the prism:

$$\frac{\sin\left(62.8^\circ\right)}{\sin b} = \frac{1.60}{1.00}$$

$$b = 33.8^\circ$$

The angle at which the radiation within the prism strikes the wall as it exits from the prism can be calculated by referring to Fig. 5-11*a*. Because angle *b* is 33.8°, angle *e* must be 90.0 – 33.8 = 56.2°. The third angle *f* in the triangle formed by the top of the prism and the path of radiation through the prism is 180.0 – 56.2 – 60.0 = 63.8°. Angle *c*, therefore, must be 90.0 – 63.8 = 26.2°. The angle at which radiation leaves the prism (angle *d*) can be calculated using Eq. (5-25):

$$\frac{\sin\left(26.2^\circ\right)}{\sin d} = \frac{1.00}{1.60}$$

$$d = 44.9^\circ$$

Diffraction Gratings

The majority of modern spectral instruments use reflective diffraction gratings to disperse polychromatic radiation. A reflective diffraction grating is constructed by carefully ruling parallel grooves into a reflective surface such as a highly polished metal. Aluminum is often used.

The master grating can be prepared mechanically or with the aid of lasers. Traditionally, a master diffraction grating was prepared by carefully cutting each parallel line in the surface of the medium used for the grating with a *ruling engine*. The process required a good deal of time and patience. Unfortunately the product often contained chips and scratches that made the grating less effective.

Since approximately 1968, lasers have been used to rule some gratings. Gratings that are ruled using lasers are *holographic gratings*. A holographic grating is prepared using two continuous-wave laser beams with slightly different emission wavelengths. The interference fringe pattern from the two lasers is projected onto a piece of glass that has a thin coating of a photosensitive plastic. The coating is less than 10 μm thick.

At the places at which the two laser beams strike the plastic in-phase with each other, the plastic is polymerized and adheres to the surface of the glass. At other places the beams arrive out-of-phase and the plastic does not polymerize. After polymerization is completed, the grating is rinsed with an appropriate solvent. The unpolymerized plastic rinses from the glass while the polymerized portion remains, leaving plastic ridges corresponding to the points at which the laser beams reenforced each other on the surface. The plastic grating is coated with aluminum, gold, or platinum in a vacuum and used as a reflective

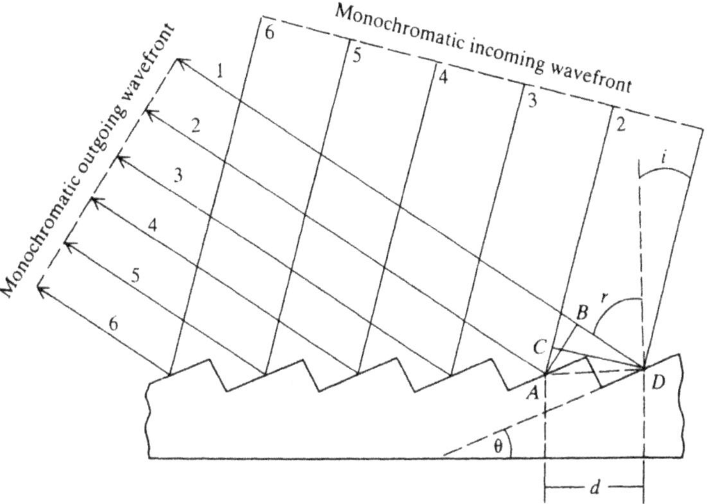

Figure 5-12 A sketch of a portion of an echelette diffraction grating.

diffraction grating. Ruling errors are considerably reduced in holographic gratings as compared to conventional gratings.

Although preparing the original master grating can be difficult, once the master grating has been prepared, duplicates can easily be formed using vacuum deposition techniques or by making plastic casts of the master. A sketch of a portion of a reflective diffraction grating is shown in Fig. 5-12.

The type of reflective diffraction grating shown in Fig. 5-12 is an *cchelette grating*. It has onc rclatively large flat surface that the incident radiation strikes, and a second smaller surface that is unused. The angle i at which incident radiation strikes the grating is measured as shown in Fig. 5-12.

Once incident radiation strikes the surface, each larger surface on the grating functions as a point source of radiation, i.e., radiation is diffracted from the grating. Constructive interference occurs when the difference in the distance that is traveled by the diffracted radiation from each surface of the groove to the wavefront is an integral number of wavelengths. In that instance all of the waves arrive at the wavefront in-phase. The in-phase radiation can be focused at a single point with a lens or mirror. If the difference in the traveled distance does not correspond to an integral number of wavelengths, destructive interference occurs. Consequently, at a particular angle of diffraction, radiation of a fixed wavelength is dispersed (Fig. 5-12).

An equation that relates the distance d between the rulings on the grating and the wavelength λ that is in-phase at the wavefront can be obtained by careful examination of Fig. 5-12. From the sketch it is apparent that beam 1 travels further than beam 2. The difference in distance traveled by the two beams must be equal to an integral number of wavelengths in order for the beams to be in-phase at the outgoing wavefront. The distance traveled by the two beams is equal between the incoming wavefront and the perpendicular *CD* between the two beams. Likewise, the distance traveled by the two diffracted beams is identical between the perpendicular *AB* between the two beams and the outgoing wavefront. Consequently, the difference in distance traveled by the two beams is $CA - DB$.

Careful examination of the locations of the perpendiculars in Fig. 5-12 makes it possible to write the following equations which relate the several angles in the sketch to each other:

$$r + BDC + CDA = 90^\circ$$

$$i + r + BDC = 90^\circ$$

Solution of the second equation for angle *BDC* and substitution into the first equation reveals that angle *i* is equivalent to angle *CDA*. The distance *CA* can be calculated using the following relationship, where *d* is the distance between rulings on the grating:

$$\text{Sin } CDA = \sin i = \frac{CA}{d}$$

$$CA = d \sin i \tag{5-27}$$

Similarly, the following equations are apparent from examination of triangle *ABD* and a perpendicular in Fig. 5-12:

$$BAD + ADB + 90 = 180^\circ$$

$$r + ADB = 90^\circ$$

Solution of the second equation for angle *ADB* and substitution into the first equation reveals that angle *BAD* is equivalent to *r*. Trigonometry can be used to solve for *DB*:

$$\sin BAD = \sin r = \frac{DB}{d}$$

$$DB = d \sin r$$

It is customary to regard the distance *DB* after reflection has occurred to be negative brcause of the change in direction. Consequently, *DB* is $-d \sin r$ rather than $+ d \sin r$. By convention angle *r* is regarded as positive if it is on the same side of the normal to the grating as angle *i*, and it is regarded as negative if it is on the opposite side of the normal. In the particular case that is shown in Fig. 5-12, the two angles are on opposite sides of the normal. Subtraction of *DB* from *CA* yields the difference in distance traveled by the two beams, which must equal an integral number of wavelengths in order for the beams to be in-phase:

$$CA - DB = n\lambda$$
$$d \sin i - (-d \sin r) = n\lambda$$
$$d(\sin i + \sin r) = n\lambda \qquad (5\text{-}28)$$

Equation (5-28) is the *grating equation*. The integer n (not to be confused with the refractive index, which has the identical symbol) in the equation corresponds to the *order* of radiation that is diffracted at angle r when the incident angle is i. It is apparent from Eq. (5-28) that more than one wavelength is diffracted at the same angle. In most cases first-order ($n = 1$) diffraction corresponds to the most intense line. Some gratings are designed so that as much as 90 per cent of the intensity of the diffracted radiation at a particular angle corresponds to first-order diffraction. Generally the distance d between rulings on the grating is approximately equal to the wavelengths that are to be dispersed with the grating.

Although radiation is emitted in all possible directions from each surface of a reflective diffraction grating, the intensity of the radiation is a maximum at the angle of reflection from the surface. Different gratings are prepared with different angles θ of the grooves so that the incident radiation is reflected at the angle that corresponds to the wavelength that is desired to be most intense. It is possible to construct gratings which yield an intense beam or *blaze* of radiation at nearly any desired angle. An advantage of the use of reflective diffraction gratings as opposed to prisms is that the dispersion is nearly linear in wavelength. With prisms the dispersion is nonlinear because of the manner in which the refractive index varies with wavelength.

Sample problem 5-7 A reflective diffraction grating contained 1750 grooves/mm. The angle of incidence of a band of polychromatic radiation was 48.2°. Determine the wavelengths that are diffracted at an angle of – 11.2°.

SOLUTION The distance d between grooves is

$$d = 1 \text{ mm}/1750 \text{ grooves} = 5.71 \times 10^{-4} \text{ mm} = 571 \text{ nm}$$

Equation (5-28) is used to solve the problem:

$$(571 \text{ nm})[\sin(48.2) + \sin(-11.2)] = n\lambda$$
$$n\lambda = 315 \text{ nm}$$
$$\lambda = \frac{315}{n} \text{ nm}$$

First order ($n = 1$):	$\lambda = 315$ nm
Second order ($n = 2$):	$\lambda = 157$ nm
Third order ($n = 3$):	$\lambda = 105$ nm, etc.

DISPERSION AND RESOLUTION

Dispersion D is the amount of separation in space between wavelengths. For a monochromator it is defined as the amount of separation between wavelengths on the focal plane. For linear dispersion it is usually defined as the distance x (mm) for each nanometer of separation:

$$D = \frac{dx}{d\lambda} \tag{5-29}$$

The *linear reciprocal dispersion* D^{-1} is the inverse of Eq. (5-29). *Angular dispersion* is the angular range $d\theta$ through which a band of wavelengths $d\lambda$ is spread ($d\theta/d\lambda$).

The *resolution* R of a monochromator is its ability to separate adjacent wavelengths of radiation. It is defined by

$$R = \frac{\lambda}{\Delta\lambda} \tag{5-30}$$

where λ is the *average* wavelength that is being used for the measurement and $\Delta\lambda$ is the difference between adjacent peaks that are just resolved. Practical definitions for "just resolved" vary from a valley between two peaks (of equal height), which is 90 per cent of the intensity of either of the peaks, to baseline separation. Spectral peaks are usually considered to be resolved when the valley between the peaks is 5 per cent of the height of either of the two equal-height peaks.

Sample problem 5-8 Determine the resolution of an instrument that is capable of separating two adjacent peaks at 207.3 and 215.1 nm. If the two peaks are separated on the focal plane of the instrument by 2.5 mm, calculate the dispersion and the linear reciprocal dispersion of the instrument.

SOLUTION Equation (5-30) is used to calculate resolution. The average wavelength is (207.3 + 215.1)/2 = 211.2 nm, and the change in the wavelength is 215.1 – 207.3 = 7.8 nm.

$$R = \frac{211.2}{7.8} = 27$$

The linear dispersion is calculated using Eq. (5-29):

$$D = \frac{2.5 \text{ mm}}{7.8 \text{ nm}} = 0.32 \text{ mm/nm}$$

Linear reciprocal dispersion is the inverse of D:

$$D^{-1} = \frac{7.8 \text{ nm}}{2.5 \text{ mm}} = 3.1 \text{ nm/mm}$$

IMPORTANT TERMS

- Absorbance
- Absorption
- Absorptive filter
- Absorptivity
- Amplitude
- Angular dispersion
- Bandpass
- Bandpass filter
- Beer's law
- Chemiluminescence
- Collisional deactivation
- Colorimeter
- Detection limit
- Dichroic filter
- Diffraction grating
- Dispersion
- Doublet state
- Electrochemiluminescence
- EMR
- Fluorescence
- Fluorometer
- Frequency
- Grating equation
- Ground state
- Holographic grating
- Intensity
- Interference filter
- Internal conversion
- Intersystem crossing
- Isobestic point
- Luminescence
- Molar absorptivity
- Monochromatic radiation
- Monochromator
- Negative deviation
- Normal fluorescence
- Percent transmittance
- Period
- Phosphorescence
- Phosphorimeter
- Photometer
- Photon
- Planck equation
- Polychromatic radiation
- Positive deviation
- Power
- Prism
- Radiationless loss
- Raman scattering
- Rayleigh scattering
- Refractive index
- Resolution
- Resonant fluorescence filter
- Scattering
- Sharp-cut filter
- Singlet state
- Spectral lines
- Spectrograph
- Spectrometer
- Spectrophotometer
- Standard-addition technique
- Stray radiation
- Transmittance
- Triplet state
- Tyndall scattering
- Wave
- Wavelength
- Working-curve method

PROBLEMS

Electromagnetic Radiation

5-1 The wavelength in a vacuum of electromagnetic radiation is 275 nm. Determine the frequency and period of the radiation. Calculate the energy that is associated with each photon of the radiation.

5-2 The velocity of radiation that has a wavelength of 589.3 nm in bauxite is 1.901×10^8 m/s. Calculate the refractive index of bauxite at 589.3 nm.

5-3 Electromagnetic radiation has a wavelength of 460.0 nm in dry air ($n = 1.00028$). Determine the frequency of the radiation and the energy possessed by one photon of the radiation.

5-4 Calculate the wavelength in a vacuum and in fused quartz ($n = 1.467$) of radiation at 15°C that has a frequency of 4.708×10^{14} Hz.

5-5 Calculate the energy per photon in air of radiation with a wavelength of 589 nm.

5-6 The index of refraction of light flint glass at 434 nm is 1.594. Calculate the energy of each photon of the radiation as it passes through the glass.

Beer's Law

5-7 The molar absorptivity of a compound in aqueous solution at 765 nm is 1.54×10^3. The percent transmittance of a solution of the compound in a cell with a 1.00-cm pathlength is 43.2. What is the concentration of the solution?

5-8 A standard 2.50×10^{-4} *M* solution of a compound was prepared and placed in a cell with a 5.00-cm pathlength. The percent transmittance of the solution at 347 nm is 58.6. Determine the molar absorptivity.

5-9 Convert the following absorbances to percent transmittance:

(*a*) 0.113 (*b*) 0.878

(*c*) 0.430 (*d*) 0.217

(*e*) 1.023

5-10 Convert the following percent transmittances to absorbances:

(*a*) 12.3 (*b*) 87.8

(*c*) 44.8 (*d*) 62.1

(*e*) 37.6

5-11 Gossypol in methanol has a molar absorptivity of 2.0×10^4 at 374 nm. If solutions between about 1×10^{-6} and 8×10^{-6} M are to be assayed spectrophotometrically, what pathlength should be chosen for the cell?

Working-Curve Method

5-12 Several standard iron(II) solutions were prepared in a solution that contained an excess of 1,10-phenanthroline. The absorbance of each iron(II)-1,10-phenanthroline (complex) solution was measured in a 1.00-cm cell at 508 nm. A 10.0-mL portion of a sample solution was added to a 25-mL volumetric flask. A 5-mL portion of 0.1 *M* sodium acetate and a 2-mL portion of 1,10-phenanthroline solution (an excess) were added to the flask and the resulting solution was diluted to the mark with distilled water. The absorbances of the standard solutions and the solution in the volumetric flask are tabulated. Determine the concentration of iron(II) in the 10-mL sample solution.

Concentration, $\times 10^{-5}$ *M*	Absorbance
Flask	0.482
2.8	0.201
5.6	0.403
8.4	0.598
11.2	0.808
14.0	1.010

5-13 The percent transmittances of a series of solutions of an organic compound were measured at 254 nm in a cell with a 1.00-cm pathlength. The results and the percent transmittance of a sample solution are listed in the table. Determine the concentration of the sample solution.

Concentration, $\times 10^{-5}$ *M*	Percent transmittance
1.25	63.1
2.50	39.7
3.75	24.9
5.00	15.7
6.25	10.1
Sample	27.5

Standard-Addition Technique

5-14 The standard-addition technique is used to assay a sample. A 10.0-mL portion of a sample solution was added to each of five 25-mL flasks that were labeled S, 1, 2, 3, and 4. A 2.00-mL portion of a standard 2.50×10^{-3} *M* solution is added to flask 1, 4.00 mL to flask 2, 6.00 mL to flask 3, and 8.00 mL to flask 4. Each solution is diluted to the mark with distilled water. The absorbance of each diluted solution is measured in a 1.00-cm cell. Determine the concentration of the 10.0-mL sample solution.

Flask	Absorbance
S	0.328
1	0.479
2	0.630
3	0.777
4	0.923

5-15 A series of solutions was prepared by adding accurately measured masses of a pure substance to 50-mL volumetric flasks and adding sufficient volumes of a sample solution to fill each flask to the mark. The percent transmittances of the solutions are tabulated. Determine the concentration of the sample solution. Assume that the solids occupied negligible volume.

Added concentration, $\times 10^{-4}$ *M*	Percent transmittance
0 (sample)	63.4
2.00	43.9
3.50	33.7
5.00	24.8
7.50	15.4

Filters

5-16 Calculate the thickness of a first-order interference filter which uses a transparent film that has a refractive index of 1.512 and that allows radiation of 528 nm to pass.

5-17 If first-order radiation of 678 nm passes through an interference filter, determine the wavelengths of the second-, third-, and fourth-order radiation that can pass through the filter.

Prisms

5-18 Monochromatic radiation enters the equilateral triangular prism that is shown in Fig. 5-11*a* at an angle of incidence that allows the radiation to travel through the prism parallel to the base. The refractive index of the prism is 1.57. Calculate the angle of incidence and the angle at which the radiation exits the prism.

5-19 Repeat Prob. 5-18 for a triangular prism that has two 45° angles at the base and a 90° angle at the apex.

5-20 If radiation from a medium with a refractive index of 1.14 strikes a second medium with a refractive index of 2.08 at an angle of incidence that is 38.2°, calculate the angle of refraction.

Diffraction Gratings

5-21 A reflective diffraction grating contained 2125 grooves/mm. The angle of incidence of a band of polychromatic radiation is 52.5°. Determine the wavelengths that are diffracted at an angle of – 10.5°.

5-22 It is necessary to design a reflective diffraction grating that disperses first-order radiation of 355 nm at an angle of – 15.0° when the incident angle is 45.0°. Determine the number of grooves per millimeter and the distance between each groove for the grating.

Dispersion and Resolution

5-23 Determine the resolution of an instrument that is capable of separating two adjacent peaks centered at 573.5 and 569.8 nm. The distance of separation between the two peaks on the focal plane of the instrument is 2.1 mm. Calculate the dispersion and the linear reciprocal dispersion of the instrument.

5-24 A particular monochromator that uses a reflective diffraction grating has a separation on the focal plane between peaks that are centered at 25.6 and 27.3 μm of 1.35 mm. Determine the linear reciprocal dispersion of the instrument.

5-25 A monochromator has a resolution of 35.00. If a spectral peak is observed at 623.5 nm, what is the wavelength of the lower energy peak that can just be resolved with the instrument?

REFERENCES

1. Winefordner, J. D. (ed.): *Spectrochemical Methods of Analysis,* Wiley-Interscience, New York, 1971.
2. Jiang, S., and G. A. Parker: *Amer. Lab.,* 13(10): 38(1981).

BIBLIOGRAPHY

Lambert, J. B., H. F. Shurvell, L. Verbit, R. G. Cooks, and G. H. Stout: *Organic Structural Analysis,* Macmillan, New York, 1976.

CHAPTER

SIX

ATOMIC ABSORPTION SPECTROPHOTOMETRY

In atoms electron transitions of outer-shell electrons correspond to the absorption or emission of electromagnetic radiation that is in the ultraviolet and visible regions. Because vibrational and rotational energy levels are not possible in atoms, the absorption and emission of radiation that occurs in molecules when an electron travels between the numerous vibrational and rotational energy states of one electron level to the numerous vibrational and rotational states of a second level are not possible. Only a single transition between each set of electron levels is possible. As a consequence the bands of emitted or absorbed radiation are narrow in atoms.

In flames, absorptive bandwidths of atoms typically vary from about 0.001 to 0.01 nm. The narrow bands of absorbed or emitted radiation that are observed with atoms are referred to as spectral *lines*. Figure 6-1 is a diagram showing the relatively few electron transitions that are observed within a potassium atom and the wavelengths of the corresponding emitted or absorbed radiation. For some elements, such as iron and uranium, hundreds of such electron transitions are observed. Potassium was chosen for inclusion in Fig. 6-1 because of the relatively small number of observed transitions.

Emission of spectral lines from atoms has been observed for well over a century. The theory behind the emission was also established over a century ago;[1] however, it was not until 1955 that Walsh[2] invented the hollow cathode lamp that made *atomic absorption spectrophotometry* (AAS) a practical method of chemical analysis. Since 1955 atomic absorption spectrophotometry has become a popular and useful method of chemical analysis. Although the widely spread adaptation of inductively coupled plasma ICP spectrometry (Chapter 8) has significantly decreased the use of AAS by analysts, absorption

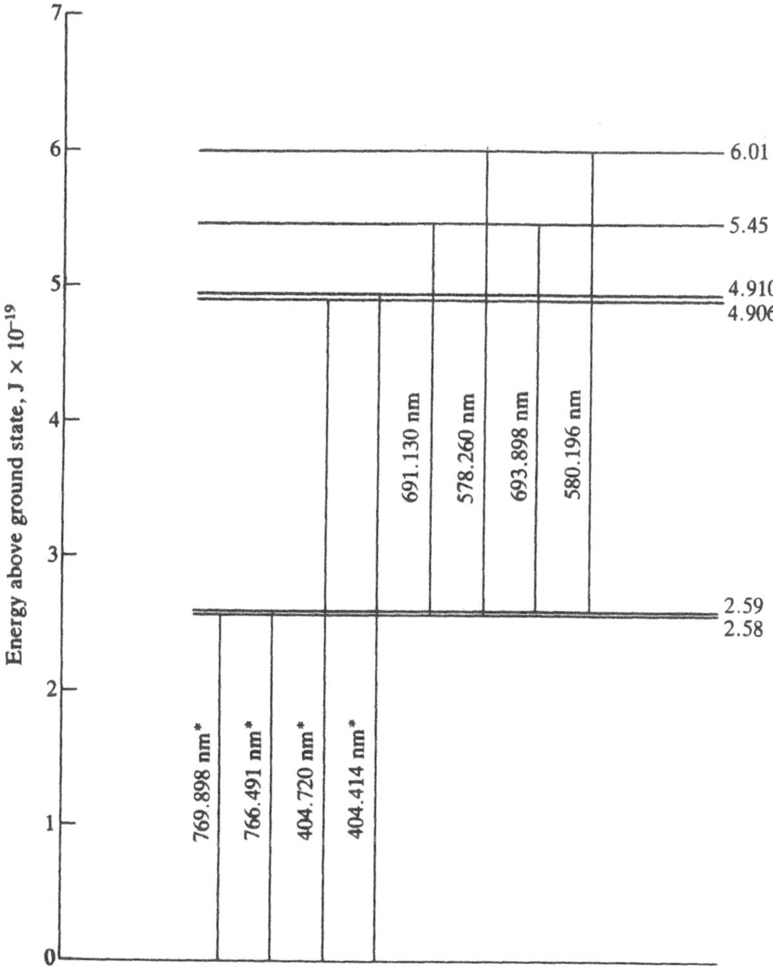

Figure 6-1 Diagram of the major energetic levels in a potassium atom. The vertical lines that connect levels represent electron transitions that occur during absorption or emission of radiation. The wavelength of the radiation is written adjacent to the vertical line that represents the transition. The transitions with asterisks are those that are normally used for AAS.

measurements remain useful and in some instances are preferred owing to lower detection limits for some elements.

SPECTRAL LINE WIDTHS

In the absence of any factor that could result in line broadening, the width of each spectral line for an atom or a monoatomic ion is approximately 10^{-5} nm. Because it is nearly impossible to avoid all of the factors that result in line broadening, atomic spectral lines are usually about two orders of magnitude broader. Of the several reasons for line broadening, two (*Doppler* and *pressure*) are nearly always encountered when atomic absorption spectrophotometry is used. Broadening owing to the presence of a magnetic field (the *Zeeman effect*) is described in the section that deals with background correction.

Doppler Broadening

The sources of radiation that are used for most measurements of atomic absorption contain energetically excited atoms that emit radiation upon returning to a lower energetic level. During atomic emission and atomic fluorescence, the radiation that is measured by the detector emanates from excited atoms in the cell. In each case (atomic absorption, atomic emission, and atomic fluorescence) the measured radiation is emitted from gaseous atoms that exhibit random motion relative to the detector. The random rapid motion of the emitting species causes Doppler broadening of the spectral line.

A basic assumption of Einstein's theory of relativity is that the speed of light is constant regardless of the direction of motion of the observer relative to the radiative source. Although the measured speed of light is fixed, the frequency and wavelength of the radiation can change as a result of the motion. The product of frequency and wavelength must always equal the speed υ of radiation in a fixed medium:

$$\upsilon = \lambda v = \frac{c}{n} \tag{5-4}$$

It can be shown[3] that the frequency of the observed radiation v_o when the radiative source is moving relative to the detector is related to the frequency v of the radiation with no motion:

$$v_o = v \frac{1 + u / \upsilon}{\left[1 - (u / \upsilon)^2\right]^{1/2}} \tag{6-1}$$

where υ (= c/n) is the velocity of the radiation in a fixed medium and u is the velocity of motion of the light source (atom) *toward* the detector. From Eq. (6-1) it is apparent that, as the velocity of the emitting atom toward the detector increases, the observed frequency of the radiation also increases. Because the emitting atoms in the radiative source or in the cell are in random motion, a series of overlapping spectral lines are observed by the detector. The result is a broadened spectral line. Doppler broadening is the major cause of broadening when atoms under low pressure emit radiation. The temperature of the source can affect Doppler broadening by changing u.

Pressure Broadening

Pressure broadening (also known as *collisional broadening*) occurs when the pressure of the atomic vapor is sufficient to cause a relatively large number of collisions that involve the emitting atoms. Usually the collisions are with molecules or atoms that are not identical to the emitting atoms. As an example, in flames the collisions can be between the emitting atoms and the molecules or atoms from the gases of the flame. Collisions that involve an emitting atom result in a shortened lifetime of the atom in an excited electron level. Because the frequency of the emission is a function of the time spent in the excited state, a change in the frequency of the emitted radiation with a consequent broadening of the spectral line occurs.

Pressure broadening that is caused by collisions between identical atoms is *resonant broadening*. Resonant broadening increases as the concentration of the assayed element increases because the probability of collisions between the atoms increases. At the temperatures typically encountered during analysis (about 2000 to 3000 K) and at a pressure of about 1 atm, the combined effect of Doppler and pressure broadening is an increase in the width of a spectral line from about 10^{-5} to about 10^{-3} nm. The usual range of observed line widths under those conditions of temperature and pressure is from 0.001 to 0.01 nm.

APPARATUS FOR ATOMIC ABSORPTION

The arrangement of the major parts of a single-beam spectrophotometer that is used for measurements of atomic absorption and a photograph of a commercial atomic absorption spectrophotometer are shown in Fig. 6-2. Electromagnetic radiation is emitted from a source at the required wavelength in the ultraviolet-visible region and is directed through a cell that contains atoms of the sample. Absorption of some of the radiation takes place in the cell. The unabsorbed radiation passes through a monochromator and into a transducer (detector) that transforms the electromagnetic radiation into an electric signal. The electric signal is amplified, if necessary, and sent to a readout device which provides the analyst with a percent transmittance, absorbance, or concentration reading. Each of the components of an atomic absorption spectrophotometer is described in a following section of the chapter.

EMR SOURCES FOR AAS

The extremely narrow spectral lines that are characteristic of atoms and monoatomic ions make it necessary to use EMR sources for AAS that are highly intense within that narrow absorptive band. Sufficient radiation to permit accurate measurement at the detector can be achieved with a narrow band of incident radiation only when the incident radiation is intense. The intensity requirement for molecular (polyatomic) spectra is fulfilled with radiation of a relatively broad bandwidth. The entire bandwidth of radiation that passes

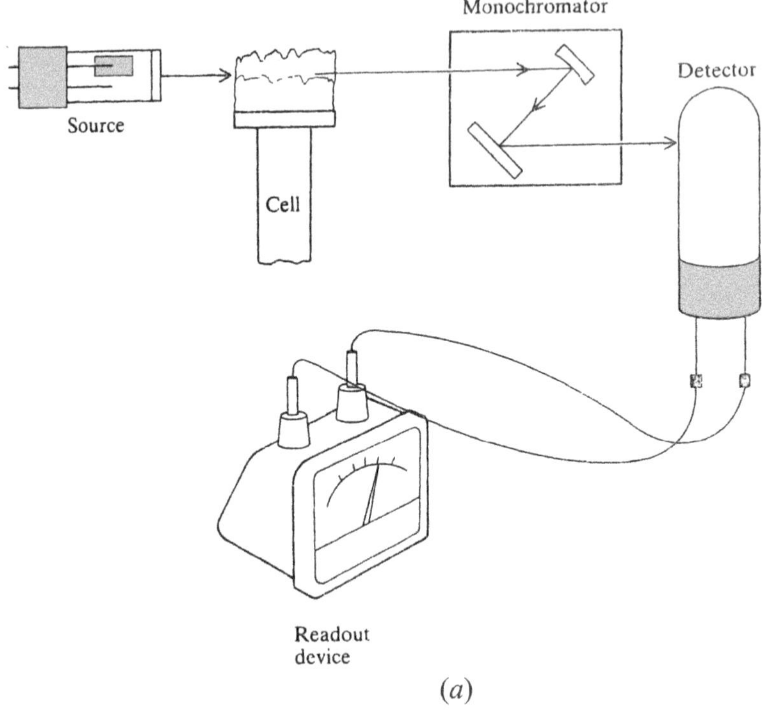

(*a*)

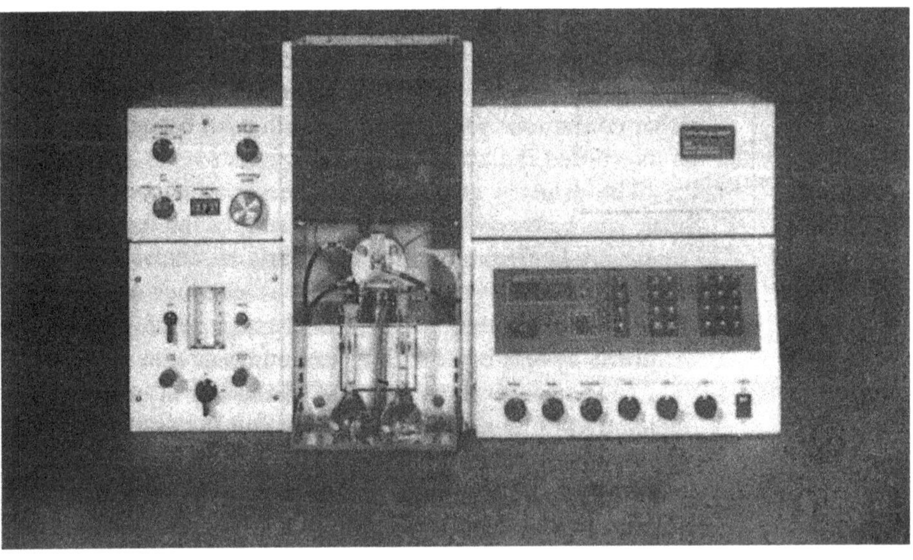

(*b*)

Figure 6-2 (*a*) Diagram of a single-beam atomic absorption spectrophotometer (*b*) Photograph of a Perkin-Elmer model 2280, single-beam atomic absorption spectrophotometer. (*Photograph courtesy of Perkin-Elmer Corporation.*)

through the monochromator easily fits within the range of wavelengths that are absorbed by the polyatomic species. The intensity within the broader wavelength range does not need to be as great because the detector integrates the rate of impinging photons within the relatively broad wavelength range within the bandpass of the monochromator. That option is not available for atomic spectra because the radiation cannot be absorbed unless it falls within the narrow bandwidth of the spectral line of the sample in the cell. Because that bandwidth is much narrower than that used to assay polyatomic species, the intensity must be greater.

The radiation coming from the source is usually interrupted at a designated rate by a chopper or by pulsing the voltage to the source. A chopper for a beam of EMR is a rotating wheel that contains one or more blades that are similar to those in a fan. The rotating wheel is placed in the path of the radiation. When a blade enters the path of radiation, the radiation strikes the blade and is prevented from continuing through the apparatus. The signal that strikes the detector is an ac square wave with a frequency that is determined by the rate of rotation of the wheel and the number of blades. By tuning the detector to respond only to the chopped signal, it is possible to eliminate steady-state signals that emanate from the cell and that might interfere with the assay. Emission of continuous radiation from the gases in the flame that often serves as the cell and emission from the assayed element are two types of steady-state signals that can interfere with the assay.

Continuous Sources

The major categories of EMR sources that are used for AAS are *continuous sources* and *line sources*. Continuous sources are those that emit broad bands of radiation from which radiation of the desired wavelength is selected with a monochromator. A relatively popular continuous source that is used for AAS is the *xenon-arc lamp*. Electricity is arced between two electrodes in an envelope that contains xenon at a reduced pressure. The electric energy causes excitation of the xenon within the lamp. When the xenon atoms and ions return to lower energetic levels, radiation is continuously emitted over the wavelength range from approximately 200 to 700 nm. The xenon-arc lamp is chosen more often than other continuous sources for use with AAS because of the relatively high intensity of its emitted radiation.

The primary advantage of the use of a continuous source is that a single lamp can be used at different wavelengths for the analysis of many elements. The xenon-arc lamp can be used for nearly all analyses that are usually done by AAS because its emissive band overlaps at least one of the commonly used absorptive lines of nearly all the analyzed elements. Automated multielement AAS can be done with a continuous source.

The major disadvantages of the use of continuous sources for AAS are the relatively low emissive intensities within the absorptive bandwidths of atoms and the requirement for a high-quality monochromator. Even the relatively high intensity of the xenon-arc lamp is usually considerably less intense within the absorptive bandwidth of a particular atom than that of an appropriate line source. A high-quality (narrow-bandpass) monochromator is required with a continuous source in order to limit the band of wavelengths that are used in the study as much as possible. Even with a good monochromator, the wavelength band that is used with a continuous source is broader than the absorptive band of the atoms in the cell. For those reasons, line rather than continuous sources are generally provided with commercial spectrophotometers.

Line Sources

The remainder of the sources that are described in the chapter are line sources. They emit highly intense radiation only in the wavelength range that corresponds to the absorptive band for a particular atom or group of atoms. They do not emit a continuum of radiation. The primary advantage of a line source is the high intensity that is achieved within a narrow bandwidth. The major disadvantage is the need to change the line source whenever the element that is to be assayed is changed. The line sources that are used most often for AAS are the *electrodeless discharge lamp* (EDL) and the *hollow-cathode* (HC) *lamp*. Other line sources such as lasers are described in the next chapter.

Electrodeless discharge lamps An electrodeless discharge lamp is prepared by placing a small amount of an element or a halide of an element into a quartz tube. A few milligrams of the iodide of the element is often used. The quartz tube is evacuated under an atmosphere of an inert gas to a pressure that is usually between 0.1 and 5 torr (13-650Pa). For most EDLs the optimal pressure is between 0.3 and 3 torr. After evacuation the tube is sealed and placed into a hollow microwave or radiofrequency cavity. The vapor pressures of the elements that are used in an EDL are sufficiently high to permit some gaseous atoms of the element to be formed in the low-pressure environment of the lamp. Application of the AC field (typically at a frequency of 2450 MHz) causes excitation of some of the gaseous atoms. As the excited atoms make the transition to the ground state or to other low-energy levels, radiation that is characteristic of the atoms is emitted and travels through the transparent end of the cavity. A sketch of a commercial EDL is shown in Fig. 6-3.

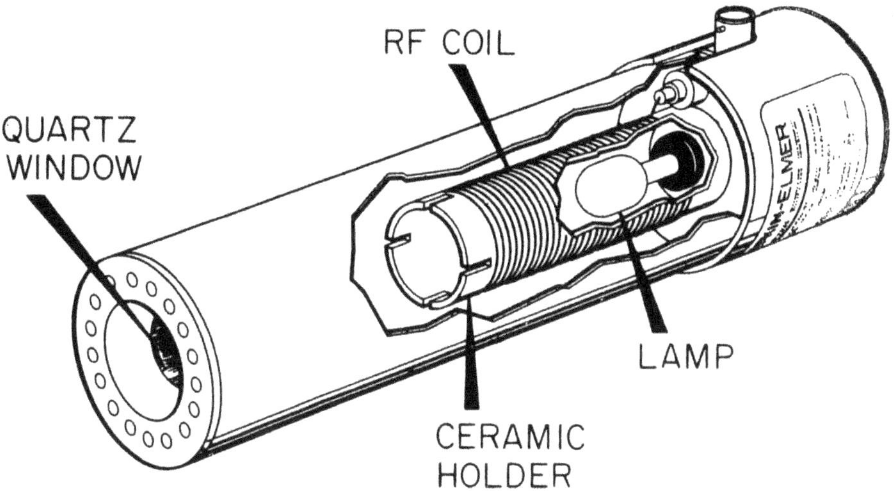

Figure 6-3 An electrodeless discharge lamp. (*Photograph courtesy of Perkin-Elmer Corporation.*)

An EDL can be operated at a power level that corresponds to any of three stages depending upon the required intensity of emitted radiation. At a low power the lamp emits a dim glow that is spread throughout the lamp. At an intermediate power level the lamp exhibits a high-intensity glow that is concentrated in a relatively small volume of the lamp. At a high power level the high-intensity glow fills the entire lamp.

Generally EDL operational lifetimes are greater than 50 h. Most EDLs emit radiation that is at least 10 times as intense as that of the corresponding HC lamp. Consequently, EDLs are more often used for fluorescent measurements than for measurements of absorption. Electrodeless discharge lamps are probably the most common source of radiation that is used for atomic fluorescence. The major disadvantage of EDLs is the lack of availability of good commercial lamps for some elements. Electrodeless discharge lamps are commercially available for As, Bi, Cd, Cs, Ge, Hg, K, P, Pb, Rb, Sb, Se, Sn, Ti, Tl, and Zn.

The intensity of the output from an EDL appears to be temperature-dependent. When using an EDL some method of temperature control is advisable. Some workers claim that the major disadvantage to the use of an EDL is the relative instability of the lamp as compared to an HC lamp. In many cases much of the instability is eliminated by minimal temperature control of the lamp.

Hollow-cathode lamps A hollow-cathode lamp consists of two electrodes that are enclosed in an evacuated glass container. Hollow-cathode lamps are the most popular sources of radiation for AAS. The material from which the anode is constructed is not particularly critical. Usually the anode is made of tungsten, although nickel and zirconium have also been used.

The cathode consists of a hollow cylinder that is constructed from the element, an alloy of the element, or a substance that is coated with the element whose spectrum is desired. The cathode is usually surrounded by a glass or porcelain shield which serves to minimize diffusion of the atoms that are formed in the lamp away from the cathode. Electrical connection to the cathode and anode is accomplished through wires or pins that are sealed through the base of the lamp.

The glass envelope that surrounds the electrodes contains an inert gas at a relatively low pressure (usually 1 or 2 torr). The inert gas, which is known as the *filler gas*, is chosen to minimize overlapping between its spectrum and the spectrum of the element on the cathode. In most lamps either neon or argon is

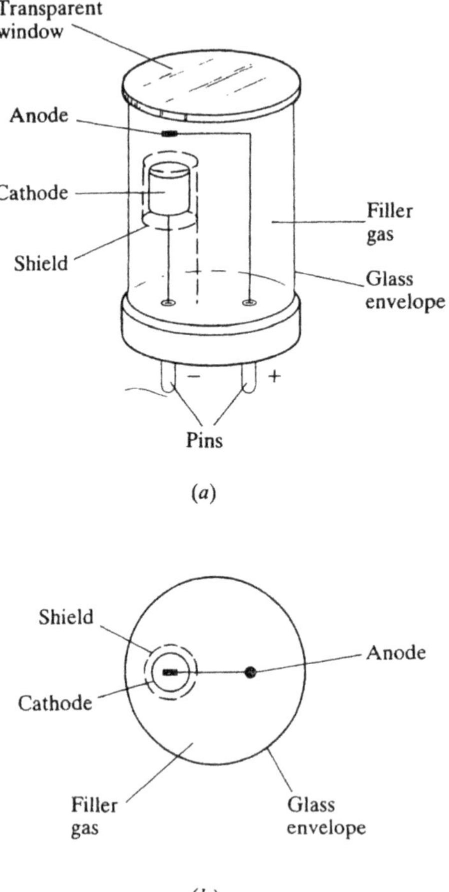

Fig 6.4 Sketches of a hollow-cathode lamp; (*a*) side view; (*b*) front view.

used as the filler gas. Helium is used in some lamps. The end of the hollow-cathode lamp is sealed with a window constructed of a material that is transparent to the radiation that is emitted by the element on the cathode. If radiation in the ultraviolet region is emitted from the lamp, the window is constructed of quartz or silica. If radiation in only the visible region is emitted, the window is constructed of glass. Sketches of a hollow-cathode lamp are shown in Fig. 6-4.

When a hollow-cathode lamp is used, a voltage of about 100 to 200 V is applied between the cathode and the anode with an external power supply. That causes a current of 1 to 25 mA to flow between the electrodes. The current is adjusted by the analyst to conform with the value recommended by the lamp manufacturer for optimal operation for the particular lamp.

Application of the potential causes formation of positive ions of the inert filler gas. The filler-gas ions are accelerated to the negative cathode, where the energy of their collision is sufficient to cause the element on the cathode to be transformed into gaseous atoms in a process that is known as *sputtering*. Absorption of energy from the applied potential and from collisions with filler-gas ions cause the gaseous atoms to become excited to higher electron-energetic states. As the electrons return to lower states, radiation that is characteristic of the excited atom is emitted. Because the radiation emitted by an HC lamp is characteristic of the element on the cathode of the lamp, it is necessary to use a lamp that has a cathode that is coated with the element that is being assayed. As an example, if it is necessary to assay for copper, a hollow cathode containing copper is used in the lamp.

The low pressure in the lamp results in minimal pressure broadening of the radiation that is emitted by the lamp. The atomic vapor of the sample in the cell has an absorptive bandwidth owing to pressure broadening that is typically four or five times greater than the bandwidth emitted from an HC lamp.

A major disadvantage to the use of HC lamps is the necessity of using a different lamp for each assayed element. The disadvantage can partially be overcome by using cathodes that are made from more than one element. The cathode in *multiple-element hollow-cathode lamps* is made from either an alloy or rings of the desired elements. The commercial multiple-element HC lamps can be used for the assay of between two and seven elements.

The elements on a multiple-element HC lamp fractionally distill from the cathode during sputtering. Because the more volatile elements are converted to vapor first, the concentration of the less volatile components on the cathode is increased as the lamp is used. Since the elements do not necessarily condense back onto the cathode as the lamp cools after use, the composition of the cathode changes with use. That leads to a change in the ratio of the intensities of the emitted radiation of the elements on the cathode. It is particularly noticeable with cathodes that are made from alloys. Although it is usually not a serious problem, the analyst should be aware of the changing lamp characteristics.

Some workers have used *demountable hollow-cathode lamps* as sources of radiation for AAS. The cathodes in such lamps are interchangeable. Consequently, a single lamp can be used for more than one element by changing to the appropriate cathode. The use of demountable hollow-cathode lamps has not become popular because of the effort and time that are required to change the cathode.

When more intense radiation is required for a particular application than a hollow-cathode lamp can provide, a *high-intensity hollow-cathode lamp* (also called a *boosted-output hollow-cathode lamp*) can be used.[4, 5] With the conventional hollow-cathode lamp that is shown in Fig. 6-4, the only control the operator has over the intensity of the emitted line is the lamp current. Unfortunately, increasing the lamp current in order to increase the intensity of the emitted radiation by increasing the number of excited atoms within the lamp also results in increased sputtering and consequently a higher atomic population. An increased atomic population causes resonant broadening and an increase in the amount of emitted radiation that is absorbed by other identical atoms within the lamp. The latter process is *self-reversal*.

The high-intensity hollow-cathode lamp is designed to minimize both resonant broadening and self-reversal while increasing the intensity of the emitted radiation. It accomplishes the goal by separating the functions of sputtering and excitation. The lamp contains two sets of electrodes, as shown in Fig. 6-5. The first set of electrodes is identical to that used in a conventional hollow-cathode lamp. The current between the two electrodes is adjusted to a relatively low value to keep sputtering low and to maintain a relatively low atomic population within the lamp.

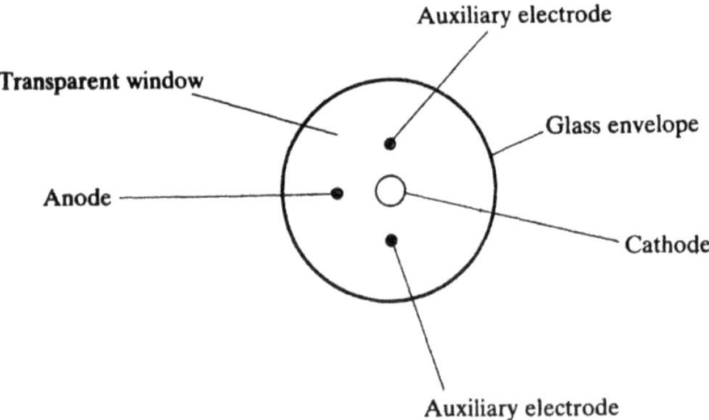

Figure 6-5 Front-view sketch of a high-intensity hollow-cathode lamp.

Excitation of the sputtered atoms is accomplished with an auxiliary pair of electrodes within the lamp. Generally the auxiliary electrodes are coated with an alkali metal salt. The auxiliary electrodes are maintained at a potential difference that is sufficiently large to cause excitation of a larger percentage of the atoms than that which occurs in the conventional HC lamp. Typically the intensity of the resulting resonant lines that are emitted from the lamp are between 30 and 100 times greater than those emitted from conventional lamps. No significant increase in line width owing to resonant broadening is observed with the lamps.

A disadvantage of the high-intensity hollow-cathode lamps is the need for a second power supply for the auxiliary electrodes. The extra intensity that is obtained with high-intensity hollow-cathode lamps is only rarely needed for AAS. The lamps are used more frequently for atomic fluorescent spectrometry.

Temperature-Gradient Lamps

A recently developed source of EMR for atomic absorption and atomic fluorescence is the *temperature-gradient lamp* (TGL). The TGL uses an electric heater to convert an element into atomic vapor. A relatively high current (about 0.5 A) which is passed through the vapor causes excitation of some of the atoms, which subsequently emit radiation at the wavelengths that are characteristic of the element in the lamp. As of this writing, the lamps have not become popular.

A sketch of a temperature-gradient lamp is shown in Fig. 6-6. The element whose emissive spectrum is desired is placed in the bulb of the lamp. The bulb is electrically heated to a temperature that produces the desired amount of atomic vapor. The temperature of the oven is carefully controlled by the electric circuit. A current is caused to flow between a filament and an anode by application of about 30 V. The filament is coated with an oxide of barium, calcium, or strontium (which have relatively low ionization energies) in order to make the filament more electron-emitting. The body of the lamp is constructed from glass but the transparent window is made of silica. By placing the window near the warm section of the lamp, condensation of hot vapor onto the window is prevented. The lamp also contains a relatively cool portion on which atomic vapor can condense and which serves to protect the window. Argon is generally used as the filler gas for the lamp at a pressure of about 1 to 5 torr.

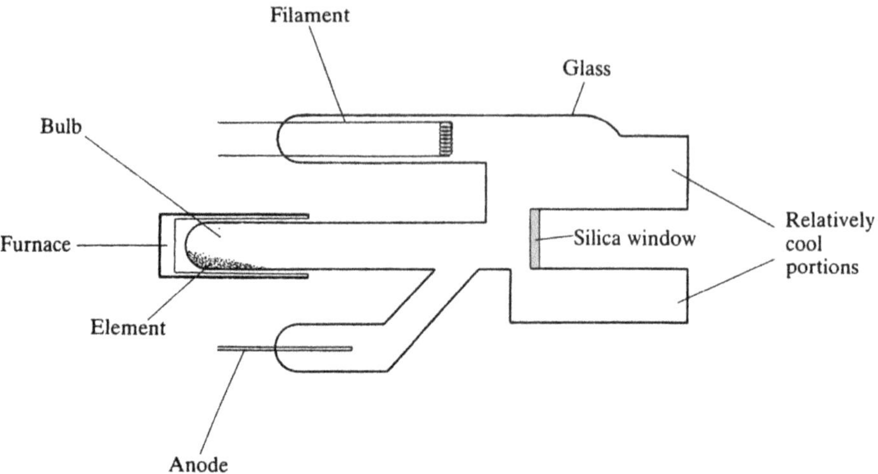

Figure 6-6 Sketch of a temperature-gradient lamp.

The intensity of the radiation that is emitted from a TGL is considerably greater than that from HC lamps and is comparable to that from EDLs. The line widths of the emitted lines from a TGL are about 0.001 nm, which is approximately the same as those from an EDL. TGLs are claimed to exhibit less self-reversal than EDLs. Both EDLs and TGLs are primarily used in AAS for studies at wavelengths that are below 200 nm for elements such as arsenic and selenium for which there are no adequate HC lamps.

CELLS

The cells that are used for AAS serve two purposes. They are used to convert polyatomic samples into atoms and to contain the atoms sufficiently long for the impinging radiation to pass through the cell and for the detector to make a measurement. In order for the cell to convert a polyatomic sample into atoms, the cell usually must be capable of supplying energy to the sample. Energy for the conversion is often supplied as heat from a flame or furnace.

Unheated Cells

Mercury is the only common metal that has an appreciable atomic vapor at room temperature and that does not require the application of energy by the cell in order to convert the analyte to atoms. The *cold-vapor method for mercury* is described because of its widespread use in environmental analysis. Mercury also can be assayed by AAS with other types of cells.

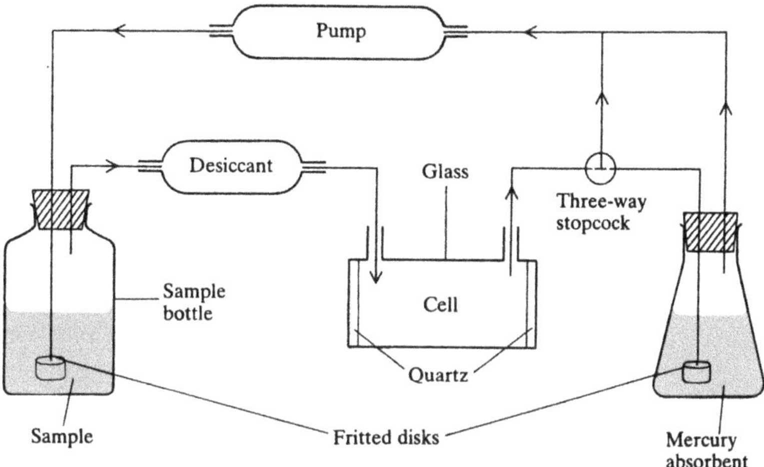

Figure 6-7 Sketch of the apparatus that is used for the cold-vapor determination of mercury by AAS.

In the usual cold-vapor method for mercury the sample solution is chemically treated with a reducing agent [usually tin(II)] which converts dissolved mercury ions to metallic mercury. Air is pumped through the solution at a constant rate of about 1 L/min and the atomic vapor of mercury is swept into a 10-cm glass cell. The cell has quartz windows that are transparent to radiation at the wavelength (253.7 nm) of the mercury line that is used for the assay. A mercury HC lamp or mercury vapor lamp is used as the source of radiation. After the vapor has passed through the cell, it is usually cycled back through the sample solution. An open system where the vapor is not pumped back into the cell can also be used. A peristaltic pump is often used in order to prevent contamination by metallic components that are found in piston pumps. After the measurement has been completed the mercury is flushed from the system by pumping the vapor through a solution (such as 0.05 *M* $KMnO_4$ and 5% H_2SO_4) that oxidizes and absorbs mercury or into a fume hood. The working-curve method is used for the analysis. A diagram of the apparatus is shown in Fig. 6-7.

Flames

Flames have been used as cells since the development of AAS. They are still more popular than other cells. The sample solution is aspirated into the flame as small drops. The solvent in the solution rapidly evaporates because of the heat from the flame. The solid particles of solute that remain after solvent evaporation melt to form a liquid, evaporate to yield a gas, and dissociate into atoms. Radiation from the lamp passes through the flame and is partially absorbed by the sample atoms. The amount of absorption is monitored with the detector.

Prior to its aspiration into the flame, the sample solution is sucked through a small-diameter tube into the stream of oxidant flowing to the flame. The flow of the oxidant past the orifice to the sample tube provides the partial vacuum (the Venturi principle) required to suck the solution into the oxidant stream. After the sample is in the oxidant stream, it is mixed with fuel gas and ignited to form the flame.

The burners used for AAS are of two types. In the *total-consumption burner* the oxidant, fuel, and sample pass directly into the flame as shown in Fig. 6-8. An advantage of a total-consumption burner is that the flame gases are not mixed prior to being burned in the flame. That minimizes the likelihood of an explosion and makes it possible to safely use gases that burn at a high velocity. A large amount of sample passes into the flame during a fixed period with a total-consumption burner. That leads to a relatively high concentration of the sample in the flame.

A disadvantage of a total-consumption burner is the relatively turbulent flame that results from the erratic cooling which occurs whenever large drops enter the flame. Large drops which are not completely vaporized in the flame can scatter radiation and cool the portion of the flame that is near the drop. The turbulence in the flame increases the amount of noise reported by the detector and leads to an unstable instrumental reading. The shape of the flame from the total-consumption burner is not ideal for AAS because the pathlength through the cell is too small for high sensitivity. For these reasons total-consumption burners are rarely used for AAS, but are used for flame emission spectrometry (Chapter 7).

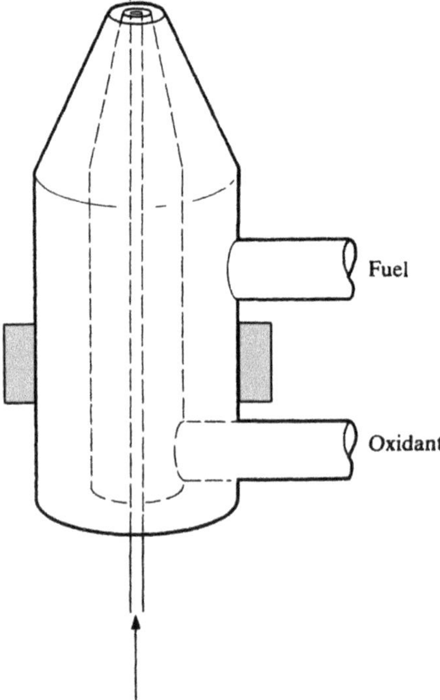

Figure 6-8 A total-consumption burner.

The turbulence that is associated with a total-consumption burner can be significantly decreased by removing the larger sample drops prior to introducing the sample into the flame. In the *premix* or *laminar-flow burner* the oxidant-sample mixture flows into a chamber located upstream from the flame, where the larger drops are separated from the mixture and discarded. The chamber is also used to mix the fuel with the oxidant and the sample. As the sample is swept through the chamber by the gaseous mixture, all except the smaller drops strike obstacles (a flow spoiler) placed in the path of the flow and fall into the bottom of the chamber. The fallen drops are channeled through a drain tube to a plastic waste-collection bottle. Small drops are swept through the chamber to the burner head and into the flame. About 90 per cent of the sample is lost in the premix chamber.

Although less sample enters the flame than in a total-consumption burner, that does not necessarily result in a significantly decreased atomic population within the flame. More sample enters the flame in a total-consumption burner, but because of the incomplete vaporization, a smaller proportion of the sample is atomized. Because the smoother burning flame in the premix burner results in a higher signal-to-noise ratio, the premix burner is preferred for most quantitative absorption analyses. A disadvantage of the premix burner is the explosive hazard when the oxidant and fuel are mixed in the premix chamber. An advantage of the burner is the ability to use burner heads that yield flames that provide a long pathlength for the radiation. A sketch of a premix burner is shown in Fig. 6-9.

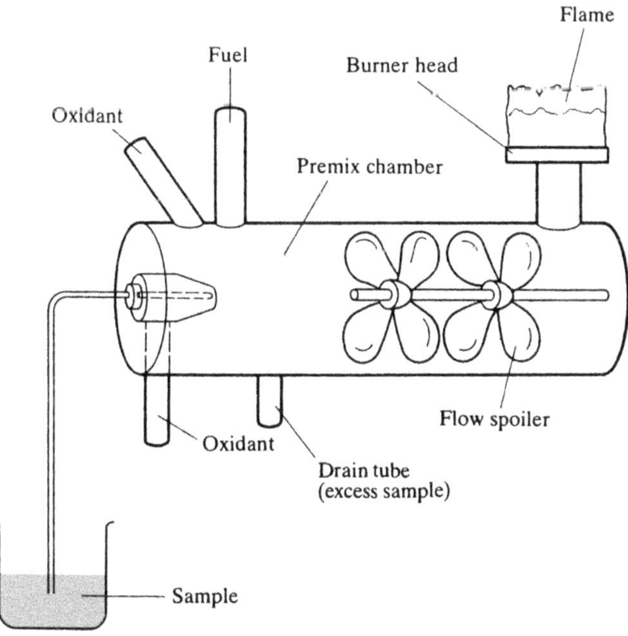

Figure 6- 9 A premix burner.

In atomic absorption spectrophotometry the absorbance is proportional to the pathlength of the cell and to the concentration of the sample within the cell (Beer's law). The concentration of atoms in a flame can be maintained at a relatively high value by keeping the volume of the flame as small as possible. The cell pathlength can be increased by increasing the length of the flame through which the EMR travels. Unfortunately, increasing the length also increases the volume of the flame. The best compromise is obtained by using a burner head that yields a relatively long narrow flame. The slot in the burner head of a premix burner is usually 5 or 10 cm in length depending upon the gases that are used as oxidant and fuel. The width of the flame perpendicular to the path of the radiation is as narrow as possible within the limitations imposed by the need for efficient atomization. A flame that is too narrow results in significant cooling from the air adjacent to the flame and a consequent decrease in atom formation within the flame. The slot in the burner head must be small enough to allow the velocity of gases flowing into the flame to be greater than the burning velocity. If the flame burns faster than the flow velocity through the slot, the flame can enter the premix chamber and explode. With fast-burning gases, slot burners cannot be used because the burning rate is too high.

Different substances require different amounts of energy to be converted from species that are found in solutions to atoms. If too little energy is used, the substance cannot be converted into atoms; and if too much energy is used, the substance can be converted into ions rather than atoms. The energy for the conversion to atoms is supplied by the heat of the flame. Ionization occurs whenever the energy that is available from the flame exceeds the ionization energy of the sample. The amount of energy that can be supplied by the flame is directly proportional to the temperature of the flame. Because different substances require different amounts of energy to be converted to atoms, it is necessary to have control over the flame temperature.

Small changes in flame temperature can be achieved by varying the ratio of the oxidant gas to the fuel gas supplying the flame. A flame in which sufficient oxidant is used to efficiently react with all of the fuel is a *lean* flame. A flame in which an excess of fuel is used is a *fuel-rich* flame. Lean flames are hotter than fuel-rich flames. Greater temperature changes are brought about by changing the oxidant, the fuel, or both the oxidant and the fuel. Table 6-1 lists the oxidant, the fuel, and the approximate maximum temperature of several flames which can be used for atomic spectroscopy. Figure 6-16 (later in the chapter) contains a list of the flames and flame conditions (lean or fuel-rich) that are used for most assays with AAS. For AAS it is generally best to choose the lowest temperature flame that will efficiently atomize the assayed element. A higher temperature could result in loss of signal owing to ionization. For most assays an air-acetylene flame is used.

Table 6-1 Several flames which are often used in atomic spectroscopy

Fuel	Oxidant	Approximate maximum temperature, °C
Natural gas	Air	1800
Propane	Air	1900
Hydrogen	Air	2000
Acetylene	Air	2300
Hydrogen	Oxygen	2700
Acetylene	Nitrous oxide	2800
Acetylene	Oxygen	3100

In addition to the fuel and oxidant, sometimes an inert gas is also used in the mixture of flame gases to help control the temperature. The two most popular examples are the argon-oxygen-acetylene flame and the helium-oxygen-acetylene flame. The flame temperature is regulated by controlling the ratio of the inert gas to oxygen. As the amount of the inert gas increases or the amount of oxygen decreases, the temperature of the flame decreases. It is possible to control the temperature up to a maximum of that of the acetylene-oxygen flame (3100°C) while using a single burner and without changing the flame gases.

For studies at wavelengths that are less than about 200 nm, special precautions are necessary to exclude from the path of the radiation atmospheric oxygen and those flame gases that absorb in that wavelength range. The entire optical path of the atomic absorption spectrophotometer is often continuously flushed with nitrogen to remove potential interferences. Nitrogen is allowed to flow over the flame in order to prevent atmospheric oxygen from entering the flame. The more popular of the flames that are protected in that manner are the *nitrogen-sheathed* air-acetylene and the nitrogen-sheathed nitrous oxide-acetylene flames. Argon sheathing can also be used although the expense is greater.

The detection limit in flame AAS is partially determined by the rate at which the sample enters the flame and partially by the efficiency of conversion of the sample to atoms. Several methods have been developed that introduce a sample into a flame with an efficiency that is greater than that normally achieved. Perhaps the most popular of-the methods is *hydride atomic absorption*. The assayed element (usually As, Bi, Ge, Hg, Sb, Se, Sn, or Te) is chemically converted to a *gaseous* metal hydride. The metal hydride is concentrated in a cold trap (often a liquid nitrogen trap), allowed to come to room temperature, and flushed into the aspirator of the burner with nitrogen. A relatively high concentration of the element enters the flame in a short time. Because no droplets of solution enter the flame, the efficiency of conversion to atoms is high and the detection limit is low. Apparently Holak[6] first used hydride atomic absorption spectrophotometry for the determination of arsenic. The method is reviewed in Reference 7.

Furnaces

Since the first reported use by L'vov[8] in 1961, furnaces have grown in popularity as cells for AAS. At present they are second in use to flames. Those made from graphite are more common and are electrically heated to temperatures as high as 3000°C. Those that are constructed from tantalum are less common and are heated to temperatures as high as 2600°C.

Furnaces are usually protected from oxidation during heating with a stream of argon or nitrogen. Nitrogen is not used at temperatures greater than 2500°C because cyanogen can be formed from the reaction with carbon furnaces. Typically the heated portion of the device is between 1 and 3 cm in length. The power supply that is used for the heating usually must be capable of delivering a current of at least 300 A at an applied potential difference of about 12 V.

A measured volume of the sample solution is placed in the furnace and electrically heated in at least three steps. A major advantage of the use of furnaces is the small sample volume that is required for an assay. Typically between 5 and 50 μL is used for each assay. In the *drying stage* the solvent is evaporated from the sample solution by holding the temperature of the furnace slightly above the boiling point of the solvent. For aqueous solutions the furnace temperature is usually adjusted to 110°C for about 30 s. The temperature in the drying stage should be sufficiently low to prevent spattering.

In the *ashing stage* the temperature is raised to a value that is sufficient to remove all volatile components of the sample but not sufficient to vaporize the analyte (assayed substance). Typically the temperature in the ashing stage is maintained between 350 and 1200°C for a period of about 45 s. In the ashing stage organic substances are volatilized and the chemical matrix of the sample is destroyed.

The final step is the *atomization stage,* in which the temperature is increased to a point at which discrete (gaseous) atoms of the analyte are formed. Typically the temperature is adjusted to between 2000 and 3000°C for a period of about 5 s. The absorption of the analyte is measured during the atomization stage. Except when no argon or nitrogen flows through the cell, the atoms rapidly diffuse out of the path of incident radiation, causing a transient absorptive signal. Usually the signal is measured with the aid of a strip-chart recorder or computer controlled data storage device, and appears as a sharp peak. Either the area or the height of the peak can be used to prepare a working curve. At each stage the analyst has control of the temperature of the furnace and the time at each temperature. With some furnaces the analyst can also control the rate of temperature increase between stages.

During the atomization stage the heating must be rapid in order to ensure a relatively high concentration of atoms within the path of radiation. Power supplies for modern furnaces are typically capable of causing an increase in temperature of at least 1000°C/s. Heating rates as high as 10^8 °C/s[9] have been achieved by using capacitive discharge to heat the furnace. The short time for the absorptive measurement requires the detector and the associated electronics to respond rapidly. In some cases the absorptive peak widths are as narrow as 0.1 s. A slow-responding detector and associated electronics leads to distortion of the absorptive peak shape and associated error in the results. A fast response is not required when a flame is used as the cell because the sample is continuously entering the flame, i.e., the signal is continuous rather than transient.

Graphite is the most popular material that is used for the construction of furnaces. *Tubular graphite furnaces* and *carbon-rod atomizers* are commercially available. The tubular furnace consists of a hollow tube of graphite into which the sample is placed. Electrical connection is made at opposite ends of the tube. Radiation from the source passes along the length and through the center of the tube. A sketch of a tubular furnace and a picture of a commercial furnace module are shown in Figs. 6-10 and 6-11. Often tubular graphite furnaces are coated with pyrolytic graphite in order to prevent diffusion of atomic species of the sample through the walls of the furnace. Pyrolytic graphite is prepared by heating graphite in an inert atmosphere. Upon cooling, the pyrolytic graphite is not as porous as graphite.

Common problems with the simple tubular furnace that is sketched in Fig. 6-10 are significant background absorption and a change in the slope of working curves with a change in the matrix (chemical environment) of the analyte. In many cases *matrix effects* can be minimized or eliminated by using a furnace that operates at a constant temperature. Several approaches have been used to allow the furnace to remain at a constant temperature. The sample can be placed on a carbon or a tungsten wire and dried and ashed outside of the furnace. The ashed sample is inserted into the furnace that is maintained at a constant atomization temperature.

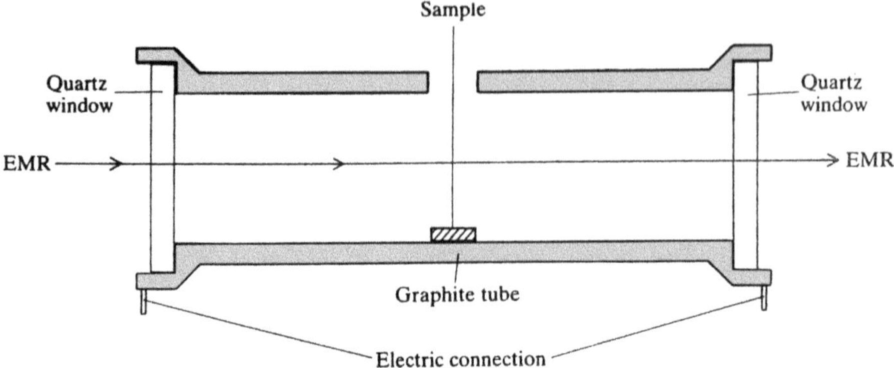

Figure 6-10 Sketch of a tubular graphite furnace. The furnace is protected from oxidation with Ar or N_2 and can have provisions for cooling.

A simpler alternative to that procedure involves the use of a *platform furnace*. A platform furnace has the same design as the conventional tubular furnace except that a small graphite platform is placed inside the furnace, as shown in Fig. 6-12. The sample is placed on the platform and the oven is operated in the normal manner (three heating stages). Because the platform is primarily heated by radiation from the tubular portion of the furnace, its temperature rises more slowly than that of the furnace walls. Consequently atomization of the sample is delayed until a time at which the temperature within the furnace becomes nearly constant. The original platform as suggested by L'vov[11] was flat. A curved platform has also been suggested.[12] Platform tubular furnaces have been shown to eliminate many of the matrix effects that are observed with conventional tubular furnaces.

The second category of graphite furnaces is the carbon-rod atomizer. A carbon-rod atomizer consists of a solid cylindrical graphite rod which has a hole for the sample cut into its center. The rod is supported by metallic terminals that provide it with electric connection to the power supply. As in the tubular graphite furnace, the carbon rod is protected from oxidation by argon or nitrogen that flows over it. The carbon rod can be enclosed in a metallic jacket with quartz windows. A sketch of a carbon-rod atomizer is shown in Fig. 6-13. A *carbon-cup atomizer* is a carbon-rod atomizer in which the sample hole is replaced with a carbon sample cup. The several types of graphite furnaces are described in References 12 through 14. Carbon-rod atomizers do not appear to be as popular as tubular furnaces.

In another type of furnace the sample is placed on a depression in an electrically heated tantalum strip. The strip is enclosed in a housing that contains quartz windows for the entrance and exit of radiation and a door to permit placing the sample in the furnace. The tantalum is protected from oxidation with argon or nitrogen. A diagram of a tantalum-strip furnace is shown in Fig. 6-14.

Because an atom of the sample in a furnace is contained within the path of EMR from the source for a longer period than an atom in a flame, furnaces can be used to assay smaller sample volumes and lower concentrations than those typically assayed by using flames. In most cases the mass of sample required to be detectable with a furnace is at least 100 times less than that required with a flame. In furnaces sample volumes in the microliter range are commonly used, whereas aspiration into a flame usually requires at least a milliliter of solution.

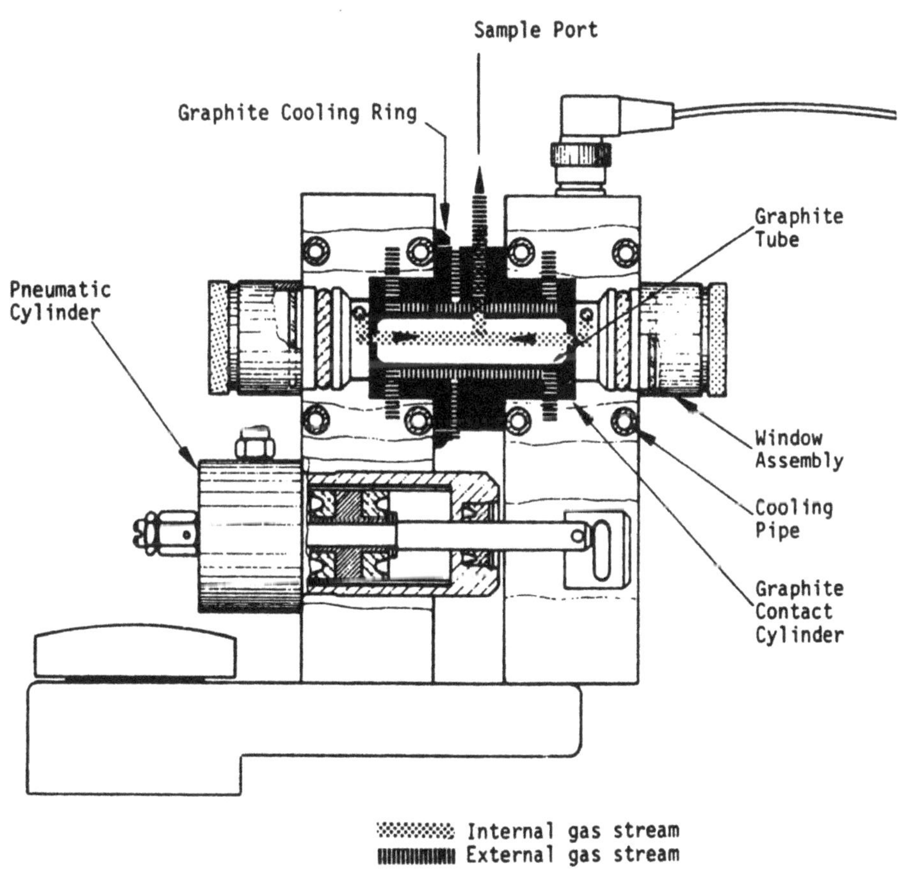

Cross-section of Furnace Assembly

(*a*)

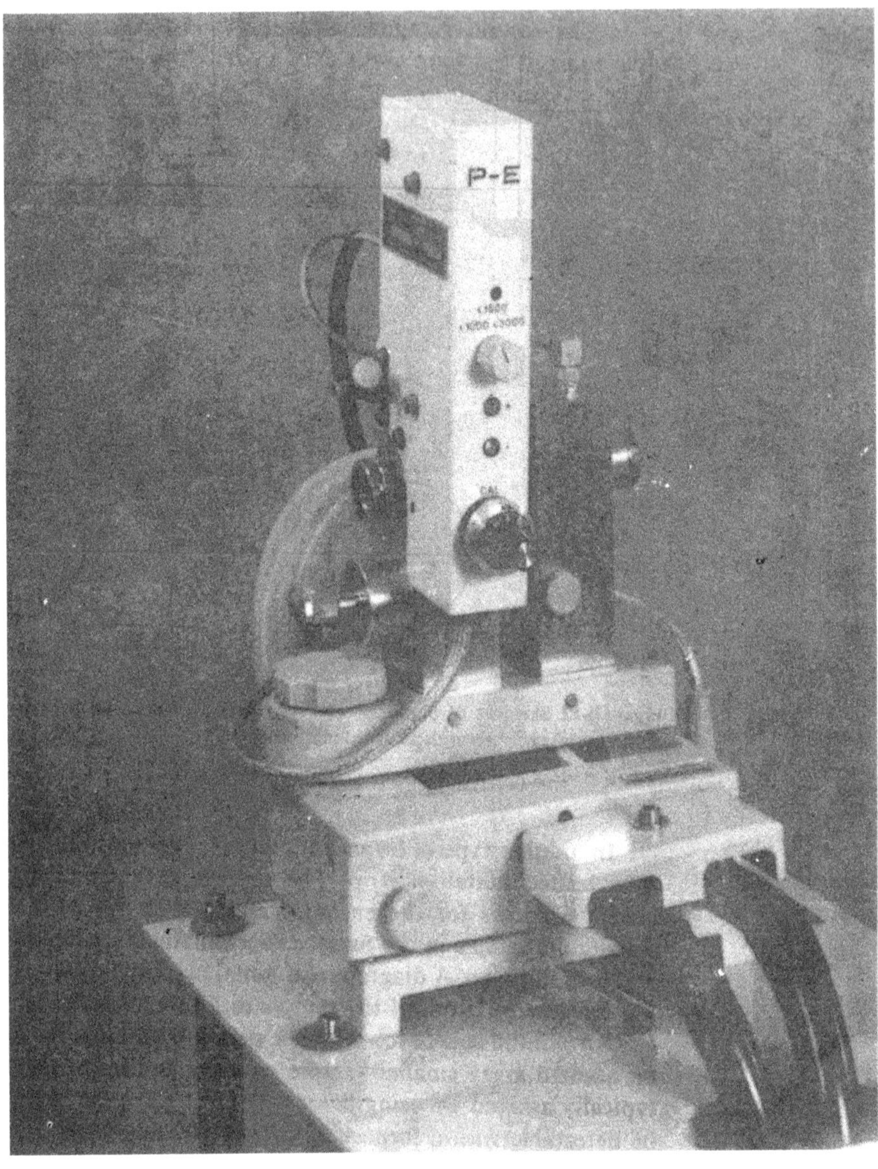

(*b*)

Figure 6-11 A commercial tubular graphite furnace. (*a*) Sketch of the furnace. (*From Beaty.*[10] *Reproduced by permission.*) (*b*) Photograph of an HGA furnace module. (*Photograph courtesy of Perkin-Elmer Corporation.*)

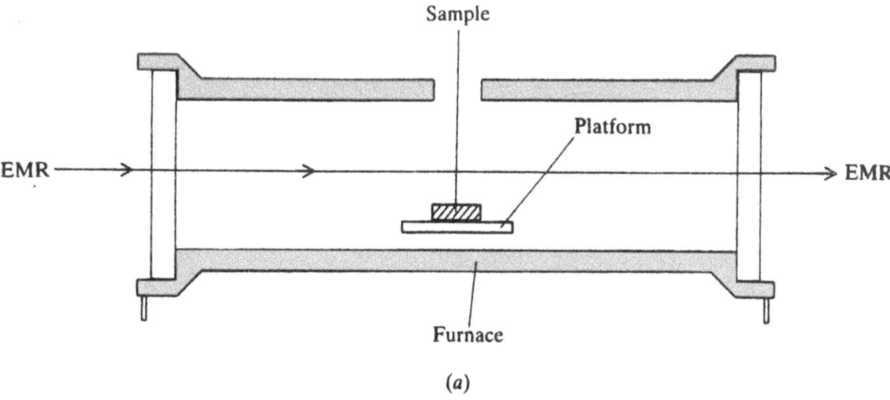

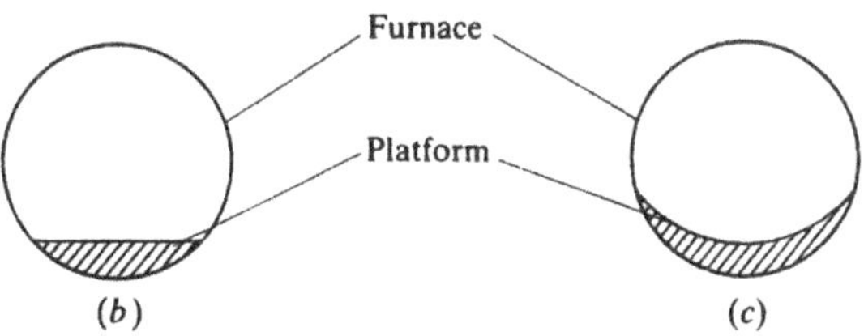

Figure 6-12 Sketches of platform tubular graphite furnaces: (*a*) side view; (*b*) end view with a flat platform; (*c*) end view with a curved platform.

Among the disadvantages of using a furnace are the additional expense of the furnace and accompanying power supply, and the extra care that must be taken to obtain reproducible results. Because the absorptive signals obtained while using furnaces are transient, i.e., not continuous as in a flame, a rapidly responding detector attached to either a computer data acquisition system or a recorder is required. Background correction is often required when furnaces are used. A description of the methods of background correction in furnaces as well as in flames is contained in a later section of the chapter.

In addition to the cells that have been described for AAS, several other types of cells have been used. Sputtering, as in an HC lamp, and plasmas have been used with some success as sources of atoms. None of the other types of cells has found the widespread use in AAS of flames and furnaces.

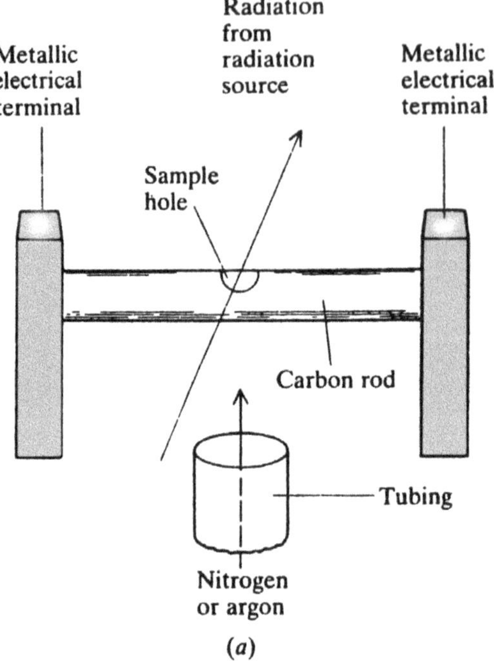

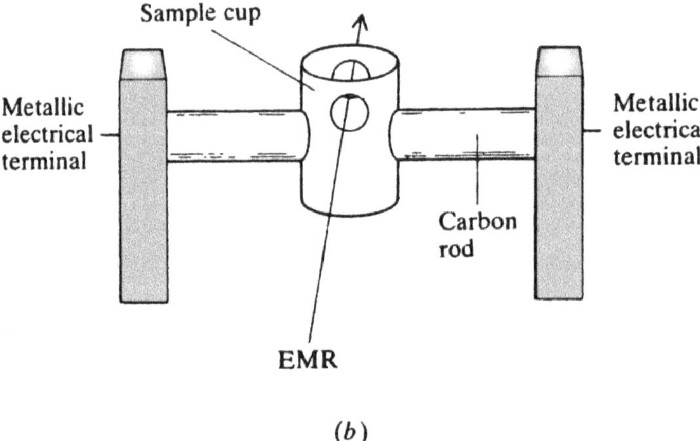

Figure 6-13 Diagrams of (*a*) a carbon-rod atomizer and (*b*) a carbon-cup atomizer.

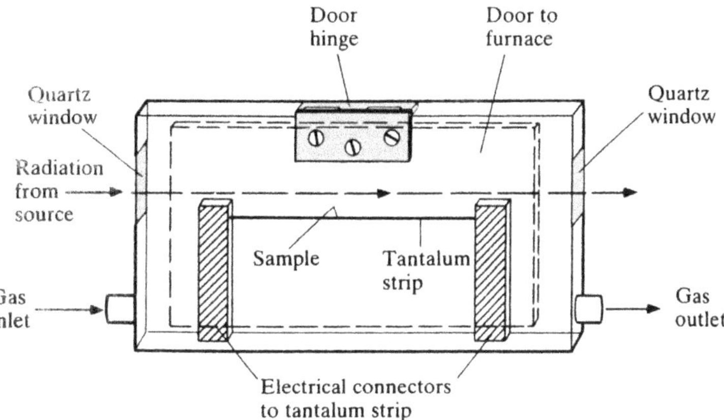

Figure 6-14 Diagram of a tantalum-strip furnace.

WAVELENGTH SELECTORS

Monochromators are used universally as the wavelength-selection devices in atomic absorption spectrophotometers. Typically the bandpass of the monochromators are 0.2, 0.5, or 1 nm. The width of the absorptive line in most cells is about 0.004 nm, which is considerably narrower than the bandpass of the monochromator.

DETECTORS

The most common detector that is used for AAS is the *photomultiplier* (PM) *tube*. A sketch of a photomultiplier tube is shown in Fig. 6-15. The tube consists of a cathode, an anode, and several additional electrodes that are called *dynodes*. All of the electrodes are enclosed in an evacuated glass envelope that is similar to an old radio tube. In the PM tube design shown in Fig. 6-15, the electrodes extend nearly the entire length of the tube. Electric connections to the electrodes are made through pins that protrude from the bottom of the tube.

If the PM tube is to be used in the uv region, a window of quartz or some other uv-transparent material is mounted in the wall of the glass tube to allow radiation to enter. EMR that enters the tube passes through a metal grill and strikes the cathode, or *photocathode* as it is often called. The cathode is coated with some substance that has a low ionization potential, i.e., with a substance that easily emits electrons. Often the cathode is coated with an oxide of cesium, although the oxides of other metals from group 1 or group 2 of the periodic table also can be used. The energy from the collision of a photon with the photocathode is sufficient to result in ejection of an electron. The photocathode serves as a transducer that converts electromagnetic radiation into electricity. The emitted electrons are attracted to dynode 1 by the relatively positive potential that is maintained on the dynode as compared to the potential on the photocathode.

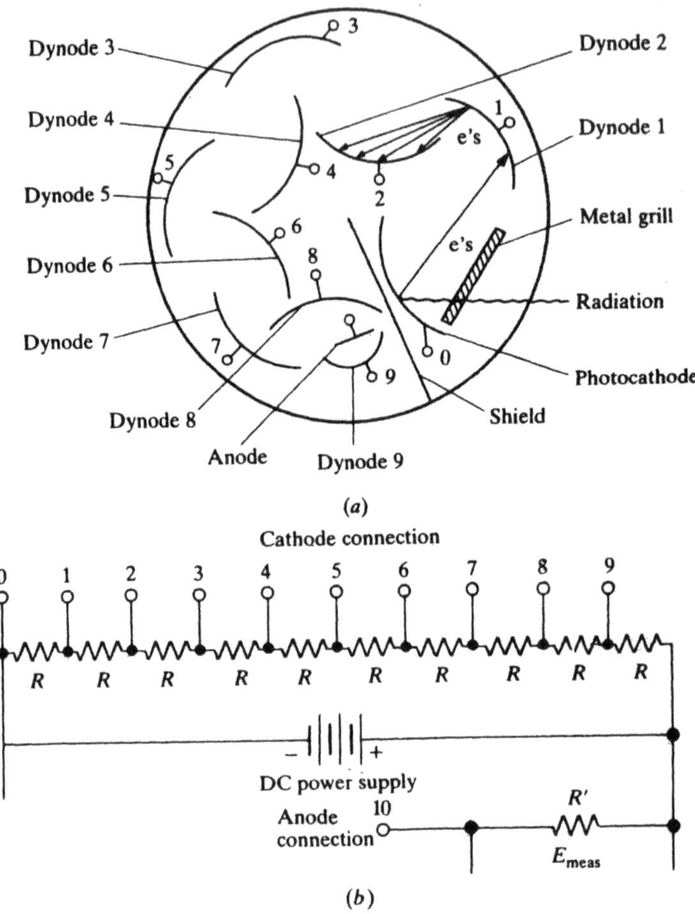

Figure 6-15 (*a*) Top-view sketch of a photomultiplier tube. The numbered circles are the connecting pins that protrude from the bottom of the tube. (*b*) The electric connections to the pins of the photomultiplier tube sketched in (*a*). A typical value for the applied potential is 900 V and a typical value for *R* is 100,000 Ω.

After acceleration by the potential difference between the photocathode and the first dynode, the electrons strike the dynode. The dynodes are coated with a compound that emits several electrons for each impinging electron from the photocathode, as illustrated in Fig. 6-15*a*. Usually the dynodes are coated with either CsSb, BeO, or GaP. Upon striking the first dynode each electron causes emission from the dynode of about four or five electrons. The result is an amplification of the electric current that flows through the PM tube.

The electrons that are emitted from dynode 1 are attracted to the more positive dynode 2. At dynode 2 each impinging electron results in emission of four or five electrons and further amplification of the electric current that flows through the PM tube. The amplification increases as the electrons that are emitted from each dynode are attracted to the next dynode by the progressively more positive potential that is maintained on each successive dynode. Eventually the electrons leave the last dynode (dynode 9 in Fig. 6-15) and are captured by the anode. The potential differences between the photocathode, the dynodes, and the anode are maintained with the voltage divider that is shown in Fig. 6-15*b*. Because the potential between the photocathode and the first dynode is particularly critical, the resistor between the two is sometimes replaced with an appropriate zener diode. The power supply of the potential divider typically supplies about 900 V and the resistors are about 100,000 Ω. Many PM tubes utilize a variable power supply so that the amount of amplification can be adjusted.

Typically the electrons that are captured by the anode flow through a standard resistor (R' in Fig. 6-15*b*). The potential across the standard resistor is measured and related to the current by Ohm's law. The current can be electronically amplified prior to measurement if necessary. Tuned amplifiers are generally used when the EMR from the source has been chopped.

Because the total amplification within a PM tube is constant under fixed conditions, the measured current is proportional to the intensity of the electromagnetic radiation that strikes the photocathode. If a particular PM tube has four dynodes and each dynode emits four electrons for each impinging electron, then four electrons are emitted at the first dynode for each electron that is emitted at the photocathode; 16 electrons (4×4) are emitted at the second dynode; 64 electrons (16×4 or $4 \times 4 \times 4$) are emitted at the third dynode; and 256 electrons (64×4 or $4 \times 4 \times 4 \times 4$) are emitted at the fourth dynode and are captured by the anode. In that particular case the amplification is a factor of 256. In general the amplification of a PM tube is the average number n of electrons emitted at each dynode raised to the power that equals the number d of dynodes in the PM tube:

$$\text{Amplification} = n^d \qquad (6\text{-}2)$$

The large amplification in a PM tube allows the use of narrow slit widths and low hollow-cathode-lamp currents.

Sample problem 6-1 If a particular photomultiplier tube contains eight dynodes and each dynode emits an average of 4.4 electrons for each impinging electron, calculate the amplification factor of the PM tube.

SOLUTION Equation (6-2) is used to calculate the amplification factor of the tube:

$$\text{Amplification} = n^d = 4.4^8 = 1.4 \times 10^5$$

In a few instruments detectors other than photomultiplier tubes are used. In instruments that are designed strictly for absorptive measurements, use of alternative detectors is rare. Among the alternative detectors that have been used are *silicon-photodiode detectors* and *resonant detectors*. There is no particular advantage in using any of those detectors for AAS when compared to PM tubes.

The operation of photodiodes was described in Chapter 3. The silicon photodiode is attached to a power supply. When EMR strikes the *n* region of the diode, an increased current flows through the device. The measured increase in current is proportional to the intensity of the impinging EMR. Photodiodes are not as sensitive as PM tubes. *Arrays* of several photodiodes can be used to increase the sensitivity of photodiode detectors. In some instruments several photodiodes or photodiode arrays are placed at different angles around the radiation-dispersing device (the monochromator) to simultaneously measure the intensity of the radiation at different wavelengths. Detectors that simultaneously monitor more than one wavelength are *multichannel detectors*.

A resonant detector consists of an evacuated envelope (similar to an HC lamp) that contains atomic vapor of the assayed element. Radiation that strikes the detector causes excitation of the atoms and subsequent resonant fluorescence. The intensity of the fluorescent radiation is measured with a PM tube. The major advantages of the resonant detector are its specificity and its lack of a need for a monochromator. Because the detector can absorb only that radiation which is characteristic of the atomic vapor, the device only responds to the wavelengths that are characteristic of the particular element within the detector. Because the detector is selective, a monochromator is not required. The major disadvantage of the device is the need to use a different detector for each element. That disadvantage has made the detector unacceptable to most analysts and manufacturers of instruments.

WAVELENGTH CHOICE

For atomic absorption spectrophotometry, the wavelength at which the assay is done usually corresponds to the wavelength at which absorption is the greatest. When a line source is used, the intensity of emitted radiation at a particular wavelength must be considered. It is necessary to choose a wavelength for the assay at which the source emits sufficient radiation to be easily measurable by the detector. It is also necessary to choose a wavelength at which the analyte strongly absorbs. The sensitivity of the analytical method depends upon the extent to which the analyte absorbs radiation. The wavelength at which absorption is the greatest is the wavelength at which the most probable electron energy-level transition takes place. It is fortunate that, in most cases, the most intense line that is emitted by a line source corresponds to the most probable absorptive electron transition. As a result, the line that is chosen for use in analysis of a particular element by AAS usually corresponds to the most intense emissive line from the hollow-cathode lamp or the electrodeless discharge lamp. Generally that corresponds to a transition to the ground electron state.

At the temperatures of the cells that are used most often for AAS, over 99 per cent of the atoms are in the ground state. Because relatively few atoms have electrons in an excited state, absorption from an excited state is considerably less likely to occur than from the ground state. Consequently, the electron transition that is most probable is from the ground state to a higher energetic state. For the purpose of quantitative analysis, it is only necessary to consider the emissive lines from the line source that result from transitions to the ground state, and which consequently can cause absorption in the analyte from the ground state.

The intensity I of an emissive line is proportional to the number of atoms N_j with electrons in a particular excited electron level at any time, the amount of energy ($E = h\nu$) that is emitted for each photon as the electrons fall from the excited state to the ground state, and the Einstein transition probability A for the electron transition:

$$I = AN_j h\nu \tag{6-3}$$

The Einstein transition probability is equal to the inverse of the average lifetime of an electron in the excited state, i.e., to the average number of transitions per second. It usually is about $10^8\ s^{-1}$.

The number of atoms in an excited energy level can be calculated from a form of the Boltzmann equation:

$$N_j = N_o(g_j / g_o)e^{-E_j/kT} \tag{6-4}$$

in which N_0 is the number of atoms in the ground state at absolute temperature T, E_j is the difference in energy between the ground state and the excited state, k is the Boltzmann constant (1.3806×10^{-23} J/K), and g_j and g_0 are the statistical weight factors for the two levels. The values of g are measures of the relative number of possible electrons in each energetic level and can be calculated using J, where J is the internal atomic quantum number for the atom in the particular energetic level:

$$g = 2J + 1 \tag{6-5}$$

By substituting the value of N_j from Eq. (6-4) into Eq. (6-3), a useful expression for the intensity of an emissive line is acquired:

$$I = Ah\nu N_o(g_j / g_o)e^{-E_j/kT} \tag{6-6}$$

Li 670.8 1,2	Be 234.9 1+,3											B 249.7 3					
Na 589.0 589.6 1,2	Mg 285.2 1+											Al 309.3 1+,3	Si 251.6 1+,3				
K 766.5 1+,2	Ca 422.7 1	Sc 391.2 3	Ti 364.3 3	V 318.4 3	Cr 357.9 1+	Mn 279.5 1,2	Fe 248.3 1	Co 240.7 1	Ni 232.0 1,2	Cu 324.8 1,2	Zn 213.9 2	Ga 287.4 1	Ge 265.2 3	As 193.7 1	Se 196.0 1		
Rb 780.0 1,2	Sr 460.7 1+	Y 407.7 3	Zr 360.1 3	Nb 405.9 3	Mo 313.3 1+		Ru 349.9 1	Rh 343.5 1,2	Pd 244.8 247.6 1,2	Ag 328.1 2	Cd 228.8 2	In 303.9 1,2	Sn 286.3 224.6 1	Sb 217.6 1,2	Te 214.3 1		
Cs 852.1 1	Ba 553.6 1+,3	La 392.8 3	Hf 307.2 3	Ta 271.5 3	W 400.8 3	Re 316.0 3		Ir 264.0 1	Pt 265.9 1,2	Au 242.8 1+,2	Hg 185.0 253.7 0,1,2	Tl 377.6 276.8 1,2	Pb 217.0 283.3 1,2	B1 223.1 1,2			

	Pr 495.1 3	Nd 463.4 3		Sm 429.7 3	Eu 459.4 3	Gd 368.4 3	Tb 432.6 3	Dy 421.2 3	Ho 410.3 3	Er 400.8 3	Tm 410.6 3	Yb 398.8 3	Lu 331.2 3
		U 351.4 3											

Figure 6-16 A periodic table showing elements that can be assayed by using atomic absorption spectroscopy. The number under the symbol for each element is the wavelength (nm) used for the assay. The number at the bottom of each square is a code for the flame type used in the assay: 0, no flame used; 1, air-acetylene; 1 +, fuel-rich air-acetylene; 2, air-propane or air-natural gas; 3, nitrous oxide-acetylene. Furnaces also can be used for most of the assays.

From Eqs. (6-4) and (6-6), it is apparent that as the temperature T increases, the number of atoms in the excited state increases and the intensity of emitted radiation increases. At the temperature of most AAS cell (2000 to 3000 K), Eq. (6-4) can be used to show that the number of atoms in the most probable excited state of the commonly assayed elements varies with the element between about 10^{-12} and 1 percent of the number of atoms in the ground state. The line that is generally used for the AAS determination of a particular element is that in which the intensity of the line as predicted by Eq. (6-6) is the greatest. For a particular element in a particular cell, the most intense line is generally that for which g_jA is the greatest. A useful listing of values of g_jA is provided in Reference 15. The wavelengths of the lines that are used most often for AAS and some information about the flames that can be used for the analyses are listed in Fig. 6-16.

USE OF AAS FOR QUALITATIVE ANALYSIS

Atomic absorption is primarily used for quantitative rather than qualitative analysis. A continuous source of radiation and a narrow-bandpass monochromator are used for qualitative AAS. The wavelength is scanned through the desired range and the spectrum recorded. The wavelengths of the absorptive lines are compared with known values for the elements. Atomic emissive spectrometry AES is more useful for qualitative analysis than atomic absorption because AES does not require use of a continuous radiative source. A description of atomic emissive spectrometry is included in the next chapter.

USE OF AAS FOR QUANTITATIVE ANALYSIS

During most measurements of atomic absorption, Beer's law

$$A = kbC \tag{6-7}$$

is obeyed. The absorbance is directly proportional to the concentration of atoms in the cell. When the cell operating parameters, the radiative source, and the wavelength used for the assay are constant, the concentration of atoms in the cell is proportional to the concentration of the element in the sample solution. Consequently, the measured absorbance is proportional to the elemental composition of the sample. Experimental variables such as the aspiration rate into a flame, the region of a flame through which EMR from the source passes, the flowrates of the gases entering the flame, and the temperature of the cell during the measurement have an effect upon the proportionality constant k.

Nevertheless, during the assay of a particular element under a fixed set of experimental conditions, the absorbance is usually directly proportional to the concentration of the element in the sample. Because the value of the proportionality constant for a particular element at a particular wavelength cannot be easily duplicated between laboratories because of variations in experimental conditions, Eq. (6-7) is never used directly to calculate concentration from a single measurement of absorbance. Instead, either the working-curve method or the standard-addition technique is used.

Most spectrophotometers contain dedicated microcomputers. The concentrations of two or more standards can be entered into the computer and the absorbances of the corresponding solutions are measured. The microcomputer uses the information to calculate the concentration of a sample solution from its absorbance. The computer uses either the working-curve method or standard-addition technique. It eliminates the need for the analyst to prepare the graphs. The elements that can be assayed with AAS and the wavelengths that normally are used for the assays are listed in Fig. 6-16.

INTERFERENCES

Interferences that are encountered while performing analyses with AAS can be classified as *chemical*, *ionization*, *spectral*, or *background interferences*. Background interferences and methods of correcting for them are described in the following sections. Chemical interferences are those that arise from chemical reactions that occur within the cell. Because the chemical reaction removes atoms of the assayed element, the atomic population within the cell and consequently the absorbance is decreased. Interfering chemical reactions occur most often in flames, in which oxygen is available to form *refractory oxides* with the assayed element.

Refractory oxides are unusually stable compounds that contain oxygen and that are not often dissociated to atoms at the temperature of the cell. Aluminum and iron form refractory oxides with the chemical formulas Al_2O_3 and Fe_2O_3 in low-temperature flames. Both strontium and magnesium can combine with aluminum and oxygen to form ternary refractory oxides. Boron, titanium, tungsten, uranium, vanadium, and zirconium also form refractory oxides. Calcium in combination with oxygen and either phosphorus or sulfur apparently forms refractory oxides with the formulas $CaO \cdot P_2O_5$ and $CaO \cdot SO_3$. Calcium phosphate is a well-known refractory compound that does not completely atomize at the temperatures of air-acetylene flames.

When a flame is used as the cell, refractory-oxide formation can be eliminated by changing to a higher temperature flame or by decreasing the concentration of available oxygen in the flame. The concentration of oxygen in the flame can be decreased by making the flame fuel-rich, i.e., by changing the ratio of gases that flow into the flame so that more of the fuel (the reductant) flows into the flame than can be consumed by the oxidant. The excess fuel reacts with most of the available oxidant and causes a decreased oxygen concentration. Changing from an air-acetylene flame to a nitrous oxide-acetylene flame sometimes eliminates chemical interference by decreasing the oxygen concentration and by increasing the temperature.

In some cases chemical interferences can be eliminated by addition of a *releasing agent* (spectroscopic buffer). A releasing agent is a chemical substance that prevents formation of a chemical interference by chemically reacting with one or more of the components of the interference to form one or more non-refractory compounds. Because the analyte cannot react to form a refractory compound, it can readily be atomized in the cell. As an example, during the assay of calcium in the presence of phosphate, a lanthanum salt is normally added. Lanthanum combines with phosphate and prevents formation of the refractory compound.

It is sometimes possible to compensate for chemical interference by using the graphical standard-addition technique. When using the standard-addition technique, the analyst must ensure that the absorbance varies linearly with the concentration of the standard additions and that the interference affects the added standards to the same extent that it affects the analyte in the original sample.

Ionization interference occurs when a significant proportion of atoms in the cell become ionized. Ionization results in a decreased concentration of atoms in the cell. Because the spectral lines of ions of an element generally do not occur at the same wavelength as the atomic lines for the element, ionization leads to low readings of absorbance.

Usually ionization is caused by a cell temperature that is too high. The energy that is required to cause ionization is supplied by the heat in the cell. Group 1 and 2 elements in the periodic table have relatively low ionization energies and consequently are more likely to be ionized at the temperatures of most AAS cells than other elements.

Ionization is dependent upon concentration. Generally as the concentration of an element increases, the relative amount of ionization of the element decreases. Apparently that is the result of increased competition between atoms of the analyte for the available energy. Working curves for analytes that are partially ionized generally curve upward as shown in Fig. 6-17 because of the decreased relative ionization with increasing concentration.

Ionization can be reduced or eliminated by decreasing the temperature in the cell. If a flame is being used, the temperature can be decreased by changing to a relatively low-temperature flame. The propane-air flame is often used for this

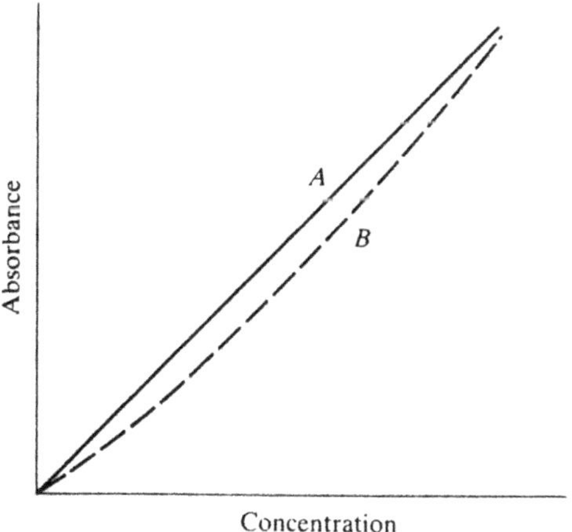

Figure 6-17 Sketch of typical absorbance working curves for an element with no ionization (curve *A*) and for the same element with ionization (curve *B*).

purpose. Alternatively, ionization can be reduced by adding a relatively large amount of an element that has a low ionization energy to the standard and sample solutions. Typically 500 to 5000 μg/mL of lithium, sodium, or potassium is added to each solution as a salt. The presence of a large concentration of an easily ionized substance results in increased competition for available energy from the cell and a large number of free electrons in the cell from the ionization process. Both processes suppress ionization of the analyte.

Spectral interferences occur when two elements or an element and a polyatomic species in the cell absorb or emit radiation at wavelengths that are nearly identical. If the interference absorbs radiation at the same wavelength as the analyte, a positive error in the assay results because the detector responds to the total absorbance in the cell. If the element emits radiation at the wavelength of the absorbance measurement, and if that radiation is measured by the detector, a negative error in the assay results. The detector "sees" more radiation than it should and consequently reports less absorbance. Spectral interferences that result from emission of radiation from elements in the cell are also classified as background interferences and are described further in the next section.

Fortunately the narrow absorptive lines of different elements rarely overlap. That is a major advantage of AAS. Because the detector is tuned to the frequency of oscillation of the EMR source, it does not respond to steady-state emissions from the cell; therefore nearly all interference owing to emission from the cell is eliminated. Among the atomic spectral lines that overlap and that can result in spectral interferences are the 285.2-nm lines of Tb and Mg, the 290.0-nm lines of Cr and Os, and the 422.7-nm lines of Ge and Ca. The easiest way to eliminate a spectral interference is to use a different spectral line of the analyte for the assay. A chemical separation prior to the assay can also be used to eliminate an interference.

Some elements form refractory oxides in the cell which can absorb or emit radiation. Because refractory oxides are polyatomic species, their absorptive and emissive bands are broad. Calcium is an example of an element that can form absorbing and emitting polyatomic species within a cell. Spectral interferences that are caused by polyatomic species can usually be eliminated either by choosing a spectral line that is beyond the wavelength range of the polyatomic substance or by adding a chemical reagent that prevents formation of the interference.

Another factor which must be considered while performing assays with flame AAS is potential differences between the viscosity of the standard and sample solutions. The viscosity of the solution affects its rate of aspiration into the flame and the atomic concentration within the flame. More viscous solutions are not aspirated into the flame as rapidly as less viscous solutions. If the viscosity of the standards is less than that of the sample, a negative error is introduced into the assay. Similarly, if the viscosity of the standards is more than that of the sample, a positive error occurs. The viscosity of a solution is determined by the components of the solution. The solution components, with the exception of the analyte, are the matrix of the analyte. Errors owing to changes in viscosity can be eliminated by *matrix matching,* i.e., by preparing the standards in exactly the same matrix as that of the sample. Whenever it is possible, matrix matching is highly recommended.

Background Correction

Background interferences (also termed *nonspecific interferences*) are usually caused by absorption by polyatomic species or scattering of radiation within the cell. The standard-addition technique cannot be used to compensate for background interferences. Generally background absorbances are considerably greater in furnaces than in flames. Background interference is usually greater at shorter wavelengths. The increased background interference at shorter wavelengths is due, at least partially, to an increase in scattering as the wavelength is shortened; refer to Chapter 14 for more details. In flames, background interference is normally insignificant at wavelengths above about 430 nm.

It is possible to correct for background interferences by measuring either the absorbance of the background or something that is proportional to the background absorbance. That quantity is subtracted from the total absorbance to yield the absorbance owing to the analyte. Several ways are available that permit estimation of the background absorbance. In a few cases it is possible to find a spectral line of an absent nonanalyte element that is near but does not overlap with the line used for the assay. The absorbance is measured with the appropriate line sources at the two wavelengths and the absorbance at the nonanalyte line is subtracted from that at the analyte line. The assumption is made that the absorbance at the nonanalyte line is completely due to the background and is identical to the background absorbance at the analyte line.

An alternative method to correct for background interferences is to use a *double-beam* spectrophotometer. Double-beam instruments are usually constructed with a variation of one of the three designs that are shown in Fig. 6-18. The design that is shown in Fig. 6-18*a* uses two radiative sources to help correct

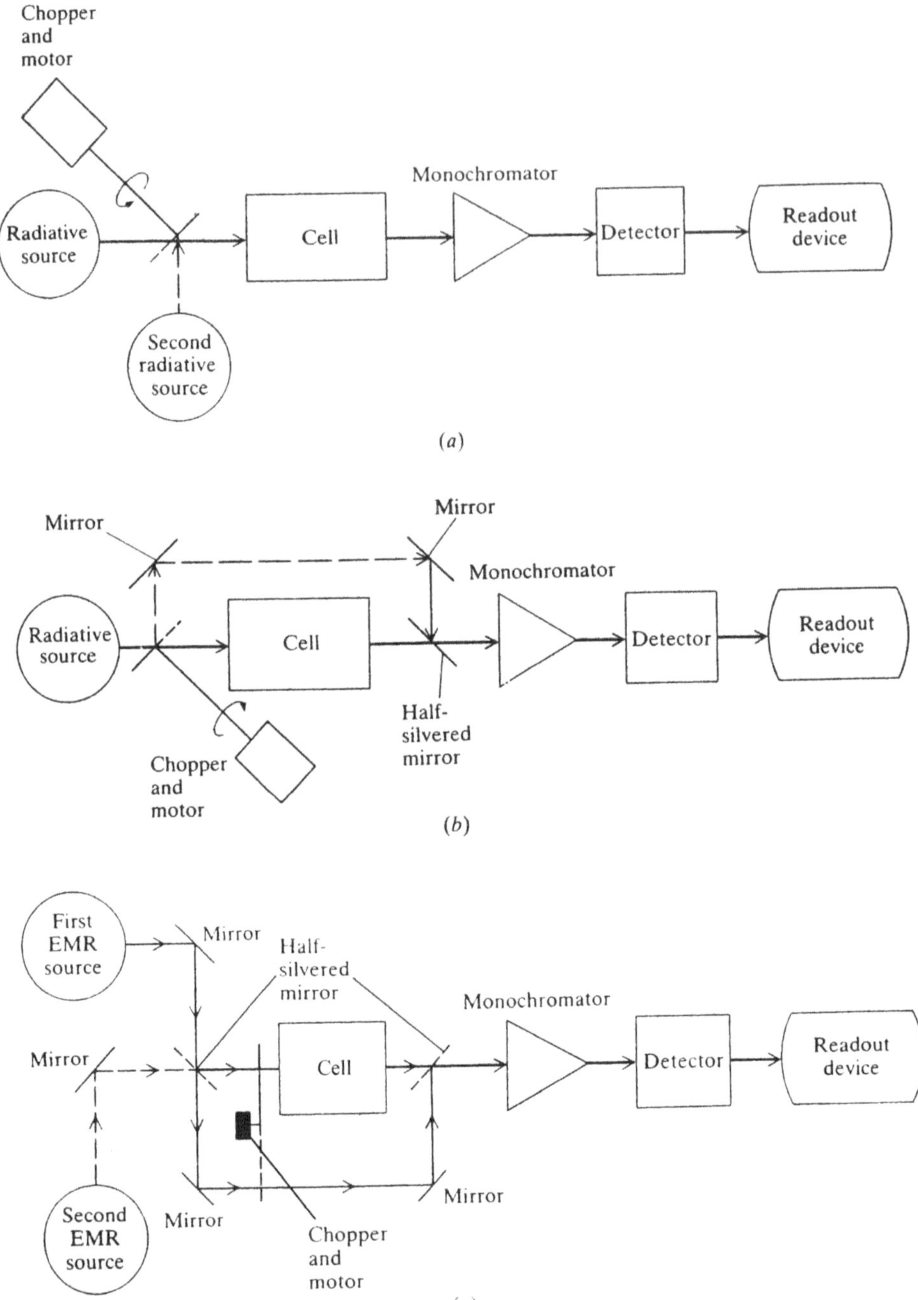

Figure 6-18 Block diagrams of double-beam atomic absorption spectrophotometers. Refer to the text for descriptions.

for background interference. The primary radiative source is a line source that is identical to that which is used in single-beam instruments. The second radiative source is a continuous source. Usually the second source is a deuterium-arc lamp for measurements that are made in the ultraviolet spectral region or a tungsten-filament lamp for measurements that are made in the visible region. Because background correction in the visible region is rarely required, many instruments do not contain a tungsten-filament lamp.

Radiation from each of the two sources alternately passes through the cell. The detector measures a square wave signal with an amplitude that is proportional to the difference in intensity of the nonabsorbed radiation from each of the two sources. The measured absorbance with the line source is due to both the analyte and background interference.

The measured band of radiation from the continuous source is much broader than that from the line source. The width of the band is determined by the bandwidth of the monochromator rather than by the width of the emitted line from the source. Because the bandwidth of the monochromator is much greater than the absorptive bandwidth of the analyte in the cell, the absorbance when the continuous beam of radiation passes through the cell is almost entirely the result of background absorption. Consequently the detector reports a background corrected readout.

The design shown in Fig. 6-18*b* uses a single source of EMR as in the single-beam instrument. The radiation is split into two portions in time with a chopper or similar device. During one portion of each cycle the radiation passes through the cell to the monochromator and detector exactly as in a single-beam instrument. During the second portion of each cycle the radiation is diverted around the cell before passing through the monochromator and striking the detector. The detector alternately measures the intensity of radiation that has passed through the cell and that has bypassed the cell. The ratio of the two intensitics (the transmittance) or the logarithm of the ratio (the absorbance) is indicated by the readout device. The apparatus automatically corrects for fluctuations in the intensity of radiation coming from the radiative source and for changes in the sensitivity of the detector. It does not correct for background interferences that occur in the cell.

The instrumental design shown in Fig. 6-18*c* combines the advantages of the designs that are shown in Fig. 6-18*a* and *b*. As in the design that is shown in Fig. 6-18*b*, radiation from the primary line source alternately passes through the cell and bypasses the cell. Consequently, the spectrophotometer automatically corrects for fluctuations in the intensity of radiation that comes from the radiative source and for changes in the sensitivity of the detector. Radiation from the second continuous source is also alternately directed through and around the cell thereby allowing the detector to automatically subtract background absorbance in the cell. The outputs from the line source and the continuous source alternately pass through the cell. Radiation from the line source alternately passes through and around the cell during one portion of each cycle and radiation from the continuous source alternately passes through and around the cell during the second portion of each cycle. Discrimination between the signals is possible because of the frequency differences. The frequency of oscillation of the radiation that passes through and around the cell from either the line or continuous source is greater than the frequency with which the line source and the continuous source are alternatively used. With either of the instrumental designs shown in Fig. 6-18*a* or *c*, the continuous source can normally be shut off when background correction is not required.

Methods of background correction that rely upon the use of a continuous source do not correct adequately for background absorbance when the broadband background absorbance contains narrow peaks owing to vibrational and rotational transitions or to spectral interferences. Use of a continuous source corrects for the average background absorption across the bandwidth of the monochromator but does not correct for narrow atomic or molecular peaks within the bandwidth that happen to overlap with the absorptive line of the analyte.

Continuous background correctors have other disadvantages. It is difficult to adjust the optics such that the beams from the continuous source and the line source follow identical paths through the cell. If the paths are not identical, the background correction might not be accurate. The half-silvered mirrors or other types of beam splitters that are used in the spectrophotometers reduce the intensity of radiation that can be used for the measurement of absorption. Finally, it is sometimes difficult to match the intensities of the continuous and line sources. Relatively recently two other methods of background correction have become available which largely eliminate the disadvantages of the use of continuous sources. Those two methods are described in the next two sections. They have found widespread acceptance, particularly for use with furnaces where significant background interference is common. Double-beam atomic absorption spectrophotometers are not in common use although they can be found in some specialized laboratories.

Zeeman Background Correctors

The *Zeeman background correctors* take advantage of the polarization of a spectral line in a magnetic field (the Zeeman effect). In the presence of a magnetic field a spectral line that is either absorbed or emitted by atoms is split into a minimum of three components, as shown in Fig. 6-19. The central component is the *pi component* which occurs at the same wavelength as the original spectral line. The remaining two *sigma components* surround the pi component. The total integrated area of the sigma components equals that of the pi component.

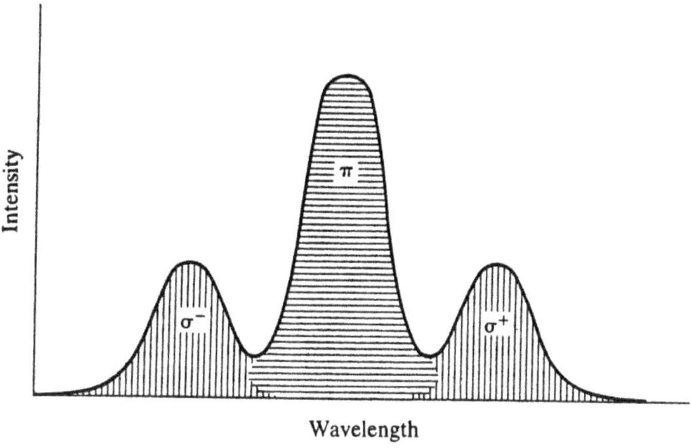

Figure 6-19 A sketch that illustrates spectral line splitting in the presence of a magnetic field (Zeeman effect). The vertical and horizontal lines represent the plane-polarized nature of the radiation.

The change in energy ΔE of EMR in the presence of a magnetic field is dependent upon the strength of the magnetic field and the energetic transition that accounts for the spectral line:

$$\Delta E = 9.274 \times 10^{-24} MB \left[1 + \frac{J(J+1) - L(L+1) + S(S+1)}{2J(J+1)} \right] \tag{6-8}$$

where M is the magnetic quantum number; ΔE is the energy shift (joules); B is the magnetic flux density (tesla); and S, L, and J respectively are the spin, orbital, and total angular momentum quantum numbers. The pi component corresponds to the transition $\Delta M = 0$ and the sigma components correspond to the transitions $\Delta M = +1$ and $\Delta M = -1$ in the magnetic field.

In addition to the wavelength separation between the sigma and pi components, the two components are also plane-polarized perpendicularly to each other. The pi component is plane-polarized parallel to the magnetic field and the sigma components are plane-polarized perpendicular to the magnetic field. Zeeman background correctors take advantage of the difference in polarity of the components. Zeeman instruments can be constructed with a magnet placed around either the radiative source or the cell. Block diagrams of the two types of instruments are shown in Fig. 6-20.

With the instrument that is shown in Fig. 6-20*a*, the magnetic field is superimposed on the radiative source causing separation of the emitted radiation into pi and sigma bands. While passing through the cell, a portion of the pi band is absorbed by the analyte and by background interferences. Because the sigma bands are displaced from the absorptive band of the analyte, they are absorbed only by background interferences. After exiting from the cell the radiation passes through a rotating polarizer that alternately allows the pi and sigma bands to pass, enter the monochromator, and strike the detector. The detector alternately responds to the pi band and the sigma bands. The difference between the two signals is background-corrected. The device automatically substracts the background signal from the signal owing to the background and the analyte to yield a signal from the analyte alone.

With the instrument that is shown in Fig. 6-20*b*, the radiation that is emitted from the line source passes through a rotating polarizer before entering the cell. The magnetic field that is superimposed on the cell causes the absorptive lines to be split into sigma and pi bands. The pi band occurs at the same wavelength as that of the emitted radiation from the source but it is plane-polarized parallel to the magnetic field, i.e., it can only absorb radiation that is polarized in the same direction. When radiation that is polarized identically to the pi band of the

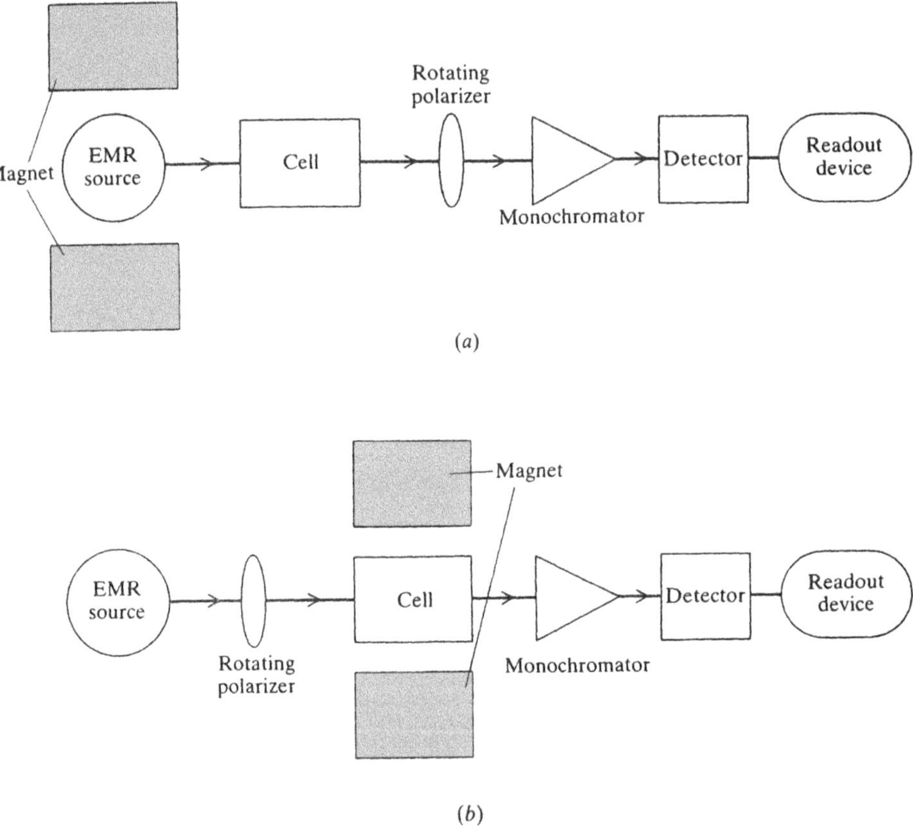

Figure 6-20 Block diagrams of the major components of Zeeman atomic absorption spectrophotometers. The magnetic field can be superimposed on either (*a*) the radiative source or (*b*) the cell.

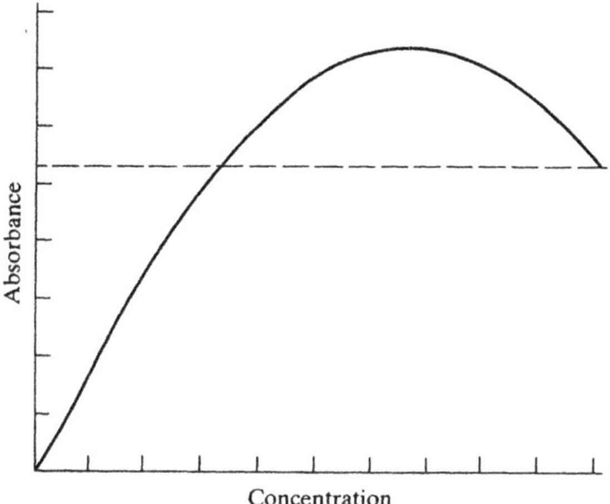

Figure 6-21 A working curve in which two concentrations can cause the same absorbance. Except at the apex, absorbances above the dashed line can be caused by either of two analyte concentrations.

analyte enters the cell, both the analyte and the background absorb. When radiation that is perpendicularly polarized enters the cell, only the background absorbs, even though the radiation is of the same wavelength as that which the analyte absorbs. The difference between the two alternating signals is due to the analyte alone.

Zeeman instruments overcome some of the disadvantages of continuous background correctors. Because only a single lamp is required, it is not necessary to match radiative sources. Problems with optical alignment of two beams of radiation are eliminated. Because correction occurs at or very near the wavelength at which the absorptive measurement is made, problems with absorption by narrow spectral lines that are superimposed on a broad band are eliminated. Generally Zeeman instruments can correct for spectral interferences if the wavelength of the interfering line is separated from that of the analyte line by at least 0.02 nm.

Zeeman instruments do have several disadvantages. With the apparatus in which the magnetic field is superimposed on the cell, the sigma bands become broader as the concentration in the cell increases. At high concentrations the sigma bands can overlap the pi band. Because the sigma bands are perpendicularly polarized to the pi band, an increase (that is caused by the analyte) in the background signal results. The resulting working curves can curve downward, resulting in two possible concentrations for a particular absorbance measurement. A working curve that has two possible concentrations that correspond to a single measurement of absorbance is shown in Fig. 6-21.

Other disadvantages include loss of radiation in the polarizer, a more bulky instrument because of the magnet, and the increased expense of the magnet. The spectral lines of some elements undergo more complex splitting than that described. The sensitivity decreases for those elements owing to background measurements that include a component that is due to the analyte. Nevertheless, Zeeman instruments have been shown[16] to be useful for 44 elements. A general description of Zeeman instruments as well as of other methods of background correction is included in Reference 17.

Smith-Hieftje Background Correctors

In 1982 the *Smith-Hieftje background-correction system* was introduced. The system takes advantage of the self-reversal of hollow-cathode lamps when operated with a high current. First the absorption owing to the analyte and the background is measured with a low lamp current as in a classical single-beam instrument. Subsequently the absorbance is measured during a short pulse with a significantly increased lamp current. At high lamp current the emissive peak from the lamp is pressure-broadened because of increased sputtering from the cathode. When the lamp is operated at high current, self-reversal occurs at the wavelength at which the analyte absorbs. Emission from the lamp at low current and at high current is illustrated in Fig. 6-22.

Absorption in the cell when the lamp is operated at high current is primarily due to the background because of the broadened band of emitted radiation and the decrease in intensity at the wavelength at which the analyte absorbs. The difference in the absorbance between that when the lamp is operated at low

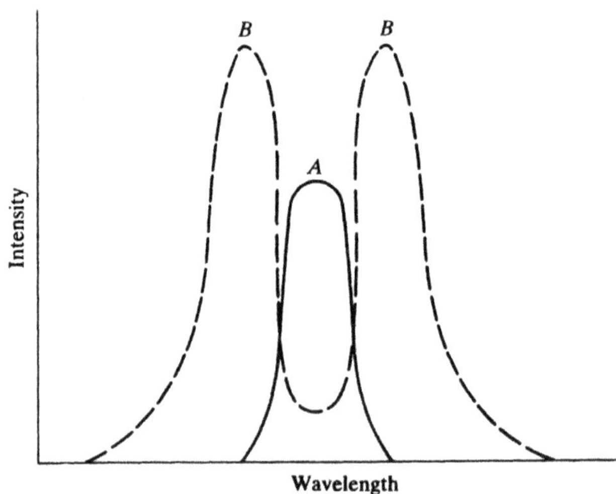

Figure 6-22 Sketch illustrating the intensity of radiation that is emitted from a hollow-cathode lamp that is operated at low current (curve *A*) and at high current (curves *B*). Self-reversal causes the valley in the broadened band at high current.

current and that when the lamp is operated at high current is the background-corrected absorbance. A commercial Smith-Hieftje instrument was introduced in 1982 by Instrumentation Laboratory, Inc.

Smith-Hieftje instruments have many of the advantages of Zeeman instruments. Only a single source is required, and good background correction is achieved when narrow-background absorptive lines are superimposed on a broad continuum. In addition, the instruments are less expensive than Zeeman instruments, do not result in double-valued working curves of the type that was shown in Fig 6-21, and do not lose radiation in a polarizer. Both Zeeman and Smith-Hieftje instruments are less sensitive than instruments that use continuous background correction.

Matrix Modification

In addition to instrumental methods to eliminate background interferences, it is sometimes possible to chemically eliminate interferences by *matrix modification*.[18] The technique is used nearly exclusively with furnaces. A substance is added that either makes the analyte less volatile or makes the interference more volatile. When the analyte is made less volatile, the sample can be ashed at a higher temperature and more background interferences are eliminated prior to measurement of the absorbance. Matrix modification of manganese solutions by the addition of magnesium nitrate decreases background interferences by that process. The addition of phosphate (H_3PO_4 or $NH_4H_2PO_4$) to many elements similarly decreases background interference. Some organic modifiers have been used.

When a matrix modifier is added that increases the volatility of interferences, the interferences are more likely to be removed during the ashing stage. Ammonium nitrate has been used as a matrix modifier of that type. It combines with some interferences to form relatively volatile products.

IMPORTANT TERMS

Ashing	Hollow-cathode lamp
Background interference	Hydride AAS
Carbon-rod atomizer	Ionization interference
Chemical interference	Lean
Continuous source	Line source
Demountable HC lamp	Matrix
Doppler broadening	Matrix effects
Dynode	Matrix matching
EDL	Matrix modification
Filler gas	Multichannel detector
Fuel rich	Multiple-element HC lamp
High-intensity HC lamp	Photocathode

Photomultiplier tube
Platform furnace
Premix burner
Pressure broadening
Refractory oxide
Releasing agent
Resonant broadening
Self-reversal
Smith-Hieftje background corrector
Spectral interference
Spectral line
Sputtering
Temperature-gradient lamp
Total-consumption burner
Xenon-arc lamp
Zeeman background corrector
Zeeman effect

PROBLEMS

Broadening

6-1 Calculate the observed wavelength of the radiation emitted from a sodium atom in a vacuum if the atom is moving toward the detector at a velocity of 8.00×10^5 m/s and if the wavelength of radiation emitted by sodium with no motion is 588.995 nm.

6-2 Calculate the observed wavelength of radiation emitted by the sodium atom in Prob. 6-1 if the atom is moving away from the detector.

6-3 Assuming a normal (gaussian) distribution of velocities of sodium atoms in a line source relative to the detector, calculate the width of the Doppler-broadened emissive line that corresponds to velocities that approach or recede from the detector at a rate that corresponds to ±1 standard deviation on the distribution-velocity graph. Assume the velocity at the ±1 standard deviation points is $\pm 7.50 \times 10^5$ m/s and that the emitted wavelength with no motion is 766.491 nm.

Photomultiplier Tubes

6-4 Calculate the amplification in a photomultiplier tube that contains 9 dynodes and in which an average of 4.3 electrons is emitted at each dynode for each impinging electron.

6-5 Calculate the amplification in a PM tube which contains 6 dynodes and in which an average of 4.7 electrons is emitted at each dynode for each impinging electron.

6-6 The amplification factor for a particular nine-dynode PM tube is 5.8×10^5. Calculate the average number of electrons that are emitted at each dynode.

6-7 An eight-dynode PM tube has an amplification factor of 4.9×10^5. Calculate the average number of electrons that are emitted at each dynode.

Wavelength Selection

6-8 Calculate the ratio in a vacuum of the intensity of the 589.0-nm sodium line at 3000 K to the same line at 2000 K if all other factors are held constant. .

6-9 Calculate the percent increase in a vacuum of potassium atoms in the excited state that leads to the 766.5-nm line as the temperature is raised from 1700 to 4500°C.

Quantitative Analysis

6-10 A sample of urine was assayed for copper with AAS at 324.8 nm. The results from the analysis by the standard-addition technique are tabulated below. What was the copper concentration in the urine?

Added copper concentration, μg/mL	Absorbance
0 (sample)	0.280
2.00	0.440
4.00	0.600
6.00	0.757
8.00	0.912

6-11 An aqueous stock solution that contained 0.100 mg/mL of calcium was prepared. Portions of the solution were added to 50-mL volumetric flasks and diluted to the mark with distilled water. A 5.00-mL portion of a natural-water sample was added to a 50-mL volumetric flask and diluted to the mark with distilled water. The absorbances of the solutions were measured. Calculate the concentration of calcium in the natural water.

Volume of stock solution	Absorbance
1.00	0.224
2.00	0.447
3.00	0.675
4.00	0.900
5.00	1.124
Diluted natural water	0.475

Background Correction

6-12 Assume that the absorbance, as measured with a line source, owing to the analyte is twice the absorbance owing to the background. Further, assume that the width of the absorptive line in the cell is 0.004 nm and the bandpass of the monochromator is 0.5 nm. Estimate the percentage of the absorbance measured with a continuous source that is due to background absorbance.

REFERENCES

1. Kirchoff, G.: *Pogg. Ann.*, **109**: 275(1860).
2. Walsh, A.: *Spectrochim. Acta,* **7**: 108(1955).
3. Halliday, D., and R. Resnick: *Physics for Students of Science and Engineering,* Wiley, New York, 1962, p. 915.
4. Sullivan, J. V., and A. Walsh: *Spectrochim. Acta,* **21**: 721(1965).
5. Van Gelder, Z.: *Appl. Spectrosc.*, **22**: 581(1968).
6. Holak, W.: *Anal. Chem.*, **41**: 1712(1969).

7. Godden, R. G., and D. R. Thomerson: *Analyst,* **105**: 1137(1980).
8. L'vov, B. V.: *Spectrochim. Acta,* **17**: 761(1961).
9. Chakrabarti, C. L., C. C. Wan, H. A. Hamed, and P. C. Bertels: *Anal. Chem.,* **53**: 444 (1981).
10. Beaty, R. D.: *Concepts, Instrumentation and Techniques in Atomic Absorption Spectrophometry,* Perkin-Elmer Corp., 1978,
11. L'vov, B. V.: *Spectrochim. Acta,* Part B, **33B**: 153(1978).
12. Koirtyohann, S. R., and M. L. Kaiser: *Anal. Chem.*, **54**: 1515A(1982).
13. Slavin, W.: *Anal. Chem.*, **54**: 685A(1982).
14. Sturgeon, R. E., and C. L. Chakrabarti: *Progr. Anal. At. Spectrosc.*, **1**: 5(1978).
15. Corliss, C. H., and W. R. Bozman: "Experimental Transition Probabilities for Spectral Lines of Seventy Elements," NBS Monograph 53, National Bureau of Standards, Washington, D.C., 1962.
16. Fernandez, F. J., S. A. Myers, and W. Slavin: *Anal. Chem.*, **52**: 741(1980).
17. Sotera, J. J., and H. L. Kahn: *Amer. Lab.*, **14**(11): 100(1982).
18. Ediger, R. D., G. E. Peterson, and J. D. Kerber: *At. Absorpt. Newsl.*, **13**: 61(1974).

BIBLIOGRAPHY

Norris, T., and J. V. Sullivan: *Amer. Lab.*, **14**(12): 67(1982).

Slavin, W.: *Anal. Chem.*, **54**: 685A(1982).

Van Loon, J. C.: *Anal. Chem.*, **53**: 332A(1981).

CHAPTER

SEVEN

FLAME EMISSION AND ATOMIC EMISSION

The analytical methods for monoatomic species that are described in Chapters 7 and 8 do not involve measurement of absorption. Emission of ultraviolet-visible radiation from atoms and monoatomic ions occurs when the atom or the ion in an excited electron state loses energy by changing to a lower energy electron state. The emitted radiation can be monitored and used for either qualitative or quantitative analysis. If the initial excitation of the atom or ion is caused by a process other than the absorption of ultraviolet-visible radiation and if emitted radiation is monitored, the analytical technique is *atomic emissive spectrometry* (AES).

AES is divided into categories according to the source of the energy that causes the initial excitation. In AES the cell is required to atomize or ionize the sample, to excite the atoms or ions, and to contain the sample for a sufficient time to obtain the analytical measurement. *Flame emissive spectrometry* (FES) uses a flame as the cell. The heat from the flame provides the energy that is required for the excitation. Other forms of AES use a plasma, electricity, or some nonflame heat source to excite the atoms.

FLAME EMISSIVE SPECTROMETRY

Flame emissive spectrometry uses a flame in a dual role to atomize the sample and to excite the atomized analyte. The radiation that is emitted when the excited atoms relax is measured and used for the assay. A block diagram of the apparatus is shown in Fig. 7-1. Often an atomic absorption spectrophotometer (Chapter 6) is used for FES. In that case the radiative source is turned off and a mechanical chopper is placed between the flame and the monochromator. The chopper is required with most instruments that were designed for use with AAS because the detectors in those instruments only respond to an ac signal.

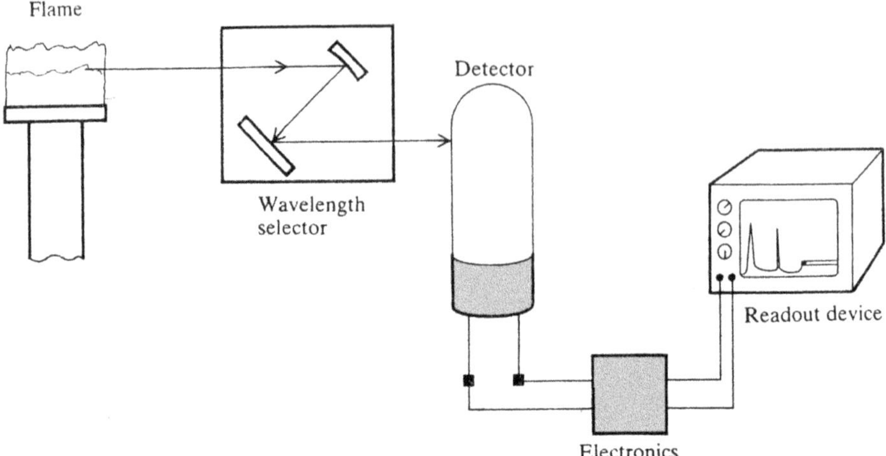

Figure 7-1 A diagram of the apparatus that is used for flame emissive spectrometry.

In instruments that are designed solely for FES, a chopper is usually not required because the detectors in those instruments are designed to respond to a dc signal.

In most instruments the wavelength selector is a monochromator. In instruments that are designed solely to assay sodium or potassium by FES, however, filters are commonly used as the wavelength selectors. The detectors are the same as those used for AAS (Chapter 6). Generally a photomultiplier tube is used as the detector. Readout devices also are identical to those used in AAS. Analog meters, digital meters, and computer-controlled devices are encountered most often.

The burners that are used for FES can be either total-consumption burners or premix burners. In the past total-consumption burners were used widely. Because of the erratic flames that result from total-consumption burners, premix burners are presently used for most analyses. The flame gases can be any of those described for AAS (Table 6-1). As the temperature increases the number of excited atoms and the intensity of the emitted radiation increases. A flame that is too hot can result in a decreased signal owing to ionization of the analyte element. Low-temperature flames can be used for easily excited group 1 elements. An air-acetylene flame can be used for the assay of the bulk of the elements. Nitrous oxide-acetylene flames can be used for refractory compounds and less easily excited elements.

The wavelength that is chosen for a quantitative analysis with FES generally corresponds to the most intense emissive line of the analyte. The most intense emissive line can be predicted with the aid of Eq. (6-6). A list of the wavelengths that are normally chosen for quantitative analysis of several of the elements that can be assayed by FES is given in Table 7-1. In some cases other

Table 7-1 A list of some of the elements that can be determined by flame emissive spectrometry and the wavelengths that are used for the assays

The elements are arranged alphabetically according to name

Element	Wavelength, nm	Element	Wavelength, nm
Al	396.2	Mg	285.2
Sb	259.8	Mn	403.1
As	235.0	Hg	253.7
Be	234.9	Mo	390.3
Bi	223.1	Ni	341.5
Ca	422.7	Pd	363.5
Cd	326.1	Rh	369.2
Co	345.4	Se	196.0
Cr	425.4	Si	251.6
Cu	327.4	Ag	328.1
Ga	417.2	Sr	460.7
Ge	265.2	Te	238.3
Au	267.6	Tl	377.6
In	451.1	Sn	284.0
Fe	372.0	V	437.9
Pb	405.8	Zn	213.9

wavelengths can be chosen for the assay. Quantitative analyses are usually done with the working-curve method.

Qualitative analyses can be done using FES by obtaining a spectrum of emitted intensity as a function of wavelength. Generally the wavelength is scanned while the relative intensity of the emitted radiation is recorded as a function of the wavelength. The spectrum is compared with spectra of known elements that have been obtained with the same apparatus or with tabulated values of the major peaks for the elements. If at least three lines from the sample can be matched with those from a known element, the identification is assumed to be positive.

The interferences that are encountered with FES are the same as those encountered in other techniques that utilize flames as sources of atoms. Chemical, spectral, and ionization interferences are essentially identical to those described for AAS. Spectral interferences are more likely to be encountered in FES than in AAS because the bandpass of the monochromator, rather than the relatively narrow emitted bandwidth of a line source, determines the wavelength range of the measured signal.

In AAS, radiation from the source was chopped and the detector was tuned to respond only to radiation at the chopping frequency. In that way some steady-state interferences emanating from the flame could be eliminated. Of course, that is not possible with FES. Generally FES is more sensitive to fluctuations of the flame than is AAS.

Difficulties associated with fluctuations in the intensity of a measured emissive line that result from fluctuating flame conditions or from a fluctuating rate of aspiration into the flame often can be overcome by use of the *internal-standard method.* A known constant concentration of a second element is added to each standard solution of the analyte. The element of known concentration is the *internal standard.* Experimentally it is advantageous to choose an internal standard which has an emission line that is near the emission line of the analyte. In that case the time required to scan between the lines is minimized and background emissions are similar at the two wavelengths. The intensities of both the analyte line and the line of the internal standard are measured. A working curve is prepared by plotting the ratio of the measured intensity of the analyte line to the intensity of the line from the internal standard. Because both elements were subject to the same environmental conditions, the assumption is made that the effects on the two elements are identical. Consequently, the resultant working curve compensates for the problems associated with erratic flames and rates of aspiration.

Sample problem 7-1 Calcium was assayed by FES. Enough strontium was added to each standard solution of calcium and to the analyte to make each solution contain 2.50 μg/mL. After aspiration of each solution into the flame the emitted intensity of the calcium line was measured at 422.7 nm and of the strontium line at 460.7 nm. Determine the concentration of calcium in the sample.

	Relative intensities	
Concentration of calcium, μg/mL	**460.7 nm**	**422.7 nm**
2.00	21.5	16.6
4.00	24.7	37.8
6.00	18.6	43.2
8.00	22.3	68.7
10.00	24.6	95.2
Sample	19.4	36.3

SOLUTION The ratio of the intensities of the 422.7-nm line to the 460.7-nm line is calculated for each solution.

Concentration, μg/mL	**I(422.7)/I(460.7)**
2.00	0.772
4.00	1.53
6.00	2.32
8.00	3.08
10.00	3.87
Sample	1.87

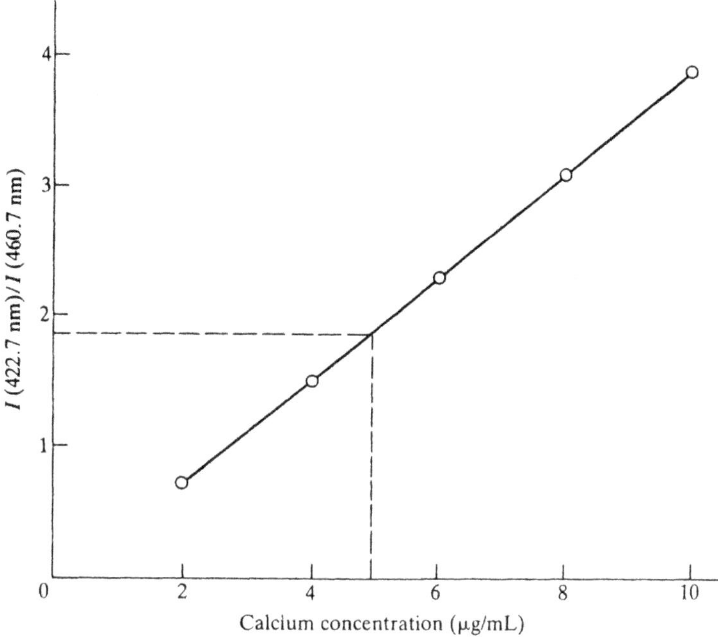

Figure 7-2 A working curve obtained by using the internal-standard method for the data in Sample problem 7-1.

A working curve (Fig. 7-2) is prepared by plotting the relative intensities of the two lines as a function of the calcium concentrations in the standard solutions. The concentration of the sample as read from the working curve is 4.85 μg/mL.

ATOMIC EMISSIVE SPECTROMETRY WITH PLASMAS

Atomic emissive spectrometry (AES) can be performed with apparatus similar to that shown in the block diagram in Fig. 7-1 where the flame is replaced with either a plasma or electrodes. Argon plasmas are used most often for nonflame AES. The high temperatures that are achieved in argon plasmas cause more efficient excitation of atoms and ions than is achieved with-flames. As a result, the intensities of the emitted lines are greater and more spectral lines are observed.

The wavelength selector for an instrument that uses a plasma is a narrow-bandpass monochromator. The wavelength of the monochromator as well as the other functions of the spectrometer are generally controlled by a microcomputer. Various detectors can be used including photomultiplier tubes and diode arrays. Several wavelengths can be simultaneously monitored or the wavelengths can be sequentially scanned. The readout devices that are used with the spectrometers include cathode-ray tubes, recorders, and line printers. A more thorough description of plasmas can be found in Chapter 8.

Qualitative analysis is done using AES in the same manner in which it is done using FES. The spectrum of the analyte is obtained and compared with the atomic and ionic spectra of possible elements in the analyte. Generally an element is considered to be in the analyte if at least three intense lines can be matched with those from the spectrum of a known element. Tabulated values of the emitted spectral lines of the elements can be found in References 1 and 2. These complement each other because Reference 1 lists the lines according to element while Reference 2 lists the lines according to wavelength.

Quantitative analysis with a plasma can be done using either an atomic or an ionic line. Ionic lines are chosen for most analyses because they are usually more intense at the temperatures of plasmas than are the atomic lines. The intensities of the spectral lines can be predicted with the aid of Eq. (6-6) and the data in the References.

Interferences that are encountered with plasmas can be grouped into the same categories as those that were encountered with AAS. Chemical interferences owing to refractory compounds are rarely a problem because plasmas have high temperatures. Spectral interferences are more plentiful when plasmas are used because an increased number of atomic and ionic lines are possible at the higher temperatures of plasmas. Plasma temperatures are in the approximate range from 6000 to 10,000 K.

AES with Electrical Discharges

An electrical discharge between two electrodes can be used to atomize or ionize a sample and to excite the resulting atoms or ions. The sample can be contained in or coated on one or both of the electrodes or the electrode(s) can be made from the analyte. The second electrode which does not contain the analyte is the *counter electrode*.

Electrical discharges can be used to assay nearly all metals and metalloids. Approximately 72 elements can be determined using electrical discharges. For analyses of solutions and gases the use of plasmas is generally preferred although electrical discharge can be used. Solid samples which have low solubility in liquid solvents are usually assayed with the aid of electrical discharges. Typically it is possible to assay about 30 elements in a single sample in less than half an hour using electrical discharges. To record the spectrum of a sample normally requires less than a minute.

Electrodes for AES

The electrodes that are used for the various forms of AES are usually constructed from graphite. Graphite is a good choice for an electrode material because it is conductive and does not spectrally interfere with the assay of most metals and metalloids. In special cases metallic electrodes (often copper) or electrodes that are fabricated from the analyte are used. Regardless of the type of electrodes that are used, a portion of each of the electrodes is consumed during the electrical discharge. The electrode material should be chosen so as not to spectrally interfere during the analysis.

Sketches of several common forms of graphite electrodes are shown in Fig. 7-3. The cylindrical graphite electrodes typically have a diameter of 6.2 mm and a length of 38 mm. Electrical discharge occurs at the pointed end of the counter electrodes (Fig. 7-3*a*, *b*, and *c*) where the strength of the electrical field is maximum. Several types of sample electrodes are available. The pointed

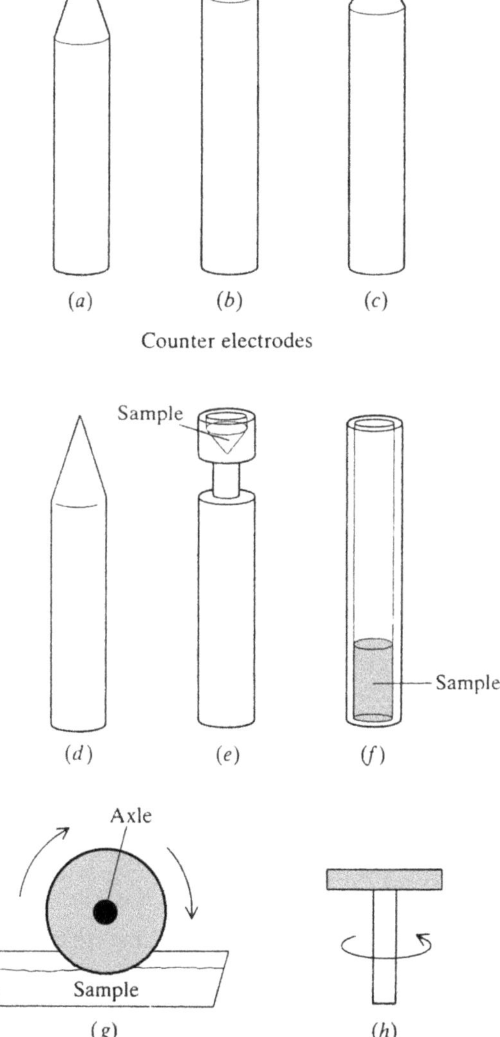

Figure 7-3 Sketches of several graphite electrodes that are used for atomic emissive spectrometry. Refer to the text for descriptions.

electrode (Fig. 7-3*d*) can be a graphite rod on which the sample solution is coated and allowed to dry before the analysis. It is also the usual design when the electrode is constructed from the analyte. Electrodes of that design are often used for steel or other metal samples.

The electrode in Fig. 7-3*e* is a *graphite-cup electrode*. The sample (usually a powder) is placed in the cup in the top of the electrode. A drill bit is used to form the cup in the electrode. Often the neck of the electrode below the cup is narrowed in order to minimize conduction of heat away from the cup during the electrical discharge. In some electrodes the neck is of the same diameter as the remainder of the electrode.

A *porous-cup electrode* is shown in Fig. 7-3*f*. It is used for solutions. Several milliliters of the solution are placed inside the electrode. The sample cavity in the electrode is prepared by drilling a hole to within about 3 mm of the end of the graphite rod. The solution slowly seeps through the bottom of the electrode. The counter electrode is placed below the porous-cup electrode.

The *rotating-disk electrode* (Fig. 7-3g) is also used for solutions. The disk, which is about 1.3 cm in diameter, is mounted on an axle and dipped into the sample solution. As the disk is rotated a film of the solution is carried to the top of the disk. The counter electrode is placed above the rotating disk at the top of the electrode. In the *rotating-platform electrode* (Fig. 7-3*h*) the sample solution is placed on the top of the disk and allowed to dry. The disk is rotated during the assay. Both forms of electrodes typically are rotated at between 5 and 30 revolutions per minute (r/min).

DC Arc

Electrical atomization/ionization and subsequent excitation of the sample can be accomplished with either spark or arc discharges. Commercial instruments often contain two or more of the electrical excitative sources. Of the several common types of arcs and sparks, the *dc arc* is the simplest. It uses a dc potential that is between 10 and 50 V to cause an electrical discharge that corresponds to a current of between 1 and 35 A to flow between the counter and the sample electrode (Fig. 7-4). The temperature generated by the electrical discharge is about 4000°C at the anode and about 3200°C at the cathode. Between the electrodes the temperature is in the 4000 to 7000°C range. The sample electrode can be either the cathode or the anode, but generally it is the anode.

Temperatures that are achieved with the dc arc are hotter than those achieved with most flames. The excitation of the sample is attributable to the combination of the high temperature and the electrical energy between the electrodes. Because different elements are vaporized (fractionally distilled) and excited at different times, it is necessary to use the arc until the entire sample has been vaporized.

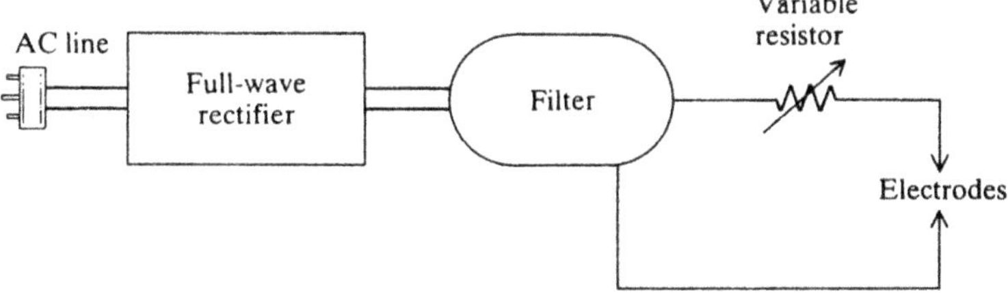

Figure 7-4 A diagram of the major components of a dc arc. The dc potential across the electrodes is usually between 10 and 50 V. The current flow between the electrodes is between 1 and 35 A.

In most instruments the dc arc is started by applying a high-potential spark across the electrodes. After the arc has been started the spark can either be shut off or allowed to continue. The dc arc yields intense emissive lines and consequently is often used for qualitative analysis. Because the dc arc wanders across the surfaces of the two electrodes and flickers, the intensities of the emissive lines are not particularly stable; i.e., the output signal from the dc arc is noisy.

Another problem that is encountered with the dc arc is the formation of gaseous cyanogen $(CN)_2$ by chemical reaction of carbon from the electrodes with nitrogen from the air. Cyanogen emits broadband radiation between about 360 and 420 nm that can interfere with many assays. The problem can be eliminated by blanketing the electrode tips and the space between the electrodes with argon or a mixture of argon (70 to 80 percent) and oxygen (20 to 30 percent). The exclusion of nitrogen prevents formation of cyanogen.

A *Stallwood jet* is a quartz enclosure that is placed around the electrodes and through which the protective gas is passed. The gas passes upward over the sample electrode. In addition to excluding nitrogen, the protective gas decreases wandering of the arc. The enclosure is constructed from quartz to permit emitted radiation to exit from the chamber.

AC Arc

An ac arc is similar to a dc arc except the discharge between the electrodes is not continuous. The cathode and anode alternate after each half-cycle of the applied ac potential. Typically, the potential supply operates at 60 Hz which results in a polarity reversal of the electrodes at a rate of 120 times each second. During the discharge in each half-cycle the current is continuous as in the dc arc.

The discharge must be restarted each time the polarity of the electrodes is switched. Because the potential that is required to start a discharge is greater than that necessary to maintain a discharge, the ac potentials that are used with ac arcs are greater than the dc potentials that are required to sustain a dc arc. The use of a potential between 2000 and 3000 V usually results in a current between 1 and 5 A.

The ac arc effectively samples the analyte during each discharge between the electrodes. Uneven sampling that is characteristic of the dc arc is prevented, with a resulting increase in reproducibility. The sensitivity of the ac arc is less than that of the dc arc. Sample solutions that are assayed using the ac arc are usually coated on the surface of the sample electrode and allowed to evaporate to dryness before the assay. Copper electrodes as well as graphite electrodes can be used with an ac arc.

Spark

The spark excitative source uses ac power, an *LC* circuit, and a spark gap that is operated by a synchronous motor to cause a spark to jump between the electrodes. The spark gap operates in a manner similar to that of the spark gap in the distributor of an automobile. Its function is to ensure that the spark jumps between the electrodes only when the potential that is stored in the capacitor in the ac circuit is at a maximum. The motor rotation is synchronized to the frequency of alternation of the current. A sketch of a simple circuit (Feussner circuit) that can be used for a spark source is shown in Fig. 7-5. Several variations of the circuit are in use in different instruments.

The potential after the step-up transformer in the circuit is between 10,000 and 50,000 V with a high-voltage source and about 1000 V with a medium-voltage source. The spark is active for periods between 10 and 100 μs and typically discharges at a rate of 120 to 180 times each second. Heating effects on the electrodes are minimized by the cooling that occurs between sparks. That leads to less fractional distillation of the sample from the electrode than is observed with the dc arc.

The time required to obtain a spectrum with a spark is about 10 s. The spark generally yields the most reproducible results and the highest precision of all of the spark and arc discharges. It is not as sensitive, however, as the dc arc. Minimum concentrations that can be assayed with a spark are about 0.01 percent for solid metallic samples and about 1 μg/mL for solutions.

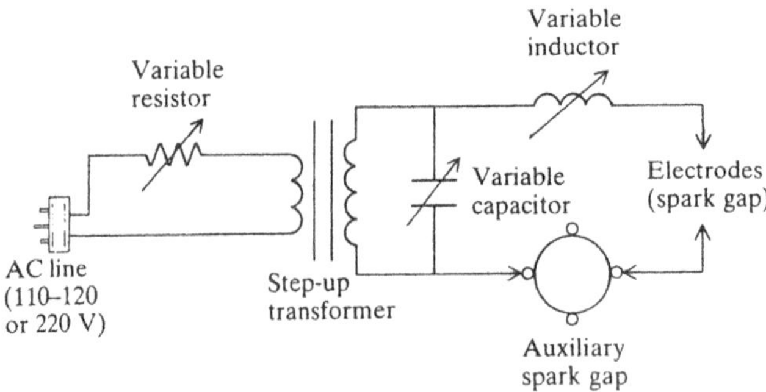

Figure 7-5 A circuit diagram of a Feussner circuit that is used to power a spark source.

Solid metallic samples are usually machined into a rod for use as the sample electrode. Normally the counter electrode is a pointed graphite or silver rod. Powders are pressed into pellets and inserted in the end of the sample electrode. Liquids often are assayed with the aid of a porous-cup electrode.

Laser Microprobe

The *laser microprobe* uses a laser (Chapter 8) to vaporize a small section on the surface of a sample. The vaporized sample passes between two ac spark electrodes that excite the sample. The resulting emissive spectrum is recorded as with the other AES methods.

The laser microprobe is ideally suited for examination of small areas on a surface. A microscope is used to focus the beam from the laser onto an area that is roughly 10 to 50 μm in diameter. Often a pulsed laser is used. The electrodes are held in place about 25 μm above the surface. The laser is fired between the electrodes. The two electrodes are sharply pointed rods that serve to control the location of the electric field during the discharges. A sketch of a laser microprobe is shown in Fig. 7-6.

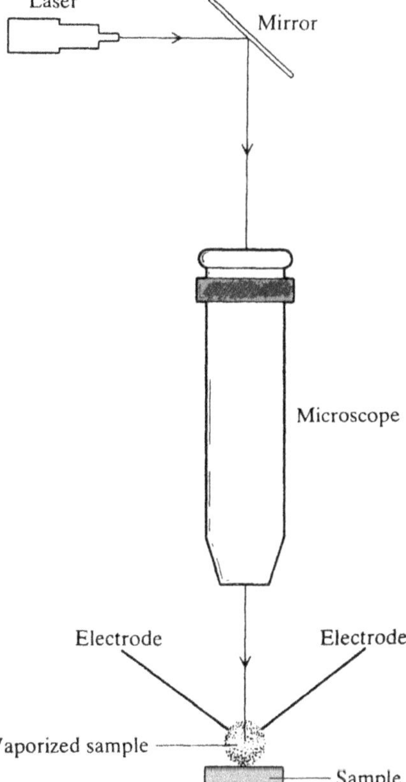

Figure 7-6 A diagram of a laser microprobe. The bottoms of the electrodes are about 25 μm above the sample.

Wavelength Selection and Detection for AES

Arc and spark instruments normally contain nonscanning monochromators. Either a series of slits is cut in the focal plane of the monochromator and a photomultiplier tube is placed behind each slit that corresponds to the wavelength of a line that is to be measured, or one or more photographic plates or pieces of film are placed on the focal plane of the monochromator. The instrument is a *spectrometer* if a photomultiplier tube or other photon detectors are used. It is a *spectrograph* if the detector is a photographic plate or film.

Commercial spectrometers can contain as many as 90 exit slits. The analyst chooses the exit slits that correspond to the spectral lines that are to be measured and places a detector behind each chosen slit. For many analyses between 20 and 35 detectors are simultaneously used at different slits to simultaneously assay one element for each detector. Each detector is termed a *channel*. A spectrometer of that design is a *direct reader* or a *direct-reading spectrometer*. If the chosen slits are too close together to permit placement of a detector behind each, mirrors can be used behind the slits to reflect the radiation to the detectors.

In a spectrograph the entire spectrum of the sample is simultaneously recorded. Measurement of the intensity of a particular spectral line is a requirement for quantitative analysis. Intensity measurements with films and plates are not as easily accomplished as they are with photomultiplier tubes. After development of the film or plate, each spectral line appears as a black image on the developed photograph or a light image on the negative. The intensity of the spectral line is proportional to the amount of darkening on the developed film or to the lack of darkening on the negative.

The amount of darkening is measured with a *densitometer*. A densitometer focuses radiation on the image of each line and uses a photomultiplier tube or other detector to measure the amount of radiation that is transmitted through or reflected by the image. The measurement is similar to the percent transmittance measurement in a spectrophotometer. The measured percent transmittance for each image is generally not directly proportional to the concentration of the assayed element. Working curves are used to determine the concentration of a particular element in a sample. About 16 spectra can be recorded on a roll of 35-mm film and 40 spectra on a 10×25-cm photographic plate.

The densitometer that is used for most measurements is a *microphotometer-comparator*. It uses a tungsten-filament lamp as the source of radiation. Microphotometer-comparators contain a slit that can be adjusted by the operator and a photon detector, such as a photomultiplier tube, that functions well in the visible region.

Qualitative Analysis with Arc and Spark AES

Qualitative analysis is performed by comparing the wavelengths of the intense lines from the sample with those for known elements. Tables [1, 2] can be used for the purpose, or spectra of known elements can be recorded and used for the comparison. It is generally agreed that at least three intense lines of a sample must be matched within a known element in order to conclude that the sample contains the element. Normally a dc arc is the source of choice for qualitative analysis because it produces intense spectral lines. Other arcs and sparks also can be used.

In order to assign wavelengths to the developed images on a photographic plate or film, it is helpful to obtain the spectrum of a reference element that has lines of known wavelengths near the lines from the sample. Iron is often used as the reference element because it emits a multitude of lines throughout the entire ultraviolet-visible region. The spectrum of the reference element is obtained on the same photographic plate as that used for the sample in order to prevent possible changes in alignment during insertion of a new plate. About 72 elements can be qualitatively and quantitatively assayed with arc or spark AES.

Quantitative Analysis with Arc and Spark AES

With direct readers quantitative analysis is straightforward. A channel is assigned for each element. The measured intensity of the spectral line is used with a working curve to quantitate the element in the sample. The wavelengths must be carefully chosen to prevent spectral interferences. Typically precisions obtained with direct readers are in the range of ± 0.3 to 3 percent.

When a photographic plate or film is used as the detector, the precision is not as good as that achieved with direct readers. In order to obtain accurate and precise results all of the experimental conditions must be carefully controlled. Variables such as exposure time, film type, and developing conditions particularly are important. Automated development of the film or plate is advisable, whenever possible, in order to minimize changes in the development process. With careful control of conditions, errors between 1 and 10 percent can be achieved using photographic detection.

Regardless of the type of detection used for the assay, the precision of the results can be improved by matrix-matching the standards with the sample. Use of the internal-standard method also improves precision. Usually a working curve is prepared by plotting the ratio or logarithm of the ratio of intensity of the standard's line to the internal standard's line as a function of the logarithm of the concentration of the standard (Prob. 7-3). The corresponding ratio for the analyte is obtained and the concentration determined from the working curve.

In many cases the precision and accuracy of an analysis of a compound that contains organic components can be increased by ashing the sample prior to the assay. Normally the sample is placed in a platinum or silica crucible and heated in a muffle furnace to 500°C. Ashing can also be done in a low-temperature oxygen plasma. The temperature should be sufficiently high to remove all traces of any organic matrix of the analyte, but it cannot be high enough to vaporize the assayed elements. The ashing process is similar to that performed in furnace cells during atomic absorption spectrophotometry.

IMPORTANT TERMS

AC arc
Atomic emissive spectrometry
Channel
Counter electrode
DC arc
Densitometer
Direct reader
Flame emissive spectrometry
Graphite-cup electrode
Internal standard
Internal-standard method
Laser microprobe
Microphotometer-comparator
Porous-cup electrode
Rotating-disk electrode
Rotating-platform electrode
Spark
Spectrograph
Spectrometer
Stallwood jet

PROBLEMS

7-1 The internal-standard method was used for the analysis of copper by FES. A series of standard solutions of copper were prepared that contained 3.00 μg/mL of cadmium. The sample solution was prepared by adding 10.0 mL of the analyte solution and 10.0 mL of a 7.50-μg/mL Cd solution to a 25-mL volumetric flask. The flask was filled to the mark with deionized water. The relative emitted intensities of copper and cadmium respectively were measured at 327.4 and 326.1 nm in each of the solutions. Determine the concentration of copper in the analyte solution.

	Relative intensities	
Copper concentration, μg/mL	**327.4 nm**	**326.1 nm**
1.20	18.7	31.5
2.40	38.6	32.7
3.60	52.7	29.8
4.80	71.7	30.4
6.00	93.9	31.3
Sample	45.3	30.2

7-2 The internal-standard method was used for the analysis of strontium by FES. A stock solution of strontium nitrate (MW 211.63) was prepared by dissolving 0.2415 g of strontium nitrate in sufficient water in a volumetric flask to prepare 1 L of solution. Varying volumes of the strontium stock solution and 10.0 mL of a 160.0-μg/mL $VOSO_4$

solution were added to 100-mL volumetric flasks with pipets. The resulting solutions were diluted to the mark with distilled water. A 10.0-mL portion of the vanadium solution and 50.0 mL of the analyte solution were added to the 100-mL volumetric flask that was labeled "sample" and the solution was diluted to the mark with distilled water. The intensities of the strontium line at 460.7 nm and the vanadium line at 437.9 nm were measured. Determine the concentration of strontium in the analyte solution.

	Relative intensities	
Strontium stock volume, mL	**460.7 nm**	**437.9 nm**
2.00	16.9	35.7
4.00	29.9	33.1
6.00	54.7	38.5
8.00	74.7	39.3
10.00	81.2	34.1
Sample	36.1	35.4

7-3 A spark was used for the determination of magnesium in a solution by the internal-standard method. Molybdenum was used as the internal standard. A series of standard magnesium solutions were prepared by dissolving $MgCl_2$ in distilled water. Each standard solution and the analyte solution contained 25.0 ng/mL of molybdenum which was prepared by dissolving ammonium molybdate in the solutions. A pipet was used to place 50 μL of each solution on copper electrodes and the solutions were evaporated to dryness. The intensities of the 279.8-nm Mg line and the 281.6-nm Mo line were measured. Determine the concentration of magnesium in the analyte solution.

	Relative intensities	
Magnesium concentration, ng/mL	**279.8 nm**	**281.6 nm**
1.05	0.67	1.8
10.5	3.4	1.6
105	18	1.5
1050	115	1.7
10,500	739	1.9
Analyte	2.5	1.8

REFERENCES

1. Corliss, C. H., and W. R. Bozeman: "Experimental Transition Probabilities for Spectral Lines of Seventy Elements," NBS Monograph 53, National Bureau of Standards, Washington, D.C., 1962.
2. Harrison, G. R.: *Massachusetts Institute of Technology Wavelength Tables,* MIT Press, Cambridge, Massachusetts, 1969.

CHAPTER

EIGHT

ATOMIC FLUORESCENCE, RESONANT IONIZATION, AND LASER-ENHANCED IONIZATION

Atomic fluorescent spectrometry (AFS) uses radiation from a line or continuous source to excite atoms to a higher electron state. The fluorescing radiation that is emitted as the excited electrons return to the ground state is measured and used for the analysis. *Resonant ionization spectroscopy* (RIS) uses the radiation from one or more lasers to successively excite an atom to an electron level above the ground state and to ionize the excited atom. *Laser-enhanced ionization* (LEI) uses a laser to excite a specific atom as in RIS. The excited atom is then thermally ionized by the heat from a flame. A potential is applied across the cell and the current flow through an external circuit is related to the degree of ionization in the cell.

ATOMIC FLUORESCENCE

The absorption of energy from a radiative source by atoms can result in atoms in an excited electron level. Atomic fluorescence occurs when the excited atoms emit radiation after initially being excited by absorption of photons. The four major categories of atomic fluorescence are illustrated in Fig. 8-1.

Energy is absorbed from a radiative source by the atoms prior to fluorescence in exactly the manner that was described in Chapter 6 for atomic absorption. In nearly every case the initial absorption takes place from the ground state of the atom. Rather than measuring the decreased intensity of the

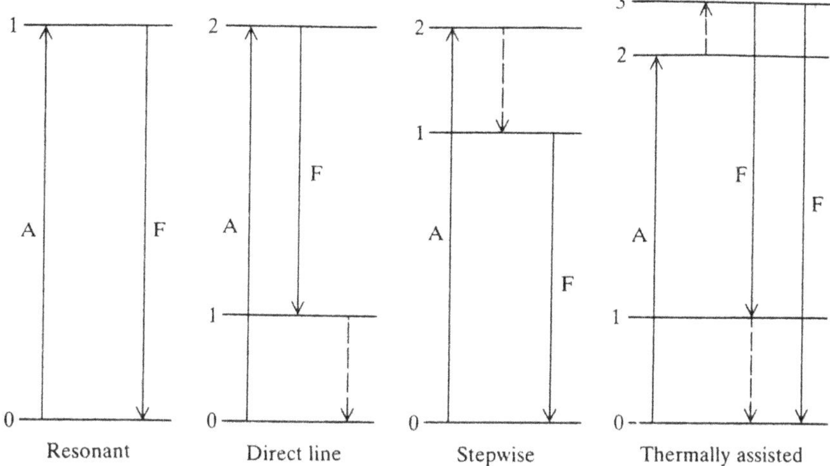

Figure 8-1 A diagram illustrating the energetic transitions that occur during atomic fluorescence in each of the four categories. A, absorption; F, fluorescence. The dashed lines represent radiationless transitions. The numbers represent the ground (0), first (1), second (2), and third (3) electron levels for the atom.

exciting radiation as a result of absorption, the intensity of the emitted (fluoresced) radiation is measured. *Resonant fluorescence* occurs when the fluoresced radiation is of the same wavelength as the absorbed radiation. Resonant fluorescence is the type that is used most often for quantitative analysis by AFS.

Direct-line fluorescence occurs when an electron in an excited state emits radiation upon falling to an electron level that is above the level from which the electron originally absorbed radiation. The wavelength of the emitted radiation is longer than the wavelength of the absorbed radiation. *Stokes fluorescence* is fluorescence in which the fluoresced radiation has a wavelength that is greater than the absorbed radiation. Further loss of energy to the ground state can occur either by emission of another photon or by collisional deactivation.

Stepwise fluorescence is preceded by absorption and collisional deactivation to a lower excited electron level. Fluorescence occurs when the atom emits a photon as the electron returns to the ground electron state. Stepwise fluorescence is a second example of Stokes fluorescence.

In some cases an atom that is in an excited electron state as a result of absorption can be further excited by a radiationless process to a higher energetic level. The enhanced excitation is usually caused by collisions of the excited atom with energetic species from the flame. Fluorescence occurs either directly from the excited state to the ground state or to a lower energy electron level that is above the ground state. In either case the emission is *thermally assisted fluorescence*. The wavelength of the emitted radiation can be either longer or shorter than that of the absorbed radiation. *Anti-Stokes fluorescence* is any form of fluorescence in which the emitted radiation has a shorter wavelength (greater energy) than the absorbed radiation.

A fifth type of fluorescence can also occur. After an atom becomes electronically excited by absorption, the excited atom transfers some or all of its energy to an atom of a different element. The atom of the second element emits radiation as its excited electron returns to the ground state (or some other lower energy, level). Atomic fluorescence of that type, which is rarely encountered, is *sensitized fluorescence.*

Apparatus for AFS

A diagram of the essential parts of an atomic fluorescent spectrometer is shown in Fig. 8-2 and a photograph of a commercial instrument is shown in Fig. 8-3. The parts are essentially identical to those used for AAS. Generally the intensity of the emitted radiation is perpendicularly measured relative to the incident radiation. In-line measurements are not made in AFS because the wavelength of the emitted radiation is often identical to the wavelength of the radiation from the source (resonant fluorescence). In order to prevent measurement of emitted radiation owing to thermal excitation within the cell (as in FES, Chapter 7), the radiative source is usually pulsed and the detector is tuned to respond only to radiation that oscillates at the pulsed frequency. Either the electric power to the source can be pulsed or a mechanical chopper can be placed in the path of the radiation between the source and the cell.

EMR Sources for AFS

Both continuous and line sources are used for AFS. In most cases the sensitivity of AFS is directly proportional to the intensity of the incident radiation, as was shown in Eq. (5-23). Consequently, high-intensity radiative sources are used in

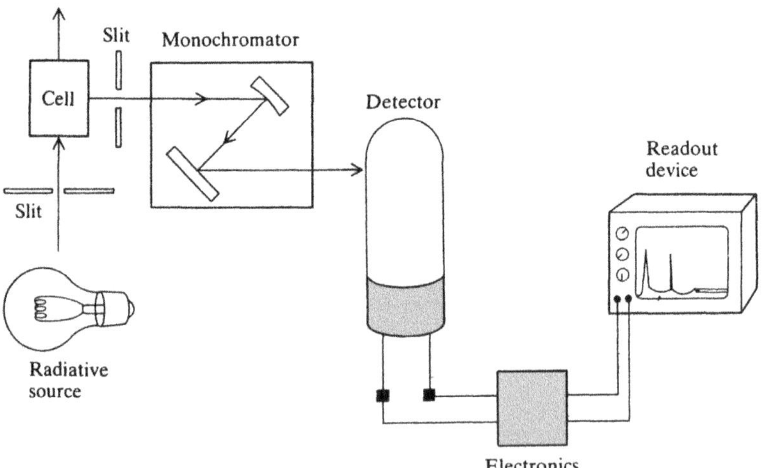

Figure 8-2 A diagram of an atomic fluorescent spectrometer.

Figure 8-3 A photograph of a commercial atomic fluorescent spectrometer. The spectrometer utilizes an inductively coupled argon plasma as the cell and a pulsed hollow-cathode lamp as the source. *(Photograph courtesy of Baird Corporation.)*

atomic fluorescent spectrometers. Generally the radiative sources that are used in AFS are more intense than those that are used in AAS. The most popular continuous source that is used in AFS is the xenon-arc lamp.

Line sources that are used for AFS include hollow-cathode lamps, high-intensity hollow-cathode lamps, electrodeless discharge lamps, and lasers. In early work vapor discharge lamps were used for Cd, Ga, Hg, In, Na, Tl, and Zn. Because of the significantly self-reversed and broadened radiation that is emitted by the vapor discharge lamps, they are rarely used in modern studies. *Inductively coupled plasmas* (ICPs) into which relatively large concentrations of the analyte element are aspirated emit intense radiation that can be used as radiative sources for AFS. ICPs at present have not found widespread acceptance as sources. A description of ICPs as cells for AFS and AES is included later in the chapter. Of the several types of line sources, the electrodeless discharge lamps have been used most frequently. A description of each of the line sources except lasers and inductively coupled plasmas was included in Chapter 6.

Lasers

Laser is an acronym for *l*ight *a*mplification by *s*timulated *e*mission of *r*adiation. A laser is a device that emits high-intensity coherent (in-phase) radiation over a narrow (typically 0.001 to 0.01 nm) bandwidth. The high intensity and narrow bandwidth of radiation that is emitted from a laser makes it a nearly ideal source for AFS at those wavelengths for which lasers are available.

The first laser was the ruby laser. It is described in order to illustrate the manner in which lasers function. Ruby consists .of crystalline Al_2O_3 into which Cr(III) has been substituted for some of the Al(III). The presence of Cr(III) imparts the red color to a ruby. The ruby laser is constructed by placing a ruby rod at one focus of a reflective elliptical cavity (Fig. 8-4). The second focus is occupied by a flashlamp that emits high-intensity continuous radiation in the region that can electronically excite chromium in the crystal. The ruby rod is the *medium*. In some ruby lasers two flashlamps are placed at the two outer foci of a double elliptical cavity and the ruby rod is placed at the center focus. The elliptical design causes radiation from the flashlamp to be focused on the ruby rod.

Radiation from the flashlamp causes chromium ions within the ruby rod to be excited to $^4F_1$ and $^4F_2$ bands. The excited ions rapidly lose energy to the excited ^2E level. The ^2E level, which is relatively stable in comparison to the two F levels, is a *metastable* state. A *population inversion* occurs when more chromium ions are in the excited ^2E level than in the ground state. A population inversion is required in order for light amplification to occur in the laser. The population inversion is created by *pumping* the laser. A flashlamp is used to pump the ruby laser. A diagram of the energy transitions that are undergone in the ruby laser is shown in Fig. 8-5.

Once the electron is in the ^2E level, radiation can be spontaneously emitted (fluoresced) when the electron returns to the ground state. The emission in the ruby laser occurs at 693.4 nm. If the radiation is emitted in a direction such that it strikes the wall of the medium (the ruby rod), it is lost. When the radiation is emitted such that it travels within the rod along its length, the radiation becomes trapped between two mirrors that are placed at opposite ends of the rod. Only parallel radiation is trapped within the optical cavity of the laser. The ends of the ruby rod can act as the mirrors if they are highly polished.

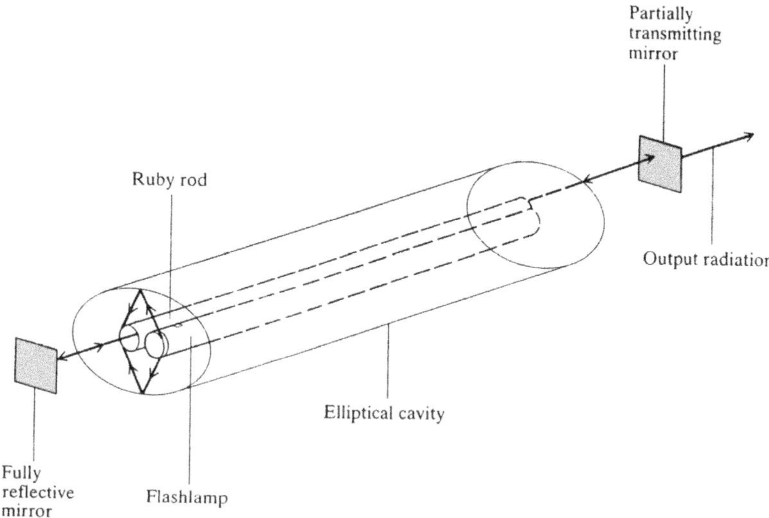

Figure 8-4 A diagram of a ruby laser that is pumped by a single flashlamp.

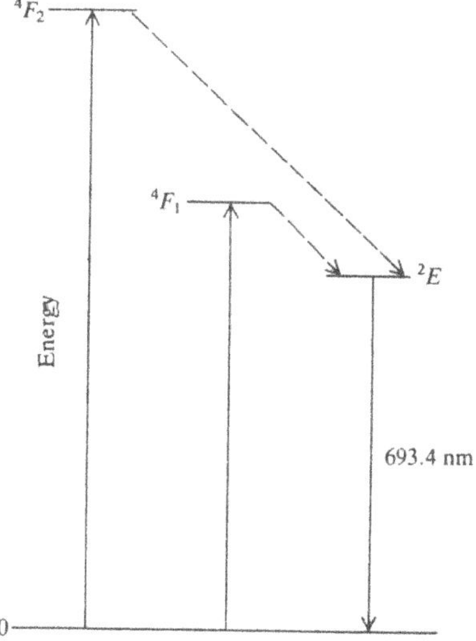

Figure 8-5 The energetic transitions in the ruby laser.

Each time a spontaneously emitted photon strikes an excited chromium ion, the ion immediately returns to the ground state with emission of radiation that is in-phase with and identical in wavelength to that of the incident radiation (693.4 nm in this case). The *stimulated emission* results in amplification of the intensity of the radiation in the optical cavity. In order for a laser to amplify it is necessary that more stimulated emission take place than absorption. That can only occur when there is a population inversion in the medium. If the medium has a high population inversion (high concentration of excited states), the gain of the laser is large. If the medium has a low population inversion, the gain is small. Lasers with high gains can better tolerate losses of radiation from the optical cavity. Lasers with low gains must have highly reflective ends to the optical cavity and can tolerate only small losses of radiation.

A portion of the radiation within the optical cavity is emitted from the laser through one of the mirrors at the end of the cavity. The mirror allows some of the radiation to pass through it while simultaneously reflecting another portion; i.e., the mirror is partially transmitting. Some configurations of the ruby laser can become hot during pumping and must be cooled.

Although ruby lasers can deliver a great deal of power, they are not particularly useful for AFS because the radiation is emitted only at a single wavelength (693.4 nm). In order to be useful as a source for AFS, lasers must be used that emit radiation at the wavelengths that are required for absorption by atoms. It is particularly desirable that one or a relatively few lasers be capable of exciting atoms of many elements. That can happen only if the wavelength emitted by the laser can be controlled. The process by which the wavelength at the output of a laser is adjusted is *tuning*. A *tunable laser* is one in which the output wavelength can be adjusted.

Lasers can be classified according to medium pumping mechanism and controlling mechanism. Lasers are available that have gaseous, liquid, or solid mediums. The ruby laser has a solid medium. The pumping mechanism is usually either optical or electrical.

Optical pumping, such as that described for the ruby laser, uses a radiative source to achieve population inversion. The radiative source can either be operated continuously or pulsed. Popular continuous pumping sources are tungsten lamps for visible excitation and arc lamps for ultraviolet excitation. Flashlamps, such as that described for the ruby laser, are used for pulsed lasers. Lasers can be used to pump other lasers. Optical pumping primarily is used for lasers that have liquid and solid mediums.

Electrical pumping is used most often for lasers that have gaseous mediums. An electric discharge is created between two electrodes in the medium. The electric energy causes excitation of the gaseous atoms or molecules and the required population inversion. Both sparks and glow discharges are used for electrical pumping.

The output from a laser can be controlled in several ways. The controlling mechanism involves maximizing loss of undesirable elements of the lasing radiation while minimizing loss of desirable elements. If a wavelength selector such as a prism or diffraction grating is placed in the optical cavity between the medium and the mirror, only those wavelengths that exit the selector so as to travel along the length and through the lasing medium are amplified. The prism or diffraction grating tunes the laser. Of course a tuning mechanism is required only when the laser can emit broadband radiation or radiation at more than one line.

Several methods are available to control the intensity or power of the beam that exits a laser. *Q-switching* allows a large population inversion to develop before emission is stimulated. Two polarizers are placed in the optical cavity so that one polarizes at a right angle to the other. Because radiation that passes through the first cannot pass through the second, no spontaneously emitted radiation is allowed to stimulate emission during the pumping process. After a large population inversion has developed, one of the polarizers is rotated thereby allowing radiation to pass through the polarizers and to stimulate emission. The large population inversion causes an emitted (lased) pulse of high intensity, Q-switching functions best for lasing mediums that have stable (long-lived) excited states.

An alternative to Q-switching is *cavity dumping*. The ends of the optical cavity are made highly reflective so that little or none of the radiation is allowed to escape. The reflecting radiation continues to stimulate emission within the medium until the desired amplification is achieved. When emission from the laser is desired, an alternating potential is applied to a material that converts the potential to a physical oscillation (the *piezoelectric effect*). The oscillating substance causes a traveling acoustic wave to be transmitted through a transparent "lens" in the path of radiation within the optical cavity. The acoustic wave within the lens acts as a diffraction grating, where the spacing between lines in the grating corresponds to the wavelength of the acoustic wave, and diffracts radiation out of the optical cavity of the laser. Cavity dumping is normally used for lasing mediums that have excited states that are stable only for short times. Cavity dumping is often used with dye lasers.

In some cases it is desirable to have pulses of in-phase radiation emitted from a laser. That can be accomplished by pulsing the pumping radiation at a frequency that corresponds to the time that is required for a photon to make a round trip through the optical cavity. Photons that arrive in the medium when the pumping radiation is emitted can stimulate emission from the excited states. Consequently, that radiation is amplified. Because photons that arrive at other times are not as likely to be able to stimulate emission, that radiation is not amplified. The process is *mode locking*.

Several of the more popular types of lasers and some of their properties are listed in Table 8-1. Gaseous mixtures of helium and fluorine with argon, krypton, or xenon are used to form *excimer lasers*. An excimer laser is one that uses a molecular medium that is stable only in the excited state. Excitation of

Table 8-1 Several popular lasers

Each excimer medium is marked with an asterisk (*)

Medium	Type	Emitted wavelength, nm
Ar	Gas	334, 351.1, 363.8, 454.5, 457.9, 465.8, 472.7, 476.5, 488.0, 496.5, 501.7, 514.5, 528.7
Kr	Gas	350.7, 356.4, 406.7, 413.1, 415.4, 468.0, 476.2, 482.5, 520.8, 530.9, 568.2, 647.1, 676.4, 752.5, 799.3
He-Ne	Gas	632.8
He-Cd	Gas	325.0, 441.6
N_2	Gas	337.1
XeF*	Gas	351
KrF*	Gas	248
ArF*	Gas	193
Ruby	Solid	693.4
Nd: YAG	Solid	266, 355, 532
Pb_{1-x} Cd_x S†	Solid	2.9×10^3 –2.6×10^4
PbS_{1-x} Se_x†	Solid	2.9×10^3 –2.6 – 10^4
Pb_{1-x} Sn_x Se†	Solid	2.9×10^3 –2.6 – 10^4
Pb_{1-x} Sn_x Te†	Solid	2.9×10^3 –2.6 – 10^4
Dyes	Liquid	217–1000

† This is a semiconductor diode laser. Lasing occurs at the *pn* junction. The emitted frequency is altered by altering the concentration of the dopant in the lead.

the rare gas to an excited electron level causes formation of the excimer molecule (XeF, KrF, or ArF) which functions as the lasing medium. Relaxation of an excimer to the ground state results in dissociation of the molecule.

The *neodymium: yttrium aluminum garnet* (Nd: YAG) laser is similar to the ruby laser except the emitted radiation corresponds to a transition from one excited state ($^4F_{3/2}$) to a lower excited state ($^4I_{11/2}$) in Nd(III). Because the transition is not to the ground state ($^4I_{9/2}$), the population inversion is achieved more readily than when the lower state is the ground state and lasing is easily accomplished. It is easier to achieve a population inversion when few ions exist in the lower excited state. When the lower state is the ground state, population inversion is more difficult to accomplish because a relatively large concentration of ions starts in the ground state.

Of the available types of lasers, the tunable dye lasers are most useful for analytical studies in the visible and near-infrared region. Many dyes can be used to prepare solutions that are used in dye lasers. Each dye solution emits lasing radiation that corresponds to the emissive fluorescent spectrum of the dye. Typically that occurs over a wavelength range of between 30 and 100 nm. A list of selected dyes that can be used in lasers is given in Table 8-2. Dye lasers are made tunable by insertion of a wavelength-selective device (grating, prism, etc.)

Table 8-2 Selected dyes that are used as mediums in lasers and the approximate wavelength ranges through which they emit radiation that is sufficiently intense to be useful

Dye	Wavelength range, nm
Stilbene 420	400-460
Coumarin 460	450-500
Coumarin 503	510-620
Rhodamine 6G	565-620
Rhodamine 610	590-660
Oxaxine 720	655-720
Nile blue 690	690-780

within the optical cavity of the laser. Typically, tunable dye lasers have an output bandwidth of about 0.01 nm.

Dye lasers can be pumped with a flashlamp or with another laser. Gas and liquid lasers such as the argon, krypton, nitrogen, excimer, and Nd: YAG lasers can be used to pump a dye laser. A diagram of a flashlamp-pumped dye laser is shown in Fig. 8-6. The dye solution is continuously pumped through the cell in the laser. As in the ruby laser, the medium and the flashlamp occupy foci within the elliptical cavity which surrounds the device. Further information related to lasers can be found in References 1 and 2, and in the reviews found in the journal *Analytical Chemistry* in even-numbered years.

Cells for AFS

The cell that is used for AFS is responsible for atomizing the sample and for holding the atoms within the path of the incident radiation sufficiently long for fluorescence to occur. In order to minimize background emission owing to

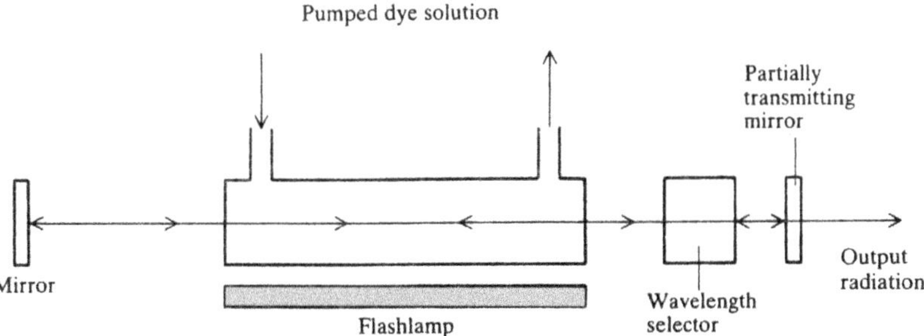

Figure 8-6 A diagram of a flashlamp-pumped dye laser.

thermal excitation within the cell, the temperature of the cell should be kept as low as possible. The temperature must be sufficiently high to result in atomization but should not be any higher than is absolutely necessary. Flames, furnaces, and plasmas have been used as cells for AFS.

Both total consumption and premix burners have been used with flames. Generally a premix chamber with a circular burner head is preferred because low noise is associated with premix burners. Circular burners are used to minimize self-reversal by minimizing the length of the flame through which an emitted photon must travel. Of course, at low analyte concentrations, self-reversal is minimized, and longer photon travel distances can be tolerated. The flames that are often used for AFS include the air-hydrogen, air-acetylene, nitrous oxide-acetylene, and the corresponding nitrogen or argon sheathed flames. The argon-oxygen-acetylene and helium-oxygen-acetylene flames can also be used.

Adaptations of the several forms of furnaces that were described in Chapter 6 can be used as cells for AFS. Laser sources combined with graphite furnace atomization provides some of the most sensitive analytical results from AFS. Among the types of furnaces that have been used are the graphite tube, carbon cup, and several other variations of carbon furnaces. Plasmas are described in the next section.

Plasmas

Inductively coupled argon plasmas (ICAPs) can be used as cells for AAS, AFS, and AES They have primarily been used for AES. The energy for the atomization or ionization of the sample is provided by an argon plasma. Plasmas formed from other media rarely have been used. A *plasma* is a stream of gaseous ions. Argon is chosen for the plasma because it is easily ionized and relatively unreactive. Although the argon plasma that is used in atomic spectroscopy can be created in any of several ways, it is most often formed electromagnetically by the action of a radiofrequency (rf) generator and an induction coil on a stream of argon.

The sample solution is either aspirated into a stream of argon by the Venturi principle or, more often, pumped into the stream with a peristaltic pump. The mixture enters a premix chamber similar to those used in premix burners. In the premix chamber the mixture passes through a series of baffles that serve to remove the larger sample drops. The smaller drops pass through the premix chamber and are guided into the plasma with a cylindrical quartz tube.

A second stream of argon that passes through a larger concentric quartz cylinder is used to create the plasma. Between two and five turns of an induction coil are wrapped around the upper end of the outer cylinder. The coil is connected to an rf generator. Often the coil is a hollow copper tube. The number of turns of the coil around the quartz cylinder determines the length of the plasma. A three-turn coil is adequate for nearly every application. The oscillating current that passes through the induction coil from the rf generator creates an oscillating electromagnetic field near the top of the outer quartz cylinder. The rf generator is usually operated at a fixed frequency between 3 and 75 MHz. A popular choice is 27 MHz. The generator normally is operated at a power of 2.5 kW or less. Often the operating power is about 1.25 kW.

A Tesla discharge (a source of electrons) provides "seed" electrons to the argon stream. The electrons are accelerated by the magnetic field from the induction coil. Upon colliding with argon atoms the electrons produce argon ions and more electrons. The produced electrons are accelerated and, after collision with other argon atoms, form more ions and electrons. After the initial Tesla discharge, the plasma is self-propagating; i.e., an external source of electrons is not required. The argon plasma absorbs energy from the oscillating magnetic field in much the same way that energy is transferred from the primary (the induction coil) to the secondary (the plasma) coil of an electric transformer (Chapter 2). The high-energy plasma created by the high-intensity magnetic field has a shape similar to that of a flame at the top opening to the outer quartz cylinder, and it reaches temperatures that are typically between 6000 and 10,000 K. The sample solution enters the center of the plasma where it is atomized or ionized. Normally the plasma is "ignited" (started) by the Tesla discharge while the quartz cylinder is vertically oriented. After ignition the plasma can be operated in either the vertical position, or it can be rotated so that it operates in the horizontal position. Either position can be utilized in AES and AFS studies. Emitted radiation from the plasma enters the detector perpendicularly to the quartz cylinder body while operated in the vertical mode, but along the length of the plasma torch while operated in the horizontal mode. Horizontal operation increases the cell pathlength and often results in lower detection limits in both AFS and AES.

Several regions exist within a plasma. The *preheating zone* is the region in which the sample first enters the plasma. Evaporation of the solvent and melting and vaporization of the salt take place in the preheating zone. In the *initial radiative zone* (IRZ) atoms are formed and excited and atomic emission

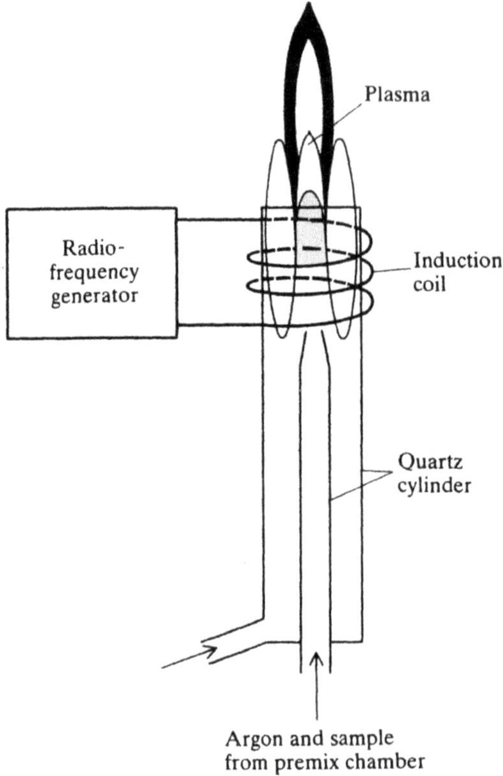

Figure 8-7 A diagram of an inductively coupled argon plasma.

takes place. In the *normal analytical zone* (NAZ) ion formation occurs. Usually + 1 and + 2 ions are formed in the NAZ. The *tail* or *plume* of the plasma is the region in which atoms can recombine to form polyatomic species. In AES studies (Chapter 7) considerable chemical interferences are observed in the tail. A yttrium salt is often used to visually locate the regions of the plasma. When the salt is aspirated into the plasma, the IRZ is red, the NAZ is blue, and the tail is red.

Because the temperature of the plasma varies with location within the plasma, the optimal sensitivity for different elements is observed at different places (heights) within the plasma. That is also true of flames. The location within the plasma at which the spectrum is observed, while operating the plasma in a vertical mode, is selected either by moving the plasma up or down with respect to the monochromator or by tilting a mirror at an angle that reflects radiation from the proper plasma location into the monochromator. When used in AES or AAS the observation is usually made at a height that is between 15 and 25 mm above the torch. For AFS the observation often is made at a height that is between 45 and 65 mm above the torch because of high-intensity emission from the plasma at lower heights. A diagram of the general structure of an ICAP is shown in Fig. 8-7. A diagram of a commercially available plasma torch and a photograph of a plasma are shown in Fig. 8-8.

In addition to the ICAP, a dc argon plasma can be used as a cell for atomic spectroscopy. It relies upon the application of less than a kilowatt of a direct current (rather than use of an rf generator and an induction coil) between two carbon anodes and a tungsten cathode. The two carbon anodes and the cathode are held at the corners of a triangle, and the sample introduction tube is placed between the two anodes at the bottom of the triangle. The tungsten cathode is located at the center in the top of the arrangement, as shown in Fig. 8-9. The dc argon plasma torch is not used in as many instruments as the ICAP torch. Details of the three-electrode dc argon plasma torch are given by Reednick.[3]

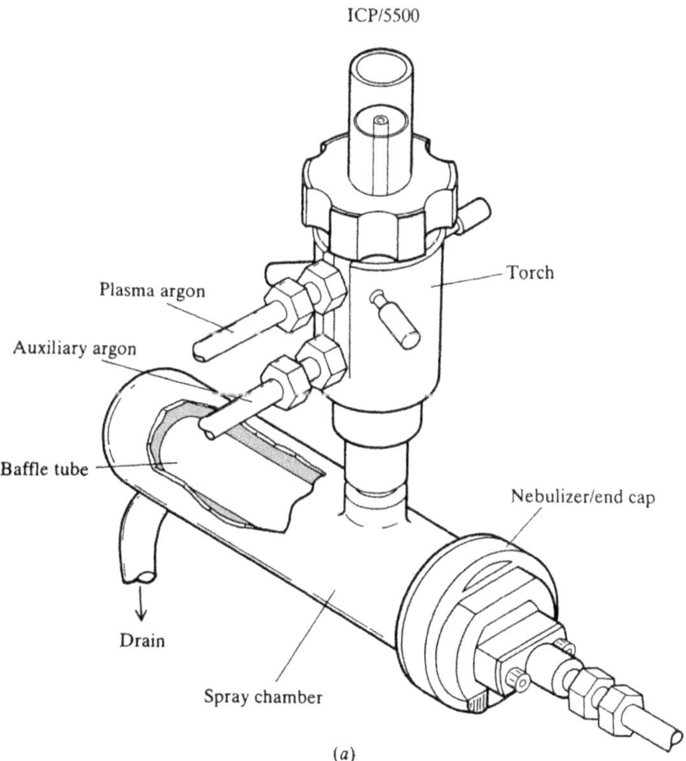

(a)

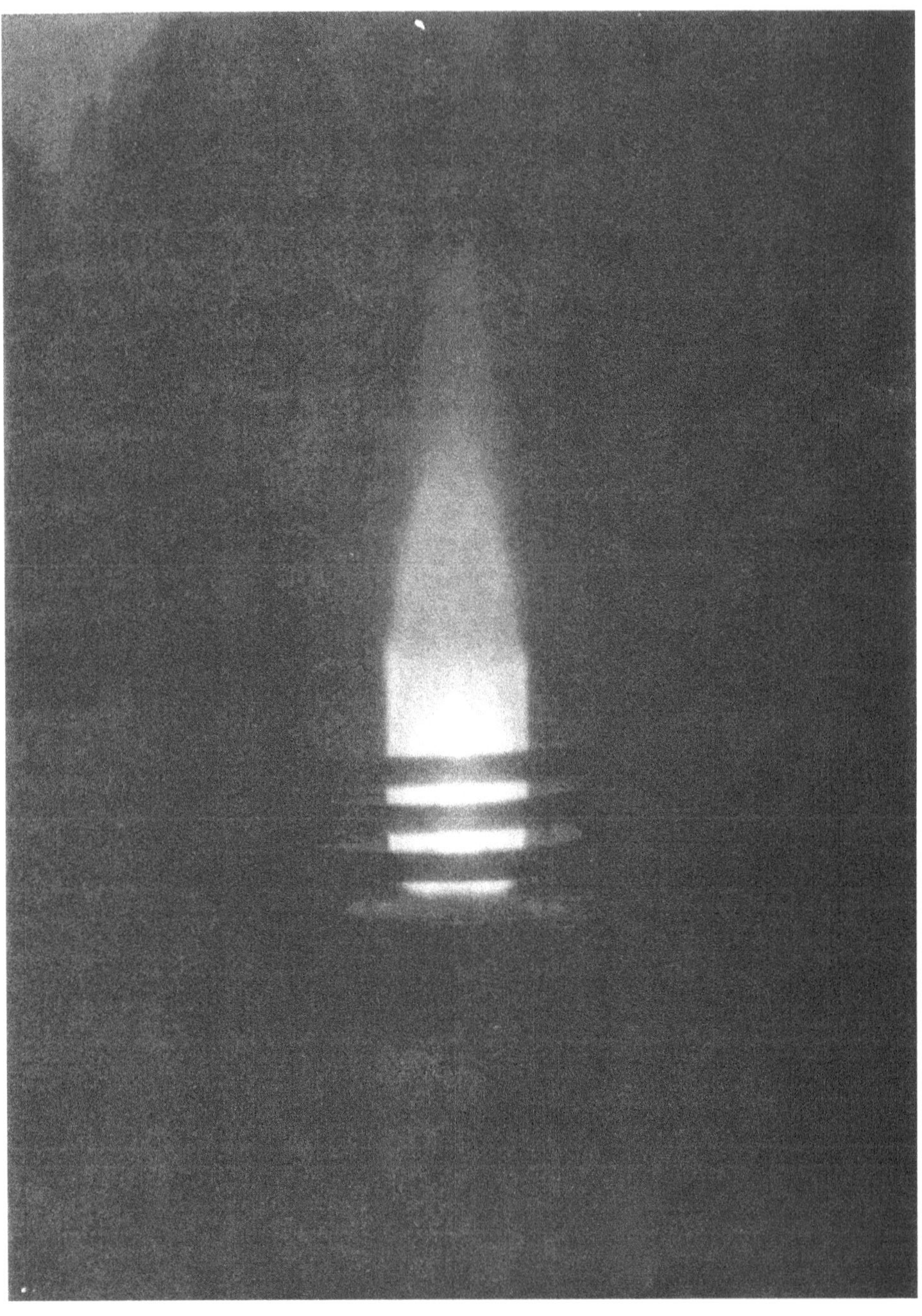

(*b*)

Figure 8-8 (*a*) Diagram of a commercial plasma torch; (*b*) photograph of the plasma. *(Photograph courtesy of Perkin-Elmer Corporation.)*

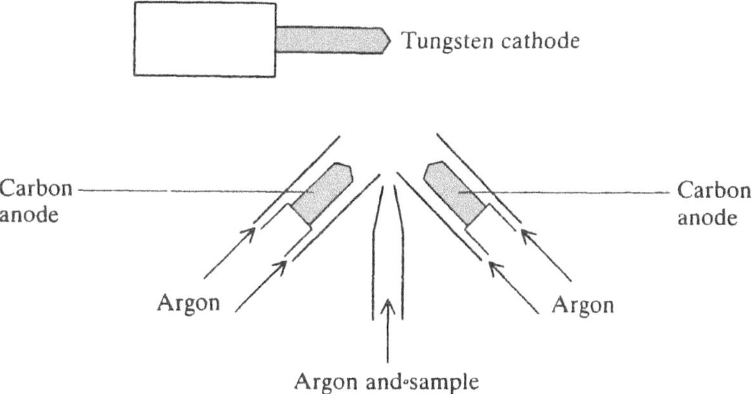

Figure 8-9 Diagram of a three-electrode dc argon plasma torch.

The high temperatures in an argon plasma provide an excellent excitative source for atomic and ionic emissive spectra. Many more spectral lines are observed with an argon plasma than with a flame, and the intensities of the observed lines are generally greater. The ICAP has been used as an intense source of exciting radiation for AFS.[4,5] Further information relating to inductively coupled plasmas can be found in Reference 6.

Because the temperature of the plasma is unusually high, it is not necessary to use only those spectral lines that involve the ground electron state. Because many spectral lines are observed with plasmas, it is desirable to use a monochromator that has a resolution of 0.02 nm or less. The commercially available ICAP instruments are generally controlled by built-in computers. The major disadvantage to the use of atomic spectroscopy with plasmas is the relatively high cost of the instrumentation. About 70 elements have been assayed (usually by AES) with the aid of plasmas.

Agilent Technologies has recently introduced a commercial AES instrument which is similar to the previously described ICP spectrometer, but which utilizes a magneticallycoupled, nitrogen plasma operating in the microwave region at 2.5 Ghz and a charged coupled device (CCD) detector. CCD detectors are described in a later chapter. An industrial magnetron is used in place of the induction coils. Nitrogen for the plasma is formed in a nitrogen generator. The nitrogen generator removes oxygen from compressed air. The result is a device that indirectly utilizes compressed air rather than argon to aspirate the sample into the torch and to form the plasma which causes excitation of the sample.

Wavelength Selectors for AFS

Most instruments that are used for AFS use monochromators to select the wavelength. The monochromator can be manually controlled or, more often, can be operated by a computer. If a line source is used the monochromator can be a low-resolution device. If a continuous source or a plasma cell is used, the monochromator must have high resolution. Filters also have been used as wavelength selectors for AFS. When resonant detectors are used, a separate wavelength selector is not required.

Detectors for AFS

Photomultiplier tubes are used most often as detectors for AFS. A description of photomultiplier tubes was included in Chapter 6. Silicon solid-state detectors occasionally are used in AFS. Because they are not as sensitive as PM tubes, they are not used as often. Resonant detectors have rarely been used for AFS. Readout devices for AFS include recorders, analog and digital meters, computer controlled devices, and video displays.

Theory of AFS

The theoretical equations that describe the intensity of emitted radiation (the *radiance*) vary depending upon the experimental conditions. The nature of the source of exciting radiation and the relative concentration of analyte atoms in the cell are two factors that must be considered when deriving the equations. If a continuous source is used, more than one excited state can be populated simultaneously and more than one fluorescent line can be observed. With most nonplasma cells, the vast majority of absorptive transitions take place from the ground state. Other transitions, however, are also possible. Several different types of fluorescence (Fig. 8-1) can occur. In addition, deexcitation can occur by radiationless transitions such as those that are caused by collisional deactivation. All of those factors complicate the theory of AFS.

The equations that are listed in this section are for the relatively simple case in which the fluorescing element contains only a single excited state. The absorptive transition is from the ground state to the excited state. Fluorescence occurs during the reverse transition; i.e., resonant fluorescence is the only considered type. Furthermore, it is assumed that a narrow line source is not used for the excitation; that the analyte atoms are uniformly distributed throughout the cell; that the cell has the same temperature throughout; that thermodynamic equilibrium is established within the cell; that the concentration of analyte atoms within the cell is small; and that the intensity of radiation from the source is uniform throughout the cell. In spite of the restrictions, the presented theory is illustrative of AFS and does apply to many actual cases. A more comprehensive treatment of the theory of AFS can be found in References 7 through 9, and in the references that are listed in those references.

Before fluorescence can occur, the element must absorb energy from the source. Consequently, the fluorescent radiance B_F is directly proportional to the total power P_a absorbed by analyte atoms within the cell. After absorption has occurred, not all of the excited atoms fluoresce. A portion of the excited atoms lose energy and return to the ground state by radiationless processes. The *fluorescent power efficiency* Y_{21} is the ratio of the power (watts) of fluorescent radiation during the energetic transition from energy state 2 to state 1 to the power of absorbed radiation. The fluorescent power efficiency is sometimes termed the quantum efficiency for the fluorescence. The fluorescent radiance is proportional to both P_a and Y_{21}:

$$B_F \propto Y_{21} P_a \tag{8-1}$$

The power absorbed by the analyte is a function of the power of radiation emitted by the source and the integrated absorptive coefficient over the width of the absorptive line:

$$P_a = E_{v12} \int k_v dv \tag{8-2}$$

In Eq. (8-2) P_a is in units of watts; E_{v12} is the spectral irradiance from the source at the frequency of absorption in units of watts per square meter-hertz; and k_v is the coefficient of absorption at the frequency v.

The proportionality (8-1) can be changed to an equation by inserting a proportionality constant. The value of the proportionality constant is $l/4\pi$ where l is the pathlength of the cell in the direction of the detector (meters) and the fluorescent radiance in the direction of the detector is the emitted power (watts) per area (square meters) per solid angle (steradians). Substituting Eq. (8-2) and the proportionality constant into the proportionality (8-1) yields

$$B_F = \frac{1}{4\pi} Y_{21} E_{v12} \int k_v dv \tag{8-3}$$

The integrated coefficient of absorption can be shown[9] to be equivalent to the terms in the following equation:

$$\int k_v \, dv = \frac{nhv_{12}B_{12}}{c} \frac{E^s_{v12}}{E^s_{v12} + E_{v12}} \tag{8-4}$$

In Eq. (8-4) n is the number of analyte atoms per cubic meter (the atomic concentration) in the cell; v_{12} is the frequency of the absorbed radiation; B_{12} is the Einstein coefficient of induced absorption (cubic meter-watt-hertz); c is the speed of EMR in a vacuum; and E^s_{v12} is the saturated spectral irradiance (watts per square meter-hertz). The saturated spectral irradiance is the irradiance from the source that results in a fluorescent radiance that is one-half of the maximal possible radiance. At saturation the number of the atoms in the ground state is equal to the number of atoms in the excited state. Substitution of the value from Eq. (8-4) into Eq. (8-3) gives

$$B_F = \frac{lY_{21}E_{\nu 12}nh\nu_{12}B_{12}}{4\pi c} \frac{E^{s}_{\nu 12}}{E^{s}_{\nu 12} + E_{\nu 12}} \tag{8-5}$$

From Eq. (8-5) it is apparent that the fluorescent radiance is directly proportional to the concentration n of analyte atoms in the cell. Under controlled conditions n is proportional to the concentration of the analyte element in the sample. The fluorescent radiance is usually proportional to concentration whenever the product of k_ν and l is at least 0.05. Consequently linear working curves are often obtained with AFS.

The working-curve method is the preferred method of quantitative analysis using AFS. Equation (8-5) was derived assuming a low concentration of analyte in the cell. At high concentrations of analyte the fluoresced radiation can be reabsorbed by atoms of analyte within the cell. At high concentrations the fluorescent radiance is proportional to the square root of concentration rather than directly proportional to the concentration.

When $E_{\nu 12}$ is much smaller than $E^{s}_{\nu 12}$ Eq. (8-5) simplifies to

$$B_F = \frac{lY_{21}\, E_{\nu 12}\, nh\nu_{12}\, B_{12}}{4\pi c} \tag{8-6}$$

From Eq. (8-6) it is apparent that the fluorescent radiance is directly proportional to the irradiance $E_{\nu 12}$ from the source. For most radiative sources (excluding some lasers) Eq. (8-6) is obeyed. Equation (8-6) can be written in the simplified form that was described in Chapter 5 [Eq. (5-23)] for generalized luminescence.

Analysis with AFS

Quantitative analysis with AFS is generally done with the working-curve method. Because the working curves are not always linear it is not safe to assume that the standard-addition technique can be used. If the linearity of the working curve has been demonstrated, the standard-addition technique can be used.

At least 58 elements have been assayed successfully with AFS. Most of the assays utilized resonant fluorescence. A list of some of the elements that have been assayed by resonant AFS and the wavelengths that were used for the assays are given in Table 8-3. Several other elements have been assayed with direct-line or stepwise AFS. A list of some of the elements that have been assayed by nonresonant AFS and the wavelengths of the observed fluorescence are given in Table 8-4. Limits of detection with AFS have been shown to be better than those with AAS in some cases but worse in other cases. The high intensity that is available from lasers should lower the limit of detection for AFS to values that are generally less than the corresponding values for AAS.

Table 8-3 A partial listing of elements that have been assayed by resonant fluorescence

Element	Wavelength, nm	Element	Wavelength, nm
Ag	328.1	Mg	285.2
As	193.7	Mn	279.5
Au	249.7, 267.6	Mo	313.3
Ba^+	455.4	Na	589.0
Be	234.0	Ni	232.0
Bi	302.5, 306.8	Pb	283.3
Ca	422.7	Rh	369.2
Co	240.7	Sb	217.6, 231.1
Cr	357.9	Se	196.0
Cu	324.8	Sr	460.7
Fe	248.3, 372.0	Te	214.3
Ge	265.2	Tl	377.6
Hg	253.7	Zn	213.9
In	410.5		

Table 8-4 A partial listing of elements that have been assayed by direct-line AFS (method 1) and by stepwise AFS (method 2)

Element	Emitted wavelength, nm	Method
Al	309.2	1
Bi	302.5	1
Fe	373.5	1
Ga	417.2	1
In	451.1	1
Pb	405.8	1
Pd	340.5	2
Se	204.0, 206.3	1
Si	251.6	2
Sn	317.5, 380.1	1
Sn	303.4	2
Ti	335.5	2
V	318.4	2

Interferences with AFS

Generally the interferences that are encountered with AFS are the same as those described in Chapter 6 for AAS. Chemical and ionization interferences are identical to those observed with AAS. Spectral interferences are most likely to occur when a continuous source is used. When a line source is used spectral interferences can occur only when the absorptive line of the interfering element or molecule is identical to that from the source and when the interference can fluoresce at the same wavelength at which the fluorescence of the analyte is observed. A few of the elements[10] that fluoresce at nearly the same wavelength include As and Cd (228.8 nm), Co and In (252.1 nm), Co and Hg (253.6 nm), Fe and Zn (213.9 nm), Ni and Sn (231.1 nm), Sb and Pb (217.0 nm), Al and V (308.2 nm), V and Si (250.7 nm), and Zn and Cu (213.9 nm).

The primary interference that is observed in AFS is caused by radiative scattering. Instruments that use continuous sources are particularly susceptible to scattering. Instruments that use furnaces also have been proven to be highly susceptible to scattering interferences. The effect of scattering can be minimized by matrix-matching the standards to the sample, by using a different spectral line for the assay, by using the Zeeman effect, and by any of several other methods. Resonant fluorescence is more susceptible to scattering interference than the other forms of AFS because scattered radiation occurs at the same wavelength as that of the incident radiation.

RESONANT IONIZATION SPECTROSCOPY

The advent of lasers has resulted in the development of several analytical techniques that utilize the high powers of the devices. The use of the laser has made it possible to determine extremely low concentrations of many elements. *Resonant ionization spectroscopy* (RIS) and *laser-enhanced ionization spectroscopy* (LEIS) are two analytical techniques that use lasers and that are capable of ultratrace analysis. In some cases RIS can be used to detect a single atom.

RIS was first used in 1974. The operating principles of RIS are relatively simple. A pulsed laser is used to excite atoms of the assayed element from the ground state to an excited electronic level. When the atom is excited from the ground state the technique is *resonant* spectroscopy. Before the atom can return to the ground state it absorbs sufficient energy from the same or a second laser to remove the excited electron from the atom, thereby resulting in an electron and a positively charged ion. Either the electron or the ion is detected with a device that responds to a single electron or ion.

The wavelength at which the laser operates is adjusted to correspond to the energetic difference between the ground and an excited state for a particular atom. Consequently the technique is selective; i.e., only atoms that are in the excited state are ionized and the only excited atoms are those that have an absorptive line that corresponds to that from the laser. RIS is more sensitive than AAS, AFS, and AES because it monitors all ions or electrons that are formed during the ionization and because all analyte atoms within the beam of the laser are ionized. AFS and AES measure only a portion of the emitted photons and only some of the excited atoms emit photons. Before the advent of lasers RIS was not possible because the radiative power was not available that could permit all excited atoms within the beam from the source to be ionized prior to loss of energy through fluorescence or radiationless deactivation.

Several ionization, mechanisms make it possible to use RIS to assay all of the elements except helium and neon. Those two elements presently cannot be assayed with RIS because of the lack of lasers at the energetic levels that are required to excite the atoms from the ground state to an excited state. The types of energetic transitions that occur with each of the RIS mechanisms that have been utilized as of the present time are summarized in Fig. 8-10.

In Fig. 8-10 the ionization energy of each atom is represented by the dashed line. In scheme 1 the same laser is used to excite the atom and to ionize the excited atom. Ionization occurs whenever sufficient energy is applied to exceed the ionization energy of the atom. Scheme 2 also requires a single laser; however the frequency of the laser must be optically doubled in order to provide sufficient energy to raise the atom to the excited state.

Schemes 3 and 4 require two lasers that are adjusted to different wavelengths. In scheme 3 one laser is used to raise the atom to a lower level excited state. The second laser raises the atom from the lower level excited state to a higher level excited state. Ionization occurs after absorption of energy from one or both of the lasers by the atom in the higher excited state. In scheme 4 the doubled frequency from one laser raises the atom to the first excited state, and a second laser that is adjusted to a different wavelength is used to further raise the atom to a second excited state. The radiation from one or both of the lasers subsequently is used to ionize the excited atom.

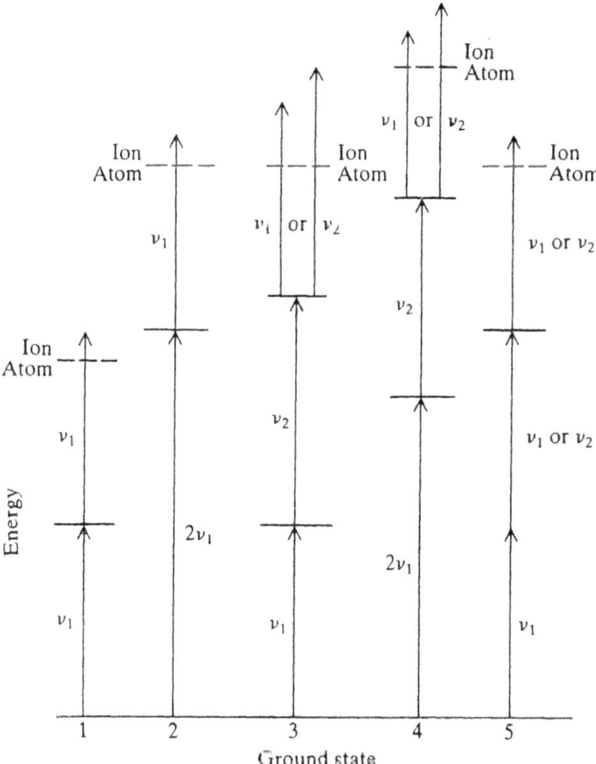

Figure 8-10 Energetic transitions during various forms of resonant ionization spectroscopy. The solid horizontal lines represent bound-electron energy levels. The dashed lines represent the ionization energies of the elements.

In scheme 5 two photons from the same or different lasers are simultaneously absorbed and result in excitation of the atom. Ionization from the excited state occurs as in scheme 3. A list of the elements and the RIS mechanism that can be used for the analysis of each with presently available lasers is given in Table 8-5.

The detectors that are used with RIS are designed to electrically measure charged particles. The particles can be either electrons or ions. Normally the detector is a *pulsed ionization chamber,* a *proportional counter,* a *Geiger-Muller counter,* or an *electron multiplier*. In each case gaseous atoms of the

Table 8-5 The elements that can be assayed with a known mechanism of resonant ionization spectroscopy

The missing elements except for He and Ne can be assayed with RIS although the mechanism is uncertain. The numbers correspond to the numbered mechanisms in Fig. 8-10. The elements are listed alphabetically according to name.

Element	Mechanism	Element	Mechanism	Element	Mechanism
Al	1	Ho	I	Rn	5
Am	2	H	5	Re	2
Sb	4	In	1	Rh	2
Ar	5	I	5	Rb	1
As	5	Ir	3	Ru	2
Ba	2	Fe	2	Sm	1
Bk	1	Kr	5	Sc	2
Be	4	La	1	Se	5
Bi	4	Pb	4	Si	4
B	4	Li	2	Ag	4
Br	5	Lu	1	Na	2
Cd	4	Mg	3	Sr	2
Ca	2	Mn	2	S	5
C	5	Hg	4	Ta	3
Ce	1	Mo	2	Tc	3
Cs	1	Nd	1	Te	4
Cl	5	Np	1	Tb	1
Cr	2	Ni	2	Tl	1
Co	2	Nb	2	Tm	1
Cu	4	N	5	Sn	2
Cm	1	Os	3	Ti	2
Dy	1	O	5	W	3
Es	2	Pd	4	U	2
Er	1	P	5	V	2
Eu	2	Pt	4	Xe	5
F	5	Pu	1	Yb	2
Gd	1	Po	4	Y	3
Ga	1	K	1	Zn	4
Ge	4	Pr	1	Zr	3
Au	4	Ra	2		

***Source*:** Data are taken from References 11 and 12.

analyte are enclosed in the detector chamber and the laser(s) is fired through the chamber to create the ions and electrons.

In the pulsed ionization chamber the laser is fired between a positive and a negative plate. The ions and electrons that are formed during the firing are attracted to the oppositely charged electrodes where they cause a current flow in an external circuit. The current flow is monitored. The pulsed ionization chamber does not amplify the current that results from the ionization. As few as 200 electrons can be measured with a pulsed ionization chamber.

In a proportional counter the electrons that are freed during the laser-induced ionization are allowed to drift into a large electric field between two electrodes. The laser is not fired between the electrodes as in the pulsed ionization chamber. Upon striking a filler gas within the electric field, ions and electrons are formed. The electrons are attracted to a large positively charged electrode and measured with external circuitry. The use of a proportional counter can result in an amplification up to 10^6. A diagram of the apparatus that is used for RIS with a proportional counter is shown in Fig. 8-11.

A Geiger-Müller counter is similar to a proportional counter except that the size of the output pulse from the detector is independent of the number of electrons that caused the pulse. An electron multiplier is another detector which functions in a manner analogous to the way in which the dynodes function in a photomultiplier tube. An electron multiplier has the advantage of being able to monitor positive ions.

A disadvantage of RIS is that the analyte must be a gas in the path of the laser. Several methods of changing an analyte to gaseous atoms have been used with RIS. The sketch in Fig. 8-12 illustrates the manner in which one group of workers[13] were able to perform an assay for sodium in a solid single crystal of silicon. The technique that they used is *laser-ablative resonant ionization spectroscopy* (LARIS). A laser is used to ablate (remove) atoms from the solid sample. A second laser passes through the cloud of atomic vapor for the RIS

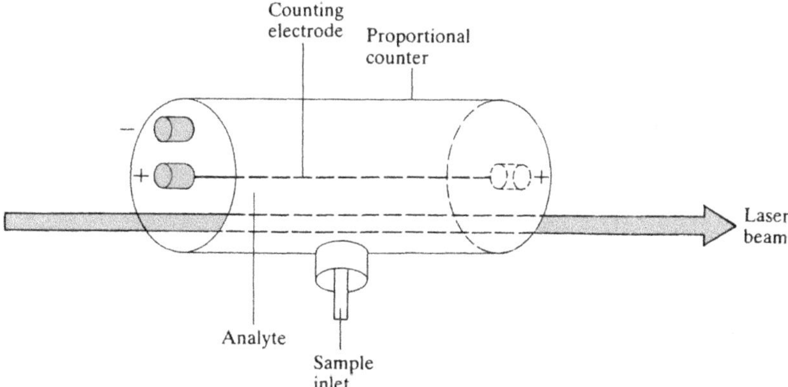

Figure 8-11 A sketch of an RIS apparatus with a proportional counter.

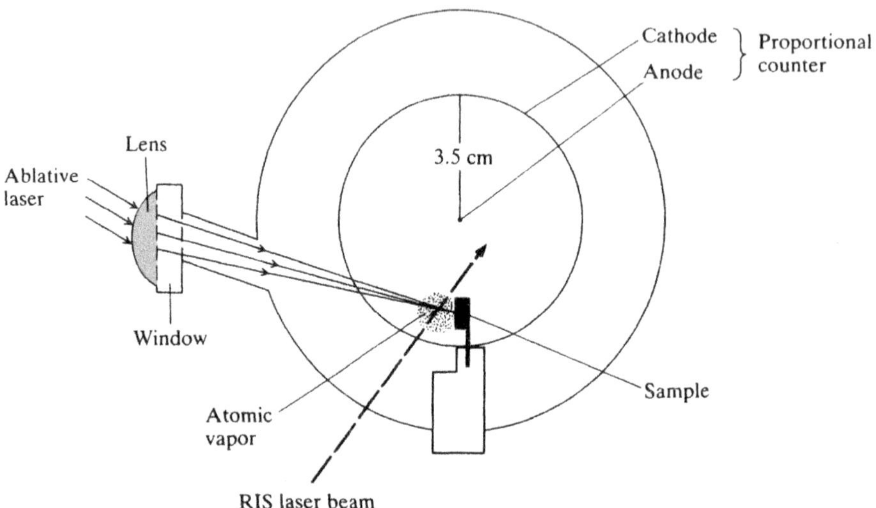

Figure 8-12 A sketch of the apparatus used for laser-ablative resonant ionization spectroscopy (LARIS). Radiation from the RIS laser enters the apparatus perpendicularly to the page through the cylinder that is the sample chamber.

assay. A proportional counter is used as the detector. RIS holds great promise for analysis at the ultratrace level. At present, however, the technique has not been sufficiently developed to be useful for routine analyses of complex samples. The detectors which are used for RIS are essentially identical to those used for radiochemical analysis. More information is available in Chapter 20.

LASER-ENHANCED IONIZATION SPECTROSCOPY

Laser-enhanced ionization spectroscopy (LEIS) uses a flame to atomize the sample. A laser serves to excite the analyte atoms to a higher electron state as in RIS. The excited atoms are ionized by the heat in the flame (collisional ionization) and the ions are electrically monitored with electrodes placed across the flame. Flame conditions are chosen that yield negligible thermal ionization from the ground state of the analyte and its matrix. The laser selectively excites analyte atoms as in RIS. Because the excited atoms require less energy to be ionized than the ground-state atoms, the excited atoms are more readily ionized by the heat from the flame. Consequently, the ions and electrons that are monitored by the electrodes primarily correspond to those from the analyte.

As with RIS, LEIS can occur through any of several mechanisms. The energy changes that are associated with the several mechanisms are shown in Fig. 8-13. Mechanism 1 corresponds to resonant (from the ground state) laser

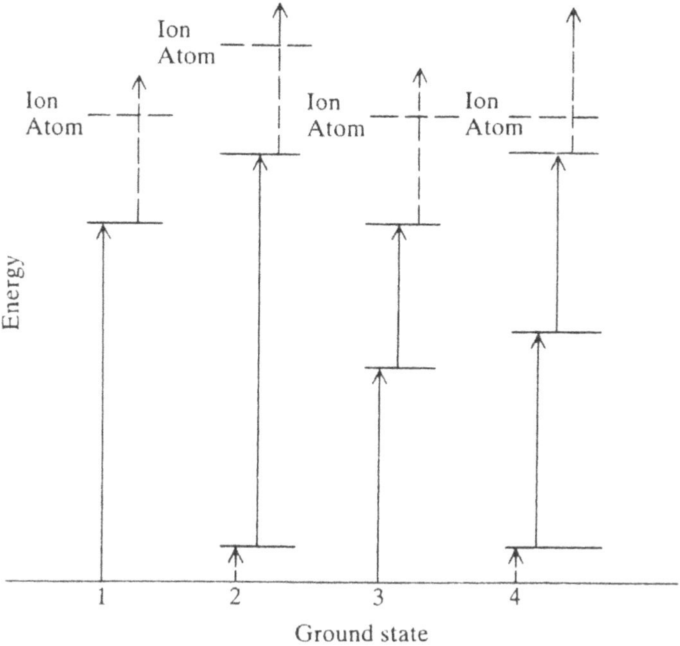

Figure 8-13 Energetic transitions during various forms of laser-enhanced ionization spectroscopy. The solid horizontal lines represent bound-electron energetic levels. The horizontal dashed lines represent the ionization energies of the elements. The solid vertical arrows represent transitions owing to absorption of radiation from a laser. The dashed vertical arrows represent thermal transitions.

excitation followed by thermal ionization. Mechanism 2 is nonresonant laser excitation after initial thermal excitation. Thermal ionization follows the nonresonant laser excitation. In scheme 3 resonant laser excitation is followed by nonresonant laser excitation and thermal ionization. The laser excitation in scheme 3 is termed *stepwise resonant laser excitation*. Mechanism 4 consists of sequential thermal excitation, stepwise laser excitation, and thermal ionization. The laser excitation in mechanism 4 is *stepwise nonresonant laser excitation*. A list of some of the approximately 70 elements that have been assayed by LEIS, the type of mechanism followed, and the wavelengths that were used for the assays are found in Table 8-6. The data in Table 8-6 were taken from Reference 14.

A diagram of the apparatus that is used for LEIS is shown in Fig. 8-14. Premix burners that are identical to those that are used in AAS are used in LEIS. Generally the flame gases are either hydrogen-air, acetylene-air, or acetylene-nitrous oxide. As in AAS the fuel and oxidant are chosen depending upon the sample. In LEIS it is particularly important to choose the coolest flame

Table 8-6 A list of some of the elements that have been assayed with LEIS

The first wavelength (nm) of each couple corresponds to the lower energy-level transition. The numbers refer to the mechanisms illustrated in Fig. 8-13. The elements are listed alphabetically according to name

Element	Mechanism	Wavelength 1	Wavelength 2
Al	1	308.2	
Ba	1	307.2	
Bi	1	306.8	
Cd	3	228.8	466.2
Ca	2	300.7	
Cr	2	298.6	
Co	3	252.1	591.7
Cu	3	324.8	453.1
Ga	1	287.4	
Au	3	242.8	479.3
In	1	303.9	
Fe	1	302.1	
Pb	3	283.3	600.2
Li	1	670.8	
Mg	1	285.2	
Mn	1	279.5	
Ni	4	300.2	576.8
K	1	294.3	
Rb	1	572.4	
Ag	1	328.1	
Na	1	285.3	
Sr	1	460.7	
Tl	2	291.8	
Sn	4	284.0	597.0

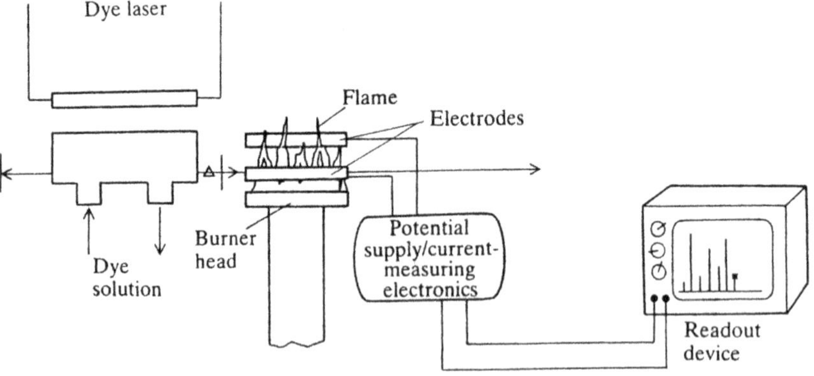

Figure 8-14 A diagram of the apparatus that is used for LEIS.

that will result in atomization of the sample. Hotter flames cause increased ionization from the ground state with a resultant increase in background signal. The hydrogen-air flame is less noisy than the other two types, but it is not sufficiently hot to atomize many samples. The acetylene-air flame is used most often. The acetylene-nitrous oxide flame is used for analytes that form refractory oxides that are difficult to atomize.

The electrodes that are used to monitor the ions in the flame can consist of either parallel anodic and cathodic plates (or rods) on opposite sides of the flame or a set of two cathodic plates on opposite sides of the flame and an anodic burner head. The electrodes are usually made of tungsten, iridium, or stainless steel. In any case a potential is applied between the cathode and the anode that is sufficient to make the electrostatic attraction between the electrodes and the ions that are formed within the flame the only significant cause for migration of the ions to the electrodes. The potential is sufficiently large to make the rate of diffusion to the electrodes, the upward flow of ions in the flame gases, and the recombination of positive and negative ions to form neutral species negligible relative to the rate of ionic migration to the electrodes. In short, the potential is sufficiently high to catch every ion that is formed before it can leave the flame or form neutral species. Typically the potential drop between the cathode and the anode is between 1000 and 2000 V.

Usually the laser is chosen to emit at a wavelength that is characteristic of a resonant absorptive line of the analyte. Tunable dye lasers are useful because of their many available emissive wavelengths. Pulsed dye lasers are used most often because they are capable of emitting at all required wavelengths above 217 nm. Continuous dye lasers also can be used but they presently are not capable of achieving emissive wavelengths that are as short as those of the pulsed lasers. The pulsed dye lasers can be pumped with a flashlamp, a nitrogen laser, or an Nd : YAG laser.

Quantitative analysis with LEIS is done with the working-curve method or the standard-addition technique. The lower limit of detection of the technique varies with the analyte from about one part per trillion to about five parts per billion. Under normal flame conditions with a typical laser, the theoretical lower limit of detection has been estimated to be about one part per trillion.[14] LEIS is normally useful over a concentration range of between 10^4 and 10^5.

Most of the interferences that are encountered with LEIS are identical to those observed with flame AAS. Chemical interferences such as the formation of refractory oxides, spectral interferences in which the laser excites an element in addition to the analyte, and ionization interferences are all possible. Refractory oxide formation can be minimized as described in Chapter 6. Spectral interferences in LEIS generally can be prevented either by changing to a different resonant absorptive line of the analyte or by changing to a two-wavelength absorptive mechanism (scheme 3 or 4 in Fig. 8-13).

The major interference in LEIS is caused by ionization from the ground state that results from the heat of the flame. Easily ionized nonanalyte elements such as the group 1 elements in the periodic table are particularly prone to ionization. Ionization of nonanalyte elements causes two problems. First, it increases the background signal in the flame and, second, the increased ions can alter the electric field strength between the electrodes. An altered electric field strength causes less attraction for ions of the analyte. Further information concerning LEIS can be found in the References.

IMPORTANT TERMS

Anti-Stokes fluorescence
Atomic fluorescent spectrometry
Cavity dumping
Direct-line fluorescence
Excimer laser
Fluorescent power efficiency
Initial radiative zone
Laser
Laser-ablative resonant ionization spectroscopy
Laser-enhanced ionization
Lasing medium
Mode locking
Normal analytical zone
Population inversion
Preheating zone
Pumping
Q-switching
Radiance
Resonant fluorescence
Resonant ionization spectroscopy
Sensitized fluorescence
Stepwise fluorescence
Stimulated emission
Stokes fluorescence
Thermally assisted fluorescence
Tunable laser

PROBLEMS

8-1 What is a population inversion and why is it a requirement for amplification in a laser?

8-2 Sketch an electrically pumped laser.

8-3 If the fluorescent power efficiency increased by 15 percent what would be the effect on the fluorescent radiance?

8-4 If the fluorescent power efficiency decreased by 11 percent, the atomic concentration in the cell increased by 25 percent, and the Einstein coefficient of induced absorption increased by 12 percent, determine the effect on the fluorescent radiance.

8-5 Several standard solutions of cadmium chloride were prepared and the fluorescent radiance of each solution was measured with a photomultiplier tube at 228.8 nm. The fluorescent radiance of a sample solution was measured under the same experimental conditions. Determine the concentration *C* of the sample.

$C \times 10^{-5}$ *M*	Relative fluorescent radiance
2.50	13.6
5.00	30.2
7.50	45.3
10.00	60.7
12.50	75.4
Sample	50.5

REFERENCES

1. Wright, J. C., and M. J. Wirth: *Anal. Chem.,* **52**: 988A(1980).
2. Wright, J. C., and M. J. Wirth: *Anal. Chem.,* **52**: 1087A(1980).
3. Reednick, J.: *Amer. Lab.,* **11**(3): 53(1979).
4. Epstein, M. S., S. Nikdel, N. Omenetto, J. Bradshaw, and J. D. Winefordner: *Anal. Chem.,* **51**: 2071(1979).
5. Kosinski, M. A., H. Uchida, and J. D. Winefordner: *Anal. Chem.,* **55**: 688(1983).
6. Faires, L. M.: *Amer. Lab.,* **14**(11): 16(1982).
7. Smith, R., in J. D. Winefordner (ed.): *Spectrochemical Methods* of *Analysis, Advances in Analytical Chemistry and Instrumentation,* vol. 9, Wiley, New York, 1971, pp. 235-282.
8. Van Loon, J. C.: *Anal. Chem.,* **53**: 332A(1981).
9. Winefordner, J. D.: *J. Chem. Educ.,* **55**: 72(1978).
10. Demers, D. R., and C. D. Allemand: *Anal. Chem.,* **53**: 1915(1981).
11. Hurst, G. S.: *J. Chem. Educ.,* **59**: 895(1982).
12. Young, J. P., G. S. Hurst, S. D. Kramer, and M. G. Payne: *Anal. Chem.,* **51** : 1050A(1979).
13. Mayo, S., T. B. Lucatorto, and G. G. Luther: *Anal. Chem.,* **54**: 553(1982).
14. Travis, J. C.: *J. Chem. Educ.,* **59**: 909(1982).

BIBLIOGRAPHY

Hurst, G. S.: *Anal. Chem.,* **53**: 1448A(1981).

Travis, J. C., G. C. Turk, and R. B. Green: *Anal. Chem.,* **54**: 1006A(1982).

CHAPTER

NINE

ULTRAVIOLET-VISIBLE SPECTROSCOPY OF POLYATOMIC SPECIES

A general description of the mechanism by which absorption in the ultraviolet uv and visible regions occurs was included in Chapter 5. A brief review of the appropriate sections of Chapter 5 prior to the study of Chapter 9 is suggested. Unlike the monoatomic species that were described in Chapters 6 through 8, polyatomic species possess vibrational and rotational energetic levels in addition to electron energetic levels (Fig. 5-7). Ultraviolet-visible absorption usually corresponds to excitation of an electron from the ground electron state to a higher electron state. Each photon of the incident radiation must possess energy that corresponds to the energetic difference between the electron levels. The energetic transition can occur from any of the rotational and vibrational levels that are associated with the lower electron state to any of the rotational and vibrational levels that are associated with the higher electron state.

Because many vibrational and rotational levels can be simultaneously occupied in different molecules and because the incident radiation that is used in most spectrophotometers contains a large number of photons, many vibrational and rotational levels that are associated with the higher electron state can be simultaneously populated in different molecules. As the wavelength of the incident radiation is altered (scanned), a molecule can be excited from the same electron, vibrational, and rotational levels to a single excited electron level, but to different vibrational and rotational levels. Because the energetic differences between the vibrational and rotational levels that are associated with a particular electron level are small, the many possible changes in energy between the vibrational and rotational levels of the two electron levels overlap, causing broad absorptive bands in the ultraviolet-visible region for polyatomic species.

Table 9-1 Divisions of the ultraviolet and visible spectral regions

In the visible region the first color is absorbed and the color in brackets is the observed (transmitted) color

Region	Wavelength range, nm
Vacuum-ultraviolet	10-200
Near-ultraviolet	200-380
Violet (yellow-green)	380-435
Blue (yellow)	435-480
Green-blue (orange-red)	480-500
Green (purple)	500-560
Yellow-green (violet)	560-580
Yellow (blue)	580-595
Orange (green-blue)	595-650
Red (blue-green)	650-780

Typically the absorptive bandwidths of polyatomic species are 50 nm or more. Molecular bandwidths are about 10,000 times broader than those of atoms and monoatomic ions because the monoatomic species do not possess vibrational and rotational levels. The wavelength of greatest absorption (the peak) for a particular electron transition corresponds to the transition from the vibrational and rotational levels in the lower electron state that are occupied by the greatest number of sample molecules to the most stable of the vibrational and rotational levels in the higher electron state.

The ultraviolet and visible regions can be divided into smaller regions as shown in Table 9-1. In the visible region the radiation is divided according to absorbed or transmitted colors. The boundaries between regions are not rigidly defined; i.e., the wavelengths listed in Table 9-1 are approximate.

ELECTRON TRANSITIONS

In most cases absorption occurs when a ground-state electron in a bonding or nonbonding molecular orbital is excited by incident radiation to a higher molecular orbital. Several types of molecular orbitals are possible. A *sigma* (σ) *orbital* is an orbital that has cylindrically symmetrical electron density around the inter-nuclear axis. Sigma orbitals are those that correspond to single bonds between atoms and to one of the bonds in double or triple bonds. Sigma orbitals usually result from the overlap of two *s* atomic orbitals, an *s* orbital with a *p* orbital, or the head-to-head overlap of two *p* atomic orbitals. If the formation of the molecular orbital increases the stability of the molecule as compared to the separated atoms, the orbital is a *bonding molecular orbital.* A molecule with a single bonding molecular orbital contains less energy than the two separated atoms that formed the molecule.

If the formation of the molecular orbital decreases the stability of the molecule, the orbital is an *antibonding molecular orbital.* Antibonding molecular orbitals are symbolically represented by placing a superscript asterisk (called a "star") on the symbol for the molecular orbital. As an example, σ^* (sigma star) is the symbol for an antibonding sigma orbital. The absence of the asterisk indicates that the orbital is a bonding molecular orbital. A σ orbital is a bonding sigma orbital.

In addition to bonding and antibonding sigma orbitals it is possible to form molecular orbitals that do not have cylindrically symmetrical electron density around the internuclear axis. That type of molecular orbital is a *pi* (π) *orbital.* Pi molecular orbitals are usually formed from side-by-side overlap of atomic *p* orbitals. A cloud of electrons exists above and below the internuclear axis, but not completely around the axis. Bonding pi orbitals are represented by π and antibonding pi orbitals by π^* (pi star). Between doubly bonded carbon atoms one of the bonds is pi and between triply bonded carbon atoms two of the bonds are pi. The remaining bond in each of the molecules is sigma.

The last category of molecular orbital neither contributes to the stability nor lack of stability of a molecule. It consists of unshared electron pairs. The energy that is associated with the electrons in the "bond" is identical to the energy of the same electrons in the individual atoms that formed the molecule. An orbital of that type is a *nonbonding molecular orbital* and is symbolized by *n*. An electron in an *n* orbital can be thought of as being in an atomic orbital that is not associated with the molecule.

Not all electron transitions between molecular orbitals are equally probable during absorption of ultraviolet and visible radiation. In order for the electric field portion of the incident EMR to be absorbed by a molecule, the energy of the EMR must correspond to the energetic difference between the electron levels in the molecule, and the dipole of the molecule must change. If the altered electron distribution is symmetrical with the original electron distribution, no change in the dipole of the molecule occurs and the electron transition is "forbidden." If a net change in the dipole of the molecule does not occur during an electron transition, no electrical work can be done on the molecule and the absorption cannot occur. More detailed descriptions of the allowed transitions in molecules and the associated selection rules can be found in the appropriate sections of References 1 through 3. Selection rules indicate the more probable transitions.

Of the possible transitions in molecules, those that result in absorption of ultraviolet or visible radiation are the n to π^*, π to π^*, n to σ^*, and σ to σ^* transitions. The σ to σ^* transitions require the greatest energy, and generally occur in the vacuum-ultraviolet at wavelength below about 150 nm. Typically C—H bonds absorb at about 125 nm and C—C bonds at about 135 nm. Because the vacuum-ultraviolet region is not accessible in most ultraviolet-visible spectrophotometers, the transitions generally are not of use for routine analysis. Because all single bonds are sigma bonds and because transitions from sigma bonds occur at energies that are beyond the range of most instruments, compounds with only single bonds do not have measurable ultraviolet-visible absorptive bands. The portion of a molecule that absorbs radiation is a *chromophore*. In the ultraviolet-visible region, strong chromophores contain some degree of unsaturation.

The n to σ^* transitions generally require less energy than σ to σ^* transitions. Typically n to σ^* transitions occur in the long-wavelength end of the vacuum-ultraviolet region and the short-wavelength end of the near-ultraviolet region (150 to 250 nm). Many of those transitions are observable with ordinary ultraviolet-visible spectrophotometers. Often an n to σ^* transition occurs when an organic molecule contains a heteroatom (N, O, S, F, Cl, Br, or I). An electron that is in a p orbital of the heteroatom is excited to a σ^* orbital of the molecule. Absorptive bands that are characteristic of alkyl halides, amines, ethers, and sulfides correspond to n to σ^* transitions. As an example, methyl iodide in hexane has an n to σ^* absorptive band at 258 nm (molar absorptivity is 380).

The π to π^* and the n to π^* transitions occur in the near-ultraviolet and visible regions (180 to 700 nm). Because those transitions occur in an easily observable spectral region, they are most useful for analysis. In both cases the chromophore must possess a degree of unsaturation in order to provide the π^* molecular orbital. Generally peaks that are associated with π to π^* transitions are highly absorptive (large absorptivities), thereby making possible quantitative assays at low concentrations. Typical molar absorptivities range from 1000 to 15,000 L/(cm·mol). Molar absorptivities of n to π^* transitions typically range from 10 to 100 L/(cm·mol).

Molecules that exhibit π to π^* transitions contain double or triple bonds or aromatic rings. An electron in a pi orbital is excited to a pi-star orbital. Molecules that exhibit n to π^* transitions include molecules that contain C=O, N=N, NO, and NO_2. An electron in a nonbonding atomic orbital that is associated with a heteroatom is excited to an antibonding pi orbital that is associated with a double or triple bond in the molecule.

QUALITATIVE ANALYSIS

Because different chromophores have uv-visible absorptive maxima at different wavelengths, it sometimes is possible to identify the absorbing molecule by comparing the wavelength of the absorptive maximum from the spectrum of a sample with the wavelengths of the absorptive maxima of known substances. Generally, a spectrum consisting of a plot of absorbance, percent transmittance, or log of absorbance as a function of wavelength is automatically obtained using a scanning spectrophotometer. The spectrum of a molecule that contains more than one chromophore usually corresponds to the sum of the expected spectra for the chromophores. If the chromophores are separated by at least two single bonds, the chromophores behave independently. If two chromophores in the same molecule are identical, a single absorptive peak is observed that has an absorbance that is approximately twice that which would be expected for either chromophore. If the chromophores are not identical, a peak (often overlapping) is observed for each.

Because absorptive bands in the uv-visible region are often quite broad, many substances have absorptive bands that overlap; and a relatively large number of compounds have absorptive maxima near each other. It usually is not safe to conclude, from the uv-visible spectrum alone, that an unknown substance even contains a particular chromophore. If the unknown substance is known to be one of a limited number of compounds, however, and if all of those compounds have absorptive maxima that differ from each other by at least 10 nm, then the spectrum of the sample can be used for qualitative analysis.

The wavelength of the absorptive maximum for a compound depends upon both the chromophore in the molecule and upon the environment of the chromophore. Functional groups that are not chromophores but that are in the same molecule as the chromophore have an effect upon the wavelength of the absorptive maximum of the chromophore. An *auxochrome* is a functional group on an absorbing molecule or ion that does not itself absorb radiation in the ultraviolet-visible region but can cause the absorptive maximum of the molecule to change wavelengths. Generally auxochromes are functional groups, such as NH, OH, SH, and halogens that contain unshared electrons.

The presence of an auxochrome in a molecule that contains a chromophore usually causes the absorptive maximum to shift to a longer wavelength (lower energy). That is a *bathochromic* or *red shift*. A *hypsochromic shift* is to a shorter wavelength. If the absorptivity of a chromophore increases with the addition of a substituent to the molecule, the increased absorbance is a *hyperchromic shift*. Hyperchromism is often caused by increased conjugation of single and double bonds within the molecule. A decrease in absorbance owing to the addition of a substituent to a molecule is a *hypochromic shift*. The effect of extended conjugation on absorptive maxima is illustrated with alkenes. The absorptive maximum increases in both wavelength and absorbance from ethylene [absorptive maximum is 162 nm and molar absorptivity is 10,000 L/(mol·cm)] to 1,3,5,7,9-pentene [335 nm and 118,000 L/(mol·cm)] to 1,3,5,7,9,11,13,15,17,19,21-undecene [470 nm and 185,000 L/(mol·cm)]. Several representative chromophores in the ultraviolet-visible region are listed in Table 9-2. More detailed information relating to organic chromophores in the ultraviolet-visible spectral region is available in References 4 and 5.

Table 9-2 Several chromophores in the ultraviolet-visible region

Chromophore	Example	Solvent	λ(max.), nm	ε(max.)
C=C (alkene)	1-Hexene	*n*-Heptane	179	12,800
C≡C (alkyne)	1-Butyne	Vapor	172	4,500
	2-Octyne	*n*-Heptane	178	10,000
			196	2,000
			225	160
CHO (aldehyde)	Acetal-dehyde	Vapor	182	10,000
			289	12.5
		n-Hexane	180	Large
			293	12
COOH (acid)	Acetic acid	Ethanol	204	41
COX	Acetyl chloride	*n*-Heptane	240	34
COOR (ester)	Ethyl acetate	Water	204	60
$CONH_2$	Acetamide	Methanol	205	160
		Water	214	60
NO_2 (nitro)	Nitro-methane	*n*-Hexane	202	4,400
			279	16
ONO (nitrite)	*n*-Butyl nitrite	Ethanol	218	1,050
			357	45
NO (nitroso)	Nitroso-butane	Ethyl ether	300	106
			665	20
ONO_2 (nitrate)	Ethyl nitrate	Dioxane	270	12
N=N (azo)	*trans*-Azomethane	Water	343	25
		Ethanol	339	5
=N=N (diazo)	Diazo-methane	Ethyl ether	417	7
	Diazo-ethane	*n*-Hexane	470	10
			490	6
C=N (azo-methine)	$Me_3CCH{=}NBu$	*n*-Hexane	244	87
	EtCH=NBu	Isooctane	238	200
Aromatic ring	Benzene	Water	203.5	7,400
			254	205
	Nitro-benzene	*n*-Heptane	252	8,700
	Chloro-benzene	*n*-Heptane	265	12,200
	Toluene	*n*-Heptane	264	10,100
	p-Xylene	*n*-Hexane	216	7,500
			269	750

Table 9-3 The wavelength of the absorptive maximum of several substituted polyenes.

Compound	Wavelength, nm	Solvent
$H_2C{=}CHCH{=}CH_2$	217	*n*-Hexane or ethanol
$CH_3CH{=}CHCH{=}CH_2$	223	Ethanol
$H_2C{=}C(CH_3)CH{=}CH_2$	222	Ethanol
$CH_3CH{=}CHCH{=}CHCH_3$	226	Ethanol
$(CH_3)_2C{=}CHCH{=}C(CH_3)_2$	242	Ethanol
$H(CH{=}CH)_3H$	268	Isooctane
$H(CH{=}CH)_4H$	304	Cyclohexane
$H(CH{=}CH)_6H$	364	Isooctane
$CH_3(CH{=}CH)_3CH_3$	275	*n*-Hexane

The effect of auxochromes on the location of the absorptive maximum of a chromophore can be seen for polyenes from the data in Table 9-3. Generally, the presence of an auxochrome substituted on the diene causes a bathochromic shift. The effect of an auxochrome on the absorptive maximum of a diene can be estimated with the Woodward-Fieser rules. The rules are summarized in Table 9-4. Each time an auxochrome is added to the diene backbone, a bathochromic shift occurs to the extent shown in the table. If two or more auxochromes are added, the shifts for the auxochromes are totaled and added to the value for the diene backbone.

A few quick calculations using the Woodward-Fieser rules and comparison with the data in Table 9-3 shows fair agreement between the calculated wavelengths and the observed wavelengths for polyenes. The rules show much better agreement for dienes in rings and typically are within 5 nm of the

Table 9-4 The Woodward-Fieser rules for conjugated dienes

Add the value for each auxochrome to that of the chromophore in order to estimate the wavelength of the absorptive maximum. R represents an alkyl or H

Chromophore	Wavelength, nm
$H_2C{=}CHCH{=}CH_2$ or heteroannular diene	214
Homoannular diene	253
Auxochrome	**Wavelength, nm**
Additional conjugated double bond	30
Exocyclic double bond	5
Alkyl	5
$OCOCH_3$	0
O-Alkyl	6
Cl or Br	5
SR	30
NR_2	60

observed values. While using Table 9-4, a *heteroannular diene* is defined as a diene in which the conjugated double bonds are in separate rings. A *homoannular diene* has the two double bonds in the same ring. Extension of the conjugation by addition of an additional conjugated double bond increases delocalization of the pi electrons. As a consequence, the energetic level of the pi-star orbitals is lowered and a bathochromic shift occurs. An exocyclic double bond is a double bond that is adjacent to a ring that contains at least one of the double bonds of the diene backbone.

Similar rules are available for α, β-unsaturated carbonyls. The rules are listed in Table 9-5. Further details of these and other rules as well as considerably more detailed descriptions of the use of ultraviolet-visible spectrophotometry for qualitative analysis can be found in References 4 and 5.

Table 9-5 Rules for predicting the wavelength of the absorptive maxima of α, β-unsaturated carbonyls

Add the value for each auxochrome to 215 nm for acyclic or six-membered rings

```
              |
β —C=C—C=O
    |   |
    β   α
```

Auxochrome	Wavelength, nm
Alpha substituents:	
Alkyl	10
OH	35
$OCOCH_3$	6
O-Alkyl	35
Cl	15
Br	25
Ring residue	10
Beta substituents:	
Alkyl	12
OH	30
$OCOCH_3$	6
O-Alkyl	30
Cl	12
Br	30
N-$(Alkyl)_2$	95
S-Alkyl	85
Ring residue	12
Conjugated double bond	30
Five-membered ring ketone	– 10
Aldehydes	– 5
Acids, esters	– 20

Sample problem 9-1 Predict the wavelength of the absorptive maximum of the following compound:

$$CH_3CH_2-\underset{|}{\overset{H}{C}}=\overset{Br}{C}-\underset{CH_3}{C}=O$$

SOLUTION The Woodward-Fieser rules (Table 9-5) can be used to predict the wavelength of the absorptive maximum:

α, β-unsaturated carbonyl backbone	215 nm
β–CH_3CH_2–	12 nm
α–Br	25 nm
Total	252 nm

The absorptive maximum should occur at about 252 nm. The CH_3 substituent on the carbon adjacent to the oxygen has no effect.

OTHER CHROMOPHORES

In addition to the primarily organic chromophores that have been described, some inorganic substances, many metal-organic complexes, and charge-transfer complexes absorb radiation in the ultraviolet-visible spectral region. Several inorganic anions absorb radiation in the ultraviolet-visible region. Among the common absorbing anions are carbonate, nitrate, and nitrite which respectively have absorptive maxima at 217, 313, and 280 and 360 nm (two maxima for nitrite). In each case absorption is caused by an n to π^* transition. If attached to organic auxochromes, the absorptive maxima generally are shifted to longer wavelengths.

Complexes of most of the transition metals absorb radiation in the ultraviolet-visible region in at least one of the oxidation states of each metal. Absorption is caused by transitions between d levels for the elements in the first two transition periods and to transitions between f levels for the elements in the lanthanide and actinide periods. Elements in the first two transition periods contain unoccupied d orbitals. When a ligand combines with the elements the d orbitals often are not involved in the bonding. In the absence of a ligand (vapor phase) the d orbitals all possess equal energy, and absorption in the ultraviolet-visible region does not take place. When a ligand is bound to the metal, electrostatic repulsion between the electron pair from the ligand and the d orbitals causes a splitting of the energetic levels of the d orbitals. Absorption occurs when an electron from a less energetic d orbital is excited to a more energetic d orbital.

Table 9-6 A list of common ligands that are arranged in approximate order of increasing electric field strength

The order of absorptive wavelength is for complexes with metals of the first two transition periods

Low strength	I^-	Long absorptive wavelength
	Br^-	
	Cl^-	
	F^-	
	OH^-	
	$C_2O_4^{2-} = H_2O$	
	SCN^-	
	NH_3	
	$H_2NCH_2CH_2NH_2$	
	o-Phenanthroline	
	NO_2^-	
High strength	CN^-	Short absorptive wavelength

The wavelength at which the absorption is observed depends upon the difference in energy between the split *d* orbitals. The energetic difference is a function of the electric field strength that is associated with the attached ligand. As the field strength of the ligand increases, the difference in energy between the *d* levels increases and the wavelength at which absorption occurs decreases. A list of several common ligands arranged according to increasing field strength is given in Table 9-6. Nearly all metallic complexes of the ligands at the top of the table absorb at relatively long wavelengths. The wavelength at which absorption occurs decreases in going from the top to the bottom of the table.

Complexes of the lanthanide elements absorb radiation during electron transitions from less energetic 4*f* orbitals to more energetic 4*f* orbitals. Complexes of the metals in the actinide period absorb radiation during electron transitions between 5*f* orbitals. In both cases the inner-shell *f* electrons are largely screened from the varying electric fields of different ligands by electrons in outer shells. Consequently, the absorptive spectrum of a particular lanthanide or actinide element is nearly independent of the ligand to which the metal is attached. The absorptive bands of the lanthanide and actinide complexes are generally narrow (typically 20 nm across) and sharp in comparison to those that are normally associated with ultraviolet-visible spectra.

The high absorptivity of some complexes is often caused by *charge transfer* within the complex. During charge transfer one of the components of a complex absorbs radiation by transferring an electron from one of its lower energetic orbitals to a higher energetic orbital of another component of the complex. That differs from absorption that occurs with the complexes that were described earlier where the electron is excited to an orbital associated with the molecule. Essentially absorbance corresponds to an internal oxidation-reduction reaction. A complex that absorbs radiation during a charge transfer is a *charge-transfer complex*.

Charge-transfer complexes are of particular interest in analytical chemistry because they generally have large absorptivities. Typically charge-transfer complexes have peak molar absorptivities in excess of 10,000. The high absorptivities make it possible to quantitatively determine low concentrations of the complexes. In many cases charge-transfer complexes are formed between a metal ion and a ligand. In most of those cases, the metal ion acts as the electron acceptor while the ligand is the electron. donor. A well-known example is the iron(III)-thiocyanate complex that can be used to quantitatively assay either iron or thiocyanate.

In a few cases the metal ion is the electron donor and the ligand is the electron acceptor. An example of that type of charge-transfer complex is the iron(II)-orthopenanthroline complex. The charge-transfer complex does not have to consist of a metallic ion and a ligand. Complexes can consist of two nonmetallic components. An example of a nonmetallic charge-transfer complex is the complex that is formed between quinone and hydroquinone.

In most cases relaxation in a charge-transfer complex occurs by the reverse of the excitative process. In some cases, however, the absorbed energy causes dissociation of the complex. The collection and storage of the photochemical decomposition products of charge-transfer complexes is a potential method of retaining solar energy. The stored reaction products are a source of chemical energy which can be used when required.

INSTRUMENTS

The instruments that are used to measure ultraviolet-visible absorption are classified according to the type of wavelength selector and detector. Colorimeters and photometers use filters while spectrophotometers use monochromators. A colorimeter uses the human eye as a detector and consequently is restricted to studies in the visible spectral region. Colorimeters and photometers are used for quantitative analysis at a fixed wavelength. In order to change the wavelength the filter must be changed. Spectrophotometers can be used at a single wavelength or can be scanned for qualitative analyses. Photometers can be used as single-wavelength detectors in liquid chromatographs.

Both *single-* and *double-beam instruments* are in common use for absorptive measurements in the ultraviolet-visible region. Although the design of the instruments varies between manufacturers, all of the instruments possess the major components that are shown in Fig. 9-1. An instrument of the single-beam type uses one beam of radiation and passes it through a single cell. A double-beam instrument divides the radiation into two beams that are passed through separate cells. Single-beam instruments are generally less expensive than double-beam instruments. Traditional, non-computer controlled, single-beam instruments are primarily used for studies performed at a single wavelength; i.e., they are not as often used to scan through a wavelength region. The wavelength at which the study is performed can be varied by adjustment of the monochromator or by changing the filter. Many modern single-beam instruments are computer controlled. Those instruments can be used to scan through a wavelength range. Generally, the solvent or "blank" solution is first scanned while the absorbance is measured at discrete wavelength intervals (often 1 or 2 nm) and recorded in computer memory. Blank solutions contain the matrix of the analyte without the assayed substance. Next, the sample is scanned in an identical manner and its spectrum recorded. Finally, the computer subtracts the solvent or blank data from the analyte data and displays the resulting spectrum (a plot of absorbance as a function of wavelength).

In double-beam instruments one of the two cells is filled with either pure solvent or a blank solution and the second cell is filled with the sample. Because the readout from the instrument is the difference between the amount of radiation that is absorbed in the two cells, the absorption in the sample is automatically corrected for absorption occurring in the blank or solvent and for changes in intensity of the incident radiation with wavelength. Double-beam instruments can be used at a fixed wavelength or they can be used to scan through a wavelength region. Because traditional single-beam instruments do not automatically correct for changes in the absorption of the blank solution and in incident radiative intensity as the wavelength changes, they cannot be used to automatically scan through a wavelength region. Double-beam instruments also have the advantage of eliminating variations in intensity of the radiative source or variations in electronic response of the detector. In that respect they are similar to double-beam atomic absorption spectrophotometers.

In some instruments the chopper that is shown in Fig. 9-1 is replaced with some other beam-splitting device. Commercial spectrophotometers are available at the present time in many designs and price ranges. The available instruments vary from portable single-beam instruments, such as the DR2800 from the Hach Company, to highly sophisticated computer-controlled instruments, such as the Lambda 950 and Lambda 1050 from Perkin-Elmer, and the CARY 5000 from Agilent Technologies.

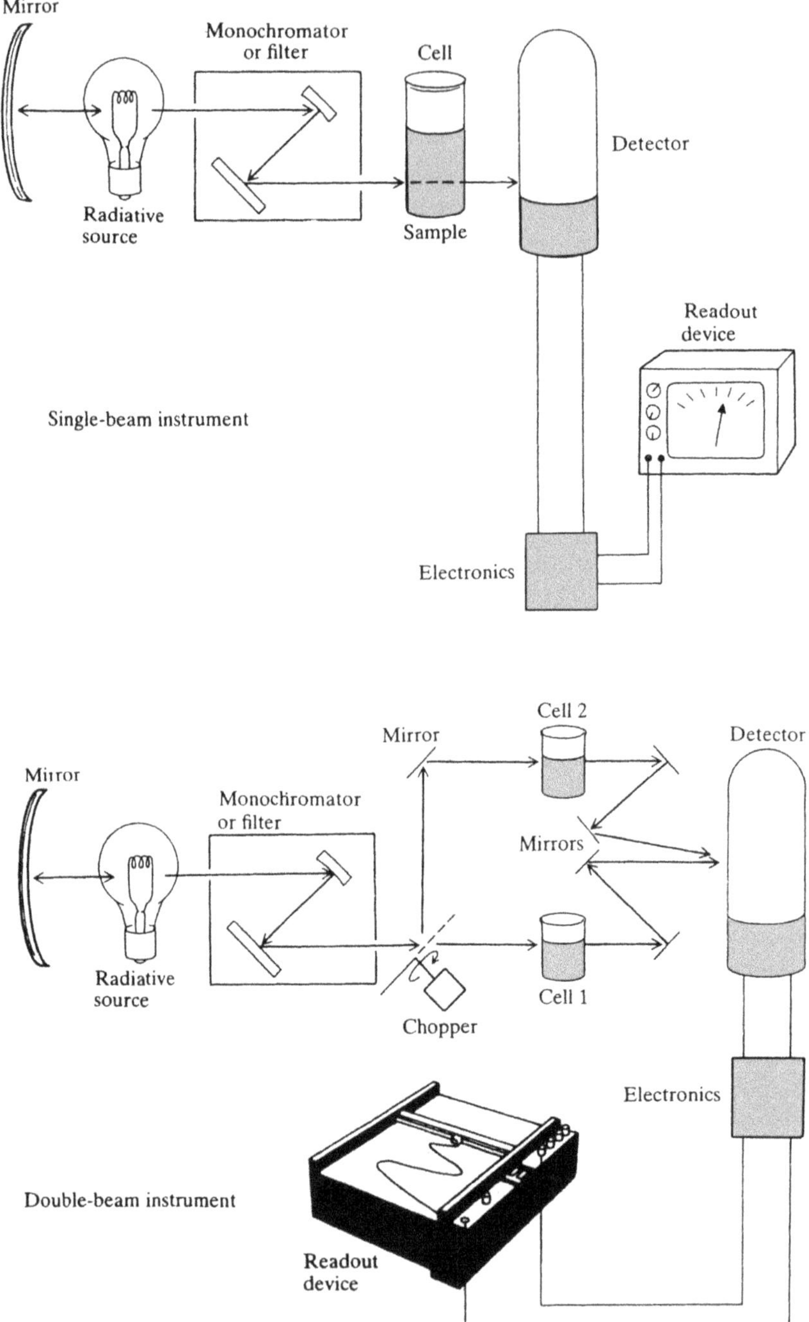

Figure 9-1 Block diagrams of two types of spectrophotometers and photometers.

Some of the computer-controlled instruments are capable of performing first- and second-derivative spectrophotometry. The first derivative of a spectrum is a plot of the slope ($dA/d\lambda$) of the spectrum as a function of wavelength. A second-derivative spectrum is a plot of the slope ($d^2A/d\lambda^2$) of the first-derivative spectrum as a function of wavelength. Some instruments have an offset control that allows the use of expanded-scale spectrophotometry (see later description). Other instruments are designed to rapidly scan the wavelength region of interest. An example of an instrument that is capable of scanning rapidly is the model 8453 from Agilent Technologies. It utilizes a photodiode array detector system (see later description) to obtain an entire spectrum in less than 1 s.

When making absorbance measurements with a single- or double-beam instrument, the usual procedure is the following:

1. Adjust the wavelength.
2. With no radiation striking the detector, set the instrument to read 0% T.
3. With the blank solution in the path of the EMR, adjust the instrument to give a 100% T reading.
4. With the sample in the path of the EMR, read the percent transmittance, absorbance, or computer-calculated concentration of the sample.

Radiative Sources and Wavelength Selectors

The source of visible radiation is usually a tungsten-filament lamp which provides radiation from about 350 to about 2500 nm. In the ultraviolet region the radiative source is often a hydrogen lamp, a deuterium lamp, or a xenon discharge lamp. Hydrogen and deuterium lamps provide continuous radiation from about 180 to 350 nm. The xenon discharge lamp provides ultraviolet radiation that is usually more intense but less stable as a function of wavelength than that from the hydrogen and deuterium lamps. When high-intensity radiation is required, lasers (Chapter 8) can be used. Lasers are rarely required for absorptive measurements in the ultraviolet-visible region.

The wavelength selector is either a filter or monochromator. Both filters and monochromators were described in Chapter 5. Filters cannot be used in scanning instruments.

Cells

In most cases the sample is a solution of the absorbing substance. The solution is held in place in the path of the radiation in a cell or cuvet. The substance from which the cell walls are made and through which the radiation passes must be transparent to radiation of the wavelength that is used for the assay. In the visible region the cell can be made of glass. However, glass absorbs ultraviolet radiation, and a cell that is to be used in the ultraviolet region is usually made of quartz or fused silica. Quartz and fused silica are also transparent in the visible region, but are more expensive than glass.

In most cases the cells are rectangular with parallel transparent windows through which the radiation passes. For most applications a cell with a 1-cm pathlength is adequate. Cells with both longer and shorter pathlengths are available for use when needed. Long-pathlength and multiple-path cells are used for analysis of unusually low-concentration samples. Gaseous samples often require use of multiple-path cells.

A multiple-path cell is one in which the incident radiation passes through the cell more than once prior to striking the detector. Usually multiple-path cells have mirrored ends that cause the radiation to reflect through the cell several times prior to exiting from the cell. In double-beam instruments the two cells should be as nearly identical as possible. Cells that have nearly identical pathlengths and that transmit radiation to nearly the same extent are *matched cells*.

Detectors

Several categories of detectors have been used in photometers and spectrophotometers. Of the several types of detectors that are available, phototubes and photomultiplier tubes are used most often. Photomultiplier tubes and diode arrays were described in Chapter 6. Diode array detectors often are used in instruments which are designed to be able to obtain an entire spectrum rapidly. An example of such an instrument is the Agilent Technologies model 8453 which was described earlier.

Barrier-layer (photovoltaic) detectors can be used where the intensity of the EMR that strikes the detector is relatively high. They are primarily found in photometers that are designed for use in the visible region. *Phototubes* contain a photocathode and an anode as in photomultiplier tubes, but do not contain dynodes. A phototube is similar in appearance to an old radio tube (Fig. 9-2). It consists of a large semicylindrical cathode and a smaller concentric anode enclosed in an evacuated glass envelope. If the phototube is to be used in the ultraviolet region, a window of quartz or some other uv-transparent material is mounted in the wall of the glass tube to allow radiation to pass into the tube. The cathode is usually coated with an oxide of cesium, but the oxides of other metals from group 1 or 2 of the periodic table also can be used.

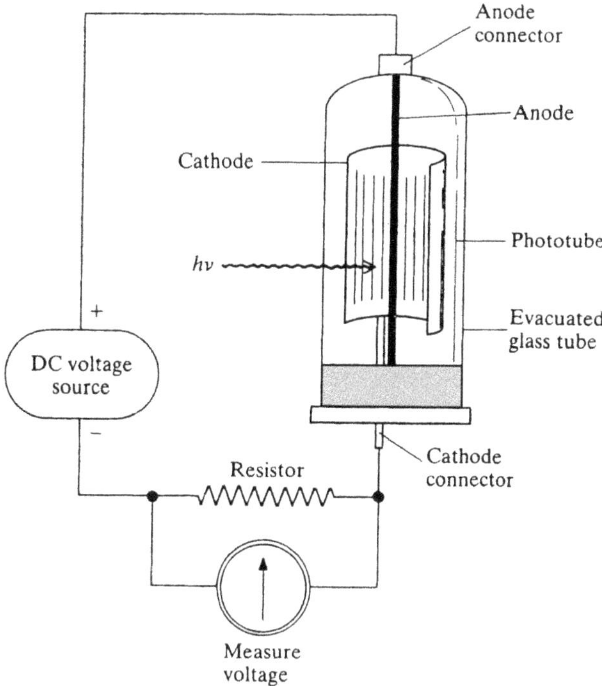

Figure 9-2 A phototube assembly.

Typically, 90 V is maintained between the cathode and the anode by an external power supply. Usually the current flowing through the circuit is monitored by measuring the voltage drop across a resistor in the circuit. The resistor is large (about 500 MΩ) in order to permit measurement of small currents. Because Ohm's law ($E = IR$) is obeyed in the circuit and because the resistance R is a constant, the measured potential E is directly proportional to the current I flowing through the circuit.

In the absence of radiation, essentially zero current flows through the circuit. As radiation strikes the cathode, electrons are emitted from the cathodic surface and are attracted to the anode. The flow of electrons causes a current flow in the circuit and an increased potential across the resistor. Because the rate of electron emission from the cathode is proportional to the intensity of the radiation striking the cathode and because the measured voltage is proportional to the current (the rate of electron emission), it follows that the measured voltage is directly proportional to the intensity of the radiation.

Because photomultiplier tubes can detect lower intensity radiation than phototubes, they are used more often than phototubes in double-beam spectrophotometers. Photomultiplier tubes cannot be used to measure relatively high radiative intensities because they exhibit instability and can be damaged by high intensities. Phototubes are often found in single-beam spectrophotometers that are intended for use in the visible region.

Readout Devices

A large variety of readout devices are available for use in photometers and spectrophotometers. Those that are encountered most often include analog meters, digital meters, recorders, printers, and computer operated devices such as video display terminals. Less-expensive instruments usually contain an analog or digital meter. Computer-controlled instruments generally have either digital meters, printers, or video displays. Scanning instruments use recorders, computer-controlled recorders, *xy* plotters, video displays or combinations of computer-controlled recorders, or plotters and printers.

QUANTITATIVE ANALYSIS

In many but not all cases, Beer's law is obeyed during ultraviolet-visible absorption by polyatomic substances. Beer's law and its use was described in Chapter 5. The most straightforward way to use Beer's law for quantitative analysis is to measure the absorbance of the sample solution at a wavelength at which the species in solution is known to absorb radiation and then directly use Beer's law ($A = abC$) to calculate concentration. Use of Beer's law requires knowledge of the cell pathlength b and the absorptivity a of the absorbing species at the wavelength of the measurement.

Cells used for absorptive measurements are prepared by the manufacturer with accurately known pathlengths which are usually listed in the manufacturer's literature accompanying the cells. If necessary, the cell pathlength can be measured with calipers or another device. The cell pathlength can also be determined by using the cell to measure the absorbance of a solution of known concentration and absorptivity. The cell pathlength can be calculated using Beer's law.

The absorptivities or molar absorptivities of many substances at specified wavelengths are listed in various tables in the literature. The values, when they are available, can be used in calculations of concentration. If a value of a is not available, it can be calculated by measuring the absorbance of solutions containing known concentrations of the absorbing species in a cell with a known pathlength. If a plot of absorbance as a function of concentration is linear and goes through the origin, the slope of the resulting straight line is ab, from which a can be calculated if b is known. Although it is possible to obtain a value of a from a single absorptive measurement of a solution of known concentration by direct substitution into Beer's law, the procedure cannot be recommended. Conclusions based upon a single measurement are not statistically sound, because it is easy to make a measurement (determinate) error that cannot be detected with a single measurement.

Sample problem 9-2 The absorbance in a 5.00-cm cell (pathlength) of aqueous standard solutions of benzotropine mesylate were measured. From the tabulated results determine the molar absorptivity of the compound at the wavelength used for the study.

Concentration × 10^{-4} *M*	Absorbance
0.50	0.109
1.00	0.219
1.50	0.330
2.00	0.441
3.00	0.654

SOLUTION The problem is best solved by preparing a working curve of absorbance as a function of concentration. If Beer's law is obeyed, it is apparent that the slope of the plot should be the product of molar

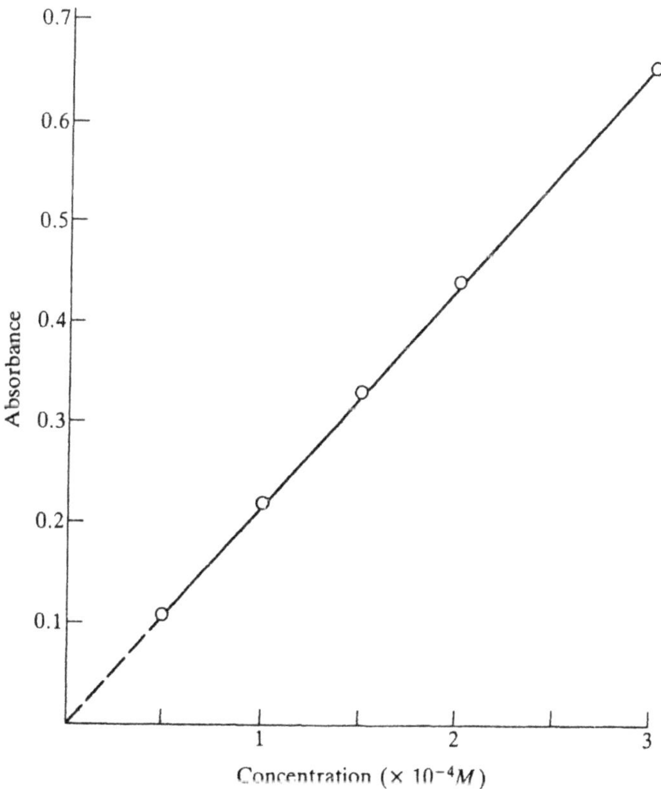

Figure 9-3 The working curve for the data of Sample problem 9-2.

absorptivity and cell pathlength. Because the pathlength is known, the molar absorptivity can be calculated. The curve is shown in Fig. 9-3.

Because the working curve is linear and can be extrapolated through the origin, Beer's law is obeyed. The slope of the straight line (least-squares fit[6]) is 2.18×10^3. The molar absorptivity can be calculated from the slope:

$$\text{Slope} = \varepsilon b$$

$$2.18 \times 10^3 = \varepsilon (5.00)$$

$$\varepsilon = 436 \text{ L/(mol. cm)}$$

In those instances in which Beer's law is not obeyed, quantitative analysis is generally performed with the working-curve method. The working-curve method is the preferred method of analysis, even in those cases in which Beer's law is obeyed, because it is more statistically sound than direct use of Beer's law with a single measurement. If Beer's law is directly used for an assay, it is important to repeat the measurement with different solutions of the sample at least thrice in order to minimize the possibility of a determinate error. If Beer's law has been demonstrated to be obyed for a particular compound and if the compound is in a matrix (solution) of unknown composition, use of the standard-addition technique (Chapter 5) is recommended in order to minimize the effect of potential interferences. Deviations from Beer's law and the reasons for those deviations were described in Chapter 5.

Wavelength Choice

When choosing a wavelength at which to make an absorbance measurement, three factors must be considered. The first factor is related to potential spectral interferences in the analyte solution. If the solution contains more than one absorbing species, the wavelength that should be chosen, whenever possible, is that at which the second species in the solution does not absorb radiation. It can be determined by using a spectrophotometer to obtain the spectra of solutions of the two pure components. If it is not possible to find a wavelength at which only the analyte absorbs radiation, then the observed absorbance is the sum of the individual absorbances of the components. That situation is described in more detail later in the chapter.

The second factor that must be considered when a wavelength is chosen is the required sensitivity of the assay. In most cases the analyst wishes to have as much sensitivity as possible and can get it by choosing a wavelength at which the absorptivity is relatively large. For a solution of a fixed concentration in a cell of constant pathlength, it is apparent from Beer's law that the absorptivity is greatest when the absorbance is greatest. Consequently, a wavelength with a relatively high absorbance or a relatively low transmittance is chosen from the spectrum.

The final factor to be considered is the sensitivity of the assay to small changes in wavelength. It is preferable to choose a wavelength at which the absorbance will not be significantly altered if the wavelength is slightly changed. Slight variation in the wavelength can result from accidentally moving the wavelength-selection knob or from a change due to instability of the instrument. By choosing a wavelength on the spectrum that is in the center of a region of relatively small change in absorbance with wavelength, the effect of minor changes in wavelength can be minimized.

The application of the three criteria for choosing a wavelength will generally lead to a successful analysis. An example of the application of the criteria to overlapping spectra is illustrated in Fig. 9-4. The solid line is the spectrum of the acid form of the indicator bromothymol blue (abbreviated to HIn). The dashed line is the spectrum of the basic form of bromothymol blue (In^-). The two species are in equilibrium in solution:

$$HIn = H^+ + In^- \tag{9-1}$$

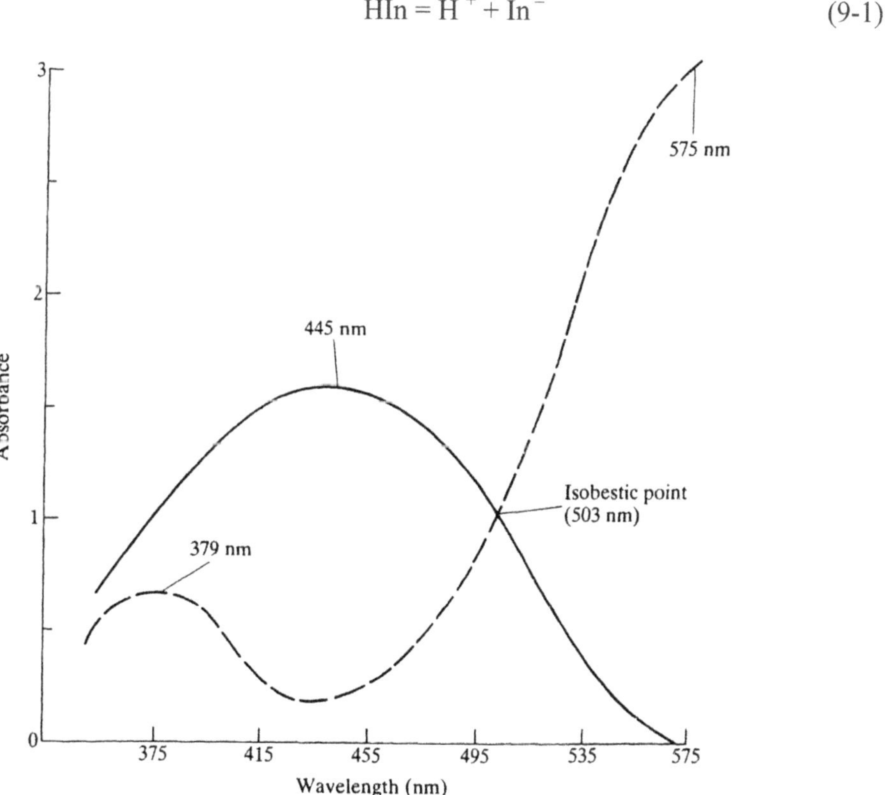

Figure 9-4 Visible spectra in 1-cm cells of the acidic (solid line) and basic (dashed line) forms of bromothymol blue at concentrations of 0.04 g/L.

If the acid form of bromothymol blue is the analyte, the choice of wavelength is relatively simple. No wavelength is available at which only the acid form of the indicator absorbs radiation; however, interference from the basic form can be minimized by choosing a wavelength at which the basic form has a low absorbance while the acid form has a high absorbance. A wavelength in the vicinity of 445 nm would probably be best. The chosen wavelength satisfies all of the criteria for selection except lack of an absorbing interference. The absorbance at 445 nm is a maximum and the absorbance does not significantly change between about 430 and 460 nm.

If the basic form of bromothymol blue is the analyte, three possible wavelengths could be chosen for the assay depending upon the circumstances. If the acid form is not present in solution (highly basic solution), the wavelength of choice would probably be about 379 nm. At 379 nm the absorbance is relatively large and little change in absorbance with wavelength is observed. A wavelength near 440 nm also could be chosen if the low sensitivity that is associated with the low absorbance could be tolerated, i.e., if the concentration of the analyte is sufficiently high.

In the presence of the interfering acid form of the indicator, it is desirable to choose a wavelength at which the acid form does not absorb radiation. A reasonable choice in that instance would be 575 nm. That wavelength also has the advantage of high sensitivity. Unfortunately a small change in wavelength in the vicinity of 575 nm could cause a significant change in absorbance, with a corresponding change in calculated concentration, because 575 nm does not occur on a plateau of the spectrum.

If the total concentration of the acidic and basic form of the indicator is desired, the isobestic point (about 503 nm) could be chosen for the assay. At the isobestic point both forms of the indicator absorb radiation to the same extent. Consequently, the form of the indicator (the pH of the solution) has no effect upon the absorbance. Because the isobestic point occurs on the falling portion of the first spectrum and the rising portion of the second spectrum, it is necessary to use a monochromator that permits radiation of a highly stable wavelength to pass. It is apparent that the choice of wavelength for a particular assay is not always easy. Several factors must be balanced and the wavelength choice is highly dependent upon actual experimental conditions.

Solvents

The choice of solvent that is used in an assay in the ultraviolet-visible region is based upon two criteria. First, of course, the sample must be soluble in the solvent. Second, the solvent must be transparent to radiation at the wavelength

Table 9-7 A list of solvents that can be used for assays in the ultraviolet or visible spectral regions

Solvent	Cutoff wavelength, nm	Solvent	Cutoff wavelength, nm
Acetic acid	260	*n*-Heptane	197
Acetone	330	Hexadecane	200
Acetonitrile	190	*n*-Hexane	195
Benzene	280	Methanol	205
1-Butanol	210	Nitromethane	380
Carbon disulfide	380	*n*-Pentane	190
Carbon tetrachloride	265	1-Propanol	205
Chloroform	245	2-Propanol	205
Cyclohexane	205	Pyridine	315
1,2-Dichloroethane	226	Tetrahydrofuran	220
Dichloromethane	239	Toluene	285
N,N-dimethylacetamide	268	2,2,2-Trifluoroethanol	190
N,N-dimethylformamide	270	2,2,4-Trimethylpentane (isooctane)	205
Dimethylsulfoxide	265		
1,4-Diosane	215	Water	192
Ethanol (95%)	204	*o*-Xylene	290
Ethyl ether	215		

at which the study is conducted. A list of several solvents that can be used for assays in the ultraviolet-visible region is given in Table 9-7. Water is the preferred choice whenever the sample is sufficiently soluble. Solubility data can be found in several reference texts such as the *Handbook of Chemistry and Physics* (Chemical Rubber Company) and the *Merck Index* (Merck and Company). The *cutoff wavelengths* that are listed in Table 9-7 are the wavelengths below which the solvent absorbs too much radiation to be useful. The solvents can be used for studies at wavelengths that are longer than the cutoff wavelengths. The cutoff wavelength can be defined in several ways. In Table 9-7 it is defined as the wavelength at which the absorbance of the solvent in a cell with a 1-cm pathlength is about 1.

Determination of Nonabsorbing Substances

In some cases it is possible to photometrically or spectrophotometrically assay a nonabsorbing substance by reacting the analyte with a chemical reagent that forms an absorbing reaction product. The chemical reagent that is chosen for the reaction must be capable of reacting with the low concentration of analyte in the solution and should be relatively selective. The reagent should either react nearly completely with the analyte or, at the very least, react to the same extent (percentage) each time. The reagent generally must be capable of reacting with analytes that are in the 10^{-5} or 10^{-6} M range. Generally the working-curve method is used for the analyses.

As with other chemical reactions, several experimental factors should be carefully controlled while performing the reaction. Identical experimental conditions must be used while reacting with the standards and the analyte. Among the factors that can affect the rate or extent of the reaction are the pH of the solution, the concentration and rate of addition of the chemical reagent, the time between the addition of the reagent and the spectral measurement, the order in which the reagents are added, the temperature of the reaction mixture, the solvent in which the reactants are dissolved, and the stability of the chemical reagent and the reaction product. Perhaps the most-often used chemical reagents are complexing agents that form absorbing (usually in the visible region) complexes with otherwise nonabsorbing metallic ions. Several common examples include the determination of iron(II) by the addition of 1,10-phenanthroline, the determination of iron(III) by the addition of thiocyanate, the determination of copper(II) or lead(II) by the addition of diphenylthiocarbazone, and the determination of copper(II) by the addition of diethyldithiocarbamate. A list of selected elements that can be assayed spectrophotometrically after addition of a chemical reagent or complexing agent is given in Table 9-8. Details relating to a specific assay can be found in Reference 7.

Mixtures of Absorbing Species

If a wavelength cannot be found at which only a single component absorbs in a mixture that contains two or more absorbing substances, it is more difficult to perform the analysis. Usually one of two methods is used in those instances: either a chemical reagent is added that, in some manner, prevents absorption by all except one substance or a wavelength is chosen for each absorbing component at which the absorbance is measured and the concentrations of the components are determined by the solution of simultaneous equations. A chemical reagent that prevents a substance from taking part in a particular reaction and therefore from absorbing radiation at a particular wavelength is a *masking agent*.[8] *Demasking* is the opposite of masking. During demasking the substance is returned to its original condition.

Table 9-8 A list of selected inorganic assays that can be performed spectrophotometrically

Element	Reagent or complexing agent	Wavelength, nm
Ag	Dithizone	450
	Rhodamine	495
Al	Ammonium aurinetricarboxylate	520
	Eriochrome cyanine R	530
As	Molybdate, hydrazine sulfate	840
Au	Rhodamine	500
B	Quinalizarin	620
	Curcumin	540
	1,1′-Dianthrimide	620
	Carminic acid	575
Bi	Iodide	460
	Dithizone	505
Br	Hypochlorite, rosaniline	570
Cd	Dithizone	540
Co	Nitroso-R-salt	420, 560
	Thiocyanate	610
Cr	Diphenylcarbazide	540
Cu	Dithizone	510
	Sodium diethyldithiocarbamate	440
	Cuproin	546
	Neocuproin	547
	Bathocuproin	479
Fe	Thiocyanate	485
	1,10-Phenanthroline	580
	Ammonium thioglycolate	535
Ge	Phenylfluorone	510
Hf	Alizarinsulfonic acid	525
Hg	Dithizone	515
In	Dithizone	515
Mn	Periodate	527
Mo	Thiocyanate	470
	Dithiol	440, 670
Ni	Dimethylglyoxime	530
P	Molybdate, hydroquinone	730
	Vanadate, molybdate	460
Pb	Dithizone	525
Pd	*p*-Nitrosodiphenylamine	510
Pt	Stannous chloride	403
Sb	Rhodamine B	565
Se	Diamino-3,3′-benzidine	340, 420
Si	Ammonium molybdate	405
Te	Hypophosphorous acid	285
Th	Thorin	545
Ti	H_2SO_4, H_2O_2	410
V	Benzohydroxamic acid	450
W	Thiocyanate	390
Zn	Dithizone	540
	Zincon	620
Zr	Alizarin red S	525

In spectroscopic studies masking usually consists of reacting the interference with a reagent that prevents formation of an absorbing substance by complexing the interference, changing the interference to a nonreactive oxidation state, or precipitating the interference. Demasking consists of reversing the process. In a mixture of two substances a typical procedure is to mask one of the substances prior to measuring the absorbance that is due solely to the remaining component. If the concentration of both substances is required, the masked substance is demasked and the absorbance is measured a second time. The difference between the two readings is that due to the masked substance. Examples of masking agents are cyanide or chloride for silver(I), EDTA for barium(II) and calcium(II), hydrogen peroxide for several elements that can change oxidation states, and ammonia for zinc(II). A detailed description of masking agents, demasking agents, and their respective uses can be found in Reference 8.

If each of the components in a mixture of absorbing substances is to be determined, it is sometimes necessary or easier to use the *method of simultaneous equations*. The method requires that Beer's law be obeyed for each of the components at each of the chosen wavelengths and that the absorbances of each of the components is additive at each wavelength. If both requirements are met for the mixture, the observed total absorbance A_T at any wavelength is the sum of the absorbances of the n individual components:

$$A_T = A_1 + A_2 + A_3 + \ldots + A_n \tag{9-2}$$

Because the absorbance of each component is given by Beer's law for that component, Eq. (9-2) can be rewritten as

$$A_T = a_1bC_1 + a_2bC_2 + a_3bC_3 + \ldots + a_n bC_n \tag{9-3}$$

The absorptivity of each component at a fixed wavelength can be calculated from absorptive measurements performed in solutions containing the single component. The cell pathlength is known or can be measured. Substitution of the values of a and b into Eq. (9-3) yields an equation with n unknowns, corresponding to the concentrations of the n absorbing species. In order to solve for n unknowns, a series of n equations must be available. Consequently, in order to determine the concentration of each absorbing species in a mixture, it is necessary to choose one wavelength for each species. An equation similar to Eq. (9-3) is written at each chosen wavelength, and the resulting n equations are solved simultaneously for the concentrations of the absorbing species.

The steps to be followed when determining the concentration of each absorbing component in a mixture are the following:

1. Obtain a spectrum (a plot of A or T as a function of wavelength) of a solution of known concentration of each component.
2. From the spectra choose a wavelength for each component at which absorptive measurements are to be made. Use the criteria described earlier in the chapter to choose each wavelength.
3. Prepare a working curve of each pure component at each of the chosen wavelengths. If the working curves are straight lines that go through the origin, Beer's law is obeyed and ab is obtained for each component at each wavelength from the slope of each working curve. If a straight line is not obtained for one or more of the components or if the line does not go through the origin, a new wavelength must be chosen.
4. Write an equation similar to Eq. (9-3) for each wavelength. Use the values of ab calculated in step 3.
5. Check to make sure the absorbances of the components are additive at each wavelength. The check can be made by comparing the calculated absorbances of a series of standard solutions of mixtures at each wavelength with the observed absorbances. The values of ab measured in step 3 can be used to obtain the calculated absorbances. If the absorbances are not additive at a particular wavelength, a new wavelength must be chosen.
6. Measure the absorbance of the analyte solution (containing all of the components) at each chosen wavelength and substitute those values into the equations written in step 4.
7. Solve the equations simultaneously for each component's concentrations. If more than three components are present in the solution, matrix algebra can be used to solve the equations.

In practice, the preceding technique is normally not useful if more than three or four components are present in the solution. That is because it is difficult to find wavelengths at which Beer's law is obeyed for all of the components and at which the absorbances are additive.

Sample problem 9-3 A solution that contained two absorbing substances was spectrophotometrically determined at two wavelengths in a cell with a 1.00-cm pathlength. The absorbance of the mixture was 0.945 at 580 nm and 0.297 at 395 nm. The molar absorptivities are listed in the table. Calculate the concentration of each component in the mixture.

	Molar absorptivity	
Component	**580 nm**	**395 nm**
1	9874	548
2	455	8374

SOLUTION At 580 nm the absorbance of the mixture is

$$A_T = \varepsilon_1 bC_1 + \varepsilon_2 bC_2$$

or

$$0.945 = 9874(1.00)C_1 + 455(1.00)C_2$$

Similarly, at 395 nm the absorbance of the mixture is

$$0.297 = 548(1.00)C_1 + 8374(1.00)C_2$$

The two simultaneous equations can be most accurately solved by solving one of the equations for the concentration of the component in that equation which has the larger coefficient. The result is substituted into the second equation. In this case the first equation is solved for C_1 and the result substituted into the second equation:

$$C_1 = \frac{0.945 - 455C_2}{9874}$$

$$0.297 = 548\left(\frac{0.945 - 455C_2}{9874}\right) + 8374C_2$$

$$C_2 = 2.93 \times 10^{-5}\ M$$

$$C_1 = \frac{0.945 - 455\left(2.93 \times 10^{-5}\right)}{9874}$$

$$C_1 = 9.44 \times 10^{-5}\ M$$

Derivative Spectrophotometry

Many modern computer-controlled spectrophotometers are capable of reporting derivative as well as conventional spectra. Many instruments are capable of plotting first- and second-derivative spectra of either transmittance or absorbance as a function of wavelength. A first-derivative spectrum is a plot of $dA/d\lambda$ or $dT/d\lambda$ as a function of λ, and a second-derivative spectrum is a plot of $d^2A/d\lambda^2$ as a function of λ. Computer-controlled instruments generally obtain the derivatives digitally. Some derivative spectrophotometers obtain derivatives mechanically or by using a dual-wavelength spectrophotometer in which the two beams of radiation are alternately passed through the cell. The two beams are displaced from each other by a constant wavelength difference during the scan.

First- and second-derivative spectra often show detail that is lacking in conventional spectra. That detail can be an aid in the qualitative or quantitative analysis of a sample. In quantitative analyses derivative spectrophotometry is particularly useful for solutions that contain an interference that absorbs at nearly the same wavelength as the analyte. In many cases the peak-to-peak height of the first-derivative spectrum is directly proportional to concentration. First- and second-derivative spectra are shown in Fig. 9-5.

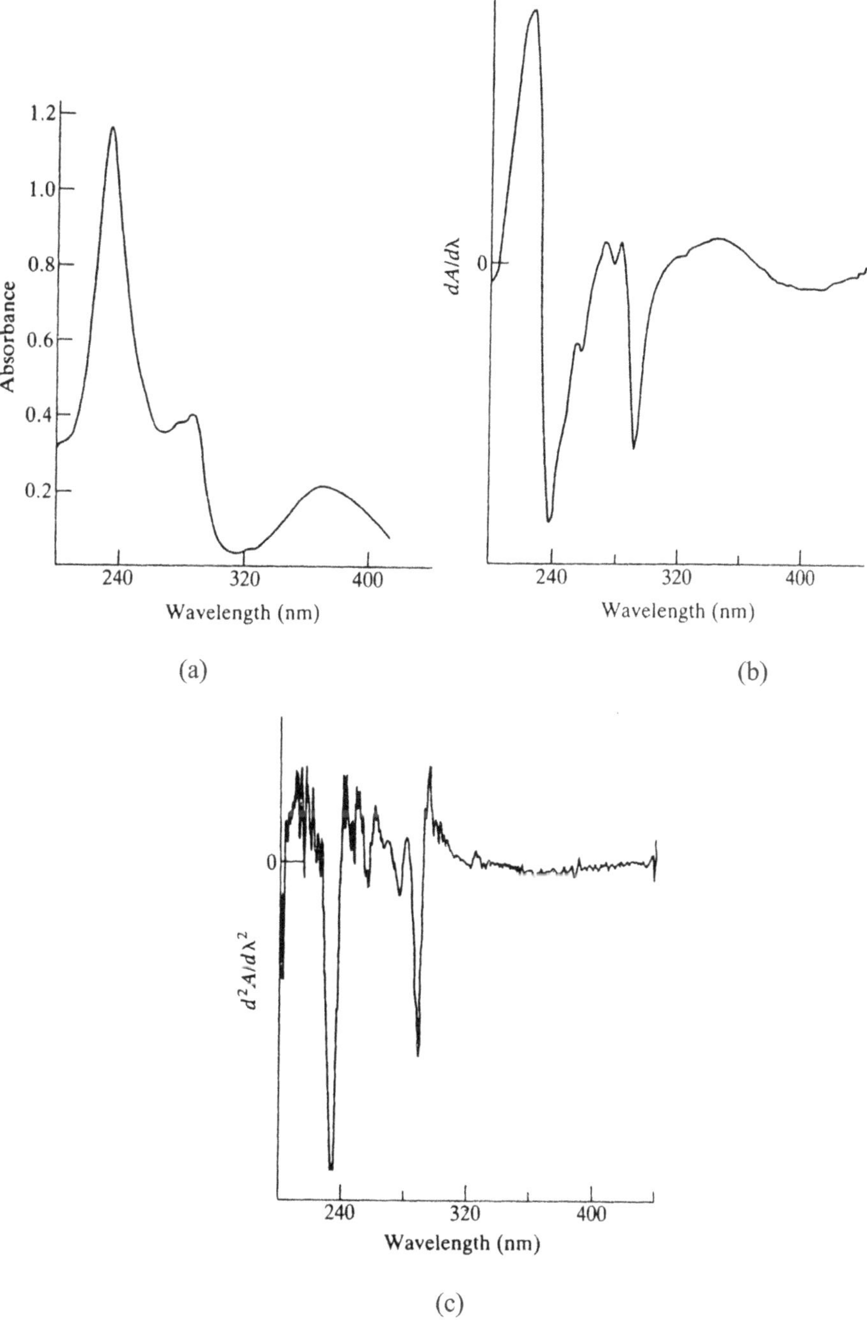

Figure 9-5 Absorbance spectra of 1.3×10^{-5} M gossypol in methanol. The spectra were recorded by using a Hitachi 100-80 spectrophotometer. (*a*) Conventional spectrum; (*b*) first-derivative spectrum; (*c*) second-derivative spectrum.

Expanded-Scale Spectrophotometry

Expanded-scale spectrophotometry is also known as *differential spectrophotometry* and *precision spectrophotometry*. It consists of expanding the readout scale using one of three procedures. It is used primarily for instruments that have analog readout devices.

If the sample has a low percent transmittance (high absorbance), the photometer or spectrophotometer is adjusted to read 0 percent transmittance in the usual manner. Rather than making the 100 percent transmittance adjustment with a blank solution or with pure solvent, the adjustment is made with a standard solution of the absorbing substance that is lower in concentration than that in the analyte. Without altering the adjustments, the percent transmittance of the analyte is measured. The effect is to expand the scale on the readout device in order to obtain a more accurate reading. If absorbance is calculated from the measured percent transmittances of several standard solutions, it is generally found that the absorbance varies linearly with concentration. Not all instruments are capable of the technique because many instruments do not have sufficient control to enable the instrument to be adjusted to 100 percent transmittance with a standard solution.

If the sample has a high percent transmittance (low absorbance), the instrument is initially adjusted to read 0 percent transmittance with a standard solution that is more concentrated than the analyte. The instrument is adjusted to read 100 percent transmittance in the usual manner with a blank solution or with pure solvent. After the initial adjustments the percent transmittance of the analyte is measured. Plots of calculated absorbance as a function of concentration using the method generally are not linear. The method requires the use of an instrument that has the capability to be adjusted to 0 percent transmittance with a standard solution in the cell.

The final procedure is useful for analytes that have percent transmittances that are somewhere within the central portion of the scale. The instrument is initially adjusted to 0 percent transmittance with a standard solution that is more concentrated than the analyte and to 100 percent transmittance with a standard solution that is less concentrated than the analyte. Without further adjustments the percent transmittance of the analyte is measured. Plots of absorbance (calculated from the percent transmittance) as a function of concentration are generally not linear with the method. With any of the three techniques, use of the working-curve method is recommended. Further details of expanded-scale spectrophotometry can be found in Reference 9.

Difference Spectrophotometry

In *difference spectrophotometry* a double-beam instrument is used to measure the difference in absorbance between two samples. Instead of a blank solution, one of the two solutions is placed in the path of the reference beam of the spectrophotometer. The other solution is placed in the usual position in the path of the sample beam. The instrument measures the difference in absorbance between the two samples. Difference spectrophotometry is particularly useful for measuring small differences between two, nearly identical solutions. Difference spectra are particularly useful for determining qualitative differences between two solutions.

TITRATIONS

If the titrant, the titrand, or the reaction product of a titration absorbs radiation in the ultraviolet-visible region, absorbance measurements, made at a fixed wavelength, can be used to locate the endpoint of the titration. Titrations that are performed with the aid of a spectrophotometer are *spectrophotometric titrations* and those performed with a photometer are *photometric titrations*. The absorbance of the titrand solution is measured after each addition of titrant and the endpoint of the titration is located from a plot of the absorbance of the titrand solution as a function of the volume of added titrant. The absorbance can be measured by removing a fraction of the titrand from the titration vessel and placing it in the absorptive cell, by continuously pumping titrand from the titration vessel into a flow-through cell, or by performing the titration inside the cell. If the titration is performed inside the cell, some provision for stirring and for making the apparatus light tight generally is required.

Before a titration curve is plotted, the absorbance reading obtained during the titration should be corrected for dilution owing to the volume of added titrant. The correction is made by multiplying the measured absorbance by the ratio of the final solution volume (original titrand volume plus added titrant volume) to the titrand volume before addition of titrant. It is often advantageous to use titrant that has a concentration that is large when compared with that of the titrand solution. In that case the volume of added titrant is negligible relative to the titrand volume and a correction for dilution is not necessary. Titrant solutions that are at least 50 times more concentrated than the titrand solution can be used for the purpose. The addition of small volumes of titrant requires the use of burets with a small capacity. Microburets with micrometer heads are commercially available that have a total capacity of 0.2 mL. Microburets with a total capacity of 2 mL are usually accurate to within 0.002 mL and are especially useful for spectrophotometric photometric titrations.

The shape of the titration curve varies with the absorbing species in the solution. The general shapes of several titration curves are shown in Fig. 9-6. Because each absorbing species in solution obeys Beer's law, the absorbance of the solution increases or decreases as the concentration of the absorbing species changes during the titration. The shapes of the titration curves shown in Fig. 9-6, as well as the several other possibilities not shown, can be easily deduced by determining the way in which the concentration of the absorbing species changes during the titration. The endpoint of a titration is located by extrapolating the linear portions of the titration curve to interception. Actual

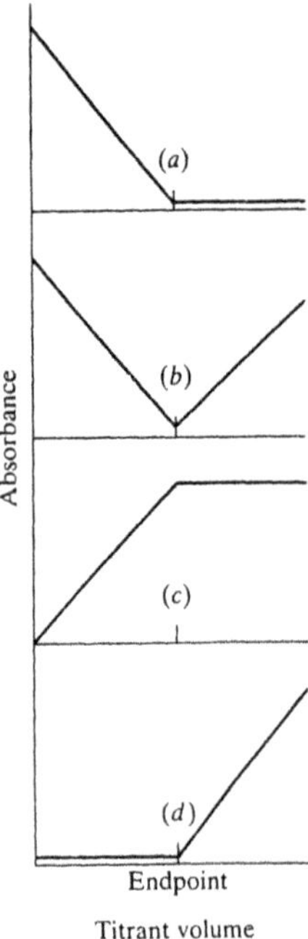

Figure 9-6 Several spectrophotometric or photometric titration curves: (*a*) only the titrand absorbs; (*b*) the titrand and the titrant absorb; (*c*) only the reaction product absorbs; (*d*) only the titrant absorbs.

titration curves are usually slightly rounded near the endpoint owing to dissociation of the reaction product.

Ligand-Metal Ratio

Some metal ions combine with some ligands to form absorbing complexes. A complex is formed when a ligand shares a pair of electrons with metal ion that uses those electrons in its vacant electron orbitals. In a complex the ligand is a Lewis base and the metal ion is a Lewis acid. Most ligands are organic compounds which contain nitrogen or oxygen atoms that are the electron donors.

Several methods can be used to determine the ratio of the ligand L to the metal M in a complex. The *mole-ratio method* and the *method of continuous variations* are the methods most often used, however, other similar methods also can be used. If the mole-ratio method (or the *molar-ratio method*) is used, a series of solutions are prepared in which the concentration of either the metal ion or the ligand is held constant and the concentration of the other component is varied. The absorbances of the solutions are measured and plotted as a function of the ratio of ligand concentration to metallic concentration or the inverse. Alternatively, the absorbances can be plotted as a function of the concentration of the varied component in each solution or as a function of some other variable (such as added volume of the varied component) that is proportional to the concentration of the varied component.

If relatively stable complexes are formed (complexes with large formation constants) and if only the complex absorbs radiation at the chosen wavelength, the absorbance of each solution increases with increased concentration of the varied component until the concentration is sufficient to cause all of the constant-concentration component to react. Further increases in concentration do not cause increased absorbance. The situation is analogous to that of a photometric or spectrophotometric titration in which the only radiation-absorbing species is the reaction product. The molar ratio of ligand to metallic ion at the extrapolated intersection of the two linear portions of the plot corresponds to the ratio in the complex. A sketch of a typical mole-ratio plot is shown in Fig. 9-7, curve *A*.

As in photometric and spectrophotometric titration curves, the plots are rounded in the vicinity of the intersection of the extrapolated linear portions. If more than one complex is formed and if each complex has a different absorptivity and is formed stepwise (not simultaneously), several linear regions are observed and the molar ratios at the extrapolated intersections of adjacent linear regions correspond to the compositions of the complexes.

In some cases both the ligand and the complex absorb radiation. If the ligand absorbs radiation at the wavelength of the study and is the constant-concentration component, the plot has the shape shown in Fig. 9-7, curve *B* or *C*. If the ligand is the variable-concentration component and both the complex and the ligand absorb radiation at the wavelength chosen for the study, the plot has the shape shown in Fig. 9-7, curve *D* or *E*.

The method of continuous variations is also known as the continuous-variation method and as Job's method. When this method is used, the total molar concentration of metallic ion plus ligand is held constant, but the ratio of the components is varied between solutions. The absorbance of each solution is plotted as a function of the mole fraction of one of the components. The wavelength chosen for the study is that at which the complex alone absorbs radiation. The absorbance curve starts at or near zero absorbance, when only component 1 is present in solution, and rises as component 2 is added and the absorbing complex is formed. Maximum absorbance is reached when the mole fractions of the metallic ion and the ligand equal the mole fractions in the complex. After the maximum, the absorbance decreases as the decreasing concentration of component 1 becomes the limiting reagent in the formation of the absorbing complex. The curve obtained by using the method of continuous variations is a *Job's plot*. An example of the plot is shown in Fig. 9-8.

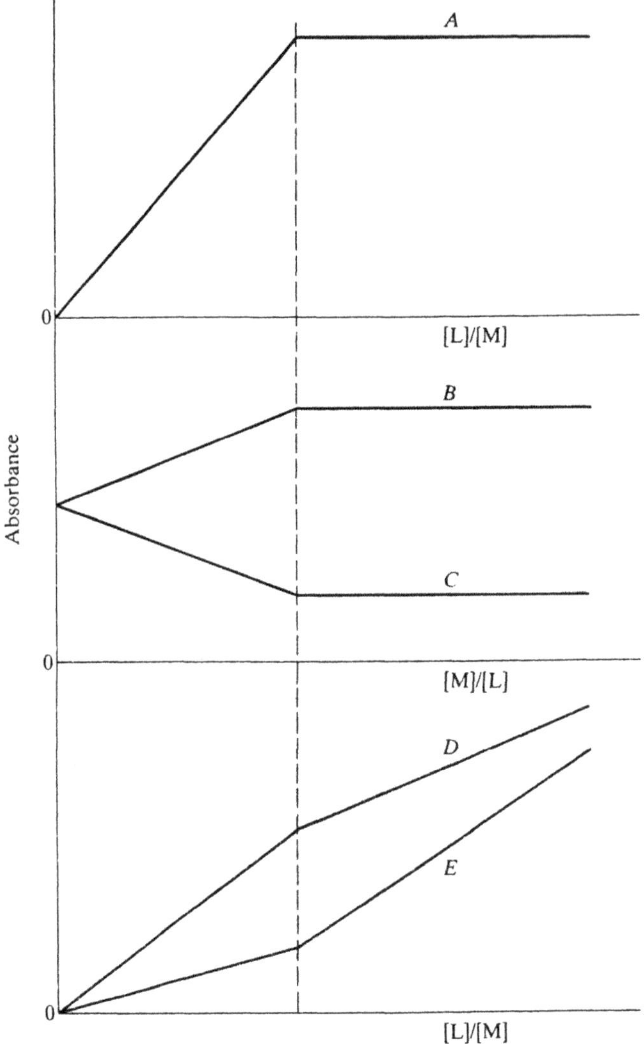

Figure 9-7 Sketches of mole-ratio plots for a complex M_xL_y. Curve *A* shows the plot where only the complex absorbs radiation; curve *B*, the ligand is the constant-concentration component and $y\varepsilon_L$ (ε_L is the molar absorptivity of the ligand and *y* is the subscript on L in the formula for the complex) is less than $\varepsilon_{M_xL_y}$; curve *C*, as in curve *B* except that $y\varepsilon_L$ is greater than $\varepsilon_{M_xL_y}$; curve *D*, the metallic ion is the constant-concentration component and $\varepsilon_{M_xL_y}$ is greater than $y\varepsilon_L$; curve *E*, as in curve *D* except that $\varepsilon_{M_xL_y}$ is less than $y\varepsilon_L$. The molar ratio at the intersection of the linear portions corresponds to the ratio in the complex.

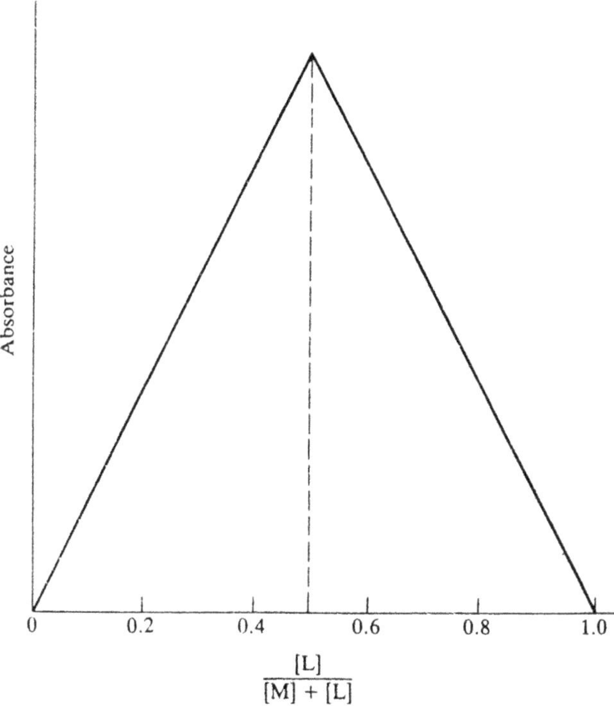

Figure 9-8 A Job's plot for a complex with a formula ML. The mole fraction (0.5) in the complex is indicated by the dashed line.

Sample problem 9-4 The total concentration of metallic ion and ligand in a series of solutions was held constant at 2.00×10^{-4} *M* (the method of continuous variations was used). The concentration of the ligand in each solution and the absorbance of each solution measured at a wavelength at which the complex absorbs are listed in the following table. Determine the ligand-metal ratio in the complex.

Ligand concentration × 10^{-4} *M*	Absorbance
0.20	0.111
0.40	0.223
0.60	0.338
0.80	0.451
1.00	0.563
1.20	0.679
1.40	0.784
1.60	0.678
1.80	0.232
2.00	0.000

SOLUTION The mole fraction of ligand in each solution is calculated.

Ligand concentration × 10^{-4} *M*	**Mole fraction**
0.20	$0.20 \times 10^{-4} / 2.00 \times 10^{-4} = 0.10$
0.40	0.20
0.60	0.30
0.80	0.40
1.00	0.50
1.20	0.60
1.40	0.70
1.60	0.80
1.80	0.90
2.00	1.00

The absorbance is plotted as a function of mole fraction (Fig. 9-9), and the mole fraction of the ligand in the complex is obtained from the point of intersection of the extrapolated linear portions of the plot. From the plot, it is evident that the mole fraction of the ligand in the complex is 0.75 $\left(\frac{3}{4}\right)$.

Consequently, three ligands are complexed to each metallic ion (4 – 3 = 1 metallic ion), and the formula of the complex is ML_3.

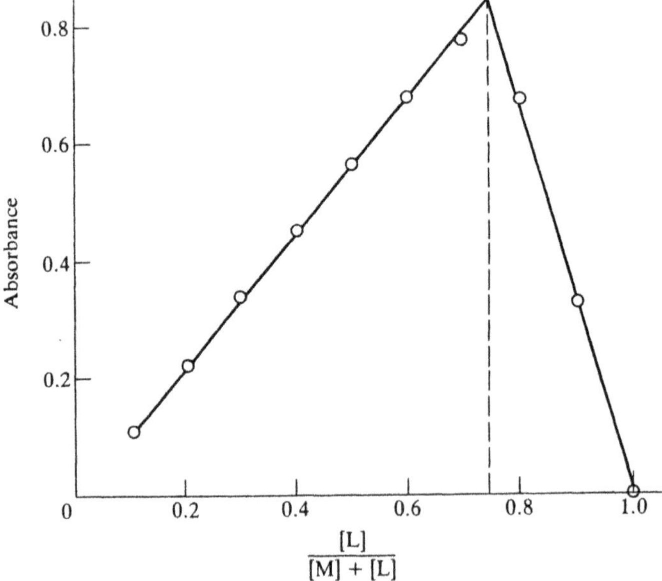

Figure 9-9 A job's plot for the data in Sample problem 9-4.

Equilibrium Constants

The dissociation constant K_a of an acid (or base) can be determined from measurements of absorbance if the acid or anion of the acid absorbs radiation, if both absorb radiation at different wavelengths, or if both absorb radiation to different extents at the same wavelength. Acid-base indicators function because the acidic and basic forms of the indicator absorb radiation at different wavelengths. They are examples of weak acids whose K_a can be determined by using measurements of absorbance.

If the dissociation of an acid (HA) is represented by

$$HA = H^{+} + A^{-} \tag{9-4}$$

the dissociation constant for the reaction is given by

$$K_a = \frac{[H^+][A^-]}{[HA]} \tag{9-5}$$

The pK_a for the acid can be calculated from Eq. (9-5) by taking the negative logarithm of both sides of the equation:

$$pK_a = pH - \log \frac{[A^-]}{[HA]} \tag{9-6}$$

The ratio $[A^-]$ / [HA] can be determined from absorbance measurements in a solution of known pH, and the pK_a of the acid can be calculated by using Eq. (9-6). Use of Eq. (9-6) at a single pH to calculate pK_a can cause a significant determinate error. The equation can be rewritten as

$$pH = \log \frac{[A^-]}{[HA]} + pK_a \tag{9-7}$$

which is in the standard form for a straight line. Log $([A^-]/[HA])$ is the independent variable and pH is the dependent variable. A plot of the pH of the solutions as a function of log $([A^-]/[HA])$ yields a straight line with an intercept on the pH axis [at log $([A^-]/[HA]) = 0$] equal to pK_a. Because the method is based upon measurements at several pH's, it usually yields a reliable value of pK_a.

Perhaps the simplest way to determine the pK_a of an absorbing weak acid or base is to plot the absorbance as a function of the pH of the solution at a wavelength at which either the acidic or basic form of the compound absorbs. Sketches of typical plots are shown in Fig. 9-10. Halfway up the curve the absorbance is one-half of the limiting absorbance (at the top of each curve) and the concentration of the absorbing species is one-half of the concentration of the species at the pH of the limiting absorbance (Beer's law). At the pH of the limiting absorbance, essentially all of the species in solution are in the form of

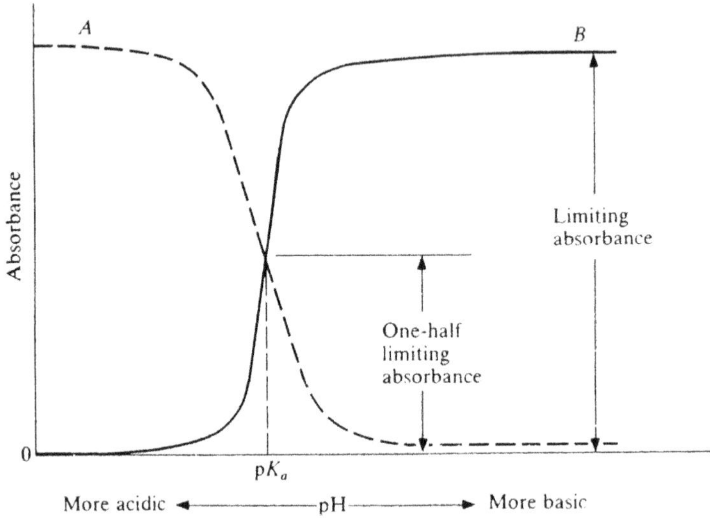

Figure 9-10 Plots of absorbance as a function of pH. Curve *A* shows measurements at a wavelength at which the acid (HA) absorbs radiation; curve *B*, measurements at a wavelength at which the anion (A^-) absorbs radiation.

the absorbing species. Consequently, halfway up the curve the concentration of the absorbing species is one-half the total concentration of HA and A^-, that is, $[A^-] = [HA]$. By making that substitution into Eq. (9-6), it is apparent that the pH at the inflection point in a plot of absorbance as a function of pH corresponds to the pK_a of the acid.

More sophisticated methods are also available for the determination of equilibrium constants from absorbance measurements. Equilibrium constants can be determined for the reactions of charge-transfer complexes and metal-ligand complexes, as well as for acid-base reactions, hydrolysis reactions, and simple polymerizations. Most of the methods involve computerized calculations of the equilibrium constants. A description of the methods is too lengthy to be included here. Further information can be found in References 10 and 11.

Kinetic Measurements

Absorbance measurements can be used to determine the rate at which a chemical reaction takes place and to determine rate constants. If a chemical reactant absorbs radiation, the rate at which the reactant disappears can be monitored by making absorbance measurements as a function of time. Likewise, if the product of a chemical reaction absorbs radiation, the rate of formation of the reaction product can be monitored by measuring the increase in absorbance with time. In either case the absorbance measurements can be made either manually, if the reaction takes place slowly, or continuously with the aid of a strip-chart recorder or a computer controlled monitoring, data storage and display system.

Generally rate constants are measured by preparing a plot of the logarithm of the absorbance of the solution or of the calculated (from the absorbance) concentration of a reactant or product as a function of time. If the absorbance is proportional to the concentration of a single reactant or product of the reaction, the slope of the plot can be used to determine a first-order or pseudo-first-order rate constant. If the logarithm of a concentration is plotted, the slope corresponds to the rate constant divided by 2.303.

Reaction-rate measurements can sometimes be used to quantitatively assay a sample. If the analyte can be reacted with a reagent in a first-order or pseudo-first- order reaction, the reaction rate can be spectrophotometrically measured with time and graphically extrapolated to time zero to yield the reaction rate at the start of the reaction. Because the initial reaction rate for a first-order reaction is directly proportional to the concentration of the reactant, the concentration of the sample can be determined. The working-curve method (plot reaction rate as a function of initial concentration), the standard-addition technique, or some other standard procedure is used.

While using kinetic measurements to perform a quantitative analysis, it is important to carefully control all experimental parameters that might affect the rate at which a reaction occurs. The experimental conditions that are used while determining the rate of reaction of standard solutions should be as nearly identical as possible to those used while measuring the reaction rate of the analyte. Experimental parameters that are of particular importance include the temperature, the concentration of the reactants, the solvent, the ionic strength of the solution, and the order in which the reactants are mixed.

IMPORTANT TERMS

Antibonding orbital
Auxochrome
Bathochromic shift
Bonding orbital
Charge-transfer complex
Chromophore
Colorimeter
Cutoff wavelength
Demasking
Derivative spectrophotometry
Difference spectrophotometry
Expanded-scale spectrophotometry
Heteroannular diene
Hyperchromic shift
Hypochromic shift
Hypsochromic shift
Job's plot
Masking agent
Matchcd cclls
Method of continuous variations
Mole-ratio method
Multiple-path cell
Nonbonding orbital
Photometer
Photometric titration
Phototube
Pi orbital
Sigma orbital
Spectrophotometer
Spectrophotometric titration

PROBLEMS

Chromophores

9-1 Predict the absorptive maximum for the following compound:

$$CH_3-\overset{\overset{\displaystyle O}{\|}}{C}-O-C{=}CHCH{=}CHCl$$

9-2 Predict the absorptive maxima for the following compounds:

9-3 Estimate the absorptive maxima of the following compounds:

9-4 Estimate the absorptive maximum of the following compound:

$$C_2H_5-\overset{\overset{\displaystyle Br}{|}}{C}{=}\underset{\underset{\displaystyle OH}{|}}{C}-CHO$$

Quantitative Analysis

9-5 A solution was prepared by dissolving 0.1235 g of a compound with a molecular weight of 214.3 in sufficient methanol to prepare 1.000 L of solution. The absorbance of the solution at 235 nm was 0.824 in a 1.00-cm cell. Calculate the molar absorptivity of the compound.

9-6 Several standard solutions of a compound were prepared and the absorbance of each solution was measured in a 5.00-cm cell. The absorbance of each solution at 534 nm was plotted as a function of its concentration. The plot was linear with a slope of 973 L/mol and went through the origin. The absorbance of the analyte solution was 0.493 in a 1.00-cm cell at 534 nm. Calculate the concentration of the absorbing compound in the analyte.

9-7 The molar absorptivity of chlorobenzene in *n*-heptane at 265 nm is 12,200 (Table 9-2). Calculate the concentration of chlorobenzene in an *n*-heptane solution that has an absorbance of 0.528 in a 1.00-cm cell at 265 nm.

9-8 Several standard solutions were prepared and the percent transmittance of each solution was measured in a 2.00-cm cell at 435 nm. The percent transmittance of an analyte solution was measured in the same cell and at the same wavelength that was used for the standard solutions. Determine the concentration of the analyte.

Concentration × 10^{-4} *M*	**%T**
0.101	85.5
0.303	65.6
0.505	49.2
0.707	36.1
0.909	27.7
Analyte	41.4

9-9 A 25.0-mL portion of an absorbing analyte solution was added with a pipet to each of five 50-mL volumetric flasks that were labeled S, 1, 2, 3, and 4. Pipets were used to add 5.00-mL of a 2.00×10^{-3} *M* standard solution of the absorbing compound to flask 1, 10.0 mL to flask 2, 15.0 mL to flask 3, and 20.0 mL to flask 4. Each flask was filled to the mark with solvent and the absorbance of each solution was measured at 580 nm in a 1.00-cm cell. Determine the concentration of the analyte solution.

Flask	**Absorbance**
S	0.343
1	0.523
2	0.695
3	0.880
4	1.056

9-10 A solution that contained two absorbing components was spectrophotometrically assayed at 225 and 510 nm. At each wavelength each component obeyed Beer's law and the absorbances were additive. The molar absorptivity of component 1 was 835 at 225 nm and 1538 at 510 nm. The molar absorptivity of component 2 was 1275 at 225 nm and 648 at 510 nm. The absorbance of the mixture in a 1.00-cm cell was 0.842 at 225 nm and 0.538 at 510 nm. Calculate the concentration of the two components in the mixture.

9-11 An analgesic tablet was assayed spectrophotometrically for acetylsalicylic acid and for caffeine. The 0.4791-g tablet was dissolved in methanol and diluted to 250.0 mL in a volumetric flask. A 1.00-mL portion of the solution was withdrawn with a pipet and added to a 100-mL volumetric flask. The flask was filled to the mark with methanol. The absorbance of the diluted solution in a 1.00-cm cell was 0.766 at 225 nm and 0.155 at 270 nm. Beer's law was obeyed for both components at both wavelengths, and the absorbances of the components were additive at both wavelengths. The molar absorptivity of acetylsalicylic acid was 8210 at 225 nm and 1090 at 270 nm. The molar absorptivity of caffeine was 5510 at 225 nm and 8790 at 270 nm. The molecular weights of acetylsalicylic acid and caffeine respectively are 180.15 and 194.19. Calculate the percentage (w/w) of acetylsalicylic acid and caffeine in the analgesic tablet.

REFERENCES

1. Atkins, P. W.: *Physical Chemistry*, W. H. Freeman, San Francisco, 1982.
2. Levine, I. N.: *Physical Chemistry*, 2d ed., McGraw-Hill, New York, 1983.
3. Gatz, C. R.: *Introduction to Quantum Chemistry*, Charles *E.* Merrill, Columbus, Ohio, 1971.
4. Lambert, J. B., H. F. Shurvell, L. Verbit, R. G. Cooks, and G. H. Stout: *Organic Structural Analysis,* Macmillan, New York, 1976.
5. Stern, E. S., and C. J. Timmons: *Electronic Absorption Spectroscopy in Organic Chemistry,* St. Martin's Press, New York, 1970.
6. Braun, R. D.: *Introduction to Chemical Analysis,* McGraw-Hill, New York, 1982, chap. 7.
7. Pinta, M.: *Detection and Determination of Trace Elements,* Ann Arbor-Humphrey Science Publishers, Ann Arbor, Michigan, 1970.
8. Cheng, K. L.: *Anal. Chem.,* **33**: 783(1961).
9. Reilley, C. N., and C. M. Crawford: *Anal. Chem.,* **27**: 716(1955).
10. Ramette, R. W.: J. *Chem. Educ.,* **44**: 647(1967).
11. Christian, S. D.: *J. Chem. Educ.,* **45**: 713(1968)

BIBLIOGRAPHY

Winefordner, J. D. (ed.): *Spectrochemical Methods* of *Analysis,* Wiley-Interscience, New York, 1971.

CHAPTER

TEN

CHEMILUMINESCENCE AND ELECTROCHEMILUMINESCENCE

A brief introduction to luminescence was included in Chapter 5. A review of the appropriate sections of Chapter 5 prior to the study of Chapter 10 is recommended. A description of the luminescence of monoatomic substances was presented in Chapter 8. Luminescence is the emission of radiation from a sample. The emission occurs when an electron in an excited electron level falls to a lower electron level while simultaneously emitting radiation with a wavelength that is characteristic of the energetic change between the levels. Because many vibrational and rotational levels are associated with each electron level in polyatomic species, the emitted radiation occurs over a broad wavelength region. The reason for broadband emission is identical to that for broadband absorption (Chapter 9) in polyatomic ions and molecules.

In analytical chemistry four types of luminescence are important. The types of luminescence vary either in the manner in which the analyte is excited or in the mechanism of the emission. *Chemiluminescence* (CL) relies upon the energy from a chemical reaction to excite the analyte. *Electrochemiluminescence* or *electrogenerated chemiluminescence* (ECL) relies upon electrical energy to cause an electrochemical reaction to occur that, in turn, causes the excitation. The reaction supplies the required excitative energy.

The forms of luminescence that are used most often for analysis are *fluorescence* and *phosphorescence*. In both cases ultraviolet-visible radiation is used to excite the molecules or ions. Fluorescence occurs when an electron in an excited level falls to a lower energy level with the simultaneous emission of radiation. During the transition the electron maintains the same spin. Normally the transition is from an excited singlet state to the ground singlet state.

Fluorescence occurs nearly instantaneously after the initial excitation and ceases when absorption ceases. Typically the electron falls to a lower electron level between 10^{-6} and 10^{-10} s after the excitation.

Phosphorescence occurs when an electron in an excited level falls to a lower energy level with the emission of radiation. During the transition the electron changes its spin state. In most cases phosphorescence occurs during a transition from an excited triplet state to the ground singlet state. Because the transition from a triplet level to a singlet level is "forbidden" the average time required for the emission of radiation is often considerably greater than that during fluorescence. Phosphorescence begins immediately after excitation but typically continues for 10^{-6} to 100 s after the absorption ceases. A more thorough description of the electron transitions that occur during fluorescence and phosphorescence is included in Chapter 5. Fluorescence and phosphorescence are described in detail in Chapter 11.

CHEMILUMINESCENCE

Chemiluminescence occurs after excitation of a molecule or ion by the energy emitted during a chemical or biochemical reaction in which the excited species is a product. In many cases the chemically excited energetic level of a molecule is identical to the energetic level that could have been attained by absorption of electromagnetic radiation. In some molecules, however, the excited levels are not identical. Chemiluminescence can occur in the ultraviolet, visible, or near-infrared regions. The majority of known chemiluminescent reactions occur in the visible region.

Bioluminescence (BL) is chemiluminescence that occurs in biological systems. Perhaps the best-known example of bioluminescence is that which occurs when fireflies emit light. The description of chemiluminescence in this chapter is brief. For more detailed descriptions refer to References 1 through 6.

Chemiluminescence generally proceeds by a variation of one of the following mechanisms :

$$\mathrm{R} \rightarrow \mathrm{P}^{*} \rightarrow \mathrm{P} + h\nu \tag{10-1}$$

$$\mathrm{R} \rightarrow \mathrm{P}^{*} \rightarrow \mathrm{P} + \text{heat} \tag{10-2}$$

$$\mathrm{R} \rightarrow \mathrm{P}^{*} \xrightarrow{\mathrm{A}} \mathrm{A}^{*} \rightarrow \mathrm{A} + h\nu \tag{10-3}$$

$$\mathrm{R} \rightarrow \mathrm{P}^{*} \xrightarrow{\mathrm{A}} \mathrm{A}^{*} \rightarrow \mathrm{A} + \text{heat} \tag{10-4}$$

In each case a reactant R undergoes a chemical reaction to form a product P* in an excited state (signified by the asterisk). The excited product can lose energy by emission of radiation [Eq. (10-1)], by emission of heat [Eq. (10-2)], or by transferring energy to a chemical acceptor A that can subsequently emit radiation [Eq. (10-3)] or heat [Eq. (10-4)]. Chemical reactions are relatively slow processes in comparison to spectral excitation. The kinetics of the reaction determines the rate of chemiluminescence. Consequently, radiative emission occurs more slowly during chemiluminescence than during fluorescence. The rate at which chemiluminescence occurs can be used to measure the kinetics of chemical reactions that precede the emission.

For simple chemiluminescent reactions it is convenient to think of the reactions occurring as illustrated in Fig. 10-1. The dark, solid curve in Fig. 10-1 represents the energetic transitions undergone during a chemical reaction in which chemiluminescence does not occur. For some reactions the activation energy E_a is greater than or equal to the energy that is associated with the excited triplet or singlet level of the products (represented by S and T, Fig. 10-1). In Fig. 10-1 the energy of the reaction (shown by the curve) is identical to the energy of the excited singlet level of the products at point 1 and to the energy of the triplet level at point 2. When the energy in the reaction is equal to that of the excited singlet or triplet level, it is possible for the molecule to cross into the excited state. Radiative emission can occur subsequently when the molecule loses energy, as shown by the vertical arrows in Fig. 10-1.

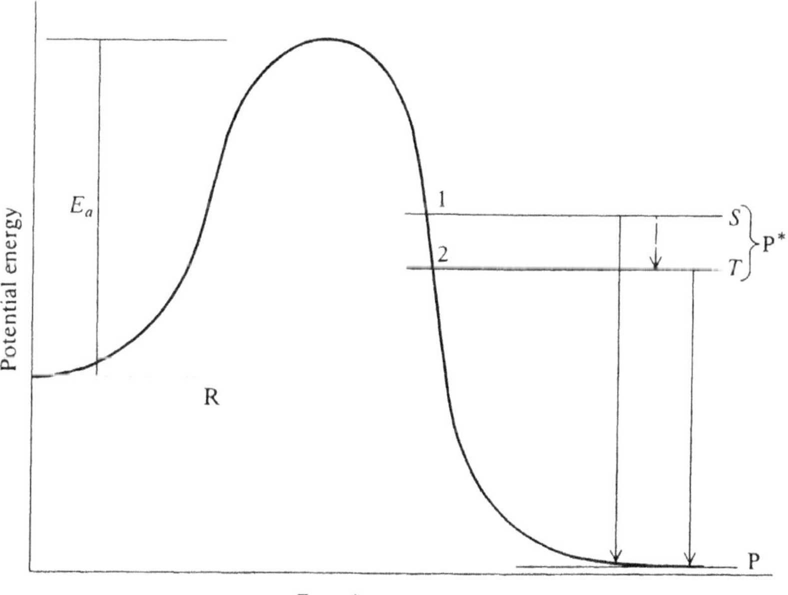

Figure 10-1 A diagram illustrating the mechanism by which a molecule can become excited during a chemical reaction. The curve shows the reaction pathway for a chemical reaction; S, an excited singlet state of the product; T, an excited triplet state of the product; R, reactants; P*, excited products; P, ground-state product. The solid vertical arrows represent radiative emission. The dashed vertical arrow represents radiationless loss of energy.

The lowest excited state for most organic molecules is a triplet state. Because crossing to the triplet state is at least 100 times slower (usually even slower) than crossing to the excited singlet state, and because the crossing must occur during that instant in the chemical reaction when the energetic level of the reaction is identical to that of the excited state, crossing to the excited singlet state is favored. Of course, that assumes that the activation energy is sufficiently large to equal or surpass the energy of the excited singlet state. In some cases the activation energy is sufficient to permit crossing only to the triplet level. When sufficient energy is available, crossing to the excited singlet level is favored.

The relatively long-lived triplet level also favors collisional quenching. Quenching occurs when luminescence ceases because relaxation occurs by a radiationless process. During collisional quenching energy is lost during collisions with other molecules or ions in the solution. In liquid solution significant emission from the triplet level is not normally observed, in part because collisional quenching occurs.

The diagram in Fig. 10-1 applies only to the reaction of diatomic molecules in which the reaction coordinate corresponds to one-dimensional bond stretching. For more complicated molecules the plot of potential energy as a function of the reaction coordinate is more complicated. The curve in Fig. 10-1 for those reactions is a surface rather than a line, and the intersection with the energetic level of the excited singlet or triplet state occurs at a line rather than a point.

Of the types of chemical reactions that can cause chemiluminescence or bioluminescence, *electron-transfer* and *fragmentation* reactions are the most important. Electron-transfer reactions often occur between organic radical cations and anions. The product of the reaction is an excited neutral molecule that potentially can emit radiation. In some cases the radical cations and anions that take part in an electron-transfer reaction can be electrogenerated. If the chemical reactants are electrogenerated, the radiative emission is electrochemiluminescence. The radicals also can be chemically generated. For example, radical anions can be formed by reducing naphthalene or diphenylanthracene with metallic potassium or sodium. Oxidation of the anion with bromine, alkyl or acyl halide, or benzoyl peroxide causes chemiluminescence from the first excited singlet state of the product.

Fragmentation of a molecule sometimes can cause chemiluminescence. As an example, chemiluminescence sometimes occurs during fragmentation of the peroxide bond in an organic molecule. The peroxide generally is initially formed by the reaction of molecular oxygen with an organic molecule. Fragmentation of peroxide which results in chemiluminescence is the basis of some biological assays.

Radiative Measurements during Chemiluminescence

For the vast majority of chemical quantitative analyses it is not necessary to make absolute measurements of the amount of emitted radiation during chemiluminescence. For other applications such as for reporting quantum yields of a reaction, however, it is necessary to be able to make absolute measurements or to be able to make relative measurements that can be mathematically converted to absolute values. Instruments that measure chemiluminescence or electro-chemiluminescence are normally calibrated using a standard lamp or a fluorescent standard.

Standard tungsten-filament and quartz-iodine lamps are available that have been calibrated with blackbody primary standards. The intensity of the emission from the lamps under controlled conditions is accurately known. The lamps are used to calibrate the response of a detector by using the detector to measure the intensity of the radiation that is emitted from the lamps. Tungsten-filament lamps are operated at a fixed temperature and used for calibrations in the visible region. Quartz-iodine lamps are operated at constant current and used for calibrations in the ultraviolet region. Other standard lamps also can be used for calibrations. The use of standard lamps is normally the most accurate way to calibrate an instrument.

Fluorescent standards are easier to use than standard lamps, but the calibrations generally are less accurate. A compound that has a known quantum yield (the ratio of the power of the fluoresced radiation to the power of the absorbed radiation) is placed in the cell of the instrument and the fluorescence from the compound is used as the standard for the calibration. Of the several fluorescent standards, solutions of quinine sulfate are used most often.

In addition to the use of standard lamps and fluorescent compounds, several other methods can be used to calibrate instruments. Among the available methods are the use of the chemiluminescent reaction of luminol[4] and the use of *actinometry*.[7] The oxidation of a known amount of luminol with hydrogen peroxide in the presence of a catalyst causes chemiluminescence of a known number of photons. In dimethylsulfoxide, luminol chemiluminesces after reaction with molecular oxygen.

Actinometry is the technique that uses photon-induced chemical reactions to measure the number of incident photons. The amount of product formed during the photochemical reaction is directly proportional to the number of incident photons. The amount of product that was formed is used to calculate the number of photons that have struck the reactant. A solution of 0.15 *M* potassium ferrioxalate in 0.05 *M* sulfuric acid is often used. The initial concentration of ferrioxalate is dependent upon the wavelength of the incident radiation and the pathlength of the cell. Upon exposure to radiation the solution reacts to form iron (II).

Chemiluminescent Apparatus

The apparatus that is used for chemiluminescent quantitative analysis generally has the components that are shown in Fig. 10-2. The entire apparatus, except for the readout device, is enclosed in a light-tight box to prevent radiation from entering the box and being measured by the detector. The instruments vary in sophistication from relatively simple battery-operated instruments for field use to highly complex computer-controlled laboratory instruments.

The cell can be similar to the cuvets that are used for spectrophotometric measurements or can be flow-through cells. In either case mirrors or *integrating spheres* are often used to reflect the radiation that is emitted from the sample to the detector. An integrating sphere is a reflecting sphere in which the cell is placed. The internal reflective surface of the sphere directs radiation to the detector which is attached to the sphere through a hole in the side. The device is designed such that regardless of the location within the sphere at which radiation is emitted, a constant fraction of the radiation strikes the detector. The detector cannot be in direct line-of-sight with the emitter because in that case the fraction of emitted radiation that strikes the detector is not easily determined. A diagram of an integrating sphere is shown in Fig. 10-3.

In some cases flames similar to those used in atomic absorption spectrophotometry are used as flow-through cells for gas-phase chemiluminescent reactions. Emission from a flame during chemiluminescence is caused by excitation during a chemical reaction rather than by thermal excitation as occurs during flame emissive spectrometry. Shock tubes have been used as cells for high-temperature kinetic studies of gas-phase reactions, but generally are not used for routine quantitative analysis.

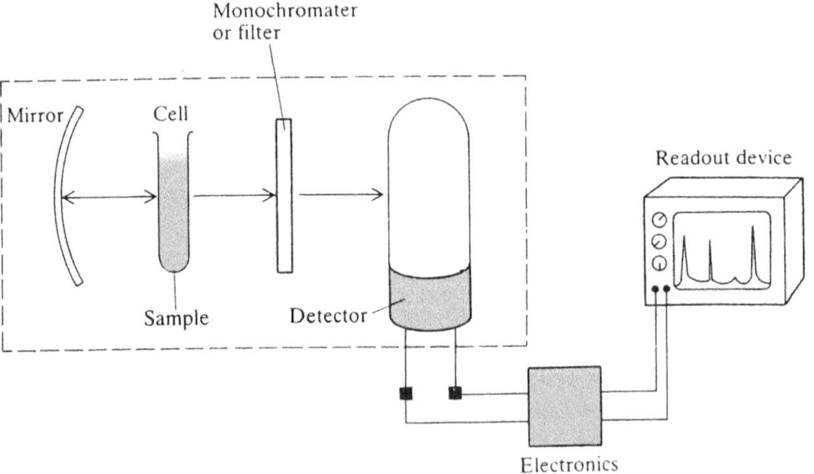

Figure 10-2 A diagram of the major components of a chemiluminescent apparatus. The dashed lines represent a radiation-tight container.

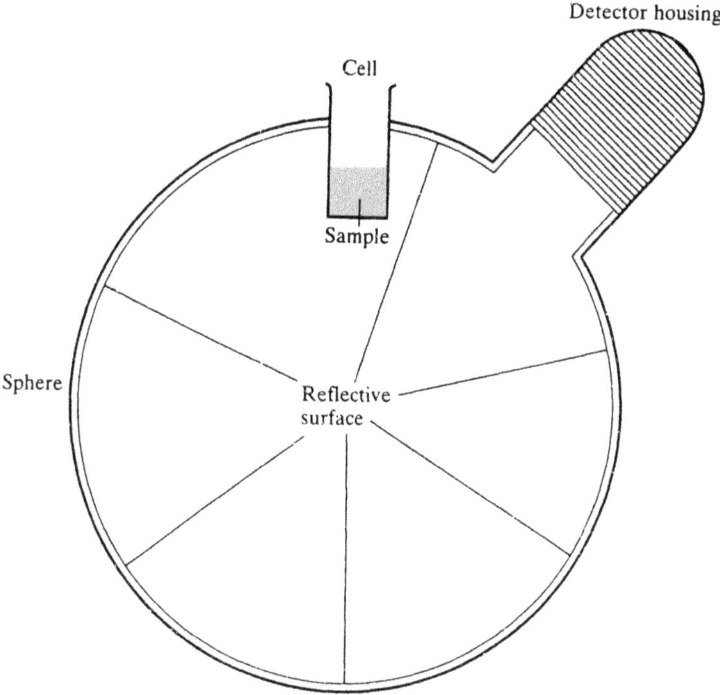

Figure 10-3 A diagram of an integrating sphere and an attached detector housing.

If discrete samples are used, the radiative measurement is started when the last chemical reagent is injected into the cell. Provision for automatic or manual injection of the reagent is required. Piston-driven pipets can be used for the purpose. The measurement is continued until the emitted radiation decays to the background level. Quantitative analysis is performed either with the standard-addition technique or with the working-curve method. Either the height of the peak or the area under the peak can be plotted as a function of concentration.

When using continuous-flow cells, the relative intensity of the emitted radiation is continuously monitored. Gas-phase chemiluminescence nearly always is measured in a continuous-flow cell. With *stopped-flow cells* the sample flow is stopped while the measurement is made. In either case the standard-addition technique or the working-curve method is normally used by plotting the relative intensity of standard solutions as a function of analyte concentration. An additional alternative is to use optical fibers (fiber optics) to transmit chemiluminescence to the detection system. An optical fiber is a flexible fiber constructed from a material, such as glass, that is transparent to the emitted luminescence. It has a reflective internal surface so that essentially all of the EMR which enters one end of the fiber is transmitted through the fiber to the opposite end which is attached to the detector assembly. For chemiluminescence measurements, typically one of the reactants is attached to the outer surface of the end of the fiber by adsorption or another method. The treated end of the fiber is dipped into the analyte solution. The second reactant is the analyte which reacts to yield chemiluminescence.

Assays can be performed either within a fixed wavelength band as in spectrophotometry or over the entire measurable spectral range of the detector. In the latter case a wavelength selector is not required. The wavelength selector can be a monochromator or filter. Filters are used most often primarily owing to the higher intensity of transmitted radiation and the reduced cost. Generally reflective diffraction gratings are used in the monochromators.

Most instruments use a photomultiplier tube as the detector. In some cases photomultiplier-tube arrays or other types of detectors can be used. A diode-array detector has been designed for chemiluminescent detection which is capable of recording 512 measurements throughout the spectral range from 200 to 840 nm in 4 ms.[8] The apparatus is computer controlled.

The readout device can be a laboratory recorder for continuous measurements and for slow individual reactions occurring in a cuvet, or a computer controlled storage and display system for all reactions. Chemiluminescent analysis is normally more reliable when the apparatus is computer controlled.[8-10]

Quantitative Chemiluminescence

Chemiluminescence has been used for quantitative analysis of gaseous reactants in reactors, for analysis of air pollutants, for atmospheric studies, and to indicate the endpoint in gas-phase titrations. In liquid solutions chemiluminescence has been used for trace metal determinations, for quantitative analysis of biochemical and organic reactants, and as a method of endpoint detection in titrations. Examples of each of the applications are presented in the following sections. More detailed information is available in References 1 through 4. Updates can be found in the reviews in the journal *Analytical Chemistry*. In nearly every instance, except for the titrations, the standard-addition technique or working-curve method is used for the analyses.

Gas-Phase Chemiluminescent Analysis

Chemiluminescence has been used extensively for the determination of gaseous air pollutants. Among the pollutants that are assayed using chemiluminescence are ozone, sulfur compounds, and nitrogen compounds. Ozone is generally determined using one of three chemiluminescent reactions. Before the early 1970s the reaction between ozone and rhodamine B that has been adsorbed on activated silica gel was used most often. The method was originally described by Regener.[11,12] An improved apparatus was later designed.[13] The method could be used to determine ozone concentrations of less than 1 ppb (part per billion).

A disadvantage of the Regener method is the need for an internal standard source of ozone for calibration. The apparatus must be frequently calibrated because the rhodamine B slowly decays.

Ozone reacts with ethene in a 1:1 molar ratio and emits a broad band of chemiluminescence that is centered at about 435 nm. It has become the basis of a standard method of analysis for ozone in air and was the Environmental Protection Agency's reference method for the 1971 federal air-quality standards. The method was proposed by Nederbragt, Van der Horst, and Van Duijn.[14]

The third chemiluminescent method for ozone utilizes the 1:1 molar reaction between ozone and NO as shown in Eq. (10-6). The reaction can be used to determine either ozone or NO. The nonanalyte is held in excess so that the amount of luminescence is determined by the amount of the analyte. Luminescence is measured at wavelengths of 600 nm and greater with an infrared-sensitive PM tube.

Atmospheric sulfur compounds such as SO_2, H_2SO_4, other sulfates, H_2S, organic sulfides, and mercaptans are of environmental concern. Of those pollutants, the compound about which the greatest concern exists is sulfur dioxide. It is a major cause of acid rain. Sulfuric acid and other atmospheric sulfates generally occur as particulates rather than as gaseous pollutants.

The sulfur compounds indiscriminately can be assayed by chemiluminescent reaction in a hydrogen-rich air-hydrogen flame. Apparently atomic sulfur in the flame reacts to form S_2 that chemiluminesces. The chemiluminescence is measured at 394 nm. If it is necessary to know the amount of each type of sulfur compound in the analyte, the chemiluminescent method can be used to detect the individual components as they are eluted from a gas chromatographic column (Chapter 26). The components are separated in the column of the chromatograph and individually enter the chemiluminescent flame detector where each component is quantitatively determined. Other substances in the air that have been determined by chemiluminescent reactions in flames include CO, P, B, and Cl_2. The Cl_2 determination requires the presence of indium in the barrel of the burner.

The atmospheric nitrogen compounds that are environmentally of the greatest interest are ammonia, amines, nitrogen monoxide, and nitrogen dioxide. Some instruments are designed to monitor NO_x, which is the sum of all mononitrogen oxides. Nitrogen monoxide and nitrogen dioxide can be determined by the chemiluminescent reaction of NO with ozone. If NO_2 is to be assayed, the NO_2 is initially converted to NO:

$$NO_2 \rightarrow NO + \frac{1}{2}O_2 \qquad (10\text{-}5)$$

which subsequently is reacted with ozone:

$$NO + O_3 \rightarrow NO_2 + O_2 + h\nu \qquad (10\text{-}6)$$

The chemiluminescent emission from Eq. (10-6) is measured at wavelengths equal to or greater than 600 nm with an infrared-sensitive PM tube and a cutoff filter that removes wavelengths shorter than 600 nm. The technique can be used for analysis in the 0.002 to 1000 ppm region. A small pump is used to mix the air sample in a reactor cell with excess ozone from an internal source.

Nitrogen monoxide is determined directly by the reaction in Eq. (10-6) without the preceding reaction shown in Eq. (10-5). If the reaction in Eq. (10-5) occurs prior to the reaction in Eq. (10-6), the result is the sum of NO and NO_2. The NO_2 concentration can be determined by subtracting the NO concentration obtained without performing the reaction in Eq. (10-5) from the NO_x concentration obtained after performing the reaction in Eq. (10-5). The concentration of NO_x in automobile exhaust and in cigarette smoke[15] is determined in the same manner.

Ammonia is a potential interference in the determination because it can react at high temperatures with oxygen to yield nitrogen dioxide:

$$NH_3 \xrightarrow{O_2\text{ (high temperature)}} NO_2 \qquad (10\text{-}7)$$

Ammonia and other basic components can be removed from the air prior to analysis by passing the sample through an acid solution in a *scrubber*. A scrubber is a vessel that contains a liquid solution and through which the gaseous sample is pumped. Usually the sample is converted with the aid of a glass frit to small bubbles that flow through the solution. If the concentration of ammonia is required, the total concentration of NO_x as determined after scrubbing is subtracted from the total concentration of NO_x and NH_3 as determined with no scrubbing.

In addition to being used to assay air pollutants, chemiluminescence has been used to monitor air flow in the upper atmosphere (above 90 km). A chemiluminescent substance is ejected into the atmosphere from a rocket and the path of the chemiluminescence is monitored as a function of time after the ejection. Usually time-lapse photography against a twilight or night sky is used to detect the chemiluminescence. Among the chemicals that have been used for the studies are NO which reacts with atomic oxygen in the upper atmosphere to yield NO_2 in an excited state, trimethylaluminum, acetylene, and $Fe(CO)_5$.

Gas-Phase Chemiluminescent Titrations

Chemiluminescence is used to locate the endpoints of some *gas-phase titrations* in fast-flow reactors. Examples include the titrations of O, H, and N. Titrations in flowing streams consist of adding the titrant to the stream at a fixed rate. The indicator in the titration is used to determine whether the titrant is added at the proper rate. As in other titrations, the reaction must proceed essentially to completion.

Atomic oxygen is titrated with nitrogen dioxide in the following reaction:

$$O + NO_2 \rightarrow NO + O_2 \qquad (10\text{-}8)$$

The chemiluminescent indicator reaction is

$$O + NO + M \rightarrow NO_2 + M + h\nu \qquad (10\text{-}9)$$

where M represents a metallic catalyst. If insufficient nitrogen dioxide is added during the titration, the chemiluminescent indicator reaction occurs and a whitish-green glow (an *afterglow*) appears downstream from the point of injection of titrant. As is typical in titrations that are performed in flowing gas streams, one of the products (NO) of the titrimetric reaction is used in the indicator reaction with the analyte.

The titrimetric reaction for the determination of H is

$$H + NO_2 \rightarrow OH + NO \qquad (10\text{-}10)$$

and the chemiluminescent indicator reaction is

$$H + NO + M \rightarrow HNO + M + h\nu \qquad (10\text{-}11)$$

Chemiluminescence occurs between 600 and 800 nm. The titrimetric reaction for N is

$$N + NO \rightarrow N_2 + O \qquad (10\text{-}12)$$

and its indicator reaction is

$$N + O + M \rightarrow NO + M + h\nu \qquad (10\text{-}13)$$

Chemiluminescence for the reaction occurs in both the ultraviolet and infrared regions.

Liquid-Phase Chemiluminescent Assays

Many chemical and biochemical species have been determined using chemiluminescence. Usually the determinations are performed in aqueous solutions. In order to illustrate the determinations, several examples will be described. Adenosine monophosphate (AMP) and adenosine diphosphate (ADP) have been determined by the chemiluminescent reaction with luciferase to yield adenosine triphosphate (ATP). This is the reaction that occurs in fireflies. Humic acid in natural waters has been determined by its chemiluminescent reaction with permanganate, and uric acid has been determined by its reaction with peroxyoxalate.

Several nitrogen compounds have been determined after conversion to NO by the ozone reaction that was described earlier [Eq. (10-6)]. Nitrite and nitrate have been determined after reduction to NO. Total bound nitrogen in organic compounds and nitrogen in petroleum have been similarly determined after pyrolysis or burning.

Many assays are based upon the chemiluminescent reaction (Fig. 10-4) of the analyte with luminol. Luminol is one of the most-used reagents for chemiluminescent assays. Among the substances that have been determined using luminol are Cl_2, HOCl, OCl^-, H_2O_2, O_2 (ferricyanide is required), and NO_2.

Luminol chemically reacts with any strong oxidizing agent (the analyte) at pH's above 8 to form a transannular peroxide that emits a blue chemiluminescence during its decomposition. A transitional free radical of oxygen (O^- or O_2^-) plays a necessary role in the reaction. Other chemiluminescent assays include the determination of O_2 by reaction with tetraaminoethylenes and the determination of S^{2-} by reaction with hypobromite.

Several trace metals (ppb region) have been determined by the highly efficient chemiluminescent reaction between luminol and hydrogen peroxide. The quantum yields (the ratio of total luminesced energy to total absorbed energy) for the reactions are unusually large. The concentrations of luminol and hydrogen peroxide and the pH of the solutions must be carefully controlled during the determinations. Typically the luminol concentration is maintained at about 1 m*M* and the hydrogen peroxide concentration at about 0.01 *M*. Inorganic ions that have been determined using luminol include Co(II), Cr(III), Cu(II), Fe(II), Mn(II), Ni(II), and V(IV). The Mn(II) assay requires the presence of a complexing amine. During the reaction the analyte usually changes oxidation state. An exception is Ni(II) which acts as a catalyst.

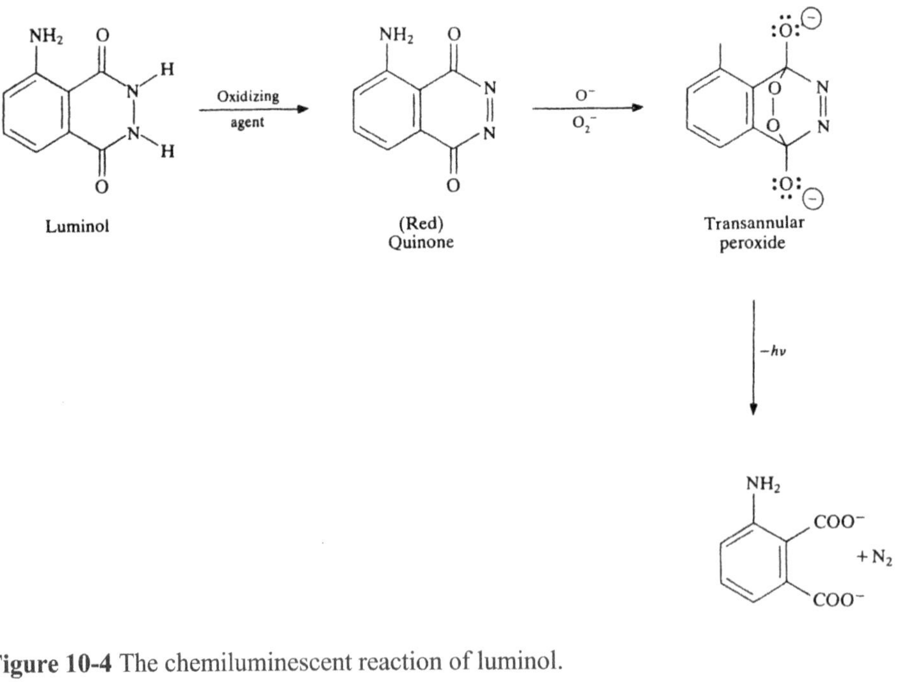

Figure 10-4 The chemiluminescent reaction of luminol.

Some analytes can be determined by chemiluminescent reaction with luminol without hydrogen peroxide. Falling into that category are OCl^-, I_2, MnO_4^-, and Fe(II). The OCl^- determination requires the presence of O_2. More detailed descriptions of specific chemiluminescent assays can be found in the first four references.

Co(II) and Ag(I) have been determined[9] by their chemiluminescent reaction with gallic acid and alkaline hydrogen peroxide. The chemiluminescent reactions of several trace metals with lophine (2,4,5-triphenylimidazole) in alkaline hydrogen peroxide solutions have been used for assays.[16] The successful reactions were those of ClO^-, Co(II), Cr(III), Cr(VI), Cu(II), Fe(III), $Fe(CN)_6{}^{3-}$, Ir(IV), $MnO_4{}^-$, and Os(IV).

Liquid-Phase Chemiluminescent Titrations

Chemiluminescent titrations in liquids are performed in a manner analogous to photometric titrations. The intensity of emitted radiation is plotted as a function of titrant volume. The endpoint is located by extrapolating to interception the linear portions of the plot prior to and after the endpoint.

The endpoints of most chemiluminescent titrations are located using the reaction of one of the chemiluminescent species (the indicator) listed in the preceding section with the analyte. Several titrations have been performed using one of the strong oxidizing or reducing agents that react with luminol to yield chemiluminescence. As examples, iodine has been used as titrant in the titrations of As(III), sulfur dioxide, and sulfite. Luminol is added to the titrand prior to each titration, and serves as the indicator. Before the endpoint little or no radiation is emitted because the iodine reacts with the analyte. After the endpoint the excess iodine reacts with luminol to chemiluminesce. Because the reaction between the titrant and the indicator at the endpoint emits considerable chemiluminescence,

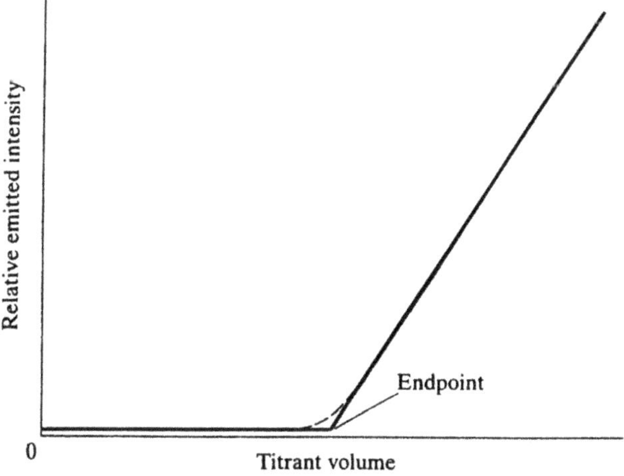

Figure 10-5 A chemiluminescent titrimetric curve for the titration of arsenic(III) with iodine.

the iodine concentration in the buret can be low (about 0.1 m*M*). The chemiluminescence increases with increasing concentration of excess titrant. A sketch of a typical chemiluminescent titrimetric curve is shown in Fig. 10-5. As with photometric titrations, the region in the vicinity of the endpoint is rounded.

ELECTROCHEMILUMINESCENCE

Electrochemiluminescence has been used primarily to study chemiluminescent electron-transfer reactions between electrogenerated radical cations and radical anions. Either two or three electrodes are placed in a solution of the precursors to the radical cations and anions. The potential that is applied to at least one of the electrodes causes generation of the radical ion. An electrogenerated radical cation reacts with an electrogenerated radical anion in an electron-transfer reaction to form an excited neutral molecule that chemiluminesces. A more thorough description of the electroanalytical apparatus that is used for formation of the radical ions is presented in Chapter 23.

The radical cation is formed when a potential is applied to the electrode that is sufficient to remove an electron from the highest, filled molecular orbital of the molecule:

$$R \rightarrow R^{\cdot +} + e^{-} \qquad (10\text{-}14)$$

The radical anion is formed when a potential is applied to the electrode that is sufficient to add an electron to the lowest, unoccupied molecular orbital of the molecule:

$$R + e^{-} \rightarrow R^{\cdot -} \qquad (10\text{-}15)$$

Chemiluminescence can occur when the radical cation reacts with the radical anion. The cation and anion can be formed from the same or different molecular species. Generally platinum electrodes are used to form the radical ions.

If the electrochemical reactions that are shown in Eqs. (10-14) and (10-15) generate radical ions that possess sufficient energy, the radical cation can react with the radical anion to form a molecule in the excited singlet state:

$$\underset{\text{Radical anion}}{R^{\cdot -}} + \underset{\text{Radical cation}}{R^{\cdot +}} \rightarrow \underset{\text{Excited singlet}}{R^{*}} + \underset{\text{Ground state}}{R} \qquad (10\text{-}16)$$

The excited molecule chemiluminesces:

$$R^{*} \rightarrow R + h\nu \qquad (10\text{-}17)$$

Not all reactions between the radical cation and the radical anion result in luminescence because energy can also be lost by radiationless processes. The mechanism by which the ions react to form a molecule in an excited singlet state that subsequently luminesces is the *S* (for singlet) route [Eqs. (10-16) and (10-17)]. The horizontal lines in brackets in the equations represent molecular-orbital energetic levels, and the arrows represent the direction of electron spin.

Many electrochemiluminescent reactions occur in which the reaction between the radical cation and the radical anion does not have sufficient energy to yield a molecule in an excited singlet state. Nevertheless, luminescence is observed. In those cases the radical ions can react to form a molecule in an excited triplet state:

$$R^{\cdot -}\left(\frac{\uparrow}{\uparrow\downarrow}\right) + R^{\cdot +}\left(\frac{}{\uparrow}\right) \rightarrow \underset{\substack{\text{Excited}\\ \text{triplet}}}{R^{*}\left(\frac{\uparrow}{\uparrow}\right)} + R\left(\frac{}{\uparrow\downarrow}\right) \tag{10-18}$$

Either luminescence can occur by relaxation from the triplet state to the ground state or two molecules in the triplet state can react [*triplet-triplet annihilation,* Eq. (10-19)] to form one molecule in an excited singlet state which can luminesce and a second molecule in the ground state. That reaction mechanism is the *T* (for triplet) route:

$$2R^{*}\left(\frac{\uparrow}{\uparrow}\right) \rightarrow R^{*}\left(\frac{\downarrow}{\uparrow}\right) + R\left(\frac{}{\uparrow\downarrow}\right) \tag{10-19}$$

$$R^{*}\left(\frac{\downarrow}{\uparrow}\right) \rightarrow R\left(\frac{}{\uparrow\downarrow}\right) + h\nu \tag{10-20}$$

An example of a compound that electrochemiluminesces by the *S* route is 9,10-diphenylanthracene (DPA). The radical cation and anion of the compound are electrogenerated. An example of a *T*-route electrochemiluminescent reaction is the reaction between the electrogenerated, radical cation of *N,N,N',N'*-tetramethyl-*p*-phenylenediamine (TMPD) with the radical anion of DPA. Electrochemiluminescence has been observed with aromatics, heterocyclics, metallic complexes, and metalloporphyrins.

Almost all electrochemiluminescent experiments are performed in organic solvents. The solvent is chosen for its ability to dissolve the luminescent compound and to stabilize the radical cation and the radical anion. Among the solvents that have been used for electrochemiluminescence are acetonitrile,

N,N-dimethylformamide, dimethylsulfoxide, methylene chloride, propylene carbonate, and tetrahydrofuran.

Electrogeneration of the radical ions can be accomplished using either transient methods in which the ions are successively generated at the same electrode, when the potential at the electrode is changed, or steady-state methods in which the cation is generated at one electrode while the anion is generated at a second electrode. In either case the ions diffuse together to react.

The common transient methods are the *triple-potential-step technique* and the *multicycle technique*. Before use of the triple-potential-step technique the electrode is adjusted to a potential E_i at which no current flows through the cell. During the first step the potential is suddenly changed from E_i to a value E_1 at which the radial anion is formed. During the second step the potential is changed from E_1 to a value E_2 at which the radical cation is electrogenerated. During the third step the potential is changed from E_2 to a value at which the anion is oxidized and the cation is simultaneously reduced (usually E_i). While the potential is E_2 the radical anions that were formed when the potential was E_1 and the radical cations that are being formed diffuse toward each other. Upon coming in contact they react and emit electrochemiluminescence. The electrochemiluminescence ceases during the third step (at E_i), when the reactants are destroyed.

The multicycle technique continuously cycles (usually a square wave) between the potential E_1 at which the anion is formed and the potential E_2 at which the cation is formed. If the cycling is done rapidly (10 Hz or faster), luminescence appears to be continuously emitted. During triple-potential-step studies a single cycle is used for each experiment and the time at each step is usually considerably longer than 0.1 s.

During use of the steady-state methods a *rotating ring-disk electrode* (RRDE) (Chapter 23) is used. One of the radical ions is formed at the planar central disk of the electrode while the oppositely charged ion is simultaneously formed at the planar concentric ring. The electrode is rotated during the experiment and centrifugal force causes the species formed at the disk to flow over the ring where it reacts with the oppositely charged radical ion.

Electrochemiluminescence has been used primarily to study reaction mechanisms. Copper(II) has been determined by measuring the intensity of the emitted radiation after electrogeneration of reactants in solutions that contain luminol.[17]

IMPORTANT TERMS

Actinometry
Afterglow
Bioluminescence
Chemiluminescence
Electrochemiluminescence
Electron-transfer reaction
Fluorescent standard
Fragmentation reaction
Gas-phase titration
Integrating sphere
Multicycle technique
Scrubber
S route
Stopped-flow cells
Triplet-triplet annihilation
Triple-potential-step technique
T route

PROBLEMS

10-1 Why is the lowest excited state for most organic molecules a triplet rather than a singlet state?

10-2 An actinometric solution initially contained 10.00 mL of 0.1478 *M* potassium ferrioxalate in 0.05 *M* sulfuric acid. After exposure to electromagnetic radiation the solution was assayed and found to contain 0.0745 *M* potassium ferrioxalate. Assuming that each incident photon reacts with one ferrioxalate ion (quantum efficiency is 1), calculate the number of photons that struck the solution. If the measurement was made over a period of 15.00 min, the surface area of the cell perpendicular to the incident radiation was 10.00 cm^2, and the incident radiation had a wavelength of 425 nm (refractive index is 1.000), calculate the average intensity of the incident radiation. Assume that the entire surface of the cell was illuminated.

10-3 A fresh portion of the 0.1478 *M* potassium ferrioxalate solution described in Prob. 10-2 was used in the cell described in Prob. 10-2 to determine the intensity of 552-nm incident radiation. After irradiating the cell for 20.0 min, the concentration of the ferrioxalate solution was 0.1013 *M*. Determine the intensity of the radiation.

10-4 Describe a chemiluminescent method for the simultaneous determination of the individual concentrations of NO, NO_2, and NH_3 in a gaseous mixture.

10-5 What is the source of the oxygen radicals in the luminol reaction (Fig. 10-4)?

10-6 Why are the steady-state ECL methods done using an RRDE rather than two stationary electrodes?

REFERENCES

1. Herring. P. J. (ed.): *Bioluminescence in Action,* Academic Press, New York, 1978.
2. Cormier, M. J., D. M. Hercules, and J. Lee (eds.): *Chemiluminescence and Bioluminescence,* Plenum, New York, 1973.
3. DeLuca, M. A., and W. D. McElroy (eds.): *Bioluminescence and Chemiluminescence,* Academic Press, New York, 1981.

4. Lee, J., A. S. Wesley, J. F. Ferguson, and H. H. Seliger, in F. H. Johnson and Y. Haneda (eds.): *Bioluminescence in Progress,* Princeton University Press, Princeton, New Jersey, 1966, p. 35.

5. Bard, A. J., and L. R. Faulkner, in A. J. Bard (ed.): *Electroanalytical Chemistry,* vol. 10, Dekker, New York, 1977, chap. 1.

6. Balzani, V., F. Bolletta, M. Ciano, and M. Maestri: *J. Chem Educ.*, **60**: 447(1983).

7. Michael, P. R., and L. R. Faulkner: *Anal. Chem.,* **48**: 1188(1976).

8. Marino, D. F., and J. D. Ingle, Jr.: *Anal Chem.,* **53**: 645(1981).

9. Stieg, S., and T. A. Nieman; *Anal. Chem.,* **52**: 800(1980).

10. Marino, D. F., and J. D. Ingle, Jr.: *Anal. Chem.,* **53**: 1175(1981).

11. Regener, V. H.: *J. Geophys. Res.,* **65**: 3975(1960).

12. Regener, V. H.: *J. Geophys. Res.,* **69**: 3795(1964).

13. Hodgeson, J. A., K. J. Frost, A. E. O'Keefe, and R. K. Stevens: *Anal. Chem.,* **42**: 1795(1970).

14. Nederbragt, G. W., A. Van der Horst, and J. Van Duijn: *Nature,* **206**: 87(1965).

15. Jenkins, R. A., and B. E. Gill: *Anal. Chem.,* **52**: 925(1980).

16. Marino, D. F., F. Wolff, and J. D. Ingle, Jr.: *Anal. Chem.,* **51**: 2051(1979).

17. Haapakka, K. E., and J. J. Kankare: *Anal. Chem. Acta,* **118**: 333(1980).

CHAPTER

ELEVEN

FLUORESCENCE AND PHOSPHORESCENCE

A brief description of fluorescence and phosphorescence was included in Chapter 5. A review of the pertinent material in Chapter 5 is recommended prior to the study of Chapter 11. Fluorescence and phosphorescence are forms of *photolumin*escence. Photoluminescence is the type of luminescence that occurs following excitation by absorption of electromagnetic radiation.

Excitation during both fluorescence and phosphorescence occurs when radiation is absorbed by an electron in the ground state of a molecule causing the electron to be excited to a higher electron state. During the absorption the electron does not reverse its spin. Generally excitation occurs from a singlet ground state to an excited singlet state.

Normally the excited molecule rapidly loses energy through radiationless processes such as vibrational relaxation and internal conversion until it arrives at the lowest vibrational level of the first excited singlet state. Fluorescence occurs if the excited electron returns to the ground state while simultaneously emitting radiation that is equivalent in energy to the energetic difference between the excited and the ground states. The electron does not change its spin during the transition.

While still in the excited state it sometimes is possible for either the unpaired ground-state electron or the excited-state electron to reverse its spin. In that instance the excited electron has a spin that is identical to that of the unpaired electron in the ground state. A molecule with an unpaired electron in an excited level that has a spin that is identical to that of another unpaired electron in the molecule is in a triplet state. In the triplet state the total electron spin for the molecule is 1, not 0 as it is in the singlet state.

According to Hund's rule, the energy of a molecule that contains two electrons with identical spins is less than that of a molecule with electrons in the same orbitals and with opposite spins. Because of the energetic difference between oppositely spinning electrons in the same molecular orbital, a molecule in a triplet state has less potential energy than the corresponding molecule in a singlet state. Nevertheless, one of the excited vibrational levels of the first excited triplet state is usually nearly identical in energy to the lowest vibrational level of the first excited singlet state. Consequently the transition from the excited singlet state to the triplet state is possible. Phosphorescence occurs when an electron in an excited triplet state relaxes to the ground singlet state while emitting radiation.

The transitions between the excited singlet state and the excited triplet state or between the excited triplet state and the ground singlet state are examples of intersystem crossing. The considerable barrier to spin reversal that exists in a molecule prevents intersystem crossing from occurring as rapidly as singlet-singlet transitions. Because of that barrier, phosphorescence occurs on a much longer time scale than fluorescence. Classical fluorescence ceases essentially immediately after the incident exciting radiation is turned off. Phosphorescence can continue to occur after removal of the exciting radiation for periods up to 100 s or more. More information on the time scales for fluorescence and phosphorescence is included in Chapter 5. Detailed information on photoluminescence can be found in References 1 and 2.

FLUORESCENCE

As previously described, fluorescence corresponds to emission of radiation during the electron transition from an excited level to a lower level (usually to the ground level) without electron-spin reversal. In most cases the fluorescent emission occurs rapidly by the mechanism described in the preceding section. In *delayed fluorescence*, however, emission occurs on the same time scale as phosphorescence.

Delayed fluorescence can occur by any of three mechanisms. *E-type delayed fluorescence* (Fig. 11-1) occurs after initial excitation to the first excited singlet state, intersystem crossing to the triplet state, thermal reexcitation to the excited singlet state, and relaxation to the ground state with simultaneous emission of radiation. The delay in the observed fluorescence is caused by the delays in the two steps involving intersystem crossing. *E*-type delayed fluorescence can occur only when the energetic difference between the excited singlet and excited triplet states is sufficiently small to permit the transition from the triplet state to the singlet state to occur by thermal excitation.

P-type delayed fluorescence occurs through triplet-triplet annihilation (Chapter 10). After excitation to the first excited singlet state by absorption, intersystem crossing occurs as in phosphorescence. Reaction of two molecules

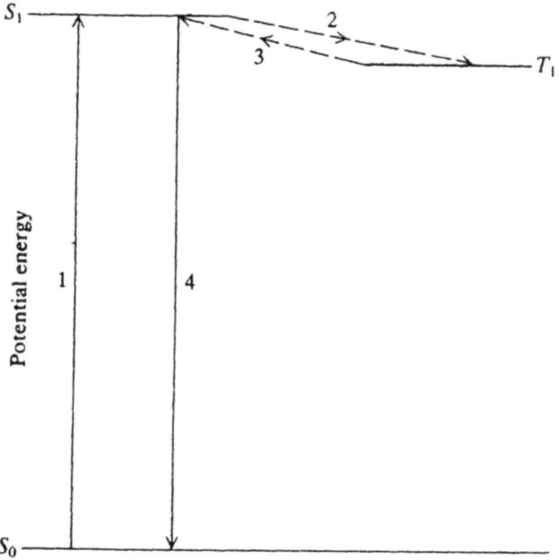

Figure 11-1 An energy-level diagram that illustrates the transitions during *E*-type delayed fluorescence. The dashed lines represent radiationless transitions. The numbers indicate the order in which the transitions occur: 1, absorption; 2, intersystem crossing to the triplet state; 3, thermal excitation to the excited singlet state; 4, fluorescence; S_o, ground singlet state; S_1, excited singlet state; T_1, first excited triplet state.

in the triplet state yields one molecule in the excited singlet state and a ground-state molecule. The excited molecule can fluoresce during relaxation to the ground state. *P*-type delayed fluorescence is summarized in the following reaction scheme:

$$^1A + h\nu_1 \rightarrow {}^1A^* \rightarrow {}^3A^* \qquad (11\text{-}1a)$$

$$2\,{}^3A^* \rightarrow {}^1A + {}^1A^* \qquad (11\text{-}1b)$$

$$^1A^* \rightarrow {}^1A + h\nu_2 \qquad (11\text{-}1c)$$

where A represents the molecule, superscript 1 represents a singlet state, superscrip 3 represents a triplet state, and the asterisk represents an excited state. The intensity of *p*-type delayed fluorescence is proportional to the square of the intensity of the incident radiation because two absorbed photons are required to produce a single emitted photon.

The final mechanism by which delayed fluorescence can occur is *recombination fluorescence*. After absorption of radiation, the molecule in the excited singlet state is photooxidized to yield a radical cation. Apparently the photooxidation occurs from an excited vibrational level of the first excited singlet state.[3] The radical cation recombines with the freed electron or with a freed electron from another molecule to yield the molecule in the first excited singlet state which subsequently fluoresces while returning to the ground state.

The mechanism by which recombination fluorescence occurs is summarized as follows:

$$\mathrm{A}\left(\underline{\uparrow\downarrow}\right) + h\nu_1 \rightarrow \mathrm{A}^*\left(\begin{matrix}\underline{\ \ \downarrow}\\ \underline{\uparrow\ \ }\end{matrix}\right) \tag{11-2a}$$

$$\mathrm{A}^*\left(\begin{matrix}\underline{\ \ \downarrow}\\ \underline{\uparrow\ \ }\end{matrix}\right) + h\nu_2 \rightarrow \mathrm{A}^{\cdot+}\left(\begin{matrix}\overline{\ \ }\\ \underline{\uparrow}\end{matrix}\right) + \mathrm{e}^- \tag{11-2b}$$

$$\mathrm{A}^{\cdot+}\left(\begin{matrix}\overline{\ \ }\\ \underline{\uparrow}\end{matrix}\right) + \mathrm{e}^- \rightarrow \mathrm{A}^*\left(\begin{matrix}\underline{\ \ \downarrow}\\ \underline{\uparrow\ \ }\end{matrix}\right) \tag{11-2c}$$

$$\mathrm{A}^*\left(\begin{matrix}\underline{\ \ \downarrow}\\ \underline{\uparrow\ \ }\end{matrix}\right) \rightarrow \mathrm{A}\left(\underline{\uparrow\downarrow}\right) + h\nu_3 \tag{11-2d}$$

The delay in the fluorescence is caused primarily by the delay during the recombination step [Eq. (11-2*c*)]. The horizontal lines in Eqs. (11-2) represent electron energetic levels and the direction of the arrows represent the spin orientations of the electrons in the energetic levels.

Another type of photoluminescence that can be classified as fluorescence because the electron does not change its spin state during emission is *excimer emission*. Excimer emission is observed at concentrations that are considerably higher than those normally associated with fluorescence. A molecule in a ground singlet state absorbs radiation and is excited to the first excited singlet state. At high concentrations of the ground-state molecule, the excited molecule reacts with a ground-state singlet to form a dimer in an excited state. The excited dimer is an *excimer*. Upon dissociation the excimer yields two ground-state monomers and emits radiation. The process is summarized as follows:

$$^1\mathrm{A} + h\nu_1 \rightarrow {}^1\mathrm{A}^* \tag{11-3a}$$

$$^1\mathrm{A}^* + {}^1\mathrm{A} \rightarrow {}^1\mathrm{A}_2^* \tag{11-3b}$$

$$^1\mathrm{A}_2^* \rightarrow 2\,{}^1\mathrm{A} + h\nu_2 \tag{11-3c}$$

The superscript 1 to the left of each A in Eqs. (11-3) is an indicator of the singlet state.

PHOTOLUMINESCENT THEORY

The amount of emitted radiation during photoluminescence is, of course, dependent upon the amount of absorbed radiation. Because it is possible for an excited molecule to return to the ground state by a radiationless loss of energy, the number of photons of emitted radiation is usually less than the number of photons of absorbed radiation. The *quantum yield, quantum efficiency,* or *luminescent efficiency* ϕ can be defined, for our purposes, as the ratio of the number of luminescent photons to the number of absorbed photons. The quantum efficiency cannot exceed unity during the normal forms of fluorescence. The quantum efficiency is constant for a particular substance under a fixed set of experimental conditions. The intensity of luminescing radiation I_L can be written in terms of the quantum efficiency and the intensity of absorbed radiation:

$$I_L = \phi(I_0 - I) \tag{11-4}$$

The intensity of absorbed radiation is equal to the difference between the intensity of the incident exciting radiation I_0 and the transmitted radiation I.

The intensity of the transmitted radiation can be calculated from Beer's law:

$$A = abC = -\log \frac{I}{I_0} \tag{11-5}$$

Solution of Beer's law for the transmitted intensity yields

$$I = I_0 \times 10^{-abC} = I_0\, e^{-2.303abC} \tag{11-6}$$

Substitution into Eq. (11-4) gives the following result:

$$I_L = \phi(I_0 - I_0\, e^{-2.303abC}) = \phi I_0(1 - e^{-2.303\, abC}) \tag{11-7}$$

The exponential term in Eq. (11-7) can be written as an infinite series using

$$e^x = 1 + x + \frac{x^2}{2!} + \frac{x^3}{3!} + \dots + \frac{x^n}{n!} \tag{11-8}$$

Substitution of the series for the exponential term in Eq. (11-7) yields

$$I_L = \phi I_0 \left[1 - 1 + 2.303abC - \frac{(2.303abC)^2}{2} + \frac{(2.303abC)^3}{6} - \dots \right]$$

$$= \phi I_0 \left[2.303abC - \frac{(2.303abC)^2}{2} + \frac{(2.303abC)^3}{6} - \dots \right] \tag{11-9}$$

For dilute solutions the absorbance (abC) is small. If abC is equal to 0.05 the second term in the series in Eq. (11-9) is 2.5 percent of the first term and the succeeding terms are even smaller. If the absorbance can be kept less than 0.05 by maintaining a low concentration of the analyte, all of the terms in the series with the exception of the first are negligible and the equation simplifies to

$$I_L = \phi I_0\, 2.303abC \tag{11-10}$$

From Eq. (11-10) it is apparent that the intensity of emitted photoluminescence is proportional to the quantum efficiency, the incident intensity of the absorbed radiation, and the concentration of the analyte. In most laboratory instruments only the fixed fraction f of the total luminescence that strikes the detector is measured and a term must be added to account for the response R of the detector. Addition of those two factors to Eq. (11-10) yields

$$I_L = 2.303 f R \phi I_0 abC \tag{11-11}$$

Combination of constants for a particular analyte, cell, and instrument gives the simplified equation

$$I_L = KI_0 C \tag{11-12}$$

When the absorbance is larger than about 0.05, the terms beyond the first term in the brackets in Eq. (11-9) become significant and a nonlinear relationship between luminescent intensity and concentration exists. When the absorbance becomes very large, Eq. (11-9) simplifies to

$$I_L = \phi I_0 \tag{11-13}$$

and the luminescent intensity becomes independent of concentration. The derivation of Eq. (11-13) assumes that the intensity of the incident radiation is the same throughout the cell. For dilute solutions the assumption is adequate, but at high concentrations significant absorption (absorbance greater than 2) decreases the intensity of the radiation as it passes through the cell and causes a decrease in the observed luminescence. Because most instruments are designed

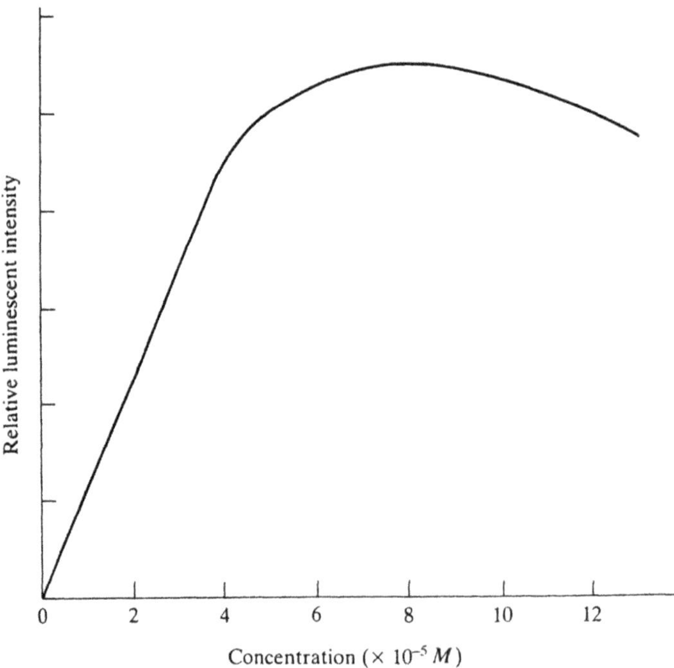

Figure 11-2 A plot of relative fluorescent intensity as a function of concentration for a substance that has a molar absorptivity of about 4×10^4.

to measure luminescent intensity in the center of the cell where the decrease in intensity can be significant at high analyte concentrations, a plot of luminescent intensity as a function of analyte concentration often goes through a maximum and actually decreases at high concentrations (Fig. 11-2).

It is possible to have two concentrations of an analyte that have identical emissive intensities. If that is suspected to be a problem in a particular assay, the analyte solution can be diluted and the luminescence remeasured. If the luminescence proportionately decreases with the dilution, the concentration is the lower of the two possible values. If the luminescence increases or does not decrease as much as expected, the analyte concentration probably is the higher of the possible values.

Electron Transitions during Photoluminescence

In most cases both fluorescence and phosphorescence are preceded by absorption and subsequent relaxation to give a molecule in the first excited singlet state. The transition between the ground state and the first excited singlet state is particularly important in determining the extent and type of photoluminescence that occurs. With most aromatic compounds and conjugated organic dienes that do not contain a heteroatom, the transition is π to π^*. The molar absorptivity for compounds that undergo a π to π^* transition is often large (in the vicinity of 10,000) which indicates that the transition is highly probable. A relatively large population of electrons in the excited singlet state results from a π to π^* transition. That favors fluorescence because the amount of absorbed radiation [$I_0 - I$ in Eq. (11-4)] is increased. Hydrocarbons that contain no heteroatoms and that luminesce always involve a π to π^* transition because nonbonding molecular orbitals are unavailable.

If a heteroatom exists in a luminescent molecule, the transition from the ground state to the first excited singlet state can be an n to π^* transition. An electron in a nonbonding orbital that is associated with the heteroatom is excited to a π^* orbital of the molecule. Molar absorptivities associated with n to π^* transitions are usually relatively small (less than 1000) in comparison to absorptivities associated with π to π^* transitions because nonbonding n orbitals do not overlap with π^* orbitals as much as bonding π orbitals do. Consequently, less fluorescence generally is observed following excitation by an n to π^* transition than is observed following excitation by a π to π^* transition.

The energetic difference between the first excited triplet state and the excited singlet state when the absorption corresponds to an n to π^* transition is smaller than that observed when the absorption corresponds to a π to π^* transition.

Consequently, although initial n to π^* transitions generally lead to less fluorescence than π to π^* transitions, they often precede increased intersystem crossing and increased phosphorescence. In many cases phosphorescence occurs when the absorption involves an n to π^* transition followed by intersystem crossing and emission during a transition from the π^* triplet state to a π ground state.

Factors That Affect Photoluminescence

Anything that can affect the energetic transitions within the molecule can affect photoluminescence. Photoluminescence is favored when the absorption is efficient (high absorptivities). Fluorescence is favored when the energetic difference between the excited singlet and triplet states is relatively large and when the energetic difference between the first excited singlet state and the ground state is sufficiently large to prevent appreciable relaxation to the ground state by radiationless processes. Phosphorescence is favored when the energetic difference between the first excited singlet state and the first excited triplet state is relatively small and when the probability of a radiationless transition from the triplet state to the ground state is low. Any physical or chemical factor that can affect any of the transitions can affect the photoluminescence. In the following sections several of the factors that can affect photoluminescence are described.

Structural rigidity Photoluminescent compounds are those compounds in which the energetic levels within the compounds favor deexcitation by emission of uv-visible radiation rather than by loss of rotational or vibrational energy. Fluorescing and phosphorescing compounds usually have a rigid planar structure. The rigidity of the molecule prevents loss of energy through rotational and vibrational energetic level changes. Any substituent on a luminescent molecule that can cause increased vibration or rotation can quench the luminescence. The planar structure of fluorescent compounds allows delocalization of the pi electrons in the molecule. That in turn increases the chance that luminescence can occur because the electrons can move to the proper location to relax into a lower energy localized orbital.

Organic compounds that contain only single bonds between the carbons do not luminesce owing to lack of absorption in the appropriate region and lack of a planar and rigid structure. Organic compounds that do luminesce generally consist of rings with alternating single and double bonds between the atoms (conjugated double bonds) in the rings. The sp^2 bonds between the carbons in the rings cause the desired planar structure, and the alternating double bonds give rigidity and provide the pi electrons necessary for luminescence.

Naphthalene

Anthracene

Fluorene

9,10-Diphenylanthracene

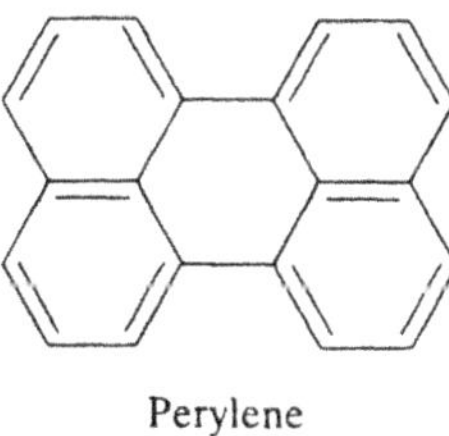

Perylene

Figure 11-3 The structural formulas of several fluorescent compounds.

In the presence of delocalized pi orbitals, absorbed energy is spread throughout a relatively large region. If extended conjugation does not exist in an absorbing molecule, the absorbed energy that is concentrated in a single bond can cause the bond to rupture. The process in which a molecule dissociates as the result of absorption of radiation is *photodecomposition*. Often relaxation occurs when absorbed energy is consumed in a photodecomposition reaction. In that case an electron no longer remains in an excited level which can lead to luminescence. Examples of fluorescing hydrocarbons with the required rigid structure are naphthalene, anthracene, fluorene, 9,10-diphenylanthracene, and perylene (Fig. 11-3).

Heteroatoms and substituents Substitutents on a luminescing compound that can delocalize the pi electrons of the compound generally increase the probability of a radiative transition occurring between the excited singlet state and the ground state. The result is enhanced fluorescence. Generally ortho-para-directing substituents donate electrons to a ring and increase the probability that fluorescence will occur. Meta-directing substituents remove electrons from a ring and reduce the fluorescent probability. The effects of several common substituents on the fluorescence of aromatic compounds are listed in Table 11-1.

Alkyl groups normally do not affect fluorescence unless they are sufficiently large to physically contact or nearly contact a portion of the molecule in addition to the bonding location. In that case internal conversion and a decrease in fluorescent intensity is relatively probable. Carbonyl substituents contain the oxygen heteroatom that adds an n to π^* transition to the molecule and causes decreased fluorescence, increased intersystem crossing, and increased phosphorescence.

The addition of cyanide also adds an n to π^* transition to a molecule, but unlike the case for carbonyl substituents, the transition is energetically greater than the π to π^* transition. Consequently, the introduction of a cyano substituent generally has little or no effect on the emitted wavelength. A small increase in fluorescent intensity is sometimes observed.

Amino substituents increase both wavelength and the intensity at which emission occurs. Amino and hydroxide substituents add weak basic and acidic functional groups to a molecule that can cause fluorescence to be pH-sensitive. Those substituents can also hydrogen-bond with some solvents causing changes in the fluorescence with the solvent.

The low-energy n to π^* transition that is introduced with nitro substituents causes fluorescent quenching in many compounds. In some cases nitro substituents cause a molecule to phosphoresce. Sulfonates usually have no effect on fluorescence. Addition of a sulfonate to a luminescent molecule can be used to increase the solubility of the molecule in aqueous solutions without altering the luminescence.

Table 11-1 A list of the effects of several substituents on the emissive wavelength and the intensity of fluorescence

Substituent	Effect on wavelength	Effect on intensity
Alkyl	None	Slight increase or decrease
COOH, CHO, COOR, CRO	Increase	Decrease
OH, OMe, OEt	Increase	Increase
CN	None	Increase
NH_2, NHR, NR_2	Increase	Increase
NO_2, NO	Large increase	Large decrease or complete quenching
SH	Increase	Decrease
SO_3H	None	None
F, Cl, Br, I	Increase	Decrease

The electromagnetic fields that are associated with relatively heavy atoms affect electron spins within a molecule more than the fields associated with lighter atoms. The addition of a relatively heavy atom to a molecule causes excited singlet and triplet electrons to become more energetically similar. That reduces the energetic difference between the singlet and triplet states and increases the probability of intersystem crossing and of phosphorescence. The probability of fluorescence is simultaneously reduced. The increased phosphorescence and decreased fluorescence with the addition of a heavy atom is the *heavy-atom effect.* If the heavy atom is a substituent on the luminescent molecule, it is the *internal heavy-atom effect.* The *external heavy-atom effect* occurs when the heavy atom is part of the solution (usually the solvent) in which the luminescent compound is dissolved rather than directly attached to the luminescent molecule.

The effect that the halides have upon a luminescent molecule is an example of the internal heavy-atom effect. Generally the fluorescent intensity decreases with the addition of a halide. As the atomic weight of the halide increases the fluorescent intensity decreases and the phosphorescent intensity increases. The effect can be illustrated using the monohalide substituents of naphthalene. The maximum (peak) relative fluorescent intensities of the compounds respectively decrease in the approximate ratio 1700 : 120 : 3 : 1 for the fluoride-, chloride-, bromide-, and iodide-substituted compounds. Simultaneously the maximum relative phosphorescent intensities increase for the same compounds in the approximate ratios 1 : 5 : 5 : 7 (see Reference 2, p. 184). The internal heavy-atom effect can be used in many cases to predict the relative fluorescent and phosphorescent intensities of different metallic complexes of the same ligand. Other factors are also involved in determining luminescent intensities of metallic complexes.

Environmental effects Several environmental factors including the solvent, the temperature, the pH, the presence of dissolved oxygen, the possibility of hydrogen bonding, and the presence of other solutes can affect the photoluminescence of a compound. Most fluorescent assays are performed in liquid solution. In most cases the solvent has the effect of broadening emitted fluorescent bands. Generally the polarity of the solvent determines the extent of interaction between the analyte and the solvent. Electrostatic attraction between the analyte and the solvent stabilizes (lowers) the energetic levels of the analyte. The effect of the solvent depends upon whether the ground state or the excited state of the analyte is stabilized more. That in turn depends upon whether the ground state or the excited state is more polar.

In order to understand the way in which the solvent affects the emitted fluorescent wavelength, it is helpful to consult the energetic diagram in Fig. 11-4. According to the *Franck-Condon principle*, the nuclear positions in a molecule do not have sufficient time to change during the short time in which electron transitions occur. Consequently, absorption of radiation occurs during an electron transition from the equilibrium geometry of the ground state to the nonequilibrium geometry (Franck-Condon state) of the excited singlet state. While in the excited state, the bonds reorient themselves to yield a more stable (lower energy) equilibrium geometry, as illustrated by the upper dashed line in Fig. 11-4.

Fluorescence occurs when the electron in the equilibrium geometry of the excited state falls to the Franck-Condon (nonequilibrium geometry) ground state. Reorientation of the bonds in the molecule returns the molecule to the lower energy geometry of the equilibrium ground state, as illustrated by the lower dashed line in Fig. 11-4. The difference in energy between the Franck-Condon excited state and the equilibrium excited state, and between the Franck-Condon ground state and the equilibrium ground state, are dependent upon the interactions between the solvent and the excited-state and ground-state molecules.

If the excited state is more polar than the ground state, then as the polarity of the solvent increases, the upper energetic state progressively is more stabilized than the ground state, and the potential energy of the equilibrium geometry in the excited state decreases while the potential energy of the ground state only slightly changes. As a result, the energy of the emitted fluorescence decreases and the central wavelength at which the fluorescence occurs increases with increasing solvent polarity. Conversely, if the ground state is more polar than

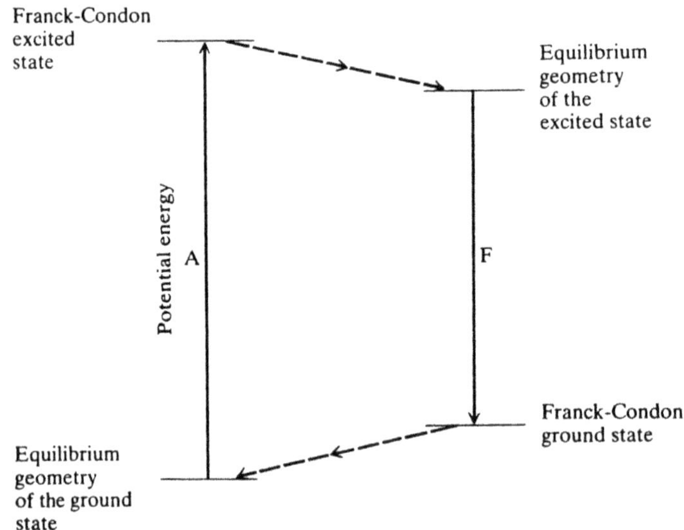

Figure 11-4 A diagram illustrating the effect of solvent on the energetic transitions that occur during fluorescence. A, absorbance; F, fluorescence. The dashed lines represent radiationless transitions. Refer to the text for a description.

the excited state, the ground state is increasingly stabilized relative to the excited state as the solvent polarity increases. Consequently, the energetic difference between the excited and ground states increases, and a shift to shorter fluorescent wavelengths occurs with increasing solvent polarity.

For n to π^* transitions the ground state of the molecule is more polar than the excited state and the wavelength of the emitted radiation becomes shorter with increasing solvent polarity. For π to π * transitions the excited state of the molecule is usually more polar than the ground state. Consequently, the central wavelength of the emitted radiation becomes longer as the solvent's polarity increases.

A solvent that contains a heavy atom can affect photoluminescence of a molecule by the external heavy-atom effect. As the mass of the atom increases the fluorescent intensity decreases and the phosphorescent intensity increases. If the phosphorescence of a compound is to be studied, a solvent that contains a heavy atom is commonly used.

Increasing temperature increases the mean velocity of the luminescent molecule in both the ground and excited states. An increased velocity causes an increased number of collisions between molecules. When a molecule in an excited state collides with another molecule, the probability of internal conversion increases causing decreased luminescence.

In those cases in which two electron energetic levels lie close to each other, an increase in temperature can have a relatively large effect on the amount of thermal excitation from one energetic level to the other. Thermally assisted intersystem crossing during *E*-type delayed fluorescence can be dramatically increased by increasing the temperature. Because the amount of luminescence can be temperature-dependent, the temperature of the cell should be controlled during luminescent studies.

Changing the pH of a solution can have an effect upon the observed fluorescence of a compound if that compound contains either an acidic or basic functional group. The molecular structure of the fluorescent species changes with pH as the weak acid or base dissociates. That change in structure can affect the observed fluorescence. When considering the effect of pH upon a particular substance, it is important to keep in mind that a molecule in an excited electron state could have a considerably different structure and electron distribution than that of the substance in its ground state. Because proton-transfer reactions are rapid, they can occur while the molecule is in the excited state. Acid-base reactions in the excited state as well as those in the ground state can affect fluorescence. Typically the difference in the pK_a of a molecule in the ground state as compared to the excited state is between 4 and 9. It is not easy to predict which state will have the higher pK_a. For assays of those compounds that are affected by pH changes, it is important to maintain the analyte and the standard solutions at the same pH.

Hydrogen bonding between the solvent and the photoluminescent compound can affect the luminescence in several ways. Hydrogen bonding normally has the greatest effect on the hydrogen-bonded heteroatom in the luminescent molecule. The nonbonding electrons on the heteroatom are stabilized by hydrogen bonding. Changes in the wavelength at which luminescence occurs are affected by pH in hydrogen-bonding solvents whenever the excited-state molecule and the ground-state molecule are stabilized to different extents. Generally molecules dissolved in hydrogen-bonding solvents that are stronger acids in the excited state than in the ground state exhibit shorter emissive wavelengths than molecules that are stronger bases in the excited state. In some cases hydrogen bonding can also cause decreased luminescent intensity owing to increased internal conversion in the hydrogen-bonded molecule.

Dissolved molecular oxygen functions as an efficient quencher of both fluorescence and phosphorescence. Quenching is caused by the interaction between the unpaired electrons in the ground triplet state of oxygen and the electrons in the excited states of the luminescent compound. The effect is enhanced in non-polar as opposed to polar solvents. While performing luminescent studies and particularly phosphorescent studies, it is important to remove dissolved oxygen prior to the measurement. Dissolved oxygen can be removed by bubbling nitrogen, helium, or argon through the solution for several minutes prior to the assay. More details relating to oxygen removal are included in Chapter 23.

The presence of a second solute in addition to the analyte can affect the luminescence of the analyte if the second solute absorbs or emits radiation at the wavelength at which either excitation from or emission by the analyte occurs. If the second solute absorbs radiation at the same wavelength at which the analyte absorbs, the intensity of the radiation that strikes the analyte can be decreased causing decreased excitation. If the second solute absorbs radiation at the wavelength at which luminescence is emitted from the analyte, some of the luminescence can be absorbed prior to exiting from the cell. The absorption of exciting or luminescing radiation by a solute is the *inner-filter effect*. Regardless of whether the exciting or the luminescing radiation is absorbed, the radiation emitted from the cell is decreased. The decrease in emitted radiation that is caused by the inner-filter effect increases as the concentration of the second solute increases. Of the several methods to compensate for the inner-filter effect, perhaps the best is the standard-addition technique. The presence of additional nonabsorbing solutes in the solution sometimes can decrease or completely quench luminescence of the analyte by collisional deactivation. Again, the standard-addition technique can be used to minimize errors in an assay that are caused by the additional solute.

LUMINESCENT APPARATUS

A block diagram of the major components of an instrument that is used to measure photoluminescence is shown in Fig. 11-5. Electromagnetic radiation from an ultraviolet-visible source passes through a wavelength selector and through the cell as in a spectrophotometer. Unlike the measurement of absorption in a spectrophotometer, however, a portion of the emitted radiation that exits from the cell is measured. Because the luminescent radiation can be emitted in broad bands that are centered at different wavelengths, a second wavelength selector is required in the path of the emitted radiation between the cell and the detector.

The emitted radiation is not usually measured in-line with the exciting radiation, as in absorptive measurements, owing to possible spectral interference from the exciting radiation. Photoluminescence has been measured at many angles relative to the incident radiation and at many locations within the cell. The most common practice is to measure the emitted radiation at 90° from the path of the exciting radiation and at the center of the cell. The signal from the detector is amplified, if required, and routed to a readout device.

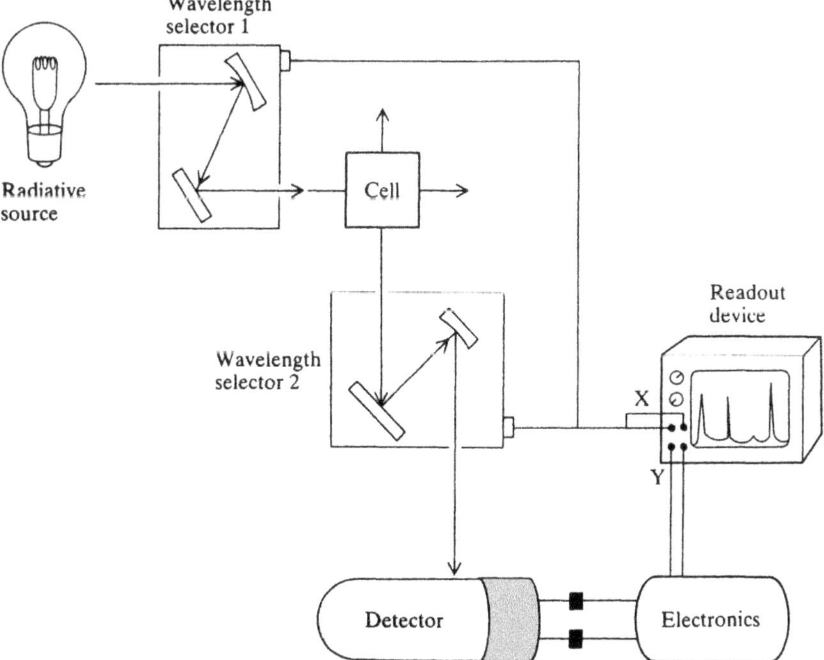

Figure 11-5 A diagram showing the major components of a single-beam instrument that is designed to measure photoluminescence.

Many instruments have an optical pathway that is more complicated than that shown in Fig. 11-5. In some instruments the radiation from the source is divided into two beams. One of the beams passes through the cell while the second beam is used as a reference. The second beam can pass through a reference solution or it can be attenuated prior to detection. Null-point detection is used in many instruments. The attenuator in the reference beam is adjusted until the response of the detector to the reference beam is identical to that of the emitted sample beam. The two beams can either be detected alternately by the same detector or simultaneously by separate detectors. Generally the ratio of the two signals (the attenuator setting) is registered on the readout device.

Optical Excitative Sources

Of the several sources of incident radiation that have been used for photoluminescent measurements, the mercury-arc lamp and the xenon gas discharge lamp have been used most often. Because the mercury-arc lamp emits line spectra rather than a continuum, it cannot be used in instruments in which the wavelength of the incident radiation is scanned. It is used nearly exclusively in instruments that use a filter as the wavelength selector between the source and the cell. The most intense line that is emitted from mercury lamps is at 253.7 nm. Other relatively intense lines occur at 296.7, 365.0, 404.7, 435.8, 546.1, 577.0, and 579.1 nm. In some mercury lamps the inside surface of the glass envelope that encloses the lamp is coated with a solid phosphorescent substance that emits radiation throughout a broad wavelength region. Emitted radiation from the mercury is absorbed by and induces emission from the phosphorescent substance. The lamp emits continuous radiation throughout the phosphorescent wavelength range of the coated compound.

High-pressure xenon lamps emit a continuum of radiation that is useful throughout the ultraviolet-visible and into the near-infrared spectral regions. Unfortunately the intensity of the emitted radiation varies with wavelength throughout the entire useful range. The lamp is often used because it emits a continuum and because it emits more intense radiation in the ultraviolet region than does a tungsten-filament lamp. In order to emit constant-intensity radiation at a particular wavelength, the lamp requires a stable power supply. During operation the lamp generates ozone which should be removed from air that is breathed by humans. Typically the lamp is operated in a hood or is vented in some other way to the atmosphere outside the laboratory.

Other radiative sources that are used in special circumstances are the tungsten-filament lamp, LEDs and lasers. The tungsten-flament lamp is restricted to use in the visible and near-ultraviolet regions. The intensity of the emitted radiation in the middle- and vacuum-ultraviolet regions generally is too low to be useful. Ultrabright LEDs have been developed that can be used as excitative sources. Often their emission is centered at 280 and 300 nm. Both pulsed and continuous dye lasers have been used as sources for specific fluorescent studies when unusual sensitivity is required. The highly intense radiation that is emitted from a laser theoretically [Eq. (11-12)] should be ideal for the determination of a luminescent analyte at low concentrations. In many cases the gain in sensitivity is not as great as hoped owing to the simultaneous increase in background noise and in Rayleigh and Raman scattering. Nevertheless, lasers have proved to be useful radiative sources for many assays.[4] In some cases it is advantageous simultaneously to use two laser sources. The outputs from the two lasers are combined to provide the required excitative energy. This combination is normally only required for very low analyte concentrations. A description of some of the ways in which radiation from two lasers can be combined is given in Chapter 8.

Wavelength Selectors

Filters, monochromators, and filter-monochromator combinations have been used as wavelength selectors. If filters are used, the instrument that is used to measure fluorescence is a *fluorometer* or *filter fluorometer*. If monochromators are used the instrument is a *spectrofluorometer*. Similarly, for measurements of phosphorescence, instruments that contain filters are *phosphorimeters* and instruments that contain monochromators are *spectrophosphorimeters*.

If filters are used, the *primary filter* is placed between the source and the cell. The *secondary filter* is placed between the cell and the detector. Generally instruments that use filters are less expensive than those instruments that use monochromators. Either absorptive or interference filters can be used in fluorometers. Fluorometers allow broader wavelength bands to strike the cell and the detector. Because more intense radiation impinges upon the sample, the use of filters usually permits more sensitivity than the use of monochromators in a quantitative analysis. Because a filter permits a broader band of radiation to pass than does a monochromator, assays performed while using filters are more prone to spectral interferences.

Spectrofluorometers are required whenever it is necessary to scan the wavelength of either the excitative radiation or the luminesced radiation. The monochromators that are used in modern instruments nearly always contain one or more reflective diffraction gratings for radiative dispersion. In general a spectrofluorometer is required when establishing a new method of fluorescent analysis because the optimal wavelengths of the excitation and emission must be established. Fluorometers are used for assays that have been done previously and for which the excitative and emissive wavelengths are known. Of course, a spectrofluorometer also can be used for routine assays.

Cells

The cells that are used for photoluminescent studies are similar to those that are used for measurements of absorption. Both test-tube-shaped and tetragonal cells are in common use. Tetragonal cells can be used only for those measurements in which the emitted radiation is measured at 0, 90, or 180° relative to the path of the exciting radiation. At other angles a portion of the radiation that is emitted in the direction of the detector can be reflected by the cell wall. Usually the test-tube cells are less expensive but are more difficult to reproducibly place in the cell holder. That can lead to small errors during assays. As during spectrophotometric measurements, the cells must have windows that are transparent to the incident and emitted radiation. Cells with glass (Pyrex) windows can be used for studies at wavelengths that are longer than about 320 nm. At shorter wavelengths it is necessary to use cells with windows that are constructed of fused silica or quartz. Fused silica is preferred because quartz slightly fluoresces.

Many phosphorescent measurements are performed in Dewar flasks that contain liquid nitrogen. In those cases the sample in the cell is frozen by the liquid nitrogen in order to minimize radiationless (vibrational and rotational) loss of energy and to maximize phosphorescence. The excitative and phosphorescent radiation passes through transparent portions of the Dewar flask.

In some circumstances flow-through cells are used. The most common use for flow-through cells is in fluorescent detectors for high-performance liquid chromatography. A description of the detectors that are used for liquid chromatography is included in Chapter 25.

In addition to the components found in a fluorometer or spectrofluorometer, an instrument designed to measure phosphorescence also permits examination of the emitted radiation while the exciting radiation is not striking the sample. Because luminescence is monitored only while exciting radiation is not striking the analyte, the luminescence must consist entirely of phosphorescence and delayed fluorescence. Conventional fluorescence essentially ceases immediately after the exciting radiation is blocked off. Because delayed fluorescence is rarely encountered, the emission usually consists solely of phosphorescence.

Several methods are available for measuring luminescence while the exciting radiation is not striking the cell. In older instruments the measurement is often accomplished with the aid of a rotating opaque cylinder that is placed over the cell. The cylinder contains one or more slits that permit radiation to enter or exit the cell. As the cylinder rotates, a slit becomes aligned with the optical source and the cell and thereby allows the radiation to enter the cell. Because a pair of slits is never simultaneously aligned with the path between the source and the cell and the path between the cell and the detector, any conventional fluorescence that is emitted from the cell in the direction of the detector strikes the inside of the cylinder and is lost. If phosphorescence occurs, however, the intensity of the emission is measured when a slit aligns with the path of emitted radiation between the cell and the detector. Sometimes a device, such as that described here, which simultaneously chops the exciting and emitting radiation at the same frequency but out-of-phase with each other is termed a *phosphoroscope*.

Another type of phosphoroscope uses rotating disks with slits in the paths of the excitative and phosphoresced radiation. The two disks are simultaneously rotated at the same rate so that the slits are never aligned. A common form of that type of phosphoroscope mounts the two disks on the same rotating shaft and is used for phosphorescent measurements that are in-line with the exciting radiation. A more detailed description of phosphoroscopes is found in Reference 5. An electronically pulsed radiative source can be used in place of mechanical chopping. The detector is tuned so that the emitted radiation is measured between 0.1 and 50 ms after the radiative source is turned off during each cycle.

Detector and Readout

The most popular detector for photoluminescent studies is the photomultiplier tube. In a few instruments phototubes are used. The response of photomultiplier tubes and of phototubes varies with the wavelength of the monitored radiation. Instruments that are designed to rapidly obtain a photoluminescent spectrum sometimes use photodiode arrays. The arrays can be designed in either one or two dimensions and can simultaneously record an entire photoluminescent spectrum. Two-dimensional arrays are sometimes used in conjunction with devices that disperse radiation according to wavelength in two dimensions.

Essentially the same readout devices that were used in spectrophotometers are available for luminescent measurements. In fluorometers the readout device is generally a digital or analog meter, null-point meter or computer controlled data acquisition and readout device. Recorders can be used for recording spectra with wavelength-scanning instruments. Many modern instruments are controlled with dedicated microcomputers that make available several computer-controlled readout devices such as video display terminals.

Photoluminescent Spectra

Scanning instruments can be used to obtain two types of spectra. If the wavelength at which the emission is observed is held constant (wavelength selector 2, Fig. 11-5) while the wavelength at which excitation occurs (wavelength selector 1, Fig. 11-5) is scanned, the spectrum is an *excitation spectrum*. If the wavelength of the exciting radiation is fixed while the wavelength at which the emission is observed is scanned, the result is an *emission spectrum*.

When studying a compound for the first time, a common practice is to use a spectrophotometer to obtain an absorption spectrum of the compound. The excitative radiation in the spectrofluorometer is adjusted (wavelength selector 1) to correspond to the wavelength at a peak in the absorption spectrum. Usually the wavelength corresponding to the center of the highest energy absorptive peak is chosen for the initial adjustment. An emission spectrum is recorded with the exciting radiation adjusted to the chosen wavelength. Subsequently the wavelength of the emitted radiation is adjusted sequentially to the values that correspond to each of the peaks in the emission spectrum and an excitation spectrum is recorded at each emissive wavelength. The wavelength of the excitation is adjusted sequentially to the value at each of the peaks in the excitation spectra and new emission spectra are obtained at each excitative wavelength.

From the results, one excitative wavelength and one emissive wavelength are chosen at which to perform the quantitative analysis. The wavelengths usually correspond to peaks in one or more of the excitation and emission spectra. While choosing the wavelengths it is necessary to keep in mind such factors as sensitivity and absence of spectral interferences at either the excitative or emissive wavelengths. If a filter fluorometer is to be used, the filters that pass radiation of the proper wavelengths can be chosen from a list such as that in Reference 6.

The intensity of radiation that is emitted by the source, the intensity of radiation that passes through each wavelength selector, and the electrical output of the photomultiplier tube at constant applied potential are all dependent upon the wavelength of the radiation. Owing to those dependencies, simple photoluminescent instruments such as that in Fig. 11-5 have a response that varies with wavelength even when the absolute intensity of the luminescence does not vary. In most cases observed peaks of identical intensity are smaller in the ultraviolet region than in the visible region owing to the decreased intensity of the xenon source in the ultraviolet region. Because the variation is generally not reproducible between different instruments, it is useful to be able to obtain *corrected spectra* that are obtained after consideration of all wavelength-variable parameters. The spectra are independent of instrumental variations.

Several methods are available that make it possible to obtain spectra in which variations in instrumental response with wavelength are minimized or eliminated. Use of a double-beam instrument can eliminate much of the problem.

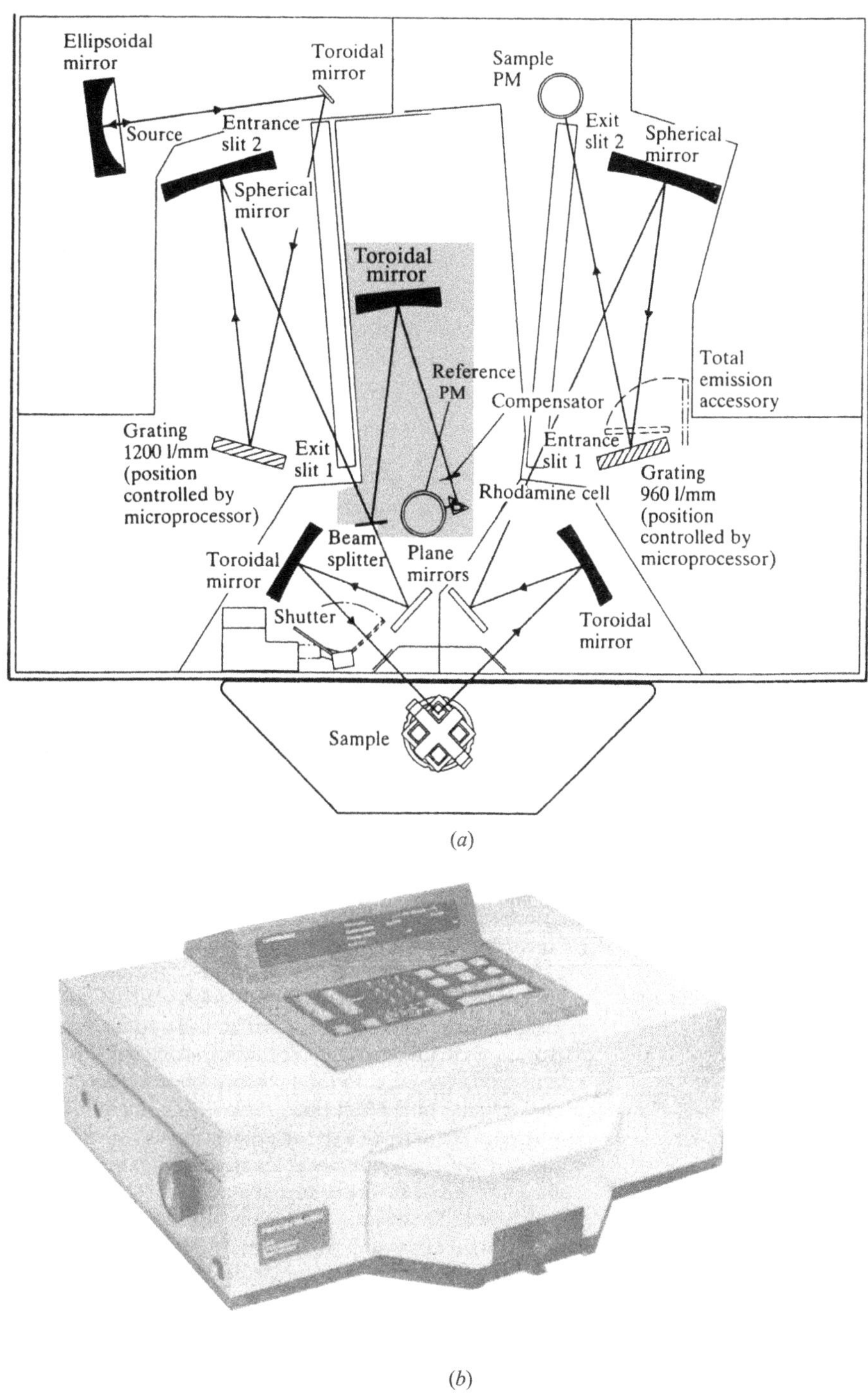

Figure 11-6 (*a*) The optical pathway and (*b*) a photograph of the LS-5 spectrofluorometer. (*Photograph courtesy* of *Perkin-Elmer Corporation.*)

The radiation from the source in some instruments is divided between two pathways after the radiation has passed through the first monochromator but before striking the cell. The division can be accomplished either with a chopper or with a beam splitter. A portion of the radiation passes through the cell and the second monochromator to the detector while the remaining radiation bypasses the cell and second monochromator and strikes the detector. If a chopper is used to divide the beam, a single detector can be used to alternately measure radiation from the two beams. If a beam splitter is used, a second detector is required to measure the intensity of the reference beam. In either case the spectrum consists of the ratio of the measured intensities of the beam emitted from the sample to the reference beam as a function of wavelength. Photographs and diagrams of optical pathways of two instruments that obtain spectra in that manner are shown in Figs. 11-6 and 11-7.

Another option is to allow the reference beam to strike a device that measures the energy of the beam. The output from the device can be used to attenuate the sample beam or to control the applied potential (the gain) of the photomultiplier tube in such a way that the output is corrected for variations in incident intensity as the wavelength is changed. Devices that can be used to measure the energy in the reference beam include thermocouples, thermopiles, and *quantum counters*.

A quantum counter is used in the instrument shown in Fig. 11-6 to compensate for the different response of the two photomultiplier tubes. A quantum counter is a device that has a response that is proportional to the rate at which photons strike it. The response is independent of the wavelength that is associated with the photons. In most cases quantum counters consist of fluorescent compounds that have excitative bands that encompass the entire possible wavelength range of the incident radiation and emit radiation with the same quantum efficiency at all possible excitative wavelengths. A solution of the fluorescing compound is placed in the reference beam before the detector. Regardless of the wavelength of the beam that strikes the compound, it emits radiation at the same wavelength and with the same efficiency. The emitted radiation is measured by the detector. The fluorescent compound that is used in the quantum counters in Figs. 11-6 and 11-7 is rhodamine B.

When the quantum efficiency of the quantum counter is known, it is possible to calculate quantum efficiencies for samples. Essentially the change in sensitivity of the monochromators, detector, and source are measured as a function of wavelength and used to calculate a series of correction factors at each wavelength. The quantum efficiency of the sample at each wavelength can be used to calculate the absolute intensity of emission and to prepare spectra of absolute emissive intensity as a function of wavelength. Many modern instruments such as those shown in Figs. 11-6 and 11-7 have provision for computer control of the instrument and for automatic plotting of absolute intensities as a function of wavelength. A multitude of companies can provide instruments for measuring fluorescence. Most of the instruments are filter fluorometers.

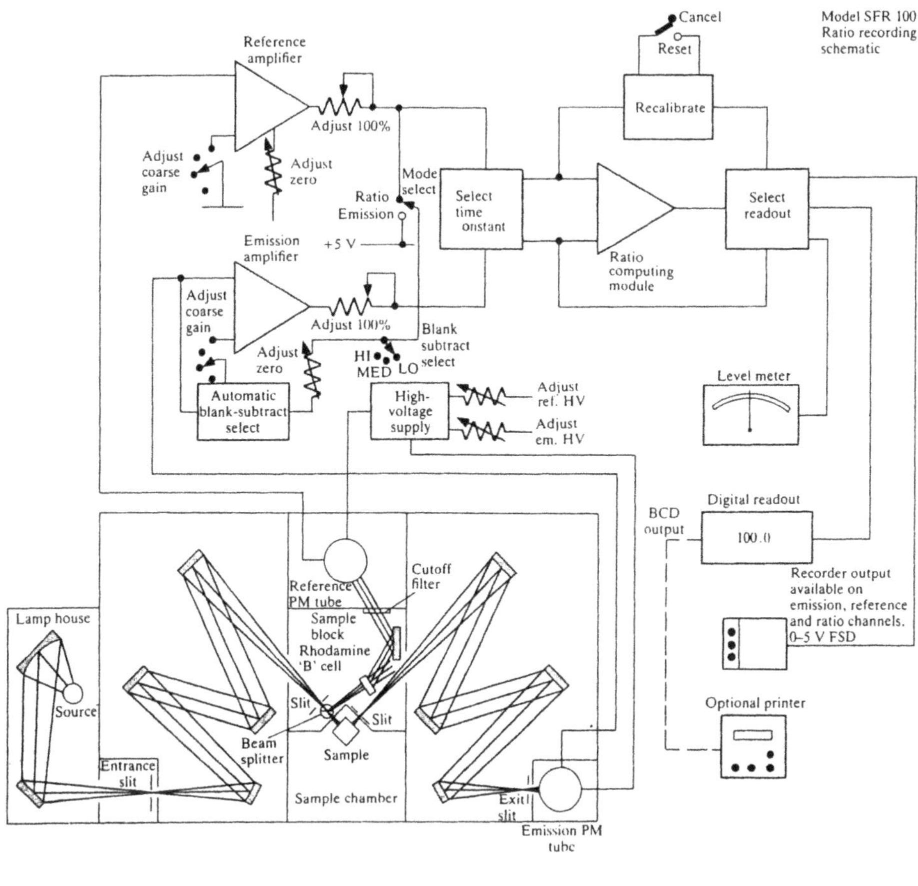

(a)

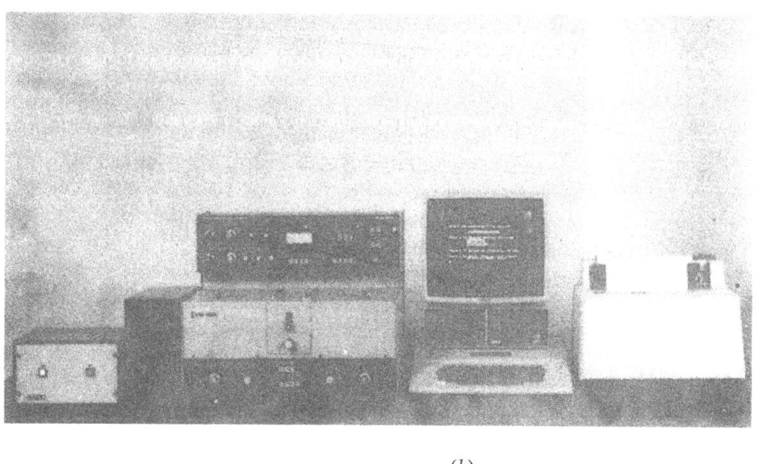

(b)

Figure 11-7 (*a*) The optical pathway and (*b*) a photograph of the SFR100 spectrofluorometer. *(Photograph courtesy* of *Baird Corporation.)*

Occasionally it is necessary to calibrate the excitative and emissive wavelengths that are used in a photoluminescent instrument. Wavelength calibration is fairly easily accomplished by placing a line source in the instrument in place of the source that is normally used. A low-pressure mercury lamp is often used. While the monochromators are adjusted to allow one of the spectral lines to pass, the wavelength reading on the instrument is adjusted to the value of the mercury line. Alternatively, a reference compound with known wavelengths of excitation and emission can be used. To obtain absolute intensity measurements or to make sensitivity adjustments, a reference compound that has a known quantum efficiency can be used. Quinine sulfate solutions are often used for that purpose. More details can be found in References 1 and 2.

PHOTOLUMINESCENT ANALYSIS

Neither fluorescence nor phosphorescence is particularly useful for qualitative analysis because many compounds photoluminesce at nearly the same wavelength. Consequently, photoluminescence is rarely used for qualitative analysis. It is useful for qualitative analysis only when the number of possible compounds in the analyte is limited and when the wavelength of the excitation or emission of each compound significantly differs from that of the others. A computer program for qualitative analysis of fluorescent compounds has been developed which compares the fluorescence of an analyte with that of 1000 fluorescent compounds.[7]

Most quantitative luminescent analyses are done by using either the working-curve method or the standard-addition technique. The two methods are also used when the sample to be assayed quantitatively quenches the fluorescence of a known amount of an added reagent. Generally the substance to be assayed falls into one of three categories. If the sample luminesces, a working curve of the luminescence of a series of standard solutions can be prepared and the concentration of the analyte can be obtained from the working curve. The standard-addition technique can also be used for luminescent samples.

In some cases the sample does not luminesce but can be converted to a luminescent species by reaction with a fluorometric or phosphorometric reagent. A fluorophore is the portion of a molecule which results in fluorescence. In studies of large (often biological) molecules, a fluorophore can be reacted with the molecule to yield a fluorescent product. Regardless of the nature of the analyte, the chemical reaction is carried out with the sample and with standards of the analyte. A working curve of luminescence as a function of concentration is prepared and the unknown concentration is obtained from the curve. Alternatively, the standard-addition technique can be used by making additions of the standard to separate portions of the sample solution prior to adding the reagent.

In some cases the sample does not luminesce and cannot be converted to a substance that does luminesce, but it does react with a luminescent substance to form a nonluminescent product. In that case the luminescence of a known, but excessive, concentration of the luminescent species is measured; a measured portion of the sample is added; and the luminescence is remeasured. The concentration is obtained from the luminescent quenching by either the working-curve method or the standard-addition technique.

Photoluminescent analysis is often used for analytes in the 10^{-5} to 10^{-8} *M* range. At those low concentrations it is necessary to take precautions to minimize interferences in the assay. The presence of radiative scattering and luminescent compounds in addition to the analyte can interfere with the assay. Contamination can come from contact with common laboratory materials, such as Tygon tubing and some detergents, that fluoresce. Standard solutions at low concentrations usually are not stable for periods of more than a few hours owing to adsorption of the standard on the walls of the container. Standard solutions should be prepared fresh daily from concentrated standard solutions. Of course, cleanliness is a prerequisite for accurate trace analysis. The solvents that are used for photoluminescent assays must be free of contaminants that can interfere with the assays. High-purity solvents should be used.

At high sensitivities Raman scattering by the solvent can interfere with fluorescent but not phosphorescent measurements. The Raman bands for a particular solvent are shifted by a fixed frequency from the frequency of the excitative radiation. The average frequency shifts for several solvents are 2.1×10^{13} (carbon tetrachloride), 9.0×10^{13} (chloroform), 8.7×10^{13} (cyclohexane), 8.7×10^{13} (ethanol), 8.1×10^{13} (methanol), and 1.0×10^{14} (water) Hz. Sometimes it is possible to choose a solvent that does not exhibit Raman scattering at the frequency of the measured emission.

Lists of fluorescent and phosphorescent compounds that can be determined by photoluminescent analysis are much too long to be given here. Lists of some of the compounds that can be determined are available in References 1, 2, and 8. The requirements for a compound to photoluminesce were described earlier in the chapter.

Analysis of Nonluminescing Compounds

Compounds that do not luminesce sometimes can be quantitatively assayed either by luminescent quenching or with the aid of a chemical reagent that reacts with the compound to form a luminescent reaction product. An example of fluorescent quenching is the quenching of the fluorescence of dichlorofluorescein in acid by nitrate. Nitrate can be determined from a working curve of the decrease in the fluorescent intensity of a fixed concentration of dichlorofluorescein as a function of nitrate concentration. Quenching can be caused by any of those factors that were described earlier or it can be caused by a chemical reaction of the luminescent compound with the analyte to form a nonluminescent product.

Many inorganic substances can be assayed by reacting the substance with a fluorometric or phosphorometric reagent to form a photoluminescent reaction product. Fluorescent assays are much more common than phosphorescent assays. Some metals can be determined by reacting the ion with a nonluminescent ligand or complexing agent to yield a luminescent product. Alternatively, the ligand or complexing agent can be determined by reaction with an appropriate metallic ion. For many fluorescent analyses the detection limit is between 0.001 and 2 μg/mL. Among the inorganic elements that have been determined by the addition of a fluorometric reagent are Al, Au, B, Be, Ca, Cd, Cu, Eu, Ga, Gd, Ge, Hf, Mg, Nb, Pd, Rh, Ru, S, Sb, Se, Si, Sm, Sn, Ta, Te, Th, W, Zn, and Zr. The structures of some of the common complexing agents that are used as fluorometric reagents are shown in Fig. 11-8. A partial listing of some of the assays that can be performed with fluorometric reagents is given in Table 11-2. More complete listings are available in References 9 and 10. In nearly every instance the complex excitation and emission correspond to π to π^* transitions.

In addition to direct determinations using fluorometric reagents, it is possible to perform indirect determinations. The usual procedure is to add an excess of a nonfluorescent complex to the analyte. The analyte, which is more strongly attracted to the metal in the complex than is the ligand, displaces the ligand in the complex. The displaced ligand fluoresces. The fluorescence is proportional to the amount of liberated ligand which is proportional to the amount of added analyte.

Luminescent determinations of the rare earth elements often result in complexes which emit radiation owing to f^* to f transitions (compare the emission with the corresponding absorption as described in Chapter 9). The emission bands for those complexes are considerably narrower than those for the π^* to π transitions of the previously described complexes.

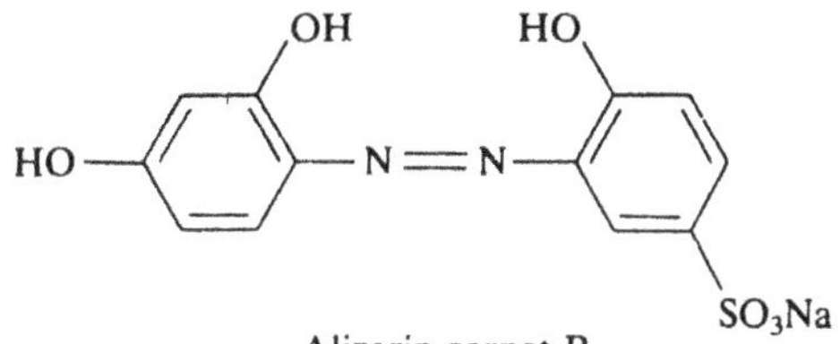

Alizarin garnet R

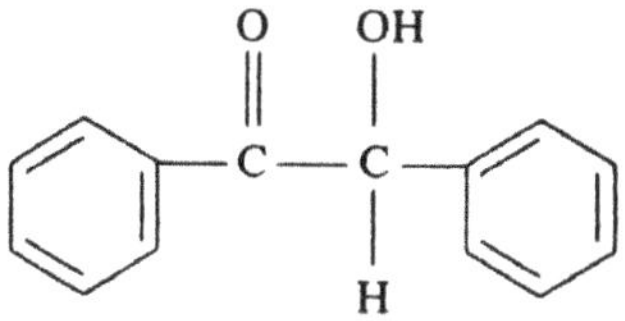

Benzoin

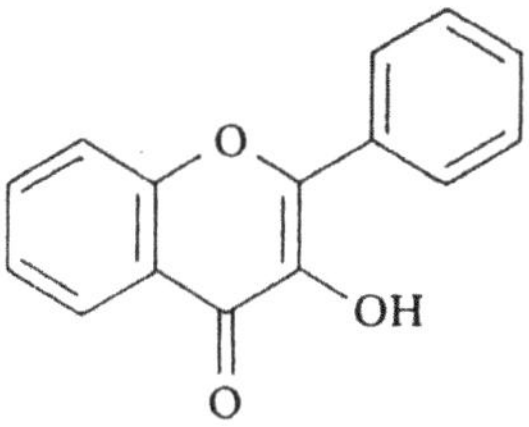

Flavanol

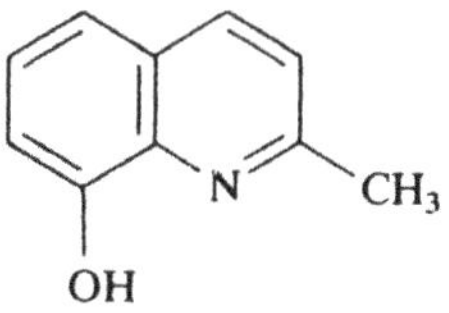

8-Hydroxyquinaldine

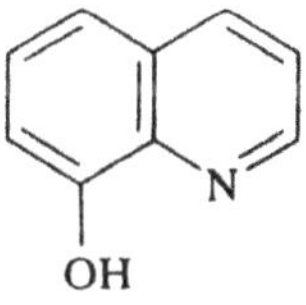

8-Hydroxyquinoline

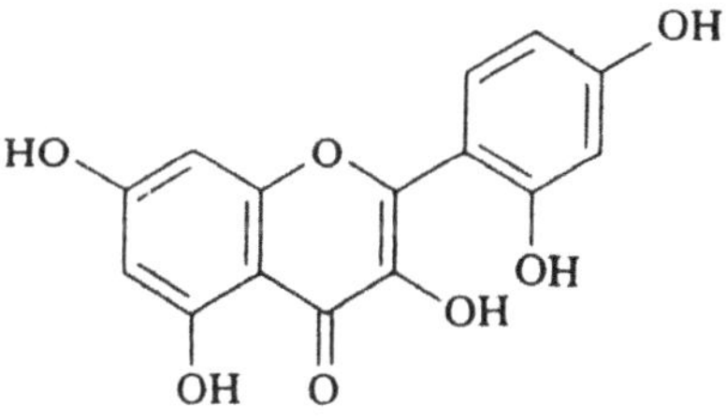

Morin

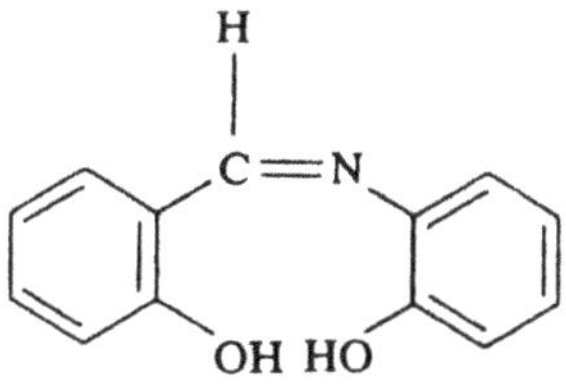

Salicylidene-*o*-aminophenol

Figure 11-8 The molecular structures of several fluorometric reagents.

Table 11-2 Several metallic ions that can be assayed using fluorometric reagents

Metal	Reagent	Fluorescent wavelengths, nm	
		Excitation	Emission
Al(III)	Alizarin garnet R	470	500
	Morin	430	500
	8-Hydroxyquinoline	365	520
	Salicylidene-*o*-amino-phenol	410	520
Be(II)	1-Amino-4-hydroxyan-thraquinone	540	620
	2,3-Hydroxynaphthoic acid	380	460
	2-(*o*-hydroxyphenyl)-benzoxazole	365	590 (broad)
Cu(II)	Tetrachlorotetraiodo-fluorescein-*o*-phenanthroline	560	570
Ga	8-Hydroxyquinaldine	–	492
	8-Hydroxyquinoline	436	470-610
	Salicylidene-*o*-amino-phenol	420	520
Ge	Benzoin	365	410 (broad)
Mg(II)	Bissalicylidene diaminobenzofuran	475	545
Sc(III)	Salicylaldehyde semicarbazone	370	455
Sn(IV)	Flavanol	400	470

Source: For details refer to References 9 and 10.

Nonluminescent organic and biochemical compounds can be determined fluorometrically or phosphorimetrically if a derivative can be quantitatively prepared that fluoresces or phosphoresces. In some cases a metallic ion can be added that reacts with the compound to form a fluorescent complex. In other cases a reaction such as a condensation can be used to extend the conjugation of an organic compound and cause luminescence.

Often a biochemical compound can be determined after reacting the compound with a *fluorescent label* (a fluorophore). The label is a compound that itself fluoresces and continues to fluoresce after the reaction. Prior to the luminescent measurement, it is necessary to separate (usually with chromatography or gel electrophoresis) unused label from the prepared derivative. In certain circumstances other types of reactions can also yield a luminescent product.

Determinations of Mixtures

If more than one luminescent compound is present in the analyte, the analysis becomes more difficult. Several methods are available that permit the simultaneous determination of more than one compound in a mixture. Of course the mixture could be separated into its components, if possible, and each component individually assayed. That procedure, performed by using extractions, distillations or other nonchromatographic methods, can be tedious and often causes increased error owing to the extra steps involved in the analysis. Generally it is preferable to perform the analysis without a prior nonchromatographic separation. The use of high performance liquid chromatography in combination with fluorometric detection of the components is a recommended method when it is possible. Liquid chromatography is described in Chapter 25.

Mixtures can be most easily assayed using most spectrofluorometers by taking advantage of differences in the excitation and emission spectra of the components. If one of the components has a wavelength in its excitation spectrum at which no other component absorbs radiation, that component can be assayed by setting the wavelength of the monochromator that is between the source and the cell to that wavelength and the other monochromator to the emissive maximum. Because only one component is excited, only that component can emit and the assay is free of interference from the other components. Use of the method assumes that energetic transfer between the analyte in the excited state and other nonanalyte molecules in the solution that are in the ground state cannot occur.

Similarly, if a wavelength in the emission spectrum of a component can be found at which only that component emits radiation, the monochromator that is between the cell and the detector can be adjusted to that value. Again the determination should be free of interference from the other components because no other component emits radiation at the chosen wavelength.

Unfortunately both the excitation spectra and the emission spectra of most substances are broad and therefore are likely to overlap with the spectra of other components in the mixture. In that case the problem becomes analogous to the simultaneous determination of absorbing substances at different wavelengths that was described in Chapter 9. A set of excitation and emission wavelengths is chosen for each component of the mixture. The change in luminescent response K' [in Eq. (11-12) $K' = KI_0$] of the instrument with changing concentration C for each component at each set of wavelengths is experimentally determined from standard solutions of the pure components, and a series of simultaneous equations is established at each set of wavelengths. The equations have the following form:

$$\begin{aligned} I\,(\text{mixture, wavelength set 1}) &= K'\,(\text{component 1, wavelength set 1})C_1 \\ &\quad + K'\,(\text{component 2, wavelength set 1})C_2 + \cdots \\ &\quad + K'\,(\text{component } n\text{, wavelength set 1})C_n \end{aligned} \qquad (11\text{-}14)$$

The luminescence of the mixture is measured at each set of wavelengths; the appropriate luminescent responses and the luminescence of the mixture is substituted into each simultaneous equation; and the equations are solved for the concentrations of each of the components.

Alternatively, an emission spectrum of a mixture can be recorded and a computer used to *deconvolute*, i.e., to separate, the portion of the spectrum owing to each component. The deconvoluted spectra are used to calculate the concentration of the components.[11] Some instruments that can be used for phosphorescent studies, e.g., the LS-5 shown in Fig. 11-6, are capable of using a pulsed source and measuring the luminescence at different times after the source has been turned off in each cycle. Pulsed lasers also can be used as sources for these studies. If the different components have different emissive lifetimes, the analysis can be performed by preparing a plot of the logarithm of the luminescent intensity as a function of time of the measurement. If the lifetimes of the components differ significantly, the plot has a linear portion for each component. The intensity that corresponds to each linear portion after extrapolation to time zero is proportional to the concentration of the corresponding component. The working-curve method is used after similar plots are prepared with standard solutions. Several other methods for the determination in a mixture are available but generally require apparatus that is not available in most laboratories.[2]

FLUORESCENCE CORRELATION SPECTROSCOPY

Fluorescence correlation spectroscopy FCS is a method in which the changing fluorescence of a sample in a limited space is monitored with time. The fluorescence changes owing to Brownian motion of the fluorescing molecules into and out of the space. The resulting fluctuations in observed fluorescence can be used to measure diffusion rates, average concentrations and the rate of chemical reactions involving the fluorescent species as well as other physical properties. The size of the observed space must be sufficiently small so that the concentration of the fluorescent species can vary but still be measurable as the substance passes into and out of the space. Consequently the optics of devices used to measure FCS normally utilize laser sources and a microscope through which the exciting and emitting radiation travel. The laser provides the high intensity required to measure the low levels of fluorescence that occur within the focus of the microscope. By using this method, it is now possible to observe a single, large fluorescing molecule, such as a protein molecule after labeling with a fluorophore.

In dual-color FCS two separate fluorescence labels are attached to different components of the reaction and each component simultaneously is observed at a different wavelength. The results can be used to measure diffusion rates, reaction rate constants, dimensions of reacting species, et al. The details of the theory and use of FCS are beyond the scope of this description but can be found in References 14 and 15. A good introduction to the method can be found in Wikipedia under the heading Fluorescence Correlation Spectroscopy.

A related technique is single molecule fluorescence spectroscopy which utilizes a fluorophore to label a biomolecule within a condensed phase sample. The labeled molecule is detected in the presence of billions or trillions of matrix molecules by using laser sources and sensitive detectors. A review of the method is found in Reference 16.

PHOSPHORESCENCE

Phosphorescence has not proven to be as popular for analysis as fluorescence. In part that is due to the smaller number of substances that phosphoresce and in part to the difficulty of making some of the measurements at the temperature of liquid nitrogen. In some cases phosphorescence occurs at room temperature. Often room-temperature phosphorescence occurs on solid surfaces. Examples of compounds that phosphoresce at room temperature include benzo [*f*] quinoline and *p*-aminobenzoic acid.[12,13] Another important consideration is phosphorescent quenching by O_2 as described earlier. A review of recent developments can be found in Reference 17.

IMPORTANT TERMS

Corrected spectra
Delayed fluorescence
Emission spectrum
E-type delayed fluorescence
Excimer emission
Excitation spectrum
External heavy-atom effect
Fluorescence
Fluorescence correlation spectroscopy
Fluorescent label
Fluorometer
Franck-Condon principle
Heavy-atom effect
Inner-filter effect
Internal heavy-atom effect
Phosphorescence
Phosphorimeter
Phosphoroscopc
Photodecomposition
Photoluminescence
Primary filter
P-type delayed fluorescence
Quantum counter
Quantum yield
Recombination fluorescence
Secondary filter
Single molecule fluorescence
Spectrofluorometer
Spectrophosphorimeter

PROBLEMS

11-1 Prepare a sketch showing the energetic levels and the orientation of the electrons for (*a*) a ground singlet state, (*b*) a ground doublet state (refer to Chapter 5), (*c*) an excited singlet state, and (*d*) an excited triplet state.

11-2 Write an equation showing the electron orientation changes [similar to Eqs. (11-2)] for the triplet-triplet annihilation reaction of Eq. (11-l*b*).

11-3 Calculate the quantum efficiency for a particular reaction in which 4.9×10^{17} photons were absorbed during the excitation and 2.9×10^{17} photons were emitted during fluorescence.

11-4 If the absorbance of a solution is 0.035, calculate the ratio of the second term in the brackets in Eq. (11-9) to the first term.

11-5 The molar absorptivity for the absorptive band of organic compound 1 was 11,100. The molar absorptivity for organic compound 2 was 655. Which of the two compounds definitely contains a heteroatom?

11-6 Which of the following compounds would be expected to have the greatest fluorescence?

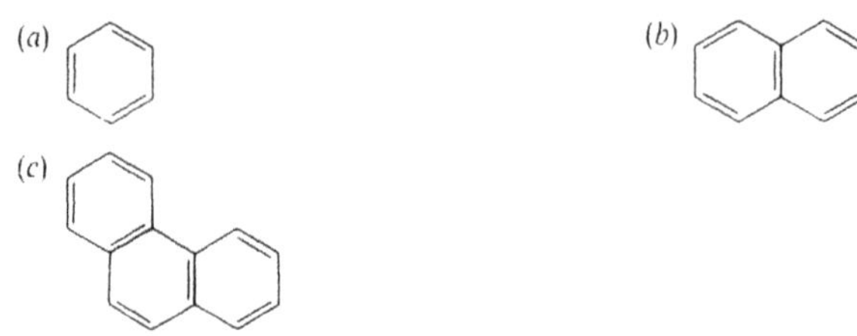

11-7 Which of the following compounds would be expected to have the greatest phosphorescence?

11-8 Arrange the following compounds in order of expected fluorescent intensity.

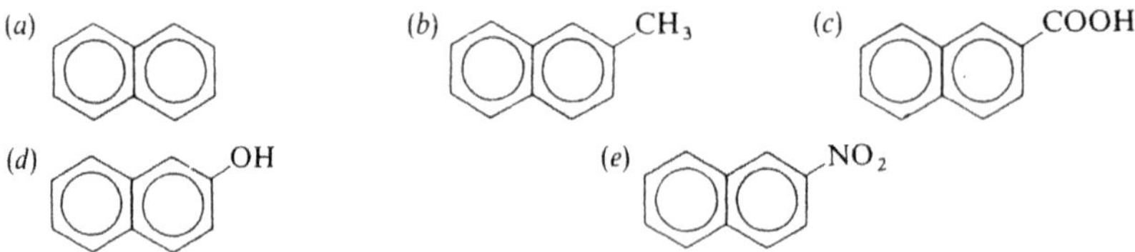

11-9 Would anthracene be expected to phosphoresce more strongly when dissolved in benzene than when dissolved in chloroform? Why or why not?

11-10 Would the wavelength at which naphthalene fluoresces be expected to become longer as the solvent was changed from benzene to diethyl ether? Why or why not?

11-11 A 5.00-mL solution that contained glycine was added to a 500-mL volumetric flask and diluted to the mark with distilled water. A 5-mL portion of the diluted solution was added with a pipet to a 10-mL volumetric flask and diluted to the mark with a pH 8.7 borate buffer solution. After reaction with 1.00 mL of a solution that contained 25.0 mg of fluorescamine dissolved in 100 mL of acetone, the fluorescence of the diluted 10-mL solution was measured. Fluorescamine reacts with glycine to form a fluorescent product. From a working curve the final solution was found to contain 1.14 mg/L of glycine. Calculate the concentration of glycine in the original 5-mL solution.

11-12 Upon excitation at 435 nm, fluorescein fluoresces at 485 nm. Nitrate quenches the fluorescence in concentrated sulfuric acid by reacting with fluorescein to form a nonfluorscent product. The concentration of fluorescein in a series of solutions was held constant at 5×10^{-7} *M* and the concentration of nitrate in the solutions was varied. The fluorescence of the standard solutions was measured. A 1.245-g soil sample was placed in an Erlenmeyer flask. The nitrate in the soil was dissolved in 10.0 mL of distilled water. A 0.500-mL portion of the solution was added to a 50-mL volumetric flask and diluted to the mark with a sulfuric acid solution that contained 5×10^{-7} *M* fluorescein. The fluorescence of the resulting solution was measured. Calculate the weight percent of nitrate in the soil.

$[NO_3^-] \times 10^{-6}$ M	Relative fluorescent intensity
0	100
0.22	94
0.44	89
1.01	70
1.99	42
Soil	64

11-13 A solution contained two fluorescent compounds. A plot of relative fluorescent intensity as a function of concentration was prepared for standard solutions of each of the pure components. Both plots were linear. At an excitative wavelength of 430 nm and an emissive wavelength of 495 nm, the slopes of the plots were 4.58×10^6 and 1.32×10^6 L/mol respectively for compounds A and B. At an excitative wavelength of 288 nm and an emissive wavelength of 385 nm, the slopes were 8.97×10^5 and 5.84×10^6 L/mol respectively for compounds A and B. The sample had a relative fluorescent intensity of 78.2 at 430/495 nm and 45.5 at 288/385 nm. Calculate the concentration of the two compounds in the sample.

REFERENCES

1. Winefordner, J. D. (ed): *Spectrochemical Methods of Analysis*, Wiley-Interscience, New York, 1971, chaps. 5, 7, and 8.
2. Seitz, W. R., in P. J. Elving, E. J. Meehan, and I. M. Kolthoff (eds.): *Treatise on Analytical Chemistry*, part I, 2d ed., vol. 7, Wiley-Interscience, New York, 1981, chap. 4.
3. Lim, E. C., and W. Y. Wen: *J. Chem. Phys.*, **39**: 847(1963).
4. Lytle, F. E.: *J. Chem. Educ.*, **59**: 915(1982).
5. O'Haver, T. C., and J. D. Winefordner: *Anal. Chem.*, **38**: 602(1966).
6. Corning Glass Works, *Glass Color Filters*, Corning, New York.
7. Miller, T. C., and L. R. Faulkner: *Anal. Chem.*, **48**: 2083(1976).
8. Williams, A. T. R.: *An Introduction to Phosphorescence Spectroscopy*, Perkin-Elmer, Beaconsfield, England, 1981.
9. West, T. S., in W. W. Meinke and B. F. Scribner (eds.): "Trace Characterization—Chemical and Physical," NBS Monograph 100, Department of Commerce, NBS, Washington, D.C., 1967.
10. Meites, L.: *Handbook of Analytical Chemistry*, McGraw-Hill, New York, 1963, p. 6-178.
11. Rechsteiner, Jr., C. E., H. S. Gold, and R. P. Buck: *Anal. Chim. Acta*, **95**: 51(1977).
12. von Wanduszka, R. M. A., and R. J. Hurtubise: *Anal. Chem.*, **48**: 1784(1976).
13. Ford, C. D., and R. J. Hurtubise: *Anal. Chem.*, **52**: 656(1980).
14. Magde, D., E. Elson, and W. W, Webb: *Phys. Rev. Lett.*, **29**: 705(1972).
15. Schwille, P., F. J. Meyer-Almes, and R. Risler: *Biophysical Journal*, **72**: 1878(1997).
16. Moerner, W. E.: J. *Phys. Chem. B*, **106**: 910(2002).
17. Hurtubise, R. J.: *Anal. Chem.*, **55**: 669A(1983).

CHAPTER

TWELVE

INFRARED SPECTROPHOTOMETRY

Infrared (ir) radiation is EMR in the wavelength range that is adjacent to and of less energy than visible radiation. The name of the region is derived from the fact that radiation in the region is less (infra) energetic than that of visible red radiation. The infrared region starts at a wavelength of about 0.7 μm and ends at a wavelength of about 500μm. Users of infrared radiation often express the energy of the radiation as a *wavenumber* $\bar{v}$ rather than as a wavelength or frequency. The wavenumber is defined as the inverse of the wavelength when the wavelength is measured in units of centimeters:

$$\bar{v} = \frac{1}{\lambda(\text{cm})} = \frac{nv}{c} \tag{12-1}$$

Units on the wavenumber scale are directly proportional to both the energy and the frequency of the radiation.

Sample problem 12-1 Convert 0.70, 2.8, and 33 μm to wavenumbers.

SOLUTION The conversions can be accomplished by application of Eq. (12-1). The wavelength must be converted to units of centimeters in order to use the equation.

$$\bar{v} = \frac{1}{0.70\ \mu\text{m} \times 1\ \text{cm}/10^4\ \mu\text{m}} = 1.4 \times 10^4 \text{cm}^{-1}$$

Similarly for 2.8 and 33 μm, the corresponding wavenumbers are

$$\bar{v} = \frac{10^4}{2.8} = 3.6 \times 10^2 \text{cm}^{-1}$$

and

$$\bar{v} = \frac{10^4}{33} = 3.0 \times 10^2 \text{cm}^{-1}$$

As was briefly described in Chapter 5, infrared absorption generally corresponds to changes in vibrational and rotational levels in a molecule, i.e., to changes in the rate or direction of vibration of a portion of a molecule in relation to the remainder of the molecule or of rotation of the molecule about its center of gravity. In most cases changes in vibrational levels require more energy than changes in rotational levels. The infrared region is divided into smaller regions according to energetic proximity to the visible spectral region. The *near-infrared region* extends from the edge of the visible region (0.7 μm or 14,000 cm^{-1}) to about 2.8 μm (3600 cm^{-1}). Spectrophotometric apparatus that is used in the near-ir region most closely resembles that used in the visible region. Some low-energy electron transitions as well as changes in vibrational and rotational levels can occur in the near-infrared region. The near-ir region is generally restricted to the study of compounds that contain OH, NH, and CH groups.

The *mid-infrared region* extends from about 2.8 μm (3600 cm^{-1}) to about 50 μm (200 cm^{-1}). Because changes in vibrational levels of most molecules occur in the mid-infrared region, it is of most use for analysis. The *far-infrared region* covers the range from 50 μm (200 cm^{-1}) to about 500 μm (20 cm^{-1}). Many purely rotational changes occur in the far-ir region. In the other ir regions, the rotational changes are superimposed upon vibrational changes. Radiation of less energy than 20 cm^{-1} is classified in the microwave or radiowave region of the spectrum.

THEORY

Infrared absorption occurs when the frequency of the alternating electric field that is associated with the incident radiation matches a possible change in a vibrational or rotational frequency of the absorbing molecule. When a match occurs, EMR can be absorbed by the molecule causing a change in the amplitude of vibration or a change in the rate of rotation.

In order for electromagnetic radiation to be absorbed by a molecule, it is necessary for the molecule to undergo a change of dipole moment during the absorption. If no change in the distribution of charge in the molecule occurs, the varying charge in the electric component of the radiation has nothing with which to interact and cannot transfer energy to the molecule. Molecules that have a completely symmetrical charge distribution and in which no change in dipole moment occurs when the molecule vibrates with a different amplitude or rotates at a different rate do not absorb infrared radiation. Substances that are transparent to infrared radiation are primarily monoatomic and homonuclear diatomic gases such as He, Ne, Cl_2, N_2, and O_2. Nearly all other substances absorb radiation in the infrared region.

An example will help to illustrate the type of change that occurs during infrared absorption that is accompanied by a change in vibrational level.

Consider a simple heteronuclear diatomic molecule such as gaseous HCl. Because the Cl is more electronegative than the H, a dipole moment exists in the molecule. Assume that the molecule is vibrating at a fixed frequency; i.e., the bond between the H and Cl periodically stretches and relaxes, as illustrated in Fig. 12-1. After absorption of energy the amplitude of the oscillations increases and the average distance between the nuclei increases. Because the separation between the nuclei is altered, the separation between the partial positive and negative charges increases and the dipole moment of the molecule is altered. The motion of vibrating nuclei is a function of the mass of the nuclei. For the case illustrated in Fig. 12-1, the H nucleus moves much farther than the Cl nucleus because the H nucleus is much lighter.

As is the case with electron levels, rotational and vibrational levels are quantized. Classical physics is often used to provide quantitative descriptions of vibrational energetic levels within simple diatomic and triatomic molecules. The nuclei are assumed to have known masses that are connected to each other by springs. The springs represent chemical bonds between the atoms. Either harmonic or anharmonic oscillations can be assumed.

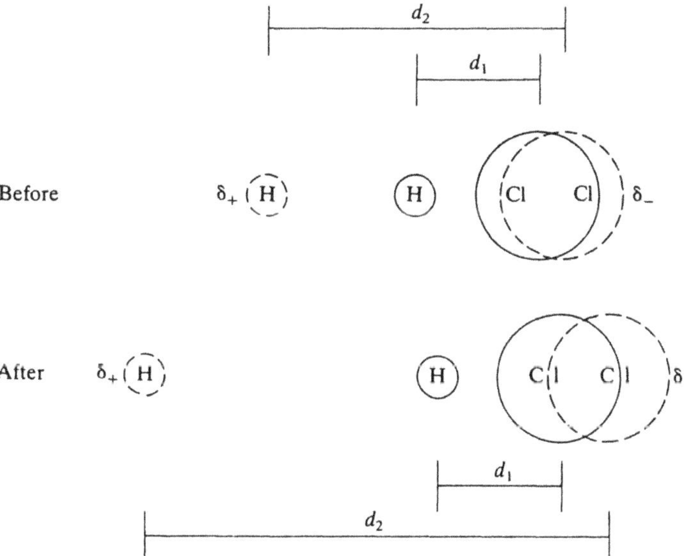

Figure 12-1 A diagram illustrating a stretching vibration of an HCl molecule before and after absorption of infrared radiation at the frequency of vibration. The circles represent nuclei. Solid circles represent the positions of the nuclei at the point of nearest approach. Dashed circles are the nuclear positions at greatest distance from each other. The space between the nuclei is compressed for clarity. d_1, internuclear distance at point of nearest approach of the nuclei; d_2, internuclear distance at point of greatest distance between the nuclei.

A better description of the possible rotational and vibrational levels is provided by quantum mechanics. Unfortunately the quantum-mechanical equations that describe vibrational and rotational changes within all except the simplest molecules are difficult, if not impossible, to solve exactly. The classical and quantum-mechanical mathematical treatments are too lengthy to be thoroughly described here. Consequently the descriptions are left to physical chemistry, physics, and spectroscopic courses. More information can be found in the first two references and the references listed therein. For most assays it is sufficient to know that the energetic levels are quantized; that vibrational level changes require more energy than rotational level changes; and that several rotational levels exist for each vibrational level in gases. In liquids and solids rotation is severely restricted and fewer rotational levels exist than in gases.

Because most analytical samples are in the liquid or solid state, the analyst is primarily concerned with changes in vibrational levels. Molecular vibrations are categorized as either *stretching* or *bending* vibrations. A stretching vibration corresponds to an oscillation along the internuclear axis such as that which was illustrated in Fig. 12-1. Stretching vibrations can be either *symmetric* or *asymmetric*.

A symmetric stretch is a stretching vibration in which the two nuclei simultaneously approach the same location or simultaneously move away from the same location. The angle between a vibrating bond and a chemical bond that is attached to one of the atoms involved in the vibration is not altered by stretching vibrations. The stretching vibration in Fig. 12-1 is a symmetric vibration in which the nuclei simultaneously approach and then simultaneously recede from the center of mass on the bond axis. Asymmetric stretching occurs when one nucleus moves toward a point while the other nucleus moves away from the point. The direction of nuclear motion during stretching vibrations is shown in Fig. 12-2 for polyatomic molecules that contain a central atom, such as C, that is bonded to four other atoms.

Bending vibrations are nuclear motions that cause a change in the angle between two vibrating bonds, and consequently that require a molecule that contains at least three atoms. Bending vibrations are classified as *rocking*, *scissoring*, *twisting*, or *wagging* depending upon the motion of the two outer nuclei in comparison to the central nucleus. The names are descriptive and can be compared to the motion of a rocking chair, the motion during operation of a pair of scissors, the twisting of two wires around a central point, and the back-and-forth motion of a wagging tail. Bending vibrations are illustrated in Fig. 12-3 for a central atom that is attached to four other atoms.

During rocking the two outer nuclei rotate in the same direction and in the same plane about a common nucleus. During scissoring the two outer nuclei rotate toward and away from each other while remaining in the same plane. The rotation is analogous to that of scissor blades while they are opened and closed.

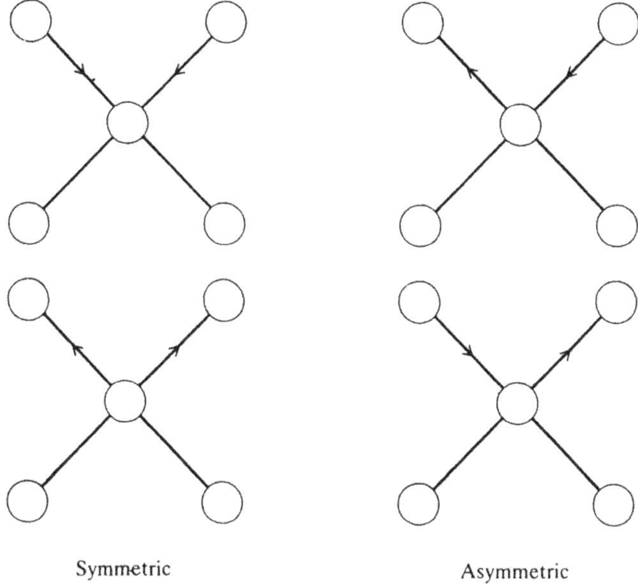

Figure 12-2 Diagrams illustrating motion during symmetric and asymmetric stretching.

Twisting occurs when the two nuclei rotate around a central nucleus but out of the initial plane defined by the nuclei. Twisting can be thought of as occurring in the same manner as that in which a person with arms stretched over the head twists first to the right and then to the left. The hands represent the outer nuclei and the body represents the central nucleus. Using the same analogy, wagging occurs when the person bends forward and backward as if to touch the floor but with bending only at the waist.

In a molecule all of the possible vibrations or rotations (if a change in dipole moment occurs) can individually be responsible for an absorptive band. Because many possibilities for a particular molecule can exist, typical infrared spectra contain many absorptive bands. That is quite different from ultraviolet-visible spectra in which few absorptive bands are observed for a single compound. The frequencies at which the absorptive bands occur are dependent upon many factors including the relative masses and polarities of the nuclei, the strengths of the bonds in the molecule, and the number of atoms in the molecule. Additionally interactions (*coupling*) between different vibrations within the same molecule can occur. Theoretically no two compounds, with the exception of optical isomers, have identical infrared spectra. Consequently, infrared spectrophotometry is particularly useful for qualitative analysis.

Vibrational coupling is most likely to occur when the interacting groups are vibrating at approximately the same frequency and when the groups are near each other. Vibrational coupling between stretching vibrations is most likely to occur when the two vibrations have an atom in common. Coupling between bending vibrations can occur when the vibrating groups share a chemical bond.

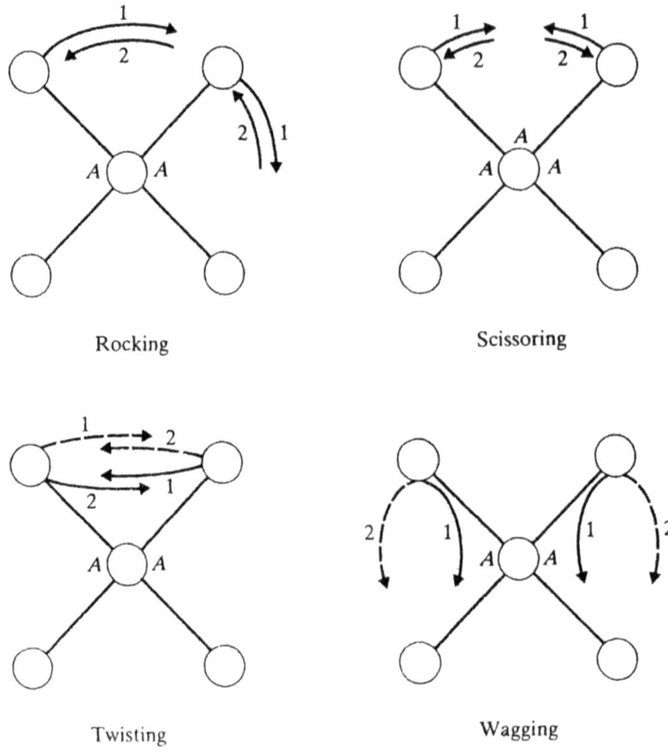

Figure 12-3 Diagrams illustrating nuclear motion during various forms of bending vibrations. The angles that are altered during the vibrations are labeled *A*. The dashed arrows during twisting and wagging represent motion behind the plane of the page and the solid arrows represent motion in front of the plane of the page. The numbers represent the orders in which the motions occur.

Coupling between a stretching vibration and a bending vibration is most likely when the bond that is stretching and shrinking is also one of the bonds involved in the bending. Vibrational coupling rarely is observed when the vibrating groups are separated by more than one bond.

It is possible to calculate the total number of possible vibrations in a molecule if the number N of atoms in the molecule is known. Each vibration can potentially result in an absorptive band in the infrared spectrum. In practice fewer absorptive bands are actually observed in most cases because some of the bands can be *degenerate* (possess equal energy) or occur beyond the spectral range of the spectrophotometer. Furthermore, some of the vibrations might not cause a change in dipole moment in the molecule and therefore are infrared-inactive.

The total number of coordinates that are required to completely specify the position and orientation in space of a molecule at a particular instant is $3N$, that is, three coordinates are required for each atom in the molecule. Each coordinate is a *degree of freedom* for the molecule. Consequently, $3N$ degrees of freedom exist for each molecule. Of course a molecule is in constant motion. The translational motion of the center of gravity can be described using three degrees of freedom. Another two degrees of freedom are required to describe the rotation of a *linear* molecule about its center of gravity or three degrees of freedom are needed to describe the rotation of a *nonlinear* molecule. The only type of motion of the molecule that has not been accounted for and that would allow a complete description of the molecular position is that owing to vibration of the individual atoms relative to each other. Because a total of $3N$ degrees of freedom are required to completely describe a molecule at any instant, the number of possible vibrational motions must be $3N - 3$ (for translational motion) $- 2$ (for rotation), or $3N - 5$, for linear molecules, and $3N - 3 - 3$, or $3N - 6$, for nonlinear molecules. It is important to understand, as described earlier, that an infrared absorption does not necessarily accompany each vibrational mode.

Sample problem 12-2 Determine the total possible number of vibrational bands for water vapor, carbon dioxide, and *n*-hexane.

SOLUTION Water has a nonlinear structure with three atoms. The number of possible vibrational modes is $3N - 6 = 3(3) - 6 = 3$. Because carbon dioxide has a linear structure consisting of three atoms, it can have $3N - 5 = 3(3) - 5 = 4$ vibrational modes. Hexane C_6H_{14} has $N = 20$ and is nonlinear. Consequently, $3N - 3 = 3(20) - 3 = 57$ vibrational modes are possible.

In addition to the factors already described, other factors that can affect the position of an infrared absorptive band include interactions between the studied solute and other solute molecules, interactions between the solute and solvent, the nature of the solvent, the concentration of the analyte, and the temperature. If the analyte can form intermolecular or intramolecular hydrogen bonds the vibrational stretching band owing to the hydrogen in the hydrogen bond is broadened and can be shifted in frequency. Substances that can take part in hydrogen bonding usually show a broad stretching band between about 2900 and 3500 cm^{-1}. Other forms of associative reactions between the analyte and other molecules in solution can have similar effects.

The solvent can affect peaks in the spectrum of an analyte if it can associate with or in some other way react with the analyte. For that reason it is generally preferable to choose a solvent for the study that is relatively inert and nonpolar. Carbon tetrachloride and carbon disulfide are recommended solvents.

The concentration of the analyte can sometimes have an effect upon the spectrum. An increase in concentration can cause increased interaction between analyte molecules. If the molecules interact the spectrum can be altered.

As the temperature increases the number of collisions and the potential interaction between molecules within the analyte increases. An increase in temperature can change the rotational levels within a molecule. That change can alter the spectrum. Generally as the temperature is increased, the absorptive bands become broader. Associative reactions such as hydrogen bonding are affected by temperature changes and consequently affect the spectral bands.

Apparatus

The apparatus that is used for infrared spectrophotometry consists of the same types of components that are used for ultraviolet-visible spectrophotometry. In most cases the order of the components is altered from that of most ultraviolet-visible spectrophotometers. A block diagram of the major components in an infrared spectrophotometer is shown in the simplified sketch in Fig. 12-4. Most infrared spectrophotometers are double-beam instruments. More information on typical instruments is included in a later section of the chapter.

Sources

In the near-infrared region a tungsten-filament lamp can be used as the source. In the important mid-infrared region the source usually is a *Globar,* a *Nernst glower,* a coil of Nichrome wire, a halogen lamp, or a tunable laser. A Globar is a bonded silicon carbide rod. Often the rod is about 7 mm in diameter and 50 mm in length. Upon electrical resistive heating the rod emits infrared radiation. It is usually operated at about 1300°C. It can be used for studies at wavenumbers as low as 200 cm^{-1}. In some cases it is necessary to water-cool the Globar.

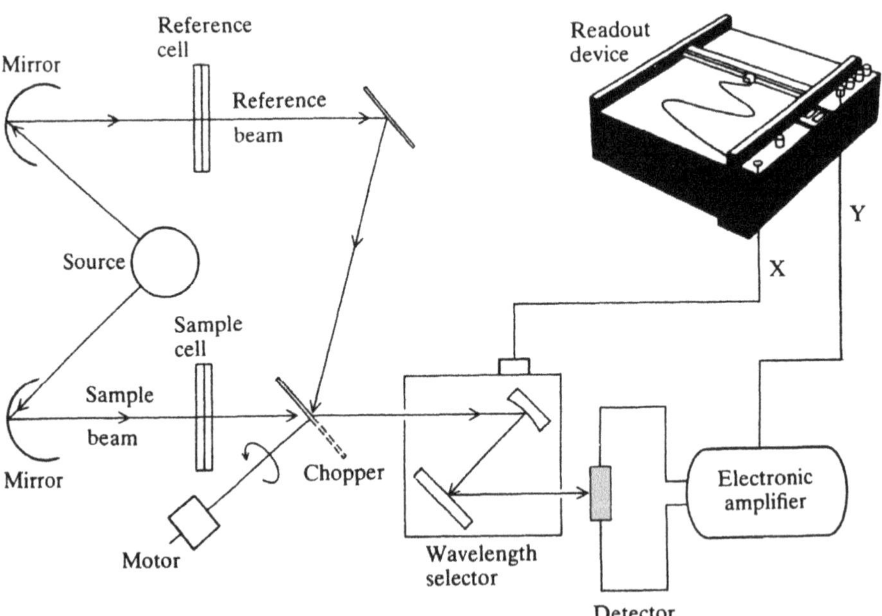

Figure 12-4 A block diagram of an infrared spectrophotometer. The arrows indicate the direction of radiative flow.

A Nernst glower is composed of a mixture of zirconium, thorium, and yttrium oxides. The mixture is formed into a hollow rod which is typically about 2 mm in diameter and between 2 and 5 cm in length. The ends of the rod are connected to the electric power supply through platinum leads. As with the Globar, passage of a current through the source results in resistive heating and consequent infrared emission. Unlike Globars, Nernst glowers must be preheated before they become conductive. A separate electrical heater is usually required. Nernst glowers are generally used in the spectral range from 1000 to 10,000 cm^{-1}. Within that range the device emits more intense radiation than a Globar.

A tight helical coil of Nichrome wire is more rugged, has a longer lifetime, and requires less care than the other sources. Unfortunately the coil also emits less intense radiation. It is electrically heated by passing a current through the wire. Upon exposure to air the wire forms an oxide coating that, when heated, emits radiation throughout most of the infrared region. Halogen lamps are not as popular as the previously described sources, but have been used in some instruments. Typically they emit useful radiation from approximately 2200 - 25,000 cm^{-1}.

Tunable lasers (Chapter 8) are available that emit radiation in the infrared region. Semiconductor diode lasers can be used for studies in the infrared region. The wavelength can be varied for each diode over a range of about 40 cm^{-1} by changing the temperature at which the device is operated. Generally the devices are operated at extremely low temperatures. Other types of lasers are also available that emit radiation in the infrared region. In most cases the high intensity radiation that is emitted by lasers is not required for infrared spectrophotometric studies. Regardless of the type of source that is used, a mirror is often used to collect the radiation and to direct it through the cell.

Cells and Sample Preparation

The type of cell that is used for a particular assay is dependent upon the physical state of the sample. In all cases it is necessary that the cell walls be transparent to radiation in the region used for the study. The entire optical system of an infrared spectrophotometer must be able to pass infrared radiation. In most instruments focusing and collimating is done with mirrors rather than with lenses in order to increase the amount of radiation that strikes the sample and to eliminate the need for constructing lenses from materials that are transparent to infrared radiation.

Solid samples Solid samples are prepared for spectrophotometric measurements using any of four methods. For quantitative studies, and for those studies in which the results have to be duplicated at a later time, it is usually preferable to prepare a solution of the analyte. Typically 1 g of sample is dissolved in sufficient solvent to prepare 10 mL of solution. The spectrum of the solution is measured in a cell with a 0.1-mm pathlength. If the sample is highly absorbant of infrared radiation, the solution can be prepared by dissolving 0.2 g of sample in sufficient solvent to prepare 10 mL of solution. Spectra obtained with carefully prepared solutions are more reproducible than those prepared using the alternative methods. The solution is injected into the cell with a syringe.

The cells that are used to contain sample solutions during spectrophotometric measurements normally have short pathlengths (typically between 0.1 and 1 mm) in order to minimize absorption by the solvent. A 0.1-mm cell is a good choice because cells that have shorter pathlengths are difficult to construct and excessive solvent absorption can occur in cells that have longer pathlengths. For trace analysis with a highly transparent solvent it is possible to use cells with pathlengths as long as 5 cm.

The windows of the cells must be constructed from a material that is transparent to infrared radiation. The most common material for cell windows is NaCl which is transparent from about 625 to 5×10^4 cm^{-1}. Other materials that can be used and the approximate spectral ranges through which they are useful include glass (5×10^3 to 2.9×10^4 cm^{-1}), quartz (2.5×10^3 to 5×10^4 cm^{-1}), fluorite (1×10^3 to 5×10^4 cm^{-1}), KCl (5×10^2 to 3×10^4 cm^{-1}), KBr (4×10^2 to 5×10^4 cm^{-1}), AgCl (4×10^2 to 1.7×10^4 cm^{-1}), CsBr (2.5×10^2 to 3×10^4 cm^{-1}), and CsI (2×10^2 to 3×10^4 cm^{-1}). A longer list of materials that can be used for cell windows can be found in Reference 1, p. 336. Because many of the materials that are used for cell windows are water-soluble, the cells must be protected from moisture in the atmosphere and in the sample. Small amounts of moisture can fog some cell windows. Particularly susceptible are windows constructed from alkali halides such as CsBr, CsI, KBr, KCl, and NaCl.

Small amounts of fogging can be eliminated relatively easily by polishing the windows either on a piece of a hard cloth, such as silk, which has been stretched over a flat surface or on a flat piece of glass. The windows should be removed from the cell prior to polishing. While using the cloth method a slurry is prepared of a commercial polishing agent (e.g., Aloxite or Barnsite) and ethanol, ethylene glycol, or water. The slurry is applied to a portion of the cloth and the windows are briskly rubbed across it. Immediately after polishing with the slurry, the window is polished on a dry portion of the cloth.

When a bare glass plate is used, the plate initially is *slightly* moistened either with an atomizer or by breathing on the plate. The window is polished on the damp glass. If neither method prepares windows with adequate transparency it might be necessary to polish the windows with an abrasive (usually 500 mesh) on a flat glass disk prior to polishing as previously described. Either kerosene or water can be used as a lubricant during the grinding. During all forms of polishing care should be taken to perform the polishing on a hard flat surface so as not to introduce a curved surface to the window.

Cell windows or entire cells should be stored either in a desiccator or in a heated compartment in order to protect them from moisture. Usually cells can be safely stored in a compartment that is heated to about 10°C above room temperature. While in place in the spectrophotometer the cell windows are generally protected by the heat from the instrument. Many instruments have heated mountings for the cells. In any case the spectrum of pure solvent should be obtained and subtracted from the spectrum of the analyte to compensate for fogging and other peculiarities of the cell windows.

Moisture in the solvent can be partially removed by shaking the solvent with a desiccant prior to solution preparation. As an alternative, moisture can be removed by the addition of an excess of 2,2-dimethoxypropane. It reacts with water to form methanol and acetone. The methanol, acetone, and excess dimethyoxypropane can be removed by vaporization during heating of the reaction mixture.

The desirable properties of a solvent that is to be used to obtain an infrared spectrum include transparency throughout the examined spectral range, a lack of interaction with the solute, a refractive index that is similar to that of the cell windows, and adequate analyte solubility. The transparency and solubility requirements are obvious. All solvents interact with the solute to a certain extent. Normally the interaction can be minimized by the use of a nonpolar solvent. Solvents that have refractive indexes that are near that of the cell windows exhibit fewer interference fringes than other solvents. When recording the spectrum of an analyte, interference fringes can be regarded as background noise that makes location of low-intensity peaks difficult. A more thorough description of interference fringes is included later in the section entitled "Quantitative Analysis."

Few of the many organic solvents possess all of the desirable properties for use in infrared spectrophotometry. No single solvent is adequate throughout the entire infrared spectral region. Consequently, at least two solvents are required for most studies. Perhaps the most useful solvent pair is carbon disulfide and carbon tetrachloride. In 0.1-mm cells, carbon disulfide is used from about 625 to 1330 cm^{-1} and carbon tetrachloride is used from about 1330 to 4000 cm^{-1}. The pair is useful throughout most of the infrared spectral range. Both solvents are toxic and should be used for solution preparation in a hood. An alternative pair of solvents that is less toxic is *n*-heptane (250 to 1000 cm^{-1}) and tetrachloroethylene (1000 to 4000 cm^{-1}). Unfortunately spectra recorded in those solvents exhibit greater background absorption than those recorded in carbon disulfide and carbon tetrachloride. Several other solvents that can be used and their approximate, useful spectral ranges are listed in Table 12-1. A longer list can be found in Reference 2, p. 171.

As an alternative to preparation of a solution the solid analyte can be ground into a fine powder and mixed with mineral oil or Nujol to form a suspension called a *mull*. The mull is pressed between two salt plates which serve as transparent windows of the cell. The mull-plates assembly is mounted in the cell holder of the spectrophotometer. The solid must be ground into particles that have a diameter that is less than the wavelength of the infrared radiation in order to prevent the particles from appearing opaque to the incident radiation. Perhaps the best way to grind the solid is with an agate mortar and pestle. A ball mill or power grinder also can be used. A single drop of mineral oil is added to no more than 20 mg of the powder to prepare the mull.

Mineral oil shows strong C—H stretching bands that can interfere with some assays. That problem can be overcome by use of a chlorinated or fluorinated oil rather than mineral oil. Perfluorokerosene or hexachlorobutadiene can be used. Often a *split mull* is used. A spilt mull consists of two mulls that have been prepared with different oils. The chlorinated or fluorinated oil is used at wavenumbers from 1340 to 4000 cm^{-1} and mineral oil is used at wavenumbers below 1340 cm^{-1}. The problems associated with solvent choice are eliminated with mulls but the spectra generally cannot be used for quantitative analysis.

Another common method of preparing solid samples is to thoroughly mix the finely powdered analyte with finely powdered KBr. The mixture is placed in a die and pressed into a transparent pellet. The pellet is mounted on a cell holder of the spectrophotometer. Often the shell of the die acts as the physical support for the pellet while the pellet is mounted in the spectrophotometer. Typically 1 mg or less of the solid analyte is mixed with about 0.1 g of KBr.

Table 12-1 A list of selected solvents and their *useless* spectral ranges for infrared spectrophotometry in 0.1-mm cells

The solvents can be used for infrared studies except in the listed regions

Solvent	Useless regions, cm^{-1}
Acetonitrile†	1300-1600, 2200-2300
Benzene†	Below 750, 3000-3100
Carbon disulfide	1400-1600, 2100-2300
Carbon tetrachloride	720-820
Chloroform	Below 820, 1200-1240, 3000-3020
Chlorotribromomethane	Below 820
Cyclohexane	850-910, 1430-1480, above 2600
Dichloromethane	Below 820, 1200-1300
n-Heptane†	1400-1500, 2800-3000
n-Hexane†	1400-1500, 2800-3000
Tetrachloroethylene	750-950

† Other low-intensity and narrow absorptive bands are observed for the solvent. A blank spectrum should be obtained.

The analyte must be ground into particles that have dimensions less than the wavelength of the radiation used for the study. That is accomplished in the same manner that was used for mull preparation. The die usually consists of a stainless steel cylinder with a bolt on each end. One bolt is inserted and serves as the bottom of the die. The powdered analyte–KBr mixture is poured into the vacant hole of the other bolt. The second bolt is inserted and tightened with a wrench to form the pellet. Unfortunately changes in the crystalline structure and in the analyte composition can occur during pressing. Furthermore, it is difficult to eliminate moisture from the pellet. Consequently the KBr-pellet method probably should not be used when a solution or mineral-oil mull can be prepared.

For those solid samples that cannot readily be prepared in one of the methods described above, it is sometimes possible to heat the sample in a pyrolysis apparatus and to spectrally examine the pyrolysis products. Often the pyrolysis products are reproducible under a fixed set of experimental conditions (temperature, heating rate, etc.). The spectra of the products can be compared with those of similarly treated known compounds for qualitative analysis. Pyrolysis is not a preferred method of sample preparation because the spectrum of the original sample is not recorded and because the process is destructive.

Liquid samples Spectra of liquid samples are usually obtained in one of three ways. For quantitative analysis and in those instances where highly reproducible results must be obtained, it is preferable to prepare a solution of the analyte. Analytes composed of aliphatic hydrocarbons are the exception to the rule. Aliphatic hydrocarbons generally yield better spectra when used as *neat* samples, i.e., as pure compounds. The same solvents and cells can be used that were described for solutions of solids.

In those cases where quantitative or semiquantitative results are required, and when an adequate solvent cannot be found, a spectrum of the neat sample can be obtained in a short-pathlength cell of the type that normally is used for solutions. When only qualitative results are required, the simplest procedure is to place a drop or two of the analyte on a flat salt plate. A second salt plate is carefully slid across the top of the sample, and the plates with the sandwiched sample are placed in a holder which is mounted in the spectrophotometer. The plates can be constructed from any of the infrared-transparent materials that were described earlier. Sodium chloride plates probably are used most often. The pathlength of a cell that is prepared in that manner is not accurately known and is not reproducible between samples. Consequently, spectra obtained with samples that are squeezed between flat salt plates cannot be used for quantitative analyses.

Gaseous samples Spectra of gaseous samples can be obtained in cylindrical cells that have relatively long pathlengths. The windows of the cells are constructed from one of the transparent salts that were described earlier. The choice of cell pathlength is based on the concentration of the analyte and the absorptivity of its absorbance bands. Spectra of highly absorbing gases can be obtained in 10-cm cells. Less strongly absorbing substances must be assayed in longer pathlength cells.

In most spectrophotometers the length of the cell cavity prevents insertion of cells that are longer than about 10 cm. Longer pathlengths are obtained by using *multipath cells*. Multipath cells use internal mirrors that are often constructed from gold (or another polished metal) to reflect the incident radiation back and forth through the gaseous sample several times prior to exiting the cell. Typical pathlengths that have been obtained with multipath cells are 1.5, 20, 40, and 120 m.

Regardless of the pathlength, the sample is pumped into the cell through a stainless steel entrance port. The valve to the entrance port and the valve to the exit port are opened and the sample is pumped into the cell. The two valves are located at opposite ends of the cell. Pumping should be maintained sufficiently long to completely flush the cell with the sample. The exit valve and then the entrance valve are closed and the cell is mounted in the spectrophotometer.

Wavelength Selection

In modern instruments wavelength selection is performed primarily with an interferometer and Fourier transforms. A description of Fourier transform infrared FTIR spectrophotometry is included later in the chapter. The wavelength selectors described in this section are used in older instruments and in instruments designed for unique assays.

In the near-infrared region a quartz prism or a diffraction grating can be used in the monochromator to disperse the radiation. Reflective diffraction gratings are most popular owing to their lack of absorption of the impinging beam of radiation. In the mid-infrared and far infrared regions, reflective diffraction gratings are nearly exclusively used. In order to eliminate radiation of several orders from simultaneously passing, some monochromators contain two diffraction gratings in series or a diffraction grating with appropriate filters. Devices that contain two serial dispersive elements are *double monochromators*.

A double monochromator is more expensive but disperses the radiation better and passes less stray radiation. Most of the advantages of a double monochromator with less expense can be attained by using a mirror to twice pass the radiation through a single dispersive device. A monochromator that contains a single dispersive device through which the radiation passes twice is a *double-pass monochromator*.

Prisms that are constructed from an alkali halide can be used for wavelength dispersion; however they have several disadvantages in comparison to reflective diffraction gratings. The resolution of prism monochromators is usually not as good as that of grating monochromators. Furthermore, the resolution in prism instruments varies with wavelength and is temperature-sensitive. Of course, alkali halide prisms must be protected from moisture. Energy passage through a prism is only a fraction of that which is reflected from a reflective diffraction grating. A single crystal of sodium chloride is used most often for prisms in the mid-infrared region; however CsBr, KBr, and LiF are also used in some instruments. KBr and CsBr are used most often between 250 and 670 cm^{-1}. LiF is used from 2000 to 10,000 cm^{-1}.

Disadvantages of grating instruments in comparison to prism instruments include possible passage of unwanted spectral orders, more radiative scattering, and a limited wavelength range for each grating. Some instruments sequentially use as many as four diffraction gratings in order to cover the entire infrared spectral region. Several popular instruments use two diffraction gratings in combination with several filters.

The energy that passes through a monochromator is related to the width of both the entrance and exit slits of the device. As long as the incident radiation completely covers the entrance slit, the energy passed through the monochromator is directly proportional to the width of the entrance slit and to the width of the exit slit. In some instruments the analyst has only a single adjustment for both slit widths and the widths are identical. In that case the energy that is transmitted through the monochromator is proportional to the square of the slit width. Resolution decreases with increased slit width. The best resolution is achieved by adjusting the slit widths to the smallest values that will permit sufficient radiation to strike the detector for accurate measurements.

Nondispersive infrared spectrophtometers do not contain a monochromator. In those instruments either the detector is designed to respond to a single substance or group of substances, or a narrow-bandwidth filter is used. The filters are usually constructed of dielectric materials. The transmitted wavelength is absorbed by an intense band of the analyte. Dielectric filter *monochromators* are formed on a disk-shaped substrate. Rotation of the disk causes different wavelengths to be transmitted through the filter. A description of the detectors that are designed to respond to specific substances is contained in the section on detectors. Wedge interference filters can be used to continuously vary the transmitted wavelength.

Detectors

Infrared detectors are classified as either *thermal detectors* or *photon* (*quantum*) *detectors*. A thermal detector is a transducer that changes thermal energy into an electric signal. The electric signal is amplified and routed to the readout device. Infrared radiation is focused on the thermal detector and alters its temperature.

Photon detectors use impinging infrared radiation to excite electrons in a semiconductor from a nonconducting to a conducting energetic level. Photon detectors respond rapidly and are among the more popular choices for use with Fourier transform infrared spectrophotometers. During the process the electric resistance of the semiconductor is lowered. Infrared radiation is not sufficiently energetic to activate photographic film or to cause a response in photoelectric detectors such as phototubes and photomultiplier tubes.

Thermal detectors Thermal detectors contain a small active element on which the radiation is focused. By blackening and insulating the element and by minimizing the size of the element, temperature change and detector response is maximized. Temperature change is approximately inversely proportional to the exposed surface area of the element. As the intensity of the impinging radiation increases, the temperature change on the element of the detector increases. Thermal detectors can be classified as *thermocouples, thermistors,* or *pneumatic devices*.

A thermocouple consists of two dissimilar metallic wires, e.g., Bi and Sb, that are joined at their ends. The surface at the junction of the wires is coated with a black metallic oxide. A change in temperature at the junction between the two wires causes a change in electric potential to develop between the wires. The potential difference between the unjoined ends of the wires is amplified and measured. The response of a thermocouple to a temperature change is relatively slow and small. Slow response is a characteristic of all of the common thermal detectors. Because the response is slow, the chopper that is used in infrared spectrophotometers that contain thermocouples must be operated at a frequency less than about 30 Hz. An advantage of a thermocouple detector is the independence of the response with changes in wavelength. If the response from a single thermocouple is inadequate, several (often six) thermocouples can be connected in series so that the output potential is the sum of the potentials across the thermocouples. A device of that design is a *thermopile*.

Thermistors are devices that have an electric resistance that is highly temperature-dependent. Sometimes a thermistor that is used in an infrared spectrophotometer is termed a *bolometer*. Many of the thermistors that are used in spectrophotometers consist of a sintered oxide of cobalt, manganese, or nickel. A Si bolometer is used for detection in the far ir region of one popular spectrophotometer, but requires liquid He cooling to operate. A liquid helium or liquid nitrogen cooled InSb bolometer is useful between 2 and 35 cm^{-1}. Because the typical response time for an uncooled thermistor is about 80 ms, choppers used in instruments that contain thermistors must operate at a frequency less than about 12 Hz. Usually a constant potential is applied across the thermistor and the difference in current flow between an illuminated thermistor and a nonilluminated thermistor is measured using a differential operational amplifier. Both thermocouples and bolometers are typically contained in evacuated glass envelopes that have infrared-transparent windows. The glass envelopes increase sensitivity and decrease noise from cooling that is caused by thermal air currents. Temperature changes in the elements of thermal detectors are on the order of a few thousandths of a degree.

Pyroelectric bolometers contain a single crystal of a pyroelectric material such as triglycine sulfate (TGS), deuterated triglycine sulfate (DTGS), lithium niobate $LiNbO_3$ or lithium tantalate $LiTaO_3$. The pyroelectric substance is placed between two metallic capacitor plates. The chopped radiation that strikes the pyroelectric substance causes an alteration in its crystalline structure that causes the surfaces of the crystal to alter their charge distribution. The change is monitored. Pyroelectric bolometers respond only to *changes* in temperature and consequently require chopped incident radiation. The response time of pyroelectric bolometers typically is 1 ms or less. The short response time makes them useful in instruments which are used to make a rapid measurement (such as FTIR spectrophotometers).

Pneumatic detectors respond to changes in volume of a nonabsorbing gas or liquid with temperature changes. Pneumatic devices that use a gas as the medium are *Golay detectors*. Diagrams of two forms of Golay detectors are given in Fig. 12-5. Infrared radiation enters the transparent window and strikes a black metallic plate. The absorbed radiation heats an inert gas (usually xenon) in a pneumatic chamber behind the plate and causes the gas to expand. As the gas expands the flexible diaphragm at the opposite end of the chamber from the metallic plate is pushed outward.

In the detector shown in Fig. 12-5*a,* visible radiation from a tungsten-filament lamp passes through one or more lenses that serve to collimate and focus the light on the mirrored outer surface of the diaphragm at the end of the pneumatic chamber. Enroute to the diaphragm the radiation passes through a grid. After reflection from the diaphragm the radiation again passes through the grid and the lens, and is reflected by a mirror to a phototube or photomultiplier (PM) tube. The phototube responds to the intensity of visible radiation that strikes it. The output from the phototube is the output from the Golay detector.

When radiation from the lamp passes through the grid, it forms the image of the grid on the mirrored surface of the diaphragm. If no infrared radiation strikes the black metallic plate, the diaphragm is in a contracted position and the reflected visible radiation strikes the opaque portion of the grid. Consequently no radiation passes through the lens and strikes the phototube. When infrared radiation strikes the black plate and warms the pneumatic chamber, the diaphragm is pushed outward and the angle of reflection of the visible radiation is altered. Because the reflected radiation no longer squarely strikes the grid, a fraction of the radiation passes through the grid and is registered by the phototube. As the gas within the pneumatic chamber expands, more radiation passes through the grid and strikes the photo-detector. The response of a Golay detector is a function of the heating of the pneumatic chamber which in turn is proportional to the intensity of the incident infrared radiation.

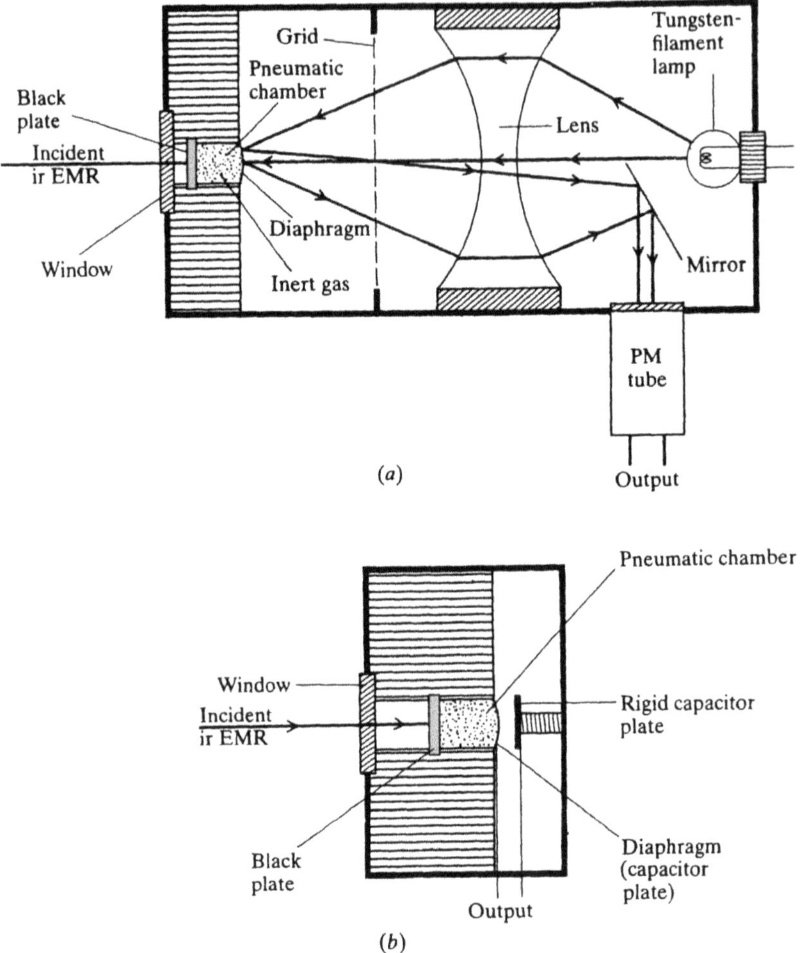

Figure 12-5 Diagrams of two forms of Golay detectors. The arrows indicate the radiative paths through the devices.

The operation of another form of Golay detector is illustrated in Fig. 12-5*b*. The visible optical system of the previously described detector is absent in the second detector. Instead, the flexible metallic diaphragm is one plate of a capacitor. A second rigid plate is held a small distance from the diaphragm and the capacitance or some electrical property that is related to the capacitance between the two plates comprises the output of the detector. Because the distance between the plates lessens as the gas in the pneumatic chamber expands, the capacitance is altered as a function of the intensity of the incident radiation.

A diagram of a nondispersive infrared photometer that is designed for the quantitative analysis of gases and that uses a thermal pneumatic detector is shown in Fig. 12-6. Chopped radiation from two identical sources is simultaneously passed through a reference cell and through a cell that contains the analyte. The chopping rate is typically about 5 Hz and the radiation passing through the two cells is in-phase. The reference cell is filled with a gaseous mixture that contains a fixed concentration of the assayed gas. The sample cell contains the analyte. The sample cell can be either continuous flow or stopped flow.

After passage through the cells, the radiation enters the detector. In some cases a filter is placed between the cell compartments and the detector. The detector contains two pneumatic chambers that are separated by a flexible metallic diaphragm. Each compartment contains a fixed concentration of the infrared-absorbing gaseous component that is assayed.

Radiation from the reference cell enters one of the compartments and radiation from the sample cell enters the other. Infrared radiation that enters the detector is absorbed by the gas in the detector, and causes the gas to expand. If the concentration of the absorbing gas in the two cells is equal, an equivalent amount of infrared radiation enters the two compartments of the detector and gaseous expansion in the compartments is identical. Consequently, the flexible diaphragm assumes an intermediate position between the two compartments.

If the concentration in the sample cell is greater than the concentration in the reference cell, more radiation is absorbed in the sample cell and less radiation

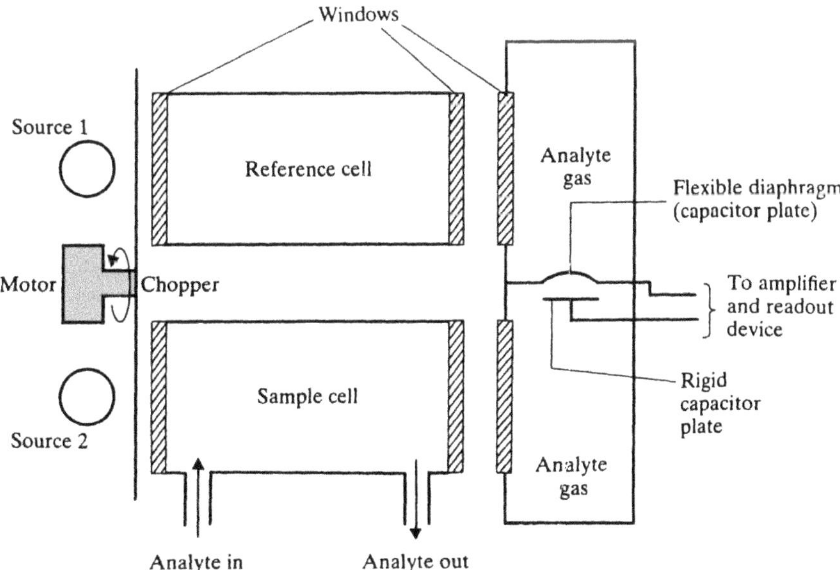

Figure 12-6 A diagram of a nondispersive infrared photometer that is used for quantitative analysis of gases.

enters the sample compartment of the detector. Consequently, the gas in the reference compartment expands more than the gas in the sample compartment of the detector and the diaphragm bulges into the sample compartment. The amount of displacement is a function of the relative absorption in the two detector compartments which is related to the amount of absorption in the two cells. If the concentration in the reference cell is greater than that in the sample cell, the diaphragm is displaced into the reference compartment of the detector.

The amount of displacement of the diaphragm is measured as in the Golay detector shown in Fig. 12-5(*b*). A rigid capacitor plate is placed in the sample compartment of the detector. The diaphragm serves as the second capacitor plate and the capacitance between the two plates is monitored. Of course, the capacitance is a function of the distance between the plates. An alternating signal is obtained using the detector because the radiation is chopped. Nondispersive instruments of the type described here often are used to monitor air pollutants such as carbon monoxide.

Photon detectors Photon detectors are constructed from both intrinsic and extrinsic semiconductors. Intrinsic semiconductors that can be used as infrared detectors include mercury cadmium telluride MCT (420 - 8,000 cm^{-1}), deuterated triglycine sulfate DTGS (10 - 700 cm^{-1}), silicon photodiodes (8,000 - 25,000 cm^{-1}) InAs, InGaAs (4,000 - 12,000 cm^{-1}), InSb (1850 - 11,700 cm^{-1}), PbS (3,000 - 5,600 cm^{-1}), and PbSe. The useful spectral ranges for some of the detectors are listed in the brackets. The detectors are usually between 10 and 100 times more sensitive than thermal detectors and respond rapidly. The range and sensitivity of intrinsic semiconductor detectors is improved, to the values listed in the brackets, by cooling the devices with liquid nitrogen or liquid helium.

Extrinsic semiconductors are useful over a broader range than intrinsic semiconductors. Typically the devices can be used as infrared detectors at all wavenumbers above 250 cm^{-1}. Doped MCT is a good choice and can be used from 450 - 12,000 cm^{-1}. Extrinsic semiconductors must be cooled to the temperature of liquid nitrogen or liquid helium in order to be useful. Because the response of semiconductor photon detectors is much more rapid than that of thermal detectors, they find use in measurements that must be made rapidly.

Readout Devices

The signal from the detector is amplified prior to being routed to the readout device. In nearly every instrument the radiation from the source is chopped. Chopped radiation is advantageous because the detector can be tuned to respond only to radiation at the chopped frequency. Background noise is significantly reduced by tuning the detector to respond to the chopped signal from the source. Most double-beam instruments use null-point or ratio detection systems.

After amplification the electric signal from the detector is directed to a readout device. In most spectrophotometers the readout device is either a recorder or a computer controlled display, such as a printer or video display terminal. Nondispersive photometers normally have analog or digital meters.

TYPICAL INSTRUMENTS

Designs of conventional infrared spectrophotometers are described in this section. Instruments that are designed to take advantage of Fourier or Hadamard transforms are described later. A description of nondispersive instruments has already been provided. Both single- and double-beam instruments are available. Double-beam instruments are much more popular than single-beam instruments.

Single-beam instruments are usually arranged as shown in Fig. 12-7. A chopper is often used in infrared spectrophotometers to prevent the detector from responding to radiation that is emitted from the cell or from some other location between the source and the detector. Any heat source is detected unless the signal is chopped and the detector is tuned to the frequency of the chopper. The chopping frequency is relatively slow owing to the slow response of most infrared detectors.

Double-beam instruments are normally of one of two types. Either the instrument utilizes an optical null measurement or it measures the ratio of the intensities of the two radiative beams. An optical diagram and a photograph of a commercial infrared spectrophotometer that uses a null measurement is shown in Fig. 12-8. In null-measuring spectrophotometers radiation from the source is split into two pathways with mirrors. Radiation in one pathway passes through the sample cell while radiation in the other pathway comprises the reference beam of the spectrophotometer. After passing through the sample compartment

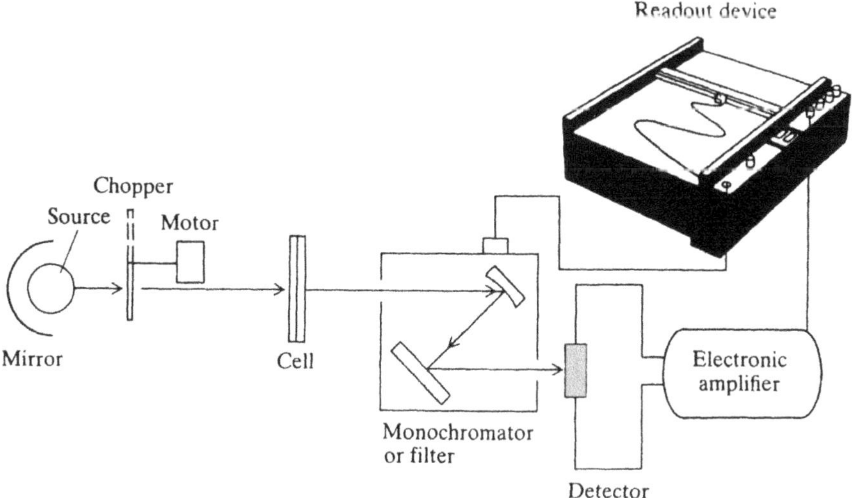

Figure 12-7 A diagram of a single-beam infrared spectrophotometer. The arrows indicate the direction of radiative flow through the instrument.

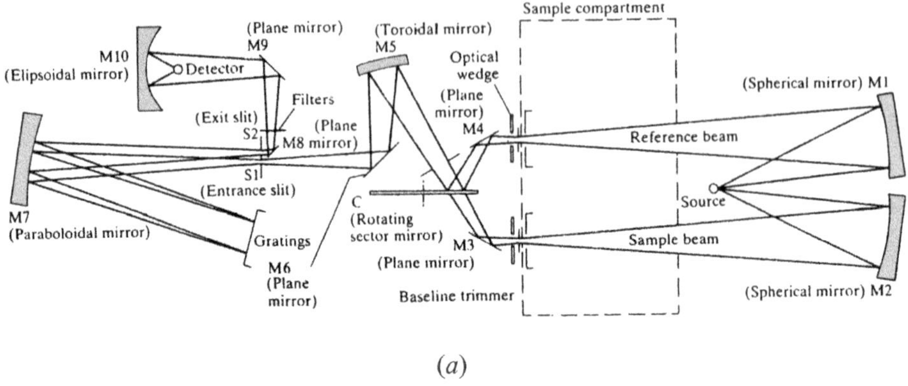

(*a*)

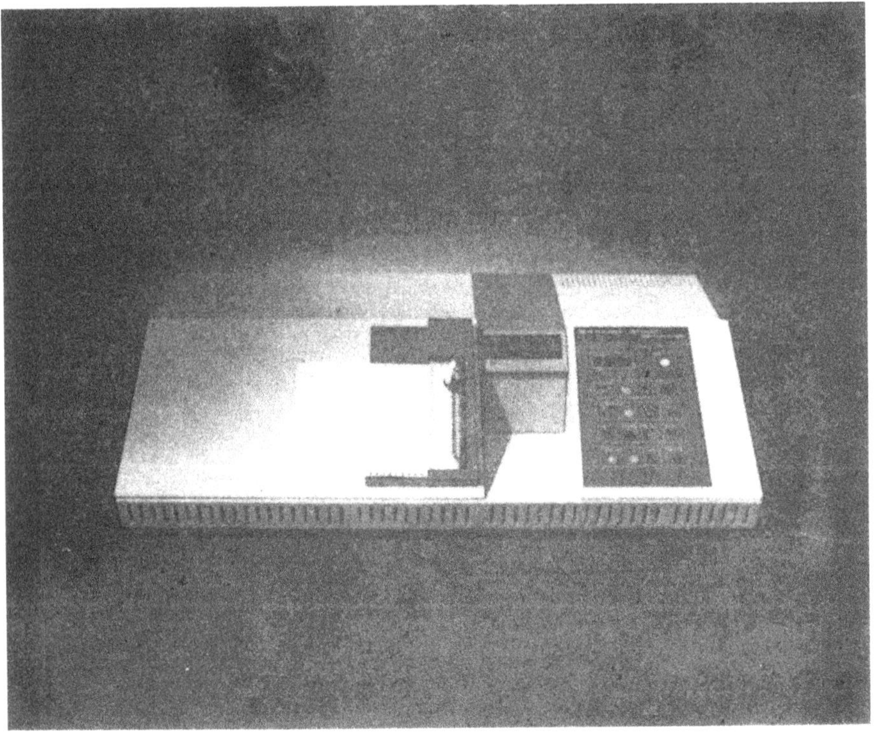

(*b*)

Figure 12-8 (*a*) An optical diagram and (*b*) a photograph of the Perkin-Elmer model 1320 infrared spectrophotometer. *(Courtesy* of *Perkin-Elmer Corporation.)*

of the instrument the reference beam passes through an *optical wedge* or a *comb*. The optical wedge (or comb) limits the amount of radiation that passes through it. As the distance the wedge is driven into the optical beam is increased, the amount of radiation which passes is decreased. A comb is a device with tapered teeth that increasingly prevents radiation from passing as it is pushed further into the beam.

The two beams are reflected to the same location and alternately pass through the monochromator and strike the detector. Usually the separation in time between the two beams is accomplished with a mirrored chopper that typically is rotated at a frequency between 5 and 15 cycles per minute. Feedback from the detector is used to drive the wedge or comb into the reference beam until the intensity of radiation that strikes the detector from the two beams is identical, i.e., until the difference between the two readings is zero (null). The readout device (usually a recorder) is driven by the deflection of the optical wedge into the reference beam. When radiation is absorbed by a sample, the wedge is driven into the reference beam until the intensity of the reference beam is decreased to the same value as that of the sample beam and the recorder pen deflects in direct proportion to the required motion of the wedge.

A disadvantage of infrared spectrophotometers that use null measurements is their inaccuracy when low-intensity radiation passes through the cell. When a strongly absorbing peak is recorded the energy that strikes the detector from both beams is too small to be accurately measured. At low radiative levels the instrument "freezes." It does not respond to small changes in intensity. Generally readings below about 30% T are inaccurate and should not be used for quantitative analysis.

An optical diagram of an instrument that measures the ratio of the intensities of the two radiative beams is shown in Fig. 12-9. In ratio-recording instruments radiation from the source is divided into two pathways as in the null-measuring instruments. In most cases the two beams are individually chopped at different frequencies prior to passing through the sample compartment of the instrument. Different chopping frequencies are required in order to permit the detector to distinguish between the two beams. After passing through the sample compartment the beams are alternately reflected to the detector by a chopper that operates at a frequency considerably less than that of the chopping

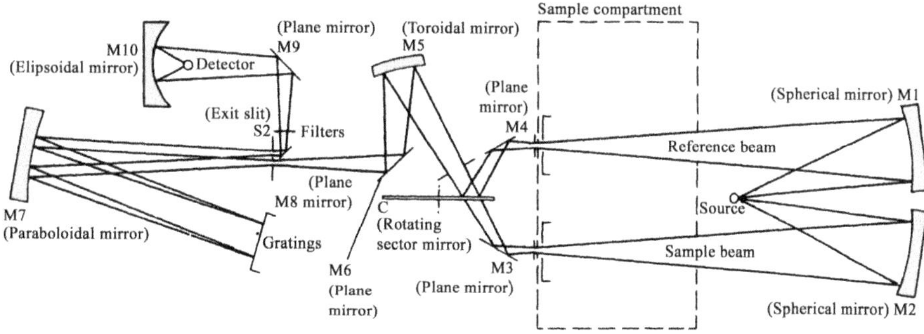

Figure 12-9 An optical diagram of the Perkin-Elmer model 1400 infrared spectrophotometer. *(Courtesy of Perkin-Elmer Corporation.)*

frequency of either individual beam. The ratio of the intensities of the two beams is obtained from the detector and is used to drive the *y* axis of the recorder or other readout device.

In addition to the instruments already described, multislit and multiple-detector instruments are occasionally used. Multislit instruments operate on the principle that high resolution can be obtained while simultaneously allowing relatively high-intensity radiation to strike the detector if many narrow slits in the monochromator are used in place of a single exit slit. The radiation from all of the slits is combined and measured by the detector. Apparently no commercial version of a multislit instrument is currently available. A summary of the operation of multislit spectrophotometers is included in Reference 1.

Rapid scans can be accomplished by using multiple detectors aligned on the focal plane of the monochromator. Typically pyroelectric detectors are used. The response of each detector is rapidly scanned in a manner analogous to the operation of a television camera and the entire spectrum can be obtained in less than a second.

ATTENUATED TOTAL REFLECTANCE

Attenuated total reflectance (ATR) is a method used for the infrared analysis of substances that cannot be easily assayed by using one of the techniques that have already been described. It is particularly useful for polymeric substances such as rubber and cured resins as well as for other substances that are difficult to handle in a conventional manner. The technique is based upon the fact that reflected radiation slightly penetrates the surface from which it is reflected.

The analyte is firmly placed against or coated on one surface of a prism or on opposite surfaces of an optically transparent element that is designed for multiple reflections (Fig. 12-10). Infrared radiation is passed through the prism or element and reflects from the surface between the prism and the sample. In the process the radiation penetrates a short distance into the sample and is *attenuated* by absorption in the sample.

Snell's law can be used to calculate the angle of refraction when radiation passes from one medium to another:

$$n_1 \sin a = n_2 \sin b \tag{12-2}$$

In Snell's law n_1 and n_2 are the refractive indexes of the first and second medium, a is the angle of the incident radiation as measured from the perpendicular to the surface between the two media, and b is the angle at which the radiation is refracted in the second medium. An example of the application of Snell's law was given in Sample problem 5-6.

From Eq. (12-2) it is apparent that as angle a increases, angle b also increases. If n_2 is less than n_1, an angle a exists for which b is 90°, that is, for

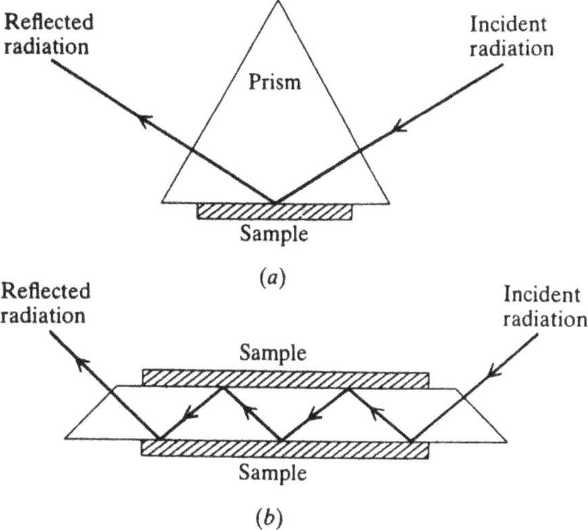

Figure 12-10 Diagrams of the cells used for attenuated total reflectance: (*a*) prism used for single reflections; (*b*) element used for multiple reflections.

which the radiation is completely reflected. That angle is the *critical angle* θ_c. Incident radiation that strikes the surface between the two media at an angle that is greater than the critical angle is completely reflected from the surface. The value of the critical angle can be calculated from Eq. (12-2) by substituting θ_c for *a* and 90° for *b*. Solving the resulting equation for θ_c gives

$$n_1 \sin \theta_c = n_2 \sin 90^\circ$$

$$\theta_c - \sin^{-1} \frac{n_2}{n_1} \tag{12-3}$$

At incident angles greater than the critical angle, complete reflection occurs at the surface. Essentially nonc of thc incident radiation is lost during the reflection although radiation can be lost due to absorption by the prism material. The critical angle is dependent upon a difference in the refractive index of the two media. Consequently, it is necessary to choose a prism material that both transmits radiation and has a refractive index that is greater than that of the sample. A substance that has a high refractive index is normally chosen. Several materials that can be used to construct the prism and the useful wavenumber range of each (in brackets) are AgBr (330 to 2.0×10^4 cm^{-1}), AgCl (500 to 2.5×10^4 cm^{-1}), CdTe (430 to 1×10^4 cm^{-1}), Ge (830 to 5000 cm^{-1}), KRS-5 (250 to 1.7×10^4 cm^{-1}), KRS-6 (310 to 2.5×10^4 cm^{-1}), and ZnSe (670 to 2×10^4 cm^{-1}).

The analyst should be aware that the refractive indexes in the two media are functions of wavclcngth. Thc angle of incidence of the radiation must be greater than that of the critical angle at all wavelengths used for the study. Because penetration into the sample decreases as the angle of incidence increases, an angle of incidence that is only slightly greater than the critical angle should be chosen. Generally the angle of incidence for most work can be estimated by using

$$a + 3 = \sin^{-1} \frac{n_2 + 0.2}{n_1} \tag{12-4}$$

where a is the angle of incidence in degrees.

Penetration depth is also related to the relative refractive indexes of the prism and the sample. As the refractive indexes approach each other, the depth of penetration increases. Prisms for use with attenuated total reflectance are usually constructed with entrance angles of 45 or 60°. The incident radiation should perpendicularly enter the prism to minimize reflection and refraction. Attachments are available for many spectrophotometers that permit use of single or multiple reflections ATR.

Sample problem 12-2 Calculate the critical angle between a prism with a refractive index of 2.03 and a sample with a refractive index of 1.34. Estimate the angle of incident radiation that should be used to obtain an ATR spectrum of the sample.

SOLUTION The critical angle can be calculated with Eq. (12-3):

$$\theta_c = \sin^{-1} \frac{1.34}{2.03} = 41.3^\circ$$

The estimated angle of incident radiation is obtained from Eq. (12-4):

$$a + 3 = \sin^{-1} \frac{1.34 + 0.2}{2.03} = 49.3^\circ$$

$$a = 46.3^\circ$$

It would be preferable to use a prism that has an entrance angle of 45° as compared to a prism that has an entrance angle of 60°. In either case the entrance angle is greater than the critical angle.

Diffuse reflectance is a method which is particularly well suited for the study of powders and other finely divided particulates. The sample is placed in a cup which is mounted inside the diffuse reflectance accessory. The accessory is placed in the sample compartment of the instrument. Normally the accessory is used with FTIR spectrophotometers. Radiation from the source enters one side of the accessory and is reflected with a mirror so that it strikes the sample. Since individual crystals of the powdered sample are oriented at different angles to the incident beam, the attenuated reflected beams exit the sample cup at multiple angles. The reflected beams are collected with a concave mirror that is placed above the sample cup. The collected resulting beam is focused on a second mirror which reflects the beam toward the detector assembly.

MULTIPLE-SCAN AND MULTIPLEX METHODS

Improvements in the signal-to-noise(S/N) ratio in a spectrum can be achieved by repeating the spectral measurements a number n times and summing or averaging the resulting signals. If a detector is used in which the noise is not a function of the signal strength, then the total signal after n scans is directly proportional to n, but the total random noise is directly proportional to the square root of n. If the signal strength for a single scan is S and the noise is N, then after n scans the total signal strength is nS and the total noise is $\sqrt{n}\,N$. After n scans the total single-to-noise ratio is $nS/\sqrt{n}\,N$ or $\sqrt{n}\,S/N$. In comparison to the single-scan signal-to-noise ratio of S/N, the gain in signal-to-noise ratio is $\sqrt{n}$ after n scans. That gain is the *Fellgett advantage*.

The improvement in S/N after multiple scans can be used to enhance a spectrum. Multiple-scan techniques are particularly useful for obtaining spectra of low-concentration analytes. Because many scans might be necessary to significantly enhance an infrared spectrum, multiple-scan instruments must rapidly scan and store the spectra in the memory of a computer.

An alternative to using multiple scans is to use a single detector to simultaneously record the oscillations of the electromagnetic waves throughout the entire spectrum as a function of time. Of course no infrared detector can respond sufficiently rapidly to measure the natural oscillations of EMR waves in the infrared region. Nevertheless, if the measurement were possible, the output from the detector would be the sum of all of the electromagnetic waves transmitted through the sample and should possess all of the information required to determine the intensity of the radiation that is transmitted at each frequency. By resolving the various frequencies from the sum reported by the detector, the entire spectrum could be constructed. A *multiplex method* is a method in which multiple pieces of data are simultaneously transmitted through a single channel. The simultaneous measurement of the oscillations throughout an entire spectrum is an example of a multiplex method. Decoding a spectrum recorded with a multiplex method is difficult, but can be accomplished within a reasonable time with the aid of a computer. Fourier transforms and, more rarely, Hadamard transforms are used to decode the information which has been recorded during the infrared measurement.

Fourier Transform Spectrophotometry

Fourier transform infrared (FTIR) spectrophotometers consist of several parts. In most instruments an interferometer such as the *Michelson interferometer* (Fig. 12-11) is used to change the frequency of the EMR from the source to a proportionately slower oscillating signal. The EMR frequency is superimposed

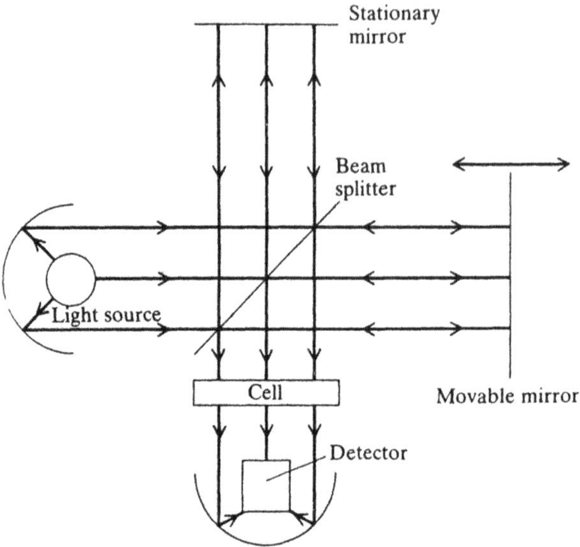

Figure 12-11 An optical diagram of a Michelson interferometer as used for infrared Fourier transform spectrophotometry. The arrows indicate the direction of radiative travel and the direction of mirror motion.

upon the slower frequency. The frequency of the oscillation must be slow enough to allow the infrared detector to accurately respond to it. With a liquid nitrogen or liquid helium cooled, MCT detector that frequency can be as high as 160 kHz. The sum of the proportionately slower oscillating signals is carried to the computer which mathematically separates the signals into individual oscillations and calculates the oscillations of the corresponding frequencies of absorbed radiation. The data are continuously recorded for a time that is sufficient to obtain the desired spectral enhancement. The method exhibits the Fellgett advantage where n is a time period (*resolution element*) rather than a number of discrete measurements. The amplitude of each resolved oscillation is a function of the intensity of the radiation.

A mathematical method called a Fourier transform is used to do the conversion from the time-domain spectrum to the conventional frequency-domain spectrum. The use of Fourier transforms and of an algorithm that considerably simplifies the computerized calculations is described in Reference 3. The frequency-domain spectrum is displayed on the readout device.

The Michelson interferometer (Fig. 12-11) functions in the following manner. Radiation from the source is emitted and directed toward a beam splitter. Half of the radiation is reflected by the beam splitter to a stationary mirror that reflects the radiation back toward the beam splitter. The remaining portion of the radiation passes through the beam splitter and strikes a mirror that is continuously moved back and forth over a distance of as much as 21 cm. After striking the movable mirror, the radiation is reflected back to the beam splitter. A portion of the radiation that was reflected from the stationary mirror and a portion of the radiation from the movable mirror combine at the beam splitter and pass through the cell. After passage through the cell the radiation strikes the detector.

If the round-trip distance through which the radiation travels between the beam splitter and the stationary mirror is identical to the round-trip distance between the beam splitter and the movable mirror, the radiation from the two mirrors arrives in-phase at the beam splitter, the cell, and the detector. As the movable mirror changes position, however, the distances between the mirrors and the beam splitter are no longer identical and radiation of a fixed wavelength will only arrive in-phase at the cell and detector when the round-trip distance between the movable mirror and the beam splitter is equal to the round-trip distance between the stationary mirror and the beam splitter plus or minus a whole-number multiple of the wavelength of the radiation. If the movable mirror moves away from the equidistant point by a factor of $\lambda/4$, the round-trip distance is altered by $\lambda/2$ and the reflected radiation is out-of-phase with that from the stationary mirror and destructively interferes. If the movable mirror is $\lambda/2$ from the equidistant point, the radiation is in-phase with that from the stationary mirror and constructively interferes. As the distance changes, different wavelengths of radiation become in-phase and out-of-phase at a frequency that is dependent upon both the wavelength of the incident radiation and the rate at which the mirror moves. By controlling the rate of mirror motion, a series of simultaneous signals that oscillate at frequencies that are directly proportional to the frequencies of the electromagnetic radiation arrive at the detector, and oscillate sufficiently slowly for the detector to measure. The detector simultaneously measures all of the frequencies that pass through the cell and routes the information to the computer which decodes the information using a Fourier transform. The decoded spectrum is directed to the readout device. Because the motion of the movable mirror must be carefully controlled in order to assure meaningful results, the motion of the movable mirror is monitored by reflection of a laser beam from a portion of the mirror, such as a corner of the mirror, that is not involved in the ir measurement. The reflected radiation from the laser is monitored. As the distance moved by the mirror changes, the reflected radiation arrives at its detector either in-phase or out-of-phase with the emitted laser wavelength. The resulting interference pattern of peaks (in-phase) and valleys (out-of-phase) is used to monitor the rate of motion of the mirror and to alter its speed if necessary.

Fourier transform infrared spectrophotometry is particularly useful in those circumstances where spectra of low-concentration samples are required and in those cases where a spectrum must be obtained rapidly. FTIR spectrophotometers can be used as detectors for chromatography.

FTIR Microscope

Spatial ir measurements on the surface or through a sample can be made with an FTIR microscope. The cell in Figure 12-11 is replaced with a microscope attachment to the FTIR spectrophotometer. The radiation from the interferometer is focused on a location on the surface of the sample. The user can choose the location of the radiation beam through a microscope which is arranged to view the ir focal point through the microscope's optics. The sample is mounted on a microscope slide which can be moved in either two or three dimensions.

Radiation from the sample is either reflected from its surface or passes through the sample. In either case the attenuated radiation is subsequently focused on the detector. Often the entire assembly is computer controlled and monitored. The FTIR spectrum of the sample can be obtained at a single location or a two or three dimensional, surface map of the analyte can be prepared showing absorption at a particular wavelength as a function of location.

Most, commercial, FTIR microscopes detect radiation in the mid ir range from about 600 - 4,000 cm^{-1}. One microscope is capable of making measurements from 80 - 25,000 cm^{-1}. Resolution with commercial instruments varies from about 1 to 3 micrometers. Sample sizes as small as a few nanograms can be studied with some instruments under ideal circumstances. Solids, liquids and gases can be assayed by using FTIR microscopy.

Hadamard Transform Spectrophotometry

Hadamard transform spectrophotometers have a source and a dispersive grating as in conventional infrared spectrophotometers. Rather than focusing a narrow-wavelength band on the detector and scanning through the wavelength region as in conventional instruments, the entire spectral range passes through an encoding mask and then is simultaneously focused on the detector. The encoding mask consists of a series of opaque and transparent slots. The width of each slot corresponds to a discrete spectral resolution element; i.e., the width of each slot corresponds to the distance on the focal plane from the grating that equals the desired bandwidth of each measurement.

The mask is stepped across the focal plane so that each dispersed wavelength band sequentially strikes or passes through each slot. A measurement by the detector is taken after each step and stored in a computer. If a total of n spectral resolution elements are required to cover the spectral range of interest, then the mask must consist of $2n - 1$ slots so that radiation always passes through or strikes a slot. After each step the mask is shifted sidewise by one slot width. A total of n measurements are made.

In order to explain the operation of the instrument consider the simple case in which only three resolution elements are required ($n = 3$). The mask would require $2n - 1 = 5$ slots. If a 1 is used to represent a transparent slot and a 0 an opaque slot, then a typical mask might be 10110. At first only the first three slots (101) on the left cover the spectral range. During the first step the total intensity I_1 measured by the detector is given by

$$I_1 = (1)I_{\lambda 1} + (0)I_{\lambda 2} + (1)I_{\lambda 3} \tag{12-5}$$

The intensity at each wavelength which corresponds to a spectral element is multiplied by 1 if the radiation passes through a slot and by 0 if the radiation strikes an opaque slot and does not pass.

After the first measurement is completed, the mask moves one position to the left. Now the three spectral elements successively see a mask that has the structure 011. An intensity measurement is made which corresponds to that shown in the following equation:

$$I_2 = (0)I_{\lambda 1} + (1)I_{\lambda 2} + (1)I_{\lambda 3} \tag{12-6}$$

Similarly, after the second measurement the mask is shifted another position to the left and the utilized portion of the mask is 110. The measured intensity during the third step is given by

$$I_3 = (1)I_{\lambda 1} + (1)I_{\lambda 2} + (0)I_{\lambda 3} \tag{12-7}$$

The three equations can be solved simultaneously by the computer for the intensity at each of the three wavelengths. A conventional spectrum is constructed from the results and displayed by the readout device. The values of 1 and 0 in the mask are used by the computer to write the simultaneous equations. For n resolution elements, n measurements are required. The resulting n simultaneous equations are solved by matrices with the computer. The calculations are relatively simple and can be rapidly accomplished by the computer because the coefficients in the equations are either 1 or 0. The mask used in the instrument shown in Fig. 12-12 contains 4093 slots and is used for 2047 spectral elements. In the instrument of Fig. 12-12, the light passes back through the dispersing device after going through the mask in order to focus the radiation on the detector.

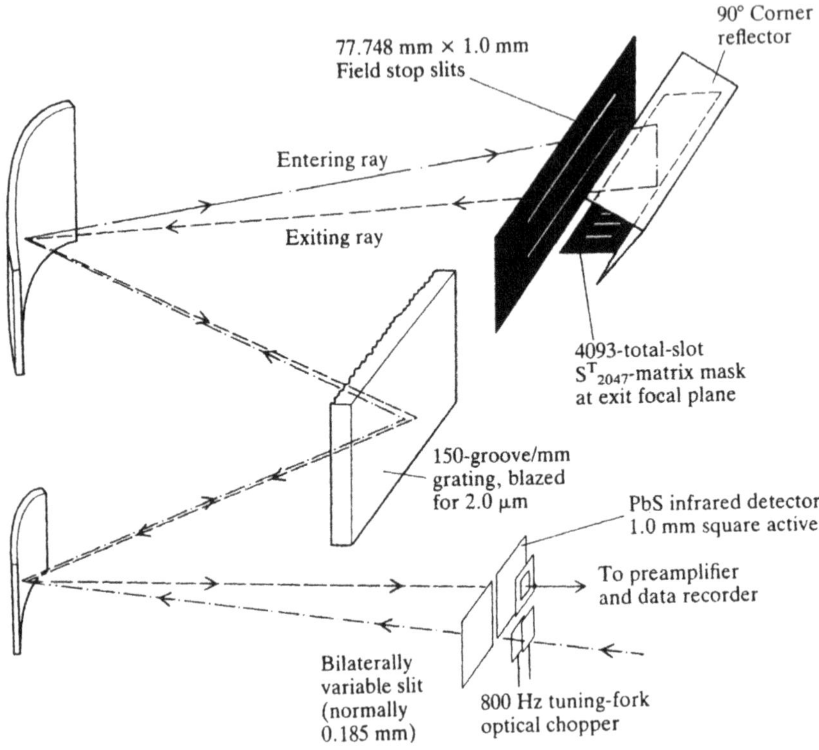

Figure 12-12 An optical diagram of one form of a Hadamard transform infrared spectrophotometer. *(Reprinted with permission from Reference 4. Copyright 1972, American Chemical Society.)*

QUALITATIVE ANALYSIS

A detailed description of the practical aspects of both qualitative and quantitative analysis by infrared spectrophotometry is provided in Reference 1. It is recommended that the reference be carefully read by all analysts who intend to routinely use infrared spectrophotometry. Infrared spectrophotometry is an excellent method for qualitative analysis because, except for optical isomers, the spectrum of a compound is unique.

Prior to recording a spectrum that is to be used for qualitative analysis, it is advisable to calibrate the wavelength settings of the instrument. Usually wavelength calibration is accomplished by using the spectrum of a compound that has peaks at known positions throughout the entire spectral region. Although any compound with known peak locations can be used, normally a polystyrene film, 1,2-dibromopropane, indene, or gaseous ammonia is used.

The best method of infrared qualitative analysis is direct comparison of the spectrum of the analyte to spectra of standards that have been obtained under identical conditions. Analysts who routinely use infrared spectrophotometry should accumulate a library of spectra of compounds and mixtures that they routinely assay and that were obtained with the same instrument that is used for the analysis.

If a reference spectrum that has been obtained in the analyst's laboratory is unavailable, the comparison can be made to spectra in any of several commercial collections of infrared spectra. A list of selected collections of infrared spectra is given in Table 12-2. A longer list containing 45 collections is given in Reference 1. Care should be taken when comparing spectra obtained with different experimental conditions because peak positions can shift and peak intensities vary, as described earlier.

If possible a computerized comparison of the analyte's spectrum with reference spectra should be completed. Computerized comparisons require little time and accurately list possible compounds in the analyte by comparison of the positions and intensities of the major peaks. Typically all bands are considered that have intensities that are 10 or 20 percent of that of the largest absorptive band.

If a reference spectrum that matches the analyte cannot be located, the analyst must rely upon *correlation charts* or tables and upon other analytical data. A correlation chart is composed of the spectral ranges throughout which spectral peaks are observed for typical functional groups. Correlation charts should never be used as the only source of information for a qualitative analysis. Conclusions that are based solely on information contained in correlation charts are often wrong. Correlation charts are generally used when computerized comparison of the spectrum with a library of reference spectra is not possible. The charts are used in a screening process to obtain a relatively small list of compounds that could be the analyte. The individual spectrum of each compound is subsequently measured or located and compared with that of the analyte.

Table 12-2 A list of selected collections of infrared spectra

C. D. Carver (ed.), *Deskbook of Infrared Spectra*, Coblentz Society, P.O. Box 9952, Kirkwood, MO 63122.
Documentation for Molecular Spectroscopy, *Working Atlas of Infrared Spectroscopy*, Butterworth, London, 1972.
R. C. Mecke and F. Lagenbucher (eds.), *Infrared Spectra of Selected Chemical Compounds*, Heyden, London.
C. J. Pouchert (ed.), The *Aldrich Library of Infrared Spectra*, Aldrich Chemical Company, Milwaukee, Wisconsin,.
Sadtler Research Laboratories, *Standard Infrared Spectra*, 3314 Spring Garden St., Philadelphia, PA 19104.

Data that could help to identify the analyte in addition to an infrared spectrum include, but are not limited to, melting points, boiling points, appearance, viscosity, solubilities, flame tests, nuclear magnetic resonance spectra, mass spectra, chromatographic data, sample history, and elemental analysis. As much information as is available should be used. Even if all of the listed information has been accumulated there are times when a positive identification is not possible. Chromatography can be used to determine the purity of a sample. If the analyte is the product of a synthesis, an infrared spectrum of the starting materials and of intermediates should be obtained and compared with that of the analyte whenever possible.

Infrared peaks can be caused by substances and circumstances unrelated to the identity of the sample. Such things as dissolved water, dissolved carbon dioxide, leached material from containers in which the sample has been stored and from plastic tubing, leaky cells, oil, and grease have absorptive bands that can mislead the analyst. In some cases it might be useful to prepare a chemical derivative of the analyte. The spectrum of the derivative could be more easily identified than that of the analyte. After identification of the derivative, the identity of the analyte can be deduced from a knowledge of the behavior of the chemical reactant that was added to the analyte.

Normally the composition of a mixture is difficult to determine with infrared spectrophotometry owing to the overlap of peaks from the components of the mixture. If the components in a mixture must be identified, any of three methods can be used. The preferred method is to separate the sample into its components and obtain a spectrum of each pure component. Perhaps the most convenient method is to perform a separation with a gas or liquid chromatograph that uses a Fourier transform infrared spectrophotometer as the detector. Other separatory methods (solvent extraction, distillation, etc.) usually require more time but can also be used. If the separated components are collected, a conventional infrared spectrophotometer can be used to obtain the spectra.

If separation is not feasible, a computer can be used to obtain potential matches with reference spectra of pure compounds. After one of the components has been identified, the computer subtracts the spectrum of the identified compound from that of the mixture to obtain a simpler spectrum that is subsequently used to determine other components in the mixture. Ideally the process can be repeated until identification of all of the components is complete.

In those cases in which one of the components has been identified, a pure sample of the identified component can be placed in a variable pathlength cell and placed in the reference beam of a double-beam spectrophotometer. The mixture is placed in the sample beam. The pathlength of the cell is adjusted to exactly cancel the contribution of the identified component from that of the mixture. An isolated peak of the identified substance can be used to locate the proper pathlength. The resulting spectrum is used to identify the remaining components. While using that method, care must be taken to ensure that the spectrophotometer does not "freeze" owing to too little radiation striking the detector. In practice this last method is the least preferable of the three methods because it does not always work.

Use of Correlation Charts

When a computerized comparison of the spectrum of the analyte with reference spectra is not possible, correlation charts can be useful in limiting the possible compounds in the analyte to those that contain certain functional groups. Spectra of compounds of likely prospects subsequently can be compared with that of the analyte. The remainder of the description in this section is restricted to the identification of pure compounds. Mixtures are identified by one of the procedures described earlier.

Several types of correlation charts or tables are available. An abbreviated version of two types are given in Tables 12-3 and 12-4. In Table 12-3, the absorptive maxima of the major peaks of several common functional groups are listed in order of decreasing wavenumber. In Table 12-4, the major peaks are listed alphabetically according to functional group. More detailed correlation charts can be found in References 1, 2, and 5.

The charts can be used in any of several ways. An author recommended method is to follow the steps listed below:

1. Carefully record the spectrum of the analyte.
2. Make a table in which each of the major peaks are listed according to wavenumber.
3. Use Table 12-3 to list possible functional groups which can be ascribed to each peak. Care should be used because the wavenumber ranges listed in Tables 12-3 and 12-4 are approximations. Occasionally peaks fall outside of the listed range for the functional group.
4. Find each possible functional group in Table 12-4 and compare the expected peaks from the table with actual peaks on the spectrum. Eliminate functional groups when peaks at the expected wavenumbers cannot be found. Make a list of those functional groups that remain. Particular attention should be paid to peaks at wavenumbers greater than about 1400 cm^{-1}. Keep in mind that the shapes and intensities of the peaks are often of as much importance as the locations.
5. Use whatever additional information is available along with the list of remaining functional groups from step 4 to postulate possible molecular structures for the analyte. The molecular weight from a mass spectrum and a nuclear magnetic resonance spectrum are particularly useful. If an elemental analysis is available, it is often useful to calculate the *degree of unsaturation* in the molecule. The degree of unsaturation for an organic molecule is the sum of the double bonds, the rings, and two times the

Table 12-3 A list of selected medium- and strong-intensity absorptive maxima arranged according to wavenumber

Wavenumber range, cm^{-1}	Functional group
3700-3600	OH (H_2O, ROH, PhOH) (dilute solution)
3530-3400	NH_2 (2 bands), NH (1 band) (dilute solution)
3500-3250	OH (polymers) (solids and liquids)
3500-3060	NH (amines, amides)
3320-3250	—C≡C—H (sharp)
3300-2400	COOH (broad)
3110-3000	C—H (C=C—H, Ph—H, CH_3X, CH_2X)
3000-2800	C—H (—CH_2—, —CH_3)
2835-2815	OCH_3
2750-2700	CHO
2260-2100	—C≡C—
2190-2130	CNS, C≡N,
2000-1650	C—H (phenyl)
1980-1950	—C=C=C—
1950-1600	C=O
1715-1630	$RCONH_2$, RCONHR
1710-1530	—COO— (broad)
1680-1630	C=C (nonconjugated, noncyclic), C=N
1680-1560	C=C (cyclic or conjugated)
1650-1590	RONO, $RONO_2$
1650-1475	$RCONH_2$, RCONHR
1615-1590	Phenyl
1615-1565	Pyridines (doublet)
1610-1560	COO^- M^+ †
1550-1490	$PhNO_2$
1515-1485	Phenyl
1475-1450	CH_2, CH_3
1440-1400	COOH
1430-1400	CO—CH_2
1420-1400	CO—NH_2
1400-1360	$(CH_3)_3C$ (two bands)
1400-1310	COO^- M^+ (broad)†
1380-1370	CH_3
1380-1360	$CH(CH_3)_2$ (two bands)
1370-1300	C—NO_2
1330-1310	Ph—CH_3
1300-1000	CF
1280-1250	$SiCH_3$
1280-1180	C—N— (aromatic)
1280-1150	—C—O—C—
1255-1240	$(CH_3)_3C$—
1275-1070	—C—O—C—
1230-1100	—C—N—
1160-1100	C=S
1200-1000	COH
1120-1030	C—NH_2
1095-1015	Si—O—Si, Si—O—C
1000-970	CH=CH_2
980-690	C=C—H
870-670	Aromatic ring
860-760	R—NH_2 (broad)

Table 12-3 *Contd...*

Wavenumber range, cm^{-1}	Functional group
835-800	CH=C (out-of-plane)
760-510	CCl
730-675	CH=CH (cis isomers)
700-550	CBr

† M represents a metal.

Table 12-4 A list of major absorptive maxima for selected functional groups in the midinfrared region
The approximate band intensities are listed in brackets: strong, s; medium, m; weak, w

Functional group	Wavenumber range, cm^{-1}	Functional group	Wavenumber range, cm^{-1}
Acetylene	3300-3250 (m or s)		1760 (s) (dilute solution)
	2250-2100 (w)		1710-1680 (s) (neat)
Alcohol (neat)	3350-3250 (s)		1440-1400 (m)
	1440-1320 (m or s)		960-910 (s)
	680-620 (m or s)	Chloro	850-650 (m)
Aldehyde	2830-2810 (m)	Cyano	2190-2130 (s)
	2740-2720 (m)	Ester	1765-1720 (s)
	1725-1695 (s)		1290-1180(s)
	1440-1320 (s)	Ether	1285-1170 (s)
Alkyl	2980-2 850 (m)		1140-1020 (s)
	1470-1450 (m)	Fluoroalkyl	1400-1000 (s)
	1400-1360 (m)	Methyl	2970-2780 (s)
Amide ($CONH_2$)	3540-3520 (m)		1475-1450 (m)
	3400-3380 (m)		1400-1365 (m)
	1680- 1660 (s)	Methylene (CH_2, alkane)	2940-2920 (m)
	1650-1610 (m)		2860-2850 (m)
(CONHR)	3440-3420 (m)		1470-1450 (m)
	1680-1640 (s)	(alkene)	3090-3070 (m)
	1560-1530 (s)		3020-2980 (m)
	1310-1290 (m)	Nitrile	2240-2220 (m)
	710-690(m)	Nitro (NO_2, aliphatic)	1570-1550(s)
($CONR_2$)	1670-1640 (s)		1380-1320 (s)
Amine (primary)	3460-3280 (m)		920-830 (m)
	2830-2810 (m)	(aromatic)	1480-1460 (s)
	1650-1590 (s)	Pyridyl (C_5H_4N)	3080-3020 (m)
(secondary)	1190-1130(m)		1620-1580 (s)
	740 700 (m)		1590-1560 (s)
Ammonium	3200 (s)		840-720 (s)
	1430-1390 (s)	Sulfate ($ROSO_3R'$)	1440-1350 (s)
Aromatic †	3100-3000(m)		1230-1150 (s)
	1630-1590 (m)	($ROSO_3M$) ‡	1260-1210 (s)
	1520-1480 (m)		810-770 (s)
	900-650(s)	Sulfonic acid (RSO_3H)	1250-1150 (s, broad)
Bromo	700-550 (m)	Thiocyanate	2175-2160 (m)
t-Butyl	2980-2850 (m)	Thiol	2590-2560 (w)
	1400-1390 (m)		700-550 (w)
	1380-1360 (s)	Vinyl (CH_2=CH—)	3095-3080 (m)
Carbonyl	1870-1650 (s, broad)		1645-1605 (m or s)
Carboxylic acid	3550 (m) (dilute solution)		1000-900 (s)
	3000-2440 (s, broad) (neat)		

† Refer to text for description.
‡ M represents a metal.

number of triple bonds in the molecule. As examples, the degree of unsaturation of benzene is 4 (three double bonds and one ring) and of acetylene is 2 (one triple bond multiplied by two). For an organic molecule of the composition $C_{\upsilon}H_{w}N_{x}O_{y}X_{z}$, where X represents a halide, the degree of unsaturation is given by

$$\text{Degree of unstaturation} = \frac{2\upsilon - w + x - z + 2}{2} \tag{12-8}$$

where the letters in Eq. (12-8) are the subscripts in the molecular formula of the compound. The number of oxygen atoms in the molecule has no effect upon the degree of unsaturation.

6. Compare the spectrum of the analyte with spectra of the postulated substances. Use spectra that are recorded on the same instrument under identical experimental conditions if possible. A commercial collection (Table 12-2) of spectra can be used for the comparison if necessary and if the variations between spectrophotometers and between experimental procedures is considered. If a match is found the analyte is identified.

Typical Infrared Spectra

A few infrared spectra can be used to illustrate the positions, shapes, and intensities of several peaks that correspond to often-encountered functional groups. The spectra that are shown throughout the remainder of the chapter were recorded with a Perkin-Elmer model 457A infrared spectrophotometer using neat samples between salt plates. The peaks that are described in the text are indicated by arrows in the figures. The shown spectra are recorded as percent transmittance as a function of wavenumber. For quantitative studies of mixtures, it is often preferable to record absorbance as a function of wavenumber.

A spectrum of pyridine that contains some dissolved water is shown in Fig. 12-13. The broad peak at about 3400 cm^{-1} is due to water. Small amounts of water are found in many samples. The location and shape of the water peak should be remembered. In most samples the water peak is much less intense than that observed in Fig. 12-13. The remainder of the peaks in the spectrum are due to pyridine. With the exception of the water peak, the peaks that are indicated by the arrows are typical of pyridyl compounds and can be used to aid in their identification.

As expected, the peak owing to an OH in the molecule occurs at about the same location and has the same broad shape (in neat samples) and high intensity as that owing to water. An example of a compound that contains an OH is 2,4-dimethylphenol (Fig. 12-14). The peak at about 3400 cm^{-1} corresponds to the OH. The remaining peaks correspond to the aromatic ring and the methyl substituents. The other peaks are described later.

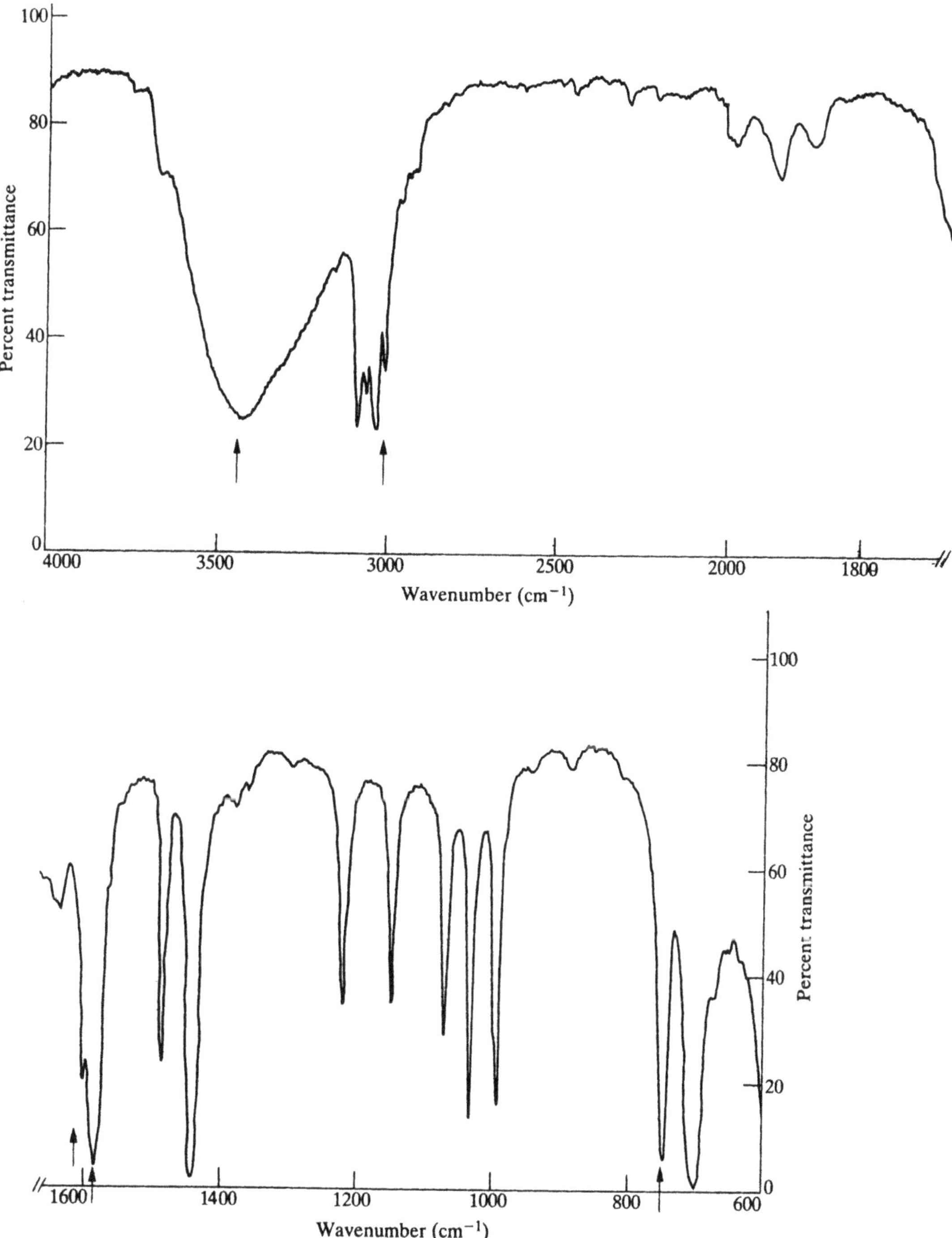

Figure 12-13 An infrared spectrum of wet pyridine. The broad peak at about 3400 cm^{-1} is caused by water. The arrows indicate peaks described in the text.

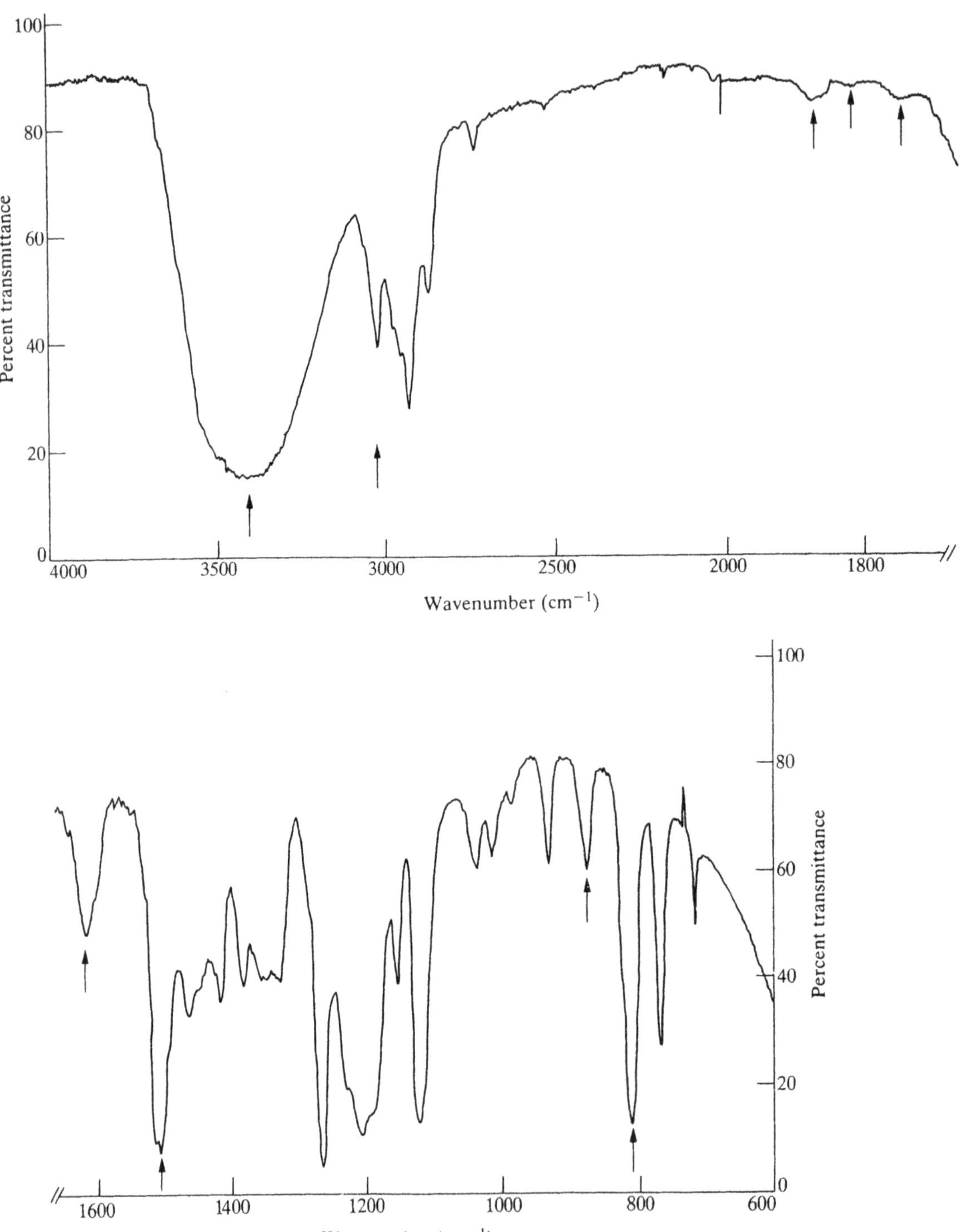

Figure 12-14 The infrared spectrum of 2,4-dimethylphenol.

A carboxylic acid has several distinctive peaks. A strong broad peak owing to OH stretching usually occurs between 2400 and 3000 cm^{-1}. A second intense but narrower peak that occurs between 1680 and 1710 cm^{-1} is caused by C=O stretching. Ketones and aldehydes also exhibit the peak. Other peaks between 1400 and 1440 cm^{-1} and between 910 and 960 cm^{-1} respectively are caused by C–O stretching and C–O–H deformation. A spectrum of 1-heptanoic acid with the characteristic peaks indicated by arrows is shown in Fig. 12-15.

Several, often overlapping, C–H stretching bands that are characteristic of alkyl functional groups occur between 2850 and 2980 cm^{-1}. The bands occur in a region of the spectrum where little else absorbs and consequently can usually be easily recognized. A spectrum of 1-bromopropane that shows the stretching bands associated with the propyl group is shown in Fig. 12-16.

The presence of substituted aromatic compounds is usually easy to detect by the presence of two or more weak overtone bands between 1660 and 2000 cm^{-1}.[6] An overtone band corresponds to excitation during absorption from the ground vibrational or rotational level to a more energetic level than the first excited level. Overtone bands are considerably less intense than fundamental bands that correspond to excitation to the first excited level. The peaks are of low intensity but are usually apparent because only C=O of the common functional groups absorbs near that region. The pattern of the peaks varies depending upon the location of the substituents on the aromatic ring. Several substituted aromatic compounds are shown in Figs. 12-14 and 12-17 through 12-19. A spectrum of benzene is shown in Fig. 12-20 for comparison.

Table 12-5 The number of weak overtone peaks between 1660 and 2000 cm^{-1} and the positions of the out-of-plane bending peaks for subtituted benzene rings

Substitution		Overtone peaks	Bending peaks, cm^{-1}
Mono		4	690, 750
Di			
	ortho	6†	750-760
	meta	4	690-710, 770-780, 810-820
	para	3	810-820
Tri			
	1, 2, 3	3	705-745, 760-780
	1, 2, 4	3	805-825, 870-885
	1, 3, 5	3‡	675-730, 810-865

† The two sets of two central peaks might overlap, giving the appearance of four rather than six peaks.

‡ The two low-frequency peaks might overlap, giving the appearance of two rather than three peaks.

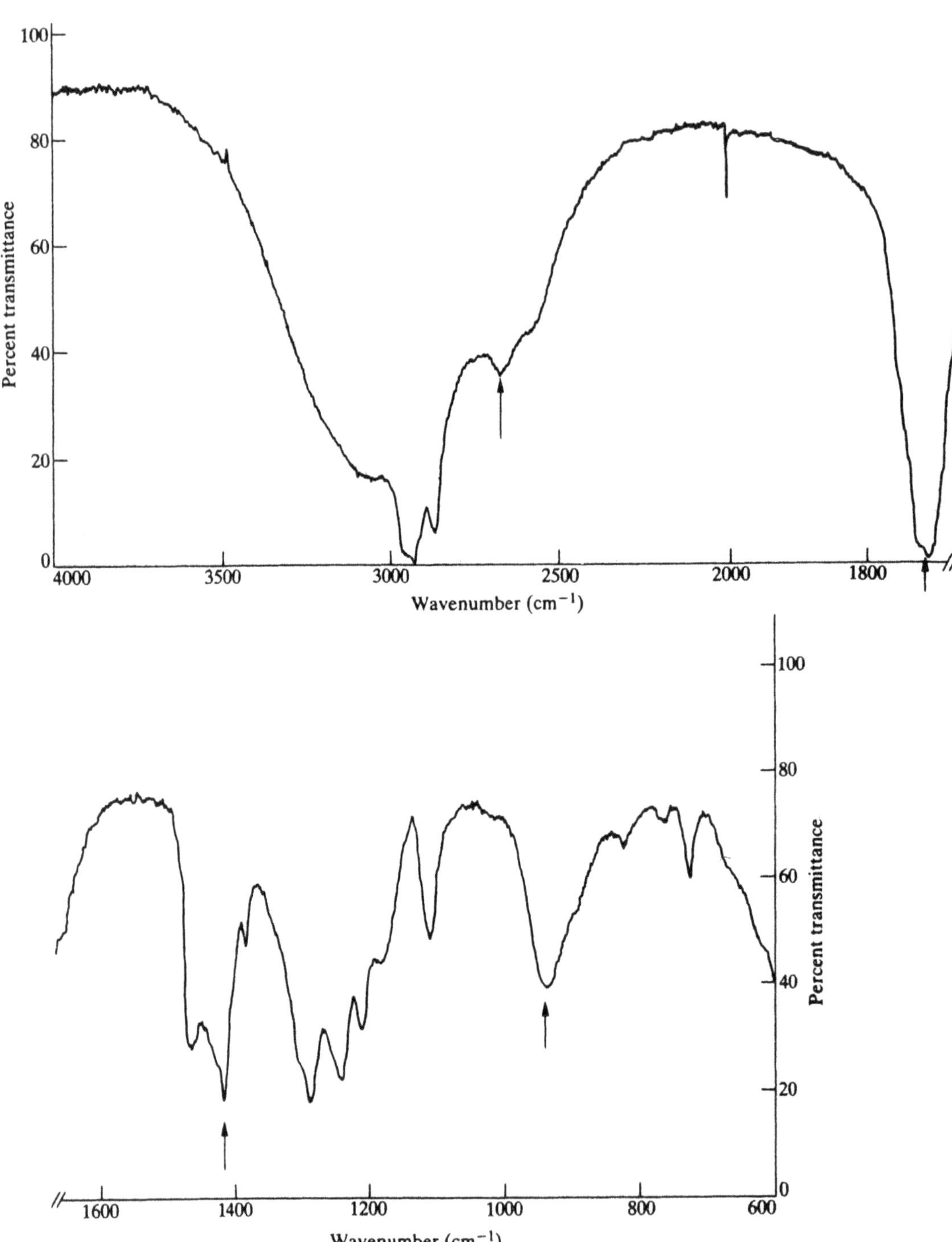

Figure 12-15 The infrared spectrum of 1-heptanoic acid.

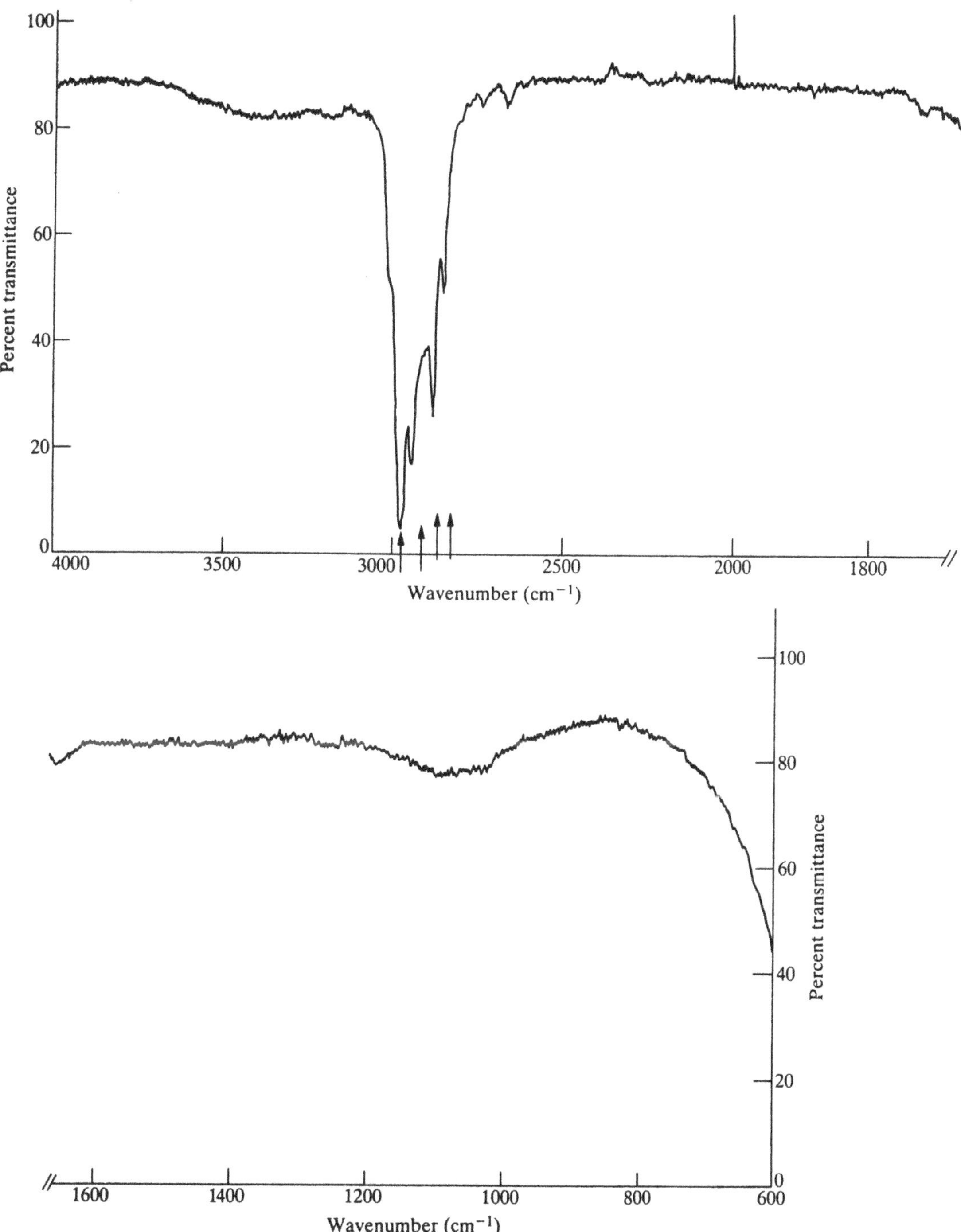

Figure 12-16 The infrared spectrum of 1-bromopropane.

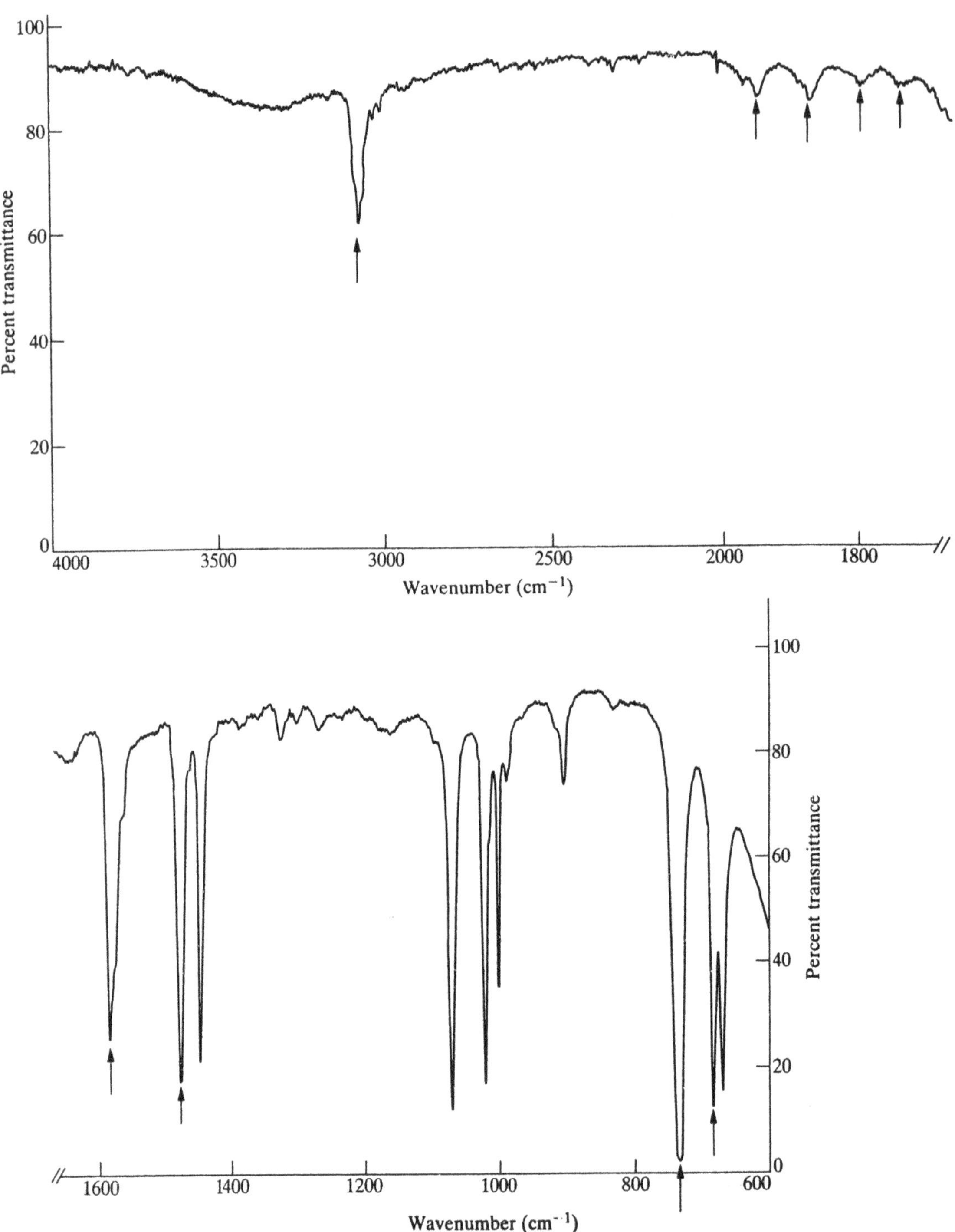

Figure 12-17 The infrared spectrum of bromobenzene.

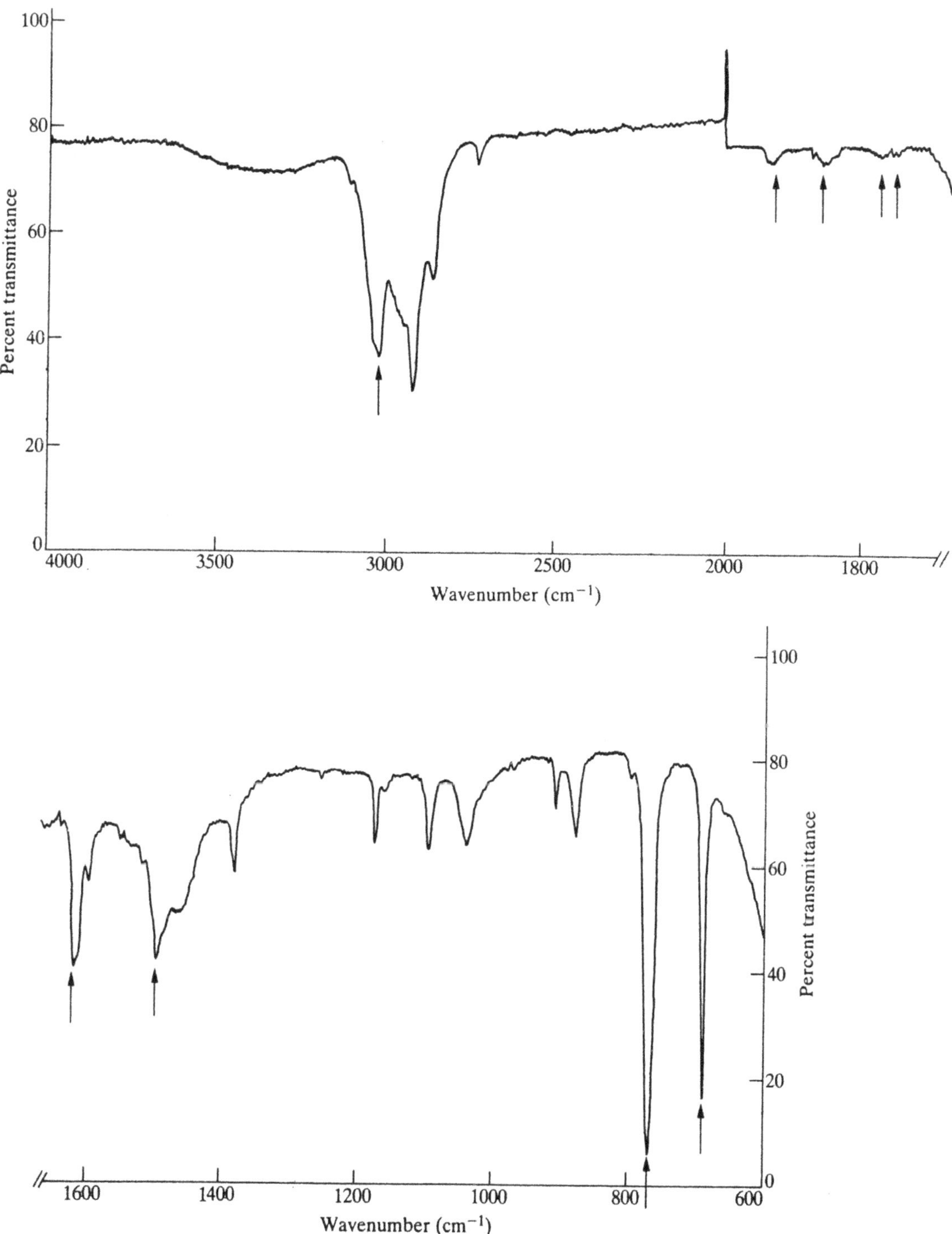

Figure 12-18 The infrared spectrum of *m*-xylene.

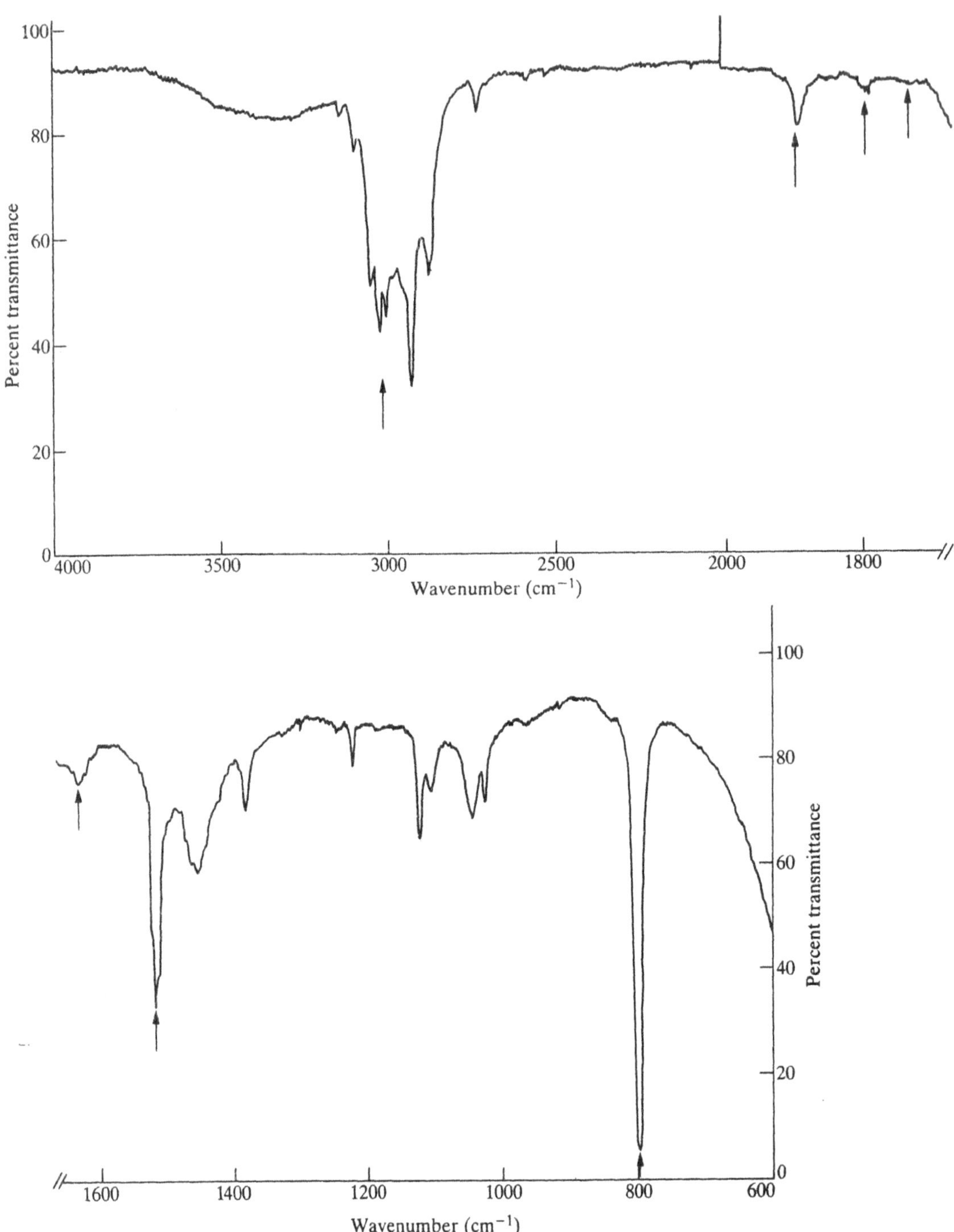

Figure 12-19 The infrared spectrum of *p*-xylene.

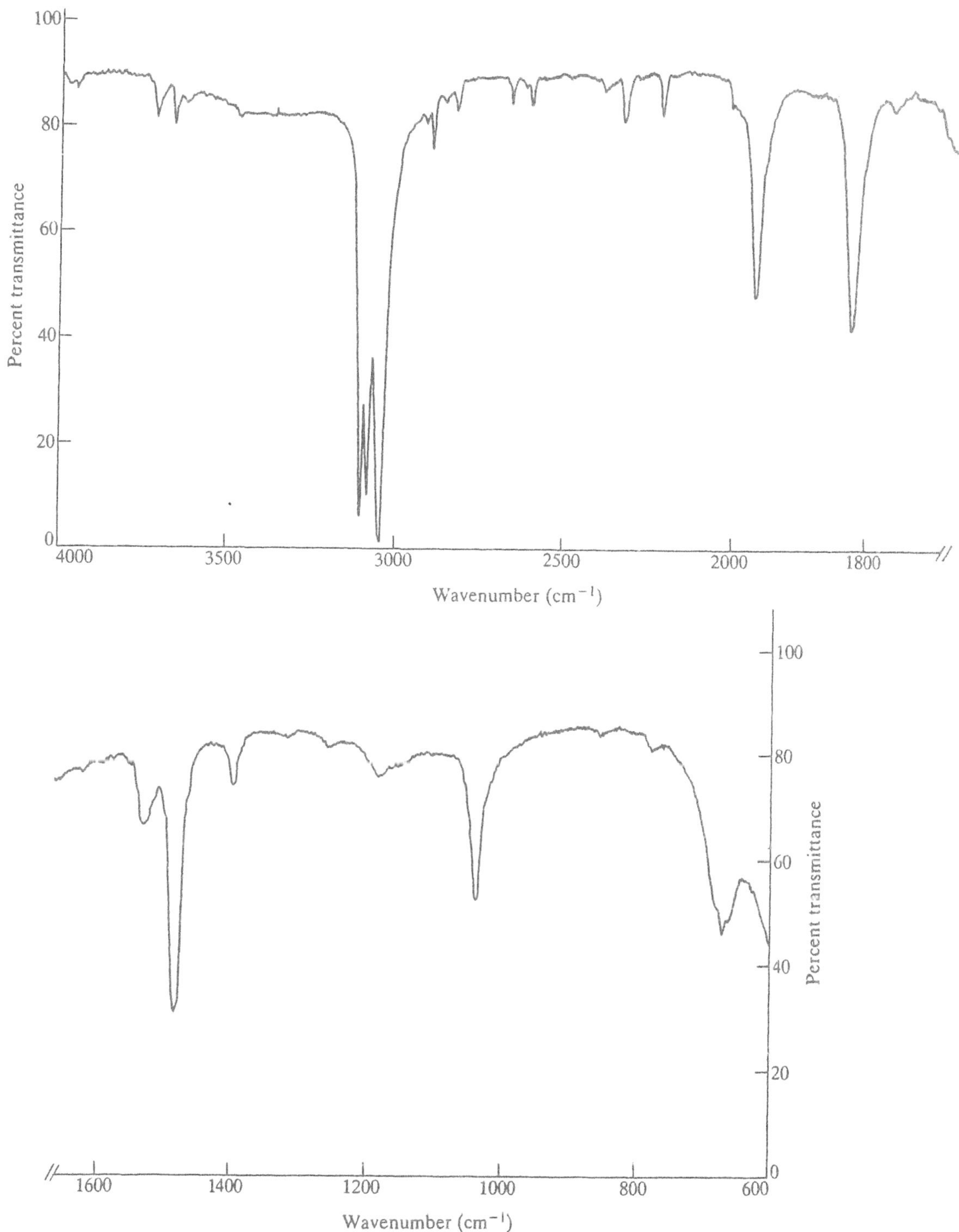

Figure 12-20 The infrared spectrum of benzene.

Assuming the aromatic compound does not have substituents at all ring positions, several C–H stretching peaks are usually observed between 3000 and 3100 cm^{-1}. A peak of variable intensity occurs between 1590 and 1630 cm^{-1} and another between 1480 and 1520 cm^{-1} owing to C=C stretching. Either one or two peaks owing to C–H out-of-plane bending occurs between 650 and 900 cm^{-1}. The location and number of the latter peaks in combination with the number and relative intensities of the overtone peaks are useful for determining the substitution pattern on the ring (Table 12-5 and Figs. 12-14 and 12-17 through 12-19).

Sample problem 12-3 The pure organic compound whose spectrum is shown in Fig. 12-21 is a liquid at room temperature with a boiling point of 131°C. An elemental analysis revealed that the compound had the empirical formula C_5H_8O. Identify the compound.

SOLUTION Equation (12-8) is used to determine the degree of unsaturation of the compound:

$$\text{Degree of unsaturation} = \frac{2(5) - 8 + 2}{2} = 2$$

The major peaks in the spectrum are listed and Table 12-3 is used to tentatively assign a functional group to each peak.

Peak location, cm^{-1}	Tentative assignment
2880-2980	CH
1750	C=O
1460	—CH_2—,—CH_3
1410	—CO—CH_2—
1155	—
960	C=C—H
835	CH=C

Because of the many possible peaks at wavenumbers less than about 1400 cm^{-1}, particular attention is paid to the peaks at higher wavenumbers. Tentative assignments at lower wavenumbers are often in error. No assignment was made to the peak at 1155 cm^{-1} because the only possible assignments (Table 12-3) contain N, S, or a single-bonded O. The presence of C=O is apparent from the strong peak at 1750 cm^{-1}.

The double bond in the carbonyl accounts for one of the two degrees of unsaturation. Another double bond or ring must be present. Twenty-one of the possible structures are (the double bond to the O is not indicated beyond structure 1):

1. $CH_3CH_2CH_2CH$=C=O
2. CH_3CH_2CH=CHCHO
3. CH_3CH=$CHCH_2CHO$
4. CH_2=$CHCH_2CHO$

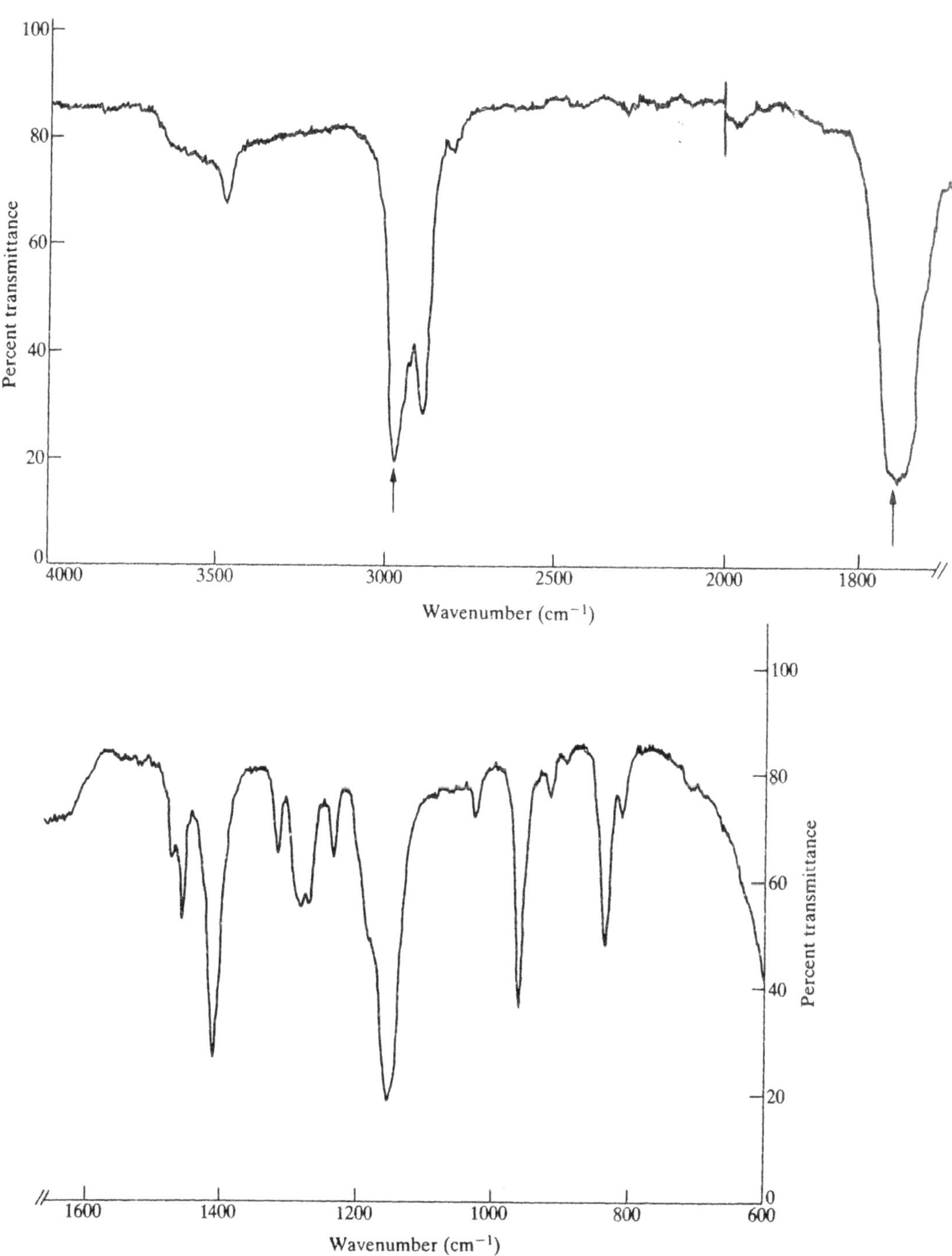

Figure 12-21 The infrared spectrum of the analyte described in Sample problem 12-3.

5. $CH_3CH{=}CHCOCH_3$
6. $CH_2{=}CHCH_2COCH_3$
7. $CH_2{=}CHCOCH_2CH_3$
8. $(CH_3)_2CHCH{=}CO$
9. $CH_3CH_2C({=}CH_2)CHO$
10. $CH_3COC({=}CH_2)CH_3$
11. $CH_3C({=}CH_2)CH_2CHO$
12. $CH_3CH{=}C(CH_3)CHO$
13. $(CH_3)_2C{=}CHCHO$
14. $CH_2CH_2CH_2CH_2CO$
15. $CH_2CH_2CH_2CHCHO$
16. $CH_2CH_2C(CH_3)HCO$
17. $CH_2CH(CH_3)CH_2CO$
18. $CH_2C(CH_3)HCHCHO$
19. $CH_2CH_2CHCH_2CHO$
20. $CH_3CHC(CH_3)HCO$
21. $CH_3CH_2CHCH_2CO$

The analyte could be any of the possibilities. Several of the compounds, e.g., those with adjacent double bonds, are unlikely but nevertheless possible. The boiling point of the analyte can be compared with literature values of the possible compounds and used to eliminate some possibilities. As an example, compound 12 has a boiling point of 116.5°C and can be eliminated. On the other hand, compound 13 has a boiling point of 133°C and should be kept as a possibility because it is close to the 131°C for the analyte. Generally compounds with boiling points greater than 5°C away from the observed value for the analyte can be eliminated.

The spectra of the remaining compounds are located and compared to that of the analyte. As is typical, not all of the compounds have spectra in the literature. If a match between the spectrum of the analyte and that of a standard is located, the compound is identified. If no match is found, a positive identification is not possible. Each time the spectrum of one of the possibilities is located and does not match that of the analyte, the compound can be eliminated as a possibility. In this case a match with Sadtler spectrum 171 is found for cyclopentanone (compound 14). It is not always possible to identify the compound from the available information.

QUANTITATIVE ANALYSIS

Quantitative analysis using infrared spectrophotometry is analogous to quantitative analysis with ultraviolet-visible spectrophotometry. In many cases Beer's law can be applied at a fixed wavelength for the assay. Because infrared peaks are narrow in comparison to those of uv-visible peaks, extra care must be taken during quantitative measurements that are made while using infrared spectrophotometry. A stable instrument is required. The temperature of the sample compartment should be controlled because changes in temperature can affect the width, position, and intensity of infrared peaks. If quantitative infrared studies are to be routinely performed, it probably would be wise to purchase a single-beam wavelength dispersive instrument or use an FTIR spectrophotometer. Caution must be exercised when using a single-beam instrument, however, owing to potential changes in peak location with concentration of the analyte. As described earlier, it is nearly impossible to set 0% T with the sample beam blocked in most null-recording instruments because the detector freezes at low energetic levels.

The precision of infrared spectrophotometric measurements increases as the care and attention to details that are taken during the assay increases. Relative precisions of about 0.5 percent are not unusual for consecutive trials with the same instrument while using identical instrumental settings. Assays of the same sample with different instruments rarely yield results that agree within better than 10 percent.

The peak that is chosen for quantitative infrared spectrophotometry should preferably be isolated and should occur at a wavelength at which no other component in the solution absorbs. If only bands at which spectral interferences occur are available, it is sometimes possible to eliminate the effect of the interference by placing a solution of the interference in a variable pathlength cell in the reference beam of the instrument. The procedure normally cannot be recommended, however, and should be used as a last resort. If an absorptive band can be found for each component of a mixture at which Beer's law is obeyed for all of the components and at which the absorbances are additive, the method of simultaneous equations can be used to determine the concentration of each component. The method is used as described in detail for ultraviolet-visible spectrophotometry in Chapter 9.

After the absorptive band has been chosen for the assay, the absorption of the sample is measured. It is necessary to use a cell with a fixed pathlength to make the measurement. Use of a sample pressed between salt plates generally will not provide accurate or reproducible results. A solution of the analyte in a solvent that does not interfere with the assay should be prepared and used in a fixed-pathlength cell. For most assays a 2% (w/v) solution is adequate. Less concentrated solutions are required in some cases. Typically a cell with a 0.1-mm pathlength is used. If a double-beam instrument is used, pure solvent is inserted into two cells that have as nearly identical pathlengths as possible. A cell is placed in the cell holder in each of the beams of the instrument. The sample beam is blocked and, if possible, the instrument is adjusted to give a reading of 0% T.

For quantitative analysis it is important that the 0% T reading be reproducible. Often a screen that causes an approximate reading of 15% T is placed in the sample beam after the initial 0% T adjustment. The instrumental reading is noted with the screen in place. In later trials the screen is reinserted and the instrument is adjusted to return the instrumental reading to the value that was initially obtained. Problems with poor detector response at low energies are eliminated with the method and the readings are reproducible.

With both beams unblocked and with pure solvent in each beam, the instrument is adjusted with the 100% T knob (if possible) to provide a reading of 90 or 95% T. Normally the instrument is not adjusted to 100% T in order to prevent the recorder trace from going off-scale during the scan. After the initial adjustment, the sample cell is removed, and flushed and filled with the analyte solution. The sample cell is replaced in the beam of the instrument and the spectrum of the sample is recorded. Scanning should be done slowly. Measurements can be made at a single wavelength with single-beam instruments without scanning if the peak location does not shift with concentration.

Upon location of the peak at which the measurement is to be made, the slit widths are adjusted to yield an analyte peak height in the middle of the percent transmittance range. Often that requires opening the slits to double or triple their normal value. The slit widths are not altered for the remaining measurements.

The percent transmittance $\%T_p$ (or transmittance) of the peak and of the baseline $\% T_b$ at the same wavenumber are measured and used to calculate the absorbance of the peak, as shown in the following equation:

$$A = \log \frac{\%T_b}{\%T_p} \tag{12-9}$$

This is illustrated in Fig. 12-22 for the peak at 1470 cm^{-1}. Normally the baseline is assumed to be the interpolated straight line across the base of the peak. A spectrum of the pure solvent can be used to obtain the percent transmittance of the baseline if desired.

Sample problem 12-4 Determine the absorbance of the 1470 cm^{-1} peak in Fig. 12-22.

SOLUTION The percent transmittance of the peak as obtained from the chart paper is 27.4. Similarly, the percent transmittance on the interpolated baseline at 1470 cm^{-1} is 81. Substituting the values into Eq. (12-9) allows calculation of the absorbance of the peak:

$$A = \log \frac{81.0}{27.4} = 0.471$$

Alternatively, if the spectrum is recorded as absorbance as a function of wavenumber, the absorbance of the baseline at the wavenumber of the peak is subtracted from the peak absorbance in order to obtain the analyte absorbance. The working-curve or the internal-standard method is preferred for most measurements. Direct application of Beer's law is possible if it has been previously demonstrated that the compound obeys Beer's law. In that case several analyte samples should be measured in order to eliminate potential determinate errors in the analysis. In order to directly apply Beer's law, the cell pathlength of the sample cell must be known.

For cells with pathlengths between 0.03 and 0.6 mm, the preferred method for measuring the pathlength is the interference-fringe method. The empty cell is placed in the sample cell holder of the instrument and the wavelength is scanned as usual with no cell in the reference beam. Owing to the difference in refractive index between the cell walls and the air inside the cell, a series of interference fringes result from constructive and destructive interference of radiation that is reflected from the internal cell walls. Interference fringes are not likely to be observed when a solution is in the cell because the difference in refractive index between the cell walls and the solution is small. Consequently, little reflection occurs. A recording of the interference fringes for a KBr cell is shown in Fig. 12-23. The number n of fringes within a wavelength region is counted and substituted into either

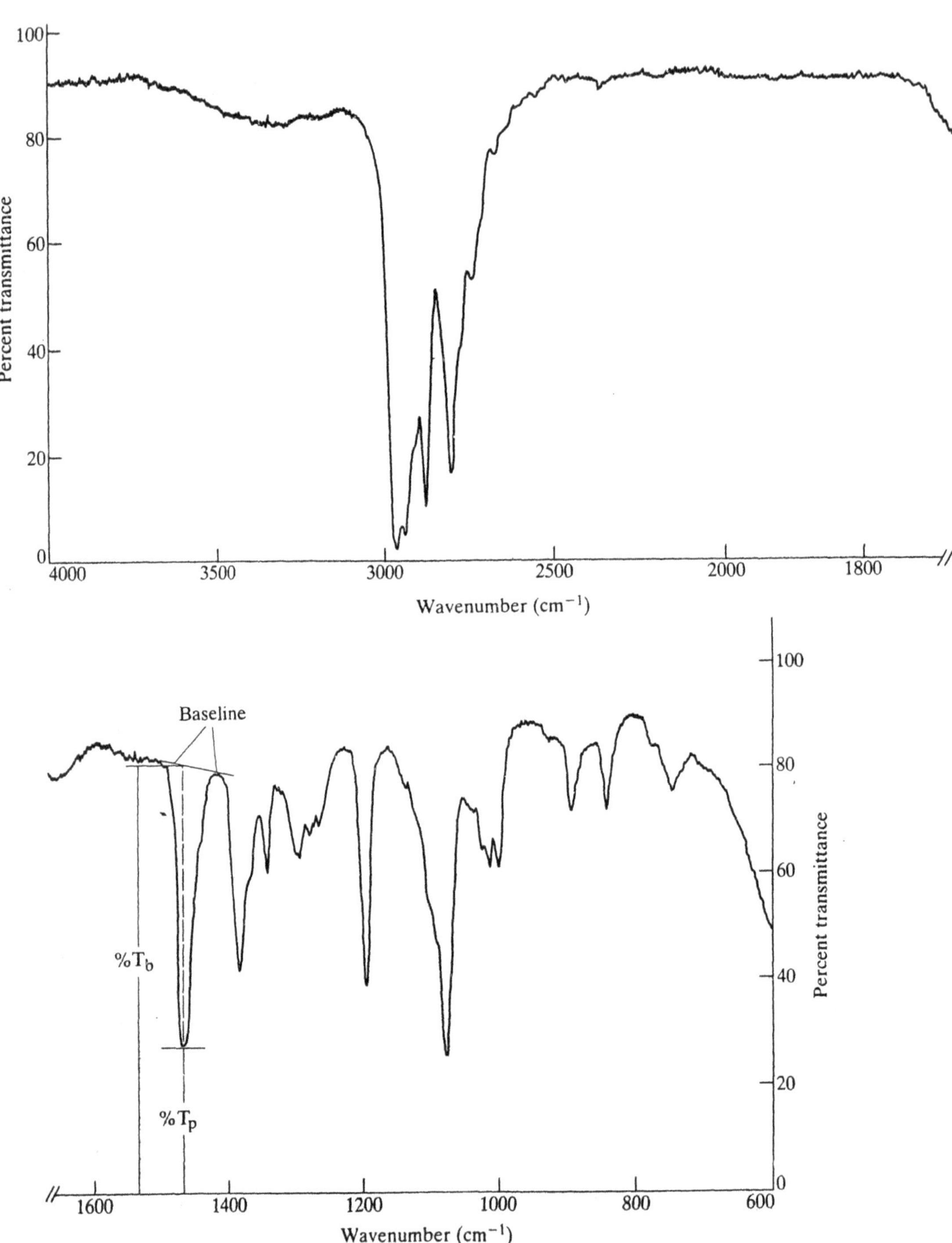

Figure-12-22 The infrared spectrum of tripropylamine which is used to illustrate absorbance measurements.

$$b\ (\text{mm}) = \frac{5n}{\bar{\nu}_1 - \bar{\nu}_2} \tag{12-10}$$

or

$$b\ (\mu\text{m}) = \frac{n\lambda_1\lambda_2}{2(\lambda_1 - \lambda_2)} \tag{12-11}$$

In Eq. (12-11) the wavelengths are measured in units of micrometers.

Sample problem 12-5 Calculate the pathlength of the cell that has the interference fringes shown in Fig. 12-23.

SOLUTION Two convenient locations on the recording are chosen and the number of fringes between the points counted. At least 10 and ideally many more fringes should be between the chosen locations. In Fig. 12-23 the two chosen wavenumbers (indicated by the arrows) are 3500 $(\bar{\nu}_1)$ and 900 $(\bar{\nu}_2)$ cm^{-1}. Between the two points are 51 fringes. The pathlength is calculated by substitution into Eq. (12-10):

$$b = \frac{5(51)}{3500 - 900} = 0.098 \text{ mm}$$

For cell pathlengths that are less than about 0.03 mm, the fringes become too far apart to permit recording of sufficient fringes within the mid-ir range to yield an accurate value for the pathlength. At cell pathlengths above about 0.6 mm, the fringes are too close together to be easily distinguished. For pathlengths greater than 0.6 mm a micrometer can be used to measure the thickness of the gasket between the cell walls. The measured distance corresponds to the cell pathlength.

An alternative method of calculating the cell pathlength is to use a standard solution of a substance that has a broad absorptive band and whose absorbance has been measured in a cell of known pathlength. The spectrum of the solution is consecutively recorded in the known pathlength cell and in the unknown pathlength cell on the same instrument with the same instrumental settings. The absorbance and concentration of the standard in the known pathlength cell are substituted into Beer's law and used to calculate the absorptivity of the substance at the wavelength of the peak. The calculated absorptivity is used with the absorbance and the concentration in the cell of unknown pathlength to calculate its pathlength. A peak should be chosen that has a peak height that is between 30 and 50% T. Often a peak of benzene, carbon tetrachloride, or *n*-hexane is used.

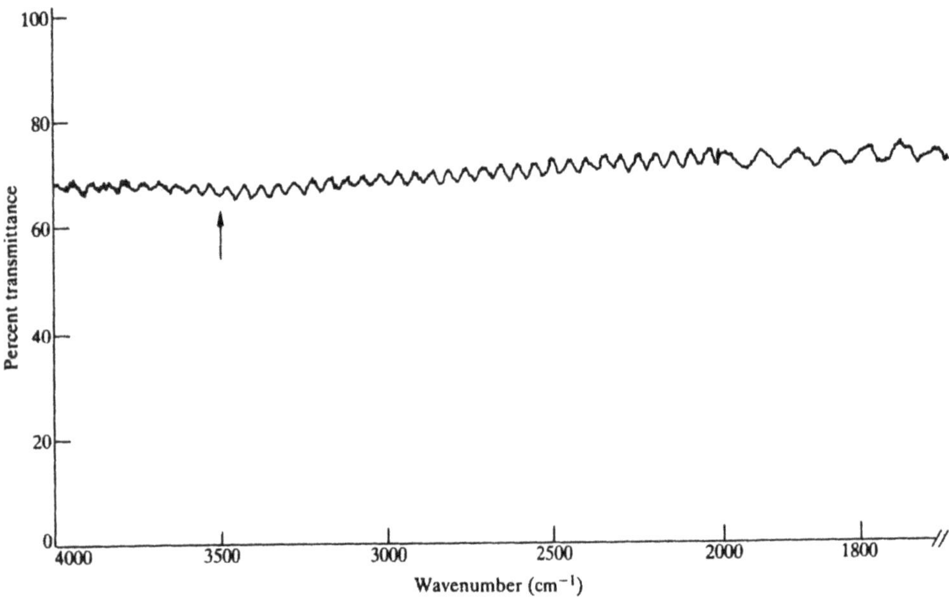

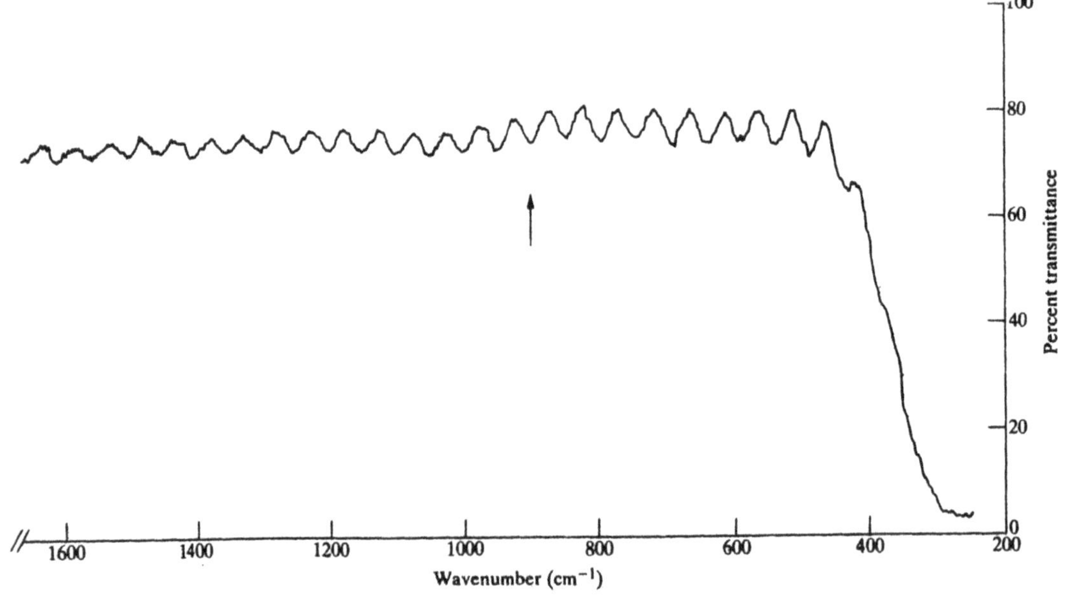

Fig- 12-23 A recording of the interference fringes for a KBr cell.

IMPORTANT TERMS

Asymmetric stretching
Attenuated total reflectance
Bending vibration
Bolometer
Correlation chart
Coupling
Critical angle
Degenerate
Degree of freedom
Degree of unsaturation
Diffuse reflectance
Double monochromator
Double-pass monochromator
Far-infrared
Fellgett advantage
Fourier transform infrared
FTIR microscope
Globar
Golay detector
Hadamard transform
Interference fringes
Michelson interferometer
Mid-infrared
Mull
Multipath cell
Multiplex method
Near-infrared
Nernst glower
Nondispersive spectrophotometer
Optical wedge
Photon detector
Pneumatic device
Pyroelectric bolometer
Rocking
Scissoring
Split mull
Stretching vibration
Symmetric stretching
Thermal detector
Thermistor
Thermocouple
Thermopile
Twisting
Wagging
Wavenumber

PROBLEMS

12-1 Convert 6.0, 12.5, and 2.7 μm to wavenumbers.

12-2 Convert 3165, 1825, and 625 cm^{-1} to micrometers.

12-3 Determine the total possible number of vibrational bands for C_6H_6.

12-4 Determine the total possible number of vibrational bands for nicothiazone $C_7H_8N_4S$.

Instrumentation

12-5 What type of cell walls can be used for quantitative analysis at 550 cm^{-1}?

12-6 If an analyte was soluble in both acetonitrile and in *n*-heptane, and if both solvents have adequate transparent ranges for a particular analysis, which solvent would probably be better? Why?

12-7 Why are mortars and pestles constructed from agate recommended for mull and pellet preparation?

12-8 What is the difference between an intrinsic and extrinsic semiconductor (see Chapter 3)?

Attenuated Total Reflectance

12-9 Calculate the critical angle between a prism with a refractive index of 2.12 and a sample with a refractive index of 1.28.

12-10 Calculate the critical angle between a prism with a refractive index of 1.95 and a sample with a refractive index of 1.31.

12-11 Estimate the proper incident angle for use with the sample and prism described in Prob. 12-9.

12-12 Estimate the proper incident angle for use with the sample and prism described in Prob. 12-10.

Multiple-Scan and Multiplex Methods

12-13 Calculate the Fellgett advantage in the signal-to-noise ratio if 558 scans of a spectrum are summed.

12-14 If the resolution element for a Fourier transform spectrophotometer is 0.50 s, calculate the Fellgett advantage if the spectrum is recorded for 10.0 min.

12-15 How many slots are required in a Hadamard transform mask used for recording from 1500 to 4000 cm^{-1} with a resolution of 2 cm^{-1}?

Qualitative Analysis

12-16 Identify the compound that has the spectrum shown in Fig. P12-16 and that has the formula $C_6H_{15}N$.

12-17 Identify the compound that has the spectrum shown in Fig. P12-17. The compound contains only C, H, and O. It has a molecular weight of 130 and a boiling point of 194°C.

12-18 Identify the compound that has the spectrum shown in Fig. P12-18 and an empirical formula C_6H_{12}.

12-19 Identify the compound that has the spectrum shown in Fig. P12-19 and the empirical formula C_4H_9I.

12-20 Identify the compound that has the spectrum shown in Fig. P12-20 and the empirical formula $C_6H_{12}O$. The boiling point of the compound is 117°C.

12-21 Identify the compound that has the spectrum shown in Fig. P12-21 and a molecular weight of 108.

12-22 Identify the compound that has the spectrum shown in Fig. P12-22 and a molecular weight of 103.

Cell Pathlength

12-23 Determine the pathlength of the cell that has the interference fringes shown in Fig. P12-23.

12-24 Determine the pathlength of the cell that has the interference fringes shown in Fig. P12-24.

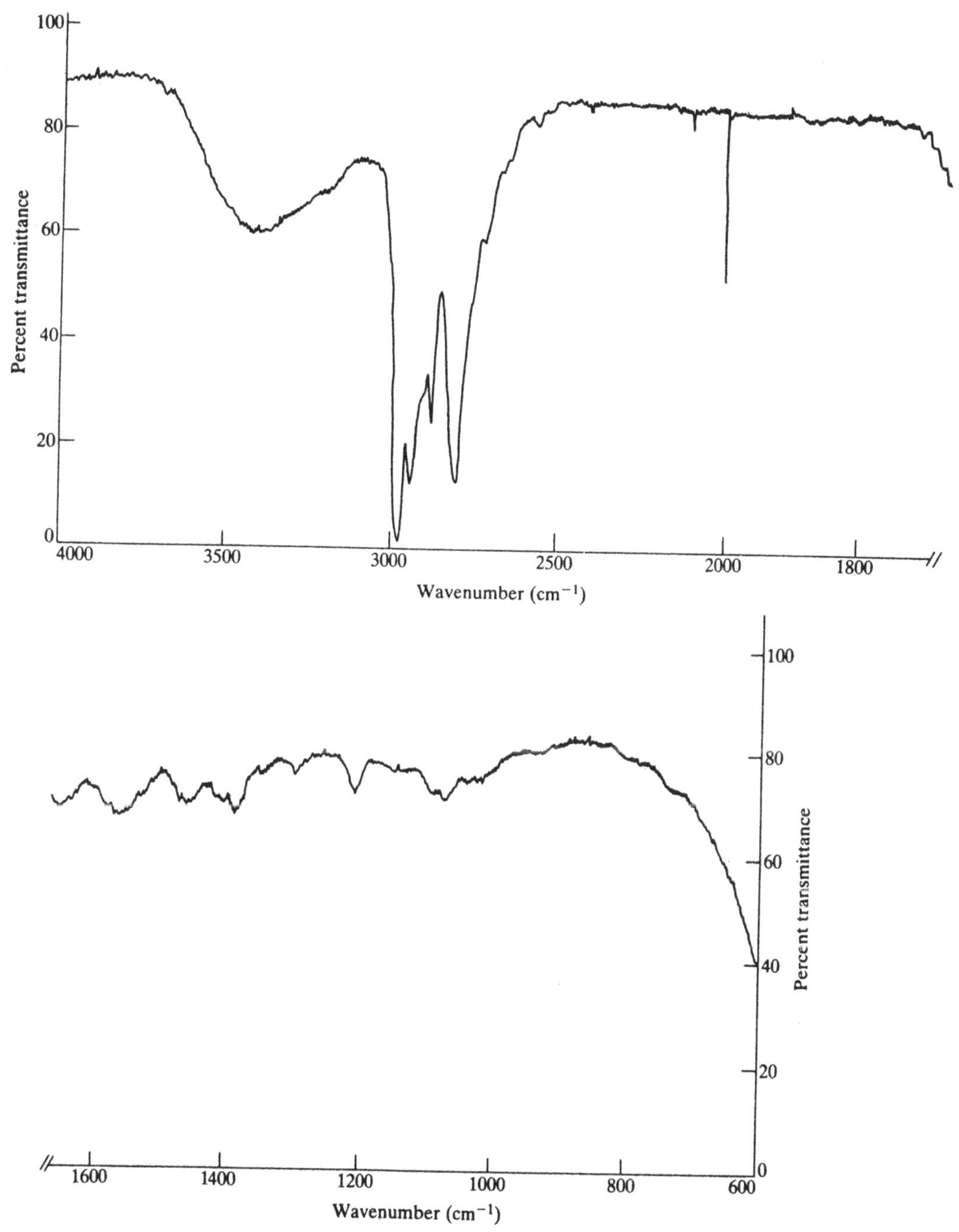

Figure P12-16

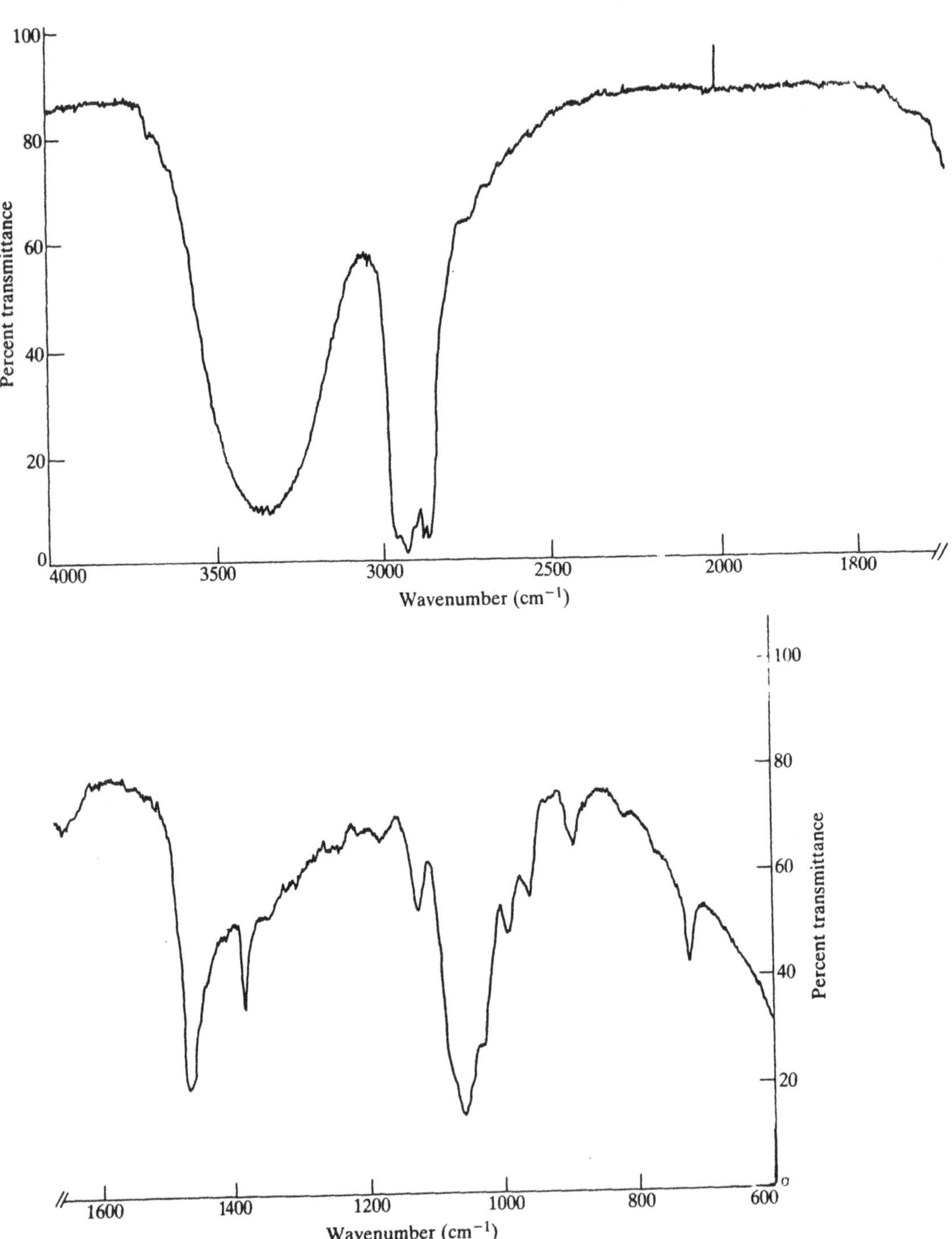

Figure P12-17

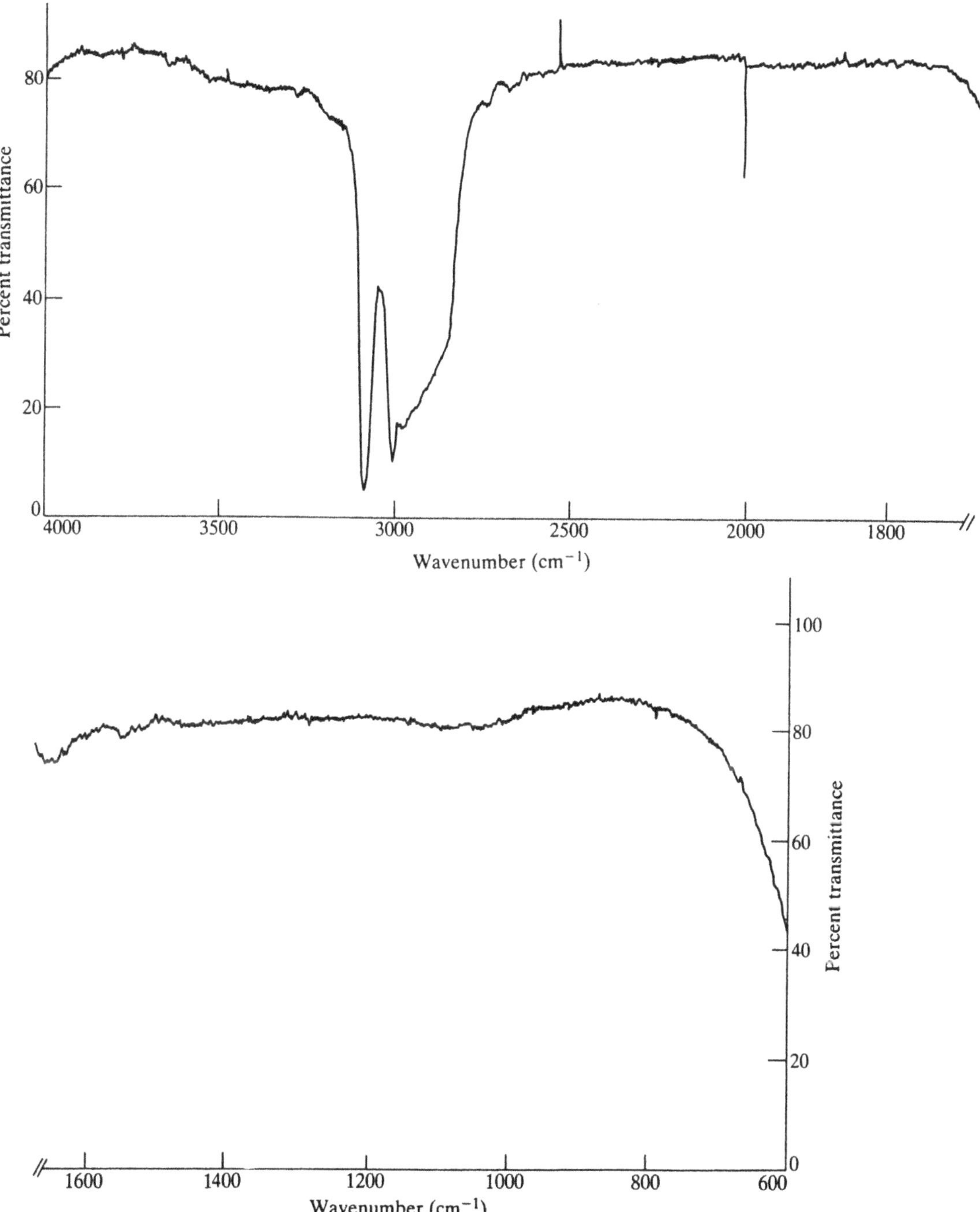

Figure P12-18

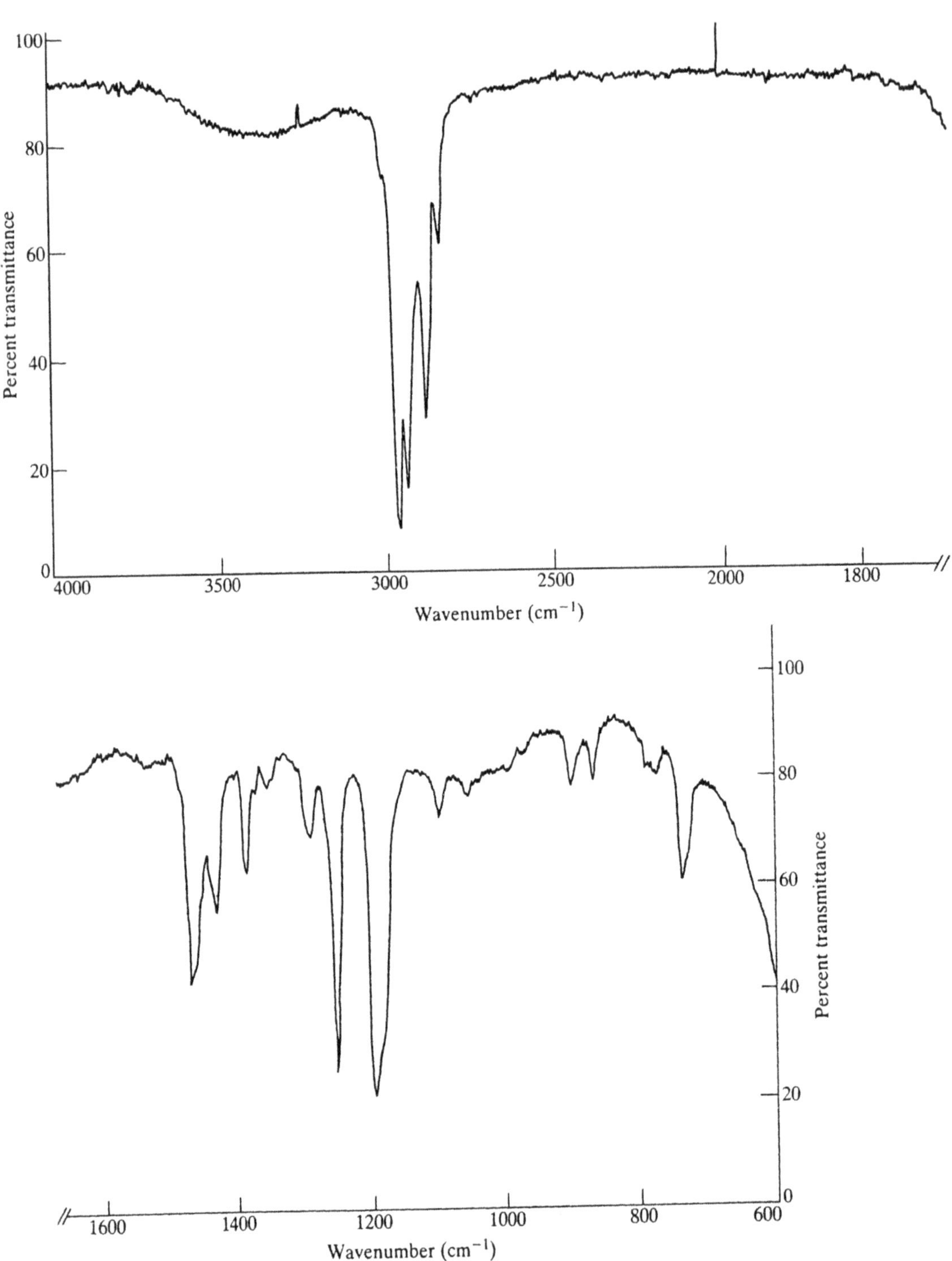

Figure Pl2-19

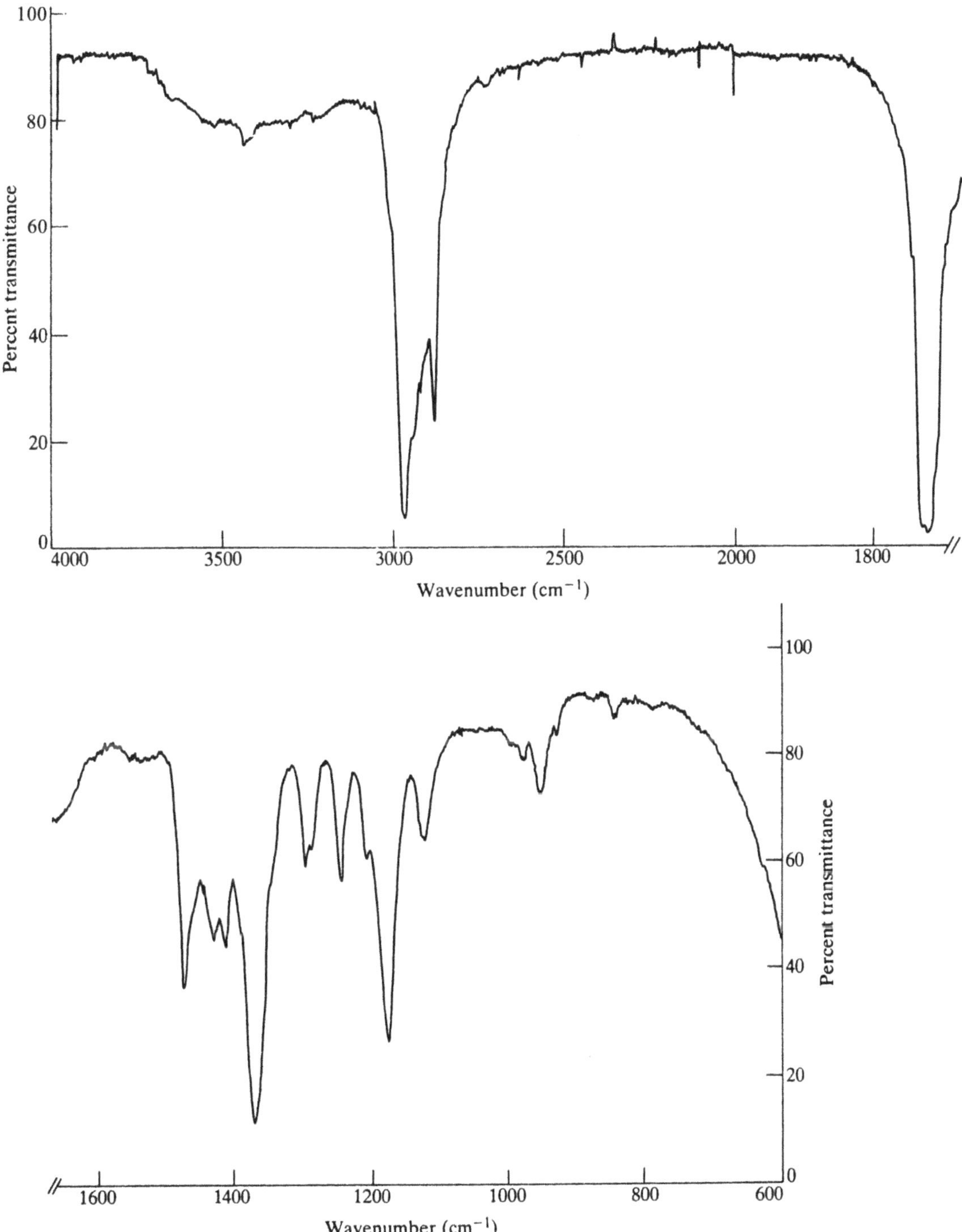

Figure Pl2-20

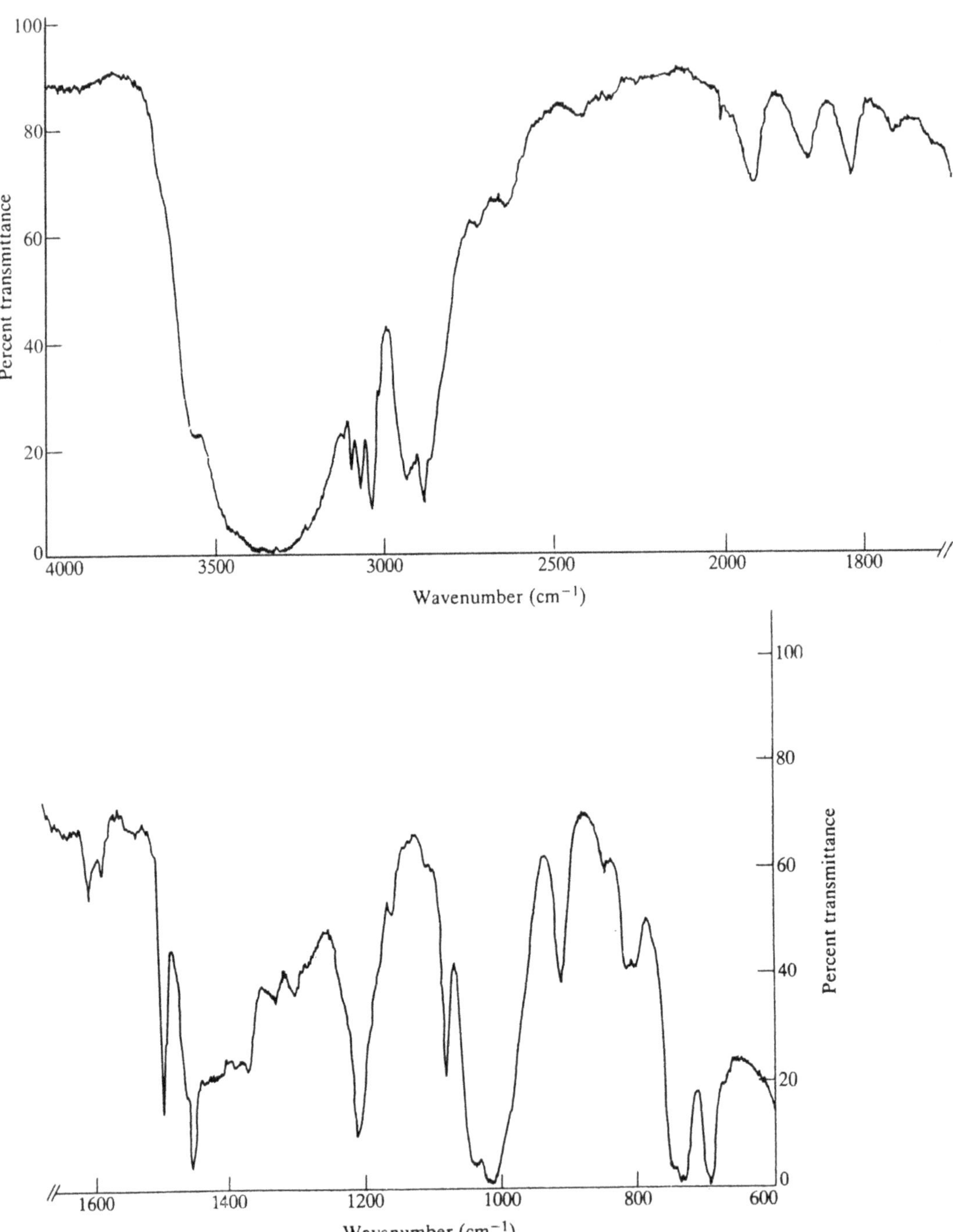

Figure P12-21

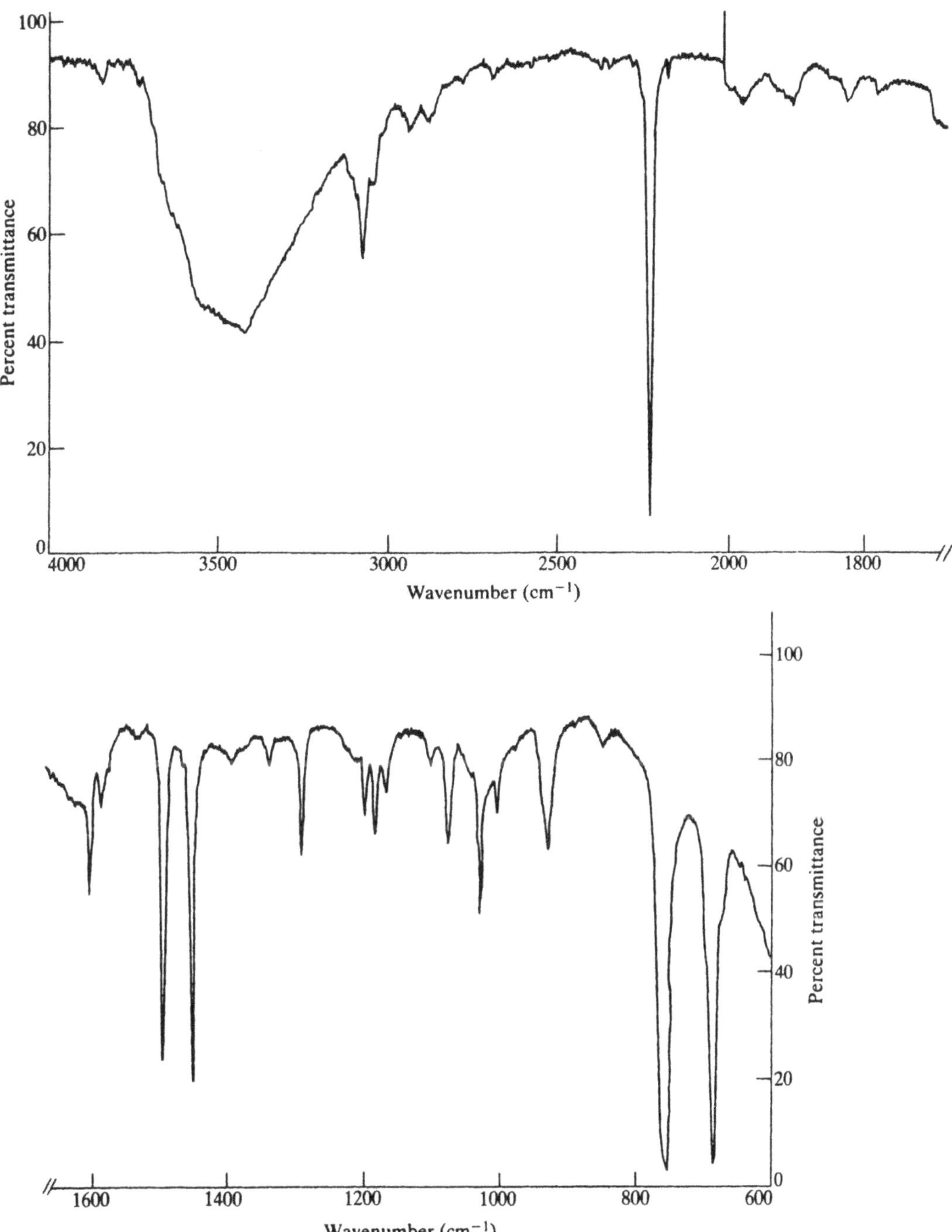

Figure P12-22

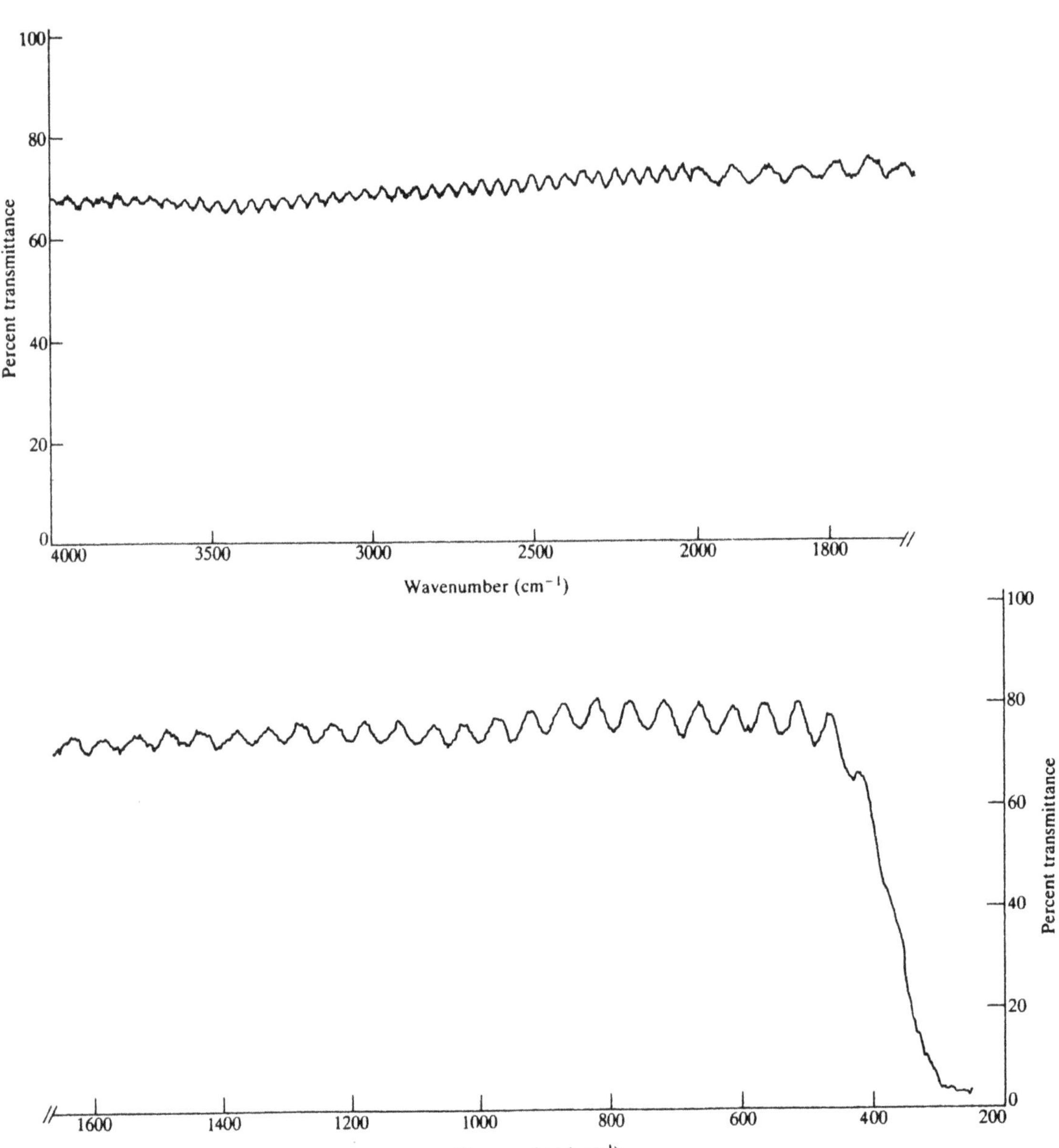

Figure P12-23

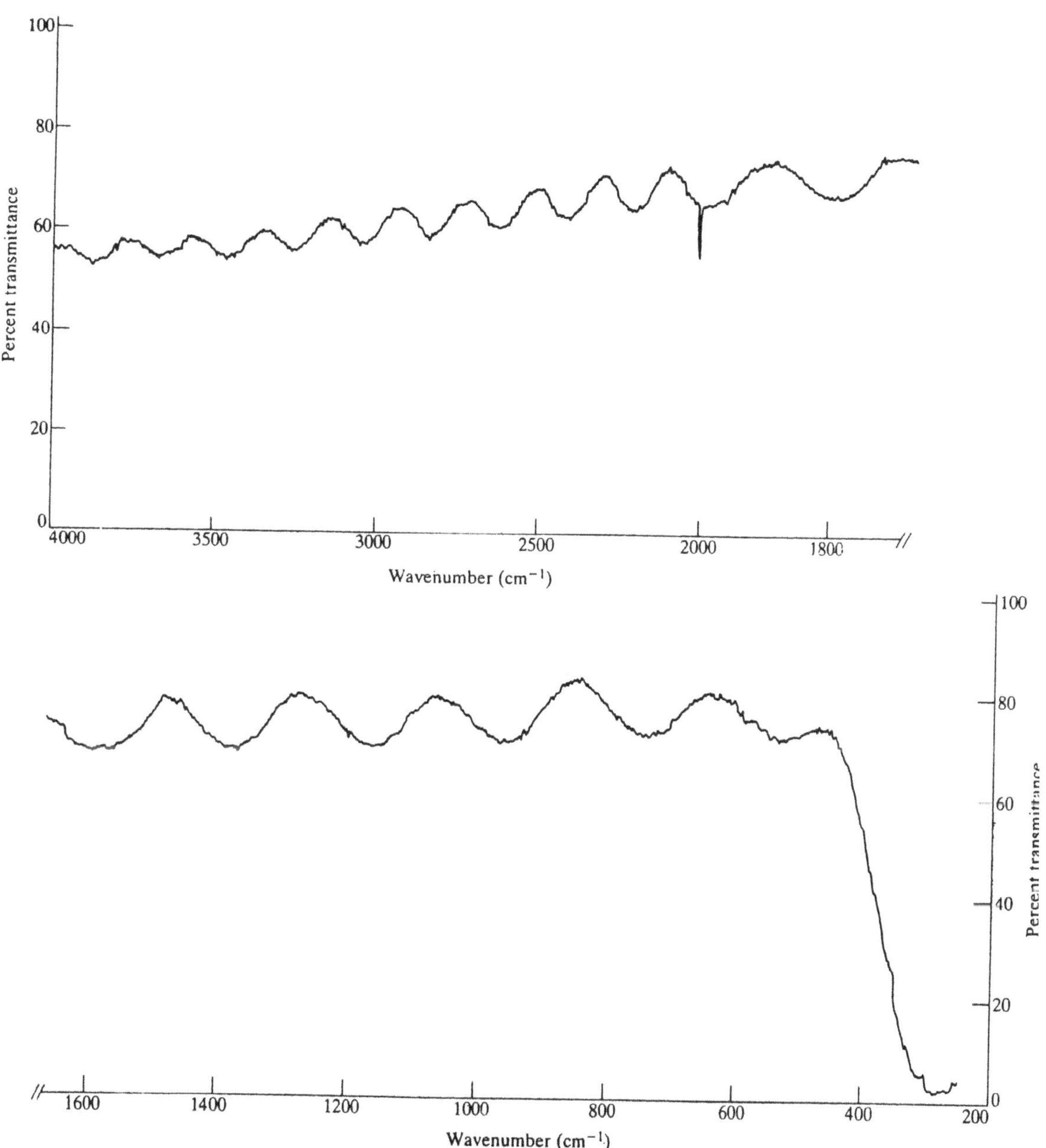

Figure P12-24

REFERENCES

1. Lee Smith, A., in P. J. Elving, E. J. Meehan, and I. M. Kolthoff (eds.): *Treatise on Analytical Chemistry,* part I, vol. 7, 2d ed., Wiley, New York, 1981, pp. 249-476.
2. Lambert, J. B., H. F. Shurvell, L. Verbit, R. G. Cooks, and G. H. Stout: *Organic Structural Analysis,* Macmillan, New York, 1976.
3. Cooper; J. W.: *The Minicomputer in the Laboratory: With Examples Using the PDP-11,* 2d ed., Wiley, New York, 1983, chap. 20.
4. Decker, Jr., J. A.: *Anal. Chem.*, **44**: 127A(1972).
5. Dolphin, D., and A. E. Wick: *Tabulation of Infrared Spectral Data,* Wiley, New York, 1977.
6. Young, C. W., R. D. Duvall, and N. Wright: *Anal. Chem.*, **23**: 709(1951).

CHAPTER

THIRTEEN

PHOTOACOUSTIC SPECTROSCOPY

Photoacoustic spectroscopy (PAS) is also referred to as *optoacoustic spectroscopy*. During the application of PAS, modulated ultraviolet-visible, infrared, or microwave radiation from a source is absorbed and spectrally excites the analyte. After excitation some of the molecules return to the ground state by radiationless processes. The thermal energy that is emitted during the relaxation causes expansion of the gaseous or liquid sample or expansion of a filler gas above the liquid or solid sample in the cell. Because the incident radiation is modulated, the gas periodically expands and contracts at the modulated frequency. The pressure waves that are generated by the relaxation process are measured by a detector that responds to the pressure waves. Photoacoustic spectra consist of a plot of the response of the detector as a function of the wavelength of the incident radiation.

Photoacoustic spectroscopy differs from the previously described spectroscopic methods in that the detector measures pressure waves (usually sound waves) that are emitted as the sample is alternately heated and cooled owing to the absorption of EMR. Because electromagnetic radiation is not measured by the detector, it is not necessary for the sample to be transparent to the incident radiation. The ability to obtain spectra of opaque samples and the high sensitivity of the technique are the major advantages of PAS. A wide array of samples, including gases, solids, liquids, smears, gels and opaque substances, can be assayed by using PAS. PAS has been used with incident radiation in the ultraviolet, visible, infrared, and microwave regions. More thorough information than presented in the chapter can be found in the first two references.

THEORY

The presented theory is brief and qualitative in nature. A more complete presentation can be found in the References. If the incident radiation is in the ultraviolet-visible spectral region, absorption by the sample causes excitation of the molecule to an excited electron state as was illustrated in Fig. 5-7. Once the molecule is in the excited state, relaxation can occur by fluorescence, phosphorescence, or a radiationless process. When relaxation occurs by emission of a photon, the energy of the emitted photon is not available as heat and does not contribute to the PAS signal. During each radiationless transition, however, energy is emitted as heat and does contribute to the PAS signal. PAS in the ultraviolet-visible spectral region is complementary to fluorescence and phosphorescence in that it provides information about radiationless relaxation whereas the luminescent methods provide information about processes that emit radiation.

The heat H which is formed during radiationless relaxation is directly proportional to the amount of radiation absorbed by the sample and to the amount of radiation which is available to be absorbed, i.e., to the intensity I of the incident beam. The amount of absorbed radiation is proportional to the absorbance of the analyte which, from Beer's law, is equal to abC, where a is the absorptivity of the analyte, b is the distance into or through the sample which is penetrated by the incident radiation, and C is the concentration of analyte in the sample. By combining the proportionality constant with a and b for a particular analyte, Equation (13-1) is obtained. This equation is analogous to Equation (11-10) which was obtained for luminescent substances.

$$H = kIC \qquad (13\text{-}1)$$

The intensity of the source determines the radiative energy that strikes the sample. From Eq. (13-1) it is apparent that high-intensity sources generate relatively large amounts of heat that subsequently cause intense PAS signals. Consequently, intense radiative sources are genrally used for PAS.

If the incident radiation is in the infrared or microwave region, the molecule can absorb radiation by excitation to a higher vibrational or rotational level as was described in Chapter 12. Relaxation is simpler than that which occurs after absorption of ultraviolet-visible radiation because the option for photon emission is eliminated. Relaxation occurs by emission of thermal energy.

Regardless of the wavelength of the incident radiation, the radiation is modulated at a frequency that causes a pressure wave to be formed that easily can be measured by the detector. Modulation is accomplished either by using a pulsed source, e.g., a pulsed laser, or by using a continuously operated source and a chopper. The frequency of the modulation must be slow relative to the rate of radiationless relaxation in the molecule in order to allow the measured PAS signal to oscillate at the same frequency as that of the modulated incident radiation. Generally modulation frequencies between 20 Hz and 100 kHz are used.

The modulation frequency is usually adjusted to a value that is different from any natural environmental oscillations that might interfere with the analysis and to a value that maximizes the intensity of the measured signal. Often an *acoustically resonant* frequency is used. The wavelength λ of an acoustically resonant signal is dependent upon the cell length L:

$$\lambda = \frac{2L}{n} \tag{13-2}$$

where n is an integer. The frequency of the oscillation is adjusted so that an integral number of half-wavelengths equals the length of the cell.

The mechanism by which heat from the sample is converted to a pressure wave depends upon the physical state of the sample. If the sample is a gas or a liquid that fills the cell, the heat that is emitted during relaxation increases the temperature of the sample. The increased temperature causes an increased pressure within the fixed-volume cell. For gases, the ideal gas law can be used to predict the increased pressure. When radiation is not striking the sample, the temperature and the pressure decrease. Pressure waves develop in the sample at the frequency of the oscillating incident radiation. Pressure waves that occur at a frequency that can be heard are sound waves. The amplitude of the waves in gaseous samples can be monitored with an acoustic detector such as a microphone. In liquid samples piezoelectric transducers are often used as detectors.

If the sample is a solid or a liquid that does not occupy the entire cell, a nonabsorbent filler gas is used to occupy the remainder of the cell. Usually the gas is either air or an inert gas. The gas is chosen so that it will not absorb radiation at any wavelength that is used for the study. Radiation from the source is focused on the sample and induces heating. The heated sample transfers some of its energy to the filler gas and the resulting pressure waves in the gas are monitored with an acoustic detector.

A combination of two mechanisms account for the wave forming in the gas. In most cases the heated sample warms the gas that is in contact with and adjacent to the sample. The heated gas expands causing a pressure wave which is measured by the detector. For samples that have low absorptivities and that are excited by incident radiation at a high modulation frequency, direct elastic coupling between the sample and the gas can account for some of the sound wave. In that case the periodic heating and cooling of the sample causes the sample to expand and contract. The expansion and contraction pushes and pulls the gas adjacent to the sample causing a pressure wave to be formed in the gas.

The intensity of the PAS signal that is obtained with a sample that does not fill the cell is related to the optical absorptive coefficient (the absorptivity) and to the *thermal diffusion length* of the sample. The thermal diffusion length is the distance in the sample through which thermal energy can be conducted during a fixed period. A *thermally thick* sample has a small thermal diffusion length. A *thermally thin* sample has a relatively large thermal diffusion length. At modulation frequencies that are greater than 10 Hz, most samples are thermally thick and have thermal diffusion lengths in the micrometer range, i.e., during the time of each oscillation thermal energy travels a relatively short distance through the sample.

For thermally thick samples that are optically transparent, only the radiation that is absorbed in the first thermal diffusion length has an effect on the acoustic signal, even though radiation can be absorbed throughout a considerably greater length. Radiation absorbed at greater distances from the surface of the sample cannot migrate to the surface sufficiently rapidly to warm the adjacent gas while the radiation is striking the sample. In that case the intensity of the acoustic signal is proportional to both the optical absorptive coefficient and to the thermal diffusion length in the sample. As the thermal diffusion length increases, more radiation that is absorbed in the interior of the sample can travel to the surface and warm the gas.

The thermal diffusion length depends upon the modulation frequency. As the frequency decreases, the thermal diffusion length increases because more time for thermal diffusion exists. Depth profiles of an absorbing solid can be acquired by changing the modulation frequency. At high frequencies (10 to 100 kHz) depths as small as 0.1 μm can be studied while at low frequencies (10 to 20 Hz) depths in the 10 to 100 μm range can be studied for samples that have a low thermal diffusivity. Samples that have high thermal diffusivities can be studied at depths of up to about 1 cm at low frequencies.

For thermally thin samples that are optically transparent, heat which is formed throughout the sample can rapidly travel to the surface and affect the PAS signal. In that case the thermal diffusion length through the sample no longer limits heating at the surface. The acoustic signal for thermally thin samples is proportional to the optical absorptive coefficient, to the length of the sample through which radiation is absorbed, and to the inverse of the modulation frequency of the radiation. As the frequency becomes lower, more heating can occur and the signal is increased. It also depends upon the thermal properties of the portion of the sample holder that is in the radiative path.

In optically opaque solids and liquids nearly all of the incident radiation is absorbed near the surface. In those cases the length of the sample has no effect upon the PAS signal. If the sample is thermally thin (the thermal diffusion length approaches or exceeds the sample length), the signal is inversely proportional to the modulation frequency and can be affected by the thermal properties of the sample holder. It is independent of the absorptive coefficient of the sample. For thermally thick solids in which the thermal diffusion length is greater than the distance of penetration of optical radiation into the sample, the PAS signal is independent of the optical absorptive coefficient, of the length of the sample, and of the thermal properties of the sample holder. It is inversely proportional, however, to the modulation frequency. Carbon black is an example of a substance that falls into that category. Carbon black is the black petroleum component that remains after removal of the lower boiling components by distillation. Because the PAS signal of carbon black (or a suspension of carbon black) is dependent only upon the modulation frequency and the intensity of the incident radiation, it is often used as a reference compound for the normalization of PAS spectra.

The final type of sample that can be encountered is the optically opaque solid or liquid that is thermally thick and for which the thermal diffusion length is less than the distance of optical penetration. In that case the acoustic signal is proportional to both the optical absorptive coefficient and to the thermal diffusion length. It is important to realize that because most opaque samples fall into this category, PAS can be used for assays of many opaque samples. Since the signal is dependent upon the modulation frequency, it is advantageous to have control of the frequency.

INSTRUMENTATION

A block diagram of a typical single-beam instrument that is used for PAS is shown in Fig. 13-1. Constant-intensity radiation from a source is chopped prior to being focused on the entrance slit of a monochromator. If the source is pulsed, the chopper is not required. Chopping can occur before or after the monochromator. After exiting the monochromator the radiation is focused on the sample in the cell. The detector is mounted in the cell wall or the cell consists of a cavity in the detector.

The signal from the acoustic detector enters a preamplifier. The preamplifier is placed close to the detector in order to minimize noise from the connecting cable. The electric signal from the preamplifier is directed to a lock-in amplifier that further amplifies that portion of the signal that oscillates at the frequency at which the chopper is operated. The output from the lock-in amplifier is displayed with a readout device. Typically the output is used to drive the *y* axis of a recorder, or is displayed on a computer controlled device.

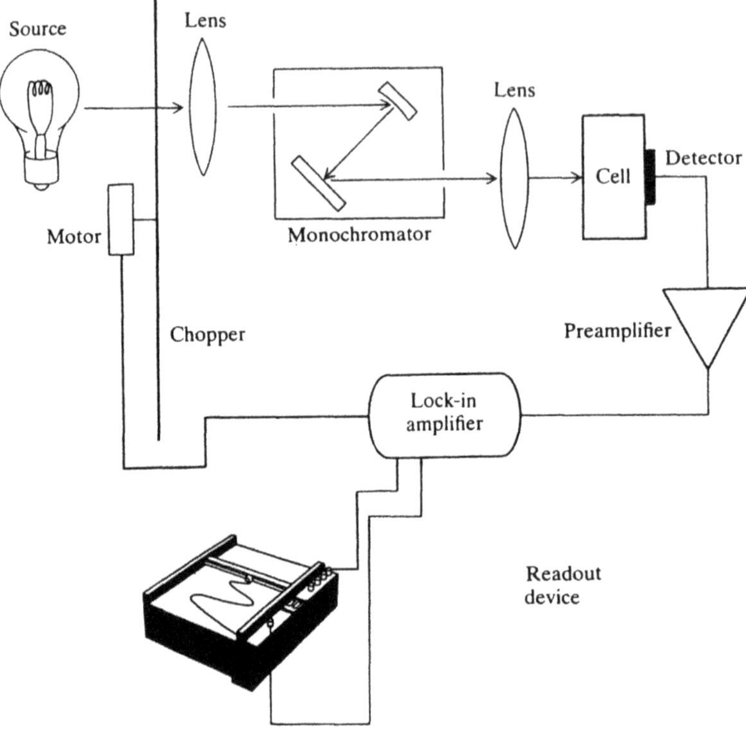

Figure 13-1 A block diagram of a single-beam photoacoustic spectrometer.

Both single- and double-beam instruments have been used. In some instruments the incident radiation is split into two portions after the monochromator but prior to the cell with a beam splitter. Half (or some other constant fraction) of the radiation enters the cell while the remainder strikes a detector that measures the radiant intensity (as opposed to an acoustic intensity). The ratio of the signal from the lock-in amplifier to the signal from the radiative detector is sent to the readout device. Instruments of that design automatically compensate for intensity of incident radiation while the wavelength is scanned. A more detailed description of typical instruments is included in a later section of the chapter. A review of PAS instruments is available in Reference 3.

Sources

The source that is used for a particular assay depends upon the wavelength range that is required for the study. Intense sources are normally used in order to maximize the output signal and to increase the signal-to-noise ratio. Most of the sources have been described in previous chapters. Tungsten-filament lamps have been used in the spectral range from about 600 to 1000 nm. Tungsten lamps are less intense at the longer wavelength end of that spectral range.

Several arc sources are available for use in PAS. The most popular is the high-pressure (50 to 70 atm) xenon-arc lamp. It has been used in the spectral range from 250 to 2500 nm. Hydrogen and deuterium arcs are useful from about 165 to 250 nm. Often hydrogen and deuterium lamps are water cooled because heat from the lamps can soften the glass casings of the lamps. Krypton lamps can be used from 125 to 165 nm. An open carbon arc can be used from about 350 to 700 nm. The carbon electrodes in the source are consumed during emission and must be periodically replaced. Some workers have used a high-pressure mercury lamp. It emits many lines between 230 and 650 nm but does not emit a continuum.

Lasers are available for use throughout most of the spectral range that is used for PAS. The lasers can be operated in either the continuous or pulsed mode. If the laser is pulsed, the lock-in amplifier is adjusted to respond to acoustic signals at the pulsed frequency. Both lasers that emit lines and lasers that emit a continuum are used. Line sources can be adjusted to a particular line by the monochromator but cannot be continuously scanned. Among the line sources that have been used for PAS are the argon ion laser, the ruby laser, the nitrogen laser, the Nd : YAG laser, the carbon dioxide laser, the carbon monoxide laser, and the nitrous oxide-carbon dioxide laser. In addition to the lines that are emitted directly by lasers, several other lines are available by addition or subtraction in a crystal of the frequency emitted by one laser to or from that emitted by another laser. Devices that combine frequencies in that manner are *difference frequency generators, sum frequency mixers,* or *optical parametric oscillators*.

Tunable dye lasers have been used as sources from 340 to 1200 nm. Typically each dye can be tuned over a wavelength range between about 50 and 100 nm. The useful wavelength range of a dye laser roughly corresponds to the fluorescent bandwidth of the dye. Unfortunately the dyes photodecompose during use and must be continuously pumped through the laser cavity and periodically replaced. Some dyes must be replaced after a few hours of use. Continuous pumping of the dye solution through the laser cavity also minimizes thermal effects during operation. The dye layer in a laser is about one-fourth of a millimeter thick. Flowrates are about 10 m/s. The dyes that are most popular for use in lasers are the coumarin dyes in the blue-green region, xanthene dyes between 500 and 700 nm, oxayine dyes from 600 to 800 nm, and carbocyanine dyes from 700 to 950 nm. Rhodamine 6G often has been used. Further details relating to dye lasers are available in Reference 4.

Other types of lasers that have been used for PAS include semiconductor diode lasers and *spin-flip Raman* (SFR) lasers. Semiconductor diode lasers are useful from 1 to 30 μm and can be continuously tuned within their useful range. They can be tuned by control of the current through the diode, the pressure of the diode, or the temperature of the diode. Among the semiconductor diode lasers that have been used are those in which the semiconductor is $GaAs_xSb_{1-x}$, $Hg_{1-x}Cd_xTe$, InAs, In_xGa_{1-x}, InSb, PbS, $PbS_{1-x}Se_x$, $Pb_{1-x}Sn_xTe$, or PbTe. Relatively inexpensive semiconductor and solid-state emitter lasers have become available for use in the near-infrared region.[5] Those lasers as well as other lasers can be used for near ir, atmospheric, trace gas assays[6] and for other environmental, space and industrial applications[7].

The CO pumped spin-flip Raman (SFR) laser is useful from about 5 to 6 μm and the CO_2 pumped SFR laser from 9 to 14 μm. The HF pumped SFR laser is used at 3 μm. SFR lasers use one laser (CO, CO_2, or HF) that emits a line to pump a semiconductor crystal that is held in a magnetic field. The photons from the pumping laser cause some electrons in the crystal to reverse (flip) their spins in the magnetic field. The photons emitted from the crystal have less energy than that associated with the pumping laser. Tuning of the emitted radiation is possible by changing the strength of the magnetic field. A popular SFR laser consists of an *n*-type InSb crystal which is pumped with either a CO or CO_2 laser. It can be operated in a continuous mode when it is pumped by the CO laser.

The intensity of the source is usually modulated, at a single wavelength, either by pulsing the source or by using a constant-intensity source with a chopper. A spectrum can be recorded by slowly changing the wavelength while the amplitude of the intensity of the incident radiation is modulated. An alternative to amplitude modulation is *wave-length modulation*. In that case the wavelength striking the sample is alternated between two values. Generally both wavelengths are scanned at the same rate and in the same direction. If the difference between the wavelengths is small in comparison to the aborptive peak width, a spectrum recorded while using wave-length modulation is the first derivative of the corresponding spectrum recorded while using amplitude (intensity) modulation. If the difference between the two wavelengths is so large that only one of the wavelengths is on a peak during the scan, the spectrum is nearly identical to that recorded with amplitude modulation. In some cases[8] the wavelength is modulated between two adjacent absorptive peaks.

In many photoacoustic spectrometers, radiation from the source is focused on the inlet slit of the monochromator. After passing through the monochromator, the radiation is focused on the sample in the cell. Consequently, two focusing elements are required. By focusing the radiation on the sample, maximum absorption and heat emission occur. A maximum PAS signal is observed.

Wavelength Selectors

In instruments that use a continuous source, such as a xenon-arc lamp, the radiation is usually made monochromatic with a reflective diffraction-grating monochromator. Sources that emit line radiation, such as some lasers, can use a monochromator or filters to select the proper line. Line sources are used in nondispersive instruments with no wavelength selector. The tuning mechanisms in tunable lasers have been described in Chapter 8.

Cells

Cell designs that are used in photoacoustic spectrometers are dependent upon the physical state of the sample. The cell must possess at least one transparent window through which incident radiation can pass. Additionally the cell must be acoustically insulated to prevent interference from external signals. Thick cell walls and good acoustic seals around the windows and other openings are used. In most cases the body of the cell is constructed from a substance, such as stainless steel or polished aluminum, that has a large *thermal mass*. A substance that has a large thermal mass requires absorption of a relatively large amount of energy before its temperature is significantly raised. In addition to stainless steel and aluminum, cells have been constructed from copper, silver, and gold. Because silver and gold are expensive, they rarely are used. Because of the large thermal mass, scattered incident radiation that strikes the cell body causes an insignificant temperature rise and PAS signal relative to the signals that occur when radiation strikes an absorbing sample.

The cell windows are constructed from a material that is transparent to the incident radiation. Windows that absorb a portion of the incident radiation not only decrease the intensity of radiation that strikes the sample but also constitute a major source of interference. Heat can be emitted by the windows into the cell if the windows have absorbed radiation. Of the commonly used window materials, sapphire is best for use in the ultraviolet-visible region. Owing to the expense of sapphire windows, quartz windows are most commonly used. In the visible region glass and plastic windows have been used. In the infrared region ZnSe windows[9] have been successfully used as well as the several salt windows described in Chapter 12.

When the analyte is a gas, the cell is most often cylindrical. In those cells that have two windows on the ends of the cylinder, radiation makes a single pass through the cell. In some *single-pass* cells acoustic baffles are used to decrease noise within the cell. In *multiple-pass* cells both ends of the cylinder are coated with a multilayer of a highly reflective dielectric material. A hole through the dielectric substance allows radiation to enter the cell. The radiation is reflected through the sample several times by the dielectric material. Multiple-pass cells can be used for assays at lower concentrations than single-pass cells.

A *differential* cell often consists of two sequentially connected cells within the same cylinder. Each cell uses a separate detector. The incident radiation sequentially passes through the cells. Sketches of the several common forms of

PAS cells that are used with gaseous samples are shown in Fig. 13-2. The sample is introduced and removed from the cells through inlet and exhaust valves or through entrance holes at the window openings. In each case the detector is attached to the cell through an opening in the wall of the cell, or the cell is a cavity within the detector. The combined cell and detector is a *spectrophone*. PAS in the infrared region has been used as a detector for gas chromatography[9] and for FTIR spectrophotometry[10].

Cells that are designed for use with solids and liquids typically use one end of the cell to support the sample or the sample holder. The distance between the sample and the window should be greater than the length of thermal diffusion in order to prevent loss of thermal energy from the sample through the window. Usually radiation passes vertically through the cell and strikes the sample. The incident radiation is not allowed to strike either the cell walls or the detector. The volume of the filler gas within the cell is kept as small as possible in order to maximize the PAS signal. In most cases the output signal is inversely proportional to the volume of the gas. The volume must be sufficiently large, however, to prevent the cell walls from significantly damping the pressure wave. The filler gas is chosen to be nonabsorbing. Often air or an inert gas is used. Use of an inert gas generally results in greater analytical sensitivity.

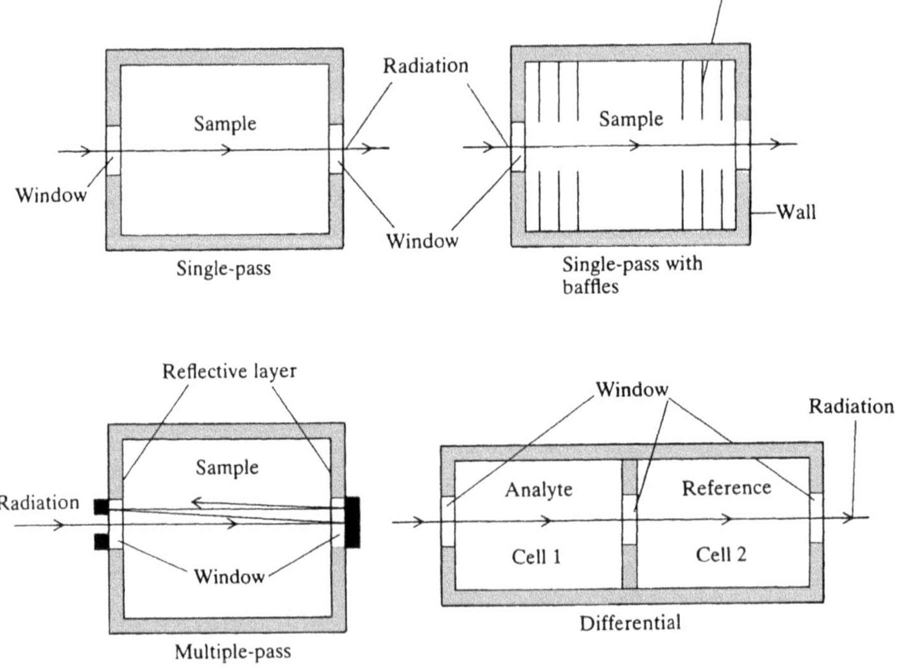

Figure 13-2 Several photoacoustic cells that are used for gaseous sample.

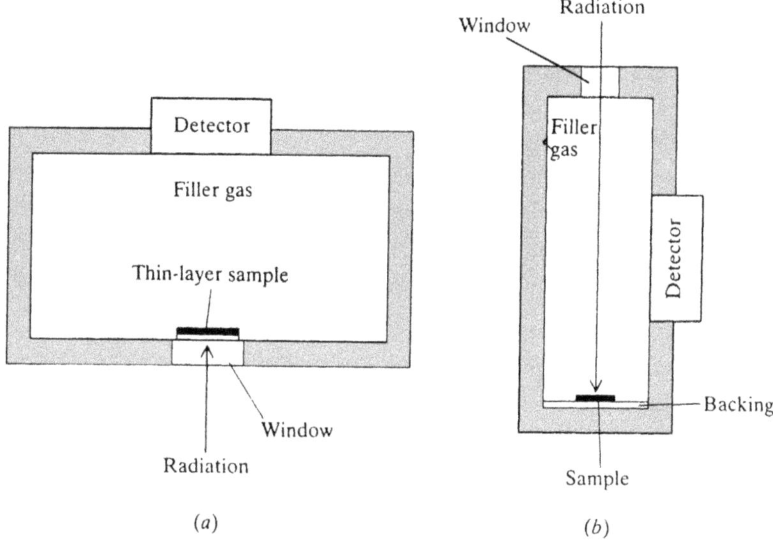

Figure 13-3 Sketches of two types of PAS cells that can be used for solids or liquids. Use of cell A requires that the sample be a thin layer of a thermally thin material.

If the sample is a thin layer of a thermally thin substance, incident radiation can strike the sample without first passing through the filler gas (Fig. 13-3*a*) because heat can rapidly travel to the inner surface of the sample. If the sample is not a thin layer of a thermally thin substance, radiation usually strikes the sample from above (Fig. 13-3*b*) after first passing through the filler gas. Samples such as crystals, powders, gels, solutions, and smears can be used.

In addition to the cells shown in Fig. 13-3, a quartz cuvet[11] has been used as a cell for liquid-phase studies, and a flow-through cell with quartz windows has been used[12] for photoacoustic detection in high-performance liquid chromatography. PAS has also been used as a method to locate and obtain spectra of sample spots on thin-layer chromatographic plates. The sample can either be removed from the plate and placed in the PAS cell [13, 14] or the open-bottom cell can be placed directly on the chromatographic plate.[15] PAS has been used to study biological materials such as mammalian cells and tree leaves.

Detectors

Microphones and piezoelectric transducers (PZT) have been used as detectors for PAS. For gaseous samples and for those solid and liquid samples that are placed in a cell such as those shown in Fig. 13-3 that contains a filler gas, the most common detector is a microphone. Of the several types of microphones, the *condenser* (or capacitor) microphone commonly is used.

A condenser microphone consists of a circular, thin metallic diaphragm that is mounted by its edge near a rigid, stationary metallic plate. The two plates serve as the plates of a capacitor in an electric circuit. The thin diaphragm is forced closer to the stationary plate whenever a pressure wave strikes it. When pressure is no longer exerted on the diaphragm by the wave, the diaphragm is returned to its initial position by the tension in the diaphragm. The result is a diaphragm that vibrates at the frequency of the modulated radiation that strikes the sample. The distance through which the diaphragm modulates is a function of the amplitude of the pressure wave that is emitted from the cell.

The plates in a condenser microphone can be *biased* in several ways. In some cases a constant charge is maintained between the two plates by an external electric circuit. In that case the change in potential between the two plates is amplified with a high-input-impedance amplifier and measured. If a constant potential difference is maintained across the two plates by an external circuit, the current flow through the circuit is amplified and measured. The current is inversely proportional to the total capacitance in the circuit. With the *electret* type of condenser microphone a permanent charge is applied to the plates. The charge can be applied with a discharging Tesla coil prior to use of the microphone. The potential difference between the two plates is amplified and monitored. In each case the amplified output from the microphone constitutes the output of the detector.

Microphones cannot be used in liquids. Piezoelectric transducers are primarily used for detecting PAS signals in solution.[16] A piezoelectric transducer is a solid material such as lead zirconate titanate which has a structure that is altered with application of pressure. Structural alteration causes a rearrangement of charge within the substance and an alteration of the potential difference across the material. A piezoelectric transducer changes a pressure wave into an electric signal which subsequently is amplified and monitored.

PZTs can be shaped into tubes which serve as cells as well as detectors (e.g., the PZT-5H piezoelectric ceramic tube from Vernitron). The sample solution is poured into the tube. Alternatively, the PZT is used in the form of a disk (e.g., the PZT-5A disk) which is attached to the wall of the cell or is dipped directly into the sample solution.

Readout Devices

Scanning instruments generally use an *xy* recorder or a computer controlled data acquisition and readout device. Computer-controlled instruments can use any of several readout devices including video display terminals, plotters, and line printers. Instruments that are used for quantitative analysis at a single wavelength can use analog or digital meters for readout, or, more often, a computer controlled device.

Typical Instruments

The most useful PAS instruments contain a continuous source and a monochromator. The monochromator can be automatically scanned to record a spectrum or can be adjusted to a single wavelength for quantitative studies. The instruments usually have a variable radiative modulation frequency. A variable modulation frequency allows examination of samples at various depths. Owing to variations with wavelength in the intensity of radiation from continuous sources, scanning instruments nearly always have a double-beam design. The radiation is usually split into two beams after passage through the monochromator. The ratio of the PAS signal from the cell to the signal in the reference beam is reported by the readout device. The detector in the reference beam responds to the energy of the radiation. Photodiode and *pyroelectric* detectors can be used in the reference beam. Pyroelectric detectors change thermal energy into an electric signal.

Some double-beam instruments use a separate PAS cell in each of the two beams. In that case a substance which absorbs radiation at all wavelengths to the same extent is placed in the cell in the reference beam. The reference PAS signal depends upon the intensity of the incident radiation but is independent of the wavelength of the incident radiation. In those instruments the ratio of the PAS signal from the sample to the signal from the reference compound is reported by the readout device. A glycerol suspension of carbon black has been extensively used as the reference compound. Parker black ink has also been used.[17]

Single-beam instruments can be used to obtain spectra; however, at least two scans are required. Most recording single-beam instruments are computer-controlled. During one scan either the intensity of the incident radiation or the PAS signal from a reference compound, such as carbon black, is recorded. The spectrum of the sample is recorded during the other scan. The ratio of the sample spectrum to the reference spectrum is calculated by the computer at each wavelength and the ratio is displayed on the readout device. A block diagram and a photograph of a commercial instrument are shown in Fig. 13-4.

QUALITATIVE ANALYSIS

Qualitative analysis that is done while using photoacoustic spectrometry is performed in the same manner as when it is performed spectrophotometrically. Spectral peak positions are compared with those of known compounds. As during spectrophotometric measurements, qualitative analysis of organic compounds is generally done while using incident radiation in the infrared region.

QUANTITATIVE ANALYSIS

In most cases the photoacoustic signal linearly increases with increasing concentration of the absorbing substance. Analyses that involve a single absorbing substance are generally performed with the working-curve method. The determination of two or more absorbing components in a mixture is usually done with the simultaneous equations method that was described in Chapter 9.

The photoacoustic signal is proportional to the intensity of the incident radiation [Eq. (13-1)]. In that regard photoacoustic spectrometry is similar to fluorometry. The intensity of the source must be stable during quantitative measurements with a single-beam instrument. By increasing the intensity of the source, the method becomes more sensitive. Because samples that lose energy by luminescing have relatively small PAS signals while samples that do not fluoresce or phosphoresce have larger PAS signals, the methods complement each other. A comparison of PAS with fluorometry and other highly sensitive spectroscopic methods can be found in Reference 18.

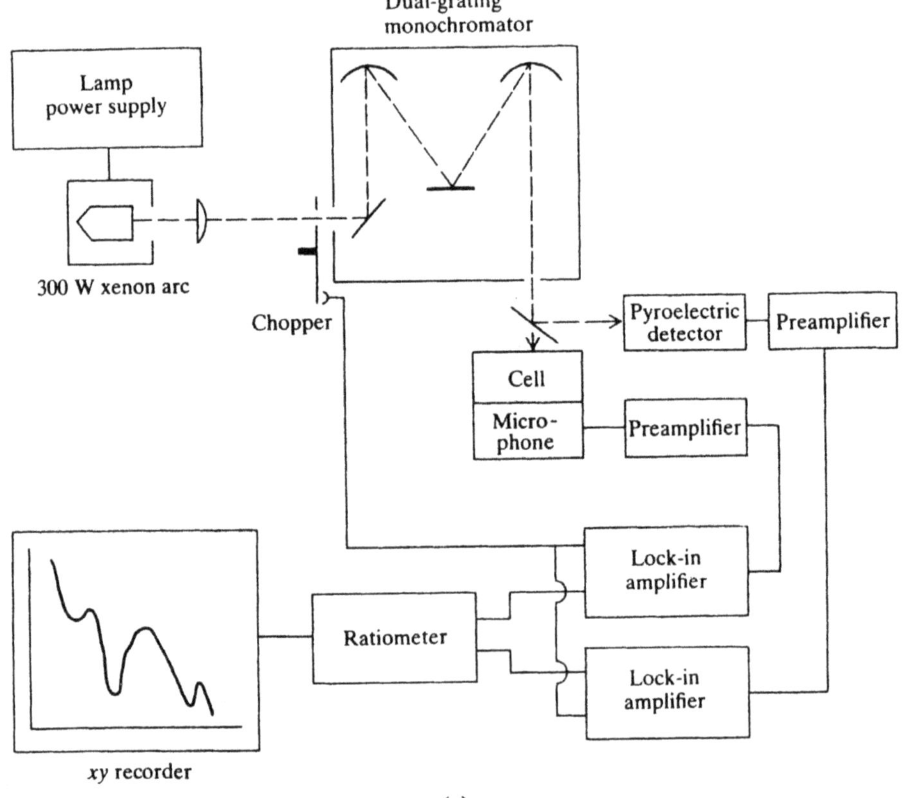

(a)

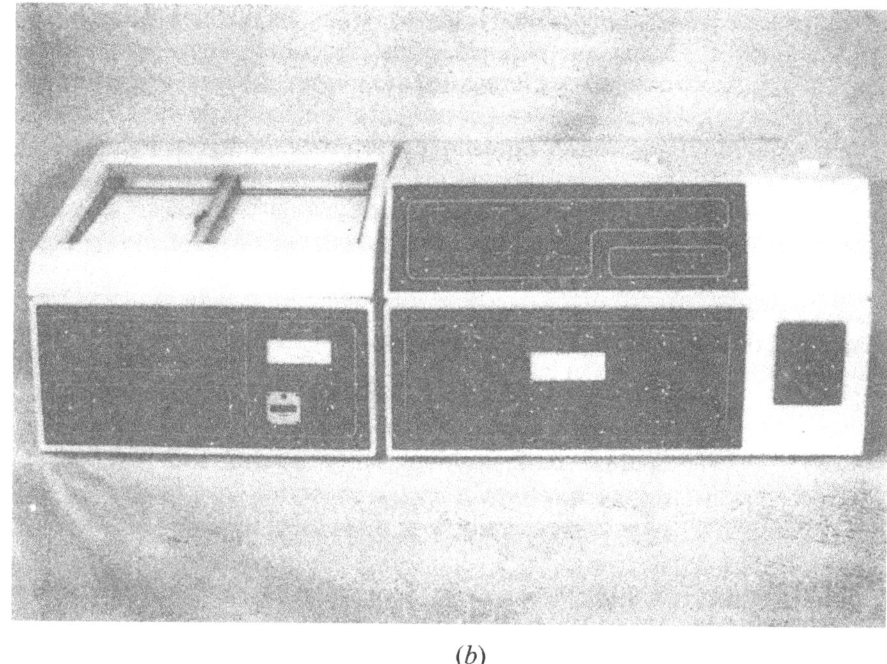

(*b*)

Figure 13-4 (*a*) A block diagram and (*b*) a photograph of the model OAS400 photoacoustic spectrometer. *(Photograph courtesy of EDT Research, 14 Trading Estate Rd., London, NW10 7LU, England.)*

IMPORTANT TERMS

Amplitude modulation	Pyroelectric detector
Condenser microphone	Single-pass cell
Difference frequency generator	Spectrophone
Differential cell	Spin-flip Raman laser
Electret microphone	Sum frequency mixer
Multiple-pass cell	Thermally thick
Optical parametric oscillator	Thermally thin
Photoacoustic spectroscopy	Thermal mass
Piezoelectric transducer	Wavelength modulation

PROBLEMS

13-1 Calculate the shortest possible cell pathlength for an acoustically resonant cell in which the modulation frequency of the source is 265 Hz if the filler gas is air in which the velocity of sound is 331.45 m/s.

13-2 What modulation frequency of the incident radiation is required in order for a cell with a 20.0-cm length to be acoustically resonant if the filler gas is air? The velocity of sound in air is 331.45 m/s.

13-3 In which of the following cases is the PAS signal proportional to the concentration of an absorbing substance in a solid? Assume that the absorptive coefficient is proportional to concentration in each case.

(*a*) Optically transparent, thermally thick sample

(*b*) Optically transparent, thermally thin sample

(*c*) Optically opaque, thermally thin sample

(*d*) Optically opaque sample with a thermal diffusion length that is less than the distance of optical penetration

13-4 The high-pressure xenon-arc lamp is often used as a source in PAS instruments. List the advantages of the source in comparison to lasers and deuterium-arc lamps.

13-5 Either photoacoustic spectrometry or fluorescent spectrometry could be chosen for quantitative analysis of solutions of each of the following compounds. Which technique is preferable in each case?

(*Hint*: Review Chapter 11.)

(*a*) Naphthalene (*b*) Butanoic acid

(*c*) Perylene (*d*) 1,3-Hexene

(*e*) Propionaldehyde (*f*) Anthracene

REFERENCES

1. Yoh-han Pao, Y. (ed.): *Optoacoustic Spectroscopy and Detection,* Academic Press, New York, 1977.
2. Rosencwaig, A.: *Photoacoustics and Photoacoustic Spectroscopy,* Wiley, New York, 1980.
3. West, G.A., J. J. Barrett, D. R. Siebert, and K. V. Reddy: *Rev. Sci. Instrum.*, **54**: 797(1983).
4. Schafer, F. P. (ed.): *Dye Lasers,* Springer-Verlag, Berlin, New York, 1973.
5. Kawabata, Y., T. Kamikabo, T. Imasaka, and N. Ishibashi: *Anal. Chem.,* **55**: 1419(1983).
6. Elia, A., C. D. Franco, P. M. Lugara and G. Scamarcio: *Sensors,* **6**: 1411(2006).
7. Miklos, A., C. H. Lim, W. W. Hsiang, G. C. Liang, A. H. Kung, A. Schmohl and P. Hess: *Appl. Opt.*, **41**: 2985(2002).
8. Yip, B. C., and E. S. Yeung: *Anal. Chem.,* **55**: 978(1983).

9. Kreuzer, L. B.: *Anal. Chem.,* **50**: 597A(1978).

10. McClelland, J. F., R. W. Jones and S. J. Bajic in *Handbook of Vibrational Spectroscopy,* J. M. Chalmers and P. R. Griffiths (eds.), John Wiley, New York, 2002.

11. Voightman, E., A. Jurgensen, and J. Winefordner: *Anal. Chem.,* **53**: 1442(1981).

12. Oda, S., and T. Sawada: *Anal. Chem.,* **53**: 471(1981).

13. Rosencwaig, A., and S. S. Hall: *Anal. Chem.,* **47**: 548(1975).

14. Castleden, S. L., C. M. Elliott, G. F. Kirkbright, and D. E. M. Spillane: *Anal. Chem.,* **51**: 2152(1979).

15. Fishman, V. A., and A. J. Bard: *Anal. Chem.,* **53**: 102(1981).

16. Lahmann, W., H. J. Ludewig, and H. Welling: *Anal. Chem.,* **49:** 549(1977).

17. Malkin, S., and D. Cahen: *Anal. Chem.,* **53**: 1426(1981).

18. Harris, T. D.: *Anal. Chem.,* **54**: 741A(1982).

CHAPTER

FOURTEEN

RADIATIVE SCATTERING

Radiative scattering occurs when an incident photon changes direction upon striking a sample particle. In most cases the change in direction is random or almost random. Several types of scattering are possible.

Tyndall scattering (also called *Mie* scattering) occurs when the diameters of the particles that cause the scattering are equal to or greater than the wavelength of the incident radiation. Colloidal and turbid solutions generally have particles with dimensions that are sufficiently large to cause Tyndall scattering when the incident radiation is in the visible region. Tyndall scattered radiation has the same wavelength as that of the incident radiation.

Rayleigh and *Raman* scattering occur when the diameters of the particles that cause the scattering are small (10 percent or less of the wavelength) in comparison to the wavelength of the incident radiation. Rayleigh and Raman scattering can occur from dissolved molecules and polyatomic ions. The wavelength of Rayleigh scattered radiation is identical to that of the incident radiation. The wavelength of the scattered radiation is either greater or less than that of the incident radiation during Raman scattering.

Tyndall scattering is nearly exclusively used for quantitative analysis. Raman scattering is primarily used for qualitative analysis but also can be used for quantitative analysis. Rayleigh scattering is rarely used for analysis.

Regardless of the type of scattering that occurs, the intensity of the scattered radiation varies with its frequency. Theoretically the intensity I is proportional to the fourth power of the radiative frequency

$$I = k\nu^4 \tag{14-1}$$

In practice the intensity varies with the frequency to some power between two and four. For Tyndall scattering the power is often closer to two whereas for Rayleigh and Raman scattering the value is usually close to four. In any case, if the ratio of the intensity of scattered radiation as a function of frequency is important to a particular study, generally it is advisable to measure that ratio rather than relying on a calculated value. As the particle size varies between that which causes Tyndall scattering and that which causes Rayleigh or Raman scattering, the power assumes values between two and four.

SCATTERING MEASUREMENTS

Quantitative analysis can be performed with scattered radiation using either *turbidimetry* or *nephelometry*. Turbidimetric measurements are analogous to spectrophotometric measurements of absorption. The detector is located after the cell and is aligned with the path of the radiation from the source as in a spectrophotometer. Turbidimetric measurements consist of measuring the decreased intensity of the incident radiation that is caused by scattering. The measurement is completely analogous to an absorptive measurement although the reason for the decreased intensity is scattering in one case and absorption in the other. Figure 14-1 is a block diagram that shows the major components of an instrument that is used for turbidimetric measurements.

At low concentrations an equation that is analogous to Beer's law describes the relationship between the incident intensity I_0, the intensity I of unscattered radiation, and the concentration C of the scattering particles:

$$S = -\log \frac{I}{I_0} = kbC \qquad (14\text{-}2)$$

This equation can be derived in the same manner used to derive Beer's law (Chapter 5). In order to avoid confusion with absorptive measurements, the

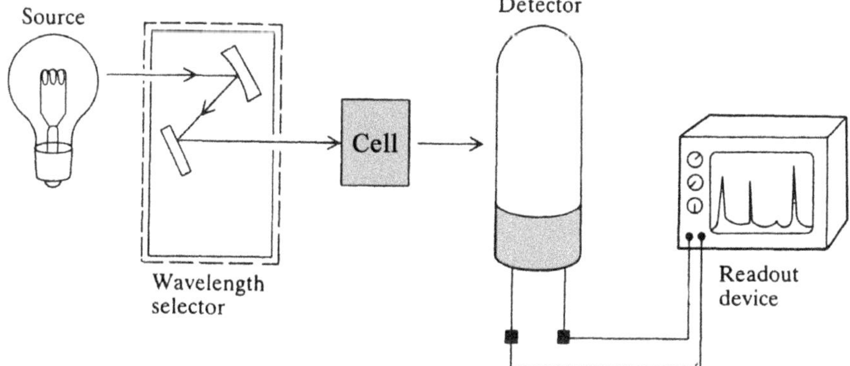

Figure 14-1 A diagram of an instrument that is used to make turbidimetric measurements. The dashed line around the wavelength selector is used to indicate that in some instruments a wavelength selector is not required.

logarithmic ratio S of the intensities is termed *turbidance* rather than absorbance. The constant k in Eq. (14-2) is the *turbidity coefficient* and the cell pathlength is b, as in Beer's law.

Turbidimetric measurements are most accurate when the ratio of I to I_0 is not near either 0 or 1. Because Rayleigh and Raman scattering generally cause relatively small decreases in intensity of the incident radiation, turbidimetric measurements are normally not used with those forms of scattering. Turbidimetric measurements are used for assays that are based upon Tyndall scattering.

Nephelometric measurements are analogous to fluorometric measurements. The detector is placed out of the path of radiation from the source. In most cases the detector is placed at 90° relative to the path of the incident radiation. Nephelometry measures the intensity of that portion of the scattered radiation that is emitted perpendicularly from the cell in the direction of the detector. Nephelometry consists of measuring the intensity of scattered radiation rather than decreased incident radiation intensity as during turbidimetric measurements. A diagram of a typical arrangement of the major components of an instrument that is used to make nephelometric measurements is shown in Fig. 14-2.

An equation similar to that listed in Chapter 11 for luminescence describes the relationship between the intensity of scattered radiation, the intensity of the

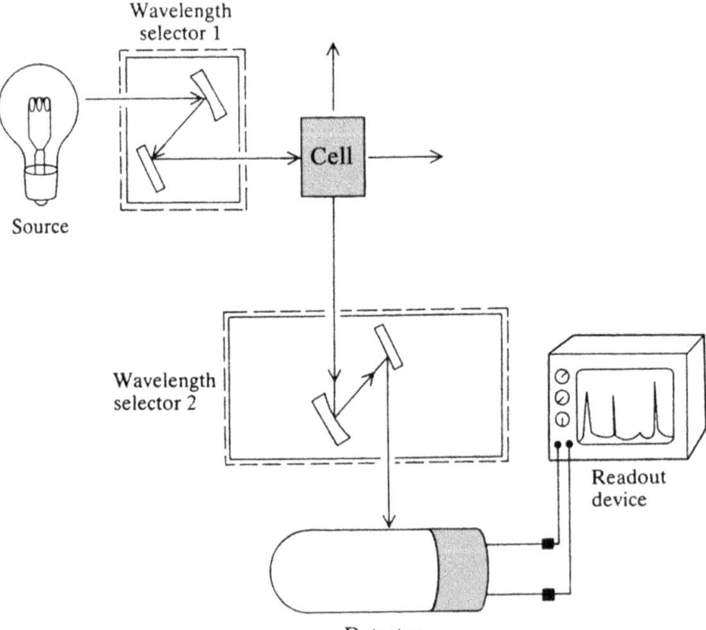

Figure 14-2 A diagram of an instrument that is used to make nephelometric measurements. The dashed lines around the wavelength selectors are used to indicate that in some instruments one or both of the wavelength selectors are absent.

incident radiation, and the concentration of particles that cause the scattering:

$$I = kI_0C \tag{14-3}$$

The value of k in Eq. (14-3) is constant only for a particular instrument and when experimental conditions are carefully controlled. As during luminescent measurements, the intensity of the scattered radiation is directly proportional to both the intensity of the incident radiation and to the concentration of the analyte. For assays of dilute solutions it is advantageous to use incident radiation that has a high intensity. Nephelometry can be used with Tyndall, Rayleigh, or Raman scattering.

TYNDALL SCATTERING

Tyndall scattering is caused by several processes that occur when incident radiation strikes particles that have dimensions equal to or greater than the wavelength of the incident radiation. Among the processes that contribute to Tyndall scattering are reflection from the surface of the particles, internal reflection from within the particles, diffraction through the particles, and refraction of the radiation as it enters and exits the particles.

Some of the factors that determine the amount and direction of Tyndall scattering are the transparency of the particles, the refractive index of the particles, the refractive index of the surrounding medium, the wavelength of the incident radiation, and the size of the particles. The theory of scattering by homogeneous spheres was presented by Mie[1] in 1908. Owing to its complexity the theory is not described here.

Because the wavelength of Tyndall scattered radiation is identical to that of the incident radiation, Tyndall scattering is not useful for qualitative analysis. The wavelength of the scattered radiation is not characteristic of a particular chemical substance. Both turbidimetry and nephelometry can be used for quantitative analysis with Tyndall scattering. In either case the working-curve method is used for the assay because the variation in either turbidance or intensity of scattered radiation with concentration is often nonlinear. For dilute solutions that have turbidances that are less than about 0.05, nephelometry is preferred. At concentrations at which the turbidance is greater than 0.05, turbidimetry generally yields more reliable results. Nephelometry is preferred at low concentrations because a small scattered intensity against a black background is easier to measure than a small change in intensity of transmitted radiation.

Turbidimetric and nephelometric measurements are usually made while using a source that emits in the visible region. Detectors which accurately and reliably respond to radiation in the visible region are readily available. Phototubes are often used in instruments designed to make turbidimetric measurements and phototubes or photomultiplier tubes for nephelometric measurements. Turbidimetric measurements are often made with an instrument that is designed for fluorometric measurements. Laboratories which routinely use Tyndall scattering for analysis sometimes use instruments that have been specifically designed for turbidimetric or nephelometric measurements. Those instruments are usually simpler in design and less expensive than spectrophotometers or fluorometers. Often they use the broad visible continuum emitted from a tungsten filament as the incident radiation, have no monochromator, and use a phototube or the human eye as the detector.

Because the intensity of Tyndall scattered radiation is often proportional to approximately the second power of the frequency of the incident radiation, the sensitivity of the assay increases as the frequency of the radiation increases. Instruments that have wavelength selectors (filters or monochromators) are typically adjusted to allow high-frequency (short-wavelength) radiation to strike the sample. Incident radiation in the blue or near-ultraviolet region is often used.

The wavelength chosen for the turbidimetric or nephelometric assay is also dependent upon the presence in solution of absorbing or fluorescing species. If substances in the solution in addition to the analyte can absorb or fluoresce, a wavelength of the incident radiation is generally chosen at which absorbance or fluorescence does not occur. If the scattering particles absorb radiation, the sensitivity of turbidimetric, but not nephelometric, determinations can be increased by choosing a wavelength at which absorbance occurs. In that case the instrument measures the sum of absorbance and turbidance. Because both absorbance and turbidance theoretically are proportional to concentration, the sum of the two should also be proportional to concentration.

Cells that are used to hold the sample during measurements of Tyndall scattering are identical to the cuvets that are used for measurements of absorbance or fluorescence. The walls must be transparent to the incident radiation and should be nonreflective. Because scattered radiation from the walls can interfere with an assay, it is sometimes advantageous to coat the exterior of the walls, except those through which radiation must pass, with a nonreflective black paint. That is particularly important for nephelometric measurements.

Quantitative Analysis with Tyndall Scattering

Owing to the frequently encountered lack of linearity, the working-curve method is most often used for quantitative analysis with either turbidimetry or nephelometry. Standard suspensions that have particle sizes that are as nearly as possible identical with those in the sample are used to prepare the working curve. Either the turbidance or the intensity of the scattered radiation at a fixed angle from the path of the incident radiation is plotted as a function of concentration. The concentration of the sample that corresponds to the instrumental reading is determined from the working curve.

Preparation of the standard suspensions is particularly critical. Car taken to control all experimental variables that can affect the size or shap suspended particles. A variation in either size or shape can significantly al turbidimetric or nephelometric measurement. The sample and stan suspensions must be prepared using identical procedures. Variables that can affect the results and that must be controlled include the concentrations of the reagents that are used to prepare the suspensions, the rate and order of mixing of the reagents, the amount of time after the reagents have been mixed and before the measurement is made, and the pH, total ionic strength, and temperature of the solution.

Measurements of the amount of Tyndall scattering are primarily used to measure the concentration of ionic substances that can react with a reagent to form a suspension. Examples include the determination of calcium(II) by the addition of an oxalic acid solution, the determination of silver(I) by the addition of a sodium chloride solution and the determination of barium(II) by the addition of a sodium sulfate solution. In the latter case a barium sulfate suspension forms from the reaction of aqueous barium(II) with sodium sulfate solution. A silver chloride suspension is formed during the reaction between silver(I) and aqueous sodium chloride; and a calcium oxalate suspension forms during the reaction between calcium(II) and aqueous oxalic acid.

Turbidimetry and nephelometry can be used to locate the endpoints of some titrations in which the titrand reacts with the titrant to form a suspension. Generally the turbidance or intensity of the scattered radiation increases before the endpoint and remains constant after the endpoint. For many titrations the portions of the titration curves before and after the endpoint are not linear. In those cases endpoint location is difficult. As the titration proceeds, the particles formed during the early portion of a titration have a chance to increase in size by combining with other particles or by precipitating on the surface of the particles that are already present. The process by which the particle size is increased is digestion. If the size of the particles changes during a titration, the titration curve is nonlinear prior to the endpoint. For titrations in which the titration curve is linear prior to the endpoint and in which the endpoint can be easily located, titrands in the 10^{-6} to 10^{-5} M range can be determined.

Nephelometric and turbidimetric measurements of Tyndall scattering have been used to locate potential precipitants in commercially prepared soft drinks and alcoholic beverages, to measure potentially equipment-clogging solids suspended in waters that are used in industrial equipment, and as an environmental analytical tool to measure suspended solids in natural waters. Nephelometric and turbidimetric measurements have also been used to measure suspended particles in gases. Smog and fog exhibit Tyndall scattering and have been quantitatively assayed using nephelometry and turbidimetry. Turbidimetric measurements that are made over a long distance, as between two buildings in a city, require sources that are highly intense. Lasers are used for such measurements.

RAYLEIGH SCATTERING

Rayleigh and Raman scattering occur when the particle dimensions are small relative to the wavelength of the incident radiation. When the particle dimensions are small and the refractive index of the particle does not greatly vary from that of the medium, the particle can behave as a radiating dipole while electromagnetic radiation passes through the medium containing the particles. The incident electromagnetic radiation has an electromagnetic field associated with it. As the wave passes the particle, the strength of the electromagnetic field at the particle varies at the frequency of the incident radiation. If the particle can be polarized by the field, a dipole is induced that oscillates at the same frequency as that of the incident radiation.

The oscillating dipole acts as a point source which emits radiation in all directions at the frequency of the oscillation. Because the frequency of the oscillating dipole is identical to the frequency of the incident radiation, the emitted (scattered) radiation from the particle has the same wavelength as that of the incident radiation. Scattering that occurs by that mechanism is Rayleigh scattering. The dipole moment μ that is induced by the electric field of the passing electromagnetic radiation is proportional to the *electric field strength E* and to the *polarizability a* of the particle:

$$\mu = aE \tag{14-4}$$

The intensity of the scattered radiation is proportional to the square of the induced dipole moment and to the fourth power of the scattered frequency:

$$I = \frac{16\pi^4 v^4 \mu^2}{3c^3} \tag{14-5}$$

Particles that are highly polarized exhibit the greatest Rayleigh scattering.

The energetic changes that occur during Rayleigh scattering are illustrated in Fig. 14-3. An induced dipole is formed in a molecule or polyatomic ion when the electric field strength of the incident radiation is relatively large at either the peak or valley of the EMR wave. That is illustrated by the upward arrow on the left in Fig. 14-3 and corresponds to excitation of the molecule. As the wave passes the particle, the electric field strength decreases, the induced dipole collapses, and radiation is emitted, as indicated by the downward arrow.

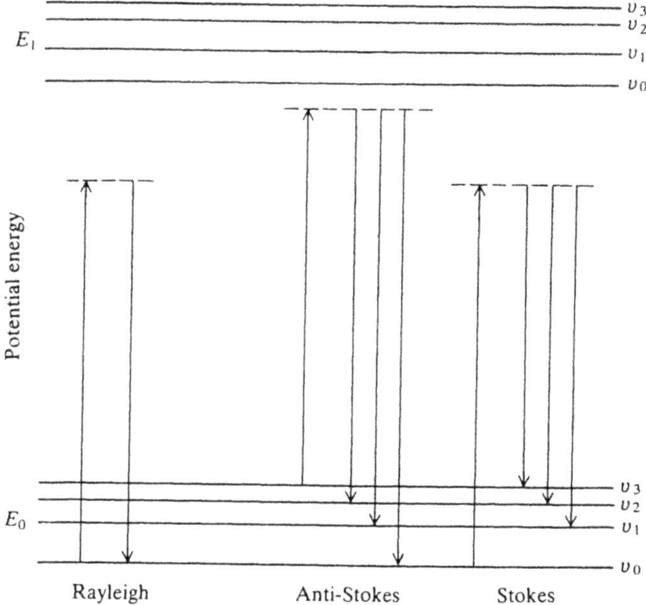

Figure 14-3 A diagram showing the energetic transitions that occur during Rayleigh and Raman scattering. E_0 and E_1 are the ground and first excited electron levels; υ_0, υ_1, υ_2, and υ_3 are vibrational levels.

Normally the molecule or polyatomic ion is initially in the lowest vibrational level of the ground electron state. The energy from each photon of the incident radiation is insufficient to cause excitation to a higher electron level as during absorption of ultraviolet-visible radiation. Because the molecule is not elevated to a higher electron level, the emission does not occur during an electron transition from an elevated electron level as in fluorescence. Rayleigh scattering should not be confused with fluorescence. The scattered radiation is emitted as the particle returns to its original energetic state. Reexcitation and scattering occurs as the next wave passes.

RAMAN SCATTERING

Raman scattering occurs by nearly the same mechanism as Rayleigh scattering except that the frequency of the scattered radiation is not identical to that of the incident radiation. The energetic changes that are associated with Raman scattering are illustrated in Fig. 14-3. Scattered radiation that has a frequency other than that of the incident radiation can be caused by either of two mechanisms. When the scattering species is initially in a vibrational level that is above the ground vibrational level, scattering can occur by return of the species to a vibrational level that is energetically lower than the initial level. The resulting, *anti-Stokes*, Raman scattered radiation has a frequency that is higher than that of the incident radiation. A major limitation of the use of Raman scattering for analysis results from the relatively low intensity of emitted radiation in relation to the incident radiation. It is typical for the ratio of the intensity of Raman scattered radiation to incident radiation to be 10^{-18} or even smaller.

If after the initial excitation the particle scatters radiation while returning to a vibrational level that is energetically above the initial level, the scattered radiation is of a lower frequency than that of the incident radiation. Raman scattering in which the scattered radiation is less energetic than the incident radiation is *Stokes scattering*. Because polyatomic species at room temperature are more likely to be found in the lowest vibrational level than in higher levels, excitation from the ground vibrational level is more likely than that from higher vibrational levels. Consequently, relaxation to a higher vibrational level after excitation is more probable than relaxation to a lower level, and Stokes scattering occurs more often than anti-Stokes scattering. Because the emitted intensity owing to Stokes scattering is generally significantly greater than that owing to anti-Stokes scattering, Stokes scattering is more often used for analysis.

The intensity of radiation that is emitted during Rayleigh scattering is typically between four and five orders of magnitude greater than that emitted during Raman scattering. The change in frequency of the scattered radiation when compared to the frequency of the incident radiation during Raman scattering is important for analysis. Because the change in frequency corresponds to the energetic difference between vibrational levels in the molecule, the technique yields information that is comparable to but not identical with that obtained from infrared absorption.

Raman scattering is primarily used for qualitative analysis. A Raman spectrum is recorded by plotting the intensity of the scattered radiation either as a function of the frequency (or wavelength or wavenumber) of the radiation or as a function of the change in frequency Δv (or wavelength or wavenumber) between the incident radiation v_i and the scattered radiation v:

$$\Delta v = v_i - v \tag{14-6}$$

Normally the latter procedure is used because it is easier for most analysts, who are accustomed to seeing infrared spectra, to interpret. The change in frequency of the radiation corresponds to the difference in energy between the vibrational levels of the scattering polyatomic species. As in infrared spectrophotometry, the difference is characteristic of particular functional groups. Consequently, Raman spectra are particularly useful for qualitative analysis. The scattered bands are *Stokes lines* if Δv is positive and *anti-Stokes lines* if Δv is negative.

Recorded values of Δv are independent of the frequency of the incident radiation. It should be remembered, however, that the intensity of the scattered radiation [Eq. (14-5)] theoretically is proportional to the fourth power of the frequency v of the scattered radiation which, in turn, is related to the frequency of the incident radiation by Eq. (14-6). The intensity of the scattered radiation can be increased by using incident radiation that has a relatively high frequency.

Resonant Raman Spectrometry

In addition to the classical forms of Raman scattering that have been described, several variations are used. Of particular importance for chemical analysis are *resonant Raman spectrometry* (RRS) and *coherent anti-Stokes Raman spectrometry* (CARS). Resonant Raman scattering occurs when the frequency of the incident radiation is that which is necessary to cause excitation of the absorbing species to an electron state above the ground state. In that case some of the Raman bands exhibit a significantly increased intensity. In most cases only the scattered frequencies that are associated with totally symmetric vibrations within the polyatomic scatterer and those vibrational modes that are associated with bonds that are affected by the change in electron state are enhanced. The enhancement can be as large as several orders of magnitude. The increased intensity permits assays to be performed on samples at much lower concentration than those used for classical Raman spectrometry. A diagram that illustrates the energetic transitions that occur during resonant Raman spectrometry is shown in Fig. 14-4. The theory for RRS is described in Reference 2.

The extent of observed enhancement in resonant Raman spectrometry is dependent upon the frequency of the incident radiation, the response of the detector, and the amount of self-absorption. The frequency of the incident radiation affects the intensity of the scattered radiation in two ways. The scattered intensity theoretically varies with the fourth power of the scattered frequency v as during classical Raman spectrometry. Additionally the intensity of the scattered radiation is dependent upon the vibrational modes that are associated with chemical bonds that are affected by the change in electron state during excitation. By altering the frequency, the enhancement is altered in much the same way as that in which fluorescent intensity is altered by scanning the excitative frequency. Owing to the variation in enhancement with incident frequency, spectrometers that are designed to be used for RRS often use a variable-frequency source rather than a fixed-frequency source like those often used for classical Raman spectrometry.

As in ultraviolet-visible spectrophotometry, the ability of the detector to respond to radiation of a fixed intensity varies with the frequency of the radiation. Because the frequency of the scattered radiation is related to the frequency of the incident radiation, it again is advantageous to have control of the frequency of the incident radiation. Self-absorption can occur when the frequency of the scattered radiation overlaps an absorptive band of the scatterer. Self-absorption decreases the measured intensity of the scattered radiation. Self-absorption can become significant at higher concentrations of the scatterer.

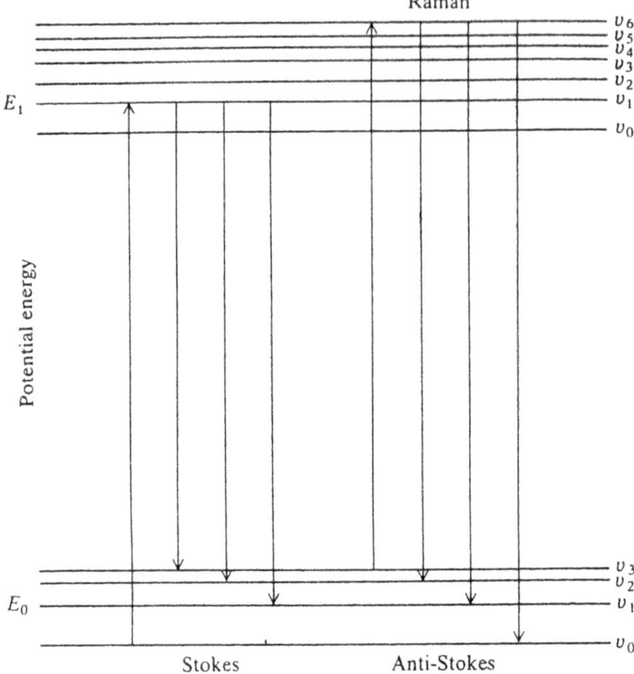

Figure 14-4 A diagram illustrating the energetic transitions that occur during resonant Raman scattering. E_0 and E_1 are the ground and first excited electron levels; υ_0 through υ_6 are vibrational levels.

If the scatterer also fluoresces, it is difficult to measure the relatively low intensity scattered radiation in the presence of the much more intense fluorescence that occurs at the same frequency. Consequently, resonant Raman spectrometry is not useful for compounds that also fluoresce. That is probably the greatest disadvantage of resonant Raman spectrometry. Its major advantage in comparison to classical Raman spectrometry is its increased sensitivity.

Nonlinear Raman Spectrometry

Rayleigh and Raman scattering depend upon the ability of the scattering particle to be polarized by the electric field that is associated with the incident radiation. The polarization causes an induced dipole moment μ. The induced dipole moment can be expressed as an expansion of the electric field strength E, as illustrated in the following equation:

$$\mu = aE + bE^2 + cE^3 + \cdots \qquad (14\text{-}7)$$

Because a is normally greater than b which normally is greater than c, etc., only the first term of the expansion is important for instruments that use low-intensity radiative sources. In those cases Eq. (14-7) simplifies to Eq. (14-4). Techniques in which the induced dipole moment (and consequently the amount of scattered radiation) is proportional to the electric field strength of the incident radiation are *linear techniques*.

With intense sources, such as lasers, the second- and higher-order terms in the expansion shown in Eq. (14-7) become important and significantly contribute to the observed scattering. If the second-order term in Eq. (14-7) is significant, trigonometry can be used[3] to show that the frequency v of the incident radiation can be doubled to yield radiation that has a new frequency $2v$. Doubling the frequency requires a relatively large value of b [Eq. (14-7)]. The value of b can be partially controlled by selecting the medium in which the doubling occurs. The newly created frequency can be used for any desired spectroscopic need including scattering inducement. A technique that relies upon one or more of the higher-order terms in Eq. (14-7) is a *nonlinear technique*.

If two lasers which emit frequencies v_1 and v_2 are simultaneously focused in a medium, in which b is large, they can combine to yield radiation with frequencies $v_1 + v_2$ and $v_1 - v_2$. Spectroscopic techniques that have taken advantage of these frequencies have been described in earlier chapters. If two laser beams are combined in an appropriate medium and the third-order term in Eq. (14-7) is significant, radiation with frequencies of $2v_1 - v_2$ and $2v_1 + v_2$ is formed.

Relatively low intensity radiation is scattered when using linear techniques. The ratio of the number of scattered photons to the number of incident photons is typically 10^{-8} or less. Because the efficiency of luminescence is much greater than that of classical and resonant Raman scattering, luminescence from either the scattering particle or some other component can overwhelm the relatively weak Raman signal. That can occur even when the fluorescent compound is present in solution at a much lower concentration than the scattering substance. Several nonlinear Raman techniques are available which are more sensitive and less prone to interference from fluorescence.

The major nonlinear techniques are *stimulated Raman scattering, hyper Raman, stimulated Raman gain* (SRG), *inverse Raman scattering* (IRS), *coherent anti-Stokes Raman spectrometry* (CARS), and *coherent Stokes Raman spectrometry* (CSRS). Of the nonlinear techniques, stimulated Raman scattering and hyper Raman have not proven to be particularly advantageous for analysis and will not be described. A description of nonlinear Raman spectrometry is available in References 4 and 5.

Coherent anti-Stokes Raman spectrometry The most popular of the nonlinear Raman methods is CARS. The energetic transitions that occur during coherent anti-Stokes Raman scattering are illustrated in Fig. 14-5. The technique requires the use of two high-powered lasers. Normally visible radiation is used. The beams from the two lasers are simultaneously focused on the sample so that the beams intersect. One of the lasers is tunable. The tunable laser is a *Stokes* *shifted probe* that emits radiation at a frequency ν_2. Typically a tunable dye laser serves as the probe. The second laser is used to *pump* the system at a frequency ν_1. The intensity of scattered anti-Stokes radiation at frequency ν_a is monitored while the frequency of the probe laser is scanned. Usually the intensity is displayed on the readout device as a function of the difference between the frequency of the pump laser and the frequency of the probe laser $\nu_1 - \nu_2$, although the intensity also can be plotted directly as a function of the frequency of the probe laser.

The two laser beams combine in the sample to form a *coherent* beam of third-order radiation at a frequency ν_3 ($\nu_3 = 2\nu_1 - \nu_2$). The combined frequency ν_3 induces polarization in the sample that corresponds to the upper energetic level shown in Fig. 14-5. Because ν_2 is continuously varied, the frequency ν_3 continuously changes. At most frequencies the intensity of the generated radiation at frequency ν_3 is low. When the difference in frequency between the pump beam and the probe beam $\nu_1 - \nu_2$ corresponds to the energetic difference between two of the vibrational levels of the scatterer, the intensity of the generated beam corresponding to the frequency ν_3 is greatly enhanced, causing excitation as illustrated in Fig. 14-5. The anti-Stokes scattered radiation of frequency ν_a is monitored. The scattered radiation at frequency ν_a is only intense when the difference between ν_1 and ν_2 corresponds to the energetic difference between the ground and a higher vibrational level of the ground electron state. The scattered radiation is termed anti-Stokes because it has greater energy than the incident radiation from the pump laser.

Unlike classical Raman spectrometry, CARS yields scattered radiation that is coherent as in a laser. CARS is typically between five and six orders of magnitude more sensitive than classical Raman spectrometry. Because the emitted radiation is coherent, it can be collected and efficiently separated from the incident radiation. CARS is a good technique to be used for samples that fluoresce because the scattered radiation occurs at a shorter wavelength

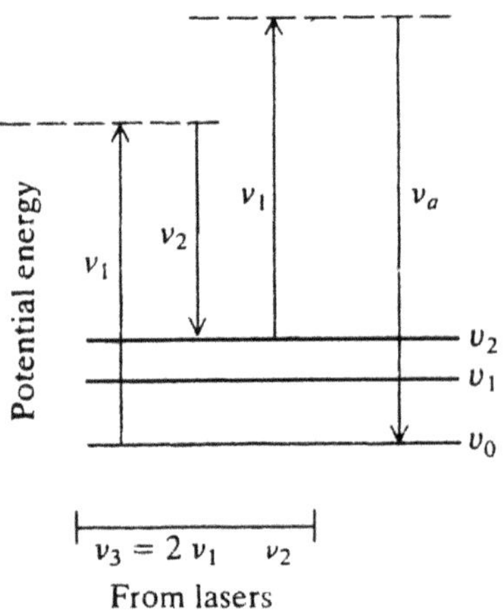

Figure 14-5 Energetic transitions that occur during coherent anti-Stokes Raman spectrometry. Frequencies ν_1 and ν_2 correspond to those of the pump and probe lasers respectively. The anti-Stokes radiation at frequency ν_a is monitored. υ_0, υ_1, and υ_2 are vibrational levels.

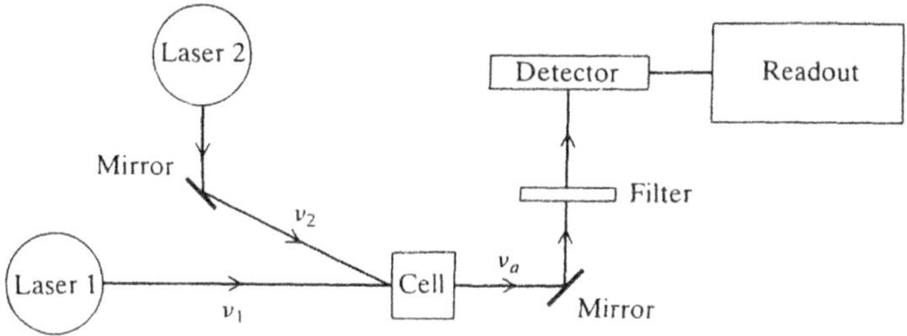

Figure 14-6 A diagram of the apparatus that is used for coherent anti-Stokes Raman spectrometry.

(higher energy) than that at which fluorescence can occur. That is a major advantage of CARS. Another advantage of CARS is the lack of a need for a monochromator in the detection system. The scattered radiation is plotted as a function of the frequency of the probe laser. Because the emitted frequency from the probe laser has a narrow bandwidth, the resolution with CARS is better than that of classical Raman spectrometry in which the limiting factor is the resolution of the monochromator that is between the sample and the detector.

The major disadvantage of CARS and most other nonlinear techniques is the presence of a weak background signal at all frequencies of the probe laser. Without some sort of corrective system, the lowest concentration in solution that can easily be measured with CARS is about 1 percent. Methods are available that allow the background signal to be decreased to a value that allows determinations at concentrations that are two or three orders of magnitude lower. A filter before the detector is often used to eliminate interfering radiation at wavelengths well removed from that of the scattered radiation. A diagram of the apparatus that is used for CARS is shown in Fig. 14-6.

A relatively recent use of CARS is found in microscopic imaging of biological and medicinal materials[6]. It is used to obtain vibrational spectra and images at a microscopic level without the need for molecular labels, such as those which are often required while making fluorescence measurements. It can be used to visualize biologic materials by observation of the contrast between resonant scattering of the stretching band associated with phosphate, OH, CH et al., and the corresponding nonresonant scattering. The result is observed by a computer through the optics of a microscope and displayed on the computer screen. Two and three dimensional images of the sample can be displayed. The technique is considerably more sensitive than Raman microscopy. It has all of the advantages of non-microscopic CARS. When near ir pump lasers are used, the technique can acquire images or spectra at depths approaching 0.4 mm into a sample.

Coherent Stokes Raman spectrometry CSRS (pronounced "scissors") is identical to CARS in all respects except that Stokes scattered radiation rather than anti-Stokes radiation is monitored. The frequency ν_2 of the probe laser is greater than that of the pump laser. The energetic transitions that occur during CSRS are illustrated in Fig. 14-7. The scattered radiation occurs at a frequency that is less than that which is emitted by the pump laser. CSRS is not as popular as CARS primarily because the scattered radiation occurs at a frequency (less than the frequency of the incident radiation) at which fluorescence from the sample can interfere.

Stimulated Raman gain and inverse Raman scattering Stimulated Raman gain and inverse Raman scattering (also called *stimulated Raman loss*) use radiation from two lasers as in CARS and CSRS. Both lasers are focused on the sample and the intensity of radiation that is used in either the pump or probe laser is monitored. As in CARS, the frequency of the pump laser is maintained constant and the frequency of the probe laser is varied. The intensity of the monitored frequency is plotted as a function of the frequency of the probe laser or as a function of the difference in frequency between the probe laser and the pump laser. When the difference in frequency between the two lasers corresponds to the energy of a vibrational level change within the sample, a change in the monitored intensity is observed.

While using stimulated Raman gain the frequency of the pump laser is greater than the frequency of the probe laser. When the difference in frequency between the two laser beams is identical to the difference between two

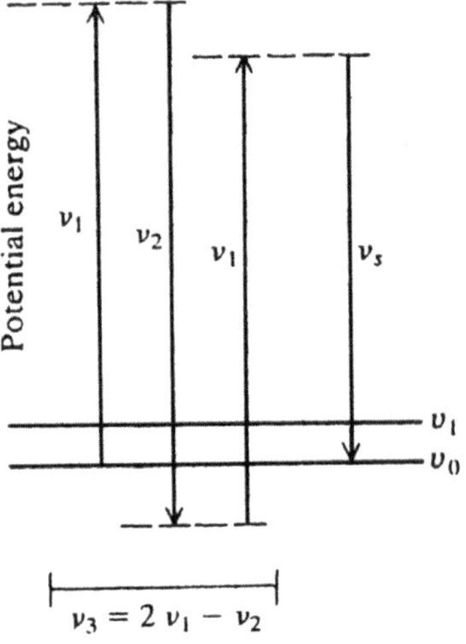

Figure 14-7 Energetic transitions that occur during coherent Stokes Raman spectrometry. Frequencies ν_1 and ν_2 correspond to those of the pump and probe lasers respectively. The Stokes scattered radiation at frequency ν_s is monitored. υ_0 and υ_1 are vibrational levels.

vibrational levels in the sample, the observed intensity at the frequency of the pump laser decreases and the intensity at the frequency of the probe laser increases. While using inverse Raman spectrometry the frequency of the probe laser is greater than that of the pump laser. When the difference in frequency between the two lasers corresponds to a vibrational level difference within the molecule, the intensity at the frequency of the probe laser decreases and the intensity at the frequency of the pump laser increases. In both cases the Raman output is coherent and directional as in CARS. Stimulated Raman gain and inverse Raman scattering have many of the same advantages that CARS and CSRS have. Nevertheless CARS remains the most popular of the nonlinear techniques for chemical analysis.

Apparatus

A block diagram of the apparatus that is used for classical Raman spectrometry is shown in Fig. 14-8. Prior to the development of useful lasers, high-intensity, low-pressure mercury-arc lamps were used as sources for Raman spectrometry. Typically the 435.8-nm emission line from the lamp was isolated with a filter and used as the incident radiation. Most modern instruments use a laser as the source. All nonlinear techniques, such as CARS, require the use of two lasers or of one laser from which two frequencies are obtained by frequency doubling or some other technique. Classical Raman spectrometry can be performed using a laser that emits a single frequency in the visible region. The helium-neon laser

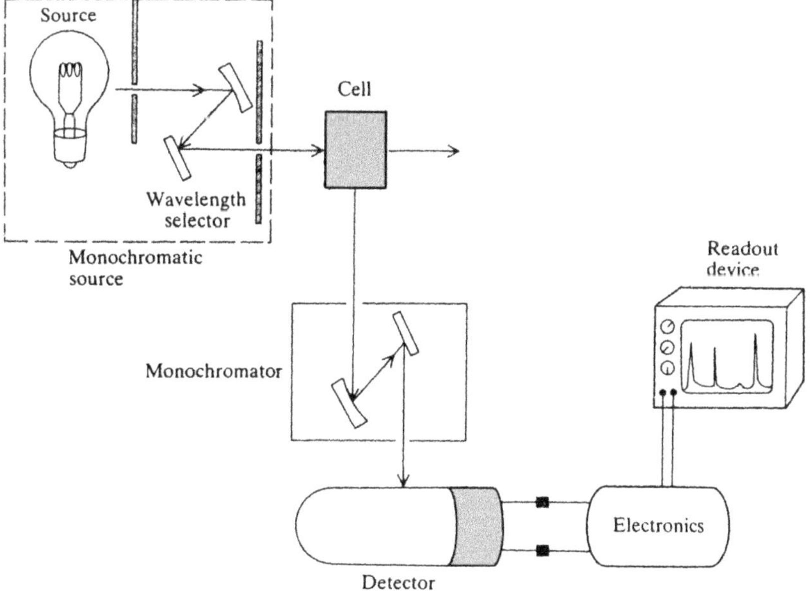

Figure 14-8 A block diagram of the apparatus that is used for classical Raman spectrometry. The dashed line is used to indicate that either a polychromatic source with a wavelength selector or a monochromatic source can be used.

which emits radiation at 632.8 nm has been a popular choice. Resonant Raman spectrometry is often done with a tunable dye laser. Of course a dye laser can also be used for classical Raman spectrometry. Both mirrors and lenses are used to direct and focus incident radiation on the sample.

The cell material must be transparent to the incident radiation and to the scattered radiation. Because the incident and scattered radiation often occur in the visible region, the cell can be constructed from glass, quartz, or silica. Often a glass capillary tube is used as a cell for liquid samples.

Wavelength selection between the cell and the detector is accomplished by using a monochromator. Both prism and high-quality diffraction grating monochromators are used. Double monochromators can be used to minimize interferences.

Normally the detector is a photomultiplier tube. Photubes are not sufficiently sensitive for scattering measurements. It is possible to use some of the other detectors that were described for use in the ultraviolet-visible region, but photomultiplier tubes are used most often. Some modern instruments use a liquid nitrogen cooled, charge-coupled device CCD as detector. The readout device and data acquisition system is often computer-controlled.

Qualitative Analysis

Raman spectrometry is most often used for qualitative analysis in a manner that is analogous to the use of infrared spectrophotometry. The recorded spectrum of the sample is compared with spectra of known compounds or solutions. If a match between the sample spectrum and a spectrum of a known substance cannot be found, it is sometimes possible to deduce the identity of the sample from correlation charts of particular group frequencies. Because the procedure is identical to that used for infrared spectrophotometry (Chapter 12), it is not described in detail.

As in infrared spectrophotometry, conclusions based solely on the use of correlation charts often are incorrect. All available analytical information should be used while identifying a sample. An analyst should never rely solely on the use of correlation charts. In large part because Raman spectrometry is not as popular as infrared spectrophotometry, the correlation charts that are used for Raman spectrometry are often not as thorough as those used for infrared spectrophotometry. Several of the major Raman scattering bands and the corresponding organic functional groups that cause the bands are listed in Table 14-1. A listing of Raman bands of inorganic compounds can be found in Reference 7.

The use of Raman spectrometry often complements the use of infrared spectrophotometry for qualitative analysis. Infrared absorption occurs during vibrational or rotational energy-level changes within the molecule. Raman scattering occurs after a dipole is induced in a molecule by the incident radiation. It is a function of the ability of the molecule to be polarized. Both techniques are indicative of vibrational levels within the sample, but owing to the different ways in which the levels are populated, the two techniques do not necessarily detect the same vibrational levels. It is possible to calculate the total number of possible vibrational modes for a molecule by the procedure described in Chapter 12 (see Sample problem 12-2). Not all vibrational levels are observed while using either Raman spectrometry or infrared spectrophotometry and often the levels that are detected by one method are not strongly detected by the other. A useful generalization is summarized by the *infrared-Raman exclusion rule* which states that for a molecule that has a *center of symmetry*, a vibrational level can be observed with infrared spectrophotometry, with Raman spectrometry, or with neither technique, but the level cannot be observed as an intense peak with both techniques. Consequently, each method provides information that might not be provided by the other technique.

Molecular Optical Laser Examiner

Raman scattering can be used to qualitatively examine the surface of samples with a *molecular optical laser examiner* (MOLE). Essentially the device is a laser microprobe that monitors Raman scattering. A laser is used as the source of incident radiation. The radiation is focused on the surface of the sample and the scattered radiation is observed by the instrument through a conventional microscope. The sample is placed on the stage of the microscope. The microscope is used to focus the incident radiation on the sample and to collect the scattered radiation. After passing through a double-grating monochromator, the scattered radiation is detected either with a photomultiplier tube or with a television-type camera. Generally the instrument is controlled with a computer. A diagram of a typical instrument is shown in Fig. 14-9.

If a photomultiplier tube is used, the Raman spectrum at a fixed location on the surface of the sample is displayed on a recorder or other appropriate readout device. If a television camera is used as the detector, a single scattered band is observed over a relatively broad region on the surface of the sample and the result is displayed on a television monitor or a computer monitor. The latter method is used to obtain a two-dimensional map of the location of a particular substance on the surface of the sample. The frequency chosen for the study is that which corresponds to an intense Raman band of the compound of interest.

The MOLE is a convenient nondestructive apparatus for the examination of polyatomic species on the surface of a sample. It differs from the laser microprobe described in an earlier chapter in that it responds to polyatomic species rather than to atoms. It has found use in archeology as a tool for examining defects in industrial materials such as glasses, for environmental analysis of dust particles, for examination of biological samples, for study of semiconductors and integrated circuits, and for many other purposes. A more detailed description of the method can be found in Reference 8.

Table 14-1 A list of Raman scattering bands for selected functional groups

The intensities of the bands are indicated by s for strong, m for medium, and w for weak

Functional group	Range of wavenumber shift, cm^{-1}	Comment
Acetylene	2300-2100 (s)	C≡C stretch
Alcohol	3650-3000 (0-w)	O—H stretch
Alkane	3000-2800 (s)	C—H stretch
	1475-1450 (s)	CH_3 antisym bend
	1350-1300 (m-s)	CH_2 bend
	400-230 (s)	C—C bend n = 3-12
Alkene	3100-3000 (m-s)	C—H stretch
	1900-1500 (m-s)	C=C stretch
	1450-1200 (s)	C—H in-plane
Amine	3550-3300 (w-m)	N—H stretch
	1380-1070 (m-s)	C—N stretch
Aromatic	1620-1580 (m-s)	C=C stretch
	1045-1015 (m)	C—H in-plane
	1010-990 (s)	Mono-, meta-, and 1,3,5 derivative
	750-700 (s)	1,2, 1,3, 1,2,4 substituted
	655-645 (s)	1,2,3 substituted
	630-610 (s)	Monosubstituted
	570-550 (s)	1,3,5 substituted
Bromo	700-490 (s)	C—Br stretch
Carbonyl	1870-1650 (w-s)	C=O stretch
Chloro	850-550 (s)	C—Cl stretch
Disulfide	550-430 (s)	S—S stretch
Ester	1100-1025 (s)	C—O—C stretch
Ether	1140-800 (m-s)	C—O—C stretch
Nitrate	1285-1260 (s)	ONO stretch
Nitrile	2260-2220 (s)	C≡N stretch
Nitrite	1660-1620 (s)	N=O stretch
Nitro (—NO_2)	1590-1530 (w-m)	N=O stretch
	1380-1340 (s)	
	920-830 (s)	C—N stretch
	650-520 (m)	NO_2 bend
Peroxide	900-845 (w-s)	O—O stretch
Pyridine	1620-1560 (m)	
	1020-980 (s)	
Sulfide	790-570 (s)	C—S stretch
Thiocyanate (—S—C≡N)	650-600 (s)	S—C stretch
Thiol	2590-2560 (s)	S—H stretch
	850-820 (s)	S—H in-plane
	790-550 (s)	C—S stretch
	340-320 (s)	S—H out-of-plane

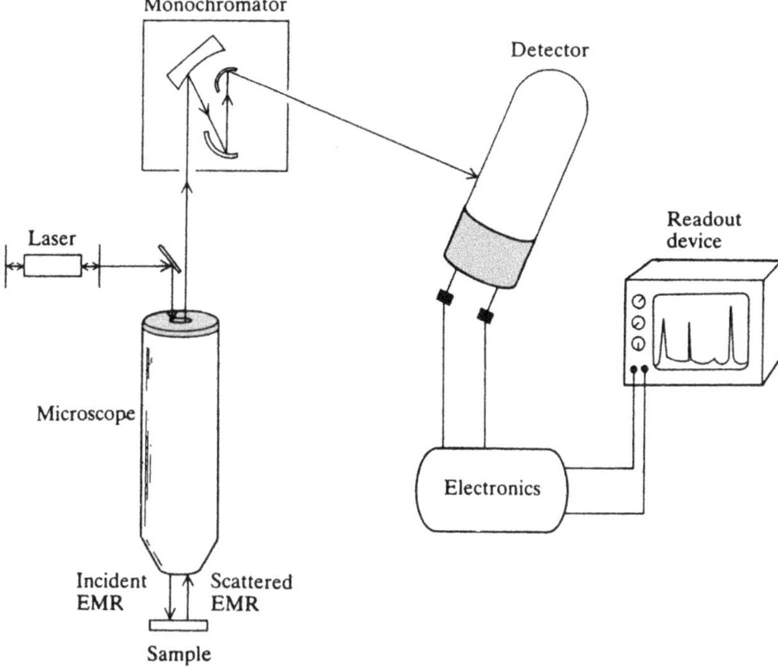

Figure 14-9 A diagram of a molecular optical laser examiner.

Surface-Enhanced Raman Scattering

An enhancement in scattered intensity of as much as six orders of magnitude relative to that observed while using classical Raman scattering can be observed if a thin layer of the sample is coated on a roughened silver, copper, or gold (the coinage metals) surface. The technique is *surface-enhanced Raman scattering* (SERS). Apparently enhancement does not occur on other surfaces. The technique was discovered in 1975.

The intensity of scattered radiation can be maximized by using nanoparticles of the coinage metal on the surface[9]. A nanoparticle is a particle with dimensions in the nanometer range. The metallic nanoparticles are deposited on a larger surface. When the particles are struck by incident radiation of a wavelength that is at least ten times longer than the dimensions of the nanoparticle, the electron cloud of the particle shifts position as the strength of the electromagnetic EM field from the wave changes while striking the nanoparticle. The result is a changing EM field within the nanoparticle which oscillates at the frequency of the incident wave. The EM field from the nanoparticle strengthens the EM field from the wave. Since the intensity of Raman scattering from analyte molecules on the surface is dependent upon the strength of the EM field from the exciting radiation, and since that field strength is heightened in the presence of nanoparticles, the intensity of the scattered radiation is significantly enhanced. At the time of the preparation of the text, it was possible to obtain scattered radiation that has an intensity which is enhanced by a factor of 10^9 - 10^{10}. The mechanism by which the surface nanoparticles resonante as the EMR wave strength changes is *localized surface plasmon resonance* (LSPR).

Quantitative Analysis

In most cases the intensity of Raman scattered radiation at a fixed frequency is directly proportional to the concentration of the scatterer. Quantitative analysis is done at a frequency that corresponds to an intense Raman peak of the sample. In nearly every case the working-curve method is used. The precautions described for quantitative analysis with infrared spectrophotometry (Chapter 12) should be exercised. Owing to the relatively narrow peaks that are observed for Raman spectrometry, care should be taken to avoid a situation that would cause the frequency of the scattered peak to shift.

IMPORTANT TERMS

Anti-Stokes scattering
Coherent anti-Stokes Raman spectrometry
Coherent Stokes Raman spectrometry
Digestion
Infrared-Raman exclusion rule
Inverse Raman scattering
Linear technique
Molecular optical laser examiner
Nanoparticles
Nephelometry
Nonlinear technique
Probe laser
Pump laser
Raman scattering
Rayleigh scattering
Resonant Raman spectrometry
Stimulated Raman gain
Stokes scattering
Surface-enhanced Raman scattering
Turbidance
Turbidimetry
Turbidity coefficient
Tyndall scattering

PROBLEMS

14-1 Determine the turbidance of a solution if the relative intensity of the incident radiation is 100.0 and the relative intensity of the transmitted radiation is 45.2.

14-2 Turbidimetry was used to examine a sample. The transmittance of the sample was 0.368; the cell pathlength was 1.00 cm; and the concentration was 112 mg/L. Determine the turbidity coefficient of the sample.

14-3 An excess of barium chloride was added to a series of standard sulfate solutions and to the analyte. The resulting barium sulfate suspensions were turbidimetrically assayed. Use the working-curve method with the tabulated results to obtain the concentration of the analyte.

Sulfate concentration, mg/L	Percent transmittance
15.0	83.8
26.0	69.2
38.7	49.0
55.4	28.3
72.6	16.1
Analyte	37.2

14-4 An excess of sodium chloride was added to a series of standard solutions containing silver(I) and to a sample solution. The resulting silver chloride solutions were nephelometrically determined. From the tabulated results determine the concentration of silver(I) in the sample.

Silver(I) concentration, mg/L	Relative intensity
1.00	16.0
3.00	37.5
5.00	51.5
7.00	63.8
9.00	79.7
Sample	56.8

14-5 Raman spectrometry was to be used for quantitative analysis of a sample. Determine the relative sensitivity of the analysis at 275 nm as compared to the sensitivity at 776 nm.

14-6 Why are Stokes Raman lines more intense than the anti-Stokes lines?

14-7 Why is CARS more generally useful than CSRS?

REFERENCES

1. Mie, G.: *Ann. Physik.*, **25**: 377(1908).
2. Strommen, D. P., and K. Nakamoto: *J. Chem. Educ.*, **54**: 474(1977).
3. Harvey, A. B.: *Anal. Chem.*, **50**: 905A(1978).
4. Borman, S. A.: *Anal. Chem.*, **54**: 1021A(1982).
5. Long, D. A.: *Raman Spectroscopy*, McGraw-Hill, New York, 1977, chap. 8.
6. Evans, C. L. and X. S. Xie in *Annual Review of Analytical Chemistry* (E. S. Yeung and R. N. Zare, eds.), vol. 1, Annual Reviews, Palo Alto, CA, 883(2008).
7. Ross, S. D.: *Inorganic Infrared and Raman Spectra*, McGraw-Hill, London, 1972.
8. Dhamelincourt, P., F. Wallart, M. Leclercq, A. T. N'Guyen, and D. O. Landon: *Anal. Chem.*, **51**: 414A(1979).
9. Stiles, P. L., J. A. Dieringer, N. C. Shah and R. P. Van Dayne in *Annual Review of Analytical Chemistry* (E. S. Yeung and R. N. Zare, eds.), vol. 1, Annual Reviews, Palo Alto, CA, 601(2008).

BIBLIOGRAPHY

Lambert, J. B., H. F. Shurvell, L. Verbit, R. G. Cooks, and G. H. Stout: *Organic Structural Analysis*, Macmillan, New York, 1976.

CHAPTER

FIFTEEN

REFRACTOMETRY

The refractive index n of a medium is defined as the ratio of the velocity c of electromagnetic radiation in a vacuum to the velocity υ of the radiation in the medium:

$$n = \frac{c}{v} \tag{5-3}$$

Because different substances have different refractive indexes, the refractive index of a substance can be used as a tool for the qualitative or quantitative analysis of the substance. Chapter 15 contains a description of the theory and use of refractometry.

Unfortunately the difference between refractive indexes of similar compounds is often small. Because most organic liquids have refractive indexes that are between 1.25 and 1.80, accurate measurements of the refractive indexes are required for analysis. Typically the refractive index must be accurately measured to at least the nearest 0.001. Some refractometers are capable of making measurements to six places to the right of the decimal.

Because the velocity of radiation is large, measurement of the radiative velocity in the sample and consequently of the refractive index cannot be directly accomplished in the sample. Refractive index measurements are more readily accomplished by measuring the change in direction (the *refraction*) of a beam of radiation as it enters the sample from a medium that has a different refractive index. The change in direction occurs in the manner that was described for the dispersion of polychromatic radiation in a prism (Chapter 5).

Although refractometry largely has been replaced as a method of organic qualitative analysis in recent years by other analytical techniques, e.g., infrared spectrophotometry, nuclear magnetic resonance spectrometry, etc., it remains a relatively rapid and useful method of analysis in specific situations. In combination with other information such as densities, boiling points, melting points, or other analytical data, refractive indexes are valuable. The ability to readily adapt refractive index measurements to industrial processes adds to the attractiveness of the technique.

THEORY

Measurements of refractive indexes rely upon the application of Snell's law:

$$\frac{\sin i}{\sin r} = \frac{v_1}{v_2} = \frac{n_2}{n_1} = n_{21} \tag{15-1}$$

Measurement of the angle of incidence (angle i as measured from the perpendicular to the surface separating the two media) in medium 1 and of the angle of refraction (angle r) in medium 2 enables calculation of the ratio of the sines of the refractive indexes in the two media. If the refractive index of one medium is known, the refractive index of the second medium (the sample) can be easily calculated. In Eq. (15-1) v_1 is the velocity of radiation in medium 1; v_2 is the velocity in medium 2; n_1 and n_2 are the refractive indexes; and n_{21} is the *relative refractive index* of medium 2 with respect to medium 1.

When monochromatic radiation passes from a medium that has one refractive index into a medium that has a different refractive index, the radiation is refracted at an angle that can be predicted with Eq. (15-1). If the radiation passes from an *optically dense* (relatively large refractive index) medium to an *optically thin* (relatively small refractive index) medium, the radiation is refracted toward the surface separating the two media. Conversely, if the radiation passes from an optically thin medium to an optically dense medium, the radiation is refracted toward the perpendicular from the surface, as shown in Fig. 15-1.

From Fig. 15-1 and Eq. (15-1) it is apparent that as the angle of incidence increases while radiation passes from an optically dense medium into a less dense medium, the angle of refraction approaches 90°. The angle in the more dense medium (the incident angle in this case) that corresponds to an angle in the less dense medium (the angle of refraction in this case) that is 90° is the *critical angle* θ_c. Most refractometers rely upon a measurement of the critical angle to calculate the refractive index of the sample. By substituting 90° for r and θ_c for i in Eq. (15-1), the following equation is obtained:

$$\sin \theta_c = \frac{n_2}{n_1} = n_{21} \tag{15-2}$$

Radiation that strikes the surface at an incident angle that is greater than the critical angle is reflected from the surface rather than refracted.

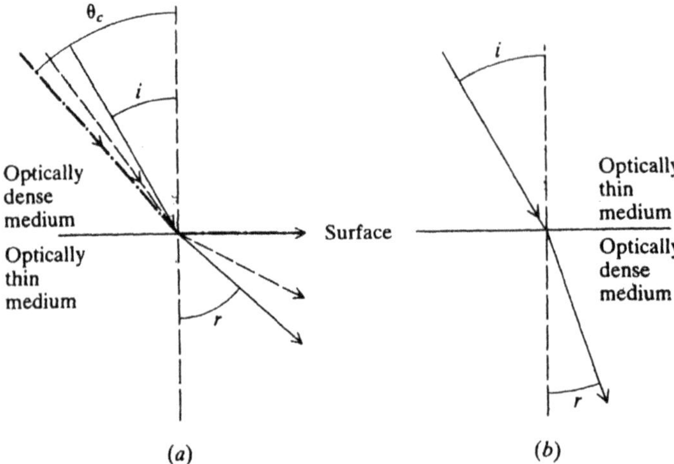

Figure 15-1 A diagram illustrating refraction as electromagnetic radiation passes between an optically dense medium that has a refractive index of 1.5 and an optically thin medium that has a refractive index of 1. The arrows indicate the direction of travel of the radiation. The vertical dashed line is perpendicular to the surface. *i*, angle of incidence; *r*, angle of refraction, θ_c, critical angle. (*a*) Radiative travel from the optically dense medium to the optically thin medium; (*b*) radiative travel from the optically thin medium to the optically dense medium.

If radiation first passes through the optically thin medium and then into the optically dense medium, the reverse process takes place. As the angle of incidence increases, the angle of refraction also increases, but the angle of incidence is always greater than the angle of refraction. When the angle of incidence becomes 90°, that is, when the incident radiation strikes the interface at a glancing angle, the radiation enters the more dense medium at the critical angle. The path that is followed by EMR at the critical angle is illustrated in Fig. 15-2.

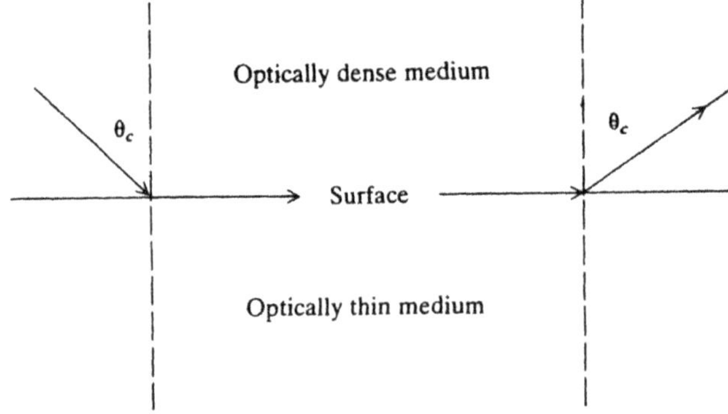

Figure 15-2 A sketch illustrating refraction at the critical angle θ_c. The dashed line is the perpendicular to the surface. The arrows indicate the direction in which electromagnetic radiation travels.

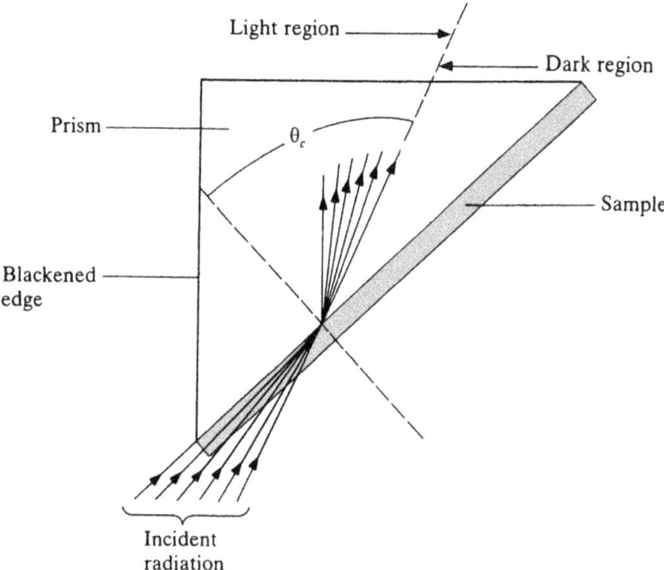

Figure 15-3 A sketch illustrating the radiative path during a measurement of refractive index.

Most of the methods for measuring the refractive index of a substance rely upon measuring the critical angle as radiation passes through the sample into an optically more dense material such as glass. The incident light from a source typically does not enter the sample and strike the surface at a single angle. Incident radiation passes through the sample and strikes the surface, separating the sample and the optically dense prism at a multitude of incident angles including 90°. The radiation refracts while passing into the optically dense prism. Because incident radiation strikes the prismatic surface at a multitude of angles, the radiation in the prism is refracted at many angles. In no case, however, is it possible for the radiation to be refracted at an angle, relative to the perpendicular, that is greater than the critical angle. Consequently, a sharp boundary exists between a light and dark region at the critical angle. The critical angle is obtained by measuring the angle of refraction at which the boundary between light and dark occurs. The radiative path during a typical measurement is shown in Fig. 15-3.

The angle at which the boundary between the light and dark regions occurs is measured with the aid of a magnifying lens assembly and a scale. The refractive index can either be directly measured on a scale within the instrument or calculated from Eq. (15-2) or a variation of Eq. (15-2). The refractive index of the prism is known. Alternatively, monochromatic radiation from the source can first pass through the prism and then impinge upon the analyte. In those instances it is more common for the boundary between the light and dark regions to be measured for the reflected radiation which passes back into the prism.

FACTORS AFFECTING THE REFRACTIVE INDEX

The refractive index of a medium is a function of the wavelength of the radiation, of the temperature, and of the concentration of the components of a mixture. The refractive index of a compressible substance is also a function of the pressure exerted upon the substance. The refractive index in a particular medium generally increases with decreasing wavelength throughout the near-infrared, visible, and near-ultraviolet regions, as illustrated in Fig. 15-4 for calcite (calcium carbonate). The change in refractive index with wavelength for a particular medium is not linear throughout the entire spectral region. Many media show a variation in refractive index with wavelength that is similar to that shown in Fig. 15-4. If the substance absorbs radiation, the plot of refractive index as a function of wavelength generally shows a discontinuity as the wavelength approaches an absorptive maximum. Typically the plots have the appearance of second-derivative plots in the vicinity of an absorptive maximum.

Owing to the variation in refractive index with wavelength, it is necessary to specify the wavelength at which measurements of refractive index are made. Most measurements are made using the two nearly equal-intensity spectral lines that are emitted by sodium vapor at 589.0 and 589.6 nm. The doublet is termed the *sodium D lines* or the sodium D line because the two lines are not resolved with some wavelength selectors. Instrument manufacturers sometimes report the measurement as being made at 589.3 nm, which is the average wavelength for the sodium D lines. When reporting a refractive index, the wavelength of the measurement is indicated as a subscript; e.g., measurements made using the sodium D lines are indicated as n_D. If a wavelength is not indicated, it is assumed that the measurements were made at the wavelength of the sodium D lines.

The change in refractive index with temperature (the *temperature coefficient* of the refractive index) varies with the medium. The expansion that occurs when a substance is heated results in a less dense medium that has a correspondingly lower refractive index. For most transparent and semitransparent liquids the change in refractive index with temperature dn/dT is between -4×10^{-4}/K and -6×10^{-4}/K. Notable exceptions to the generalization are carbon disulfide and water which respectively have refractive index temperature coefficients of -8×10^{-4}/K and -1×10^{-4}/K. For solids the variation in refractive index with temperature is generally appreciably smaller than that in liquids. Typical values vary from -10^{-6}/K to -10^{-4}/K. Most solids have a refractive index temperature coefficient of about -1×10^{-5}/K.

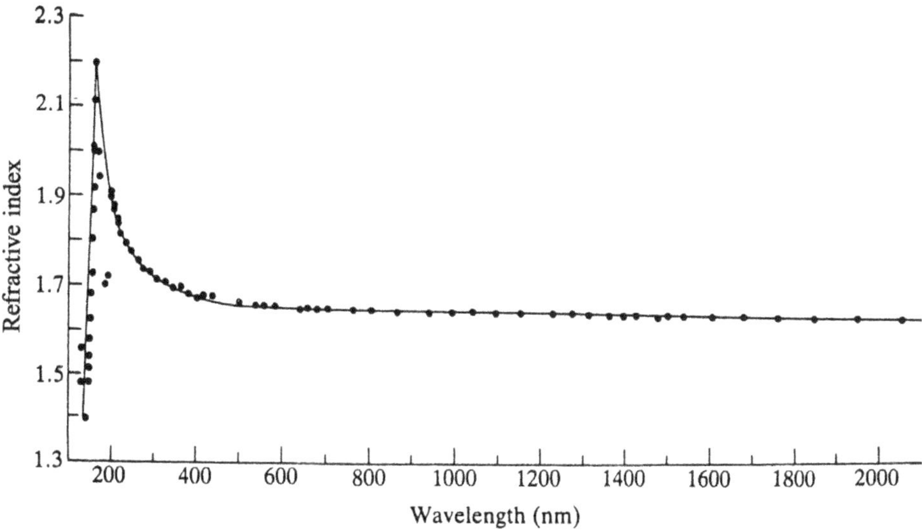

Figure 15-4 A plot showing the variation in refractive index with wavelength for calcite at 20°C.

In addition to density changes, temperature changes can also cause changes in the vibrational frequencies of a molecule. Generally ultraviolet absorptive bands move toward longer wavelengths as the temperature increases. The refractive index can be significantly altered near an absorptive band. The effect which temperature has upon density is the major cause for changes in refractive index with temperature. Because the refractive index varies with temperature, it is necessary to regulate the temperature of the sample and of temperature-sensitive portions of the apparatus. In order to obtain measurements that are accurate to the nearest 0.00001, it is necessary to control the temperature to within ± 0.02°C. Less accuracy, of course, requires less rigid control. The temperature at which a refractive index is measured is indicated as a superscript on the symbol for the refractive index; for example, n_D^{20} means the refractive index was measured at 20°C using the sodium D lines. Most measurements are made at either 20 or 25°C.

Variations in the refractive index with pressure are caused by the increased density of the medium as the pressure is increased. For liquids, dn/dP is usually about 3×10^{-5}/atm. For solids the effect is even less. For both solids and liquids the change in refractive index with pressure is negligible for pressure changes encountered in typical laboratory environments. For gases the effect is greater, as expected from the ideal gas law, and the pressure must be regulated during measurements of refractive indexes.

The refractive index of a solution is dependent upon the concentration of the solution components. Advantage can be taken of the variation for quantitative analysis. A description of the effect of concentration upon refractive index is included later in the chapter in the section that deals with quantitative analysis. More information on the theory of refraction and on the factors that affect it can be found in References 1 to 6.

APPARATUS

A variety of instruments are used to measure refraction. The more common laboratory instruments include the *Abbe*, *precision Abbe*, *immersion*, *Pulfrich*, and *differential refractometers*. Other instruments use interference patterns to make refractive measurements. The instruments can have either analog or digital displays. The readout devices described in the following sections are analog, but can easily be altered to provide digital readings with LCD or LED displays. Modern laboratory instruments often have the capability to be computer controlled and have computerized data acquisition.

Abbe Refractometer

Abbe refractometers are probably used more often for laboratory measurements of refractive indexes than any other type of instrument. The original version of the instrument was designed by Ernst Abbe in 1869. A diagram of one version of an Abbe refractometer is shown in Fig. 15-5.

Two prisms are mounted on the front of the instrument near the operator. The upper prism is hinged. At least 0.03 mL of a liquid sample is required for a measurement. The upper prism is rotated away from the lower prism and an eye dropper is used to place the sample on the exposed horizontal surface of the lower prism. After the upper prism is swung to the closed position and latched into place, the sample is sandwiched between the two prisms.

White light from a source, which is often directly attached to the instrument, is directed through the illuminating prism, the sample, and the refracting prism. If the sample is opaque, incident radiation enters through the refracting prism (rather than the illuminating prism), reflects from the sample, and repasses through the refracting prism. The temperature of the sample compartment is controlled by a flow of thermostated water through a water jacket that surrounds the prisms. Sometimes water flows through a hollowed portion of a prism.

After the radiation exits from the sample compartment, it passes through *Amici prisms* that allow only the sodium D lines to enter the field-of-view. The wavelength at which the measurement is made can be altered by rotating the Amici prisms. An Amici prism consists of three regions made from two different types of glass. Typically the two outer regions are composed of crown glass and the center region of flint glass. Radiation that passes through an Amici prism is highly dispersed. The only radiation that exits the Amici prisms and is focused in the field-of-view of the instrument corresponds to that of the sodium D lines. The Amici prisms compensate for the dispersed radiation in such a way that the results are the same as those that would have been obtained if a sodium-arc source had been used in place of the continuous source of visible radiation. The prisms are rotated until any colored fringe disappears and a sharp contrast occurs between the dark and light regions during measurement of the critical angle.

If a sodium arc lamp is used as the source, the Amici prisms are not needed. If the instrument contains Amici prisms and a sodium-arc lamp is used, the prisms are set for zero dispersion. An Amici prism can be rotated to either of two positions that are 180° apart and that provide the correct amount of compensating dispersion. For greatest accuracy several measurements are made at each position of the Amici prism and the results are averaged.

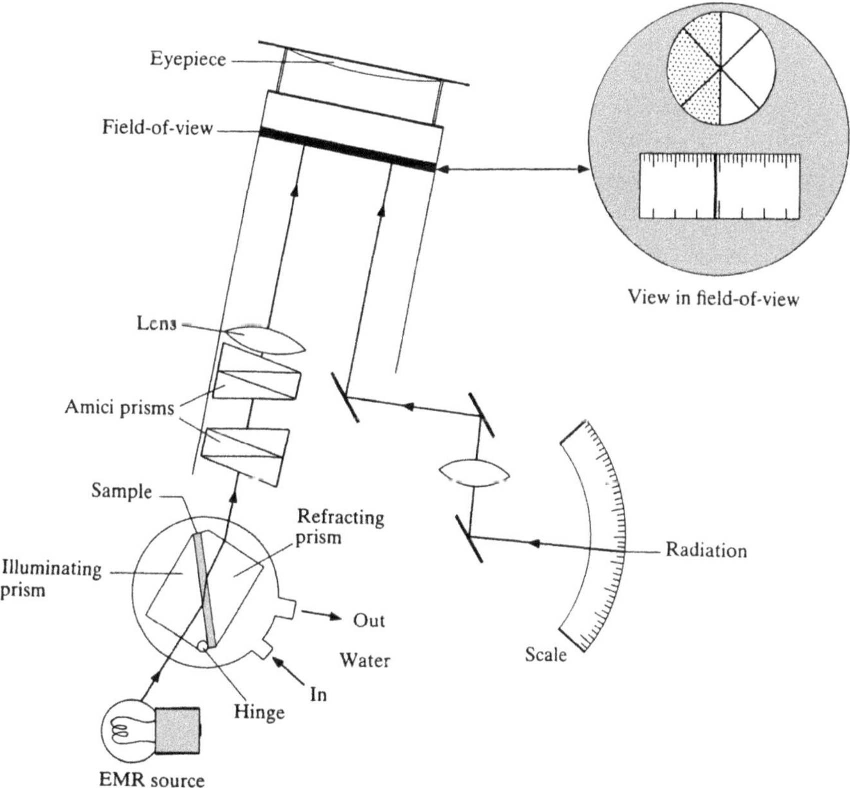

Figure 15-5 A diagram of an Abbe refractometer. The light path is indicated by the arrows. The view as seen through the eyepiece is shown on the right.

An eyepiece similar to that in a microscope is affixed to the top of the instrument. Often the lens in the eyepiece slightly magnifies (typically by a factor of 1.3) the image in the field-of-view. The refracting prism is rotated until the boundary between the light and dark regions, which occurs at the critical angle, appears in the field-of-view of the instrument. The boundary is aligned in the cross-hairs of the eyepiece. As the refractive prism is rotated, a calibrated scale which is coupled to the prism simultaneously revolves. Radiation passes through or reflects from the calibrated scale and is focused in the field-of-view of the instrument. The value on the illuminated scale is observed through the eyepiece. The refractive index of a substance is related to its critical angle as measured with an Abbe refractometer as follows:

$$n = \sin\theta_c \cos A + \sin A\left(n_p^2 - \sin^2\theta_c\right)^{1/2} \quad (15\text{-}3)$$

where A is the prism angle and n_p is the refractive index of the prism.

In some instruments the boundary and the scale reading can be viewed simultaneously through a single eyepiece, as illustrated in Fig. 15-5. In other instruments the boundary is aligned and then a switch must be thrown that illuminates the scale while removing the boundary from view. A few instruments provide a separate eyepiece for the boundary and the scale. Generally the scale is calibrated to directly provide the refractive index. In some instruments the scale is also calibrated in percent sucrose at 20°C. The Abbe refractometer can measure refractive index with an accuracy of ± 0.0001 and dissolved sucrose with an accuracy of ± 0.05 percent. The range of measurements with the device is relatively large. Typically the refractive index can be measured between 1.30 and 1.71 and dissolved sucrose can be measured between 0 and 85 percent. The temperature must be controlled within ± 0.2°C.

Precision Abbe Refractometer

A precision Abbe refractometer is similar in design to the Abbe refractometer. A line source is used with the instrument. Usually either the D lines from a sodium-arc lamp or the 435.8- or 546.1-nm line from a mercury lamp is used. No Amici compensating prisms are used, but the illuminating and refracting prisms are larger and more precisely manufactured than those in the Abbe refractometer. Because larger prisms and a monochromatic source is used, the accuracy of measurements that are made with a precision Abbe refractometer is ± 0.00002. Temperature control within ± 0.02°C is required.

Immersion Refractometer

Immersion or *dipping refractometers* are similar in construction to Abbe refractometers. The major differences are that a single prism is used for the measurement, water jackets are not used, and the boundary between the light and dark region in the field-of-view is superimposed on a fixed scale. A diagram of an immersion refractometer is shown in Fig. 15-6.

The prism portion of the device is dipped into the solution which is to be measured. The sample can be contained in a cup which attaches to the end of the refractometer or the prism can be dipped into the sample solution in any container that is sufficiently large. The device is particularly useful for making measurements in large vats used in industrial processes. Radiation must be directed toward the prism from below, as indicated in Fig. 15-6. That can be accomplished either by placing the radiative source below the prism or, more often, by using a mirror to direct radiation toward the prism. Normally white radiation is used.

An Amici prism and a lens are used as in the Abbe refractometer to compensate for the use of polychromatic incident radiation. Because a fixed scale is used, the range of the device with a single prism is more limited than the range of the Abbe refractometer. Typically refractive index measurements can be made throughout a range of about 0.04 with a single prism; e.g., the range for an L1 prism is from 1.325 to 1.366. Measurements are accurate to the fifth place to the right of the decimal point if the sample is well thermostated. In order to make measurements over a broader range, a series of prisms are supplied. The prism in the refractometer can easily be replaced because it slides into or out of the end of the device and snaps into or out of place. With a series of prisms, immersion refractometers can be used for refractive index measurements over the range from 1.325 to 1.647.

The refractive index is directly measured on the scale in the field-of-view of the instrument at the boundary between the light and dark regions, as illustrated in Fig. 15-6*b*. Often the immersion refractometer is used for measurement of the percent sucrose in a solution. The measuring ranges of the 10 prisms that accompany the Zeiss immersion refractometer are listed in Table 15-1.

Another popular refractometer which is similar to the immersion refractometer is the *hand-held Abbe refractometer*. It has a fixed scale and can only be used over a narrow range of refractive indexes (typically about 0.1 unit). The hand-held Abbe refractometer allows measurements to the nearest 0.001. Many hand-held refractometers have scales that are calibrated in the specific units required for a particular quantitative analysis; i.e., the devices are designed for a specific analysis rather than for general use. The sample is placed on the surface of the prism with an eye dropper or similar device and white light from an overhead incandescent lamp or the sun is used. The refractometer is sufficiently small to be able to be carried in a pocket. Instruments of this design are particularly popular for measurements that are made in the field, e.g., for making measurements of sugar content from fruits in an orchard or vineyard.

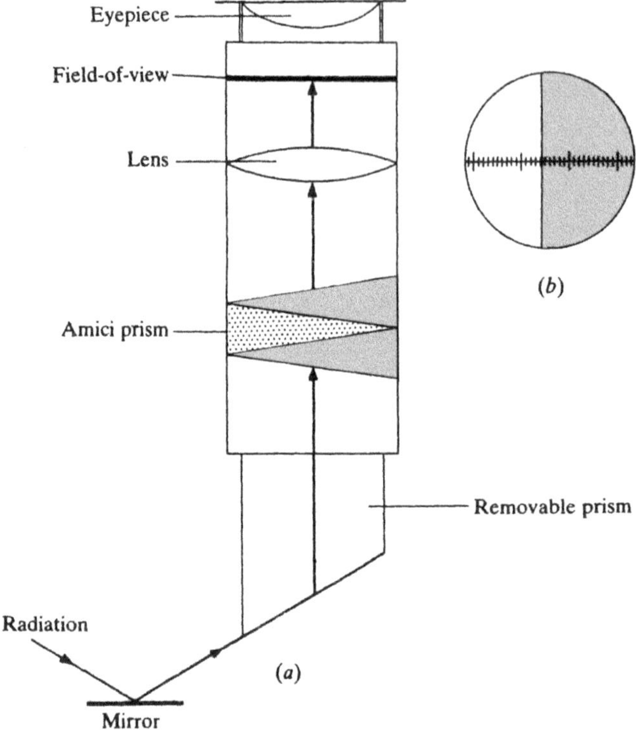

Figure 15-6 (*a*) Diagram of an immersion refractometer. The path of radiation through the device is indicated by the arrows. (*b*) View through the eyepiece during a measurement.

Table 15-1 The measuring range for several prisms, assuming the sodium D lines are used for the measurement

Prism type	Range, n_D
1 † or L1	1.3254-1.3664
L2	1.3642-1.3999
L3	1.3989-1.4360
L4	1.4350-1.4678
L5	1.4668-1.5021
L6	1.5011-1.5322
L7	1.5312-1.5631
L8	1.5621-1.5899
L9	1.5889-1.6205
L10	1.6195-1.6470

† The symbol 1 refers to a nontempered prism with the lowest range of refractive indexes. A prefix L refers to a tempered prism.

Pulfrich Refractometer

The Pulfrich refractometer uses radiation from a monochromatic source. Use of monochromatic EMR eliminates the need for dispersive devices such as Amici prisms. A square prism is used. The critical angle is viewed with an eyepiece that rotates on a scale that is calibrated to the nearest 0.5 degree of arc. The critical angle is measured on the scale when the boundary between the light and dark region is in the cross-hairs in the field-of-view. The measured critical angle is converted to an actual critical angle by subtraction of a zero-point correction. The actual critical angle is used to calculate the refractive index of the sample as follows:

$$n = \left(\mathrm{n}_p^2 - \sin^2 \theta_c\right)^{1/2} \tag{15-4}$$

where n_p is the refractive index of the prism. Eq. (15-4) is identical to Eq. (15-3) when the prism angle is 90°. Alternatively a table that is supplied by the manufacturer can be used to convert the measured critical angle to the refractive index.

A diagram of the important parts of a Pulfrich refractometer is shown in Fig. 15-7. With most Pulfrich refractometers refractive indexes that are accurate to ±0.0001 can be obtained. The devices can be used over a range that is at least as large as that of the Abbe refractometer. Typically between 2 and 4 mL of a liquid sample are required for an assay. Solid samples can be measured by placing the flat surface of the sample on the upper surface of the prism. A contact liquid, such as a light oil, which has a refractive index that is slightly greater than that of the sample is used.

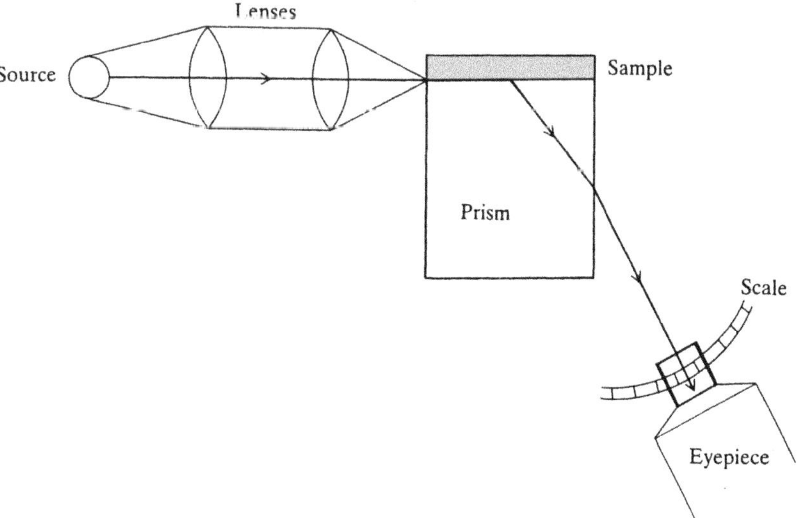

Figure 15-7 A diagram of a Pulfrich refractometer.

Differential Refractometers

Differential refractometers measure the difference in refraction between a sample and a reference. In most cases the radiation successively passes through the two media and the displacement of the refracted radiation is monitored. Of the several forms of differential refractometers, the *Debye refractometer* and the *Brice-Phoenix refractometer* are most popular. The major advantage of differential refractometers is their ability to measure refractive indexes with an accuracy of $\pm 3 \times 10^{-6}$. Temperature control of the sample is not as critical as might be expected for the obtainable accuracy because the difference in the temperature coefficient between two media is relatively small in comparison to the temperature coefficient of either medium. The temperatures of the two media must be within 0.01°C of each other.

The sample solution in the Debye refractometer is used as a filling solution inside a prism. The filled prism is dipped into a vessel that contains a reference substance (usually the solvent). Monochromatic radiation from the source passes through a slit, the solvent, and a prism containing the sample, as illustrated in Fig. 15-8. After having passed through the sample and solvent, the image of the slit is focused on a scale. First the prism and the vessel are both filled with solvent and the location of the slit image on the scale is recorded. Then the prism is filled with the sample solution and the new location of the slit image is noted. For small displacements d (the distance between the first and second readings) of the image, the refractive index n of the sample solution is related to that of the solvent n_s, the refractive angle A of the prism (typically 125 degrees), and the focal length f of the lens between the cell and the scale as follows:

$$d = 2f \tan\left(A\frac{n - n_s}{2}\right) \tag{15-5}$$

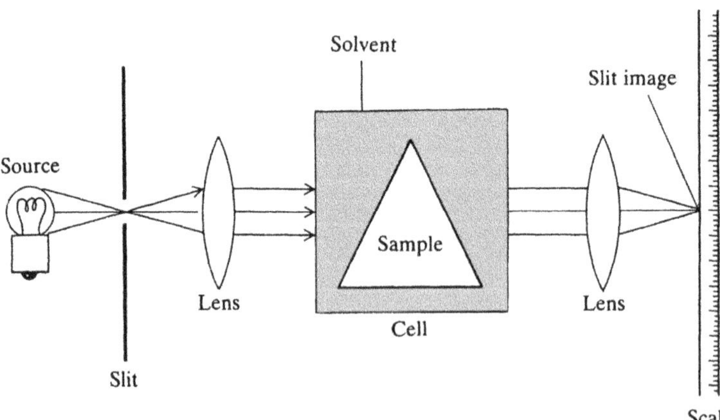

Figure 15-8 A diagram of a Debye differential refractometer.

The Brice-Phoenix refractometer is commercially available from the Phoenix Precision Instrument Company. The differential cell that is used in the device consists of two prisms. The prisms are respectively filled with a reference liquid and the sample (which is the reference liquid during the first measurement). Each prism holds 1 mL of liquid and is filled through the top. Radiation from a line source passes through a filter to isolate the desired spectral line and subsequently through a slit. A mercury lamp is normally used as the source. The instrument manufactured by the Phoenix Precision Instrument Company also provides for a second lamp which can be used in place of the mercury lamp.

After passing through the slit, the radiation passes through the cell which is held in a water-jacketed housing for temperature control. A lens is used to focus radiation from the cell onto the graduated field-of-view of a microscope. The location of the slit image can be measured to the nearest micrometer. As with the Debye instrument, the displacement of the image is related to the refractive index of the two liquids. For the Brice-Phoenix instrument the displacement is directly proportional to the difference in refractive index between the two liquids. Diagrams of the essential parts of the instrument and of the cell are shown in Fig. 15-9. Several automated versions of the refractometer have also been developed.

Double Beam Refractometers

A few double beam refractometers are in use. Typically a single source provides the radiation for the two beams. The radiation passes through the prism and is split into two beams at the interface of the prism and the analyte. One portion is reflected back through the prism and strikes a detector. That beam is the reference beam. The second portion strikes the sample solution. The reflected radiation at the critical angle passes back through the prism and strikes a second detector. The difference in reading from the two detectors is used to determine the refractive index or concentration of analyte.

Interferometric Refractometers

Some refractometers use an interferometer, similar to that described in Chapter 12, to make the measurement. Radiation from the source is divided into two portions. One portion of the radiation passes through the sample while the second portion passes through a reference medium. The distance the beams travel is nearly identical, but the beam which passes through the medium with the higher refractive index is slowed relative to the other beam. Upon recombining, the two beams are out-of-phase with each other owing to the different velocities of radiation in the two media. The combined beam is focused on a scale on which a series of constructive and destructive interference bands are observed.

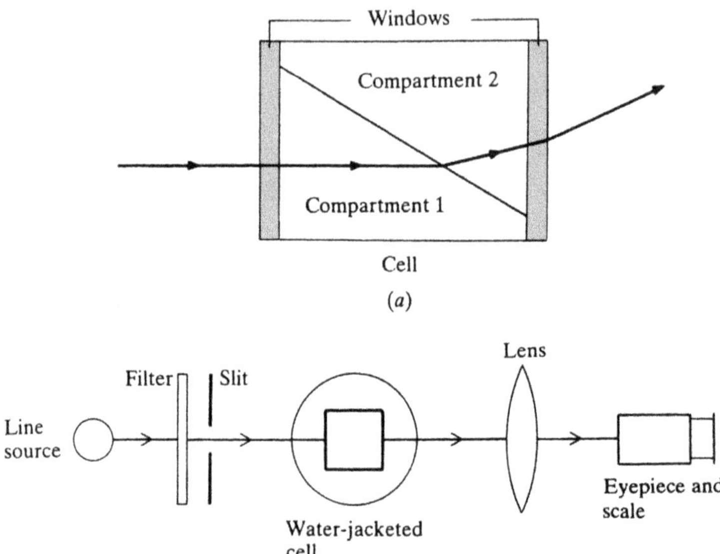

Figure 15-9 (*a*) Top-view sketch of the cell that is used in the Brice-Phoenix differential refractometer. (*b*) Sketch of the refractometer. The arrows indicate the radiative path.

The change in the location of the interference bands between that observed when both beams pass through the reference medium and that observed when one of the beams passes through the sample is related to the difference in refractive index between the reference and the sample. Typically the change in refractive index between the two media can be measured to an accuracy of $\pm 1 \times 10^{-7}$ when a monochromatic source is used. Even greater accuracy can be obtained with modified instruments. In any case the sample and reference must have refractive indexes that are close to each other. Because interferometric refractometers are differential instruments, temperature control is not as critical as that of nondifferential instruments. It is necessary, however, that the sample and the reference be at the same temperature.

Several types of interferometric refractometers are available. The Rayleigh-Haber-Löwe refractometer is shown in Fig. 15-10. Monochromatic radiation from the source passes through a slit and is converted to a parallel beam by the lens. A set of slits divides the beam into two portions which respectively pass through the cell that contains the reference medium of known refractive index and the sample cell.

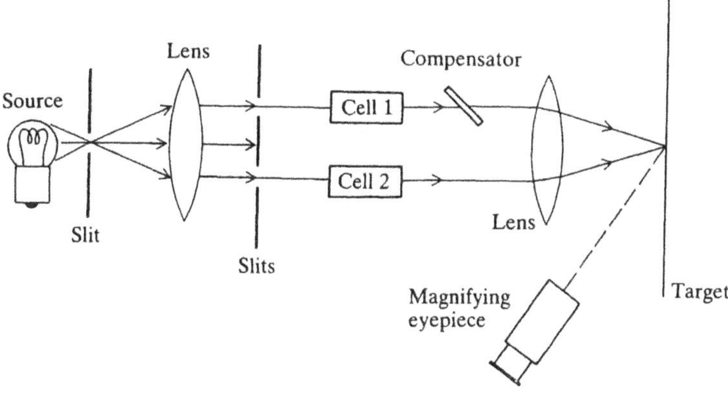

Figure 15-10 A diagram of a Rayleigh-Haber-Löwe interferometer. Cell 1 contains a medium of lower refractive index than that contained in cell 2.

A glass plate is used as a *compensator* in the path of the radiation that passes through the cell that contains the medium with the smaller refractive index. The two beams pass through a second lens which combines and focuses the radiation on the target. The image on the target consists of a series of light and dark bands that are caused by constructive and destructive interference from the two beams. Initially the two cells are filled with the same medium and the location of the bright zero-order band of radiation is observed through the eyepiece and noted. After replacement of the medium in one of the two cells with the sample, the difference in refractive indexes of the two media in the cells causes the position of the zero-order band to shift from its original position.

The compensator is rotated until the zero-order band returns to its initial position. The amount of rotation of the compensator plate is a function of the difference in refractive index between the two media. Rotation of the compensator increases the distance through the plate through which radiation passes. Owing to the relatively high refractive index of the glass, the radiation that passes through the compensator is slowed until the zero-order band returns to the starting position.

Instrumental Components

Some of the individual components of refractometers have been described in previous sections. Radiative sources can be tungsten filament (incandescent) lamps, sodium lamps, light emitting diodes LEDs, or sunlight. LEDs are often found in digital, hand held refractometers.

Prisms can be constructed from various glasses, sapphire, yttrium aluminum garnet YAG or other materials. The sample compartment and prism assemblies typically are thermostated with water jackets in laboratory instruments which are designed to make accurate measurements. Hand held refractometers normally are not water jacketed, but are used in environments at or near room temperature. Results obtained from those devices are less accurate than those obtained from thermostated laboratory instruments. Some laboratory and field instruments have a temperature probe which allows for automatic temperature compensation ATC of the results.

The readout devices can be the analog devices described in the preceding sections, or digital, seven-segment displays that utilized either LCDs or LEDs. Digital field instruments utilize batteries as their power supply. Regardless of whether analog or digital displays are used, the readout display can be calibrated to provide refractive indexes, Brix units, Baume units, or any of multiple concentration units for different substances. Instruments are in use which monitor plasma in blood, salinity, specific gravity, sucrose, fructose, inverted sugars, alcohol, ethylene glycol, propylene glycol and multiple other materials. Additionally, some instruments are designed to provide automated feedback to control various industrial products and processes. Those instruments often use dip cells or flow through cells to continuously monitor refraction.

QUALITATIVE ANALYSIS

Qualitative analysis of pure substances can be done by comparing the measured refractive index of the sample with that of known compounds. It is important to measure the refractive indexes of samples and standards by identical procedures. In all cases it is necessary to control the temperature and the wavelength at which the measurements are made. If the sample and standards are gases, the pressure must be controlled. If an Abbe refractometer is used, the boundary at the critical angle should be alternately approached from the light side and from the dark side. The set of measurements typically is repeated twice and averaged. The four measurements are repeated at the second setting of the Amici prisms (180° rotation) and averaged. The average of the means of the data sets at the two prism settings is taken to be the refractive index of the medium.

In those cases in which a match cannot be found between the refractive index of the sample and a standard, it is sometimes possible to perform a qualitative analysis[2, 4] by comparing the *Lorentz-Lorenz molar refraction R* of the sample with a calculated value for a hypothesized substance. The Lorentz-Lorenz (or Lorenz-Lorentz) molar refraction (in cubic centimeters per mole) is defined by

$$R = \frac{n^2 - 1}{n^2 + 2}\frac{M}{d} \tag{15-6}$$

where M is the molecular weight of the sample and d is its density (g/cm^3).

While obtaining the molar refraction of the sample, it is preferable to make the refractive index and density measurements on the same portion of the sample and at the same temperature. The measured refractive index and density of the sample and the molecular weight of the compound which is hypothesized to be the sample are substituted into Eq. (15-6), and the molar refraction is calculated.

Table 15-2 Increments of molar refraction for various organic components, assuming refractive index measurements are made using the sodium D lines

Component	Increment of *R*
C	2.418
H	1.100
C=C	1.733
C≡C	2.398
F	0.95
Cl	5.967
Br	8.865
I	13.900
O (hydroxyl)	1.525
O (ether)	1.643
O (carbonyl)	2.211
O (ester)	1.64
Aliphatic amines	
primary	2.322
secondary	2.502
tertiary	2.840
Aromatic amines	
primary	3.21
secondary	3.59
tertiary	4.36
Three-membered ring	0.71
Four-membered ring	0.48

The calculated molar refraction of the sample is compared to estimated values of known substances.

The estimated value of the hypothesized substance is obtained by adding the molar refractions of the atoms in the substance to the molar refractions associated with unsaturated bonds and with three- or four-membered rings:

$$R = (\text{contributions of double bonds, triple bonds, rings}) + (\text{contributions of atoms}) \qquad (15\text{-}7)$$

For an homologous series the difference between the molar refractions of the adjacent members of the series is approximately constant. Consequently, the molar refraction of an organic compound can be estimated by adding the contribution of each of the parts of the compound. The contributions of each of several common atoms and groups is listed in Table 15-2. Measured and estimated values of *R* are generally in good agreement, except for substances that contain conjugated double bonds or rings.

Sample problem 15-1 Estimate the molar refraction at the sodium D lines for 1,5-hexadiene.

SOLUTION 1, 5-hexadiene has the structure

$$CH_2=CHCH_2CH_2CH=CH_2$$

From Table 15-1 the following increments are obtained.

Component		Increment
2	C=C	1.733 × 2 = 3.466
10	H	1.100 × 10 = 11.00
6	C	2.418 × 6 = 14.51
		Total 28.98

[The calculated value for the compound using Eq. (15-6) is 28.99.] While performing the calculation it is important to understand that the contribution of the double bonds is added to that of all six carbon atoms including those involved in the double bonding.

QUANTITATIVE ANALYSIS

Quantitative analysis of a solute is most often performed with the working-curve method by plotting the refractive index as a function of concentration. The working curve is linear (Probs. 15-7 and 15-8) for some solutes but more often is curved. While performing a quantitative analysis the concentration is typically expressed as molarity or a weight percentage. Aqueous protein solutions are sometimes assayed using *specific refractive increment* k' as the concentration unit:

$$k' = \frac{n - n_0}{C} \tag{15-8}$$

In Eq. (15-8) C is the concentration in grams per milliliter, n is the refractive index of the solution, and n_0 is the refractive index of the solvent. The measurements are performed with a differential instrument using pure solvent as the reference medium.

In those cases where the working curve prepared by plotting n as a function of C is nonlinear, a plot of n^2 as a function of C is often linear. A maximum or a minimum in a working curve can occur when the solution components associate or when the intermolecular forces between the solute and the solvent are significantly different from the forces between the pure components. As an example, a maximum is observed in the working curve prepared by plotting the refractive index as a function of the concentration of ethanol in aqueous solutions.

APPLICATIONS

In addition to the use of refractometry for general qualitative and quantitative analysis, a few specific applications are worthy of particular attention. Hand-held refractometers have found widespread use in specialized assays. In those instruments that are designed for a specialized application, the scale in the refractometer is calibrated in concentration units of the analyte. Some of those specialized applications were previously described. Hand-held refractometers are used to determined the percent sucrose in aqueous solutions, the specific gravity of urine, and the concentration of protein. Common uses of Abbe and Pulfrich refractometers include the assay of NaCl or KCl in aqueous solutions, the assay of globulins or albumens in serum, and determination of the weight percentage of ethanol in aqueous solutions.

Some liquid chromatographic detectors rely upon refractometry, and refractometry is used as a technique to locate boundaries formed during electrophoresis. A flow-through cell is used to make differential measurements in liquid chromatographic detectors. The change in refractive index is converted to an electric signal and used to drive the y axis of a strip-chart recorder. Digitization and storage of refractive indexes that are measured by chromatographic detectors is possible with some detectors. The results can be used to recreate the chromatogram which is displayed by a computer-controlled device. A description of liquid chromatographic detectors is included in Chapter 25.

Refractometry is used to monitor the degree of separation that takes place during distillation or during separation by continuous or batch extraction. Refractometry is particularly useful in industry as a method of continuously monitoring flowing streams. Often a refractometer is permanently placed in a pipe through which a product of the industrial process flows. The monitored refractive index can be used in process control apparatus to help determine the operating conditions of the industrial plant. A description of process control apparatus is included in Chapter 28.

IMPORTANT TERMS

Abbe refractometer
Amici prism
Compensator
Critical angle
Differential refractometer
Immersion refractometer
Lorentz-Lorenz molar refraction
Optically dense
Optically thin
Precision Abbe refractometer
Pulfrich refractometer
Refraction
Sodium D lines
Specific refractive increment
Temperature coefficient of the refractive index

PROBLEMS

15-1 Determine the change in temperature that is required to cause an increase of 1×10^{-4} in the refractive index of a compound which has a refractive index temperature coefficient of -5×10^{-4} K.

15-2 Determine the change in temperature that is required to decrease the refractive index of carbon disulfide and of water by 1×10^{-4}.

15-3 What is the maximum temperature change that can be tolerated with a hand-held Abbe refractometer if an error of ±0.001 can be tolerated and if the temperature coefficient of the sample is -5×10^{-4} K?

15-4 Estimate the Lorentz-Lorenz molar refraction of diethyl ether.

15-5 The refractive index of acetone at 20°C as measured with the sodium D lines is 1.3571. The density at 20°C is 0.7908 g/cm^3. Calculate the molar refraction of acetone with Eq. (15-6) and compare the value to that estimated with Eq. (15-7).

15-6 The refractive index of a sample at 20°C measured with the sodium D lines is 1.3721. The density of the sample at 20°C is 1.0491 g/cm^3. The sample is known to be carbon tetrachloride, acetone, ethanol, chloroform, acetic acid, or ethyl acetate. Compare the molar refraction of the sample with that estimated for each compound and use the results to determine the identity of the sample.

15-7 The refractive indexes of several standard acetone-chloroform solutions and of a sample are listed in the accompanying table. Use the working-curve method to determine the percentage (w/w) of acetone in the sample.

Mole fraction acetone	**Refractive index** $\left(n_D^{20}\right)$
0.0000	1.3562
0.1030	1.3660
0.2020	1.3750
0.2995	1.3840
0.4085	1.3940
0.4980	1.4020
Sample	1.3760

15-8 The nitrogen content in hydrotreated shale oil can be determined using refractometry.[7] The refractive index of a series of standards and of a sample are listed in the table. Determine the percent nitrogen in the hydrotreated shale-oil sample.

Percent nitrogen (w/w)	n_D^{20}
0.20	1.4598
0.40	1.4629
0.60	1.4658
0.80	1.4689
1.00	1.4720
Sample	1.4673

REFERENCES

1. Degenhard, W. E., in F. D. Snell and C. L. Hilton (eds.): *Encyclopedia of Industrial Chemical Analysis, General Techniques P-Z*, vol. 3, Interscience, New York, 1966, pp. 392-407.

2. Schwartz, B. D., in F. J. Welcher (ed): *Standard Methods of Chemical Analysis*, vol. 3, part A, 6th ed., D. Van Nostrand, Princeton, New Jersey, 1966, pp. 250-257.

3. Niemczyk, T. M., in T. Kuwana (ed.): *Physical Methods in Modern Chemical Analysis*, vol. 2, Academic Press, New York, 1980, pp. 337-400.

4. Bauer, N., K. Fajans, and S. Z. Lewin, in A. Weissberger (ed.): *Physical Methods of Organic Chemistry*, vol. 1, part II, 3d ed., Interscience, New York, 1959, pp. 1139-1281.

5. Lewin, S. Z., and N. Bauer, in I. M. Kolthoff and P. J. Elving (eds.): *Treatise on Analytical Chemistry*, part I, vol. 6, Interscience, New York, 1965, pp. 3895-3931.

6. Schwartz, R. D., in L. Meites (ed.): *Handbook of Analytical Chemistry*, McGraw-Hill, New York, 1963, pp. 6-271-6-275.

7. Saint-Just, J., and O. A. Larson: *Anal. Chem.*, **51**: 1097(1979).

CHAPTER

SIXTEEN

NUCLEAR MAGNETIC RESONANCE SPECTROSCOPY

Some nuclei spin about their axes in a manner that is analogous to that in which electrons spin. In the presence of an externally applied magnetic field, a spinning nucleus can only assume a limited number of stable orientations. *Nuclear* mag*netic resonance* (nmr) occurs when the spinning nucleus in a lower energetic orientation in a magnetic field absorbs sufficient electromagnetic radiation to be excited to a higher energetic orientation. Because the energy that is required for the excitation varies with the type and environment of the nucleus, nmr spectroscopy can be used for qualitative chemical analysis. During nmr studies the sample is placed between the poles of a strong magnet. In much the same manner in which a spinning top or a gyroscope precesses about its axis of rotation under the influence of the gravitational force of the earth, the spinning nucleus precesses under the influence of the applied magnetic field. Only certain stable orientations (energetic levels) of the precessing nucleus relative to the applied magnetic field are possible. Nuclear magnetic resonance spectroscopy consists of measuring the energy that is required to change a spinning nucleus from a stable (lower energy) orientation to a less stable (higher energy) orientation in the magnetic field.

Because different spinning nuclei precess at different frequencies in the magnetic field, a different frequency of absorbed radiation is required to cause different spinning nuclei to change their orientations. The frequency at which absorption occurs can be used for qualitative analysis. The decreased intensity of incident radiation owing to absorption during a particular transition is related to the number of nuclei in the sample that undergo the transition and can be used for quantitative analysis. Nuclear magnetic resonance was first demonstrated in 1945 by Felix Bloch of Stanford University and by Edward Purcell of Harvard University while working independently of each other. They shared the Nobel Prize in physics in 1952 for their work.

THEORY

The spin angular momentum which a nucleus possesses is specified by a *spin quantum number I* for the nucleus. The spin quantum number can be any integer or half-integer. Nuclei, such as ^{12}C and ^{16}O which do not spin, have a spin quantum number that is 0. Nuclei which do not spin, and thus have no spin angular momentum, cannot be detected by nmr spectroscopy. Nuclei which have nonzero spin quantum numbers can be detected by nmr spectroscopy.

A spinning nucleus can assume one orientation in the magnetic field for each allowed *magnetic quantum number* m_I of the nucleus. The allowed magnetic quantum numbers for a nucleus with a spin quantum number I are $I, I - 1, I - 2, \ldots, -I + 2, -I + 1, -I$. Consequently, $2I + 1$ orientations (the number of terms in the series) of a nucleus in the magnetic field are possible. As the magnetic quantum number decreases, the energetic level increases, i.e., the energetic level that corresponds to $m_I = +\frac{1}{2}$ is lower than the level that corresponds to $m_I = -\frac{1}{2}$

A nucleus with a spin quantum number of $\frac{1}{2}$ can have $2\left(\frac{1}{2}\right) + 1 = 2$ orientations (energetic levels) in a magnetic field. The more stable (lower energy) orientation corresponds to orientation of the magnetic field generated by the spinning nucleus with the external magnetic field. The less stable orientation corresponds to opposite orientations of the two magnetic fields. The two possible orientations for a nucleus with $I = \frac{1}{2}$ are illustrated in Fig. 16-1. If the energy that corresponds to the difference between the two levels ($E_B - E_A$) is applied to a nucleus that is in a magnetic field, the nucleus can absorb the energy and change to an alignment that is opposite to that of the external magnetic field. The change in orientation of the spinning nucleus relative to the externally applied magnetic field is a nuclear magnetic resonance.

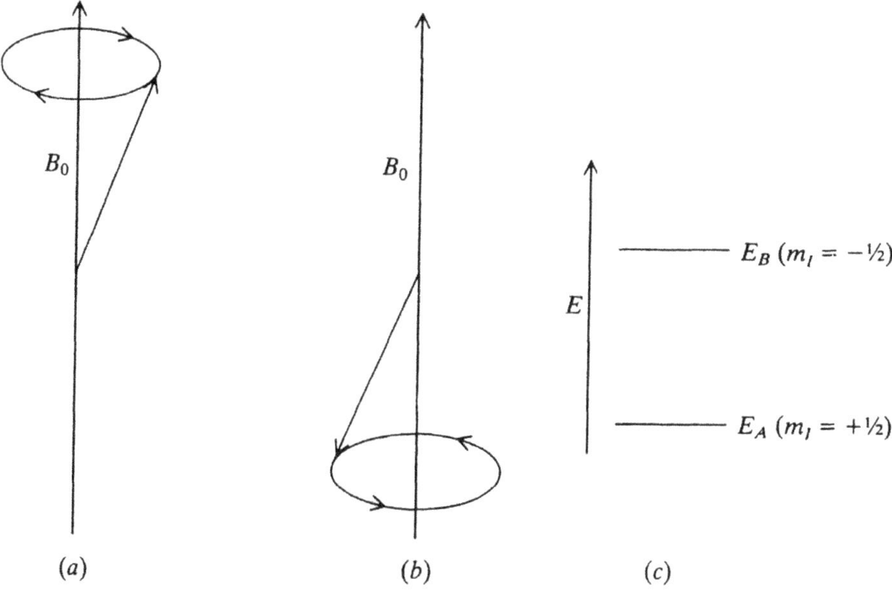

Figure 16-1 A diagram showing the possible orientations of a precessing nucleus with a spin quantum number of $\frac{1}{2}$ relative to an externally applied magnetic field B_0. (*a*) Nuclear alignment with the field $\left(m_I = +\frac{1}{2}\right)$; (*b*) nuclear alignment against the field $\left(m_I = -\frac{1}{2}\right)$; (*c*) an energy-level diagram for the two alignments.

A nucleus with $I = 1$ can have $2(1) + 1 = 3$ orientations in a magnetic field. The lowest energy orientation is with the applied field. The highest energy orientation is against the field. The third orientation is perpendicular to the field.

Spinning nuclei that have a spherical charge distribution have a spin quantum number of $\frac{1}{2}$. Examples of nuclei that have a spin quantum number of $\frac{1}{2}$ include ^{13}C, ^{19}F, ^{1}H, ^{3}H, ^{15}N, and ^{31}P. A nucleus that has a spin quantum number that is $\frac{1}{2}$ has a *magnetic moment* μ but does not have an electric quadrupole moment. Magnetic moment is a measure of the torque experienced by a magnet that is oriented at 90° to an external magnetic field of unit intensity.

Nuclei that do not have a spherical charge distribution are *quadrupolar nuclei*. They have both a magnetic moment and an electric quadrupole moment. Nuclei in that category have a spin quantum number that is greater than $\frac{1}{2}$.

In order for a nucleus to highly absorb electromagnetic radiation in a magnetic field, the nucleus must be relatively abundant in the sample and must possess a relatively large magnetic moment μ. Nuclei which exhibit both of these properties in some samples include ^{1}H, ^{19}F, and ^{31}P. Most nmr measurements are concerned solely with the ^{1}H nucleus. Measurements of other nuclei often require the use of a signal-enhancement technique in order to observe the nmr spectrum. Of the less abundant nuclear isotopes that exhibit nuclear magnetic resonance, ^{13}C, ^{15}N, and ^{17}O are generally of most interest to the chemist. A list of some of the common nuclei, their spin quantum numbers, and their magnetic moments is given in Table 16- 1.

The energy of a nucleus in the presence of an externally applied magnetic field is given by

$$E = \frac{-m_I \mu \beta B_0}{I} \tag{16-1}$$

where μ is the magnetic moment of the nucleus measured in *nuclear magnetons*, β is the value of a single nuclear magneton (5.0505×10^{-31} J/G), and B_0 is the magnetic flux density (tesla or gauss) of the externally applied magnetic field. The product $\mu\beta$ is the magnetic moment of a nucleus in units of joules per gauss. The magnetic flux density is often referred to as the magnetic field strength. Values of μ for several nuclei are listed in Table 16-1.

The energy that is required to cause a nucleus to go from a lower energetic level to a higher level can be determined with the aid of Eq. (16-1). As an example, for a ^{1}H nucleus the spin quantum number I is $\frac{1}{2}$ and the allowed values of m_I are $+\frac{1}{2}$ and $-\frac{1}{2}$. For all nuclei the selection rule for nuclear transitions is $\Delta m_I = \pm 1$; that is, only those transitions are allowed in which the magnetic quantum number changes by 1. Other transitions are observed in some circumstances although those transitions are forbidden. The lower energetic level E_L

Table 16-1 The spin quantum number *I* and magnetic moment μ (nuclear magnetons) of several nuclei

I	Nucleus	Mass Number	μ	*I*	Nucleus	Mass number	μ
0	C	12			Ba	135	0.8323
	O	16			Ba	137	0.9311
$\frac{1}{2}$	Ag	107	−0.1130		Be	9	−1.1774
	Ag	109	−0.1299		Br	79	2.0990
	C	13	0.70216		Br	81	2.2626
	Cd	111	−0.5922		Cl	35	0.82088
	Cd	113	−0.6195		Cl	37	0.68328
	F	19	2.6272		Cr	53	−0.47354
	H	1	2.79268		Cu	63	2.2206
	H	2	0.85738		Ga	71	2.5549
	H	3	2.9787		Gd	155	−0.24
	Hg	199	0.499		Gd	157	−0.32
	N	15	−0.28304		Hg	201	−0.607
	Os	187	0.1		K	39	0.39094
	P	31	1.1305		K	41	0.21453
	Pb	207	0.5837		Li	7	3.2560
	Pt	195	0.6004		Na	23	2.2161
					Rb	87	2.7414
	Se	77	0.5333		S	33	0.64274
	Si	29	−0.55477	$\frac{5}{2}$	Al	27	3.6385
	Sn	115	−0.9132		I	127	2.7940
	Sn	117	−0.9949		Mg	25	−0.85470
	Sn	119	1.0408		Mn	55	3.4610
	W	183	0.115		O	17	−1.8930
1	H	2	0.85738		Ti	47	−0.78711
	Li	6	0.82191		Zn	67	0.8735
	N	14	0.40357	3	B	10	1.8006
$\frac{3}{2}$	Au	197	0.136	$\frac{7}{2}$	Ca	20	−1.3153
	As	75	1.4348		Co	59	4.6388
$\frac{3}{2}$	B	11	2.6880		Cs	133	2.5642
					Ti	49	−1.1023

$\left(m_I = +\frac{1}{2}\right)$ is given by

$$E_L = \frac{-\left(+\frac{1}{2}\right)\mu\beta B_0}{\frac{1}{2}} = -\mu\beta B_0 \tag{16-2}$$

and the higher energetic level E_H by

$$E_H = \frac{-\left(-\frac{1}{2}\right)\mu\beta B_0}{\frac{1}{2}} = \mu\beta B_0 \tag{16-3}$$

The difference in energy ΔE between the two levels is determined by subtracting the lower level from the higher level:

$$\Delta E = E_H - E_L = \mu\beta B_0 - (-\mu\beta B_0) = 2\mu\beta B_0 \tag{16-4}$$

The difference in energy is the amount which must be absorbed by a proton with $m_I = +\frac{1}{2}$ in order to invert its spin relative to the externally applied magnetic field.

For the general case in which the nucleus has a spin quantum number I and for which the lower energetic orientation in a magnetic field corresponds to m_{IL} and the higher energetic orientation corresponds to m_{IH}, the difference in energy ΔE between the two levels is given by

$$\begin{aligned}\Delta E = E_H - E_L &= \frac{-m_{IH}\mu\beta B_0}{I} - \frac{-m_{IL}\mu\beta B_0}{I} \\ &= \frac{\mu\beta B_0\left(m_{IL} - m_{IH}\right)}{I}\end{aligned} \tag{16-5}$$

It is apparent from Eq. (16-5) that the energy that must be absorbed by a nucleus in a lower energetic level in order to cause the nucleus to become excited to a higher level is directly proportional to the magnetic moment ($\mu\beta$) of the particular nucleus and to the magnetic flux density B_0 of the applied magnetic field. The frequency of electromagnetic radiation that corresponds to the difference in energy between the two levels can be calculated by setting ΔE from Eq. (16-5) equal to the value from the Planck equation and solving for the frequency v. The result is

$$\Delta E = hv = \frac{\mu\beta B_0\left(m_{IL} - m_{IH}\right)}{I}$$

$$v = \frac{\mu\beta B_0\left(m_{IL} - m_{IH}\right)}{hI} \tag{16-6}$$

Sample problem 16-1 Calculate the energy and the frequency of radiation that is required to excite a proton from the lower to the upper energetic level if the applied field has a magnetic flux density of 14,092 G. For a proton μ is 2.7927 nuclear magnetons (Table 16-1).

SOLUTION The spin quantum number for a proton (Table 16-1) is $\frac{1}{2}$. Consequently, the allowed orientations in the magnetic field correspond to $m_{IL} = +\frac{1}{2}$ and $m_{IH} = -\frac{1}{2}$. ΔE is calculated using Eq. (16-5):

$$\Delta E = \frac{\mu\beta B_0(m_{IL} - m_{IH})}{I} = \frac{2.7927(5.0505 \times 10^{-31}\,\text{J/G})(14{,}092\,\text{G})\left(\frac{1}{2} - -\frac{1}{2}\right)}{\frac{1}{2}}$$

$$= 3.98 \times 10^{-26}\ \text{J}$$

The corresponding frequency can be obtained from Planck's equation:

$$\Delta E = hv$$

$$3.98 \times 10^{-26}\ \text{J} = (6.6256 \times 10^{-34}\ \text{J} \cdot \text{s})\ v$$

$$v = 6.00 \times 10^{7}\ \text{s}^{-1} = 60.0\ \text{MHz}$$

Nuclear magnetic resonance occurs when energy corresponding to the frequency that is required for absorption is applied to the nucleus in the magnetic field. Magnets that have a flux density of 14,092 G often are used for studies of protons. For magnets of that strength, protons resonate (absorb energy) at a radiofrequency of 60 MHz (Sample problem 16-1). Consequently proton nmr spectrophotometers which use a magnetic flux density of about 14 kG contain a radiofrequency source which emits radiation at 60 MHz.

Of course, if the magnetic field is altered then the frequency at which a particular nucleus resonates also changes [Eq. (16-6)]. Because each different type of nucleus has unique values of m_I, I, and μ, different nuclei resonate at different frequencies. It is possible to study more than one type of nucleus with the same instrument either by scanning the frequency in a fixed magnetic field or by scanning the magnetic flux density while holding the frequency at a fixed value. Absorption is recorded as a function of the variable parameter. One modern desktop instrument is capable of varying the frequency from 2 - 65 MHz. Another manufacturer offers more sophisticated instruments with operating frequencies between 300 MHz and 1 GHz.

Relaxation

After a nucleus has absorbed radiation, it can return (relax) to the lower energetic level by several processes. If relaxation did not occur, all of the nuclei would soon be in the upper energetic level and further absorption would become impossible. *Saturation* of an upper energetic level occurs when no further excitation to the level can occur until a nucleus in an upper level relaxes to a lower level. At equilibrium in a magnetic field the relative number of nuclei in each of the energetic levels in the absence of an applied radiofrequency field can be calculated using a Boltzmann distribution as follows:

$$\frac{n_H}{n_L} = e^{-\mu\beta B_0 / IkT} \tag{16-7}$$

where n_L is the number of nuclei in the lower energetic level, n_H is the number in the higher level, and k is the Boltzmann constant (1.381×10^{-23} J/K).

Sample problem 16-2 Determine the ratio of the number of hydrogen nuclei in the upper energetic level to those in the lower energetic level at 25°C in a magnetic field with a flux density of 14,092 G.

SOLUTION The appropriate values are substituted into Eq. (16-7). In this instance μ is 2.7927 nuclear magnetons (Table 16-1); β is 5.0505×10^{-31} J/G; I is $\frac{1}{2}$; k is 1.381×10^{-23} J/K; and T is 298 K.

$$\frac{n_H}{n_L} = e^{-2.7927\left(5.0505 \times 10^{-31}\right)(14,092)/\left[\left(1.381 \times 10^{-23}\right)298/2\right]}$$

$$= e^{-9.659 \times 10^{-6}} = 0.99999$$

From the result of Sample problem 16-2, it is apparent that the population of nuclei in the lower energetic level is only slightly greater than that in the upper level. The situation is quite different from that which was described in earlier chapters which dealt with other forms of spectroscopy. Nuclear relaxation after absorption must readily occur in order to prevent saturation of the upper energetic level.

Unlike the case of molecules or atoms that absorb in the ultraviolet-visible spectral region, nuclear relaxation generally cannot occur by emission of radiation. Nuclear relaxation normally occurs by *spin-lattice relaxation*, *spin-spin relaxation*, or *quadrupolar relaxation*. Spin-lattice relaxation occurs when the precessional frequency of the excited nucleus is identical in frequency and phase to the fluctuating magnetic field of a different magnetic nucleus in the sample. Because the nuclei surrounding the excited nucleus are in random motion, it is likely that at least some of the nuclei have the required magnetic field to interact with the excited nucleus. All nuclei in the sample except the excited nucleus are part of the lattice regardless of the physical state of the sample. The excited nucleus relaxes by losing energy to the lattice. The transferred energy appears as vibrational or rotational energy in the lattice.

Spin-lattice relaxation is a first-order process. The *spin-lattice relaxation time* T_1 is the inverse of the first-order rate constant for the process. The spin-lattice relaxation time is a measure of the lifetime of a nucleus in the excited state. It is the time required to return 63 percent of the excited nuclei to the lower energetic level. Because spin-lattice relaxation depends upon the motion of magnetic nuclei in the sample, T_1 increases as the mobility decreases. In gases and low-viscosity liquids, the spin-lattice relaxation time is short while in high-viscosity liquids and solids it is long. In solids and high-viscosity liquids the spin-lattice relaxation time varies between 0.01 and 100 s. It is generally advisable to have relaxation times that are between 0.3 and 3 s. At slower relaxation times the steady-state signal is not sufficiently large owing to near saturation of the upper energetic level. At faster relaxation times spectral line broadening occurs causing decreased resolution of the spectral peaks. Line widths in nmr spectra are approximately inversely proportional to the relaxation time.

Other factors that can affect spin-lattice relaxation times include the presence in the vicinity of the nucleus of an unpaired electron and the presence of nuclei with asymmetrical charge distributions (I greater than $\frac{1}{2}$). Unpaired electrons cause relatively strong fluctuating magnetic fields which can dramatically shorten the spin-lattice relaxation time. The presence of a magnetic nucleus with an asymmetrical charge distribution also causes a fluctuating magnetic field that provides an alternative mechanism by which spin-lattice relaxation can occur. That can shorten the spin-lattice relaxation time.

Spin-spin relaxation occurs when two adjacent nuclei are precessing at the same frequency but are in different energetic levels. In that instance the high-energy nucleus can transfer energy to the low-energy nucleus, causing the two nuclei to exchange spin states. Although the total number of nuclei in each spin state remains constant during spin-spin relaxation, the lifetime of a particular nucleus in an excited state is decreased. Because the *spin-spin relaxation time* T_2 is decreased for a particular nucleus, the line width in the nmr spectrum is broadened. Spin-spin relaxation is most likely to occur when the frequency corresponding to the applied energy does not exactly correspond to the precessional frequency of the nucleus of interest. Because spin-spin relaxation times are generally short for viscous liquids and crystalline solids, the line widths are broad and of little use. Generally spin-spin relaxation times predominate over spin-lattice relaxation times in determining nmr spectral line widths; i.e., viscous liquids and solids yield spectra with broad peaks.

Quadrupolar relaxation only occurs when quadrupolar nuclei are surrounded by asymmetrical electron clouds. A quadrupolar nucleus has a nuclear charge distribution that is ellipsoidal. Nuclei that have spin quantum numbers that are greater than $\frac{1}{2}$ are quadrupolar, while nuclei that have spin quantum numbers of 0 or $\frac{1}{2}$ have a spherical charge distribution. An asymmetrical electron cloud has an electron distribution that is not spherical, tetrahedral, or in any other symmetrical geometry. If the electron distribution is symmetrical, the electric field surrounding the electron cloud is constant as the molecule tumbles through the solution. If the electron distribution is asymmetrical, the tumbling molecule causes a fluctuating electric field around the molecule. The fluctuating field exerts a force on a quadrupolar nucleus that can change the orientation of the nucleus in the magnetic field.

It is important to understand that the interaction between the nucleus and the electric field that surrounds the electron cloud only occurs when both the nucleus and the electron cloud have asymmetrical charge distributions. Otherwise any interaction that occurs does not cause an alteration of the nuclear orientation relative to the magnetic field. The interaction between the asymmetrical electron cloud around the tumbling molecule and the quadrupolar nucleus is completely electrostatic; i.e., it depends upon the fluctuating attraction of the negative electron cloud for an end of the ellipsoidal positive nucleus. If the interaction causes a change in orientation of the magnetic field of the nucleus from a higher to a lower energetic level, quadrupolar relaxation has occurred.

Quadrupolar relaxation does not occur with nuclei such as ^{1}H and ^{13}C which have a spin quantum number of $\frac{1}{2}$. Nuclei which have larger spin quantum numbers (Table 16-1) such as ^{2}H, ^{14}N, and ^{35}Cl, however, can undergo quadrupolar relaxation. The presence of a quadrupolar nucleus in a sample can affect the spectral splitting pattern (see later section of the chapter) observed for other nuclei with spin quantum numbers of $\frac{1}{2}$. If the quadrupolar relaxation time is short, a nucleus near the quadrupolar nucleus only experiences the effect of the average magnetic moment associated with the upper and lower energetic levels of the quadrupolar nucleus. If the quadrupolar relaxation time is long, the neighboring nucleus experiences the magnetic moments of both the upper and lower energetic levels and the spectral peak is split as described in the section titled "Spin-Spin Splitting."

Chemical Shifts

The electron cloud around a particular nucleus can partially shield the nucleus from the externally applied magnetic field. In the presence of a magnetic field, electrons around a nucleus move in such a way as to form a small magnetic field that normally opposes the external field. The magnetic field that is experienced by a particular nucleus is slightly less than that of the applied field. Because the magnetic flux density B_0 at the nucleus slightly varies as the density of the electron cloud around the nucleus changes, nuclei of the same element that have different electron-cloud densities absorb EMR at slightly different frequencies [Eq. (16-61)]. As the density of the electron cloud increases, the magnetic flux density at the nucleus decreases causing a decrease in the frequency at which absorption occurs.

The density of the electron cloud around a particular nucleus is a function of the environment in the vicinity of the nucleus. Of particular importance in determining the density of the electron cloud is the electronegativity of chemical groups that are attached to the nucleus of interest. As the electronegativity of an attached group increases, electrons are drawn toward the electronegative group and away from the nucleus. Consequently, the amount of shielding decreases; the field strength experienced by the nucleus increases; and a spectral shift to a higher frequency occurs. The change in spectral peak frequency between that observed for a nucleus in a reference compound and that observed for the nucleus of interest is the *chemical shift* and is characteristic of the chemical environment of the nucleus. Chemical shifts that are caused by electronegative groups are rarely important over distances greater than one or two bond lengths.

The magnetic flux density β_e experienced by a particular nucleus is related to the applied magnetic flux density B_0 by

$$B_e = B_0(1 - \sigma) \tag{16-8}$$

where the *shielding constant* σ is a dimensionless constant that is characteristic of a particular nucleus and environment. The shielding constant can be either positive or negative. It is not possible to accurately predict a value of the shielding constant for a particular nucleus in a particular environment although, as described earlier, it is possible to predict trends in the way that the shielding constant varies between nuclei in different environments.

In addition to the electronegativity of groups that are near the studied nucleus, *magnetically anisotropic* groups in the vicinity of the nucleus influence the field strength that is experienced by the nucleus. An anisotropic group is a group of atoms, a bond, or a collection of electrons in which the orientation of the group in the magnetic field defines the direction of motion of the electrons in the group. An example of an anisotropic group is an aromatic ring such as is found in benzene. If the ring is oriented so that the plane of the ring is perpendicular to the applied magnetic field, the pi electrons in the ring are relatively free to move and circulate in doughnut-shaped clouds above and below the plane of the ring, as illustrated in Fig. 16-2. The electrons circulate in the direction predicted by the left-hand rule; i.e., in the direction in which the fingers point in a fist made from a left hand when the extended thumb points in the direction of the applied magnetic field.

The circular motion of the electrons in a ring is a *ring current* and causes formation of a circular magnetic field around the edges of the ring as predicted

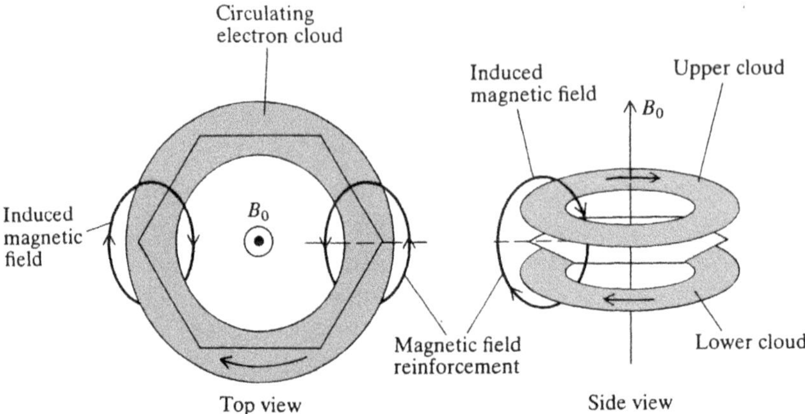

Figure 16-2 A sketch illustrating the circulating pi electrons and the induced magnetic field in an external magnetic field. In the top view the external field is directed from behind the page toward the viewer. In the side view the direction of application of the field is indicated by the vertical arrow.

by the right-hand rule. Because the magnetic field that is caused by the electrons is circular, it opposes the external magnetic field at the interior of the ring, reinforces the field in the external plane of the ring, and is perpendicular to the applied field directly above and below the ring. The strength of the induced magnetic field varies with the distance from the edge of the ring. Because most chemical substituents on an aromatic ring are in the external plane of the ring, the magnetic field experienced by those nuclei is greater than that of the external field, and the frequency at which absorption occurs is relatively large.

If the ring is placed in the field so that the plane of the ring is parallel to the magnetic field, little circulation of the electrons is possible owing to the orientation of the pi electrons; and no enhancement or decrease in the applied magnetic field is experienced by a nucleus that is attached to the ring. Other magnetically anisotropic groups include C═C, C═C, C═O, and to a slight extent even C—C. The electron circulation in unsaturated and to a lesser extent in saturated bonds is completely analogous to that described for aromatic rings, although not over as large a distance. The chemical shift observed for a particular nucleus is the sum of the effects resulting from the electronegativity of neighboring groups and the effect of magnetic anisotropy.

It is apparent from Eq. (16-8) that the magnetic field strength experienced by a nucleus is dependent upon both the shielding constant of the nucleus in a particular environment and upon the strength of the externally applied magnetic field. Absorption by a particular nucleus as observed with instruments which use different field strengths occurs at different frequencies. It is advantageous to have a term for measuring the spectral position of peaks which is independent of the strength of the applied magnetic field and consequently of the instrument.

It is conventional to express the chemical shift of an absorptive band in *parts per million* δ of the magnetic field:

$$\delta = \frac{B_{\text{ref}} - B_{\text{samp}}}{\nu} \times 10^6 \tag{16-9}$$

The difference between the magnetic field strength B_{ref} at which the reference absorbs and the magnetic field strength B_{samp} at which a sample compound absorbs is divided by the radiofrequency ν at which the instrument is operated. The result is multiplied by 10^6 to yield a dimensionless term in which the chemical shift is expressed in parts per million of the applied radiofrequency. The magnetic flux density of the reference and the sample, and the frequency must all be expressed in identical units in order to yield a dimensionless result.

Typically the magnetic flux densities and the frequency are expressed in frequency units (hertz). Conversion from units of magnetic flux density to frequency units is accomplished by equating the magnetic flux density to the radiofrequency at which a nucleus absorbs (Sample problem 16-3). Values of the chemical shift in parts per million are independent of the strength of the applied magnetic field and of the radiofrequency used for the study. A particular nucleus appears at the same chemical shift regardless of whether the instrument uses 60, 100, 500 MHz, or some other frequency. The major advantage of using instruments that operate at higher frequencies is increased spectral resolution. Peaks are better separated from each other at higher frequencies. The sensitivity of the measurement also increases with approximately the $\frac{3}{2}$ power of the field strength.

Sample problem 16-3 Calculate the chemical shift of a particular nucleus in a 60-MHz instrument if the reference nucleus absorbs at a magnetic flux density that is 0.063 G greater than that at which the sample nucleus absorbs.

SOLUTION For a hydrogen nucleus a magnetic flux density of 14×10^3 G is required to cause absorption at 60 MHz (Sample problem 16-1). That relationship can be used to convert the chemical shift in gauss to hertz:

$$B_{\text{ref}} - B_{\text{samp}} = 0.063 \text{ G} \times \frac{60 \times 10^6 \text{ Hz}}{14 \times 10^3 \text{ G}}$$
$$= 2.7 \times 10^2 \text{ Hz}$$

Substitution into Eq. (16-9) yields the desired result:

$$\delta = \frac{2.7 \times 10^2 \text{ Hz}}{60 \times 10^6 \text{ Hz}} \times 10^6 = 4.5 \text{ ppm}$$

A positive value of the chemical shift indicates that the sample nucleus experiences a greater magnetic field strength than the reference nucleus in the same externally applied magnetic field. By convention positive values of δ correspond to displacement to the left on nmr spectra, i.e., downfield from the reference nucleus. Positive chemical shifts are associated with anisotropic effects which increase the field strength at the studied nucleus, e.g., for the protons on benzene, and with decreased electron shielding owing to an electronegative group in the vicinity of the nucleus.

Reference Compounds

Reference compounds are used in nmr measurements [Eq. (16-9)] because the changes in magnetic flux density that are required to cause absorption of identical nuclei in different chemical environments are small. For the case described in Sample problem 16-3, the change in magnetic flux density was 0.063 G. Because the applied density was about 14×10^3 G, the change in the magnetic flux density was 4.5×10^{-4} percent. It is difficult to make absolute measurements of the magnetic flux density and to control the absolute field density with that accuracy. Differential measurements are considerably easier. Modern instruments typically can make differential measurements with an accuracy of about 1 in 10^{10}, which is well below the ability of nmr spectrometers to make absolute measurements.

For studies of protons the universally accepted reference compound is tetramethylsilane (TMS) [$Si(CH_3)_4$]. TMS is chosen as the reference because all of the protons in the compound are identical and absorb at an unusually high field strength. The relatively large amount of electron shielding in TMS is caused by the presence of the electropositive silicon in the compound. Reference compounds that are used for nmr spectroscopy of several nuclei are listed in Table 16-2. In each case positive chemical shifts correspond to absorption downfield (at lower applied field strengths) from the reference compound.

Proton magnetic resonance (pmr) was the first type of nmr to gain wide acceptance for chemical analysis. It is still the most widely used form of nmr. Table 16-3 is a list of chemical shifts of protons (^1H nuclei) in several chemical environments relative to TMS. Nuclear magnetic resonance with ^{13}C nuclei, while being used less often than pmr, is used more often than nmr with other nuclei. Generally it is not as easy to correlate carbon-13 chemical shifts with substituent polarity as it is to correlate proton shifts. The chemical shifts relative to TMS of ^{13}C nuclei in selected environments are listed in Table 16-4.

Table 16-2 Reference compounds that can be used for the nmr study of selected nuclei

Nucleus	Mass number	References compound
B	11	$F_3B: O(CH_2CH_3)_2$
C	13	$(CH_3)_4Si$
F	19	CCl_3F
H	1	$(CH_3)_4Si$
N	14	$(CH_3)_4N^+$
N	15	$(CH_3)_4N^+$
P	31	H_3PO_4 (85% aq soln)

Table 16-3 Typical ranges of chemical shifts δ for protons in selected environments

Substituents on bonds that are not shown are generally not important in determining the chemical shift

Group†	Chemical shift, ppm
$(CH_3)_4Si$	0
R_2NH	0.4-5.0
ROH (monomer, dil soln)	0.5
RNH_2	0.5-2.0
CH_3C	0.7-1.3
$HCCNR_2$	1.0-1.8
CH_3CX (X = F, Cl, Br, I OH, OR, OAr, N, SH)	1.0-2.0
RCH_2R	1.2-1.4
$RCHR_2$	1.5-1.8
$CH_3C{=}C$	1.6-1.9
$CH_3C{=}O$	1.9-2.6
$HC\equiv C$	2.0-3.1
CH_3Ar	2.1-2.5
CH_3S^-	2.1-2.8
CH_3N	2.1-3.0
ArSH	2.8-4.0
ROH (polymeric)	3.0-5.2
CH_3O^-	3.3-4.0
RCH_2X (X = Cl, Br, OR)	3.4-3.8
ArOH (polymeric)	4.5-7.7
$H_2C{=}C$	4.6-7.7
$RCH{=}CR_2$	5.0-6.0
$HNC{=}O$	5.5-8.5
ArH	6.0-9.5
RHN^+ ‡	7.1-7.7
Benzene	7.27
HCOO	8.0-8.2
$ArHN^+$	8.5-9.5
ArCHO	9.0-10.5
RCHO	9.4-12.0
RCOOH (dimer, nonpolar solvent)	9.7-13.2
ArOH (intramolecularly bonded)	10.5-15.5
HO_3S^-	11.0-13.0
RCOOH (monomer)	12.0
Enol	15.0-16.0

† R represents a saturated substituent or H.

‡ Dissolved in trifluoroacetic acid.

Table 16-4 Typical ranges of chemical shifts for ^{13}C nuclei in selected environments

When more than one C is shown, the shift is for the underlined nucleus

Group †	Chemical shift, ppm
CI	–295-45
CBr	–30-50
R_4C	5-60
CSR	5-70
CCl	20-110
CSOR	35-55
CSO_2R	35-55
COH	40-90
COC	55-90
CNO_2	60-80
CF	70-135
HC≡CR	60-75
RC≡CR	65-90
$O—\underline{C}—O$	85-115
Aromatics	90-170
RSCN	95-120
$H_2\underline{C}=CR_2$	105-125
RC≡N	105-130
$H_2C=\underline{C}R_2$	110-150
$(RCO)_2O$ (anhydrides)	145-175
$RC=C\underline{C}OOR'$	150-175
$RCONH_2$ (amides)	150-175
$RC=C\underline{C}OOH$	155-175
RCOOR′	155-175
RCOOH	165-190
$RCOO^-$	170-195
$RC=C\underline{C}HO$	175-195
$R_2C=O$	175-225
$(RC=C)_2\underline{C}O$	180-215
CS_2	193

† R and R ′ represent saturated substituents or H.

The values in the tables are normal ranges of chemical shifts. Because a particular substituent is occasionally outside the listed range, chemical shift tables should be used cautiously. Nuclear magnetic resonance of ^{13}C nuclei is *carbon magnetic resonance* (cmr). Longer listings of proton and carbon chemical shifts can be found in References 1 through 7. A particularly thorough listing is found in Reference 7. Representative chemical shifts for ^{19}F, ^{15}N, ^{31}P, and ^{11}B can be found in Reference 6. Tables of chemical shift ranges can be used as an aid in identifying the environment of each examined nucleus. They can be useful for qualitative analysis.

Spin-Spin Splitting

The average magnetic field strength that is experienced by a particular nucleus is determined by the strength of the externally applied magnetic field and by the amount of electron shielding. If two nuclei with nonzero spin quantum numbers are near each other, the magnetic moment of one nucleus can affect the magnetic field strength that is felt by the other nucleus. The weak magnetic field that is generated by the second spinning nucleus is sufficiently strong to interact with the first monitored nucleus only when the nuclei are near each other. A particular second nucleus assumes one of its allowed orientations in the externally applied magnetic field. The magnetic field that is experienced by the first nucleus is the sum of that owing to the externally applied field after electron shielding and that owing to the second nucleus. The second nucleus can only exist in allowed orientations in the external field; and the orientation determines the portion of the magnetic field strength from the spinning nucleus that is exerted on the first nucleus. Therefore, the first nucleus can experience a slightly different total magnetic field strength for each of the allowed orientations of the second nucleus. The single nmr spectral peak that would have been observed if the second nucleus was not present can be split into two or more peaks corresponding to each of the allowed orientations in the external magnetic field of the second nucleus. The interaction between spinning magnetic nuclei is *spin-spin coupling* and the resulting splitting of the nmr absorptive peak is *spin-spin splitting*.

Spin-spin coupling occurs through the electrons that bond the nuclei to the remainder of the molecule. The spin orientation of the second nucleus causes a bonding electron near the second nucleus to alter its spin orientation. The spin orientation of the bonding electron affects the spin orientation of neighboring bonding electrons which can affect the first monitored nucleus that is in the vicinity of the latter electrons. For protons, spin-spin coupling normally does not occur through distances greater than three bond lengths, although in some circumstances coupling is possible through as many as five bond lengths.

Spin-spin coupling can be illustrated by examining the relatively simple case of coupling between two nuclei that have identical spin quantum numbers I of $\frac{1}{2}$.

The spin-spin splitting patterns in the pmr spectrum of ethyl acetate (Fig. 16-3) can serve as an example. From the structure of ethyl acetate it is apparent that protons in the molecule exist in three different environments. Owing to different electron shielding in each environment, a separate peak is observed for each of the different types of H, as indicated by the numbers in Fig. 16-3 and by the chemical shifts in Table 16-3. In order to observe the splitting patterns that are described, it is necessary for the magnetic fields that are experienced by the coupled nuclei to be significantly different. A nucleus that experiences an identical magnetic field to that of a second nucleus, i.e., the two nuclei have identical chemical shifts, does not split the spectral peak of the second nucleus. Both nuclei appear as a single peak. Consequently, a proton of type 1 (Fig. 16-3) will not result in peak splitting of another type 1 proton.

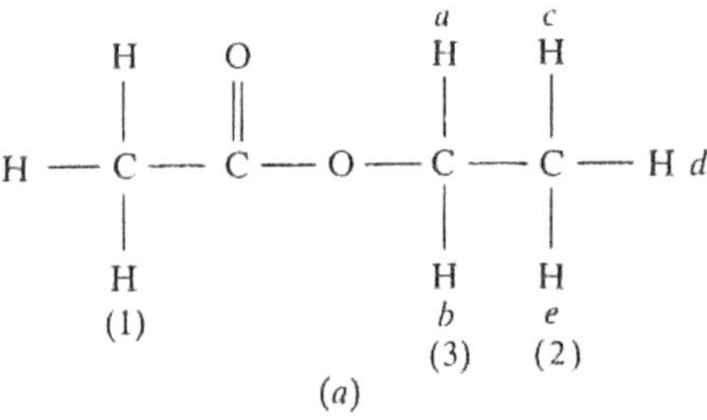

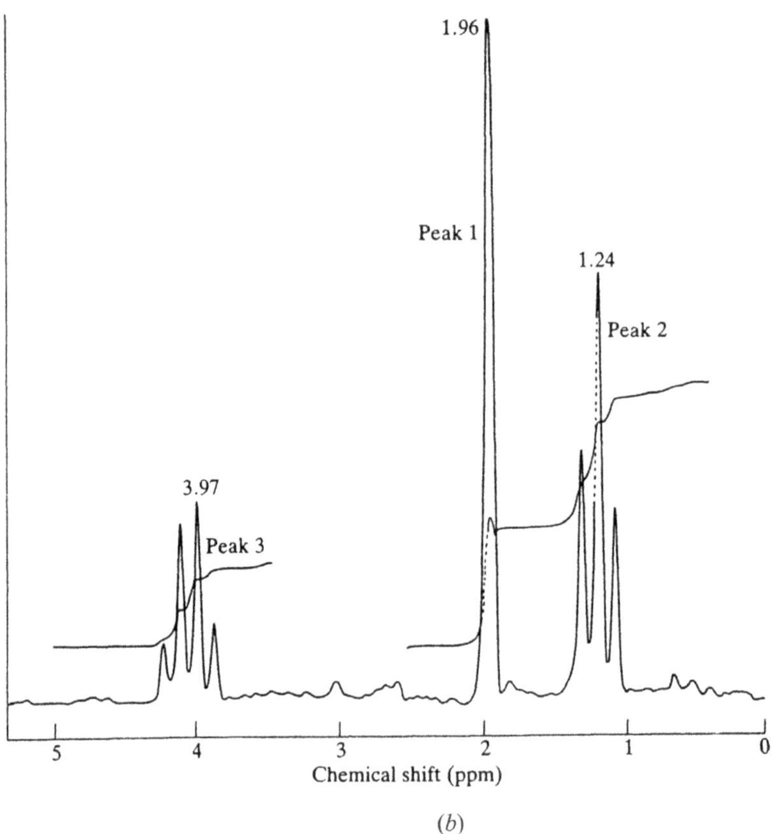

Figure 16-3 (*a*) The structure and (*b*) the nmr spectrum of ethyl acetate. The spectrum was recorded with a Varian T-60 spectrometer. The numbers near the tops of some of the peaks are the chemical shifts of the peaks.

The nearest magnetic nuclei to the type 1 protons of ethyl acetate are the methylene (type 3) protons which are five bond lengths removed. Keep in mind that ^{12}C and ^{16}O are not magnetic (I equals 0) and do not magnetically couple to the examined protons. Because coupling over distances greater than three bonds is usually not significant, the type 1 protons appear as a single unsplit peak (peak 1, Fig. 16-3*b*).

Because the methyl protons on the ethyl group (type 2 in Fig. 16-3) are separated from the methylene protons on the adjacent carbon by three bond lengths, spin-spin coupling can occur. The effect of the coupling depends upon the magnetic fields exhibited by the methylene protons which in turn is determined by the orientation of the methylene protons in the external field. For protons (I equals $\frac{1}{2}$) two orientations are possible in the magnetic field corresponding to $m_I = +\frac{1}{2}$ and $m_I = -\frac{1}{2}$. The two orientations are nearly equally probable (Sample problem 16-2). The total effect of all of the methylene protons in the sample on the methyl protons is reflected by the nmr spectrum.

Because the probabilities of each proton being in either of the two orientations is nearly equal, it is relatively simple to treat the problem statistically. First label the two type 3 protons as *a* and *b* for convenience. The two protons can both be oriented with the external field (m_{Ia} and m_{Ib} are both $+\frac{1}{2}$); the two protons can both be oriented against the external field (m_{Ia} and m_{Ib} are $-\frac{1}{2}$); or one proton can be oriented with the field while the other is oriented against the flied (m_{Ia} is $+\frac{1}{2}$ while m_{Ib} is $-\frac{1}{2}$ or m_{Ia} is $-\frac{1}{2}$ while m_{Ib} is $+\frac{1}{2}$). The possibilities are illustrated in Fig. 16-4. Each of the four possibilities is equally probable.

When the orientations are identical and in the direction of the applied magnetic field, the combination of the magnetic fields from the protons cause the methyl protons to experience a magnetic field that is slightly greater than the external field. Consequently, a slightly lower external field strength is required to cause resonance. Conversely, when the two orientations are identical and opposed to the magnetic field, a slightly higher external field strength is required for resonance. When the two nuclear orientations are opposite, the magnetic field of one nucleus cancels that of the other and no change of the resonant frequency is observed.

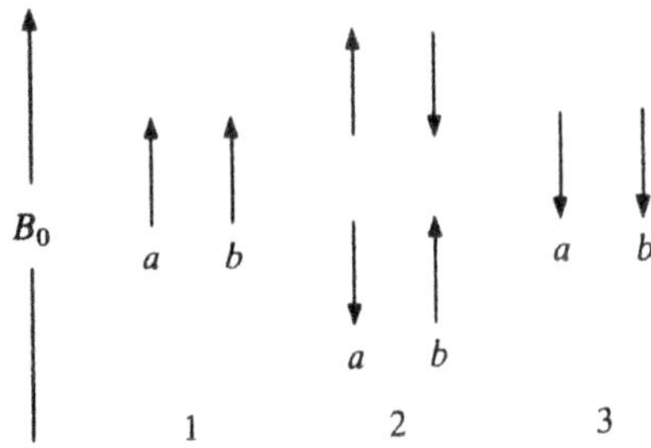

Figure 16-4 The possible orientations relative to an external magnetic field of methylene protons. The letters correspond to the labeled type 3 nuclei in Fig. 16-3. In case 1 the strength of the external field B_0 that is required for resonance of the methyl protons is less than that required in case 2 which is less than that required in case 3.

From the qualitative description presented here it is apparent that the absorptive peak that corresponds to the methyl hydrogens should appear as a triplet. The relative intensities of each portion of the triplet correspond to the probabilities of finding the two methylene protons in each of the three possible combinations. The probabilities (Fig. 16-4) of finding the two protons simultaneously oriented either with or against the applied field are equal. The probability of finding one proton oriented with the field while the other is against the field is twice that of either of the other two combinations because two combinations of orientations result in identical net magnetic fields. Consequently, the areas under the three peaks corresponding to the methyl protons should be in the ratio of 1 : 2 : 1.

Similarly a treatment of the splitting patterns observed for the methylene protons (type 3, Fig. 16-3) depends upon the possible orientations of the three methyl protons (type 2, Fig. 16-3) in the magnetic field. The possible orientations are illustrated in Fig. 16-5. The three protons can be simultaneously aligned with the field, simultaneously aligned against the field, two protons can be aligned with the field while the third is against the field, or two protons can be aligned against the field while the third is with the field. Because the possible combinations of orientations cause four different magnetic field strengths to be experienced by the methylene protons, the spectral peak owing to the protons is split into a quartet with peak areas in the ratio of 1 : 3 : 3 : 1. Splitting patterns provide information related to the number of magnetic nuclei near the studied nucleus.

By now it is apparent that the number of peaks that are observed for a hydrogen nucleus in a particular environment is dependent upon the number of hydrogen atoms that are attached to adjacent carbon atoms. In most cases hydrogens that are further from the studied nucleus do not cause spin-spin splitting. The number of peaks p that are observed for a particular hydrogen nucleus is one greater than the number n of *equivalent* hydrogen atoms that are attached to adjacent carbon atoms:

$$p = n + 1 \tag{16-10}$$

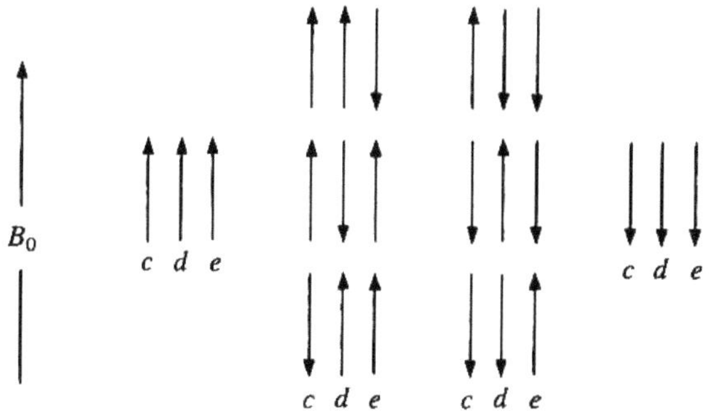

Figure 16-5 The possible orientations of the three methyl protons (type 2, Fig. 16-3) relative to the applied magnetic field.

If the hydrogen nuclei that cause the splitting are not in identical environments, the situation is more complicated because the strengths of the magnetic fields that are associated with the nuclei are not identical. In that case the number of expected peaks can be determined by sequentially treating the studied nucleus as if it were split first by all of the coupled nuclei in a single environment and then treating each split peak as if it were further split by nuclei in the second environment, etc. If splitting is caused by n_1 nuclei in environment 1 and n_2 nuclei in environment 2, a total of $(n_1 + 1)(n_2 + 1)$ peaks should be observed. Typically some of the peaks either overlap or are too small to be seen. More complex spectra of that and other types are described in the References.

The relative areas under each of the split peaks varies in the ratio of the coefficients of the binomial expansion

$$(a + b)^n \tag{16-11}$$

where n is the number of coupled protons that cause the spin-spin splitting. A binomial expansion has coefficients of the proper magnitude for nuclei with spin quantum numbers of $\frac{1}{2}$ because each term (a or b) in the expansion represents a stable orientation (m_I equals $+\frac{1}{2}$ or $-\frac{1}{2}$) relative to the applied magnetic field of the nuclei that cause the splitting.

As an example, for the triplet corresponding to the methyl protons on the ethyl group of ethyl acetate, the ratio of the peaks can be obtained by substituting 2 (for the two methylene protons) for n in Eq. (16-11) and performing the expansion. The coefficients of the terms in the expansion are the relative peak areas (1 : 2 : 1):

$$(a + b)^2 = 1a^2 + 2ab + 1b^2 \tag{16-12}$$

A simple way to remember the terms in a binomial expansion is with the aid of Pascal's triangle (Fig. 16-6). The triangle is constructed by placing a 1 at the

n	Relative peak area/Binomial coefficient
0	1
1	1 1
2	1 2 1
3	1 3 3 1
4	1 4 6 4 1
5	1 5 10 10 5 1
6	1 6 15 20 15 6 1
7	1 7 21 35 35 21 7 1
8	1 8 28 56 70 56 28 8 1
9	1 9 36 84 126 126 84 36 9 1
10	1 10 45 120 210 252 210 120 45 10 1

Figure 16-6 Pascal's triangle. The numbers in the triangle are the coefficients in a binomial expansion and correspond to relative peak areas of multiplets that are observed for nuclei with spin quantum numbers of $\frac{1}{2}$. The number n of identical neighboring magnetic nuclei are listed in the column on the left.

top and at the two outer edges of each horizontal row of numbers. The intermediate numbers in the triangle are obtained by adding the two adjacent numbers in the series immediately above the desired number in the triangle. For example, in order to obtain the coefficient of the second term in the expansion when n is 4, the two numbers immediately above the desired number in the triangle when n is 3 are added ($1 + 3 = 4$). The answer is the desired coefficient.

For nuclei with a spin quantum number that is not equal to $\frac{1}{2}$, a similar expansion can be used which has a term that corresponds to each of the stable nuclear orientations. For example, the relative peak heights that are expected when a proton is split by n equivalent nuclei that have a spin quantum number of 1 is given by the coefficients in a trinomial expansion of the form $(a + b + c)^n$. The letters (a, b, and c) represent the three stable orientations of nuclei in the external magnetic field. The lowest energetic level (when m_I equals + 1) is represented by a; the intermediate energetic level (when m_I equals 0) is represented by b; and the highest energetic level (when m_I equals –1) is represented by c.

Some of the terms in the expansion are energetically equivalent and are added together to yield the proper peak areas. For example, when n is 2, $(a + b + c)^2$ is $a^2 + 2ab + 2ac + b^2 + 2bc + c^2$. Because ac (m_I is + 1 and m_I is –1) and b^2 (m_I is 0 and m_I is 0) are energetically equivalent, the two terms are added together and the peak areas of the multiplet are in the ratio 1 (a^2 term) to 2 (ab term) to 3 (ac and b^2 terms) to 2 (bc term) to 1 (c^2 term). The terms are arranged in order of increasing energy.

Most nuclei that are studied using nmr spectroscopy have a spin quantum number of $\frac{1}{2}$ and can be treated using Eqs. (16-10) and (16-11). If the nucleus of interest is coupled to a nucleus with a spin quantum number that is greater than $\frac{1}{2}$, more splitting is observed than described for coupled hydrogen nuclei. In general the number of peaks in the multiplet is equal to $2nI + 1$ where n is the number of equivalent nuclei that have a spin quantum number I and that are coupled to the studied nucleus.

If a peak is split by two or more sets of nuclei that have spin quantum numbers of $\frac{1}{2}$ and that are in different environments, each set of coupled nuclei acts independently upon the peak. Equation (16-11) can be used to determine the peak areas owing to one set of nuclei and then each of the split peaks is further split according to Eq. (16-11) by the second set of coupled nuclei. If a third set of coupled nuclei are involved, further splitting occurs. Typically not all of the split peaks are observed because some of the peaks overlap or are too small. A complex multiplet is observed.

The relatively simple spectra that have been described in this section are *first-order spectra*. Interpretation of more complex spectra is too difficult to be described in this chapter. Further information is available in the References. First-order spectra are observed when the ratio of the difference in central frequencies $\Delta\nu$ between the coupled nuclei to the *coupling constant* J of the split peaks is 20 or greater. Spectra that resemble first-order spectra are observed for values of $\Delta\nu/J$ that are greater than about 10. The coupling constant is the frequency difference between adjacent peaks in a multiplet. The coupling constant is independent of the strength of the external magnetic field. It is a function of the ability of the nuclei to couple. If the difference in the mean chemical shifts between the coupled nuclei is not large relative to the separation between the split peaks, the coupled nuclei do not act equally on the studied nuclei and the spectrum becomes more complex.

The coupling constant for a pair of coupled nuclei is independent of the strength of the external magnetic field. Owing to its independence, it is usually measured in absolute units (hertz) rather than in relative units (parts per million). The difference in chemical shift between the coupled nuclei increases as the strength of the magnetic field increases. Consequently, $\Delta\nu/J$ increases and first-order spectra are more likely to be observed when instruments with larger magnetic fields are used. As might be expected, the increased likelihood of observing first order spectra, the increased resolution and increased chemical shifts are advantages of using instruments which have greater magnetic field strengths. Unfortunately the expense of nmr spectrometers generally increases as the strength of the magnetic field that is used in the instruments increases. Coupling constants can sometimes provide information that aids in the qualitative analysis of a sample.

Peak Areas

The area under an nmr peak is directly proportional to the amount of radiofrequency energy that is absorbed by a particular type of nucleus in a particular environment. Absorption requires that the nucleus initially be in a lower energetic level. At equilibrium the fraction of the nuclear population that is in a lower energetic level is constant (Sample problem 16-2). The amount of absorption that can occur and the spectral peak area are directly proportional to the number of nuclei in the lower level. Because the number of nuclei in the lower level is a constant fraction of the total number of nuclei in that environment, the peak area is proportional to the total nuclear population in the environment. If high-resolution nmr spectroscopy is used, i.e., if all of the nuclei that are studied are of the same element, the energy absorbed by a nucleus is nearly constant regardless of its environment. Consequently, the proportionality constant that relates peak area to nuclear population is the same for all of the peaks in the nmr spectrum.

That information is useful because it allows the integrated peak areas to be used to determine the ratio of nuclei at each chemical shift in a sample. The point can be illustrated with the nmr spectrum of ethyl acetate that was shown in Fig. 16-3. The molecule contains a total of eight protons which appear at three chemical shifts. The ratio of the peak areas is identical to the ratio of the number of protons of each type in the molecule; i.e., the ratio of the areas under peaks 1, 2, and 3 is 3 : 3 : 2, which is identical to the ratio of the number of protons in each environment in the molecule. An integrator trace from an operational amplifier integrator has been recorded over the spectrum in Fig. 16-3. The vertical deflection of the pen as it passes over a spectral peak is directly proportional to the area under the peak.

Most nmr spectrometers are equipped with an integrator. In some instruments the integrator is a variation of the operational amplifier devices that were described in Chapter 4. In other instruments digital integration is used. Typically the integrators that accompany nmr instruments have a relative accuracy of about ± 1.5 percent. Conclusions concerning the number of nuclei at each chemical shift that require greater accuracy cannot be made with those integrators.

APPARATUS

Instruments that are used to obtain nmr spectra are usually named nmr *spectrometers* in spite of the fact that absorption rather than emission is measured. Nmr spectrometers contain a stable source of the applied magnetic field, a radio-frequency transmitter, a radiofrequency receiver, a recorder, an integrator, and a method for rotating the sample. Either the magnetic field or the radiofrequency energy can be scanned. In addition to the sample compartment, some instruments contain a compartment for a reference material that is used to adjust the zero chemical-shift setting on the spectrum. In those instruments the difference in chemical shift between the sample and the reference is displayed on the x axis of the recorder. Some instruments contain additional components that are designed for specific applications. Nearly all modern instruments are computer controlled and utilize computerized data acquisition systems. A diagram of the major components, excluding the computer and computer interface, of an nmr spectrometer in which the magnetic field flux density is scanned is shown in Fig. 16-7.

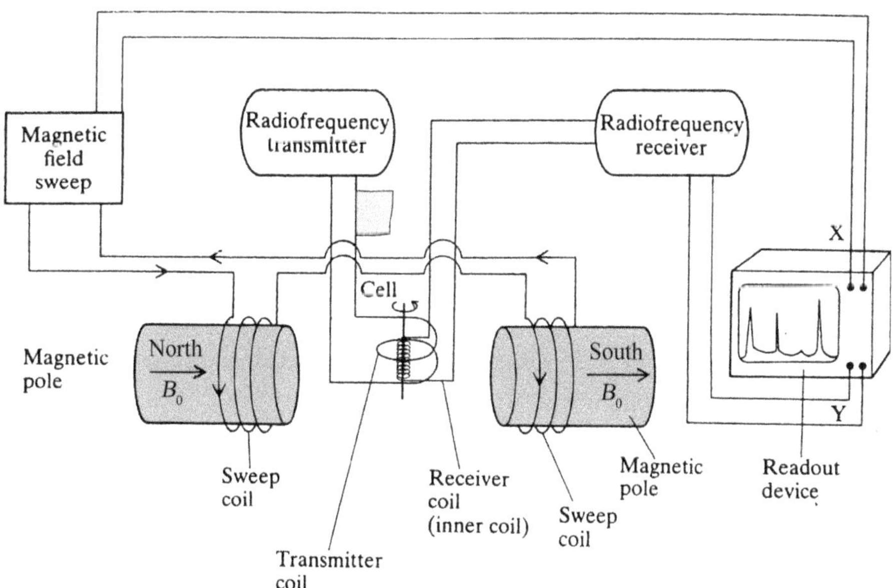

Figure 16-7 A diagram of an nmr spectrometer. The arrows on the sweep coils indicate the direction of current flow through the coils so that the applied magnetic field is in the indicated direction.

The sample is placed between the poles of a magnet. If the magnetic field in the instrument is to be varied, sweep coils are usually wrapped around or placed between the magnetic poles in such a way that the magnetic field that is associated with a current that flows through the sweep coils adds to the magnetic field from the poles of the magnet. By changing the current, the flux density of the magnetic field that passes through the sample is altered. The magnetic field should be both stable and homogeneous. Because spectral peak widths depend upon the homogeneity of the magnetic flux density throughout the sample, the magnetic field must be highly homogeneous. An additional magnetic coil called a *shim coil* is often used in each of the three axes to adjust the magnetic field until the width of an observed spectral peak is minimized.

The magnets used in nmr spectrometers can be permanent magnets, superconducting magnets, or electromagnets. Permanent magnets are least expensive and provide an ease of operation that is not available with the other types. The major disadvantage of permanent magnets is their lack of flexibility. Superconducting magnets are relatively compact and can be used to achieve magnetic flux densities that are greater than those that are possible with other magnets. Their disadvantages include a higher expense and the requirement to cool the magnets with liquid nitrogen or liquid helium. Because resolution and chemical shift (measured in hertz) increase with magnetic flux density, superconducting magnets are used in instruments in which high resolution is a requirement. Peak areas are also a function of magnetic flux density. Normally peak areas are directly proportional to $B_0^{3/2}$. The major disadvantage of superconducting magnets is their expense. In electromagnets the flux density can be altered by changing the current that passes through the coils of the magnet. Electromagnets are relatively insensitive to temperature changes whereas the temperature of both permanent magnets and superconducting magnets must be controlled.

Radiofrequency radiation is applied to the sample perpendicularly to the direction of the magnetic field. In Fig. 16-7 the radiofrequency coils are wrapped around the cell. The radiofrequency transmitter contains an oscillator. The operational frequency of the oscillator can be controlled either by the resonate frequency of a crystal or by some other means.

The radiofrequency signal is detected by the radiofrequency receiver. Generally the receiver coils are also wrapped around the cell. In some instruments the receiver and the transmitter share the same coils. When radiation from the transmitter is absorbed by the sample, the receiver coils detect a decreased electric potential in the coils. The signal from the radiofrequency receiver is used to drive the y axis of the recorder or other readout device, or is monitored by the computerized data acquisition system.

The sweep generator is used to drive the *x* axis of the readout device. If the instrument is to be used to record spectra owing to more than one elemental type of nucleus (for example, ^{1}H and ^{13}C), a separate transmitter and set of coils is required for each nuclear type. Those coils and transmitter are enclosed in the *probe* of the instrument. Generally when a different nucleus is studied, the probe of the instrument must be changed.

As described earlier, a spectrum can be obtained either by holding the radiofrequency constant while varying the magnetic field (*field-sweep method*) or by varying the radiofrequency while keeping the magnetic flux density fixed (*frequency-sweep method*). Although the field-sweep method is used in many instruments, instruments that are designed for highly accurate measurements of chemical shifts and instruments that are designed for decoupling experiments (described later in the chapter) generally utilize the frequency-sweep method. If the frequency-sweep method is used, the radiofrequency energy that is emitted in the instrument must be highly stable.

In either case the chemical-shift setting on the instrument is adjusted either with an internal standard, such as TMS, that is added to the sample or with a reference sample, such as water, that is placed in a separate cell within the instrument. In either case the instrument records the difference in chemical shift between the two chemicals on the *x* axis of the readout device. A differential method such as those described here is required because it is difficult to control either the magnetic field or the radiofrequency with the accuracy required for nmr measurements. Typically control to 1 part in 10^8 or 10^9 is required.

Cells that are used for nmr spectral studies are constructed from a material that does not absorb radiofrequency radiation within the range in which the sample is expected to absorb. For pmr studies the cells are often constructed of borosilicate glass. Most instruments are designed to be used with 5-mm OD (outer diameter) cylindrical cells; however, some instruments can accommodate cells up to 1.5 cm in diameter. Larger cells are used when the sample is dilute, the sensitivity of the instrument to the nucleus is low, or the natural abundance of the nucleus is low. Owing to the low abundance of ^{13}C, 10-mm cells sometimes are used for cmr studies. Cylindrical cells that have a spherical cavity within the cell can be used for samples of limited volume. Cells that contain a spherical cavity require about 0.035 mL of sample to obtain a spectrum while the more common cylindrical cells require about 0.35 mL. The cell length is usually between 15 and 20 cm. Typically the cell is filled to a depth of between one-eighth and one-sixth of the cell length and carefully lowered into the space between the poles of the magnet.

Generally samples are rotated around the *y* axis corresponding to the length of the cylindrical cell. By spinning the sample, small inhomogeneities in the *xz* plane of the magnetic field are averaged throughout the sample. Spinning causes better resolution between adjacent spectral peaks. Because averaging does not occur along the vertical (y) axis, the y-axis shim control must be carefully adjusted to create a uniform field and minimize peak width. Rotation of the sample is generally accomplished by inserting the cell into a turbine. The turbine-cell assembly is lowered into the instrument and rotated at 30 to 50 Hz by a stream of air from a compressor or a high-pressure air tank.

Solvents

If the sample is a nonviscous liquid, it can be placed directly into the cell for analysis. The spectral peaks of high-viscosity liquids and of solids are normally broad and poorly resolved. Better spectral resolution can be obtained by using a solution of the viscous sample in a nonviscous solvent. Spectra of solids which are not dissolved in a solvent are described in a later section.

The solvent that is chosen for a particular sample must be nonviscous, capable of dissolving the sample, and have no peaks within the spectral range of the study. For pmr the last requirement can be met by using a solvent that contains no H. Often carbon tetrachloride, carbon disulfide, or a deuterated solvent is used. Carbon tetrachloride and carbon disulfide are less expensive than the deuterated solvents, but the solubility of many compounds in those solvents is insufficient. Among the deuterated solvents, deuterated chloroform $CDCl_3$ and deuterated benzene C_6D_6 are used most often. Other deuterated solvents which can be used for pmr studies include CD_3CN, CD_3OD, CD_3SOCD_3, CD_3COCD_3, $C_6H_5CD_3$, and D_2O. Solvents that do not contain protons and that can be used for pmr include CF_2Cl_2, SO_2FCl, and SO_2. Solvents that contain protons and can be used throughout restricted spectral ranges are dimethyl ether, ethylene chloride, propylene, cyclopropane, cyclopentane, dichloromethane, and chlorofluoromethane. Most of the solvents that contain protons are used for studies at temperatures below 25°C. Dimethylformamide and dimethylsulfoxide are useful solvents at temperatures above 30°C.

QUALITATIVE ANALYSIS

An analyst has a great deal of latitude in approaching a qualitative analysis. In many cases the procedure outlined in the following steps will yield satisfactory results while using nmr spectroscopy. In some cases the steps in the analysis are taken in a different order. In other cases it might be possible to eliminate one or more of the steps. In all cases it is best to have additional analytical information from some source other than nmr spectra prior to reaching any conclusions.

1. The chemical shift of each environmentally different nucleus is measured and used in conjunction with correlation tables such as those in Tables 16-3 and 16-4 to obtain a list of possible groups that could have caused each of the spectral singlets or multiplets.
2. The integrations of the spectral peak areas are used to determine the relative number of nuclei in the sample that are associated with each group.
3. The splitting patterns of the first-order spectral peaks are used as described in the section entitled "Spin-Spin Splitting" to determine the number of magnetic nuclei in the vicinity of the nucleus. The integrated areas are used in conjunction with the splitting patterns to determine the groups that are adjacent to each other in the sample molecule.
4. From the information that was obtained in the first three steps and from information that is available from other analytical methods, a list is compiled of possible sample compounds. Other analytical information which can be particularly useful includes elemental analysis, infrared spectra, Raman spectra, uv-visible spectra, mass spectra, boiling points, melting points, and the results of functional group analyses. Any other available analytical data also can be useful.
5. Whenever possible, a sample of each of the hypothesized sample components is obtained. The nmr spectrum of each hypothesized compound is obtained and compared with that of the sample. Comparisons using other spectral methods are also helpful. If the comparisons yield identical results for a hypothesized compound and the sample, the sample can normally be concluded to be the hypothesized compound. If a hypothesized compound cannot be located, the spectrum of the sample can sometimes be compared to that found in a published series of spectra. The nmr spectra published by Sadtler Research Laboratories are particularly useful.

The use of a pmr spectrum in a qualitative analysis is illustrated in Sample problem 16-4. The pmr spectra that are shown throughout the chapter were recorded on neat liquid samples.

Sample problem 16-4 The pmr spectrum of a pure compound is shown in Fig. 16-8. The molecular weight of the compound was determined from mass spectral data to be 106. Identify the compound.

SOLUTION The spectrum consists of three groups of peaks. A singlet occurs at a chemical shift of 10 ppm, a multiplet at about 7.9 ppm, and a multiplet at about 7.6 ppm. From Table 16-3 it is apparent that the singlet at 10 ppm corresponds either to an aldehydic proton or to a proton on a carboxylic acid. Similarly, the other two peaks can be assigned either to protons on an aromatic ring or to a proton that is attached to a N.

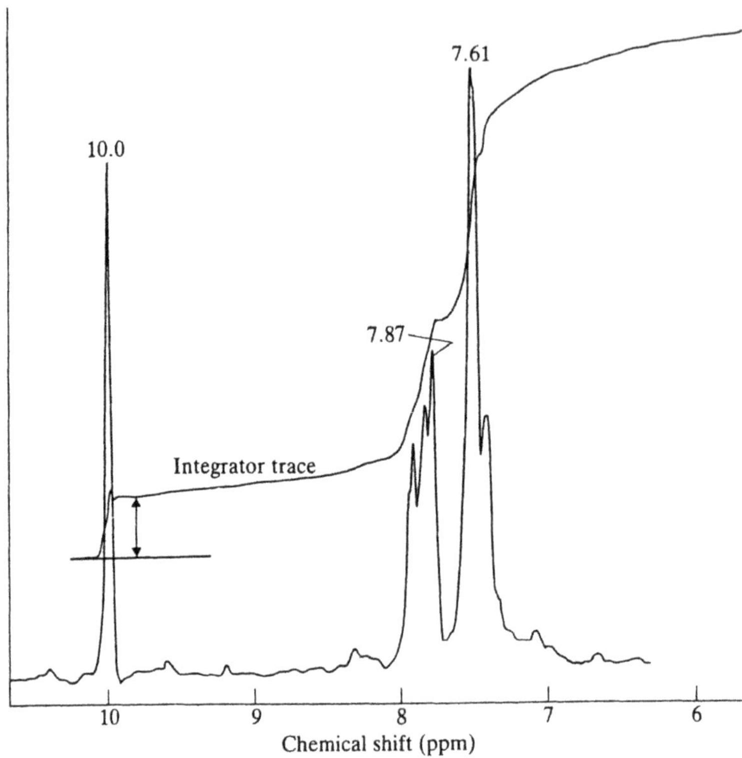

Figure 16-8 The pmr spectrum of the compound described in Sample problem 16-4. The chemical shifts of some of the major peaks are listed near the tops of the peaks.

The trace of the integrator is superimposed over the pmr spectrum in Fig. 16-8. The vertical deflection of the trace as the recorder pen passes over a peak is directly proportional to the number of protons in the sample compound that corresponds to the integrated peak. The measurement of the deflection for the peak at 10 ppm is illustrated by the double-headed arrow in Fig. 16-8. The integrator trace reveals that the number of protons that correspond to the three groups of peaks is in the ratio 1: 2 : 3 for the 10, 7.9, and 7.6 ppm peaks respectively.

The lack of splitting of the peak at 10 ppm indicates that the proton is not within three bond lengths of another proton. The splitting patterns of the remaining two multiplets appear to be more complex than that expected for first-order spectra. Consequently it is probably not safe to apply the rules for first-order splitting to the two multiplets.

The readily available pmr spectral information has now been exhausted and it is necessary to use other analytical data in combination with the pmr data to propose possible chemical structures of the compound. The integrations of the two multiplets indicate that the multiplet at 7.6 ppm is caused by

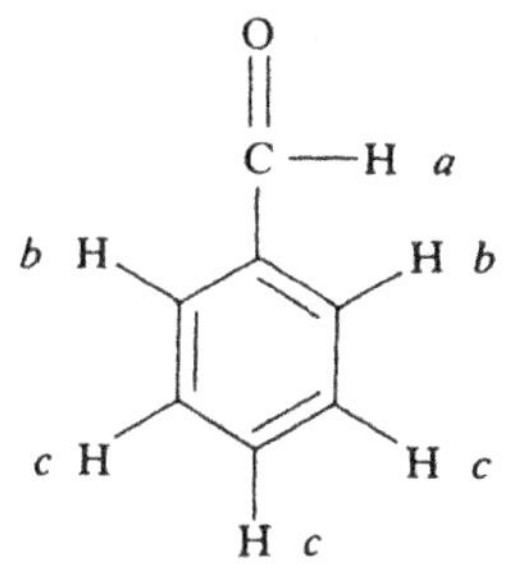

Figure 16-9 The molecular structure of benzaldehyde. The small letters are used to indicate protons in different chemical environments. Refer to Sample problem 16-4 for a description.

an integral multiple of three protons while that at 7.9 ppm is caused by an integral multiple of two protons. The combination of three and two protons suggests a singly substituted aromatic ring. It should be noted that monosubstituted aromatic rings which contain an electronegative substituent nearly always exhibit a set of two multiplets that have an appearance and chemical shift that is similar to that shown in Fig. 16-8.

If the two multiplets are caused by the five protons on an aromatic ring, 77 amu (atomic mass unit) of the 106 molecular weight is accounted for by C_6H_5, leaving 29 amu. Because —COOH has a mass of 45 amu, it can be eliminated as causing the 10 ppm peak. An aldehydic group —CHO accounts for 29 amu. Consequently, the compound can be concluded to be benzaldehyde (Fig. 16-9). The aldehydic proton can be assigned to the peak at 10 ppm. The two ring protons that are ortho to the aldehyde (the *b* protons in Fig. 16-9) are in a different chemical environment from the ring protons (the *c* protons in Fig. 16-9) that are meta or para to the aldehyde. The integration is used to assign the multiplet at 7.9 ppm to the *b* protons and the multiplet at 7.6 ppm to the *c* protons.

The final step of the analysis is to obtain a pmr spectrum of benzaldehyde and to compare it to the sample spectrum. In this instance the two spectra were identical, and the compound can be identified as benzaldehyde.

While using splitting patterns and chemical shifts to interpret spectra, it is important to be aware of potential chemical reactions that can occur in the sample. An example is observed with samples that contain hydrogen that is bonded to an oxygen. A pmr spectrum of neat ethanol is shown in Fig. 16-10. The compound contains protons in three chemical environments, as illustrated in Fig. 16-11. The triplet at about 1.2 ppm corresponds to the three methyl protons (labeled *c* in Fig. 16-11). The peak is split into a triplet by the two protons on the adjacent carbon. The quadruplet at about 3.7 ppm corresponds to the two methylene protons (labeled *b*). The peak is split into a quadruplet by the three methyl protons. The singlet at 5.3 ppm is caused by the hydroxyl proton (labeled *a*).

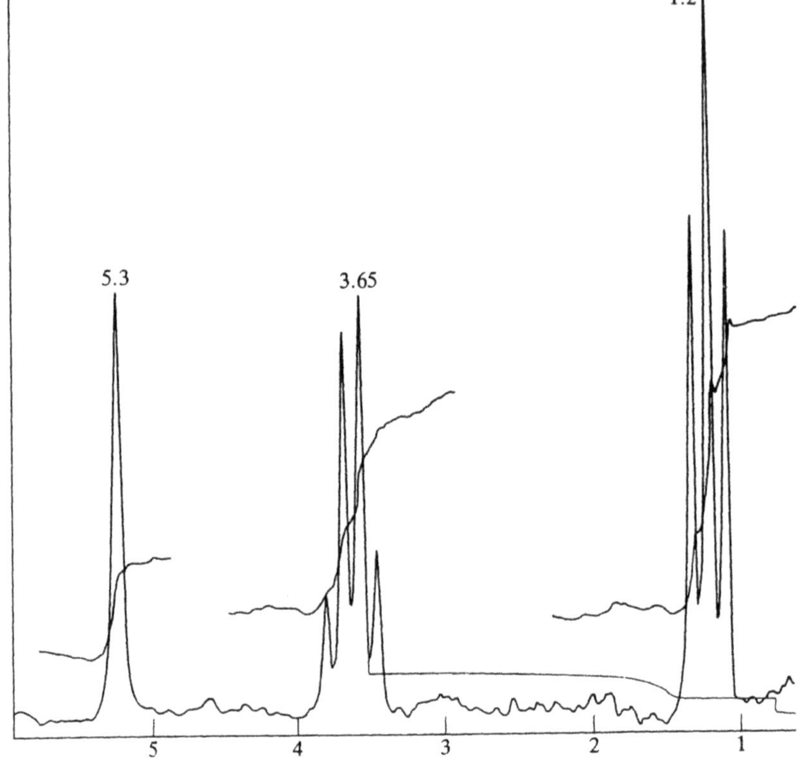

Figure 16-10 The pmr spectrum of neat ethanol. The chemical shifts of some of the peaks are listed near the tops of the peaks.

Because the hydroxyl proton is within three bond lengths of the two methylene protons, coupling between the protons is expected. Each of the four peaks in the quadruplet should have been split into a doublet ($n + 1$ rule) by the single hydroxyl proton, thereby forming eight peaks. Further, the peak owing to the hydroxyl proton should have been split into a triplet by the two methylene protons.

Coupling between the methylene protons and the hydroxyl proton generally does not occur in neat ethanol because acid impurities in the sample catalyze the

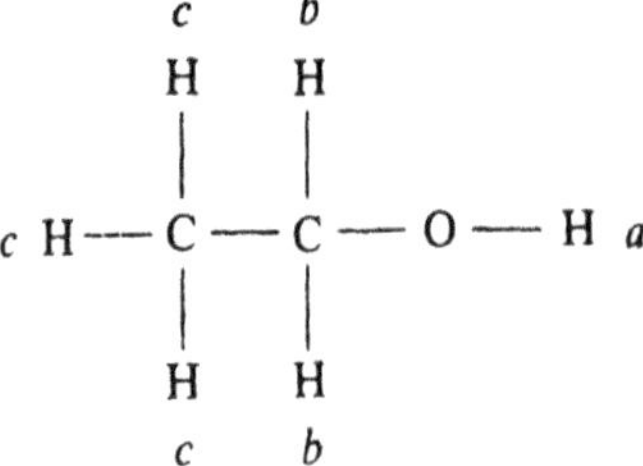

Figure 16-11 The molecular structure of ethanol. The small letters indicate the chemical environments of the protons.

rapid acid-base exchange of the hydroxylic proton. Because the proton does not reside continuously on the oxygen for a significant period, it is not coupled to the methylene protons. The expected splitting patterns can be observed if the rate of exchange of the hydroxylic proton can be decreased. That can be accomplished by adding a base, such as anhydrous sodium carbonate or alumina, to react with the acid impurities that catalyze the exchange. Use of a deuterated solvent also slows the rate of proton exchange. Not all alcohols exhibit rapid exchange of the hydroxylic proton. Those alcohols in which the rapid exchange is not observed have spectra with the anticipated splitting patterns.

The position of the pmr peak owing to the hydroxyl proton varies with concentration and with the solvent. Neat samples of ethanol (Fig. 16-11) usually exhibit a peak at about 5.3 ppm. As the concentration in a nonpolar solvent is decreased, the chemical shift of the peak decreases. The peak corresponding to a 5 percent solution of ethanol in a nonpolar solvent typically has a chemical shift of about 2 ppm.

The peak position is also affected by the temperature of the solution. The variation in peak position with temperature is attributable to intermolecular hydrogen bonding. When the hydroxyl proton forms a hydrogen bond with an electronegative group on a different molecule, the electron density around the proton is decreased owing to the attraction for electrons of the electronegative element. The consequent deshielding causes the peak to appear at a greater chemical shift. As the amount of hydrogen bonding decreases, such as when the concentration is decreased in a nonpolar solvent, the electron shielding increases and the peak position moves upfield to a smaller chemical shift. As might be expected, compounds which can form intramolecular hydrogen bonds are less affected by concentration changes than those which form intermolecular bonds.

Shift Reagents

Spectra that contain overlapping peaks can be difficult to interpret. Splitting patterns are difficult, if not impossible, to detect and to interpret, and peak areas cannot be readily obtained. When severely overlapping peaks are encountered, a shift reagent can sometimes be used to separate the peaks. In most cases useful shift reagents consist of an organometallic complex of one of the lanthanide elements. Addition of a *lanthanide shift reagent* to an organic compound that contains a basic functional group produces a Lewis acid-base product. It is generally believed that the metal exerts an electrostatic force through space on magnetic nuclei of the sample in the vicinity of the lanthanide element. The electrostatic force causes a change in the extent of electron shielding around the nucleus and a consequent change in the chemical shift associated with the nucleus. The chemical shift is generally large and can be as much as 20 ppm.

Only nuclei near the bonding site with the shift reagent are affected. A chemical shift can be observed for nuclei other than protons although the reagents are primarily used in pmr studies.

Shift reagents are available which contain any of several of the lanthanide elements including dysprosium, europium, holmium, praseodymium, and ytterbium. Of the available reagents tris(dipivalomethanato)europium(III) [$Eu(dpm)_3$] and 1,1,1,2,2,3,3-heptafluoro-7,7-dimethyloctanedionato-europium (III) [$Eu(fod)_3$] are particularly useful. Europium compounds cause downfield shifts whereas praseodymium compounds cause upfield shifts. Upfield shifts cannot be accounted for by the mechanism described in the preceding paragraph. Lanthanide shift reagents have been used to change the chemical shift of protons on several types of functional groups including alcohols, aldehydes, amines, esters, ethers, ketones, lactones, nitriles, oximes, sulfoxides, and thioethers. The shift reagents are not useful when reacted with acids, alkenes, halides, nitro compounds, or phenols.

QUANTITATIVE ANALYSIS

Although nmr spectrometry has been used primarily for organic qualitative analysis, it also can be used as a tool for quantitative analysis. Integrated peak areas are used for quantitative analysis of a component in the sample. A graphical internal-standard method is preferred. Use of an internal standard minimizes errors owing to instrumental variations during an assay. A series of standard solutions in the same solvent are prepared. Each solution is prepared to contain a fixed concentration of an internal standard which has a spectral peak that does not overlap with any peak of the analyte. A spectrum of each solution is recorded and integrated. The integrated area of a peak of the analyte is divided by the integrated area of the peak of the internal standard for each solution. A working curve is prepared by plotting the area ratios as a function of concentration. The working curves are nearly always linear.

Sufficient internal standard is added to the analyte solution to yield a concentration that is identical to that in the standard solutions. The spectrum is recorded and integrated exactly as was done for the standard solutions, and the ratio of the integrated analyte peak area to that of the internal standard is determined. The concentration of analyte in the sample is determined from the working curve. As usual the assay should be performed several times in order to determine the precision and to estimate the accuracy of the assay.

While performing quantitative nmr analyses, the same precautions that are used with other spectral methods should be taken. An analyte peak for the analysis should be chosen which is separated from those of potential interferences and from that of the internal standard. Because nmr peaks are often narrow, it is not normally difficult to meet that requirement. The possibility that a spectral peak of the analyte will overlap with a spectral peak of the standard can be minimized by choosing an internal standard that has a single peak. Popular choices of internal standards are methylene bromide and 1,3,5-trinitrobenzene in nonaqueous solutions, and a salt of terephthalic acid in aqueous solutions.

NONPROTONIC NMR SPECTRA

Because pmr is used more frequently than nmr of other nuclei, most of the descriptions and examples in the chapter are related to pmr. Spectra obtained of other nuclei can also provide valuable qualitative and quantitative analytical information. The nonprotonic nuclei that have attracted the greatest interest are ^{13}C, ^{19}F, and ^{31}P. Because all of those nuclei have a spin quantum number of $\frac{1}{2}$ (Table 16-1), the splitting patterns observed in nmr spectra with those nuclei are similar to those observed with protons.

A nonprotonic nucleus in a molecule can couple with a proton. The coupling constants between magnetic nonprotonic nuclei and protons are generally greater than those between protons. Coupling constants between ^{1}H and ^{19}F which are separated by two bond lengths (H—C—F) typically range from 40 to 80 Hz while coupling constants for the nuclei which are separated by three bond lengths are usually less than 30 Hz. Coupling constants between ^{31}P and ^{1}H range from a maximum of about 700 Hz over one bond length to less than 30 Hz over three bond lengths.

In many cases the magnetic sensitivity of nonprotonic nuclei is lower than that of protons. Additionally, in most compounds the natural abundance of nonprotonic magnetic nuclei is considerably less than that of protons. Those two factors combine to yield nmr spectra of nonprotonic nuclei that have relatively low signal-to-noise ratios. The spectral peaks are small and often cannot be located when apparatus similar to that which was used for pmr is used to obtain the spectrum. Because the signal-to-noise ratio is low for nonprotonic

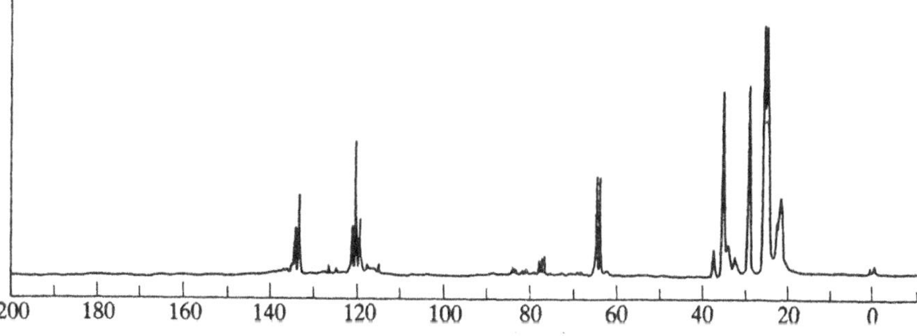

Figure 16-12 A cmr spectrum of 2-vinyl-1-oxaspiro{5.2}octane

O
H
$CH{=}CH_2$

that was obtained using a 50.3-MHz spectrometer. The reference compound was TMS and the solvent was $CDCl_3$. *(Courtesy of Dr. August A. Gallo, Department of Chemistry, University of Louisiana at Lafayette.)*

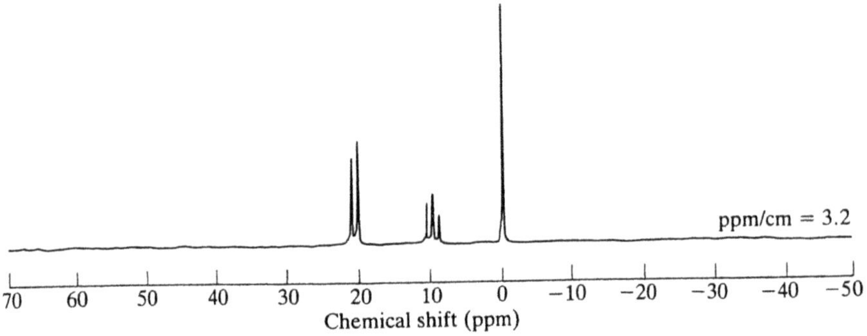

Figure 16-13 A ^{31}P spectrum of sodium pentakis(trifluoroethoxyl)-1-oxo-cyclotriphosphazenate monodiglymate, $N_3P_3(OCH_2CF_3)_5ONa\cdot([CH_3OCH_2CH_2]_2O)$. The spectrum was obtained using an 81-MHz spectrometer. The singlet at 0 ppm corresponds to the reference compound (H_3PO_4). The solvent was $CDCl_3$. *(Courtesy of Dr. Sigred Lanoux, Department of Chemistry, University of Louisiana at Lafayette.)*

nuclei, most instruments that are designed to record nmr spectra of those nuclei use multiple scans in combination with a signal-averaging technique. Many instruments use a Fourier transform to extract the spectral peaks from the background noise. Fourier transform instruments are also used to obtain pmr spectra of dilute solutions and of complex molecules, such as proteins, in which the amount of a particular proton in the molecule is small.

Another difference between pmr spectra and other types of nmr spectra is the observed range of chemical shifts. A perusal of Table 16-3 reveals that most pmr chemical shifts occur within a range of 10 ppm. Table 16-4 indicates that the chemical shifts for ^{13}C nuclei cover a range of about 200 ppm. Similarly ^{19}F and ^{13}P spectra cover a range of about 300 and 400 ppm respectively. Normally cmr spectrometry is not used for quantitative analysis owing to the nonlinear working curves that are obtained as a result of the nuclear Overhauser effect (described in a later section). A cmr spectrum is shown in Fig. 16-12 and a ^{13}P spectrum is shown in Fig. 16-13. More detailed information on nonprotonic nmr spectra can be found in References 1, 2, 4, and 6.

MULTIPLE RESONANCE

Spectral interpretation often can be simplified by the use of an instrument which has the capability of simultaneously exposing the sample to more than one wavelength of radiofrequency radiation. If two wavelengths are used, the technique is *double resonance.* The usual procedure is to first record the spectrum in the normal manner while using a single source of radiofrequency radiation. After the spectrum has been recorded, one of the radiofrequency sources is adjusted to a frequency corresponding to one of the spectral peaks and the spectrum is recorded using the second source. The first source is adjusted to a frequency at which a particular nucleus absorbs while the second source is used in the normal manner to record the spectrum.

The set of nuclei that are continuously irradiated rapidly absorb and emit radiation. Because the irradiated nuclei rapidly alternate between energetic levels, nuclei in the vicinity of the irradiated nuclei no longer are capable of distinguishing between the two levels. The nuclei experience the average magnetic moment of the two levels of the irradiated nuclei, and consequently the spectral peaks are not split by the irradiated nuclei. Multiplets which resulted from coupling to the irradiated nuclei collapse to singlets. The resulting spectrum not only is simplified but also allows the analyst to determine which nuclei are coupled to the irradiated nuclei. Double resonance which is accomplished in the described manner is termed *spin decoupling. Homonuclear double resonance and homonuclear spin decoupling* occur when both radiative sources are used to irradiate nuclei of the same elemental type. *Heteronuclear double resonance* uses one radiative source to irradiate one type of nucleus, for example, ^{31}P, while the second source records the spectrum of another nucleus, for example, ^{1}H. A spin-decoupled cmr spectrum is shown in Fig. 16-14 of the same compound whose cmr spectrum is shown in Fig. 16-12.

Double-resonant spectral studies can be done either by using a constant magnetic field strength while scanning one of the frequencies or by holding both frequencies fixed while scanning the magnetic field. The frequency-sweep method is generally preferred. Decoupling occurs with the field-sweep method only when the difference between the two fixed frequencies corresponds to the difference between resonant frequencies of two nuclei at the particular magnetic flux density at a particular instant during the scan. Spectra obtained with the field-sweep method are more difficult to interpret than those obtained with the frequency-sweep method. That has led to relatively little use of the field-sweep method.

The ability to control the bandwidth of the constant-frequency radiation that is used in a double-resonant study is advantageous. A broad bandwidth is normally used in heteronuclear double resonance in order to eliminate coupling to all nuclei of a single element. The bandwidth must be broad because the nuclei of the irradiated element could exist in more than one environment and thereby absorb at different frequencies. That technique is *noise decoupling.* Noise decoupling is routinely used to decouple ^{13}C nuclei from protons.

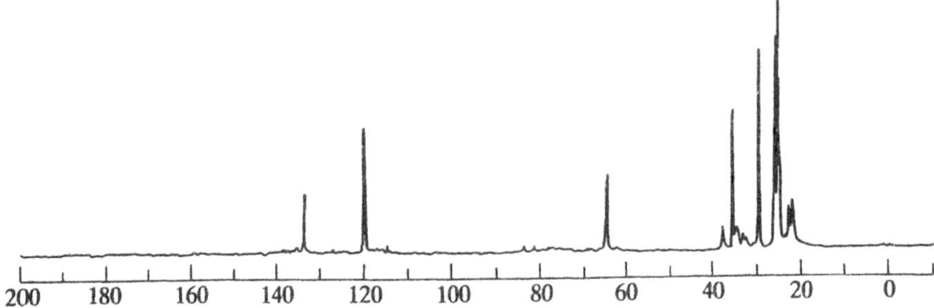

Figure 16-14 A heteronuclear double-resonance spectrum of 2-vinyl-l-oxaspiro[5.2]octane (ct Fig. 16-12). The sample was continuously irradiated at the frequency corresponding to protonic absorption while the cmr spectrum was recorded. The spectrum is hydrogen decoupled. *(Courtesy of Dr. August A. Gallo, Department of Chemistry, University of Louisiana at Lafayette.)*

Selective spin decoupling is primarily used with homonuclear double resonance. The bandwidth of the constant-frequency radiation is only sufficiently broad to decouple from neighboring nuclei those nuclei that absorb at a single-resonant frequency. Selective spin decoupling is the procedure described in the first paragraph of this section. A further decrease in the intensity of the radiation can cause decoupling of only the more weakly coupled nuclei in the vicinity of the irradiated nuclei. The more strongly coupled nuclei remain coupled to the irradiated nuclei. That technique is *spin tickling*. It can be used to locate hidden lines in a complex spectrum in which many spectral peaks overlap.

In the previously described procedures, the intensity of each peak at the frequency of the scanned radiofrequency was recorded as a function of the chemical shift that corresponded to the varied radiofrequency. *Internuclear double resonance* (INDOR) is a technique in which the intensity of the peak corresponding to the nuclei that are continuously irradiated at a fixed frequency is recorded as a function of the chemical shift corresponding to the varied frequency. A change in peak height is observed whenever the scanned frequency becomes identical to that of a band of a coupled nucleus. A change in peak height is observed only for nuclei in the sample that are coupled to the irradiated nuclei. INDOR can be used for both homonuclear and heteronuclear studies.

INDOR is a convenient and relatively inexpensive method to examine nonprotonic nuclei. A high expense is normally associated with the use of nonprotonic nmr instruments. If a nonprotonic nucleus is coupled to a proton, the fixed-frequency radiation can be adjusted to the frequency of one of the peaks in the multiplet of the coupled proton. The second frequency is scanned throughout the expected range of the nonprotonic nucleus. The intensity of the irradiated protonic peak is recorded as a function of the scanned frequency of the nonprotonic nuclei. Each time the scanned frequency passes through a spectral band of the nucleus, decoupling occurs between the proton and the nonprotonic nucleus. Owing to the decoupling, the multiplet corresponding to the coupled proton collapses to a singlet. If the fixed frequency at which the proton is irradiated is not identical to the frequency of the collapsed singlet, e.g., if the multiplet is a doublet, a decrease in intensity occurs and a negative peak is recorded. A peak is observed at a pmr frequency each time the scanned frequency equals the frequency at which a coupled nonprotonic nucleus absorbs radiation. Spectra of nonprotonic nuclei can be recorded in that manner without the use of Fourier transform instruments. Both positive and negative peaks can be observed with INDOR.

NUCLEAR OVERHAUSER EFFECT

Advantage can be taken of double-resonant studies and nuclear relaxation through dipole-dipole interactions to increase the sensitivity of an nmr analysis and to gain structural information about the analyte. Because the irradiating frequency in nmr spectrometers is relatively intense, relaxation by the methods described earlier is sometimes insufficient to return an irradiated nucleus to the lower energetic level after its initial excitation. The intensity of the observed peak is, of course, related to the number of nuclei in the lower energetic level. As the number of nuclei in the lower level increases, more radiation can be absorbed and the spectral peak height increases.

If a magnetic nucleus is rigidly held a short distance from the irradiated nucleus, an additional relaxation mechanism is available in which the irradiated nucleus transfers its excess energy to the neighboring nucleus through a dipole-dipole interaction. It should be emphasized that because relaxation by that mechanism occurs through space rather than through chemical bonds, the nuclei involved in the relaxation process do not have to be spin-coupled. The dipole-dipole relaxation mechanism, which results in increased peak height, is the *nuclear Overhauser effect* (NOE). It is named for the first person who observed the effect between electrons and nuclei.

Suppose two nuclei, *A* and *B*, that both have spin quantum numbers equal to $\frac{1}{2}$ are sufficiently near each other to allow dipole-dipole interactions to occur. In the usual study nmr absorption at one of the two nuclei (nucleus *B*) is observed. At any instant the two nuclei can exist in any of four possible combinations of spin states, (as indicated by the numbers to the right of each energy level) as illustrated in Fig. 16-15 where *H* represents the higher energetic level ($m_I = -\frac{1}{2}$) and *L* represents the lower level ($m_I = +\frac{1}{2}$) for each nucleus. Without irradiation of nucleus *A*, the ratio of the number *n* of *B* nuclei that are in the upper level to the *B* nuclei that are in the lower level is equal to the ratio of the probability p of finding nuclei in level 2 to finding nuclei in level 1 and to the ratio of the probability of finding nuclei in level 4 to finding nuclei in level 3:

$$\frac{n(B, H)}{n(B, L)} = \frac{p_2}{p_1} = \frac{p_4}{p_3} \tag{16-13}$$

The probability of finding the nuclei in any of the four allowed combinations of energetic levels can be determined from a Boltzmann distribution as described earlier in the chapter. Combined energetic levels are used because the nuclei undergo dipole-dipole interactions. Substitution of a Boltzmann term for each level into Eq. (16-13) yields

$$\frac{n(B, H)}{n(B, L)} = \frac{e^{-E_2/kT}}{e^{-E_1/kT}} = \frac{e^{-E_4/kT}}{e^{-E_3/kT}} = e^{-(E_2-E_1)/kT} = e^{-(E_4-E_3)/kT} \tag{16-14}$$

If nucleus A is irradiated at its resonant frequency, the nucleus is rapidly converted between the lower and higher energetic levels, i.e., between levels 1 and 3 and between levels 2 and 4. If the nucleus is sufficiently irradiated to cause saturation, the probability of finding an A nucleus in an upper level is identical to the probability of finding it at a lower level. Consequently the probability of finding the nucleus in level 1 at any instant is identical to the probability of finding it in level 3 ($p_1 = p_3$) and the probability of finding it in level 2 is identical to that of finding it in level 4 ($p_2 = p_4$). It is important to understand that those equalities are only true when nucleus A is irradiated and when nuclei A and B undergo dipole-dipole interactions. Substitution of p_1 for p_3 in the last term in Eq. (16-13) followed by introduction of the appropriate Boltzmann terms yields

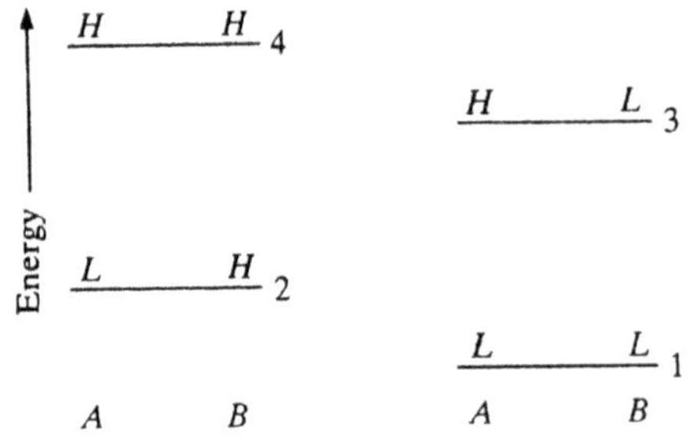

Figure 16-15 A representation of the possible combinations of energetic levels for two nuclei with spin quantum numbers of $\frac{1}{2}$.

$$\frac{n(B,\, H)}{n(B,\, L)} = \frac{p_4}{p_3} = \frac{p_4}{p_1} = \frac{e^{-E_4/kT}}{e^{-E_1/kT}}$$

$$= e^{-(E_4 - E_1)/kT} \tag{16-15}$$

It is apparent from Eq. (16-15) that the ratio of the number of B nuclei in the upper level to the number in the lower level is a function of the energetic difference between level 4, where both nuclei A and B are in upper levels, and level 1, where both nuclei are in lower levels. Because that energetic difference is greater than that between levels 1 and 2 or between levels 3 and 4 [Eq. (16-14)], more of the B nuclei reside in the lower energetic level when the A nuclei are irradiated than when the irradiation does not occur. Consequently, more absorption can occur and the peak area corresponding to the B nuclei increases.

In the usual study which takes advantage of the nuclear Overhauser effect, one of the two nuclei is continuously irradiated at its resonant frequency while an nmr spectrum is recorded throughout the spectral range which encompasses the second nucleus. Sensitivity for the second nucleus is generally significantly enhanced. The method is often used to enhance the sensitivity of ^{13}C assays. Irradiation of ^{1}H nuclei permits increased ^{13}C peak areas. An increase in peak area of about 200 percent can be achieved. If both nuclei are protons, the increased sensitivity is about 50 percent.

If relaxation can occur by a process other than dipole-dipole interaction, some of the increased sensitivity is lost. The best results are obtained when nucleus B cannot easily interact with nuclei other than nucleus A. The two nuclei should be closer to each other than to other nuclei and a solvent should be used which does not contain magnetic nuclei. The NOE can be used as a method of signal enhancement to determine the structure of a sample compound and to locate hidden spectral lines. The effect is only observed for nuclei which physically are close to each other in the sample.

SOLID SAMPLES

Most nmr studies are performed on relatively nonviscous liquid samples (neat or solution) which are rotated in the magnetic field to average variations in the magnetic flux density. As the sample becomes more viscous, the spectral peaks become broader and less useful. Spectra of solids that are obtained while using conventional nmr apparatus have very broad peaks owing to many overlapping spectral bands. Such spectra are nearly useless. In some cases either it is inconvenient to prepare a solution of a solid sample or it is feared that solution preparation will cause a change in the molecular structure. In those cases it is advantageous to be able to obtain nmr spectra of solid samples. Several reviews[8-10,12] of high-resolution nmr spectrometry of solids have been published and can be consulted for more details than those presented in this section.

Among the causes of the broad peaks that are observed with solid samples are shielding owing to anisotropic effects, dipole-dipole coupling, and quadrupolar interactions with nuclei that have a spin quantum number that is greater than $\frac{1}{2}$. Anisotropic shielding causes band broadening owing to the infinite number of possible orientations of anisotropic groups in the magnetic field. Each orientation causes a different amount of chemical shielding, and a broad spectral peak is observed corresponding to the series of overlapping absorptive maxima. As described in an earlier section of the chapter, anisotropic effects are most common in unsaturated groups. Aromatic rings are particularly affected.

The effect of dipole-dipole coupling in solids is generally greater than that in liquids. Its effect upon the location of an nmr peak in solids is large in comparison to electron shielding. The effect of dipole-dipole coupling upon the chemical shift of a nucleus is a function of both the magnetic flux density from the externally applied field and of the angle, relative to the magnetic field, of the vector that connects the dipolar nuclei. Again, the infinite number of orientations of the nuclei in the magnetic field causes band broadening.

If a quadrupolar nucleus is present, the effect of the interaction between the nonspherical charge distribution in the nucleus and the electric fields from the electrons in the vicinity of the nucleus is considerably greater than that owing to either chemical shielding or dipolar coupling. The interaction is dependent upon the quadrupolar moment of the nucleus and the electric field strength caused by the electrons. If the environment of the quadrupolar nucleus is cubically symmetric, the effect does not occur.

Of course, all of the factors which cause line broadening in the solid state are also present in liquids. In liquids, however, the individual molecules are not held rigidly in place and tumble through the liquid. Tumbling averages dipolar interactions to zero and, owing to the rapid changes of orientation in the magnetic field, eliminates anisotropic effects. Tumbling in liquids also provides a spin-lattice relaxation mechanism which considerably shortens the spin-lattice relaxation time.

Several methods are available that can decrease band broadening in solids. Use of one or more of the methods can result in nmr spectra of solids that are nearly as well resolved as those of liquids. Nearly all anisotropic effects can be eliminated by rapidly rotating the sample in the magnetic field. Dipolar interactions between two nuclei are proportional to $3 \cos^2 \theta - 1$. Consequently, anisotropic effects can be eliminated by spinning the sample at an angle θ, relative to the magnetic field, at which $3 \cos^2 \theta - 1 = 0$. That angle is 54.74° and is termed the *magic angle*. The technique by which the sample is rotated in the magnetic field at the magic angle is *magic-angle spinning* (MAS).

In order for MAS to be effective, the spinning rate must be of the same magnitude as the anisotropic chemical shift that is to be eliminated. For aromatic compounds the chemical shift owing to anisotropy is between 100 and 200 ppm. If a 15 MHz instrument is used for the study, the corresponding rotation rate should be between 100 and 200 ppm of 15 MHz which corresponds to between 1500 and 3000 Hz. Of course, if the operating frequency of the instrument is larger, the rotation rate must also be larger. Air-driven turbines are generally used to perform the rotation. Careful attention must be paid to the design of the sample cell and the turbine in order to allow rotation at those relatively high rates.

Double resonance can be used to cause spin decoupling in solids. The procedure is similar to that described earlier for use in liquids. For studies of nuclei in solids at low concentration or of weakly magnetic nuclei, *cross-polarization* (CP) is valuable. Cross-polarization is a technique for transferring magnetism from a relatively abundant nucleus, such as protons, to a less abundant nucleus, such as ^{13}C. It is often used in combination with MAS to provide structural information from cmr data on solids. A description of MAS/CP is provided in Reference 13.

With ^{13}C CP, two radiofrequency pulses are used at the resonant frequencies of ^1H and ^{13}C in the particular magnetic field to cause the two types of nuclei to precess at the same frequency along the axis corresponding to thc magnetic field. When the nuclei precess at the same frequency, magnetism efficiently can be transferred from protons to ^{13}C nuclei. After the ^{13}C pulse is terminated, the free-induction decay of the ^{13}C nuclei are observed as a function of time using the Fourier transform method; i.e., the decay in ^{13}C signal is observed as the nuclei return to the lower energetic level. After the Fourier transform has been performed, the results are displayed in the usual manner as a function of frequency.

Cross-polarization shortens the spin-lattice relaxation time in solids and allows multiple Fourier transform spectra to be obtained in a reasonable time. CP differs from most nmr methods that have been described earlier in this chapter in that it uses a pulse rather than a continuous wave (CW). Thc cntirc spectral range is simultaneously recorded as a function of time after each pulse instead of recording the spectrum by scanning the magnetic field or the radiofrequency separately.

MAS can also be used to obtain meaningful spectra of solids at rotation rates lower than those normally used, if the spectrum is recorded in two dimensions rather than one, and if double resonance is simultaneously used to eliminate homogeneous dipolar broadening. That method is described in Reference 10. A description of the use of two-dimensional carbon magnetic resonance for obtaining a map of the carbon backbone in a compound can be found in Reference 14.

MAS of solid samples has proven to be of particular value in the study of semiconductors, inorganic solids, glasses, polymers and biological solids. As an example, Rb-87, Sn-119, F-19, Pt-195, Se-77, P-31 and Mn-55 nmr spectra have been used to study semiconductors and solid ionic conductors. Double resonance methods as well as other multiple resonance methods are popular in solid nmr studies. For biological materials pmr, cmr and multiple resonance methods have been used most often.

KINETIC STUDIES

Rate constants can be obtained for some chemical reactions from nmr data. The technique which uses nmr measurements to determine kinetic parameters is *dynamic nuclear magnetic resonance* (dnmr). When a nucleus A on a molecule is coupled to another nucleus B on the same molecule, nucleus A causes spin-spin splitting of the spectral peak for nucleus B. The splitting pattern is observed in the nmr spectrum of the compound. If nucleus A, however, undergoes a rapid chemical reaction which causes the nucleus to become decoupled from nucleus B, the multiplet corresponding to nucleus B collapses to a singlet. That is the reason that the spectral peak corresponding to the hydroxylic proton in neat samples of ethanol (Fig. 16-10) is a singlet rather than a multiplet. The hydroxylic proton undergoes rapid intermolecular exchange and consequently is not on the molecule for a sufficient period to cause the anticipated peak splitting to occur.

Because no coupling occurs when the rate of exchange is rapid, a singlet is observed for the hydroxyl proton and the methylene protons are not split by the hydroxyl proton. At slow rates of exchange the anticipated splitting patterns are observed. The splitting pattern in the nmr spectrum of a sample, therefore, can be used to determine whether a reaction is rapid or slow.

Generally the rate of a chemical reaction is a function of the concentration of the reactants and the temperature at which the reaction is conducted. Typically one of the two variables is changed while the second is held constant. The nmr spectrum is recorded after each variation and the splitting pattern is observed.

In order for coupling to be observed, the coupling constant (hertz) between the peaks corresponding to the two nuclei must be smaller than the rate constant of the first-order or pseudo first-order reaction. The spectra are recorded as a function of concentration or temperature until the multiplet collapses to a singlet. The rate constant is determined from the spectral data at that concentration and temperature.

First-order unimolecular reactions are the simplest to study. When nmr spectra are observed as a function of temperature, the temperature at which the two sets of peaks collapse to form a single peak is the *coalescence temperature* T_c. The coalescence temperature is a function of the magnetic flux density of the externally applied field. For the case in which the two nuclei are coupled to each other, the rate constant at the coalescence temperature is given by

$$k = \pi\left(\frac{\Delta v^2 + 6J^2}{2}\right)^{1/2} \tag{16-16}$$

where J is the coupling constant of a multiplet. The rate constant can be used to calculate the Gibbs free energy of activation at the coalescence temperature by using

$$G = 2.303T_c\left(10.32 + \log\frac{T_c}{k}\right) \tag{16-17}$$

An example of a unimolecular reaction that can be observed with pmr, and which does not involve spin-spin coupling, is nonaromatic ring reversal. Separate peaks are observed for an axial and an equatorial proton on a single ring site when the rate of ring reversal, e.g., conversion between boat and chair configurations, is slow. When the rate is rapid the two configurations rapidly interchange and a single peak is observed at an intermediate chemical shift. In those cases the pmr spectra are typically observed as a function of temperature. For the special case in which two uncoupled singlets that correspond to two sets of nuclei at equal concentrations collapse to a single peak, the rate constant at the coalescence temperature is given by

$$k = \frac{\pi\Delta v}{\sqrt{2}} \tag{16-18}$$

In Eq. (16-18) Δv is the separation (hertz) between the two peaks when the rate of ring reversal is slow; i.e., before the two peaks coalesce. Take note that Eq. (16-18) is identical to Eq. (16-16) when $J = 0$.

An alternative approach to determining rate constants is to use a theoretically derived equation to calculate the spectrum that would be obtained at a particular rate constant. Owing to the complexity and length of the calculations, computers are used to obtain theoretical spectra. A series of spectra that have been theoretically obtained at different rate constants is compared to the observed spectrum of the sample. The rate constant of the sample corresponds to the rate constant of the theoretical spectrum which is identical to that of the sample. The spectra can be compared either visually or with the aid of a computer. That method is more generally useful than that which relies upon a single measurement at the coalescence temperature because the rate constant can be determined at many temperatures. The method by which rate constants are determined in that manner from an entire spectrum is the *complete lineshape* (cls) method. Further information relating to nmr kinetic studies is available in Reference 6 and in the references listed there.

Plots of log k obtained with the cls method as a function of the inverse of the absolute temperature at which the rate constant was determined can be used with the Arrhenius equation:

$$k = Ae^{-E_a/RT} \tag{16-19}$$

to determine A and the *activation energy* E_a for a reaction. Rearrangement of the Arrhenius equation results in

$$\log k = \frac{-E_a}{2.303RT} + \log A \tag{16-20}$$

from which it is apparent that the slope of a plot of log k as a function of $1/T$ is $-E_a/2.303R$, and the intercept on the log k axis is log A. Similarly, values of the enthalpy and entropy of activation can be obtained. Most physical chemistry texts can be consulted for the appropriate equations.

In some cases, during a study of nmr spectra as a function of time, a peak is observed to appear and then disappear. In those cases the peak corresponds to an intermediate in a chemical reaction. The identification of an intermediate is a valuable aid in determining the mechanism of the reaction.

In a few cases a negative nmr peak is observed during the recording of a series of nmr spectra as a function of time. Generally the peak is not caused by the reactants but disappears as the reaction proceeds. Negative peaks correspond to emission rather than absorption and can be attributed to formation of a free radical. The presence of a negative peak is normally caused by *chemically induced dynamic nuclear polarization* (CIDNP). CIDNP results from spin interactions between a nucleus and the unpaired electron of the free radical. A description of the phenomenon is included in Reference 6, chap. 7.

IMPORTANT TERMS

- Anisotropy
- Carbon magnetic resonance
- CIDNP
- Chemical shift
- Coalescence temperature
- Complete lineshape (cls) method
- Coupling constant
- Cross-polarization
- Double resonance
- Dynamic nmr
- Electron shielding
- Field sweep
- First-order spectrum
- Frequency sweep
- Heteronuclear double resonance
- Homonuclear double resonance
- Internuclear double resonance
- Lanthanide shift reagent
- Magic angle
- Magic-angle spinning
- Magnetic flux density
- Magnetic moment
- Magnetic quantum number
- Noise decoupling
- Nuclear magnetic resonance (nmr)
- Nuclear magneton
- Nuclear Overhauser effect
- Nuclear relaxation
- Parts per million
- Proton magnetic resonance
- Quadrupolar nuclei
- Quadrupolar relaxation
- Ring current
- Selective spin decoupling
- Shielding constant
- Shift reagent
- Shim coil
- Spin decoupling
- Spin-lattice relaxation
- Spin-lattice relaxation time
- Spin quantum number
- Spin-spin coupling
- Spin-spin relaxation
- Spin-spin relaxation time
- Spin-spin splitting
- Spin tickling
- TMS
- Tumbling

PROBLEMS

Theory

16-1 Calculate the energy of radiation that is required to excite a ^{13}C nucleus from the lower to the upper energetic level if the applied field has a flux density of 1.000×10^4 G.

16-2 Calculate the energy of radiation that is required for excitation in each of the two allowed transitions for a ^{14}N nucleus if the applied field has a magnetic flux density of 1.000×10^4 G.

16-3 Calculate the frequency required for each of the transitions described in Probs. 16-1 and 16- 2.

16-4 Determine the ratio of ^{13}C nuclei in the upper energetic level to the lower energetic level at 30°C in a magnetic field that has a flux density of 25,100 G.

16-5 Determine the ratio of ^{19}F nuclei in the upper energetic level relative to the lower level at 25°C in a magnetic field that has a flux density of 1.500×10^4 G.

16-6 Calculate the pmr chemical shift in a 100 MHz instrument if the reference nucleus absorbs at a magnetic flux density that is 0.063 G greater than that at which the sample nucleus absorbs. A 100-MHz instrument operates at 23.48 kG. Compare the answer to that in Sample problem 16-3 and comment on the relative advantage of a 100-MHz instrument as compared to a 60-MHz instrument.

16-7 Determine the relative area under each of the peaks of the multiplet corresponding to the methine proton in $(CH_3)_2CHCl$.

16-8 Predict the splitting patterns and the ratio of the peak areas in each multiplet as well as between the multiplets in the pmr spectrum of $CH_3CH_2OC(CH_3)_3$.

Qualitative Analysis

The chemical shifts of some of the major peaks in each of the following pmr spectra are listed near the tops of the peaks. Identify the compound that corresponds to each spectrum from the spectrum and any additional information that is found in each problem. The compounds can contain C, H, N, O, and Cl.

16-9 The molecular weight of the compound is 123.

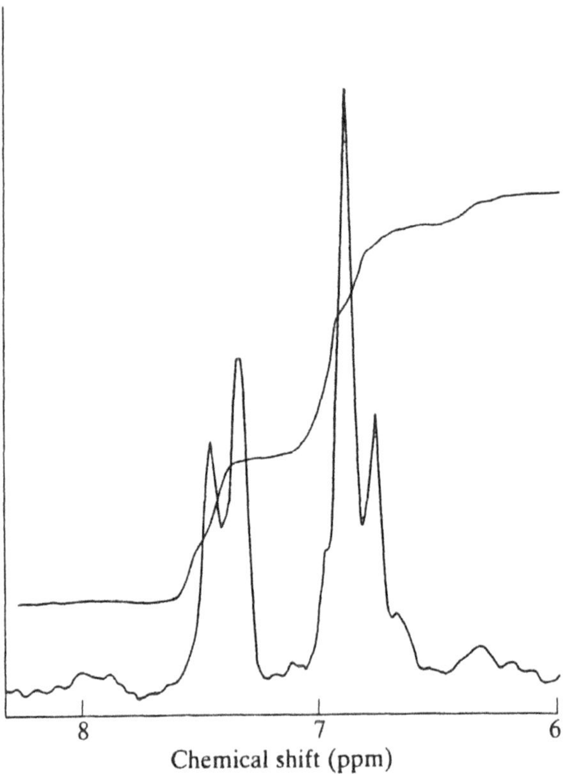

Figure P16-9

16-10 The molecular weight is 92.

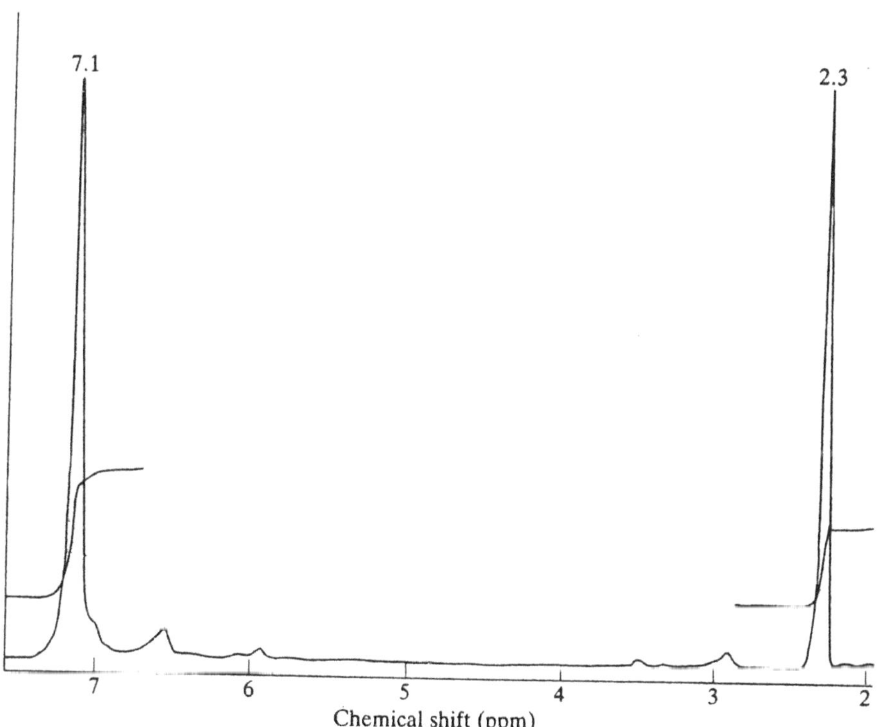

Figure P16-10

16-11 The molecular weight is 136.

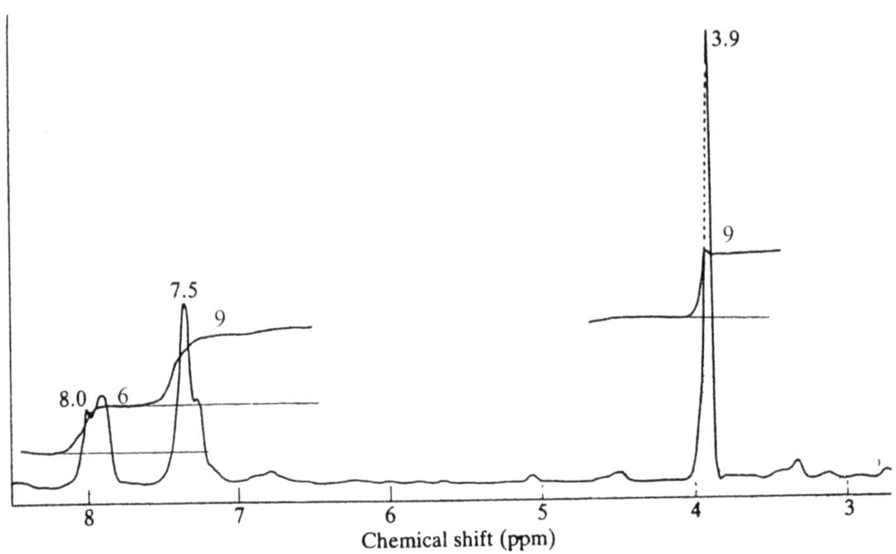

Figure P16-11

16-12 The formula is $C_5H_{10}O_2$.

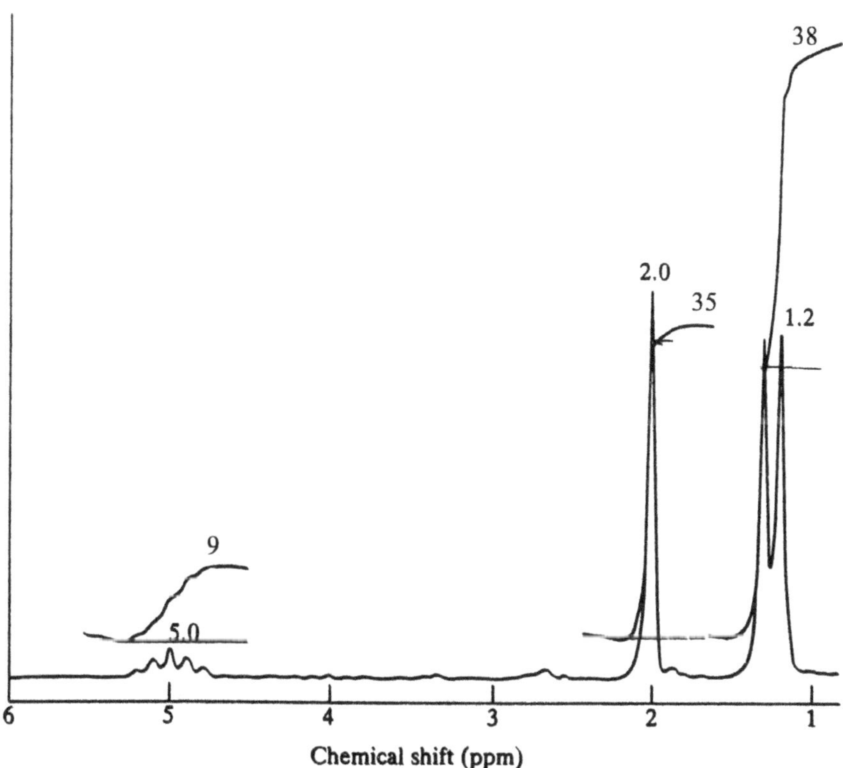

Figure P16-12

16-13 The formula is C_6H_{12}.

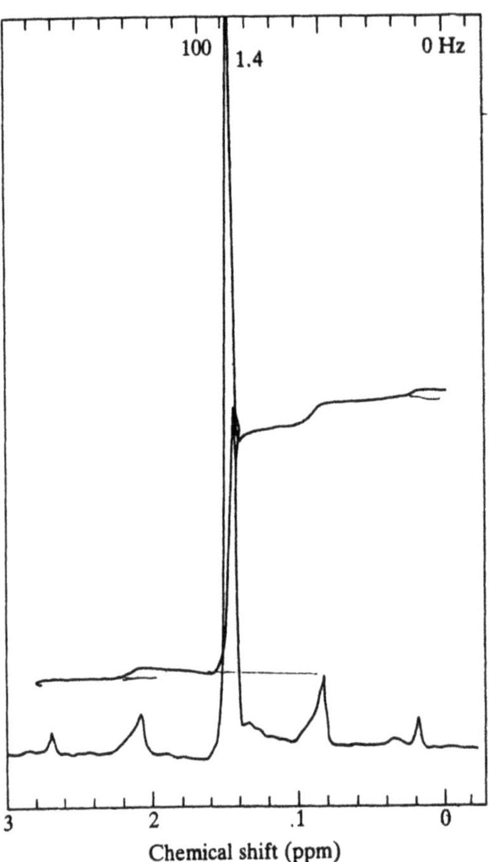

Figure P16-13

16-14 The molecular weight is 85.

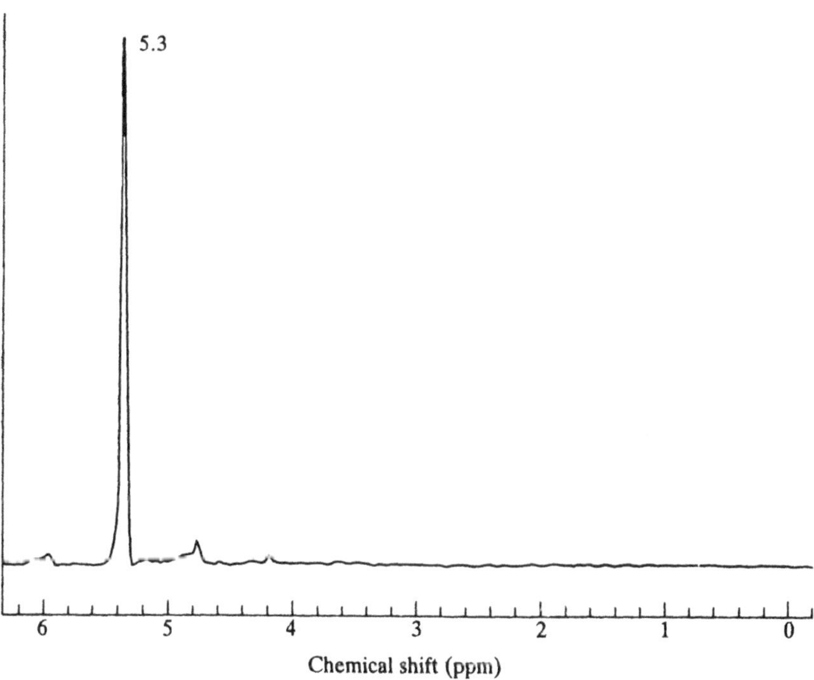

Figure P16-14

16-15 The molecular weight is 62.

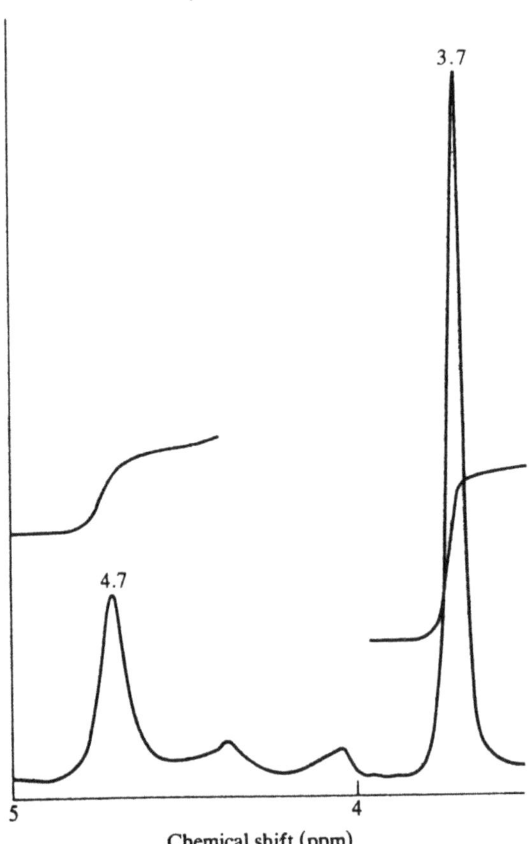

Figure P16-15

16-16 The formula is $C_6H_{12}O$.

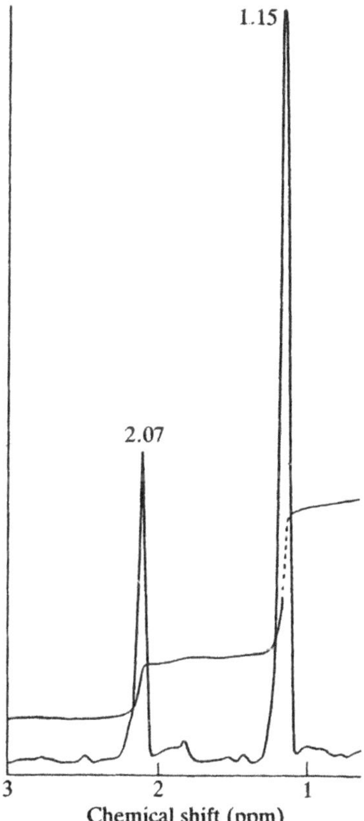

Figure P16-16

16-17 The molecular weight is 120.

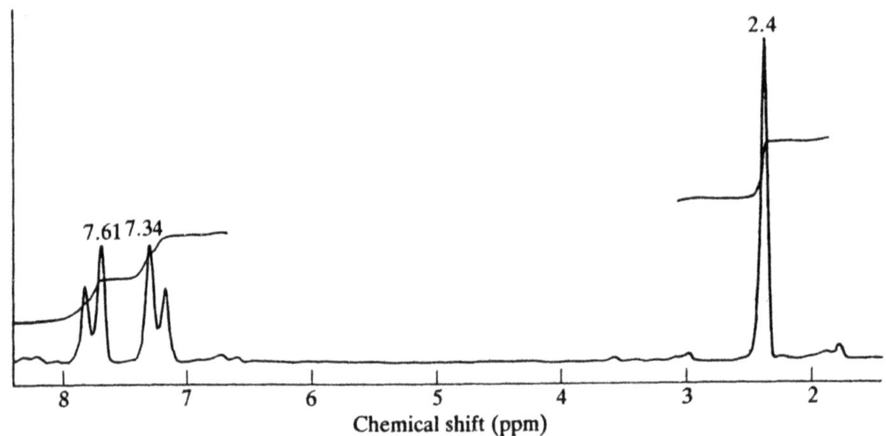

Figure P16- 17

16-18 The molecular weight is 79.

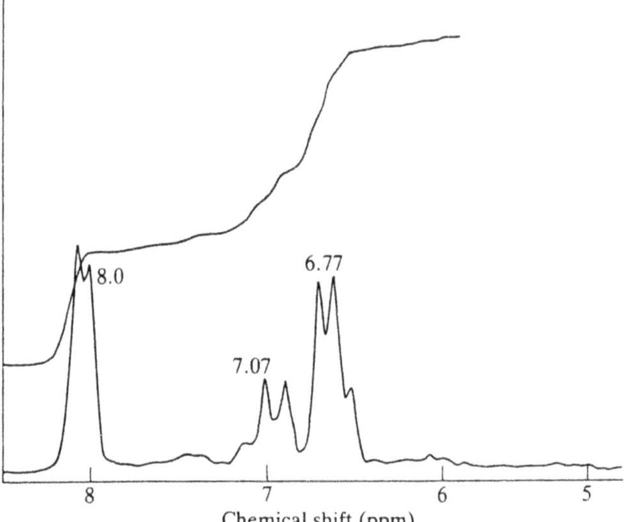

Figure P16-18

Quantitative Analysis

16-19 Methylene bromide was chosen as the internal standard for the quantitative analysis of propionaldehyde in a solution. A series of standard solutions were prepared by using pipets to successively add 0.5, 1.0, 1.5, 2.0, and 2.5 mL of propionaldehyde to each of five labeled 10-mL volumetric flasks. A pipet was used to add 1 mL of methylene bromide to each flask and the flasks were filled to the mark with solvent. An 8-mL portion of the analyte was added by pipet to each of three 10-mL volumetric flasks. A 1-mL portion of methylene bromide was added to each flask and the solutions were filled to the mark with solvent. The integrated peak areas corresponding to the methylene bromide protons and the aldehydic proton of propionaldehyde were measured for each solution. Determine the percentage (v/v) of propionaldehyde in each sample solution, the mean percentage, and the standard deviation of the results.

Volume of aldehyde, mL	Relative aldehydic peak area	Relative CH_2Br_2 peak area
0.5	5.0	12.0
1.0	8.7	10.0
1.5	18.6	14.0
2.0	19.5	11.0
2.5	26.8	12.0
Sample 1	19.4	12.1
Sample 2	18.5	11.2
Sample 3	19.5	12.4

Kinetic Studies

16-20 At –85°C the rate of ring reversal for a particular six-membered ring is slow and two singlets are observed corresponding to an axial and an equatorial proton. At – 85°C the separation between the two singlets is 25 Hz. At –50°C the peaks coalesce to a single peak. Estimate the rate constant for the conversion at the coalescence temperature.

REFERENCES

1. Silverstein, R. M., G. C. Bassler, and T. C. Morrill: *Spectrometric Identification of Organic Compounds*, 4th ed., Wiley, New York, 1981.
2. Levy, G. C., and G. L. Nelson: *Carbon-13 Nuclear Magnetic Resonance for Organic Chemists*, Wiley-Interscience, New York, 1972.
3. Ionin, B. I., and B. A. Ershov: *NMR Spectroscopy in Organic Chemistry*, Plenum, New York, 1970
4. Johnson, L. F., and W. C. Jankowski: *Carbon-13 NMR Spectra, A Collection of Assigned, Coded and Indexed Spectra*, Wiley, New York, 1972.
5. "Nuclear Magnetic Resonance Spectra," Sadtler Research Laboratories, Philadelphia, Pennsylvania.

6. Lambert, J. B., H. F. Shurvell, L. Verbit, R. G. Cooks, and G. H. Stout: *Organic Structural Analysis*, Macmillan, New York, 1976.
7. "NMR Chemical Shift Index," Sadtler Research Laboratories, Philadelphia, Pennsylvania.
8. Reinhold, M.: *Amer. Laboratory*, 14(2): 164(1982).
9. Gerstein, B. C.: *Anal. Chem.,* 55: 781A(1983).
10. Gerstein, B. C.: *Anal. Chem.,* 55: 899A(1983).
11. Dybowski, C., S. Bai and S. van Bramer: *Anal. Chem*., 76:3263(2004).
12. Dybowski, C. and S. Bai: *Anal. Chem*., 78:3853(2006).
13. Miknis, F. P., V. J. Bartuska, and G.E. Maciel: *Amer. Laboratory,* 11(11): 19(1979).
14. Shoolery, J. N.: *Ind. Res. Dev.,* 25(11): 90(1983).

CHAPTER

SEVENTEEN

ELECTRON SPIN RESONANCE SPECTROMETRY

Electron spin resonance (esr) occurs when a spinning electron in an externally applied magnetic field absorbs sufficient electromagnetic radiation to cause inversion of the spin state of the electron. Electron spin resonance is also known as *electron paramagnetic resonance* (epr) and as *electron magnetic resonance.* A substance that contains at least one unpaired electron is *paramagnetic.* Information relating to electron spin resonance in addition to that described in the chapter can be found in References 1 to 4.

All electrons have a spin quantum number I (also often given the symbol S for electrons) that is $\frac{1}{2}$. Electron spin resonance spectra cannot be obtained of substances that contain only paired electrons, because the opposing spin states of paired electrons yield a substance that has no net magnetic moment. Only unpaired electrons can be observed while using esr spectroscopy. Examples of substances which can be studied with esr spectroscopy include organic free radicals, substances in a triplet state (two unpaired electrons with a total spin quantum number of 1), various compounds or complexes of transition metals which contain unpaired *d*-shell electrons, compounds or complexes of lanthanide or actinide elements which contain unpaired electrons in 4*f*, 5*f*, or 6*d* orbitals, compounds which have a singlet ground state but a low-energy easily populated triplet state, and metals or semiconductors which have electrons in a conduction band.

Electron spin resonance spectra are obtained in a manner that is analogous to the manner in which nmr spectra are obtained. Normally the strength of the applied magnetic field is varied while the frequency of the electromagnetic radiation is held fixed. Some instruments, however, have been designed to vary the EMR frequency while maintaining a constant field strength. The field-sweep method is used in most instruments because the dimensions of the sample cavity permit only a single resonant frequency. Most esr spectra consist of a plot of the first derivative of the absorbed energy as a function of frequency. First-derivative spectra are recorded in order to improve resolution, to decrease background noise, and because the detectors that are normally used in esr spectrometers respond to changes in absorption.

THEORY

The theory of esr spectroscopy closely parallels that of nmr spectroscopy. It is advisable to review the appropriate sections of Chapter 16 prior to reading this section. In some cases it is slightly more complicated than nmr theory because both the spinning electron and the motion of the electron in its orbital about an atom or molecule contribute to the magnetic moment. In most cases the contribution to the magnetic moment from orbital motion is negligible in comparison to that from the spin of the electron. Because the spin quantum number of an electron is $\frac{1}{2}$, only two stable orientations corresponding to m_s values of $-\frac{1}{2}$ and $+\frac{1}{2}$ exist in a magnetic field. The magnetic quantum number of an electron is given the symbol m_s in order to distinguish it from the magnetic quantum number m_I of a nucleus.

As with spinning nuclei, the change in energy between the two stable energetic levels of an electron in a magnetic field are given by

$$\Delta E = E_H - E_L = 2\mu B_0 \tag{17-1}$$

where μ is the magnetic moment of the electron measured in units of joules per tesla or joules per gauss. A tesla T is the SI unit of magnetic flux density; a gauss G is the corresponding cgs unit: 1 G is equivalent to 10^{-4} T. The value of μ in Eq. (17-1) is identical to $\mu\beta$ in Eq. (16-4) owing to the change in units. In Eq. (16-4) μ was in units of nuclear magnetons and β was the conversion factor required to convert nuclear magnetons to joules per gauss. The magnetic flux density (magnetic field strength) of the applied magnetic field is B_0. The magnetic moment of an electron is usually several orders of magnitude greater than that of a nucleus. The larger magnetic moment of electrons results in increased analytical sensitivity when compared to nmr analytical results.

The frequency that is required to cause resonance in an electron can be calculated by equating ΔE from Eq. (17-1) to hv from the Planck equation. The result is shown in the following:

$$\Delta E = hv = 2\mu B_0$$

$$v = \frac{2\,\mu B_0}{h} \tag{17-2}$$

The resonant frequency of electrons is directly proportional to the magnetic flux density of the magnetic field. For a free electron, the magnetic moment μ_e is 9.285×10^{-24} J/T.

A typical frequency used for esr studies is 9500 MHz (9.5 GHz). The corresponding magnetic flux density at which a free electron resonates is 0.339 T (3390 G), as calculated by substituting the frequency into Eq. (17-2) and solving the equation for B_0. For electrons that are associated with an atom or molecule, the magnetic moment at a resonant frequency of 9500 MHz can range from about 0.005 to 0.55 T. Radiation at a frequency of 9500 MHz is the *X band*, and is used often for esr studies. Other frequencies which are used for esr studies include, but are not limited to, 1-2 GHz (*L band*), 2-4 GHz (*S band*), 23 GHz (*K band*) and 35 GHz (*Q band*). The corresponding magnetic flux densities for a free electron are 0.0357-0.0714, 0.0714-0.143, 0.821 and 1.249 T. Some instruments have been designed to operate at very low frequencies from approximately 200 - 600 MHz. They are typically used for *in vivo* and other biological studies. The two extremes of instrumental frequencies are illustrated by the instruments described in References 5 and 6. The instrument described in Reference 5 operates at 95 GHz while the instrument in Reference 6 operates at 283 MHz.

The magnetic moment of an electron in a magnetic field is related to *I* by

$$\mu = g\mu_B I \tag{17-3}$$

where μ_B is the Bohr magneton (9.2732×10^{-28} J/G or 9.2732×10^{-24} J/T). The *g-factor* or *g-value* g in Eq. (17-3) is a dimensionless constant for an electron in a particular environment. Some spectroscopists refer to g as the spectroscopic splitting factor. For a free electron the *g-value* g_e is 2.002319. Electrons which are associated with a chemical substance have g-factors and magnetic moments that vary from that of a free electron, owing to the effect of the orbital in which the electron resides and the environment of the electron. Although most g-factors are between 2 and 3, the values can range from about 1.5 to about 6.

The g-factor in esr is analogous to the chemical shift in nmr. Because the g-factors of many compounds are similar, their utility for qualitative analysis is normally more limited than that of the chemical shift in nmr spectroscopy. Substitution of the magnetic moment from Eq. (17-3) and the spin quantum number for an electron into Eqs. (17-1) and (17-2) yields

$$\Delta E = 2\frac{g\mu_B}{2}B_0 = g\mu_B\,B_0 \tag{17-4}$$

and

$$\nu = \frac{2(g\mu_B/2)B_0}{h} = \frac{g\mu_B\,B_0}{h} \tag{17-5}$$

Relaxation

Relaxation is necessary in esr in order to prevent saturation of the upper energetic level during absorption. The two major mechanisms by which relaxation can occur during esr studies are *spin-lattice relaxation* and *spin-spin relaxation*. Spin-lattice relaxation results from interactions between an electron in an excited energetic level and the surroundings. Dipole-dipole interactions constitute one path by which spin-lattice relaxation can occur. Because spin-lattice relaxation is caused by the interaction between the unpaired electron and thermal energetic levels of the lattice, relaxation times can be increased by lowering the temperature. In some cases the sample must be cooled to the temperature of liquid helium, hydrogen, or nitrogen in order to observe spectra with relatively narrow line widths.

Spin-spin relaxation occurs when a spinning electron in the upper energetic level transfers its energy to an electron in the lower level. The electrons exchange spin states. The effect of spin-spin relaxation is to shorten the relaxation time. In solution, exchange of electron spins generally occurs through bimolecular processes. Two molecules or ions which each contain an unpaired electron collide and the electrons exchange spins. The number of electrons undergoing relaxation by that process can be decreased by decreasing the concentration of free radicals in the solution. As the concentration decreases, the number of collisions between molecules with unpaired electrons decreases, and the exchange is less frequent. The consequent narrowing of the esr spectral lines is *exchange narrowing*.

As in other forms of spectroscopy, the spectral line width is related to the relaxation time. As the relaxation time increases, the spectral line narrows. It is desirable to have a relaxation time which is sufficiently rapid to prevent saturation of the upper energetic level, but sufficiently slow to yield narrow spectral peaks. A typical sharp line width of an esr spectral peak at half the peak height for an organic radical is about 1×10^{-5} T or 0.1 G. In some cases the change in spectral line width with concentration or with temperature can be used to obtain kinetic information about a particular relaxation process. The use of esr in kinetic studies is similar to the use of nmr in kinetic studies.

The ratio of the number of electrons in the upper energetic level n_H to those in the lower level n_L can be calculated by using

$$\frac{n_H}{n_L} = e^{-2\mu B_0/kT} = e^{-h\nu/kT} \qquad (17\text{-}6)$$

Instrumentation

It is customary to name an instrument that is used to make esr measurements an esr spectrometer. The instrumental components that are required in a typical esr spectrometer include a source of electromagnetic radiation, a source of a stable, but variable, magnetic flux density, a sample holder, a detection system, and a readout device. A diagram of the major components of an esr spectrometer is shown in Fig. 17-1.

Many, but not all, instruments are designed to operate at constant frequency. The common frequencies of operation were listed in the Theory section. Electromagnetic radiation that oscillates at a frequency of 9.5 GHz is used often. An example of a radiative source that is used in esr spectrometers is a *reflex klystron*. A diagram of a reflex klystron is shown in Fig. 17-2. A heated cathode emits a beam of electrons which are accelerated by a positive potential on an accelerator grid toward a resonant cavity. The beam of electrons enters the resonant cavity through grid 1 and is accelerated or decelerated by an alternating potential that is applied to grid 2. The potential of the grid is alternated at a frequency that is related to the desired output frequency of the klystron. When the potential is positive, the electrons are accelerated. When the potential is negative, the electrons are slowed. Of course, the effect upon the electrons depends upon their distance from the grid at any instant.

Because the grid potential alternates, the electrons that leave the cavity are velocity-modulated. After passing through grid 2, the more rapid electrons that left the cavity when the grid potential was positive overtake the slower electrons that left the cavity earlier when the grid potential was less positive. With proper design the electrons accumulate in groups prior to returning to the cavity. The electrons are repelled by a negative reflector filament and directed back toward the resonant cavity. The frequency at which the groups of electrons reenter the cavity is adjusted by varying the potential on the reflector filament.

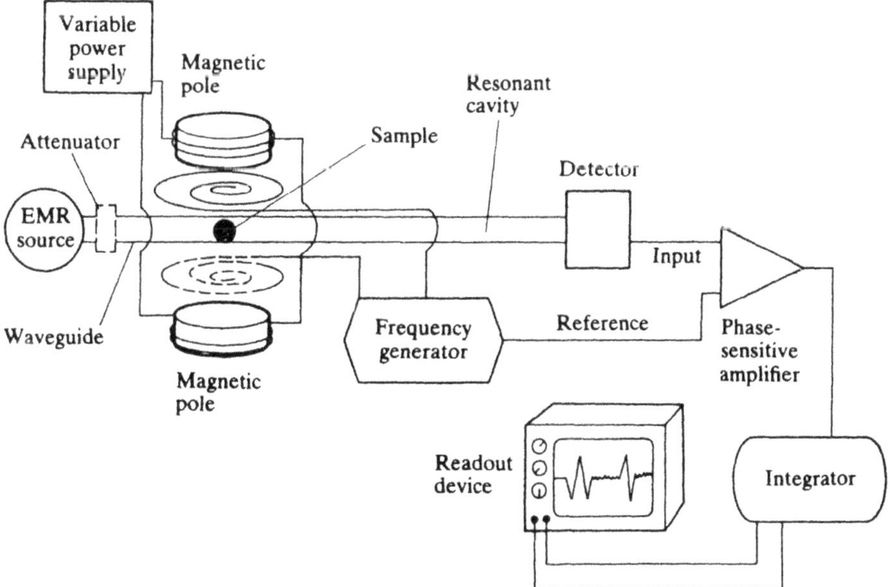

Figure 17-1 A diagram of an esr spectrometer.

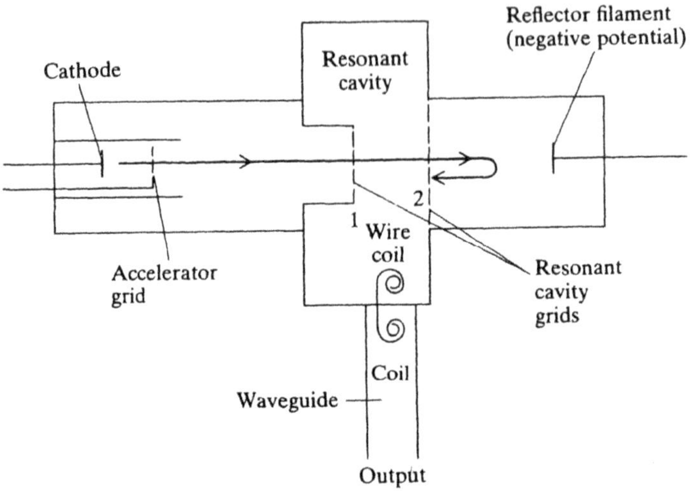

Figure 17-2 A diagram of a klystron.

When the frequency at which the returning groups of electrons strike grid 2 is identical to the resonant frequency of the cavity, a resonating magnetic field is established in the resonant cavity. The resonant frequency of the cavity can be altered by deforming the cavity walls or by using plungers to alter the length of the cavity. Power is removed from the resonant cavity at the resonant frequency through a coil of wire. The alternating magnetic field in the coil establishes an oscillating magnetic field in a *waveguide* which carries the signal to the sample. A waveguide is a hollow rectangular tube. Generally the waveguide is constructed from either copper or brass and is coated on the interior walls with a highly conductive element such as silver or gold. The output frequency of a klystron can be monitored with a frequency meter or a device that proportionately decreases the output frequency and measures the decreased frequency with a frequency counter. Further information on the design and use of klystrons as well as on other aspects of esr instrumentation is available in Reference 7.

After leaving the klystron, the radiation passes through a waveguide and strikes the sample which is held in place in a cavity in the waveguide. The magnetic field is applied to the sample with an electromagnet. The poles of the electromagnet are placed on opposite sides of the sample holder in the waveguide. The flux density of the magnetic field can be scanned by varying the voltage from the power supply that is attached to the magnet. In order to scan through the region of greatest interest in esr spectroscopy at a frequency of 9.5 GHz, the magnetic flux density should be capable of being altered throughout a range from about 0.005 to 0.55 T or 50 to 5500 G. In some instruments more than one set of magnetic coils are used. The instrument described in Reference 6 has a cylindrical yoke with 16 poles. The magnetic field is generally stable to 1 part in 10^6 or 10^7. When temperature controlled magnets are used, greater stability is possible. The flux density of the magnetic field is monitored either with a gaussmeter or by comparison with a standard compound of known g-factor.

In addition to the electromagnetic field, an alternating magnetic field is applied to the sample. The frequency of the alternating field is 100 kHz in many spectrometers, but can be other values. The amplitude of the alternating field is small in comparison to that of the field from the electromagnet. The sample experiences the effect of the sum of the two fields. Use of an alternating field allows the detector to discriminate between the decreased signal at the detector owing to absorption and the decreased signal owing to dispersion. The spectrometer is operated in a manner that causes the two signals to be 90° out-of-phase. A detection system that is phase-sensitive permits measurement of the signal owing to absorption while minimizing measurement of the signal owing to dispersion. An increase in the signal-to-noise ratio results.

If the frequency of the alternating magnetic field is 400 Hz or less, the coils that are used to apply the field can be mounted outside the waveguide. At higher frequencies the coils must be mounted on the walls of the cell holder within the waveguide because higher frequencies do not sufficiently penetrate the metallic walls of the waveguide. Application of an alternating field to the sample causes the detector to measure the first derivative of the signal owing to absorption.

The resonant cavity can have several designs. A popular resonant cavity is the *magic T* that is illustrated in Fig. 17-3. The output from the klystron is perpendicularly applied through a waveguide to the center of a T that is constructed from waveguides. Because radiation enters the T perpendicularly, it is equally divided between arms 2 and 3 as long as the two arms absorb the radiation equally. No radiation enters arm 4 which contains the detector. Arm 3 contains the sample cell and arm 2 contains something that has the same impedance as a nonabsorbing sample. A tuner which is included either in arm 2 or arm 3 permits the impedance of the two arms to be matched in the absence of an absorbing sample so that no radiation strikes the detector. When the sample absorbs radiation, the two arms are no longer matched and some of the radiation from the klystron enters arm 4 and is measured by the detector. Other resonant cavities are described in Reference 7.

The cell is usually a cylindrical quartz or glass tube that is about 3 mm in diameter. In some cases capillaries or thin rectangular tubes are used. Unlike nmr cells, esr cells are not rotated by a turbine in the cell cavity. In some instruments provision is made for orienting the cell at different angles relative to the magnetic field in order to study anisotropic effects in solids. Multiple instruments have been constructed which have cell dimensions which are sufficiently large to accommodate small laboratory animals such as rats. *Surface coil resonantors* are designed to be placed on the surface of a living animal or person, in a manner similar to that in which a stethoscope is used, rather than placing the animal in a cell cavity.

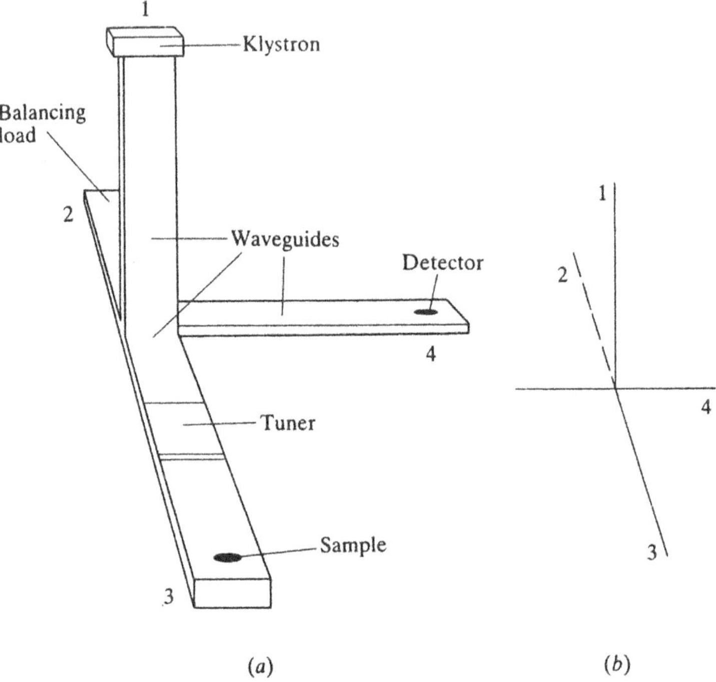

Figure 17-3 A diagram of a magic T. (*a*) Diagram illustrating the relative locations of the klystron, detector, and sample; (*b*) the spatial relationship between the arms of the device.

The cell is constructed from a material that does not absorb radiation in the region in which the sample absorbs. Studies of free radicals and triplet-state electrons are usually performed in quartz cells. Generally esr spectra that are obtained in quartz cells have better signal-to-noise ratios than those obtained in glass cells. Glass cannot be used for studies of triplet-state electrons because most glasses contain magnetic impurities such as Fe(III) which absorb in the same region as the sample. Glass is often used for studies of organic free radicals in polar solvents.

At low temperatures the cell can be a miniature Dewar flask. Dual cells in separate cavities can be used to simultaneously obtain the spectrum of a reference and a sample. The two signals are distinguished from each other by using different magnetic field frequencies in the two cavities and frequency-selective detectors.

The detector is a device that responds to radiation in the gigahertz region. Often the detector is constructed from a crystalline silicon-tungsten semiconductor. The crystal functions as a rectifier by changing alternating electromagnetic radiation that strikes it from the source into a direct current. The current is amplified, electronically manipulated, and directed to the readout device. The readout device is a recorder, video display terminal, or other computer-controlled device. Because the magnetic field that is applied to the sample has an alternating component, the signal that is sent to the readout device is the first derivative of the peak owing to absorption. A sketch of a single esr peak is shown in Fig. 17-4. The *peak-to-peak line width* ΔB of an esr peak is the separation on the magnetic flux-density axis between the minimum and maximum of the first-derivative curve.

Detection limits are dependent upon the type of sample, sample size, cell design, detector sensitivity, peak shape, frequency of incident radiation, and the electronics of the instrument. For most instruments the detection limit corresponds to about 10^{10} or 10^{11} unpaired electrons. That corresponds to a concentration of about 10^{-10} or 10^{-9} *M*. In water or other polar solvents, the lower concentration limit at which an esr peak can be observed is about 10^{-7} *M*. Quantitative assays typically can be done at concentrations of about 10^{-6} *M* or greater. The sensitivity increases as the frequency of the incident radiation increases. If 35 GHz radiation (*Q* band) is used rather than 9.5 GHz radiation (*X* band), the sensitivity is increased by a factor of about 20.

Bench top and considerably larger commercial instruments are available. The size of the instrument increases with the size of the magnet. The resolution and sensitivity increases with the field strength of the magnet and the frequency of the radiation. Many modern instruments take advantage of signal averaging methods such as Fourier transforms. Instruments which utilize Fourier transforms are Fourier transform electron spin resonance FTESR spectrometers.

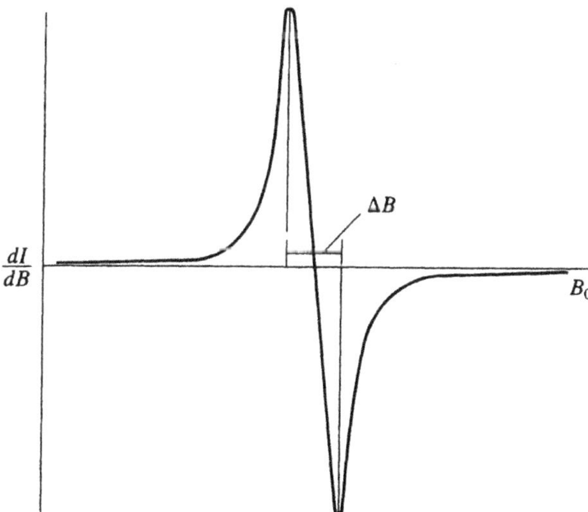

Figure 17-4 A sketch of an esr spectral peak. The first derivative of the intensity of the radiation that strikes the detector is plotted as a function of the applied, steady-state magnetic flux density. ΔB is the peak-to-peak line width of the peak.

Spin-Spin Splitting

Magnetic nuclei in the vicinity of an unpaired electron can couple with the spinning electron in much the same way as two nuclei can couple during nmr studies. Consequently, the spectral peak owing to the electron can be split into a multiplet. Because spin-spin splitting in esr spectrometry is similar to that in nmr spectrometry, it is not described in detail.

A single magnetic nucleus with a spin quantum number I in the vicinity of the electron causes spin-spin splitting of the esr signal into $2I + 1$ equally spaced lines. A set of n, equivalent, magnetic nuclei cause splitting into $2nI + 1$ lines. Spin-spin splitting that is caused by a magnetic nucleus is *hyperfine splitting*. In order for spin-spin coupling to occur, it is necessary that the unpaired electron have a finite probability of being within the range of the magnetic field of the nucleus. Because free radical electrons are often delocalized, the electrons can approach many magnetic nuclei within a molecule or ion. All of the magnetic nuclei can spin-spin couple with the electron. Consequently, the splitting patterns that are observed in esr spectra are often more complex than those observed in nmr spectra.

In the presence of a magnetic field, the energetic level corresponding to each value of m_s for the unpaired electron in a hydrogen atom is given by

$$E = g\mu_B B_0 m_s + Am_I m_s \tag{17-7}$$

Similar expressions can be written for more complex atoms and molecules. In Eq. (17-7), g is the g-factor; μ_B is the Bohr magneton; m_s is the quantum number for the electron; m_I is the quantum number for the magnetic ion; and A is the hyperfine splitting constant. Normally the applied magnetic field strength B_0 is much greater than the magnetic field strength of the spinning nucleus, and the splitting between the lines is independent of the applied field.

Equation (17-7) ignores the effects of nuclear interactions. The selection rules for the allowed transitions in esr spectrometry are $\Delta m_s = \pm 1$ and $\Delta m_I = 0$; that is, the quantum number for the electron can change by one unit but that of the nucleus is not allowed to change. If the electron is coupled to more than one nucleus, the second selection rule is $\Delta M_I = 0$ where M_I is the total nuclear spin moment. In practice it is sometimes possible to observe spectral lines that correspond to forbidden transitions. Generally these lines are of much lower intensity than those corresponding to allowed transitions.

In addition to their use in chemical analysis, hyperfine splitting patterns have been used to assign spin quantum numbers to isotopes. The spin quantum numbers of ^{60}CO, ^{57}Fe, ^{99}Ru, ^{101}Ru, and ^{50}V have been assigned using that method.

Figure 17-5 shows the energetic levels of an unpaired electron that is coupled to a single magnetic nucleus which has a spin quantum number equal to 2. In the absence of the magnetic nucleus, the electron can reside in either of two energetic levels corresponding to m_s values of $-\frac{1}{2}$ or $+\frac{1}{2}$ in a magnetic field. The lower energetic level is assigned the negative quantum number. The lower energetic level (aligned with the field) has an m_s value $(-\frac{1}{2})$ whose sign is opposite to that of a nucleus which is in the lower level, owing to the difference in sign of the charge on the two bodies. The electron is negative while a nucleus is positive. The two levels are illustrated in Fig. 17-5*a*.

In the presence of a single magnetic nucleus, each energetic level is split into $2I + 1$ levels by the $2I + 1$ stable orientations of the nucleus in the external magnetic field. For a nucleus with a spin quantum number equal to 2, each electron level is split into five sublevels corresponding to m_I values of $-2, -1, 0, +1$, and $+2$ as shown in Fig. 17-5*b*. The allowed transitions can be predicted using the selection rules $\Delta m_s = \pm 1$ and $\Delta m_I = 0$. It is important to realize when applying the selection rules that in the upper energetic levels (when m_s is $+\frac{1}{2}$), the order of increasing energy of the m_I values is opposite to that in the lower energetic levels (when m_s is $-\frac{1}{2}$). In this instance five transitions are allowed as predicted using the $2I + 1$ rule. Each transition corresponds to an observed spectral line.

The relative intensities of esr hyperfine lines can be calculated in the same manner that was used to calculate the intensities of nmr lines that were split owing to spin-spin coupling. If the unpaired electron interacts with a single magnetic nucleus, each of the $2I + 1$ energetic levels that are associated with each electron level are nearly equally populated and each of the lines are of equal intensity. As an example, if the esr line is split into a quartet by ^{35}Cl (I is $\frac{3}{2}$, $2I + 1 = 4$), each of the four hyperfine lines is of equal intensity.

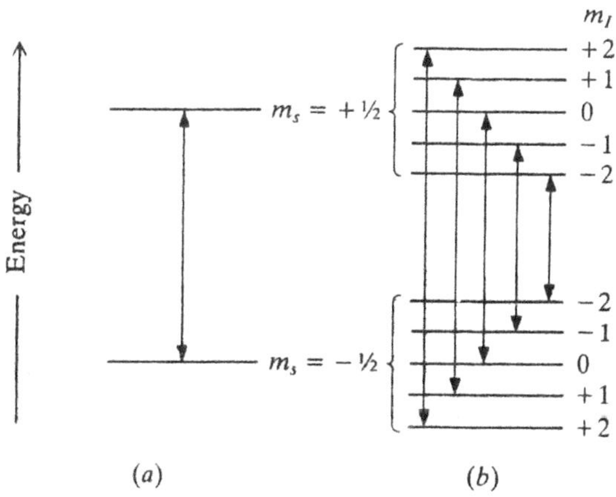

Figure 17-5 Energy-level diagrams illustrating the stable levels in which an unpaired electron can exist in a magnetic field. The allowed energetic transitions are indicated by the arrows. (*a*) No magnetic nucleus is present; (*b*) a magnetic nucleus that has a spin quantum of 2 is coupled to the electron.

An esr peak is split into $2nI + 1$ hyperfine lines of unequal intensity by n equivalent magnetic nuclei. A peak which is coupled to n_1 nuclei of one type and n_2 nuclei of a second type is split into $(2n_1 I_1 + 1)(2n_2 I_2 + 1)$ hyperfine lines where I_1 is the spin quantum number of the n_1 nuclei and I_2 is that of the n_2 nuclei. Similarly, a spectrum is split into $(2n_1 I_1 + 1)(2n_2 I_2 + 1) \ldots (2n_x I_x + 1)$ hyperfine lines by a set of x types of magnetic nuclei.

If the esr line is coupled to n equivalent nuclei with a spin quantum number of $\frac{1}{2}$ (such as ^{1}H nuclei), the intensity of the hyperfine lines are in the ratio of the coefficients of a binomial expansion. The intensities of the hyperfine lines in the esr spectrum of a methyl radical $\cdot CH_3$ can serve to illustrate the point. The peak is split by the three protons into $2nI + 1 = 2(3)(\frac{1}{2}) + 1 = 4$ peaks. The relative intensities of the peaks are in the ratio of the coefficients in the binomial expansion $(a + b)^3 = 1a^3 + 3a^2b + 3ab^2 + 1b^3$, that is, in the ratio 1: 3 : 3 : 1. The ratios can easily be determined with the aid of Pascal's triangle (Fig. 16-6).

The spectrum of an unpaired electron which is coupled to n nonequivalent protons is split into 2^n hyperfine lines. Experimentally, esr peak areas are obtained by instrumentally measuring the second integral of the first-derivative spectrum. The second integral is obtained by performing two consecutive integrations of the spectrum.

If the studied chemical species exists in a triplet state, i.e., if it contains two unpaired electrons, the situation becomes more complicated. In that case each electron couples to the other unpaired electron as well as to any magnetic nuclei. For a triplet state the total spin quantum number is 1 and the values of M_s are -1, 0, and $+1$. Consequently, transitions in a magnetic field between $M_s = -1$ and $M_s = 0$, and between $M_s = 0$ and $M_s = +1$ are permitted even in the absence of a magnetic nucleus. The situation is further complicated, in some cases, by magnetic dipole-dipole interactions between the unpaired electrons. Dipole-dipole interactions can cause energetic levels corresponding to each of the M_s values to be observed without application of a magnetic field. The result is *zero-field splitting*. Zero-field splitting can be sufficiently large to prevent observation of absorption in the microwave region. Further information that relates to the esr spectra of triplet-state compounds can be found in Reference 4.

QUALITATIVE ANALYSIS

The hyperfine structure in an esr spectrum can be used as a tool for qualitative analysis by using a procedure that is similar to that used in nmr spectrometry. If an unpaired electron is delocalized, all of the magnetic nuclei that it approaches can spin-spin couple to the electron. As an example, the unpaired electron in the benzene anion radical $C_6H_6^{\cdot}$ is delocalized throughout the entire molecule.

Because the electron can couple to six identical protons, the esr spectrum contains $2nI + 1 = 7$ hyperfine lines.

The coupling of a localized free-radical electron to a proton is generally only significant over distances corresponding to one or two bond lengths. In some cases, however, coupling can be observed over at least four bond lengths. Significant coupling normally occurs only when the unpaired electron and the proton are attached either to the same atom or to adjacent atoms. Coupling with ^{1}H is greatest when the molecular orbital of the free electron and the orbital of the proton are coplanar.

Standard compounds which have known g-factors are used in esr spectral studies for standardization. The standard that is used most often is the 1,1-diphenyl-2-picrylhydrazyl (DPPH) free radical. The molecular structure of the free radical is shown in Fig. 17-6. The g-factor for the DPPH radical is 2.0036. Other standards include solutions of nitrosodisulfonate [$ON(SO_3)_2^{2-}$] and synthetic ruby. Synthetic ruby is generally attached to the wall of the cell during use. Potassium nitrosodisulfonate is Fremy's salt. DPPH can be used as a quantitative as well as a qualitative standard. It contains 1.53×10^{21} unpaired electrons for each gram of the compound. Fremy's salt can also be used as a quantitative standard.

The g-factor for the standard is often not very different from that of the sample. Consequently, the standard is rarely added to the sample solution. In a single-cavity instrument, spectra are consecutively obtained of the standard and the sample. It is more convenient to obtain the two spectra simultaneously in a dual-cavity instrument. The first spectrum is superimposed on the second by the readout device. The two spectra are distinguished from each other by the frequency of the applied alternating field.

The coupling constants observed in esr spectra vary between compounds and can sometimes be used for qualitative analyses. The coupling constant is the distance between the two central peaks of a multiplet; i.e., the distance on the magnetic flux axis between the locations at which the first-derivative peaks cross the axis. It is a function of the fraction of time that the unpaired electron spends in an s electron orbital. The coupling constant increases as the electron nears a magnetic nucleus. Protons which are not magnetically equivalent, as observed by the unpaired electron, cause different coupling constants. Those coupling constants can be used to help determine the structure of an analyte. The hyperfine splitting constant A for hydrocarbon radicals can be calculated from[8]

$$A = \rho Q \tag{17-8}$$

where ρ is the pi density of the unpaired electron and Q is a constant which is equal to 0.00224 T or 22.4 G.

Spectra that have overlapping peaks are difficult to interpret. Sometimes a computer can be used to interpret the spectra. The normal approach is to use

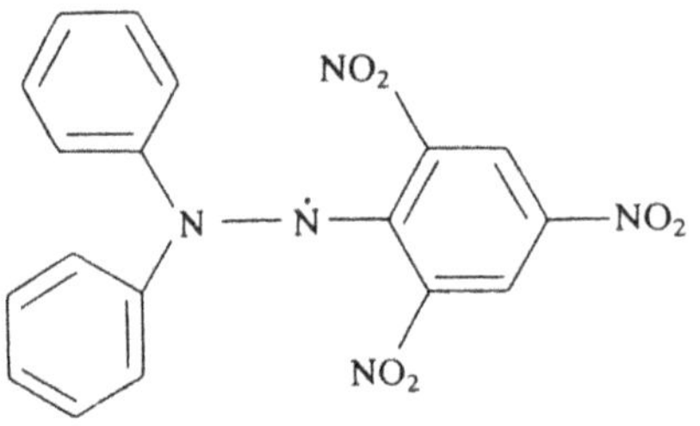

Figure 17-6 The molecular structure of the 1,1-diphenyl-2-picrylhydrazyl (DPPH) radical.

measured or theoretical coupling constants to obtain a computer-calculated spectrum which is compared to the actual spectrum. If the spectra match, the sample is assumed to correspond to the single substance or to the mixture whose coupling constants were entered into the computer. Spin decoupling and spin labeling studies can also provide qualitative information. Each of these techniques are described later in the chapter.

QUANTITATIVE ANALYSIS

Quantitative esr assays can be performed by the working-curve method using either peak heights or peak areas, or by comparing the area or height of a sample peak with that of a peak of a reference material. The reference must contain a known number of unpaired electrons. When comparing peak heights the reference should have a line shape and line width that is similar to that of the sample, and should possess the other desired properties of standard substances (stability, purity, etc.). If the peak shapes of the sample and standard are identical, the ratio of the first-derivative peak heights h are directly proportional to the concentrations of unpaired electrons:

$$\frac{C_{\text{sample}}}{C_{\text{ref}}} = \frac{h_{\text{sample}}}{h_{\text{ref}}} \qquad (17\text{-}9)$$

If the line shapes are not the same, the peak areas A can be used for the calculation:

$$\frac{C_{\text{sample}}}{C_{\text{ref}}} = \frac{A_{\text{sample}}}{A_{\text{ref}}} \qquad (17\text{-}10)$$

If integrated peak areas are not available, the areas can be estimated from the peak heights and the peak-to-peak line widths ΔB of the first-derivative peaks:

$$A = h(\Delta B)^2 \qquad (17\text{–}11)$$

Substitution of A from Eq. (17-11) into Eq. (17-10) yields

$$\frac{C_{\text{sample}}}{C_{\text{ref}}} = \frac{h_{\text{sample}}(\Delta B_{\text{sample}})^2}{h_{\text{ref}}(\Delta B_{\text{ref}})^2} \qquad (17\text{-}12)$$

The use of Eq. (17-12) typically gives results that are accurate to within about 10 percent at the parts per billion level. Substances which can be used for quantitative standards include DPPH and Fremy's salt. A dual-cavity instrument is advantageous when using a standard because both the sample and the standard can be placed simultaneously in the instrument.

The working-curve method usually gives more accurate results than the use of Eq. (17-12) with a single measurement, but it is experimentally more difficult. The second integration of the first-derivative peak is plotted as a function of the concentration of a series of standard solutions that are prepared from the same substance as that in the analyte. Care should be taken while recording the spectra and obtaining the integrations to ensure identical instrumental operating conditions during each scan. The spectra used to prepare the working curve should be obtained just prior to or just after obtaining the sample spectrum. If a double-cavity instrument is available, the spectrum of a reference compound can be simultaneously recorded with that of each standard solution, and the ratio of the area under the peak of each standard solution to the area under the peak of the reference compound can be plotted as a function of the concentration of the standard solution.

If an analyte does not exhibit an esr spectrum, sometimes it is possible to create one or more unpaired electrons in the substance by labeling the substance with a reactant which contains a free radical or by irradiating the analyte with the emissions from a radioactive material. Radiochemical methods of analysis are described in Chapter 20. The interaction between the emission and the analyte results in one or more unpaired electrons in the compound. A similar approach can be used to study living animals or humans. In those instances either radiation can be used to generate a free radical or sometimes a paramagnetic material at low levels (to prevent poisoning) can be administered in order to study a particular organ. Two dimensional and three dimensional images of the organ can be created from the esr results.

An additional use for esr spectrometry is in dating archeological materials of minerals. Among the items dated in this manner are fossilized teeth, sedimentary quartz, flint and limestone. The sample is irradiated with the emissions from a source of radioactivity causing electrons to be removed from the analyte. The esr spectrum of the resulting material is obtained. The change in the esr spectrum is related to the age of the materials.

MULTIPLE RESONANCE

Multiple resonance can be used in esr spectrometry as well as in nmr spectrometry. The procedures used in esr spectrometry are similar to those used in nmr spectrometry. Double resonance is typically used in esr spectroscopy as a method to improve the resolution in a spectrum and to assign coupled spectral lines. Although several variations of double resonance can be used, *electron nuclear double resonance* (ENDOR) and *electron double resonance* (ELDOR) are used most often.

ENDOR is the technique in which the sample is simultaneously irradiated in the gigahertz region and in the radiofrequency region. The gigahertz radiation is absorbed by an unpaired electron in the sample and the radiofrequency radiation can be absorbed by magnetic nuclei. The radiofrequency radiation is scanned throughout the frequency range in which a nucleus that is coupled to the unpaired electron is expected to absorb. Electron spin resonance absorption at a fixed frequency is observed. The esr signal is displayed on the y axis of the readout device as a function of the frequency of the scanned nmr radiation. When the scanned frequency becomes identical to the absorptive frequency of a nucleus, the nucleus resonates between allowed energetic levels. Because the coupled electron is no longer capable of distinguishing between the energetic levels of the nucleus, spin-spin splitting no longer occurs. The height of the observed peak (y-axis display) either becomes zero, if the peak is not in the center of the multiplet, or increases, if the observed peak is in the center. Either way it is possible to use the scanned frequency to determine the nucleus that is coupled to the unpaired electron.

Interpretation of the esr spectrum also can be simplified by keeping the radiofrequency fixed at a value that is absorbed while obtaining the esr spectrum. Such decoupling experiments can be useful in the interpretation of complex spectra. ENDOR has also been used to determine g-values and hyperfine coupling constants. A description of those applications can be found in Reference 2.

ELDOR is the technique in which the sample simultaneously is irradiated with two beams of radiation in the gigahertz region. Observations are made at one frequency while the second frequency is scanned. As in ENDOR the y axis of the readout device is used to display the absorptive signal at the monitored frequency as a function of the varied frequency or the difference between the two frequencies. ELDOR is primarily used to identify overlapping resonant esr peaks and to aid in assigning the peaks. A more thorough description of both ENDOR and ELDOR can be found in Reference 2.

SPIN LABELING

It is often possible to obtain structural information about macromolecules by selectively attaching a small compound with an unpaired electron to a specific site on the macromolecule. The product of the reaction can be studied using esr spectrometry. Hyperfine splitting in the recorded spectrum can be used to arrive at conclusions concerning the chemical structure and types of nuclei in the vicinity of the label. Among other things, information can be obtained concerning relative polarities, conformational changes, chemical reactions, and phase transitions in the portion of the macromolecule near the labeled site. The technique is *spin labeling*, and the compound containing the unpaired electron which is to be attached to the larger molecule is a *spin label*. The technique has found widespread use for examination of biological molecules such as proteins and cell membranes.

The spin label is chosen to react with a specific functional group or protein. Often the label reacts with a specific amino acid. Generally the label contains a nitroxide ring or a complex of a transition metal which provides the unpaired electron. The label must contain both a chemically active site that can bond to a specific site on a macromolecule and a group of atoms that form a bridge between the free radical and the macromolecule. The bridging group determines the portions of the macromolecule that are near the unpaired electron. As the distance between the unpaired electron and a studied portion of the macromolecule decreases, the effect upon the spectrum increases. The length, polarity, flexibility, and hydrophobic nature of the bridging group must be considered while choosing spin labels. A list that contains a few spin labels and the groups to which those labels selectively bond is given in Table 17-1. More information and a list containing many more spin labels can be found in Reference 9.

METALLIC COMPLEXES

Electron spin resonance spectrometry can be used to study complexes of the transition metals. In most cases the unpaired electron is associated with the metal in the complex. Owing to changes in polarity between different atoms, ligands have electromagnetic fields associated with them. The fields are present regardless of the presence of a magnetic nucleus in the ligands. Often the fields are relatively strong in the vicinity of the metal and the unpaired electron, and can couple with the unpaired electron to cause splitting of the esr spectral line into a multiplet. The splitting that is caused by the field of a ligand is *fine splitting* (cf. hyperfine splitting), and the spectrum consists of a multiplet of lines that are termed *fine structure*.

The amount of splitting depends upon the initial energetic level of the unpaired electron, the strength of the electric field from the ligand, and the symmetry of the electric field that surrounds the metal which has the unpaired electron. Splitting occurs only when the electric field is not symmetrical around the unpaired electron. Asymmetrical electric fields from ligands cause spectral splitting and anisotropy because the field is stronger in some orientations than in others.

If the ligand contains a magnetic nucleus and if the unpaired electron is delocalized throughout the complex, further splitting can be observed owing to spin-spin coupling with the magnetic nucleus. That splitting is *superhyperfine splitting*. Fine splitting and superhyperfine splitting can cause highly complex esr spectra. Further information can be obtained from the References.

Table 17-1 Selected spin labels and the group on a macromolecule to which each label bonds

The groups in brackets are the locations on the main group to which chemical bonding occurs

Spin label †	Bonded group
$ClHg-C_6H_4-C(=O)NHR$	Cysteine(SH)
$ICH_2C(=O)NHR$	Cysteine(SH), histidine, imidazole
2,4-dichloro-6-(NHR)-1,3,5-triazine (Cl, Cl, NHR on triazine ring)	Histidine, imidazole, lysine
$O_2N-C_6H_4-C(=O)R'$	Serine(OH)
$BrCH_2C(=O)NHR$	Methionine(SCH_3)
$N_2CHC(=O)R'$	Asparagine(C=O)
ATP-R	Myosin
$CH_3C(=O)NHR$	Lysozyme
RC≡N	Heme

† R is 2,2,6,6-tetramethylpiperidine-N-O' (ring with CH_3, CH_3, N–O', CH_3, CH_3)

R' is 2,2,5,5-tetramethylpyrrolidine-N (ring with CH_3, CH_3, N, CH_3, CH_3)

ATP is adenosine triphosphate.

OTHER USES OF ESR SPECTROMETRY

Electron spin resonance spectrometry can be used for several purposes in addition to its use as a qualitative analytical tool. Some of the alternative uses are described here. Esr spectrometry can be used to demonstrate the presence of an unpaired electron. Generally the technique is used to study intermediates which are formed during chemical, electrochemical, or spectrochemical reactions. If possible, the reaction is performed in the cell within the cavity of an esr spectrometer. The presence of an esr signal indicates that the reaction proceeds by a free-radical mechanism. Sometimes it is possible to use the spectrum to identify the location of the unpaired electron in the intermediate. While using esr spectrometry to study chemical reactions, the chemist should be aware that some side reactions can form free radicals that can be detected and mistaken for products of the primary reaction.

Because esr spectrometry is highly sensitive, quantitative analysis of trace quantities of substances that contain a metallic ion with an unpaired electron are possible. Typically the assays can be performed at concentrations in the micromolar or higher range. Some of the ions that have been quantitatively determined by using esr spectrometry are Cr(III), Cu(II), Gd(III), Fe(III), Mn(II), and Ti(III). The number of peaks observed in each case is, of course, determined by the spin quantum number of the nucleus to which the unpaired electron is spin-spin coupled.

Esr spectrometry has also been used to measure surface areas. A substance which contains an unpaired electron is adsorbed on a surface to be measured. The spectrum of the sample and the adsorbed paramagnetic substance is obtained and doubly integrated. The integrated area is directly proportional to the amount of the adsorbed paramagnetic substance and consequently to the surface area. Paramagnetic substances used in those studies must contain a known number of unpaired electrons.

IMPORTANT TERMS

- Electron double resonance
- Electron nuclear double resonance
- Electron spin resonance
- Exchange narrowing
- Fine splitting
- FTESR spectrometer
- *g*-Factor
- Hyperfine splitting
- Hyperfine splitting constant
- *K* band
- Magic T
- Paramagnetic
- Peak-to-peak line width
- *Q* band
- Reflex klystron
- Spin label
- Spin labeling
- Spin-lattice relaxation
- Spin-spin relaxation
- Superhyperfine splitting
- Surface coil resonator
- Waveguide
- *X* band
- Zero-field splitting

PROBLEMS

Theory

17-1 Calculate the magnetic flux density at which a free electron (μ_e is 9.285×10^{-24} J/T) resonates when the spectrometer is operated with the *X* band, the *K* band, and the *Q* band.

17-2 Calculate the energetic changes for an electron during resonance when the *X*, *K*, and *Q* bands are used in the esr spectrometer.

17-3 If an unpaired electron in a particular environment has a g-factor of 2.015, calculate the magnetic flux density required to cause the electron to resonate at a frequency of 9.500 GHz.

17-4 If resonance was observed for an unpaired electron at a magnetic flux density of 0.3157 T and a frequency of 9.500 GHz, calculate the g-factor for the electron.

17-5 Calculate the ratio of free electrons in the upper energetic level relative to the number in the lower level at – 35.0°C in magnetic fields with flux densities of 0.339, 0.821, and 1.249 T.

Spectra

17-6 Define fine splitting, hyperfine splitting, and superhyperfine splitting. Compare the causes of each type of splitting.

17-7 Determine the number of esr spectral peaks that would be expected for each of the following ions. Assume the unpaired electron is coupled only to the single nucleus.

(*a*) $^{53}Cr^{3+}$ (*b*) $^{63}Cu^{2+}$ (*c*) $^{157}Gd^{3+}$

(*d*) $^{55}Mn^{2+}$ (*e*) $^{47}Ti^{3+}$

17-8 Prepare energy-level diagrams similar to that in Fig. 17-5 illustrating the allowed transitions for a single electron which is coupled to nuclei with spin quantum numbers of $\frac{1}{2}$, 1, $\frac{3}{2}$, $\frac{5}{2}$, and 3.

17-9 Calculate the number of peaks and the relative area under each peak in the esr spectrum of $(CH_3)_3C\cdot$.

17-10 Calculate the number of peaks in the esr spectrum of $CH_3\dot{C}H_2$.

17-11 The esr spectrum of Fig. P17-11 was obtained of a sample which was known to contain one of the following free radicals. Use the spectrum to identify the free radical in the sample.

(a) (b) (c) (d)

(Me represents methyl, CH_3—)

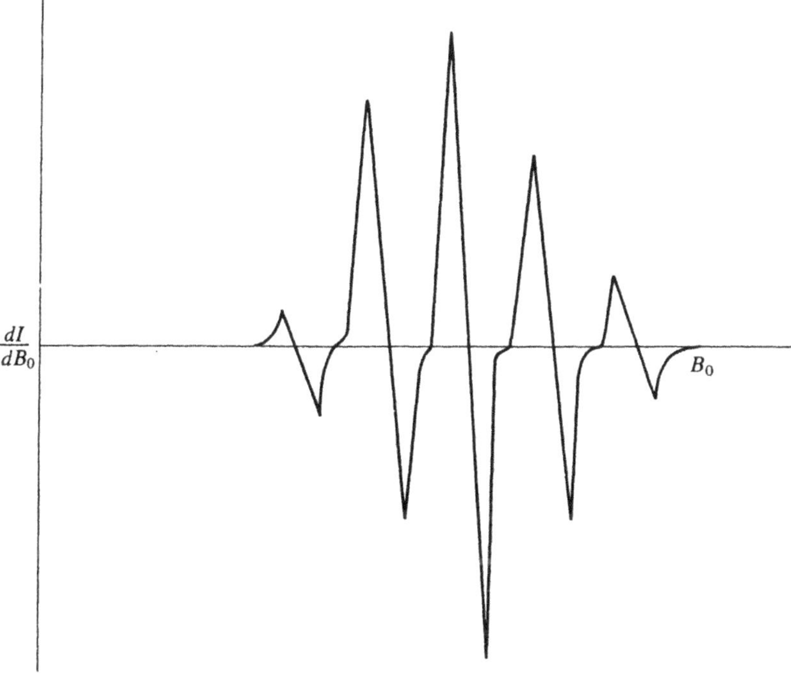

Figure P17-11

Quantitative Analysis

17-12 A 0.25-mL sample containing an organic free radical which contained one unpaired electron for each radical yielded an esr spectrum that contained a singlet with a peak-to-peak line width of 0.15 G and a relative peak height of 24.3. The spectrum of 0.25 mL of a 2.50×10^{-5} *M* DPPH sample was recorded with identical instrumental settings. The peak-to-peak line width of the DPPH spectrum was 0.12 G and the relative peak height was 31.6. Estimate the concentration of free radical in the sample.

17-13 The esr spectra of a series of standard solutions and of a sample were recorded. The same peak in each spectrum was doubly integrated and the values tabulated. Determine the concentration in the sample.

Concentration, × 10^{-5} M	**Relative area**
1.00	19.5
2.00	39.1
3.00	58.7
4.00	78.2
5.00	97.8
Sample	42.7

REFERENCES

1. Alger, R. S.: *Electron Paramagnetic Resonance: Techniques and Applications*, Interscience, New York, 1968.
2. Wertz, J. E., and J. R. Bolton: *Electron Spin Resonance: Elementary Theory and Practical Applications*, McGraw-Hill, New York, 1972.
3. Hill, H. A. O., and P. Day: *Physical Methods in Advanced Inorganic Chemistry*, Interscience, London, 1968, chap. 7.
4. Drago, R. S.: *Physical Methods in Chemistry*, Saunders, Philadelphia, 1977, chap. 9.
5. Hofbauer, W., K. A. Earle, C. R. Dunnam, J. F. Moscicki and J. H. Freed: *Rev. Sci. Instrum.*, **75**: 1194(2004).
6. Alecci, M., S. D. Penna, A. Sotgiu, L. Testa and I. Vannucci: *Rev. Sci. Instrum.*, **63**: 4263(1992).
7. Price, L. W.: *Electronic Laboratory Techniques*, J. and A. Churchill, London, 1969, pp. 22-25, 161-168.
8. McConnell, H. M.: *J. Chem. Phys.*, **24**: 632, 764(1956).
9. Likhtenshtein, G. I.: *Spin Labeling Methods in Molecular Biology*, Wiley, New York, 1976.

CHAPTER

EIGHTEEN

X-RAY METHODS

X-rays were first discovered in 1895 by W. C. Röentgen while working with a discharge tube. He termed the radiation x-rays because the nature of the radiation was not understood. Röentgen received a Nobel prize for the discovery in 1901. X-ray radiation falls in the wavelength range from about 0.01 to 10 nm. The most useful portion of the x-ray region for chemical analysis lies between 0.07 and 0.2 nm.

X-rays are created when an electron in an occupied outer shell of an atom or ion falls into a vacant inner shell. The outer-shell electron falls to the inner shell within 10^{-12} to 10^{-14} s after the vacancy in the inner shell is created. The emitted radiation has an energy that is identical to the energetic difference between the two shells. Because the energetic difference between inner shells is greater than that between outer shells, the emitted photon is in the x-ray region rather than in the lower energy ultraviolet-visible region.

The electron vacancy in the inner shell can be created in several ways. When the atom is bombarded with a particle that possesses sufficient energy to knock an inner-shell electron completely out of the atom, an ion with a vacant inner shell is formed. Electrons, protons, and alpha particles, as well as other particles, can be used to bombard the substance. The atom can also be irradiated with x-rays which have sufficient energy to remove an inner-shell electron. Additionally, x-rays can be formed by nuclear capture of an inner-shell electron as the electron nears the nucleus. Such processes can occur in radioactive elements and form an element with an atomic number that is one unit less than that of the original element, and an electron vacancy in an inner shell.

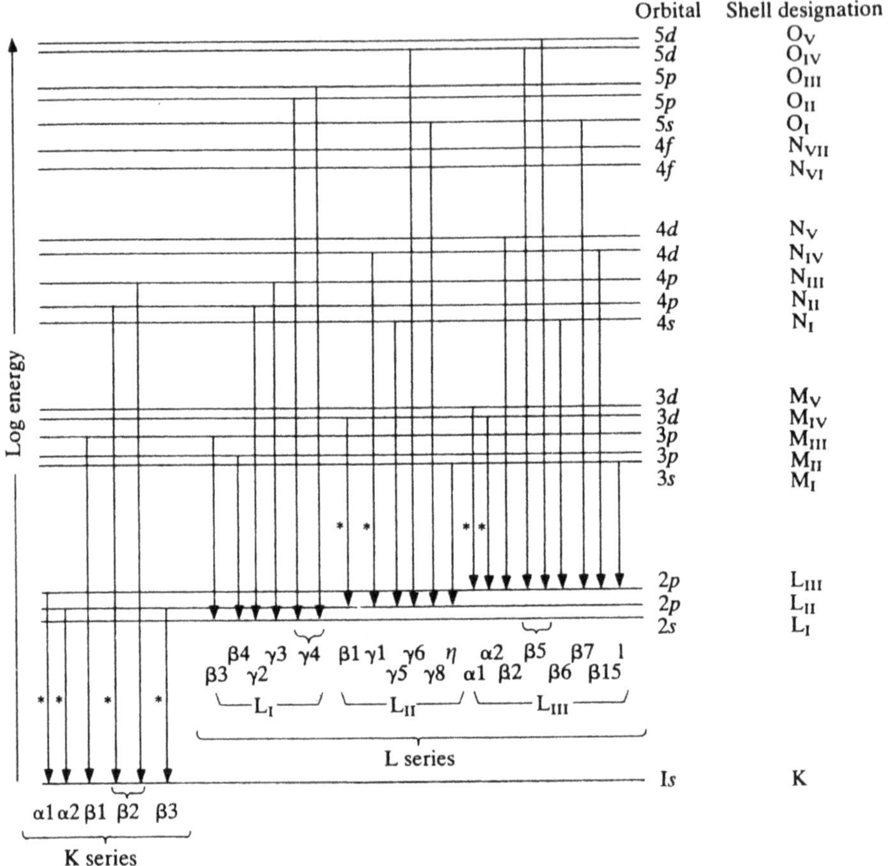

Figure 18-1 An energy-level diagram for uranium showing the transitions that cause the more intense K- and L-series emissive lines. The M-series lines are of relatively low intensity and are not shown. The lines of greatest intensity are labeled with an asterisk.

A diagram illustrating the energy-level transitions that occur during emission of x-rays from uranium is shown in Fig. 18-1. It is informative to compare the transitions shown in Fig. 18-1 with the transitions for ultraviolet-visible radiation that are illustrated in Fig. 5-7. The major x-ray techniques which are used for analysis are absorption, emission, fluorescence, diffraction, electron probe analysis, and electron microscopy.

THEORY

X-Ray Spectral Lines

An advantage of the use of x-rays for analysis is the relatively small number of spectral lines that are observed during absorption or emission. The major emitted lines of uranium are illustrated in the energy-level diagram shown in Fig. 18-1. The energy axis on the diagram is logarithmic for clarity. The energetic differences are greater between inner-shell levels than between outer levels. Similar diagrams can be drawn for other elements although lighter elements, of course, have fewer possible transitions owing to fewer electrons in outer shells. Only the more intense lines are shown in Fig. 18-1. The lines which are most often used for analysis are marked with an asterisk.

The atomic orbital (1*s*, 2*s*, etc.) corresponding to each energetic level is noted on the right of Fig. 18-1. The quantum mechanical selection rule for allowed transitions is $\Delta l = \pm 1$; that is, allowed transitions correspond to *s* to *p*, *p* to *s* or *d*, *d* to *p* or *f* orbitals, etc. The emitted lines are grouped into series according to the destination of the electron. The K series of lines corresponds to the emission of an x-ray which accompanies the transition of an electron to the K shell (principal quantum number *n* is 1). Similarly, the L series of lines result from transitions to the L shell (*n* is 2), and the M series from transitions to the M shell (*n* is 3).

The distinguishable electron levels are labeled with the letter associated with the shell and a Roman numeral which corresponds to the relative energetic level within the shell. As an example, for the L shell, three energetic sublevels can be distinguished. The lowest level is labeled L_I, the intermediate level L_{II}, and the highest level L_{III}. It is important to understand that many of the lines are nearly equal in energy and cannot be distinguished by all analytical instruments.

The observed spectral lines are further labeled with a Greek letter and an Arabic numeral. The Greek letters and Arabic numerals were used to label the lines prior to a complete acceptance of quantum mechanics. Consequently, the labels do not always appear logical based upon current knowledge. The labeling was originally based upon relative line intensities, that is, α lines were more intense than β lines, and an $\alpha 1$ line was more intense than an $\alpha 2$ line. Although that relationship is not always true for all elements, generally the most intense line in a spectrum is the $K_{\alpha 1}$ or $K_{\alpha 2}$ line. Note that the line series is written as an upper case letter, and the Greek letter and Arabic numeral are smaller and often written as subscripts. Elements with atomic numbers of 23 or less only emit the K series of x-ray lines. Elements with larger atomic numbers emit at least the K- and L-series lines. A more thorough description can be found in the References.

X-Ray Tubes

X-rays are often generated in a partially evacuated tube termed an *x-ray tube* by bombarding an appropriate element (the anode) with a beam of electrons. The electrons can be generated either by bombarding a cathode with positive ions of a gas left in the x-ray tube or by thermionic emission from a heated filament. In many modern tubes, a heated filament is used as the source of electrons. The number of electrons emitted each second (the *electron beam current*) is controlled by adjusting the temperature of the filament.

The emitted electrons are accelerated toward the target by a large electric potential that is maintained between the filament and the anodic target. The kinetic energy possessed by the electrons prior to striking the target can be controlled by altering the potential between the target and the filament.

X-ray tubes are designed with permanent or demountable targets. Tubes which have demountable targets are used whenever the target is the sample which is to be analyzed, or when it is necessary to use filtered x-rays which have monochromatic lines at different wavelengths. Different emitted lines can be obtained by changing the target material. X-ray tubes which contain a permanent target are permanently sealed. Those with demountable targets are continuously evacuated with a diffusion vacuum pump and a mechanical rotary pump. Vacuum pumps are not required with the tubes which have a permanent target because the tubes are permanently sealed under a vacuum.

Generally the target must be water or air cooled because more than 99 percent of the kinetic energy of the incident electrons is converted to thermal energy in the target. Less than 1 percent of the kinetic energy is converted to x-ray photons. Without cooling the target could melt. Water cooling is accomplished by pumping cold water through the tube and over or near the anode. Some more recently developed tubes do not generate as much heat and can be air cooled with the aid of a fan. A diagram of one form of an x-ray tube is shown in Fig. 18-2.

When the electron beam strikes the target, the electrons can either cause expulsion of an inner-shell electron from the target material or the beam can lose some or all of its kinetic energy by other interactions with the atoms in the target. In the first case the incident electron transfers sufficient energy to the atom to expel an electron and form an ion with a vacant inner shell. As an outer-shell electron falls into the vacancy, an x-ray with a characteristic line frequency is emitted.

In the second case, insufficient energy is transferred from the electron to cause ionization; however, excitation of an inner-shell electron to the valence shell is possible. The valence-shell orbitals are collectively termed the *valence band*. Because the difference in energy between valence-shell orbitals is small relative to that between energetic levels associated with x-ray emission, electrons which fall from adjacent energetic orbitals in the valence band to vacant inner shells emit x-rays throughout a broad spectral band. The width of the emitted continuum of radiation equals the width of the valence band in the target.

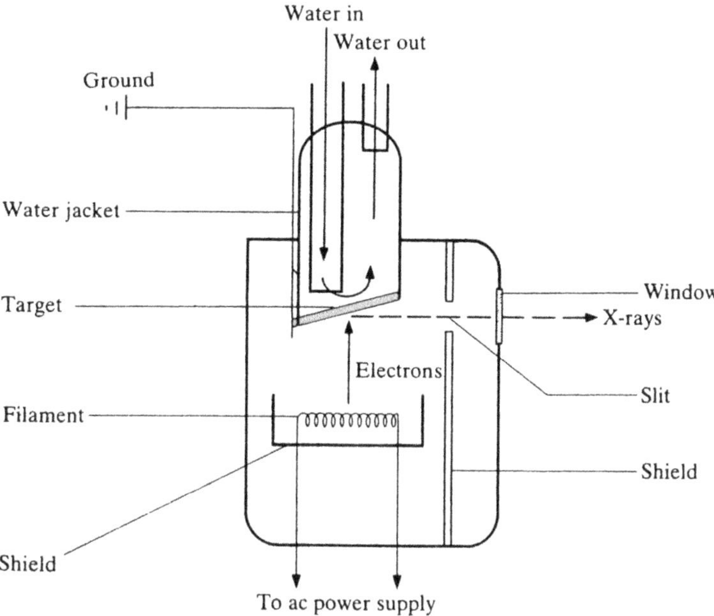

Figure 18-2 A diagram of an x-ray tube.

The energy of the emitted x-ray is identical to the loss in kinetic energy of the incident electron during the interaction between the bombarding electron and the target. Because the incident electron can lose its energy gradually, by successive collisions with several atoms in the target, the emitted x-rays can be of any energy that is equal to or less than the initial kinetic energy possessed by the electrons. Consequently, a continuum of x-ray radiation is emitted by the target. The characteristic line spectrum that is caused by electron expulsion from the target is superimposed upon the emitted continuum. The continuum is referred to as *white* radiation or by the German term *Bremsstrahlung*.

The high-energy limit of the Bremsstrahlung corresponds to emitted x-ray photons which have the same energy as that of the incident electrons. The wavelength at which all of the electron energy is converted to an x-ray is the *short-wavelength cutoff* λ_0. It is independent of the target material, but does depend upon the *accelerating potential* V in the x-ray tube. The short-wavelength cutoff can be calculated by setting the kinetic energy of each incident electron equal to the energy of the emitted x-ray. The energy of the incident electron is Ve, where e is the charge on the electron. The energy of the emitted photon at the short-wavelength cutoff is hc/λ_0 (remember that the refractive index is 1 in a vacuum). Equating the two and solving for λ_0 gives

$$Ve = \frac{hc}{\lambda_0}$$

$$\lambda_0 = \frac{hc}{Ve} \tag{18-1}$$

Combining the constants in Eq. (18-1) gives the *Duane-Hunt* equation:

$$\lambda_0 = \frac{1239.8}{V} \tag{18-2}$$

where λ_0 is in units of nanometers and V is in volts.

The absolute intensity I of emitted x-rays at any applied voltage approximately increases with the atomic number Z of the target element:

$$I = aZ(E_p - E) + bZ^2 \tag{18-3}$$

where a and b are constants; E_p is the energy (volts) at the peak of the continuum; and E is the energy (volts) that corresponds to the particular emitted x-ray. The shape of the intensity-wavelength plot for the continuum is nearly independent of the target, but the height of the plot increases with the atomic number of the target. The area A under the intensity-wavelength curve (the integrated intensity) of the continuum is proportional to the electron-beam current i (milliamperes), the atomic number, and the square of the accelerating potential (kilovolts):

$$A = KiZV^2 \tag{18-4}$$

where K is the proportionality constant. Often the lamps are operated between 10 and 4000 W with an accelerating potential between 30 and 160 kV. Electron beam currents typically vary from 1 - 1000 microamperes.

To a first approximation the wavelength of maximum intensity in a continuum is $1.5\lambda_0$. Several continua for different target elements at the same accelerating potential are shown in Fig. 18-3. The short-wavelength cutoff of all four elements is identical and can be calculated by using Eq. (18-2). The change in the short-wavelength cutoff for a particular element at several accelerating potentials is illustrated by the continua in Fig. 18-4. The area under each continuum increases with accelerating potential as predicted in Eq. (18-4).

If x-rays are produced by a process other than electron bombardment, the emitted continuous spectrum is either of low intensity or nonexistent. Bombardment with protons yields a continuum which is about 3 million times less intense than that observed after electron bombardment. Proton-induced x-ray emission is described later in the chapter. Fluorescence caused by absorption of x-rays is monochromatic.

The metals that are used as targets in x-ray tubes which are used as sources generally have a relatively large atomic number, a good thermal conductivity, and a high melting point. A large atomic number ensures an intense emitted

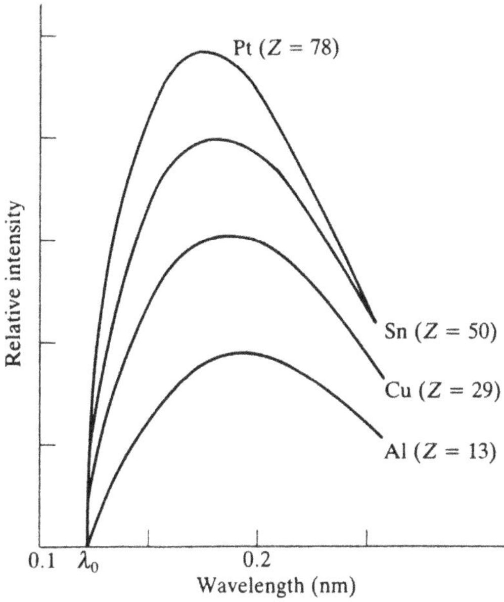

Figure 18-3 Several continua that are observed during x-ray emission from x-ray tubes that have different targets at an accelerating potential of 10 kV.

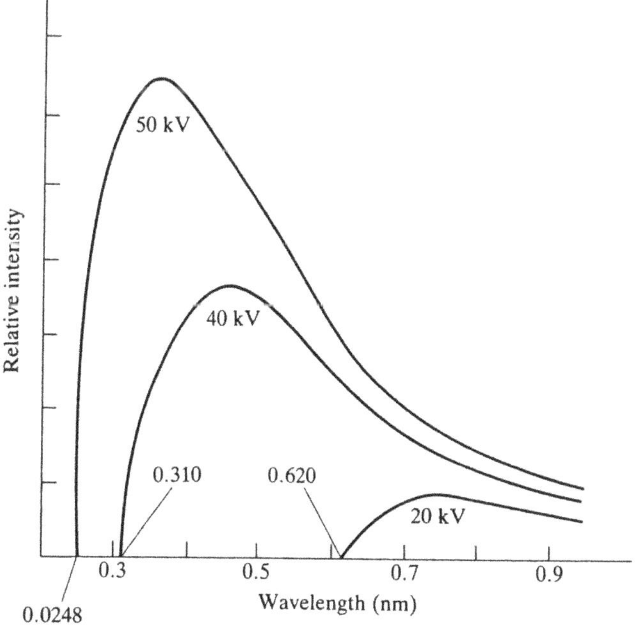

Figure 18-4 X-ray continua at different accelerating potentials at a fixed electron beam current. The short-wavelength cutoff at each accelerating potential is indicated.

continuum [Eq. (18-4)]. A high thermal conductivity helps to prevent the target from overheating. Some of the elements that are used as targets include Ag, Co, Cr, Cu, Fe, Mn, Mo, Ni, Pt, Rh, Sc, Ti, W, and Y. For general use tungsten is often preferred. Tubes which contain the other elements cannot be used at power levels as high as those at which tungsten-targets tubes can be used. Other elements are used whenever a desirable specific line is emitted by those elements or when one of the tungsten lines interferes with the analysis. Tubes with chromium targets are preferred by many analysts for elements with atomic numbers less than 22 (titanium) because the characteristic line spectrum emitted by chromium contains high-intensity lines which efficiently excite low-atomic-number elements such as aluminum and chlorine. At least one instrument maker uses Rh anodes in their equipment. Often a Be window is used in x-ray tubes. For studies of elements with low atomic numbers, it usually is necessary to use tubes which have no windows.

Line spectra can be emitted from x-ray tubes when the energy of the incident electrons is sufficient to ionize the target atom during expulsion of an electron from an inner shell. If the kinetic energy of the incident electrons is sufficient to expel an inner-shell electron, an electron from an outer shell can fall into the vacancy while emitting an x-ray. The energy of the x-ray, of course, must always be less than that of the incident electron. The energy of the incident electrons is determined by the accelerating potential between the filament and the target. If the accelerating potential is insufficient to cause electron emission from an inner shell of the atoms in the target, only a continuum is emitted. At higher accelerating potentials, the line-spectrum characteristic of the target element is superimposed upon the Bremsstrahlung. The emitted spectrum from

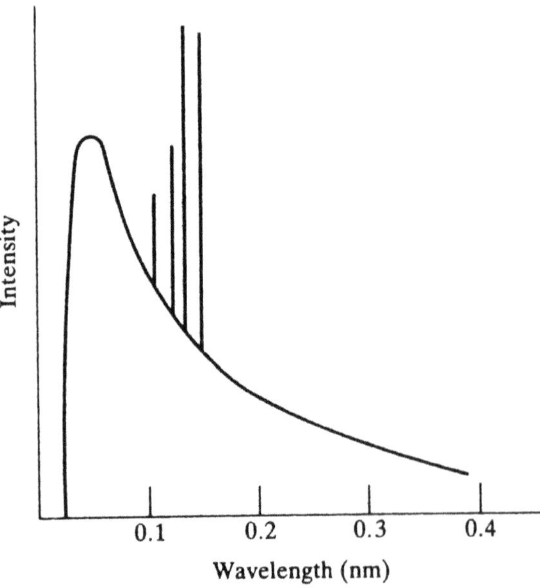

Figure 18-5 The spectrum from an x-ray tube with a tungsten target. The tube was operated at 50 kV.

an x-ray tube with a tungsten target is shown in Fig. 18-5. X-ray tubes are typically operated at accelerating potentials that are between 30 and 160 kV.

The minimal accelerating potential that is required to produce line spectra increases with increasing atomic number of the target element. In work that was published in 1913 and 1914, H. G. J. Moseley showed that the frequency of the K_α line was related to the atomic number of the emitting element by

$$\nu = 2.48 \times 10^{15} (Z - 1)^2 \tag{18-5}$$

It is apparent from Eq. (18-5) that the atomic number is proportional to the square root of the emitted frequency. The observation was regarded as evidence of the validity of Bohr atomic theory. Equations similar to Eq. (18-5) can be written for other series of emitted x-rays.

X –Ray Emission

It is possible to use spectra of emitted x-rays for analysis in much the same way as atomic emissive spectra are used. The sample is placed on the end of a holder made from an element such as copper, and the assembly is used as the target in an x-ray tube with a demountable target. Alternatively, metallic samples can be machined to fit the target socket in the tube. The emitted spectrum is either wavelength dispersed WD or energy dispersed ED and used for qualitative analysis. The method is rarely used for analysis owing to the experimental difficulties involved in preparing the target. It is easier and the results are more reproducible when x-ray fluorescence is used for the assay.

ABSORPTION

Measurements of absorption that are made in the x-ray region are analogous to measurements that are made in the ultraviolet-visible region. Radiation from an x-ray source is collimated and passed through a thin sample. Some of the x-ray photons are absorbed while passing through the sample. The intensity or power of the attenuated signal is monitored after appropriate collimation and dispersion. The samples used during absorptive measurements must not be so thick as to completely absorb the x-ray radiation.

During absorption the energy of the incident x-ray generally is sufficient to eject an electron from an inner shell of a sample atom, leaving a positive ion with a vacant inner shell. The process also causes fluorescence when an outer-shell electron emits an x-ray photon as it falls into the inner shell. X-ray fluorescence is described in a later section.

If the energy of the incident x-ray photon is absorbed by an atom of the sample, then the energy of the incident x-ray must equal the sum of the kinetic energy of the ejected electron and the potential energy of the ion with the vacant inner shell. In that regard, x-ray absorption differs from atomic absorption in the ultraviolet-visible region. In the latter case the entire energy of the absorbed incident photon is used to increase the potential energy of the atom by causing an outer-shell electron to move to a higher orbital.

In the case of x-ray absorption, nearly all of the energy from the incident photon that is in excess of that required to cause ionization is converted into kinetic energy of the ejected electron. A small proportion of the excess energy can be absorbed by the nucleus of the bombarded atom. The shapes of the spectral bands of absorbed x-rays differ from those of atomic absorptive bands. Atomic absorption yields narrow absorptive bands at energies that are identical to the energetic differences between the lower and higher energy atomic orbitals. X-ray absorptive bands are broad. The minimal energy in an x-ray spectral band corresponds to the energy that is just sufficient to cause ejection of an inner-shell electron. Any photon energy in excess of that minimum can also cause x-ray absorption.

The probability of x-ray absorption is greatest when the energy of the incident photon is identical to that which is required to just remove an electron from the absorbing atom, i.e., when the kinetic energy of the ejected electron equals zero. The probability of absorption of x-rays by expulsion of an electron in a particular inner shell decreases as the energy increases above the minimal value. A sketch of a typical x-ray absorptive spectrum is shown in Fig. 18-6. Each sharp decrease in absorptive coefficient occurs at a wavelength that corresponds to the minimal energy required to remove an electron from a specific inner shell and to form an ion. The wavelength at which each sharp decrease occurs is an *absorptive edge*. An absorptive edge exists for each set of inner-shell electrons. In Fig. 18-6 the locations of the K, L_I, L_{II}, and L_{III} absorptive edges are indicated. Absorptive edges can be used for qualitative analysis in much the same way that absorptive maxima are used in atomic absorption or emitted spectral lines are used in atomic emission.

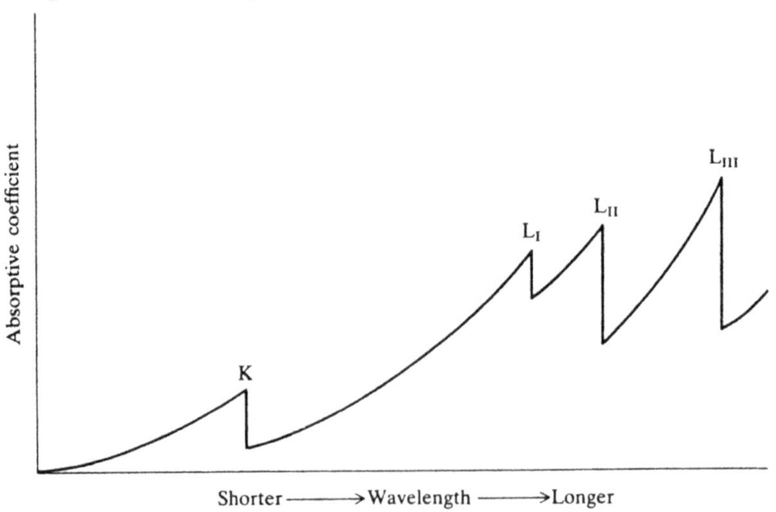

Figure 18-6 A sketch of a typical x-ray absorptive spectrum. The K and L absorptive edges are marked.

Beer's law is obeyed during measurements of x-ray absorption at a specified wavelength. In order for Beer's law to be obeyed, the incident x-ray radiation must be a narrow collimated beam. Generally Beer's law for x-ray absorption is written as follows:

$$\ln \frac{I_0}{I} = \ln \frac{P_0}{P} = \mu x \tag{18-6}$$

where I_0 and P_0 are the intensity and power respectively of the incident, radiation; I and P are the intensity and power after the beam has passed a distance x through the sample; and μ is the *linear absorptive coefficient.*

The linear absorptive coefficient is characteristic of both the concentration of the absorbing elements in the path of the x-rays and the identity of the absorbing elements. The linear absorptive coefficient is a combination of the absorptivity coefficient and the concentration which were used in the form of Beer's law used for ultraviolet-visible spectrophotometry. Another useful form of Beer's law can be obtained by substituting the following relationship

$$\mu = \mu_m \rho \tag{18-7}$$

into Eq. (18-6) to yield

$$\ln \frac{I_0}{I} = \ln \frac{P_0}{P} = \mu_m \rho x \tag{18-8}$$

In Eqs. (18-7) and (18-8), μ_m is the *mass absorptive coefficient* and ρ is the sample density (grams per cubic centimeter).

The advantage of using Eq. (18-8) rather than Eq. (18-6) is that the mass absorptive coefficient is constant for a particular element at a particular wavelength regardless of the physical (solid, liquid, or gas) or chemical (type of compound, etc.) state in which the element exists. Values of mass absorptive coefficients are listed in standard reference texts. The mass absorptive coefficient at a specified wavelength is approximately proportional to the fourth power of the atomic number of the absorbing element. For a particular element, the coefficient increases (Fig. 18-6) with wavelength until an absorptive edge is reached. For the K- and L-series bands, the coefficient exponentially increases with wavelength to between the 2.5 and 3.0 power. The coefficient at a specified wavelength is approximately inversely proportional to the atomic weight of the absorbing element.

Owing to the broad bands in x-ray absorptive spectra, mixtures such as alloys or chemical compounds which contain more than one heavy element often have more than one absorbing element at a particular wavelength. Generally the absorbance of the mixture or compound is the sum of the absorbances of each of the absorbing elements. The mass absorptive coefficient of a mixture of absorbing species at a particular wavelength can be calculated by

$$\mu_m = \sum \mu_{mi} W_i \tag{18-9}$$

where μ_{mi} and W_i respectively are the mass absorptive coefficient and the weight fraction of the *i*th component in the mixture. Eq. (18-9) is analogous to Eq. (9-2) which was used for uv-visible spectra of polyatomic species.

Sample problem 18-1 Calculate the mass absorptive coefficient of an alloy which consists of 39 percent nickel and 61 percent copper at the wavelength corresponding to K_α radiation of copper. The mass absorptive coefficient at that wavelength is 49.3 cm^2/g for nickel and 52.7 cm^2/g for copper.

SOLUTION Substitution into Eq. (18-9) provides the desired result:

$$\mu_m = 0.39 \times 49.3 + 0.61 \times 52.7 = 51 \text{ cm}^2/\text{g}$$

In addition to absorption, scattering can reduce the intensity of x-ray radiation that passes through a sample. Absorption and scattering are additive:

$$\mu = \tau + \sigma \tag{18-10}$$

where τ is the *photoelectric* (true) *absorptive coefficient* and σ is the *scattering coefficient*. The absorptive coefficients in Eqs. (18-6), (18-8), and (18-9) incorporate contributions from absorption and scattering. The relative amount of scattering decreases with increasing atomic mass and increases with decreasing wavelength. Reference 1 provides more details.

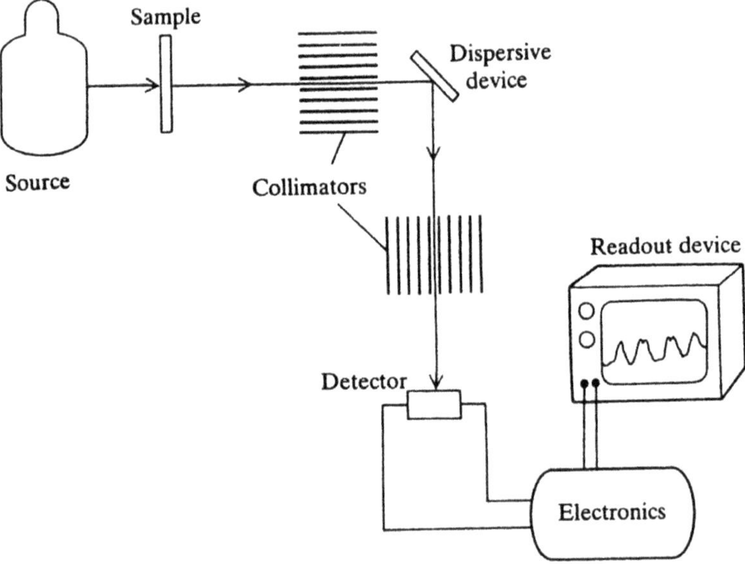

Figure 18-7 Diagram showing the major components of the apparatus used to measure x-ray absorption. The arrows indicate the radiative path. The dispersive device is not required if the detector can distinguish between radiation of different energies.

Absorptive Apparatus

The apparatus used for x-ray absorption includes a source of x-rays, one or more collimators, a sample holder, a method for dispersing the absorbed radiation, and a detector. Often the components are arranged in the order shown in the block diagram in Fig. 18-7. Most of the individual components used for measurements of absorption are identical to those used for other spectroscopic methods in the x-ray region. Consequently, much of the description in the following sections is applicable to x-ray fluorescence and diffraction as well as to absorption. Throughout a portion of the x-ray region, atmospheric components absorb. In that spectral region the apparatus either is evacuated by vacuum pumping or a continuous flow of a low-molecular-weight gas such as helium is continuously flushed through the apparatus.

Sources

The most common source for x-ray analysis is the x-ray tube. Modern tubes typically are operated at an accelerating potential between 30 and 160 kV, and a power of between 10 W and 4 kW. The tubes are cooled to prevent melting and emit about 10^{14} photons/(sr · s). The tubes were described earlier. Sealed high-vacuum tubes are preferred to those tubes that have demountable targets.

Both the portion of the continuum and the line spectrum which is emitted at wavelengths less than a particular absorptive edge of the sample can be absorbed. Even though the Bremsstrahlung which is emitted by the x-ray tube can cause excitation of the sample, it does not cause emission of a continuum. Only the line spectrum that is characteristic of the fluorescing element is emitted.

In addition to x-ray tubes, radioactive sources can be used during measurements of absorption and fluorescence. X-ray emission can be induced by bombardment of the sample with electrons or protons. Radiation in the x-ray region can be emitted from radioactive sources by *electron capture* (EC), gamma (γ) emission, or beta (β) emission. The radiochemical methods of analysis are described in Chapter 20.

An electron in an orbital surrounding a nucleus has a finite probability of existing in or near the nucleus. Electron capture occurs when the electron enters the nucleus and combines with a proton to yield a neutron. The resulting atom has an atomic number which is one unit less than that of the original atom. The atomic weight does not change. If a K-shell electron is captured, the process is *K capture.* Similarly, capture of an electron from the L or M shell is *L capture* or *M capture.* K capture occurs most often owing to the greater probability of a K-shell electron existing in the nucleus.

Upon electron capture, a vacancy is created in an inner electron shell. When an electron in an outer shell falls into the vacancy, an x-ray photon is emitted that is characteristic of the newly formed element. An example of the emission of an x-ray photon during K capture is

$$^{55}_{26}Fe \rightarrow ^{55}_{25}Mn + \text{x-ray (Mn } K_\alpha \text{ x-ray photon, 210.3 pm)} \qquad (18\text{-}11)$$

Gamma emission also occurs in the x-ray region. Gamma emission, however, is caused by transitions between *nuclear* energetic levels rather than between electron levels. Gamma radiation is spectrally indistinguishable from x-ray radiation.

Beta emitters are radioactive nuclei which decay by emission of electrons (β particles). The emitted particles are used without electrical acceleration to bombard an element which subsequently emits x-rays. The spectrum of the emitted x-rays can contain both a continuum and lines (if the energy of the electrons is sufficient). The process is similar to that which occurs in x-ray tubes.

In addition to the individual emissive pathways already described, some radioactive sources emit radiation in the x-ray region by a combination of two or more processes. Some radioactive nuclei simultaneously emit radiation and an alpha particle. An alpha particle is a helium nucleus. If the element remaining after the emission of an alpha particle is in an excited level, x-rays and/or gamma rays can be emitted with the alpha particle. Alpha emitters are health hazards and consequently are used only for special assays. A list of some of the common radioactive sources which are used for x-ray analysis is provided in Table 18-1.

Radioactive sources do not require use of a power supply and x-ray tube. A disadvantage of radioactive sources is the relatively low intensity of the radiation which is emitted. The intensity of x-ray radiation that is emitted from

Table 18.1 A list of selected radioactive sources which can be used for x-ray analysis

Source	Half-life	Emitter type	Emitted radiation	Wavelength, pm	Typical activity, Bq†
^{241}Am	458 y	α, γ	Np L_α	88.9	4×10^7
			NP L_β	69.8	
			NP L_γ	59.7	
				20.8, 47	
^{109}Cd	1.3 y	EC, γ	Ag K	56.4	7×10^7
			γ	14.1	
^{57}Co	270 d	EC, γ	Fe K_α	193.7	2×10^7
			Fe K_β	175.7	
			γ	86.1, 10.17	
^{55}Fe	2.7 y	EC	Mn K_α	210.3	7×10^7
^{153}Gd	236 d	EC, γ	Eu K	29.5	4×10^7
			γ	12.0, 12.8	
$^{3}H/Zr$	12.4 y		Continuum	103-620	4×10^{10} to 1.1×10^{11}
^{125}I	60 d	EC, γ	Te K	8.3	2×10^7
				35.2	
$^{147}Pm/Al$	2.6 y	β	Continuum	12-124	2×10^{10}

† A becquerel Bq is the quantity of a radioactive substance which undergoes a single disintegration each second.

radioactive sources is generally at least six orders of magnitude less than that emitted from x-ray tubes. That necessitates an instrumental design in which the source is placed as near the sample as possible. The detector must also be placed near the sample to minimize radiative loss.

A second disadvantage of radioactive sources is that radiation is continuously emitted. An x-ray tube can be turned off, but a radioactive source continuously emits potentially harmful radiation. Radioactive sources find the greatest use for the relatively simple assays in which only a single heavy element is present to absorb radiation and for those assays which must be performed away from the laboratory and in the absence of a power supply capable of operating an x-ray tube. An example of the first type of assay is the determination of lead in gasoline. Radioactive sources are particularly useful for on-site assays and for some process-control applications. Instruments that use radioactive sources are relatively small and can be used with battery-operated power supplies for the detector and readout device. No power supply is required for the source.

Collimation

Collimation is necessary when monochromatic radiation from a dispersive crystalline grating is used because the angle of reflection from the crystal is dependent upon the incident angle. The angle of incidence must be known in order to use the Bragg equation. Collimation can be accomplished in several ways. If collimation into a single plane of radiation is adequate, the collimator consists of a series of equally spaced, parallel metallic plates. Typically the separation between the plates is between 0.13 and 1.3 mm and the length is about 10 cm. Radiation which strikes a metallic plate is absorbed. Only radiation which is parallel or nearly parallel to the plates passes between the plates. Depending upon the spacing between the plates, the divergence of the emerging beam is typically between 0.07 and 0.7° from parallel. Collimation can also be accomplished with a bundle of identical metallic hollow cylindrical tubes. Typically the tubes have a diameter that is less than 0.5 mm. Because a collimated beam should enter the detector, a collimator is often placed between the dispersing crystal and the detector.

Sample Handling

For absorptive measurements, the sample must be sufficiently thin so as not to absorb all of the incident x-ray radiation. If a metal is to be examined, a thin piece of metallic foil can be used. Solutions can be held in a cell which has transparent windows that are constructed from a material which only contains elements with low atomic numbers. Often thin sheets of Mylar are used as windows. Mylar is the registered trademark of Dupont Tejin Films. It is a polyethylene terephthalate PET plastic sheet.

Wavelength-Dispersive Devices

Absorptive and fluorescent measurements should be made while using monochromatic incident radiation. The radiation which comes from the cell can be made monochromatic prior to being measured with the detector, or polychromatic radiation can be used in combination with a detector that distinguishes between radiation of different energies (an energy dispersive ED device). Filters and wavelength-dispersive WD devices (monochromators) can be used to make the radiation monochromatic.

The filters which are used for x-ray analyses are thin metallic foils. The metal is chosen so that it has a K edge at a wavelength that is either just slightly longer or slightly shorter than the wavelength of the x-ray radiation which is used for the assay. Metallic filters can be used in either of two ways. The *balanced filter method* (Fig. 18-8) uses two filters to isolate radiation of the desired wavelength. The first filter has an absorptive edge at a wavelength which is slightly longer than the wavelength at which the measurement is to be made. Normally absorption is measured at or near the absorptive edge of the analyte. Because the first filter absorbs radiation at wavelengths below its absorptive edge, it does not permit radiation at the wavelength at which the analyte absorbs to pass. After an absorbance measurement is made with the first filter, a second filter, which has an absorptive edge at a wavelength which is slightly less than that of the absorptive edge of the analyte is used in place of the first filter. A second absorbance measurement is made and the difference between the two measurements is proportional to the concentration of the analyte. The thicknesses of the two filters must be such that the two filters absorb radiation to the same extent in all regions in which both filters absorb.

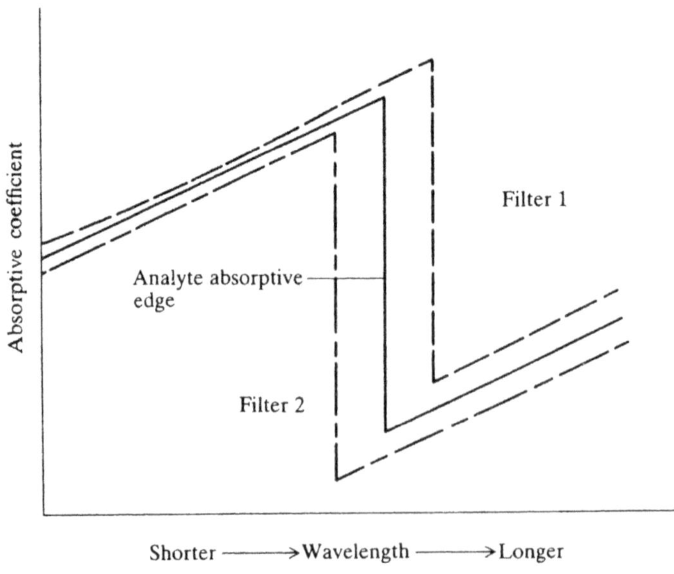

Figure 18-8 Illustrative sketch of absorptive spectra of an analyte and two filters that are used to isolate the wavelength at the absorptive edge of the analyte. The spectra are vertically displaced for clarity.

The analytical technique in which measurements are made before and after an absorptive edge of an analyte is the *absorptive edge method*. The method can utilize filters to isolate the absorptive edge of the analyte as described for the balanced filter method, or it can use a monochromator to isolate the absorptive edge. Monochromators are used more often than filters.

A second method of radiative filtration takes advantage of the differences in mass absorptive coefficients of a single filter at two wavelengths. The method is normally used during measurements of x-ray fluorescence or x-ray diffraction. As an example, a nickel filter can be used to isolate the copper K_α line from the copper K_β line during x-ray crystallographic studies when the source is an x-ray tube that contains a copper target. The mass absorptive coefficients for nickel are 4.57 and 27.5 m^2/kg for the K_α and K_β lines at 0.154 and 0.139 nm respectively. Owing to the difference in absorptivity coefficient, a nickel foil which is 20 μm thick transmits 44 percent of the K_α radiation but only 0.75 percent of the K_β radiation, thereby effectively isolating the K_α radiation. The K absorptive edge for nickel is at 0.149 nm which is between the copper K_α and K_β lines. Filter thicknesses typically vary from 10 to 70 μm. A list of selected filters which primarily have been used to isolate emissive lines from x-ray tubes is provided in Table 18-2.

Monochromators that disperse radiation in the x-ray region contain single-crystal diffraction gratings. Because the spacing between nuclei in crystals is approximately equal to the wavelength in the x-ray region, the crystals can be used as reflective diffraction gratings. The crystal which is used for the dispersion is the *analyzing crystal*.

Block diagrams of two monochromators which are used for x-ray analysis are shown in Figs.18-9 and 18-10. A monochromator consists of an analyzing crystal between entrance and exit slits or collimators. The monochromator

Table 18-2 A list of selected x-ray filters and their K edges

Wavelengths greater than the K edge are transmitted while shorter wavelengths are absorbed

Filter	K edge, pm
Co	160.8
Fe	174.3
Mn	189.6
Ni	148.8
Pd	50.9
V	226.9
Zr	68.9

shown in Fig. 18-9 uses a flat crystal and consequently is easier to construct than the monochromator shown in Fig. 18-10. The optical system shown in Fig. 18-10 uses a curved crystal at the outer edge of a Rowland circle. Flat reflective crystals are used more often than curved crystals. Several less-popular optical systems can also be used for x-ray analysis and are described in Reference 2, chap. 4.

The reflective wavelength-dispersive crystals which are used for x-ray analysis obey the Bragg equation:

$$n\lambda = 2d \sin \theta \tag{18-12}$$

Diffraction was described in Chapter 5. The Bragg equation is Eq. (5-28) (the grating equation) applied to the special case where the angle of incidence equals the angle of reflection. In Eq. (18-12), n is the order of the diffracted (reflected) radiation, λ is the wavelength that is diffracted at angle θ relative to the crystalline surface, and d is the spacing between adjacent layers of nuclei in the analyzing crystal. The factor 2 in Eq. (18-12) is the result of the reflected radiation successively passing twice across the space between adjacent planes in the crystal.

Incident radiation passes between atoms in the top layer of the crystal and is reflected from atoms in the second layer. The radiation again passes through the space between layers and exits from the crystal. Measurement of the angle of incidence is illustrated in Fig. 18-9. For a particular angle of incidence and a fixed distance between layers in the crystal, the only radiation that is reflected in-phase from successive layers of the analyzing crystal corresponds to that with an integral multiple n of the wavelength λ.

The radiative path through a monochromator which utilizes a flat crystal is shown in Fig. 18-9. The radiation strikes and is reflected from the crystal at an angle θ. The wavelength of radiation which strikes the detector is altered by simultaneously rotating the crystal, the exit collimator, and the detector relative to the incident radiation. The rotatable platform upon which the crystal, exit collimator, and detector are mounted is a *goniometer*. Goniometers are carefully made to permit accurate measurement of the angle at which the incident radiation strikes the analyzing crystal.

Reflection, of course, occurs from the analyzing crystal when the angle of incidence equals the angle of reflection. If the angle of incidence relative to the surface of the crystal is θ, the angle of the reflected radiation relative to the crystalline surface also must be θ. Because the analyzing crystal was already rotated θ degrees relative to the incident radiation, the reflected radiation must be at an angle 2θ *relative to the incident radiation* in order to make the angle of reflection equal to the angle of incidence. Consequently, whenever the analyzing crystal is rotated relative to the incident radiation, the exit collimator

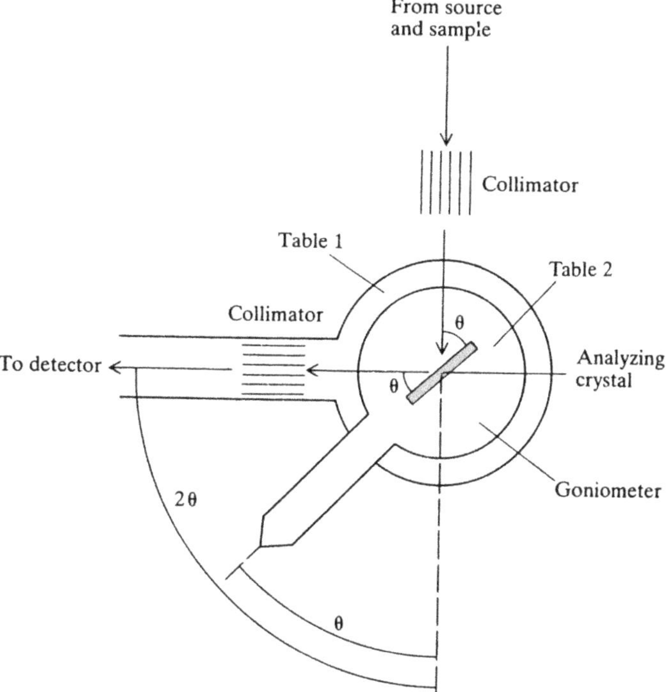

Figure 18-9 A diagram of a flat-crystal monochromator that is used for x-ray analysis.

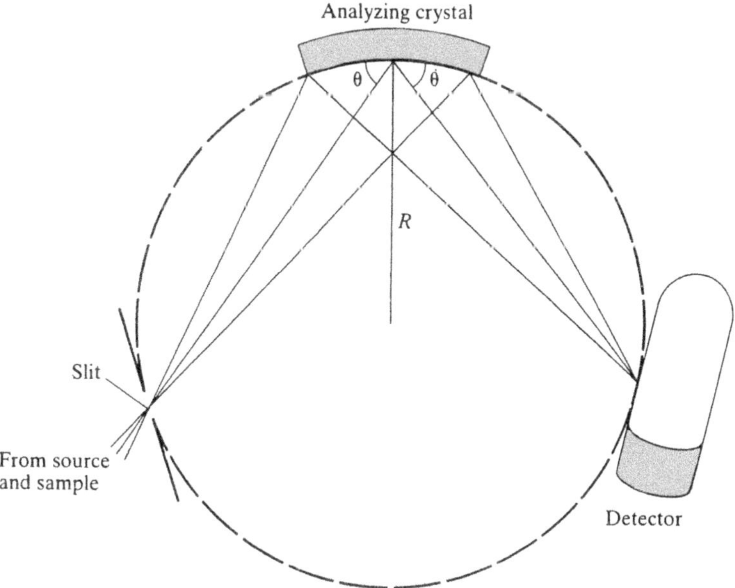

Figure 18-10 Diagram of a curved-crystal monochromator that is used for x-ray analysis. The dashed line indicates the circumference of a Rowland circle.

and detector must be rotated twice as much in order to keep the detector aligned with the reflected radiation. For that reason, it is necessary to mount the analyzing crystal on a separate platform of the goniometer from the exit collimator and detector.

The curved-crystal monochromator shown in Fig. 18-10 functions in a slightly different manner from the flat-crystal monochromator. Divergent radiation which enters the circle as a point source through a slit is focused on the circumference of a Rowland circle. The analyzing crystal must have a radius of curvature $2R$ which is twice the radius of the Rowland circle. For best results the inner surface of the crystal is ground to a radius of curvature R which is identical to that of the Rowland circle. The ground surface is placed on a portion of the circumference of the circle. Some monochromators do not use a ground crystal. If the crystal is not ground the diffracted radiation does not converge to produce a line image at the detector.

Radiation from the source and sample passes through a slit at the circumference of the circle. After striking the analyzing crystal, the reflected radiation passes through a second slit on the opposite edge of the circle from the entrance slit and enters the detector. By altering the angle at which the incident radiation strikes the crystal, the wavelength of the reflected radiation is varied as predicted by the Bragg equation. In order to keep the angle of incidence equal to the angle of reflection, the distances on the circumference of the circle between the crystal and the entrance and exit slits must be equal.

The major advantage of a curved-crystal monochromator is that, in addition to diffraction, the radiation is focused on the exit slit by the curved crystal. That allows incident radiation to be used which is more divergent than that which can be tolerated by a flat-crystal monochromator. As a consequence, slits rather than collimators can be used, and the monochromator can function with radiation that is less intense than that required when a flat-crystal monochromator is used.

Table 18-3 Selected analyzing crystals which can be used for x-ray analysis

Crystal	*d*, nm
Ammonium dihydrogen phosphate (ADP)	0.5325
Barium stearate	5.00
Ethylenediamine-*d*-tartrate (EDDT)	0.4404
Lithium fluoride	0.2014
Mica	0.9965
Pentaerythritol	0.437
Potassium hydrogen phthalate	1.3315
Silicon	0.3135
Sodium chloride	0.2820
Topaz	0.1356

Analyzing crystals normally consist of elements of low atomic number in order to minimize absorption of x-rays by the crystal. Because no single crystal can be used throughout the entire x-ray region, many instruments are equipped with two or three interchangeable crystals. The useful spectral range of a particular crystal is determined by the spacing between adjacent layers in the crystal as predicted by the Bragg equation. X-ray monochromators are usually designed so that the angle θ can be varied between about 5 and 80°. Substitution of those values and the d spacing in a particular crystal into Eq. (18-12) permits calculation of the range of wavelengths which can be studied with the crystal.

Crystals with large internuclear spacings can be used throughout larger spectral ranges than crystals with small spacings. The disadvantage of using crystals with large lattice spacings is the relatively low *dispersion* $d\theta/d\lambda$ which is attainable with the crystals. Dispersion can be determined by differentiating Eq. (18-12). The result is as follows:

$$\frac{d\theta}{d\lambda} = \frac{n}{2d\cos\theta} \tag{18-13}$$

It is apparent from Eq. (18-13) that dispersion is inversely proportional to the crystalline spacing and is a function of the angle of incidence. Several analyzing crystals which can be used for x-ray analysis are listed in Table 18-3.

Energy-Dispersive Devices

Rather than using a monochromator to disperse x-ray radiation according to wavelength, a detector can be used which responds differently to radiation of different energies. In those detectors the output voltage or current is proportional to the energy or the frequency of the radiation which strikes the detector. By using appropriate electronic gates in the output circuit of the detector it is possible to discriminate between incident radiation of different wavelengths without using a monochromator. Typically a multichannel analyzer separates the incident radiation according to energy ($E = h\nu$) into between approximately 100 and 1000 groups. Each group or channel of the multichannel analyzer measures a distinct energy range of about 20 eV across the x-ray spectrum. An instrument which relies upon the change in response of the detector as a function of frequency and which does not use a wavelength-dispersive monochromator is *energy-dispersive* ED. Often energy-dispersive instruments have lower resolution than wavelength-dispersive instruments, but can simultaneously monitor a wide range of frequencies and can more efficiently detect x-ray radiation. An x-ray spectrum that was recorded by using an energy-dispersive detector is shown in Fig. 18-11. Energy-dispersive detectors in modern instruments often are constructed from semiconductors, and are described in following sections of the chapter. Gas-ionization detectors and some other devices also can be used as ED detectors.

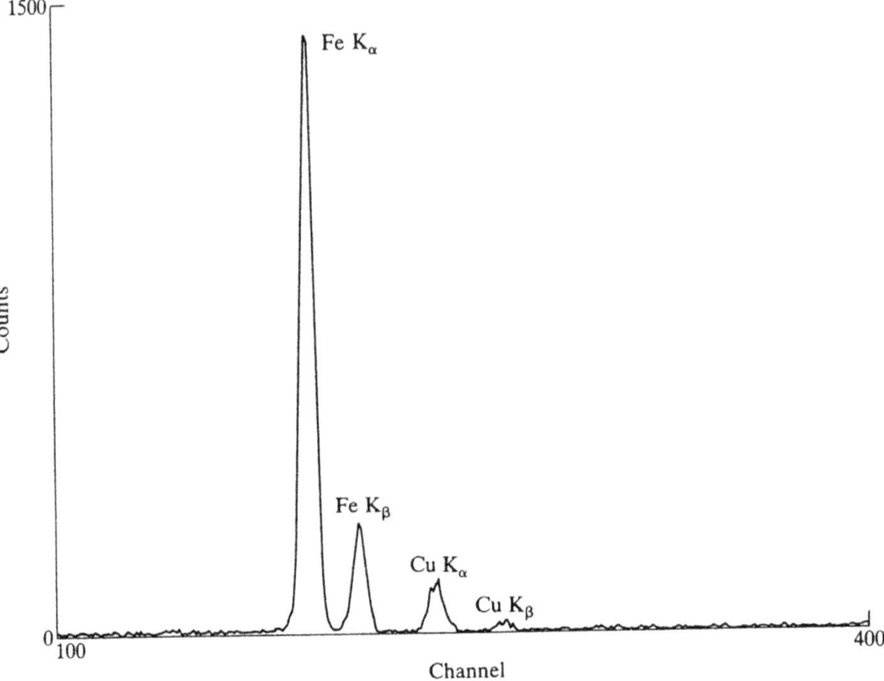

Figure 18-11 A spectrum of emitted x-rays obtained by using an energy-dispersive detector. *(Courtesy of Dr. John R. Meriwether, Department of Physics, University of Louisiana at Lafayette.)*

Detectors

The detectors that are used for x-ray absorption, emission, and fluorescence can be classified as *gas-ionization detectors, scintillation detectors,* or *semiconductor detectors*. Each type of detector has particular advantages and disadvantages which should be considered while choosing a detector for a particular analysis.

Gas-ionization detectors Gas-ionization detectors contain an easily ionized gas and two electrodes. Gases which can be used in the detectors include argon, helium, krypton, methane, and xenon. Generally one of the electrodes is the metallic case of the detector which is grounded for protection of the analyst. The second electrode is normally placed in the center of the cylindrical detector and is maintained at a positive potential relative to the ground.

The window through which x-rays enter the detector is often constructed of mica or beryllium when the detector is permanently sealed. An exit window is provided in some detectors to prevent unabsorbed x-ray photons from striking the metallic case and generating false pulses. In those detectors the windows are located on opposite sides of the cylinder, as shown later in Fig. 18-14.

Upon entering the detector, an x-ray photon can interact with the filler gas in either of two ways. In a few cases the energy of the incident photon can be absorbed by an inner-shell electron of the gas causing formation of a positive ion of the gas and a free electron. In those cases the ion fluoresces x-rays as outer-shell electrons fall into inner-shell vacancies. It is more probable that the energy of the incident x-ray photon is absorbed by a valence electron and forms a positive ion which has filled inner shells and a free electron. The emitted electron possesses a kinetic energy that equals the difference in kinetic energy between that of the incident photon and the energy that is required to ionize the molecule of gas.

The ejected electron can collide with other gaseous molecules to produce more positive ions and electrons. The combination of positive ions and electrons that are formed by the process are *ion pairs*. The positive ions and electrons are accelerated by the potential difference between the two electrodes, and, upon colliding with filler-gas molecules, can cause further ion-pair formation before striking the electrodes. Because the electrons are smaller, they are accelerated more than the positive ions. Their collisions with other molecules are responsible for most further ion-pair formation.

The positive ions flow toward the grounded casing of the detector while the electrons flow toward the central electrode. Upon striking the casing, the positive ions are neutralized. The electrons striking the central electrode cause a current to flow in an external circuit which is amplified and appears as an electric pulse in the readout circuit. Either the number of pulses is counted or an average output current is measured. A diagram of a gas-ionization detector is shown in Fig. 18-12.

As the potential difference between the electrodes in the gas-ionization detector increases, the acceleration of the ion pairs prior to striking the electrodes increases and the number of ion pairs that can be formed increases. Increased ion-pair formation causes an increased electric pulse when the electrons strike the anode and a consequent increased sensitivity of the detector. The number of electrons that are produced for each incident x-ray photon approximately varies with potential difference between the electrodes, as shown in Fig. 18-13. Note that the ordinate in Fig. 18-13 is logarithmic. The height of the curve in regions A and B varies with the energy of the incident x-rays.

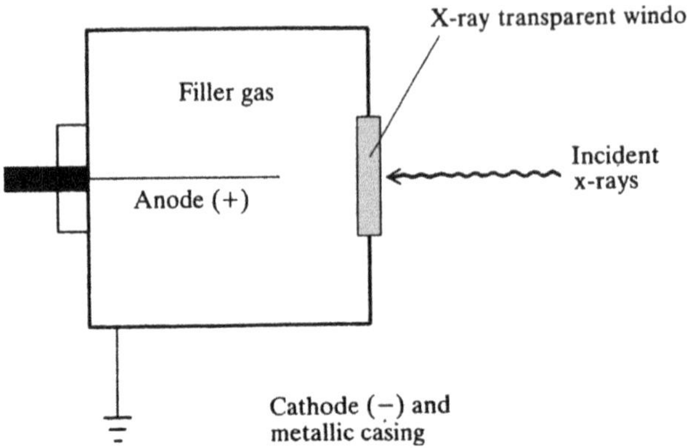

Figure 18-12 A diagram of a gas-ionization detector.

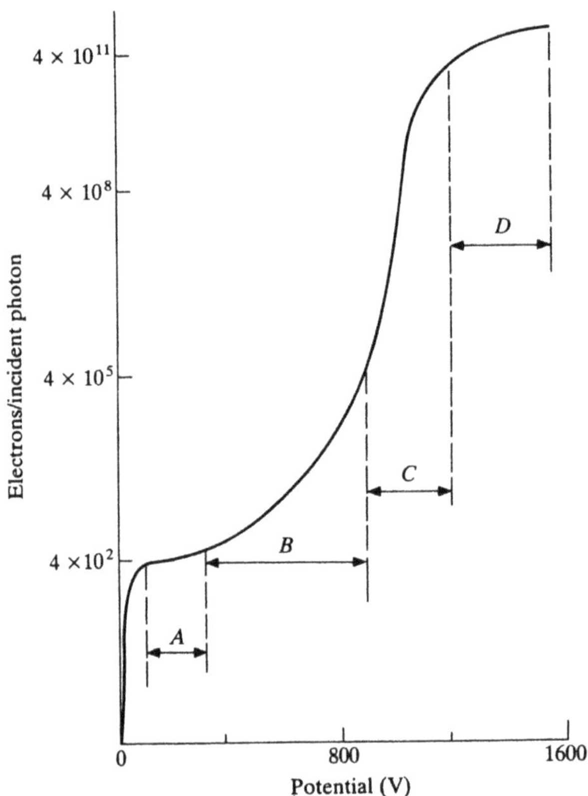

Figure 18-13 The variation in number of electrons formed in an argon gas-ionization detector as a function of potential difference between the electrodes. The plot is for incident x-ray radiation that has a wavelength of 0.1 nm. The ordinate is logarithmic.

The number of ion pairs which can be formed solely by collision between the incident x-ray photon and molecules of the filler gas increases with the energy of each photon. As an example, if an incident photon that has a wavelength of 0.10 nm enters a detector which is filled with argon, the number of ion pairs p which can be formed by direct interaction of the photon with the gas, assuming 100 percent efficiency, equals the energy E of the photon divided by the energy required to form each ion pair. The energy required to produce an ion pair in argon is about 4.8×10^{-18} J. That is roughly twice the ionization energy E_I of argon because about half of the energy from the photon is lost in other processes. Energy transfer to other filler gases is also approximately 50 percent efficient. The number of ion pairs which are formed by direct interaction with a photon is given by

$$p = \frac{E}{2E_I} \tag{18-14}$$

where the 2 in the denominator reflects the 50 percent efficiency of the process. The energy possessed by a photon which has a wavelength of 0.10 nm is 2.0×10^{-15} J, as shown in the following calculations:

$$E = \frac{hc}{\lambda} = \frac{6.626 \times 10^{-34}\ \text{J} \cdot \text{s} \times 2.9979 \times 10^{8}\ \text{m/s}}{0.10 \times 10^{-9}\ \text{m}} = 2.0 \times 10^{-15}\,\text{J}$$

The number of ion pairs which can be formed for each photon is 2.0×10^{-15} J/4.8×10^{-18} J, which equals 4.2×10^{2} ion pairs. Figure 18-13 was constructed for incident radiation which has a wavelength of 0.10 nm in an argon gas-ionization detector.

Plots of electrons formed for each incident photon as a function of potential difference between the electrodes can be divided into several regions, as illustrated in Fig. 18-13. At potentials below that in region A, the number of electrons which strike the anode is less than the number which was formed in the detector by interaction with the incident x-ray radiation. In that case the electric field is too low to attract the electrons to the anode before some of them recombine with positive ions. Potentials in that region are not used in gas-ionization detectors.

In region A (the *saturation region*) the number of electrons which strike the anode for each photon that enters the detector is approximately constant and equals the number formed solely by interaction of the photon with the filler gas. A gas-ionization detector which operates with a potential difference in that region is an *ionization chamber*. Although the ionization chamber has the disadvantage of not amplifying the signal, it has the advantage of having an output pulse that is a function of the energy of the incident photon. For example, an incident photon that has a wavelength of 1.0 nm produces 42 ion pairs in argon rather than the 420 pairs that are produced when the radiation has a wavelength of 0.10 nm. An increase in wavelength by a factor of 10 decreases the size of the output pulse from the detector by a factor of 10.

Ionization chambers can be used in energy-dispersive instruments owing to the relationship between pulse height and energy of each incident photon.

Ionization chambers are rarely used, however, for x-ray analysis because the sensitivity is low. The relatively low potential in ionization chambers causes a relatively slow response because the attraction of charged particles to an electrode is proportional to the potential of the electrode. A slow response prevents individual x-ray photons from being counted when the intensity of the incident beam is high. In such cases, the output is an averaged steady-state signal rather than a rapid series of pulses. When the intensity of the incident beam is low, individual pulses can be counted after electronic amplification.

In the *proportional region* (region *B* in Fig. 18-13) the pulse height from the detector increases with increasing potential owing to further ion-pair formation caused by collisions between the filler-gas molecules and the electrons formed during interaction with the incident x-ray photon. In that region the number of ion pairs which are formed at any potential is proportional to the number which was formed during the initial interaction between the incident x-ray photons and the filler gas. A gas-ionization detector which is operated in the proportional region is a *proportional counter*. The amplification which is achieved in the proportional region increases with potential difference, as illustrated in Fig. 18-13. It can vary from slightly more than that in the saturation region to over 10^4 at the upper potentials. In the proportional region the pulse height consists of the *amplification factor A* multiplied by the number of electrons formed during the interaction between each x-ray photon and the filler gas. For the case described earlier in which the incident photon had a wavelength of 0.10 nm and the filler gas was argon, the pulse height for each incident photon would be $420A$. The detector can be used in energy-dispersive instruments.

The proportional counter can handle counting rates that approach 1 MHz. If the counter is used in an energy-dispersive instrument, a highly stable power supply and a pulse-height selector must be used. Proportional counters can be classified either as *sealed proportional counters* or as *flow proportional counters*.

Sealed proportional counters contain a filler gas which is permanently sealed inside the detector. The windows of the detector are made of beryllium or mica and do not permit the filler gas to leak from the detector. Flow proportional counters often have Mylar windows and a filler gas which is different from that which is used in sealed proportional counters. Typically the Mylar windows are 6.35 μm thick for studies in the spectral range from 0.2 to 1 nm. At wavelengths longer than 1 nm, the windows can be constructed from polypropylene, Formvar, or nitrocellulose.

At wavelengths less than 1 nm, the filler gas normally is 90 percent argon and 10 percent methane. At longer wavelengths the filler gas contains greater amounts of methane and can even consist of 100 percent methane. Helium or argon can also be used at wavelengths longer than 1 nm. Because the windows in flow proportional counters are porous, the filler gas must be continuously flushed through the detector. A diagram of a flow proportional counter is shown in Fig. 18-14.

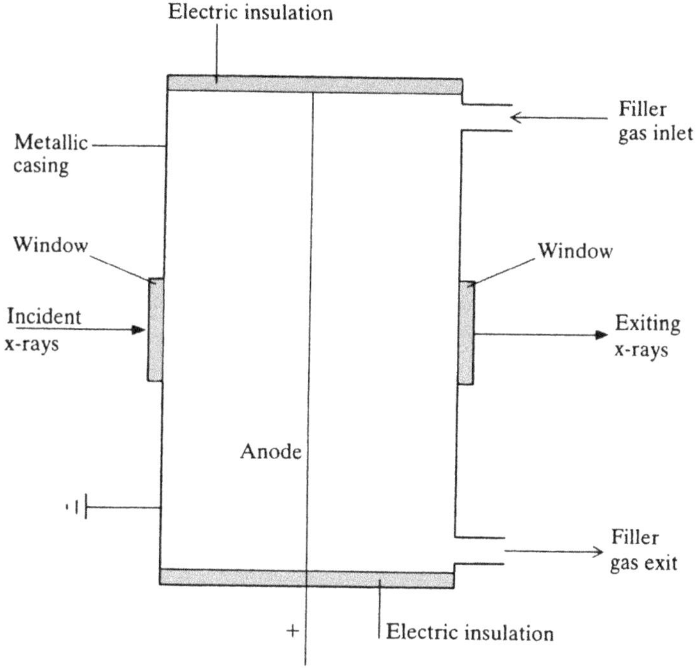

Figure 18-14 A diagram of a flow proportional counter.

In region *C* of Fig. 18-13, the response of a gas-ionization detector is changing from that observed in the proportional region to that observed in the *Geiger counter region.* Region *C* normally is not used for chemical analysis. The final potential region which is useful for chemical analysis is the Geiger counter region (region *D* in Fig. 18-13).

At the high potential differences which are used in the Geiger counter region, the amplification factor is large and independent of the energy of the incident x-ray photons. Typically the amplification factor is 1 billion or greater. At applied potentials that are greater than those in the Geiger counter region, the filler gas can be ionized by the electrical energy applied between the electrodes, causing an output signal without incident photons. That region is of no use for chemical analysis.

A detector which uses applied potential differences in the Geiger counter region is a *Geiger counter* or a *Geiger-Müller* (GM) *counter.* The Geiger counter is the most popular x-ray detector for use in wavelength-dispersive instruments. At pulse rates up to about 500 counts/s, the detector outputs a single pulse of fixed magnitude for each incident x-ray photon regardless of wavelength. At higher incident intensities, the outputted pulse rate is less than the rate at which x-ray photons enter the device.

The decreased counting rate is caused by the *dead time* of the detector. The dead time of a gas-ionization detector is the minimum time after a pulse before a second pulse can be recorded. The dead time is a function of the rate at which the positive ions can migrate to the cathode of the detector. It limits the upper counting rate for a detector. Typically the dead time in a Geiger counter is less than 0.5 ms and often it is near 0.2 ms. That compares to a dead time of about 1 μs in a proportional counter. The upper counting rate of a Geiger counter is restricted to about 20,000 counts/s. Because the fraction of measured pulses for each incident photon varies with the counting rate, it is important to calibrate a Geiger counter at counting rates near those expected for the analyte.

In addition to the normal filler gas, a Geiger counter contains a small amount of a gas that is electronegative and that acts as a trap for emitted electrons. Typically a halogen is used. The halogen *quenching gas* captures the secondary electrons that are emitted by the positive ions and the electrons that are emitted when the ions strike the cathode. Secondary emissions cause undesirable output pulses from the detector. Argon is used as the quenching gas in some Geiger counters. Quenching gases in addition to halogens which can be used include alcohols and methane. The major advantage of Geiger counters is the large amplification which often makes it unnecessary to use electronic amplification. The major disadvantages are the inability to use the detectors in energy-dispersive systems and the inability to use the detectors at high counting rates.

Scintillation detectors Upon striking a phosphor such as zinc sulfide or calcium tungstate, x-rays cause the excitation of ground-state electrons in the phosphor to higher electron levels. As the electrons return to the ground state, visible radiation is emitted. A detector which monitors the amount of emitted visible radiation from a phosphor after excitation by incident x-rays is a *scintillation detector*.

Early scintillation detectors consisted of a phosphor powder which was attached to the envelope of a photomultiplier tube. The incident x-rays struck the powder causing visible radiation to be emitted. The visible radiation was amplified and measured with a photomultiplier tube and the accompanying electric circuit. Scintillation detectors of that type are *phosphor-photoelectric detectors*. Phosphor-photoelectric detectors require relatively intense incident radiation which causes nearly continuous emission of visible radiation. The detector emits a steady-state signal. It is impossible to count individual x-ray photons with a phosphor-photoelectric detector.

Modern scintillation detectors generally use a single crystal rather than a powder to change absorbed x-ray radiation into visible radiation. In many detectors a thallium-doped crystal of sodium iodide is used. The crystal usually contains about 1 mol% thallium. The transparent crystal is attached to the window or is used as the window of the photomultiplier tube. The outer edge of the crystal is covered with a reflective coating, which is often made of aluminum. The coating minimizes the loss of emitted visible radiation, minimizes the entrance of extraneous radiation, and directs emitted radiation

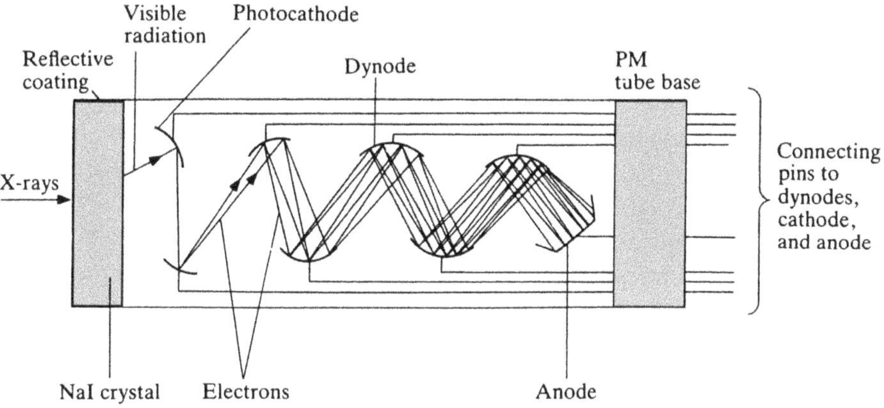

Figure 18-15 A diagram of a scintillation counter.

toward the photomultiplier tube The coating also serves to prevent atmospheric moisture from damaging the crystal. A diagram of a scintillation detector is shown in Fig. 18-15.

X-rays which strike the crystal are absorbed by iodine atoms in the crystal. The excited iodine atoms transfer some of their excess energy to thallium atoms which subsequently emit radiation at 410 nm. The photons strike the photocathode of the photomultiplier tube causing emission of electrons. The electrons emitted by the photocathode are accelerated and amplified by the dynodes in the tube and the resulting current which flows in an external circuit is measured. Photomultiplier tubes were described in Chapter 6.

Each incident x-ray photon causes an output pulse of electrons from the photomultiplier tube. Unlike earlier scintillation detectors, this type can be used with counting circuits that monitor the number of output pulses. Consequently it is termed a *scintillation counter*. If the incident x-ray beam is highly intense, the output pulses become so close together that a steady-state output signal rather than pulses is observed.

Scintillation counters are one of the more generally useful types of x-ray detectors. They have a low dead time that is about 0.25 μs and a high sensitivity. The amplification factor is typically several thousand and is determined, in part, by the number of dynodes in the photomultiplier tube. At x-ray wavelengths below about 0.2 nm, the output of the detector (either counts per second or current) is proportional to the energy of the incident x-ray photon. In those instances the detector can be used with a pulse-height analyzer in an energy-dispersive instrument. The peak width obtained from a plot of pulse amplitude as a function of incident x-ray energy is roughly two or three times broader than that obtained with a proportional counter. Consequently the resolution obtainable with a scintillation counter is lower than that with a proportional counter. Some detectors contain a continuous-strip magnetic electron multiplier or a curved-channel electron multiplier, similar to those used in mass spectrometric detectors, rather than dynodes. Mass spectrometric detectors are described in Chapter 21.

Semiconductors The most common semiconductor detector that is used for x-ray analysis is the *lithium-drifted silicon* Si(Li) *detector. Lithium drifted germanium* Ge(Li) *detectors* and *intrinsic semiconductor detectors*, such as the intrinsic silicon and germanium detectors, also are used. In order to obtain optimal results, most of the semiconductor detectors must be cooled with liquid nitrogen. Since the Si(Li) detector remains popular, it will be described in more detail. Other semiconductor detectors are prepared and operate in a similar manner. The lithium-drifted silicon detector is formed by allowing lithium to diffuse into a silicon semiconductor. Lithium is deposited on one surface of a piece of *p*-doped silicon. Upon heating to a temperature of between 400 and 500°C, the lithium diffuses into the silicon. The diffusion process is stopped well before the lithium has diffused throughout the entire crystal.

Because lithium is electropositive, it easily can donate electrons to the crystalline structure, thereby changing that portion of the *p*-type silicon to an *n*-type silicon. An *np* junction is formed at the interface of the lithium-drifted region with the remainder of the silicon crystal. A dc potential is applied across the silicon while the crystal is at 400 to 500°C. The positive lead is attached to the lithium-drifted region and the negative lead to the *p*-type silicon region. Application of the potential difference across the semiconductor causes electrons to be withdrawn from the lithium region and positive holes to be withdrawn from the *p*-type silicon region.

Some of the lithium ions which form upon removal of electrons from the *n* region migrate ("drift") across the *np* junction to fill the vacancies left by removal of the holes from the *p* region. After some of the lithium ions have drifted into the silicon, the crystal is cooled and the lithium ions become relatively immobile. The crystal contains three regions. The *n*-type region contains lithium atoms in silicon, the *p*-type region is similar to that in a silicon diode, and the central region contains lithium ions in silicon. The central intrinsic region has a high resistance relative to the other two regions, and is the portion of the device which is sensitive to x-rays. A diagram of one form of a lithium-drifted silicon detector is shown in Fig. 18-16.

In use, the positive lead is attached to the *n* region with a thin metallic foil which is often made from aluminum. The *p* region is attached through a thin, transparent metallic film, that is typically composed of gold, to the negative lead. X-rays enter the intrinsic region after passing through the *p* region and excite thousands of electrons to a conduction band. The detector is normally used either with no window or with a beryllium window. The decreased resistance in the intrinsic region when the electrons enter the conduction band permits a flow of current through the device. A large current pulse is generated from each incident x-ray photon.

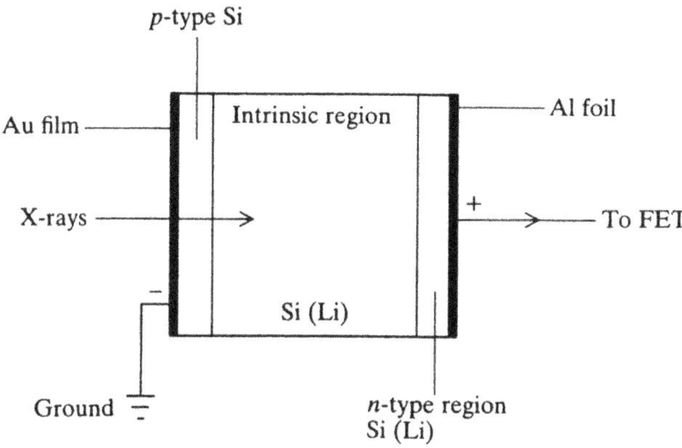

Figure 18-16 A diagram of one form of a lithium-drifted silicon detector.

The output from the detector is usually amplified by a factor of about 10 with a field-effect transistor (FET) which is either attached to the lead coming from the *n* region of the detector or which is on the same block and directly attached to the detector. The output from the FET is further amplified prior to being routed to the readout device.

The output of the detector is directly proportional to the energy of the incident x-ray photon. Normally the detector is operated at 77 K (in a liquid nitrogen bath) in order to decrease background thermal noise, to increase resolution, and to minimize further lithium drifting. Liquid helium rather than liquid nitrogen can be used for cooling. In some modern detectors it is only necessary to use liquid nitrogen or liquid helium during maintenance. Generally the detectors are operated in a vacuum to prevent moisture condensation at the low operating temperature. A major advantage of the detector is the large number of electrons in the intrinsic region which are promoted to a conduction band for each incident x-ray photon. Because that number is greater than the number of ion pairs which are formed in proportional gas-ionization counters for each incident x-ray photon, the detector has better energetic resolution than proportional counters. The resolution is at least an order of magnitude better than that of scintillation counters. At present the device is probably the most widely used detector in energy-dispersive instruments.

Lithium-drifted germanium detectors are prepared and operated in a manner that is analogous to that of the lithium-drifted silicon detectors. Research is being done to develop semiconductor detectors which can be used at room temperature. Detectors constructed of mercury(II) iodide are useful at room temperature. It is likely that research of semiconductor detectors which can be operated at room temperature will continue owing to their convenience of use.

Readout Devices

Readout devices which can be used in equipment that measure x-ray emission, absorption, or fluorescence are generally electronic digital counters. Modern instruments are computer-controlled, thereby making it possible to use the usual computer-operated readout devices such as video display terminals and printers. Older instruments and some less expensive modern instruments use strip-chart recorders.

CHEMICAL ANALYSIS BY X-RAY ABSORPTION

A good description of the use of x-ray absorption for analysis is contained in several chapters in Reference 3. The method has been used for both qualitative and quantitative analysis although x-ray fluorescence is preferred for most assays.

Qualitative analysis is generally performed by comparing the observed absorptive edges in the spectrum of the sample with tabulated absorptive edges for known elements. A list of the K and L critical absorptive edges for the elements with atomic numbers between 12 and 100 is given in Table 18-4. The fine structure on absorptive edges can be related to the structure of an analyte. The analytical technique which measures the fine spectral structure and uses it for analysis is *extended x-ray absorption fine-structure* (EXAFS) *spectroscopy*.

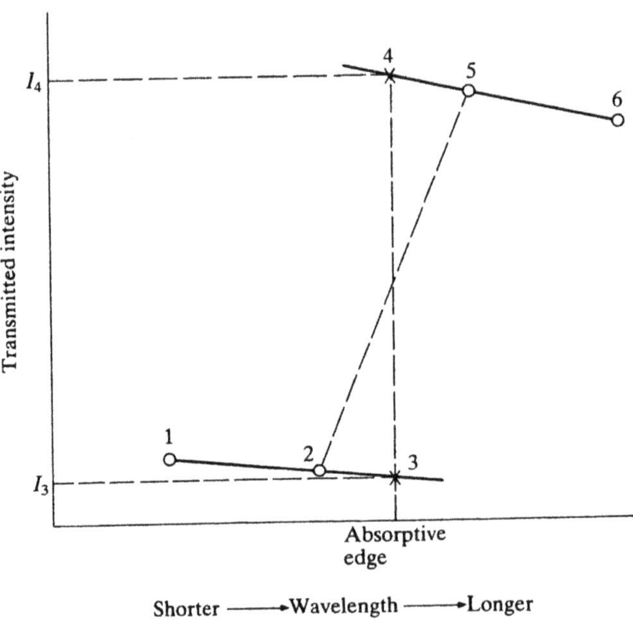

Figure 18-17 An illustration of the use of the absorptive edge method for x-ray absorption.

Table 18-4 The K and L critical absorptive edges of the elements with atomic numbers Z between 12 and100

Z	Element	Critical absorptive edge, pm†			
		K	L_I	L_{II}	L_{III}
12	Mg	951.17			
13	Al	795.11			
14	Si	674.46			
15	P	578.66	8102	9684.3	
16	S	501.82	6422.8	7604.9	7651.9
17	Cl	439.69	5208.4	6136.6	6167.2
18	Ar	387.07	4319.2	5039.0	5080.3
19	K	343.65	3635.2	4202.0	4245.2
20	Ca	307.02	3106.8	3541.7	3582.7
21	Sc	275.73	2683.1	3016.1	3045.7
22	Ti	249.73	2338.9	2683.1	2718.4
23	V	226.90	2052.3	2370.2	2406.99
24	Cr	207.01	1825.6	2122.6	2159.58
25	Mn	189.64	1626.8	1889.6	1924.84
26	Fe	174.33	1460.1	1716.9	1748.38
27	Co	160.81	1334.3	1553.4	1583.14
28	Ni	148.80	1226.7	1413.5	1444.76
29	Cu	138.04	1126.9	1299.4	1325.78
30	Zn	128.33	1033.0	1183.95	1210.55
31	Ga	119.57	953.5	1061.30	1085.46
32	Ge	111.65	872.2	996.46	1022.77
33	As	104.50	810.71	912.81	937.67
34	Se	97.98	746.7	842.12	866.24
35	Br	92.00	692.5	775.23	798.71
36	Kr	86.55	645.6	716.53	742.27
37	Rb	81.55	599.75	665.38	687.52
38	Sr	76.97	558.26	617.23	638.68
39	Y	72.76	523.22	575.51	596.18
40	Zr	68.88	486.72	537.78	558.29
41	Nb	65.30	458.09	522.97	522.26
42	Mo	61.98	429.84	471.80	491.25
43	Tc	58.91	407.46	443.83	463.11
44	Ru	56.05	384.13	417.95	436.89
45	Rh	53.38	362.59	394.23	412.96
46	Pd	50.92	342.75	372.38	390.81
47	Ag	48.58	325.40	351.38	369.83
48	Cd	46.41	308.45	332.57	350.38
49	In	44.39	292.57	314.69	332.44
50	Sn	42.47	277.71	298.23	315.59
51	Sb	40.66	263.94	282.97	300.00
52	Te	38.97	251.05	268.74	285.54
53	I	37.38	238.87	255.27	271.94
54	Xe	35.85	227.37	242.90	259.24
55	Cs	34.47	216.72	231.38	247.39
56	Ba	33.14	206.77	220.44	236.28
57	La	31.84	197.29	210.31	225.83
58	Ce	30.65	188.94	201.08	216.39
59	Pr	29.52	181.08	192.40	207.70

Table 18.4 *Contd...*

Z	Element	Critical absorptive edge, pm†			
		K	L_I	L_{II}	L_{III}
60	Nd	28.45	173.52	184.28	199.47
61	Pm	27.43	166.84	176.58	191.89
62	Sm	26.46	159.86	170.25	184.45
63	Eu	25.55	153.64	162.61	177.53
64	Gd	24.68	147.70	156.13	170.95
65	Tb	23.84	142.10	150.11	164.86
66	Dy	23.05	136.48	143.79	157.92
67	Ho	22.29	131.73	138.97	153.53
68	Er	21.57	126.81	133.83	148.22
69	Tm	20.89	122.21	128.75	143.28
70	Yb	20.22	118.18	124.27	138.61
71	Lu	19.58	114.01	119.85	134.14
72	Hf	18.98	109.96	115.48	129.71
73	Ta	18.39	106.10	111.37	125.51
74	W	17.84	102.45	107.46	121.51
75	Re	17.31	98.95	103.68	117.70
76	Os	16.78	95.57	100.10	114.04
77	Ir	16.29	92.30	96.76	110.57
78	Pt	15.82	89.32	93.42	107.24
79	Au	15.34	86.34	90.26	103.99
80	Hg	14.92	83.53	87.22	100.92
81	TI	14.47	80.80	84.34	97.93
82	Pb	14.08	78.15	81.51	95.03
83	Bi	13.71	75.68	78.87	92.34
84	Po	13.31	73.22	76.38	89.76
85	At	12.95	70.92	73.87	87.23
86	Rn	12.59	68.68	71.53	84.85
87	Fr	12.26	66.54	69.29	82.53
88	Ra	11.93	64.46	67.09	80.28
89	Ac	11.62	62.48	65.00	78.16
90	Th	11.29	60.60	63.00	76.11
91	Pa	11.03	58.75	61.06	74.14
92	U	10.78	56.96	59.19	72.22
93	Np	10.49	55.31	57.42	70.39
94	Pu	10.23	53.66	55.71	68.64
95	Am	9.97	52.08	54.04	66.93
96	Cm	9.67	50.60	52.46	65.28
97	Bk	9.51	49.13	50.93	63.67
98	Cf	9.28	47.71	49.45	62.14
99	Es	9.07	46.36	48.01	60.65
100	Fm	8.86	45.06	46.65	59.23

† Absorptive edges that occur at wavelengths longer than 10 nm are not listed because they are outside of the range of most x-ray instruments.

Quantitative analysis is generally performed by using the absorptive edge method. Normally the K or the L_{III} edge of an element is used for the assay because those edges exhibit relatively large changes in x-ray intensity and because interference from neighboring absorptive edges is less likely than when other edges are chosen.

The transmitted x-ray intensity is measured at a minimum of two wavelengths that are shorter than the edge and at two wavelengths that are longer than the absorptive edge in the spectrum, as illustrated in Fig. 18-17. Fig. 18-17 does not look like Fig. 18-8 because the absorptive coefficient (in Fig. 18-8) rather than transmitted intensity was used as the ordinate. For the case shown in Fig. 18-17, measurements were made at points 1 and 2 and at points 5 and 6. The two sets of points are used to mathematically or graphically extrapolate the data to the absorptive edge. The data taken at the shorter wavelengths are extrapolated to point 3 and the data at the longer wavelengths to point 4. Points 3 and 4 are both at the wavelength of the absorptive edge. The intensities at points 3 and 4 are obtained from the extrapolation.

For most quantitative analyses, the distance through the sample is fixed during the four measurements, and it is convenient to restate Beer's law [Eq. (18-8)] for x-ray analysis:

$$\ln\frac{I_0}{I} = \mu_m m \tag{18-15}$$

In Eq. (18-15) the sample density and distance have been combined into a single term m which has units of grams per square centimeter. It is the mass of the sample within a 1-cm^2 cross-sectional area.

Because the sample is not composed of a single element, the mass absorptive coefficient is given by Eq. (18-9). If it is assumed that the sample consists of an analyte and a matrix, Eq. (18-9) becomes

$$\mu_m = \mu_s W_s + \mu_M W_M \tag{18-16}$$

where μ_s is the mass absorptive coefficient of the pure analyte and μ_M is the coefficient of the matrix. Substitution from Eq. (18-16) into Eq. (18-15) yields

$$\ln\frac{I_0}{I} = \mu_s W_s m + \mu_M W_M m \tag{18-17}$$

Substitution of the intensities at points 3 and 4 in Fig. 18-17 into Eq. (18-17) produces two equations:

$$\ln\frac{I_0}{I_3} = \mu_{s3} W_s m + \mu_M W_M m$$

$$\ln\frac{I_0}{I_4} = \mu_{s4} W_s m + \mu_M W_M m$$

Subtraction of one of the equations from the other yields

$$\ln\frac{I_0}{I_3} - \ln\frac{I_0}{I_4} = \mu_{s3} W_s m + \mu_M W_M m - \mu_{s4} W_s m - \mu_M W_M m$$

$$\ln\frac{I_4}{I_3} = (\mu_{s3} - \mu_{s4}) W_s m \qquad (18\text{-}18)$$

Equation (18-18) is used to calculate $W_s m$, which is the mass of the analyzed element in a 1-cm^2 cross-sectional area of the sample. In order to directly apply Eq. (18-18), it is necessary to know the two values of the mass absorptive coefficient for the pure analyzed element at the absorptive edge. Those values can be obtained by extrapolating tabulated values to the desired wavelength or from reference tables in the literature.

While using Eq. (18-18) the two intensities can be recorded in any units which are proportional to intensity as long as both intensities are recorded with the same units. As an example, the number of pulses from a detector in a fixed period or the counting rate can be used. As an alternative to direct application of Eq. (18-18), the working-curve method can be used by plotting ln (I_4/I_3) as a function of concentration for a series of standards. That method has the advantage of not requiring knowledge of the mass absorptive coefficients. Another approach is to use a single intensity measurement near the top and the bottom of the absorptive edge, rather than extrapolated values. That method is less accurate than the method of extrapolation. Absorptive methods do have the advantage of using a ratio of intensities which, at least partially, compensates for matrix effects.

Sample problem 18-2 Iron in a solution was assayed using the absorptive edge method across the K edge. The sample was placed in the beam of the instrument and the number of pulses from the detector during a period of 15 s was recorded at each of four wavelengths. The results are tabulated. The mass absorptive coefficient at the lower transmitted-intensity portion of the absorptive edge is 465 cm^2/g and at the higher transmitted-intensity portion is 54 cm^2/g. Determine the amount of iron in the sample.

Number	Wavelength, pm	Counts/ 15 s
1	153.9	219
2	165.6	163
5	175.0	2852
6	193.0	2261

SOLUTION The K edge of iron occurs at 174.33 pm (Table 18-4). Linear extrapolation of the tabulated data to 174.33 pm can be done using proportions and the differences between readings. The numbers on the left of the tabulated data correspond to the numbered points in Fig. 18-17. The lower intensity at 174.3 pm can be obtained using the following proportion:

$$\frac{I_1 - I_2}{\lambda_1 - \lambda_2} = \frac{I_1 - I_3}{\lambda_1 - \lambda_3}$$

$$\frac{219 - 163}{153.9 - 165.6} = \frac{219 - I_3}{153.9 - 174.3}$$

$$I_3 = 121 \text{ counts/15 s}$$

Similarly, the higher intensity is obtained by extrapolation of the data at points 5 and 6:

$$\frac{I_5 - I_6}{\lambda_5 - \lambda_6} = \frac{I_4 - I_6}{\lambda_4 - \lambda_6}$$

$$\frac{2852 - 2261}{175.0 - 193.0} = \frac{I_4 - 2261}{174.3 - 193.0}$$

$$I_4 = 2.88 \times 10^3 \text{ counts/15 s}$$

Substitution of the appropriate values into Eq. (18-18) and solution of the equation yields the desired result:

$$\ln \frac{2.88 \times 10^3}{121} = \left(465 - 54 \text{ cm}^2/\text{g}\right) W_s m$$

$$W_s m = 7.71 \times 10^{-3} \text{ g/cm}^2$$

Absorptiometry with x-rays is not used as often as x-ray fluorescence for qualitative and quantitative analysis. Nevertheless, absorption is particularly useful for certain applications such as measuring the thickness of steel or other metallic strips and performing selected quantitative analyses. The thickness of metallic strips can be continuously monitored during production of the strips and the results used for process control. If the strip is of uniform composition, the amount of absorbed radiation increases with thickness according to Beer's law.

Quantitative analyses which are preferentially performed by using x-ray absorption are those analyses in which portable instrumentation is required and those analyses in which a single heavy element exists in the sample. Portable x-ray apparatus generally use small radioactive sources.

X-RAY FLUORESCENCE

Prior to x-ray fluorescence an x-ray photon from a source is absorbed by the sample producing a positive ion with a vacancy in an inner electron shell. As an electron from an outer shell falls into the vacancy, an x-ray photon is emitted. The emission is x-ray fluorescence only when the initial excitation occurs by absorption of an x-ray photon.

The incident photon possesses the relatively large energy that is required to completely remove the electron from the element while the fluoresced photon possesses the smaller energy corresponding to an energetic transition from an outer electron shell to the vacant inner shell. Consequently radiation is fluoresced at a wavelength which is greater than that of the absorbed radiation. In order for fluorescence to occur, the x-ray source must emit radiation that is sufficiently energetic to expel an electron from an inner electron shell. If the source is an x-ray tube, the operating voltage must be sufficient to produce a short-wavelength cutoff [Eq. (18-2)] which is less than the absorptive edge of the analyzed element. The bandwidth of the emitted radiation is small in comparison to absorptive bandwidths. A continuum of radiation, such as that emitted from x-ray tubes, is not emitted from the sample during fluorescence.

In order to measure absorption it is necessary for the sample to be sufficiently thin to permit x-ray radiation to pass through it without being completely absorbed. Measurements of fluorescence do not have the same requirement. The x-ray radiation only must penetrate the surface of the sample in order for fluorescence to occur. As in the ultraviolet-visible region, the intensity of the fluorescent radiation is directly proportional to the concentration of the fluorescing substance in the sample. The intensity is also a function of the amount of incident radiation absorbed by the sample and of the portion of the fluorescent radiation which can be self-absorbed by the sample.

For the typical case in which the incident radiation only penetrates a small distance relative to the thickness of the sample, the intensity I of the fluoresced radiation is given by

$$I = \frac{kW}{\mu_i + \mu_f} \qquad (18\text{-}19)$$

where k is a proportionality constant, W is the weight fraction of the analyzed element, and μ_i and μ_f are the linear absorptive coefficients for the entire sample (not the analyte alone) at the wavelength of the incident and the fluorescent radiation respectively. Because all of the incident radiation is absorbed, the intensity is independent of distance through the sample.

Because the linear absorptive coefficients for the entire sample appear in the denominator of Eq. (18-19), the components of the sample in addition to the analyte can affect the intensity of the emitted radiation. The change in intensity of the fluoresced radiation which is caused by components in the sample other than the analyte is the *matrix effect*. While measuring x-ray fluorescence, it is usually necessary to consider the matrix effect. The matrix effect can be responsible for major errors in quantitative x-ray fluorescence. The matrix can both absorb incident radiation and fluoresce. Fluoresced radiation from the matrix contributes to the matrix effect if it falls within the bandpass of the detector or if it is absorbed by the analyte and induces fluorescence.

Apparatus for X-Ray Fluorescence

The apparatus that is used to induce and monitor x-ray fluorescence is similar to that which is used to induce and monitor x-ray absorption. Owing to the lack of commercial apparatus, measurements of x-ray absorption are sometimes made with modified fluorescence apparatus. Instruments that are used for fluorescence can contain either wavelength- or energy-dispersive detection systems. Block diagrams of wavelength- and energy-dispersive instruments are shown in Fig. 18-18.

The individual components of the apparatus are identical to those described earlier. The source is generally an x-ray tube for laboratory measurements. Often chromium targets are used for long-wavelength studies and tungsten targets for shorter wavelength studies, however, any of the previously listed targets can be used. Radioactive sources can be used for measurements made outside the laboratory.

In energy-dispersive instruments, it is not necessary for the beam of photons which strikes the sample to be monochromatic. Consequently, the output from the x-ray tube can be used directly. Energy-dispersive instruments normally have less resolution than wavelength-dispersive instruments. The limitations owing to the relatively low dispersion in energy-dispersive instruments often can be overcome by using monochromatic incident radiation. Because monochromatic incident radiation selectively excites the sample, much unwanted fluorescence and the need for high resolution can be eliminated. Consequently, some energy-dispersive instruments contain a filter or a second target between the source and the sample. Incident radiation which strikes the second target induces fluorescence of those lines which are characteristic of the target. The fluoresced radiation is used as exciting radiation for the sample. The wavelengths striking the sample can be controlled by changing the second target. The second target is termed the *first fluorescer*.

Solids, liquids, and gases can be assayed by using x-ray fluorescence. If the solid is a single object, the surface should be ground and polished prior to being placed in the x-ray beam. The particle size of powdered samples should be carefully controlled and the powder should be homogeneous. Often powders are mixed with a binder and pressed into a single pellet. Alternatively, powders can be fused with an appropriate flux, such as sodium tetraborate, to form a glass disk.

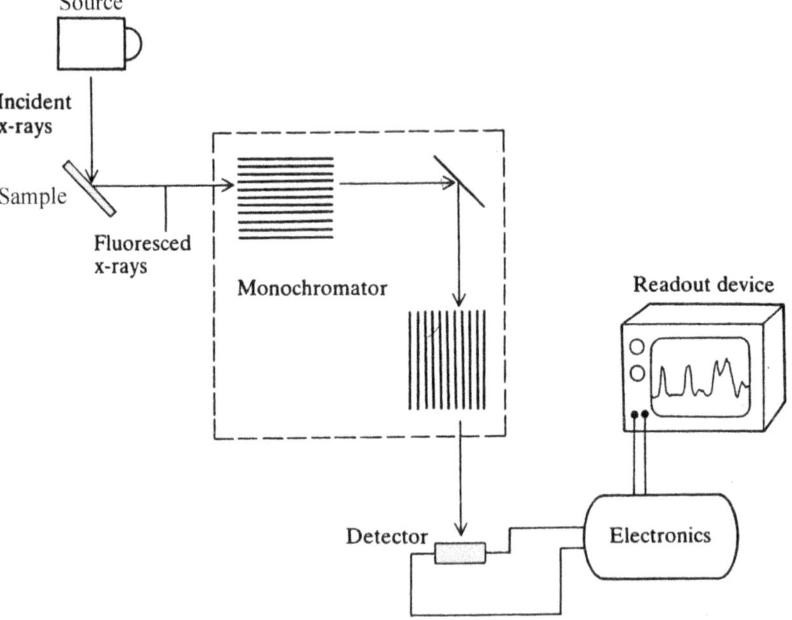

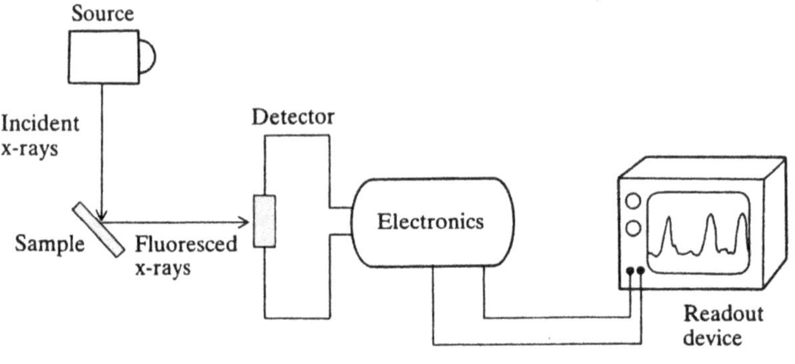

Figure 18-18 Diagrams of wavelength-dispersive and energy-dispersive x-ray fluorometers.

Solutions are placed in plastic or metallic cells that are constructed from low atomic number elements. Aluminum cells are often used. The windows in the cells typically consist of thin stretched sheets of Mylar. The analyst stretches the sheet over an open end of the sample cup and slips a ring over the Mylar film to hold the sheet in place. Gaseous samples are contained in vessels which are capable of withstanding the high pressures within the cell. The windows must be constructed from a material that is transparent to x-rays and is structurally strong. Mylar sheets cannot be used because they lack the required strength. Beryllium windows can be used. If it is necessary to assay elements which have atomic numbers less than 11 (Na), windowless optics are used. Under ideal conditions, windowless optics instruments can be used to assay elements which have atomic numbers of 4 (Be) or greater.

Collimation is accomplished with the devices described for x-ray absorption. X-ray radiation is normally dispersed with a crystalline grating which is mounted on a goniometer as shown in Figs. 18-9 and 18-10. In Fig. 18-18 the collimators are considered part of the monochromator.

The detectors can be any of those devices which were described earlier. Semiconductor detectors are used most often in energy-dispersive instruments. Energy-dispersive instruments generally have less resolution than wavelength-dispersive instruments but are more sensitive. Typically energy-dispersive instruments are two orders of magnitude more sensitive than wavelength-dispersive instruments. Energy-dispersive instruments yield large signal-to-noise ratios owing to the Fellgett advantage that is associated with the relatively large number of counts that are recorded by the instruments on a single sample.

In addition to the detectors which previously have been described, some x-ray fluorometers and other x-ray instruments utilize *charged coupled device* detectors. CCD detectors have found multiple uses including the acquisition of digital images. In that case the devices are termed CCD cameras. The incident x-ray radiation from the fluorometer strikes a photoactive capacitor array resulting in an accumulation of electric charge on the illuminated region of the array. Each capacitor in the array normally contains a Si region which has been doped with B to create a p-type semiconductor. This region constitutes the photosensitive element. On top of the p-doped Si is a layer of silicon dioxide which serves as the insulator between the plates in each capacitor. Electrodes are attached to the top of the silicon dioxide layer and serve as the plates of the capacitor. Each IGFET is isolated from adjoining IGFETS in the array by channel stops consisting of thermally grown silicon oxide layers. Upon irradiation with x-rays, a charge accumulates in the p-doped region under an irradiated electrode.

After the charge has accumulated on the irradiated portions of the array, an electric circuit is used to apply charges to the capacitor plates which cause each array component to transfer its charge stepwise to the adjacent array component. The device functions as a shift register (Chapter 4). The last array component transfers its charge to an amplifier which converts the charge to a potential and amplifies it. When the devices are used in spectrometry, the potential is digitized and stored in memory. The array is contained in an integrated circuit chip with an upper surface which is exposed to incident x-rays. The devices were originally developed in 1969 by Boyle and Smith (Reference 7) while working at AT&T Bell Labs. One instrument maker uses a CCD sensor which has a 1024 × 122 pixel array. Each pixel corresponds to a component of the array and occupies a space of 24 × 24 micrometers. Readout devices are identical to those described for measurements of absorption.

The National Institute of Standards and Technology NIST in the United States has recently developed a microcalorimeter detector which shows great promise. It is an ED detector with a 5 eV resolution throughout the entire x-ray range. It uses a bismuth block to monitor heat changes. The heat changes are converted to electrical pulses which are amplified and monitored. The bismuth sensor is operated at a temperature of 0.15 K with cooling supplied by a combination of liquid nitrogen, liquid helium and an adiabatic demagnetization refrigerator. At the low operating temperature the detector is a superconductor which is approximately 100 times more sensitive than the previously described Si detectors. An example of the use of the detector is described in Reference 8.

Chemical Analysis by X-Ray Fluorescence

Qualitative analysis that is performed by using x-ray fluorescence is relatively straightforward. A spectrum is recorded of intensity or logarithm of intensity as a function of wavelength, energy, or channel number (energy-dispersive instruments), and the positions of the spectral peaks are compared to those of known elements. The use of x-ray fluorescence for qualitative and quantitative analysis of environmental samples is described in Reference 4.

Quantitative analysis is similar to that in the ultraviolet-visible region. If the sample is sufficiently large, if long counting times are used, and if other conditions are favorable, assays can be performed at concentrations that are 0.1 ppm or lower. Assays of trace constituents can be preceded by concentrating the analyte. Among the methods that can be used to concentrate the analyte are precipitation, electrochemical plating, and chromatography.

Quantitative analysis can be performed in several ways. Regardless of the method, the matrix effect can be important. Perhaps the most popular method uses a single standard in a matrix which is as similar as possible to that of the sample. If the matrices are identical, the matrix effects in the analyte and the standard are identical. In that case the fluorescent intensity of the standard can be used to determine the proportionality constant in the equation that relates fluorescent intensity to concentration of the analyte:

$$C_u = C_s \left(\frac{I_u}{I_s} \right) \tag{18-20}$$

Equation (18-20) can be used to calculate the concentration C_u of the analyte from the concentration C_s of the standard and the corresponding fluorescent intensities I_u and I_s.

The working-curve method in which all of the standards are prepared in matrices similar to that of the analyte normally yields more accurate results than those obtainable with a single standard. Unfortunately the working-curve method is more time-consuming. The standard-addition technique can also be used to correct for the matrix effect. Often the standard-addition technique is performed by using a single standard addition. The graphical method in which a series of standard additions are used is preferable because it minimizes the probability of a determinate error in the analysis.

In some cases the internal-standard method is used. The internal standard should have a fluorescent line which is as near that of the analyte as possible. The ratio of the intensity of the line owing to the analyte to that owing to the internal standard is measured and used with the working-curve method or with Eq. (18-20) in place of I_u and I_s. The assumption made while using the internal-standard method is that the effect of the matrix upon the analyte is identical to that upon the internal standard. The effects cancel when the ratio of the intensities is obtained. While using that method it is important to be aware of any absorptive edge that might exist at a wavelength between that of the analyte and standard lines, and of the ability of an emitted line from the matrix to excite the element (either analyte or standard) which has the lower atomic number. The internal standard is usually chosen to be an element which has an atomic number that is one unit higher or lower than that of the analyte.

When the composition of the matrix is unknown or cannot be duplicated in the standard solutions, and when it is impractical to use the standard-addition technique, it sometimes is possible to reduce matrix effects by dissolving the sample and standard in a relatively large amount of the same matrix. The matrix can be either a solid or liquid solvent. In a related method, thin films of the analyte and the standards are used during the analysis. The matrix effects are minimized because the number of x-ray photons that can be absorbed or fluoresced by the thin layer of the matrix is relatively small. Another technique relies upon measurements of the ratio of the intensity of the fluorescent radiation from the analyte to the intensity of scattered radiation. The assumption is that both fluorescent radiation and scattered radiation are absorbed to the same extent by the matrix.

The detection limit during quantitative analyses can be controlled by the amount of self-absorption which occurs. Typically the detection limit is about 0.01 percent. As the amount of self-absorption increases, the intensity of the measurable emitted radiation decreases and the detection limit increases. The accuracy and precision of a quantitative analysis are a function of the number of counts taken by the detector during the assay. The counting error is inversely proportional to the square root of the number of pulses. Consequently, as the number of counted pulses increases by a factor of 100, the counting error decreases by a factor of 10. Of course, improved accuracy and precision must be balanced against longer analytical times when deciding upon a procedure.

X-ray fluorescence can be used for qualitative and quantitative analyses of all elements with an atomic number of 11 (Na) or greater. It can be used with less accuracy for elements with lower atomic numbers. In favorable circumstances the detection limit can be as low as 0.1 parts per million. Relative precision at high concentrations is often less than 0.1 percent.

PARTICLE-INDUCED X-RAY EMISSION

Earlier in the chapter methods were described in which incident electrons and incident x-ray photons caused x-ray emission. It is also possible to induce x-ray emission by bombarding a sample with positive particles from an accelerator. That technique is *particle-induced x-ray emission* (PIXE). Although alpha particles and other positively charged particles can be used, protons are used most often to bombard the sample, and the technique is *proton-induced x-ray emission* (also abbreviated as PIXE).

After bombardment the sample emits x-rays at the frequencies corresponding to the allowed energetic transitions, as during x-ray fluorescence. The kinetic energy possessed by the incident protons is determined by the particular accelerator that is used. Normally protons with energies between 2×10^{-14} and 3×10^{-12} J are used. Many instruments contain an energy-dispersive Si(Li) or intrinsic Ge detector. The particle beam is focused on the desired area on the surface of the sample and the emitted x-rays are monitored. Depending upon exposure time, the sample can be assayed at depths between about 1 nm and about 1 μm. The diameter of the beam used for many assays is about 5 mm. Beams with diameters of about 1 mm are available for work on smaller areas.

If the beam can be maintained at a specific location for varying periods, a depth profile of the sample can be obtained. By moving the beam from location to location across the surface, an elemental surface map is obtained. A PIXE instrument that is capable of scanning across a surface is a *proton microprobe.* The major advantages of the proton microprobe for surface analysis are its relatively deep penetration and its high sensitivity. The detection limit for PIXE varies with sample thickness and can be as low as 1 pg. The major disadvantages of a proton microprobe are the high cost associated with the apparatus and the loss of volatile components during irradiation. The electron microprobe is presently used more often for surface analysis. A good description of several PIXE instruments, the procedures used for analyses, and applications can be found in Reference 4.

ELECTRON MICROPROBE

An instrument which has proven to be particularly useful for the analysis of surfaces is the *electron microprobe. Electron probe x-ray microanalysis* (EPMA) is the analytical technique that relies upon the use of an electron microprobe. Magnetic fields are used to focus a beam of electrons on the surface of a sample. The electron bombardment causes emission from the sample, as in an x-ray tube, and the emitted x-rays are monitored. An optical microscope is attached to the device in such a manner that the portion of the surface that is observed in the optical microscope is identical to the portion upon which the electron beam is focused. Figure 18-19 is a diagram that shows the essential parts of a wavelength-dispersive electron microprobe.

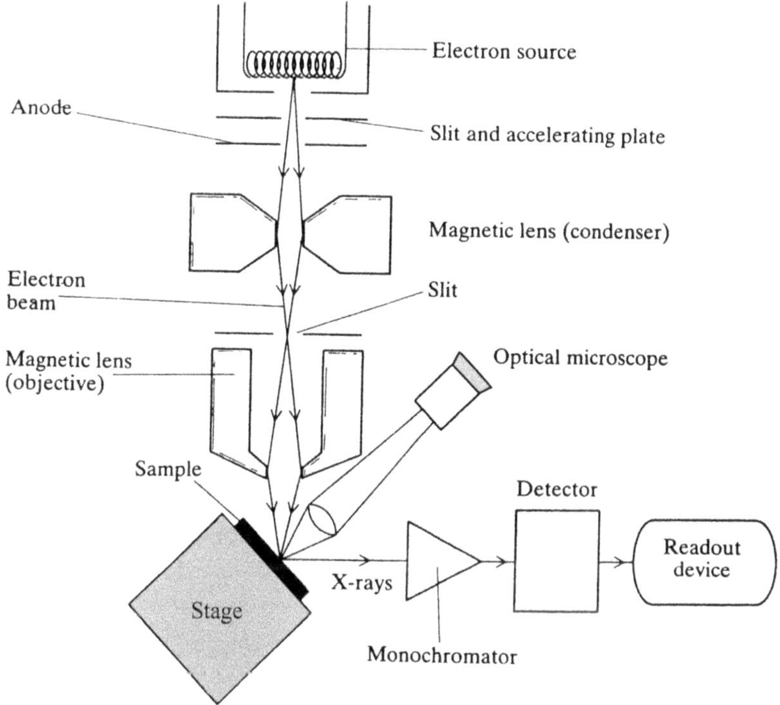

Figure 18-19 A diagram of an electron microprobe. The path of electrons and x-rays through the device is indicated by the arrows.

The electron source usually is a heated tungsten cathode. The electrons are accelerated by an anode through one or more slits and focused by two magnetic lenses onto the sample. Typical beam diameters are in the range from 0.1 to 5 μm. Owing to the small beam diameter, the spatial resolution on the surface of the sample is better than that of the proton microprobe. The electron beam, the optical microscope, and the entrance to the x-ray spectrometer are all focused on the same spot on the sample.

The sample is placed on a stage similar to a microscope stage. The stage is accurately moved while the sample is observed through the microscope until the sampling site is located. Movement is stopped at the chosen site and the electron beam is discharged. The energy of the electron beam is determined by the change in potential between the cathode and the anode. Typically the accelerating potential is between 10 and 50 kV which applies kinetic energies between 1.6×10^{-15} and 8×10^{-15} J to the electrons. The apparatus is operated in a vacuum at pressures below 10^{-5} torr. Typical detection limits for EPMA vary from 10^{-14} to 10^{-12} g. Relative accuracy during quantitative analyses is generally in the range from 1 to 5 percent at concentrations that are greater than about 1 percent.

The large number of electrons that strike the sample can cause an accumulation of charge which must be disseminated. If the sample is a good electrical conductor, dissemination occurs automatically. Poor conductors are mounted on a thin metallic strip which facilitates dispersal of the charge. As with the proton microprobe, the electron microprobe can be used to map the surface of a sample and to obtain a depth profile at a fixed location. Both energy-dispersive and wavelength-dispersive spectrometers are used in conjunction with an electron microprobe. In some instruments both types of detection are available simultaneously.

SCANNING ELECTRON MICROSCOPE

An instrument that is related to the electron microprobe is the *scanning electron microscope* (SEM). A beam of electrons is focused on the surface of a sample as with EPMA. An image of the surface of a sample can be obtained from the intensity of backscattered electrons. A *scanning transmission electron microscope* (STEM) obtains an image of the sample from the electrons that are transmitted through the sample. With either the SEM or the STEM, a fixed-energy electron beam is scanned across the surface of the sample. The results appear on the readout device as a visual image of the surface of the sample. The results are similar to those obtained from an optical microscope except that the magnification is considerably greater. STEMs have been developed which have beam diameters as small as 2 nm. They can be used for studies of particulates and thin foils.

In addition to providing a visual image of a sample, scanning electron microscopes are used in conjunction with x-ray spectrometers to obtain the elemental composition of the sample. The process is similar to that used with the electron microprobe. The elemental composition can be obtained across the surface of the sample during a scan or at various depths at a single location. Scanning electron microscopes generally use energy-dispersive x-ray spectrometers.

X-RAY DIFFRACTION

X-ray diffraction by the analyzing crystal in a wavelength-dispersive instrument was described earlier in the chapter. The incident angle, relative to a crystal which has a fixed internuclear spacing, is related to the wavelength of the reflected radiation. If monochromatic incident radiation is used, the angle at which radiation is diffracted from the crystal can be related to the atomic spacing in a crystal with the Bragg equation [Eq. (18-12)]. That is the basis of *x-ray crystallography*.

X-ray crystallography is used to determine the spacing between atoms in crystals. The interatomic spacings are used to deduce crystalline structures. In addition to use in structural determinations, x-ray diffraction can be used in qualitative or quantitative analysis. Descriptions of the uses of x-ray diffraction can be found in References 5 and 6.

One instrument which traditionally has been used for x-ray crystallography is the *x-ray powder camera*. A block diagram of an x-ray powder camera is shown in Fig. 18-20. Radiation from the source is made monochromatic by passage through a filter. X-ray tubes are used as sources in most instruments. The monochromatic beam of x-rays passes through a collimator and strikes the sample. The collimator is a cylinder with entrance and exit holes.

A powder is used in a powder camera rather than a single crystal. The powder can be placed in a cylindrical plastic, glass, or lithium borate container, or it can be mixed with a binder and dried into a cylindrical shape. Typically a 1 mm^3 portion of the powder contains about 10^{12} crystals. Each of the crystals has a volume of about 10^{-21} m^3. Owing to the large number of randomly oriented crystals in the sample, a significant number of crystals are oriented at

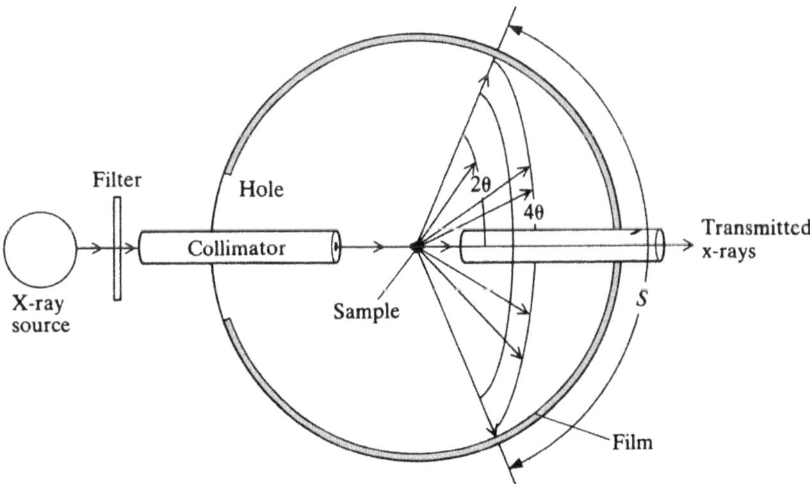

Figure 18-20 A diagram of an x-ray powder camera. The angle between the path of the transmitted x-rays and any portion of the diffracted cone is 2θ.

all angles relative to the x-ray beam. Consequently, regardless of the atomic spacing within the crystalline powder, some of the crystals are properly oriented to reflect radiation as predicted by the Bragg equation.

Owing to the nearly infinite number of orientations of the crystals in the powder, a cone of x-rays with an apex at the sample is reflected from the crystal for a particular interatomic spacing, as illustrated in Fig. 18-20. The terms reflected radiation and diffracted radiation are used interchangeably in the description. The detector for the x-ray beam in older instruments is a strip of photographic film that is wrapped around the inner edge of the circular casing that surrounds the sample. The center of the circle formed by the film is occupied by the sample. Holes exist in the film for the collimator and for the transmitted beam of unreflected x-rays. A portion of the reflected cone of x-rays from the sample exposes an arc on the film at the points of intersection. In some instruments the film forms a half-circle rather than a complete circle around the sample. A separate arc is exposed on the film for each interatomic spacing within the crystal.

After the film has been exposed, the source is turned off and the film is removed and developed. The developed film strip has a series of darkened lines as illustrated in Fig. 18-21. Each line is an exposed arc on the film which is characteristic of a single internuclear distance. The angle between the beam of x-rays and the line that connects the sample with an arc is 2θ, as indicated in Figs. 18-9 and 18-20. Because the radius of the circle which had the film around its circumference is known, the distance on the film from the entrance (collimator) or exit (transmitted beam) holes that corresponds to a particular angle 2θ can be calculated. Substitution of the values into Eq. (18-12) makes it possible to calculate the interatomic spacing d within the crystal.

In practice, the distance S on the film is measured between the two arcs of the same cone on opposite sides of the exit hole. The angle between the two lines which connect the arcs with the sample is 4θ, as indicated in Fig. 18-20, that is, 2θ between the transmitted beam and each line. The angle 4θ can be calculated in radians by dividing the distance S on the circle that encompasses the angle by the radius r of the circle. For a powder camera the result is

$$4\theta_{\text{rad}} = \frac{S}{r} \tag{18-21}$$

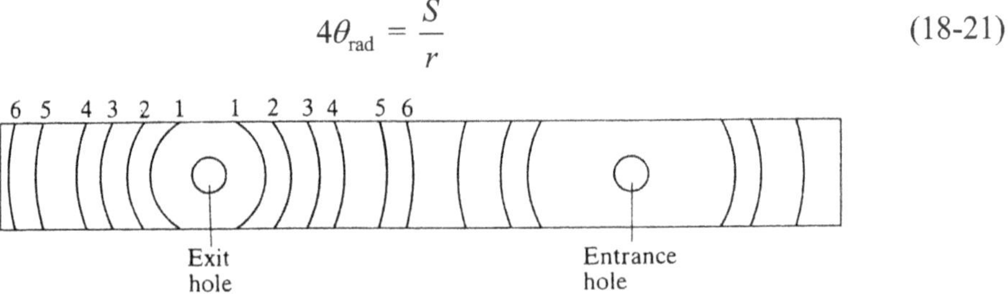

Figure 18-21 A diagram of a developed powder photograph. The numbers around the exit hole correspond to developed arcs of diffracted radiation.

The angle which is substituted into the Bragg equation must be measured in degrees. Because 1 rad is 57.295779°, the equation

$$\theta_{\text{rad}} \text{ radians} \times 57.296 \text{ degrees/radian} = \theta_{\text{deg}} \tag{18-22}$$

can be used to convert θ_{rad} to θ_{deg}. Substitution from Eq. (18-22) into Eq. (18-21) gives

$$\frac{4\theta_{\text{deg}}}{57.296} = \frac{S}{r}$$

$$\theta_{\text{deg}} = \frac{57.296S}{4r} \tag{18-23}$$

From Eq. (18-23) it is apparent that calculations would be simplified if the radius of the powder camera was constructed to be a multiple of 57.296. The radius of many cameras is 57.296 mm, which makes θ_{deg} equivalent to one-fourth of the distance S (mm) between the two exposed arcs on the film.

The distance on the film that corresponds to one degree can be calculated for a camera with any particular radius r by setting 360°, which corresponds to one revolution around a circle, equal to the circumference $2\pi r$ of the circle. The result is

$$360^{\circ} = 2\pi r$$

$$1^{\circ} = \frac{2\pi r}{360} = 0.017453r \tag{18-24}$$

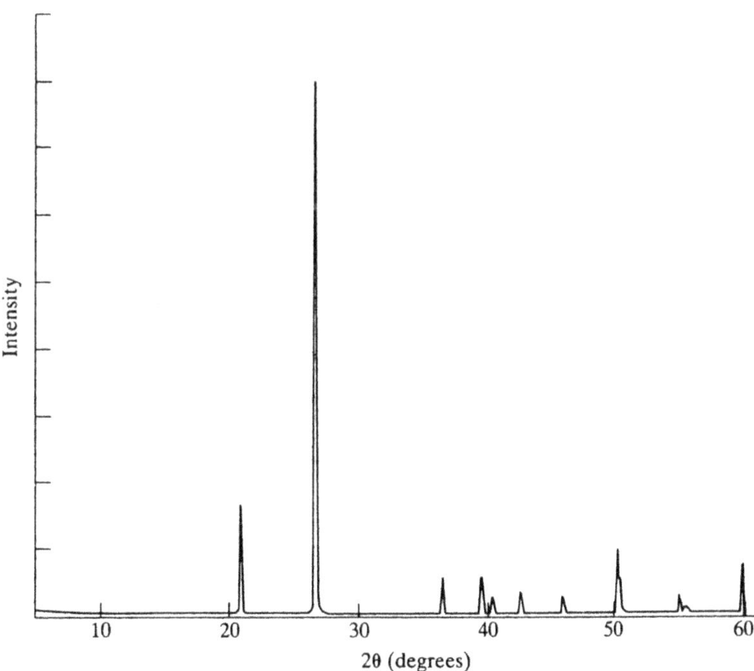

Figure 18-22 The x-ray diffraction spectrum of α-quartz obtained by using a goniometer. (*Courtesy of Dr. Alan Bailey, Department of Geology, University of Louisiana at Lafayette.*)

For the case in which the radius is 57.296 mm, substitution into Eq. (18-24) reveals that 1° corresponds to 1 mm.

The incident beam of x-rays used for x-ray crystallography should be monochromatic. In many cases the copper K_α line at 0.154 nm is used. Some powder cameras have a motor attached to the sample holder which is used to rotate the sample during the measurement. Rotation is particularly important when the particle size in the sample is relatively large. Without rotation the diffraction pattern in coarse crystals is a function of the physical orientation of the crystals in the sample.

A difficulty which is sometimes encountered when using a powder camera is overlapping diffracted beams. In those instances it is preferable to perform structural determinations on single crystals rather than on powders. Each crystal is exposed to x-rays at many angles relative to the surfaces of the crystal and each exposure is individually used to determine a particular spacing within the crystal.

Detectors other than photographic film can be used to measure diffraction. The detector can be mounted on a goniometer and slowly rotated around the sample while the intensity of radiation is recorded as a function of angle. Gas-ionization and scintillation detectors also can be used. An x-ray diffraction spectrum that was obtained by using a goniometer is shown in Fig. 18-22. Multichannel detection also can be used. Multiple detectors are mounted at different angles but at the same distance from the sample and are used to simultaneously measure diffracted radiation. Some modern instruments utilize CCD cameras to obtain the digital image of the diffracted x-ray radiation. The devices appear to be ideally suited for use in x-ray diffractometers and likely will see increased use.

Sample problem 18-3 A powder camera was used to determine the spacing in a powdered crystal. The radius of the camera was 57.296 mm and the incident radiation was the Cu K_α line. The distance S on the developed film between the two opposite arcs of one of the diffracted cones of x-rays was 58.88 mm. Determine the internuclear spacing d that caused the diffraction.

SOLUTION For a camera with a radius of 57.296 mm, each millimeter on the film corresponds to 1°. Consequently, the angle 4θ that encompasses the distance S is 58.88°. The angle θ is therefore

$$\frac{58.88^\circ}{4} = 14.72^\circ$$

Substitution into the Bragg equation for first-order radiation and solution for d provides the internuclear spacing:

$$n\lambda = 2d \sin \theta$$

$$(1)(0.154 \text{ nm}) = 2d \sin (14.72^\circ)$$

$$d = 0.303 \text{ nm}$$

Chemical Analysis by X-Ray Diffraction

Both qualitative and quantitative chemical analysis can be performed by using x-ray diffraction. Qualitative analysis is generally performed with a powder camera. The image caused by x-ray diffraction in the sample is compared to images caused by diffraction in standards. All images should be obtained by using the same instrument. Alternatively, the crystalline spacing corresponding to the three or four most-intense developed arcs can be calculated and compared with those in a library file. Files are available from the American Society for Testing and Materials (ASTM), the Bureau of Mines, the Geological Society, and several other organizations.

Quantitative analysis is rarely performed using diffraction measurements. Quantitative measurements can be performed by measuring the intensity of a particular diffracted line at a fixed angle relative to the incident radiation. The measured intensity is compared to that of a standard or of several standards. Qualitative and quantitative analyses with x-ray diffraction have found numerous industrial and laboratory applications.

IMPORTANT TERMS

Absorptive edge
Absorptive edge method
Analyzing crystal
Balanced filter method
Beta emission
Bremsstrahlung
CCD detector
Collimator
Dead time
Dispersion
Duane-Hunt equation
Electron beam current
Electron capture
Electron microprobe
Energy-dispersive detector
EPMA
EXAFS
Filter
Gamma emission
Gas-ionization detector
Geiger counter region
Geiger-Müller counter
Goniometer
Ion pairs
Ionization chamber
K capture
K series
L capture
L series
Line spectra
Linear absorptive coefficient
M series
Mass absorptive coefficient
Matrix effect
Microcalorimeter detector
Particle-induced x-ray emission
Photoelectric absorptive coefficient
Powder camera
Proportional counter
Proportional region
Proton microprobe
Quenching gas
Saturation region
Scanning electron microscope
Scattering coefficient
Scintillation counter
Scintillation detector
Semiconductor detector
Short-wavelength cutoff
Valence band
X-ray
X-ray crystallography
X-ray tube

PROBLEMS

Theory

18-1 Calculate the potential difference between the filament and the target in an x-ray tube which has a short-wavelength cutoff of 0.110 nm.

18-2 The short-wavelength cutoff for an x-ray tube was 0.152 nm. Determine the accelerating potential used in the tube.

18-3 The accelerating potential in an x-ray tube was 25.0 kV. Calculate the short-wavelength cutoff of the lamp.

18-4 Estimate the wavelength of maximum intensity for the x-ray tubes described in Probs. 18-1 through 18-3.

18-5 Estimate the relative integrated areas of the emitted spectra of four x-ray tubes that are operated with identical electron beam currents and accelerating potentials when the targets in the lamps respectively are Pt, Sn, Cu, and Al (see Fig. 18-3).

18-6 Estimate the frequencies and the wavelengths of the K lines of Al, Cu, Sn, and Pt.

Absorption

18-7 Calculate the mass absorptive coefficient at 0.436 nm of an alloy consisting of 85.0 percent Fe, 5.0 percent Ni, 9.0 percent Cu, and 1.0 percent Zn. The mass absorptive coefficients for the pure elements at 0.436 nm are 610, 715, 760, and 910 cm^2/g respectively for Fe, Ni, Cu, and Zn.

18-8 Calculate the mass absorptive coefficient at 0.193 nm of a homogeneous mixture that contains 60.0 percent $CaCl_2$ and 40.0 percent $CuCl_2$. The mass absorptive coefficients at 0.193 nm of Ca, Cl, and Cu respectively are 306, 198, and 96.2 cm^2/g.

18-9 Estimate the thickness of a foil composed of the alloy described in Prob. 18-7 if the detector recorded 10,848 counts/min of transmitted x-rays when the foil was not in the path of the x-rays, and 1023 counts/min when the foil was placed in the path. The density of pure Fe, Ni, Cu, and Zn respectively are 7.87, 8.90, 8.96, and 7.13 g/cm^3.

18-10 The L_{III} critical absorptive edge of silver was used with the absorptive edge method for quantitative analysis of silver in a solution. The mass absorptive coefficients of silver at the absorptive edge are 354 and 1410 cm^2/g. Determine the amount of silver in the solution from the tabulated data.

Wavelength, pm	Detector response, counts/min
360.0	1483
365.0	1276
375.0	8784
380.0	8106

Analyzing Crystals

18-11 Calculate the incident angle at which x-rays from a source must strike a LiF analyzing crystal in order to reflect primary radiation with a wavelength of 0.284 nm.

18-12 If the angle at which an analyzing crystal can be varied relative to the incident radiation is from 5.0 to 80°, calculate the wavelength regions at which a barium stearate, an ADP, a LiF, and a topaz crystal respectively can be used for diffracting primary radiation.

Detectors

18-13 Estimate the number of ion pairs formed in a gas-ionization detector which is filled with argon if the incident radiation has a wavelength of 0.307 nm.

18-14 If the detector described in Prob. 18-13 is operated in the proportional region at a potential at which the amplification factor is 1.5×10^3, calculate the number of electrons that strike the anode for each incident x-ray with a wavelength of 0.307 nm.

Fluorescence

18-15 The K_α line of copper at 0.154 nm was used to perform a quantitative analysis for copper in an aqueous solution. With the analyte in the instrument, the detector response was 2585 counts in 15 s. A standard solution which contained 1.00 percent (w/w) of copper caused a detector response of 3842 counts in 15 s. Determine the concentration of copper in the analyte.

18-16 The copper analysis described in Prob. 18-15 was repeated for a second analyte. The time was recorded for the analyte and the standard to yield a detector count of 7500. The 1 percent standard reached 7500 counts in 29.3 s and the analyte reached the count in 15.8 s. Determine the concentration of copper in the analyte.

Diffraction

18-17 A powdered silicon sample was exposed to Cu K_α x-rays in a powder camera. The radius of the camera was 57.3 mm. The distance on the exposed film between the two opposite arcs of the most-intense diffracted cone of radiation was 32.73 mm. Determine the spacing in silicon.

18-18 Powdered zinc blende was exposed to copper K_α x-rays in a powder camera with a radius of 114.6 mm. The distance on the exposed film between the two arcs of the same radiative cone was 60.00 mm. Determine the spacing in zinc blende.

REFERENCES

1. Brown, J. G.: *X-Rays and Their Applications*, Plenum, New York, 1966.
2. Birks, L. S.: *X-Ray Spectrochemical Analysis*, 2d ed., Interscience, New York, 1969.
3. Liebhafsky, H. A., H. G. Pfeiffer, E. H. Winslow, and P. D. Zemany: *X-Ray Absorption and Emission in Analytical Chemistry*, Wiley, New York, 1960.
4. Dzubay, T. G. (ed.): *X-Ray Fluorescence Analysis of Environmental Samples*, Ann Arbor Science, Ann Arbor, Michigan, 1977.

5. Bragg, L.: *The Development of X-Ray Analysis*, Hafner Press. New York, 1975.

6. Woolfson, M. M.: *An Introduction to X-Ray Crystallography*, Cambridge University Press, London, 1970.

7. Boyle, W. S. and G. E. Smith: *Bell Sys. Tech.* J, **49**:585(1970).

8. Jach, T. J.: *Surf. Interfac. Anal.*, **42**:1646(2010).

BIBLIOGRAPHY

Azaroff, L. V. (ed): *X-Ray Spectroscopy*, McGraw-Hill, New York, 1974.

Goldstein, J. I., D. E. Newbury, P. Echlin, D. C. Joy, C. Fiori, and E. Lifshin: *Scanning Electron Microscopy and X-ray Microanalysis*, Plenum, New York, 1981.

Herglotz, H. K., and L. S. Birks (eds.): *X-Ray Spectrometry*, Marcel Dekker, New York, 1978.

CHAPTER

NINETEEN

ELECTRON SPECTROSCOPY

Chapter 18 contained a description of the ionization process which occurs when a sample is bombarded either with electrons or with x-rays. The emphasis in the chapter was on measurement of the x-rays emitted after the bombardment and on measurement of x-ray absorption. During the ionization process one or more electrons are expelled from the sample. Measurement of the kinetic energy of the expelled electrons and of the number of expelled electrons can provide qualitative and quantitative information. *Electron spectroscopy* characterizes the electrons emitted from a sample.

Ionization can be caused by bombarding the sample with x-rays, ultraviolet radiation, electrons or ions. If the incident radiation is x-rays, the spectroscopic technique is either *electron spectroscopy for chemical analysis* (ESCA) or *x-ray photoelectron spectroscopy* (XPS). The two terms are used interchangeably. If ionizing radiation in the ultraviolet region is used, the analytical method is either *ultraviolet photoelectric spectroscopy* (UPS) or *photoelectric spectroscopy* (PES). Again, the terms are interchangeable. Some workers use *photoelectric spectroscopy of the inner shell* (PESIS) to indicate XPS and *photoelectric spectroscopy of the outer shell* (PESOS) to indicate UPS, for reasons which will become apparent later in the chapter. When the sample is bombarded with electrons and the emission of other electrons is monitored, the technique is *Auger* (pronounced "Oh Jay") *electron spectroscopy* (AES). Although several other forms of electron spectroscopy are possible, ESCA, UPS, and AES, or a variation of one of those methods, are used more often than other forms of electron spectroscopy.

The ionization process that occurs during ESCA is

$$S + h\nu_1(\text{x-rays}) \rightarrow S^{+*} + e^- \qquad (19\text{-}1)$$

The incident x-ray radiation is sufficiently energetic to cause an inner-shell electron to be ejected from the sample atoms yielding an ion $S^{+}*$ in an excited state. The excited-state ion has a vacancy in an inner shell. An asterisk is used to indicate an excited state. The kinetic energy of the ejected electrons is monitored during ESCA studies.

The ionization that occurs during UPS is similar to that which occurs during ESCA. The major difference is that the energy of the incident radiation is in the vacuum ultraviolet spectral region rather than in the x-ray region. The energy of each incident photon is insufficient to eject an inner-shell electron, but is sufficient to expel an electron from a valence orbital. ESCA can be used to study electrons in both core and valence orbitals. UPS can only be used to study electrons in valence orbitals. The energetic resolution in UPS is better than that in ESCA. It is possible to use UPS to study vibrational and some rotational energetic levels as well as valence electron levels. Generally vibrational and rotational levels cannot be resolved with ESCA.

After an inner-shell electron has been expelled by an incident x-ray [Eq. (19-1)], the excited ion can relax in either of two ways. One of the ways, x-ray fluorescence, was described in Chapter 18. An electron in an outer shell falls into the vacancy in the inner shell with the simultaneous emission of an x-ray photon. The process is as follows:

$$S^{+*} \rightarrow S^{+} + h\nu_2 \text{ (x-ray fluorescence)} \qquad (19\text{-}2)$$

Alternatively, the potential energy that is lost by the outer-shell electron when it falls into the vacant orbital can be absorbed by a second electron. If the energy is sufficient, the second electron can be ejected from the sample yielding a doubly charged ion. The process is the *Auger effect*, and the electron that is ejected as the result of the effect is an *Auger electron*. The process takes its name from M. Pierre Auger who first observed the emission while working with a cloud chamber. His observations were published in 1925. The Auger effect is as follows:

$$S^{+*} \rightarrow S^{2+} + e^- \text{(Auger electron)} \qquad (19\text{-}3)$$

The Auger effect and x-ray fluorescence are competing processes. The probability that x-ray fluorescence occurs increases with increasing atomic number of the emitting element. Conversely, the probability of emission of an Auger electron decreases with increasing atomic number. If the electron vacancy is in the K shell, the two processes are approximately equally probable for arsenic (Z is 33). At atomic numbers of 11 or less, Auger emission occurs nearly exclusively and x-ray fluorescence rarely occurs. X-ray fluorescence is less useful for elements with low atomic numbers whereas Auger emission occurs most readily for those same elements. To put it in another way, for those processes in which the loss of potential energy by the electron as it falls into the vacant inner shell is roughly 1.6×10^{-16} J or less, the Auger effect dominates. For transitions in which the energetic change is greater than 1.6×10^{-15} J, x-ray fluorescence dominates.

Ionization can be caused by bombarding the sample with electrons or positive ions as well as with photons. When the sample is bombarded with electrons, each incident electron can transfer all or any portion of its kinetic energy to the sample. Because the transferred energy is not quantized, the electrons that are emitted during the process can possess any kinetic energy. In effect a continuum of electron kinetic energies is emitted from the sample. That is considerably different from the process which occurs during ESCA. In the latter case the entire energy of the incident photon is absorbed and the emitted electrons possess discrete kinetic energies. Ionization by incident electrons is illustrated by

$$S + e^{-} \rightarrow S^{+}* + 2e^{-} \text{(continuum)} \qquad (19\text{-}4)$$

After the initial ionization by impinging electrons, Auger emission can occur as shown in Eq. (19-3). Auger electrons have discrete kinetic energies because the Auger effect is caused by absorption of discrete amounts of energy during the fall of an electron from a specific outer shell to a specific inner shell. An Auger spectrum consists of discrete electron-energy bands superimposed on a background continuum. Although Auger emission can occur after excitation by x-rays, it occurs more often after excitation by incident electrons. Generally electron bombardment is used during Auger electron spectroscopy.

Although x-rays can penetrate a relatively large distance into some samples, the electrons that are emitted during the ionization or during Auger emission normally do not have sufficient energy to penetrate distances greater than about 10 nm. Typical depths from which an electron can be emitted from a sample are between 1 and 5 nm. Because only those electrons that are emitted from the sample can be detected, electron spectroscopy is used for studies at depths of less than 10 nm. Electron spectroscopy, therefore, is a method of surface analysis.

Because the kinetic energy of each electron emitted during ESCA, UPS, or AES is related to the energy of the orbital from which the electron was ejected, and because orbital energies are characteristic of the atom or molecule, electron spectroscopy can be used for qualitative analysis. Since the number of emitted electrons under a specified set of experimental conditions is usually proportional to the concentration of the emitter, electron spectroscopy can be used for quantitative analysis. Detailed descriptions of the major forms of electron spectroscopy can be found in References 1 through 3.

ELECTRON SPECTROSCOPY FOR CHEMICAL ANALYSIS

The energy-level diagram shown in Fig. 19-1 illustrates the process that occurs when an x-ray photon is used to ionize an atom of a solid. An incident x-ray with an energy E_0 strikes the atom. If an inner-shell electron absorbs the photon energy, the electron is ejected from the sample. In Fig. 19-1 a K-shell electron is ejected. Electrons from other shells also can absorb the energy.

Because electron spectroscopy is often used to study solids, it is useful to briefly describe the energetic levels within the solid. The lower energetic levels are electron levels associated with a particular atom. The lower levels are the 1*s*, 2*s*, 2*p*, 3*s*, 3*p*, 3*d*, etc., atomic orbitals. The higher energy atomic orbitals are relatively closely spaced and cumulatively make up the valence band of the atom.

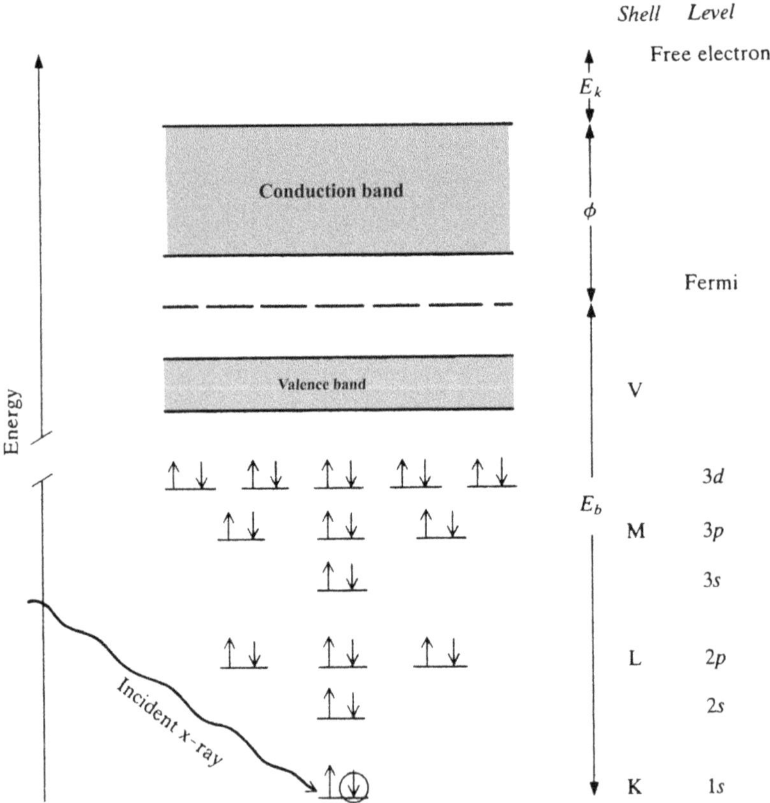

Figure 19-1 An energy-level diagram illustrating the changes that occur during x-ray-induced electron emission from the K shell in a solid electric insulator. The ejected electron is circled. The diagram is not drawn to scale.

An electron which possesses sufficient energy to pass through the valence band is no longer in an atomic orbital and consequently is no longer controlled solely by the attraction to a single nucleus. The energetic level at which the attraction to a particular atom becomes identical to its attraction to other atoms in the solid is the *Fermi level*. The Fermi level is the point of zero potential energy on the energetic axis of energy-level diagrams. It is the reference point for energetic measurements in solids.

After passing through the Fermi level the electron enters the conduction band of the solid. In a metallic conductor the valence band and conduction band are adjacent to each other on the energetic scale. In electric insulators the two bands are energetically separated from each other as illustrated in Fig. 19-1. In the conduction band an electron is free to move throughout the solid. The electron is no longer required to stay within the influence of a single atom. In order to be ejected from a solid, the electron must have sufficient energy to pass through the conduction band.

After the electron has passed through the conduction band, it is free. Any portion of the energy that was absorbed from the photon that is in excess of the energetic difference between the outer edge of the conduction band and the inner shell in which the electron originally resided appears as kinetic energy of the ejected electron. The energy which is required to move the electron to the Fermi level is the *binding energy* E_b of the electron. The *work function* ϕ of the substance is the energy that is required to move the electron from the Fermi level to the free-electron level with no kinetic energy.

It is apparent that the kinetic energy E_k of the ejected electron is a function of the energy of the incident x-ray photon and of the difference in energetic levels within the sample. Measurements of the binding energy of a substance can provide qualitative analytical data as well as information on electron levels within a sample.

Recoil energy E_r is the energy associated with the recoil of the atom as the electron is ejected. Recoil energy is that fraction of the total kinetic energy that is equal to the ratio of the mass of the ejected electron to the mass of the remaining ion. Because the electron mass is small in comparison to the ionic mass, recoil energy is important only when resolution between rotational levels is required. It is not shown in Fig. 19-1.

The spectrometer that is used to measure the kinetic energies of emitted electrons is a solid, like the sample, and consequently possesses energetic levels that are similar to those of the sample. Because the energetic level at which an electron can be emitted from the spectrometer is usually not identical to that at which an electron is emitted from the sample, an energetic difference exists between the spectrometer and the sample. That difference can accelerate (or decelerate) the electron that was emitted from the sample as it travels between the sample and the spectrometer. The kinetic energy of the accelerated electron is not identical to the kinetic energy of the electron that was emitted from the sample. Consequently, the kinetic energy that is measured by the spectrometer is not equal to that possessed by the electron immediately after it was ejected from the sample.

Figure 19-2 illustrates the relative energetic levels in a solid sample and in a spectrometer that is used to measure binding energies. The portion of the figure on the left is a shortened version of Fig. 19-1. It was drawn assuming the sample to be an electric insulator. The portion on the right shows the important energetic levels within the spectrometer. It should be noted that the conduction and valence bands within the spectrometer are adjacent, thereby indicating that the spectrometer is an electric conductor, i.e., metallic. The subscript "s" in the figure refers to the sample and the subscript "sp" to the spectrometer. It also should be noted that the Fermi levels in the sample and spectrometer are, by definition, equivalent on the energetic scale.

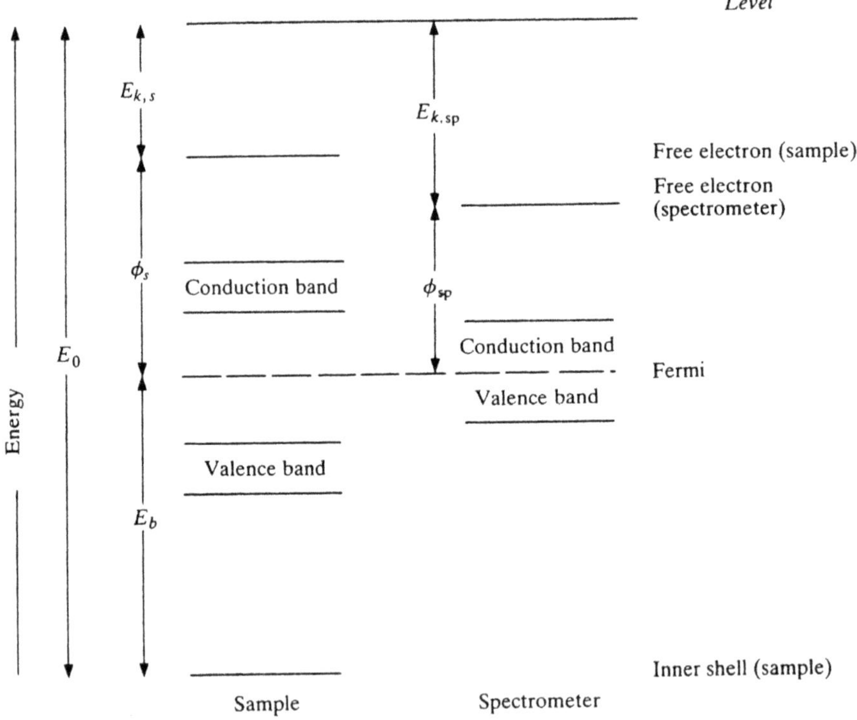

Figure 19-2 An energy-level diagram illustrating the changes that occur during measurements of binding energy. The diagram is not drawn to scale.

The energy E_0 absorbed by the electron from the incident x-ray photon is partitioned between E_b, ϕ_s, and $E_{k,s}$. Because the work function of the sample ϕ_s is generally not identical to that ϕ_{sp} of the spectrometer, a potential difference exists between the two surfaces, and the ejected electron is either accelerated or decelerated as it moves toward the spectrometer. If the work function of the spectrometer is less than that of the sample (as shown in Fig. 19-2), the free electron is accelerated by the potential difference between the sample and the spectrometer. When the electron strikes the detector of the spectrometer it possesses a kinetic energy $E_{k,sp}$ which is not identical to the kinetic energy $E_{k,s}$ of the ejected electron.

From Fig. 19-2 it is apparent that the energy of the incident photon is equal to the sum of the binding energy in the sample, the kinetic energy of the free electron when it strikes the spectrometer, and the work function of the spectrometer:

$$E_0 = E_b + \phi_{sp} + E_{k,\,sp} \tag{19-5}$$

Substitution from the Planck equation for E_b and solution of the equation for the binding energy yields

$$E_b = h\nu(x\text{-ray}) - E_{k,\,sp} - \phi_{sp} \tag{19-6}$$

Because the work function of the spectrometer appears in Eq. (19-6), absolute measurements of the binding energy can only be made after the work function has been determined.

The work function varies between instruments, but is constant for a particular instrument. The kinetic energy of the electrons that strike the detector is measured by the spectrometer. Either the electrons can be separated using electric or magnetic fields according to energy prior to detection or an energy-dispersive detector can be used. The first approach is analogous to the use of a monochromator in an uv-visible spectrophotometer. Binding energies are unique for a particular atom in a particular compound and can be used for qualitative analysis. Because the binding energies are affected by the number and position of electrons in the valence band, the binding energy is a function of the oxidation state of the element and of the chemical groups to which the element is bonded.

The work function of a spectrometer can be determined by measuring the kinetic energy of the electrons that were ejected from a sample which has a known binding energy. Often the binding energy (1.34×10^{-17} J; 83.8 eV) of the $4f_{7/2}$ electron in gold is used for the calibration. Alternatively, relative changes in binding energy rather than absolute values can be measured. A description of binding energy measurements can be found in Reference 4.

ESCA Satellite Peaks

From the preceding description a single ESCA peak would be expected for each electron level from which an electron is ejected. In addition to those peaks, *satellite* peaks are observed in some spectra. Several mechanisms can lead to satellite peaks.

After an inner-shell electron is ejected, the outer-shell electrons are exposed to a greater nuclear charge because the amount of electron shielding has been decreased. The sudden change in effective charge of the nucleus makes it possible for the ion to undergo either monopole excitation or further ionization. If an outer-shell electron is excited to a higher orbital, a satellite peak is observed. That process is *electron shakeup* or monopole excitation. Generally shakeup consists of simultaneous excitation of a valence-shell electron to an unoccupied bound level and ejection of an inner-shell electron.

During shakeup the electron is excited to an orbital in which only the principal quantum number has been increased. All other quantum numbers remain at the same values which they possessed prior to the excitation. The resulting satellite peak is observed at a kinetic energy which is less than that of the peak caused by ejection of the inner-shell electron. The energetic difference ΔE_s between the peak owing to ejection of an inner-shell electron with no shakeup and the shakeup peak is equivalent to the difference in energy between the excited state of the ion after the electron has gone into the higher orbital and the ground state of the ion with the inner-shell vacancy. The kinetic energy of the ejected electron is decreased by the amount required to excite the second electron to a higher orbital.

A second related process is *electron shakeoff* during which a second electron is simultaneously ejected from the atom with the inner-shell electron. The change in kinetic energy between the inner-shell electron that is ejected during shakeoff and the kinetic energy of the same electron without shakeoff is equal to the sum of the energy required to ionize the ground-state ion with the inner-shell vacancy and the kinetic energy of the second ejected electron. The process is similar to the Auger effect; however, the kinetic energy of the ejected inner-shell electron is monitored rather than that of the second electron. Ejection of the second electron is not dependent upon an electron falling into the inner-shell vacancy.

The maximum kinetic energy which the ejected inner-shell electron can have corresponds to the special case in which the second electron is ejected with zero kinetic energy. That limiting kinetic energy is the *ionization limit*. The ionization limit is also the most probable kinetic energy. Consequently, shakeoff peaks are characterized by a broad rising continuum. The continuum terminates in a maximum at a kinetic energy which is separated from the primary ESCA peak by the ionization energy of the ground-state ion with the vacant inner shell.

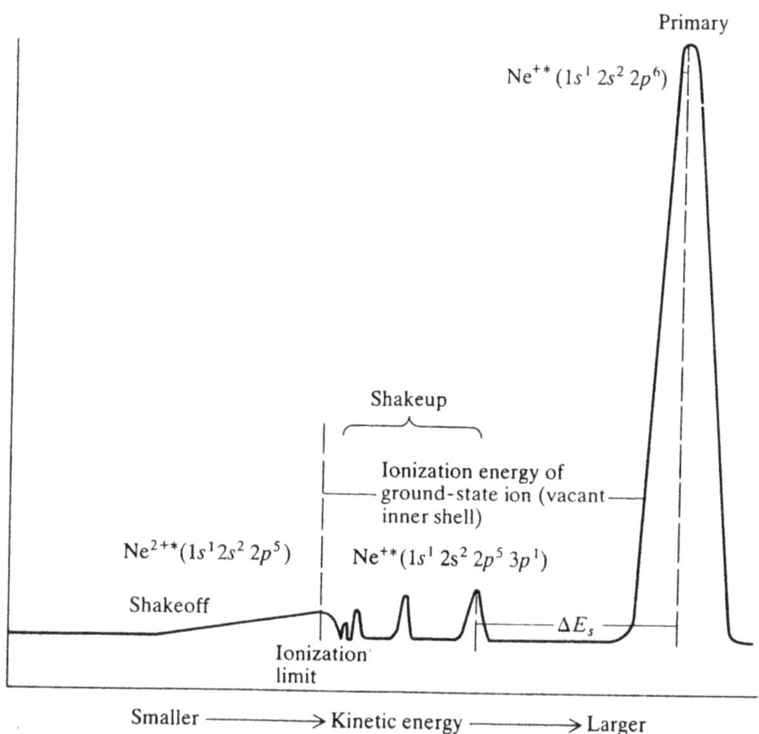

Figure 19-3 Sketch illustrating the ESCA primary and satellite peaks for neon. The arrangements of the electrons in the excited ions which are formed after electron ejections are printed near the tops of the primary peak, the first shakeup peak, and the shakeoff continuum. The height of the shakeoff continuum is exaggerated for clarity.

Electron shakeoff is another form of monopole ionization. Although electron shakeoff occurs often, it is not usually recognized because the broad continuum owing to shakeoff is obscured by the background signal. A sketch illustrating the spectra observed during electron shakeup and shakeoff is shown in Fig. 19-3.

Shakeup peaks are generally smaller than the primary peak. In some cases, however, the satellite peaks can be as large as 80 percent of the height of the primary peak. If the intensity of the shakeup peak increases, it follows that the intensity of the primary peak must decrease. Satellite peaks are generally not used for quantitative analysis, but can provide qualitative information and information about relative energetic levels within the sample.

Spectral Splitting

Some samples contain more than one unpaired electron in the valence shell. In those cases expulsion of an inner-shell electron can produce an ion in one of two or more excited states. The different excited states are caused by different ways in which the unpaired electrons in the valence shells can couple with each other. The spectrum contains a multiplet rather than a singlet. As an example, a solid which contains Fe^{3+} has an electron structure in which the five 3*d* electrons are unpaired and have identical spins. If one of the paired 3*s* electrons is ejected by absorption of the energy from an x-ray photon, the remaining unpaired electron in the 3*s* level can either be aligned with or against the spin of the 3*d* electrons. Because the two arrangements correspond to different energetic states, a doublet can be observed in the ESCA spectrum. Further details can be found in the References 1 through 3.

ESCA Chemical Shifts

The binding energy of an electron is partially determined by the number and location of electrons in the valence shell of the atom. Anything which can alter the electron density around the atom can affect the kinetic energy of the ejected electron. As was mentioned earlier, the oxidation state of the atom can affect the kinetic energy of the ejected electron. Similarly, the chemical groups which are bound to the atom can change the electron density and cause a *chemical shift* in the ESCA peak position. Such chemical shifts can be used, in much the same way that chemical shifts are used in nmr spectrometry, for qualitative analysis.

Chemical shifts in various compounds in which an electron is ejected from carbon, nitrogen, phosphorus, sulfur, oxygen, and many other elements have been studied. Correlation charts can be used to relate the chemical shift to the functional group that is attached to the atom from which the electron was ejected. Chemical shifts corresponding to ejection of a 1*s* electron from carbon have been more extensively studied than those of the other elements. Unfortunately, carbon is also one of the more difficult elements to study because the sample can be easily contaminated by compounds which contain carbon such as hydrocarbons from vacuum pump oil, stopcock grease in the apparatus, and hydrocarbons from the O rings used for sealing the apparatus.

Chemical shifts can be measured relative to a reference compound, e.g., N_2 is often used as the reference for nitrogen compounds, or the actual binding energy for the compound can be calculated. Knowledge of the work function of the instrument [Eq. (19-6)] is required to calculate absolute values of binding energies. Table 19-1 is a correlation chart containing selected chemical shifts. Chemical shifts can be measured for all elements which have an inner-shell electron in a chemical compound which can be ejected. i.e., for all elements except H and He.

Table 19-1 A correlation chart indicating the approximate binding energies for ejection of a 1*s* electron from nitrogen or oxygen and a 2*s* electron from sulfur in selected compounds

Compound	Binding energy, eV
N compounds	
RNH	397.1-398.3
RC—NR	398-399
$RCONH_2$	398-399.7
RC≡N	398.5
RN═NR	399-399.5
RNH^+	400.3-400.5
R_4N^+	401-402.3
$ArNH^+$	401.1
ArNOH	402.9-404
RONO	403.9
$NO2^-$	404
RNO_2	404.9-405.6
NO_3^-	407.4
O compounds	
$P(CH_3)_3O$	535.9
$S(CH_3)_2O_2$	537.7
$POCl_3$	537.8
$VOCl_3$	538.7
CrO_2Cl_2	538.9
POF_3	538.9
SO_2Cl_2	539.3
SO_2F_2	540.3
S compounds †	
S^{2-}	160.9
$S_2O_3^{2-}$	160.9, 166.9
RSH	161.8-162.2
RSR	161.9-162.4
ArSR	162.5-162.9
RSSR	162.7-162.9
SO_3^{2-}	165.4
RSO_2R'	166.7-167.3
SO_4^{2-}	167.8-168.1

† The 2*p* electron binding energies for sulfur vary from about 161 to 171 eV. A correlation chart can be found in Reference 2(p. 200).

The spectral peak associated with ejection of a particular electron from a particular compound can vary throughout a range of about 10 eV. An electronvolt (eV) is 1.60×10^{-19} J. Often units of electronvolts are used in x-ray and electron spectroscopy because they are easy to relate to instrumental parameters and because they are of a convenient magnitude.

Binding energies for electrons in the elements range from nearly 0 to about 1500 eV. The chemical shift for a particular element covers a small portion of that range. Because electron spectral peaks typically have a width between 1 and 3 eV, considerable overlap of spectral peaks for the same element that is chemically bonded to different functional groups can occur. Computers can be used to deconvolute overlapping gaussian-shaped peaks. More information is available in References 1, 2, and 4.

APPARATUS USED FOR ESCA

An electron spectrometer must contain a source of monochromatic x-rays, a sample holder, a means of resolving the kinetic energies of the emitted electrons, and a detector. A diagram showing the essential components of an electron spectrometer that is used for ESCA is shown in Fig. 19-4.

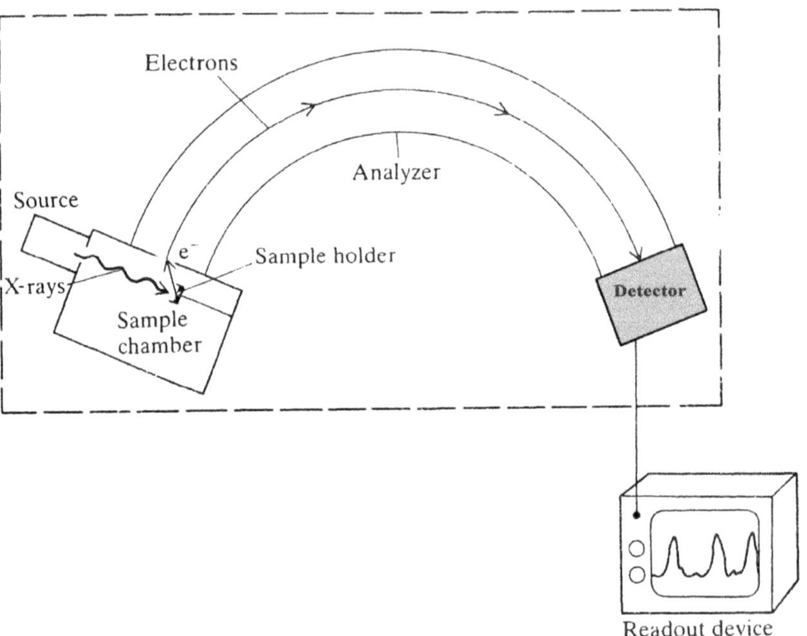

Figure 19-4 A diagram showing the essential parts of an ESCA spectrometer.

X-Ray Sources

The source used in ESCA instruments is generally an x-ray tube which has either an aluminum or a magnesium target. Many modern x-ray tubes contain both targets. The K lines from the tubes are used. Among the other target materials and the emitted spectral lines which have been used for ESCA are Cu(K_{α}), Na($K_{\alpha 1,2}$) Ti($K_{\alpha 1}$), Y(M), and Zr(M). Advantages of the aluminum and magnesium tubes are the relatively intense emitted radiation and the relatively narrow spectral width of the emitted K_{α} lines. A sodium target is spectrally acceptable but difficult to fabricate. Band broadening occurs for elements with higher atomic numbers. Some analysts prefer to use x-ray tubes with demountable targets so that the target can be exchanged easily for one of another material when emitted radiation of a different energy is required.

X-ray tubes that contain aluminum or magnesium targets emit $K_{\alpha 1}$, $K_{\alpha 2}$, $K_{\alpha'}$, $K_{\alpha 3}$, $K_{\alpha 4}$, $K_{\alpha 5}$, $K_{\alpha 6}$, and K_{β} radiation at respective wavelengths of 833.93, 834.17, 830.73, 828.51, 827.46, 822.90, 820.94, and 796.0 pm for aluminum, and 988.83, 989.07, 985.30, 982.25, 981.01, 975.37, 972.92, and 952.1 pm for magnesium. The $K_{\alpha 1}$ and $K_{\alpha 2}$ x-rays are generally not resolved and together are at least 10 times more intense than the other emitted lines. The tubes also emit Bremsstrahlung radiation. The use of beryllium or aluminum windows in x-ray tubes with magnesium or aluminum targets filters out much of the continuum and line spectra other than the K_{α} radiation. Roughly half of the x-rays passing through the windows is K_{α} radiation.

For many applications further resolution of the radiation emitted from the x-ray tubes is not required, and the radiation directly strikes the sample. In those instances when unusually low background radiation or when more monochromatic incident radiation is required, a crystalline diffraction-grating monochromator can be placed between the source and the sample. Some commercial instruments utilize a silicon dioxide crystal. Often the monochromator is based upon a Rowland circle similar to that described in Chapter 18. Monochromators are generally used when studying conductive bands in solids.

Samples

The sample is mounted as near to the source and to the entrance slit of the electron analyzer as possible. That permits the emitted electrons to enter the remainder of the spectrometer with the greatest efficiency. Modern instruments can monitor emitted electrons within several nanometers of the surface. A requirement of all forms of electron spectrometry is that the emitted electrons strike the detector without colliding with any other particles between the sample and the detector. Collisions, of course, alter the kinetic energy of the ejected electrons. Because the measured kinetic energy is used for qualitative analysis as well as for studies of atomic or molecular orbitals, collisions must be eliminated. That requires operation of most of the spectrometer in a high vacuum. Pressures as low as 7×10^{-8} Pa are achievable with commercially available pumping systems.

The use of a vacuum places restrictions on the types of samples which can be examined with electron spectroscopy. Sufficient sample must exist at the relatively low pressure within the instrument to emit enough electrons to be easily measurable. Most liquids cannot be studied because, at the low pressure in the instrument, liquids boil. Of course, a liquid can be frozen and studied as a solid. Gases can be studied by keeping the pressure in the sample compartment relatively high and the pressure in the dispersive device and the detector relatively low. The higher pressure in the sample compartment permits sufficient electrons to be ejected to yield a fairly strong signal and the lower pressure in the remainder of the apparatus minimizes the probability of an electron collision. The two pressure levels are accomplished by differential pumping across the slit that separates the sample compartment from the analyzer. Typically the pressure within the sample compartment is maintained at about 1 Pa (1 torr = 133.32 Pa) while that in the remainder of the apparatus is about 10^{-3} Pa.

Electron spectroscopy is used most often for surface studies of solids. The solid is placed as near to the source and the slit between the sample chamber and the remainder of the apparatus as possible. The apparatus is evacuated to at least 10^{-4} Pa. Some studies are done at pressures as low as 10^{-8} Pa. The ultrahigh vacuums that are used in the apparatus are accomplished by use of diffusion pumps, ion pumps, turbomolecular pumps, or cryostatic pumps. Sometimes a combination of pumps is used. Often the listed pump is preceded by a mechanical pump. If an oil-diffusion pump is used, contamination of the sample by oil from the pump is sometimes observed. Differential pumping can be used.

Solid samples are attached to the sample holder in any of several ways. Perhaps the most convenient is to use tape which has adhesive on both sides. The sample is mounted on one side and the opposite side is mounted to the sample holder. Several samples can be mounted simultaneously on a sample holder which can be rotated or which can be raised or lowered from outside the instrument. After one sample has been studied, the next sample is moved into the path of the x-ray beam. That process eliminates the need to break and reestablish the vacuum each time a new sample is examined.

Because electron spectrometry is a surface technique, unusual care must be taken to ensure that the surface of the sample is not contaminated. The vacuum must be clean. Even then, contamination by hydrocarbons from vacuum grease and O rings is possible. The surface of the sample is generally contaminated prior to study. Many times the composition of the surface layer is not representative of the bulk composition of the sample.

Several processes can be used to clean the surface prior to a study. Sometimes surface contamination is removed by mechanical scraping. Oxides on metallic samples can be removed by submerging the sample in a gaseous reductant. A popular cleaning method is to sputter the contaminants from the surface of the sample by bombardment with a beam of gaseous ions. Argon ions are often used. A description of the use of ion bombardment is presented later in the chapter in the section titled Chemical Analysis Using ESCA.

During an analysis the sample emits electrons. That can lead to an accumulation of a positive charge on the sample which could hinder the escape of further electrons or decrease the kinetic energy which they possess upon leaving the sample. It is necessary to prevent the deleterious effects of charge accumulation. If the sample is an electric conductor, electric connection can be maintained between the sample and the metallic portions of the spectrometer. In that case the charge is dissipated throughout the spectrometer. If the sample is an insulator, charge accumulation can be prevented by flooding the sample with low-energy electrons from an external source.

Analyzers

The portion of an electron spectrometer that resolves the ejected electrons into components of different kinetic energy is the *analyzer*. The analyzer is analogous to the monochromator that is used with photons. Generally analyzers use electric or magnetic fields to separate electrons of different kinetic energies.

Any magnetic field can affect the path of an electron that has been ejected from the sample. It is critical that the only field which influences an ejected electron is the field imposed by the analyzer. Such things as the earth's magnetic field, automobile traffic, subway trains, and the presence of electric motors or other electric components can be sources of electric or magnetic fields which can influence the path of an ejected electron. It is necessary to protect the ejected electron from all sources of electromagnetic fields except those imposed by the analyzer. It is desirable to protect the entire spectrometer from extraneous fields, but it is critical that the analyzer portion of the apparatus be protected.

Protection from extraneous fields typically is accomplished either with the use of Helmholtz coils or with the use of ferromagnetic shielding. Helmholtz coils consist of two equivalent circular coils of wire. The first coil is placed directly over the second coil so that the coils are parallel. The coils are wound in opposite directions. The x-ray tube, sample compartment, analyzer, and detector are placed between the coils. When an equal current passes through the coils, the spectrometer is exposed to a region of constant magnetic field. The current through the coils is adjusted to cancel the extraneous magnetic field.

Some workers use square rather than circular coils. Sometimes a combination of square and circular coils is used. If it is necessary to cancel extraneous fields in three dimensions, a set of three coils is used. Some workers use a vertical set of circular coils and two horizontal sets of square coils. Cancellation of the vertical portion of the earth's magnetic field is more important than cancellation of the horizontal component. Because an extraneous magnetic field can fluctuate with time, it is generally advisable to use an automated feedback system to control the current through the coils. The probe of a magnetometer is placed near the spectrometer and measures the magnetic field. The deviation of the measured field strength from zero is used to control the current through the coils. Generally the field is controlled within $\pm 2 \times 10^{-9}$ T of zero.

An advantage of using Helmholtz coils is the accessibility to the spectrometer which they permit. The coils can be sufficiently large to permit the operator to step between the coils to work with the instrument. With a shielded instrument it might be necessary to remove a portion or all of the shielding to access the instrument. A disadvantage of the use of Helmholtz coils is their inability to compensate for a rapidly oscillating extraneous field such as that which surrounds a wire carrying a large ac current.

As an alternative to the use of Helmholtz coils, the apparatus can be surrounded by a highly paramagnetic material. The material shields the spectrometer from extraneous fields. A ferromagnetic alloy is often used. Typically the shielding material is formed into a hollow cylinder which has a diameter about twice the length of the spectrometer and a length about thrice that of the spectrometer. The instrument is placed inside the cylinder. Shielding is most useful for instruments that have electrostatic, rather than magnetic, analyzers because the shielding disrupts the magnetic field imposed by magnetic analyzers. Sometimes a combination of shielding and relatively small Helmholtz coils is used.

Three types of analyzers are available. The *retarding potential analyzer* applies a potential difference between the sample and the detector that opposes the electron flow. The electrons which reach the detector are restricted to those which have a kinetic energy that is greater than that of the applied potential difference. The mean kinetic energy of the electrons that strike the detector increases with the applied potential. The remaining two types of analyzers use either a magnetic field or an electrostatic field to spatially separate the electrons into bands corresponding to different kinetic energies. Only that band which strikes the entrance slit to the detector is measured.

The retarding potential analyzer has the simplest design. The analyzer can be cylindrical, spherical, or linear. The linear analyzer uses parallel retarding plates with one or more slits. Block diagrams of three types of retarding potential analyzers are shown in Fig. 19-5.

Of the retarding potential analyzers, the spherical type apparently has the best combination of high sensitivity and good resolution. Regardless of the type, the electrons which are emitted from the sample pass through the first plate and are exposed to the retarding potential between the plates. Those electrons which have sufficient kinetic energy to overcome the effect of the applied potential pass through the second plate and strike the collector. The signal from the collector is electronically manipulated, as required, and sent to the readout device. Approximately 70 percent of the electron beam can pass through the metallic plates without being absorbed.

The resolution in cylindrical analyzers is limited because not all electrons are ejected from the sample at an angle that allows them to pass through the cylindrical plates at right angles. Regardless of the direction of electron ejection within a spherical analyzer, the electrons pass perpendicularly through the plates. The analyzer type shown in Fig. 19-5*b* uses a slit to limit the emitted electrons to those which perpendicularly pass through the plates.

Those electrons which do not possess sufficient energy to overcome the applied potential are retarded and do not strike the collector. A spectrum is recorded by varying the retardation potential with time. Each time the potential is decreased below the value at which retardation of a particular band of electrons occurs, the intensity of the spectrum increases stepwise as shown in Fig. 19-6*a*.

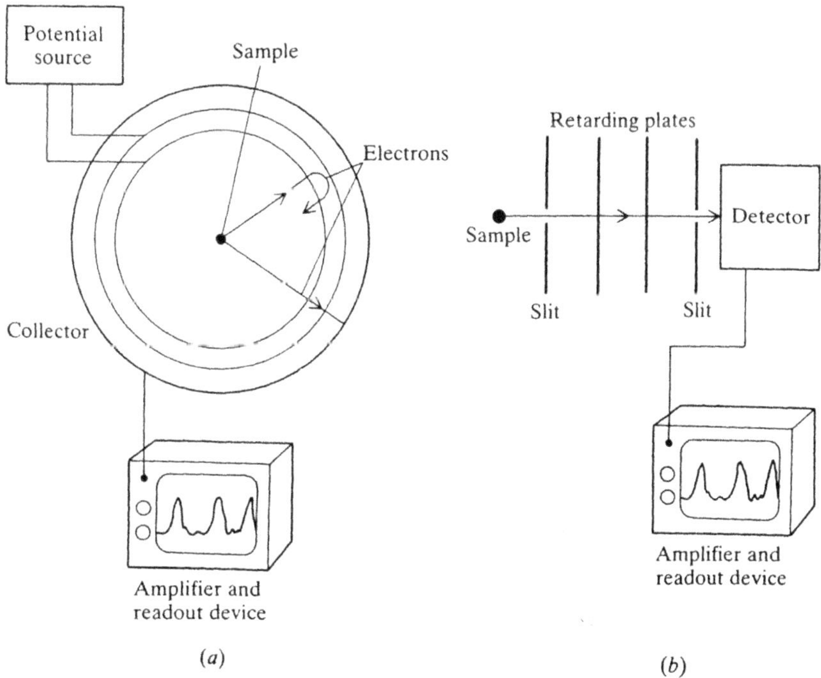

Figure 19-5 Top-view diagrams of the major types of retarding potential analyzers: (*a*) the cylindrical or spherical type; (*b*) the linear type.

The background signal is large with retarding potential analyzers because all electrons that have more than the kinetic energy required to overcome the applied potential contribute. That is particularly troublesome while using low retarding potentials. Normally the first derivative of the spectrum is electronically obtained and displayed on the readout device as illustrated in Fig. 19-6*b*.

Magnetic field analyzers use a magnetic field to focus ejected electrons of a particular kinetic energy onto the detector. By altering the magnetic field strength, electrons of different kinetic energies strike the detector. The most common magnetic field analyzer uses double focusing by four cylindrical coils above the trajectory of the electrons to generate an inhomogeneous magnetic field. Electrons entering a magnetic field B that has cylindrical symmetry assume a circular path with a radius r that is given by

$$r = \frac{mv}{Be} \tag{19-7}$$

where m and v are the mass and velocity respectively of the electron.

As the magnetic field is altered, electrons with differing velocities strike the detector. In the double-focusing analyzer the angle through which the electrons

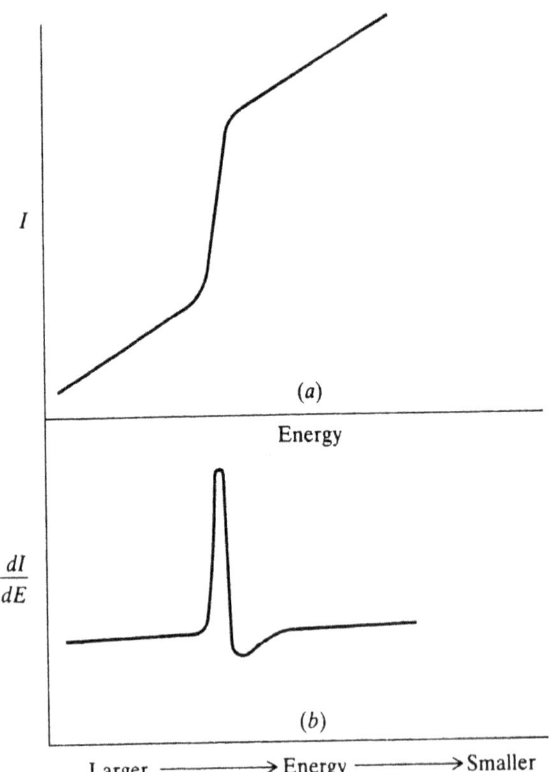

Figure 19-6 Sketches of the spectra obtained for a single energetic band of electrons with a retarding potential analyzer. (*a*) Electron intensity I as a function of decreasing retardation potential; (*b*) the first-derivative spectrum.

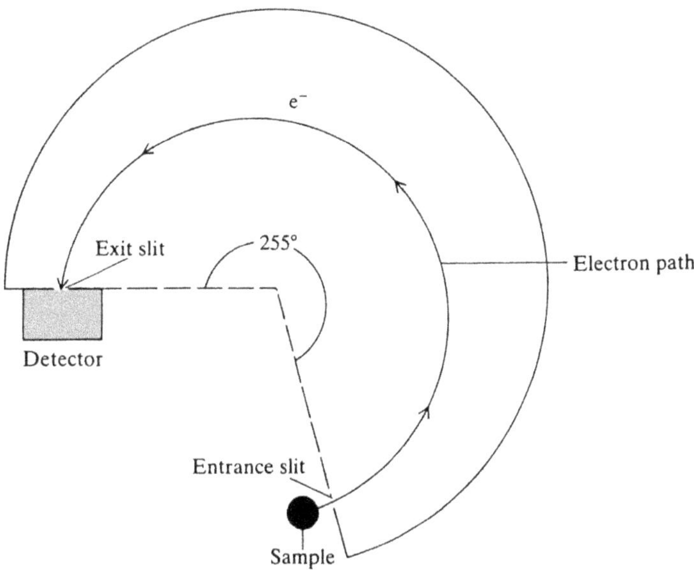

Figure 19-7 Diagram illustrating the electron path through a double-focusing magnetic field analyzer.

pass between the source and the detector is 254.56°, as illustrated in Fig. 19-7. The analyzers are manufactured from a noniron metal such as aluminum or brass and often have a radius of about 30 cm. The major disadvantages of the magnetic field analyzer are the expense, which is roughly 10 times that of an electrostatic field analyzer of approximately equal resolution, and the need for good extraneous field compensation. A description of double-focusing analyzers is included in Chapter 21.

Most of the electron spectrometers which are used for chemical analysis contain an *electrostatic field analyzer*. The two most common electrostatic field analyzers are the *spherical electrostatic field analyzer* and the *cylindrical mirror analyzer* (CMA). The spherical analyzer consists of two concentric spheres or hemispheres. Typically the radius of the outer sphere is 20 or 40 cm. Electrons from the sample enter the gap between the two spheres and are forced by the potential difference between the spheres to assume a curved path. Because the outer sphere is maintained at a negative potential relative to the inner sphere, the path of the electrons bends toward the center of the spheres, as illustrated in Fig. 19-8. Only those electrons which possess the proper kinetic energy E_k pass between the spheres and strike the detector. The detector is placed at 180° from the sample.

To a first approximation the potential difference ΔE that is required to focus electrons of a particular kinetic energy on the detector, when the radii of the two spheres are R_1 and R_2, is given by

$$\Delta E = E_k \frac{R_2^2 - R_1^2}{eR_1R_2} \tag{19-8}$$

This equation only applies to electrons traveling at nonrelativistic velocities, i.e., to electrons that have kinetic energies that are less than 3×10^{-16} J. At

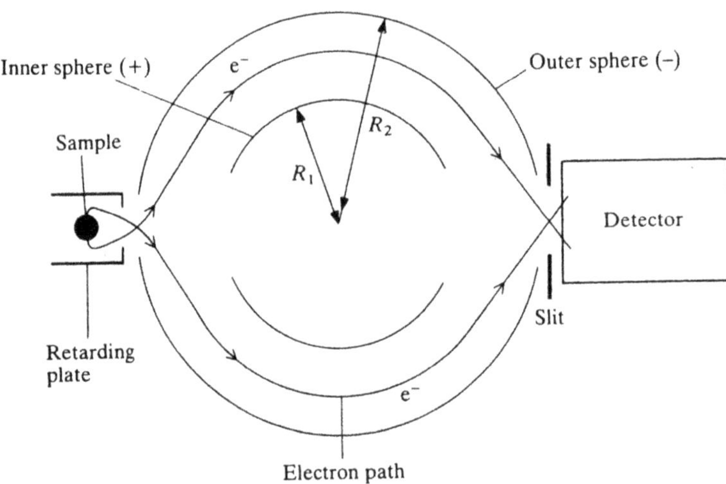

Figure 19-8 A diagram of a spherical electrostatic field analyzer that has a preceding retarding potential plate.

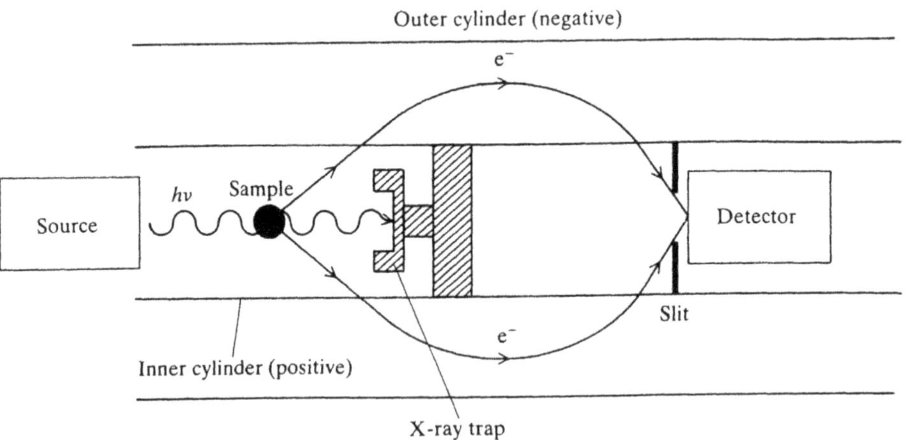

Figure 19-9 A diagram of an electron spectrometer that utilizes a cylindrical mirror analyzer.

higher kinetic energies the required potential difference does not quite vary linearly with kinetic energy as predicted by Eq. (19-8). Sometimes a retarding potential is applied to a plate between the sample and the spherical analyzer. Because the potential slows the electrons entering the analyzer by a factor of about 10, it allows the analyzer to be smaller.

The cylindrical mirror analyzer is an electrostatic field analyzer which uses concentric cylindrical plates rather than spherical plates. The solid angle around the sample from which electrons enter the analyzer is usually greater in the CMA than in the spherical design, and the sample size is smaller. A disadvantage of the CMA is the lack of a sharp focal plane upon which the electrons are focused. That is particularly troublesome for those detectors which require focusing on their entrance slits. A diagram of a CMA is shown in Fig. 19-9. Several analyzers in addition to those described have been used. Further details are available in Reference 2 and in the references cited there.

Detectors

Some spectrometers use an electrometer to measure the current that flows through a detective element which captures the electrons. The detective element is a cup or trap upon which the electron beam falls. The sensitivity of the detector is relatively low. About 10,000 electrons/s must strike the detector in order to produce a useful output signal. Counting individual electrons is not possible with the detector.

Geiger-Müller counters have been used and offer higher sensitivity as described in Chapter 18. Proportional counters have also been used as detectors. Generally the use of proportional counters requires that low-energy electrons be accelerated before entering the detector. Some workers have used energy-dispersive detectors which are similar to those that are used for x-ray analysis.

Perhaps the most common detector is the *electron multiplier*. Essentially an electron multiplier is equivalent to the dynodes in the photomultiplier tube which was described in Chapter 6. An electron multiplier can consist of a series of separate elements comparable to the dynodes found in the photomultiplier tube or it can consist of a single element. If separate elements are used, an incident electron which strikes the first stage of the device induces the emission of several electrons which are accelerated to the second stage by a potential difference between the two stages. Upon striking the second stage, each electron causes emission of several more electrons which are subsequently accelerated to the third stage. The electron multiplier normally has between 8 and 15 stages. The electrons ejected from the last stage are captured by an anode, and the current flow through the anode is monitored.

Continuous-channel analyzers operate in a similar manner but contain only a single detective element. Often the element is a small-diameter glass tube which is coated on its interior surface with a poor electric conductor. Because the resistance of the coating is large, a potential difference of between 2 and 3 kV can be applied across the length of the device. Essentially the device acts as a voltage divider where the potential drop relative to the entrance hole increases with the distance from the entrance hole.

Electrons strike the inner wall near the entrance of the detector and induce the emission of several electrons which are accelerated into the tube by the increasingly positive potential. Upon striking the wall farther into the device, further amplification occurs as several electrons are emitted for each impinging electron. As the electrons continue through the channel, the amplification increases. Eventually the electrons reach the exit end of the tube where they are captured by an anode and cause a current to flow in an external circuit. The current is monitored. The gain in a continuous-channel detector is between 10^4 and 10^8 depending upon the design. Continuous-channel electron multipliers are more rugged than multistage electron multipliers, and consequently are used more often. Discrete electrons can be counted with the devices at counting rates up to about 20,000 counts/s. Variations in the physical design of the continuous-channel analyzers are found in instruments obtained from different manufacturers, however, the operating principles of the analyzers are similar.

Traditionally an electron multiplier is placed behind the exit slit from the analyzer. Only those electrons with the proper kinetic energy to strike the slit are measured by the detector. The analyzer focuses electrons of different kinetic energies upon the slit during the spectral scan. An alternative to the traditional procedure is to use a *position-sensitive detector*.

A position-sensitive detector is used either with no entrance slit or with a wide slit. It monitors the number of electrons which strike it as a function of location. Electrons that have multiple kinetic energies simultaneously strike the detector. An entire spectrum or portion of a spectrum is obtained without altering the electric or magnetic field in the analyzer.

A position-sensitive detector can consist of an array of electron multipliers. The arrays are available on a single plate. Each electron multiplier in the array occupies a different location in the output beam from the analyzer and responds to electrons of a different kinetic energy. The output from each electron multiplier can be either continuously monitored or monitored in rapid succession.

Another position-sensitive detector uses a single detective element which is sufficiently large to cover the entire dispersed electron beam from the analyzer. A mask that is placed in front of the detective element is used with a Hadamard transform to obtain spectra. The use of the mask and the Hadamard transform for electron spectroscopy is identical to their use in infrared spectrophotometry (Chapter 12).

Readout Devices

Modern electron spectrometers are computer-controlled. Readout devices of computer-controlled instruments include video display terminals, digital plotters, and printers. Other readout devices which are used include strip-chart and *xy* recorders. Computer data storage is advantageous because it permits data manipulations, such as signal averaging and background subtraction, to be performed automatically. A description of older commercial instruments is provided in Reference 2 (pp. 58-63).

CHEMICAL ANALYSIS USING ESCA

Qualitative analysis is performed by comparing absolute or relative binding energies of the sample with those of known substances. During the comparisons it is important to understand that the binding energies for a particular element are affected by the oxidation state and the chemical environment of the element. Qualitative analyses usually yield good results because few of the elemental spectral peaks overlap. ESCA can be used to assay all elements except H and He.

It also is important to realize that ejected electrons can rarely penetrate more than 5 nm of the surface. Even though the incident x-rays can penetrate the sample to depths of several hundred nanometers, only that portion of the sample from which electrons can escape can be studied. Consequently, chemical analyses are limited to the several nanometers adjacent to the surface.

Quantitative assays are performed by using spectral peak areas. If appropriate standards can be prepared, the working-curve method can be used. Such assays yield results that can be within 2 percent of the correct value. Use of a single standard typically yields results that are within 5 percent of the correct value.

The diameter of the incident x-ray beam in modern instruments often is several millimeters, but in some instruments it is as small as 0.03 mm. In older instruments the beam diameter was about 1 cm. Those instruments which have small beam diameters can be scanned across the analyte surface to obtain a surface map. Of course, as the beam diameter decreases, the resolution in the map increases.

If it is necessary to use ESCA to perform an analysis that is representative of the bulk of a sample rather than of the surface, it is advisable to remove the surface layer prior to performing the analysis. Although surface etching or scraping with a mechanical device can be used to remove the layer, it is more acceptable to bombard the surface with positively charged ions. Many spectrometers are equipped with an argon ion source which provides ions that bombard the sample. Ions other than those of argon are used less frequently.

The ionic beam is focused at the same location on the sample as that at which the x-ray beam is to be focused. Upon striking the sample some of the kinetic energy of the ions is transferred to the sample, and the sample atoms are *sputtered* from the surface. The depth of the crater or *etching* which is formed during sputtering is a function of the time during which the ionic beam is focused on the sample. Greater bombardment times produce deeper craters. After the surface layer is removed, the analysis is performed within the crater. Ionic sputtering can be used in combination with ESCA or other forms of electron spectroscopy to obtain a depth profile of the sample. A spectrum is obtained after etching to the desired depth. More spectra can be obtained after etching to deeper depths if desired.

Applications

ESCA as well as other forms of electron spectroscopy have been used for many types of surface studies. Surface studies of low atomic weight elements are difficult to perform by most other techniques. An example of the use of electron spectroscopy has been to study the surface of industrial catalysts and compounds which adhere to the surfaces of catalysts. Active sites on catalysts can also be studied.

The ability of ESCA to distinguish between oxidation states can be used to advantage. ESCA has been used to distinguish between different compounds of the same element on surfaces. Other applications include the examination of oxide layers on metals, polymers on surfaces, and contaminants on semiconductors. Electron spectroscopy even has been used to study compounds on human skin.

By observing the change in area of a spectral peak with time, electron spectroscopy can be used for kinetic studies. ESCA also has been used to obtain differential rates of reaction of two substances on a single surface. Some of the applications of electron spectroscopy are summarized in References 5 through 7.

AUGER ELECTRON SPECTROSCOPY

A brief description of the Auger process was included in the introduction to the chapter. After initial ejection of an inner-shell electron, an outer-shell electron falls into the vacancy. The energy which is lost when the electron falls into the vacancy is used to eject a second outer-shell electron from the atom. Because a total of two electrons are ejected, a doubly charged positive ion is created. The initial ejection of the inner-shell electron can be caused by absorption of energy from an x-ray photon, an electron, or a positive ion.

A shorthand notation is used to indicate the electron transitions during the Auger process. Thus a KLL Auger process is one in which the initial electron was ejected from the K shell; the second electron fell from the L shell to the K shell; and simultaneously a third electron was ejected from the L shell. The KLL process consists of initial ejection from the 1*s* level followed by relaxation of a 2*s* electron into the 1*s* vacancy and emission of a 2*p* electron.

If one or more of the electrons was originally in a valence orbital, the process is sometimes indicated using a V rather than the letter that corresponds to the particular valence orbital. The V collectively represents all the valence orbitals. As an example, an MMV Auger process is one in which an electron is initially ejected from the M shell. Subsequently, an electron in a more energetic orbital in the M shell falls into the vacancy with simultaneous emission of an electron from a valence orbital. The product has vacancies in the M shell and in a valence orbital. A diagram indicating the energetic transitions that occur during a KLL Auger process is shown in Fig. 19-10.

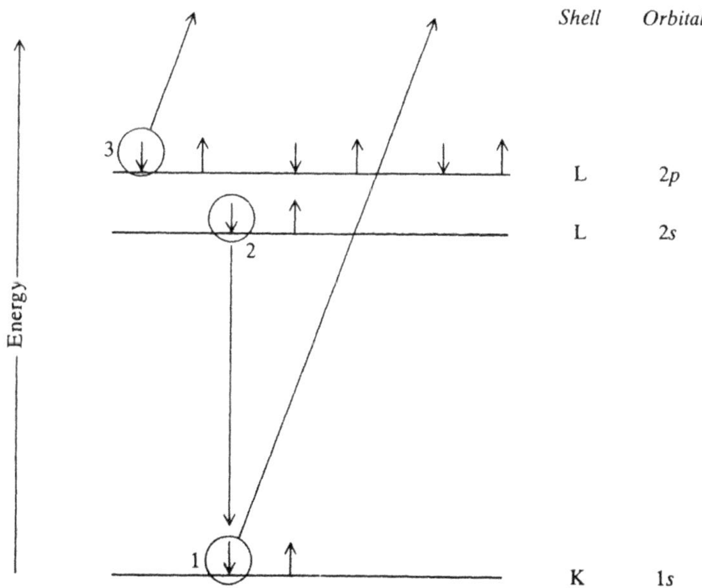

Figure 19-10 A KLL Auger process in neon. Each arrowhead indicates an electron. The electron spin is indicated by the direction in which the arrow is pointing. The electrons that undergo transitions are circled. Electron 1 is initially ejected and electron 3 is simultaneously ejected as electron 2 falls to the K shell.

For the simple KLL Auger process described above, the kinetic energy of the ejected Auger electron corresponds to the difference between the energy lost by electron 2 (Fig. 19-10) as it falls into the inner shell and the energy required to remove the Auger electron (electron 3) from the atom. The assumption is that all of the energy that is lost when electron 2 falls to the lower shell is absorbed by the Auger electron. A portion of the absorbed energy is used to remove the Auger electron from the atom and the remainder appears as kinetic energy of the ejected electron. Mathematically the kinetic energy E_A of the ejected Auger electron is given by

$$E_A = (E_1 - E_2) - E_3 = E_1 - E_2 - E_3 \tag{19-9}$$

where E_1, E_2, and E_3 respectively are the binding energies of electrons 1, 2, and 3 in Fig. 19-10. The requirement for Auger emission is that E_1 is greater than the sum of E_2 and E_3.

More complicated Auger processes can also occur. Electron shakeoff can take place in AES as well as in ESCA, resulting in simultaneous ejection of two or more Auger electrons. Simultaneous ejection of two Auger electrons is a *double Auger* process. A *radiative* or *semi-Auger* process occurs when a portion of the energy emitted by the outer-shell electron which falls into the vacancy is used to eject an Auger electron while the remainder is used to simultaneously eject a photon, as during x-ray fluorescence.

A *vacancy cascade* can occur after the initial Auger electron is ejected from an inner shell. The ejection creates a vacancy which is filled when an electron from a higher shell falls into the vacancy. That process can be accompanied by emission of a second Auger electron and formation of a triply positive ion. For elements that have large atomic numbers (many orbitals), multiple Auger emissions can occur following the initial ejection of an inner-shell electron. The average ionic charge on xenon following formation of an initial K-shell vacancy is +8. Vacancy cascades in some heavy elements have led to ionic charges as large as +22, corresponding to 21 Auger emissions.

Although Auger emissions are observed after bombardment with x-rays or positive ions, electron bombardment is used most often for AES. Electron bombardment is normally preferred to x-ray bombardment because electron beams are available that are orders of magnitude more intense than x-ray beams and because electron beams can be focused onto a smaller spot on the sample. An incident beam of greater intensity causes greater Auger emission, i.e., a stronger signal. A small beam diameter improves spatial resolution on the surface of a sample. Commercial instruments have beam diameters as small as 8 nm.

A disadvantage of electron bombardment is the large background signal caused by scattered incident electrons. The signal-to-noise ratio in the spectrum is high in comparison to that observed during x-ray bombardment, even though the background signal is large. Because the background signal is large, Auger spectra are often recorded using the first derivative of the emitted electron intensity as a function of kinetic energy. A second disadvantage of electron bombardment is the relatively large amount of damage done to the sample by the electron beam in comparison to that done by an x-ray beam. The primary advantage of proton, alpha particle, or heavy ion bombardment is the low background signal which is observed in the spectra.

As in ESCA, the depth at which a sample can be studied is determined by the *inelastic mean free path* (IMFP) of the ejected electrons. The IMFP is the mean distance in the sample through which an electron can travel before undergoing an inelastic collision. Incident electrons that are used to create the initial inner-shell vacancy have sufficient energy to cause ionization, even after several collisions. Consequently, the mean free path of the incident electrons is not the limiting factor in determining the depth at which studies can be performed. The depth at which measurements can be made using AES is about 2 nm.

AES Apparatus

Except for the source, the electron spectrometers that are used to record Auger spectra are identical to those used for ESCA. Often the same instrument is used to record both types of spectra. The source used for AES is usually an *electron gun*.

Electron guns can use an electrically heated tungsten filament as the electron source or an LaB_6 emitter. The intensity of the electron beam can be controlled by adjusting the current that flows through the filament. The electrons pass through a slit and are accelerated to the desired kinetic energy by a potential difference between two or more electrodes. Altering the potential between the electrodes alters the energy of the electron beam. The operator can control both the beam intensity and the beam energy. The diameter of the beam that is used for AES on solid samples is usually less than 1 mm and is often about 25 μm. Beam diameters as low as 8 nm are available in instruments from some manufacturers. In *scanning Auger microprobes* (SAMs) the beam diameter determines the spatial resolution obtained while preparing surface maps.

The SAM is similar to the microprobes described in preceding chapters. A small-diameter electron beam is scanned across the surface of a sample and the spectral peak corresponding to one or more elements is monitored as the position of the beam changes. The result is a highly resolved map of the surface for the chosen elements.

AES instruments are usually equipped with an ion-beam source which is used to clean the surface and to obtain depth profiles. Etching by bombardment with argon ions is done with ionic beam energies between 0.5 and 5 keV. Calibration of the etching rate is difficult. Typical etching rates are between 1 and 10 nm/s and vary with the sample. When ion-beam etching is used to obtain depth profiles, the electron beam is centered in the much broader ion beam. Typically the ion beam has a diameter of several millimeters.

Because the many scattered electrons that accompany use of electron bombardment cause a large background signal, the output from AES is often displayed as a first derivative. The differentiation can be accomplished by superimposing a low-amplitude ac potential on the dc potential that is applied to the sweep source. The in-phase portion of the output current from the detector is measured with a lock-in amplifier.

Chemical Analysis with AES

Chemical shifts in the position of Auger spectral peaks do occur, e.g., chemical shifts are observed between Sn, SnO, and SnO_2. They are not as easily recognized as in ESCA because the spectral peaks are broad and sometimes complex and because the chemical shifts are not as large. It is not normally possible to use AES chemical shifts to obtain information related to the oxidation state or the chemical environment of an element. Qualitative analysis with AES is limited to elemental analysis.

Quantitative measurements are made, as in ESCA, by monitoring the area under spectral peaks. For quantitative analysis the area under the undifferentiated peak, after background subtraction, must be measured. Either an undifferentiated spectrum can be recorded or a double integration of the first-derivative spectrum can be obtained. In either case computerized data collection and manipulation is common. Sometimes the peak-to-peak height of the first-derivative spectral peak can be used for semiquantitative analysis.

All elements except H and He produce Auger spectra. Of course, H and He cannot produce the spectra because they have only a single electron shell. Both gases and solids can be studied with AES.

A scanning Auger microprobe is an excellent instrument for qualitatively and quantitatively mapping the surface of a sample. Depth profiles are obtained after etching the sample by ionic bombardment to the desired depth. Multiplexing devices permit simultaneous determinations of six to eight elements during a single scan. More details and examples of specific applications can be found in the References. Applications and advances in electron spectroscopy can be found in Reference 8, and in updated versions of that review paper which are published in the journal Analytical Chemistry.

ULTRAVIOLET PHOTOELECTRON SPECTROSCOPY

UPS is similar to ESCA in that a photon is used to bombard a sample. Energy from the photon is absorbed and the kinetic energy of the ejected electron is measured. UPS is used most often to study gaseous molecules. The photons used for UPS have energies in the ultraviolet or vacuum-ultraviolet spectral region.

The energy of the uv photons is insufficient to eject inner-shell electrons from most elements but is sufficient to eject valence-shell electrons from some molecules and atoms. The kinetic energy possessed by each ejected electron is the difference between the energy of the incident photon and the ionization energy of the sample. The ionization energy as defined for UPS is the difference between the energy possessed by the ion after electron ejection and the energy possessed by the ground-state molecule.

UPS cannot be used to study inner-shell orbitals, but it can provide information about valence orbitals. The high resolution that is obtainable with UPS allows it to be used to study vibrational and, in some cases, rotational transitions. Because vibrational and rotational spectra are related to bonding in a molecule as well as to the oxidation state of a particular atom in the molecule, UPS can be used for qualitative analysis.

UPS is often used to study electrons in the valence orbitals of gaseous molecules. The ejected electrons come from molecular orbitals rather than from atomic orbitals. Often those electrons are delocalized throughout a portion of the molecule and cannot be assigned to a particular element. UPS can be used to study ionization energies, photoionization processes, spin-orbit coupling in polyatomic ions, and instability of polyatomic ions, as well as to perform chemical analysis. More information than that which can be provided here can be found in References 3 and 9.

Table 19-2 A list of selected filler gases that can be used in UPS source lamps

Filler gas	Resonant line, nm
Ar	106.6659
	104.82 19
He	58.43340
	53. 70296
	52.22128
	30.3781 †
	25.63 17†
	24.3027†
	23.7331†
Kr	123.5838
	116.4867
Ne	74.3718
	73.5895
Xe	146.9610
	129.5586

† The resonant line is caused by emission of a photon from He^+.

UPS Apparatus

The apparatus used for UPS is essentially the same as that used for gas-phase studies with ESCA except for the radiative source. Sources used for UPS are microwave lamps or discharge lamps that contain a rare gas or hydrogen. The radiation that strikes the sample must be monochromatic. Resonant radiation which has been emitted from the lamp is used. Often a high-energy resonant line of helium is chosen for use with UPS. Several filler gases which can be used in sources and their emitted resonant lines are listed in Table 19-2. In addition to the filler gases listed in Table 19-2, hydrogen, nitrogen, and oxygen have been used.

Chemical Analysis with UPS

UPS can be used for qualitative analysis of gases or volatile liquids. In either case absorption of ultraviolet radiation occurs in the gaseous phase. If a mixture of volatile liquids is examined, the liquid that has the higher vapor pressure produces the more intense peaks because it has a higher concentration in the gaseous phase. Qualitative analysis is accomplished by comparing the spectrum of the sample with spectra of standards. Ultraviolet photoelectron spectra are used for qualitative analysis in much the same way that infrared spectra are used. Correlation charts are unavailable for UPS. Semiquantitative analyses can be performed using spectral peak areas that are obtained from gaseous samples.

IMPORTANT TERMS

AES
Analyzer
Auger effect
Auger electron
Binding energy
Chemical shift
Cylindrical mirror analyzer
Double Auger process
Electron gun
Electron multiplier
Electron shakeoff
Electron shakeup
Electron spectroscopy
Electrostatic field analyzer
ESCA
Fermi level
Helmholtz coils
Inelastic mean free path
Ionization limit
Magnetic field analyzer
Photoelectric spectroscopy
Position-sensitive detector
Recoil energy
Retarding potential analyzer
SAM
Satellite peaks
Semi-Auger process
Sputtering
UPS
Vacancy cascade
Work function
XPS

PROBLEMS

19-1 Why does the kinetic energy of an electron change between the time at which it is ejected from the sample and the time at which it strikes the detector?

19-2 Estimate the ratio of the recoil energy to the kinetic energy of an ejected electron from C.

19-3 Estimate the ratio of the recoil energy to the kinetic energy of an ejected electron from Li.

19-4 Calculate the binding energy for F if the incident x-ray photon that was used to create the inner-shell vacancy had a wavelength of 834 pm (Al, K_{α}); the work function of the spectrometer was 4.71 eV; and the kinetic energy of the measured electron was 799 eV.

19-5 The 1*s* electron of Na has a binding energy of 1072 eV. Estimate the work function of the electron spectrometer if the incident radiation is the $K_{\alpha 1}$ line of magnesium and the kinetic energy of the measured electron is 176.7 eV.

19-6 A prominent ESCA peak of a sample occurred at a binding energy of 168 eV. The sample was known to contain either $Ba(NO_3)_2$ or $BaSO_4$. Which compound was in the sample?

19-7 What is the difference between electron shakeup and electron shakeoff?

19-8 List the major advantages and disadvantages of the three categories of analyzers.

19-9 Why must the analyzer be evacuated in an ESCA spectrometer?

19-10 If an average electron strikes the wall of a continuous-channel detector 12 times as it passes through the device and if an average of 4 electrons is emitted for each incident electron each time the electron strikes the wall, determine the gain of the detector.

19-11 What is meant by an LMM Auger process?

19-12 An Auger peak was observed at 548 eV. Estimate the difference in energy between the inner shell from which the electron was ejected and the outer shell from which the second electron fell. The binding energy of the Auger electron is 575 eV.

19-13 Why is UPS not used to study inner-shell orbitals?

19-14 Why is it possible to resolve peaks that correspond to vibrational level changes with UPS but not with AES?

REFERENCES

1. Brundle, C. R., and A. D. Baker (eds.): *Electron Spectroscopy: Theory, Techniques and Applications*, vol. 1, Academic Press, London, 1977.
2. Carlson, T. A.: *Photoelectric and Auger Spectroscopy*, Plenum, New York, 1975.
3. Rabalais, J. W.: *Principles of Ultraviolet Photoelectron Spectroscopy*, Wiley-Interscience, New York, 1977.
4. Hercules, D. M.: *Anal. Chem.*, **42**: 20A(1970).
5. Hercules, D. M., and S. H. Hercules: *J. Chem. Educ.*, **61**: 402(1984).
6. Hercules, D. M., and S. H. Hercules: *J. Chem. Educ.*, **61**: 483(1984).
7. Hercules, D. M.: *Anal. Chem.*, **50**: 734A(1978).
8. Turner, N. H., B. I. Dunlap, and R. J. Colton: *Anal. Chem.*, **56**: 373R(1984).
9. Betteridge, D., and A. D. Baker: *Anal. Chem.*, **42**: 43A(1970).

CHAPTER

TWENTY

RADIOCHEMICAL METHODS

The nuclei of some atoms are unstable and emit particles or radiation. The branch of chemistry which is concerned with the decay of unstable nuclei and the products of the decay is *radiochemistry*. In many instances radiochemical measurements can be used for qualitative or quantitative chemical analysis. As with the other spectroanalytical methods, the energy of the emitted particle or radiation can be used for qualitative analysis and the magnitude of the emission for quantitative analysis. Radiochemical analysis can be used to assay nuclei which spontaneously emit and stable nuclei from which emission can be induced by any of several mechanisms.

An isotope of an element that emits a particle or electromagnetic radiation is a *radioisotope*. A nucleus that emits is a *radionuclide*. Over 1000 radionuclides are known. A substance which spontaneously emits radiation or a particle or which undergoes *fission* is *radioactive*. Fission is the process by which a radionuclide decays into at least two pieces which each have atomic numbers greater than 2.

The nuclear emissions that are important for chemical analysis are listed in Table 20-1. Other emissions are possible, but generally are not important for chemical analysis. The ability of a nuclide to spontaneously emit is a function of the ratio of protons to neutrons in the nuclide. For elements with atomic numbers of about 20 or less, a nuclide is stable (does not emit) if the ratio of protons to neutrons is approximately 1. As the atomic number increases beyond 20, the ratio of protons to neutrons in a stable nucleus becomes less than 1. Emission continues until a stable nucleus is achieved.

Table 20-1 Nuclear emissions of interest for chemical analysis

Emission	Symbol	Charge†	Mass, amu ‡
Alpha	α, ${}^{4}_{2}He$	+2	4.003
Electron (beta)			
Negatron	e^- (β^-)	−1	5.486×10^{-4}
Positron	e^+ (β^+)	+1	5.486×10^{-4}
Neutron	n	0	1.009
Photon (gamma)	γ	—	—
Proton	p	+1	1.008

† Charges are given in units of elementary charge. An elementary charge is the charge of the proton which corresponds to 1.6021×10^{-19} C.

‡ An atomic mass unit (amu) is 1.6606×10^{-27} kg.

NUCLEAR EMISSIONS

Alpha Particles

An alpha particle α is a helium nucleus. It has a charge of + 2 owing to the two protons and a mass of 4 amu. Because alpha particles are heavier than other nuclear emissions, their mean velocity is less than those of other nuclear emissions that have the same kinetic energy. The size of alpha particles prevents them from penetrating as far into a substance as other nuclear emissions. Alpha particles are primarily emitted from heavy nuclides. After emission the nuclide contains two less neutrons and protons. The radionuclide is the *parent* and the product of the disintegration is the *daughter*. A general reaction for a spontaneous decay of nuclide X to yield nuclide Y is as follows:

$$\underset{\text{Mother}}{{}^{A}_{Z}X} \rightarrow \underset{\text{Daughter}}{{}^{A-4}_{Z-2}Y} + \underset{\text{Alpha particle}}{{}^{4}_{2}He^{2+}} \qquad (20\text{-}1)$$

The subscript Z on the left of the symbol for an element represents the atomic number of the element. The superscript A represents the mass number. Subscripts and superscripts on the right of an elemental symbol respectively represent the number of atoms in a compound or ion and the ionic charge, as in other chemical equations. The charge in Eq. (20-1) is not balanced, for simplicity. Many nuclides with atomic numbers greater than 82 decay by alpha emission.

Alpha particles that are emitted from a particular nuclide possess discrete kinetic energies. Alpha particles rapidly lose kinetic energy as they penetrate into matter. Owing to the rapid loss, alpha particles are highly effective in ionizing matter through which they pass; however, they only penetrate a relatively short distance. Typical penetrations by alpha particles are in the micrometer range. Alpha particles interact with electrons in matter causing ionization. Eventually alpha particles capture electrons and are converted to helium atoms.

Beta Particles

Both negatively and positively charged electrons can be emitted from nuclei. Emitted electrons are *beta particles* β. A negative beta particle is a *negatron* and a positive beta particle is a *positron*. The absence of a sign on the symbol for an emitted electron is assumed to mean that the particle has a negative charge. In the nucleus electrons are formed either by decay of a neutron to yield a proton and a negatron or by decay of a proton to yield a neutron and a positron:

$$n \rightarrow p + e^- \qquad (20\text{-}2a)$$

$$p \rightarrow n + e^+ \qquad (20\text{-}3a)$$

In either case the electron is ejected from the nucleus.

Alternatively the reactions can be written as Eqs. (20-2b) and (20-3b) where the subscripts to the left of the symbols for the particles are the elementary charges. When nuclear reactions are written in this manner, the sum of the subscripts of the products must equal the sum of the subscripts of the reactants and the sum of the superscripts of the products must equal the sum of the superscripts of the reactants in a balanced reaction.

$${}^{1}_{0}n \rightarrow {}^{1}_{1}p + {}^{0}_{-1}e \qquad (20\text{-}2b)$$

$${}^{1}_{1}p \rightarrow {}^{1}_{0}n + {}^{0}_{1}e \qquad (20\text{-}3b)$$

Electrons can also be ejected from an atom as described in Chapter 19 by the Auger process following electron capture by the nucleus. If a negatron is ejected from the nucleus, the daughter nuclide has the same mass number as the parent, but an atomic number that is one unit greater. Similarly, if a positron is ejected, the daughter has an atomic number that is one unit less than that of the parent. Examples of radiochemical reactions which involve emission of positrons and negatrons are as follows:

$${}^{176}_{71}Lu \rightarrow {}^{176}_{72}Hf + e^- \qquad (20\text{-}4)$$

$${}^{40}_{19}K \rightarrow {}^{40}_{18}Ar + e^+ \qquad (20\text{-}5)$$

Because electrons have a much lower mass than that of alpha particles, beta particles are less effective than alpha particles at ionization of matter through which they pass. They do penetrate, however, to considerably greater depths than alpha particles. Beta particles can penetrate about 500 times the distance penetrated by alpha particles which have equivalent energy. The kinetic energy of beta particles, unlike that of alpha particles, is continuous. The maximum energy E_{max} which a beta particle can possess is determined by the energetic loss by the emitting atom. It is characteristic of a particular nuclear reaction and can be used for qualitative analysis.

While passing through matter, electrons partially lose their kinetic energy by ionization or excitation of the target atoms, and by inducing Bremsstrahlung emission. Beta particles primarily lose their kinetic energy through interactions with electrons in the target atoms. Bremsstrahlung emission is continuous as described in Chapter 18. It is caused by the rapid deceleration of electrons in matter. Bremsstrahlung emission can be thought of as a shock wave of electromagnetic radiation that occurs as a result of the deceleration.

After positrons are slowed by interaction with matter, they are eventually *annihilated* by reacting with a negative electron in the matter to form two equivalent gamma rays:

$$e^- + e^+ \rightarrow 2\gamma \text{ (0.511 MeV)} \tag{20-6}$$

The entire mass of the electrons is converted to the energy that is predicted by Einstein's equation:

$$E = mc^2 \tag{20-7}$$

The equation can be used to show that each gamma ray must possess 0.511 MeV (Sample problem 20-1). Gamma ray emission at 0.511 MeV is indicative of positron emission.

Sample problem 20-1 Calculate the energy of the gamma rays that are formed during positron annihilation as shown in Eq. (20-6).

SOLUTION The mass of each electron in the reaction is 5.486×10^{-4} amu (Table 20-1). The mass can be converted to kilograms with the conversion factor listed in the second footnote of the table:

$$2e^- \times 5.486 \times 10^{-4} \text{ amu/e}^- \times 1.6606 \times 10^{-27} \text{ kg/amu} = 1.822 \times 10^{-30} \text{ kg}$$

Substitution into Eq. (20-7) provides the energy:

$$E = 1.822 \times 10^{-30} \text{ kg } (2.9979 \times 10^8 \text{ m/s})^2 = 1.638 \times 10^{-13} \text{ J}$$

Conversion from joules to electronvolts gives the total energy possessed by the two gamma ray photons:

$$1.638 \times 10^{-13} \text{ J} \times 1 \text{ eV}/1.60219 \times 10^{-19} \text{ J} = 1.022 \times 10^6 \text{ eV}$$

The energy possessed by each photon is

$$1.022 \text{ MeV}/2 = 0.511 \text{ MeV}$$

Neutrons

Neutrons are primarily used during chemical analysis to induce radioactivity in a bombarded sample. Because neutrons have no charge, they can penetrate matter to a greater depth than other nuclear particles. Neutrons can penetrate matter to a depth that is at least an order of magnitude greater than that penetrated by electrons. Bombarding neutrons can interact with a sample either by transferring energy to the target nucleus or by being captured by the nucleus. In the former case the neutrons are scattered and the nucleus is energetically excited. In the latter case the energy that is lost during the collision and the change in proton-neutron ratio can cause emission of nuclear particles from the unstable daughter nucleus. Measurement of the emitted particles or of the gamma rays which are often simultaneously emitted can be used for chemical analysis. The analytical technique in which neutrons are used to produce an unstable nucleus is *neutron activation analysis* (NAA). It is described later in the chapter.

Gamma Rays

Any excess energy created during a radiochemical reaction can be emitted as photons. If the photons are emitted from the nucleus, they are gamma rays. Gamma rays are identical to x-rays except for their source. X-rays are formed during fall of an outer orbital electron to a vacant inner orbital whereas gamma rays are formed during relaxation of a nucleus. Gamma radiation can be simultaneously emitted with alpha, beta, or other particles if the particulate emission produces a daughter nuclide in an excited state. Photoemission occurs during relaxation.

The energy of emitted gamma rays is characteristic of the change in nuclear energetic levels for a particular nuclide. Consequently, the emissions occur at discrete energies which can be used to identify the emitting nuclide. The gamma ray spectrum of a sample can be used for qualitative analysis by comparison with the spectra of known nuclides. A gamma ray spectrum is shown in Fig. 20-1.

The penetration of gamma rays into matter is identical to that of x-rays (Chapter 18). The penetration that is required to halve the intensity of the radiation can be calculated from the absorptivity coefficient of the sample and Beer's law. Generally penetration decreases as the atomic number of the sample increases. Shielding materials that are used for protection from gamma radiation normally contain heavy elements such as lead. Penetration into matter of gamma rays is greater than that of alpha or beta particles and is of about the same magnitude as that of neutrons.

The major forms of gamma ray interaction with matter involve the *photoelectric effect*, *Compton scattering*, and *pair production*. The photoelectric effect occurs when the energy of the photon is absorbed by an electron which subsequently is ejected from the atom. Absorption of x-rays, which is identical to the absorption of gamma rays, was described in Chapter 18. Electron ejection after absorption was described in Chapter 19. The kinetic energy of the ejected electron is discrete, if the incident radiation is monochromatic, because it corresponds to the difference in energy between that of the incident photon and the binding energy of the ejected electron. Of course, the emitted electron can undergo further interaction with the sample as described earlier.

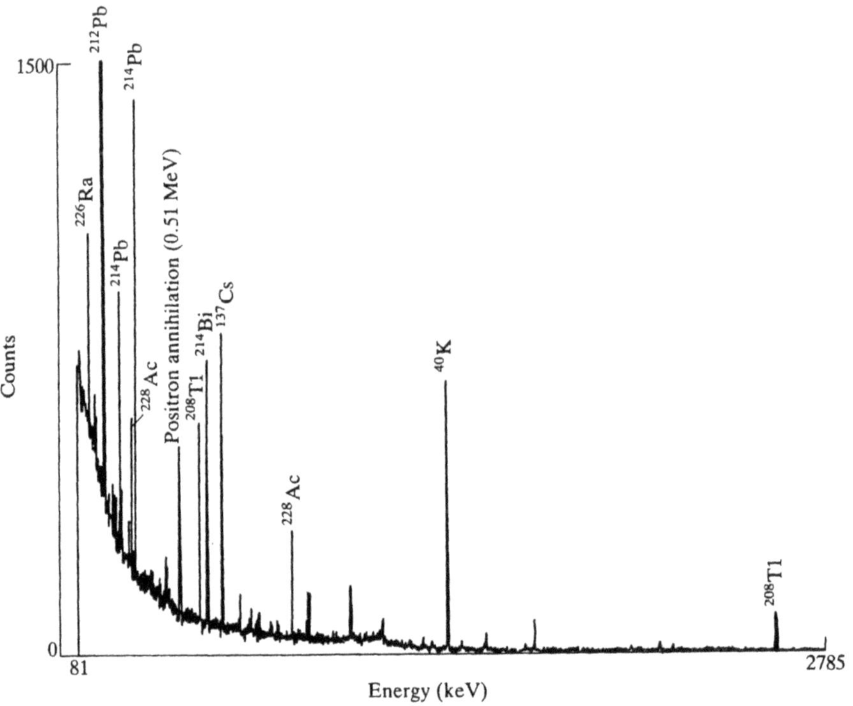

Figure 20-1 Spectrum of gamma rays that were emitted from surface soil near Grand Coteau, Louisiana. *(Courtesy of Dr. John Meriwether, Department of Physics, University of Louisiana at Lafayette.)*

Compton scattering occurs when the energy of the incident photon is partially absorbed by an electron in an atom rather than completely absorbed as occurs during the photoelectric effect. If sufficient energy is transferred, a bound electron can be ejected from the atom as during the photoelectric effect. In that case the kinetic energy of the ejected electron is the difference in energy between the energy that is lost by the gamma ray photon and the binding energy of the electron. A single photon can sequentially interact with more than one electron during Compton scattering.

Pair production can occur whenever the mass equivalent of the energy of the incident photon is greater than that of two electrons (Sample problem 20-1). If that is the case, the photon energy, upon striking an absorber (a heavy nucleus), can be converted into one negatron and one positron:

$$\gamma(1.02 \text{ MeV or more}) \rightarrow e^- + e^+ \qquad (20\text{-}8)$$

The minimal energy which the photon must possess in order for the reaction to occur is 1.02 MeV. Any photon energy in excess of that required to induce the reaction is converted to kinetic energy of the ejected positive and negative electrons.

Miscellaneous Nuclear Particles

In addition to the nuclear particles which have already been described, a number of other particles also exist and can be emitted during nuclear reactions. Among these are the lambda particle $\wedge$, the sigma particle Σ, the xi particle Ξ, the omega particle Ω, the helion h $\left({}_{2}^{3}He^{2+}\right)$, the pion Π, the k meson K, the muon μ, the neutrino v, the deuteron d ($^{2}H^{+}$), and the triton t ($^{3}H^{+}$). A bar or tilde over the symbol for a particle indicates the corresponding antiparticle. As an example, $\bar{v}$ is the, symbol for an antineutrino. Because none of those particles are presently useful for chemical analysis, they will not be described in greater detail. More information can be obtained from the first ten References and other texts dedicated to radiochemistry.

NUCLEAR REACTIONS

Nuclear reactions can be written either in the way in which chemical reactions are written with the reactants on the left of an arrow and the products on the right or by using a shorthand notation. In the latter case the parent nuclide is written on the left followed by parentheses, within which the bombarding and emitted particles or radiation are sequentially listed. To the right of the bracket is written the daughter nuclide. As an example, the following nuclear reaction:

$$ {}_{1}^{3}H^{+} + {}_{1}^{2}H^{+} \rightarrow {}_{2}^{4}He^{2+} + {}_{0}^{1}n \qquad (20\text{-}9) $$

can be written in the shorthand notation as follows:

$$ {}_{1}^{3}H(d,\ n)\ {}_{2}^{4}He \qquad (20\text{-}10) $$

The subscript on the left of the symbol for a nuclide is the atomic number of the nuclide. Because the atomic number defines the element, use of a subscript and the symbol for an element is redundant. Consequently, nuclear reactions can be written without the subscripts. An equivalent way of expressing (20-10) is

$$ {}^{3}H(d,\ n){}^{4}He \qquad (20\text{-}11) $$

Radiochemical Decay and Activity

Because the radiochemical decomposition of an unstable nuclide does not require a collision with any other body, the process occurs by a first-order mechanism, and first-order kinetics can be used to describe the decay. The rate at which radiochemical decay occurs is the *activity A* of the nuclide. The activity is given by the following first-order rate law:

$$A = -\frac{dN}{dt} = \lambda N \qquad (20\text{-}12)$$

where N is the number of parent nuclei present at time t and λ is the first-order rate constant (1/s). The rate constant for the decay is the *nuclear decay constant.*

Rearrangement of Eq. (20-12) and integration between the limits of N_0 at time t_0 and N at time t gives

$$-\ln\frac{N}{N_0} = \lambda(t - t_0) \qquad (20\text{-}13)$$

If the time t_0 at the start of the decay is assumed to be zero, Eq. (20-13) reduces to

$$\ln\frac{N}{N_0} = -\lambda t \qquad (20\text{-}14)$$

Solution of Eq. (20-14) for the number of parent nuclides remaining after time t yields

$$N = N_0\, e^{-\lambda t} \qquad (20\text{-}15)$$

The *half-life* $t_{1/2}$ of a radiochemical reaction is defined as the time at which one half of the nuclides have decayed; that is, t is $t_{1/2}$ when N is $N_0/2$. Substitution into Eq. (20-14) and solution for $t_{1/2}$ produces the result

$$\ln\frac{N_0/2}{N_0} = -\lambda t_{1/2}$$

$$t_{1/2} = \frac{0.693}{\lambda} \qquad (20\text{-}16)$$

Equation (20-16) can be used to determine the nuclear decay constant for a reaction from a measurement of the half-life. Better accuracy can be achieved, however, from a plot of ln N as a function of t [Eq. (20-14)]. The slope of the plot is the negative of the nuclear decay constant.

The half-life is a constant for a particular radionuclide. Because half-lives vary between radionuclides, they can sometimes be used to identify the nuclides.

Half-lives vary from fractions of a second to billions of years. Radionuclides that have half-lives between a fraction of a minute and several thousand years can be used for chemical analysis. Table 20-2 is a list of selected radionuclides which are used in chemical analysis and the corresponding half-lives. The radionuclides listed in Table 20-2 were chosen from among those that are used as sources or as tracers, and among those that are artificially produced after bombardment of a sample with neutrons.

Table 20-2 A list of selected radio nuclides that have analytical utility

Nuclide	Half-life	Particle emissions
^{110}Ag	24.4s	e^-
^{28}Al	2.31 min	e^-
^{198}Au	2.693 d	e^-
^{82}Br	35.5 h	e^-
^{14}C	5730 y	e^-
^{45}Ca	165 d	e^-
^{49}Ca	8.8 min	e^-
^{38}Cl	37.3 min	e^-
^{57}Co	270d	†
^{60}Co	5.26 y	e^-
^{51}Cr	27.8 d	†
^{64}Cu	12.9 h	e^-
^{66}Cu	5.1 min	e^-
^{55}Fe	2.6 y	†
^{59}Fe	45 d	e^-
^{3}H	12.26 y	e^-
^{203}Hg	46.57 d	e^-
^{128}I	25.1 min	e^-, e^+†
^{131}I	8.07d	e^-
^{42}K	12.4 h	e^-
^{27}Mg	9.5 min	e^-
^{56}Mn	2.576 h	e^-
^{22}Na	2.602 y	e^+
^{24}Na	15 h	e^-
^{32}P	14.3 d	e^-
^{36}S	88d	e^-
^{65}Zn	243.6 d	e^+†

† Decays by electron capture.

The SI unit of radionuclidic activity is the *becquerel* (Bq) which corresponds to decay of a single nuclide each second. Another unit which often is used is the *curie* (Ci). A curie equals 3.7×10^{10} Bq. The *specific activity* is the activity corresponding to a specified amount of the substance. A mass, volume, or molar unit can be used. Common units of specific activity are Bq/g, Bq/mL, Bq/mmol, μCi/g, μCi/mL, mCi/mL, mCi/mmol, and μCi/mmol.

Instrumentation

The simplest apparatus that is used for radiochemical analysis consists solely of a detector to monitor emissions from the sample. Some instruments also contain a cell which is used to hold the sample and a source of nuclear particles which are used to induce radioactivity in the sample. Often the apparatus must be properly shielded to prevent human exposure to harmful emissions and to minimize measurement of background activity.

Detectors The detectors that are used for radiochemical analysis are identical to those that were described in Chapter 18 for x-ray analysis. If a photographic plate is used as a detector, the analytical technique is *autoradiography* or *radioautography*. More commonly used detectors are the gas-ionization, scintillation, and semiconductor detectors. All of those detectors were described in Chapter 18.

The gas-ionization detectors rely upon ionization of a filler gas within the detector by radiation that is emitted from the sample. Ionization chambers, proportional counters, and Geiger-Muller (GM) counters have been used. Proportional and GM counters are used more often than ionization chambers because they are more sensitive. Often lead shielding, which is about 5 cm thick, is placed around the apparatus to lower the background count from environmental alpha, beta, and gamma radiation as well as from cosmic rays.

It is important to understand that gas-ionization detectors do not respond equally to alpha and beta particles. Because an alpha particle produces about 100 times more primary ionization than that produced by a beta particle, the detector is more sensitive to alpha particles when operated in the saturation or proportional region. In the Geiger region the response is nearly identical for both alpha and beta particles. An ionization chamber or a proportional counter that is used to exclusively monitor alpha particles generally requires no shielding from the environment owing to the magnitude of ionization caused by the particles. A typical plot of pulse height from the detector as a function of potential difference between the electrodes of the detector is shown in Fig. 20-2 for alpha and beta particles. It might be useful to compare Fig. 20-2 with Fig. 18-13 for x-ray radiation.

Ionization chambers are primarily used to measure alpha particles. Proportional counters are used for both alpha and beta particles. GM counters are relatively inexpensive. They are normally used in portable instruments and to measure total beta activity from radiochemical tracers. Neutrons cannot be directly detected with ionization chambers owing to their lack of charge. They can, however, be indirectly detected after undergoing a nuclear reaction that yields a charged particle. Often the detectors that are used to measure neutrons either utilize the nuclear reaction with ^{10}B which yields an alpha particle, or the reaction with tritium (^{3}H) nuclei which yields a proton. The number of resulting charged particles is proportional to the number of neutrons and the charged particles can be detected with gas-ionization and scintillation detectors.

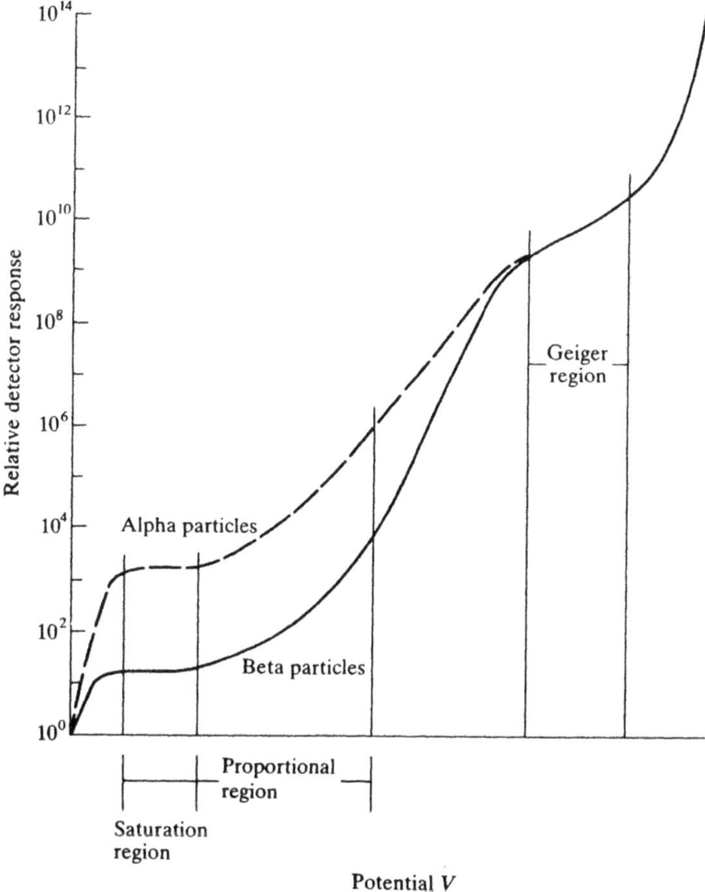

Figure 20-2 Variation in response of gas-ionization detectors with potential difference between the electrodes of the detector.

Scintillation detectors consist of a scintillator and one or two photomultiplier tubes. The radioactive emission strikes the scintillator and induces emission of visible radiation which is monitored by the photomultiplier tube. Scintillation detectors are used most often to monitor beta particles and gamma radiation. Beta particles with energies above 0.2 MeV are generally measured with proportional or GM counters. Lower energy beta particles are monitored with scintillation counters.

The solid NaI(Tl) detector which was described in Chapter 18 is a common scintillation detector. Some detectors contain scintillant solutions. In those cases both the sample and the scintillator are usually dissolved in the same portion of solvent. Liquid scintillators that have been used include anthracene, 2,5-diphenyloxazole (PPO), phenylbiphenyloxadiazole (PBD), *p*-terphenyl, and αnaphthylphenoxazole (NPO). A detector which uses a liquid scintillator is a liquid scintillation detector. Radiation that is emitted from the scintillators is monitored with one or more photomultiplier (PM) tubes. Often two PM tubes, in conjunction with an appropriate electric circuit which utilizes AND gates (Chapter 4), are used to minimize noise. Only those pulses are counted which simultaneously are observed by both tubes. A spurious pulse in either tube is ignored. Noise can be decreased by cooling the tubes and amplifier to about −10°C. In some instruments a reference light source, such as a pulsed LED, is used to check the detector output.

Semiconductor detectors also were described in Chapter 18. They are used to monitor γ- or x-ray radiation. Si(Li), Ge(Li), and intrinsic Ge and CdZnTe (CZT) detectors are used. The detectors become more conductive upon being struck by radiation. The CZT detector has not been previously described. A description of the device can be found in Reference 11 and the references found therein. Generally the semiconductor detectors work best when cooled to liquid nitrogen temperature (77 K) in order to minimize thermal conduction.

Regardless of the type of detector, it is essential that the geometrical relationship between the sample or standard and the detector be constant during all measurements. Proportional, scintillation, and semiconductor detectors are energy-dispersive when operated with a multichannel analyzer. More information related to radiochemical detectors can be found in the first six References.

Sample Handling and Safety

Care must be taken when handling radioactive substances in order to prevent human exposure to dangerous doses of radiation. Special care should be used while working with gamma ray emitters. Even while working with low levels of radiation, care must be used to prevent radioactive contamination of the laboratory which could cause a significant increase in the background signal that is monitored by the detector.

The harmful effects of radiation are caused by chemical reactions, such as ionizations, excitations, and dissociations of bodily chemicals that are induced by the radioactive emissions. The SI unit of radiative dosage is the *sievert* (Sv). A sievert is an energetic dose of 1 J/kg of exposed material. A more commonly used unit is the *roentgen equivalent man* (rem) which is equivalent to 10^{-2} Sv.

People who work with radioactive substances should have whole-body exposures of no more than 5 rem/y, although short-term hand exposure can be as large as 75 rem/y. A person who is not occupationally involved with radiation should not be exposed to more than 0.5 rem/y. Typical background radiation in the United States is 0.12 rem/y at sea level and 0.25 rem/y at

1500 m above sea level. The short-term dose which causes death in humans 50 percent of the time is about 500 rem. Analysts who use radiochemical substances should carefully monitor their exposure with a film badge, pocket ion chamber, portable counter, survey meter, thermoluminescence dosimeter, or other device. Every precaution should be taken to prevent inhalation or ingestion of radioactive materials.

When working with radioactive substances several routine practices should be followed. Rubber, surgical, or disposable gloves should be used. Whenever possible, the radioactive substances should be handled in glove boxes or hoods. Paper tissues can be used to absorb spills. Work is best done inside trays to contain spills. Pipeting should *never* be done by mouth, particularly when working with radiochemical compounds.

Compounds that emit gamma rays should be handled with particular care. It might be necessary to perform manipulations behind lead shields or by remote control from a different room. Tongs should be used in addition to gloves. Descriptions of the safety techniques that are used while manipulating radioactive substances can be found in References 1 and 2.

Sources

Radiochemical methods can be used to chemically assay nonradioactive samples after bombarding the samples with particles or EMR from a source. The induced radioactivity in the bombarded sample is monitored and used to obtain the desired analytical information. The sample can be bombarded with alpha particles, neutrons, protons, gamma rays, or some other particle. An analytical method in which radiation is induced by bombardment is *activation analysis*. The method which utilizes gamma rays is *photon activation analysis* and the method which utilizes neutrons is *neutron activation analysis* (NAA). NAA is the most popular type of activation analysis. A thorough description of sources is available in Reference 1.

Several forms of charged particle accelerators can be used as sources. The simplest type accelerates the particle between two electrodes. The sample is placed in the path of the accelerated particles. J. D. Cockroft and E. T. S. Walton built the first such device in 1932 for use with protons. *Cockroft-Walton accelerators* use accelerating potentials as large as 4 MV with a proton current of 10 mA or less.

Electrostatic generators were first used for acceleration of positive ions by R. J. Van de Graaff in 1929. Often the devices are termed *Van de Graaff generators*. A high potential source transfers positive or negative charges to a rotating belt from which the charges flow through the pointed conductor of a conducting comb to a reservoir where the charges accumulate. Typically the reservoir is a hollow metallic sphere and the belt runs through the center of the sphere. The belt can be replaced with a steel chain in which the steel portions are separated by nylon spacers. The chain is termed either a *pelletron* or a *laddertron*. Commercial generators are capable of achieving potentials up to 6 MeV with proton currents of about 100 μA. Electrostatic generators can be used to obtain accelerated particles with energies that can be controlled with a relative accuracy of about 0.1 percent. They are rarely used for chemical analysis.

An electrostatic accelerator that contains a positive ionic source in one end of a tube followed by multiple pairs of electrodes of progressively more negative potential is an *accelerating tube*. The ions enter the tube and are accelerated between each pair of electrodes to the following more negative pair. The sample is placed in the path of the particles at the opposite end of the tube from the ionic source.

Other accelerating devices which rely upon static or alternating electric fields for acceleration are *tandem Van de Graaff generators* and several forms of *linear accelerators* (*linacs*). Because neither of these devices are routinely used for chemical analysis, they are not described. A device which is sometimes used as a source for chemical analysis is the *cyclotron* (Fig. 20-3).

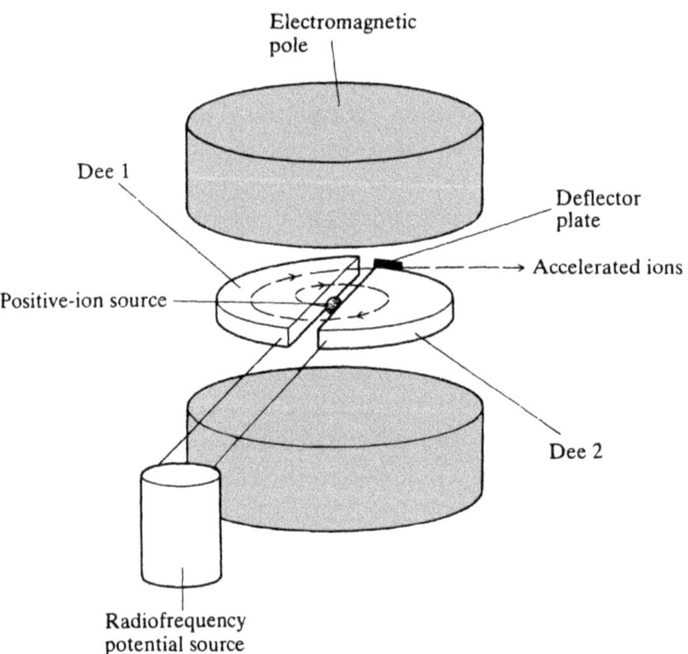

Figure 20-3 Side-view diagram of a cyclotron.

A cyclotron contains two hollow semicylindrical boxes which function as electrodes. Owing to the shape of the boxes, each box is termed a *dee*. The dees are placed between the circular poles of an electromagnet. A positive ionic source is located in the center of the gap between the two dees, as shown in Fig. 20-3. A positive ion that is emitted from the ionic source is accelerated toward the negatively charged dee.

Upon entering the dee the ion is no longer accelerated by the electric field, but is forced into a semicircular path by the magnetic field. As the ion leaves the first dee, the electric field between the dees is reversed and the ion is accelerated toward the second dee. The magnetic field forces the ion into a semicircular path within the second dee. When the ion enters the gap between the dees the electric field is again reversed and further acceleration toward the first dee occurs.

The potential between the dees is alternated at a radiofrequency. As the ion is accelerated it gains momentum and consequently assumes an outwardly spiraling path through the dees. Eventually the ion becomes sufficiently accelerated to reach the outer edge of one of the dees, where it leaves the cyclotron and strikes the target. Cyclotrons have been constructed which are capable of accelerating protons to energies as large as 700 MeV. Considerably less energy is required for chemical analysis.

Other accelerators that use electromagnetic fields and that can be used as sources of high-energy positive ions or electrons are synchrotrons and betatrons (for electrons). Descriptions of those devices can be found in Reference 1. Gamma rays are normally obtained from selected decaying radionuclides such as ^{2}H or ^{9}Be, light-element reactions such as the $^{3}H(p, \gamma)^{4}He$ reaction which emits gamma rays at 19.8 MeV, or as a byproduct during electron acceleration.

Bremsstrahlung x-ray radiation, which can be used as the source radiation for some radiochemical reactions, is produced when electrons are rapidly decelerated upon passing through matter. A disadvantage of the use of Bremsstrahlung radiation in some studies is its lack of monochromaticity; i.e., Bremsstrahlung radiation is a continuum. Bremsstrahlung radiation can be passed through a monochromator if monochromatic radiation is required. Monochromatic gamma radiation can be obtained as a byproduct of negatron-positron annihilation. Electromagnetic radiation produced in a synchrotron can also be used to bombard a sample.

Neutron sources are particularly important because they are used in neutron activation analysis. Some radiochemical reactions and disintegrations produce neutrons and can be used in sources. A typical source contains a mixture of a radionuclide, which emits alpha particles or gamma rays, and beryllium or deuterium. Sometimes another light element, such as boron, is used. The alpha or gamma radiation induces neutron emission from the beryllium or deuterium. Alpha emitters are mixed with the light element because alpha particles have a short pathlength owing to their large size. Gamma emitters can be either separated or mixed with the light element. If the two are not mixed, the gamma emitter is usually placed in a capsule that is surrounded by the light element. Alpha particles react with ^{9}Be to yield ^{12}C and a neutron:

$$^{9}_{4}\text{Be} + {}^{4}_{2}\text{He} \rightarrow {}^{12}_{6}\text{C} + {}^{1}_{0}\text{n} + Q \qquad (20\text{-}17)$$

The Q in Eq. (20-17) represents the energy emitted for each nucleus of the target during the reaction. Gamma rays react with beryllium or deuterium as indicated:

$$^{9}_{4}\text{Be} + \gamma \rightarrow {}^{8}_{4}\text{Be} + {}^{1}_{0}\text{n} + Q \qquad (20\text{-}18)$$

$$^{2}_{1}\text{H} + \gamma \rightarrow {}^{1}_{1}\text{H} + {}^{1}_{0}\text{n} + Q \qquad (20\text{-}19)$$

A reaction in which energy is produced has a positive value of Q and is *exoergic*. A reaction in which energy is absorbed has a negative value of Q and is *endoergic*. Exoergic reactions emit energy either as gamma radiation or as kinetic energy of the emitted particles. Endoergic reactions absorb kinetic energy from the incident particles. Reactions (20-18) and (20-19) are endoergic. Several neutron sources are listed in Table 20-3.

Table 20-3 Selected neutron sources

Source †	Reaction
Ra/Be	$^{9}Be(\alpha, n)^{12}C$
Ra + Be	$^{9}Be(\gamma, n)^{8}Be \rightarrow 2\ ^{4}He$
Po/Be	$^{9}Be(\alpha, n)^{12}C$
$^{239}Pu/Be$	$^{9}Be(\alpha, n)^{12}C$
^{124}Sb, Be	$^{9}Be(\gamma, n)^{8}Be \rightarrow 2\ ^{4}He$
^{24}Na, Be	$^{9}Be(\gamma, n)^{8}Be \rightarrow 2\ ^{4}He$
$^{241}Am/Be$	$^{9}Be(\alpha, n)^{12}C$
^{24}Na, D_2O	$^{2}H(\gamma, n)^{1}H$
^{88}Y, D_2O	$^{2}H(\gamma, n)^{1}H$
^{140}La, D_2O	$^{2}H(\gamma, n)^{1}H$

† A slash between two nuclides indicates that the nuclides are mixed. An addition sign indicates that the nuclides are separated. A comma between the nuclides indicates that the nuclides can be mixed or separated.

Fission reactions are another source of neutrons. Fission is the spontaneous decomposition of a radionuclide into at least two relatively heavy fragments. The fission reaction of ^{252}Cf is used often as a source of neutrons. Approximately 97 percent of the ^{252}Cf decays by alpha emission. The remaining 3 percent decays by fission. During each fission reaction nearly four neutrons are emitted. The source emits neutrons at a rate of 2×10^9 n/mg · s and has a half-life of 2.65 y.

Nuclear reactions of accelerated particles and nuclear chain reactions can produce greater quantities of neutrons than those available from fission reactions and nuclear reactions with unaccelerated particles. As an example, deuterons that are accelerated to energies of 100 keV or greater can be used to form neutrons by the $^2H(d, n)^3He$ reaction where the target is generally solid D_2O. As the energy of the accelerated particles increase, the neutron yield dramatically increases. If 100-keV deuterons are used for the reaction described above, about 0.7 neutrons are produced for 10^7 deuterons; however, if 1-MeV deuterons are used the yield is 80 neutrons for 10^7 deuterons. Several other neutron sources which use accelerated particles are listed in Table 20-4.

Nuclear chain reactors provide greater quantities of neutrons than other sources. Generally the reactors are fueled with uranium which is enriched in ^{235}U or with plutonium. *Moderators* are used to slow the fast neutrons emitted during fisson reactions. A moderator is a substance, such as H_2O or D_2O, that absorbs much of the kinetic energy of the fast neutrons which are emitted in the reactor. The slowed neutrons are *thermal neutrons.* Low atomic weight nuclei are used in moderators because they absorb a greater portion of the kinetic energy of the incident neutrons than that absorbed by larger nuclei.

The process by which a portion of the energy of an incident particle is lost to a target nucleus in the form of kinetic energy is *elastic scattering. Inelastic scattering* occurs when the incident particle transfers a portion of its energy to the target nucleus causing excitation of the nucleus. Subsequent gamma ray emission occurs from the nucleus during relaxation. A description of nuclear reactors is available in Reference 1.

If monoenergetic neutrons are required, either a crystalline monochromator or a time-of-flight monochromator can be used. Slow neutrons can be diffracted from crystalline diffraction gratings such as those used for x-rays. Time-of-flight monochromators use a chopper after the source or a modulated accelerator ionic beam to emit electrons in bursts. The chopper contains a neutron-absorbing blade. After passing the chopper, neutrons travel through a fixed distance prior to striking the sample. The detector responds only to emissions from the sample that occur at a fixed time after the neutrons left the source. Because the velocity of the neutrons defines their kinetic energy and because the pathlength is fixed, only neutrons of a defined kinetic energy are measured in the study.

Table 20-4 A list of selected reactions which utilize accelerated particles and which can be used as neutron sources

Reaction	
$^{9}Be(\gamma, n)^{8}Be$	$^{3}H(d, n)^{4}He$
$^{9}Be(d, n)^{10}Be$	$^{7}Li(p, n)^{7}Be$
$^{2}H(\gamma, n)^{1}H$	$^{7}Li(d, n)^{8}Be$
$^{2}H(d, n)^{3}He$	

Statistical Considerations during Measurements

Each radioactive nucleus in a radionuclide decays independently of the remaining nuclei in the sample. Although it is not possible to predict when a particular nucleus will decay, statistics can be used to predict when decay is most probable in a large number of nuclei. A brief introduction to statistics and its relationship to chemical analysis can be found in some introductory texts dedicated to chemical analysis.[8] A description of the use of statistics as applied to radiochemical analysis as well as details of the derivations of the equations can be found in Reference 1 (Chap. 9). Most of the statistics that are described in this section also apply to other analytical techniques, such as x-ray and electron spectroscopy, that use readout devices that count pulses from a detector.

The binomial distribution law

$$W(m) = \frac{N_0!\,p^m(1-p)^{N_0-m}}{(N_0-m)!\,m!} \tag{20-20}$$

is used to calculate the probability $W(m)$ of obtaining m radiochemical disintegrations during a time t when the number of radioactive nuclei originally present (at time 0) is N_0 and the probability of a particular nucleus decaying is p. Use of Eq. (20-20) and the equation for first-order decay [Eq. (20-15)] allows derivation of an expression for the average number M of nuclei which decay in time t:

$$M = N_0(1 - e^{-\lambda t}) \tag{20-21}$$

Equation (20-21) can be used to estimate the average number of counts for a particular decay that is registered by the detector if the fraction f of the total emissions from the decay which are observed by the detector is known.

Further mathematical manipulations can be used to obtain

$$\sigma = (Me^{-\lambda t})^{1/2} \tag{20-22}$$

which describes the standard deviation σ of the binomial distribution as applied to radiochemical decay. Equation (20-22) can be simplified to

$$\sigma = M^{1/2} \tag{20-23}$$

if the time t of the measurement is small in comparison to the half-life [Eq. (20-16)] of the radionuclide. In practice measurements normally are made during a period which is small in comparison to the half-life of the decaying radionuclide and Eq. (20-23) can be assumed to be correct.

Equation (20-23) can be used to calculate the number of counts that must be acquired during a radiochemical analysis in order to obtain results that have a desired precision. If it is assumed that the number n of counts measured by the detector is proportional to the average number of nuclei which decay, when t is small n can be substituted for M in Eq. (20-23) to yield

$$\sigma = n^{1/2} \tag{20-24}$$

Equation (20-24) can be used to estimate the standard deviation of the results of a measurement if a relatively large number (50 or more) of counts is made during the measurement.

While using Eq. (20-24) n is the total number of counts that are made within the measuring period and that correspond to a particular radiochemical decay. If a multichannel analyzer is used, n corresponds to the total number of counts from all of the channels associated with a particular peak. It is apparent from Eq. (20-24) that the absolute value of the standard deviation of the number of counts increases with the number of counts. As an example, if 100 counts are made, the standard deviation of the results would be $100^{1/2} = 10$ counts. If 10,000 counts were taken, however, the standard deviation would be 100 counts. While using detectors which output counted pulses, it can be advantageous to measure the time required to obtain a constant count for each sample and standard because that assures an identical precision for each of the results.

Sometimes it is more convenient to measure the *counting rate* R rather than the total number of counts. The counting rate is the number of counts divided by the time during which the count was obtained:

$$R = \frac{n}{t} \tag{20-25}$$

It is proportional to the activity of the decaying nuclide. The standard deviation σ_R of the counting rate is the standard deviation of the number of counts [Eq. (20-24)] divided by the time of the measurement:

$$\sigma_R = \frac{n^{1/2}}{t} \tag{20-26}$$

Solution of Eq. (20-25) for n and substitution into Eq. (20-26) gives an expression for the standard deviation of the counting rate:

$$\sigma_R = \frac{(Rt)^{1/2}}{t} = \left(\frac{R}{t}\right)^{1/2} \tag{20-27}$$

From Eq. (20-27) it is apparent that as the time increases for a measurement, the standard deviation of the measurement of counting rate decreases and the precision increases. Percentage relative standard deviations are calculated by dividing the standard deviation by the number of counts or by the counting rate and multiplying the result by 100.

Sample problem 20-2 During a particular measurement 225 counts were registered during 75 s. Estimate the standard deviation of the number of counts and of the counting rate. Determine the counting rate.

SOLUTION Equation (20-24) is used to estimate the standard deviation of the number of counts:

$$\sigma = n^{1/2} = 225^{1/2} = 15.0 \text{ counts}$$

The counting rate is obtained from Eq. (20-25) and the standard deviation of the counting rate from Eq. (20-27):

$$R = \frac{n}{t} = \frac{225}{75} = 3.0 \text{ counts/s}$$

$$\sigma_R = \left(\frac{R}{t}\right)^{1/2} = \left(\frac{3.0}{75}\right)^{1/2} = 0.20 \text{ counts/s}$$

Confidence limits[8] can be calculated from tabulated values of t at a particular confidence level if a further indication of the precision of a radiochemical measurement is required. For a fairly large number of radioactive nuclei that are observed throughout a period of time that is short relative to their half-lives, a Poisson distribution can be used to approximate the distribution of the results. If more than approximately 100 counts are measured, a gaussian distribution can be used to approximate the distribution of the results. The standard deviation of a multistep analytical procedure, assuming a gaussian distribution, can be estimated by obtaining the square root of the sum of the squares of the standard deviations for each step, i.e., by obtaining the square root of the total variance.

Background Corrections

The activity of the sample, as measured with a detector, normally includes a contribution from the background. Subtraction of the background contribution to the activity from the total activity yields the value owing to the analyte. The background contribution can be estimated by linearly interpolating the baseline that is recorded before and after the peak, as illustrated in Fig. 20-4. The area under the peak, as measured from the interpolated background, is used for the chemical analysis. Alternatively, the background count below the peak can be estimated from the sum of the areas of the triangle and the rectangle below the peak (Fig. 20-4). If A and B are the points at which the peak intersects the background, the background area a_b is given by

$$a_b = (C_B - C_{A-1})n_B + \frac{(C_B - C_{A-1})(n_A - n_B)}{2} \qquad (20\text{-}28)$$

where C_A and C_B are the channel numbers at points A and B, and n_A and n_B are the corresponding detector counts. Channel number C_{A-1} rather than number C_A is used in the calculations to ensure that the point at channel C_A is included in the calculated area. Upon rearrangement Eq. (20-28) yields

$$a_b = \frac{(n_A + n_B)(C_B - C_{A-1})}{2} \qquad (20\text{-}29)$$

The corrected peak area a_c is the total area a_t between channels A and B after subtraction of the background area a_b. The total area is the sum of the counts at each channel between A and B. The corrected peak area is given by

$$a_c = \sum_{i=A}^{B} n_i - \frac{(n_A + n_B)(C_B - C_{A-1})}{2} \qquad (20\text{-}30)$$

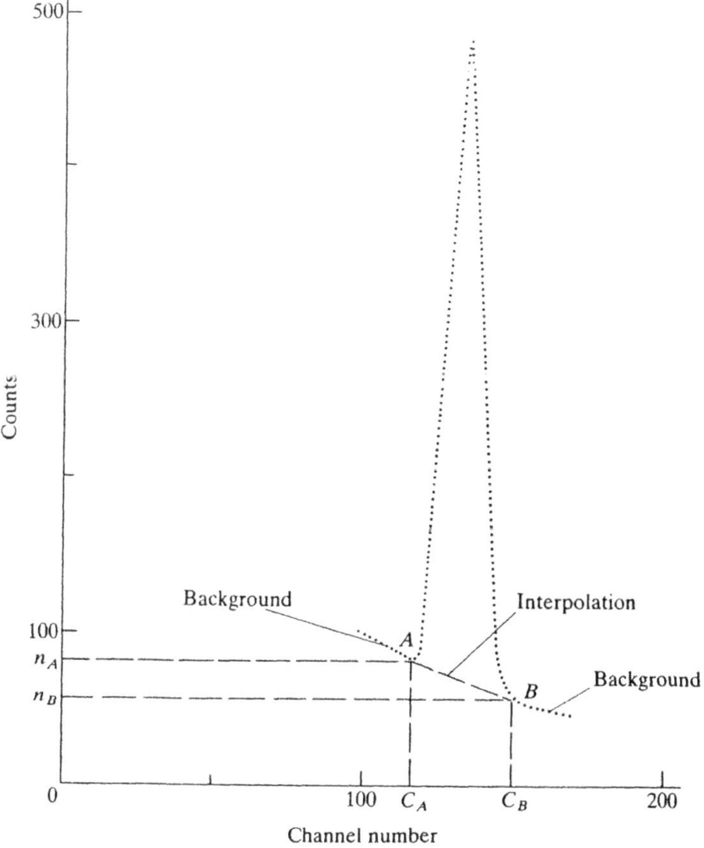

Figure 20-4 Diagram illustrating a method of background compensation. A and B are the points at which the peak intersects the background.

Random error occurs during measurement of the total count at each channel and during estimation of the background. If the measurements obey the gaussian distribution law, the random error σ_c in the calculated peak area is the square root of the sum of the standard deviations of the total measurement σ_t and the background measurement σ_b:

$$\sigma_c = \left(\sigma_t^2 + \sigma_b^2\right)^{1/2} \tag{20-31}$$

RADIOCHEMICAL ANALYSIS

In the following sections several forms of radiochemical analysis are described. Perhaps the most popular of the methods for chemical analysis is neutron activation analysis. Those techniques in which the measured emissions are resolved into energetic bands can generally be used for both qualitative and quantitative analysis. Those techniques in which the total activity of the sample is monitored are used for quantitative analysis or to track a particular component.

Activation Analysis

A nonradioactive sample can sometimes become radioactive after bombardment with particles or gamma radiation. Bombardment with neutrons is particularly advantageous because the lack of charge and the mass of a neutron allows efficient penetration and energetic transfer to the target. The analytical technique in which radioactive emissions are monitored from a sample that has been bombarded with neutrons is neutron activation analysis.

Normally the sample is bombarded with neutrons for a period that equals at least one and sometimes several half-lives of the radionuclide product. The bombarded sample is allowed to decay for a fixed period before measurements are made in order to allow emissions from interferences with short half-lives to decrease to a negligible or small value, and in order to make the sample less hazardous. Either beta particles or gamma radiation is monitored. Beta particle emission is energetically continuous whereas gamma ray emission is discrete. Consequently, gamma ray emissions are often preferentially measured even though measurements of beta particle emissions are more sensitive. In most cases relatively slow *thermal* neutrons are used for the bombardment. Thermal neutrons are those that have energies less than about 0.2 eV. At room temperature thermal neutrons have an energy of about 0.04 eV. Neutrons that are emitted in reactors have energies in the megaelectronvolt region. They are slowed by collisions with a moderator to thermal velocities. It is estimated that about 20 collisions are required to properly reduce the energy. The slowing process requires less than 1 ms.

The most important type of reaction that occurs during neutron bombardment is a *capture reaction*, during which the neutron bombards and is captured by the analyte nucleus. The resulting excited radionuclide has a mass number that is one unit larger than that of the original nuclide. The energy imparted to the product nuclide by the neutron equals the sum of the kinetic energy of the neutron and the binding energy of the neutron in the produced nucleus. The imparted energy excites the nucleus to a higher energetic level. Relaxation occurs by the emission of alpha particles, beta particles, neutrons, protons, or gamma radiation. If sufficient energy is available more than one emission can occur.

If radiochemical decay of the product radionuclide is exoergic, the decay occurs spontaneously. If the reaction is endoergic, however, the decay can only occur if the incident neutron can transfer sufficient energy to the nuclide. Because a portion of the energy of the incident neutron is used to cause the target nucleus to physically move, not all of the energy is available for nuclear excitation. Furthermore, if a charged particle, such as a proton, is emitted during the ensuing reaction, a portion of the energy must be used to overcome the electrostatic attraction between the particle and the nucleus. The energy that is required to overcome that attraction is the *coulombic barrier energy*. For an endoergic reaction, the incident electron must have sufficient energy to initiate the radiochemical reaction, to cause the target nucleus to move, and to overcome the coulombic barrier energy.

During bombardment with fast neutrons, the probability that a nuclear reaction will occur is nearly proportional to the cross-sectional area of the target nucleus. Consequently, the reaction probability during nuclear bombardment can be expressed as a *cross section* σ which has area units (generally square centimeters). The cross section is associated with a single nucleus or atom. Even though the reaction probability is not directly proportional to the cross-sectional area during bombardment with thermal neutrons or charged particles, the practice of expressing reaction probabilities as cross sections persists. A cross section that is 10^{-24} cm^2 is considered large and consequently is termed a *barn* (b).

The net rate r at which a target reacts to produce radioactive atoms while being bombarded on all sides (e.g., in a reactor) with neutrons is given by

$$r = \frac{dN^*}{dt} = \phi N\sigma - \lambda N^* \qquad (20\text{-}32)$$

where N is the number of nuclei in the sample; N^* is the number of radionuclei formed by the reaction during time t; ϕ is the neutron flux (particles per square centimeter-second); σ is the cross section (square centimeter); and λ is the nuclear decay constant (per second). The number of target nuclei can be calculated from the atomic weight of the target and Avogadro's number. The first term on the far right of Eq. (20-32) is the rate at which the radionuclei are formed by the bombardment process, and the negative term is the rate at which the produced radionuclei decay.

Substitution of $Ne^{-\lambda t}$ [Eq. (20-15)] for N^* in Eq. (20-32) followed by integration over the time interval from 0 to t during which irradiation takes place results in

$$N^* = \frac{\phi\sigma N\left(1-e^{-\lambda t}\right)}{\lambda} \tag{20-33}$$

Equation (20-33) can be used to calculate the number of radionuclei present at the end of the irradiation. Substitution of $0.693/t_{1/2}$ [Eq. (20-16)] for λ in Eq. (20-33) followed by substitution of the resulting value of N^* into Eq. (20-12) yields

$$N^* = \frac{\phi\sigma N\left(1-e^{-0.693t/t_{1/2}}\right)}{\lambda}$$

$$A = \lambda N^* = \phi\sigma N\left(1-e^{-0.693t/t_{1/2}}\right) \tag{20-34}$$

The activity A in Eq. (20-34) is that of the radionuclide produced during the bombardment at the moment when the bombardment is stopped. Equation (20-34) assumes that the sample is sufficiently thin to allow attenuation of the neutron flux to be ignored.

From Eq. (20-34) it is apparent that the activity increases as the bombardment time increases. Typical bombardment times range from one to six half-lives of the produced radionuclide. Bombardment times beyond six half-lives do not significantly increase the activity because the relationship in Eq. (20-34) is exponential. Although neutron bombardment is most popular, bombardment with other particles, such as protons, also can be used. The theory for bombardment with other particles is nearly identical to that described for nuclear bombardment. A description of irradiation with gamma rays can be found in References 1 and 3 to 6 along with other helpful information.

Qualitative and Quantitative NAA

Instruments used for NAA generally monitor the gamma rays emitted from the sample after neutron bombardment. The sample is bombarded for a fixed period. After bombardment the sample is allowed to decay during a second period and the spectrum is recorded during a third period. A Ge(Li), CZT. Si(Li) or other, high-resolution, semiconductor detector is often used to measure the ejected photons and to resolve them into spectral peaks. Rather than using a multichannel analyzer, the pulse height can be digitized and stored in computer memory. The peak energy can be used for qualitative analysis and the peak area for quantitative analysis.

The standard(s) and the sample are simultaneously bombarded with neutrons during NAA. Normally a single standard is used for quantitative analysis. The peak areas of the sample and the standard are used to calculate the number of nuclei of a particular type in the sample. A direct proportionality can be assumed between the peak area and number of nuclei when equal irradiation times and neutron fluxes are used; i.e., when the sample and standard are simultaneously bombarded. The direct method of NAA which is described here is simple and nondestructive. It is termed *instrumental neutron activation analysis* (INAA).

Prior to the development of high-resolution Ge(Li) detectors, the detectors that were used for NAA generally were incapable of resolving the peaks in the gamma ray spectra. Consequently, it was necessary to use another method to distinguish between radionuclides in the sample. Some workers continue to use one of the other methods. When the sample is chemically manipulated after bombardment but before counting, the technique is *radiochemical neutron activation analysis* (RNAA).

Any of several separatory methods have been used to remove potentially interfering radionuclides prior to measuring the total particulate or gamma ray emission. A typical procedure consists of dissolving sufficient bombarded sample to contain about 5 μg of the assayed element and adding a relatively large known amount (often 10 or 15 mg) of a carrier. A carrier is a pure nonbombarded portion of the analyte. It is added to make the total concentration of the analyzed element in the sample solution sufficiently large to permit a simple separation.

After addition of the carrier, the analyte is separated from radionuclides that could interfere with the assay. The separation can be accomplished by performing a precipitation, a chromatographic separation, an ion exchange, an electrochemical separation, an extraction, or by using another procedure. After the separation is completed, the radionuclide of interest is counted and the results are used to calculate the amount of the radionuclide in the original sample.

In another method the decay curve owing to a particular emission is recorded after neutron bombardment. Either the total intensity of the emission or the logarithm of the activity is recorded as a function of time. The decay curve is the sum of the individual decay curves for each of the radionuclides in the sample. Each radionuclide has a characteristic half-life and decays to insignificance at a different time. At long times only the radionuclide that has the greatest half-life remains and the logarithmic activity plot becomes linear. Extrapolation of that linear portion to zero time allows estimation of that portion of the activity of the irradiated sample that is caused by that nuclide. The activity can be used for quantitative analysis. The identity of the radionuclide can sometimes be deduced from its measured half-life [Eq. (20-16)]. The half-life is the time on the extrapolated portion of the curve at which the activity is half of the initial activity.

After identification of the component that has the greatest half-life, its extrapolated decay curve is subtracted ('stripped') from the remainder of the decay curve and the process is repeated for the component with the next longest half-life. A linear portion should exist for each nuclide if the half-lives of the nuclides are sufficiently different. The procedure is continued until all components are identified and quantitatively determined. Subtraction of the curve corresponding to the long-lived radionuclides from the remainder of the decay curve can be done manually, but is more often accomplished with the aid of an appropriate computer program.

Simple quantitative analyses in which the sample contains two or three known radionuclides that have significantly different half-lives can be performed by yet another procedure. At a particular time the activity of each component in the mixture can be calculated by substituting the half-life of the component [Eq. (20-16)] into Eqs. (20-15) and (20-12) followed by substituting Eq. (20-15) into Eq. (20-12). The total activity A_T of a mixture of n components at time t is the sum of the individual activities:

$$A_T = \sum_{i=0}^{n} A_i = \sum_{i=0}^{n} \frac{0.693 N_{0,\,i} e^{-0.693t/t_{1/2,\,i}}}{t_{1/2,\,i}} \tag{20-35}$$

A series of simultaneous equations are written using Eq. (20-35) for the total activity of the sample at different times. A separate equation is written for each time at which a measurement is made. A total activity measurement is made at each time; the half-lives of the radionuclides are substituted into the equations; and the equations are simultaneously solved for the number of radionuclides of each type. The number of measurements and simultaneous equations must equal the number of radionuclides. The procedure makes the reasonable assumption that the activities of the radionuclides are additive. The process is similar to that used for spectrophotometric multicomponent analysis in the ultraviolet-visible spectral region.

Generally INAA is preferred for chemical assays because the measurements are simple and because mathematical and chemical manipulations are minimal. INAA can be used for chemical assays at levels as low as 10^{-10} g/g for selected nuclides such as Dy, Eu, In, and Mn; however, it has a detection limit which is no better than that of spectrophotometric and electroanalytical methods for about 43 other elements. Because the other analytical methods are less expensive, they are preferred for those assays. INAA has a better detection limit than standard spectrophotometric and electroanalytical methods for about 32 elements.

Isotopic Dilution

Isotopic dilution is the technique in which a known amount of a radioactive material is homogeneously mixed with a nonradioactive analyte. The added radioactive component is as nearly identical to the analyte as possible. It must behave identically to the analyte during the separatory portion of the assay. The radioactive component is usually identical to the analyte except that it contains a radionuclide.

The analyte and the radioactive component are chemically separated, as pure compounds, from the remainder of the sample. Not all of the analyte must be removed from the sample, but the product of the separation must be pure. A weighed portion of the separated components is measured for its radioactivity and the result is used to calculate the amount of the analyte in the original sample. The method of quantitative analysis which relies upon addition of a radioactive component to the sample as described here is *direct isotopic dilution analysis* (DIDA).

During DIDA it is convenient to use the *specific activity* (SA) of the components. The specific activity, as defined for this technique, is the ratio of the number of radioactive atoms of a specified element in a sample to the total number of atoms in the sample. During DIDA the ratio of two specific activities are used. Because the units in the numerator and denominator of the ratio cancel, the specific activities can be replaced by anything which is proportional to the specific activities. For isotopic dilution analyses, the specific activity is generally replaced with the ratio of the activity to a unit mass of the substance. That ratio as well as other ratios are also termed specific activity. If the latter definition of specific activity is used, the specific activity SA_s of the radioactive component which is added to the sample is given by

$$SA_s = \frac{A_s}{m^*} \tag{20-36}$$

where m^* is the mass of the added radioactive component and A_s is its activity. Similarly, the specific activity SA_M of the separated mixture of the radioactive component and the analyte is given by

$$\mathrm{SA_M} = \frac{A_M}{m_\mathrm{M}} \tag{20-37}$$

where m_M is the mass of the mixture and A_M is its activity.

Because the mass of the mixture is the sum of the mass of the radioactive component and the mass of the analyte, Eq. (20-37) can be rewritten as

$$\mathrm{SA_M} = \frac{A_M}{m + m^*} \tag{20-38}$$

The number of radionuclides in the pure radioactive component is identical to the number in the mixture. Consequently, the activity A_s of the pure radionuclide and the activity A_M of the mixture must be identical. By setting A_s from Eq. (20-36) equal to A_M from Eq. (20-38), m is found to equal the following result:

$$\mathrm{SA}_S m^* = \mathrm{SA}_M(m + m^*)$$

$$m = \frac{m^*(\mathrm{SA}_S - \mathrm{SA}_M)}{\mathrm{SA}_M} \tag{20-39}$$

Equation (20-39) can be used to calculate the mass of the analyte in the sample. Normally the mass m^* of the added radioactive substance is approximately equal to the mass m of the analyte so that the difference in the bracketed term in Eq. (20-39) is neither unusually large nor unusually small.

Sample problem 20-3 Dissolved hydrogen in metals can be determined using isotopic dilution. A 100.0-g sample of aluminum was placed in an evacuated chamber at 25°C. Tritiated hydrogen which had been stored as uranium hydride was released into a calibrated 1.00-mL vessel. The pressure in the vessel was 0.452 atm at 25°C. The tritiated hydrogen was emptied into the chamber containing the aluminum and an internal gas-ionization detector was used to measure the activity of the added tritium. The activity was 749 counts/min. After the measurement the sample was heated to 620°C for 20 min and a second measurement was made. During the heating the hydrogen that is dissolved in the aluminum equilibrates with the tritiated hydrogen. The activity after equilibration was 447 counts/min. Determine the volume of dissolved hydrogen in the metal at 25°C and 1.00 atm.

SOLUTION Because gases are being measured, and because tritiated hydrogen has a molecular weight that significantly differs from that of normal hydrogen, it is convenient to use Eq. (20-39) written in terms of moles of gas rather than mass. The form of the equation which is used for the calculation is

$$n = \frac{n^*(\mathrm{SA}_S - \mathrm{SA}_M)}{\mathrm{SA}_M} \tag{20-40}$$

The number of moles n* of tritiated hydrogen that was added to the 1.00-mL vessel can be calculated using the ideal gas law:

$$PV = nRT$$

$$(0.452\ \mathrm{atm})(1.00 \times 10^{-3}\ \mathrm{L}) = n^*(0.0821\ \mathrm{L \cdot atm/K \cdot mol})(298\ \mathrm{K})$$

$$n^* = 1.85 \times 10^{-5}\ \mathrm{mol}$$

Substitution of the specific activities and n^* into Eq. (20-40) yields the total number of moles n of normal hydrogen in the gaseous mixture and in the aluminum. Because all of that hydrogen was originally in the aluminum, it corresponds to the number of moles of dissolved hydrogen in the aluminum at 25°C:

$$n = \frac{1.85 \times 10^{-5}(749 - 447)}{447} = 1.25 \times 10^{-5}\ \mathrm{mol}$$

The volume of dissolved hydrogen at 25°C and 1 atm in the aluminum is

$$(1.00\ \mathrm{atm})V = (1.25 \times 10^{-5}\ \mathrm{mol})(0.0821\ \mathrm{L \cdot atm/K \cdot mol})(298\ \mathrm{K})$$

$$V = 3.06 \times 10^{-4}\ \mathrm{L} = 0.306\ \mathrm{mL}$$

A variation of the isotopic dilution method can be used to determine the mass of a radionuclide after dilution with a nonradioactive nuclide of the same element. That method is *inverse isotopic dilution analysis* (IIDA). The principle is the same as that used for DIDA and the governing equation is shown as follows:

$$A_M = A^*$$

$$\mathrm{SA}_M(m + m^*) = A^*$$

$$\frac{\mathrm{A}_M(m + m^*)}{m_M} = A^*$$

$$m^* = \frac{A^* m_{\mathrm{M}}}{A_M} - m \tag{20-41}$$

Radiometric Titrations

A *radiometric analysis* is one in which a radioactive substance is used in a chemical reaction with a nonradioactive substance for the purpose of quantitatively assaying the nonradioactive substance. The radioactive substance is normally used in a titration with the nonradioactive substance. The endpoint of the titration is located from a titration curve which is a plot of radioactivity as a function of titrant volume.

Precipitation titrations are most easily adapted to the radiometric method because a separation occurs during the titration. Complexometric and redox titrations can also be performed by using the radiometric method. The radioactive compound is used to prepare a standard solution which can be used either as the titrand or the titrant. An example of a *radiometric titration* is the determination of chloride by titration with a standard solution of radioactive $^{110}Ag^+$. The activity of the solution is plotted as a function of the volume of added Ag^+. Prior to the endpoint all Ag^+ added to the solution precipitates as AgCl and the activity of the solution is low. After the endpoint, however, excess Ag^+ is added to the solution and the activity rises. The titration curve has two linear portions which are extrapolated to intersection at the endpoint. Other examples of radiometric titrations include the determination of chromate by a precipitation reaction with radioactive $^{110}Ag^+$, the determination of fluoride by titration with radioactive Ca^{2+} to yield a precipitate of CaF_2, the titration of Mg^{2+} or Zn^{2+} by titration with radioactive $^{32}PO_4^{3-}$ to yield a precipitate of the phosphate, and determination of Ag^+ by titration with radioactive Br^- to yield insoluble AgBr. The primary advantage of radiometric titrations is the high sensitivity that cannot be attained relative to most other titrations. The mass of analyte that can be assayed by using a radiometric precipitation titration is usually limited to about 1 mg by the solubility product of the precipitate. More information is available in Reference 3.

Radiorelease Methods

Another category of radiometric methods of analysis is the radiorelease methods. A radioactive chemical reagent is reacted with the sample to yield a radioactive product which is "released" in an easily separable form. Often the radioactive product is a gas. The radioactivity of the separated reaction product is measured and the result is used to calculate the amount of the reactive species in the sample. As an example, the amount of OH in air can be determined by reacting the OH in a Teflon bag with radioactive ^{14}CO. The rapid reaction produces radioactive CO_2 and H. The product $^{14}CO_2$ is separated from the mixture by freezing and its radioactivity measured.

The determination of $SO_2(g)$ in air is another example. The air is pumped through an aqueous solution of radioactive potassium iodate (the radionuclide is I). The radioactive I_2 that is formed during the reaction is extracted into chloroform and its activity is measured.

Radioactive Tracers

A radionuclide can be used in place of a nonradioactive nuclide of the same element to label a site on a molecule or an entire molecule. By monitoring the radioactivity, the physical location of the molecule can be followed. Such studies have been useful for following flow in pipes and for locating blocked blood vessels in humans. Radioactive tracers can sometimes be used to identify reaction mechanisms and to measure reaction rate constants.

Mössbauer Spectroscopy

Mössbauer spectroscopy is a technique in which resonant absorption of gamma rays by nuclei is measured. During absorption the nucleus is excited from its ground state to a higher energetic level. It should be emphasized that the energetic transitions that are monitored correspond to nuclear rather than electron transitions. Because it is presently not used by most analytical chemists, Mössbauer spectroscopy is only briefly described. Further information can be found in Reference 9 and the references listed there, as well as in the reviews published in *Analytical Chemistry*.[10]

Sources of gamma rays that are used for Mössbauer spectroscopy are radionuclides that initially decay by emission of a particle or by electron capture. The resulting excited nucleus emits a gamma ray as it relaxes to the ground level. The emitted gamma ray is characteristic of the radionuclide.

The gamma ray from the source is absorbed by a nucleus in the sample. As a radionuclide in the source relaxes, energy ΔE is emitted which corresponds to the difference between the energy E_e of the excited level and the energy E_g of the ground level:

$$\Delta E = E_e - E_g = h\nu \text{ (gamma ray)} + E_r \qquad (20\text{-}42)$$

Part of the energetic difference is emitted as a gamma ray and, the remainder is the recoil energy E_r of the radionuclide.

During the emission, momentum must be conserved. The momentum of the gamma ray photon must equal that of the recoiling nucleus. In order for resonant absorption to occur in the sample nucleus, the recoil energy must be small in comparison to that of the emitted photon. That can be accomplished by using a source and a sample that are solids. The tightly bound nucleus in the source behaves as a small portion of a heavy body. Vibrational excitations of the lattice must be minimized. Because the mass of the solid is relatively large, the velocity imparted to the body during recoil is negligible and the process is termed a *recoilless emission*. For a recoilless emission, E_r in Eq. (20-42) is small and the gamma ray has the energy which is characteristic of the difference in energetic levels in the nucleus. The emitted photon can be absorbed by a similar nucleus in the sample.

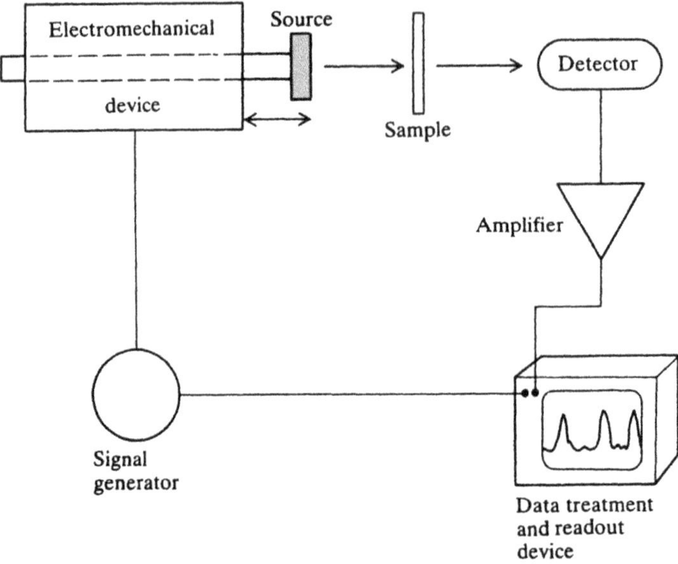

Figure 20-5 Diagram of a Mössbauer spectrometer

The detectors used for Mössbauer spectroscopy are the same as those used for x-rays. Scintillation, proportional, and semiconductor detectors are used. The energy at which absorption occurs in the sample can be affected by the chemical environment of the sample nucleus. Small spectral shifts and splitting patterns are measured and related to chemical structure or other interactions.

Because the energetic shifts that are observed during Mössbauer spectroscopy are very small, the resolution of the instrument must be large. The energy of the incident radiation is changed by using first-order Doppler shifts; i.e., the sample and the source are moved relative to each other during a measurement. The velocity of the motion determines the amount of the Doppler shift and consequently the frequency of the radiation which strikes the sample. The energetic axis on the spectrum is normally graduated in velocity units (millimeters per second) corresponding to the rate of motion. A positive velocity corresponds to motion of the source toward the sample. A Mössbauer spectrum is a plot of gamma ray intensity as a function of Doppler velocity. A block diagram of a typical Mössbauer spectrometer is shown in Fig. 20-5.

Movement of the source relative to the detector, in modern instruments, is normally accomplished by using an electromechanical device such as the drive coil in a loudspeaker. The velocity is continuously changed in a regular pattern while the spectrum is recorded. Multiple scans are used to create a spectrum. Computerized control of the instrument and computerized data acquisition are used. Often it requires hours or even days to record a spectrum. Generally the apparatus is cooled with liquid helium or liquid nitrogen to decrease recoil energy. Sometimes an external magnetic field is applied to the sample in order to aid in spectral splitting studies.

Nuclear hyperfine interactions can be investigated using Mössbauer spectrometry. The interactions are classified as electric monopole interactions, magnetic dipole interactions, and electric quadrupole interactions. Hyperfine interactions cause spectral splitting which is similar to that observed during nmr (Chapter 16) and esr (Chapter 17) studies. Shifts in peak positions can be caused by changes in electron density, changes in the ligands attached to the atom that is studied, and changes in the electronegativity of attached groups. Mössbauer spectrometry can be used to study those changes as well as to identify oxidation states of the Mössbauer atom. The spectra exhibit chemical shifts similar to those observed in nuclear magnetic resonance spectrometry. Those chemical shifts can be used for qualitative analysis. The spectra can be used as a fingerprint to identify specific substances. Spectral shifts can also be used to observe phase changes.

Miscellaneous Methods

Radiochemical methods which are used less often than those that were described in the chapter include absorptive measurements of alpha particles, beta particles, neutrons, or gamma rays, measurement of backscattered alpha particles, beta particles, or gamma rays, and autoradiography. Autoradiography is the technique in which a photographic film is used as the detector. It is primarily used for qualitative studies. Radiochemical methods have been used to detect effluents from chromatographic columns. Details of these as well as other uses of the radiochemical methods can be found in Reference 3.

Carbon-14, which is produced by the $^{14}N(n, p)^{14}C$ reaction in the atmosphere is often used to estimate the age of carbon-containing materials. The half-life of ^{14}C is 5730 y and its specific activity is 13.6 disintegrations/min · g. Assuming a constant flux of cosmic rays in the atmosphere, the natural ^{14}C concentration is constant. The ^{14}C becomes incorporated into CO_2 which is used by plants and animals. When an organism which contains ^{14}C dies, but is preserved, the ^{14}C decays. The specific activity of the organism can be used to calculate the period during which radioactive disintegration has occurred.

IMPORTANT TERMS

Accelerating tube
Activation analysis
Activity
Alpha particles
Annihilation
Autoradiography
Barn
Becquerel
Beta particles
Capture reaction
Carrier
Cockroft-Walton accelerator
Compton scattering
Coulombic barrier energy
Counting rate
Cross section
Curie
Cyclotron
CZT detector
Daughter
Dee
DIDA
Elastic scattering
Endoergic
Exoergic
Fission
Gamma ray
Half-life
IIDA
INAA
Inelastic scattering
Isotopic dilution
Moderator
Mössbauer spectroscopy
Negatron
Neutron activation analysis
Nuclear decay constant
Pair production
Parent
Pelletron
Photoelectric effect
Photon activation analysis
Positron
Radiochemistry
Radioisotope
Radiometric analysis
Radiometric titration
Radionuclide
Radiorelease
Recoilless emission
RNAA
Roentgen equivalent man
Sievert
Specific activity
Van de Graaff generator

PROBLEMS

Reactions

20-1 Write the following radiochemical reactions in shorthand notation:

(*a*) $^{6}Li + n \rightarrow {}^{3}H + {}^{4}He$

(*b*) $^{9}Be + \gamma \rightarrow {}^{1}n + {}^{8}Be$

(*c*) $^{54}Fe + {}^{4}He \rightarrow {}^{57}Ni + {}^{1}n$

(*d*) $^{37}Ar + e^{-} \rightarrow {}^{37}Cl + \nu$

Kinetics

20-2 The nuclear decay constant for beta decay of ^{20}F is 0.0608 s^{-1}. Calculate the activity of a sample which contains 1.4 mg of ^{20}F.

20-3 The half-life of ^{24}Na, which undergoes beta decay, is 15.0 h. If a sample initially contains 0.842 mg of the radionuclide, what time is required for the mass of the nuclide to be reduced to 0.347 mg?

20-4 A sample which contained ^{210}Po underwent alpha decay. After 45.0 d the activity of the sample was reduced from 1.78×10^{15} d^{-1} to 1.42×10^{15} d^{-1}. Determine the nuclear decay constant for the reaction and the half-life of ^{210}Po.

Statistics

20-5 Estimate the standard deviation and the percentage relative standard deviation for a sample in which (*a*) 2000 and (*b*) 20,000 counts were acquired.

20-6 During a period of 15.0 min, 1148 counts were obtained. Determine the counting rate and the standard deviation of the counting rate.

20-7 During 145 s, 453 counts were registered. Estimate the standard deviation of the number of counts and the counting rate.

Background Correction

20-8 The following tabulated data were obtained. Estimate the number of background counts between channels 22 and 35. Estimate the background-corrected number of counts under the corresponding peak.

Channel number	Counts	Channel number	Counts
20	330	29	935
21	325	30	890
22	320	31	500
23	325	32	395
24	398	33	295
25	507	34	245
26	895	35	239
27	941	36	237
28	955	37	235

20-9 Estimate the standard deviation of the corrected peak area determined in Prob. 20-8.

Quantitative Analysis

20-10 Estimate the activity of ^{90}Y formed from 0.050 mg of ^{89}Y during the $^{89}Y(n, \gamma)^{90}Y$ reaction after an irradiation period of 72.0 h. The cross section for the reaction is 1.31 b; the half-life of ^{90}Y is 64.3 h; and the neutron flux is 2.0×10^{7} n/cm^2 · s.

20-11 Calculate the time delay between irradiation and counting for the sample described in Prob. 20-10 if the activity at the start of the measurement must be 125 min^{-1}.

20-12 A 1.000-g sample of an alloy containing cobalt was dissolved in acid. A 5.00-mL portion of a solution that contained ^{60}Co was added to the solution and stirred. The resulting solution was diluted to 250.0 mL. After removal of the iron in the sample by extraction, the cobalt was anodically deposited on a platinum electrode as Co_2O_3. The activity of the deposit was 847 counts/min. The process was repeated with 1.000 g of an alloy that contained 0.0105 percent ^{60}Co, and the activity of the deposit was 758 counts/min. Calculate the percentage of Co in the alloy.

20-13 A 25.0-mL sample containing silver was to be assayed by DIDA. The sample was transferred to a beaker and 50.0 mg of $^{110}AgClO_4$ was added. After addition of an excess of a solution of sodium perchlorate, the silver perchlorate was extracted into toluene. The toluene portion was removed and the solvent evaporated under vacuum. After drying, the separated silver perchlorate weighed 0.2187 g. A GM counter was used to measure the counting rates. The 50.0 mg of $^{110}AgClO_4$ had a counting rate of 2545 counts/min and the separated $AgClO_4$ had a counting rate of 852 counts/min. Both counting rates were corrected for background. Determine the molarity of silver(I) in the 25.0-mL sample.

20-14 The following tabulated data were obtained from the radiometric titration of a 25.0-mL Ag^+ solution with 1.49×10^{-3} M $^{36}Cl^-$. After each addition of titrant the counting rate in the solution above the precipitate was measured. Determine the silver concentration in the sample.

Volume	Counting rate, min^{-1}
0.00	105
1.00	115
2.00	127
3.00	137
4.00	148
5.00	155
6.00	180
7.00	240
8.00	340
9.00	470
10.00	620
11.00	780

REFERENCES

1. Friedlander, G., J. W. Kennedy, E. S. Macias, and J. M. Miller: *Nuclear Radiochemistry*, 3d ed., Wiley, New York, 1981.
2. Faires, R. A., and G. G. J. Boswell: *Radioisotope Laboratory Techniques*, 4th ed. Butterworths, London, 1981.
3. Coomber, D. I. (ed.): *Radiochemical Methods in Analysis*, Plenum, New York, 1975.
4. Moses, A. J.: *Nuclear Techniques in Analytical Chemistry*, Macmillan, New York, 1964.
5. Lenihan, J. M. A., S. J. Thomson, and V. P. Guinn (eds.): *Advances in Activation Analysis*, vol. 2, Academic, New York, 1972.
6. Hoste, J., J. Op De Beeck, R. Gijbels, F. Adams, P. Van Den Winkel, and D. De Soete: *Activation Analysis*, CRC Press, Cleveland, Ohio, 1971.
7. Aliev, A. I., V. I. Drynkin, D. I. Leipunskaya, and V. A. Kasatkin: *Handbook of Nuclear Data for Neutron Activation Analysis*, Halsted Press, New York, 1970.
8. Braun, R. D.: *Introduction to Chemical Analysis*, McGraw-Hill, New York, 1982.
9. Stevens, J. G., and M. J. Ruiz, in P. J. Elving and I. M. Kolthoff (eds.): *Treatise on Analytical Chemistry*, part I, vol. 10, 2d ed., Wiley, New York, 1983, pp. 439-522.
10. Stevens, J. G., and L. H. Bowen: *Anal. Chem.*, **56**: 199R(1984).
11. Chen, H., S. A. Awadalla, F. Harris, P. Lu, R. Redden, G. Bindley, A. Copete, J. Hong, J. Grindlay, M. Amman, J. S. Lee, P. Luke, I. Kuvvetli and C. Bultz-Jorgensen: IEEE Trans. Nucl. Sci., 55: 1567(2008).

CHAPTER

TWENTY-ONE

MASS SPECTROMETRY

Mass spectrometry is the analytical technique in which a mixture of gaseous ions are separated according to their mass-charge *m/z* ratios. Some workers use *m/e* to represent the mass-charge ratio. A *mass spectrum* is a plot of relative pressure or concentration of the gaseous components as a function of their mass-charge ratios. Both positive and negative ions can be studied. Although use of negative-ion mass spectrometry is possible, positive-ion mass spectrometry is used most often. Construction of the first mass spectrometer was reported by Dempster in 1918.[1]

Because mass spectrometry can only be used to separate gaseous ions, non-gaseous samples are first converted to a gas and then to ions prior to being separated. The conversion to a gas is accomplished in the *inlet system* of the mass spectrometer. A portion of the gaseous sample is admitted to the *ionic source* where the gaseous molecules are converted to gaseous ions. Only those samples which can be converted to a gas can be studied using mass spectrometry. In the past that requirement restricted the samples, which could be examined, to relatively volatile inorganic and organic compounds. With modern instrumentation and techniques that restriction has largely been overcome. It is now possible to study polymeric and biochemical compounds with molecular weights up to nearly 60,000 and to assay most elements and many inorganic compounds.

The vast majority of the gaseous ions that are created in the mass spectrometer are singly charged. The mass-charge ratio of those ions is equal to the mass of the ions. Care must be exercised while attempting to interpret mass spectra, however, because some multiply charged ions can be formed. A doubly charged ion has a mass spectral peak at a mass-charge ratio that is one-half the value for a singly charged ion with the same mass.

The singly charged ion which has a mass that is equivalent to the mass of the unionized sample molecule is the *molecular or parent ion*. The molecular ion is formed either by removing a single electron from the molecule or by adding an electron to the molecule. Often the ionizing process causes a molecule to break into several charged fragments, each of which yields a mass spectral peak. The ion which is present in the greatest amount in the mixture and which consequently corresponds to the most intense spectral peak is the *base ion*. An ion which splits into two or more fragments while in the *mass analyzer* portion of a mass spectrometer is a *metastable ion*. The mass analyzer is the portion of the mass spectrometer in which the ions are separated according to their mass-charge ratios.

Mass spectrometry is primarily used to assay organic compounds. Both qualitative and quantitative analyses can be performed. Mass spectrometry can be used to study unimolecular gas-phase reactions of ions, to obtain rate constants for gas-phase reactions, to study the ionic products formed during photoelectron spectroscopy, to measure thermodynamic properties of ions, to study ion-molecule reactions, to determine the chemical structure of biochemical compounds (amino acids, peptides, proteins, carbohydrates, antibiotics, lipids, steroids, vitamins, coenzymes, and nucleic acids), and to perform pharmacological studies.

Qualitative analysis with mass spectrometry is accomplished by comparing the mass spectrum of the sample with those of standards which were recorded under similar experimental conditions. Alternatively, the spectrum can be used to hypothesize the identity of the fragments which formed from the analyte molecule during ionization. A knowledge of relative ionic stabilities in combination with the observed fragmentations can be used to hypothesize a chemical structure of the analyte. The molecular weight (molar mass) of the sample is obtained from the m/z of the parent-ion peak.

Quantitative analysis is accomplished by comparing the intensity of a single peak from the spectrum of the sample with that of a series of standards. More information related to mass spectrometry than that which can be presented here can be obtained by consulting References 2 through 5. Although those readings do not contain descriptions of the most recent advances in mass spectrometry, they do provide a great deal of basic information which is useful in understanding mass spectrometry.

INSTRUMENTATION

Many variations of mass spectrometers are in use. Each instrument consists of the five categories of components shown in Fig. 21-1. The inlet system is responsible for accepting the sample and converting it to a gas. The ionic source converts the gaseous molecules to gaseous ions. The mass analyzer separates the gaseous ions according to m/z. The detector measures the relative

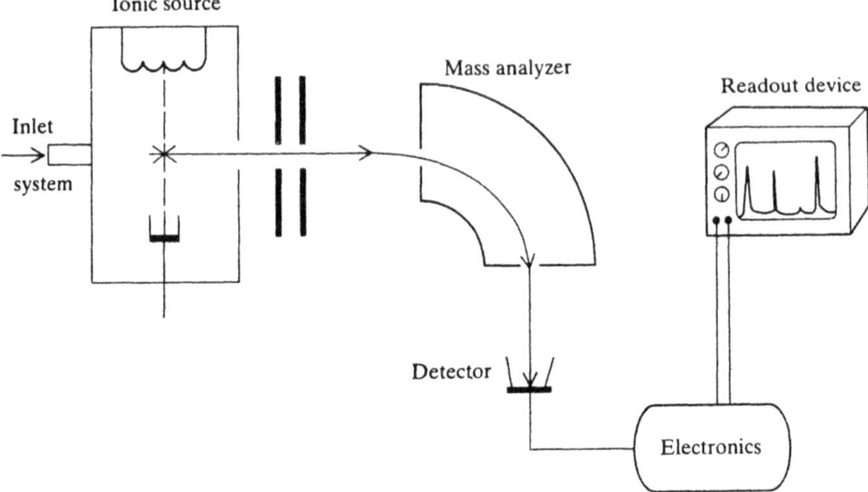

Figure 21-1 Diagram of a mass spectrometer. The arrows indicate the direction in which the sample passes through the instrument.

concentrations of the separated gaseous ions. The readout device accepts the signal from the detector and converts it to a form that can be used by the analyst. Details of each of the components of a mass spectrometer are contained in the following sections.

Inlet Systems

The type of inlet system that is used depends upon the physical state and the history of the sample. Solids, liquids, and gases can all be studied by using mass spectrometry. Inlet systems that are designed to introduce gaseous samples into the mass spectrometer are the simplest. Often they consist of a slightly porous sintered glass plug through which the gaseous sample leaks from the vessel in which it is contained into tubing. The opposite end of the tubing is attached to the ionic source. The pressure of the gaseous sample is reduced by expanding the sample into a relatively large-volume reservoir between the plug and the ionic source. Typically the pressure at the inlet to the ionic source is between 0.13 and 13 Pa (0.001 – 0.1 torr). Many instruments reduce the pressure to about 1 Pa although in some cases the pressure is reduced to less than 0.1 Pa.

Liquids and solids that have an appreciable vapor pressure (1 Pa) at room temperature can be introduced using essentially the same inlet system. The pressure differential across the plug causes the vapor above the sample to flow through the plug and into the tubing. Inlet systems which are operated at room temperature or below are *cold inlets*.

Solid and liquid samples that have relatively low vapor pressures at room temperature can sometimes be studied in a mass spectrometer by using a heated inlet system. Several types of *hot inlets* are used. In some instruments the sample reservoir, glass plug, and tubing leading to the ionic source are all heated to a temperature at which the sample has an adequate vapor pressure. Typically the temperature is maintained between 200 and 300°C. Often the entire inlet system is constructed from glass in order to prevent metal-catalyzed reactions of the sample from occurring on the walls of the inlet system.

A second type of hot inlet is useful for liquid samples. The sample is injected through a septum into a heated reservoir. The vaporized sample diffuses through several valves into the ionic source. Normally the reservoir is constructed from stainless steel when the samples do not interact with the metal.

Either solids or liquids can be placed in a capillary tube which is inserted in the end of a metallic rod. The rod is termed a *probe*. The end of the probe which contains the sample is pushed into a chamber near the ionic source. The chamber is sealed, evacuated, and heated. Vapor from the sample diffuses through a small hole into the ionic chamber. The pinhole between the inlet system and the ionic source has a diameter that is 50 μm or less. The hole is in gold foil in some instruments. Several variations of the inlet systems which are described here are available. Inlet systems for use when a mass spectrometer is interfaced to a gas or liquid chromatograph are described later in this chapter and in Chapters 25 and 26.

Ionic Sources

The function of the ionic source is to convert the gaseous sample molecules to ions which can be separated in the mass analyzer. Because the energy that is required for the conversion significantly differs with the molecule, no single method of ionization can be used for all samples. The major types of ionic sources are described in the following sections. Those methods of ionization which can impart a relatively large amount of energy to a sample are the *hard* methods. The *soft* methods impart less energy and consequently are used for samples which are fragile, easily ionized or when fragmentation might prevent formation of a parent peak and consequently make it difficult or impossible to determine the molecular weight of a molecule.

Electron-bombardment ionization The ionic source which is used most often relies upon bombardment of the gaseous sample molecules with electrons that have a selected kinetic energy. The source that uses electron bombardment to produce ions is the *electron-bombardment* or electron ionization EI source. A diagram of an EI source is shown in Fig. 21-2.

The sample normally leaks into the ionic source through a slit or hole in the metallic repeller plate or through a hole in the ionic source that is on the wall and is perpendicular to the repeller plate and grid G2. A relatively small potential difference is maintained between the repeller plate and grid G3. The potential difference is established to repel ions formed in the source from the repeller plate and to direct them toward grid G3. If positive ions are studied, the repeller plate is maintained at a positive potential relative to grid G3. Generally the potential difference is several volts per centimeter.

Electrons which are used to bombard the sample molecules are emitted from the filament shown on the left of Fig. 21-2. Many instruments contain tungsten filaments; however, rhenium, tantalum, other metals, carbon, and coated filaments can be used. Some instruments use filaments composed of LaB_6 coated on tungsten.

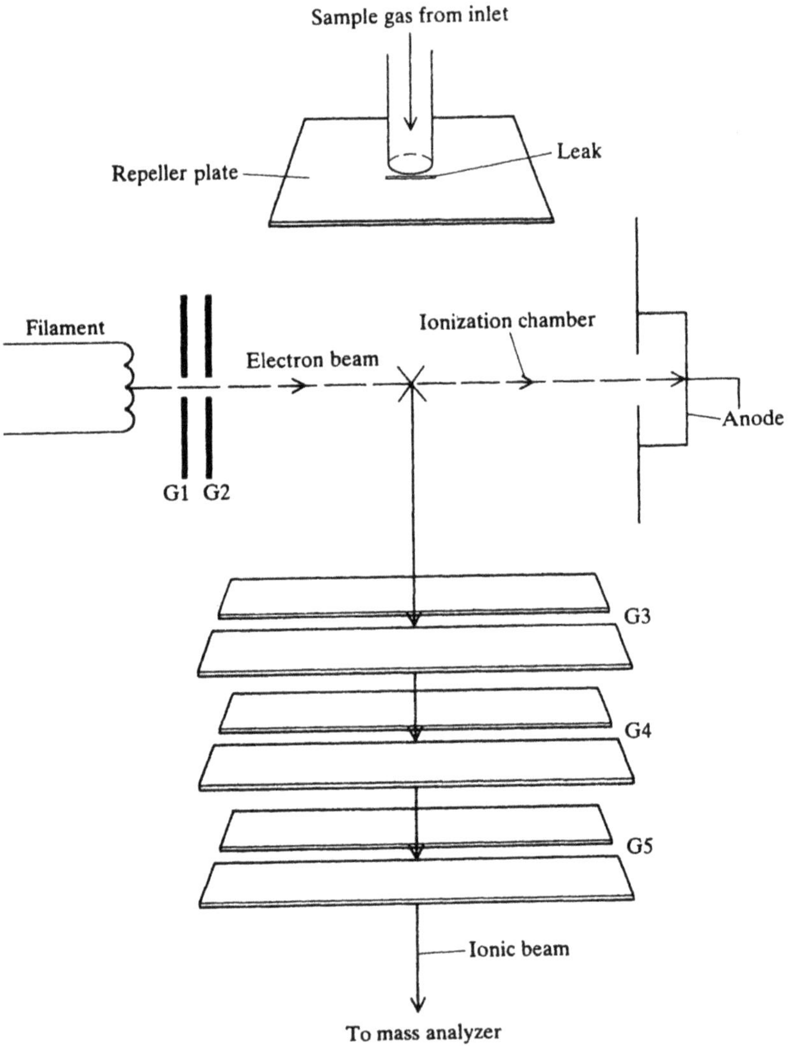

Figure 21-2 Diagram of an electron-bombardment ionic source. Grids G1 and G2 are used to accelerate the electrons. Grids G3, G4, and G5 are used to accelerate the ions that are formed in the source.

The electrons emitted from the filament are accelerated toward the sample molecules by a potential difference between grids G1 and G2. Grid G2 is maintained at a positive potential relative to grid G1. In many instruments the potential difference between the grids can be continuously varied from about 5 to about 100 V. In some instruments the potential difference can be adjusted to 1000 V or greater. Many mass spectra are recorded using a potential difference of 70 V which, of course, imparts a kinetic energy of 70 eV to the electrons. By varying the energy of the bombarding electrons, different amounts of energy are imparted to the sample molecules during the bombardment, causing differing extents of ionization.

Upon being struck by an accelerated electron, a molecule can be converted to one or more positive ions, negative ions, or to a molecule in an excited state. If the bombarding electrons possess sufficient energy, the molecule can be broken into several ionic fragments. A potential difference of 70 V between grids G1 and G2 is generally sufficient to rupture any of the chemical bonds in an organic molecule. The more stable fragments are present in the larger amounts and are responsible for the more intense spectral peaks. Electrons which do not strike a molecule are captured by the anode. The anode and grid G2 are maintained at approximately the same potential.

The cations or the anions (not both) that are formed by electron bombardment are repelled by the repeller plate toward grid G3. Grids G4 and G5 are progressively more negative if positive ions are examined or are progressively more positive if negative ions are studied. The potential difference between grids G3 and G4 and between grids G4 and G5 are used to impart a specific kinetic energy to the ions. Some instruments use a different number of grids for the acceleration. At least two grids are required for the acceleration, however, some EI sources contain more than three grids. The grids also serve to collimate the ionic beam. The grids can be circular, rather than rectangular as shown in Figure 21-1, with a hole in the center of each grid which allows the ions to pass. Typically the total potential difference between the grids is between 1000 and 8000 V. The beam current owing to the accelerated ions which leave grid G5 and enter the mass analyzer is usually between 10^{-15} and 10^{-10} A.

Arc and spark ionization Both arcs and sparks through a gaseous sample can be used to initiate ionization. *Arc discharge* ionic sources have primarily been used to produce relatively intense ionic beams such as those that are required during separation of large quantities of isotopes. The arc is caused by emission of electrons from a filament. As in the EI source, the electrons flow through the sample gas to an anode. The gaseous sample normally has a pressure between 0.01 and 1 Pa. The ions generally pass through several collimating slits prior to entering the mass analyzer.

A *spark* ionic source is used to ionize a solid sample which has been formed into an electrode or which has been packed into a hole in an electrode. An rf oscillator circuit is used to generate a spark discharge between the sample electrode and a second electrode. The second electrode is sometimes also constructed from the sample. Typically a potential difference between 5 and 25 kV is applied between the electrodes. The oscillator circuit causes between 25 and 1000 pulses/s. Each pulse has a duration of between 20 and 200 μs. The discharge causes ionization of some of the solid sample.

Because the energetic spread of the ions formed by using a spark ionic source is relatively large, a double-focusing mass analyzer is used. The source produces a relatively high concentration of multiply charged ions. Many instruments that use a spark ionic source also use a photographic plate as the detector. The source can be used to assay substances such as metals, elements, inorganic solids, and organic solids. Organic solids are generally packed into holes in aluminum electrodes. The ions generated in the source are accelerated and collimated as they pass through holes in two or more electrodes similar to those used in the EI source. Detection limits below 0.1 ppm are typical.

Photoionization Some organic compounds have ionization energies that are less than 13 eV. Irradiating those compounds with photons that have an energy of at least 13 eV causes ionization. A suitable lamp, such as a helium discharge lamp (21.21 eV), can be used to irradiate gaseous sample molecules in the ionization chamber of a mass spectrometer. The ions that are produced during the irradiation are withdrawn from the chamber as in the EI source.

Depending on the sample, radiation that has an energy between 5 and 50 eV can be used for ionization. That corresponds to wavelengths between 248 and 24.8 nm. Because the radiation is in the ultraviolet and vacuum ultraviolet regions, those uv sources that were described in Chapter 9 can be used in the ionic source. High-intensity incident radiation is required to produce ionic currents that are competitive with those from the EI source. Because photoionization is gentler than EI at 70 eV, the mass spectra that are obtained after photoionization are less complex than those that are obtained after electron bombardment. Photoionization with lasers is described in a following section.

Thermal ionization For some elements thermal energy is sufficient to vaporize and partially ionize a sample. The sample is placed on a filament which is electrically heated. The emitted ions are withdrawn from the ionic source, accelerated toward the mass analyzer, and collimated through slits in parallel electrodes similar to grids G3 through G5 in Fig. 21-2. An ionic source which relies upon application of thermal energy to a sample on an electric filament is a *surface-emissive* source. Sometimes thermal ionization is termed *direct chemical ionization* (DCI). A diagram of a simple surface-emissive source is shown in Fig. 21-3.

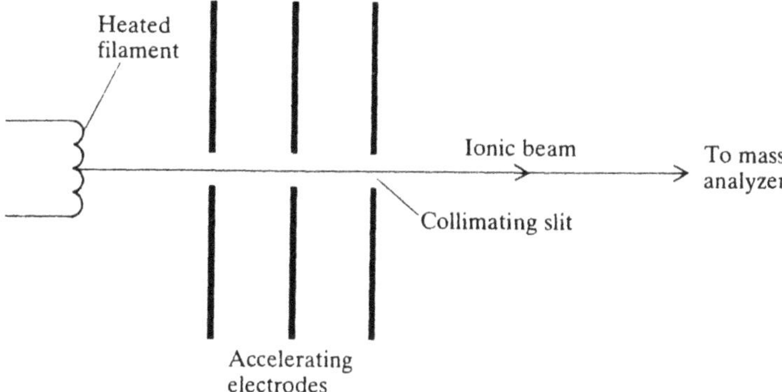

Figure 21-3 Diagram of a surface-emissive ionic source.

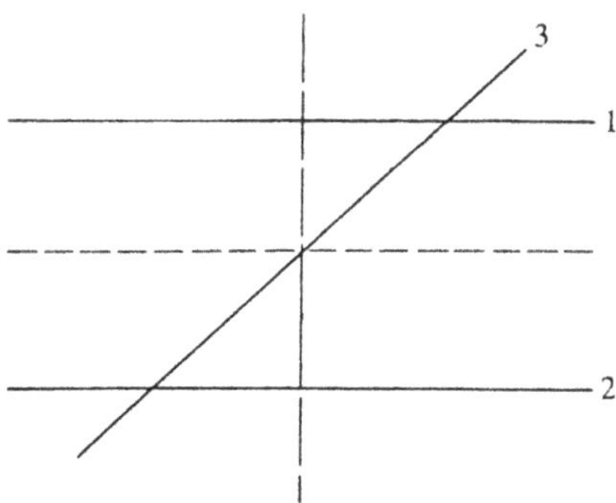

Figure 21-4 The arrangement of the filaments in one form of a three-filament surface-emissive source.

The filament in the surface-emissive source is generally constructed from tungsten; however, tantalum and rhenium have been used at higher temperatures and platinum has been used at lower temperatures. Wires that have been coated with a polymeric material, e.g., polyimide, also can be used as filaments. The sample either can be condensed onto a relatively cool filament from the gaseous phase or added to a cavity in the filament. At least three filament designs have been used. The single wire filament is the simplest. The gaseous sample is condensed onto that filament.

A canoe-shaped filament is used in the second design. The ends of the canoe are used for electrical connection and the cavity in the canoe is filled with the sample. Often that design is easier to use than the single-wire filament.

The third design uses three wire filaments. Two of the filaments are parallel and the third filament is perpendicularly oriented to the plane of the other two. The third filament is placed between the two parallel filaments as shown in Fig. 21-4. The sample is placed on one or both of the parallel filaments. The current through the filament that supports the sample causes sufficient heating to vaporize the sample. The perpendicular filament is heated to a higher temperature in order to efficiently ionize the vaporized sample. A variation of the three-filament source uses three parallel filaments. Surface emission can be used for the mass spectral analysis of at least 29 elements. A more detailed description of the source can be found in Reference 2.

Chemical ionization Gas-phase collisional reactions between a gaseous ionic species and the sample molecule can yield an ionic product that is related to the sample molecule. The mass spectrum of the product can be used for qualitative and quantitative analysis of the sample. The method of ionization in which a chemical reaction is used to form an ionic species is *chemical ionization* (CI).

Chemical ionization is more gentle than EI primarily because the reactive ionic species loses energy gradually during multiple collisions with different gaseous sample molecules; i.e., not all of the energy is transferred to a single sample molecule. CI is particularly useful for the study of compounds, such as alcohols, amines, esters, amino acids, etc., which are highly fragmented or which do not yield a molecular ion during EI. Rearrangement reactions prior to fragmentation occur less often during CI than during EI. Generally that makes CI spectra easier to interpret and has led to its popularity.

The ions that are used to collide with sample molecules are normally formed by electron bombardment of a *reagent gas* in an EI source. The reagent gas and the sample are mixed in the ionization chamber of an EI source. In order to minimize ionization of the sample by electron bombardment, it is necessary to keep the molar ratio (the pressure ratio) of the reagent gas to the sample equal to or greater than 1000. If the ratio is large nearly all of the bombarding electrons strike the reagent gas. Collisions of the ionized reagent gas with the sample molecules are highly probable while ionization of the sample molecules by electron bombardment is statistically insignificant. The choice of a reagent gas is based upon its availability, its ease of ionization, and its chemical reactions with the sample. Reagent gases should be highly volatile so that they can be easily pumped out of the ionic source, and should not harm the filament or other parts of the ionic source.

Because it is desirable to maximize the collisions between the ionized reagent gas and sample molecules, CI sources are usually operated at pressures between 25 and 300 Pa. That compares to operating pressures in the EI source of about 10^{-4} Pa. At the high pressures in the CI source the mean free path of the ionized reagent-gas molecules is relatively short (about 0.2 μm), and many collisions occur between the ions and molecules.

The ionization process in a CI source is summarized as follows:

$$RH + e^- \rightarrow RH^+ + 2e^- \qquad (21\text{-}1)$$

$$RH^+ + RH \rightarrow RH_2^+ + R \qquad (21\text{-}2)$$

$$RH^+ + RH \rightarrow R^+ + RH_2 \qquad (21\text{-}3)$$

$$RH_2^+ + SH \rightarrow RH + SH_2^+ \qquad (21\text{-}4)$$

$$R^+ + SH \rightarrow RH + S^+ \qquad (21\text{-}5)$$

$$RH_2^+ + SH \rightarrow SH\text{—}RH_2^+ \qquad (21\text{-}6)$$

In the equations RH is the reagent gas and SH is the sample molecule. Reaction (21-1) represents ionization of the reagent gas by electron bombardment. Equations (21-2) and (21-3) are two of the more common secondary ionization reactions between the ion formed during electron bombardment and the reagent gas molecule. Other reactions are also possible. The ionic products formed in Eqs. (21-2) and (21-3) can react with sample molecules as shown in Eqs. (21-4) through (21-6) to yield ionic products. The predominant reactions are those which yield the most stable products.

All of the ions formed in the source eventually enter the mass analyzer. That portion of the mass spectrum which is attributable to the ions that contain a portion or all of the sample molecule (SH_2^+, S^+, and $SH—RH_2^+$) are used for chemical analysis. Normally the ionized reagent gas has a relatively low mass and only that portion of the spectrum which occurs at greater m/z than that of the reagent ions is used for an analysis. The ionic products in Eqs. (21-4) and (21-5) were formed during protonic transfer reactions. In some cases acid-catalyzed isomerization reactions can occur during the transfer of a proton. The ionic product in Eq. (2 1-6) is an *adduct ion*. The m/z of the ions formed in Eqs. (21-4), (21-5), and (21-6) respectively are equal to $M + 1$, $M - 1$, and $M + m(\text{RH}) + 1$, where M is the molecular weight of the sample molecule and $m(\text{RH})$ is the molecular weight of the reagent gas.

Reagent gases which are commonly used during CI are methane, isobutane, and ammonia. Other less common reagent gases include methyl vinyl ether, methyl propane, tetramethylsilane, nitric oxide, ethane, propane, and argon. Methane and ammonia are primarily ionized by Eq. (21-2) to yield CH_5^+ and NH_4^+. Isobutane is primarily ionized by Eq. (21-3) to yield $C_4H_9^+$. Because CH_5^+ is acidic, it often reacts by transferring a proton to the sample, which acts as a base, as shown in Eq. (21-4). As an example, the CI of acetone in methane forms $(CH_3)_2COH^+$ which has a mass spectral peak one unit higher than the molecular weight of acetone.

If the ionic reagent gas can form a stable Lewis acid-base product with the sample molecule, the sample ionization can occur by Eq. (21-6). An example is the reaction of ammonium (ammonia reagent gas) with an amine to yield an adduct ion with a molecular weight of M +18. Ionization occurs by Eq. (21-5) whenever the sample molecule is a stronger acid than the ionized reagent gas. In many spectra the ionic products formed in Eqs. (21-4), (21-5), and (21-6) are all observed, although not in equal intensity

Fragmentation also occurs during CI. Generally the fragmentation in organic compounds occurs by loss of neutral molecules to yield relatively stable fragment ions with an even number of electrons. Free radical ions are relatively unstable. Fragmentation during CI produces fewer ionic fragments than during EI. The fragments correspond to the ions that have the greatest stability and consequently are particularly useful for qualitative analysis. The extent of fragmentation is dependent upon the energy of the ionic reagent gas and the relative stability of the product ion. For reagent gases in a homologous series, the greatest fragmentation generally occurs when the reagent gas of lowest molecular weight is used. Low-molecular-weight reagent gases more efficiently transfer energy to the sample molecule. Some instruments are designed to utilize both EI and CI. In some instruments the source can be changed from one form of ionization to the other in less than 1 s. More information can be found in References 3 through 5.

Negative chemical ionization *Negative chemical ionization* (NCI) occurs by chemical reaction of the sample molecule with an ionized reagent gas in a manner similar to that described in the preceding section. The product of the reaction, however, is a negative ion whose mass spectrum is subsequently recorded. NCI is useful for the study of molecules which are highly electronegative. Examples of the many organic compounds which can be assayed using NCI include halides, some oxygenated compounds such as quinones, and aromatic nitro compounds. Because CI, which yields positive ions, is more useful for compounds with low electronegativity, CI and NCI complement each other.

At low pressures, such as those at which the EI source is operated, negative ions are not formed efficiently. At the relatively high pressures (at least 100 Pa) at which NCI sources are operated, negative ions can be formed with relatively high efficiency. Often NCI is caused by the collision of a negative reagent gas ion with the sample molecule.

Electron bombardment of the reagent gas can produce negative ions:

$$XY + e^- \rightarrow X + Y^- \tag{21-7}$$

where XY represents the reagent gas and Y^- the negative product ion. As during CI, which yields positive ions, the concentration of the reagent gas is large in comparison to that of the sample molecules in the ionization chamber. Collision of the negative ion with a sample molecule can yield a negative product ion by any of several mechanisms. The following are some of the more common mechanisms which involve hydrogen transfer:

$$Y^- + SH_2 \rightarrow SH\cdot + HY^- \tag{21-8}$$

$$Y^- + SH_2 \rightarrow SH^- + HY \tag{21-9}$$

where SH_2 represents the sample molecule.

In Eq. (21-8) a hydrogen atom is abstracted and in Eq. (21-9) a proton is abstracted. Other relatively common reactions that occur during collision of the negative ion with the sample molecule include H_2 abstraction from the sample, hydrogen atom displacement by the negative ion, alkyl displacement by the negative ion, nucleophilic displacement on the sample molecule to yield a displaced negative ion, and base-induced eliminations.

Several of the more common reagent gases used during NCI are dichloromethane, nitrous oxide, and a mixture of nitrous oxide and methane. The reagent gas must have an electronegative component in order to form a negative ion during electron bombardment. Electron bombardment of dichloromethane yields chloride ion which reacts upon collision with sample molecules. Nitrous oxide forms the highly reactive $O\cdot^-$ ion. It can undergo hydrogen atom abstraction reactions [Eq.(21-8)] with sample molecules to yield hydroxide ions, proton abstraction to yield neutral OH· and a negatively charged sample ion with mass $M-1$, H_2 abstraction to yield a negative ion with mass $M-2$, hydrogen atom displacement to yield a negative ion with mass $M+15$, and alkyl displacement.

A reactive negative ion can be formed by the addition of methane to nitrous oxide. Nearly all of the $O\cdot^-$ formed during electron bombardment is rapidly converted to OH^- by hydrogen atom extraction during collisions with methane [Eq. (21-8)]. The hydroxide formed during the reaction subsequently collides with sample molecules to produce negative product ions. Much of the energy that was initially possessed by the $O\cdot^-$ is lost during collisions with methane. The hydroxide consequently undergoes fewer reactions than those undergone by $O\cdot^-$. In most cases it undergoes proton abstraction to form water and a negative ion with a mass $M-1$:

$$OH^- + SH \rightarrow H_2O + S^- \qquad (21\text{-}10)$$

where SH represents the sample. Generally little fragmentation occurs.

Another process by which negative ions of the sample molecule can be formed in an NCI source is *electron capture*. An electron is captured by a molecular orbital in the sample during electron bombardment. The sample molecule must have a vacancy in an orbital which is relatively stable (low energy). Unsaturated organic compounds as well as compounds which contain phosphorus, sulfur, chlorine, bromine, or iodine can capture electrons. Normally electron capture requires the presence of a third body to remove the excess kinetic energy possessed by the electron prior to the collision. Otherwise fragmentation occurs. Often methane is used as the third body. Methane or any other gas which is used to dissipate the excess energy during the three-body collision (electron, sample, and methane) is a *buffer* or *moderating gas*. Electron capture, as described here, should not be confused with electron capture by a nucleus which was described in Chapter 20.

Because CI complements NCI, some mass spectrometers are designed to use both processes. An instrument which alternatively withdraws positive and negative ions from the chemical ionization chamber at a rate of about 10 kHz into a quadrupolar mass analyzer uses *pulsed positive-ion/negative-ion chemical ionization* (PPINICI). The positive ions and negative ions are measured with separate detectors. The positive-ion spectrum and the negative-ion spectrum are simultaneously recorded.

Field ionization and field desorption Sample ionization can be induced by application of a strong electric field. A *field ionization* (FI) source consists of two pointed electrodes between which a potential gradient of about 10^7 or 10^8 V/cm is applied. Gaseous sample molecules are ionized upon passing between the electrodes. If the source is used to ionize samples that are adsorbed on the anode, the technique is *field desorption* (FD). Ionization occurs by quantum tunneling of a valence electron from a sample molecule. The electron is attracted to the anode while the cation is repelled from the anode. The positive sample ion is extracted from the ionic chamber and accelerated through slits in several electrodes similar to the grids used in the EI source.

The tip of the anode must be highly pointed in order to achieve the required potential gradient. The efficiency of the electrode can be increased by growing microneedles of carbon or another electrode material on the surface of the electrode tip. The array of microscopic needles constitutes a *multipoint array electrode*. Its efficiency can be several orders of magnitude greater than that of a conventional electrode.

Relatively little fragmentation occurs with the FI source; that which does occur is usually caused by thermal decomposition or by collisional reactions between molecular and ionic species in the vicinity of the electrode. Protonic transfer can occur during those reactions yielding ionic species with molecular weights of $M - 1$ and $M + 1$ where M is the molecular weight of the analyte. The initial ionization produces ions that have a molecular weight of M. FI causes intense molecular ion peaks.

Sample molecules that have high electron affinities can capture electrons during FI to yield negative ions. That process is *negative-ion field ionization*. It is used to generate negative ions whose mass spectra subsequently can be obtained.

Prior to FD a layer of a dilute solution of the sample is placed on the electrode with a syringe. The electrode upon which the sample is placed is the *emitter*. After evaporation of the solvent the electrode is placed in the ionic chamber and ionization is induced as during FI.

Laser-induced ionization Lasers that emit ultraviolet or visible radiation sometimes can be used to induce ionization in gaseous molecules. Often the energy possessed by a single photon in the near-ultraviolet or the visible spectral region is insufficient to induce ionization in a molecule. The high intensity of the radiation that is emitted from a laser makes it possible for a molecule to simultaneously absorb multiple photons, as described in Chapter 8. Ionization that is caused by simultaneous absorption of more than one photon is *multiple photon ionization* (MPI) (or multiphoton ionization). Both resonant and nonresonant transitions can be used. Because absorption of the first photon excites the molecule to an excited electron state, the initial absorptive process is wavelength-dependent; i.e., the energy of the photon must equal the energetic difference between electron levels. By varying the wavelength of the incident photon, different species can be excited. Absorption of subsequent photons provides a total absorbed energy that is sufficient to remove the electron from the sample, thereby forming a positive ion.

Lasers can also be used to ionize sample molecules that are coated on a solid substrate. The laser provides the energy that is required for both the vaporization and the ionization. The method by which a sample on a solid is vaporized and ionized by a laser is *laser desorption* (LD) ionization. After ionization the ions are accelerated and collimated through slits in electrodes identical to those used in other ionic sources. Further details relating to LD ionization can be found in Reference 6.

Matrix-assisted laser desorption/ionization Matrix-assisted Laser Desorption/Ionization MALDI is a soft laser ionization technique that has proved to be particularly useful for ionization of macromolecules (large organic molecules, biomolecules such as proteins and peptides, organic polymers, etc.). It was proposed in 1985 (Reference 7). A solution containing the analyte and typically a larger amount of a matrix compound are prepared in an aqueous, organic or mixed aqueous-organic solvent. A drop of the solution is placed on a MALDI plate, which often is metallic, and the solvent is removed by vaporization. The resulting spot is a mixed solid crystal containing the analyte and the matrix.

The matrix molecule typically is a relatively low molecular weight compound, such as picolinic acid, a substituted cinnamic acid or 2,5-dihydroxybenzoic acid, which easily can absorb radiation from a uv or ir laser and which can react with the analyte to form ionized analyte ions. The EMR absorption requirement is met by choosing a matrix which contains conjugated double bonds or is an aromatic compound. The chemical reactivity requirement is often fulfilled by use of an acid matrix molecule which can transfer a proton to the analyte.

The mixed solid crystal is irradiated with a laser which preferentially transfers energy to the matrix molecules to form matrix ions. The matrix thereby protects the fragile analyte molecules from directly absorbing radiation. Macromolecules often are highly fragmented with harder ionic sources and usually have spectra with no parent peak. Interactions of matrix ions with analyte molecules yield analyte ions which typically have molecular weights of M, M+1 and M–1 corresponding to the parent ion, the gain of a proton from the matrix and the loss of a proton to the matrix.

Although multiple types of lasers can be used in the ionic source, a relatively inexpensive nitrogen laser (emission at 337 nm) or a Nd:YAG laser (266 and 355 nm) is used in many commercial sources. Since a pulsed laser often is used, the source typically is used with time-of-flight mass analyzers (see later section in this chapter) because the laser pulse can form ion pulses which meet the requirement for a pulse of entering ions in the analyzer. It is possible, however, to use MALDI sources with other mass analyzers. Either ionization can occur in a vacuum, typically at about 1 Pa, as is done with most other types of ionic sources, or it can be accomplished at atmospheric pressure. In the latter case the technique is atmospheric pressure MALDI AP-MALDI. MALDI has become a popular and important method for ionizing macromolecules.

Photoelectric ionization The ionic products formed during photoelectric spectroscopy (PES) (Chapter 19) can be studied in a mass spectrometer. The mass spectrum of the ionized sample that remains after photonic bombardment is measured rather than the spectrum of the emitted electrons. The ionization is identical to that described in Chapter 19. Considerable research has been done using photoelectric ionization, as indicated in Reference 8.

Ionization by ionic bombardment If a beam of primary ions strikes the surface of a solid with sufficient force, the ions can become imbedded in the solid. The kinetic energy that is lost by the ions during the collision cause sputtering from the surface of neutral fragments and secondary positive and negative ions. The technique in which the mass spectrum of the secondary ions that are formed during the collision is obtained is *secondary ion mass spectrometry* (SIMS). Both positive and negative ions can be studied.

Often ions of an inert gas, such as argon or neon, or oxygen ions are used for the bombardment. Other ions can also be used. Bombardment with ions of an inert gas or oxygen produces greater sputtering of positive ions than of negative ions. Sputtering of negative ions can be enhanced by bombardment with ions such as Cs^+ of electropositive elements. In any case the ionic yield from the sputtering is generally less than 1 percent. The remainder of the sputtered particles are neutral species.

Figure 21-5 is a diagram that shows the important parts of a secondary ion mass spectrometer. The ionic gun supplies primary ions that have a kinetic energy which is normally between 0.3 and 5 keV. After the primary ions strike the sample surface, secondary ions are emitted and pass through a lens and filtering system. A quadrupolar mass analyzer is normally used to resolve the ionic mixture according to the mass-charge ratio.

Sputtering can be accomplished by using either of two methods. *Dynamic* sputtering occurs when the current of the primary ionic beam is maintained at about 1 $\mu A/cm^2$. That beam current disrupts the surface to a depth of about 4 nm. Consequently, dynamic sputtering from the surface is used when it is not necessary to distinguish between molecules that are within a depth of 4 nm from each other.

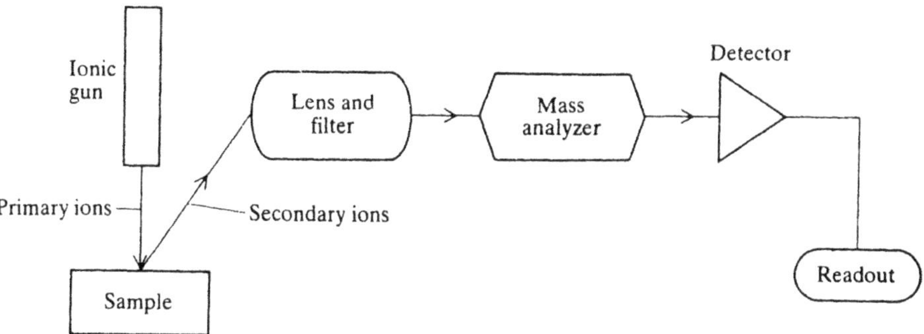

Figure 21-5 Diagram of a secondary ion mass spectrometer. The arrows indicate the direction of ionic flow.

Static sputtering uses an ionic beam current that is about three orders of magnitude smaller than that used during dynamic sputtering. When a beam current of about 1 nA/cm^2 is used, the bombarding particles penetrate the surface to smaller depths than that penetrated during dynamic sputtering. Consequently, the mass spectrum is representative of the upper several atomic layers of the surface.

The sensitivity that is achieved while using SIMS varies considerably between different elements within a specified chemical matrix and for the same element between matrices. Because the variations are large and difficult to control, SIMS is primarily used for qualitative analysis of surfaces. Sensitivity variations that are as large as five orders of magnitude have been observed. Even though quantitative analyses are rarely possible with SIMS, the technique is highly sensitive. In some cases it is possible to detect ions at a concentration that corresponds to as little as one-millionth of a monolayer of the surface. It can be used to qualitatively determine all elements.

Depth profiling can be rapidly done using SIMS in the dynamic mode. The ionic beam can be used in the dynamic mode as an *ionic microprobe* to study composition across a surface. Spatial resolution with an ionic microprobe is about 1 μm on a surface. Spatial resolution is not as good as that obtained using an Auger microprobe. SIMS has been used to assay many elements on many types of surfaces. Secondary ions also have been sputtered from a liquid surface rather than from a solid. The form of SIMS in which sputtering occurs from a liquid is *liquid secondary ion mass spectrometry* (LSIMS). A description of the technique can be found in Reference 9.

Another technique which also uses ionic bombardment of a surface is *ion scattering spectroscopy* (ISS). ISS is briefly described in this section, because it is similar to SIMS, although it is not a form of mass spectrometry. The incident ionic beam used during ISS is identical to that used during SIMS. During ion-scattering studies, however, the kinetic energies of the ions are measured after scattering from the surface. The energy of each scattered ion is a function of the energy of the incident ion and of the mass of the atom on the surface from which the ion was scattered.

A block diagram of the apparatus that is used for ISS is shown in Fig. 21-6. Ions that are emitted from the ionic source strike the sample surface and are scattered. The scattered ions are usually energetically resolved with a cylindrical mirror analyzer (CMA) (Chapter 17) which is coaxial with the ionic source. As during SIMS, ionic scattering is performed at a pressure of about 0.001 Pa. ISS and SIMS are often performed with a single instrument which has both a mass analyzer and a CMA.

Either dynamic or static sputtering can be used during ISS, as during SIMS. Use in the dynamic mode disrupts the surface during the measurement. ISS is truly a surface technique because ions which penetrate beyond the top atomic layer of the surface are not likely to be scattered. The incident ionic beam typically has an energy of about 1 keV. Spectral resolution and peak location are a function of the mass of the incident ions. Resolution is best when the incident ions and the sample atoms have nearly equal masses. ISS is used for elemental qualitative and semiquantitative analysis. Considerable effort is being expended to make it more useful for quantitative analysis. Further details are available in Reference 9.

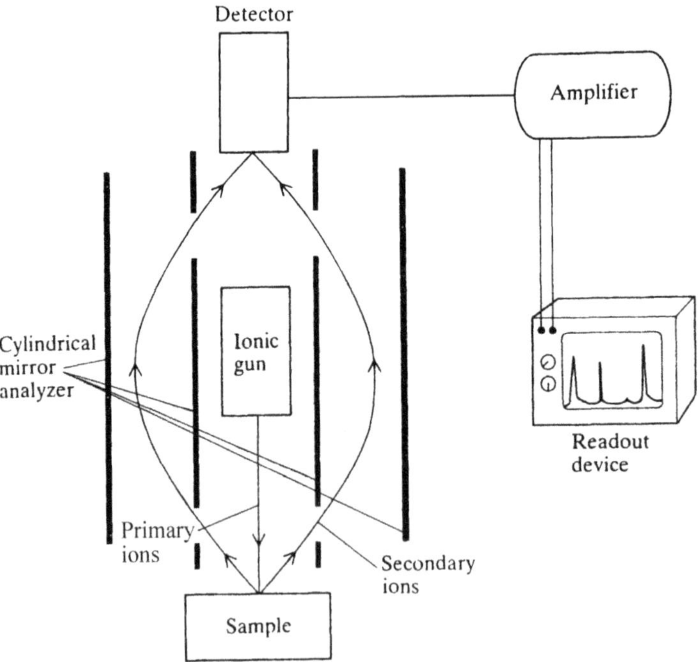

Figure 21-6 Diagram of the apparatus that is used for ion-scattering spectroscopy.

Ionization by atomic bombardment Atomic bombardment can be used in the same way that ionic bombardment was used to sputter ionic particles from the surface of a solid or liquid. Normally ionic bombardment is used for the analysis of solid surfaces and atomic bombardment for liquid surfaces. The use of incident atoms to induce ionization is *fast-atom bombardment* (FAB). The form of mass spectrometry which relies upon FAB for ionization is *fast-atom bombardment mass spectrometry* (FABMS). The apparatus is similar to that shown in Fig. 21-5 where the primary ionic source is replaced with an atomic source. Often Xe, Ne, He, or Ar atoms are used to bombard the surface. The efficiency of the ionization increases with increasing mass of the bombarding atoms. The mass analyzer that is used to monitor the sputtered ions can be magnetic, double focusing, or of some other type.

Fast atoms are generated from ions of the desired atoms. First, atoms of the desired atoms, e.g., Xe, are ionized and are accelerated to the desired energy in an electric field. The accelerated ions subsequently pass through a xenon (or other gas) chamber in which the ions undergo charge exchange and are converted to fast atoms. The fast atoms impinge upon the sample causing ionization. The sample ions that are produced during the bombardment are repelled from the sample and toward the mass analyzer by maintaining an appropriate charge on the metallic probe on which the sample is placed.

FABMS has been used to study most of the substances studied by SIMS as well as liquid samples. Compounds which can only be studied by EI or CI after chemical derivatization can often be studied without chemical derivatization by FABMS. Often a small volume (about $0.3 \mu L$) of a solution of the sample is placed on a solid probe and inserted into the mass spectrometer in the path of the atomic beam. The sample is generally prepared in a viscous fluid matrix such as glycerol. A description of the experimental parameters that are used during FABMS is available in Reference 10.

Electrospray ionization Electrospray ionization ESI has become a popular method to interface high performance liquid chromatography (Chapter 25) to mass spectrometry while simultaneously serving as a soft ionic source. The technique utilizes a metallic capillary nozzle, with an internal diameter of about 0.1 mm, through which the analyte enters, and a sampling cone which contains a pin hole in its center and through which the created ions pass while entering the high vacuum region of the mass analyzer. The nozzle and cone are separated by a gaseous space, in which desolvation and ionization of the analyte occurs. A potential difference of typically 2-6 kV is applied between the nozzle and cone. As the liquid sample passes through the metallic capillary, it acquires a charge which depends upon the polarity of the capillary. If the capillary is the anode, the entering solution forms positive ions. Similarly, if the capillary is negatively charged, the components of the entering solution form negative ions.

As the solution is aspirated as a mist into the region between the nozzle and cone, the ions within the mist are repelled by the electric charge on the nozzle and attracted toward the opposite charge on the sampling cone. In the space between the nozzle and cone, the sample is converted to gaseous ions and the solvent is removed. The analyte ions apparently are formed by a combination of charge transfer and acid-base reactions which occur primarily in the liquid phase, but which also can occur in the gas phase. The molecular weights of the generated ions can be M, M + 1 (H^+ addition), M + 23 (Na^+ addition), M + 39 (K^+ addition), M + 18 (NH_4^+ addition), M + 18 (H_2O addition), M + 18n (gain of nH_2O), M + 19 (H_3O^+ addition), as well as masses corresponding to other reactions which form positive ions. When negative ions are monitored, the ionic masses are primarily M, M-1 (proton loss), M-1-18n (loss of a proton and n water molecules), M + 35 (chloride addition) and other masses corresponding to other chemical reactions.

The ionic source has been used to ionize analytes with molecular weights as large as 100,000. The method is most effective when used for the analysis of acids, bases and polarizable molecules which relatively easily can be ionized in solution. Compounds which contain O, N and other electronegative atoms as well as molecules which have a relatively large dipole moment are good candidates for ionization by ESI. Of course, the analyte ion also must be vaporized relatively easily.

Inductively coupled plasma ionization Unlike most of the other ionic sources, inductively coupled plasma ICP sources are used to form elemental ions rather than polyatomic ions. When ICP is used as an ionic source for mass spectrometry the method is inductively coupled plasma mass spectrometry ICPMS. ICPs were described in Chapter 8. ICPMS can be used to assay nearly all elements which have atomic masses between 7 (Li) and 238 (U).

Essentially the process in ICPMS systems consists of the following steps. First, the sample solution, which can be the effluent from a chromatographic column, is aspirated into the spray chamber of the ICP where the larger drops are removed. The resulting aerosol mist consisting of small droplets enters the plasma torch and is ionized. The intensity of emitted emission from the plasma can be measured at one or more wavelengths, however, that is not a requirement. The ions formed in the plasma are pulled through a small entrance hole in an interface cone that is similar to the sampling cone which is utilized during ESI, and into the mass analyzer of the mass spectrometer. Typically, the ionizer is combined with a magnetic sector, quadrupolar or time-of-flight mass analyzer. Reviews of the development and uses of ICPMS can be found in References 11 and 12.

Mass Analyzers and Resolution

The mass analyzer in a mass spectrometer is positioned between the ionic source and the detector. It is the portion of the mass spectrometer in which the sample ions or the fragmented sample ions are separated according to *m/z*. The major types of mass analyzers are the *magnetic analyzer*, the *electrostatic analyzer*, the *time-of-flight analyzer, the quadrupolar analyzer*, and the *ion cyclotron resonance analyzer*. Several variations of those analyzers are available.

The *mass resolving power R* (also termed the *resolution*) that is achieved by the mass analyzer in a mass spectrometer is a measure of the ability of the analyzer to separate ions which have different mass-charge ratios. The mass resolving power in mass spectrometry is defined as the ratio of the average mass M (amu) of two adjacent peaks that correspond to "just resolved" singly charged ions to the change in mass ΔM between the peaks:

$$R = \frac{M}{\Delta M} \tag{21-11}$$

Several methods are used to define "just resolved." In most cases the definition is related to the ratio of the height ΔH of the valley between two peaks of equal height H to the height of either peak. Definitions of "just resolved" usually correspond to values of $\Delta H/H$ which are between 0.001 and 0.1, i.e., to a valley which is between 0.1 and 10 percent of the peak height. The 10 percent definition is used most often. Some analysts, however, define resolving power by using a valley which is 50% of the peak height. The manner in which ΔH and H are measured is illustrated in Fig. 21-7. The usual definition of "just resolved" is as follows:

$$\frac{\Delta H}{H} = 0.1 \tag{21-12}$$

In some cases it is difficult to locate two adjacent peaks that are of equal height and that have a valley between the peaks that is 10 percent of the peak height. In that case the peak width of any single peak at 5 percent of the height of the peak $w_{0.05}$ can be used as the value of ΔM in Eq. (21-11). The resolution in that case is defined as

$$R = \frac{M}{w_{0.05}} \tag{21-13}$$

where M is the mass (amu) at the center of the peak. Equation (21-13) and the combination of Eqs. (21-11) and (21-12) yield identical results if the peaks are gaussian. Other analysts calculate resolving power from a single peak by using Equation (21-13) where the width of the peak measured at 50% of its height is used rather than the width at 5% of the peak height. In either case the m/z at which the resolution is calculated should be specified.

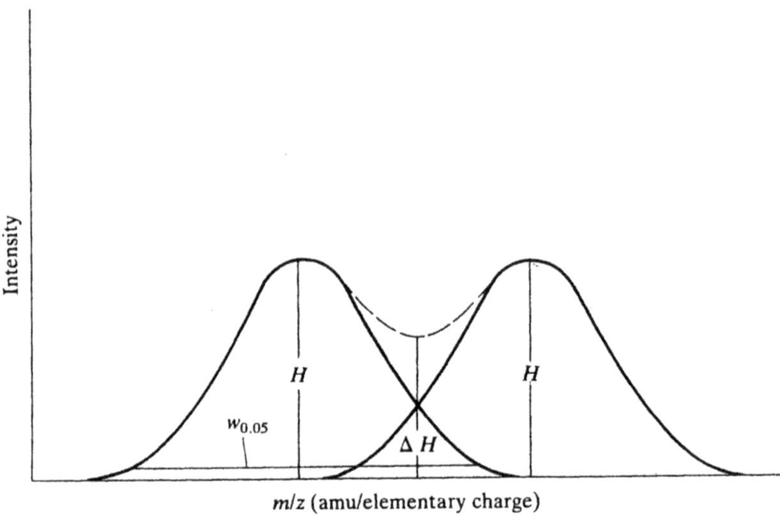

Figure 21-7 Overlapping mass spectral peaks of equal height H. The solid lines represent the separate peaks. The dashed line represents the combined spectrum.

Sample problem 21-1 Calculate the resolution in a mass spectrum if a peak that is centered at 245 amu has a peak width at 5 percent of the peak height that is 0.52 amu.

SOLUTION The resolution can be calculated using Eq. (21-13):

$$R = \frac{245}{0.52} = 4.7 \times 10^2$$

Magnetic analyzers Magnetic mass analyzers rely upon the effect of a magnetic field on moving ions to separate the ions. The accelerated ions that leave the ionic source possess nearly identical kinetic energies regardless of their masses. The ionic beam is directed between the poles of an electromagnet. The magnetic field causes the ionic beam to assume a curved path. Because the amount of curvature for a particular ion is dependent upon m/z of the ion, separation is achieved. A diagram illustrating one form of a magnetic analyzer is shown in Fig. 21-8.

If an ion which is formed in the ionic source initially has no kinetic energy, the energy that is imposed by the potential drop E between the accelerating electrodes is zE, where z is the charge on the ion in units of coulombs. That kinetic energy is possessed by every ion that enters the magnetic field. The kinetic energy that is possessed by an ion with mass m and velocity v as it enters the magnetic field is also equivalent to $mv^2/2$. Equating the two ways of expressing the kinetic energy gives:

$$\frac{mv^2}{2} = zE \tag{12-14}$$

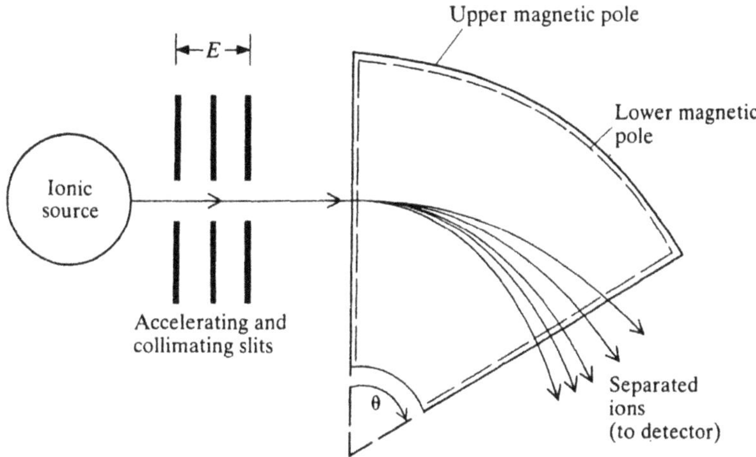

Figure 21-8 Diagram of a magnetic analyzer. The arrows indicate the direction of ionic flow through the analyzer. The ions pass between the two poles of a magnet.

The magnetic field exerts a force F_B on the ions which pass between the poles of the magnet:

$$F_B = Bzv \tag{21-15}$$

where B is the magnetic flux density. The force exerted by the magnetic field must equal the centrifugal force F_c:

$$F_c = \frac{mv^2}{r} \tag{21-16}$$

that is experienced by the ions as they assume a curved path with a radius r in the magnetic field. Setting F_B from Eq. (21-15) equal to F_c from Eq. (21-16) yields

$$Bzv = \frac{mv^2}{r} \tag{21-17}$$

Equation (21-17) can be solved for the velocity of an ion as it passes through the magnetic field:

$$v = \frac{Bzr}{m} \tag{21-18}$$

Substitution of the velocity into Eq. (21-14) yields

$$\frac{m\left(Bzr/m\right)^2}{2} = zE \tag{21-19}$$

Solution of Eq. (21-19) for m/z of the ion gives

$$\frac{m}{z} = \frac{B^2 r^2}{2E} \tag{21-20}$$

If m is converted from mass (kilograms) to amu and z is converted to units of elementary charge from coulombs, Eq. (21-17) can be written as

$$\frac{\text{amu}}{z} = 4.824 \times 10^7 \frac{B^2 r^2}{E} \tag{21-21}$$

where amu is the mass of the ion in amu, z is its elementary charge, and all other quantities are in SI units.

From Eq. (21-20) it is apparent that for a fixed set of instrumental parameters, the radius of curvature of an ion in the magnetic field increases with m/z. Heavier ions of a fixed charge are deflected less by the magnetic field than lighter ions. Consequently, the ions emerging from the magnetic field are separated in space according to their m/z. In many mass spectrometers the detector is placed at a fixed angle relative to the ionic beam that enters the magnetic field and at a fixed radius within the magnetic field. In that case combination of all of the constants in Eq. (21-20) into a single constant k gives

$$\frac{m}{z} = \frac{kB^2}{E} \tag{21-22}$$

For a fixed radius of curvature, the m/z of the ions which strike the detector can be controlled either by varying the magnetic field strength B or the accelerating potential E. Varying either of those quantities with time while holding the remaining variable fixed permits scanning the m/z of the ions that strike the detector. A mass spectrum is recorded while the parameter is varied. In most instruments the magnetic field is varied. The angle of deflection of the ionic beam in a mass analyzer is normally fixed at 60, 90, or 180 degrees, although other angles can be used. Modern magnetic analyzers can resolve integral values of m/z (amu per elementary charge) up to about 2000. This can be compared with mass resolving powers which are significantly higher with some time-of-flight, ion cyclotron resonance and orbitrap mass analyzers.

Sample problem 21-2 Calculate the magnetic flux density that is required to focus a $C_4H_9^+$ ion on the detector in a mass spectrometer in which the accelerating potential is maintained at 2.00 kV and the radius of curvature of the focused ionic beam at the exit slit is 30.0 cm.

SOLUTION The value of m/z for the ion is 57 amu/elementary charge, which can be converted to SI units as follows (Table 20-1, footnote):

$$57 \text{ amu/elem. chg.} \times 1.66 \times 10^{-27} \text{ kg/amu} \times 1 \text{ elem. chg.}/1.60 \times 10^{-19} \text{ C}$$

$$= 5.9 \times 10^{-7} \text{ kg/C} = m/z$$

Substituting the appropriate values into Eq. (21-20) and solution for B^2 yields

$$5.9 \times 10^{-7} \text{ kg/C} = B^2 \frac{(0.300 \text{ m})^2}{2\left(2.00 \times 10^3 \text{ V}\right)}$$

$$B^2 = 2.6 \times 10^{-2} \text{ kg} \cdot \text{V/m}^2.\text{C}$$

Substituting the definitions of the volt and coulomb yield

$$B^2 = 2.6 \times 10^{-2} \text{ kg}(\text{kg} \cdot \text{m}^2/\text{s}^3 \cdot \text{A})/(\text{A} \cdot \text{s})\text{m}^2$$

$$= 2.6 \times 10^{-2} \text{ kg}^2/\text{s}^4 \cdot \text{A}^2$$

from which B can be calculated (a tesla T is a $\text{kg/s}^2 \cdot \text{A}$):

$$B = 0.16 \text{ kg/s}^2 \cdot \text{A} = 0.16 \text{ T}$$

Alternatively, Eq. (21-21) can be used without performing the initial conversion step in the calculation.

Double-focusing analyzers Not all of the ions are formed at the same location in the ionic source. Consequently, the ions do not all experience identical accelerating potentials from the accelerating electrodes of the ionic source. Furthermore, the ions do not all possess identical kinetic energies at the time of their formation. The combination of the two factors causes ions to enter the mass analyzer with slightly different kinetic energies. That, of course, leads to peak broadening because the value of E in Eqs. (21-19) through (21-22) slightly varies between ions.

A double-focusing analyzer uses a cylindrical electric sector (an *electrostatic analyzer*) between the ionic source and the magnetic mass analyzer to restrict the energies of the ions that enter the magnetic analyzer to a relatively narrow band. The electric sector focuses the ions according to energy and the magnetic sector focuses the ions according to direction. Consequently, the combined apparatus is a double-focusing analyzer. Double-focusing analyzers have resolutions that are approximately an order of magnitude greater than that of magnetic analyzers. Modern double-focusing analyzers can typically scan to m/z of about 2500 with a mass resolving power of about 20,000. A diagram of a double-focusing analyzer is shown in Fig. 21-9.

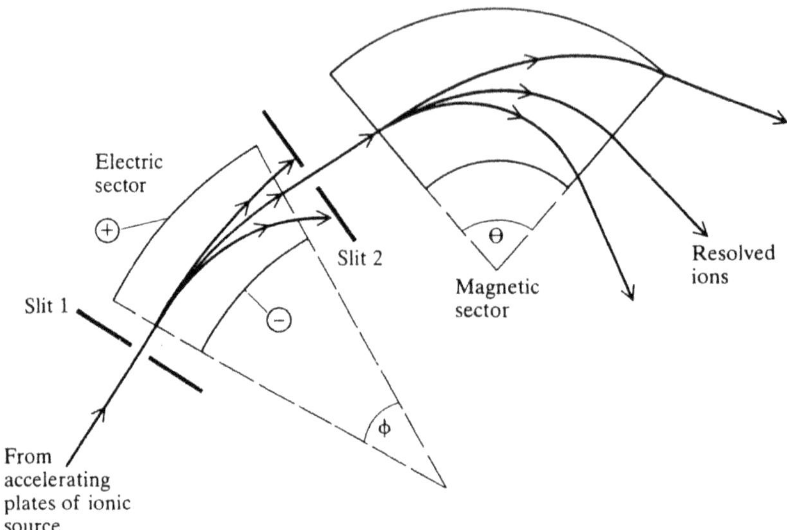

Figure 21-9 Diagram of a double-focusing mass analyzer which is used for resolution of positive ions. The arrows indicate the flow direction of positive ions through the analyzer.

The ions from the source pass through a slit that collimates the beam prior to entering the electric sector. An electric potential is applied between the two plates of the spherical electric sector such that the entering ions are repelled by the outer electrode and attracted toward the inner electrode. If positive ions are studied, the outer plate is positive relative to the inner plate. The radius of curvature r_e of the ions as they pass through the electric sector is given by

$$r_e = \frac{2E}{V} \tag{21-23}$$

where V is the potential difference between the two plates and E is the accelerating potential imparted to the ions by the accelerating grids of the ionic source.

Only those ions which have nearly identical kinetic energies are focused on slit 2 between the electric and magnetic sectors. After passing through slit 2 the ions enter the magnetic sector. The equations that are obeyed in the magnetic sector are identical to those described in the section devoted to magnetic analyzers. An ion which strikes the detector must simultaneously obey Eqs. (21-20) and (21-23).

Several variations of the double-focusing analyzer are in use. In *reverse-geometry* instruments the positions of the electric and magnetic sectors shown in Fig. 21-9 are reversed. That geometry is often used to study metastable ions. A *triple-sector* (*triple-focusing*) mass analyzer contains a second electric sector after the magnetic sector in the double-focusing analyzer; i.e., the magnetic sector is sandwiched between two electric sectors. The triple-sector mass analyzer is also used to study metastable ions.

Time-of-flight analyzers The operation of a time-of-flight TOF analyzer is relatively easy to understand. Ions that exit from the ionic source have essentially identical kinetic energies. Because the masses of the ions differ, the velocities must differ. If a group of ions that have different masses simultaneously enter the mass analyzer, the heavier ions have a velocity that is less than that of the lighter ions. Consequently, the time that is required for an ion to travel a fixed distance in the analyzer varies with the mass of the ion.

A diagram of a time-of-flight mass spectrometer is shown in Fig. 21-10. The ions from the source enter the evacuated flight tube in carefully timed pulses. The flight tube must be evacuated to a pressure at which the ions are unlikely to strike another body between the last accelerating plate and the detector. Lighter ions strike the detector before heavier ions. The electronics of the apparatus must be fast because the time between entrance of the ionic pulse and the arrival of the ions at the detector is short. The separation in time of different ions which strike the detector is generally less than a microsecond. The readout device normally is a computer-controlled device.

The potential between the last accelerating grid and the slit is zero during a measurement. A repelling potential is often placed on the slit after an ion of interest has passed and before ions of the next m/z enter. That allows a detector and readout device that respond relatively slowly to measure the intensity of a single ion. Ions are ejected from the source in pulses at a frequency of about 10 kHz. When MALDI ionic sources are used, the laser pulse corresponds to the rate of ion ejection into the TOF mass analyzer.

The kinetic energy of the ions as they enter the flight tube is given by Eq. (21-14). Solution of Eq. (21-14) for m/z yields

$$\frac{m}{z} = \frac{2E}{v^2} \tag{21-24}$$

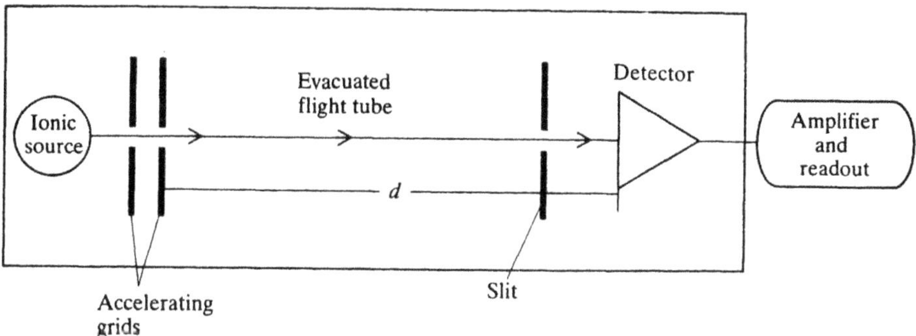

Figure 21-10 Diagram of a time-of-flight mass spectrometer.

The velocity which a particular ion has is equal to the distance d between the last accelerating grid and the detector divided by the time t of flight between the two:

$$v = \frac{d}{t} \tag{21-25}$$

Substitution for v in Eq. (21-24) gives

$$\frac{m}{z} = \frac{2Et^2}{d^2} \tag{21-26}$$

For a mass spectrometer that has a fixed accelerating potential and a fixed distance through the mass analyzer, the simplified version of Eq. (21-26) can be shown as follows:

$$\frac{m}{z} = kt^2 \tag{21-27}$$

where k is a constant ($k = 2E/d^2$). From Eqs. (21-26) and (21-27) it is apparent that m/z of the ions that strike the detector varies with the square of the time required to traverse the distance between the last accelerating grid and the detector.

Sample problem 21-3 A time-of-flight mass spectrometer has a flight path of 100.0 cm and uses an accelerating potential of 2500 V. Calculate the time required for ionic fragments with m/z 100 and 101 to strike the detector. Calculate the difference in time of arrival of the two ions at the detector.

SOLUTION The SI values for the m/z 100 and 101 fragments can be calculated as in Sample problem 21-2.

m/z 100:

$$100 \text{ amu/elem. chg.} \times 1.66 \times 10^{-27} \text{ kg/amu} \times 1 \text{ elem. chg.}/1.60 \times 10^{-19} \text{ C} = 1.04 \times 10^{-6} \text{ kg/C}$$

m/z 101:

$$101 \times 1.66 \times 10^{-27} \times 1/1.60 \times 10^{-19} = 1.05 \times 10^{-6} \text{ kg/C}$$

Substitution into Eq. (21-26) for the two ions yields

m/z 100:

$$1.04 \times 10^{-6} \text{kg / C} = \frac{2(2500 \text{ V})t^2}{(1.000 \text{ m})^2}$$

$$t^2 = 2.08 \times 10^{-10} \text{ kg} \cdot \text{m}^2 / \text{C} \cdot \text{V}$$

$$(1 \text{ C} = 1 \text{ A} \cdot \text{s}; 1 \text{ V} = 1 \text{ kg} \cdot \text{m}^2 / \text{s}^3 \cdot \text{A})$$

$$t^2 = 2.08 \times 10^{-10} \text{ s}^2$$

$$t = 1.44 \times 10^{-5} \text{ s} = 14.4 \ \mu\text{s}$$

m/z 101:

$$1.05 \times 10^{-6} = \frac{2(2500)t^2}{(1.000)^2}$$

$$t = 14.5 \ \mu\text{s}$$

The difference in time between arrival of the two ions is 14.5 – 14.4 = 0.1 μs = 100 ns.

It is apparent that for an ion with a particular m/z, the time required to reach the detector increases with the distance through the analyzer. It follows that the resolving power in TOF analyzers can be increased by increasing the flight path in the analyzers. Since restrictions on space and vacuum limit the dimensions of TOF analyzers, longer flight paths often are achieved by using multiple pass mass analyzers. In multiple pass analyzers the ions are forced to take nonlinear paths through the analyzer by means of applied electric fields.

A reflectron TOF analyzer applies an electrostatic field between two electrodes to the incident beam from the source in a manner similar to that which was used to apply the field in the electric sector of a double-focusing mass analyzer. The applied potential causes the ions to assume a curved, rather than linear, path through the flight tube. Near the end of the flight tube the ions enter a *reflectron* which consists of a series of electrodes of potentials which are sufficient to cause the ion beam to be returned ("reflected") to the flight tube while simultaneously displacing it in space. The detector is placed behind a slit in the path of the reflected beam.

Some multiple pass TOF analyzers use toroidal reflectron sectors to force the ions to pass multiple times through the flight tube. Total distances traveled by the ions can approach 700 m after more than 500 cycles through the flight tube. Mass resolving powers achieved with toroidal reflectrons have surpassed 350,000 when measured at m/z 28. A description of multiple-pass TOF mass analyzers as well as a description of other high resolving power instruments can be found in Reference 13.

Quadrupolar analyzer A quadrupolar mass analyzer consists of four (hence the name) parallel metallic rods arranged as shown in Fig. 21-11. In most instruments the rods are cylindrical, although rods that have a hyperbolic shape are also used. The accelerated ionic beam from the ionic source passes through a collimating hole that is aligned with the space between the four rods.

Diagonally opposite rods are electrically connected i.e., rod 1 is connected to rod 3 and rod 2 is connected to rod 4, as illustrated in Fig. 21-11. A dc potential difference and a radiofrequency ac potential are simultaneously applied between the two groups of connected rods. Positive ions that enter the space between the electrodes are repelled by the rods that are momentarily positively charged and attracted to the rods that are negatively charged. Because the relative charge on the sets of rods is continuously changing, the ions follow an irregular oscillating

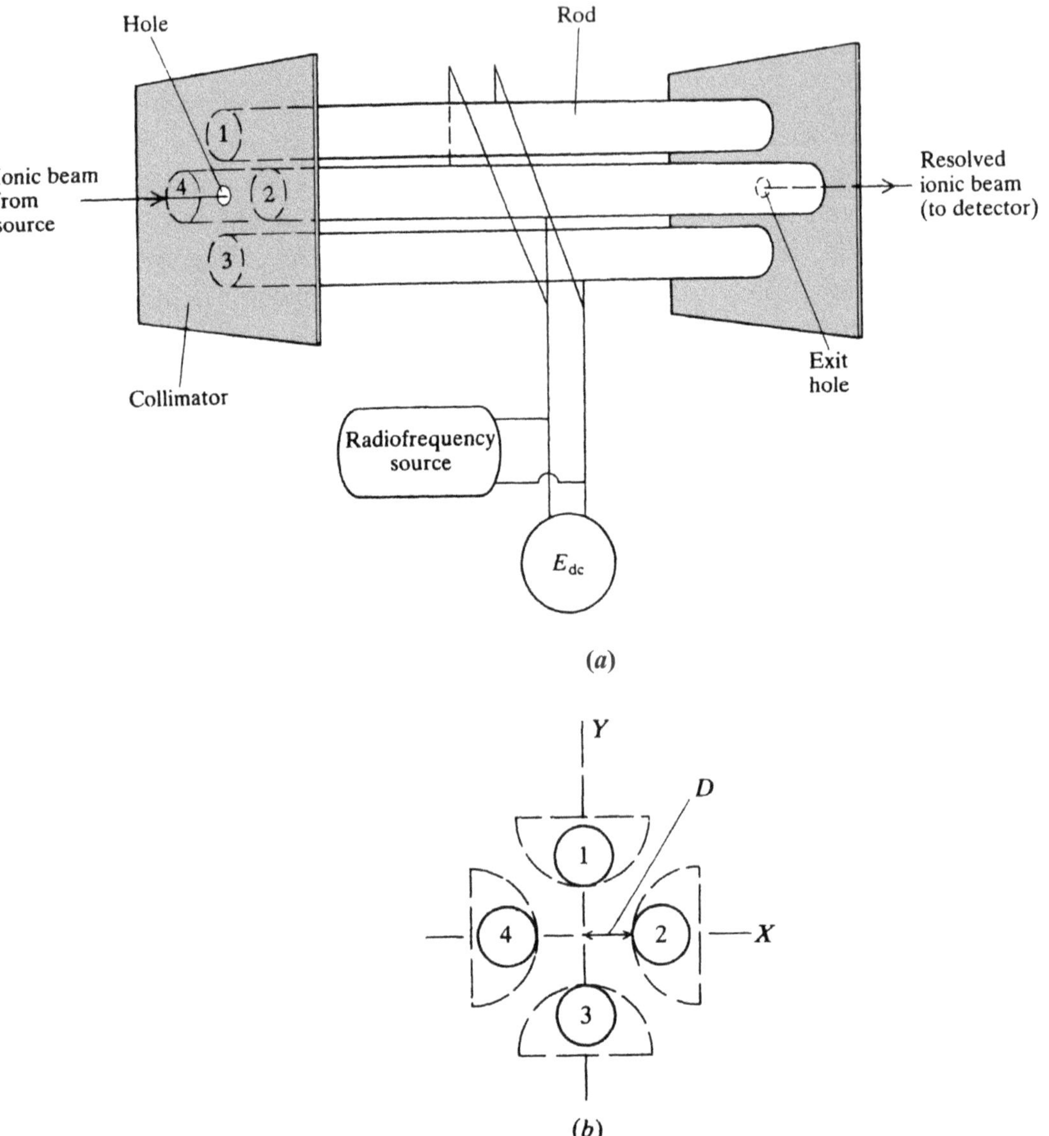

Figure 21-11 (*a*) Diagram of a quadrupolar mass analyzer; (*b*) end view of a quadrupolar mass analyzer that uses cylindrical rods. The dashed lines in (*b*) indicate the shape of hyperbolic rods that are used in some instruments.

path between the rods. Only those ions that can pass through the space between the rods strike the exit hole and are measured by the detector. Other ions strike one of the roads and are not detected.

Because the electric fields are perpendicularly applied to the direction of travel of the ions, the fields have no effect upon ionic motion that is parallel to the lengths of the rods. If the rods are hyperbolically shaped (dashed lines, Fig. 21-11*b*), the potential E that is experienced by an ion at any point between the rods at time t after entering the analyzer is the sum of the applied ac and dc potentials:

$$E = \frac{(x^2 - y^2)(E_{dc} + E_{ac} \cos \omega t)}{D^2} \tag{21-28}$$

In the equation x and y are the coordinates of the ion in the plane that is perpendicular to the rods (Fig. 21-11*b*); D is the distance between the inner edge of the rod and the center of the space between the rods; ω is the frequency of the applied alternating potential (radians per second); E_{dC} is the applied dc potential; and E_{ac} is the amplitude of the applied ac potential.

Equation (21-28) strictly applies only to the case in which hyperbolic rods are used; however, it is approximately true when cylindrical rods are used. Because hyperbolic rods are more difficult to prepare than cylindrical rods, instruments that contain hyperbolic rods are generally more expensive than those that contain cylindrical rods. The force exerted on a charged particle by an electric field is the product of the potential and the charge divided by the distance through which the particle moved while the potential was applied. The forces in the x and y directions (F_x and F_y) exerted on a particle with a charge z are given by

$$F_x = -z\frac{dE}{dx} \tag{21-29}$$

$$F_y = -z\frac{dE}{dy} \tag{21-30}$$

Substitution into Eqs. (21-29) and (21-30) of the value in Eq. (21-28) and differentiation leads to

$$F_x = -z\frac{(E_{dc} + E_{ac} \cos \omega t)2x}{D^2} \tag{21-31}$$

$$F_y = -z\frac{(E_{dc} + E_{ac} \cos \omega t)2y}{D^2} \tag{21-32}$$

Force is defined as the product of mass and acceleration. Acceleration is the second derivative of distance with time. By substituting those definitions into Eqs. (21-31) and (21-32) and rearranging, the following equations are obtained:

$$F_x = ma_x = m\frac{d^2x}{dt^2}$$

$$F_y = ma_y = m\frac{d^2y}{dt^2}$$

$$\frac{d^2x}{dt^2} + \left(\frac{z}{m}\right)\frac{2x(E_{dc} + E_{ac}\cos\omega t)}{D^2} = 0 \qquad (21\text{-}33)$$

$$\frac{d^2y}{dt^2} + \left(\frac{z}{m}\right)\frac{2y(E_{dc} + E_{ac}\cos\omega t)}{D^2} = 0 \qquad (21\text{-}34)$$

Equations (21-33) and (21-34) describe the motion of an ion through the analyzer. It is apparent from the equations that the motion is a function of z/m of the ions. A spectrum can be obtained either by maintaining E_{dc}/E_{ac} constant while varying ω or by varying E_{dc}/E_{ac} while holding ω fixed. While using the first method, the best resolution is achieved when E_{dc}/E_{ac} is slightly less than 0.168.

The mass resolving power and m/z range that are available with a quadrupolar analyzer are about the same as that of a magnetic analyzer. The maximum obtainable resolving power is about 2000 but the typical resolving power is about 700. The upper m/z that can be detected is about 1200. The major advantages of the quadrupolar analyzer are its ability to rapidly obtain spectra and its relatively inexpensive price. An entire mass spectrum can be obtained in a few milliseconds. That makes the analyzer ideal for use as a detector with a chromatograph. The mass spectrum of the separated components are obtained as they are eluted from the column. Quadrupolar analyzers are relatively inexpensive because they do not require an expensive magnet. Mass analyzers that contain more or fewer than four rods but which operate on the same principle as the quadrupolar analyzer are available but are used less often. Triple quadrupolar mass analyzers recently have been used in a few instruments.

Ion cyclotron resonance analyzer An EI source that has no exit slit is used with the ion cyclotron resonance ICR analyzer. Separation of the ions is accomplished by simultaneously applying a magnetic flux density B perpendicular to the direction in which the ions leave the source and an oscillating electric field perpendicular to both the direction of flow of the ions and the magnetic field as shown in Fig. 21-12. The *drift velocity* v_d of the ions in the analyzer is the velocity exhibited by the ions along the length of the analyzer. It is given by

$$v_d = \frac{E}{B} \qquad (21\text{-}35)$$

where E is the static electric field (the *drift potential*) applied to the ions in the ionic source.

An ion with a mass m in a magnetic field that is perpendicular to the direction of drift experiences a force F given by

$$F = Bzv \tag{21-36}$$

which is perpendicular to the field direction and to the direction of drift. The force causes the ion to move in a circular path perpendicular to the direction of drift at a velocity v. Both positive and negative ions assume circular paths although the direction of motion of negative ions is opposite to that of positive ions. The mass analyzer is termed an ion *cyclotron* resonance analyzer because the ions assume a cycloidal path through the device.

By setting the force from Eq. (21-36) equal to the centrifugal force of the ion in its circular path, the following equation is obtained:

$$\frac{mv^2}{r} = Bzv \tag{21-37}$$

The frequency ω_c with which the ion completes a circle is v/r (radians per second) and can be obtained by rearranging Eq. (21-37) to yield

$$\omega_c = \frac{v}{r} = \frac{Bz}{m} \tag{21-38}$$

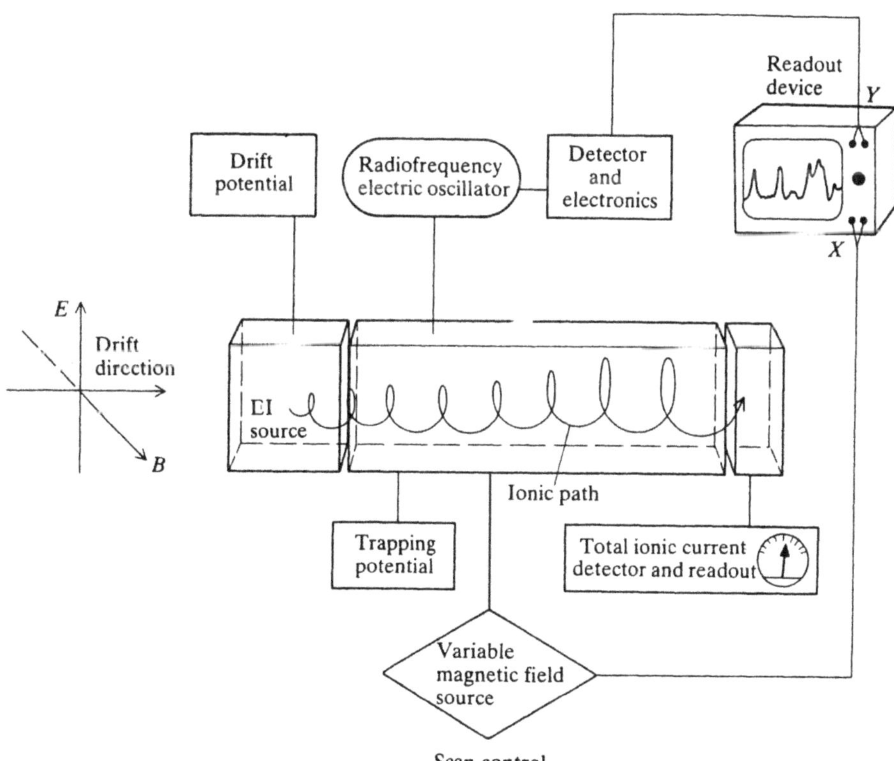

Figure 21-12 Diagram of an ion cyclotron resonance mass spectrometer.

The ions are prevented from striking the walls of the analyzer by application of a *trapping potential* to the walls.

From Eq. (21-38) it is apparent that the frequency of the oscillation of an ion in a fixed magnetic field is inversely proportional to m/z for the ion. The frequency of the oscillation is independent of the velocity of the ion in its cycloidal path; however, the radius of the circle traced by the ion increases as the velocity increases [Eq. (21-38)]. Energy from an oscillating electric field can be absorbed by the ions when the frequency ω of oscillation of the magnetic field equals the frequency of oscillation of the ion in its circular path. The absorption that occurs when the two frequencies are identical is *resonant* absorption or *resonance*.

When resonance occurs, the kinetic energy, circular velocity, and radius of motion of the ion increase. The increasing radius of motion of the ion is illustrated in Fig. 21-13. The electric frequency at which resonance occurs can be determined by monitoring the power required by the electric field oscillator.

When resonance occurs, energy is transferred to ions of a particular m/z and the power used by the oscillator increases. Normally the magnetic field in the icr analyzer is scanned and the frequency of the electric field is held constant at a radiofrequency. The detector monitors the amount of the rf radiation absorbed by the ions at resonance. The detector is similar to that used for nmr spectroscopy as described in Chapter 16. Normally the total ionic current is also monitored as shown in Figure 21-12. From Eq. (21-38) it is apparent that the magnetic field strength at the resonant frequency for an ion is directly proportional to m/z for the ion.

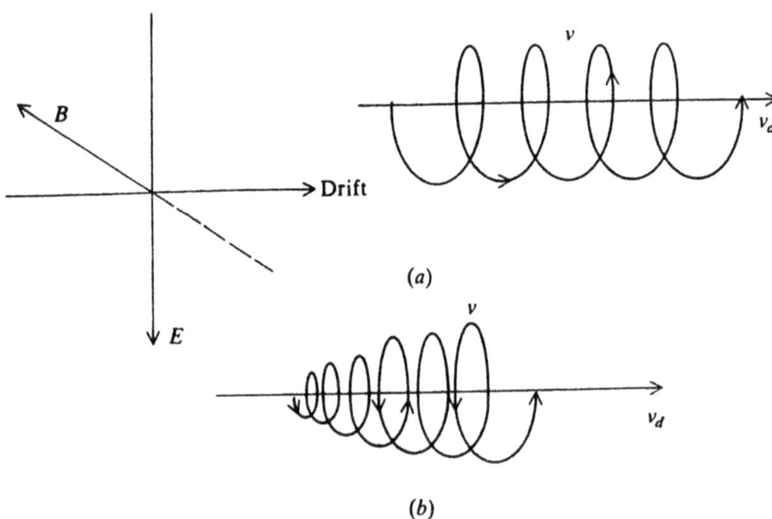

Figure 21-13 The path of ionic particles under the influence of a magnetic field of strength B in an ICR analyzer in (a) the absence of and (b) during application of an electric field E that is alternating at the resonant frequency.

Because the ions take a cycloidal path and because the drift velocity is relatively small, a typical ion spends between 1 and 10 ms in the analyzer rather than the approximately 1 μs spent in a magnetic analyzer. In some cases ions can be trapped in the analyzer for periods of up to about 1 s. The analyzer has a length of several centimeters. The relatively long time spent in the analyzer by the ions considerably increases the probability that ion-molecule collisions and reactions will occur. That is a major advantage of the ICR analyzer. It can be used to study gas-phase chemical reactions and the structure of neutral molecules by examining the ionic reaction products. Fourier transform methods often are used with the analyzer, in which case the instrument is a Fourier transform ion cyclotron resonance FTICR mass spectrometer. Normally the ionic masses that can be studied with ICR mass spectrometry are less than about 400 amu. ICR analyzers can take cylindrical and rectilinear shapes as well as the cubic shape shown in Figure 21-12.

The frequency of the applied electric field can be maintained at one value while the frequency at which absorption is observed is fixed at a second value. The resulting double-resonant studies are similar to those used for nmr spectrometry and can be used to study ion-molecule reactions. Normally irradiation is done at the frequency at which the reactant ion absorbs, and absorption is measured at the resonant frequency of a product ion.

Orbitrap mass analyzers An *orbitrap* mass analyzer (Reference14) utilizes two, concentric coaxial electrodes to capture ions in orbits between the electrodes. Each electrode is shaped so that its diameter varies with distance along its axis in a manner that changes with the radius of the rotating ions between the electrodes. The diameter is greatest at the center of the electrode and smallest at the two ends.

Pulses of ions enter the space between the two electrodes and rotate around the smaller, central electrode at axial frequencies that are typically between 50 and 150 kHz. Detection is done in a manner similar to that used in ICR analyzers.

Detectors

The common detectors that are used in mass spectrometers are the *Faraday cup*, the electron multiplier, and the scintillation counter. The Faraday cup is the simplest of the detectors. It consists of a metallic cup that is held at a potential relative to the remainder of the spectrometer that allows ions to be captured. Often the cup is maintained at virtual ground potential. Ions exiting the mass analyzer pass through a collimating slit and, in some instruments, through one or more suppressor grids or electrodes prior to striking the cup. The suppressor electrode is maintained at a potential that causes any secondary ions that are emitted during bombardment of the cup by sample ions to be returned to the cup.

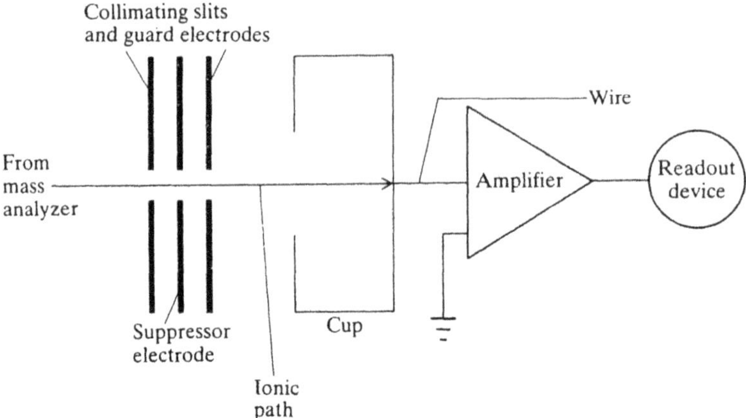

Figure 21-14 Diagram of a Faraday cup detector.

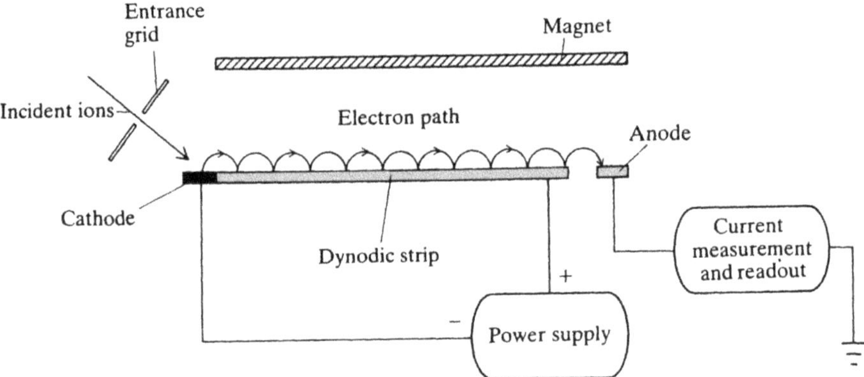

Figure 21-15 Diagram of a dynodic strip electron multiplier.

The current caused by ionic bombardment of the cup is converted to a potential, amplified electronically and displayed on the readout device. The Faraday cup has the advantages of being simple and dependable. With proper amplification it can be used to monitor ionic currents of about 10^{-15}A or greater. The detector is shown in Fig. 21-14.

The most popular of the several forms of detectors are the electron multipliers. Several types of electron multipliers are used. Early versions of the electron multiplier resembled photomultiplier tubes (see Fig. 6-15). Incident ions (rather than photons) from the mass analyzer strike the cathode. Each ion induces the emission of several electrons. The electrons are attracted to a dynode. Each electron that strikes a dynode induces the emission of several more electrons which subsequently are attracted to a second dynode. The current is increased at each dynode in the device and eventually is measured in an external circuit that is electrically connected to the anode. Typically electron multipliers of that type contain between 10 and 20 dynodes and amplify the incident ionic current by between five and seven orders of magnitude. The dynodes are usually constructed from 2% Be-Cu, Mg-Ag, or brass alloys. Sometimes a magnetic field is used to help direct the electrons from one dynode to the next and to protect the electrons from the effects of external magnetic fields.

Another type of electron multiplier uses a *dynodic strip* in place of separate dynodes (Fig. 21-15). A potential difference is maintained across the strip by a regulated power supply. The potential drop across the strip is typically between 1 and 2 kV. Sometimes the strip is placed on a glass support.

Ions from the analyzer bombard the cathode at one end of the strip inducing emission of electrons. A magnetic field or crossed magnetic and electric fields are used to direct the emitted electrons in a cycloidal path back toward the strip. Upon striking the strip, the current is further amplified by the emission of several electrons for each impinging electron. The process is repeated many times as the electrons travel along the length of the strip toward the anode. After reaching the end of the strip the electrons strike the anode. The resulting current flow in an external circuit is monitored and displayed on the readout device.

A channel electron multiplier is similar to a dynodic strip electron multiplier. The dynode consists of a small-diameter tube. The interior of the tube is coated with an electron-emitting substance and functions as the amplifying device. Incident ions enter the tube from one end and strike an interior wall. The electrons that are emitted when the ions strike the wall are accelerated down the length of the tube by the potential difference maintained across the length of the tube. Each time the electrons strike an interior wall further amplification occurs. Eventually the electrons are captured by the anode which is located at the end of the tube. The current flow through the anode is measured as in other types of electron multipliers. A diagram of a single-channel electron multiplier is shown in Fig. 21-16.

Often an array of channel electron multipliers are used. Each channel is close packed parallel to the other channels so that the assembly looks like a container filled with straws. The length of each channel is about 1 mm and the internal diameter of each channel is typically between 10 and 30 μm. A channel electron multiplier array can be used to simultaneously monitor ions of all masses in the sample because ions with different masses strike different channels. The array eliminates the need to perform an *m/z* scan in the mass analyzer. The output from each channel can either be rapidly and sequentially monitored or all of the outputs can be monitored simultaneously. If further amplification is required, a second channel electron multiplier array can be serially attached to the end of the first array.

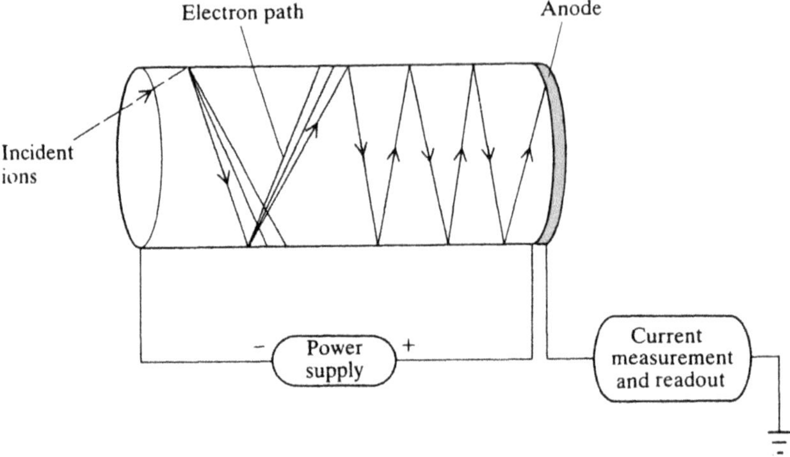

Figure 21-16 Diagram of a single channel in a channel electron multiplier

Photographic plates can be used to simultaneously measure ions throughout a broad *m/z* range as they exit the mass analyzer. The plates offer the advantages of high resolution and a low detection limit when used with long exposure times. The major disadvantages are the delay and the effort that is required to develop the plates. Photographic plates are primarily used for qualitative mass spectral studies because quantitative analyses require measurements of the density of the developed spectral lines. Those measurements are not as accurate as intensity measurements obtained by using other detectors. Photographic plate detection is rarely used in modern instruments and is mentioned only because of its historical importance.

The last common type of detector is the scintillation counter. The detector uses a photomultiplier tube to count the number of radiative emissions that are induced by bombardment of a phosphor screen with ions from the mass analyzer. The output from the PM tube is monitored and displayed by the readout device. The devices are similar to those described in Chapter 18 for the detection of x-rays.

Readout Devices

The readout devices that can be used for mass spectrometry, include strip-chart recorders, mirror galvanometer recorders, oscilloscopes, and various computer-controlled devices. Strip-chart recorders can only be used with those instruments that scan slowly. Mirror galvanometer recorders use a beam of light in place of the pen in strip-chart recorders. The light deflects across light-sensitive recorder paper leaving a visible trace. Those recorders can be used for more rapid scans, but hardly are used in instruments constructed after 1980. Computer-controlled data acquisition and readout devices are used most often.

CHEMICAL ANALYSIS

Mass spectrometry can be used for qualitative and quantitative chemical analysis. The mass that corresponds to the molecular ion is the molecular weight of the compound. Accurate molecular weight determinations with high-resolution spectrometers are particularly useful for qualitative analysis.

By changing the electron accelerating potential that is used to produce ions in the source, it is possible to measure the *ionization potential* and *appearance potential* of the sample. The appearance potential is the minimum energy that is required to produce an ion and any accompanying neutral fragments from the sample. The ionization potential is the minimum energy that is required to remove a single electron from a neutral atom or molecule. The ionization potential is a special case of the appearance potential. The molecular weight, the ionization potential, and the appearance potential can be used to help identify a molecule.

An examination of the fragmentation patterns that occur during ionization is generally most useful for qualitative analysis. Because the series of mass spectral peaks that are observed for a particular compound while using fixed experimental parameters is often unique, qualitative analysis can be performed by comparing the sample spectrum to spectra of known compounds. Alternatively, the m/z at which the major peaks in a spectrum are observed can sometimes be used to deduce the identity of a compound. In either case analytical information from other techniques (ir, nmr, uv spectra, etc.) should be used to aid in the analysis whenever that information is available.

The procedure that is used when performing a mass spectral qualitative analysis of a compound whose spectrum is not on file involves at least three steps. First, the molecular weight of the compound is determined from the spectral position of the molecular ion. Next, the spectral fragmentation pattern is checked for the presence of elements with unusual isotopic abundances. Finally, the fragmentation pattern is used to deduce the stable ionic fragments in the molecule. The combined information from the three steps is used to postulate the identity of the sample. If the pure postulated compound is available a mass spectrum is obtained and compared to that of the sample.

Quantitative analysis can be accomplished by comparing the intensity of a particular mass spectral peak, which is unique to the analyte, with peak intensities corresponding to a series of standards of varying concentration. Simultaneous determinations of several components in a mixture can be accomplished, as in spectrophotometry, by using the simultaneous equations method. A peak corresponding to each component which is to be assayed is chosen. A set of simultaneous equations is established at each chosen m/z and the equations are solved simultaneously. The more common techniques that can be used for chemical analysis with mass spectrometry are described in the following sections.

Appearance Potential

Appearance and ionization potentials are determined experimentally. Generally an *ionization efficiency curve* is prepared by plotting the observed ionic current from an EI source as a function of the energy (between G1 and G2 in Fig. 21-2) of the electron beam in the source. A typical ionization efficiency curve is shown in Fig. 21-17. The efficiency of the ionization increases with the accelerating potential of the electron beam as shown by the increasing ionic current in the figure. At electron energies above about 50 or 60 eV, ionization of organic compounds is nearly complete and the beam current ceases to rapidly increase with increasing electron energy. That is typical of ionization efficiency curves and is one reason for routinely using 70-eV beam energies while assaying organic species.

At low electron energies nearly all of the kinetic energy of the impinging electron is used to ionize the sample molecule or atom. Consequently, the appearance or ionization potential of a sample equals the electron energy at which the ionic current becomes a nonzero value. Because the ionization efficiency curve is not linear at low ionic currents, the energy at which the ionic current becomes greater than zero is difficult to locate.

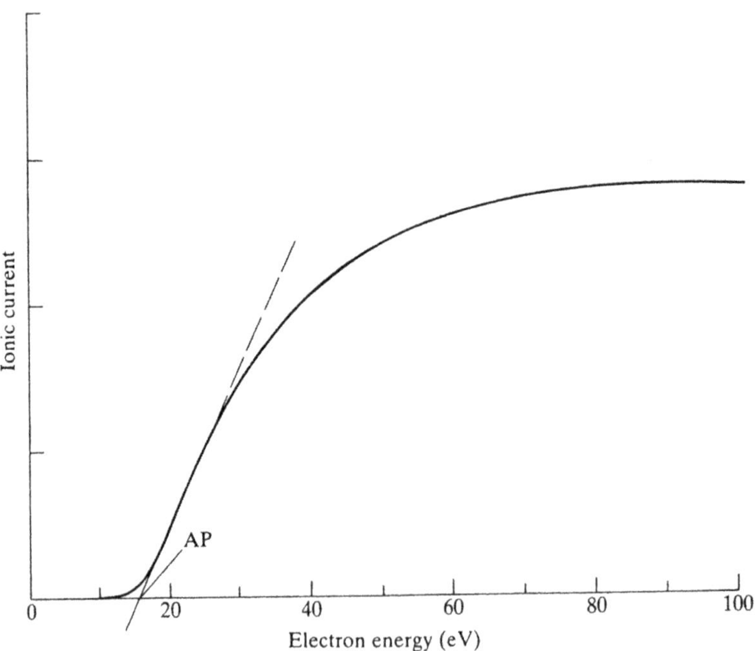

Figure 21-17 An ionization efficiency curve for the formation of O_2^+ ions with an EI source. The dashed line on the linear rising portion of the curve is used to estimate the upper limit of the appearance potential (AP).

The simplest way to estimate the appearance potential (AP) is to extrapolate the linear portion of the curve, as shown in Fig. 21-17, to the electron-energy axis. The energy at the intersection of the extrapolated line and the axis is assumed to be the appearance potential. The linear extrapolation often yields an appearance potential that is too high, because ionization (and the ionic current) actually occurs in the curve at lower electron energies. Nevertheless, the method can be used to determine the upper limit of the appearance potential. Other more complicated methods sometimes yield more reliable results. Several methods for obtaining appearance and ionization potentials from mass spectrometric data are described in Reference 2 (chap. 8).

Molecular Weight

A molecular weight determination can be particularly helpful during a qualitative analysis. Molecular weights can be directly measured with mass spectrometry. Often an EI source, which is operated at an electron energy of 70 eV, is used for the measurement. The mass spectral peak that appears at the greatest *m/z,* which is not an isotopic peak, normally corresponds to the molecular weight of the compound. As an example, for the compound whose mass spectrum is shown in Fig. 21-18, the molecular weight is 98. Unfortunately not all compounds have a molecular ion when bombarded with electrons at 70 eV. A well-known example is carbon tetrachloride in which the ion of greatest *m/z* corresponds to CCl_3^+ rather than CCl_4^{+}. At an electron energy of 70 eV, most organic compounds are highly fragmented and the peak that is caused by the molecular ion can be small relative to the peaks of more stable fragments.

For those reasons it is often advantageous either to use an EI source at a lower electron energy or to use a more gentle ionic source for molecular weight determinations. As the electron energy in the EI source is decreased, the extent of sample fragmentation decreases and the relative abundance (and peak height) of the molecular ion increases. A CI source can be used to measure molecular weights. In the latter case the ion with the largest *m/z* is the product of a chemical reaction between the sample and the reagent gas, as described earlier in the chapter. The molecular weight is determined by subtracting the appropriate mass from the observed *m/z*. For example, if methane is used as the reagent gas, the product of the chemical reaction [Eqs. (21-2) and (21-4)] normally has a molecular weight that is one unit greater than that of the sample. Consequently, the molecular weight of the sample can be calculated by subtracting one unit from the largest nonisotopic *m/z* observed in the spectrum. As an example, the CI mass spectrum of *trans*-2-ethylidene-cycloheptanone is shown in Fig. 21-19. The molecular ion appears at *m/z* 139 and the molecular weight of the analyte is 139 – 1 = 138.

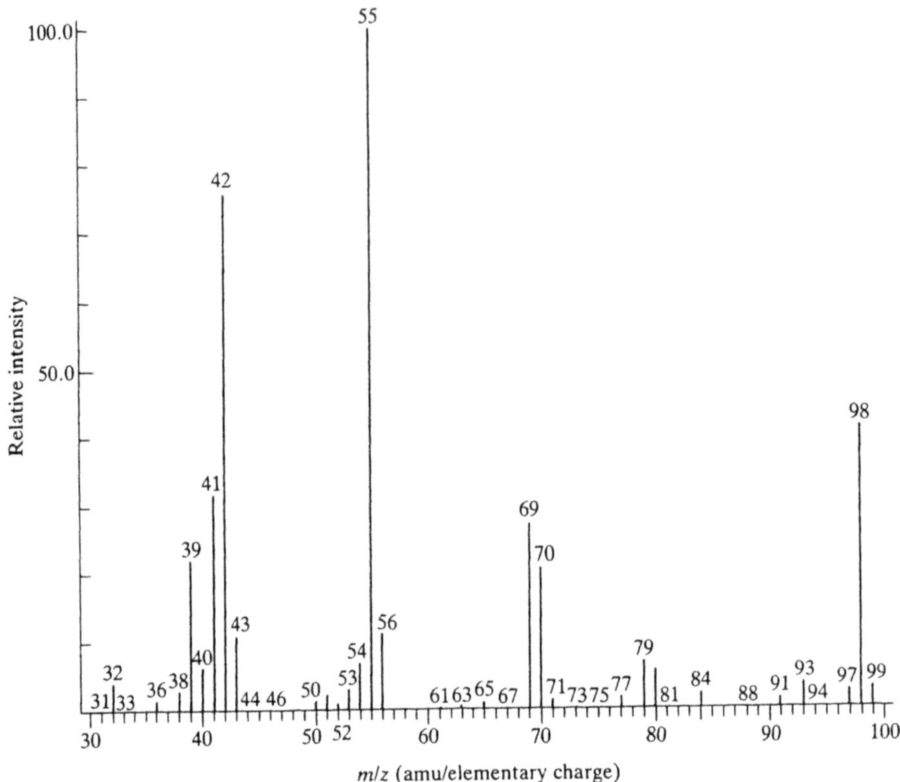

Figure 21-18 The mass spectrum of cyclohexanone. (*Courtesy of Dr. August A. Gallo, Department of Chemistry, University of Louisiana at Lafayette.)*

If a mass spectrometer is available which is capable of sufficient resolution to allow the molecular weight to be determined to the nearest 0.001 or 0.0001 amu, the measured molecular weight can be used to obtain the molecular formula of an organic sample. Compounds rarely have identical molecular weights when measured with the high accuracy achievable with high-resolution mass spectrometers. Accurate compilations of molecular weights and the corresponding molecular formulas can be found in References 15 and 16. Once the formula of an organic compound has been determined, the number of rings and double bonds in the compound can be calculated using

$$\text{Rings + double bonds} = \frac{2w - x + y + 2}{2} \tag{21-39}$$

where w, x, and y are the subscripts in the formula $C_w H_x N_y O_Z$. A triple bond counts as two double bonds or rings.

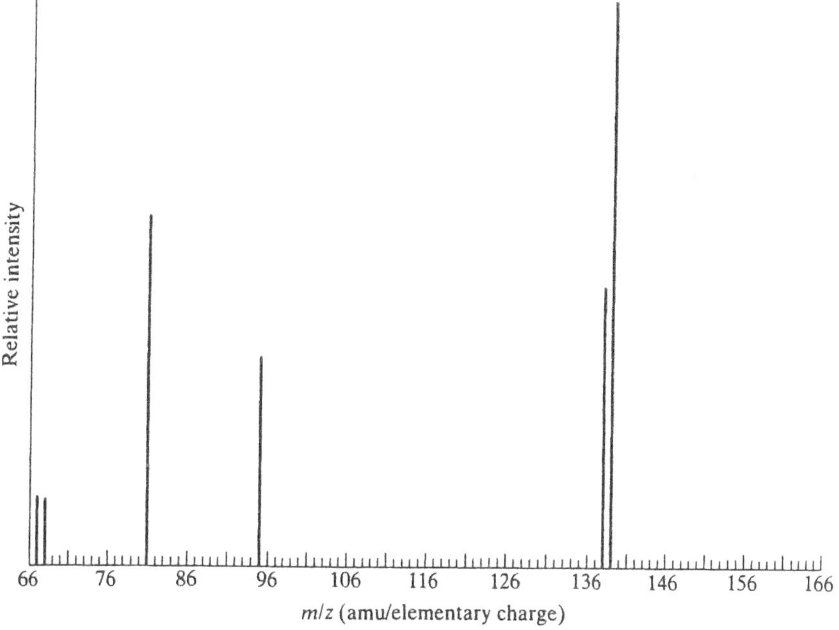

Figure 21-19 The CI mass spectrum of trans-2-ethylidene-cycloheptanone.

The reagent gas was methane. (*Courtesy of Dr. August A. Gallo, Department of Chemistry, University of Louisiana at Lafayette.*)

Mass Spectra

An examination of the ionic fragmentation pattern that is observed in a mass spectrum can be useful for qualitative analysis. Generally an EI source is used with a 70 eV electron beam energy to break the sample molecule into a series of ionic fragments. The mass spectrum is a plot of the intensity of the fragments as a function of *m/z*. The more stable fragments yield the more intense spectral peaks. An examination of the *m/z* of the relatively intense peaks, therefore, provides information that can be used to deduce the identity of the fragments, and consequently the identity of the sample.

Because the fragmentation patterns of dissimilar molecules considerably differ from each other, the identity of a sample molecule can be determined by comparing the mass spectrum of the sample with spectra of known compounds. Most modern mass spectrometers are computer-controlled devices that automatically compare the recorded spectrum with the spectra of known compounds in a reference library. Generally the reference library is stored on a hard drive of the computer. The *m/z* values and relative intensities of the major peaks in the spectrum of the sample are compared with those of the compounds in the reference library. The computer lists the compounds in the library that have spectra that most nearly match that of the sample. The sample spectrum must be recorded by using experimental conditions that are nearly identical to those used to record the spectrum of the reference compound in the library.

If the number of possible sample compounds is limited, the comparison can be made visually, as is often done with infrared spectra. In that case it is useful to create a library of mass spectra of reference compounds that have been recorded by using the same instrument and the same instrumental parameters that are used to record the spectrum of the sample. The sample and reference spectra should be identical or nearly identical in order for the two spectra to be assumed to be attributable to the same compound.

In some cases a computerized library search cannot be performed or does not yield a match between the spectrum of the sample and that of the reference compounds. In that case the *m/z* of some or all of the major peaks and the change in *m/z* between major peaks sometimes can be used to deduce the chemical structure of the sample. A brief description of the manner in which that can be accomplished is described later. More information can be found in References 3, 4, 5, and 17.

Isotopic Abundances

Several elements have isotopic abundances that are unusual. Normally it is easy to recognize the presence of one of those elements in the sample. The isotopic abundances of the stable elements are listed in Table 21-1.

Although Table 21-1 is extensive, it should be emphasized that the more common sources are incapable of ionizing most salts. Many salts cannot be examined using an EI source because the vapor pressures of the salts are insufficient. Consequently, spectra of metallic elements are rarely observed when a common source is used. An ICP ionic source is a good choice to use during the analysis of those analytes. The spectra of some organometallic compounds can be observed with an EI source.

An organic compound which contains an element that has a distinctive isotopic abundance can be easily recognized from the mass spectrum of the substance. An example of an element that has a distinctive isotopic abundance is bromine. The 79 and 81 amu isotopes are nearly equally abundant (Table 21-1). Consequently, approximately half of the ionic fragments that contain a single bromine atom contain ^{79}Br while the remainder contain ^{81}Br. The portion of the mass spectrum that is attributable to an ionic fragment which contains a single bromine atom consists of a series of peaks of nearly equal intensity that are spaced two units apart on the *m/z* axis, as shown in Fig. 21-20 and as marked by the asterisks. Any ionic fragment that does not exhibit that distinctive spectral pattern does not contain one atom of bromine.

Table 21-1 The isotopic abundances of the stable elements

Relative abundances of less than 1 percent are not listed except for hydrogen. Unlisted stable elements consist of over 99 percent of a single isotope

Element	Isotopic mass	Relative abundance, %
H	1	99.985
	2	0.015
Li	6	7.5
	7	92.5
B	10	20
	11	80
C	12	98.89
	13	1.11
Ne	20	90.51
	22	9.22
Mg	24	78.99
	25	10.00
	26	11.01
Si	28	92.23
	29	4.67
	30	3.10
S	32	95.0
	34	4.22
Cl	35	75.77
	37	24.23
K	39	93.26
	41	6.73
Ca	40	96.941
	44	2.086
Ti	46	8.0
	47	7.5
	48	73.7
	49	5.5
	50	5.3
Cr	50	4.35
	52	83.79
	53	9.50
	54	2.36
Fe	54	5.8
	56	91.8
	57	2.1
Ni	58	68.27
	60	26.10
	61	1.13
	62	3.59
Cu	63	69.2
	65	30.8
Zn	64	48.6
	66	27.9
	67	4.1
	68	18.8
Ga	69	60
	71	40
Ge	70	20.5
	72	27.4
	73	7.8
	74	36.5
	76	7.8
Se	76	9.0
	77	7.6
	78	23.5
	80	49.8
	82	9.2
Br	79	50.69
	81	49.31
Kr	80	2.25
	82	11.6
	83	11.5
	84	57.0
	86	17.3
Rb	85	72.17
	87	27.83
Sr	86	9.9
	87	7.0
	88	82.6
Zr	90	51.4
	91	11.2
	92	17.1
	94	17.5
	96	2.8
Mo	92	14.8
	94	9.3
	95	15.9
	96	16.7
	97	9.6
	98	24.1
	100	9.6
Ru	96	5.5
	98	1.9
	99	12.7
	100	12.6
	101	17.1
	102	31.6
	104	18.6
Pd	102	1.0
	104	11.0
	105	22.2
	106	27.3
	108	26.7
	110	11.8
Ag	107	51.83
	109	48.17
Cd	106	1.2
	110	12.4
	111	12.8
	112	24.0
	113	12.3

Element	Isotopic mass	Relative abundance, %
	114	28.8
	116	7.6
In	113	4.3
	115	95.7
Sn	112	1.0
	116	14.7
	117	7.7
	118	24.3
	119	8.6
	120	32.4
	122	4.6
	124	5.6
Sb	121	57.3
	123	42.7
Te	122	2.5
	124	4.6
	125	7.0
	126	18.7
	128	31.7
	130	34.5
Xe	128	1.9
	129	26.4
	130	4.1
	131	21.2
	132	26.9
	134	10.4
	136	8.9
Ba	134	2.4
	135	6.6
	136	7.9
	137	11.2
	138	71.7
Ce	140	88.4
	142	11.1
Nd	142	27.2
	143	12.2
	144	23.8
	145	8.3
	146	17.2
	148	5.7
	150	5.6
Sm	144	3.1
	147	15.1
	148	11.3
	149	13.9
	150	7.4
	152	26.6
	154	22.6
Eu	151	47.8
	153	52.2
Gd	154	2.2
	155	14.8
	156	20.5
	157	15.7
	158	24.8

Element	Isotopic mass	Relative abundance, %
	160	21.8
Dy	160	2.34
	161	18.9
	162	25.5
	163	24.9
	164	28.2
Er	164	1.6
	166	33.4
	167	22.9
	168	27.0
	170	15.0
Yb	170	3.1
	171	14.3
	172	21.9
	173	16.2
	174	31.7
	176	12.7
Lu	175	97.4
	176	2.6
Hf	176	5.2
	177	18.5
	178	27.1
	179	13.8
	180	35.2
W	182	26.3
	183	14.3
	184	30.7
	186	28.6
Re	185	37.4
	187	62.6
Os	186	1.6
	187	1.6
	188	13.3
	189	16.1
	190	26.4
	192	41.0
Ir	191	37.3
	193	62.7
Pt	194	32.9
	195	33.8
	196	25.3
	198	7.2
Hg	198	10.1
	199	16.9
	200	23.1
	201	13.2
	202	29.7
	204	6.8
Tl	203	29.5
	205	70.5
Pb	204	1.4
	206	24.1
	207	22.1
	208	52.4

Similarly, a compound that contains a single atom of chlorine has a mass spectrum consisting of a series of two peaks with relative intensities in the ratio of 3 to 1. The peaks are separated by two units on the *m/z* axis. Compounds that contain a single sulfur atom (Fig. 21-21) have fragmentation patterns that consist of peaks with a ratio of relative intensities of 19 to 1 separated by two units on the *m/z* axis. Organic compounds that contain B or Si can similarly be recognized by the characteristic isotopic pattern in the mass spectra of the compounds. The relative isotopic abundances of C and H are such that the major isotopic peaks are separated by one unit from peaks with appreciably lower intensities.

If an ionic fragment contains more than one atom of an element that has more than one isotope, the relative heights of the peaks are determined by the probability of finding the various combinations of isotopes in the ionic fragment. As an example, if a particular ionic fragment contains two atoms of

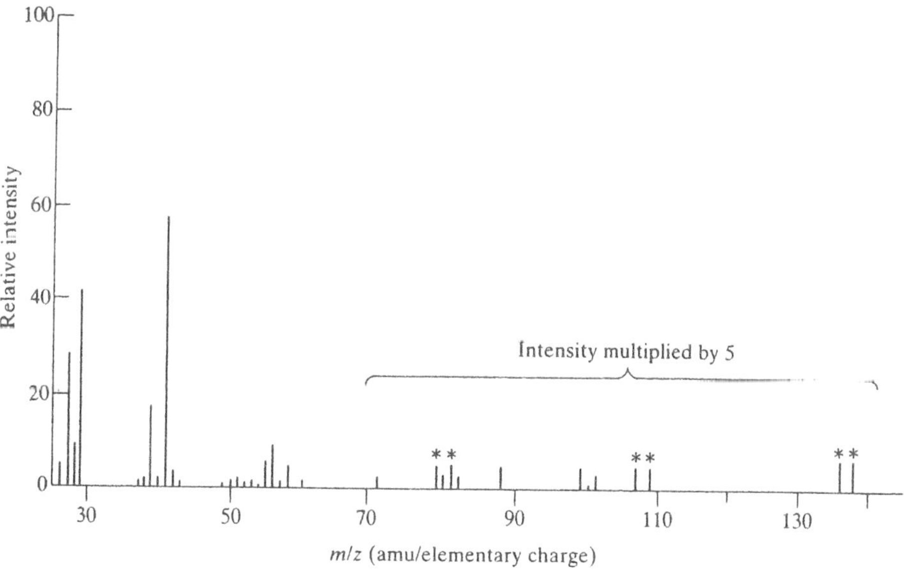

Figure 21-20 The mass spectrum of 2-bromobutane. The bromine isotopic peaks are indicated by asterisks. The spectrum was recorded with an EI source at an accelerating electron beam potential of 70 eV.

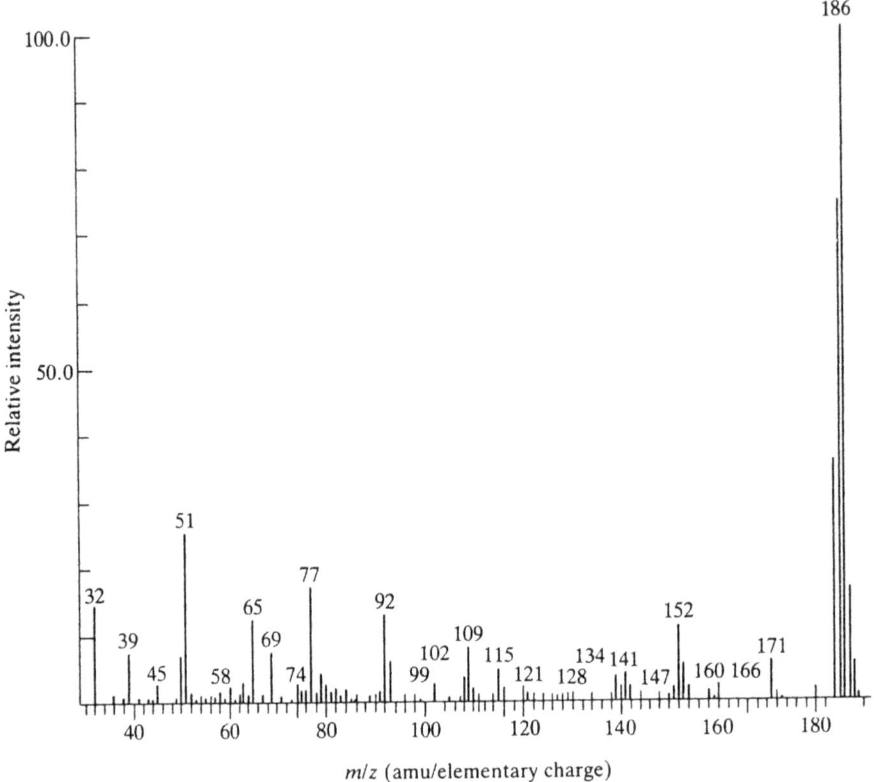

Figure 21-21 A mass spectrum of phenyl sulfide $(C_6H_5)_2S$ (MW 186). *(Courtesy of Dr. August A. Gallo, Department of Chemistry, University of Louisiana at Lafayette.)*

chlorine, three possible combinations of the two isotopes exist. The fragment could contain ^{35}Cl ^{35}Cl, ^{35}Cl ^{37}Cl, or ^{37}Cl ^{37}Cl. Of course, not all of the possibilities are equally probable. The probability of having ^{35}Cl ^{35}Cl is equal to the number of possible combinations (1) that yield the result multiplied by the chance (relative abundance, Table 21-1) of choosing the isotope. For $^{35}Cl^{35}Cl$ the probability is (1)(0.7577)(0.7577) = 0.5741.

Similarly, the probability of one ^{35}Cl and one ^{37}Cl appearing in the same fragment is the number of possible combinations (2: $^{35}Cl^{37}Cl$ and $^{37}Cl^{35}Cl$) multiplied by the chance of picking a ^{35}Cl and the chance of picking a ^{37}Cl. For $^{35}Cl^{37}Cl$ the probability is (2)(0.7577)(0.2423) = 0.3672. The probability of $^{37}Cl^{37}Cl$ is (1) (0.2423)(0.2423) = 0.0587. Consequently, three isotopic peaks should appear in the mass spectrum of a compound that contains two chlorine atoms. The peaks should appear at m/z of $M + 70$, $M + 72$, and $M + 74$, where M is the mass of the nonchlorine portion of the ionic fragment. The respective relative intensities of the peaks should be 0.5741/0.3672/0.0587 (9.78/6.26/1.00). Similar calculations can be used to obtain the relative abundances of other isotopic peaks.

Fragmentation Patterns

The extent of fragmentation that is observed in a mass spectrum is primarily determined by the energy that is transferred to the molecules in the ionic source. Ionic sources that transfer large amounts of energy to the sample cause more thorough fragmentation than the sources which transfer small amounts. An EI source that is operated at an electron beam energy of 70 eV generally transfers sufficient energy to break any bond in an organic compound.

The mass spectral peaks that have the greatest intensity correspond to the ionic fragments that are most stable. Consequently, a knowledge of the relative stabilities of the possible ionic fragments can be used to estimate relative peak heights in a mass spectrum. Of course, the best way to use fragmentation patterns for qualitative analysis is by direct comparison of the sample spectrum to spectra of standards. If that method cannot be used or does not produce a result, it is sometimes possible to use the major peaks in the mass spectrum to determine the most stable ionic fragments. The fragments can be reassembled, as in a puzzle, to reveal the identity of the sample.

A few rules are useful while trying to determine the identity of an organic compound from its mass spectrum. The rules should be used cautiously because they do not apply to all compounds. Carbon-carbon bonds are weaker than carbon-hydrogen bonds. Consequently, in saturated hydrocarbons fragmentation is more likely to occur between carbon atoms. Fragmentation is more likely to occur at a branch in the carbon chain. Generally the positive charge remains with the fragment that contains the greater degree of branching. Tertiary carbonium ions are more stable than secondary carbonium ions, which in turn are more stable than primary carbonium ions. For the compound shown in Fig. 21-22*a*, cleavage is expected around the branched carbon and relatively intense peaks are expected at *m/z* 57 and 71. Less intense peaks are observed for cleavage at nonbranched portions of the chain. For straight-chain alkanes the relative intensities of the peaks generally decrease with increasing *m/z* for fragments with molecular weights that are greater than about 57. Often groups of peaks are separated by *m/z* 14, corresponding to fragmentation around adjacent —CH_2— groups.

In alkenes cleavage is favored at the second bond away from the double bond, as illustrated in Fig. 21-22*b*. Apparently the fragment containing the double bond is stabilized by the allylic resonant structure. The charge remains with the unsaturated fragment. Similar fragmentation occurs in a ring that contains a single double bond.

Aromatic compounds often have unusually intense molecular ions. Aromatic compounds that have a carbon side chain have an intense peak at *m/z* 91, corresponding to cleavage at the beta bond from the ring. The actual fragment ion is probably a rearrangement product of that shown in Fig. 21-22*c*.

Organic compounds that contain a terminal noncarbon group generally rupture at the bond adjacent to the noncarbon group. The positive charge remains with the organic fragment (Fig. 21-22*d*). Compounds that contain a metallic element usually cleave at the bond to the metal. The positive charge remains with the metallic fragment (Fig. 21-22*e*). Compounds that contain a C═O generally fragment adjacent to the C═O. The positive charge remains with the fragment containing the oxygen (Fig. 21-22*f*). More information can be found in Reference 5.

Occasionally an ion with a mass m_0 that is formed in the source is accelerated out of the ionic source, but decomposes into an ion of smaller mass m before entering the magnetic mass analyzer. The mass spectral peak of the *metastable ion* is broader than other peaks and is observed at a mass m^* that is given by

$$m^* = \frac{m^2}{m_0} \tag{21-40}$$

Sometimes the presence of a metastable peak can be used to aid in identifying an analyte. More often, however, metastable peaks provide information that can be used to help determine gas-phase reaction mechanisms.

Because many organic ions undergo chemical rearrangements while in a mass spectrometer, mass spectral peaks sometimes occur at *m/z* values that are not anticipated. The organic chemistry of the many possible rearrangements is beyond the scope of the present chapter. Descriptions of some of the rearrangement reactions can be found in several of the References.

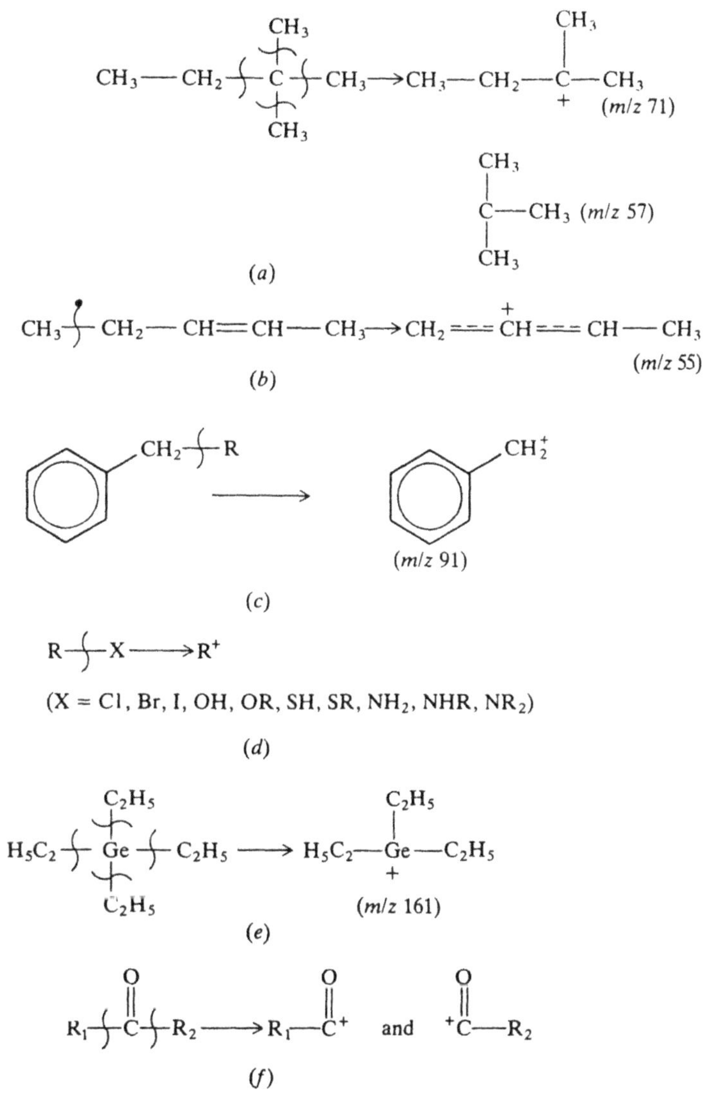

$$CH_3-CH_2-C(CH_3)_2-CH_3 \rightarrow CH_3-CH_2-\overset{+}{C}(CH_3)-CH_3 \ (m/z\ 71)$$

$$\overset{+}{C}(CH_3)_3 \ (m/z\ 57)$$

(*a*)

$$CH_3-CH_2-CH{=}CH-CH_3 \rightarrow CH_2{=}\overset{+}{C}H{=}CH-CH_3 \ (m/z\ 55)$$

(*b*)

$$C_6H_5-CH_2-R \longrightarrow C_6H_5-CH_2^+ \ (m/z\ 91)$$

(*c*)

$$R-X \longrightarrow R^+$$

$(X = Cl, Br, I, OH, OR, SH, SR, NH_2, NHR, NR_2)$

(*d*)

$$H_5C_2-Ge(C_2H_5)_2-C_2H_5 \longrightarrow H_5C_2-\overset{+}{Ge}(C_2H_5)-C_2H_5 \ (m/z\ 161)$$

(*e*)

$$R_1-C(=O)-R_2 \longrightarrow R_1-C^+(=O) \quad \text{and} \quad {}^+C(=O)-R_2$$

(*f*)

Figure 21-22 Typical fragmentations of organic compounds that occur in a mass spectrometer.

QUANTITATIVE ANALYSIS

Quantitative analysis can be performed using mass spectral peak heights. The peak heights are directly proportional to the concentrations of the components that are responsible for the peaks. Mass spectra are recorded of the sample and standards of all of the components in the sample. The molar quantity of the sample or standard that is used to obtain each spectrum is generally maintained at a fixed value by keeping the total ionic current through the mass spectrometer constant. A peak is chosen for each component in the mixture and the peak height in the sample is assumed to be the sum of the peak heights attributable to each of the components at the particular *m/z*. The peak height of each component at a particular *m/z* is determined from the spectra of the pure standards. The several simultaneous equations are solved for the concentration of each component. The process is essentially identical to that used for the quantitative determination of a mixture by ultraviolet-visible spectrophotometry as described in Chapter 9.

Quantitative analysis of complex mixtures by using the simultaneous equations method is often difficult and time consuming. Furthermore, mathematical and measurement errors can easily be made. Consequently, complex mixtures are rarely assayed using the method. An attractive alternative is to separate the components of the mixture prior to introduction into the mass spectrometer. Of the several separatory methods, chromatography (Chapters 24 through 26) is most often used in conjunction with mass spectrometry.

Both gas and liquid chromatographs have been coupled to mass spectrometers. The complex sample is inserted into the chromatograph. The sample is separated by the chromatograph into its components and the chromatogram is recorded. The chromatographic peak heights or areas are generally used for the quantitative analysis.

The outlet from the chromatograph is coupled to the inlet from the mass spectrometer. A mass spectrum of each component is obtained as that component leaves the chromatograph. Because the mass spectrum must be obtained rapidly, a quadrupolar mass analyzer often is used. The mass spectrum of each component is used to qualitatively analyze the component. The combined chromatographic-mass spectral technique performs both quantitative and qualitative chemical analysis of the sample.

A gas chromatograph (GC) is relatively easy to interface to a mass spectrometer (MS) because the sample component as it exits the GC is gaseous. A portion of the component flows directly into the inlet system of the mass spectrometer. If a CI source is used, the carrier gas in the GC often serves as the reagent gas in the MS. Often a capillary column is used in the GC. An introduction to the use of GC/MS is presented in Reference 18, with a further description in Chapter 26.

Before insertion of a solid sample into the GC it is usually necessary to chemically convert all or part of the sample into a relatively volatile liquid. That can be accomplished by reacting the sample with a reagent that yields a volatile product. Acetylation reactions have been used with some compounds. Sugars can be reduced with sodium borohydride to relatively volatile compounds; acids are often converted to esters before insertion into the GC; and some nucleosides can be assayed after conversion to the trimethylsilyl derivatives.

The coupling of high-performance liquid chromatography (HPLC) to mass spectrometry produces another useful combined technique liquid chromatography/mass spectrometry LC/MS. Interfacing a liquid chromatograph to a mass spectrometer is more difficult than interfacing a gas chromatograph to a mass spectrometer. If the HPLC solvent is carefully chosen, sometimes the LC outlet can be directly connected to the MS inlet. Only a portion of the effluent from the LC enters the MS. The vaporized solvent in the MS can serve as the reagent gas in a CI source. Another alternative is to physically collect the separated components from the LC and manually inject them into the MS. LC/MS is described in Chapter 25. Electrospray ionization has recently become one of the most popular methods to interface an LC to a mass spectrometer.

Tandem Mass Spectrometry

In addition to its use as an analytical tool, mass spectrometry can be used to study fragmentation mechanisms. A particularly useful technique for that purpose is *tandem mass spectrometry* (MS/MS) in which two mass spectrometers are coupled to each other. The first mass spectrometer uses a "soft" ionization technique to form a relatively high concentration of molecular ions that have not been highly fragmented. The molecular ions subsequently are fragmented by collision with a neutral reagent gas in a process termed *collision-activated decomposition* (CAD). The daughter ions formed during the decomposition are examined in the second mass spectrometer. The technique is thoroughly described in Reference 19.

IMPORTANT TERMS

Adduct ion
Appearance potential
Arc discharge source
Base ion
Buffer gas
CAD
Channel electron multiplier
Chemical ionization
Cold inlet
Double-focusing analyzer
Dynamic sputtering
Dynodic strip
Electron capture
Electron multiplier
Electron-bombardment ionization
Electrospray ionization
Electrostatic analyzer
Emitter
Faraday cup
Fast-atom bombardment
Field desorption
Field ionization
Hot inlet
ICPMS
Ion cyclotron resonance
Ionic microprobe
Ionic source
Ionization efficiency curve
Ionization potential
Ion-scattering spectroscopy
Isotopic abundance
Laser desorption ionization
LSIMS
Magnetic analyzer
MALDI
Mass analyzer
Mass resolving power
Mass spectrometry
Metastable ion
Molecular ion
Multiple photon ionization
Negative chemical ionization

Negative-ion field ionization
Orbitrap analyzer
Parent ion
Photoionization
PPINICI
Quadrupolar analyzer
Reagent gas
Reflectron
Resolution
Resonant absorption
Scintillation counter
Secondary ion mass spectrometry
Spark ionic source
Static sputtering
Surface-emissive ionic source
Tandem mass spectrometry
Thermal ionization
Time-of-flight analyzer

PROBLEMS

Resolution

21-1 Mass spectral peaks at *m/z* 1257 and 1259 were just resolved. The two peaks were of equal height. Calculate the resolution of the mass spectrometer.

21-2 Identical intensity, mass spectral peaks at *m/z* 748 and 749 exhibited a 10 percent valley between the peaks. Calculate the resolution of the mass spectrometer.

21-3 A mass spectral peak centered at *m/z* 447 had a peak width at 5 percent of the peak height of 0.34. Estimate the resolution of the mass spectrometer.

21-4 A mass spectral peak centered at *m/z* 352 had a peak width at 5 percent of the peak height of 0.41. Determine the resolution of the mass spectrometer.

Mass Analyzers

21-5 Calculate the magnetic flux density that is required to focus an ion with *m/z* 245 on the exit slit of a magnetic analyzer in which the accelerating potential is 7500 V and the radius of curvature of the ionic beam at the exit slit is 25.0 cm.

21-6 What accelerating potential is required to focus an ion with *m/z* 356 on the entrance slit of a detector if the magnetic analyzer has a radius of curvature at the location of the detector of 22.5 cm, and if the applied magnetic flux density is 1.00 T?

21-7 In a particular time-of-flight mass spectrometer an ion with *m/z* 298 required 1.75 μs to strike the detector after emission from the source. Determine the time needed for an ion with *m/z* 425 to strike the detector.

21-8 A time-of-flight mass spectrometer has a flight path of 1.250 m and an accelerating potential of 2875 V. Calculate the time required for ions with *m/z* 100, 150, 200, and 250 to strike the detector after emission from the source.

21-9 Calculate the frequency of oscillation of an ion with *m/z* 100 in an icr analyzer with a magnetic flux density of 5.0 T.

Analysis

21-10 The following tabulated data were obtained. Determine the upper limit of the appearance potential for the sample in the mass spectrometer.

Electron energy, eV	Relative ionic current
20	0.5
24	2.6
28	6.8
32	14.0
34	18.3
36	24.3
38	30.0
40	34.4
44	41.7
50	48.2
58	51.3
66	51.6

21-11 The quadrupolar mass spectrum of coumarin 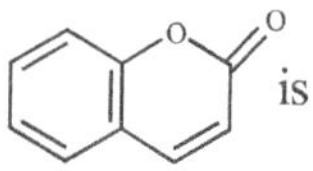 is shown in Fig. P21-11. Assign ionic fragments to the four largest peaks.

21-12 Some of the major peaks in the mass spectrum of 1-bromo-2-methyl-3-cyclohexyl-1-propene [$BrCH{=}C(CH_3)CH_2C_6H_{11}$] are listed below. Assign an ionic fragment to each peak.

m/z	Relative intensity
218	10
216	10
137	6
136	8
134	8
83	87
55	100
41	51

21-13 In the mass spectrum of *n*-butane a metastable ion is observed at *m/z* 31.9. Propose a reaction mechanism that would yield the observed metastable peak.

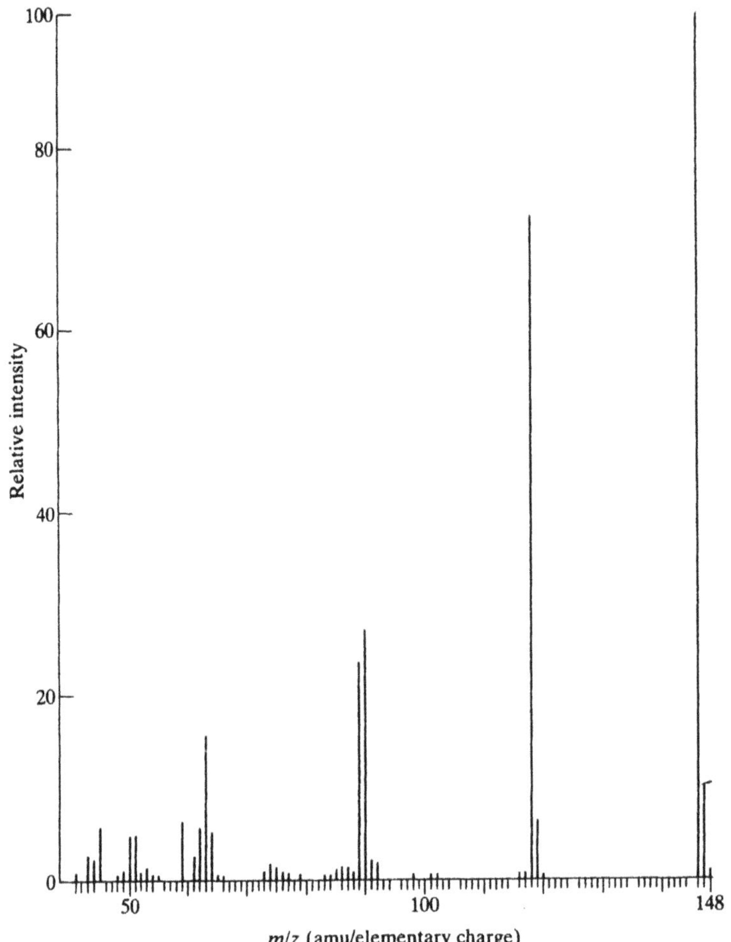

Figure P21-11

21-14 A mass spectrum was obtained of equal molar quantities (equal total ionic current) of pure *n*-butanol, pure *t*-butanol, and a mixture of the two compounds. The relative peak intensities at *m/z* 43 and 57 were chosen for the quantitative analysis. The spectrum amplitude was maintained at a fixed value during recording of the spectra. From the following tabulated data, determine the percentage of *n*-butanol and *t*-butanol in the mixture.

Sample	*m/z*	Relative intensity
n-Butanol	43	87.2
	57	18.5
t-Butanol	43	3.5
	57	92.0
Mixture	43	62.5
	57	71.4

REFERENCES

1. Dempster, A. J.: *Phys. Rev.*, 11:316(1918).
2. Kiser, R. W.: *Introduction to Mass Spectrometry and Its Applications,* Prentice-Hall, Englewood Cliffs, New Jersey, 1965.
3. McLafferty, F. W.: *Interpretation of Mass Spectra,* 3d ed., University Science Books, Mill Valley, California, 1980.
4. Howe, I., D. H. Williams, and R. D. Bowen: *Mass Spectrometry: Principles and Applications,* 2d ed., McGraw-Hill, New York, 1981.
5. Rose, M. E., and R. A. W. Johnstone: *Mass Spectrometry for Chemists and Biochemists,* Cambridge University Press, Cambridge, 1982.
6. Hillenkamp, F., in A. Benningbove (ed.): *Proceedings of 2nd International Conference of Ion Formation from Organic Solids,* Springer-Verlag, New York, 1983, p. 190.
7. Karas, M., D. Bachmann and F. Hillenkamp: *Anal. Chem.*, 57: 2935(1985).
8. Burlingame, A. L., J. O. Whitney, and D. H. Russell: *Anal. Chem.,* **56**: 417R(1984).
9. Hercules, D. M., and S. H. Hercules: *J. Chem. Educ.,* **61**: 592(1984).
10. Martin, S. A., C. E. Costello, and K. Blemann: *Anal. Chem.,* **54**: 2362(1982).
11. Beauchemin, D.: *Anal. Chem.*, **76**: 3395(2004).
12. Beauchemin, D.: *Anal. Chem.*, **78**: 4111(2006).
13. Marshall, A. G. and C. L. Hendrickson in Annual Review of Analytical Chemistry, E. S. Yeung and R. N. Zare (eds.), Annual Reviews, Palo Alto, California, USA: 579(2008).
14. Markov, A.: *Anal. Chem.*, 72: 1156(2000).
15. Benyon, J. H., and A. E. Williams: *Mass and Abundance Tables for Use in Mass Spectrometry,* Elsevier, Amsterdam, 1963.
16. Lederberg, J.: *Compilation of Molecular Formulas for Mass Spectrometry,* Holden-Day, San Francisco, 1964.
17. Lambert, J. B., H. F. Shurvell, L. Verbit, R. G. Cooks, and G. H. Stout: *Organic Structural Analysis, Macmillan,* New York, 1976.
18. Karasek, F. W., and A. C. Viau: *J. Chem. Educ.*, **61**: A233(1984).
19. McLafferty, F. W. (ed.): *Tandem Mass Spectrometry,* Wiley, New York, 1983.

CHAPTER

TWENTY-TWO

POTENTIOMETRY

Potentiometry is an analytical technique in which the amount of a substance in solution is determined, either directly or indirectly, from measurement of the electromotive force (emf) between two probes (electrodes) that are dipped into the solution. The emf is measured in units of volts or millivolts. The electrode at which reduction takes place is the *cathode,* and that at which oxidation occurs is the *anode*.

The measured emf between the electrodes varies in a predictable fashion with changes in the concentration of the species in solution. An electrode at which the emf varies as a function of solution concentration is used to indicate, i.e., determine, the concentration of a solution species, and consequently is termed an *indicator electrode*. In some cases the indicator electrode is an inert-metal wire such as a platinum wire. Such an electrode acquires an emf which is characteristic of both the identity of the redox couple in solution and the ratio of the amounts of the dissolved oxidant and reductant. It responds to changes in concentration of any component of any electroactive redox couple. Inert-metal indicator electrodes are generally used to follow the progress of oxidation-reduction titrations.

Another type of potentiometric indicator electrode preferentially responds to a single solution species, for example, Ca^{2+} or Cl^{-}, or a limited number of solution species. It is usually called an *ion-selective electrode*; however, if the solution species being measured is not an ion, it is sometimes called a *molecule-selective electrode*. Ion-selective electrodes are used to assay directly or indirectly a sample solution containing the species for which the electrode is selective.

If an inert-metal indicator electrode is dipped into a solution containing an unchanging ratio of the oxidized and reduced forms of a redox couple, the electrode has a constant emf. As long as the ratio of the oxidant to reductant is constant, the emf of the electrode is constant. An electrode with a constant emf which is normally used in conjunction with an indicator electrode is a *reference electrode*. It is usually either an inert-metal wire dipping into a solution of an oxidant and a reductant of a redox couple at fixed concentrations or a metal that is a constituent of a redox couple dipping into a solution of a fixed concentration of the ion of the oxidized form of the metal. The emf between two reference electrodes is constant.

A single indicator electrode is used in combination with a single reference electrode for most potentiometric measurements. The emf between an indicator electrode and a reference electrode is a *potential* or *potential difference*. A potential measurement implies the use of a reference electrode. A measurement of electromotive force does not imply that one of the leads from the measuring instrument is connected to a reference electrode. The use of potential and potential difference in Chapters 22 and 23 is slightly different from that used in Chapters 2 through 4 where the terms were used to mean emf as defined in this chapter.

Potentiometry is highly versatile. If the concentration of a solution species is determined from a single potential measurement in the sample solution, the technique is *direct potentiometry*. Direct potentiometry is usually done with an ion-selective electrode. A titration in which potentiometry is used to locate the endpoint is a *potentiometric titration*. Potentiometry can be used to follow redox, acid-base, complexometric, and precipitation titrations.

THEORY

A simple oxidation-reduction half-reaction can be written in the general form as follows:

$$a\mathrm{O_X} + n\mathrm{e}^- = b\mathrm{Red} \tag{22-1}$$

where Ox is the oxidized form of the species, Red is the reduced form, n is the number of electrons e^- involved in the half-reaction, and a and b are the coefficients of Ox and Red in the balanced half-reaction. When writing the formulas for reversible chemical reactions in this chapter and the following chapters, equal signs (=) are used, for the sake of simplicity, in place of two opposing arrows. The thermodynamic feasibility of the reaction taking place can be determined from the change in the Gibbs free energy ΔG for the reaction. As the change in the Gibbs free energy (also called the free energy) of the reaction becomes more negative, the reaction becomes more thermodynamically feasible; i.e., the occurrence of the reaction becomes more likely. Of course, no half-reaction can ocur alone; every reduction must be accompanied by a simultaneous oxidation of some other species. Nevertheless, the change in the free energy for the half-reaction is an indication of the thermodynamic feasibility of the reaction occurring.

It can be shown from thermodynamics (see References 1 and 2) that the change in G for Eq. (22-1) is given by

$$\Delta G = \Delta G^{\circ} + RT \ln \frac{a_{\text{Red}}^{b}}{a_{\text{Ox}}^{a}} \tag{22-2}$$

where R is the gas constant (8.314 V · C/K · mol), T is the absolute temperature, a_{Red} is the activity of Red, a_{O_x} is the activity of Ox, and $G°$ is the Gibbs free energy of the substance in its standard state.

In general, a superscript ° above any thermodynamic term means the thermodynamic term is for the substance in its standard state. The standard state for all substances is that which occurs at one atmosphere of pressure and, unless otherwise specified, at 25°C. For a gas the standard state is the pure gas; for a liquid it is the pure liquid; and for a solid it is the pure solid in a specified crystalline state.

The free energy change for a reaction can be related to the potential E of a reaction by

$$\Delta G = -nFE \tag{22-3}$$

$$\Delta G^{\circ} = -nFE^{\circ} \tag{22-4}$$

where F is the Faraday constant[3] (96,485 C/mol) and $E°$ is the *standard potential* for the particular reaction. The standard potential is the potential for the reaction at 25°C when the activities (or fugacities for gases) of the reaction components are unity. A list of selected standard potentials is given in Table 22-1. A considerably more detailed list can be found in Reference 4.

Substitution of the values of ΔG and $\Delta G°$ from Eqs. (22-3) and (22-4) into Eq. (22-2) yields

$$-nFE = -nFE^{\circ} + RT \ln \frac{a_{\text{Red}}^{b}}{a_{\text{Ox}}^{a}} \tag{22-5}$$

Equation (22-5) can be solved for the potential of the half-reaction as follows:

$$E = E^{\circ} - \frac{RT}{nF} \ln \frac{a_{\text{Red}}^{b}}{a_{\text{Ox}}^{a}} \tag{22-6}$$

Equation (22-6) is one form of the *Nernst equation*. Whenever one of the components of a half-reaction is a gas, the activity in Eq. (22-6) is replaced by the fugacity of the gas. If the gas is an ideal gas, the fugacity is replaced by the partial pressure (atm) of the gas. It can be seen from Eq. (22-6) that the potential of the half-reaction equals the standard potential of the half-reaction when the logarithmic term becomes zero. That occurs for the reaction shown in Eq. (22-1)

Table 22-1 Selected standard potentials at 25°C in aqueous solutions relative to the standard hydrogen electrode

Half-reaction	$E°$, V
$F_2(g) + 2H^+ + 2e^- = 2HF$	3.06
$O_3 + 2H^+ + 2e^- = O_2 + H_2O$	2.07
$S_2O_8^{2-} + 2e^- = 2SO_4^{2-}$	2.01
$Ag^{2+} + e^- = Ag^+$	2.00
$Co^{3+} + e^- = Co^{2+}$	1.84
$H_2O_2 + 2H^+ + 2e^- = 2H_2O$	1.77
$MnO_4^- + 4H^+ + 3e^- = MnO_2(s) + 2H_2O$	1.70
$Ce^{4+} + e^- = Ce^{3+}$ (1 M $HClO_4$)	1.70
$Ce^{4+} + e^- = Ce^{3+}$	1.61
$H_5IO_6 + H^+ + 2e^- = IO_3^- + 3H_2O$	1.60
$Bi_2O_4 + 4H^+ + 2e^- = 2BiO^+ + 2H_2O$	1.59
$2BrO_3^- + 12H^+ + 10e^- = Br_2 + 6H_2O$	1.52
$MnO_4^- + 8H^+ + 5e^- = Mn^{2+} + 4H_2O$	1.51
$PbO_2 + 4H^+ + 2e^- = Pb^{2+} + 2H_2O$	1.455
$Cl_2 + 2e^- = 2Cl^-$	1.36
$Cr_2O_7^{2-} + 14H^+ + 6e^- = 2Cr^{3+} + 7H_2O$	1.33
$Tl^{3+} + 2e^- = Tl^+$	1.28
$MnO_2(s) + 4H^+ + 2e^- = Mn^{2+} + 2H_2O$	1.23
$O_2(g) + 4H^+ + 4e^- = 2H_2O$	1.229
$2IO_3^- + 12H^+ + 10e^- = I_2 + 6H_2O$	1.20
$3Br_2(l) + 2e^- = 2Br_3^-$	1.096
$2ICl_2^- + 2e^- = I_2 + 4Cl^-$	1.06
$VO_2^+ + 2H^+ + e^- = VO^{2+} + H_2O$	1.00
$HNO_2 + H^+ + e^- = NO(g) + H_2O$	1.00
$NO_3^- + 3H^+ + 2e^- = HNO_2 + H_2O$	0.94
$2Hg^{2+} + 2e^- = Hg_2^{2+}$	0.92
$Cu^{2+} + I^- + e^- = CuI$	0.86
$Ag^+ + e^- = Ag$	0.799
$Hg_2^{2+} + 2e^- = 2Hg$	0.789
$Fe^{3+} + e^- = Fe^{2+}$	0.771
$C_6H_4O_2$(quinone) $+ 2H^+ + 2e^- = C_6H_4(OH)_2$	0.699
$O_2(g) + 2H^+ + 2e^- = H_2O_2$	0.695
$2HgCl_2 + 2e^- = Hg_2Cl_2(s) + 2Cl^-$	0.63
$Hg_2SO_4(s) + 2e^- = 2Hg + SO_4^{2-}$	0.615
$Sb_2O_5 + 6H^+ + 4e^- = 2SbO^+ + 3H_2O$	0.581
$H_3AsO_4 + 2H^+ + 2e^- = HAsO_2 + 2H_2O$	0.559
$I_3^- + 2e^- = 3I^-$	0.536
$I_2 + 2e^- = 2I^-$	0.5355
$Cu^+ + e^- = Cu$	0.521
$H_2SO_3 + 4H^+ + 4e^- = S + 3H_2O$	0.45
$Fe(CN)_6^{3-} + e^- = Fe(CN)_6^{4-}$	0.36

Half-reaction	$E°$, V
$Cu^{2+} + 2e^- = Cu$	0.337
$VO^{2+} + 2H^+ + 2e^- = V^{3+} + H_2O$	0.337
$UO_2^{2+} + 4H^+ + 2e^- = U^{4+} + 2H_2O$	0.334
$BiO^+ + 2H^+ + 3e^- = Bi + H_2O$	0.32
$Hg_2Cl_2(s) + 2e^- = 2Hg + 2Cl^-$	0.268
$AgCl(s) + e^- = Ag + Cl^-$	0.222
$SbO^+ + 2H^+ + 3e^- = Sb + H_2O$	0.212
$CuCl_3^{2-} + e^- = Cu + 3Cl^-$	0.178
$SO_4^{2-} + 4H^+ + 2e^- = SO_2(aq) + 2H_2O$	0.17
$Cu^{2+} + e^- = Cu^+$	0.153
$Sn^{4+} + 2e^- = Sn^{2+}$	0.151
$S + 2H^+ + 2e^- = H_2S(g)$	0.141
$TiO^{2+} + 2H^+ + e^- = Ti^{3+} + H_2O$	0.10
$S_4O_6^{2-} + 2e^- = 2S_2O_3^{2-}$	0.08
$AgBr(s) + e^- = Ag + Br^-$	0.0713
$2H^+ + 2e^- = H_2(g)$	0.0000
$Pb^{2+} + 2e^- = Pb$	–0.126
$Sn^{2+} + 2e^- = Sn$	–0.136
$AgI(s) + e^- = Ag + I^-$	–0.151
$Ni^{2+} + 2e^- = Ni$	–0.25
$V^{3+} + e- = V^{2+}$	–0.255
$Co^{2+} + 2e^- = Co$	–0.277
$Ag(CN)_2^- + e- = Ag + 2CN^-$	–0.31
$PbSO_4(s) + 2e^- = Pb + SO_4^{2-}$	–0.356
$Cd^{2+} + 2e^- = Cd$	–0.403
$Cr^{3+} + e^- = Cr^{2+}$	–0.41
$Fe^{2+} + 2e^- = Fe$	–0.440
$2CO_2 + 2H^+ + 2e^- = H_2C_2O_4$	–0.49
$H_3PO_3 + 2H^+ + 2e^- = H_3PO_2 + H_2O$	–0.499
$U^{4+} + e^- = U^{3+}$	–0.607
$Zn^{2+} + 2e^- = Zn$	–0.763
$Cr^{2+} + 2e^- = Cr$	–0.913
$Mn^{2+} + 2e^- = Mn$	–1.18
$Zr^{4+} + 4e^- = Zr$	–1.539
$Al^{3+} + 3e^- = Al$	–1.66
$Th^{4+} + 4e^- = Th$	–1.899
$Mg^{2+} + 2e^- = Mg$	–2.37
$La^{3+} + 3e^- = La$	–2.522
$Na^+ + e^- = Na$	–2.714
$Ca^{2+} + 2e^- = Ca$	–2.87
$Sr^{2+} + 2e^- = Sr$	–2.9
$K^+ + e^- = K$	–2.925
$Li^+ + e^- = Li$	–3.045

when $a_{\mathrm{Red}}^{b} = a_{\mathrm{OX}}^{a}$. It is often more convenient to rewrite Eq. (22-6) using base 10 logarithms:

$$E = E^{\circ} - \frac{2.303RT}{nF} \log \frac{a_{\mathrm{Red}}^{b}}{a_{\mathrm{OX}}^{a}} \tag{22-7}$$

At 25.0°C Eq. (22-7) can be further simplified by substitution and combination of the constants preceding the logarithmic term to yield

$$E = E^{\circ} - \frac{0.05917}{n} \log \frac{a_{\mathrm{Red}}^{b}}{a_{\mathrm{OX}}^{a}} \tag{22-8}$$

The equation from which the Nernst equation was derived [Eq. (22-2)] is valid only for thermodynamically reversible reactions. Consequently, the Nernst equation is valid only for reversible reactions. A reversible reaction is one which can be made to reverse its direction of reaction [Eq. (22-1)] by an infinitesimal change, in the appropriate direction, of the potential of the reaction. The reversal can be accomplished by changing the potential of the electrode in a solution containing the components of the half-reaction.

A working definition of reversibility takes advantage of the fact that the Nernst equation is valid only for reversible reactions. If a half-reaction obeys the Nernst equation, it is apparently reversible, but if it does not obey the Nernst equation, it definitely is not reversible. Reversibility can usually be assumed if a plot of E as a function of log ($a_{\mathrm{Red}}^{b} / a_{\mathrm{OX}}^{a}$) is a straight line with a slope of $-2.303RT/nF$.

The activities used in the preceding equations are related to the concentrations of the species in a solution by

$$a = fC \tag{22-9}$$

where f is the dimensionless *activity coefficient* and C is the concentration. In analytical chemistry the concentrations are usually expressed in molarity, although molality is often used by workers in other disciplines. While molarity is sensitive to temperature changes because the volume of a solution often varies with temperature, molality is independent of temperature. If studies are conducted at more than one temperature, molality is the preferred unit.

For aqueous solutions with ionic strengths I less than 0.001 M, the activity coefficient can be estimated from Debye-Hückel theory:

$$-\log f = Az^2 I^{1/2} \tag{22-10}$$

Equation (22-10) can be used only for dilute solutions. An extended form of the equation:

$$-\log f = \frac{Az^2 I^{1/2}}{1 + 1.5I^{1/2}} \tag{22-11}$$

can be used at ionic strengths up to about 0.01 M.

Table 22-2 Values of the constant A in Eqs. (22-10) and (22-11) for aqueous solutions at different temperatures when the concentration is expressed in molarity

Temperature, °C	A
5	0.495
10	0.499
15	0.503
20	0.507
25	0.512
30	0.516
35	0.521
40	0.526

Debye-Hückel theory is based on the assumption that each ion in solution is surrounded by ions of the opposite charge. Debye and Hückel used the Boltzmann distribution law to estimate the charge density of the oppositely charged ions around each ion. In Eqs. (22-10) and (22-11), A is a temperature-dependent constant and z is the charge on the ion whose activity coefficient is estimated. Values of A at several temperatures are listed in Table 22-2. More information on Debye-Hückel theory as it applies to potentiometry as well as a more thorough listing of the values of the constant in Eq. (22-10) and several extended versions of Eq. (22-10) can be found in Reference 5. The ionic strength of a solution is dependent upon all of the ions in the solution and can be calculated using

$$I = \frac{1}{2} \sum C_i z_i^2 \tag{22-12}$$

where C_i is the concentration of each individual ion i in solution and z_i is the charge of each ion.

In solutions of low ionic strength, f is nearly 1 and the activity [Eq. (22-9)] is nearly equal to the concentration of the species in solution. Consequently, if a dilute solution is used, the Nernst equation can be rewritten as

$$E = E^{\circ} - \frac{RT}{nF} \ln \frac{[\text{Red}]^b}{[\text{Ox}]^a} \tag{22-13}$$

where the square brackets indicate the molarity of the substance within the brackets.

Sample problem 20-1 Calculate the activity of calcium ion in a solution containing $1.50 \times 10^{-4} M$ $CaCl_2$ and $5.00 \times 10^{-4} M$ NaCl at 25°C.

SOLUTION The ionic strength of the solution can be calculated by using Eq. (22-12):

$$I = \frac{1}{2}\,[(1.50 \times 10^{-4})\,(2)^2 + (3.00 \times 10^{-4})\,(-1)^2 + (5.00 \times 10^{-4})\,(+\,1)^2$$

$$+\,(5.00 \times 10^{-4})(-1)^2] = 9.50 \times 10^{-4}\ M$$

Since the ionic strength is less than 0.001 M, Eq. (22-10) can be used to calculate the activity coefficient. After substitution of the value of I, the value of A from Table 22-2 at 25°C, and the charge of the calcium ion, Eq. (22-10) is solved for the activity coefficient:

$$-\log f = 0.512(2)^2(9.50 \times 10^{-4})^{1/2} = 6.31 \times 10^{-2}$$

$$f = 0.865$$

The activity of the calcium ion is calculated from the activity coefficient and the calcium ion concentration with the aid of Eq. (22-9):

$$a = fC = 0.865(1.50 \times 10^{-4}) = 1.30 \times 10^{-4}\ M$$

Potentials and Standard Potentials

It is impossible to measure the absolute potential of a single electrode dipped into a solution. At least two electrodes must be in electrical contact with the sample solution in order to complete the circuit and allow the measurement. Because it is possible to measure the emf only between two electrodes, it is impossible to determine the absolute value of E associated with a single electrode; i.e., it is possible only to make relative measurements of the emf or potential of one electrode as compared with a second electrode. By international agreement the standard potential $E°$ of the hydrogen half-reaction has been assigned a value of exactly zero volts:

$$2H^+ + 2e^-\ H_2(g), \qquad E° = 0\ V$$

The potentials or standard potentials for other half-reactions can be determined by using the hydrogen half-reaction as the reference electrode and measuring the potential of the half-reaction with respect to that *hydrogen electrode.* A hydrogen electrode at 25°C in which the activity of H^+ and the fugacity of H_2 are 1 is a *standard hydrogen electrode* and has a potential of 0 V. Experimentally measurements relative to the hydrogen electrode can be accomplished by attaching the hydrogen electrode to the negative terminal of the potential-measuring instrument. The positive terminal is attached to the indicator electrode which is dipped into a solution of the half-reaction whose potential is to be measured. The "negative" or "positive" terminal of a voltmeter, as used here, can be a misnomer, because the terminals can be either positive or negative depending upon the circumstances. The "negative" terminal as used in this connection corresponds to the "common" (black) terminal. The "positive" terminal is the red terminal. On a pH meter the "negative" terminal usually corresponds to the reference electrode socket. If the measured potential is positive, the electrode attached to the red terminal is positive relative to the electrode attached to the negative terminal. If the measured potential is negative, the electrode attached to the red terminal is negative relative to the electrode attached to the black terminal. Both electrodes must make electrical contact with the solution whose potential is measured.

The electrical circuit between the hydrogen electrode and the test solution is usually completed with a *salt bridge* containing an electrolyte solution, as shown in Fig. 22-1. The electrolyte used in the solution is often potassium chloride (see the description of liquid-junction potentials later in the chapter). The electrolyte solution can be held in place in the salt bridge with slightly porous plugs made from glass frits which have been inserted into the ends of the filled tube, with asbestos fibers, or with some other type of porous plug. As an alternative, about 5 percent by weight of agar can be dissolved in the heated electrolyte solution and poured into the salt bridge. Upon cooling, the solution gels and holds the electrolyte solution inside the tube.

An electrochemical cell can be conveniently represented in shorthand notation for the hydrogen electrode and platinum electrode pair shown in Fig. 22-1:

$$\text{Pt, } H_2(g) \mid H^+(aq) \| \text{Ox}(aq), \text{Red}(aq) \mid \text{Pt} \tag{22-14}$$

In the notation, a single vertical line indicates a phase boundary and a double vertical line represents a physical separation of components as by a salt bridge. The portion of Eq. (22-14) to the left of the double vertical line is the hydrogen electrode and the portion to the right represents the platinum indicator electrode dipping into a solution containing dissolved Ox and Red. Generally the portion of the cell that is attached to the negative terminal of the potential-measuring instrument is written on the left.

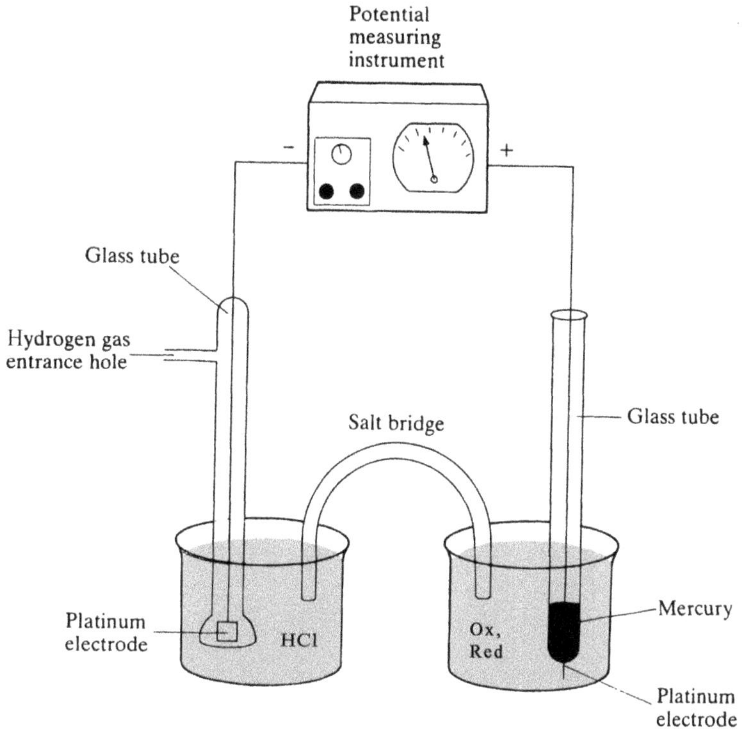

Figure 22-1 Apparatus for measuring the potential of a half-reaction. The mercury is used to make electric contact between the Pt indicator electrode and the lead to the potential measuring instrument.

Formal Potentials

When one or more of the components of a redox couple take part in a chemical reaction within the cell which occurs to an unknown extent, such as a complexation reaction, it is impossible to determine the standard potential of the half-reaction. Because the activities of the redox species in solution are unknown, the standard potential for the half-reaction cannot be calculated using Eq. (22-12). In those instances it is convenient to measure and use the *formal potential* $E°'$. Measurement of the formal potential does not require an exact knowledge of the activities or concentrations of the species which actually exist in the solution, but only of the concentrations of the species added to the solution. For the generalized half-reaction shown in Eq. (22-1), the formal potential is defined as follows:

$$E = E^{\circ\prime} - \frac{RT}{nF} \ln \frac{C^b_{\text{Red}}}{C^a_{\text{Ox}}} \tag{22-15}$$

where C_{Red} and C_{O_x} are added concentrations (*M*). Compare Eq. (22-15) to Eqs. (22-6) and (22-13). It is important to note that *added* concentrations rather than actual activities are used in the logarithmic term of Eq. (22-15). When a formal potential is reported, it is necessary to specify the exact solution composition, because the formal potential can vary with changes in the solution components and their concentrations. Care must be taken when using formal potentials to ensure that the solution components are identical with those used during the measurement of the formal potential.

Sample problem 22-2 Enough ferricyanide and ferrocyanide were added to a 0.1 *M* hydrochloric acid solution to make the apparent concentrations of ferricyanide and ferrocyanide equal to 0.010 *M* and 0.020 *M* respectively at 25°C. A platinum indicator electrode connected to the positive terminal of a voltmeter and a reference electrode connected to the negative terminal were inserted, and the measured potential difference between the electrodes was 0.30 V. If the reference electrode has a constant potential of 0.24 V, calculate the formal potential for the reduction of ferricyanide to ferrocyanide:

$$\text{Fe (CN)}_6^{3-} + e^- = \text{Fe (CN)}_6^{4-}$$

SOLUTION The potential difference of the cell is the difference between the potential of the ferricyanide-ferrocyanide half-reaction E (attached to the positive terminal) and the potential of the reference electrode:

$$0.30 = E - 0.24$$

The potential of the ferricyanide-ferrocyanide half-reaction can be calculated by substituting E from Eq. (22-15) into the above equation:

$$0.30 = E^{o\prime} - \frac{0.059}{1} \log \frac{0.020}{0.010} - 0.24$$

Solution of the equation for the formal potential yields

$$E^{o\prime} = 0.30 + 0.24 + 0.059(0.30) = 0.56 \text{ V}$$

Liquid-Junction Potentials

The potential difference measured by the electronic instrument between two electrodes is the sum of all of the potential differences between the electrical connections of the device. If a potential difference, in addition to that owing to the reference and indicator electrodes, occurs in the circuit in the electrochemical cell, the measuring device reports the sum of the potential differences between the two electrodes and the additional potential difference. The total potential difference in the cell differs from the potential difference attributable solely to the redox reactions. If an unrecognized potential does occur in the circuit, incorrect analytical results can be obtained from direct potentiometric measurements.

A common example of an unwanted potential, which can be found in electrochemical cells, is the *liquid-junction potential*. It occurs at the interface between solutions containing different electrolytes or between solutions containing different concentrations of the same electrolyte. At the interface between those solutions, a diffusion of ionic species from regions of relatively high concentration to regions of lower concentration occurs. In most cases the positive and negative ions of an electrolyte do not diffuse at the same rate, The partially separated charged ions cause the liquid-junction potential.

Although liquid-junction potentials are normally not the major contributor to the measured potential of an electrochemical cell, they can constitute a significant portion of that potential. The liquid-junction potential between solutions containing positive and negative ions that diffuse at vastly different rates is considerably greater than that for solutions containing ions that diffuse at similar rates. The liquid-junction potential between a solution containing 0.1 *M* HCl and a solution containing 0.01 *M* HCl has been calculated to be 40 mV, whereas the liquid-junction potential between a solution containing 0.1 *M* and 0.01 *M* KCl is only −1.2 mV.[6] In the case of the HCl solution, the H^+ diffuses more rapidly from the 0.1 *M* solution to the 0.01 *M* solution than does the Cl^-, thereby causing a net positive charge in the more dilute solution and a negative charge in the more concentrated solution. In the KCl solutions the diffusion rates are nearly equal, although the Cl^- does diffuse slightly more rapidly into the more dilute solution than does the K^+. That leads to less separation of charged particles across the interface and a smaller liquid-junction potential.

Liquid-junction potentials often occur at the interface between a salt bridge and a half-cell solution. They are commonly minimized in either of two ways. The preferable method is to use the same concentration of the same electrolyte in the salt bridge and the half-cell solutions. Unfortunately, that is often experimentally impossible. As an alternative, the electrolyte in the salt bridge can be chosen to contain positive and negative ions which diffuse at nearly the same rate. The electrolyte most often chosen for salt bridges is potassium chloride. However, because ammonium and nitrate also diffuse at nearly the same rate, ammonium nitrate is sometimes used. Ammonium ion has the disadvantage of being able to alter the pH of solutions into which it diffuses and of being capable of taking part in acid-base reactions. Other electrolyte solutions are used in special circumstances.

Theory of Ion-Selective Electrodes

Most ion-selective electrodes ISEs have the general structure shown in Fig. 22-2. The electrodes consist of an internal reference electrode dipping into an internal reference solution which is enclosed in a cylinder of a nonreactive material such as plastic or glass. The solution inside the ISE is held in place by an ion-selective membrane, which is sealed across the lower end of the cylinder. The physical nature of the membrane varies with the type of ISE. The various ISE types are described later in the chapter.

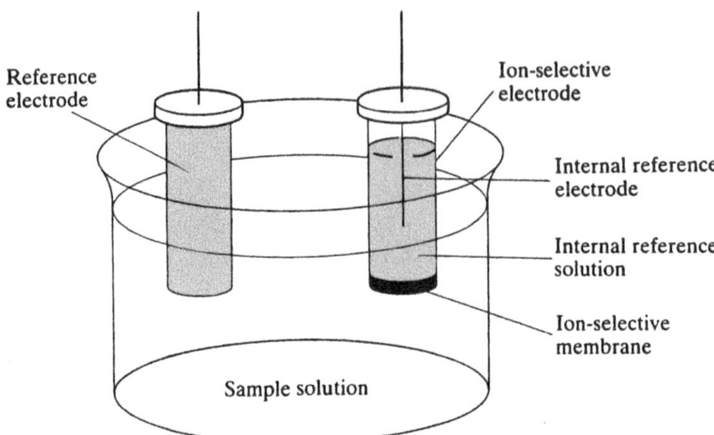

Figure 22-2 Diagram of the electrodes used for analysis with an ion-selective electrode.

The selectivity of an ISE results from the ion-selective membrane. When the ISE is dipped into a solution containing ions to which the electrode responds, a potential is developed across the ion-selective membrane. The potential is dependent upon the activity of the ion in the sample solution and the activity of the ion in the internal reference solution:

$$E_m = \frac{2.303RT}{zF}\log\frac{a_i}{a_j} = \frac{2.303RT}{zF}\log a_i - \frac{2.303RT}{zF}\log a_j \quad (22\text{-}16)$$

In Eq. (22-16) E_m is the potential across the membrane relative to the internal reference electrode; z is the elementary charge on the analyzed ion; a_i is its constant activity on the sample side of the membrane; and a_j is its activity in the internal reference solution on the other side of the membrane.

The potential difference E measured by a pH meter, digital voltmeter (DVM), digital multimeter (DMM), or other high-input-impedance potential measuring device is the difference in potential between that of the ISE (E_m) and that of the external reference electrode E_r:

$$E = E_m - E_r \quad (22\text{-}17)$$

Substitution of E_m from Eq. (22-16) into Eq. (22-17) and combination of the constants yield

$$E = \frac{2.303RT}{zF}\log a_i - \frac{2.303RT}{zF}\log a_j - E_r$$

$$= \text{constant} + \frac{2.303RT}{zF}\log a_i \quad (22\text{-}18)$$

It is apparent from Eq. (22-18) that the measured potential difference is related to the activity a_i of the assayed ion in the sample.

An electrode which obeys Eq. (22-18) is said to exhibit nernstian behavior. The interested reader can find a detailed derivation of Eq. (22-18) in Reference 7 (chap. 1). At 25°C Eq. (22-18) can be rewritten as

$$E = \text{constant} + \frac{0.05917}{z}\log a_i \tag{22-19}$$

From Eq. (22-19) it can be seen that a change in activity of a univalent ion by a factor of 10 should result in a potential change of 0.05917 V at 25°C. For a divalent ion, the potential change corresponding to a tenfold activity change should be 0.02959 V. Because the voltage changes are relatively small, some analysts prefer to use units of millivolts rather than volts. If the values of R and F are substituted into Eq. (22-18) and E and the constants are expressed in millivolts rather than volts, the following equation is obtained:

$$E = \text{constant} + \frac{0.1984T}{z}\log a_i \tag{22-20}$$

Equations (22-18) to (22-20) have been written on the assumption that the electrode responds only to ion i. Ion-selective electrodes often respond to other ions as well. If one or more of those other ions are in solution with i, the electrode response is given by

$$E = \text{constant} + \frac{2.303RT}{zF}\log\left(a_i + k_{i,j}a_j^{z/a} + b_{i,k}\, a_k^{z/b} + \cdots\right) \tag{22-21}$$

The constants $k_{i,j}$ and $k_{i,k}$ in Eq. (22-21) are the *selectivity coefficients* for ions j and k respectively relative to ion i. Selectivity coefficients are a measure of the relative response of the electrode to each interfering ion (*j, k, etc.)* as compared with the measured ion i. As the electrode's response to the interfering ion decreases, the selectivity coefficient also decreases. In Equation (22-21) the charges on ions j and k are a and b respectively. Actually, the theory of ISEs is somewhat more involved than presented here. Further details can be obtained from the appropriate sections of References 7 and 8.

Sample problem 22-3 A sodium-selective electrode and a reference electrode are dipped into a solution of sodium ion at an activity of 1.25×10^{-3} *M*. The potential of the sodium-selective electrode in the solution was – 0.2034 V. Calculate the potential of the same electrode pair in a solution containing sodium ion at an activity of 1.50×10^{-3} *M* and potassium ion at an activity of 1.20×10^{-3} *M* if the selectivity coefficient of the sodium-selective electrode for potassium ion $k_{Na,\,K}$ is 0.24. The temperatures of both solutions are 25.0°C.

SOLUTION Equation (22-19) and the potential of the electrode in the sodium ion solution can be used to calculate the value of the constant for the electrode:

$$E = \text{constant} + \frac{0.05917}{z}\log a_i$$

$$-0.2034 = \text{constant} + \frac{0.05917}{1}\log\left(1.25\times10^{-3}\right)$$

$$\text{Constant} = -0.0316$$

Equation (22-21) is used to calculate the potential of the solution containing both sodium ion and potassium ion:

$$E = \text{constant} + \frac{0.05917}{z}\log\left(a_i + k_{i,j}a_j^{z/a}\right)$$

$$= -0.0316 + 0.05917\log\left[1.50\times10^{-3} + 0.24\left(1.20\times10^{-3}\right)\right]$$

$$= -0.1942\ \text{V}$$

The selectivity coefficient of an ISE for a particular interference can be determined in several ways. The easiest way requires two potential measurements. First, the potential is measured in a solution containing a known activity of the ion for which the electrode is selective, but no interfering ions. Equation (22-18) is used to determine the value of the constant. A second solution is prepared which contains a known activity of both the assayed ion and an interfering ion. The potential of the second solution is measured using the same electrodes that were used for the first measurement. The result and the constant determined from the previous measurement are substituted into Eq. (22-21) which is then solved for the selectivity coefficient. While using that method it is important that sufficient interference be added prior to the second measurement to significantly alter the measured potential. The method suffers the usual disadvantages associated with small numbers of measurements; i.e., an undetected error can easily be introduced into one or both of the measurements. The measurements should be repeated several times with different solutions to allow estimation of random errors and detection of determinate errors.

A second way to determine a selectivity coefficient is the *fixed interference method*. A series of standard solutions are prepared which contain a fixed activity of the interference and varying activities of the analyte. The potential of each solution is measured and the results plotted as a function of the logarithm (or negative logarithm) of the activity of the analyte.

In the absence of an interference the potential should vary linearly with the logarithm of the activity as predicted by Eq. (22-18). At relatively high activities of the analyte, the contribution to the cell potential of the interference is negligible and the potential does vary linearly with log a_i as predicted. As the

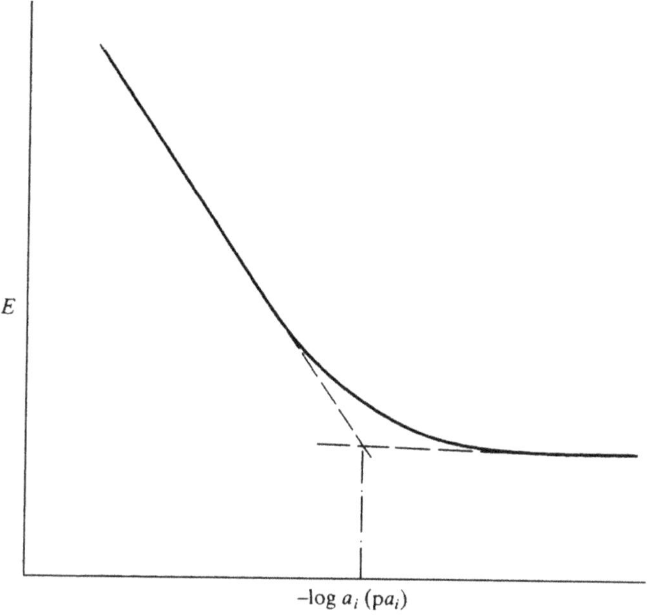

Figure 22-3 Change in potential of an ion-selective electrode with activity a_i of an analyte in the presence of a fixed concentration of an interference.

analyte activity is decreased, however, the relative contribution to the potential by the interference increases and eventually predominates. Because the activity of the interference is fixed, the potential eventually ceases to significantly change, and a plot of E as a function of log a_i changes slope as illustrated in Fig. 22-3.

The two linear portions of the plotted curve are extrapolated to the intersection as shown in Fig. 22-3. The activity of the analyte at the point of intersection corresponds to the activity at which the contributions to the cell potential of the analyte and the interference are equal. At that point the following equation is obeyed:

$$k_{i,j} = \frac{a_i}{a_j^{z/a}} \tag{22-22}$$

Substitution of the appropriate values into Eq. (22-22) allows calculation of the selectivity coefficient.

Sample problem 22-4 A series of standard solution were prepared which each contained nitrate at an activity of 1.0×10^{-3} M and varying activities of perchlorate as indicated in the following table. The potential of each solution was measured with a perchlorate-selective electrode. Determine the selectivity coefficient for nitrate with the perchlorate-selective electrode.

a_{ClO4^-}	*E*, V
1.0×10^{-2}	0.5143
5.0×10^{-3}	0.5321
1.0×10^{-3}	0.5735
5.0×10^{-4}	0.5885
1.0×10^{-4}	0.6282
5.0×10^{-5}	0.6293
1.0×10^{-5}	0.6315

SOLUTION The logarithm of the activity of perchlorate is calculated and the potential is plotted as a function of $-\log a_{ClO4^-}$. The plot is shown in Fig.22-4. The perchlorate activity at the intersection of the two extrapolated linear portions of the curve occurs at $-\log a_{ClO4^-} = 3.92$. The corresponding perchlorate activity is

$$-\log a_{ClO4^-} = 3.92$$

$$a_{ClO4^-} = 1.2 \times 10^{-4}$$

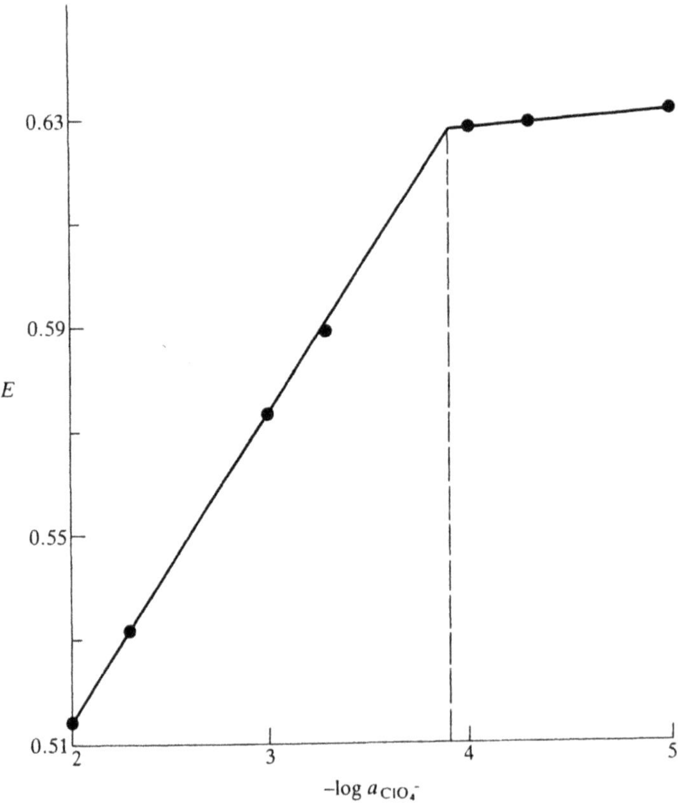

Figure 22-4 The plotted data from Sample problem 22-4 for a perchlorate-selective electrode.

Substitution of the appropriate values into Eq. (22-22) permits calculation of the selectivity coefficient:

$$k_{ClO_4^-, NO_3^-} = \frac{1.2 \times 10^{-4}}{\left(1.0 \times 10^{-3}\right)^{-1/-1}}$$

$$k_{ClO_4^-, NO_3^-} = 0.12$$

APPARATUS FOR POTENTIOMETRY

Direct potentiometry and potentiometric titrations require the use of two electrodes and a potential-measuring device. Usually one of the electrodes is an indicator electrode and the other is a reference electrode. Although the potential-measuring device can be a digital voltmeter (DVM), a digital multimeter (DMM), or, in some cases, a potentiometer, it is usually a pH meter.

Electrodes

Regardless of whether an electrode is used as a reference or an indicator electrode, it can often be classified into one of three categories. *Electrodes of the first kind* consist of a metal dipping into a solution of the ions of the metal. The reversible reaction at the electrode surface involves only the metal and its ions and is of the general form as follows:

$$M^{n+} + ne^- = M \tag{22-23}$$

Examples of electrodes of the first kind are silver dipping into silver nitrate solution, and mercury in contact with a solution of mercury(I) ions.

Electrodes of the second kind consist of a metallic wire or strip dipping into a solution that both contains metallic ions and is saturated with a sparingly soluble salt of the metallic ion. The electron transfer at the metallic surface must be reversible. The reaction taking place in that electrode can be regarded as occurring in two steps as shown in the following two examples:

$$Ag^+ + e^- = Ag \tag{22-24}$$

$$Ag^+ + Cl^- = AgCl\ (satd) \tag{22-25}$$

$$Hg_2^{2+} + 2e^- = 2Hg(l) \tag{22-26}$$

$$Hg_2^{2+} + 2Cl^- = Hg_2Cl_2(satd) \tag{22-27}$$

Alternatively, it can be thought of as taking place in a single step by adding the appropriate equations to give

$$AgCl\ (satd) + e^- = Ag + Cl- \tag{22-28}$$

$$Hg_2Cl_2\ (satd) + 2e^- = 2Hg(l) + 2Cl^- \tag{22-29}$$

In each case, an electrode of the second kind responds to changes in the activity of the anion of the sparingly soluble salt. The electrode corresponding to the reduction of mercury(I) to mercury in the presence of chloride will be used as an example. The Nernst equation written for the reaction in Eq. (22-26) is given by

$$E = E^{o}_{Hg_2^{2+}} + \frac{RT}{2F}\ln a_{Hg_2^{2+}} \tag{22-30}$$

The activity of mercury(I) can be calculated from the solubility product K_{sp} of the salt and substituted into Eq. (22-30) to give

$$K_{sp} = (a_{Hg_2^{2+}})(a_{Cl^-})^2 \tag{22-31}$$

$$a_{Hg_2^{2+}} = \frac{K_{sp}}{a^2_{Cl^-}} \tag{22-32}$$

$$E = E^{o}_{Hg_2^{2+}} + \frac{RT}{2F}\ln\frac{K_{sp}}{a^2_{Cl^-}} \tag{22-33}$$

Rearrangement of Eq. (22-33) yields

$$E = E^{o}_{Hg_2^{2+}} + \frac{RT}{2F}\ln K_{sp} - \frac{RT}{F}\ln a_{Cl^-} \tag{22-34}$$

Another way to obtain the potential of the electrode is to write the Nernst equation directly for Eq. (22-29) as shown:

$$E = E^{o}_{Hg_2Cl_2} - \frac{RT}{F}\ln a_{Cl^-} \tag{22-35}$$

From a comparison of Eq. (22-34) with Eq. (22-35), it is apparent that $E^{o}_{Hg_2Cl_2}$ is equal to $E^{o}_{Hg_2^{2+}}$ + $RT/2F$ in K_{sp}. Both Eqs. (22-34) and (22-35) show the dependence of the electrode potential on the activity of chloride in the solution.

An *electrode of the third kind* uses a saturated solution of two sparingly soluble salts with a common anion or complexes of two cations containing the same chelating agent. All chemical equilibria and the electron transfer must be reversible. One of the two salts contains cations of the electrode metal and the second contains cations to which the electrode responds. An example of an electrode of the third kind is the mercury-mercury oxalate-calcium oxalate-calcium ion electrode. An electrode of that kind can be represented in shorthand notation as shown:

$$Hg(l) \mid Hg_2C_2O_4(s) \mid CaC_2O_4(s) \mid Ca^{2+} \tag{22-36}$$

The chemical equilibria are listed as follows:

$$Hg_2^{2+} + 2e^- = 2Hg \tag{22-37}$$

$$Hg_2C_2O_4(s) = Hg_2^{2+} + C_2O_4^{2-} \tag{22-38}$$

$$C_2O_4^{2-} + Ca^{2+} = CaC_2O_4(s) \tag{22-39}$$

the sum of the equilibria being given by

$$Hg_2C_2O_4(s) + Ca^{2+} + 2e^- = 2Hg + CaC_2O_4(s) \qquad (22\text{-}40)$$

The Nernst equation for the overall reaction can be developed in a manner analogous to that in which Eq. (22-35) was developed, as follows:

$$E = E^o_{Hg_2C_2O_4/CaC_2O_4} + \frac{RT}{2F}\ln a_{Ca^{2+}} \qquad (22\text{-}41)$$

or it can be developed in a manner analogous to the way in which Eq. (22-34) was developed, as follows:

$$E = E^o_{Hg_2{}^{2+}} + \frac{RT}{2F}\ln K_{sp,Hg_2C_2O_4} - \frac{RT}{2F}\ln K_{sp,CaC_2O_4} + \frac{RT}{2F}\ln a_{Ca^{2+}} \qquad (22\text{-}42)$$

In either case it can be seen that the electrode responds to changes in activity of a cation which is not identical to that of the electrode metal.

Reference Electrodes

As mentioned earlier in the chapter, the hydrogen electrode (Fig. 22-1) can be used as a reference electrode if the hydrogen pressure and the H^+ activity are held constant. Generally the advantages associated with the zero standard potential of the hydrogen electrode are more than offset by the disadvantages of the electrode. The hydrogen electrode is cumbersome, and it responds slowly to changes in H^+ activity. The latter, of course, is not a disadvantage if the hydrogen electrode is used strictly as a reference electrode. Because H_2 can chemically reduce some organic compounds, it is necessary to use a salt bridge to separate the hydrogen electrode from a sample solution containing the organic compounds. Hydrogen reductions are sometimes catalyzed by platinum from the electrode. For those reasons the hydrogen electrode is rarely used in potentiometric investigations.

Several *secondary reference electrodes* are used more often than the hydrogen electrode for potentiometric measurements. A secondary reference electrode has a potential that has been accurately measured relative to the hydrogen electrode. Several of the more popular secondary reference electrodes are described in the following paragraphs. A more thorough description of reference electrodes is given in Reference 9.

The *silver-silver chloride electrode* is an electrode of the second kind [Eqs. (22-24), (22-25), and (22-28)] which serves as a reference electrode if it is placed in a solution that is saturated in AgCl and has a constant concentration of chloride. It is often used as the internal reference electrode in ion-selective electrodes (Fig. 22-2). A silver-silver chloride reference electrode can be prepared by dipping a silver wire into a solution that has a fixed chloride concentration (often a saturated or a 1-*M* KCl solution) and to which enough solid silver chloride had been added to saturate the solution.

Alternatively, the silver chloride can be formed in the solution by adding a small amount of silver nitrate to a solution containing chloride. Electrodes with more reproducible potentials are normally prepared by placing the silver wire in direct contact with silver chloride. That can be accomplished by making the silver wire the anode and passing electric current through the wire until sufficient Ag^+ has been generated to react with Cl^- in the solution to form a thin layer of AgCl on the wire or by dipping the silver wire in a melt of AgCl. Upon cooling the melt, the AgCl solidifies on the wire. The wire-salt assembly is dipped into the reference chloride solution. Some electrodes are prepared by initially plating a layer of silver on a platinum electrode. Subsequently, a portion of the deposited silver is electrolytically converted to AgCl on the surface of the electrode. At 25°C the potential of the silver-silver chloride electrode is 0.222 V in 1 *M* potassium chloride, 0.205 V in 3.5 *M* potassium chloride, and 0.199 V in saturated potassium chloride.

Calomel electrodes constitute another category of reference electrodes. They contain a saturated solution of calomel (a common name for Hg_2Cl_2), as shown in the half-cell in the following equation:

$$Hg(l) \mid Hg_2Cl_2(satd), KCl(aq)\| \qquad (22\text{-}43)$$

The overall half-reaction which takes place at a calomel electrode is that given in Eq. (22-29). The potential of a calomel electrode depends upon the activity of chloride in the half-cell. A calomel electrode containing a saturated solution of potassium chloride is a *saturated calomel electrode* (sce); an electrode which contains 1 *M* potassium chloride is a *normal calomel electrode* (nce); and an electrode which contains 0.1 *M* potassium is a *decinormal calomel electrode* (dce). Normal, as used in this sense, refers to the non-SI concentration unit which is now rarely used. For a +1 ion, normality is identical to molarity. The sce is the most widely used of the calomel electrodes.

A bottle-type sce is diagrammed in Fig. 22-5. The bottom of a wide-mouth bottle is filled to a depth of about 1 cm with liquid mercury. After a solid layer of a mercury (I) chloride-potassium chloride mixture is placed on top of the mercury, the bottle is filled nearly to the top with saturated potassium chloride solution. The solid potassium chloride and mercury (I) chloride are added to ensure that the solution is saturated in both salts. Electrical connection to the mercury pool is made either with a platinum electrode dipping into the mercury or with a platinum wire sealed through the wall of the bottle and into the mercury pool. Electrical connection to the other half-cell is completed through a potassium chloride salt bridge. The platinum electrode and salt bridge are held in place with a rubber stopper placed in the top of the bottle.

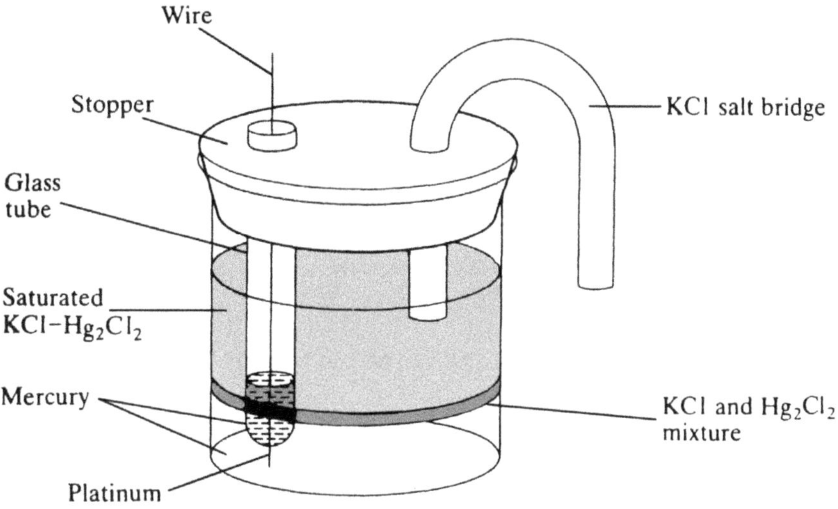

Figure 22-5 Diagram of a bottle-type saturated calomel electrode.

Commercial versions of calomel electrodes are prepared in glass or plastic tubes as shown in Fig. 22-6. An inner tube containing the mercury, mercury (I) chloride, and potassium chloride dips into a potassium chloride and saturated mercury (I) chloride solution contained in an outer tube. Electrical connection to the other half-cell is made through a fiber junction, a crack, a glass sleeve, a ceramic plug, or by some other means at the right most end in Figure 22-6 of the outer tube. At 25°C the potentials of the calomel electrodes which contain potassium chloride solutions of the indicated concentrations are 0.244 V (saturated), 0.246 V (4.0 *M*), 0.250 V (3.5 *M*), 0.280 V (1.0 *M*), and 0.336 V (0.1 *M*).

In a few cases, as when an ion-selective electrode is used to determine chloride, neither a silver-silver chloride nor a calomel electrode can be used as the reference electrode because the sample could be contaminated with chloride from the electrode. In those cases a reference electrode that often is used is the mercury-mercury (I) sulfate electrode. The half-reaction taking place in that reference electrode is shown by

$$Hg_2SO_4(satd) + 2e^- = 2Hg(l) + SO_4^{2-} \qquad (22\text{-}44)$$

The mercury-mercury (I) sulfate electrode is an electrode of the second kind whose potential is dependent upon the sulfate concentration of the solution. The standard potential of the half-reaction in Eq. (22-44) is 0.6125 V. In a saturated solution of potassium sulfate the electrode has a potential of 0.658 V. In one version of the electrode, 1 *M* Na_2SO_4 rather than saturated K_2SO_4 is used, and the potential is 0.648 V. The construction of the electrode is similar to that of the calomel electrodes.

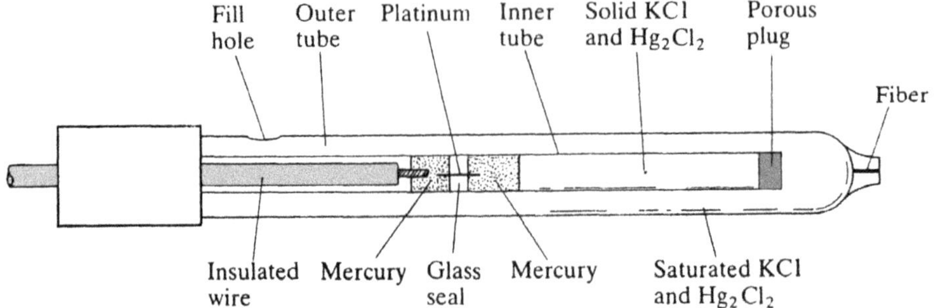

Figure 22-6 Diagram of a commercial version of a saturated calomel electrode.

If potentiometry is to be used to follow the course of a titration rather than for direct potentiometry, relative changes in potential are of more interest than absolute potentials. Consequently, the requirements for reference electrodes are considerably less strict than the requirements for electrodes which are used for direct potentiometry. The reference electrode used in a potentiometric titration must have a constant potential only during the titration. Furthermore, the potential of the electrode does not have to be known as long as it is constant.

A widely applicable and simple reference electrode which can be used for potentiometric redox titrations is the *titrant-stream reference electrode*. It consists of an inert indicator electrode (usually platinum) sealed through the wall of the buret tip so that the electrode is constantly immersed in the titrant. Because the potential of the electrode is determined by the concentration of the redox titrant components, and because those components do not change during the course of the titration, the potential of the electrode remains constant. Electrical connection to the titrand solution is completed by dipping the buret tip into the titrand. The titrant-stream electrode has the advantages of being simple, rugged, and widely applicable. A diagram of a titrant-stream electrode is shown in Fig. 22-7.

All of the previously described electrodes have been used for studies in aqueous solutions. Sometimes it is advantageous or necessary to use a non-aqueous solvent. In those cases it is generally not possible to use a reference electrode prepared with an aqueous internal solution. The use of a reference electrode containing an aqueous solution in a nonaqueous solvent can result in salt bridges plugged with salts soluble in one solvent but not the other, relatively large liquid-junction potentials, and contamination of the nonaqueous solvent with water from the reference electrode.

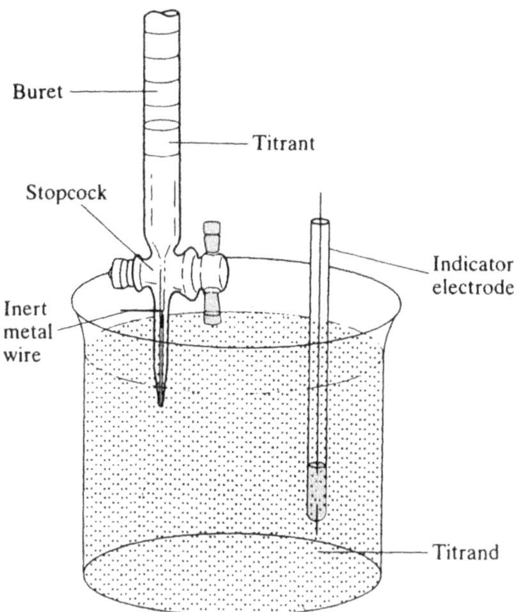

Figure 22-7 Diagram of the apparatus used for potentiometric titration with a titrant-stream reference electrode.

Table 22-3 Selected reference electrodes used for non-aqueous potentiometric measurements

Reference electrode †	Solvent
Ag | AgCl(*satd*), Cl^-	Acetic acid, formamide, dimethylformamide, dimethylacetamide, methylformamide
Ag | Ag^+(0.01 *M*)	Acetonitrile, dimethylformamide
Glass ‡	Acetonitrile
Antimony	Acetonitrile
Hg | Hg_2SO_4, H_2SO_4	Acetic acid
sce	Dimethylsulfoxide
Tl(Hg) | TlCl(*satd*), Cl^-	Liquid ammonia, dimethylsulfoxide
Cd(Hg) | $CdCl_2$(*satd*), Cl^-	Dimethylformamide, dimethylsulfoxide
Cu | Cu(II)(2.5 m*M*)	Pyridine

† Prepared using the nonaqueous solvent.

‡ The glass electrode is described later in the chapter. It can be used as a reference electrode in solutions in which the pH is constant.

Some workers have partially surmounted those problems by using an aqueous reference electrode with a nonaqueous salt bridge. Although that approach eliminates the problem of solvent contamination with water, the other problems remain. In general, use of an aqueous reference electrode and a non-aqueous salt bridge in a nonaqueous solvent cannot be encouraged. Whenever possible, it is preferable to use a reference electrode prepared in the same solvent as found in the sample.

The silver-silver chloride reference electrode has found the greatest use in nonaqueous solvents. Titrant-stream reference electrodes are generally useful for nonaqueous as well as aqueous titrations. Some of the reference electrodes and the solvents in which they have been used are listed in Table 22-3.

Indicator Electrodes

The most popular indicator electrode for potentiometric oxidation-reduction studies is a platinum wire. A typical platinum indicator electrode is shown as part of the sce in Fig. 22-5. It usually consists of a platinum wire sealed through the end of a glass tube. Electrical connection to the platinum is achieved by dipping a bare wire (often copper, iron, or nickel) into a mercury pool in contact with the platinum on the inside of the glass tube. The use of mercury can sometimes be avoided by inserting the wire in a small bundle of metallic foil which is in contact with the platinum wire on the inside of the tube. Generally a platinum wire which is sufficiently long to be used by directly dipping it into the solution is too expensive to be used as an indicator electrode. Other indicator electrodes which are sometimes used for potentiometric studies are constructed from mercury, gold, silver, "glassy" carbon, or graphite. The indicator-reference electrode pair can be made quite small for use in biological systems or in other situations requiring small electrodes. Although ion-selective electrodes are normally used as indicator electrodes in direct potentiometry, they can also be used in potentiometric titrations involving ions for which the electrodes are selective.

Ion-Selective Electrodes

A general description of ion-selective electrodes and the theory of operation of those electrodes was presented in an earlier section of the chapter. The major components of most ion-selective electrodes are shown in Fig. 22-2. Ion-selective electrodes can be classified according to the type of membrane used in the electrode. The major categories of ion-selective electrodes are glass electrodes, liquid-ion-exchanger membrane electrodes, solid-state membrane electrodes, neutral-carrier membrane electrodes, coated-wire electrodes, field-effect-transistor electrodes, gas-sensing electrodes, air-gap electrodes, and biomembrane electrodes.

Glass electrodes The composition of the glass ion-selective membrane in a glass electrode is carefully controlled. By altering the composition of the glass, it is possible to make the electrode selective for different ions. The glasses used in glass membranes are composed of mixtures of oxides of elements which have an oxidation number of at least 3 and elements with an oxidation number of 1 or 2. Usually the glasses contain 60 to 75 mole percent SiO_2, 2 to 20 percent Al_2O_3 or La_2O_3, 0 to 6 percent BaO or CaO, and a variable amount of a group 1A oxide. The mixture of the oxides is melted and cooled to form the glass. Monovalent cations in the three-dimensional glass structure are relatively mobile. Consequently, monovalent cations from a solution into which the glass is dipped can penetrate into the surface of the glass and be cation-exchanged at negatively charged sites in the glass. Because the concentration of the analyzed ion in the sample solution differs from that in the internal reference solution, a potential difference develops across the membrane as indicated in Eq. (22-16). Glass membranes are selective for monovalent cations because polyvalent ions cannot easily penetrate the surface of the membrane. Evidently the selectivity of glass electrodes is related both to the ability of the various monovalent cations to penetrate into the glass membrane and to the degree of attraction of the cations to the negative sites within the glass.

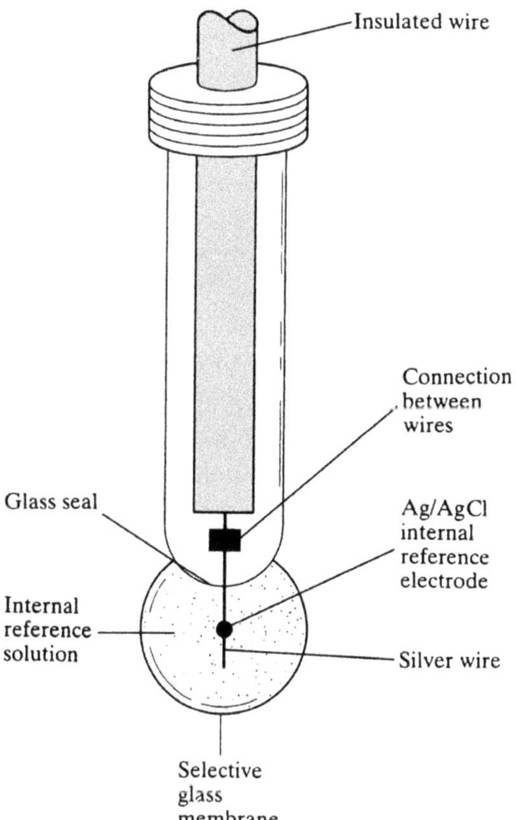

Figure 22-8 Diagram of a glass electrode.

A diagram of a glass electrode is shown in Fig. 22-8. The electrical resistance across glass membranes is typically between 50 and 500 MΩ and requires the use of a high-input-impedance potential-measuring device such as a pH meter. Glass electrodes which are selective for H^+ (the pH electrode), Li^+, Na^+, K^+, Cs^+, Ag^+, Tl^+, and NH_4^+ are available. The electrodes are used most often to measure hydrogen and sodium ions.

Interferences with glass electrodes It has been known for quite some time that the glass pH electrode gives erroneous results in highly alkaline solutions. That type of error in the determination of pH is *alkaline error*. Actually the term is a misnomer, because the error is caused by interference from relatively high concentrations of the univalent cation of the base rather than from the base itself. For example, the pH error in sodium hydroxide solutions is caused by Na^+ to which the electrode slightly responds. In alkaline solutions the hydronium concentration is low and the concentration of the univalent cation of the base is high. As an example, in a pH 13 NaOH solution, the hydronium concentration is 10^{-13} M and the sodium concentration is 0.1 M. Even though the selectivity coefficient [Eq. (22-21)] for sodium is small with the pH electrode, at a ratio of Na^+ to H^+ that is 10^{12}, the response of the electrode to Na^+ becomes significant relative to the response to H^+.

Some glasses also give erroneous results in highly acid (pH less than zero) solutions. Although that type of error is termed *acid error,* it is not necessarily caused by an acid. In the dilute solutions which are normally assayed with glass electrodes, the activity of water is nearly unity. In solutions containing high solute concentrations, the activity of the water in equilibrium with the glass membrane is less than unity. Because the response of the glass electrode differs with the activity of the water which is in equilibrium with the glass membrane, aqueous solutions containing high solute concentrations cause erroneous electrode responses. That type of error is often encountered in highly acidic solutions; consequently, it is referred to as acid error.

A glass electrode which is selective for a specific cation is subject to interference from other monovalent cations. Because nearly all glass electrodes are subject to interference from H^+ , it is advisable, when using non-pH glass electrodes, to prepare the working curve and the sample with buffered solutions. Of course, careful attention must be paid to the cations used to prepare the buffered solutions so as not to introduce more potential interferences.

Liquid-ion-exchanger membrane electrodes A diagram of a typical liquid-ion-exchanger membrane electrode is shown in Fig. 22-9. The inner compartment of the electrode contains a reference electrode and an aqueous reference solution. The outer compartment contains an organic liquid ion exchanger. In the commercial version of the electrode the two compartments are filled through holes under the screw cap.

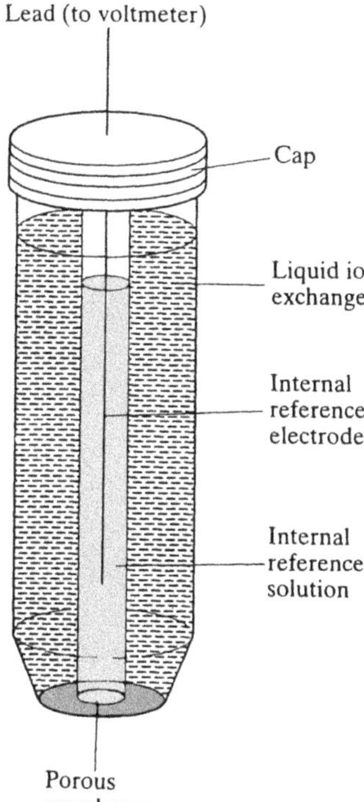

Figure 22-9 Diagram of a liquid-ion-exchanger membrane electrode.

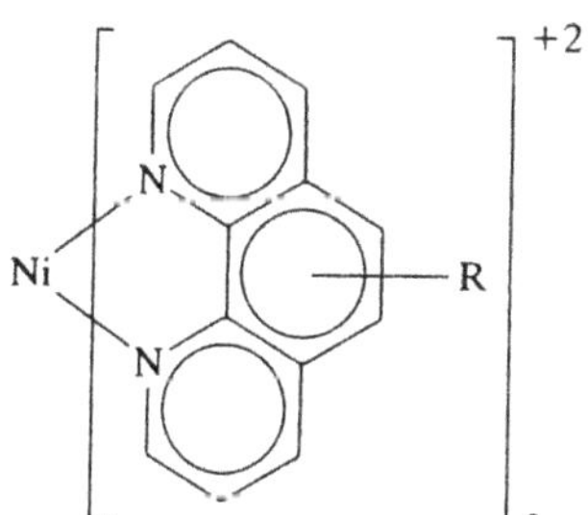

Figure 22-10 Chemical structure of a liquid anion exchanger which has been used for NO_3^- and BF_4^-.

The liquid ion exchanger is insoluble in the solvent in which the electrode is to be used (water) and is nonvolatile at room temperature. The ion exchanger is dissolved in a relatively high-molecular-weight solvent such as dioctylphenylphosphonate. Liquid ion exchangers consist of polar ionic sites attached to relatively large nonpolar organic molecules. The ionic sites are negative in a cation exchanger and positive in an anion exchanger. Typical liquid ion exchangers are $(RO)_2PO_2^-$ (for Ca^{2+} and Mg^{2+}) and $RSCH_2COO^-$ (for Cu^{2+} and Pb^{2+}). R in the ion exchangers can be any of several organic groups, for example *p*-(1,1,3,3-tetramethylbutyl)phenyl, *p*-(*n*-*octyl*)phenyl, and decyl.

The S and O in ion exchangers of the $RSCH_2COO^-$ type selectively form chelate rings with certain ions, e.g., with Cu^{2+} or Pb^{2+}. Liquid anion exchangers are usually complexes formed between a transition metal and an organic ligand. The structure of a typical liquid anion exchanger is shown in Fig. 22-10.

The ion exchanger and reference solutions are held in place inside the electrodes by a porous membrane. Although the membrane can be made from polyvinyl chloride (PVC), it is usually constructed from some form of cellulose, e.g., from cellulose acetate. The membrane is prepared to have a pore diameter of about 100 nm. Chemical treatment makes the membrane hydrophobic. The membrane is in physical contact with the liquid ion exchanger and becomes permeated with it. Because the membrane is hydrophobic, water from the internal reference solution and from the sample solution is prevented from mixing with the liquid ion exchanger. A second type of liquid-ion-exchanger membrane utlizes a polymeric membrane to permanently hold the ion-exchanger solution in place within the membrane. That type of membrane does not need to be in direct contact with a solution of the liquid ion exchanger. In either case the shell of the electrode is made from an inert material such as glass or an organic polymer.

Liquid-ion-exchanger membrane electrodes owe their selectivity to their ability to selectively exchange ions. Upon contacting the membrane, an ion from the aqueous solution exchanges with an ion on a polar site in the ion exchanger. The newly created ion and ion-exchanger combination can freely diffuse throughout the membrane. The ionic conductivity of the membrane results from the mobility of the ion within the membrane. The potential across the membrane is related to the ionic conductivity within the membrane. Liquid-ion-exchanger membrane electrodes have been used to assay for Ca^{2+}, K^+, Li^+, Na^+, Mg^{2+}, Ni^{2+}, Zn^{2+}, Tl^+, Ag^+, Hg^{2+}, water hardness ($Ca^{2+} + Mg^{2+}$) Cu^{2+}, Pb^{2+}, Cl^-, BF_4^-, NO_3^-, ClO_4^-, $Cr_2O_7^{2-}$, benzoate, SCN^-, and other ions. The useful concentration range of liquid-ion-exchanger membrane electrodes varies with the assayed ion, but typically is from about 1×10^{-6} to 1 M.

Solid-state membrane electrodes A solid-state membrane electrode has the general structure shown in Fig. 22-11. The solid-state membrane can be a single crystal, a pellet made from a sparingly soluble salt, or a sparingly soluble salt embedded in an inert matrix, e.g., rubber. Because the single-crystal and pellet membranes are homogeneous, electrodes containing them are referred to as *homogeneous membrane electrodes*. The membrane consisting of the sparingly soluble salt in the inert binding material is a *heterogeneous membrane electrode*.

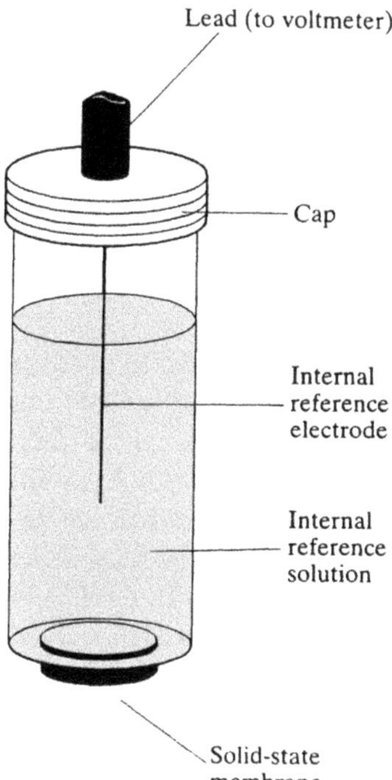

Figure 22-11 Diagram of a solid-state membrane electrode.

The lanthanum fluoride (LaF_3) membrane is the only single-crystal membrane that is widely used in ion-selective electrodes. As is done during the preparation of extrinsic semiconductors by doping (Chapter 3), the resistance of the LaF_3 crystal is decreased by replacing a relatively small number of La^{3+} ions in the crystal with Eu^{2+} ions. Fluoride ions migrate from vacancy to vacancy in the doped LaF_3 crystal. As a fluoride ion abandons one position in the crystalline structure, it leaves a hole into which another fluoride can migrate. The result is a crystal which exhibits ionic conductivity rather than the electron mobility that is observed in extrinsic semiconductors. The conductance (Chapter 23) through the membrane, as well as the potential across the membrane, can be related to analyte concentration for many solid-state membrane electrodes. Vacancies in the crystalline structure have exactly the proper size, charge, and shape to hold a fluoride ion. Because fluoride can selectively migrate through the crystal, the lanthanum fluoride membrane is selective for fluoride. If no fluoride is present in the sample solution, the LaF_3 membrane electrode can be used to assay for La^{3+}. Because the K_{sp} of LaF_3 is small, the activity of F^- in the solution (from dissolved membrane) is determined by the activity of the La^{3+} in the sample, and the electrode responds to changes in La^{3+} activity. The fluoride selective electrode is of particular interest owing to its popular use for the determination of fluoride in drinking and environmental waters.

Some sparingly soluble salts can be pressed into polycrystalline membranes that have a mobile ion and consequently exhibit ionic conductivity. Both silver chloride and silver bromide can be pressed into membranes in which the silver ion is mobile. Those membranes can be used in electrodes which are selective for either Ag^+ or the halide. Because the behavior of AgCl and AgBr membranes is affected by light, the membranes are of little practical use in most ion-selective electrodes. Membranes made from silver sulfide also have Ag^+ mobility, but they are not affected by light. Electrodes which have Ag_2S membranes can be used to assay either Ag^+ or S^{2-}.

The good membrane properties of Ag_2S can be used to advantage by preparing membranes from a mixture of Ag_2S and a sparingly soluble salt that has an ion in common with Ag_2S. Halide-selective membranes can be prepared by mixing a silver halide, for example, AgCl, AgBr, or AgI, with Ag_2S and compressing the mixture into a membrane. Because the K_{sp} for the silver halide is greater than the K_{sp}, for Ag_2S (about 10^{-51}), the halide activity in the sample determines the silver ion activity in the solution and the electrode responds to changes in activity of the halide. The activity of the halide in the solution fixes the silver ion activity (from dissolved membrane) at the value required by the solubility product of the silver halide. That activity determines the ionic conductivity in the membrane and consequently the potential across the membrane.

Membranes which are made from a mixture of Ag_2S and a sparingly soluble metal sulfide are selective for the metallic ion. The solubility product of the metallic sulfide must be larger than that of Ag_2S, so that the activity of Ag^+ in the sample solution will be determined by the metal ion activity rather than by the activity of dissolved Ag_2S from the membrane. Ion-selective electrodes containing Ag_2S-MS membranes which are selective for Cu^{2+}, Pb^{2+}, and Cd^{2+} can be prepared. All electrodes prepared from membranes containing Ag_2S respond to changes in the activity of Ag^+, S^{2-}, and any ion which forms a salt with Ag^+ or S^{2-} that has a K_{sp} which is comparable with or larger than that of the Ag_2S salt in the membrane. Some of the homogeneous membrane electrodes which have been used and their membrane compositions are listed in Table 22-4.

Table 22-4 Selected pellets used as homogeneous membranes in ion-selective electrodes

Pellet composition	Analyte
Ag_2S	Ag^+, S^{2-}
Ag_2S-AgBr	Br^-
Ag_2S-AgCl	Cl^-
Ag_2S-AgCN	CN^-
Ag_2S-AgI	I^-, CN^-
Ag_2S-AgSCN	SCN^-
Ag_2S-CdS	Cd^{2+}
Ag_2S-CuS	Cu^{2+}
Ag_2S-PbS	Pb^{2+}

In some cases the salt mixture can be melted rather than pressed into a homogeneous membrane. In several ion-selective electrodes mercury (I) sulfide (Hg_2S) or a mixture of Ag_2S and a sparingly soluble salt of Hg_2^{2+} or S^{2-} is used in the membrane. Such an electrode behaves like the Ag_2S electrodes (Table 22-4); e.g., the Hg_2S–Hg_2Br_2 electrode is selective for Br^-.

A heterogeneous membrane consists of an active ingredient dispersed throughout an inert binding material. The inert binder provides the physical properties that are required of the membrane, and the active ingredient provides the membrane's selectivity. Wax, silicone rubber, polyvinyl chloride PVC, and several other polymeric substances have been used as inert binders in ISEs. Silicone rubber and PVC are the most popular binders. Generally, a heterogeneous membrane is prepared by mixing the active ingredient with a binder prior to polymerization of the binder. After the polymerizing agent has been added, the membrane is formed into a thin sheet (a fraction of a millimeter thick). An electrode membrane of the appropriate size is cut from the sheet and attached to the end of the electrode body.

The active ingredient in heterogeneous membranes is often a sparingly soluble substance similar to the substances in homogeneous membranes. Heterogeneous membrane electrodes behave similarly to homogeneous membrane electrodes. As an example, the use of a silver halide as the active ingredient in a heterogeneous membrane electrode results in selectivity for both the halide and silver. Likewise, a barium sulfate-impregnated membrane is selective for sulfate and barium, and a silver sulfide-impregnated membrane is selective for sulfide and silver. As with pressed homogeneous membrane electrodes, an interfering ion is one which combines with the oppositely charged ion of the active ingredient to form a sparingly soluble salt. Heterogeneous membranes used in ISEs have also been prepared from ion exchangers similar to those used in the liquid-ion-exchanger electrodes described earlier.

Neutral-carrier membrane electrodes Neutral-carrier membrane electrodes have essentially the same design as liquid-ion-exchanger membrane electrodes. The liquid ion exchanger is replaced in neutral-carrier membranes with a neutral complexing agent (a neutral carrier), such as a crown ether, which is dissolved in a highly water-insoluble organic solvent. The neutral carrier complexes with the analyte at the membrane-sample interface to form a charged complex which is extracted from the aqueous solution into the organic solvent in the membrane. The selectivity of the membrane for a particular ion depends upon the ability to extract the ion into the membrane, which in turn depends upon the ability of the ion to form a complex with the neutral carrier. The major difference between a liquid-ion-exchanger membrane and a neutral-carrier membrane is the charge on the mobile species in the liquid membrane. After complexation and extraction, the species in the neutral-carrier membrane has the same charge as the extracted ion. After ion exchange and extraction, however, the species in the ion-exchanger membrane is neutral.

The neutral carriers originally used to prepare neutral-carrier membrane electrodes were naturally occurring substances such as valinomycin (used in K^+– selective electrodes). Synthetically prepared neutral carriers have also been used. Several neutral carriers and the ions for which they are used are listed in Table 22-5.

Table 22-5 Selected neutral carriers that are used in neutral-carrier membrane electrodes

Analyzed ion	Neutral carrier
K^+	Valinomycin
Ca^{2+}	
Na^+	

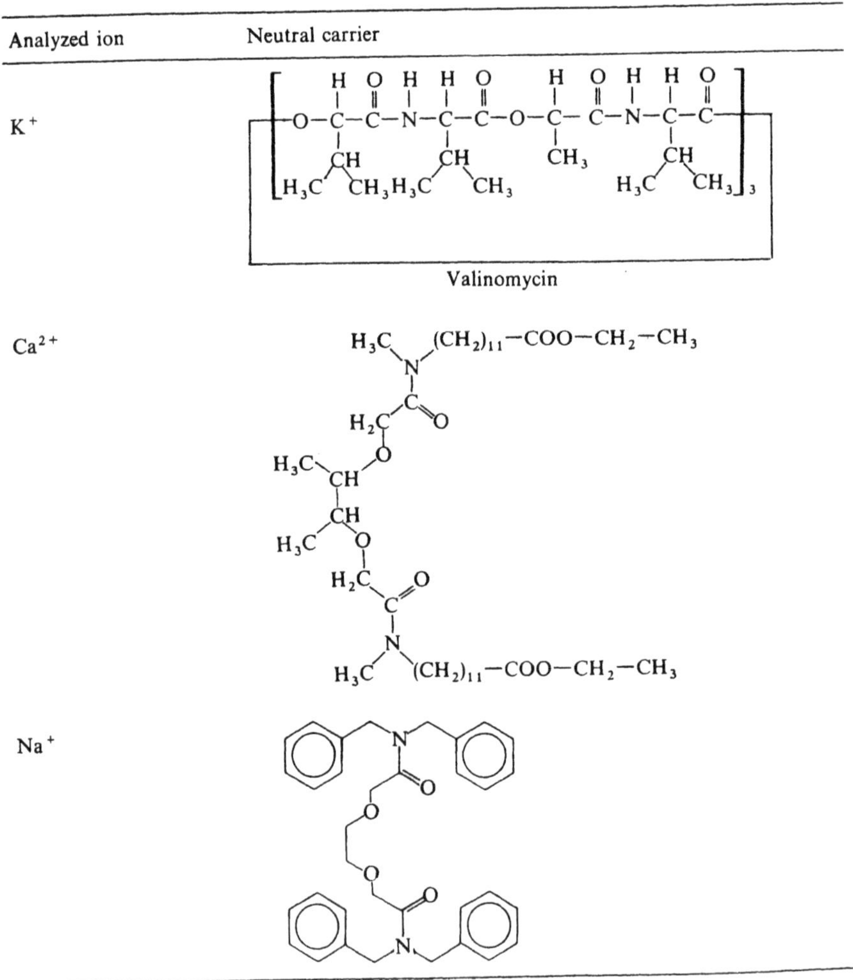

The solvent in which the neutral carrier is dissolved is usually a high-boiling organic compound such as nitrobenzene (used in a Ba^{2+} -selective electrode), dibutylsebacate (used in a K^{+}-selective electrode), and *o*-nitrophenyl-*n*-octyl-ether (used in a Ca^{2+}- selective electrode). The physical support for the neutral carrier and solvent is often either a cellulose membrane (a cellulose ester membrane is used with one Ba^{2+}-selective electrode) or, more commonly, a PVC membrane. In some electrodes the solution is immobilized in a polymeric matrix such as PVC. In addition to the K^{+}-,Ca^{2+}-, Ba^{2+}-, and Na^{+}-selective electrodes already mentioned, neutral-carrier membrane electrodes that are selective for Li^{+} ,H^{+}, Mg^{2+}, NH_4^{+}, Sr^{2+}, and other ions have been prepared. Advances in theory and applications of ion-selective electrodes of all kinds can be found in the biennial reviews in *Analytical Chemistry.*[10]

Coated-wire electrodes The size of the ion-selective electrodes described earlier is too large (tip diameters between 3 and 15 mm) to permit their use in applications in which the sample volume is small. Ion-selective electrodes with tip diameters of less than about 10 μm can be used as detective devices for chromatography and flow-injection analysis (refer to later chapters) as well as for numerous biological applications such as *in vivo* monitoring of a particular ion or molecule. Coated-wire electrodes can be considerably smaller than other forms of ion-selective electrodes because the internal filling solution is eliminated and the ion-selective membrane is coated directly on the internal electrode wire. A diagram of a coated-wire electrode is shown in Fig. 22-12.

The ion-selective membranes utilized in coated-wire electrodes usually consist of either an ion exchanger or neutral carrier immobilized in a polymeric material that is coated on the electrode. Other types of membranes, such as a sparingly soluble salt in a polymeric matrix, also can be used. The usual method of preparation of the electrodes is as follows. First, the metal on the interior of the electrode is sealed into glass or some other suitable material so that several millimeters or less of the wire is exposed. Subsequently the exposed wire is successively dipped into a solution of the polymeric material and then into a solution of the ion exchanger or neutral carrier. After the electrode has air-dried, the dipping procedure is repeated, if necessary, until the membrane coating on the wire is the desired thickness. Alternatively, the wire can be dipped into a single solution containing both the membrane material and the polymerizer. The polymeric matrix can be any of several materials including PVC, polymethyl acrylate (PMM), or epoxy.

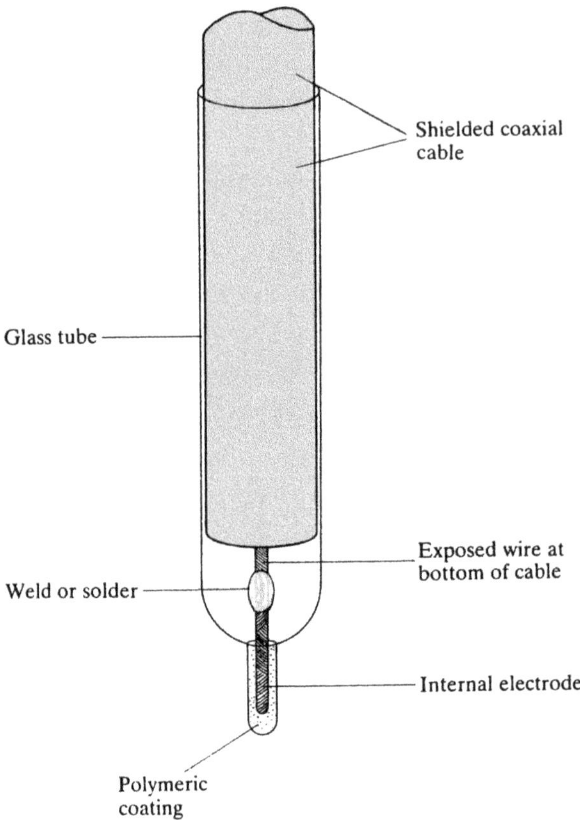

Figure 22-12 Diagram of a coated-wire electrode.

The internal electrode can be constructed from any of several metals. If the metal is platinum, the electrode responds to any redox couple which is capable of passing through the membrane. If platinum or vitreous ("glassy") carbon is coated with a pure polymeric material, which contains no neutral carrier or ion exchanger, the electrodes preferentially respond to H^+ because most other chemical substances are hindered by the membrane from contacting the electrode. Other materials that have been used as the internal electrode include copper and silver wire, and graphite rods.

Coated-wire electrodes are more sturdy than other ISEs and can be constructed with small tips. The electrodes can be thought of as electrodes of the second kind in which the precipitate has been replaced with a polymeric coating. Essentially the electrodes respond to whatever chemical species can penetrate into the coating. A more detailed description of the electrodes can be found in References 11 and 12.

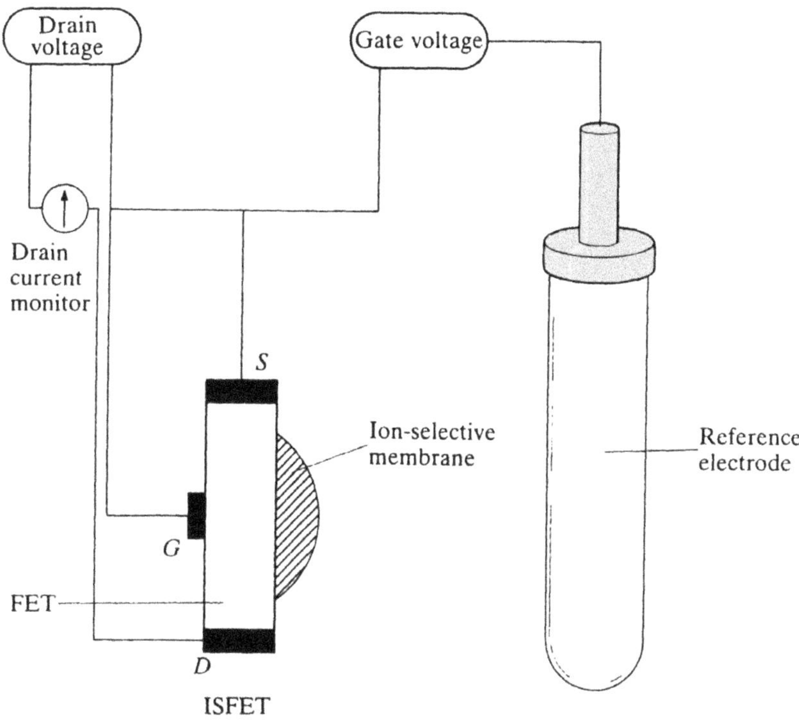

Figure 22-13 Diagram showing the electric connections between an ISFET and a reference electrode. The drain current varies with the activity of the analyte. The size of the ISFET is exaggerated for clarity.

Ion-selective field-effect transistors The ion-selective field-effect transistors (ISFETs) comprise a relatively new class of ISEs. The electrode consists of an ion-selective membrane deposited or coated on the gate of a field-effect transistor (FET) (Chapter 3, Fig. 3-11). The membrane can be a sparingly soluble compound such as silver bromide (solid-state membrane) or some other type of membrane such as an ion exchanger or neutral carrier in a PVC matrix. Often membranes in a PVC matrix are used. Membranes in a PVC matrix can be forced to adhere to the gate of the FET by placing a polyimide mesh over the gate prior to coating it with the membrane.

The potential at the membrane is partially determined by the activity of the analyte in solution. That potential determines the flow of current through the drain of the FET. The drain current consequently varies with the activity of the analyte and is the monitored factor. Both n- and p-type FETs can be used. A diagram illustrating an ISFET and its use is shown in Fig. 22-13. A more thorough description of ISFETs is provided by R. P. Buck in Reference 11 (chapter 1). ISFET electrodes, as well as some other semiconductor based electrodes, can be miniaturized, and consequently have many of the same advantages displayed by coated wire electrodes.

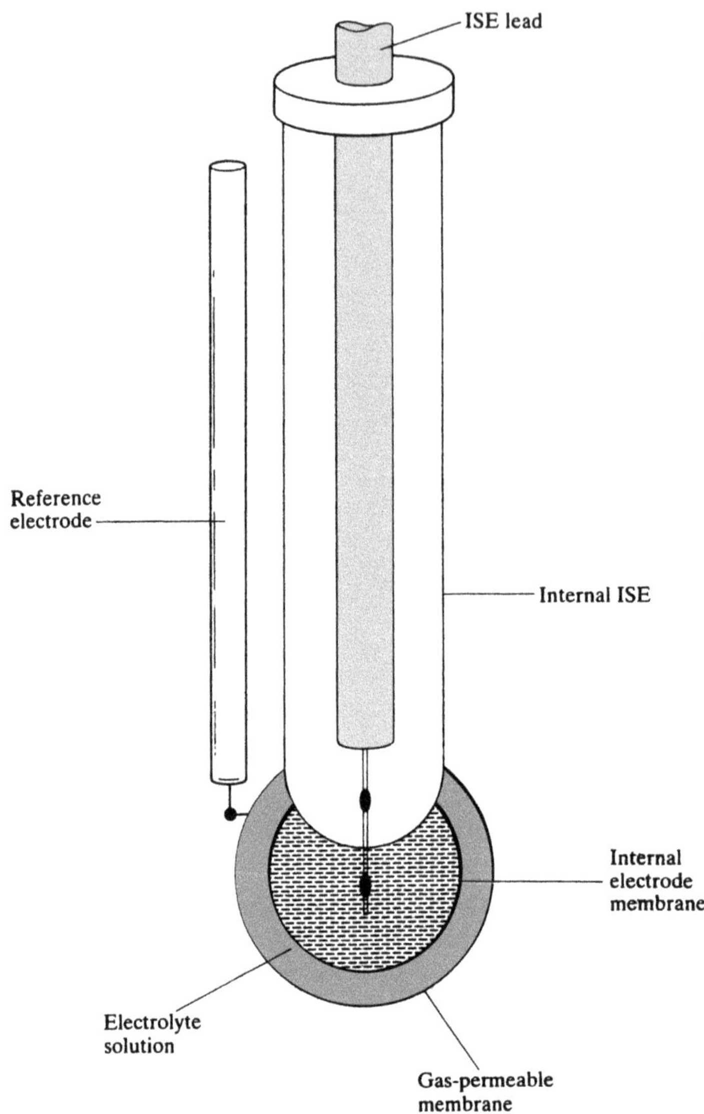

Figure 22-14 Diagram of a gas-sensing electrode.

Gas-sensing electrodes Gas-sensing electrodes are used to assay gases dissolved in aqueous solutions. A gas-sensing electrode is constructed by enclosing the ion-selective membrane of one of the ISEs that were described earlier (usually a glass pH membrane) in a second, gas-permeable hydrophobic membrane (Fig. 22-14). A thin layer of an electrolyte solution is held between the two membranes. Gas-sensing electrodes also have a small reference electrode enclosed within the gas-permeable membrane. The potential between the internal ISE and the reference electrode within the outer membrane is monitored. The gas-permeable membrane holds a constant volume of solution around the internal ISE into which the gaseous analyte can diffuse. The hydrophobic gas-permeable membrane can be composed of any substance which allows passage of dissolved gas but prevents the solution within the membrane from escaping.

Gas from the sample solution passes through the submerged gas-permeable membrane and equilibrates in the electrolyte solution between the two membranes. The gas reacts reversibly with the electrolyte solution to form an ion to which the ion-selective electrode responds. Because the activity of the ion that is formed between the two membranes is proportional to the amount of gas dissolved in the sample, the electrode response is directly related to the activity of the gas in the sample.

The gases (primarily NH_3, SO_2, and CO_2) which are detected by gas-sensing electrodes based on the pH electrode equilibrate with the electrolyte solution to alter its pH:

$$NH_3 + H_2O = NH_4^+ + OH^- \tag{22-45}$$

$$SO_2 + H_2O = HSO_3^- + H^+ \tag{22-46}$$

$$CO_2 + H_2O = HCO_3^- + H^+ \tag{22-47}$$

The electrolyte solution for each electrode contains the ion shown on the right-hand side of the equations. The ammonia electrode contains an ammonium chloride electrolyte solution; the sulfur dioxide electrode contains sodium hydrogen sulfite; and the carbon dioxide electrode contains sodium hydrogen carbonate. By controlling the activity of the electrolyte solution, the gaseous activity becomes proportional to the activity of either OH^- [Eq. (22-45)] or H^+ [Eqs. (22-46) and (22-47)], as can readily be seen by substitution into the appropriate expression for the equilibrium constant. The glass electrode monitors the activity of H^+ and consequently of the gas.

Other gases can be assayed with gas-sensing electrodes based on other internal ion-selective electrodes. Hydrogen sulfide, hydrogen cyanide, hydrogen fluoride, and chlorine can be assayed by using internal homogeneous membrane electrodes containing the appropriate silver salt. The selectivity of gas-sensing electrodes is determined by the internal ISE, the electrolyte solution, and the type of gas-permeable membrane. The gas-permeable membrane can be made from silicone rubber, Teflon (registered trademark of Du Pont), polypropylene, fluorinated ethylene propylene, polyvinylidene fluoride (a microporous filter), or some other substance.

Gas-sensing electrodes have the disadvantage of possessing a relatively long response time. Typically they require from 1 to 7 min after insertion into a sample solution to reach equilibrium. Commercial gas-sensing electrodes for NH_3, CO_2, H_2S, NO_x, SO_2, and Cl_2 are available. Gas-sensing electrodes which are selective for $(C_2H_5)_2NH$, HCN, HF, NO_2, and other gases have also been used.

Air-gap electrodes Another form of gas-sensing electrode is the air-gap electrode (Fig. 22-15), invented by Ruzicka and Hansen.[13] Because the air-gap electrode does not come in direct contact with the sample, the need for a gas-permeable membrane is eliminated. A very thin layer of an appropriate electrolyte solution is adsorbed on the surface of the membrane of the glass electrode. The electrolyte solution is adsorbed on the glass membrane when the membrane comes in contact with a sponge containing the electrolyte solution and a wetting agent. The reference electrode makes contact with the adsorbed electrolyte layer through a small, porous, ceramic salt bridge.

The air-gap electrode is used to assay ionic species which can be chemically converted to gases. The analysis of HCO_3^- will serve as an example. The HCO_3^- solution is placed in the sample holder (Fig. 22-15) and an acid is added to convert the HCO_3^- (*aq*) to CO_2(*g*). The sample holder is placed in position under the electrode and stirred with a magnetic stirrer and stirrer bar. Carbon

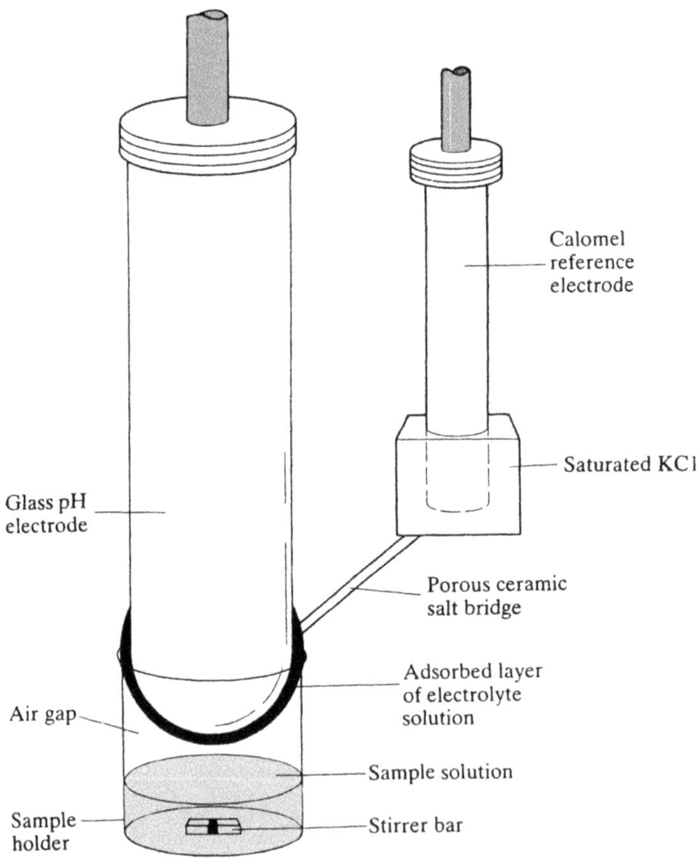

Figure 22-15 Diagram of an air-gap electrode. The reference and glass electrodes are enclosed in a single tube. The sample compartment attaches with a gastight seal to the bottom of the electrode tube.

dioxide, which is emitted during the chemical reaction, equilibrates with the electrolyte solution on the glass membrane and alters the pH of the solution. The glass electrode measures the pH of the resulting solution. The electrolyte solutions used with the air-gap electrode are the same as those used with other gas-sensing electrodes. The air-gap electrode has a faster response time (owing to the thinner layer of electrolyte solution) and a longer lifetime than most of the other types of gas-sensing electrodes. A typical response time for an air-gap electrode is less than a minute. Air-gap electrodes have primarily been used for analysis of NH_4^+, HSO_3^-, and substances which can be converted into NH_4^+ or HSO_3^-. As an example, an air-gap electrode can be used for the determination of urea in blood. The urea is enzymatically converted to NH_4^+ and subsequently assayed as previously described.

Biomembrane electrodes A biomembrane electrode is an ion-selective electrode of one of the types described earlier which has a membrane that is coated with an enzyme-containing acrylamide gel. The gel and enzyme are held in place on the surface of the ion-selective electrode by an inert physical support. The design is similar to that of the gas-sensing electrode (Fig. 22-14). Often the support is a sheet of cellophane or a piece of gauze made from dacron or nylon. The physical support is wrapped around the electrode membrane and tied in place. The acrylamide gel containing the enzyme is coagulated on the support-electrode combination. In an early paper[14] describing the urea-selective electrode, the solid support was a strip of nylon from a stocking.

Enzymes are highly selective biochemical catalysts. Generally, an enzyme catalyzes only a small number of reactions. The selectivity of biomembrane electrodes is attributable to the selectivity of the enzymes that are used in the electrodes. In biomembrane electrodes the enzyme-catalyzed reaction of the analyte yields an ionic reaction product which is monitored by the internal ion-selective electrode. It should be possible to construct a biomembrane electrode which is selective for any species for which an enzyme reaction can be found if the product of the reaction is an ion for which an ISE exists.

The major credit for the development of biomembrane electrodes belongs to G. G. Guilbault and his associates at the University of New Orleans.[14, 15] The operation of the urea-selective electrode will serve to illustrate the operation of biomembrane electrodes. The glass membrane of an ammonium-sensitive glass electrode is coated with an acrylamide gel layer containing the enzyme urease. When the electrode is dipped into a solution containing urea, the following reaction occurs to yield NH_4^+:

$$\underset{\text{Urea}}{CO(NH_2)_2} + H_2O \xrightarrow{\text{Urease}} 2NH_4^+ + CO_2 \qquad (22\text{-}48)$$

The NH_4^+ formed during the reaction is measured at the ammonium selective electrode. A working curve is prepared by plotting the potential of the electrode in standard urea solutions as a function of the logarithm of urea concentration. The urea concentration in the sample is obtained from the working curve. Unfortunately the enzymes used in biomembrane electrodes gradually decay and the enzyme-containing gel must be periodically replaced. The biomembrane of the urea electrode lasts about 2 weeks. Biomembrane electrodes usually have relatively long response times of 5 or more minutes.

A *biological electrode* uses living bacteria, which excrete the desired enzymes, in place of pure enzymes. Biological electrodes last longer than electrodes containing pure enzymes because the living bacteria can be replenished by storing the electrodes in a proper nutrient. A disadvantage of biological electrodes is that the bacteria can produce undesired enzymes that catalyze other reactions as well as those required for the assay.

Biomembrane electrodes which are selective for enzymes have also been designed. In that case the substance whose reaction is catalyzed (the *substrate*) is placed in the biomembrane in place of the enzyme. The substrate becomes at least partially depleted during the reaction and must be replaced regularly.

A closely related electrode system is the *enzyme-reactor electrode*. An enzyme is immobilized on an inert solid support (such as porous glass beads) which is packed into a reactor or column. The substrate solution is pumped through the reactor at a rate that allows conversion of the sample to ions. The ionic solution flows through a chamber containing an ISE which measures their activity. As with biomembrane electrodes, a working curve is used for the assay. Biomembrane or enzyme-reactor electrodes that are selective for acetylcholine, adenosine monophosphate (AMP), adenosine deaminase, amino acids, amygdalin, arginine, asparagine, cholesterol, creatinine, glucose, glutamine, guanine, histidine, lactose, maltose, penicillin, pyruvate, sucrose, tyrosine, urea, and mercury have been used.

QUALITATIVE ANALYSIS

Because an indicator electrode that is dipped into a sample solution has a potential that is a function of both the identity and the amount of the species in solution, it seems reasonable to expect potentiometry to be useful for qualitative as well as quantitative analysis. In practice potentiometry at an inert-metal electrode can rarely be used for qualitative analysis because considerable potential overlap occurs for different chemical species at different concentrations; i.e., the potentials of many species are identical at some concentration of each of the species. Even when an ion-selective electrode is used, it is difficult to conclude with any certainty that a particular electrode response is caused by the ion for which the electrode is selective. An ISE is normally not specific for a single substance. Potentiometry is much better suited to quantitative analysis, to which the remainder of the chapter is devoted.

QUANTITATIVE ANALYSIS

Differential Potentiometry

In *differential potentiometry* two identical indicator electrodes are placed in separate solutions, as shown in Fig. 22-16. The two solutions are joined together by a salt bridge and the two indicator electrodes are connected to the terminals of a high-precision ion-selective electrode meter (a pI meter), DVM or DMM. A known-concentration solution of the substance to be assayed and to which the indicator electrode responds is placed in one container and a known volume of the sample is placed in the other container. The electromotive force between the two solutions is dependent upon the ratio of the activities of the species in the two solutions:

$$E = \frac{RT}{nF}\ln a_1 - \frac{RT}{nF}\ln a_2 = \frac{RT}{nF}\ln\frac{a_1}{a_2} \tag{22-49}$$

where a_1 and a_2 are the activities of the substance in the two solutions. If the concentrations of the substance in the two solutions are equal, $a_1 = a_2$ and the emf is zero.

A solution that contains a relatively large known concentration of the analyte molecule or ion is added from a buret to the half-cell containing the more dilute solution.

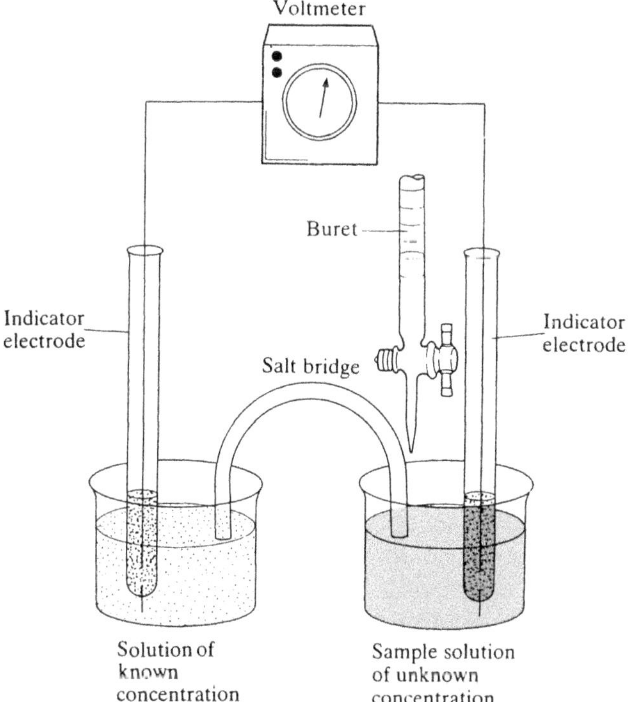

Figure 22-16 Diagram of the apparatus used for differential potentiometry.

The volume and concentration of the added solution when the emf between the two cells is zero is used to calculate the sample concentration (Sample problem 22-5). Either the "titrant" can be added until the emf becomes zero, at which time the volume of the titrant can be read from the buret, or a titration curve can be plotted and the endpoint can be located at zero emf. The terms "titrant" and "titration curve" are used loosely here because a titration is actually not being performed. That is, the substance in the buret does not chemically react with a substance in the half-cell to which it is added.

Sample problem 22-5 A 0.0500-*M* solution of silver nitrate is placed in one container and 50.0 mL of the sample solution is placed in another container. The two containers are electrically connected with a potassium nitrate salt bridge. Identical silver-wire indicator electrodes are inserted into the solutions and connected to a voltmeter. The addition of 5.24 mL of 0.100 *M* silver nitrate to the sample solution is required to make the emf between the two solutions zero. Calculate the concentration of Ag^+ in the original sample solution.

SOLUTION A 5.24-mL portion of 0.100 *M* Ag^+ was added to 50.0 mL of the sample solution to yield a solution that has a total volume of 55.2 mL and a concentration equal to that in the other container (0.0500 *M*). The total amount of Ag^+ in the solution after the addition was

$$0.0500 \text{ mmol/mL} \times 55.2 \text{ mL} = 2.76 \text{ mmol}$$

The amount of added Ag^+ was

$$0.100 \text{ mmol/mL} \times 5.24 \text{ mL} = 0.524 \text{ mmol}$$

Consequently, the amount of Ag^+ originally present in the sample solution was

$$2.76 \text{ mmol} - 0.524 \text{ mmol} = 2.24 \text{ mmol}$$

The original concentration of the sample solution is

$$[Ag^+] = \frac{2.24 \text{ mmol}}{50.0 \text{ mL}} = 0.0447\ M$$

Potentiometric Titrations

The Nernst equation [Eq. (22-6)) can be used to predict that the variation in potential during a redox titration results in a potentiometric titration curve of either the shape shown in Fig. 22-17 (Sample problem 22-6) or the mirror image of the curve shown in Fig. 22-17. The endpoint of the titration is usually considered to correspond to the inflection point in the curve. Actually the equivalence point corresponds to the inflection point only for titrations in which the molar quantities of titrand and added titrant at the equivalence point are equal (when the coefficients for the titrand and added titrant in the balanced chemical equation are identical) and for titrations in which the titrand is not significantly diluted by the addition of titrant during the titration.

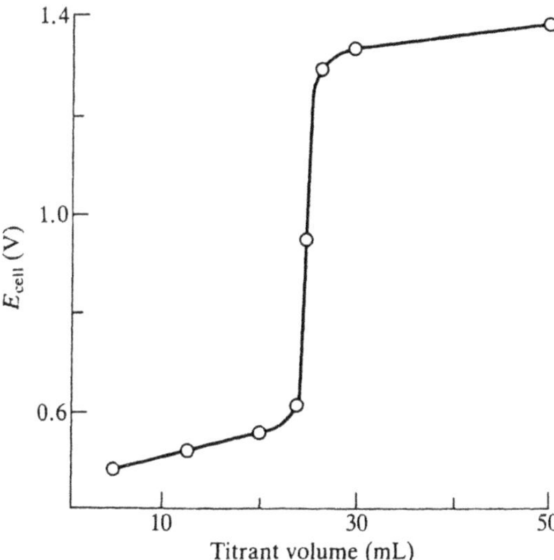

Figure 22-17 The calculated potentiometric titration curve for the titration of 25.0 mL of 0.0100 *M* Fe^{2+} with 0.0100 *M* Ce^{4+}. The equivalence point and the endpoint are located at a titrant volume of 25.0 mL.

If the titrand and titrant do not react in a 1:1 molar ratio, the equivalence point of the titration does not correspond to the inflection point. Because the titration curve is steep in the vicinity of the inflection point, the difference in titrant volume between that corresponding to the inflection point and the equivalence point is usually small. Consequently, little error is introduced by using the volume at the inflection point as the endpoint. Because the endpoint of potentiometric titrations is usually easy to locate and because potentiometry can be used to follow many types of titrations, potentiometric titrimetry is a popular method of quantitative analysis. Most potentiometric titrations are done using either a platinum indicator electrode (for oxidation-reduction titrations) or a glass pH electrode (for acid-base titrations). Calculation of the indicator electrode potential throughout a redox titration is illustrated in Sample problem 22-6. Calculation of the equivalence-point potential for a titration in which the inflection point of the titration curve does not coincide with the equivalence point is illustrated in Sample problem 22-7.

Sample problem 22-6 Calculate the potential at 25°C of a cell consisting of an sce and a platinum-wire indicator electrode dipping into a titration vessel that initially contains 25.00 mL of 0.0100 *M* Fe^{2+} after the addition of 5.00, 12.50, 20.00, 24.00, 25.00, 26.00, 30.00, and 50.00 mL of 0.0100 *M* Ce^{4+}. The sce is attached to the negative terminal of the voltmeter.

SOLUTION The half-reactions and standard potentials (Table 22-1) are

$$Fe^{3+} + e^- = Fe^2, \quad E° = 0.771 \text{ V}$$
$$Ce^{4+} + e^- = Ce^{3+}, \quad E° = 1.61 \text{ V}$$

The balanced chemical reaction for the titration is

$$Fe^{2+} + Ce^{4+} = Fe^{3+} + Ce^{3+}$$

Although the potential of the cell can be calculated by using either half-reaction, it is only possible to determine the concentrations of the oxidized and reduced forms of the titrand (Fe^{3+}, Fe^{2+}) before the equivalence point and of the titrant (Ce^{4+},Ce^{3+}) after the equivalence point. Concentrations are normally used in the calculations rather than activities with little loss of accuracy. Before the equivalence point, the following equations can be used to calculate the cell potential at 25°C:

$$E_{cell} = E - E_{sce}$$

$$E = E^{o} - 0.05917 \log \frac{[Fe^{2+}]}{[Fe^{3+}]}$$

$$= 0.771 - 0.05917 \log \frac{[Fe^{2+}]}{[Fe^{3+}]}$$

$$E_{cell} = 0.771 - 0.05917 \log \frac{[Fe^{2+}]}{[Fe^{3+}]} - 0.244$$

$$= 0.527 - 0.05917 \log \frac{[Fe^{2+}]}{[Fe^{3+}]}$$

After the addition of 5.00 mL of Ce^{4+}, the concentrations are

$$[Fe^{3+}] = \frac{5.00 \text{ mL} \times 0.0100 \text{ mmol/mL}}{25.00 + 5.00 \text{ mL}}$$

$$= 1.67 \times 10^{-3}\, M$$

$$[Fe^{2+}] = \frac{(25.00 \times 0.100) - (5.00 \times 0.0100)}{25.00 + 5.00}$$

$$= 6.67 \times 10^{-3}\, M$$

The cell potential is

$$E_{cell} = 0.527 - 0.05917 \log \frac{6.67 \times 10^{-3}}{1.67 \times 10^{-3}} = 0.491 \text{ V}$$

Likewise, after the addition of 12.50, 20.00, and 24.00 mL of titrant, the concentrations and cell potentials are as follows:

12.50 mL:

$$[Fe^{3+}] = \frac{12.50 \times 0.0100}{25.00 + 12.50}$$

$$= 3.33 \times 10^{-3}\, M$$

$$[Fe^{2+}] = \frac{25.00 \times 0.0100 - 12.50 \times 0.0100}{25.00 + 12.50}$$

$$= 3.33 \times 10^{-3}\, M$$

$E_{cell} = 0.527$ V (logarithmic term is 0)

20.00 mL:

$$[Fe^{3+}] = \frac{20.00 \times 0.0100}{25.00 + 20.00}$$

$$= 4.44 \times 10^{-3}\, M$$

$$[Fe^{2+}] = \frac{25.00 \times 0.0100 - 20.00 \times 0.0100}{25.00 + 20.00}$$

$$= 1.11 \times 10^{-3}\, M$$

$$E_{cell} = 0.527 - 0.05917 \log \frac{1.11 \times 10^{-3}}{4.44 \times 10^{3}}$$

$$= 0.536 \text{ V}$$

24.00 mL:

$$[Fe^{3+}] = \frac{24.00 \times 0.0100}{25.00 + 24.00}$$

$$= 4.90 \times 10^{-3}\, M$$

$$[Fe^{2+}] = \frac{25.00 \times 0.0100 - 24.00 \times 0.0100}{25.00 + 24.00}$$

$$= 2.04 \times 10^{-4}\, M$$

$$E_{cell} = 0.527 - 0.05917 \log \frac{2.04 \times 10^{-4}}{4.90 \times 10^{-3}}$$

$$= 0.609 \text{ V}$$

The equivalence point of the titration is reached when 25.00 mL of titrant is added. At the equivalence point it is necessary to use both half-reactions to calculate the potential:

$$E = 0.771 - 0.05917 \log \frac{[Fe^{2+}]}{[Fe^{3+}]}$$

and

$$E = 1.61 - 0.05917 \log \frac{[Ce^{3+}]}{[Ce^{4+}]}$$

Addition of the two equations yields

$$2E = 2.38 - 0.05917 \log \frac{[Fe^{2+}][Ce^{3+}]}{[Fe^{3+}][Ce^{4+}]}$$

At the equivalence point it is apparent from the balanced chemical equation that

$$[Fe^{3+}] = [Ce^{3+}] \text{ and } [Fe^{2+}] = [Ce^{4+}]$$

When those substitutions are made into the equation for E and the equation is solved, the logarithmic term becomes zero and

$$2E = 2.38$$

$$E = 1.19 \text{ V}$$

$$E_{cell} = E - E_{sce} = 1.19 - 0.24 = 0.95 \text{ V}$$

After the equivalence point, the following equations are used to calculate the cell potential:

$$E_{cell} = E - E_{sce}$$

$$E = 1.61 - 0.05917 \log \frac{[Ce^{3+}]}{[Ce^{4+}]}$$

$$E_{cell} = 1.61 - 0.05917 \log \frac{[Ce^{3+}]}{[Ce^{4+}]} - 0.244$$

$$= 1.37 - 0.05917 \log \frac{[Ce^{3+}]}{[Ce^{4+}]}$$

After the addition of 26.00, 30.00, and 50.00 mL of titrant, the concentration and cell potentials are as follows. The concentrations of Ce^{4+} are determined from the excess Ce^{4+} added beyond the equivalence point.

$$26.00\ \text{mL:} \qquad [Ce^{4+}] = \frac{(26.00 - 25.00)0.0100}{25.00 + 26.00}$$

$$= 1.96 \times 10^{-4}\ M$$

$$[Ce^{3+}] = \frac{25.00 \times 0.0100}{25.00 + 26.00}$$

$$= 4.90 \times 10^{-3}\ M$$

$$E_{cell} = 1.37 - 0.05917 \log \frac{4.90 \times 10^{-3}}{1.96 \times 10^{-3}}$$

$$= 1.29\ \text{V}$$

$$30.00\ \text{mL:} \qquad [Ce^{4+}] = \frac{(30.00 - 25.00)0.0100}{25.00 + 30.00}$$

$$= 9.09 \times 10^{-3}\ M$$

$$[Ce^{3+}] = \frac{25.00 \times 0.0100}{25.00 + 30.00}$$

$$= 4.55 \times 10^{-4}\ M$$

$$E_{cell} = 1.37 - 0.05917 \log \frac{4.55 \times 10^{-3}}{9.09 \times 10^{-4}}$$

$$= 1.33\ \text{V}$$

$$50.00\ \text{mL:} \qquad [Ce^{4+}] = \frac{(50.00 - 25.00)0.0100}{25.00 + 50.00}$$

$$= 3.33 \times 10^{-3}\ M$$

$$[Ce^{3+}] = \frac{25.00 \times 0.0100}{25.00 + 50.00}$$

$$= 3.33 \times 10^{-3}\ M$$

$$E_{cell} = 1.37 - 0.05917 \log \frac{3.33 \times 10^{-3}}{3.33 \times 10^{-3}}$$

$$= 1.37\ \text{V}$$

The results of the calculations are plotted as the titration curve shown in Fig. 22-17.

Sample problem 22-7 Calculate the potential at 25°C relative to the saturated calomel electrode at the equivalence point of the titration of 25.0 mL of 0.010 *M* Fe^{2+} with 0.010 *M* $Cr_2O_7^{2-}$ in a solution that is buffered at pH 3.50.

SOLUTION The half-reactions and standard potentials (Table 22-1) are

$$Fe^{3+} + e^- = Fe^{2+}, \qquad E° = 0.771 \text{ V}$$

$$Cr_2O_7^{2-} + 14H^+ + 6e^- = 2Cr^{3+} + 7H_2O, \qquad E° = 1.33 \text{ V}$$

The balanced chemical reaction for the titration is

$$6Fe^{2+} + Cr_2O_7^{2-} + 14H^+ = 6Fe^{3+} + 2Cr^{3+} + 7H_2O$$

The potential E of the indicator electrode, relative to the standard hydrogen electrode, is given by

$$E = 0.771 - 0.05917 \log \frac{[Fe^{2+}]}{[Fe^{3+}]}$$

and

$$E = 1.33 - \frac{0.05917}{6} \log \frac{[Cr^{3+}]^2}{[Cr_2O_7^{2-}][H^+]^{14}}$$

The two equations should be combined in such a manner that it is possible to eliminate concentrations of reactants that cannot be calculated at the equivalence point. For this titration neither the concentration of Fe^{2+} nor the concentration of $Cr_2O_7^{2-}$ can be calculated at the equivalence point without knowledge of the equilibrium constant for the reaction. If the second equation is multiplied by 6 prior to addition of the two equations, $[Fe^{2+}]$ and $[Cr_2O_7^{2-}]$ can be written in the same fraction. Thus

$$E = 0.771 - 0.05917 \log \frac{[Fe^{2+}]}{[Fe^{3+}]}$$

and

$$6E = 7.98 - 0.05917 \log \frac{[Cr^{3+}]^2}{[Cr_2O_7^{2-}][H^+]^{14}}$$

added together give

$$7E - 8.75 - 0.05917 \log \frac{[Fe^{2+}][Cr^{3+}]^2}{[Fe^{3+}][Cr_2O_7^{2-}][H^+]^{14}}$$

At the equivalence point

$$[Fe^{2+}] = 6[Cr_2O_7^{2-}]$$

and the equation simplifies to

$$7E = 8.75 - 0.05917 \log \frac{6[Cr_2O_7^{2-}][Cr^{3+}]^2}{[Fe^{3+}][Cr_2O_7^{2-}][H^+]^{14}}$$

or

$$7E = 8.75 - 0.05917 \log \frac{6[Cr^{3+}]^2}{[Fe^{3+}][H^+]^{14}}$$

The volume of added dichromate at the equivalence point is

$$0.010 \text{ mmol } Fe^{2+}/\text{mL} \times 25.0 \text{ mL} \times 1 \text{ mmol } Cr_2O_7^{2-}/6 \text{ mmol } Fe^{2+} \times 1 \text{ mL}/0.010 \text{ mmol } Cr_2O_7^{2-} = 4.2 \text{ mL}$$

The concentration of Fe^{3+} can be calculated as in Sample problem 22-6:

$$[Fe^{3+}] = \frac{0.010 \text{ mmol/mL} \times 25.0 \text{ mL}}{25.0 + 4.2 \text{ mL}} = 8.6 \times 10^{-3}\ M$$

At the equivalence point the concentration of Cr^{3+} can be calculated by using the balanced chemical equation:

$$[Cr^{3+}] = [Fe^{3+}] \times \frac{2}{6} = 8.6 \times 10^{-3} \times \frac{2}{6} = 2.9 \times 10^{-3} M$$

The concentration of H^+ is obtained from the pH:

$$\text{pH} = -\log [H^+] = 3.50; \qquad [H^+] = 3.2 \times 10^{-4}\ M$$

Substitution of the values into the equation for E makes it possible to calculate E:

$$7E = 8.75 - 0.05917 \log \frac{6[2.9 \times 10^{-3}]^2}{[8.6 \times 10^{-3}][3.2 \times 10^{-4}]^{14}}$$

$$E = 0.855 \text{ V}$$

The potential relative to the see is

$$E = 0.855 - 0.244 = 0.611 \text{ V}$$

If an ISE is used as the indicator electrode in a potentiometric titration, the electrode potential can be predicted by using Eq. (22-18). Titration curves obtained with ISEs have the same shape as the curves obtained with inert-metal indicator electrodes. The glass pH electrode has been widely used to locate the endpoints of acid-base titrations. When a pH electrode is used, the pH, which varies linearly with the cell potential, rather than the cell potential is plotted as a function of the titrant volume. Acid-base titration curves have the same shapes as redox titration curves, and the endpoints of the titrations are located in the same manner. With other ISEs, p*I* (*I* is the ion) is usually plotted against added titrant volume in the same manner in which pH is plotted for acid-base titrations.

Potentiometric titrations can be performed in some nonaqueous solvents as well as in water. Generally, inert-metal indicator electrodes are used in those titrations because most ISEs cannot be used in nonaqueous solvents. The glass pH electrode and some solid-state electrodes are exceptions. Among the non-aqueous solvents in which potentiometric titrations have been performed are acetic acid, acetonitrile, *N,N*-dimethylformamide, dimethylsulfoxide, ethanol, liquid ammonia, methanol, nitrobenzene, nitromethane, and pyridine.

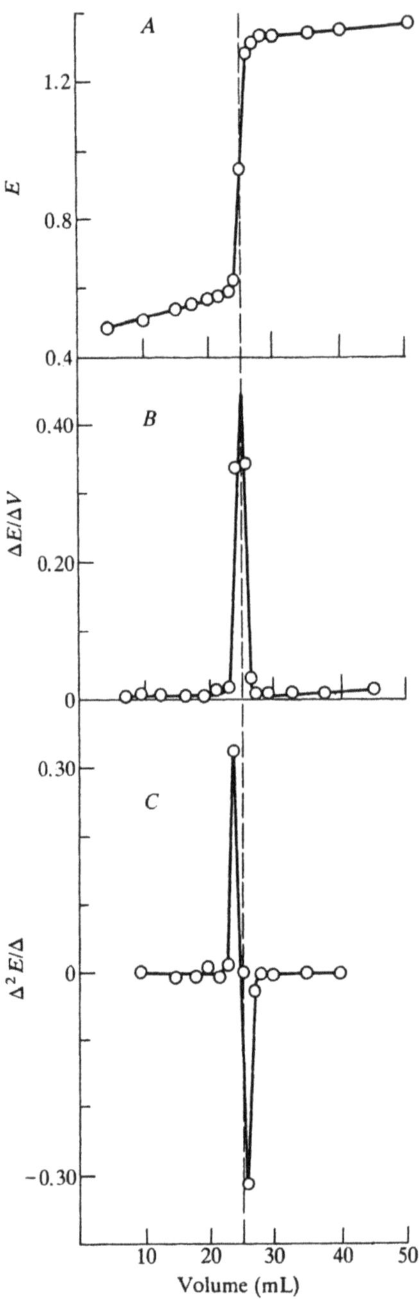

Figure 22-18 A series of potentiometric titration curves for the data given in Sample problem 22-8. The endpoint of each titration is indicated by the dashed line at 25 mL: curve *A*, regular potentiometric titration curve; curve *B*, first-derivative potentiometric titration curve; curve *C*, second-derivative potentiometric titration curve.

In some cases it is easier to locate the endpoint of a potentiometric titration from a first- or second-derivative potentiometric titration curve. A *first-derivative potentiometric titration curve* is prepared from potential and volume data for the titration by plotting the change in potential ΔE for consecutive additions (n and $n + 1$) of titrant divided by the change in volume ΔV for the same additions as a function of the average volume of titrant added between consecutive additions:

$$\frac{\Delta E}{\Delta V} = \frac{E_{n+1} - E_n}{V_{n+1} - V_n} \tag{22-50}$$

A first-derivative potentiometric curve reaches a maximum at the volume corresponding to the inflection point in a normal potentiometric titration curve. A first-derivative potentiometric titration curve is shown in Fig. 22-18, curve B.

A *second-derivative potentiometric titration curve* is a plot of the change in the change of potential [$\Delta(\Delta E)$ is given the symbol $\Delta^2 E$ in the same manner that a second derivative is given the symbol d^2E] divided by the change in the average volumes used previously to calculate ΔV (i.e., given the symbol ΔV^2) as a function of the titrant volume:

$$\frac{\Delta^2 E}{\Delta V^2} = \frac{\Delta E_{m+1} - \Delta E_m}{V_{\text{avg},\,m+1} - V_{\text{avg},\,m}} \tag{22-51}$$

The second-derivative potentiometric titration passes through the volume axis at the endpoint, that is, $\Delta^2E/\Delta V^2 = 0$ at the endpoint (Fig. 22-18, curve C). In most cases it is not worth the effort to prepare first- and second-derivative potentiometric titration curves manually, but it is worthwhile to learn how to prepare the curves. Some instruments that are used for potentiometric titrations automatically plot first- or second-derivative titration curves.

Sample problem 22-8 The following data were obtained during a potentiometric titration. Plot the normal, first-, and second-derivative potentiometric titration curves.

Data		Solution				
V, mL	E, V	ΔE	ΔV	$\Delta E/\Delta V$	V_{avg}	$\Delta^2E/\Delta V^2$
5.00	0.489					
		0.019	5.00	0.004	7.50	
10.00	0.508					0.003
		0.032	5.00	0.006	12.50	
15.00	0.540					–0.004
		0.015	3.00	0.005	16.50	
18.00	0.555					–0.002
		0.007	2.00	0.004	19.00	
20.00	0.562					0.006
		0.018	2.00	0.009	21.00	
22.00	0.580					–0.003

		0.014	1.00	0.014	22.50	
23.00	0.594					0.002
		0.016	1.00	0.016	23.50	
24.00	0.610					0.324
		0.340	1.00	0.340	24.50	
25.00	0.950					0
		0.340	1.00	0.340	25.50	
26.00	1.290					–0.309
		0.031	1.00	0.031	26.50	
27.00	1.321					–0.026
		0.005	1.00	0.005	27.50	
28.00	1.326					–0.001
		0.007	2.00	0.004	29.00	
30.00	1.333					0
		0.007	5.00	0.001	32.50	
35.00	1.340					0.002
		0.010	5.00	0.002	37.50	
40.00	1.350					0.001
		0.020	10.00	0.002	45.00	
50.00	1.370					

SOLUTION The data are listed in the first two vertical columns in the table. The results of the calculations used to determine the first- and second-derivative potentiometric titration curves are given in the last five columns of the table. ΔE was calculated for each pair of consecutive values of E by subtracting the first value from the second; e.g., between 23.00 and 24.00 mL ,ΔE = 0.610 – 0.594 = 0.016 V. ΔV was calculated by subtracting the first volume of each pair of consecutive volumes from the second volume; thus between 23.00 and 24.00 mL, V = 24.00 – 23.00 = 1.00 mL. $\Delta E / \Delta V$ was calculated by dividing the value of ΔE calculated above by ΔV. V_{avg} is the average amount of total titrant added for each pair of data points; e.g., between 10.00 and 15.00 mL, V_{avg} = (10.00 + 15.00)/2 = 12.50 mL. The first-derivative titration curve is a plot of $\Delta E / \Delta V$ as a function of V_{avg}.

$\Delta^2 E$ is obtained by subtracting the first ΔE of a consecutive pair from the second. As an example, for the values of ΔE at V_{avg} 19.00 and 21.00 mL, $\Delta^2 E$ = 0.018 – 0.007 = 0.011. ΔV^2 is calculated by subtracting adjacent values of V_{avg}. Between V_{avg} of 19.00 and 21.00 mL, ΔV^2 = 21.00 – 19.00 = 2.00. $\Delta^2 E/\Delta V^2$ = 0.011/2.00 = 0.006. A second-derivative potentiometric titration curve is a plot of $\Delta^2 E/\Delta V^2$ as a function of the mean of V_{avg} for the two points used to calculate $\Delta^2 E/\Delta V^2$. The regular first-derivative and second-derivative potentiometric titration curves are plotted in Fig. 22-18. The endpoint of the titration is indicated by the dashed line at 25 mL.

Controlled-Current Potentiometry with Two Indicator Electrodes

The change in potential between two metallic indicator electrodes of approximately equal size can be used to locate the endpoint of a titration while a constant current is applied between the electrodes. That method of analysis is *controlled-current potentiometry with two indicator electrodes*. Some analysts prefer to apply the term *bipotentiometry* to the technique. Normally, small platinum-wire electrodes are used. The constant current passed between the two electrodes depends upon the size of the electrodes and other experimental parameters. The current is usually between 1 and 5 μA.

The constant-current source can be a battery in series with a resistor as shown in Fig. 22-19, a constant-current supply which is built into some pH meters for the purpose, or some other source. If a battery-resistor current source is used, the resistor must have a resistance that is large relative to the resistance of the solution. If the resistance of the resistor is not large, the total resistance in the circuit (the sum of the resistances owing to the resistor and the solution) and consequently the current flowing between the electrodes can significantly vary during the titration owing to resistance variation in the solution. A change in the resistance of the solution is expected during a titration because the ions in the solution and the volume of the solution change. If the resistance of the resistor is large in comparison to the greatest possible resistance through the solution, the total resistance R_T in the circuit is equal to that of the resistor:

$$R_T = R_{\text{resistor}} + R_{\text{soln}} = R_{\text{resistor}} \tag{22-52}$$

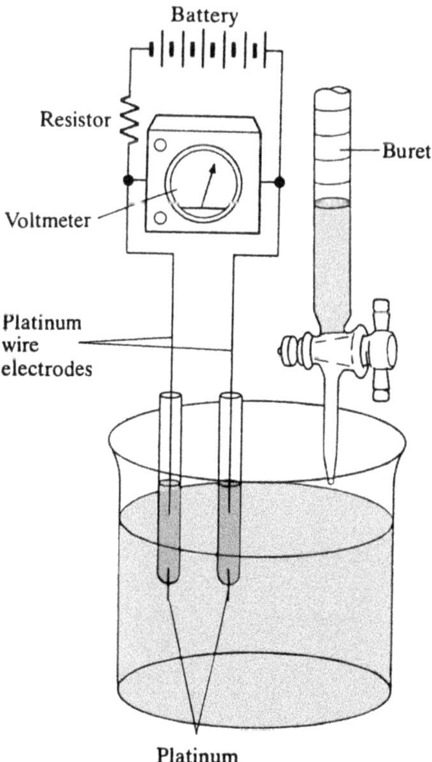

Figure 22-19 Diagram or the apparatus used for controlled-current potentiometric titrations with two indicator electrodes.

In that case the current is determined by the emf applied between the two electrodes. In any case a high-input impedance voltmeter must be used to prevent providing an alternative path for the constant current.

Generally the resistance of the resistor is at least 10 MΩ. The emf applied by the battery or other constant-voltage source can be determined by using Ohm's law ($E = iR$). If a 1 μA current is required and a resistor of 100 MΩ is used, the battery or power supply must apply 100 V as shown in the following calculation:

$$E = (1 \times 10^{-6}\ \text{A})\ (100 \times 10^{6}\ \Omega) = 100\ \text{V}$$

The shape of the titration curves obtained by using controlled-current potentiometry with two indicator electrodes is dependent upon the nature of the titrand and titrant (Reference 6, pp. 153-157). If both the titrand and the titrant are reversible redox couples, the titration curve peaks at the endpoint. If the titrant is reversible but the titrand is irreversible, the emf between the electrodes remains relatively constant prior to the endpoint and drops starting at the endpoint. If the titrand is reversible but the titrant is not, the first half of the titration curve is the same as when titrand and titrant are reversible, but at the endpoint the emf between the electrodes reaches a maximum and stays near that value for the remainder of the titration. The shapes of the titration curves are shown in Fig. 22-20. Bipotentiometry often is used to determine the end point of Karl Fischer titrations of water in methanol solutions. Karl Fischer titrations are described in Chapter 23.

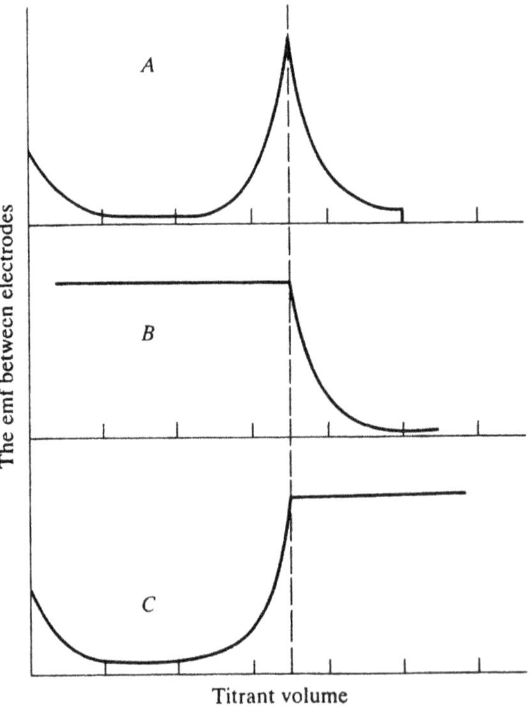

Figure 22-20 Diagram of controlled current potentiometric titration curves with two indicator electrodes: curve *A*, both titrand and titrant are reversible redox couples; curve *B*, only titrant is reversible; curve *C*, only titrand is reversible. The endpoints of the titrations are indicated by the dashed line.

Quantitative Analysis with Ion-Selective Electrodes

In most cases quantitative analysis by direct potentiometry is done with an ion-selective electrode and one of the reference electrodes described in an earlier section of the chapter. An electroactive species is generally quantitatively analyzed by using a working curve, by directly reading the pI of the species I from a pH meter (or pI meter), or with a variation of the standard-addition technique. A plot of cell potential as a function of the logarithm of the activity of an ion for which an electrode is selective is a straight line [Eq. (22-18)]. Because the activity coefficient in the relatively dilute solutions analyzed with ISEs is often nearly unity, a plot of cell potential as a function of the logarithm of the concentration of the ion is nearly linear. A working curve encompassing the concentration region of the sample solution can be prepared by plotting the cell potential as a function of the logarithm of either the activity or the concentration of the ion of interest. A nearly identical method is to plot the cell potential as a function of pI of the species (p$I = -\log a_I$). The cell potential of the sample solution is measured, and the activity or concentration of the ion is read from the working curve.

If the direct-reading method is used, two standard solutions which bracket the activity or concentration of the analyte are used to adjust the pI meter so that the meter simultaneously yields the proper readings for both solutions. Without further adjustment, the electrodes are placed in the sample and the pI of the sample is measured. The assumption made while using the method is that the instrument accurately measures any value of pI between the values of the standards. That method is used most often for pH measurements, but can also be used for pI measurements of other chemical species.

With either method of analysis, it is advantageous to keep the concentration of interferences at least roughly constant in the standards and sample. Because the concentrations of interferences in many samples are unknown, it is sometimes impossible to adjust the concentrations in the standards properly. If the interference is H^+ or OH^-, a buffer can be added to the solution to keep the concentration of the interference constant. An example of the application of the procedure was mentioned earlier. Assays with glass electrodes for non-H^+ monovalent cations are often made by using solutions that have been buffered in order to maintain the concentration of the interfering H^+ at a constant value. Of course, none of the buffer components should interfere with the analysis.

Similarly metal-ion buffers can sometimes be prepared to control the concentration of a particular metallic ion at a fixed value. Those buffers rely upon complexation reactions between the buffered metallic ion and an appropriate ligand. A difficulty that is encountered while trying to prepare metal-ion buffers is a lack of accurate dissociation constants for the complexes. The dissociation constants are used to calculate the metal-ion concentrations in the buffers.

In the presence of an interference of unknown identity and concentration, the standard-addition technique or one of its variations can be used for the assay. The standard-addition (or the *known-addition*) technique, as utilized with ion-selective electrodes, varies somewhat from that described for other analytical techniques. With other analytical methods, it is assumed that the instrumental response is proportional to concentration and is zero at zero concentration. The potential of an ion-selective electrode generally is not zero when the activity of the analyte is zero. Use of the known-addition method or one of its variations has the advantage of minimizing the effect of potential interferences in the solution upon the assay. In addition to minimizing the effect of interferences to which the electrode responds, the method also minimizes the effect of chemical species that form complexes with the analyte.

If an ISE is dipped into a solution containing an analyte ion with an activity a, the potential of the electrode as predicted by Eq. (22-18) is given by

$$E = \text{constant} + 2.303\,\frac{RT}{zF}\log a \tag{22-53}$$

The activity of the analyte ion in the solution is equivalent to the product of the activity coefficient of the ion and the concentration of free ion in the solution [Eq. (22-9)]. In many cases the known-addition method is used in solutions in which a complexing agent is present that partially reacts with the analyte ion. If that is the case, the activity of the ion is the product of the activity coefficient, concentration, and the fraction F of uncomplexed ion:

$$a = \mathrm{F}fC \tag{22-54}$$

Substitution of the activity from Eq. (22-54) into Eq. (22-53) gives

$$E = \text{constant} + 2.303\,\frac{RT}{zF}\,\log \mathrm{F}f\mathrm{C} \tag{22-55}$$

If a standard solution of the analyte ion with a concentration C_s and volume v is added to the analyte which has a volume V, the total activity a_T of the ion in the solution is given by

$$a_T = F_T f_T\left(\frac{VC + vC_s}{V + v}\right) \tag{22-56}$$

where F_T is the fraction of uncomplexed ion in the solution to which the standard has been added and f_T is its activity coefficient. The term in the brackets is the total concentration of the ion in the solution. The potential E_T of the ISE when it is dipped into the solution is given by

$$E_T = \text{constant} + 2.303\,\frac{RT}{zF}\log\left(F_T f_T \frac{VC + vC_s}{V + v}\right) \tag{22-57}$$

The change in the potential ΔE of the ISE between that measured before and after addition of the standard can be obtained by subtracting E [Eq. (22-55)] from E_T [Eq. (22-57)] to yield

$$\Delta E = 2.303\,\frac{RT}{zF}\log\left(F_T f_T \frac{VC + vC_s}{Ff\left(V + v\right)C}\right) \tag{22-58}$$

In most cases it is possible to assume that F is equal to F_T and f is equal to f_T. If that is the case, the equation simplifies to

$$\Delta E = 2.303\,\frac{RT}{zF}\log\frac{VC + vC_s}{\left(V + v\right)C} \tag{22-59}$$

Substitution of known and measured values into Eq. (22-59) permits calculation of the analyte concentration C. If the ISE does not exhibit nernstian behavior, Eq. (22-59) can be replaced with

$$\Delta E = S\log\frac{VC + vC_s}{\left(V + v\right)C} \tag{22-60}$$

where S is the slope obtained from a plot of the electrode potential as a function of the logarithm of the analyte activity.

A similar analytical technique that is sometimes applied to measurements made with ISEs is the *standard-subtraction technique*. A portion of the analyte ion is removed from the solution by chemical reaction with an added chemical reagent. The potential of the ISE is measured in the solution before and after addition of the chemical reagent. In that case the following equation is obeyed if the chemical reagent reacts in a 1:1 molar ratio with the analyte ion:

$$\Delta E = 2.303\,\frac{RT}{zF}\log\frac{VC - vC_s}{\left(V + v\right)C} \tag{22-61}$$

Another related technique is the *dilution method* in which the potential of the sample solution is measured before and after dilution of the solution.

Equations (22-59) and (22-61) assumed that Eq. (22-18) is obeyed; i.e., that the slope of a plot of potential as a function of log a is $2.303RT/zF$. Sometimes the slope S significantly varies from that predicted by Eq. (22-18). In that case the potential of the ISE is given by

$$E = \text{constant} + S \log f FC \qquad (22\text{-}62)$$

If S is known, the known-addition technique can be used with Eq. (22-60). If S is not known and cannot be measured, the standard-addition technique cannot be used.

If Eq. (22-62) is obeyed and the value of S is unknown, it is advantageous to use an analytical method that is independent of S. The known-addition technique required making two measurements in order to determine the analyte concentration because Eq. (22-53) contained two unknown terms (constant and a). Consequently, two independent simultaneous equations [Eqs. (22-55) and (22-57)] were solved to determine the concentration. Similar reasoning suggests that measurement of the potential of a solution after each of two consecutive additions would result in three independent equations which could be simultaneously solved to yield the analyte concentration. The known-addition technique that requires a potential measurement in the sample solution as well as after each of two additions of a standard solution is the *double known-addition method* (DKAM). In order to use the DKAM, S must be constant throughout the concentration range used for the study.

The potential of the ISE in the sample solution is given by Eq. (22-62). After addition of a volume V_S of a standard solution of concentration C_S to the sample, the potential E_1 of the ISE is given by

$$E_1 = \text{constant} + S \log\left(F_1 f_1 \frac{CV + C_s V_s}{V + V_s} \right) \qquad (22\text{-}63)$$

where V is the original volume of the analyte solution. For simplicity the volume of the second addition of standard solution is generally identical to the volume of the first addition. If that is the case, the potential E_2 of the electrode after the second addition of the standard solution is given by

$$E_2 = \text{constant} + S \log\left(F_2 f_2 \frac{CV + 2C_s V_s}{V + 2V_s} \right) \qquad (22\text{-}64)$$

The constant can be eliminated, from the equations by subtracting one equation from another. The change in potential ΔE_1 between that measured after the first addition and that in the analyte solution can be obtained by subtracting Eq. (22-62) from Eq. (22-63). The result is

$$\Delta E_1 = S\log\left(f_1F_1\frac{CV + C_s\ V_s}{fFC\left(V + V_s\right)}\right) \tag{22-65}$$

Similarly, the change in potential between that measured after the second addition and that in the analyte solution can be obtained by subtracting Eq. (22-62) from Eq. (22-64). That result is

$$\Delta E_2 = S\log\left(f_2F_2\frac{CV + 2C_s\ V_s}{fFC\left(V + 2V_s\right)}\right) \tag{22-66}$$

If experimental conditions are chosen such that the activity coefficients and fraction of uncomplexed ion is constant, the equations simplify to

$$\Delta E_1 = S\log\frac{CV + C_s\ V_s}{C\left(V + V_s\right)} \tag{22-67}$$

$$\Delta E_2 = S\log\frac{CV + 2C_s\ V_s}{C\left(V + 2V_s\right)} \tag{22-68}$$

The slope S can be eliminated from the equations by dividing one of the equations by the other. The resulting ratio R is given as

$$R = \frac{\Delta E_2}{\Delta E_1} = \frac{\log[(CV + 2C_s\ V_s)/C(V + 2V_s)]}{\log[(CV + C_s\ V_s)/C\left(V + V_s\right)} \tag{22-69}$$

Equation (22-69) can be used with the appropriate measured and known parameters to determine the concentration C of the analyte ion in the sample solution.

Unfortunately the direct solution of Eq. (22-69) for C is impossible. It is possible, however, to solve the problem by successive approximations or some other method using a computer. The ratio R corresponding to several values of CV/C_SV_S are provided in Table 22-6. The three potential measurements are made and the value of R calculated. Reference is made to Table 22-6 from which the corresponding value of CV/C_SV_S is obtained. Substitution for the appropriate values allows calculation of the analyte concentration. The major drawback to the DKAM is the relatively large error that can be introduced into the analysis by a small error in a potential measurement. A description of the error associated with the DKAM and methods for minimizing that error can be found in Reference 16.

Table 22-6 Values of $CV/C_S V_S$ for corresponding values of R for use with the double known-addition method

R	$CV/C_s V_s$	R	$CV/C_s V_s$
1.270	0.100	1.595	1.056
1.280	0.113	1.600	1.086
1.290	0.126	1.605	1.116
1.300	0.140	1.610	1.147
1.310	0.154	1.615	1.179
1.320	0.170	1.620	1.213
1.330	0.186	1.625	1.245
1.340	0.203	1.630	1.280
1.350	0.221	1.635	1.415
1.360	0.240	1.640	1.353
1.370	0.260	1.645	1.391
1.380	0.280	1.650	1.430
1.390	0.302	1.655	1.469
1.400	0.325	1.660	1.510
1.410	0.349	1.665	1.554
1.420	0.373	1.670	1.598
1.430	0.399	1.675	1.643
1.440	0.427	1.680	1.691
1.450	0.455	1.685	1.738
1.460	0.485	1.690	1.787
1.470	0.516	1.695	1.840
1.475	0.532	1.700	1.894
1.480	0.548	1.705	1.948
1.485	0.565	1.710	2.006
1.490	0.582	1.7 15	2.066
1.495	0.600	1.720	2.126
1.500	0.618	1.725	2.190
1.505	0.637	1.730	2.256
1.510	0.655	1.735	2.326
1.515	0.675	1.740	2.397
1.520	0.694	1.745	2.470
1.525	0.714	1.750	2.549
1.530	0.735	1.755	2.629
1.535	0.756	1.760	2.711
1.540	0.778	1.765	2.801
1.545	0.801	1.770	2.892
1.550	0.823	1.775	2.985
1.555	0.847	1.780	3.088
1.560	0.870	1.785	3.193
1.565	0.896	1.790	3.301
1.570	0.920	1.795	3.416
1.575	0.946	1.800	3.536
1.580	0.973	1.805	3.664
1.585	1.000	1.810	3.797
1.590	1.029	1.815	3.939

Sample problem 22-9 An ion-selective electrode was dipped into a 50.0 mL sample solution containing the ion to which the electrode responds. The potential of the electrode was 113.5 mV. After addition of 5.00 mL of a 5.00×10^{-3} *M* solution of the analyte ion, the electrode potential was 132.0 V. After a second 5.00 mL addition the potential was 144.4 mV. Determine the concentration of analyte ion in the sample.

SOLUTION Values are obtained for ΔE_1 and ΔE_2 by subtracting the potential of the electrode in the sample solution from the measured potentials after each addition of standard:

$$\Delta E_1 = 132.0 - 113.5 = 18.5 \text{ mV}$$

$$\Delta E_2 = 144.4 - 113.5 = 30.9 \text{ mV}$$

R is calculated using Eq. (22-69):

$$R = \frac{30.9 \text{ mV}}{18.5 \text{ mV}} = 1.67$$

Reference to Table 22-6 indicates that the corresponding value of $CV/\ Cs\ V_s$ is 1.598. Substitution into the expression provides the desired analyte concentration:

$$1.598 = \frac{CV}{C_s\ V_s} = \frac{C(50.0 \text{ mL})}{(5.00 \times 10^{-3}\ M)(5.00 \text{ mL})}$$

$$C = 7.99 \times 10^{-4} M$$

When ISEs are used to determine concentrations, rather than activities, it is necessary to maintain a constant ionic strength in the standards and the sample. The ionic strength is adjusted by adding an excess (usually enough to make the solution 0.1 *M* or greater in the electrolyte) of an inert electrolyte. Because its concentration is high, the inert electrolyte makes the ionic strength [Eq. (22-12)] of the solution constant regardless of the concentration of other species in solution. That, in turn, keeps the activity coefficient [Eqs. (22-10) and (22-11)] of the analyte constant and consequently makes the activity of the analyte directly proportional only to the concentration of the ion. That is, $a = fC$, but f is a constant, so a is directly proportional only to C. The net result is that the electrode potential responds to the logarithm of the concentration of the analyte:

$$E = \text{constant} + 2.303 \frac{RT}{zF} \log C \qquad (22\text{-}70)$$

The ionic strength is routinely adjusted to a constant value when natural-water samples are assayed for F^- with the LaF_3 solid-state electrode. A fixed amount of a *total ionic strength adjustment buffer* (TISAB) is added to all standards and the sample. For fluoride the TISAB usually contains 1 *M* sodium chloride, 0.25 *M* acetic acid, 0.75 *M* sodium acetate, and 1 m*M* sodium citrate or 1,2-diaminocyclohexane-N,N,N',N'-tetraacetic acid. The high sodium chloride concentration controls the ionic strength; the acetic acid and sodium acetate buffer the solution at a constant pH to eliminate the effect of interference from hydroxide; and the 1 m*M* component is used as a complexing agent for metallic ions which could react with fluoride. Similar TISAB solutions can be designed and used with other ISEs to eliminate the effect of interfering ions as well as to keep the ionic strength constant.

IMPORTANT TERMS

Acid error
Activity
Activity coefficient
Air-gap electrode
Alkaline error
Anode
Biological electrode
Biomembrane electrode
Calomel electrode
Cathode
Coated-wire electrode
Bipotentiometry
Differential potentiometry
Direct potentiometry
DKAM
Electrode of the first kind
Electrode of the second kind
Electrode of the third kind
Electromotive force
Enzyme-reactor electrode
First-derivative potentiometric titration curve
Fixed interference method
Formal potential
Gas-sensing electrode
Glass electrode
Heterogeneous membrane electrode
Homogeneous membrane electrode
Indicator electrode
Ionic strength
Ion-selective electrode
ISFET
Known-addition method
Liquid-ion-exchanger membrane electrode
Liquid-junction potential
Molecule-selective electrode
Nernst equation
Neutral-carrier membrane electrode
Potential
Potentiometric titration
Reference electrode
Secondary reference electrode
Second-derivative potentiometric titration curve
Selectivity coefficient
Solid-state membrane electrode
Standard potential
Titrant-stream reference electrode
Total ionic strength adjustment buffer

PROBLEMS

Activity and Ionic Strength

22-1 Calculate the ionic strength of a solution which contains 0.015 *M* $CaCl_2$ and 0.010 *M* $Cu(NO_3)_2$.

22-2 Calculate the ionic strength of a solution which contains 1.00 *M* acetic acid ($K_a = 1.76 \times 10^{-5}$).

22-3 Calculate the ionic strength of a solution which contains 0.0100 *M* NaCl, 0.010 *M* Na_2SO_4, and 0.050 *M* Na_3PO_4.

22-4 Calculate the ionic strength of a solution which contains 0.0250 *M* NaCl, 0.0035 *M* Na_2SO_4, and 0.0100*M* HCl.

22-5 Calculate the activity coefficient of K^+ at 20°C in a solution with an ionic strength of 0.0081 *M*.

22-6 Calculate the activity coefficient of Ca^{2+} in a solution which contains 1.5×10^{-3} *M* $CaCl_2$ at 30°C.

22-7 Calculate the activity of Na^+ in a solution which contains 1.00×10^{-4} *M* NaCl at 25°C.

22-8 Calculate the activity of Cl^- in a solution which contains 1.00×10^{-4} *M* NaCl and 1.00×10^{-3} *M* KCl at 25°C.

22-9 Calculate the activity of K^+ in a solution which contains 1.00×10^{-4} *M* KCl and 1.00×10^{-2} *M* HNO_3 at 30°C.

22-10 Calculate the activity of Mg^{2+} at 25°C in a solution which contains 1.20×10^{-4} *M* magnesium nitrate and 2.10×10^{-4} *M* sodium nitrate.

Nernst Equation

22-11 Calculate the molar ratio of Cu^+ to Cu^{2+} in a solution into which a platinum indicator electrode and a saturated calomel electrode are dipped at 25°C if the cell potential is 0.000 V.

22-12 Calculate the cell potential of a solution containing 1.00×10^{-3} *M* $Cr_2O_7^{2-}$ and 2.00×10^{-3} *M* Cr^{3+} at pH 4.00 and 25.0°C. The Pt indicator electrode is connected to the positive terminal of the DVM and the sce to the negative terminal.

22-13 The potential of the cell listed below is 0.200 V at 25°C. Calculate the concentration of chloride. The platinum electrode is attached to the negative terminal of the voltmeter:

$$Pt \mid Hg \mid Hg_2Cl_2\,(satd),\ KCl\,(satd) \parallel Cl^-,\ AgCl\,(satd) \mid Ag$$

22-14 Calculate the potential at 25°C of a cell consisting of an sce attached to the negative terminal of a voltmeter and a silver wire electrode attached to the positive terminal. The silver wire electrode is dipping into a solution which contains 1.29×10^{-3} *M* Ag^+.

22-15 Given the following half-reactions and standard potentials, calculate K_{sp} for AgBr at 25°C:

$$AgBr(s) + e^- = Ag(s) + Br^-, \qquad E° = 0.0713 \text{ V}$$
$$Ag^+ + e^- = Ag(s), \qquad E° = 0.799 \text{ V}$$

22-16 Calculate the potential at 25°C of a hydrogen electrode in a solution in which pH = 3.685 and the $H_2(g)$ partial pressure is 0.600 atm.

22-17 Calculate the potential of a cell consisting of an sce and a Pt indicator electrode dipping into a solution containing 2.00 m*M* Mn^{2+} and 5.00 m*M* MnO_4^- at pH 4.00 and 25°C. The sce is attached to the negative terminal of the voltmeter.

22-18 Calculate the cell potential for a cell containing a platinum indicator electrode and an sce reference electrode dipping into a solution containing 0.0100 *M* quinone ($C_6H_4O_2$) and 0.0125 *M* hydroquinone [$C_6H_4(OH)_2$] at pH 5.000. Assume all activity coefficients are 1.00 and the sce is attached to the negative terminal of the pH meter.

22-19 What is the potential difference of a cell consisting of an sce and a platinum indicator electrode dipping into a solution containing 0.0100 *M* Sn^{4+} and 0.050 *M* Sn^{2+}? Assume all activity coefficients are unity and the sce is attached to the negative terminal of the voltmeter.

22-20 Calculate the equilibrium constant for the reaction

$$6Ti^{3+} + Cr_2O_7^{2-} + 2H^+ = 6TiO^{2+} + 2Cr^{3+} + H_2O$$

from the standard potentials for the two half-reactions.

22-21 Calculate the formal potential at 25°C for the reduction of the iron(II)-2-cysteine complex to metallic iron if the concentrations of cysteine ion and the complex placed in solution were 6.00×10^{-3} and 7.00×10^{-3} *M* respectively, and the cell potential was –0.969 V when measured with respect to the sce (negative terminal).

Selectivity Coefficients

22-22 A sodium-selective glass electrode when dipped into a solution containing Na^+ with an activity of 1.00×10^{-4} *M* resulted in a cell potential of 0.587 V. When the same electrode was dipped into a solution which contained both Na^+ and H^+ with activities of 1.00×10^{-4} *M*, the cell potential was 0.625 V. Calculate the selectivity coefficient $K_{Na,H}$ for the electrode.

22-23 A nitrate-selective electrode and a reference electrode were dipped into a solution of nitrate with an activity of 1.00×10^{-3} *M*, and the potential of the nitrate-selective electrode was – 122.4 mV. The same electrode couple was dipped into a solution containing both nitrate and chloride at an activity of 1.00×10^{-3} *M* each. The potential of the nitrate-selective electrode in that solution was –124.8 mV. The temperatures of both solutions were 25.0°C. Calculate the selectivity coefficient of the nitrate-selective electrode for chloride.

22-24 The selectivity coefficient of a coated-wire chloride-selective electrode for bromide is 1.2. The potential of the electrode at 25°C in a solution containing chloride at an activity of 1.20×10^{-4} *M* is –148.5 mV relative to an sce. Determine the potential of the electrode in a solution that contains chloride at an activity of 2.4×10^{-4} *M* and bromide at an activity of 3.0×10^{-4} *M*.

22-25 A series of standard solutions were prepared which each contained Al^{3+} at an activity of 1.00×10^{-3} *M*, and varying activities of Cd^{2+} as indicated in the following table. The potential of each solution was measured with a solid-state cadmium-selective electrode. Determine the selectivity coefficient for aluminum with the cadmium-selective electrode.

$a_{Cd^{2+}}$, *M*	*E*, mV
3.0×10^{-3}	249.1
1.0×10^{-3}	233.7
3.0×10^{-4}	216.8
1.0×10^{-4}	204.7
3.0×10^{-5}	204.3
1.0×10^{-5}	204.0

Differential Potentiometry

22-26 Two identical silver wire electrodes were dipped into separate solutions of silver nitrate which were connected with a potassium nitrate salt bridge. The solution on the left contained 1.00×10^{-4} *M* Ag^{+} and the sample solution was in the container on the right. The electrode in the left-hand solution was connected to the negative terminal of the DVM and the electrode in the solution on the right was connected to the positive terminal. The emf between the two electrodes was – 0.0015 V at 25°C. Calculate the concentration of the sample solution.

22-27 Identical Ag/AgCl electrodes were placed in separate solutions connected by an appropriate salt bridge. The first solution contained 1.000×10^{-3} *M* KCl and the second contained 100.0 mL of a sample. After addition of 4.75 mL of 0.0100 *M* KCl to the sample solution, the emf between the electrodes was zero. Calculate the concentration of the sample.

Potentiometric Titrations

22-28 Calculate the potential at 25°C of a cell consisting of an sce and an Ag/AgCl indicator electrode dipping into a titration vessel, which initially contained 50.0 mL of 0.0150 *M* NaCl, after the addition of 5.00 mL of 0.100 *M* $AgNO_3$. The sce is attached to the negative terminal of the DVM.

22-29 Calculate the potential (relative to the standard hydrogen electrode) at the equivalence point for the titration of Fe^{2+} with MnO_4^- at pH 4.00 and 25°C.

22-30 Calculate the potential at 25°C of a cell consisting of an sce and a platinum wire indicator electrode dipping into a titration vessel, which initially contained 20.00 mL of 0.0100 *M* MnO_4^- buffered at pH 4.00, after the addition of 5.00, 10.00, 15.00, 17.00, 19.00, 20.00, 21.00, 23.00, 25.00, 30.00, and 40.00 mL of 0.0500 *M* Fe^{2+}. Plot the titration curve and indicate the equivalence point. The sce is attached to the negative terminal of the DVM.

22-31 The following data were obtained from a potentiometric titration of 50.0 mL of hydrochloric acid with 0.100 *M* sodium hydroxide by using a glass pH electrode. Calculate the original concentration of hydrochloric acid.

NaOH volume, mL	pH
4.00	1.12
9.00	1.30
14.00	1.50
19.00	1.88
21.09	2.14
23.00	2.50
24.00	6.97
25.00	11.01
27.00	11.55
29.00	11.80
35.00	12.16
40.00	12.33
50.00	12.61

22-32 Calculate the potential at 25°C of a cell consisting of an sce (negative terminal) and a platinum wire indicator electrode dipping into a titration vessel, which initially contained a pH 2.00 solution of 20.00 mL of 0.0105 *M* Fe^{2+}, after the addition of 4.00 mL of 0.0105 *M* potassium permanganate.

22-33 A 1.000-g sample of a brass was dissolved in nitric acid and boiled to near dryness. The resulting solution was dissolved in about 25 mL of water and neutralized by the addition of aqueous ammonia. The neutralized solution was reacted with a large excess of potassium iodide as shown in the following equation:

$$2Cu^{2+} + 4I^{-} = 2CuI + I_2$$

The iodine liberated from the reaction required 24.30 mL of 0.1000 *M* thiosulfate solution to reach the endpoint of the potentiometric titration. Calculate the percentage by weight of copper in the brass.

22-34 Plot the first- and second-derivative potentiometric titration curves for the data given in Prob. 22-31. The first-derivative pH titration curve is a plot of Δ pH/ΔV as a function of V_{avg} and the second-derivative curve is a plot of Δ^2 pH/ΔV^2 as a function of V.

Quantitative Analysis with ISEs

22-35 The following data were obtained with a cadmium-selective solid-state electrode. The ionic strength of each solution was 0.3 *M* and the reference electrode was an sce. Determine the concentration of Cd^{2+} in the sample.

$[Cd^{2+}]$, *M*	*E*, V
1.00×10^{-1}	– 0.115
8.75×10^{-3}	– 0.150
1.00×10^{-3}	– 0.175
3.16×10^{-5}	– 0.213
3.16×10^{-6}	– 0.242
Sample	– 0.200

22-36 The following data were obtained at pH 7 and 25°C with a sodium-selective glass electrode and a silver-silver chloride (1 *M* KCl) reference electrode connected to the sample compartment with a salt bridge. Determine the activity of Na^+ in the sample.

pNa	*E*, V
0.00	0.122
1.00	0.062
2.00	0.000
3.00	– 0.059
4.00	– 0.117
5.00	– 0.148
6.00	– 0.165
Sample	– 0.050

22-37 A 50.0 mL sample solution containing fluoride at 25°C was assayed by the known-addition technique with a fluoride-selective electrode. The potential of the electrode in the sample solution was –102.5 mV. After the addition of 5.00 mL of 1.00×10^{-2} *M* fluoride, the measured potential was – 112.7 mV. Calculate the concentration of fluoride in the sample.

22-38 A copper (II)-selective electrode was used to perform an analysis of a sample solution by the DKAM. The potential of a 25.0 mL portion of the sample solution was 247.2 mV. After the first addition of a 5.00 mL aliquot of a 2.00×10^{-3} *M* Cu^{2+} solution, the potential was 261.8 mV. After a second addition of a 5.00 mL aliquot, the potential was 271.9 mV. Determine the concentration of copper (II) in the sample.

REFERENCES

1. Daniels, F., and R. A. Alberty: *Physical Chemistry*, 2d ed., Wiley, New York, 1962.
2. Klotz, I. M.: *Chemical Thermodynamics, Benjamin*, New York, 1964, pp. 358-359.
3. Diehl, H.: *Anal. Chem.*, **51**: 318A(1979).
4. Milazzo, G., and S. Caroli: *Tables of Standard Electrode Potentials*, Wiley, New York, 1978.
5. Serjeant, E. P.: *Potentiometry and Potentiometric Titrations*, Wiley, New York, 1984.
6. Lingane, J. J.: *Electroanalytical Chemistry*, 2d ed., Interscience, New York, 1966.
7. Durst, R. A. (ed.): Ion-Selective Electrodes, NBS Special Publication 314, National Bureau of Standards, Washington, D.C., 1969.
8. Pungor, E. (ed.): *Ion-Selective Electrodes*, Elsevier, New York, 1978.

9. Ives, D. J. G., and G. J. Janz (eds.): *Reference Electrodes — Theory and Practice, Academic*, New York, 1961.
10. Bakker, E. and Y. Qin: *Anal. Chem.*, **78**: 3965(2006)
11. Freiser, H. (ed): *Ion-Selective Electrodes in Analytical Chemistry*, vol. 1, Plenum, New York, 1978.
12. Freiser, H. (ed.): *Ion-Selective Electrodes in Analytical Chemistry*, vol. 2, Plenum, New York, 1980.
13. Ruzicka, J., and E. H. Hansen: *Anal. Chim. Acta*, **69**: 129(1974).
14. Montalvo, Jr., J. G., and G. G. Guilbault, *Anal. Chem.*, **41**: 187(1969).
15. Guilbault, G. G., and J. Montalvo, Jr.: *J. Amer. Chem. Soc.*, **91**: 2164(1969).
16. Longhi, P., T. Mussini, and S. Rondinini: *Anal. Lett.*, **15**: 1601(1982).

CHAPTER

TWENTY-THREE

NONPOTENTIOMETRIC ELECTROANALYSIS

This chapter is devoted to a brief description of several nonpotentiometric electroanalytical methods. In order to keep the chapter's length reasonable, several of the less-used techniques have been omitted. Further information on a specific technique can be obtained from the References for the chapter.

VOLTAMMETRY AND POLAROGRAPHY

Voltammetry is a technique in which a varying potential is applied to an electrode in a sample solution. The electrode at which the potential is applied is the *indicator* or *working* electrode. The current flowing through the electrode is monitored as a function of the applied potential. The results are displayed by a recorder, computer controlled device, or other readout device as a current-potential plot. *Polarography* is that type of voltammetry that uses an electrode whose surface is renewed as the potential is varied. A mercury electrode in which mercury is continuously flowing into the mercury drop is nearly exclusively used for polarography. Other types of voltammetry use a solid electrode or a stationary mercury drop into which no mercury flows.

During an analysis, the potential at the indicator electrode is applied in a predictable manner. Sometimes the variation in applied potential with time is termed the *excitation waveform*. Voltammetry and polarography can be contrasted with potentiometry (Chapter 22), in which the potential difference between two electrodes is measured while the current flowing between the electrodes is held constant (usually at nearly zero).

Several subdivisions of voltammetry are possible which differ in the manner in which the applied potential varies with time. Regardless of the manner in which the potential varies with time, when the potential becomes sufficient to cause reduction or oxidation of a redox species in the solution, a current flows through the indicator electrode. That current is measured and plotted as a function of the applied potential by the readout device. The portion of the analytical instrument that applies the potential to the indicator electrode is a *potentiostat*. The combined potentiostat and current-monitoring apparatus is sometimes termed a *voltammetric analyzer*. A voltammetric analyzer can be used for polarographic and voltammetric measurements. The plot of current as a function of potential is a *polarogram* or *voltammogram*.

The potential at which the current starts to flow corresponds to the energy that is required to initiate an oxidation or reduction of a chemical species at or near the indicator electrode surface. Because that energy varies with the chemical species, it sometimes is possible to use the potential for qualitative analysis. The current flow is proportional to the concentration of the redox species and is used for quantitative analysis. Because current can only flow when an oxidation or reduction occurs at the indicator electrode, voltammetry and polarography are only useful for the determination of chemical species that can be oxidized or reduced.

Electrodes

Voltammetric indicator electrodes are usually constructed from platinum, mercury, gold, or some form of carbon. Other electrode materials are used in specific cases. For polarography the indicator electrode is nearly always a *dropping mercury electrode* (dme). A dme consists of an elevated mercury reservoir connected through glass or flexible tubing (transparent plastic tubing is often used) to a glass capillary tube with a small-diameter bore (about 0.06 mm). The open end of the glass capillary is dipped into the solution to be assayed. Mercury flows through the capillary and forms small spherical drops at the lower open end. The surface of the mercury drop is exposed to the sample and serves as the site at which the electrochemical reaction takes place.

Because mercury continuously flows from the tip of the capillary, the mercury drop eventually becomes sufficiently large to fall from the capillary and a new drop forms. The flow of fresh mercury to the electrode surface continuously renews the surface. Renewal of the electrode surface prevents the surface from becoming permanently contaminated with electrode reaction products or with adsorbed species in solution. A major advantage of a dme is the highly reproducible results that are possible because the electrode surface is continuously renewed. The major disadvantage to the use of mercury electrodes is the danger associated with mercury contamination of the environment. Normally used mercury from the electrodes is captured, treated to remove contaminants and recycled.

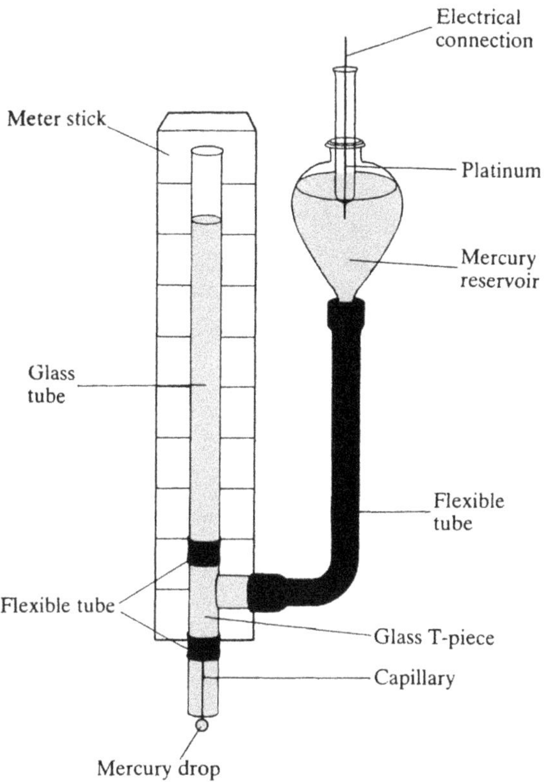

Figure 23-1 Diagram of one form of a dropping mercury electrode.

Electrical connection to a dme is made through a platinum wire inserted into the mercury somewhere above the capillary. The time between drops is the *drop time t* (seconds) and the mass of mercury which flows from the capillary tip in a 1-s period is the *mercury flowrate m* (milligrams per second). Both the drop time and the mercury flowrate can be adjusted by altering the height of the mercury reservoir and thereby changing the pressure of mercury on the capillary. In some electrodes the drop time is controlled with a mechanical device that physically dislodges the drops from the electrode at chosen time intervals. Often the electrode is mounted on a meter stick to make it possible to easily measure the height of the mercury reservoir above the bottom of the capillary. For classical polarography t is adjusted to a value between 2 and 8 s. Shorter drop times can be used when the drop is physically knocked from the electrode. A diagram of a classical dme is shown in Fig. 23-1.

A variation of a dme is the *static mercury drop electrode* (smde).[1] The smde uses mercury drops as the electrode material, as does the dme; however, the mechanism of drop formation differs from that used by the dme. A measured quantity of mercury from a reservoir flows through a solenoid-activated valve and into the capillary. Because the internal diameter of the capillary is about thrice that used for a dme, the mercury rapidly flows through the capillary and forms a drop at the end of the capillary. After the desired quantity of mercury has passed, the valve is automatically closed and the drop ceases to grow. The current measurement with an smde is made while the drop area is constant rather than during the entire life of the drop as with the dme. After the measurement is made, the drop is automatically dislodged from the capillary and the procedure is repeated. A scan can be obtained at an smde either over the lifetime of many drops, as during polarographic studies at the dme, or on a single drop (cf. the hanging mercury drop electrode that is described later).

When either voltammetry or polarography is used, the potential of the indicator electrode is varied relative to a reference electrode. In aqueous solutions the reference electrode is usually an sce, although other reference electrodes also can be used. In a nonaqueous solvent the reference electrode is often one of those nonaqueous reference electrodes described in Chapter 22.

Older voltammetric analyzers were designed to be used with two electrodes (indicator and reference). When two electrodes are used, all of the current flowing through the cell must flow through the reference electrode as well as through the indicator electrode. As will become apparent, that can be a disadvantage. Modern instruments normally use three electrodes in order to prevent significant amounts of current from flowing through the reference electrode. A high input impedance connection to the voltammetric analyzer is used to minimize current flow through the reference electrode.

In any solution a potential drop occurs between the reference electrode and the indicator electrode as predicted by Ohm's law ($E = iR$), where i is the current flowing between the indicator and reference electrodes and R is the total resistance in the cell between the electrodes. The potential at the indicator electrode is the difference between the potential applied by the potentiostat between the indicator and reference electrodes and the iR drop:

$$E_{\text{ind}} = E_{\text{appl}} - iR \tag{23-1}$$

If the iR drop is negligible relative to the applied potential, the indicator electrode potential equals the applied potential. Because it is usually advantageous to know the potential at the indicator electrode, it is desirable to keep the iR drop negligibly small. That can be done by keeping i, R, or both small.

The usual approach is to try to minimize the current flowing between the reference and indicator electrodes. That can be accomplished by using a three-electrode arrangement, in which the bulk of the cell current flows between the indicator electrode and a third electrode. Because the reference electrode is used only to control the potential of the indicator electrode, the current flow between the reference and indicator electrodes (and the resulting iR drop) can be small.

The third electrode in a three-electrode arrangement is an *auxiliary* or *counter electrode*. The auxiliary electrode is usually constructed from an inert conductor with a relatively large surface area. A platinum foil or gauze is often used to construct the electrode. The auxiliary electrode is placed in a separate compartment of the electrochemical cell in order to prevent mixing of the reaction products formed at the auxiliary electrode with the components of the analyte solution. Electrical connection between the compartments is made through a salt bridge.

The flow of a significant current through a reference electrode can also be deleterious because it can alter the concentrations of the redox species inside the reference electrode. A variation in the concentrations of the redox species within the electrode causes an alteration of the potential of the reference electrode. That can sometimes be avoided in a two-electrode system by increasing the size of the reference electrode, i.e., by increasing the amount of redox species in the electrode and thereby increasing the amount of current necessary to significantly alter the ratio of the activities of the reduced to oxidized forms. It is normally more convenient to use a smaller reference electrode in a three-electrode system. Commercially prepared reference electrodes that are designed for potentiometric use can be used for voltammetry or polarography with a three-electrode system, but they cannot be used with a two-electrode system. The assembled apparatus for voltammetry with three electrodes is shown in Fig. 23-2.

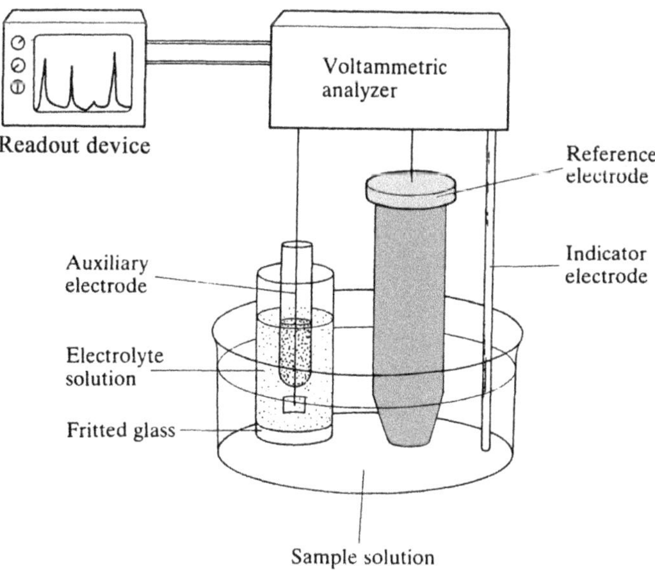

Figure 23-2 Diagram of the apparatus used for voltammetry with a three-electrode system.

POLAROGRAPHIC PRINCIPLES

The theory of polarography is described in the following sections in more detail than that of the other voltammetric methods in spite of the fact that some of the other methods are more useful for some types of assays than classical polarography. Polarography is described in more detail because it is historically more important than the other techniques. The theory and terminology of the other techniques is similar to that of polarography. The description of polarography is used as a vehicle to introduce terms and concepts that apply to all forms of voltammetry. Details of the theory of the other techniques can be found in References 2 and 3.

Polarography is almost exclusively performed with a dme. The variation in the potential applied at the indicator electrode with time is a *voltage* or *potential ramp*. The voltage ramp for polarography is linear with a slope (*scan rate*) of about 1 to 5 mV/s for classical polarography and with a drop time of between 2 and 8 s. If a mechanical device is used to control the drop time at a smaller value (usually between 0.01 and 1 s), the scan rate can be as high as 200 or 300 mV/s. The latter technique is *rapid dc polarography.*[2]

For rapid dc polarography a useful generalization is that no more than about 5 mV should be scanned during the lifetime of each drop. In aqueous solution the useful potential range of the dme is from 0 to about –2 V (versus sce). At potentials more positive than about 0 V, the mercury on the electrode surface is oxidized causing a large anodic current. By definition *anodic current* is a current resulting from an oxidation at the indicator electrode and a *cathodic current* results from a reduction at the electrode.

At potentials more negative than about – 2 V, water is reduced on the mercury electrode to H_2:

$$2H_2O + 2e^- = H_2 + 2OH^-, \qquad E^\circ - 0.828\ \text{V} \qquad (23\text{-}2)$$

From the standard reduction potential given in Eq. (23-2), it is apparent that water should be reduced to H_2 at a potential which is more positive than the – 2 V (at about pH 7) at which the reduction is first noticed at the dme. When comparing these potentials it is important to remember that the standard potential is measured relative to the standard hydrogen electrode. The pH has an effect because hydroxide is a product of the reduction. The difference between the electrode potential at which a reaction starts to occur and the value at which, from Nernst equation calculations, it should occur is the *overpotential* or *overvoltage*.

Overpotential is caused by a kinetically slow electrochemical reaction. The overpotential for the reduction of water to H_2 on mercury is about – 1 V, but the overpotential for the same reduction on a platinum, gold, or carbon electrode is – 0.2 V or less. Consequently, a mercury electrode can be used to study electrochemical reactions that occur at potentials more negative than those of the reactions that can be studied by using other electrodes. The increased cathodic range which is available with mercury electrodes is another advantage to the use of the electrodes.

If the sample contains a reducible species (Ox), the initial potential of the electrode is adjusted to a value at which the reduction does not occur, and the polarogram is obtained by scanning toward more negative potentials. By convention cathodic currents usually are recorded as upward deflections and are considered to be positive. Anodic currents usually correspond to downward deflections and are considered to be negative. The potential axis normally becomes more negative upon progressing to the right. When a potential at which the reduction of Ox to Red can occur is reached, a current flows through the cell:

$$\mathrm{Ox} + ne^{-} = \mathrm{Red} \tag{23-3}$$

Initially the current increases linearly with increasing potential, as predicted by Ohm's law. Eventually, however, the potential becomes sufficiently negative to completely reduce all of the Ox which is at the surface of the electrode. When that occurs the current flowing through the cell is dependent upon the rate at which Ox can travel through the solution to the electrode surface. The solution is not stirred and experimental conditions are adjusted such that the electric field from the electrode has negligible effect on movement of analyte ions. Typically that is accomplished by adding a relatively large concentration of an electroinactive electrolyte, termed the supporting electrolyte, to the solution. The supporting electrolyte often is potassium chloride. Consequently, the only means of transport of Ox to the electrode surface is by diffusion. The current flowing through the cell that is limited by the rate of diffusion to the indicator electrode is the *diffusion current* i_d. Because the rate of diffusion of Ox to the electrode surface is dependent upon the differential between the concentration in the bulk solution and that at the electrode surface, and because the concentration differential is constant, the rate of diffusion to the electrode surface and the diffusion current are constant for a fixed concentration of Ox. A typical polarogram is shown in Fig. 23-3.

Tast polarography is a variation of polarography in which the current is measured only during the last fraction of a second before the drop falls from the capillary. A mechanical device is used to strike or shake the capillary at fixed time intervals in order to dislodge the drop and to control the drop time. The current is measured just prior to its dislodgement. A Tast polarogram is similar to the classical polarogram shown in Fig. 23-3 except that the oscillations owing to changing drop surface are not observed and the polarogram traces the tops of the oscillations observed with a classical polarogram. A Tast polarogram is shown in Fig. 23-4.

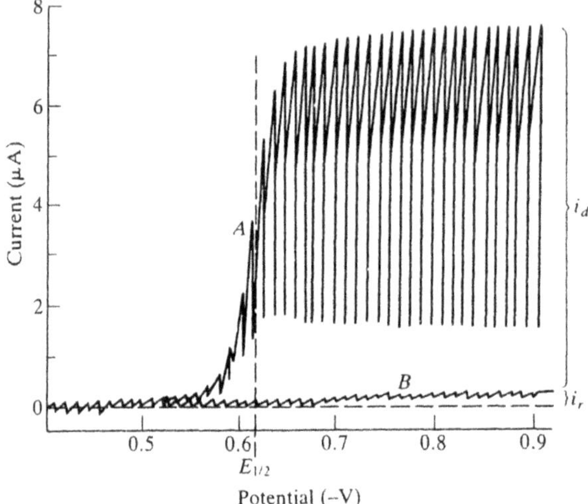

Figure 23-3 Polarograms of 1.2 m*M* cadmium(II) chloride obtained at a scan rate of 2 mV/s in an aqueous 0.1-*M* potassium chloride solution (sce reference electrode) (curve *A*) and 0.1-*M* potassium chloride solution (curve *B*). i_d, diffusion current; i_r, residual current; $E_{1/2}$, half-wave potential.

Polarographic Current

It is apparent from Fig. 23-3 that the current oscillates throughout the recording of a polarogram. That is caused by the constantly changing size of the mercury drop. As the drop grows, the electrode area increases (and expands into the electric double layer surrounding the drop) with a corresponding increase in the amount of current which flows through the electrode. When the drop falls from the capillary and a new drop forms, a sharp decrease in the current occurs owing to the decreased electrode area.

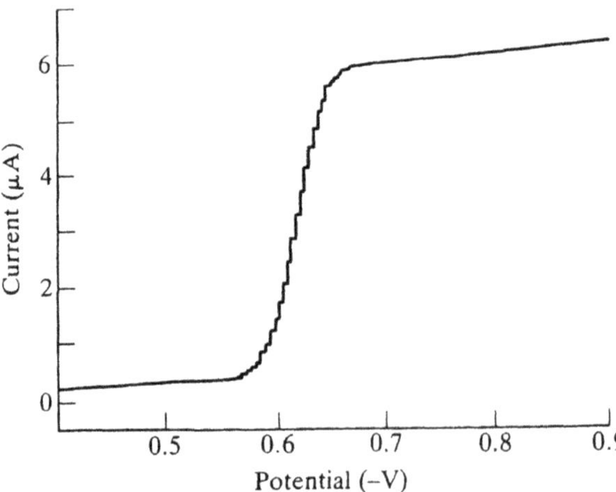

Figure 23-4 Tast polarogram of 1.2 m*M* cadmium chloride in an aqueous 0.1-*M* potassium chloride solution. The reference electrode is an sce and the drop time is 2 s.

In the absence of an electroactive species (Fig. 23-3, curve *B*), a small current, known as the *residual current* i_r, flows through the cell at most potentials. Residual currents are also observed when using other voltammetric techniques. The residual current is the combination of the current flowing as a result of the reduction or oxidation of any impurities in solution and the *capacitive current* (also called the *charging current*).

A charged electrode attracts ions of the opposite charge; e.g., in a potassium chloride solution, a negatively charged dme will attract a layer containing more potassium ions than chloride ions to the electrode surface. The potassium ions, in turn, attract a layer that contains more chloride ions. The result is an *electric double layer* of partially separated charged particles. The current that is required to form the double layer is the capacitive current. The capacitive current is greatest at the initiation of each drop and rapidly decays as the electric double layer forms. A further description of the capacitive current is contained in the sections relating to pulse polarography and differential pulse polarography.

The diffusion current is the portion of the total current that is controlled by the rate of diffusion of the electroactive species to the electrode surface. It is caused by the concentration gradient between the bulk of the solution and the electrode surface. Diffusion currents occur whenever a concentration gradient occurs between any electrode surface and the bulk of the solution. They can occur while using any electrochemical technique. The limiting current is the total current which flows on the plateau of the polarographic wave as illustrated in Fig. 23-3. The diffusion current is the difference between the *limiting current* flowing through the cell on the plateau of the wave and the residual current. Because it is sometimes difficult to obtain a polarogram of the background electrolyte solution without the electroactive species, the residual current often is estimated by extrapolating the current prior to the polarographic wave to the potential of the i_d measurement, as shown in Fig. 23-5.

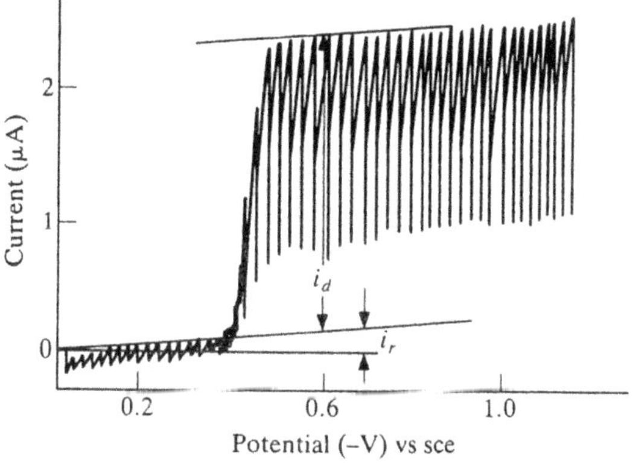

Figure 23-5 Polarogram of 0.41 m*M* flavin adenine dinucleotide (FAD) in pH 8.2 Tris buffer. i_d, the instantaneous diffusion current measured at – 0.6 V versus sce; i_r, the residual current measured at– 0.7 V.

The diffusion current can be measured in either of two ways. If the device recording the polarogram is undamped and responds rapidly to changes in current, i_d can be measured from the top of the oscillations of the residual current to the top of the oscillations on the plateau of the polarographic wave. The diffusion current measured in that way (Fig. 23-5) is the *instantaneous diffusion current*. It is related to the concentration of the electroactive species through the form of the *Ilkovic equation*:

$$i_d = 708nD^{1/2}m^{2/3}t^{1/6}C \tag{23-4}$$

where i_d is the diffusion current (μA), C is the concentration of the electroactive species (mM), t is the drop time (s), m is the mercury flowrate (mg/s) in the capillary, D is the diffusion coefficient (cm^2/s) of the electroactive species, and n is the number of electrons transferred in the half-reaction [Eq. (23-3)].

If the readout device does not respond rapidly enough to measure the instantaneous current, or if the device is damped, the diffusion current is measured from the center of the oscillations of the residual current to the center of the oscillations on the plateau of the polarographic wave. The center of the oscillations is used as an estimate of the time-averaged current that flows through the dme. The diffusion current measured in that way is the *average diffusion current* (Fig. 23-6). Nearly all modern recorders are capable of accurately recording the instantaneous current at the end of the mercury drop life. The form of the Ilkovic equation obeyed by the average diffusion current is

$$i_d = 607nD^{1/2}m^{2/3}t^{1/6}C \tag{23-5}$$

Derivations of Eqs. (23-4) and (23-5) can be found in References 3 and 4.

In addition to the residual and diffusion currents, a *migration current* can flow in an electrochemical cell. The migration current is caused by the attraction or repulsion of the electroactive ion by the charged indicator electrode. It corresponds to the movement of ions owing to the electric field between the electrodes. The attraction or repulsion can cause an increase or decrease in the rate at which the ion arrives at the electrode surface and therefore can cause an additional positive or negative current. That type of current does not occur when the electroactive species has no electric charge. Migration currents can be observed regardless of the voltammetric technique that is used.

The migration current can be virtually eliminated by adding an excess of a supporting electrolyte to the sample solution. Aqueous solutions are routinely made 0.1 to 0.5 M in potassium chloride or some other electrolyte prior to recording the polarograms of the solutions. In nonaqueous solvents sodium perchlorate, tetrabutylammonium perchlorate, or some other tetraalkylammonium perchlorate is often used as the supporting electrolyte.

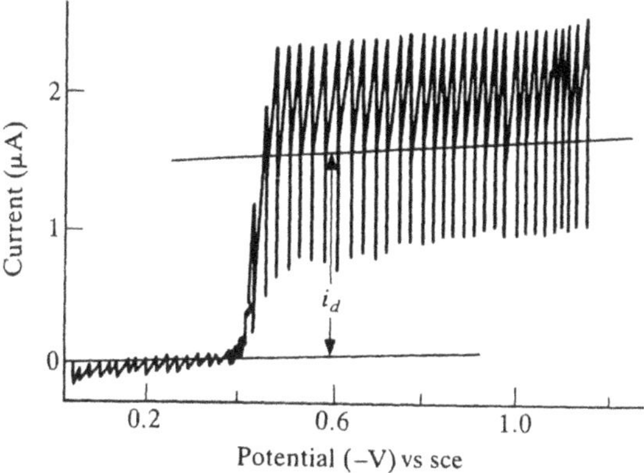

Figure 23-6 Polarogram of 0.4 m*M* flavin adenine dinucleotide (FAD) showing the average diffusion current i_d as measured at − 0.6 V versus sce.

Sometimes the polarographic wave does not have a constant limiting current as the potential is changed. A current peak in the plateau region of the polarogram is a *current maximum*. A current maximum can be found at the start of the plateau or at a potential beyond the start of the plateau. Current maxima are caused by convection around the mercury drop. Normally convection is caused by a change in surface tension at the electrode surface owing to adsorption of an electroactive solution species on the dme.

If the diffusion current cannot be measured either before or after the current maximum, it becomes necessary to remove the current maximum prior to using the wave for quantitative analysis. That can be accomplished in most cases by the addition of a small amount of a *maximum suppressor*. A maximum suppressor is an electroinactive surface-active substance, such as a detergent, which is adsorbed on the electrode surface and prevents adsorption of other species which can cause the maximum. Gelatin, indicators, and detergents, among other substances, have been used as maximum suppressors. Triton X-100 (a detergent) is probably the most widely used maximum suppressor. About 6 drops of a 0.1% (v/v) aqueous solution of Triton X-100 is usually sufficient to suppress a current maximum in a 10-mL sample solution. Care should be taken to add no more of a maximum suppressor than is necessary, because the addition of too much can cause a shift in the potential at which the polarographic wave occurs, a change in the shape of the wave, or a decrease in the height of the wave.

Polarographic Potential

The equations that are derived in this section apply to reversible electrochemical reactions in which the reaction rate is limited by the rate of mass transport (by diffusion for polarography) to the electrode. The rate at which the electrochemical reaction occurs at the electrode surface is assumed to be fast relative to the rate of mass transport. The approach is general and applies to other voltammetric techniques as well as to polarography, e.g., to hydrodynamic voltammetry where mass transport occurs by convection.

The potential of the dme can be related to the concentrations of the soluble oxidized and reduced species [Eq. (23-3)] at the electrode surface by the Nernst equation:

$$E = E^{\circ} - \frac{RT}{nF} \ln \frac{[\mathrm{Red}]_o}{[\mathrm{OX}]_o} \tag{23-6}$$

In Eq. (23-6) the subscript o's are used to indicate that the concentrations are at the electrode surface. Concentrations are substituted for activities because, at the low concentrations at which voltammetric measurements are made, activity coefficients are nearly 1. For the reversible reduction shown in Eq. (23-3), the cathodic current flowing through the cell is directly proportional to the difference between the concentration in the bulk solution and that at the electrode surface:

$$i = K_{O_x} ([\mathrm{Ox}] - [\mathrm{Ox}]_o) \tag{23-7}$$

i.e., the current is governed by the rate of mass transport by diffusion to the electrode surface.

On the plateau of the polarographic wave, the concentration of Ox at the electrode surface is zero owing to the immediate reduction of Ox upon its arrival at the electrode surface (see earlier description) and the current is the cathodic diffusion current $i_{d,c}$. Substitution of $i_{d,c}$ for i and 0 for $[\mathrm{Ox}]_o$ in Eq. (23-7) yields

$$i_{d,c} = K_{O_x} [\mathrm{Ox}] \tag{23-8}$$

By simultaneously solving Eqs. (23-7) and (23-8) for $[\mathrm{Ox}]_o$, the following result is obtained:

$$[\mathrm{O_X}]_o = \frac{i_{d,c} - i}{K_{\mathrm{OX}}} \tag{23-9}$$

For the special case in which Red is soluble, the current flowing through the cell is proportional to the rate of diffusion of Red away from the working electrode:

$$i = K_{\mathrm{Red}}([\mathrm{Red}]_o - [\mathrm{Red}]) \tag{23-10}$$

Because the concentration of Red in the bulk solution is essentially zero, Eq. (23-10) simplifies to

$$i = K_{\mathrm{Red}}[\mathrm{Red}]_o \tag{23-11}$$

which can be solved for $[\text{Red}]_o$ with the following result:

$$[\text{Red}]_o = \frac{i}{K_{\text{Red}}} \tag{23-12}$$

Upon substitution of the values of $[\text{Ox}]_o$ and $[\text{Red}]_o$ from Eqs. (23-9) and (23-12) into Eq. (23-6), the following equation is obtained:

$$E = E^o - \frac{RT}{nF} \ln \frac{iK_{\text{Ox}}}{K_{\text{Red}}(i_{d,c} - i)} \tag{23-13}$$

That equation relates the potential of the indicator electrode to the current flowing through the cell. Rearrangement of Eq. (23-13) yields

$$E = E^o - \frac{RT}{nF} \ln \frac{K_{\text{Ox}}}{K_{\text{Red}}} - \frac{RT}{nF} \ln \frac{i}{i_{d,c} - i} \tag{23-14}$$

It is important to understand that the Nernst equation, from which all of the other equations in this section are derived, is valid only at chemical equilibrium. Consequently polarographic scan rates must be sufficiently slow to ensure that chemical equilibrium is maintained. The *half-wave potential* $E_{1/2}$ is used to describe a voltammogram that exhibits a plateau such as that which occurs in polarograms. It is defined as the potential when the current is one-half of the diffusion current, i.e., the potential halfway up the rising portion of the polarographic wave (see Fig. 23-3):

$$E = E_{1/2} \text{ when } i = \frac{i_{d,c}}{2} \tag{23-15}$$

Upon substitution of the equalities in Eq. (23-15) into Eq. (23-14), the following equation is obtained:

$$\begin{aligned} E_{1/2} &= E^o - \frac{RT}{nF} \ln \frac{K_{\text{Ox}}}{K_{\text{Red}}} - \frac{RT}{nF} \ln \frac{i_{d,c}/2}{i_{d,c} - i_{d,c}/2} \\ &= E^o - \frac{RT}{nF} \ln \frac{K_{\text{Ox}}}{K_{\text{Red}}} \end{aligned} \tag{23-16}$$

If that value is substituted into Eq. (23-14), the following equation is obtained:

$$E = E_{1/2} - \frac{RT}{nF} \ln \frac{i}{i_{d,c} - i} \tag{23-17}$$

It relates the indicator electrode potential to the half-wave potential and the current flowing through the cell.

From Eq. (23-16) it is apparent that the half-wave potential for an electroactive species is a constant related to the standard potential for the electrochemical reduction. Under fixed experimental conditions (constant pH, ionic strength, etc.) the half-wave potential sometimes can be used as a tool for qualitative analysis. The disadvantage of using $E_{1/2}$ for qualitative analysis is the fairly large number of electroactive substances which have half-wave potentials near each other. If a limited number of possibilities exist, and if all of the possible electroactive species have $E_{1/2}$ values which are separated by at least 0.1 V under specific experimental conditions, comparison of the $E_{1/2}$ of the sample with those of known substances is an acceptable method of qualitative analysis.

Equations similar to Eq. (23-17) can be derived for the case of the reversible oxidation of Red to Ox:

$$E = E_{1/2} - \frac{RT}{nF} \ln \frac{i_{d,a} - i}{i} \tag{23-18}$$

and the reversible case in which Ox and Red are simultaneously present in the bulk solution:

$$E = E_{1/2} - \frac{RT}{nF} \ln \frac{i - i_{d,a}}{i_{d,c} - i} \tag{23-19}$$

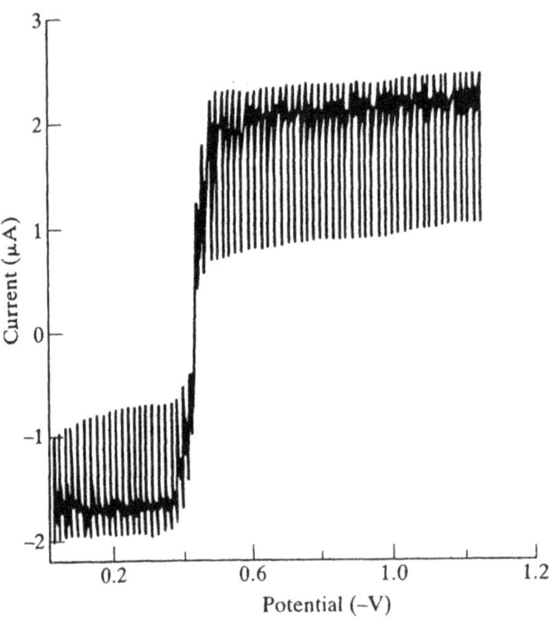

Figure 23-7 Polarogram of a solution containing 0.4 mM flavin adenine dinucleotide (FAD) and 0.4 mM reduced flavin adenine dinucleotide ($FADH_2$) in pH 8.2 Tris buffer. The reference electrode is an sce.

In Eqs. (23-18) and (23-19), $i_{d,\,a}$ is the anodic diffusion current, $i_{d,\,c}$ is the cathodic diffusion current, and both Ox and Red are assumed to be soluble. A reversible polarographic wave corresponding to a solution which contains both Ox and Red is called a combination oxidation and reduction wave (Fig. 23-7).

Equations (23-17) through (23-19) apply only for the reversible electrochemical reaction shown in Eq. (23-3). Although the reaction shown in Eq. (23-3) is probably the most often encountered type of reaction of inorganic substances, many organic compounds and some inorganic compounds obey different equations. If the coefficients of Ox and Red in the balanced electrochemical half-reaction are not identical, other equations relating potential to current are obtained. As an example, for reduction of the dimer R—R to form the monomer R—H (R—R + $2H^+ + 2e^- = 2$R—H), the equation derived by following the procedure outlined in this section is

$$E = E^{\circ} + \frac{RT}{F}\ln\,[H^+] - \frac{RT}{2F}\ln\frac{K_{Ox}}{K_{Red}^2} - \frac{RT}{2F}\ln\frac{i_d^2}{i_{d,c} - i} \tag{23-20}$$

For that particular case, $E_{1/2}$ is dependent upon the diffusion current (i.e., upon the concentration of R—R) and the pH of the solution as follows:

$$E_{1/2} = E^{\circ} + \frac{RT}{F}\ln\,[H^+] - \frac{RT}{2F}\ln\frac{K_{Ox}}{K_{Red}^2} - \frac{RT}{2F}\ln\frac{i_d}{2} \tag{23-21}$$

Oxygen Removal

Regardless of whether polarography or voltammetry is used, the presence of O_2 from air dissolved in the solution can interfere with the analysis of an electroactive species. Dissolved oxygen exhibits two reduction waves corresponding to the following half-reactions:

$$O_2 + 2H^+ + 2e^- = H_2O_2 \tag{23-22}$$

$$H_2O_2 + 2H^+ + 2e^- = 2H_2O \tag{23-23}$$

The reduction of O_2 to H_2O_2 occurs at a dme between 0 and – 0.1 V (versus sce) and the reduction of H_2O_2 occurs at about – 0.9 V. Because the reductions of many other substances also occur in those regions, it is desirable to remove oxygen from the solution prior to a voltammetric or polarographic study.

The method used most often for removing O_2 from a solution consists of bubbling an electroinactive gas through the solution. The solution becomes saturated with the electroinactive gas and the dissolved O_2 is swept from the solution. Either nitrogen or argon can be used for the purpose. Nitrogen is used more often than argon because it is usually less expensive. The time necessary to remove all oxygen from the solution depends upon the volume of solution, the rate of introduction of the electroinactive gas to the solution, and the size of the bubbles forced through the solution. Between 2 and 5 min is usually required to deaerate a 10- to 25-mL solution if the nitrogen or argon tank is attached through tubing to a submerged Pasteur pipet or to a coarse-porosity glass frit, and if the gas is bubbled through the solution at a brisk pace.

Polarographic Analysis

Qualitative polarographic analysis is done by comparing the half-wave potential of the sample with the half-wave potential of known substances under as nearly identical experimental conditions as possible. For the general redox reaction shown in Eq. (23-3), the half-wave potential can be accurately measured by preparing a *log plot* of applied potential E as a function of log $[i/(i_{d,\,c} - i)]$ [Eq. (23-17)] for a reduction, log $[(i_{d,a} - i)/i]$ [Eq. (23-18)] for an oxidation, or log $[(i - i_{d,a})/(i_{d,\,c} - i)]$ [Eq. (23-19)] for a combined oxidation and reduction wave. The resulting linear plot intercepts the potential axis at the half-wave potential.

Quantitative polarographic analysis is usually done by using a working curve or the standard-addition technique. The Ilkovic equation [Eqs. (23-4) and (23-5)] also can be used to calculate concentrations if the quantity $nD^{1/2}m^{2/3}t^{1/6}$ can be determined. The *capillary characteristics* m and t can be obtained by accurately measuring the drop time in the sample solution at a potential on the plateau of the polarographic wave and by weighing the mercury which corresponds to a known number of drops that were collected when the potential was a value on the plateau of the wave.

If the redox reaction is reversible, a plot of E as a function of the natural logarithmic term in Eqs. (23-17), (23-18), or (23-19) yields a straight line with a slope of $-RT/nF$ from which n can be determined for use in the Ilkovic equation. Unfortunately the method cannot be used to calculate n for an irreversible process. Because irreversible reactions do not obey the Nernst equation, they do not yield logarithmic-plot slopes of $-RT/nF$. The diffusion coefficient D must be obtained from some other source if the Ilkovic equation is to be used to determine concentrations. A list of the substances which can be assayed by each type of voltammetry and polarography is much too long to be given here. Nearly every biochemical, inorganic, and organic species which can be assayed by chemical oxidation or reduction, as well as many other species, can be assayed by polarography or voltammetry.

Sample problem 23-1 The following data were obtained on the rising portion of a polarographic wave for the reduction of 4.20 mM nitrate in dimethylformamide (solvent) at 25°C. On the plateau of the polarographic wave the drop time was 34.9 s for 10 drops, and 20 drops had a mass of 0.1296 g. The diffusion current was 14.25 μA. Determine $E_{1/2}$, n, and D for nitrate in dimethylformamide. All current measurements were made at the top of the undamped recorder trace.

E versus Ag/AgNO$_3$(0.010 *M*), V	i_c, μA
–2.400	1.00
– 2.420	2.28
–2.440	3.50
– 2.460	6.00
–2.480	9.00
–2.500	11.17

SOLUTION Initially a logarithmic plot is prepared in order to obtain $E_{1/2}$ and n. Substitution of the value of the constants into Eq. (23-17) and conversion to base 10 logarithms yields the equation shown below:

$$E = E_{1/2} - \frac{0.059}{n} \log \frac{i}{i_{d,c} - i}$$

The values of $i/(i_{d,c} - i)$ can be calculated as follows:

–2.400 V: $\frac{1.00}{14.25 - 1.00} = 0.0755$

–2.420 V: $\frac{2.28}{14.25 - 2.28} = 0.190$

–2.440 V: $\frac{3.50}{14.25 - 3.50} = 0.325$

–2.460 V: $\frac{6.00}{14.25 - 6.00} = 0.727$

–2.480 V: $\frac{9.00}{14.25 - 9.00} = 1.71$

–2.500 V: $\frac{11.17}{14.25 - 11.17} = 3.63$

The base 10 logarithm of each of those values is obtained from a logarithmic table or a calculator:

E, –V	log ($i/i_{d,c} - i$)
2.400	– 1.122
2.420	– 0.721
2.440	– 0.488
2.460	– 0.138
2.480	0.233
2.500	0.560

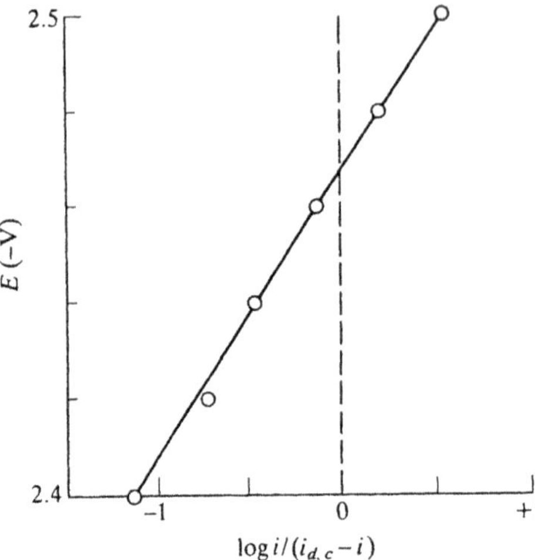

Figure 23-8 The log plot of the polarographic data presented in Sample problem 23-1.

The potentials in the table are plotted as a function of the corresponding logarithmic terms to give the log plot shown in Fig. 23-8. At the point on the plotted line at which the logarithmic term is zero, the half-wave potential is determined to be − 2.466 V relative to the Ag/$AgNO_3$(0.01 *M*) reference electrode.

Any two convenient points on the line (the points at − 2.400 V and − 2.500 V were chosen for this calculation) can be used to calculate the slope of the line:

$$\text{Slope} = \frac{-2.500 - (-2.400)}{0.560 - (-1.122)}$$

$$= \frac{-0.100}{1.682} = -0.0594 \text{ V}$$

The theoretic slope is − 0.059/*n* V and the value of *n* can be determined by setting the theoretical slope equal to the actual slope:

$$-0.059/n \text{ V} = -0.0594 \text{ V}$$

$$n = 1$$

Because the current measurements were made at the top of the undamped recorder trace, the form of the Ilkovic equation listed in Eq. (23-4) can be used to calculate *D*. The drop time is t = 34.9 s/10 drops = 3.49 s/drop, that is, t = 3.49 s. The mercury flowrate is

$$m = 129.6 \text{ mg/20 drops} \times 1 \text{ drop/3.49 s} = 1.86 \text{ mg/s}$$

From those values $m^{2/3}$ and $t^{1/6}$ can be calculated:

$$m^{2/3} = 1.86^{2/3} = 1.51 \text{ mg}^{2/3}/\text{s}^{2/3}$$

$$t^{1/6} = 3.49^{1/6} = 1.23 \text{ s}^{1/6}$$

Substitution of the appropriate values into the Ilkovic equation allows calculation of the diffusion coefficient:

$$i_d = 708nD^{1/2}m^{2/3}t^{1/6}C$$
$$14.25 = 708(1)D^{1/2}(1.51)(1.23)(4.20)$$
$$D^{1/2} = 2.58 \times 10^{-3}$$
$$D = 6.66 \times 10^{-6}\ \text{cm}^2/\text{s}$$

AC Polarography

The most popular voltage ramp used in ac polarography consists of a small-amplitude, sinusoidally alternating potential applied on top of (or added to) the normal polarographic ramp. Typically, the amplitude of the applied alternating potential is between 5 and 50 mV and the frequency of the ac potential is between 10 and 100 Hz. A diagram of the ac polarographic ramp is shown in Fig. 23-9. The applied alternating potential causes an alternating current to flow through the cell. Only the ac component of the current is measured and the ac polarogram is obtained by plotting the ac current as a function of the dc potential.

For a reversible redox reaction, the ac polarogram is a symmetrical peak which is essentially the first derivative of the classical polarogram of the sample. The peak height i_p is directly proportional to the concentration of the electroactive species and to the square root of the frequency:

$$i_p = \frac{2.5n^2F^2AD^{1/2}\Delta EN^{1/2}C}{4RT} \tag{23-24}$$

In Eq. (23-24), ΔE is the amplitude (V) of the superimposed alternating potential, A is the electrode area (cm^2), N is the frequency (Hz) of the alternating potential, D is the diffusion coefficient (cm^2/s), C is the concentration (mM), i_p is the peak current (μA), and the other terms have their usual meanings.

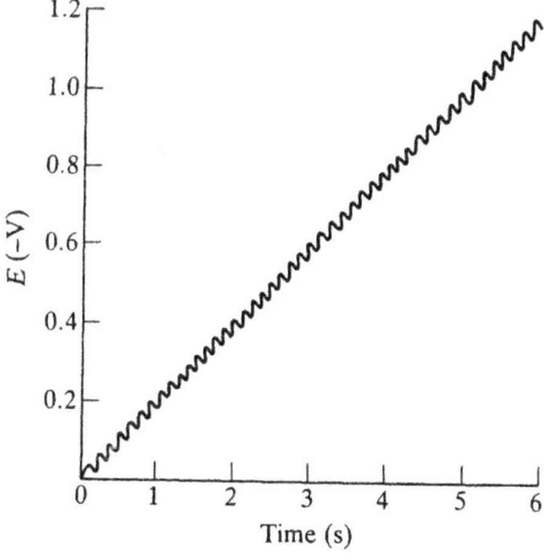

Figure 23-9 An ac polarographic voltage ramp. The actual frequency of the ac potential is greater than that shown.

The applied ac potential alternately drives a reversible electrode reaction to the right and then to the left. If the reaction is not completely reversible, the peak height is less and the peak width is greater than in the reversible case. The peak width at half the peak height for a reversible reaction is $90/n$ mV. For completely irreversible reactions, a small broad peak is observed at a potential (for reductions) significantly negative of the expected potential for a reversible reaction. Because in many cases the reversibility of the electrochemical reaction depends upon the frequency of the applied ac potential, it is possible to obtain kinetic information about the reaction by altering the frequency of the applied ac potential and observing the effect on the shape of the ac polarogram.

For a reversible reaction, the *faradaic current* is out-of-phase with the applied potential and leads the applied potential by 45°. The faradaic current is that portion of the total current that is caused by the redox half-reaction occurring at the indicator electrode. The charging current leads the applied potential by 90°. Essentially the characteristics of the charging current are identical to the current associated with charging a capacitor as described in Chapter 2. The phase relationship between the applied potential (plotted on the horizontal axis), the faradaic current, and the charging current (plotted on the vertical axis) are illustrated in Fig. 23-10.

An advantage in using ac polarography as a quantitative analytical tool is achieved by using a phase-sensitive lock-in amplifier to measure the ac current. If the current is measured at 0° from the applied potential (in-phase with the applied potential), the amplitude of the measured faradaic current is decreased to 70.7 percent of its value at 45° from the applied potential, as illustrated in Fig. 23-10. The charging current is not related to the concentration of the electroactive species and consequently is a component of the "noise" of the total current. The charging current, when measured at 0° in relation to the applied potential, is decreased to essentially zero. Consequently, by measuring the current in-phase with the applied potential, the ratio of the faradaic current to the charging current is increased with a resulting increase in the signal-to-noise (S/N) ratio of the technique. That, in turn, can result in a detection limit about 10 to 20 times lower than that observed without a phase-sensitive lock-in amplifier. Typical detection limits when using total current (not phase-selective) ac polarography are about 10^{-4} *M*. A typical ac polarogram is shown in Fig. 23-11. Further information on ac polarography can be found in References 2, 3, and 5.

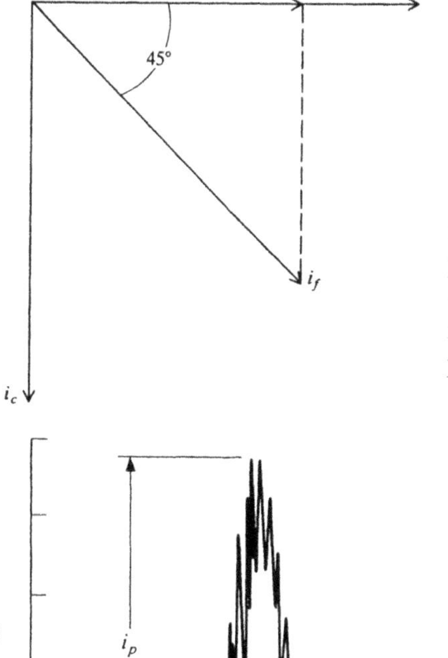

Figure 23-10 A diagram illustrating the relationship between the applied alternating potential E_{appl}, the faradaic current i_f, and the charging current i_c. The faradaic current measured at 0° relative to the applied potential is indicated by $i_{f,0}$.

Figure 23-11 Diagram of an ac polarogram for a reversible electrochemical reaction.

Pulse Polarography

Under ideal circumstances classical polarography is useful over the concentration range from about 10^{-2} to 10^{-5} *M*. The lower concentration limit is determined by the signal-to-noise ratio for a particular analysis. In classical polarography a major source of noise at low concentrations is the capacitive current resulting from charging of the electric double layer at the electrode. By using pulse or differential pulse polarography, most of the capacitive current can be eliminated with a resultant increase in the S/N ratio of about 100.

In pulse and differential pulse polarography advantage is taken of the relatively rapid decrease in the capacitive current as compared with the faradaic current after application of a potential to an electrode. In pulse polarography a potential pulse is typically applied to the mercury drop about 57 ms prior to the drop's fall from the capillary. The capacitive current exponentially, decays to nearly zero during the first 40 ms of the pulse, and the remaining faradaic current is measured during the last 17 ms (that corresponds to 1 cycle at the 60-Hz ac line frequency used in the United States) of the pulse.

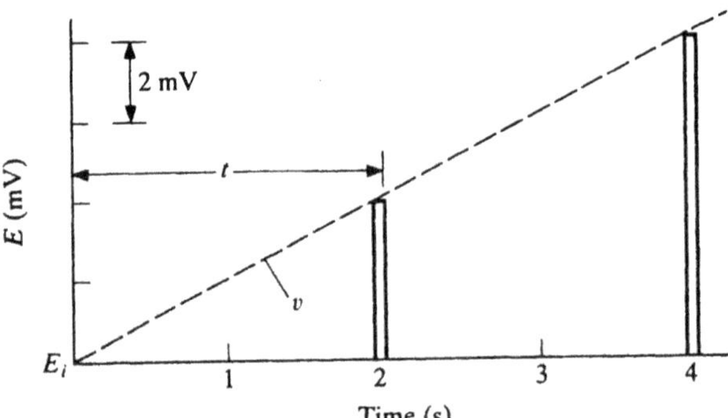

Figure 23-12 A typical voltage ramp for pulse polarography. The potential pulse is typically applied for 57 ms just before the drop falls from the capillary. The current is typically measured during the last 17 ms of the pulse. E_i, the initial potential; t, drop time (2 s in this case); v, the scan rate, which corresponds to the slope of the dashed line (2 mV/s in this case).

The pulse is applied just before the drop is removed from the capillary. The drop time must be controlled with a device which dislodges each drop at a selected time interval. A sketch of the voltage ramp used during pulse polarography is shown in Fig. 23-12. In addition to control over the drop time, the analyst can control the size of the potential pulses by controlling the scan rate. As the scan rate is increased, the change in pulse size is necessarily increased at a selected drop time.

By making the current measurement at the end of the drop when the electrode area is at a maximum and the current flowing through the electrode is at its highest, the sensitivity is maximized. By applying the potential for a short time during the life of the drop, the layer of depleted electroactive substance (i.e., the layer adjacent to the electrode in which the electroactive substance has reacted) is kept small and electroactive substances can diffuse to the electrode more rapidly than during classical polarography. The combination of the decreased capacitive current, measurement while the electrode area is at a maximum, and the small depleted layer during the measurement causes a significant increase in sensitivity relative to that obtained with classical polarogtaphy.

A pulse polarogram is a plot of the current which is typically measured during the 17 ms period prior to the fall of the drop as a function of the dc potential corresponding to the dashed line in Fig. 23-12. The current measured during a pulse is continuously plotted until the next pulse is applied and a new current is measured. A pulse polarogram (Fig. 23-13) has the same shape as a classical polarogram and as a Tast polarogram. It is instructive to compare Fig. 23-13 with Figs. 23-3 and 23-4.

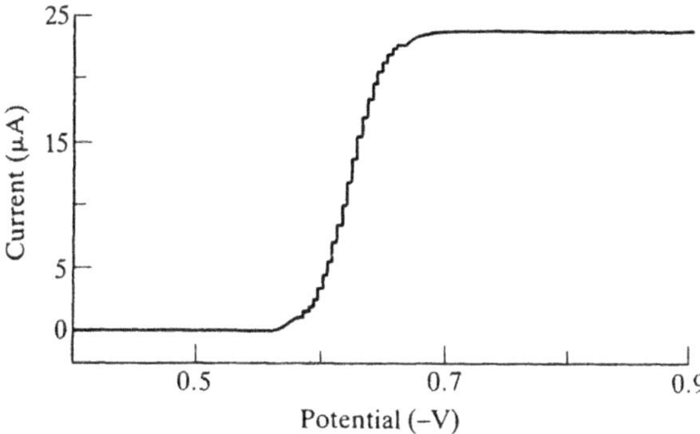

Figure 23-13 Pulse polarogram of 1.2 m*M* cadmium chloride in an aqueous 0.1-*M* potassium chloride solution. $v = 2$ mV/s; $t = 2$ s. The reference electrode is an sce.

The diffusion current measured from a pulse polarogram for a reversible reaction is given by[6]

$$i_d = \frac{nF\, AD^{1/2}C}{\pi^{1/2}\delta^{1/2}} \tag{23-25}$$

In Eq. (23-25), δ is the length of time the pulse is applied excluding the time during the current measurement (i.e., typically 57 – 17 = 40 ms), and the other terms have the same meanings as before. Generally either the working-curve method or the standard-addition technique is used for quantitative analysis with pulse polarography. The diffusion current is plotted either as a function of the concentration of a series of standard solutions or as a function of the added concentration of analyte to the sample as described in previous chapters.

The pulsed technique can be applied to solid electrodes as well as to the dme. In those instances the technique is *pulse voltammetry*. An introduction to pulse polarography and pulse voltammetry can be found in Reference 7.

Differential Pulse Polarography

The voltage ramp for differential pulse polarography consists of a classical polarographic ramp upon which a potential pulse with an amplitude which is usually between 5 and 100 mV is applied during the last 57 ms of the drop. In differential pulse polarography, the current is typically measured during the 17 ms prior to application of the pulse and during the last 17 ms during application of the pulse and after decay of the capacitive current. The polarogram is a plot of the difference between the two currents as a function of the applied potential. It approximates the first derivative of a classical polarogram. A typical voltage ramp for differential pulse polarography is shown in Fig. 23-14, and a typical differential pulse polarogram is shown in Fig. 23-15 (cf. Figs. 23-3 and 23-12).

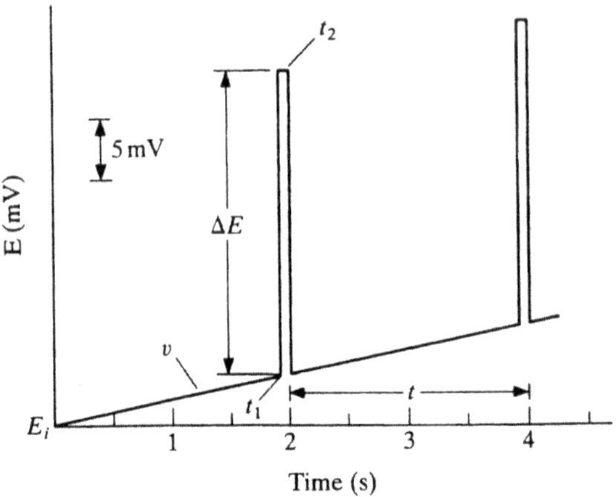

Figure 23-14 A voltage ramp for differential pulse polarography. The current is measured just prior to application of the pulse at t_1 and typically during the last 17 ms of the pulse at t_2. The difference $i_2 - i_1$ is plotted in the polarogram. ΔE, potential pulse amplitude (25 mV in this case); E_i, initial potential; t, drop time (2 s in this case); v, slope of the line or the scan rate (2 mV/s in this case).

The height of the peak (Δi_p in Fig. 23-15) in a pulse polarogram for a reversible system is related to concentration and the other experimental parameters by[6,8]

$$i_p = \frac{nF\,AD^{1/2}C[\exp(nF\Delta E/2RT) - 1]}{\pi^{1/2}\delta^{1/2}[\exp(nF\Delta E/2RT) + 1]} \tag{23-26}$$

In Eq. (23-26) ΔE is the amplitude of the applied pulse and the other terms have their usual meanings. ΔE is considered to be positive for a cathodic pulse and negative for an anodic pulse.

Quantitative analysis is done by using the working-curve method or the standard-addition technique. The operator has control over the drop time, the scan rate, and the amplitude of the applied potential pulse. By increasing the amplitude of the potential pulse, the technique is made more sensitive to concentration but is less capable of resolving adjacent peaks. If a single peak is present in the polarogram or if more than one peak is present, but the peaks are well separated on the potential axis, a potential pulse of perhaps 50 or 100 mV can be used to obtain maximum sensitivity. If two or more peaks occur at nearly the same potential in the polarogram, it is preferable to sacrifice sensitivity by decreasing the pulse amplitude to perhaps 5 or 10 mV in order to achieve better resolution of the peaks.

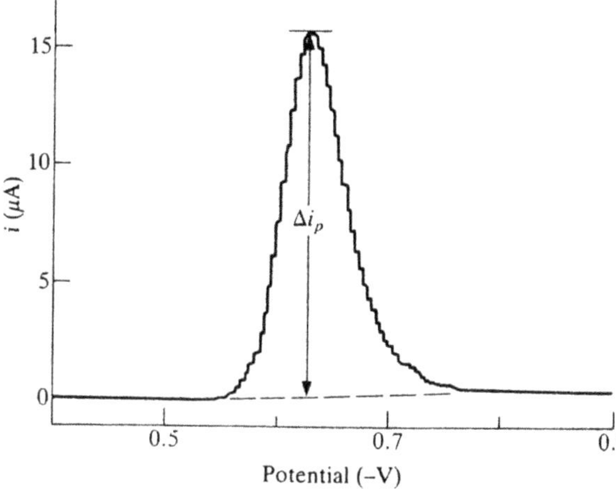

Figure 23-15 A differential pulse polarogram of 1.2 m*M* cadmium chloride in an aqueous 0.1-*M* potassium chloride solution. $v = 2$ mV/s; $t = 2$ s; $\Delta E = 5$ mV. The reference electrode is an sce.

Square Wave Polarography

The potential ramp that is used in square wave polarography is a square wave superimposed on a staircase ramp as shown in Fig. 23-16. The staircase ramp is indicated by the dashed line. The height of each "step" ΔE_s in the staircase ramp is typically 10 mV. The amplitude E_{sw} of the square wave is often adjusted to about $50/n$ mV which can be shown to yield the optimum combination of resolution and sensitivity. Consequently, if the analyte undergoes a one-electron oxidation or reduction, E_{sw} is adjusted to 50 mV, but if the oxidation or reduction is a two-electron process, E_{sw} is adjusted to 25 mV. The period τ of each step on the staircase is twice the time t_p of application of the square wave pulse, as illustrated in Fig. 23-16.

The current is measured at the end of each square wave pulse at time t_1 (Fig. 23-16) and just before application of the next pulse at t_2. A pulse polarogram is a plot of the change in current Δi between the two measurements $(i_{t_1} - i_{t_2})$ as a function of the average applied potential during each pulse. As with differential pulse polarography, the polarogram has the shape of the first derivative of a classical polarogram, as shown in Fig. 23-17.

Unlike pulse polarography and differential pulse polarography, the entire scan is obtained on a single drop during square wave polarography. Typically a relatively long drop time (5 s or more) is used for the study. A physical method is used to control the drop time and the entire apparatus is computer-controlled. The scan rate is controlled by adjusting ΔE_s and the frequency of the square wave. The square wave frequency is usually in the range between 100 and 1000 Hz. A typical value is 200 Hz. If the indicator electrode potential is to be scanned throughout a 1-V potential range while applying a 200-Hz square wave with a staircase amplitude of 10 mV, the scan requires 0.50 s as shown in the following calculation:

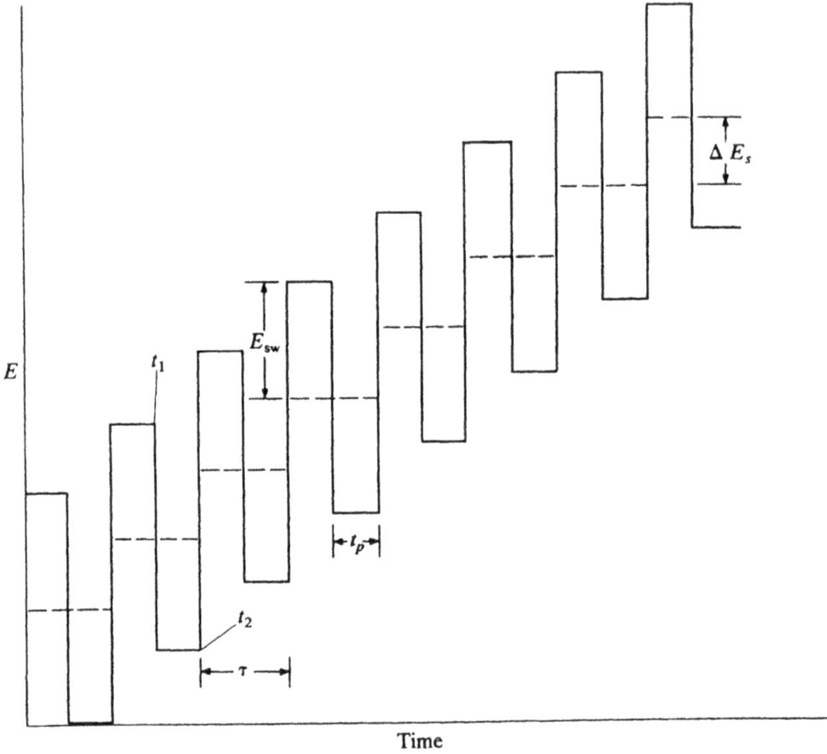

Figure 23-16 The voltage ramp that is applied while obtaining a square wave polarogram. E_{sw}, square wave potential; ΔE_s, staircase amplitude; τ, period of the staircase potential; t_p, period of the square wave. The current is measured at t_1 and t_2 during each square wave pulse.

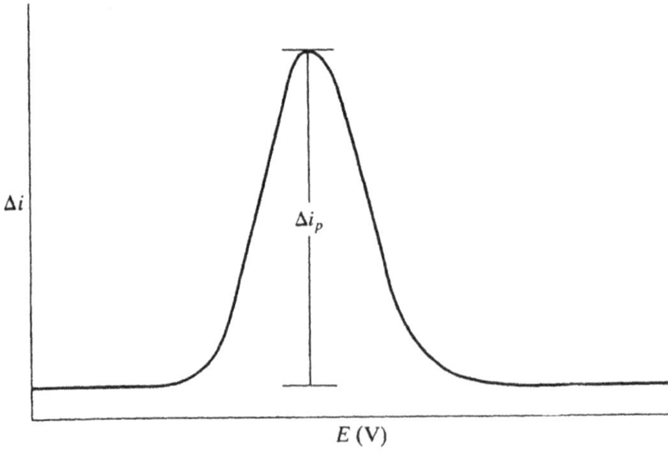

Figure 23-17 A square wave polarogram.

$$1\text{ V} \times \frac{1\text{ cycle}}{0.010\text{ V}} \times \frac{1\text{ s}}{200\text{ cycles}} = 0.50\text{ s}$$

The polarogram is obtained, in this case, during the last 0.50 s of the drop life. A major advantage of square wave polarography is the speed with which data can be obtained. The peak current Δi_p is proportional to concentration, as shown for a reversible electrochemical reaction:

$$\Delta i_p = \frac{nF\,AD^{1/2}\;C\Delta\Psi(E_s, E_{sw})}{\left(\pi t_p\right)^{1/2}} \qquad (23\text{-}27)$$

All of the terms in Eq. (23-17) have been defined previously except for $\Delta\Psi$ (E_s, E_{sw}), which is a dimensionless function (the change in the *current function*) of ΔE_s and E_{sw}.

Square wave polarography can be done using a dme or smde. If a solid electrode is used, the technique is termed square wave voltammetry. More detailed descriptions of the operation and theory of square wave polarography and voltammetry can be found in References 9 through 12.

TRACE ANALYSIS

Whenever assays are performed on low-concentration (less than 10^{-5} *M*) samples, some difficulties are encountered regardless of the analytical method used, and other difficulties are more often encountered while using polarographic or voltammetric techniques. The problem of accurately preparing standard solutions for use with the working-curve method is common to all analytical techniques. At low concentrations the sample often adsorbs on the wall of the vessel in which it is placed and thereby decreases the concentration in solution. The problem is often encountered with metallic cations in glass containers, and it can sometimes be alleviated by switching to polyethylene or polypropylene containers and rinsing the containers with nitric acid prior to use. A problem encountered with polarography is keeping small concentrations of electroactive impurities in the supporting electrolyte from interfering with the analysis. Small concentrations of impurities in the solvent used for an assay can also sometimes interfere with the analysis.

VOLTAMMETRIC PRINCIPLES

Voltammetry at Solid Electrodes

The voltage ramps and current-measuring techniques applicable in pulse polarography, differential pulse polarography, and square wave polarography are not restricted to use at a dme. If the same techniques are used at a stationary or rotating solid electrode, the techniques are respectively called *pulse voltammetry*, *differential pulse voltammetry*, and *square wave voltammetry*. Many types of polarographic waveforms can be used in addition to those described; however, they have not been used as often.

In some cases it is necessary to examine an electrochemical reaction which occurs at a potential more positive than that which can be obtained at a mercury electrode. As was mentioned earlier, the most positive potential at which an electrochemical reaction can be examined on a mercury electrode is determined by the potential at which mercury starts to be oxidized. That potential depends upon the components of the solution into which the electrode is dipped. In a solution containing potassium chloride, mercury oxidation starts at about 0 V relative to an sce. At an electrode constructed from some other substance such as platinum, graphite, glassy carbon (also called vitreous carbon), or gold, the positive potential at which an electrochemical reaction can be examined is limited by the oxidation of the solvent or the supporting electrolyte rather than by oxidation of the electrode. In aqueous solutions the limiting potential is often determined by the oxidation of water, as shown by

$$O_2 + 4H^+ + 4e^- = 2H_2O, \qquad E^o = 1.229 \text{ V} \tag{23-28}$$

As is customary when writing standard potential reactions, the half reaction is written as a reduction rather than an oxidation. As expected from Eq. (23-28), the limiting positive potential is dependent upon pH. At pH 7 electrochemical measurements can be made to a potential of about + 1 V relative to the sce.

The limiting negative potential in aqueous solutions with platinum, gold, and carbon electrodes is often determined by the reduction of water [Eq. (23-2)] Because the hydrogen overvoltage on those electrodes is relatively small, the limiting negative potential is about – 1 V relative to the sce (cf. –2 V on mercury). That value is also pH-dependent. An advantage in using a solid electrode is the extended positive potential range within which electrochemical reactions can be examined. A disadvantage is the decreased negative potential range. Measurements made with solid electrodes generally are not as reproducible as those made with a dme. A more thorough description of the theory and applications of solid electrodes is given in Reference 13.

Hydrodynamic Voltammetry

Voltammetry at a rotating electrode is probably the most popular form of solid electrode voltammetry which is used for quantitative analysis. The electrode can be a disk or a wire. Because the theory of voltammetry for the disk is best established, use of the disk is often advantageous. If a wire electrode is used, it can be rotated along its axis, perpendicular to its axis, or at some other angle relative to its axis. Wire electrodes which are vibrated rather than rotated have also been used for analysis, although usually not as successfully as the rotated electrodes. Several rotated electrodes are sketched in Fig. 23-18.

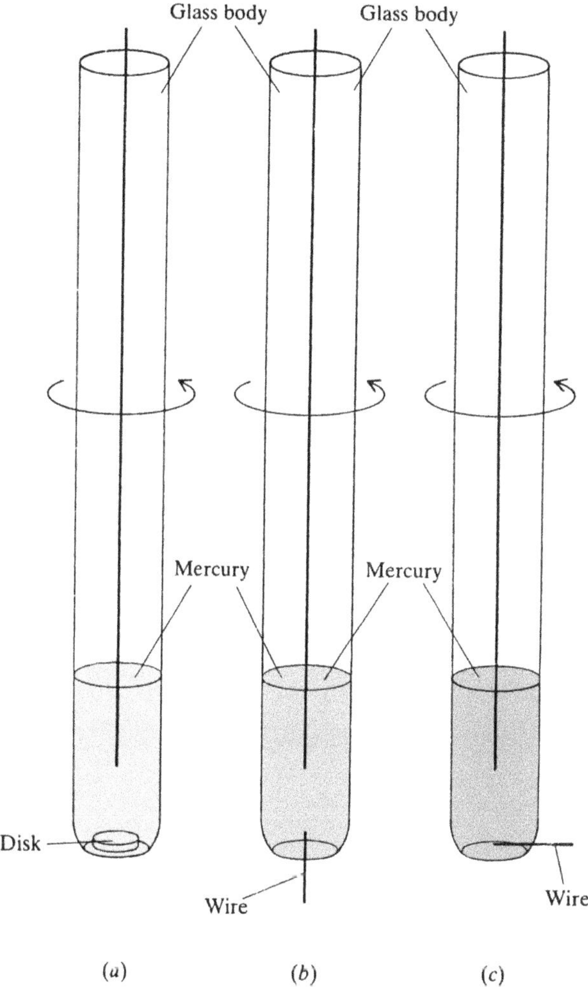

Figure 23-18 Several types of rotating electrodes used for hydrodynamic voltammetry. All electrodes are rotated around the axis of the electrode body. (*a*) Disk electrode; (*b*) and (*c*), wire electrodes. The arrows indicate the direction of rotation.

If the electrode is rotated, the electroactive species is brought to the electrode by convection, i.e., by the stirring owing to the rotation. Because the theory of convective flow to a rotated electrode of electroactive species is identical with the theory of flowing fluids (hydrodynamics), that type of voltammetry is termed *hydrodynamic voltammetry*. The voltage ramp used for hydrodynamic voltammetry is identical with that used for polarography, i.e., the potential is scanned at a rate of about 1 to 5 mV/s. The potential can be scanned in either a positive or negative direction. Usually the potential is initially adjusted to a value at which the electroactive species does not react and then scanned in the direction which will cause the electrochemical reaction to occur. For a reduction the potential is normally scanned toward more negative potentials and for an oxidation it is usually scanned toward more positive potentials.

The voltammogram obtained by using hydrodynamic voltammetry has a shape (Fig. 23-19) similar to that obtained with polarography. The current flowing on the plateau of the wave is controlled by convection (rather than diffusion) of the electroactive species to the electrode surface. The height of the wave measured from the extrapolated residual current is usually called the *limiting current* i_{lim} rather than the diffusion current. The potential at which the current is one-half of the limiting current is the half-wave potential $E_{1/2}$ as in polarography.

Regardless of the type of rotating electrode used for an analysis, the limiting current is directly proportional to the concentration of the electroactive species. For a *rotated disk electrode* (rde) in which only unidirectional flow (upward to the electrode surface as illustrated in Fig. 23-20) of the electroactive species can occur, the limiting current is proportional to the square root of the rotation rate of the electrode as well as to the concentration of the electroactive species. The relationship between limiting current, concentration, electrode rotation rate, and several other parameters is given by

$$i_{lim} = 1.55nFAD^{2/3}\,\nu^{-1/6}\pi^{1/2}N^{1/2}C \tag{23-29}$$

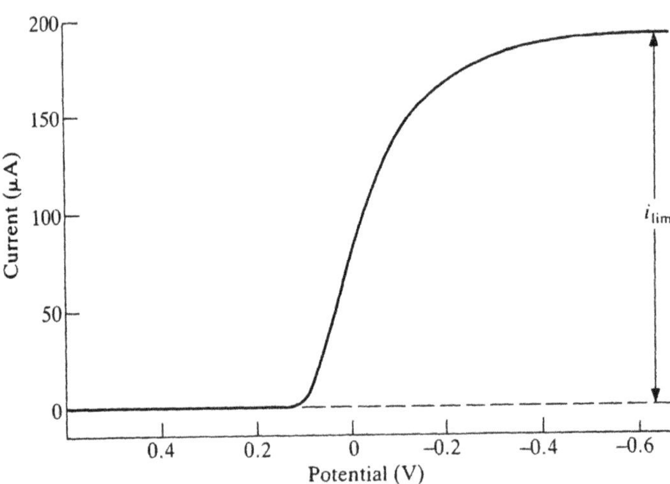

Figure 23-19 A hydrodynamic voltammogram of 4.6 m*M* mercury(II) acetate in pH 7.5 Tris buffer at a rotated glassy carbon disk electrode. N = 30 Hz; v = 5 mV/s. The reference electrode is an sce.

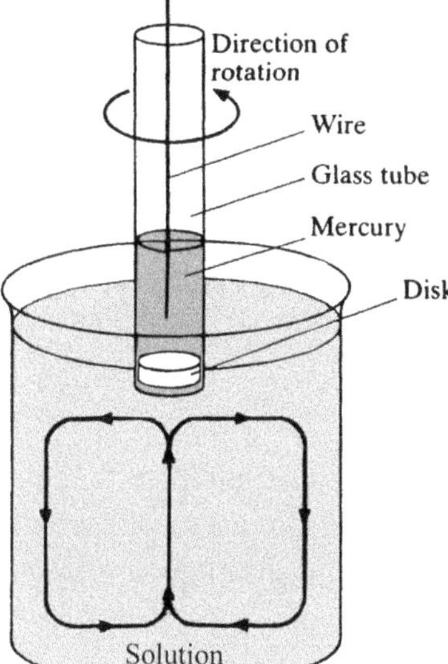

Figure 23-20 Diagram of a rotated disk electrode showing the direction of solution flow to the electrode during rotation.

All of the terms in Eq. (23-29) have the same meanings as reported earlier except for v (the kinematic viscosity of the solvent, cm^2/s) and N (the rotation rate of the electrode, r/s). The kinematic viscosity is the viscosity of the solution divided by the solution's density.

The limiting current can be used for quantitative analysis in the same way as the polarographic diffusion current is used. Both the working-curve method and the standard-addition technique are often used. The half-wave potential is a measure of the amount of energy required to reduce or oxidize an electroactive species. Under favorable conditions, it can be used as a tool for qualitative analysis. (See the earlier description concerning the polarographic half-wave potential.)

Information other than qualitative and quantitative analytical data can sometimes be obtained by using rotating electrodes. Sometimes the rate and mechanism of an electrochemical reaction can be deduced from data obtained by using a *rotated ring-disk electrode* (rrde). An rrde is a disk electrode that contains a concentric ring of electrode material which has a radius larger than that of the disk. The ring and disk are separated from each other by an insulating material. Separate electrical connections to the ring and the disk are maintained. A sketch of the bottom of an rrde is shown in Fig. 23-21.

An electrochemical reaction product, which is formed at the disk, travels outward over the ring as the electrode rotates. The rate of rotation of the electrode determines the rate of transfer of material from the disk to the ring.

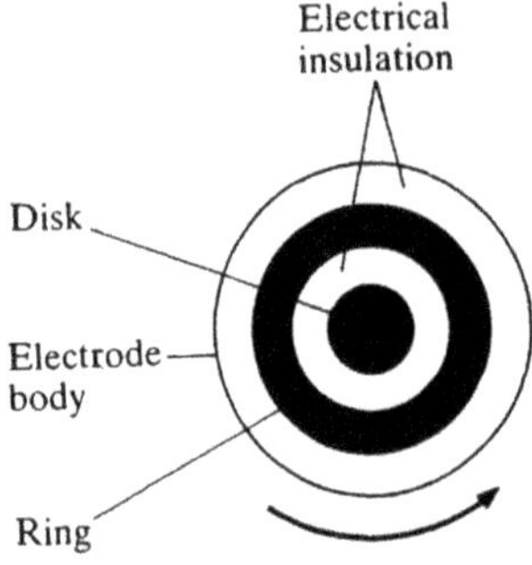

Figure 23-21 Diagram of the bottom of a rotated ring-disk electrode (rrde). The arrow indicates the direction of rotation.

As the rotation rate is increased, the rate of transfer to the ring is increased. The potential at the disk is scanned or is adjusted to a potential at which an electrochemical reaction takes place, and the ring is adjusted to a potential at which the disk reaction products react. By observing the amount of reaction product which reaches the ring as a function of electrode rotation rate or as a function of the disk potential, it is sometimes possible to determine the rate constant for the reaction or the reaction mechanism. That and other applications of rotated electrodes are described by Adams.[13]

Triangular Wave Voltammetry

Triangular wave voltammetry utilizes a triangular voltage ramp (Fig. 23-22). The potential is scanned at a fixed rate from the initial potential to a maximum (or minimum) potential where the scan direction is reversed and the potential is returned at the same scan rate to the initial potential. The initial direction of potential scan can be either negative or positive. The indicator electrode can be a *hanging mercury drop electrode* (hmde), a solid wire or disk electrode (normally constructed from carbon, platinum, or gold), or some other solid or liquid electrode. In some cases a chemically modified electrode[14] can be used. An hmde consists of a mercury drop hanging from a capillary, platinum-wire electrode, or some other device. The hmde differs from the dme in that a single stationary drop is used with the hmde, whereas the mercury is continuously flowing in the dme. Sketches of two hmde's are shown in Fig. 23-23. The static mercury drop electrode that was described earlier can be used as an hmde if a single drop is used for the study.

If the dme is used for triangular wave polarography, the entire scan is performed during a fraction of the drop time. Usually the scan is performed near the end of the life of the drop but before the drop falls. Large drop times (about 10 s) are generally used with the procedure. Because the scan is rapid, the electrode area does not significantly increase during the scan.

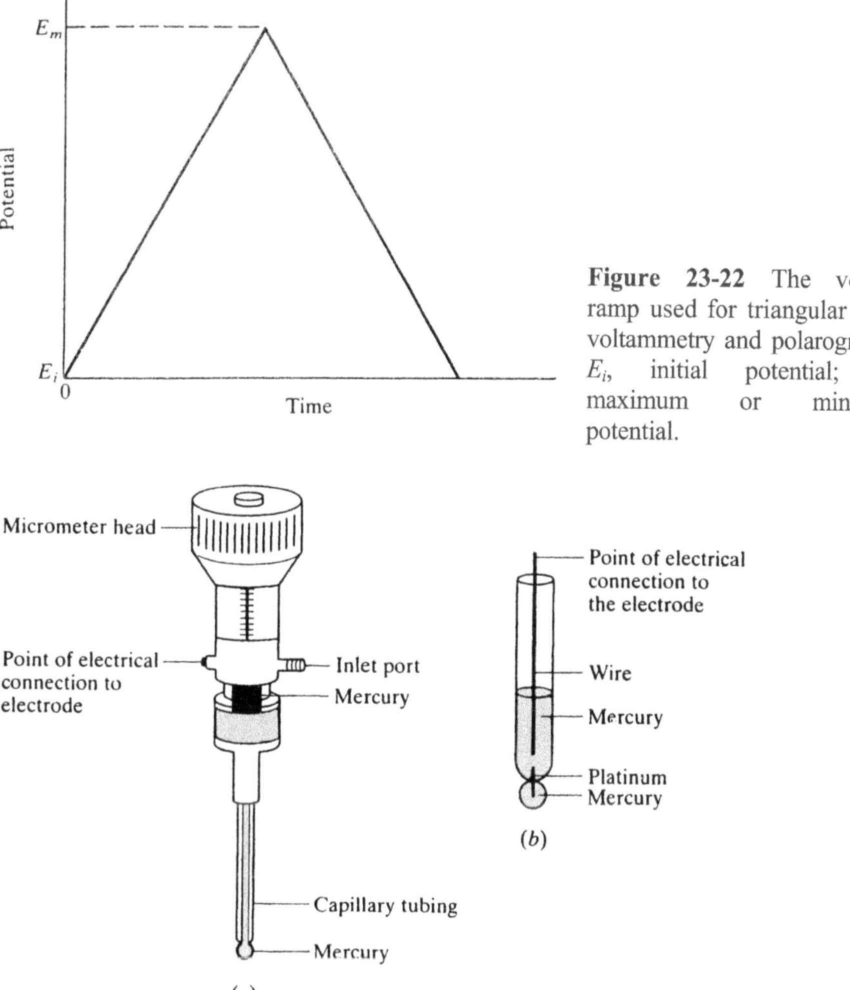

Figure 23-22 The voltage ramp used for triangular wave voltammetry and polarography. E_i, initial potential; E_m, maximum or minimum potential.

Figure 23-23 Two forms of hanging mercury drop electrodes (hmdes): (*a*) micrometer head hmde; (*b*) hmde with platinum wire connector. The inlet port is used to attach a vacuum line to fill the electrode with mercury. The wire is for electric connection to the electrode.

The electrodes used for triangular wave voltammetry are assumed to be stationary, unlike those used for polarography and voltammetry at a rotating electrode. As a consequence of the lack of motion of the electrode, the shape of the voltammogram differs somewhat from the polarographic and voltammetric shapes already described. Because neither the solution nor the electrode is in motion, the rate of transfer of electroactive species to the electrode is controlled by diffusion.

As the potential at which the electroactive species starts to react is reached, a current begins to flow through the cell. Increased potential leads to an increased current as in classical polarography. Eventually the electroactive species reacts as rapidly as it arrives at the electrode surface and the current flowing through the cell becomes diffusion-controlled. With a further increase in the applied potential, i.e., as time increases, the depleted layer of electroactive species around the electrode grows. The width of the depleted layer through which the electroactive species must diffuse increases with the square root of time. The result is a larger average distance through which the electroactive species must diffuse to reach the electrode and a consequent decrease in the current. The current is inversely proportional to the square root of time after all of the electroactive material initially present at the electrode surface has reacted.

The decreasing current with increasing potential causes the recorded voltammogram to exhibit a peak rather than a plateau as observed during classical polarography. Eventually the current becomes nearly constant because a relatively constant thickness diffusion layer is achieved. The peak current i_p at a stationary disk electrode for a reversible electrochemical reaction which involves only electron transfer between the electrode and the electroactive species is given by

$$i_p = 0.27n^{3/2}AD^{1/2}Cv^{1/2} \tag{23-30}$$

where i_p is the peak current (A), A is the electrode area (cm^2), D is the diffusion coefficient (cm^2/s), C is the concentration (mM), and v is the scan rate (V/s).

When the potential reaches the top of the triangle in the voltage ramp (Fig. 23-22), the direction of the scan is reversed and the potential returns to its initial value. During the forward portion of the scan, the electroactive species reacts at the electrode surface. The reaction product is a reduced species as shown in Eq. (23-31) if the initial potential scan is toward more negative potentials or an oxidized species if the initial scan is toward more positive potentials:

$$\text{Ox} + ne^- \rightarrow \text{Red} \tag{23-31}$$

If the reaction product is also electroactive, i.e., if the initial electrochemical reaction is reversible, the product can react during the reverse scan to yield the original electroactive species:

$$\text{Red} \rightarrow \text{Ox} + ne^- \tag{23-32}$$

which is the reverse of Eq. (23-31). As a consequence, a current flows through the cell in the opposite direction to that of the current flow which yielded the original peak.

For a reversible reaction, the peak observed during the reverse scan is the same size as the peak observed during the forward scan, and the two peaks are separated by $0.058/n$ V at 25°C. If the electrochemical reaction is not reversible, the two peaks are separated by a greater potential difference. If the reverse reaction does not occur at all, the second peak, corresponding to the reverse reaction, is not observed. If the reaction product is electroactive but does not react to yield the initial electroactive species, a second peak is observed, but not at the expected potential for the reverse reaction. A triangular wave voltammogram is shown in Fig. 23-24.

The peak potential of an electroactive species is dependent upon the chemical identity of the species and the scan rate used during the study. Generally, the peak potential is not as useful for qualitative analysis as was the polarographic or hydrodynamic voltammetric half-wave potential. The peak current, as measured from the extrapolated baseline prior to the peak (Fig. 23-24), is directly proportional to the concentration of the electroactive species at a fixed scan rate. It generally increases with increasing scan rate. For a reversible electrochemical reaction, the peak current is proportional to the square root of the scan rate.

The peak current during the forward scan of a triangular wave voltammogram can be used for quantitative analysis if the scan rate is held constant. The working-curve method is preferred for analysis with that technique, although the standard-addition technique also can be used. Scan rates used for triangular wave voltammetry are usually in the range of 0.02 to 10 V/s. For quantitative analysis, scan rates between 0.05 and 0.2 V/s are preferred. At scan rates above 0.5 V/s, most recorders are unable to respond sufficiently rapidly to accurately display the voltammogram. In those instances computer controlled data acquisition and display are used most often.

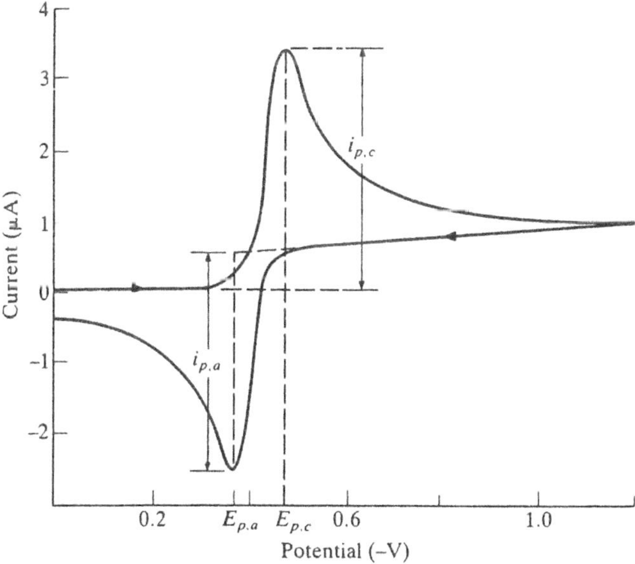

Figure 23-24 Triangular wave polarogram of 0.72 m*M* flavin adenine dinucleotide in pH 8.2 Tris buffer. The reference electrode was an sce. The arrows indicate the direction of the scan. $v = 0.100$ V/s.

Although triangular wave voltammetry is useful for quantitative analysis, it is even more useful as a method for determining electrochemical reaction mechanisms and reaction kinetics. A description of those uses is beyond the scope of the text. The interested reader is referred to References 3 and 13.

If repetitive forward-reverse scans are performed sequentially without a pause between the scans, the technique is termed *cyclic voltammetry*. Unfortunately some confusion arises because some workers use the term cyclic voltammetry to refer to triangular wave voltammetry (a single forward-reverse scan). In any case the shape of the voltammogram changes slightly between that observed during the initial cycle and subsequent cycles. Most studies are performed using the initial forward-reverse cycle. An introduction to triangular wave voltammetry can be found in Reference 15. A review of the use of electrochemical methods for the study of biological systems is available in Reference 16.

Tafel Plots

The voltammetric methods which previously were described are used to study substances that are dissolved in solution. In some instances it is desirable to study the working electrode material rather than dissolved species. This often is the case when examining the corrosion characteristics of metals. The working electrode in those cases is constructed from the studied metal. Tafel plots of potential as a function of the logarithm of measured current density (Fig. 23-25) can be used to obtain the desired corrosion characteristics. The sign of the current is ignored while preparing the plots. The current density is the current flowing through the electrode divided by the electrode area which is exposed to the solution. During the study the electrodes typically are submerged in a corrosive solution which often consists of synthetic sea water.

While obtaining the plots, a slow linear potential scan rate, which is typically in the 0.1-1 mV/s range, is used to ensure equilibrium conditions at the test electrode. Scans often require 45 min or longer for completion owing to the slow scan rate. The corrosion potential E_{corr} is the potential at which the extrapolated linear cathodic portion of the plot becomes equal to the extrapolated linear anodic portion of the plot. At that point no net current flows through the cell because the anodic and cathodic currents are equal. The corrosion current I_{corr} is estimated as the current density at the intersection of the two linear potions. The corrosion current is used with the half reaction for the oxidation of the metal and the density of the metal to calculate a corrosion rate for the metal. Corrosion rates are normally calculated in units of depth of corrosion (mm, cm, inches, etc.) per year. A more thorough description can be found in Reference 17.

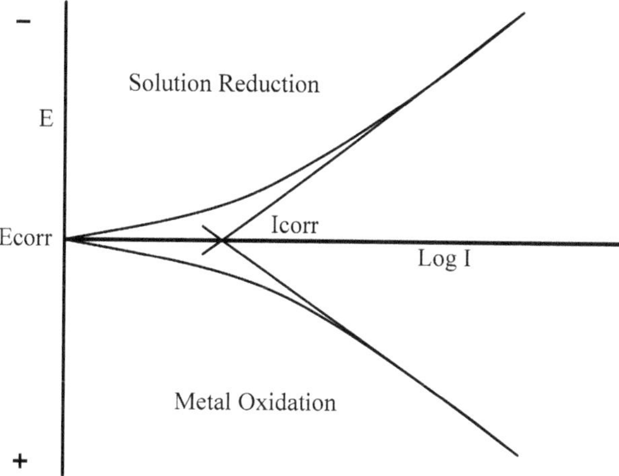

Figure 23-25 A Tafel plot. E_{corr}, corrosion potential; I_{corr}, corrosion current density; I_{corr}, corrosion current.

Sample problem 23-2 A Tafel plot was prepared for an iron electrode. The density of iron is 7.86 g/cm^3 and the molar mass of iron is 55.85. The corrosion current density obtained from the Tafel plot was 0.0875 mA/cm^2. Calculate the corrosion rate of iron in mm/year in the test solution.

SOLUTION Essentially the problem is one of units conversion. The half reaction for the corrosion of iron is

$$Fe \rightarrow Fe^{2+} + 2e^-$$

0.0875 mA/cm^2 × 1 cm^3/7.86 g × 55.85 g/mol Fe × 1 mol Fe/2 mol e's × 1 mol e's/96,485 C × 1 C/A·s × 1 A/10^3 mA × 60 s/min × 60 min/h × 24 h/d × 365 d/y ×10 mm/cm – 1.02 mm/y

POLAROGRAPHIC AND VOLTAMMETRIC INSTRUMENTATION

Although many types of polarography and voltammetry exist, it is not usually necessary to use a different instrument for each method. Multipurpose instruments that are commercially available from several manufacturers are either capable of being used for all of the polarographic and voltammetric techniques or can be made useful for a particular technique by the addition of a module to the basic instrument. Among the more popular of such instruments are those manufactured by Princeton Applied Research, Astra Scientific International, ECO, eDAQ, Uniscan Instruments, Solartron Analytical, and Bioanalytical Systems. Most instruments can be used for amperometry, controlled-potential coulometry, controlled-potential electrogravimetry, and anodic stripping voltammetry, as well as for the polarographic and voltammetric methods.

AMPEROMETRY

As has already been described, the polarographic diffusion current and the limiting current obtained in hydrodynamic voltammetry are proportional to the concentration of the electroactive species. *Amperometry* is the electroanalytical technique in which the potential of the indicator electrode is adjusted to a fixed value on the plateau of the polarographic or voltammetric wave of an electroactive species, and the current flowing through the electrode is measured as a function of concentration. Amperometry is a measure of the limiting current at a fixed applied potential as a function of concentration; it is analogous to spectrophotometric absorbance measurements. The potential, rather than the wavelength, is maintained constant, and the current, rather than the absorbance, is monitored. The current is directly proportional to the concentration of the electroactive species, and can be used for quantitative analysis of the electroactive species by either the working-curve method or the standard-addition technique.

Amperometry has primarily been used as a method for locating the endpoints of titrations involving an electroactive species. The current is directly proportional to the concentration of electroactive species in the solution and it decreases or increases during the titration as the concentration of the electroactive species changes. The endpoint of an *amperometric titration* is located by extrapolation of the straight-line portions of the titration curve, before and after the endpoint, to interception at the endpoint. The shapes of amperometric titration curves are identical to the shapes of spectrophotometric titration curves. The shapes of several amperometric titration curves are shown in Fig. 23-26. As with the absorbance in spectrophotometric titration curves, the current measured during amperometric titrations should be corrected for dilution of the titrand by the titrant prior to plotting of the curves, and the titration curves are usually rounded near the endpoint.

Amperometric titration with two indicator electrodes is the electroanalytical technique in which the current is measured between two indicator electrodes while an emf is imposed between the two electrodes. Usually the indicator electrodes are identical platinum electrodes which are rigidly held a fixed distance from each other in the cell. The emf applied between the two electrodes is normally between 0.01 and 0.25 V. Generally the sensitivity of the technique increases with increasing emf; however, if the emf is too large, the accuracy of the titration can decrease. For most purposes an emf of about 50 mV works well.

As during amperometric titrations with a single indicator electrode, it is necessary to keep mental track of the species in the titration vessel during the titration in order to understand the shapes of the titration curves obtained with two indicator electrodes. Three situations are often encountered during amperometric titrations which utilize two indicator electrodes. The titrand can

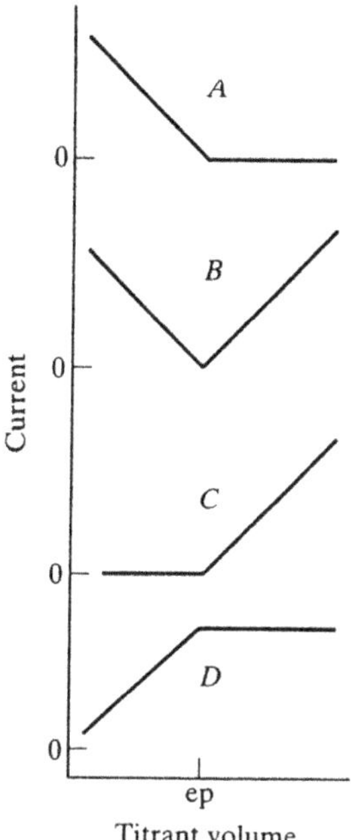

Figure 23-26 The shapes of amperometric titration curves. The end-points occur at ep. Curve *A*, titration of an electroactive titrand with a nonelectroactive titrant to yield a nonelectroactive product; curve *B*, titration of an electroactive titrand with an electroactive titrant to yield a nonelectroactive product; curve *C*, titration of a nonelectroactive titrand with an electroactive titrant to yield a nonelectroactive product; curve *D*, titration of a nonelectroactive titrand with a nonelectroactive titrant to yield an electroactive product.

be reversibly electroactive while the titrant is not; the titrant can be reversibly electroactive while the titrand is not; and both titrand and titrant can be reversibly electroactive. For the purpose of the description, "reversibly electroactive" is taken to mean that the forward and reverse reactions occur at potentials that are separated from each other by no more than the applied emf between the two indicator electrodes (typically 50 mV).

To simplify the description, it will be assumed that the titrand is originally in the oxidized form and that it is being titrated with a titrant in the reduced form; i.e., Eq. (23-3) is going from left to right for the substance being titrated:

$$\text{Ox} + ne^{-} = \text{Red} \qquad (23\text{-}3)$$

If it is assumed that the titrand is originally in the reduced form rather than the oxidized form, the shape of the titration curve is the same except that the direction of current flow is reversed through the cell.

At the start of the titration only the oxidized form of the titrant [Ox in Eq. (23-3)] is present in the titration vessel. Ox could be reduced at one of the indicator electrodes if some other species in solution were simultaneously oxidized at the other electrode; i.e., electrochemical reactions must occur simultaneously at both electrodes if a current is to flow through the cell. Because no reduced species is present in the solution, an oxidation cannot occur; no reaction occurs at either electrode; and no current flows through the external circuit. Because the potential difference between the electrodes is relatively small, it is impossible to simultaneously reduce Ox at the anode while the solvent or electrolyte is oxidized at the cathode. As titrant is added to the titration vessel, some of the oxidized form of the titrand is converted to the reduced form (Red). If the titrand is reversibly electroactive, the oxidized form of the titrand can be reduced at one indicator electrode while the reduced form of the titrand, which was formed by addition of titrant, can be oxidized at the other indicator electrode. The result is a current flow through the circuit.

The current flowing through each electrode is dependent upon the concentrations and the diffusion coefficients of the electroactive species in the solution (as in amperometry with a single indicator electrode). If the indicator electrodes are rotated, the current is also dependent upon the rotation rate as during hydrodynamic voltammetry. For the purposes of the present description, the diffusion coefficients of Red and Ox are assumed to be identical, and if the identical electrodes are rotated, the rotation rates are assumed to be identical. Consequently, the current flow is dependent only upon the concentrations. In the early part of the titration, the oxidized form of the titrand has a greater concentration than the reduced form. Because the current flowing through both electrodes is necessarily identical, the current flowing through the system is limited by the amount of the electrochemical oxidation (Red $\rightarrow$ Ox + ne^-) which can take place at the anode.

As the titration proceeds, the current increases because the concentration of Red increases until the titration is halfway to the equivalence point. At the halfway mark, the concentrations of the oxidized and reduced forms of the titrand are equal and the current reaches a maximum. After the halfway mark, the concentration of the oxidized form of the titrand becomes less than the concentration of the reduced form of the titrand and consequently limits current flow through the cell. As more Ox is removed during the titration, the current decreases until it reaches zero at the equivalence point. If the titrand were not reversibly electroactive, no current flow would have been observed prior to the equivalence point of the titration.

After the equivalence point, the concentration of the oxidized form of the original titrand is essentially zero and no current owing to the original oxidized titrand can flow. Prior to the equivalence point, the reduced form of the titrant species was chemically reacted to yield the oxidized form of the titrant species;

i.e., only the oxidized form of the titrant species was present in the solution. After the equivalence point, the reduced form of the titrant is added to the solution. The result is a current flow through the cell if the titrant is reversibly electroactive or no current flow if the titrant is not reversibly electroactive. If the titrant is reversibly electroactive, the current increases as more titrant is added. It becomes approximately constant when the concentration of the oxidized form of the titrant species becomes equal to or less than the concentration of the reduced form of the titrant species in the titration vessel. Sketches of the shapes of the titration curves obtained during amperometric titrations with two indicator electrodes are shown in Fig. 23-27.

Amperometric titrations and amperometric titrations with two indicator electrodes are widely used. The endpoints of the important *Karl Fischer titrations* are usually located using either amperometry with two indicator electrodes or controlled-current potentiometry with two indicator electrodes. Karl Fischer titrimetry is the titration of water in a nonaqueous solvent (usually methanol) according to the following overall reaction:

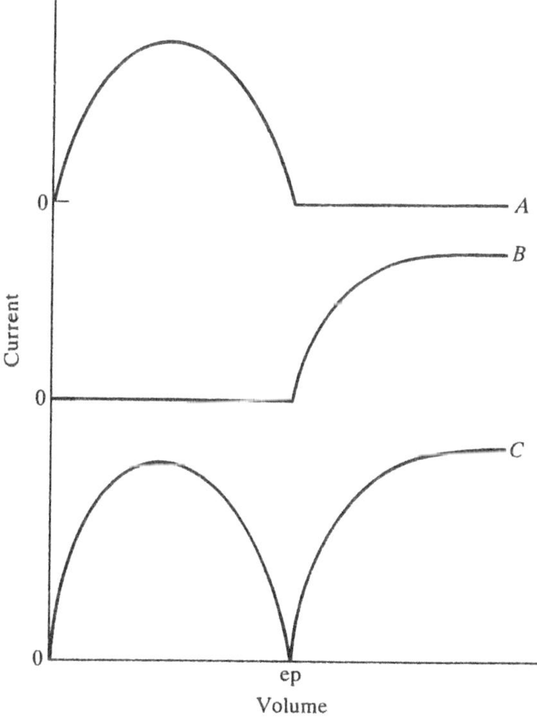

Figure 23-27 The shapes of titration curves obtained for amperometric titrations with two indicator electrodes. The endpoints are marked with ep. Curve *A*, only titrand is reversibly electroactive; curve *B*, only titrant is reversibly electroactive; curve *C*, both titrand and titrant are reversibly electroactive.

$$I_2 + SO_2 + 3C_5H_5N(\text{pyridine}) + CH_3OH + H_2O \rightarrow$$
$$2C_5H_5NHI + C_5H_5NSO_3OCH_3 \qquad (23\text{-}33)$$

A description of the theory and many uses of amperometric titrations is given by Stock.[18]

ANODIC STRIPPING VOLTAMMETRY

Anodic stripping voltammetry (ASV) is useful for determining small concentrations of metallic ions in solution. The technique is primarily used for assays at concentrations that are between about 10^{-5} and 10^{-10} *M*. Assays cannot be performed in that region by classical polarography or voltammetry although the pulsed techniques can be used at concentrations down to about 10^{-6} or 10^{-7} *M*. ASV is performed in two steps. In the first step the potential of the indicator electrode is adjusted to a value at which a metal ion M^{n+} is reduced to the corresponding metal M on the electrode. During the first step the metal is concentrated on the electrode.

The plating of the metal on the electrode (or the formation of an amalgam if the electrode is mercury) takes place for a controlled period of time. Typically between 10 and 30 min is used for plating, although the time can be either shorter or longer. The amount of plated metal increases as the time increases; consequently, the sensitivity of the technique increases with increased plating time. After the plating, the potential of the electrode is scanned from the initial potential toward more positive potentials. The voltage ramp used during the scan can be linear or it can be any of the ramps used for ac polarography, pulse polarography, differential pulse polarography, or another technique.

A "stripping" peak is observed at the potential at which the deposited metal is oxidized. At a fixed scan rate and plating time, the height of the stripping peak is proportional to the original concentration of the metallic ion in solution. Because the concentrations that are measured with stripping voltammetry are low, it is important to make a blank correction to the results to allow for any interferences in the solution.

The electrodes that are most often used for ASV are the hmde, the *mercury thin-film electrode* (mtfe), and the platinum electrode. The mtfe is usually prepared by simultaneously plating a thin film of mercury from a mercury(II) nitrate solution onto a glassy carbon electrode with the sample; i.e., mercury(II) nitrate is added to the sample solution. An advantage of the mtfe over the hmde is that the mtfe can be rotated during the deposition step without loss of the mercury film. That can speed sample deposition. An anodic stripping voltammogram is shown in Fig. 23-28.

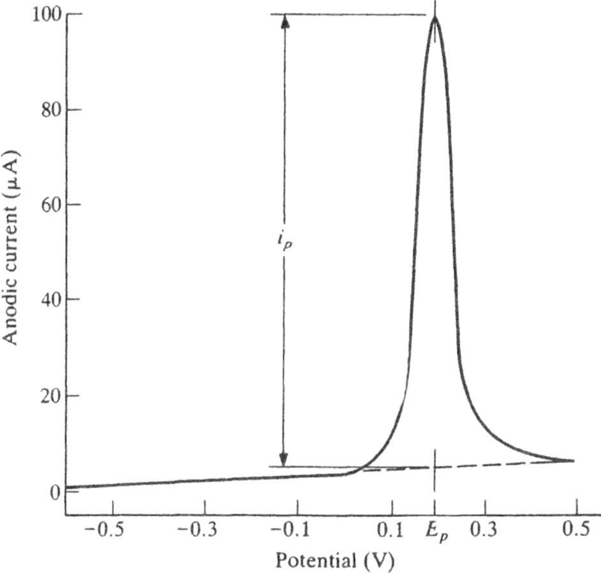

Figure 23-28 An anodic stripping voltammogram of an aqueous solution containing mercury(II). Plating potential, – 0.9 V (versus sce); v, 0.02 V/s; electrolysis time, 5 min; i_p, peak current; E_p, peak potential. The indicator electrode is a glassy carbon electrode.

Anodic stripping voltammetry is used for quantitative analysis by either the working-curve method or the standard-addition technique. The plating time, plating potential, and scan rate must be held constant for the sample and standards. In a few cases a deposit can be formed on the electrode by oxidation, rather than reduction, of a chemical species. In those instances the electrode is the anode during the plating step and the cathode during the stripping step. That method is *cathodic stripping voltammetry*.

ELECTROGRAVIMETRY

Electrogravimetry is an electroanalytical technique in which a substance is electrochemically reacted to form a solid plate on one of the electrodes in the cell. The electrode is weighed with and without the plate, and the original concentration of the substance in solution is calculated from the mass of the plated electrochemical reaction product. In most cases electrogravimetry is used to assay metallic ions which are exhaustively reduced to form metallic plates on the cathode. The electrode at which the electrochemical reaction of interest occurs is the *working electrode* (rather than the indicator electrode) for those techniques (e.g., electrogravimetry, coulometry, and other forms of exhaustive electrolysis) in which a relatively large quantity of electricity flows through the cell.

Platinum electrodes are normally used for electrogravimetry. The cathode is usually a hollow cylindrical electrode constructed from platinum gauze. Gauze is used rather than platinum foil because the metallic plate adheres better to the gauze. The anode is also usually a hollow, cylindrical platinum gauze electrode. The diameter of the anode is less than that of the cathode and the anode is usually placed inside, i.e., concentric with, the cathode during the electrolysis. The electrode at which the deposition occurs is the larger electrode. A sketch of a cathode that is used for electrogravimetry is shown in Fig. 23-29. During the electrolysis, mass transfer to the electrode occurs by convection; i.e., the solution is stirred with a mechanical stirrer, magnetic stirrer, or by rotating the anode (the inner electrode).

Either the current or the potential can be controlled during an electrogravimetric assay. If only one species in solution can possibly plate on the electrode, controlled-current electrogravimetry is preferred because the current can be adjusted to a sufficiently high value to complete the plating in a relatively short time. The emf between the two electrodes increases during the study. If more than a single species can be plated onto the electrode, controlled-potential electrogravimetry is the method of choice because controlled-current electrogravimetry can lead to the simultaneous plating of more than one metal.

During controlled-potential electrogravimetric studies, the working electrode is adjusted to a potential on the plateau of the voltammetric wave corresponding to reduction of the metallic ion. Because the potential of the working electrode must be controlled, it is necessary to use a reference electrode during controlled-potential assays. Usually a three-electrode arrangement is used.

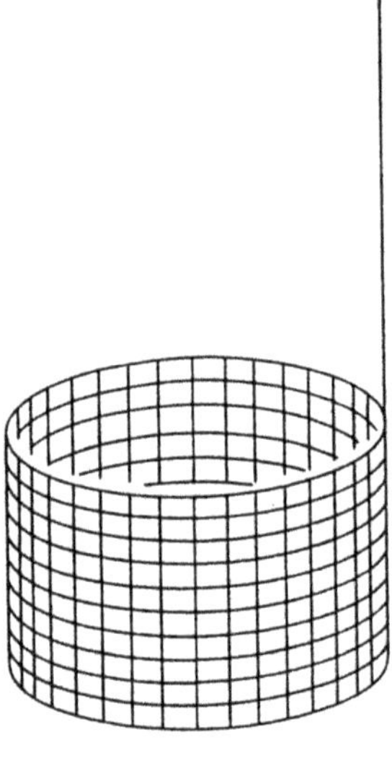

Figure 23-29 Sketch of a platinum gauze cathode used for electrogravimetry. Typical dimensions are 4.5 cm (diameter) × 4 cm (height of gauze) × 11.5 cm (total height).

At a fixed stirring rate, the electrolytic current is dependent upon the concentration of the electroactive species. As the concentration decreases during the electrolysis, the current also decreases. The decreasing current usually results in a longer plating time during controlled-potential electrogravimetry than during controlled-current electrogravimetry. By adjusting the potential of the working electrode to a value at which only a single species can plate on the working electrode, it is possible to eliminate interferences with the assay owing to more than one deposited metal. That is the primary advantage of controlled-potential electrogravimetry as compared to controlled-current electrogravimetry.

COULOMETRY

Coulometry is an electroanalytical technique in which the total quantity of electric charge which is used for an exhaustive electrolysis of an electroactive analyte is measured. Because the units of electric charge are coulombs (C) or some multiple of coulombs, the technique is named coulometry. The number of coulombs that are required to react with all of an electroactive analyte is mathematically converted to moles of electrons (faradays) by using the Faraday constant (96,485 C/F; 1 F = 1 mol electrons). The balanced half-reaction for the electrochemical reaction can be used to calculate the number of moles of the electroactive species which reacted and consequently the original concentration of the electroactive analyte.

The device which measures the quantity of electric charge used during an electrochemical reaction is a *coulometer*. Much of the early coulometric work was done using chemical coulometers. A chemical coulometer consists of an electrolyte solution of an electroactive species that is placed in series in the circuit between the source of electricity and the working electrode. As current flows through the coulometer, an electrochemical reaction occurs in the solution which generates an easily measurable product (typically a gas or electrode plate). Measurement of the amount of product formed in the coulometer and knowledge of the coulometer's reaction make it possible to calculate the quantity of electric charge passed through the coulometer and the cell.

Some controlled-current coulometers accurately measure the time of flow of the constant current through the electrochemical cell, and use the following equation to calculate the quantity of electric charge that is used:

$$Q = It \tag{23-34}$$

The quantity of electric charge, measured in coulombs, is calculated by multiplying the constant current (amperes) by the time (seconds). When Eq. (23-34) is used, it is necessary that all of the current going into the cell be used in the electrochemical reaction of the analyte. The percentage of the total cell current that is used for a specific electrochemical reaction is the *current efficiency* of the reaction. To use Eq. (23-34), it is necessary to have 100 percent current efficiency.

At the start of an assay by controlled-current coulometry, 100 percent current efficiency can be obtained by adjusting the current flowing through the cell to a value that is less than the total current which could flow through the cell as a result of mass transfer of the electroactive species to the electrode. If a voltammogram of the analyte under the experimental conditions to be used for the coulometric analysis (same stirring rate, electrodes, etc.) were recorded, a voltammetric wave with a limiting current corresponding to the reaction of the electroactive species would be obtained, as shown in Fig. 23-30, curve *A*. The desired 100 percent current efficiency can be obtained by applying a constant current i_{appl} that is less than the current which could flow through the cell owing to the electrochemical reaction of the analyte.

Of course, the voltammetric wave height is proportional to the concentration of the electroactive species. As the concentration of the electroactive species is reduced during the coulometric assay, the wave height decreases as shown in Fig. 23-30, curve *B*. Eventually the wave height becomes less than the constant applied current (Fig. 23-30, curve *C*). When that occurs, some of the current (i_1 in Fig. 23-30) is used for the electrolysis of the analyte, and the remainder of the current i_2 must be used in some other electrochemical reaction. The excess current can be used to oxidize or reduce other solution components including the supporting electrolyte and the solvent. With the loss of 100 percent current efficiency, Eq. (23-34) cannot be used to calculate Q.

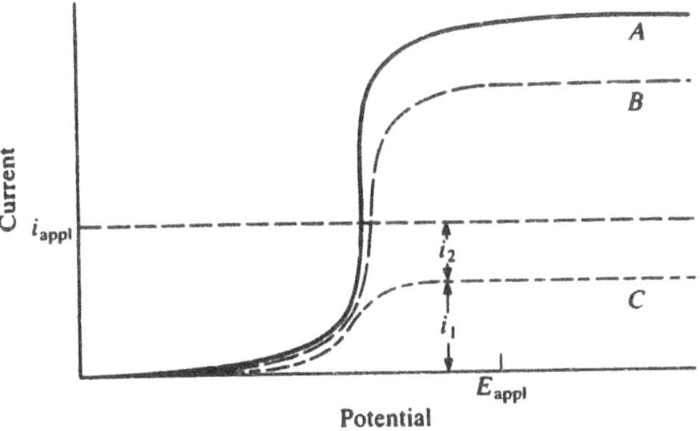

Figure 23-30 Sketches of voltammograms obtained during a coulometric study. The constant current applied during the hypothetical controlled-current coulometric experiment described in the text is indicated by i_{appl}. Curve *A*, before the study; curve *B*, after start of the coulometric study but before curve *C* was recorded; curve *C*, after 100 percent current efficiency is no longer attained.

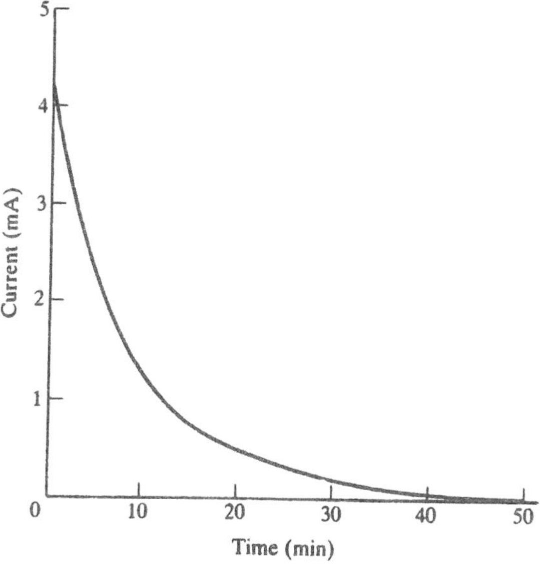

Figure 23-31 The decay in current with time during a controlled-potential coulometric study of 10 mL of 1 m*M* mercury(II) acetate. The applied potential at the platinum-foil working electrode is – 0.2 V relative to an sce.

Two methods can be used to avoid an analytical error owing to less than 100 percent current efficiency. As in electrogravimetry, the potential rather than the current of the working electrode can be controlled. By adjusting the potential to a value on the plateau of the voltammetric wave (E_{appl} in Fig. 23-30), 100 percent current efficiency during the electrochemical reaction of the most easily oxidized or reduced species is assured. The current obtained during controlled-potential coulometry decreases with time, as shown in Fig. 23-31, owing to the decreasing concentration of the electroactive species. The plot of current as a function of time is usually automatically recorded during controlled-potential coulometry by the electroanalytical instrument.

A related technique in which the current at a stationary electrode in an unstirred solution is monitored as a function of time is *chronoamperometry*. The current is monitored only for the first portion of the electrochemical reaction rather than until the electrochemical reactant is exhausted as in controlled-potential coulometry; Because the technique is normally not used for chemical analysis, it is not described further. More information can be found in Reference 3.

To obtain the quantity of electricity consumed during controlled-potential coulometry, it is necessary to integrate (i.e., to measure the area under) the *i-t* curve rather than to apply Eq. (23-34), which is only applicable to constant-current coulometry. The integration can be accomplished in several ways, e.g., with a planimeter, with an electronic integrator, or by weighing the recorder paper under the curve. Modern instruments nearly always accomplish the integration electronically. Although analysis by controlled-potential coulometry usually requires more time than analysis by controlled-current coulometry, controlled-potential coulometry does have the advantage of assuring 100 percent current efficiency. Controlled-potential coulometry is used to assay samples that can be oxidized or reduced.

Sample problem 23-3 A controlled-potential coulometric assay was performed at a potential on the plateau of the voltammetric wave of Cu^{2+}. A 25.0-mL sample solution was assayed with the results shown in Fig. 23-32. Calculate the concentration of the Cu^{2+} in the solution.

SOLUTION One way to determine the area under the *i-t* curve is by approximation of the area with various geometric figures of known area. The area under the curve in Fig. 23-32 is divided into triangles and rectangles, and the areas of all the portions are added to find the total area. To clarify the calculations, each triangle and rectangle is numbered in Fig. 23-32 and the areas are added sequentially starting with number 1.

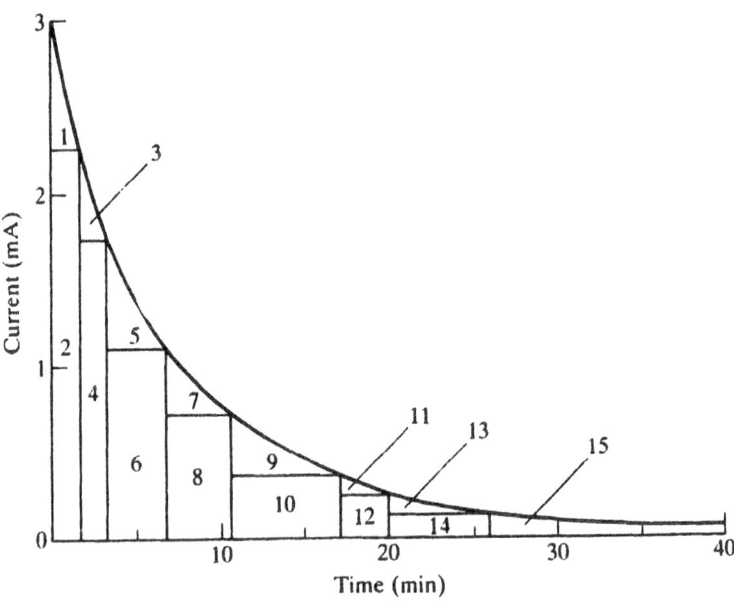

Figure 23-32 Plot of current as a function of time for the controlled-potential coulometric analysis described in Sample problem 23-3.

$$
\begin{aligned}
Q = it = {} & \frac{1}{2}(1.5\ \text{min})(0.75\ \text{mA}) + (1.5\ \text{min})(2.25\ \text{mA}) \\
& + \frac{1}{2}(1.5)(0.55) + (1.5)(1.70) \\
& + \frac{1}{2}(3.5)(0.60) + (3.5)(1.10) \\
& + \frac{1}{2}(4.0)(0.40) + (4.0)(0.70) \\
& + \frac{1}{2}(6.5)(0.35) + (6.5)(0.35) \\
& + \frac{1}{2}(3.0)(0.10) + (3.0)(0.25) \\
& + \frac{1}{2}(6.0)(0.10) + (6.0)(0.15) \\
& + \frac{1}{2}(14.0)(0.15) \\
= {} & 0.56 + 3.4 + 0.41 + 2.6 + 1.1 + 3.9 + 0.80 + 0.28 + 1.1 \\
& + 2.3 + 0.15 + 0.75 + 0.30 + 0.90 + 1.1 \\
= {} & 19.6\ \text{mA} \cdot \text{min}
\end{aligned}
$$

The units of Q must be coulombs to be useful:

$$19.6\ \text{mA} \cdot \text{min} \times \frac{1\text{A}}{1000\ \text{mA}} \times 60\ \text{s/min} = 1.18\ \text{A} \cdot \text{s} = 1.18\ \text{C}$$

The Faraday constant is used to convert the number of coulombs to faradays:

$$1.18\ \text{C} \times \frac{1\text{F}}{96{,}485\text{C}} = 1.22 \times 10^{-5}\ \text{F}$$

The balanced electrochemical reaction

$$Cu^{2+} + 2e^{-} \rightarrow Cu$$

is used to calculate the number of moles of Cu^{2+} reacted:

$$1.22 \times 10^{-5}\ \text{F} \times \frac{1\ \text{mol}\ Cu^{2+}}{2\text{F}} = 6.10 \times 10^{-6}\ \text{mol}\ Cu^{2+}$$

The original concentration can be obtained by dividing the number of moles by the volume:

$$
\begin{aligned}
[Cu^{2+}] &= \frac{6.10 \times 10^{-6}\ \text{mol}}{25.0\ \text{mL} \times 1\text{L}/1000\ \text{mL}} \\
&= 2.44 \times 10^{-4}\ M\ Cu^{2+}
\end{aligned}
$$

The second coulometric method by which 100 percent current efficiency can be maintained is utilized during *coulometric titrations*. A titrant is generated (usually in the titration vessel) at controlled current. The concentration of the electroactive substance from which the titrant is generated is maintained at a relatively high value by adding a large excess (in comparison to that necessary to generate sufficient titrant to react with all of the titrand) of the substance prior to starting the coulometric titration. By maintaining a high concentration of the electroactive species, it is possible to achieve 100 percent current efficiency during generation of the titrant. The electrogenerated titrant reacts with the titrand as in a normal titration. The endpoint of the titration is located in any of the usual ways, e.g., by using a chemical indicator, potentiometry, amperometry, or spectrophotometry.

The titrant does not have to be electrogenerated in the titration vessel, although that is the usual approach. If it is generated in the vessel, the working electrode is dipped into the titrand solution and the second electrode is placed in a compartment separated from the titrand solution by a salt bridge. That prevents the electrochemical reaction products that are formed at the second electrode from reacting with either the titrand or the electrogenerated titrant. Typically, the second electrode is placed in an electrolyte solution inside a glass

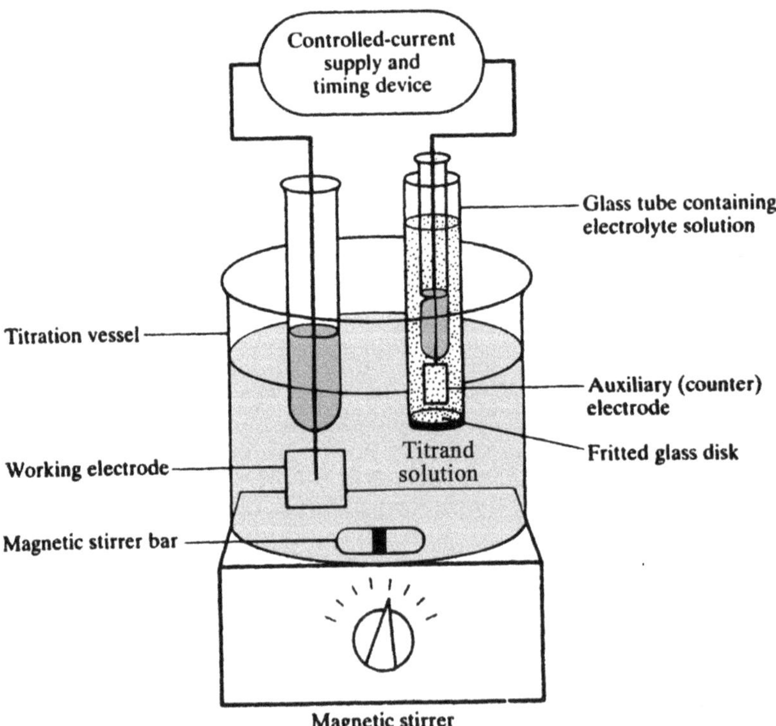

Figure 23-33 Diagram of the apparatus used for coulometric titrations.

tube with a sintered glass disk in the bottom, as shown in Fig 23-33. The glass tube is dipped into the titrand solution.

Electrogenerated titrants can be used in many types of titrations. Cerium(IV) which is electrogenerated from cerium(III) sulfate can be used for redox titrations. Silver(I) can be generated from a silver anode for use in precipitation titrations, e.g., for the precipitation of chloride. Both H^+ and OH^- can be electrogenerated from aqueous electrolyte solutions for use in acid-base titrations. Other titrants which are commonly electrogenerated for use in coulometric titrations include copper(I), iron(II), chlorine (Cl_2), bromine (Br_2), and ferrocyanide.

Sample problem 23-4 Constant-current coulometry was used to assay a solution containing iron(II). To ensure 100 percent current efficiency, the assay was performed in a 0.1-*M* cerium(III) sulfate-sulfuric acid solution. The electrode reaction was a combination of the reaction of iron(II) to form iron(III) and the reaction of cerium(III) to form cerium(IV):

$$Fe^{2+} \rightarrow Fe^{3+} + e^-$$

$$Ce^{3+} \rightarrow Ce^{4+} + e^-$$

Cerium(IV) which formed at the electrode reacted chemically with iron(II) to yield cerium(III) and iron(III):

$$Ce^{4+} + Fe^{2+} \rightarrow Ce^{3+} + Fe^{3+}$$

At the endpoint of the titration of a 25.0-mL sample, a controlled current of 6.43 mA had flowed for 3 mm 43 s. Calculate the concentration of Fe^{2+} in the sample.

SOLUTION The quantity of electricity which flowed through the solution prior to the endpoint can be calculated by using Eq. (23-34):

$$Q = It = (6.43 \times 10^{-3}\ \text{A})(223\ \text{s}) = 1.43\ \text{C}$$

That value is used to calculate the moles of electrons (faradays) used in the formation of iron(III) either directly or indirectly through the chemical reaction with cerium(IV). In either case the number of coulombs used is the same:

$$1.43\ \text{C} \times \frac{1\text{F}}{96{,}485\ \text{C}} = 1.48 \times 10^{-5}\ \text{F}$$

The number of faradays is used with the balanced electrochemical equations to calculate the number of moles of Fe^{2+} in the original solution:

$$1.48 \times 10^{-5}\ \text{F} \times \frac{1\ \text{mol Fe}^{2+}}{1\ \text{F}} = 1.48 \times 10^{-5}\ \text{mol Fe}^{2+}$$

The concentration is calculated by dividing the number of moles of Fe^{2+} by the sample volume:

$$[Fe^{2+}] = \frac{1.48 \times 10^{-5} \text{ mol}}{0.0250 \text{ L}} = 5.92 \times 10^{-4}\ M$$

Another technique which involves measurement of the quantity of electric charge is *chronocoulometry*. The potential of the indicator electrode is stepped to a value at which the chemical sample becomes electroactive, and Q is recorded as a function of time. Sometimes the potential is changed twice while Q is monitored as a function of time. The technique has primarily been used to study the adsorption of electroactive substances on the electrode rather than for chemical analysis. More information can be found in References 3 and 19.

SPECTROELECTROCHEMISTRY

In some cases it is advantageous to combine electrolysis or coulometry (usually at controlled potential) with an optical method of analysis. The combined analytical method is *spectroelectrochemistry*. The incident beam of radiation can be passed through an *optically transparent electrode* (OTE), reflected from the electrode surface, or passed parallel to and along the surface of the electrode. In any case, the concentration of the reactants or products of the electrochemical reaction can be monitored at or near the electrode surface during the electrolysis.

Optically transparent electrodes are generally constructed by coating a thin layer (typically between 10 and 500 nm) of the electrode material on a substrate that is transparent to the radiation used to monitor the progress of the reaction. The coating is transparent because the thickness of the coating is comparable to or smaller than the wavelength of the radiation used for the study. Typical electrode materials that are used in spectroelectrochemical studies include carbon, gold, platinum, and tin(IV) oxide. The substrate upon which the materials are coated can be glass or a transparent plastic for studies in the visible region, quartz for studies in the ultraviolet region, and germanium for studies in the infrared region.

As an alternative to a thin coating of electrically conductive material on a transparent substrate, an OTE can be constructed from a fine grid of the conductive (electrode) material or from a porous piece of glassy carbon. In the first instance the radiation passes through the holes in the grid. The holes must be sufficiently small so that the radiation can only pass through a layer of solution that is adjacent to the electrode surface. Porous glassy carbon can be used as an OTE as long as the pores are small and permit radiation to travel through the entire width of the electrode.

Most spectroelectrochemical studies use ultraviolet, visible, or infrared radiation to obtain an absorptive spectrum or to monitor the absorption at a fixed wavelength as a function of the electrolysis time. Such studies are often used to measure the rate constant of the electrochemical reaction or a following chemical reaction that occurs at the electrode surface. Other spectral techniques that have been used to follow the progress of electrochemical reactions at the surfaces of electrodes include fluorescence, specular reflectance, internal reflectance, and nmr spectrometry. During reflectance studies the radiation is reflected (often multiple times) from the surface of the electrode prior to striking the detector. A brief introduction to spectroelectrochemistry can be found in Reference 20.

CONDUCTOMETRY

The electrical *conductance* G (Chapter 2) of a solution is a measure of the solution's ability to conduct electricity. Because the ability of a solution to conduct an electric current decreases as the resistance R of the solution increases, the conductance is defined as the inverse of the resistance:

$$G = \frac{1}{R} \tag{23-35}$$

The unit of conductance is the siemens (S). The conductance of a solution which has a resistance of 1 Ω is 1 S.

Conductivity measurements are usually made by using an ac emf of about 6 V applied across two identical platinum foil electrodes upon which a thin layer of platinum has been plated. Plantinized platinum electrodes, i.e., those on which platinum has been plated, have a larger surface area than shiny platinum electrodes. That makes it possible to measure the conductance more accurately. An ac emf rather than a dc emf is used for the measurement in order to prevent a decrease in the measured conductance with time owing to a partial separation of the positive and negative ions as they are attracted to the charged electrodes. Often the ac frequency used for the measurement is 60, 100, or 1000 Hz. The conductivity bridge used to measure conductance is often a modified Wheatstone bridge such as that described in Reference 21 and in Chapter 2.

Electricity is conducted in solution by ions. The resistance of a solution containing a fixed concentration of an electrolyte at constant temperature is directly proportional to the distance l between the two identical electrodes used for the measurement and inversely proportional to the cross-sectional area A between the conducting portions of the electrodes:

$$R = \frac{\rho l}{A} \tag{23-36}$$

The proportionality constant ρ in Eq. (23-36) is the *specific resistance*. By substituting Eq. (23-36) into Eq. (23-35), an expression which relates conductance to l and A is obtained:

$$G = \frac{A}{\rho l} = \frac{kA}{l} \tag{23-37}$$

The *specific conductance* k in Eq. (23-37) is the inverse of the specific resistance. As is apparent from Eq. (23-37), the specific conductance is equal to the conductance when A equals l.

Both R and G vary with temperature and electrolyte concentration. The conductance of most electrolyte solutions increases by about 1 to 2 percent for each degree of temperature rise. As the concentration of the electrolyte solution increases, the resistance of the solution decreases and the conductance of the solution increases. The conductance of dilute solutions is directly proportional to the concentration of the electrolyte solution. That relationship enables conductometry to be used for quantitative analysis of electrolyte solutions. The *molar conductance* Λ_m of a solution is the specific conductance of a solution containing one mole of the electrolyte in each cubic centimeter of solution. Consequently, if the concentration C of the conducting species is given in units of mol/1000 cm^3, i.e., as molarity, the molar conductance is defined according to

$$\Lambda_m = \frac{1000k}{C} \tag{23-38}$$

The conductance of an electrolyte solution can be expressed as a function of the molar conductance by solving Eq. (23-38) for k and substituting the value into Eq. (23-37). The result is

$$G = \frac{\Lambda_m CA}{1000\,l} \tag{23-39}$$

Although in an actual cell it is usually difficult to measure A and l, the ratio l/A can be calculated from a measurement of G for a solution which has known values of Λ_m and C. The ratio l/A is the *cell constant* θ. Equations (23-37) and (23-39) can be expressed as functions of the cell constant by substituting θ for l/A:

$$G = \frac{k}{\theta} \tag{23-40}$$

$$G = \frac{\Lambda_m C}{1000\,\theta} \tag{23-41}$$

Table 23-1 The specific conductances k of several potassium chloride solutions at 25°C

Concentration, *M*	*k*, S/cm
1.000×10^{-3}	1.469×10^{-4}
1.000×10^{-2}	1.413×10^{-3}
1.000×10^{-1}	1.288×10^{-2}
1.000	0.1118

The cell constant is usually determined by measuring the conductance of a solution containing a known concentration of an electrolyte with a known specific conductance. The appropriate values are substituted into Eq. (23-40), and the equation is solved for the cell constant. The specific conductances of several potassium chloride solutions (Table 23-1) have been accurately measured and are often used for the calculation.

Table 23-2 Molar conductances at infinite dilution $\left(\Lambda_m^0\right)$ for several ions at 25°C

Cation	Λ_m^0, S · cm²/mol	Anion	Λ_m^0, S · cm²/mol
H^+	349.8	OH^-	198.6
Li^+	38.6	F^-	55
Na^+	50.1	Cl^-	76.4
K^+	73.5	Br^-	78.1
Rb^+	77.8	I^-	76.8
NH_4^+	73.4	NO_3^-	71.4
Ag^+	61.9	ClO_4^-	67
		IO_4^-	55
Mg^{2+}	106	Formate	55
Ca^{2+}	119	Acetate	41
Sr^{2+}	119	HCO_3^-	45
Ba^{2+}	127	Benzoate	32
Fe^{2+}	108	SCN^-	66
Co^{2+}	106		
Cu^{2+}	107	CO_3^{2-}	139
Zn^{2+}	106	SO_4^{2-}	160
Hg^{2+}	106	$C_2O_4^{2-}$	148
Pb^{2+}	139	CrO_4^{2-}	170
Fe^{3+}	204	PO_4^{3-}	210
La^{3+}	209	$Fe(CN)_6^{3-}$	303
Ce^{3+}	209		
		$Fe(CN)_6^{4-}$	444

The conductance of a solution is the sum of the conductances of all of the ions in the solution. The conductance of an ion in solution is related to the charge, size, and concentration of the ion. A comparison of ions of about the same size and concentration reveals that the conductance is greater for ions of greater charge. As the size of the ion and any solvent molecules attached to it in solution decreases, the ability of the ion to move through the solution increases with a consequent increase in the conductivity of the ion. As the concentration of a particular ion increases, the conductance owing to the ion increases. The conductance of a mixture of ions is the sum of the individual conductances [Eq. 23-41] for all of the individual ions in the solution:

$$G = \frac{\sum \Lambda_{m,\,i}\, C_i}{1000\theta} \tag{23-42}$$

To use Eq. (23-42) to calculate the conductance of a solution, it is necessary to know the value of Λ_m for each ion in the solution. Values of Λ_m have been measured for various concentrations of different ions, and the results have been extrapolated to give a value of the molar conductance at zero concentration (Λ_m^0). At infinite dilution (zero concentration) the ions act completely independently of each other. As the concentration increases, the ions act less independently and conductance is somewhat less than would have been expected by using the molar conductances at zero concentration in Eq. (23-42). Because the molar conductance decreases with increasing concentration, Λ_m / Λ_m^0 decreases with increasing concentration. In dilute solutions, Λ_m^0 can be used as an estimate of Λ_m in Eqs. (23-41) and (23-42). Values of Λ_m^0 for several ions are listed in Table 23-2.

Sample problem 23-5 A 1.000×10^{-2} *M* KCl solution was added to a cell and the measured conductance was 283×10^{-6} S. Estimate the conductance of a solution containing 1.00×10^{-4} *M* HCl and 2.00×10^{-4} *M* $Fe_2(C_2O_4)_3$ in the same cell.

SOLUTION The cell constant θ can be obtained by using the specific conductance of the KCl solution from Table 23-1 and Eq. (23-40):

$$G = \frac{k}{\theta}$$

$$2.83 \times 10^{-4}\ \text{S} = \frac{1.413 \times 10^{-3}\ \text{S/cm}}{\theta}$$

$$\theta = 4.99\ \text{cm}^{-1}$$

The cell constant and the values, of Λ_m^0 from Table 23-2 are used with Eq. (23-42) to estimate the conductance of the solution:

$$[H^+] = 1.00 \times 10^{-4}, \qquad \Lambda^0_{m,\,H^+} = 349.8$$

$$[Cl^-] = 1.00 \times 10^{-4}, \qquad \Lambda^0_{m,\,Cl^-} = 76.4$$

$$[Fe^{3+}] = 4.00 \times 10^{-4}, \qquad \Lambda^0_{m,\,Fe^{3+}} = 204$$

$$[C_2O_4{}^{2-}] = 6.00 \times 10^{-4}, \qquad \Lambda^0_{m,C_2O_4{}^{2-}} = 148$$

$$G = \frac{349.8\left(1.00 \times 10^{-4}\right) + 76.4\left(1.00 \times 10^{-4}\right) + 204\left(4.00 \times 10^{-4}\right) + 148\left(6.00 \times 10^{-4}\right)}{1000(4.99)}$$

$$= 4.27 \times 10^{-5}\ \mathrm{S}$$

The ratio of the conductance owing to a particular ion j in a solution to the total conductance of the solution is the *transference number* N of the ion:

$$N = \frac{\Lambda_{m,\,j}\, C_j}{\sum \Lambda_{m,\,i}\, C_i} \tag{23-43}$$

Equation (23-43) and the values in Table 23-2 can be used to estimate transference numbers for dilute solutions.

Sample problem 23-6 Estimate the transference number of Cl^- in the solution described in Sample problem 23-5.

SOLUTION The values listed in Sample problem 23-5 are substituted into Eq. (23-43):

$$N = \frac{76.4\left(1.00 \times 10^{-4}\right)}{349.8\left(1.00 \times 10^{-4}\right) + 76.4\left(1.00 \times 10^{-4}\right) + 204\left(4.00 \times 10^{-4}\right) + 148\left(6.00 \times 10^{-4}\right)}$$

$$= 3.59 \times 10^{-2}$$

In solutions which contain a limited number of ionic species, it is possible to use conductometry to estimate values of solubility products K_{sp}, dissociation constants, and various other equilibrium constants, and to measure dielectric constants. Conductometry is also used as a method for determining the purity of water. Conductivity meters can be calibrated in units used to measure water hardness or in any other convenient units. Conductivity measurements are widely used to check the purity of distilled and deionized water. Many deionizing columns are equipped with a conductivity meter which measures the conductance of the water coming from the column. When the conductance exceeds a predetermined value, the meter indicates that the column should be changed or regenerated. A more thorough description of some of the applications of conductivity measurements is given in Reference 21 (chap. 9).

Sample problem 23-7 The cell constant of a conductivity cell at 25°C is 1.20 cm^{-1}. The cell was filled with a saturated solution of AgCl at 25°C. The conductance of the solution was 1.44 × 1.0^{-6} S. Estimate the solubility product of silver chloride.

SOLUTION The concentration of Ag^+ and Cl^- in the saturated solution can be calculated from Eq. (23-42) and the molar conductances in Table 23-2 (in dilute solutions Λ_m nearly equals Λ_m^0):

$$G = \frac{\sum \Lambda_{m,i} C_i}{1000\ \theta}$$

$$G = \frac{\Lambda^0_{m,\,Ag+} C_{Ag+} + \Lambda^0_{m,Cl^-} C_{Cl^-}}{1000\ \theta}$$

$$1.44 \times 10^{-6} = \frac{61.9 C_{Ag^+} + 76.4 C_{Cl^-}}{1000(1.20)}$$

For this particular case, $C_{Ag+} = C_{Cl^-} = C$ because one ion of Ag^+ is formed during the dissolution each time one ion of Cl^- is formed:

$$AgCl(s) = Ag^+ + Cl^-$$

Consequently,

$$1.44 \times 10^{-6} = \frac{61.9C + 76.4C}{1.20 \times 10^3}$$

$$C = C_{Ag}^{\ +} = C_{Cl}^{\ -} = 1.25 \times 10^{-5}\ M$$

The solubility product constant can be estimated by using those values:

$$K_{sp} = [Ag^+][Cl^-] = (1.25 \times 10^{-5})(1.25 \times 10^{-5}) = 1.56 \times 10^{-10}$$

Conductometric Titrations

If a change in the conductance of a titrand solution occurs during a titration, conductometry can be used to locate the endpoint of the titration. The conductance of the titrand is measured after each addition of titrant, and the titration curve, consisting of conductance as a function of added titrant volume, is plotted. If necessary the conductivity readings are corrected for dilution of the titrand by the titrant during the titration. Often the titrant is at least 20 times as concentrated as the titrand, so that dilution during the titration is negligible. Extrapolation to interception of the linear portions of the titration curve yields the endpoint. Conductometric titrations are useful for acid-base, precipitation, and complexation titrations. Redox conductometric titrations are not usually successful, owing to the relatively small change in total conductance during those titrations.

The shape of a conductometric titration curve can be predicted by keeping track of ionic concentration changes in the titration vessel during the titration.

The molar conductances in Table 23-2 can be used to estimate changes in conductivity. The titration of a hydrochloric acid solution with sodium hydroxide will serve as an example. At the start of the titration, the titrand contains a relatively high concentration of H^+ (Λ_m^0 = 349.8) and Cl^- (Λ_m^0 = 76.4). As sodium hydroxide is added, the following reaction occurs:

$$H^+ + Cl^- \xrightarrow{Na^+,\ OH^-} Na^+ + Cl^- + H_2O \tag{23-44}$$

The components to the left of the arrow are the ions present in the titration vessel prior to reaction with sodium hydroxide and the species to the right of the arrow are present after reaction. The components above the arrow are in the titrant. It can be seen that Cl^- is not involved in the reaction; i.e., it is a spectator ion. Consequently, neglecting dilution, the portion of the conductance owing to Cl^- does not change as the titration proceeds. The only change with the addition of titrant is the replacement of H^+ (Λ_m^0 = 349.8) with Na^+ (Λ_m^0 = 50.1). Because Na^+ contributes less (smaller molar conductance) to the conductance of the solution than H^+, the conductance decreases with the addition of titrant prior to the endpoint.

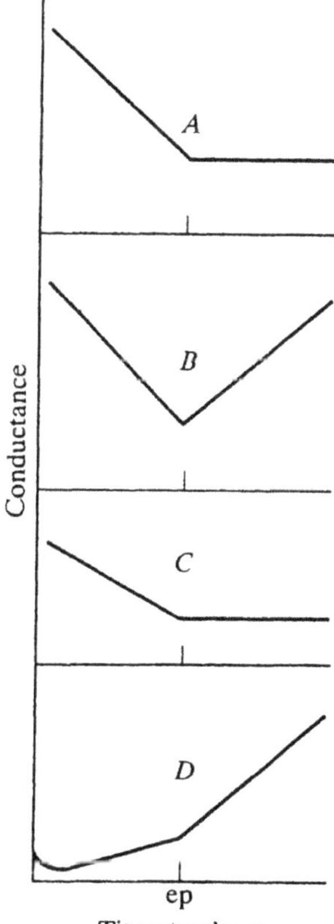

Figure 23-34 Diagrams of acid-base conductometric titration curves. The endpoint of each titration is located at ep. Curve *A*, the titration of HCl with NH_3, curve *B*, the titration of HCl with NaOH; curve *C*, the titration of NaOH with acetic acid; curve *D*, the titration of acetic acid with NaOH. The slight decrease in conductance in curve *D* at the start of the titration is caused by suppression of the dissociation of acetic acid by the common-ion effect.

After the endpoint the conductance no longer decreases because no more H^+ is removed. Because both Na^+ and the highly conducting OH^- (Λ_m^0 198.6) are added after the endpoint, the conductance increases resulting in a V-shaped titration curve. The shapes of other conductometric titration curves can be deduced in the same manner. Sketches of several conductometric titration curves are shown in Fig. 23-34.

Electrochemical Impedance Spectroscopy

During electrochemical impedance spectroscopy (EIS) measurements, the electrical impedance (Chapter 2) of a system is monitored as a function of the frequency of a small amplitude ac potential. Impedance in electrochemical systems consists of a resistance and a capacitance component as shown in Eqs. (2-34) and (2-35). Generally inductance does not contribute to impedance in electrochemical systems. At the metal electrode-solution interface, the charging of the electric double layer or a contribution from an adsorbed layer on the metallic surface constitutes the capacitance component. The electron transfer makes up the resistance component.

The normal approach that is used during EIS measurements is to vary the frequency of the applied ac potential over a wide range while recording the impedance. Electrical "equivalent" circuits consisting of resistors and capacitors are constructed and the impedance of those circuits is measured or calculated at varying applied frequencies. The plot of the electrochemical system is compared to those of the equivalent circuits. The equivalent circuits are varied until a match is achieved. Upon acquiring matching spectra, the equivalent circuit is assumed to be a good model for the electrochemical system. The corresponding arrangement of capacitors and resistors is subsequently used to explain the actual electrochemical process that occurs at the electrode-solution interface. For the interested reader, more thorough descriptions are available in References 22 and 23.

Localized electrochemical impedance spectroscopy (LEIS) uses a small electrode probe which is held near the surface of the studied material. The impedance measurements are obtained over a wide frequency range of the applied small sinusoidal potential as in EIS. In this manner the EIS is recorded at a single location rather than averaged over the entire surface of the studied material. Surface maps can be acquired. Both EIS and LEIS typically are used to study factors which have an effect upon the capacitance at the electrode-solution interface. Examples include the study of the degradation of coatings on the surface, corrosion properties such as passivation, and the study of biofilms on metal surfaces.

Scanning Electrochemical Microscopy

Another scanning electroanalytical method is Scanning Electrochemical Microscopy (SECM). It utilizes ultramicroelectrodes which have electrode radii in the nanometer region to scan above an examined surface while monitoring an electric property. Typically a current is monitored at a controlled potential in order to obtain a surface map of current as a function of surface location, or a voltammogram at a particular location above the surface is recorded. Descriptions of SECM are available in References 24 and 25.

IMPORTANT TERMS

AC polarography
Amperometric titration
Amperometric titration with two indicator electrodes
Amperometry
Anodic stripping voltammetry
Auxiliary electrode
Capacitive current
Cathodic stripping voltammetry
Cell constant
Conductance
Conductometric titration
Conductometry
Controlled-current coulometry
Controlled-current electrogravimetry
Controlled-potential coulometry
Controlled-potential electrogravimetry
Coulometer
Coulometric titration
Coulometry
Current efficiency
Differential pulse polarography
Diffusion coefficient
Diffusion current
Diffusion layer
Dropping mercury electrode
Electric double layer
Electrochemical impedance spectroscopy
Electrogravimetry
Faradaic current
Half-wave potential
Hanging mercury drop electrode
Hydrodynamic voltammetry
Ilkovic equation
iR drop
Karl Fischer titrimetry
LEIS
Limiting current
Maximum suppressor
Mercury thin-film electrode
Migration current
Molar conductance
Overpotential (overvoltage)
Polarogram
Polarography
Pulse polarography
Rapid dc polarography
Residual current
Rotated disk electrode
Scanning electrochemical microscopy
Specific conductance
Specific resistance
Square wave polarography
Static mercury drop electrode
Supporting electrolyte
Tafel plots
Tast polarography
Transference number
Triangular wave voltammetry
Voltage ramp
Voltammetric analyzer
Voltammetry
Voltammogram
Working electrode

PROBLEMS

Polarography

23-1 The diffusion currents of a series of standard solutions of lead(II) and a sample solution containing lead(II) are listed in the following table. Calculate the concentration of lead(II) in the sample solution.

Concentration, m*M*	Diffusion current μA
0.52	2.2
1.16	5.5
2.00	7.8
3.11	12.8
4.19	17.7
5.32	22.0
Sample	14.7

23-2 The mass of 20 drops of mercury was found to be 0.1320 g and the drop time was 4.94 s. Calculate the flowrate of mercury from the capillary (mg/s).

23-3 In the solvent *N,N*-dimethylformamide (DMF), nitrate is reduced in a one-electron step to NO_2. A polarogram of a 8.4-m*M* KNO_3 solution obtained by using a dme with $m^{2/3}\,t^{1/6}$ equal to 1.86 $mg^{2/3}\,s^{-1/2}$ was recorded. The instantaneous diffusion current was 24 μA. Estimate the diffusion coefficient of nitrate in DMF.

23-4 The standard-addition technique was used for the polarographic assay of a Cd^{2+} sample. From the data listed in the table, calculate the concentration of Cd^{2+} in the sample.

Added concentration, m*M*	Diffusion current, μA
0 (sample)	3.8
1.20	5.5
2.50	7.2
4.04	9.5
6.00	12.2
7.50	14.3

23-5 At a potential on the plateau of a polarographic wave for the two-electron reduction of a metal ion M^{2+} to the metal M, *m* and *t* were respectively equal to 1.46 mg/s and 4.29 s. Polarograms of a series of standard M^{2+} solutions were recorded and the average diffusion currents were plotted as a function of concentration. The slope of the plot was 4.92 μA/m*M*. Calculate the diffusion coefficient of M^{2+}.

23-6 The polarogram of a 1.25-m*M* solution of zinc(II) had an instantaneous diffusion current of 7.12 μA. The capillary characteristics were t = 3.47 s and m = 1.42 mg/s. Determine the diffusion coefficient of zinc(II) in the solution.

23-7 The instantaneous diffusion current at the dme of a 2.10-m*M* solution of lead(II) (n = 2) was 15.2 μA. If $m^{2/3}\ t^{1/6}$ for the capillary used in the study was 1.66 $mg^{2/3}s^{-1/2}$, determine the diffusion coefficient of lead(II) in the solution.

23-8 The diffusion coefficient of oxygen at 25°C in aqueous solutions is 2.65 × 10^{-5} cm^2/s. A dme with $m^{2/3}t^{1/6}$ of 1.86 $mg^{2/3}s^{-1/2}$ was used to assay a natural-water sample. The instantaneous diffusion current of the first oxygen wave was 2.3 μA. Calculate the concentration of dissolved oxygen in the water.

23-9 The data in the following table were obtained at 25°C on the rising portion of a reversible cathodic wave which obeys the general equation Ox + ne^- = Red. The diffusion current of the wave was 10.0 μA. Determine the number of electrons transferred in the electrochemical reaction and the half-wave potential.

***E*, –V, versus see**	***i*, μA**
0.600	0.91
0.612	2.01
0.620	3.20
0.627	4.44
0.639	6.66
0.650	8.25

23-10 The following data were obtained on an anodic polarographic wave at 25°C. The electroactive compound obeyed the general equation

$$\text{Red} \rightarrow \text{Ox} + ne^-$$

1s the electrochemical reaction reversible?

***E*, – V, versus sce**	**log [($i_{d,a}$ – i)/i]**
0.100	– 1.000
0.120	– 0.759
0.140	– 0.516
0.160	– 0.278
0.180	– 0.040
0.200	0.200
0.220	0.440

AC and Pulse Polarography

23-11 The following data were obtained from ac polarograms with a series of Cd^{2+} standards and a sample. Determine the cadmium concentration in the sample.

Concentration, m*M*	Peak current, μA
0.50	1.20
1.00	2.42
2.00	4.86
3.00	7.28
3.75	9.11
Sample	4.73

23-12 If the time during one complete cycle at a line frequency of 50.0 Hz is used to measure the current during pulse polarography, calculate the time of the measurement.

23-13 Differential pulse polarography was used with the standard-addition technique to assay an aqueous solution of nitrobenzene which is a reducible organic pollutant. From the results recorded in the following table, determine the pollutant concentration in the sample.

Added concentration, m*M*	Peak current, μA
0 (sample)	2.51
0.100	4.16
0.200	5.75
0.300	7.42
0.400	9.10

Voltammetry

23-14 An electroactive species yielded a wave with a limiting current of 15.2 μA at an rde which was rotated at 10.0 r/s. What limiting current would be expected at 30.0 r/s?

23-15 During the forward scan of a triangular wave voltammogram at a disk electrode, a peak current of 25.4 μA was observed at a scan rate of 0.250 V/s. Estimate the peak current at a scan rate of 50.0 mV/s assuming a reversible electrochemical reaction.

23-16 A triangular wave voltammogram was obtained on an oxidizable electroactive species ($n = 2$). During the forward scan the peak potential was 0.368 V and during the reverse scan the peak potential was 0.319 V. Was the electrochemical reaction reversible at that scan rate?

23-17 Potassium ferrocyanide ($n = 1$) has a diffusion coefficient of 6.5×10^{-6} cm^2/s during its oxidation in 0.1 *M* KCl. It was used to measure the area of a stationary disk electrode. At a scan rate of 100 mV/s the anodic peak current for the oxidation of a 1.0-m*M* solution of potassium ferrocyanide was 32 μA. Estimate the electrode area.

Tafel Plots

23-18 An iron electrode had a corrosion current density of 0.125 mA/cm^2 as determined from a Tafel plot. Calculate the corrosion rate in cm/y.

Amperometry

23-19 The tabulated data were obtained during the amperometric titration of 50.0 mL of iron(II) with 0.100-*M* cerium(IV) at a potential at which only the iron(II) is electroactive:

$$Fe^{2+} + Ce^{4+} = Fe^{3+} + Ce^{3+}$$

Calculate the concentration of iron(II) in the original sample.

Volume of Ce^{4+}, mL	Anodic current, μA
0	29.2
0.5	23.4
1.0	17.8
1.5	12.2
2.0	6.5
2.5	1.5
3.0	0.7
3.5	0.5
4.0	0.3
5.0	−0.1

23-20 The following data were obtained during the amperometric titration with two indicator electrodes of a 50.0-mL solution of iodine (I_2) with 0.112-*M* thiosulfate ($S_2O_3^{2-}$):

$$I_2 + 2S_2O_3^{2-} = 2I^- + S_4O_6^{2-}$$

Calculate the concentration of iodine in the original sample solution.

Volume, mL	Current, μA
1.00	11.0
1.50	11.5
2.00	11.0
2.50	9.0
2.90	6.3
3.10	3.7
3.20	2.1
3.50	0.4
4.00	0.3
4.50	0.3
5.00	0.3
6.00	0.3

Anodic Stripping Voltammetry

23-21 Anodic stripping voltammetry at a platinum disk electrode was used to assay a water sample for Cu^{2+}. The potential was adjusted to –1.3 V (versus sce) for 5 min prior to scanning anodically at 50 mV/s. The height of the stripping peak was recorded for a series of standard solutions and the sample. Determine the concentration of Cu^{2+} in the sample solution.

$[Cu^{2+}]$, × 10^{-5} *M*	i_p, μA
1.00	1.20
2.00	2.37
4.00	4.82
6.00	7.27
8.00	9.67
10.00	12.14
Sample	5.36

Coulometry

23-22 A 10.0-mL solution containing phenol (C_6H_5OH) was added to a 100-mL coulometric cell. Hydrochloric acid and sodium bromide were added to the cell, along with sufficient water to make the total volume about 50 mL. Bromine (Br_2) was electrogenerated from the bromide in the solution and was used as titrant for the phenol. The endpoint of the coulometric titration was located amperometrically. The time required to reach the endpoint with a 6.43-mA source was 112 s. Calculate the concentration of phenol in the sample.

$$2Br^- \rightarrow Br_2 + 2e^-$$

$$C_6H_5OH + 3Br_2 \rightarrow C_6H_2Br_3OH + 3HBr$$

23-23 A 10.0-mL aqueous solution of the biochemical compound flavin adenine dinucleotide (FAD) was assayed by controlled-potential coulometry at – 0.8 V versus sce:

$$FAD + 2H^+ + 2e^- = FADH_2$$

The area under the current-time curve was 33.7 mA · min. Determine the concentration of FAD in the sample.

23-24 Nicotinamide adenine dinucleotide (NAD^+) can be coulometrically assayed at constant potential by reduction to the dimer $(NAD)_2$:

$$2NAD^+ + 2e^- = (NAD)_2$$

If the area under the current-time curve for the reduction of a 15.0-mL aqueous solution of NAD^+ is 54.3 mA · min, calculate the concentration of NAD^+ in the sample.

Conductometry

23-25 The conductance of a 0.100-*M* KCl solution was 1.58 mS at 25°C. Calculate the cell constant.

23-26 Estimate the conductance of a solution containing 1.15×10^{-4} *M* NaCl and 5.65×10^{-5} *M* K_2CrO_4 in the cell described in Prob. 23-25.

23-27 Calculate the transference number of Fe^{2+} in a solution containing 1.00×10^{-3} *M* $FeCl_2$ and 0.100 *M* HCl.

23-28 The cell constant of a conductance cell at 25°C was 2.38 cm^{-1}. The conductance of a saturated solution of $BaSO_4$ placed in the cell was 1.27 μS. Estimate the solubility product of $BaSO_4$.

23-29 Exactly 6.00 mL of acetic acid (density = 1.05 g/mL, MW 60.05) was added to a 1-L volumetric flask, and the flask was filled to the mark with distilled water. A portion of the resulting solution was added to a conductance cell which had a cell constant of 1.25 cm^{-1}. The measured conductance was 416 μS. Estimate the acid dissociation constant of acetic acid.

23-30 The data listed in the following table were obtained during the conductometric titration of 50.0 mL of acetic acid with 0.100 *M* sodium hydroxide. Determine the concentration of the acetic acid.

Volume of NaOH, mL	G, μS
0	55
0.50	30
1.00	35
1.50	49
2.00	62
2.50	76
3.00	91
3.50	129
4.00	170
4.50	209
5.00	252
6.00	337

REFERENCES

1. Peterson, W. M.: *Amer. Lab.*, 11(12): 69(1979).
2. Bond, A. M.: *Modern Polarographic Methods in Analytical Chemistry*, Dekker, New York,1980.
3. Bard, A. J., and L. R. Faulkner: *Electrochemical Methods*, Wiley, New York, 1980.
4. Meites, L.: *Polarographic Techniques*, 2d ed., Wiley, New York, 1965.

5. Breyer, B., and H. H. Bauer: *Alternating Current Polarography and Tensammetry*, Interscience, New York, 1963.
6. Parry, E. P., and R. A. Osteryoung: *Anal. Chem.*, **37**: 1634(1965).
7. Osteryoung, J. : *J. Chem. Educ.*, **60**: 296(1983).
8. Birke, R. L.: *Anal. Chem.*, 50: 1489(1978).
9. Osteryoung, J. G., and R. A. Osteryoung: *Anal. Chem.*, 57: 101A(l985).
10. Christie, J. H., J. A. Turner, and R. A. Osteryoung: *Anal. Chem.*, **49**: 1899(1977).
11. Turner, J. A., J. H. Christie, M. Vukovic, and R. A. Osteryoung: *Anal. Chem.*, **49**: 1904(1977).
12. O'Dea, J. J., J. Osteryoung, and R. A Osteryoung: *Anal. Chem.*, **53**: 695(1981).
13. Adams, R. N.: *Electrochemistry at Solid Electrodes*, Dekker, New York, 1969.
14. Bard, A. J.: *J. Chem. Educ.*, **60**: 302(1983).
15. Evans, D. H., K. M. O'Connell, R. A. Petersen, and M. J. Kelly: *J. Chem. Educ.*, **60**: 290(1983).
16. Adams, K. L., M. Puchades and A. G. Ewing in Annual Review of Analytical Chemistry, E. S. Yeung and R. N. Zare (eds), Annual Reviews, Palo Alto, California, **1**: 329(2008).
17. Bockris, J. O'M. and S. U. M. Khan: Surface Electrochemistry - A Molecular Level Approach, Plenum Press, New York, 1993.
18. Stock, J. T.: *Amperometric Titrations*, Interscience, New York, 1965.
19. Anson, F. C., and R. A. Osteryoung: *J. Chem. Educ.*, **60**: 293(1983).
20. Heineman, W. R.: *J. Chem. Educ.*, **60**: 305(1983).
21. Lingane, J. J.: *Electroanalytical Chemistry*, Interscience, New York, 1958.
22. MacDonald, J. R.:Electrochim. Acta, 35:1483(1990).
23. Barsoukov, E. and J. R. MacDonald:Impedance Spectroscopy-Theory, Experimental Applications, Wiley-Interscience, New York (2005).
24. Amemiya, S., A. J. Bard, F. F. Fan, M. V Mirkin and P. R. Unwin in Annual Reviews of Analytical Chemistry, E. S. Yeung and R. N. Zare(eds), Annual Reviews, Palo Alto, California, 1:91(2008).
25. Poggi, M. A., E. D. Gadsby and L. A. Bottomley: Anal. Chem., 76:3429(2004).

CHAPTER

TWENTY-FOUR

INTRODUCTION TO CHROMATOGRAPHY

Chromatography comprises the group of techniques that are used to separate mixture components by the relative attraction of each component to a stationary phase while a mobile phase passes over or through the stationary phase. After the mixture is added to the stationary phase, the mobile phase is allowed to flow continuously over the stationary phase, starting from the point at which the sample was added. The mixture is partitioned (divided) between the stationary and mobile phases and moves along with the mobile phase at a rate which is dependent upon the relative attraction of each component for the two phases. A species which is more strongly attracted to the mobile phase than to the stationary phase is swept along with the mobile phase more rapidly than a species which is more strongly attracted to the stationary phase.

Because different substances adhere to the stationary phase to different extents, a separation of the components can be effected. The mixture components can be separated in space at a fixed time after the start of the separation or in time at a fixed distance away from the initial location of the sample. The time required for a component to move a fixed distance or the distance a component travels in a fixed time can be used for qualitative analysis. The measured amount of each separated component can be used for quantitative analysis.

Chromatography is divided into categories depending upon the nature of the stationary and mobile phases. Some of the more common forms of chromatography are listed in Table 24-1. *Liquid chromatography* is any form of chromatography in which the mobile phase is a liquid. *Gas chromatography* is any form of chromatography in which the mobile phase is a gas. If the mobile phase is a gas and the stationary phase is a liquid, the technique is *gas-liquid chromatography*. If the mobile phase is a gas and the stationary phase is a solid,

Table 24-1 Several categories of chromatography

Type	Mobile phase	Stationary phase
Gas-liquid	Gas	Liquid
Gas-solid	Gas	Solid
Ion pair	Liquid	Liquid
Ion exchange	Liquid	Gel
Liquid-liquid	Liquid	Liquid
Liquid-solid	Liquid	Solid
Thin layer	Liquid	Solid on flat plate
Paper	Liquid	Liquid on paper
Size exclusion	Liquid	Gel

the method is *gas-solid chromatography. Liquid-liquid chromatography* and *liquid-solid chromatography* have liquid mobile phases with liquid and solid stationary phases respectively. Several other types of chromatography are described in Chapters 25 and 26. The present chapter describes some of the characteristics that are common to nearly all forms of chromatography. Specific characteristics of a particular type of chromatography are described in the appropriate section in Chapter 25 or Chapter 26.

A *chromatogram* is a plot of the detector response as a function of time, mobile phase volume, or distance. A chromatogram is normally *developed* by following one of three procedures. The chromatographic process by which sample components are separated is termed *developing* the chromatogram. If the analyte solution is continuously added to the stationary phase and the total concentration of the mixture components eluted from the stationary phase is plotted as a function of time or eluate volume, the chromatographic method is

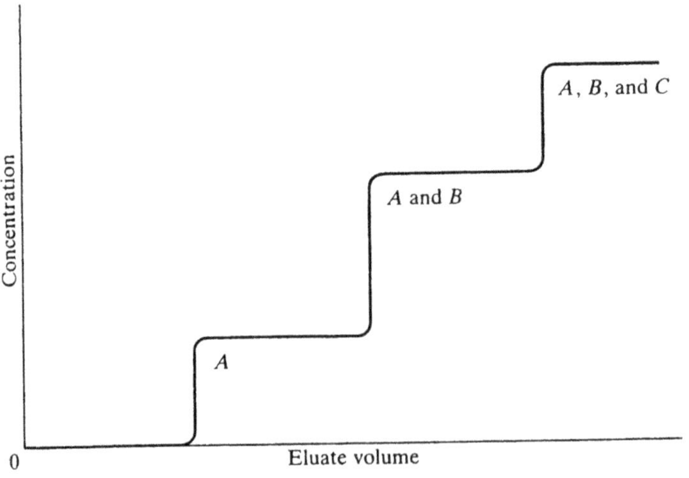

Figure 24-1 A liquid chromatogram of a three-component (*A*, *B*, and *C*) mixture that was obtained by frontal analysis.

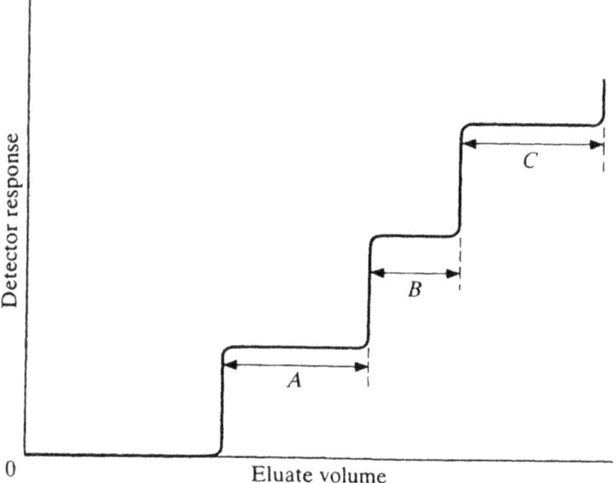

Figure 24-2 A liquid chromatogram of a three-component (*A*, *B*, and *C*) mixture that was obtained by displacement analysis. The eluate volumes (indicated by the arrows) are proportional to the concentrations of the components.

frontal analysis. A diagram of a liquid chromatogram that was obtained by using frontal analysis is shown in Fig. 24-1. The *eluent* is the mobile phase; the *eluate* is the solution of sample and mobile phase which has passed through or over the stationary phase.

Prior to *displacement analysis* a fixed volume (block) of the analyte solution is added to the stationary phase. It is subsequently forced from the stationary phase by continuous addition of a solution of a substance that is more strongly attracted to the stationary phase than are the sample components. The chromatogram consists of a plot of total analyte concentration in the eluate as a function of eluate volume or time of elution, as shown in Fig. 24-2.

Elution analysis is the most popular chromatographic technique. The sample is placed on the stationary phase in a single block and is eluted from the stationary phase with pure solvent (the mobile phase). The chromatogram is a plot of solute concentration in the eluate as a function of eluted volume or time. *Gradient-elution analysis* is a category of elution analysis in which the composition of the mobile phase is gradually changed during development of the chromatogram. A gas chromatogram obtained by using elution analysis is shown in Fig. 24-3.

Chromatography is a valuable analytical tool because it can separate the components of a mixture while simultaneously providing both qualitative and quantitative analytical information about each component. Often that is possible even though the physical characteristics of the sample components (and cones-quently the relative attractions to the stationary and mobile phases) differ only slightly. Because elution analysis is used much more often than the other methods, except where otherwise specified, the remainder of Chapters 24, 25, and 26 are restricted to descriptions of its use.

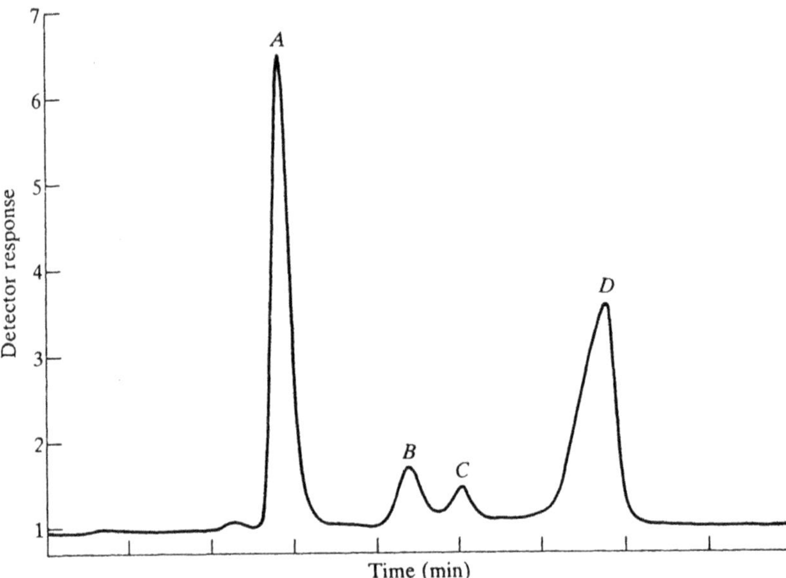

Figure 24-3 A gas chromatogram of several components in a corrosion inhibitor. The chromatogram was developed by using elution analysis. The area under each peak is proportional to the concentration of that component.

THEORY

The relative concentration of a particular analyte component in or on the stationary phase as compared with the concentration in the mobile phase during a chromatographic study is the *distribution* or *partition coefficient* K for that species. The distribution coefficient is defined as the ratio of the concentration of the species in the stationary phase C_s to the concentration of the species in the mobile phase C_m:

$$K = \frac{C_s}{C_m} \tag{24-1}$$

The distribution coefficient is related to the velocity with which a component moves along the stationary phase during development. As K increases, the average velocity of motion decreases. Because different components have different distribution coefficients for a particular set of experimental conditions, the distribution coefficient of a component could be compared to the coefficients of known species for qualitative analysis. Unfortunately, it is rarely possible to obtain values of K easily, and use of distribution coefficients for qualitative analysis is only occasionally possible.

A more easily measured parameter, which can also be used for qualitative analysis, is the *retention time* t of a chemical species. The retention time is the time required to elute a sample component from the stationary phase; i.e., it is the time required for the mobile phase to sweep the component from the stationary phase. A similar and also useful parameter is the *retention volume V*. The retention volume is the volume of mobile phase required to elute the sample component from the stationary phase. The retention volume and the retention time are related to each other by

$$V = Ft \tag{24-2}$$

where F is the flowrate of the mobile phase (units are those of volume per time).

The time required for the mobile phase to flow over or through the stationary phase, i.e., the time for an average mobile-phase molecule to flow from one end of the stationary phase to the other, is t_m. The corresponding mobile-phase volume V_m is the *dead volume*, the *interstitial volume*, or the *void volume*. V_m and t_m are related to each other by

$$V_m = Ft_m \tag{24-3}$$

The most commonly used method of chromatographic qualitative analysis consists of comparing retention times or retention volumes of sample components with those obtained for known substances under identical conditions. Because a change in any of several experimental conditions (stationary phase, mobile phase, flowrate, temperature, and so on) can alter V and t, it is not possible to prepare a list of generally useful values of V and t. If the stationary and mobile phases are not changed, the *relative retention* α can be used to eliminite the effect of other experimental variables:

$$\alpha = \frac{t - t_m}{t_{\text{ref}} - t_m} = \frac{V - V_m}{V_{\text{ref}} - V_m} \tag{24-4}$$

This is the ratio of the retention time or volume of the substance, after correction for t_m and V_m to the corrected retention time or volume of a reference compound.

Because the retention volume at a constant flowrate is proportional to the retention time, the two definitions of relative retention are equivalent. The direct proportionality between the distribution coefficient K for a substance and the corrected retention time or retention volume of the substance makes it possible to express the relative retention in terms of distribution coefficients:

$$\alpha = \frac{K}{K_{\text{ref}}} \tag{24-5}$$

In ion-exchange chromatography, the relative retention [Eq. (24-5)] sometimes is called the *separation factor*.

Because a distribution coefficient cannot be directly measured from a chromatogram, a need exists for a term which is related to the distribution coefficient but which can be directly determined from the chromatogram. That need is filled by the *capacity factor* k'. If the definition of concentration is substituted into Eq. (24-1), the following result is obtained:

$$K = \frac{m_s / V_s}{m_m / V_m} = \frac{m_s}{m_m}\frac{V_m}{V_s} = k' \frac{V_m}{V_s} \tag{24-6}$$

where m_s and m_m are the number of moles (or mass depending upon the chosen concentration unit) of a species in the stationary and mobile phases respectively and V_s and V_m are the volume of the stationary and mobile phases inside the *column* (the tube holding the stationary phase). It is apparent from Eq. (24-6) that the capacity factor is equivalent to the ratio of the amount of the species in the stationary phase to the amount in the mobile phase. Rearrangement of Eq. (24-6) yields

$$k' = \frac{KV_s}{V_m} \tag{24-7}$$

The capacity factor can be determined directly from the chromatogram with the aid of

$$t = t_m\left(1 + k'\right) \tag{24-8}$$

or

$$V = V_m\left(1 + k'\right) \tag{24-9}$$

Another equation which is sometimes useful is that for the *retention ratio r*. The retention ratio is the ratio of t_m to t:

$$r = \frac{t_m}{t} = \frac{t_m}{t_m\left(1 + k'\right)} = \frac{1}{1 + k'} = \frac{1}{1 + K(V_s / V_m)} = \frac{V_m}{V_m + KV_s} \tag{24-10}$$

That ratio is important because it gives the fraction of time that an average component molecule spends in the mobile phase.

Sample problem 24-1 A gas chromatogram of a mixture that contained benzene, anthracene, and air (not retarded on the column) was obtained. The retention time of each component was measured and recorded. Assuming the column is a cylindrical tube with a length of 50.0 cm and an internal diameter of 1.00 cm, and the flowrate is 30 cm^3/min, calculate k' for benzene, V_m, V_s (assuming the total volume of the column is $V_m + V_s$), K for benzene, the relative retention of anthracene with respect to benzene, and the fraction of time that an average molecule of benzene spends in the mobile phase.

Compound	***t*, min**
Benzene	3.24
Anthracene	5.73
Air	0.25

SOLUTION Equation (24-8) can be used to calculate k' for benzene. Because air is not retarded in the column, $t_m = 0.25$ min.

$$t = t_m(1 + k')$$

$$3.24 = 0.25(1 + k')$$

$$0.25k' = 3.24 - 0.25$$

$$k' = 12$$

V_m can be calculated from the flowrate and t_m with the aid of Eq. (24-3):

$$V_m = Ft_m$$
$$= (30 \text{ cm}^3/\text{min})(0.25 \text{ min}) = 7.5 \text{ cm}^3$$

The total volume of the column can be calculated from the column's dimensions:

$$V = \pi r^2 L$$
$$= 3.14(0.500 \text{ cm})^2(50.0 \text{ cm}) = 39.3 \text{ cm}^3$$

Because for *this* column $V_m + V_s$ = total volume (that is not normally the case), V_s can be calculated:

$$7.5 \text{ cm}^3 + V_s = 39.3 \text{ cm}^3$$
$$V_s = 31.8 \text{ cm}^3$$

The distribution coefficient for benzene can be calculated by using Eq. (24-6):

$$K = \frac{k'V_m}{V_s}$$
$$= \frac{12(7.5)}{31.8} = 2.8$$

The relative retention of anthracene with respect to benzene is determined with Eq. (24-4):

$$\alpha = \frac{t - t_m}{t_{\text{ref}} - t_m}$$
$$= \frac{5.73 - 0.25}{3.24 - 0.25} = 1.83$$

The fraction of time that an average molecule of benzene is in the mobile phase is obtained by using Eq. (24-10):

$$r = \frac{t_m}{t}$$
$$= \frac{0.25 \text{ min}}{3.24 \text{ min}} = 0.077$$

An average benzene molecule spends 7.7 percent of its time within the column in the mobile phase.

CHROMATOGRAPHIC BAND BROADENING

It is unfortunate that not all molecules of a particular chemical species pass through a chromatographic column (or over a chromatographic plate, etc.) at exactly the same rate. Some molecules move more rapidly through the column than do other molecules of the same substance. As a consequence, a chromatogram consists of a series of broadened bands as shown in Fig. 24-4, rather than a series of narrow vertical lines. The time or volume corresponding to the apex of each peak is the most probable retention time or volume for that particular chemical species and is used as the measured retention time or retention volume for the species.

Many factors can broaden chromatographic bands. Among the factors are: (1) sample introduction to the column in a band of appreciable width, (2) diffusion within the column from regions of higher concentration to regions of lower concentration, (3) flow of different molecules through the column in nonequivalent paths of different length, and (4) failure of the solute to establish equilibrium between the stationary and mobile phases.

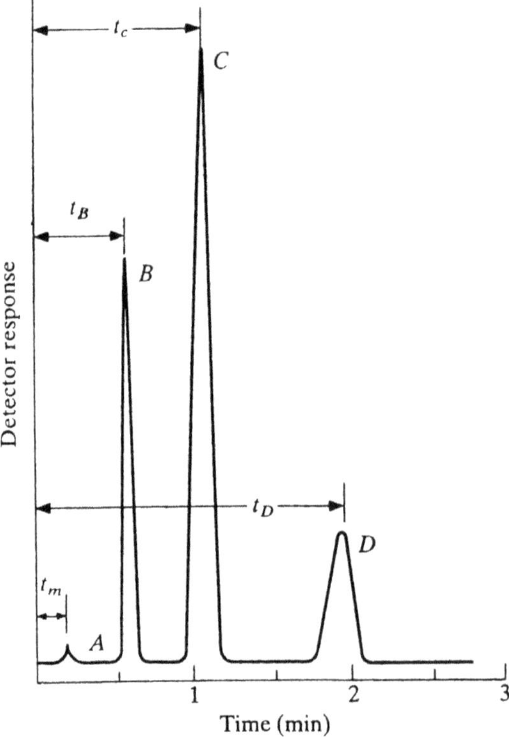

Figure 24-4 A gas chromatogram of a sample that contains air (peak *A*), heptane (peak *B*), octane (peak *C*), and nonane (peak *D*). Air is a nonretarded substance that is used to measure t_m. The retention times for the components are measured horizontally to the tops of the peaks as indicated.

It is impossible to introduce an entire sample into a column at the same instant; i.e., the sample enters the column in a band of finite thickness. Generally, the width of the sample band which enters the column can be directly related to the width of the eluted band. An increase in the bandwidth of the introduced sample causes increased bandwidths of the separated components. For that reason, it is advisable to keep the width of the sample band introduced into a column or added to a thin-layer chromatographic plate as small as possible.

It is common knowledge that a dissolved species in a stationary solvent diffuses from regions of higher concentration to those of lower concentration. If a spoonful of sugar is added to a cup of tea or coffee, it will eventually dissolve; if the concentration of the sugar is allowed to equilibrate, it will be identical in all parts of the solution. That is caused by diffusion. Although diffusion occurs in chromatography also, the relatively fast motion of the mobile phase prevents the concentration of a sample component from becoming identical at all points within the column. The velocity of the component as it is swept through the column by the mobile phase is greater than the velocity of the component caused by diffusion.

During development of a chromatogram, some sample molecules diffuse into the lower concentration region in front of the center of the band while other molecules diffuse into the lower concentration region behind the center of the band. The net result is band broadening. Generally, the amount of band broadening owing to diffusion increases as the time for diffusion within the column increases, i.e., as the retention time of the component increases. Because the retention time of a component decreases with increasing mobile-phase flow-rate, the amount of band broadening owing to diffusion decreases with increasing flowrate. Relatively narrow bands can be expected at higher flowrates. Unfortunately, peak resolution (separation) usually decreases with increasing flowrate. Often it is necessary to use a flowrate that is a compromise between that which yields large peak separation and that which yields narrow bandwidths.

An analyte molecule or ion can follow many paths while traveling through a column. Not all of them are of equal length. If a molecule follows a relatively long path through the column, it will exit from the column later than a molecule following a path of average distance. Likewise, if a molecule follows a relatively short path through the column, it will exit from the column earlier than the average molecule. As a consequence, the band broadens during chromatographic development. Band broadening owing to different pathlengths can be decreased by making all pathlengths as nearly equal as possible. That is accomplished by making the particles of stationary phase as nearly identical in size as possible.

If a solute molecule is unable to equilibrate between the stationary and mobile phases, it spends more time in one of the phases than it otherwise would have, with a resultant broadening of the chromatographic band. That is particularly problematic when the degree to which equilibrium is established varies between identical component molecules. As the flowrate of the mobile phase increases, less time exists for the solute to establish equilibrium. Consequently, band broadening that is caused by nonequilibrium increases as the flowrate increases. The causes of band broadening in gas chromatography are described in more detail in Chapter 26.

EFFICIENCY

The *theoretical plate* concept which is used to measure the efficiency of a distillation column has been adapted for use in chromatography. A chromatographic theoretical plate can be thought of as the portion of a column which is required to establish a single equilibration of the sample component between the stationary and mobile phases. If both the stationary and mobile phases are liquids, a theoretical plate in a chromatographic column is the portion of the column required to yield the same concentrations in the stationary and mobile phases as in the two liquid phases after an extraction. Within a chromatographic column many equilibrations occur, and separations can generally be made more efficiently than with extractive apparatus. As the number of theoretical plates n in a column increases, the efficiency of the column increases. In this section and the following section dealing with resolution, the equations are written while using retention times. Similar identical equations can be written by using retention volumes in place of retention times.

Another measure of column efficiency is the height equivalent of a theoretical plate (H). The height equivalent of a theoretical plate is the length of column corresponding to a single theoretical plate. It can be calculated from the length L of the column and the number of theoretical plates within the column:

$$H = \frac{L}{n} \tag{24-11}$$

An efficient column is one for which n is large or H is small.

The efficiency of a chromatographic column is a measure of the ability of a column to separate components of a mixture. Because a complete separation of mixture components is most likely to occur when each chromatographic peak has a narrow bandwidth, the efficiency of a column increases as the bandwidth of a peak decreases. Under identical experimental conditions, diffusion should cause a peak with a long retention time to be broader than a second peak with a short retention time. Because the efficiency of the column has not changed, efficiency must be related to retention time as well as to peak width. If the peak widths for a sample component on two separate columns are identical, but the retention time of the component on the first column is greater than that on the second column, the efficiency of the first column is greater than that of the second column. The efficiency of a chromatographic separation is defined as the square of the retention time t for a component divided by the *variance* V of the chromatographic peak:

$$n = \frac{t^2}{V} \tag{24-12}$$

If the chromatographic peak is gaussian (normal) in shape, the variance is the square of the *standard deviation* σ of the peak, and Eq. (24-12) can be rewritten as

$$n = \frac{t^2}{\sigma^2} \tag{24-13}$$

It is statistically possible to show for a gaussian peak that Eq. (24-13) is equivalent to either

$$n = \left(\frac{4t}{w}\right)^2 \tag{24-14}$$

in which the peak width is w, or

$$n = 5.54\left(\frac{t}{w_{1/2}}\right)^2 \tag{24-15}$$

in which the peak width at half the peak height is $w_{1/2}$.

Either Eq. (24-14) or Eq. (24-15) can be used to calculate n. The peak width w and the half-peak width $w_{1/2}$ must be in the same units as the retention time when either equation is used. The peak width is the distance along the baseline between the extrapolated tangents to the rising portions of the peak. Many chromatographers prefer to use Eq. (24-15) because the use of that equation does not require the analyst to draw tangents to the chromatographic peak to estimate w. Measurements of w and $w_{1/2}$ are illustrated in Sample problem 24-2.

Sample problem 24-2 Calculate the number of theoretical plates and the height equivalent of a theoretical plate for the 50.0-cm column used to obtain the chromatogram shown in Fig. 24-5.

SOLUTION Eq. (24-14) or Eq. (24-15) can be used to calculate n. If Eq. (24-14) is used, a straight line is drawn tangent to each edge of the peak and the width w of the peak is measured as the distance between the intersections of the two tangent lines with the baseline. In that manner w is found to be 8.3 s. The retention time t is 59 s. Substitution into Eq. (24-14) permits calculation of the efficiency of the column:

$$n = \left(\frac{4 \times 59}{8.3}\right)^2 = 8.0 \times 10^2 \text{ theoretical plates}$$

Equation (24-11) is used to calculate H:

$$H = \frac{L}{n} = \frac{50.0 \text{ cm}}{8.0 \times 10^2 \text{ plates}} = 6.3 \times 10^{-2} \text{ cm/plate}$$

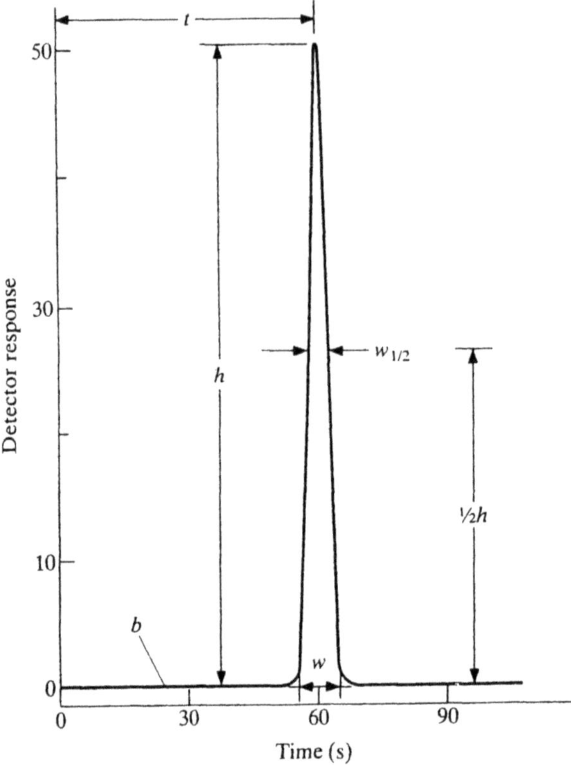

Figure 24-5 A gas chromatogram of heptane that was obtained with a 50-cm column. *b*, baseline; *t*, retention time; *h*, peak height; $w_{1/2}$, half-peak width; *w*, peak width.

To use Eq. (24-15) to calculate *n*, it is necessary to measure the half-peak width $w_{1/2}$. The peak height as measured from the baseline is 50.2 units, and half the peak height is 25.1 units. The width of the peak at a height of 25.1 units is 4.9 s. Substitution for $w_{1/2}$ into Eq. (24-15) allows calculation of *n*:

$$n = 5.54\left(\frac{59}{4.9}\right)^2 = 8.0 \times 10^2 \text{ plate}$$

Then

$$H = \frac{L}{n} = \frac{50.0 \text{ cm}}{8.0 \times 10^2 \text{ plates}} = 6.3 \times 10^{-2} \text{ cm/plate}$$

If the chromatographic peak is asymmetrical or is symmetrical but not gaussian, determination of the efficiency of a chromatographic separation is slightly more complex. The amount of asymmetry in a peak is generally measured with the *asymmetry factor* a_f:

$$a_f = \frac{B}{A} \tag{24-16}$$

The asymmetry factor is the ratio of the portion of the peak width to the right B of the retention time to the portion of the peak width to the left A of the retention time. The widths are measured at a fixed percentage of the height of the peak. Typically the widths are measured at 5, 10, 15, 30, or 50 percent of the peak height. Measurement of the widths at 10 percent of the peak height is generally preferred because the result can be used in equations for efficiency. Measurements of A and B at 10 percent of the peak height are illustrated in Fig. 24-6.

If a chromatographic peak exhibits tailing, i.e., is wider on the right than the left, a_f is greater than unity. Conversely, if the peak is wider on the left (shorter times) than on the right, the asymmetry factor is less than unity. An asymmetry factor of unity indicates a symmetrical peak.

Several methods have been proposed to determine the efficiency of a chromatographic separation which produces an asymmetrical peak. Of the methods, the graphical method suggested by Barber and Carr[1] and the computational method suggested by Foley and Dorsey[2] do not require use of a computer and consequently are suitable for use in all laboratories. Both methods assume that the asymmetrical chromatographic peak has an exponentially modified gaussian shape. The equation derived in Reference 2 for the calculation of theoretical plates is generally easiest to use and is shown as follows:

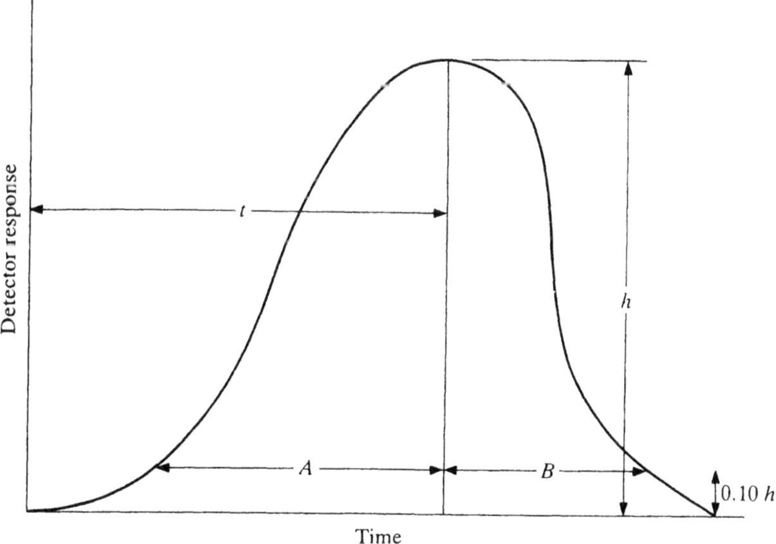

Figure 24-6 A diagram of an asymmetrical chromatographic peak. t, retention time; h, peak height; $0.10h$, 10 percent of the peak height; A, portion of the peak width at 10 percent of the peak height prior to the retention time; B, portion of the peak width at 10 percent of the peak height after the retention time.

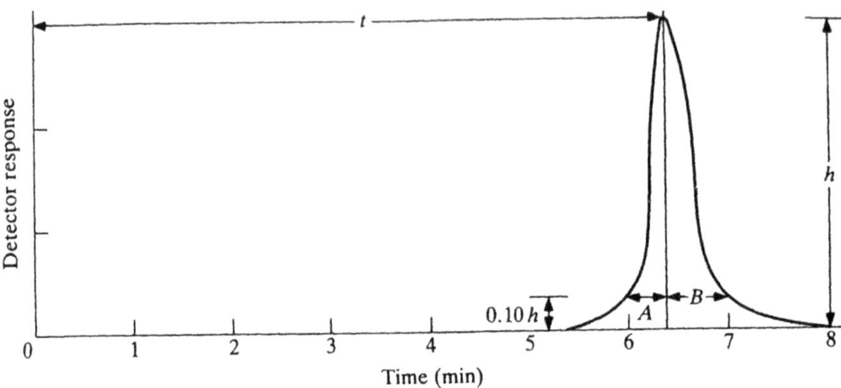

Figure 24-7 The chromatogram described in Sample problem 24-3.

$$n = \frac{41.7(t / w_{0.1})^2}{a_{f,0.1} + 1.25} \tag{24-17}$$

where $a_{f,\,0.1}$ and $w_{0.1}$ respectively are the asymmetry factor and peak width at 10 percent of the peak height. The equation yields results that are within 1.5 percent of the correct value for asymmetry factors that are between 1.09 and 2.76. Most skewed chromatographic peaks are within that range.

Sample problem 24-3 Calculate the asymmetry factor at 10 percent of the peak height of the chromatographic peak shown in Fig. 24-7. Determine the number of theoretical plates and the height equivalent of a theoretical plate for the 180-cm column.

SOLUTION The retention time of the peak is 6.40 min and the peak height is 3.00 units. At 10 percent of the peak height (0.30 units), the portion of the peak width to the left of (before) the retention time is 0.42 min. The corresponding portion of the peak width to the right of the retention time is 0.64 min, and $w_{0.1}$ is 1.06 min. The asymmetry factor at 10 percent of the peak height is calculated using Eq. (24-16):

$$a_f = \frac{B}{A} = \frac{0.64}{0.42} = 1.5$$

The number of theoretical plates can be determined using Eq. (24-17):

$$n = \frac{41.7(6.40\ \text{min}/1.06\ \text{min})^2}{1.5 + 1.25} = 5.5 \times 10^2 \text{ plates}$$

The height equivalent of a theoretical plate can be calculated by using the column length:

$$H = \frac{L}{n} = \frac{180\ \text{cm}}{5.5 \times 10^2 \text{ plates}} = 0.33\ \text{cm/plate}$$

RESOLUTION

Resolution is a measure of the separation between adjacent peaks in a chromatogram. The amount of separation between two peaks is a function of two parameters. As the difference between the retention times of the peaks increases, the separation increases; i.e., resolution is directly proportional to the difference in retention times of the peaks. The amount of separation is also dependent upon that portion of the width of each peak which is nearest to the adjacent peak. As the half-widths of the peaks increase, the amount of separation between the peaks decreases; i.e., the resolution decreases. For symmetrical peaks which have a gaussian shape and which are of about the same size, the resolution R between two peaks is given by

$$R = \frac{t_2 - t_1}{4\sigma} = \frac{t_2 - t_1}{w_1/2 + w_2/2} = \frac{2(t_2 - t_1)}{w_1 + w_2} \tag{24-18}$$

where t_1 and t_2 are the retention times of the peaks (t_2 is greater than t_1), σ is the standard deviation associated with each peak (4σ is the peak width for a gaussian peak), and w_1 and w_2 are the widths of the peaks ($w_1/2$ and $w_2/2$ are half the peak widths). The peak widths are measured by using tangents to the sides of the peaks as was done for calculations of theoretical plates. For gaussian shaped peaks, a resolution of 1.0 corresponds to 2 percent peak overlap. That is adequate for most quantitative analyses. If the peaks are of approximately equal size, a resolution of 1.5 corresponds to essentially complete separation between the peaks (less than 1 percent overlap).

Although Eq. (24-18) is often applied to peaks of all shapes, it works best for gaussian shaped peaks. A better working definition of resolution between a pair of peaks, at least one of which is not symmetrical, is given by

$$R = \frac{t_2 - t_1}{w_{1R} + w_{2L}} \tag{24-19}$$

where w_{1R} is the portion of the peak 1 width which is to the right of t_1 (i.e., adjacent to peak 2) and w_{2L} is the portion of the peak 2 width which is to the left of t_2 (Fig. 24-8). Equations (24-18) and (24-19) can be written in terms of retention volumes by replacing t_1 and t_2 by V_1 and V_2 respectively. To use Eq. (24-18) or Eq. (24-19), the peak widths must be measured with the same units as are used to measure the retention times or retention volumes. The resolution between gaussian peaks can be related to the relative retention α, the number of theoretical plates n, and the capacity factor k' by

$$R = \frac{n^{1/2}(\alpha - 1)k'}{4(1 + k')} \tag{24-20}$$

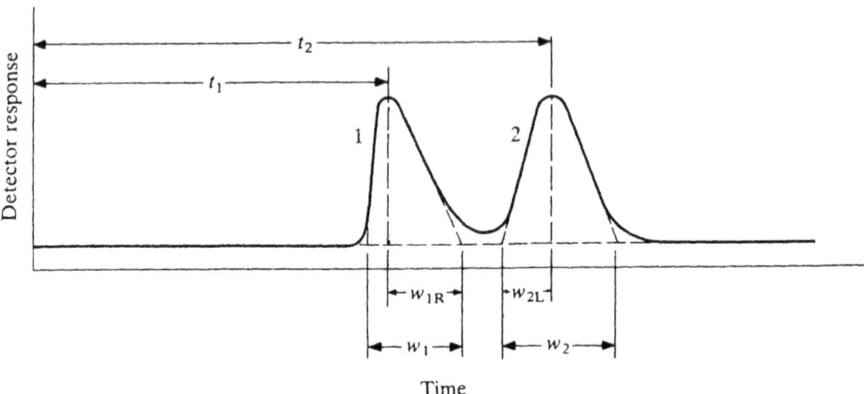

Figure 24-8 Slightly overlapping chromatographic peaks. Peak 1 is not symmetrical around its retention time. t_1, retention time of peak 1; t_2, retention time of peak 2; w_1, width of peak 1; w_2, width of peak 2; w_{1R}, the portion of the width of peak 1 to the right of t_1; w_{2L}, the portion of the width of peak 2 to the left of t_2.

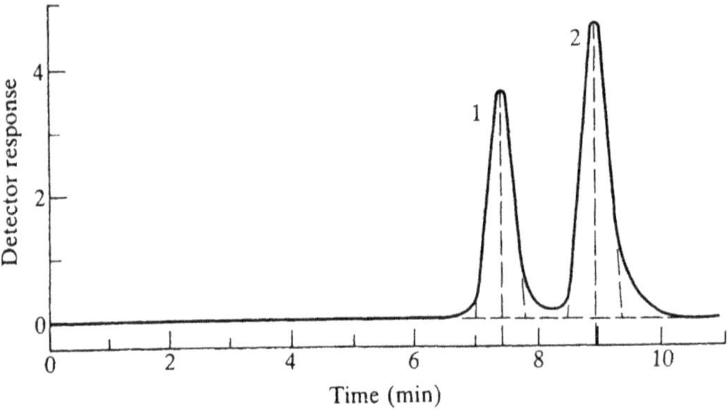

Figure 24-9 A liquid chromatogram of a mixture of acetylsalicylic acid (peak 1) and salicylic acid (peak 2).

Sample problem 24-4 Determine the resolution between the acetylsalicylic acid peak and the salicylic acid peak in the liquid chromatogram shown in Fig. 24-9. Is that resolution adequate for quantitative analysis?

SOLUTION The retention times of peaks 1 and 2 are 7.42 and 8.92 min respectively. The widths of the two peaks (measured as in Sample problem 24-2) are 0.87 and 0.91 min respectively. The resolution is determined by substitution of those values into Eq. (24-18):

$$R = \frac{2(8.92 - 7.42)}{0.87 + 0.91} = 1.69$$

Because the resolution is greater than unity, the separation is adequate for analysis.

IMPORTANT TERMS

Asymmetry factor
Capacity factor
Chromatogram
Chromatographic development
Chromatography
Dead volume
Distribution (partition) coefficient
Efficiency
Eluate
Eluent
Elution analysis
Frontal analysis
Gas chromatography
Gradient elution analysis
Height equivalent of a theoretical plate
Liquid chromatography
Relative retention
Resolution
Retention ratio
Retention time
Retention volume
Separation factor
Theoretical plate

PROBLEMS

Relative Retention

24-1 Compounds A, B, and C had respective retention volumes of 1.25, 15.0, and 22.5 mL. If compound A is not retarded on the column, calculate the relative retention of C with respect to B.

24-2 The following data were obtained on a reversed-phase HPLC column. All retention times were measured from the peak of a nonretarded component. Calculate the relative retention of each component with respect to 3-methoxytyramine. Determine the fraction of time that an average molecule of metanephrine spends in the mobile phase if t_m is 33 s.

Component	***t*, min**
Vanillinemandelic acid	3.23
Normetanephrine	3.87
Metanephrine	5.81
3-Methoxytyramine	7.31
Homovanillic acid	11.70

Capacity Factor

24-3 Calculate the capacity factor for components B and C on the column described in Prob. 24-1.

24-4 If t_m for the compounds listed in Prob. 24-2 is 0.15 min, calculate the capacity factor for normetanephrine.

Flowrate

24-5 The following retention times were recorded for the separation of a series of pesticides on a reversed-phase HPLC column. Assuming a flowrate of 1.15 cm^3/min, calculate the retention volume for each pesticide.

Pesticide	***t*, min**
Baygon	5.25
Carbamate	6.03
Barban	12.41

Efficiency

24-6 On a particular 25.0-cm-long column, benzene had a retention time of 33.0 min and a peak width of 15.1 min. Calculate the number of theoretical plates in the column and the height equivalent of a theoretical plate.

24-7 Calculate H and n for a 25.0-cm column for which didodecylphthalate was found to have a retention time of 9.59 min and a peak width of 1.20 min.

24-8 Calculate H and n for a 25.0-cm column if methylbenzyl alcohol has a retention time of 17.6 min and a half-peak width of 0.59 min.

24-9 The width of a chromatographic peak is 33 s and the retention time of the peak is 7.52 min. Calculate the H for the 1.50-m column.

24-10 The retention time of an organic compound on a 100-cm liquid chromatographic column was10.0 min. The width of the chromatographic peak was 24 s. Calculate the height equivalent of a theoretical plate for the column.

24-11 An asymmetrical chromatographic peak has an asymmetry factor, measured at 10 percent of the peak height, that is 1.58. The retention time of the peak is 4.57 min and the peak width at 10 percent of the peak height is 34 s. Determine the number of theoretical plates in the column.

Resolution

24-12 Calculate the resolution between two gaussian peaks which have respective retention times of 18.5 and 20.9 min and peak widths of 130 and 182 s.

24-13 Calculate the resolution between two adjacent chromatographic peaks which have retention times of 22.9 and 27.5 min and peak widths of 5.8 and 6.1 min.

24-14 Calculate the resolution between two chromatographic peaks if the retention time corresponding to the first peak is 10.52 min, the retention time corresponding to the second peak is 11.36 min, and the widths of the two peaks are 0.38 and 0.48 min respectively. Is the resolution adequate for analysis?

24-15 Calculate the resolution between two chromatographic peaks which have retention times (measured in recorder chart-paper divisions) of 16.25 and 17.00 units. The width of the first peak was 0.50 unit and the width of the second peak was 0.80 unit.

24-16 An asymmetrical chromatographic peak had a retention time of 15.27 min and a peak width after (to the right of) the retention time of 0.37 min. An adjacent peak had a retention time of 16.43 min and a peak width before (to the left of) the retention time of 0.58 min. Determine the resolution between the peaks.

Asymmetry Factor

24-17 Calculate the asymmetry factor at 10 percent of the peak height for a chromatographic peak for which the peak width at 10 percent of the peak height is 0.41 min to the left of the retention time and 0.83 min to the right of the retention time.

24-18 Calculate the asymmetry factor at 15 percent of the peak height for a chromatographic peak for which $w_{0.15L}$ is 28 s and $w_{0.15R}$ is 43 s.

REFERENCES

1. Barber, W. E., and P. W. Carr: *Anal. Chem.*, **53**: 1939(1981).
2. Foley, J. P., and J. G. Dorsey: *Anal. Chem.*, **55**: 730(1983).

CHAPTER

TWENTY-FIVE

LIQUID CHROMATOGRAPHY

Liquid chromatography (LC) is the general name for the several types of chromatography which use a liquid mobile phase. Several categories of liquid chromatography are used. *Liquid-solid chromatography* (LSC) uses a liquid mobile phase and a solid stationary phase. The rate at which a chemical species travels through an LSC column is determined by the relative amount of time it spends dissolved in the mobile phase as compared to the amount of time that the species spends adsorbed on the surface of the solid stationary phase. As the time of adsorption increases, the retention time also increases. Because LSC depends upon adsorption to the stationary phase, the technique is sometimes termed *adsorption chromatography*.

Liquid-liquid chromatography (LLC) uses immiscible liquids for the stationary and mobile phases. The stationary-phase liquid is held in place either as a coating on solid particles which are packed into the column or as a coating on the interior of the column wall. The rate at which a chemical species travels through an LLC column depends upon the relative solubility of the species in the stationary phase as compared with its solubility in the mobile phase. As the relative solubility in the stationary phase increases, the retention time of the species lengthens. Because separation in LLC depends upon the ability of the mixture components to partition, i.e., divide, themselves between two liquid phases, LLC is sometimes termed *partition chromatography*. The partition coefficient [Eq. (24-1)] is used to quantitatively describe the separation that occurs between the two liquid phases.

Ion-exchange chromatography uses a cation or anion exchanger as the stationary phase. The mobile phase is normally water, although in some circumstances mixtures of water and organic solvents are used. Sample component ions are attracted to exchange sites on the ion exchanger. Separation of the sample into ionic groups is based on the relative attraction of the ion-exchange sites for the individual ionic types. If an ion is strongly attracted to an ion-exchange site that is already occupied by a weakly attracted ion, the strongly attracted ion displaces the weakly attracted ion. The displacement of one ion on a stationary phase by a second ion is *ion exchange*. As the attraction between an ion and the stationary ion-exchange site increases, the retention time of the ion in the column increases. Ion-exchange chromatography differs from the forms of liquid chromatography described previously in that ions rather than neutral molecules are separated.

A particularly useful form of LC is actually a subdivision of each of the previously described types. By decreasing the particle size of the stationary packing material, the surface area of the stationary phase that is exposed to the mobile phase is increased. The result is a tightly packed column that is more efficient. To achieve the same mobile-phase flowrate through a tightly packed column as through a less tightly packed column, it is necessary to use a higher mobile-phase inlet pressure. The type of liquid chromatography which uses these high-efficiency columns and relatively high mobile-phase inlet pressures is *high-performance liquid chromatography* (HPLC). HPLC is the most popular form of chromatography that is used for chemical analysis.

Plane chromatography is a form of liquid chromatography in which the stationary phase is held on or in a plane rather than in a column. The plane can be a glass or plastic plate (*thin-layer chromatography*) or a piece of filter paper or cellulose (*paper chromatography*). In either case the mobile phase is a liquid. Separation is caused by adsorption, relative solubility, or both adsorption and relative solubility on or in the stationary phase. In *size-exclusion chromatography* (SEC), a gel is used as the stationary phase. Separation is based on molecular size. In *affinity chromatography* a stationary phase preferentially binds a specific sample component to it chemically. Separation in that case results from the chemical reactivity of the sample components with the stationary phase. *Electrophoresis* utilizes charged electrodes to aid the separation. An ion migrates toward the oppositely charged electrode at a rate that is dependent upon the ionic charge and the mobility of the ion in or on the stationary phase. Ions are separated based upon their charge and mobility.

LIQUID-SOLID CHROMATOGRAPHY

The descriptive material in the sections that are devoted to LSC, LLC, and ion-exchange chromatography apply both to traditional and high-performance chromatography. The chromatographic principles that are introduced are identical for traditional and high-performance liquid chromatography. Several later sections in the chapter are devoted exclusively to HPLC. The high-performance methods are generally more useful for chemical analysis than the traditional chromatographic methods.

The assembled apparatus that is used for HPLC is described in a later section. The column used for traditional LSC is usually a glass tube with a valve at the bottom. Although columns are manufactured specifically for LSC, a 50-mL buret can be used. Either a plug of glass wool or a porous glass frit is placed above the valve in the column to hold in the stationary packing material. While the valve is closed, the mobile phase is poured into the column and a slurry of the stationary phase is added. The walls of the column are tapped to ensure uniform packing and the absence of air pockets, and a thin plug of glass wool is placed on the top of the packing material to hold the stationary phase in place.

The valve is opened and the mobile phase is drained out of the column until the level of the mobile phase is slightly higher than the level of the stationary phase. The column should never be allowed to dry; i.e., the stationary phase must always be covered with the mobile phase. Otherwise, air pockets can develop within the column and impede the separation. A small volume of the sample solution is added to the packed column and drained onto the stationary phase by opening the valve. After the sample has been added to the stationary phase, additional mobile phase is added and the chromatogram allowed to develop. Chromatographic development is the separation process which takes place as the mobile phase flows over the column packing material.

Chromatographic separation is based upon the relative abilities of the sample components to be adsorbed on the stationary phase. As a molecule of a sample component enters an LSC column, it can either be adsorbed on the stationary phase or remain in the mobile phase. A strongly adsorbed sample component spends a greater proportion of its time within the column on the stationary phase than does a weakly adsorbed component. Consequently, the retention time or volume increases as the amount of adsorption on the stationary phase increases.

Adsorption on the stationary phase is usually caused by the attraction between a polar or charged group on the stationary phase and a group of opposite polarity on the analyte component. In some cases adsorption of a sample component on the stationary phase results from attraction to the stationary phase of a dipole that has been induced in the sample component by the stationary phase. Many other adsorption mechanisms are possible, but generally are less important than those two.

The amount of adsorption of a chemical species on the stationary phase usually increases as the concentration of the species in the mobile phase increases. The amount of adsorption is also temperature-dependent. A plot of the equilibrium concentration of the species adsorbed on the stationary phase (usually expressed as moles or mass of species per mass of stationary phase) as a function of the concentration of the species in the mobile phase at a fixed temperature is an *adsorptive isotherm*. Of the several possible shapes of adsorptive isotherms, the two shown in Fig. 25-1 are the most common.

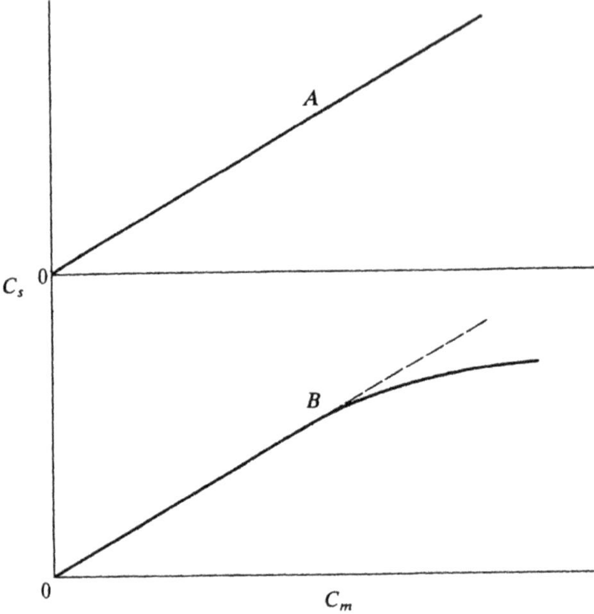

Figure 25-1 Common adsorptive isotherms: curve *A*, linear adsorptive isotherm; curve *B*, Freundlich adsorptive isotherm.

Generally the amount of adsorption on the stationary phase increases linearly with concentration in the mobile phase until the adsorptive sites on the stationary phase become nearly completely occupied. When the number of available sites becomes small compared with the number of solute molecules that are strongly attracted to those sites, the slope of the adsorptive isotherm decreases. A linear adsorptive isotherm (Fig. 25-1, curve *A*) indicates the presence of available sites on the stationary phase throughout the studied concentration range. A Freundlich adsorptive isotherm (Fig. 25-1, curve *B*) indicates near saturation of the available sites. A column on which a Freundlich isotherm is observed has chromatographic peaks for those components which exhibit tailing, i.e., which are asymmetrical. The asymmetry factors [Eq. (24-16)] for those peaks are greater than unity.

The slope of a linear adsorptive isotherm equals the distribution coefficient ($K = C_s/C_m$) for the species between the stationary and mobile phases. The distribution coefficient for a species which has a Freundlich isotherm is given by

$$K = \frac{C_s}{C_m^{1/n}} \tag{25-1}$$

where n is a constant for a particular set of experimental conditions. The value of n is greater than unity.

LSC Stationary Phases

The solid stationary phase is usually composed of fine granular particles of nearly equal size. The stationary phase must be insoluble in the mobile phase and chemically inert to the mobile phase and the sample components. It must have particle dimensions which permit separation in a reasonable time. Additionally, the stationary phase must either be capable of being regenerated after use (all adsorbed species removed) or be sufficiently inexpensive that it can be discarded after each use. HPLC stationary phases are reused many times and generally are too expensive to discard after a single use. To minimize differences in separating capability between different batches of the stationary phase, it is necessary that the stationary phase be available in a reproducible form.

Adsorption is usually caused by the attraction of a sample component to polar sites on the stationary phase. Because the amount of adsorption generally increases as the polarity of the stationary phase increases, the analyst can control the retention time of a component by careful choice of the stationary phase. Several stationary phases which have been used for LSC are listed in order of polarity in Table 25-1. By increasing the polarity of the stationary phase, the retention time of polar components is increased.

Of the several adsorbents listed in Table 25-1, alumina and silica gel are used most often for traditional LSC and for HPLC with a solid stationary phase. Alumina is available in several grades (activities) based upon the amount of adsorbed water which the alumina contains. Adsorbed water occupies adsorptive sites on the stationary phase and decreases the capacity of the alumina to adsorb. Alumina is usually categorized according to the Brockmann scale of activity shown in Table 25-2. As the Brockmann number increases, the amount of adsorbed water increases and the ability of the alumina to adsorb other chemical species decreases. Silica gel is a solid polymer of silicic acid (H_2SiO_3). It is a porous amorphous (noncrystalline) solid which contains polar SiOH adsorptive sites.

Table 25-1 Several stationary phases used for traditional LSC

The materials are listed from most polar at the top of the table to least polar at the bottom

High polarity	Alumina (Al_2O_3)†	Greatest adsorption
	Magnesium oxide	
	Charcoal	
	Silica gel	
	Calcium oxide	
	Magnesium carbonate	
	Calcium carbonate	
	Potassium carbonate	
	Sodium carbonate	
	Starch	
Low polarity	Cellulose	Least adsorption

† Several grades of alumina are used. See the text for a description.

Table 25-2 The Brockmann scale of activity for alumina

Brockmann scale number	Percent (w/w) water
I	0
II	3
III	6
IV	10
V	15

LSC Mobile Phases

The mobile phase competes with sample components for adsorptive sites on the stationary phase. Increasing the polarity of the mobile phase results in greater adsorption of mobile-phase molecules on the stationary phase and a resultant decrease in the number of adsorptive sites which are available for the sample components. Consequently, increasing the polarity of the mobile phase decreases the retention times of the sample components. Alternatively, the sample components can be thought of as partitioning themselves between the stationary and mobile phases. As the polarity of the mobile phase increases, the solubility in the mobile phase and consequently the concentrations of the components in the mobile phase increase and the retention times decrease.

Table 25-3 Selected solvents used as mobile phases in LSC and HPLC

The solvents are listed in order of polarity. Use of the low-polarity solvents at the top of the table results in relatively long analyte retention times. The mobile phases which are in bracket are all nonpolar.

Low polarity	[pentane, hexane, heptane, petroleum ether, carbon tetrachloride cyclohexane]	Long retention time
	toluene	
	benzene	
	esters	
	chloroform	
	ethyl ether	
	dichloromethane	
	chlorobenzene	
	methanol, ethanol, 2-propanol, tert-butanol	
	ethyl acetate	
	water	
	chlorobutane	
	dichloroethane	
	methyl ethyl ketone, acetone	
High polarity	nitromethane	Short retention time

In *gradient-elution analysis* two, three, or more solvents of differing polarity are combined to form the mobile phase and are used to elute the sample through the column. Initially the mobile phase is composed entirely or mostly of the less polar solvent. The initial low polarity of the mobile phase allows separation of those low-polarity sample components that have short retention times. During development of the chromatogram, the relative amount of the more polar solvent(s) in the mobile phase is increased while the amount of the less polar solvent is decreased. In that way the components which are more strongly adsorbed on the stationary phase can be eluted in a shorter time. The solvents used for gradient elution analysis must be completely soluble in each other.

The analyst can exert some control over the degree of separation during LSC by choosing the solvent or mixture of solvents to be used as the mobile phase. A list of several solvents used for mobile phases in LSC and HPLC is given in Table 25-3. When several mobile phases have similar polarities, such as occurs with the solvents in brackets at the top of Table 25-3, mobile phase choice is based upon analyte component solubilities in the solvent, the miscibility of the solvent with other mobile phases that are used during gradient elution development and expense. The use of binary and ternary solvents in the mobile phase is common. More complex mixtures are rarely required although some advantage is claimed[1] for the use of mixtures of four solvents. The use of ternary and quaternary solvent systems in HPLC by gradient-elution analysis is described in Reference 2.

LSC Detectors

The concentrations of the sample components in the eluate can be measured in any of several ways. Many analysts prefer to collect fractions (constant-volume portions) of eluate as it leaves the column and individually assay each fraction. The method is tedious but accurate. Often automated fraction collectors are used to collect the fractions. In many cases the concentration in the eluate can be continuously monitored with a detector. The electrical signal from the detector is plotted with the aid of a strip-chart recorder or other readout device. Many of the analytical instruments that were described in preceding chapters have been adapted to permit their use as LC detectors. Instrumental detectors are particularly useful for HPLC. The detectors described in this section are primarily used in HPLC. A brief description of some of the LC detectors that are in use can be found in Reference 3.

The detectors used most often for LSC (primarily for HPLC) are those which utilize a flow-through uv-visible absorbance or fluorescence cell and the evaporative light scattering detector (ELSD). The uv-visible or fluorescence cell is connected to the exit side of the LSC column and the absorbance or fluorescence is continuously monitored at a particular wavelength. A wavelength in the ultraviolet or visible region is used. Detectors which operate with either fixed- or variable-wavelength radiation are available. Diode array detectors are commercially available which can acquire an entire effluent spectrum at frequencies up to 160 Hz. Of course, absorption and fluorescence detectors are useful only for sample components which exhibit absorbance or fluorescence at the wavelength at which the detector is adjusted. Fortunately, many organic compounds absorb and/or fluoresce, at least to a small degree, at the same wavelength. Many aromatic compounds absorb between about 220 and 260 nm.

Sometimes the addition of a second reactive column to the apparatus permits chemical conversion of the nonabsorbing or nonfluorescing component to a compound that absorbs or fluoresces. The reactive column can be placed either before or after the chromatographic column. It contains an appropriate chemical reactant. A description of precolumns and postcolumns that are used for amino acid assays is available in Reference 4. Both conventional sources of electromagnetic radiation and lasers have been used as EMR sources in LC detectors.

Refractive index detectors rely upon measurement of the refractive index of the eluate as it leaves the column. Changes in the refractive index of the mobile phase occur whenever a sample component is eluted from the column. Refractive index detectors can be used to detect more compounds than detectors that rely upon absorbance and fluorescence; however, generally the sensitivity of the refractive index detectors is less. Conventional radiative sources as well as lasers have been used in refractive index detectors.

Other optical methods that have been used in LC detectors include Rayleigh scattering, Raman scattering, and flame photometry. With the scattering methods lasers are often used as sources. A description of the uses of lasers in LC detectors can be found in Reference 5. The flame photometric detector is similar to the detector that is more commonly used with gas chromatography (Chapter 26). The eluate from an LC column is sprayed into a flame. The intensity of radiation that is emitted from the flame at a fixed wavelength is monitored.

Fourier transform infrared (FTIR) spectrophotometry is another optical technique that has been used to detect sample components in the LC eluate. A disadvantage of the use of FTIR detectors is absorption in the infrared region by many mobile phases. Computerized background subtraction has helped to solve the problem.

Nuclear magnetic resonance (nmr) spectrometers have been used as detectors for LC. Generally Fourier transform instrumentation must be used to achieve the speed necessary for use with HPLC. A major disadvantage of the use of the detector is the large expense associated with FT-nmr instrumentation. A description of the use of nmr detectors for LC is contained in Reference 6.

Electrochemical detectors have found use in liquid chromatographs in part owing to their rugged construction and ease of miniaturization. The eluate from the column is directed through a flow-through electrochemical cell. Amperometric detective systems most often are used. The potential between the electrodes in the cell is adjusted to a value at which the sample components are electroactive, and the mobile phase is electroinactive. The current flowing through the cell is monitored. Other electrochemical detectors have been designed to use differential pulse voltammetry, triangular wave voltammetry, conductometry, and other methods. The combination of liquid chromatography with electrochemical detection is *liquid chromatography/ electrochemistry* (LCEC).

A single indicator electrode or multiple indicator electrodes can be used with LCEC. If two indicator electrodes are used, the electrodes are generally maintained at different potentials. If the electrodes are placed abreast of each other in the eluate, the chromatogram often consists of a plot of the ratio of the current flowing through the two electrodes as a function of time or volume.

If the two indicator electrodes are serially ordered in the eluate, the potential of the first electrode can be adjusted to a value at which the sample component electrochemically reacts to form a product that is monitored at the second electrode. The chromatogram consists of a plot of current flowing through the second electrode as a function of time or volume. The two major advantages of electrochemical detection are the selectivity that is possible by control of the electrode potential and high sensitivity. The use of LCEC is described in References 7 through 9. A commercial LCEC detector is described in Reference 10. The factors that should be considered when selecting the indicator electrode for use in LCEC are listed in Reference 11.

Mass spectrometry is another instrumental method that has been used to detect components as they exit from LC columns. LC/MS was briefly described in Chapter 21. A good deal of attention has been directed toward design of interfaces between LC columns and mass spectrometers. Although many interfaces have been designed, only three are in common use.[12] In one type, the effluent from the chromatograph is split into two portions. The fraction that enters the ionization chamber of the mass spectrometer is small in comparison to the total volume of eluate. Alternatively, the entire eluate from a microbore column can be directed into the mass spectrometer. The *direct liquid inlet system* can only be used if the eluate volume is small.

The second interface consists of a moving wire or belt upon which the eluate is deposited or sprayed. The wire or belt is used to transport the sample components into the mass spectrometer. The wire is heated and the solvent evaporated from the eluate prior to entering the ionization chamber of the mass spectrometer. Within the ionization chamber the sample components are removed from the wire and ionized at reduced pressure and elevated temperature.

The third interface is a *thermospray device*. The eluate from the column passes through a heated capillary and is directly sprayed into the ionization chamber of the mass spectrometer. Regardless of the type of interface, a quadrupolar mass analyzer is generally used because the mass spectral scans must be obtained rapidly. The major disadvantage of using mass spectrometry to detect LC sample components is the expense associated with the mass spectrometer. The major advantage is the great deal of qualitative and quantitative information that can be obtained from mass spectra of the separated sample components. LC/MS applications in the pharmaceutical industry have been reviewed in Reference 13.

A relatively recent addition to the list of HPLC detectors is the evaporative light scattering detector. The column effluent, or a portion of the effluent, is mixed with a nebulizer gas, which often is nitrogen, and is sprayed into a detection chamber as droplets similar to those formed during aspiration into the torches that are used for atomic absorption spectrophotometry and inductively coupled plasma spectrometry. In the detection chamber the mobile phase solvent evaporates leaving small, suspended, analyte particulates. Laser radiation passes through the gas-phase suspension and is scattered by the particulates to a radiation detector. The detector responds to the intensity of the scattered radiation. Often the radiation detector is a photomultiplier tube. The lasers used in the detector are relatively inexpensive, durable lasers which emit in the visible spectral region. The detector has good sensitivity and is useful when the analyte neither absorbs nor fluoresces. It can be used for nearly all molecular analytes and when analyte absorption is masked by absorption of the mobile phase.

Many types of detectors in addition to those previously described have been used with liquid chromatography. Among the parameters that have been monitored by the detectors are atomic absorption, chemiluminescence, inductively coupled plasma emission, photoacoustic absorption, radioactivity, and the response of ion-selective electrodes (particularly the response of the pH electrode). The detectors that have been described for use with LSC have also been used with other forms of LC. They are particularly useful for the several forms of HPLC.

Functional Groups Adsorbed on LSC Columns

In general the degree of attraction of molecules to the LSC stationary phase increases with increasing polarity of the molecule. A functional group that increases the polarity of a molecule also increases the amount of adsorption on the stationary phase and the retention time of the molecule on the column. Several categories of functional groups are listed in Table 25-4 in their approximate order of polarity. Because the degree of adsorption on the LSC stationary phase is related to the polarity of the sample components, Table 25-4 can be used as a guide to predict the order of elution of otherwise similar compounds which contain different functional groups.

Table 25-4 Several functional groups in their approximate order of elution from LSC columns

High polarity (long retention time)	Organic acids (RCOOH), bases (NH_3, OH^-)
	Alcohols, thiols
	Amines, nitro group
	Aldehydes, ketones
	Esters
	Halides
Low polarity (short retention time)	Unsaturated hydrocarbons
	Saturated hydrocarbons
	Perfluorocarbons (C_nF_{2n+2})

Sample problem 25-1 Predict the order of elution of the following compounds from an LSC column:

$$CH_3CH_2CH_3,\ CH_3COOCH_3,\ CH_3CH(NH_2)CH_3,\ CH_3CH_2COOH,$$
$$CH_3CH_2CH_2OH,\ CH_3CH{=}CH_2,\ CH_3CH_2CH_2Cl$$

SOLUTION From Table 25-4 it is apparent that the order of elution should be:

Shortest retention time: $CH_3CH_2CH_3$, $CH_3CH = CH_2$, $CH_3CH_2CH_2Cl$, CH_3COOCH_3, $CH_3CH(NH_2)CH_3$, $CH_3CH_2CH_2OH$, CH_3CH_2COOH: longest retention time

The position of a functional group on a molecule can affect the polarity of the molecule and the molecule's retention time on an LSC column. As the polarity increases, the retention time also increases. The difference between the retention times on an LSC column of the cis and trans isomers of azobenzene (Fig. 25-2) will serve to illustrate the point. The only difference between the two isomers is the relative positions of the benzene rings within the molecules. *cis*-Azobenzene has the electronegative nitrogens on the same side of the molecule and the less electronegative benzene rings on the opposite side.

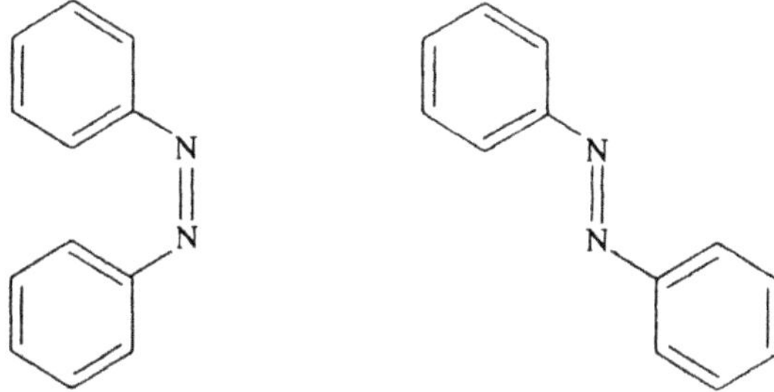

Figure 25-2 The chemical structures of *cis*- and *trans*-azobenzene.

trans-Azobenzene has the nitrogens in the center of the molecule and benzene rings on each side. Because the electronegative groups in *cis*-azobenzene are concentrated on one side of the molecule, *cis*-azobenzene is more polar and consequently is held on an LSC column longer than *trans*-azobenzene.

The mass of a sample component also can have an effect upon the extent of adsorption of the component on an LSC stationary phase. If the polarity of a series of compounds is nearly identical, i.e., if the compounds contain the same functional groups oriented in the same direction, heavier compounds are held on a column longer than lighter compounds. That is a statement of *Traub's rule*. An equivalent statement of Traub's rule is that the adsorption on an LSC column increases with increasing molecular weight for a homologous series of compounds.

LIQUID-LIQUID CHROMATOGRAPHY

The procedure followed in LLC and the detectors and the columns used in LLC are identical to those used in LSC. The difference between LSC and LLC is in the stationary phase which is used to pack the column. An LLC stationary phase is a liquid which often is coated on a solid support. The solid support is usually a polar material that strongly adsorbs the polar stationary liquid.

When the mobile phase is the less polar of the two liquids, the technique is *normal-phase chromatography*. If the mobile phase is the more polar of the two liquid phases, the technique is *reversed-phase chromatography* (RPC). In many instances the stationary phase is water-coated on silica gel or some other adsorbent. The mobile phase can be any one of the solvents listed in Table 25-3 that is immiscible in the stationary phase. If the stationary phase is water, the mobile phase is usually a less polar solvent (a solvent above water in Table 25-3). For obvious reasons, it is necessary to choose a mobile phase that is immiscible with the stationary phase.

For reversed-phase chromatography the solid support is generally a relatively nonpolar substance, such as powdered rubber, on which the nonpolar stationary phase, e.g., benzene or carbon tetrachloride, is coated. The mobile phase is usually water in reversed-phase chromatography, although it could be any solvent which is immiscible with the stationary phase and which has a higher polarity. A table listing several adsorbents, stationary phases, and mobile phases is given in Reference 14 (p. 104).

The proportion of time spent on the LLC stationary phase by a sample component depends on the relative solubility of the component in the stationary phase as compared with the solubility in the mobile phase. As the relative solubility in the stationary phase increases, the time in the stationary phase increases and the retention time in the column increases. The order of elution of components in a mixture can be predicted if the distribution coefficient for each component in the two solvents is known. As the distribution coefficient increases, the retention time for the component increases.

ION-EXCHANGE CHROMATOGRAPHY

The ion-exchange resins used as stationary phases in ion-exchange chromatography consist of synthetic cross-linked polymers containing covalently bonded, ionizable functional groups. Anionic groups are used in cation-exchange resins and cationic groups are used in anion-exchange resins. The *counterion* (oppositely charged ion) of the bonded ion is relatively free to dissolve in the aqueous mobile phase as it flows through the column; i.e., the ion-exchange resin behaves as a polyelectrolyte.

Sample ions that have been added to an ion-exchange column are attracted to the oppositely charged, stationary ionic sites on the resin. If the affinity of the sample ions for the stationary ions on the ion exchanger is greater than that of the counterions of the ion exchanger for the stationary ions, the sample ions ion-exchange by attaching to the ion exchanger and replacing the counterions. The locations on the resin at which ion exchange takes place are *ion-exchange sites.*

Separation of an ionic mixture on an ion-exchange resin results from the different affinities of the different sample ions for the stationary ion-exchange sites. As the force of attraction between the ion-exchange site and the ion increases, the relative time spent on the stationary phase and the retention time (or volume) of the ion increases. At all times during ion exchange, electroneutrality is maintained in the column; e.g., if a resin initially contains attached H^+, a + 1 ion would exchange with one H^+, a + 2 ion with two H^+, etc.

Ion-Exchange Resins

The polymeric backbone of ion-exchange resins is usually a styrene-divinylbenzene polymer, a polymethacrylate polymer, a methacrylic aciddivinylbenzene polymer, or an acrylic acid-divinylbenzene polymer. The formation of the styrene-divinylbenzene polymer is as follows:

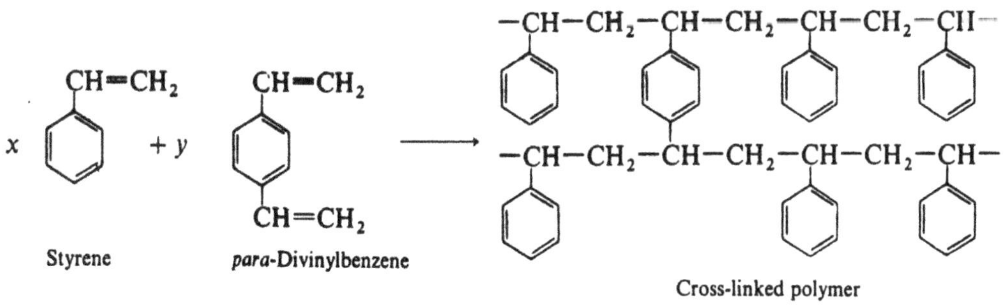

x Styrene ($CH{=}CH_2$ on benzene) $+\ y$ *para*-Divinylbenzene ($CH{=}CH_2$, $CH{=}CH_2$) ⟶ Cross-linked polymer (25-2)

$-CH-CH_2-CH-CH_2-CH-CH_2-CH-$

$-CH-CH_2-CH-CH_2-CH-CH_2-CH-$

Styrene *para*-Divinylbenzene Cross-linked polymer

The amount of cross-linking (connection between the chains) in the polymer is controlled by the relative amount of divinylbenzene used during polymeric formation [ratio of y/x in Eq. (25-2)]. Often about 1 mole of divinylbenzene is

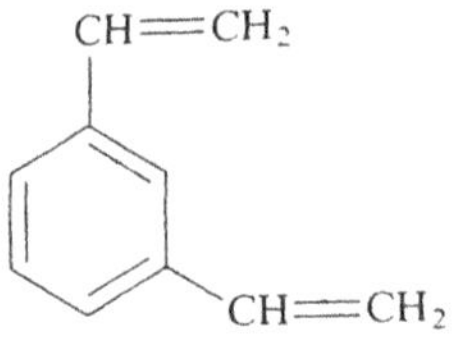

Figure 25-3 *meta*-Divinylbenzene.

used for every 11 moles of styrene. Although *para*-divinylbenzene is most often used to cross-link polymeric chains, *meta*-divinylbenzene (Fig. 25-3) can also be used. The polymer as supplied by the manufacturers is generally in the form of beads.

Cation exchangers generally are formed by addition of acid functional groups to the polymer and anion exchangers are generally formed by addition of basic functional groups. A styrene-divinylbenzene polymer, when reacted with sulfuric acid, forms a polymerized cation exchanger with substituted —SO_3H groups as shown in Fig. 25-4. Other cation- and anion-exchange resins are formed by chemical reactions to yield other substituted polymers. An ion-exchange resin which contains a single kind of functional group is a *mono-functional* resin and a resin containing more than one kind of functional group is a *polyfunctional* resin. *Mixed-bed* ion-exchange resins contain both acidic and basic functional groups and can be used for both cation and anion exchange.

Ion-exchange resins are divided into strongly basic (strong anion exchangers), moderately basic, weakly basic, strongly acidic (strong cation exchangers), and weakly acidic categories, depending upon the strength of the acidic or basic functional groups on the polymer. The greater the basicity or acidity of the functional group, the greater the attraction to oppositely charged sample ions. Strongly basic anion exchangers have a greater attraction for anions than weakly basic anion exchangers and strongly acidic cation exchangers have a greater attraction for cations than weakly acidic cation

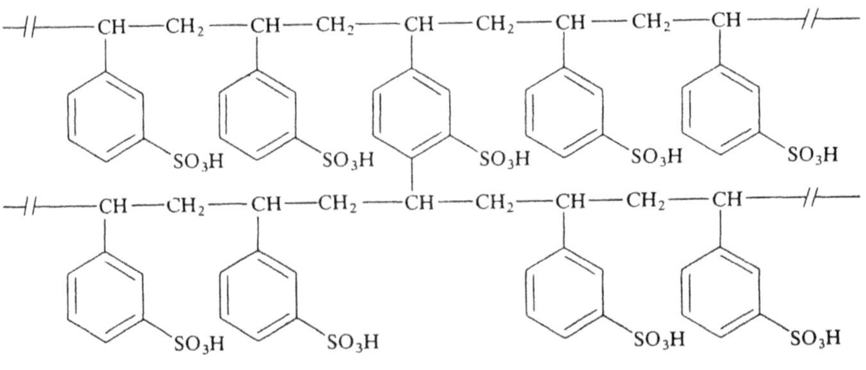

Figure 25-4 A cation-exchange resin that contains sulfonate (SO_3H) functional groups.

Table 25-5 Several gel ion-exchange resins

Classification	Functional group	Polymeric support†	Useful pH range	Trade name	Source‡
Strongly basic (strong anion exchanger) (Q type)	Tetraalkyl-ammonium hydroxide	S-DVB	0-14	ANGA-542	Baker
				REXYN 201 OH	Fisher
	—$CH_2N(CH_3)_3{}^+Cl^-$	S-DVB		Amberlite IRA 400C	R and H
				Dowex-1	Dow
	—$CH_2N(CH_3)_3{}^+OH^-$	S-DVB		Amberlite IRA 400 OH	R and H
	Tetraalkyl-ammonium chloride	S-DVB		IONAC A-540	MCB
				REXYN 201 C	Fisher
Moderately basic	—$N(CH_3)_2$	S-DVB		CGA-301	Baker
	DEAE	S-DVB	0-14	301VHP	Vydac
Weakly basic	—NH_2	S-DVB	0-9	Amberlite IRA 93	R and H
			0-7	Amberlite IRA 45	R and H
			0-12	ANGA-316	Baker
	—NH_2	A-DVB	0-7	Amberlite IRA 68	R and H
Strongly acidic (strong cation exchanger) (S type)	—$SO_3{}^-H^+$	S-DVB	0-14	Dowex-50	Dow
				IONAC C-242	MCB
				400 VHP	Vydac
	—$SO_3{}^-Na^+$	S-DVB	0-14	Amberlite IR 120P	R and H
				Amberlite 200	R and H
				Amberlite 252	R and H
				CGC-241	Baker
Weakly acidic	—COO^-H^+	S-DVB	0-12	CGC-270	Baker
				Amberlite CG 50	R and H
	—COO^-H^+	MA-DVB	5-14	Amberlite IRC 50	R and H

† S-DVB is styrene-divinylbenzene; MA-DVB is methacrylate-divinylbenzene; A-DVB is acrylate-divinylbenzene.

‡ Baker is J. T. Baker Chemical Co.; Fisher is Fisher Scientific Co.; MCB is Matheson, Coleman, and Bell Manufacturing Chemists; Dow is Dow Chemical Co.; R and H is Rohm and Haas Co.

exchangers. Several cation- and anion-exchange resins are listed in Table 25-5. Most of the ion-exchange resins listed in Table 25-5 form gels in water. Some ion-exchange resins are prepared by chemically bonding the functional group to porous macroreticular (i.e., netlike) or macroporous particles. Those substances do not swell in solution to yield gels.

As is apparent from Table 25-5, strong anion exchangers usually contain a tetraalkylammonium group bonded to the polymeric support; moderately strong anion exchangers contain bonded dialkylammonium groups; and weak anion exchangers contain bonded amines. Strong anion exchangers which contain tetraalkylammonium groups are termed "Q" type anion exchangers. Many moderately strong anion exchangers contain diethylaminoethane (DEAE) [$-CH_2CH_2-NH^+(CH_2CH_3)_2$]. Strong cation exchangers sometimes are termed "S" type cation exchangers and usually contain $-SO_3^-$ groups. Although weak cation exchangers most often contain bonded $-COO^-$ groups, some weak cation exchangers contain –OH, –SH, or $-PO_3H_2$ groups. The ion-exchange resin is supplied with a single kind of counterion. Generally, cation exchangers are supplied with Na^+ or H^+ counterions. An ion-exchange resin containing Na^+ is said to be in the *sodium form*. Likewise, an H^+ –containing cation-exchange resin is in the *hydrogen form*. Anion-exchange resins are most often in the chloride or hydroxide form except for those containing amines ($-NH_2$), which are supplied as the free base (not ionic).

Attraction to Ion-Exchange Resins

To a first approximation Coulomb's law can be used to predict the force F of attraction between an ion-exchange site of charge q(resin) and an oppositely charged ion with a charge q(ion) which are separated by distance r:

$$\mathrm{F} \propto \frac{q(\text{ion}) \times q(\text{resin})}{r^2} = \frac{q(\text{ion}) \times q(\text{resin})}{r(\text{hydrated ion})^2} \tag{25-3}$$

If the charge of the ion is assumed to be located in the center of the hydrated ion in solution, then the distance of the closest approach between the ion-exchange site and the ion is approximately equal to the radius of the *hydrated* ion r(hydrated ion). It is apparent from Eq. (25-3) that the force of attraction between a specific ion-exchange resin and an ion increases as the charge of the ion increases, but decreases with increasing radius of the hydrated ion. Consequently, an ion-exchange resin has a greater attraction for ions with greater charge and smaller diameter. As the charge increases and/or as the radius of the hydrated ion decreases, the retention time of the ion on an ion-exchange column increases.

The effect of charge on retention on an ion-exchange column is demonstrated by the relative retentions of Na^+, Ca^{2+}, La^{3+}, and Th^{4+} on cation-exchange columns. All four ions have nearly the same radius, but the order of attraction to a resin is $Th^{4+} > La^{3+} > Ca^{2+} > Na^+$. That is, the ion with greatest charge is most strongly attracted, and the attraction decreases with decreasing charge. The approximate order of attraction of ions of the same charge to cation-exchange columns is given in Table 25-6. The approximate order of attraction of monovalent anions to an anion-exchange column is given in Table 25-7. In some specific cases, the orders of attraction can be other than those listed.

Table 25-6 The approximate order of attraction of several cations to cation-exchange columns

	Monovalent	Divalent
Least strongly attracted (short retention time)	Li^+	Be^{2+}
	H^+	Mn^{2+}
	Na^+	Mg^{2+}
	NH_4^+	Zn^{2+}
	K^+	Co^{2+}
	Rb^+	Cu^{2+}
	Cs^+	Cd^{2+}
	Ag^+	Ni^{2+}
Most strongly attracted (long retention time)	Tl^+	Ca^{2+}
		Sr^{2+}
		Pb^{2+}
		Ba^{2+}

Table 25-7 The approximate order of attraction of several monovalent anions to anion-exchange columns

Least strongly attracted (short retention time)	OH^-, F^-
	CH_3COO^-
	$HCOO^-$
	$H_2PO_4^-$
	HCO_3^-
	Cl^-
	NO_2^-
	HSO_3^-
	CN^-
	Br^-
	NO_3^-
Most strongly attracted	HSO_3^-
(long retention time)	I^-

From the listings given in Tables 25-6 and 25-7, it can be seen that generally the degree of attraction of an ion to an ion-exchange resin increases in going down in a group in the periodic table. For the group 1 elements the order of attraction is $Li^+ < Na^+ < K^+ < Rb^+ < Cs^+$; for the group 2 elements the order of attraction is $Be^{2+} < Mg^{2+} < Ca^{2+} < Sr^{2+} < Ba^{2+}$; and for the group 17 elements the order of attraction is $F^- < Cl^- < Br^- < I^-$. No similar trend is generally observable in proceeding across a period in the periodic table. In each case the degree of attraction to the ion-exchange resin apparently decreases with increasing radius of the hydrated ion.

Effect of pH on Attraction to Ion-Exchange Columns

If one or more of the analyte ions that are passing through an ion-exchange column is the conjugate acid or base of a weak base or weak acid, the pH of the solution can have an important effect on the degree of dissociation of the weak acid or base. Because ions alone can be retained on ion-exchange columns and because the pH of the solution can affect the relative number of ions in the solution, the pH can have a large effect on the retention of the species on the column.

A weak acid reaction is shown as follows:

$$HA = H^{+} + A^{-} \tag{25-4}$$

As the pH of the mobile phase passing through an anion-exchange column decreases, the concentration of H^{+} in the solution increases and the equilibrium of Eq. (25-4) is shifted to the left. That causes a decrease in the number of A^{-} ions in the column and a decrease in the amount of A^{-} that is retained on the resin. Similarly, as the pH is decreased in a cation-exchange column that is used to separate the conjugate acid of a weak base from other components, the relative number of cations increases and consequently the degree of retention on the column increases. It is clear that whenever a weak acid or weak base is involved, control of the pH of the mobile phase is desirable. Control of the pH can sometimes be used to advantage in separating ions which have significantly different acidic or basic equilibrium constants. In many cases the ion cannot be easily separated by other methods. Advantage is taken of the effect of pH in order to separate biochemical compounds, such as weak biologic acids on anion exchange columns and proteins on cation exchange columns. Essentially, weak acids are separated based upon their pK_a values while the bases, such as proteins, are separated based upon the pK_a of the protonated compounds.

Ion-Exchange Apparatus

Ion-exchange columns are similar to the columns used for LSC and LLC. If the ion-exchange resin is a gel, a slurry of the resin is packed into the column in much the same way that LLC and LSC columns are packed. The level of the mobile phase, which is added to the top of the column, should never be allowed to fall below the top of the resin. The mobile phase is usually an aqueous solution of an ion that has a relatively weak affinity for the ion-exchange sites on the stationary phase. Dilute solutions of a strong acid, for example, HCl, are often used in cation-exchange columns, and dilute hydroxide solutions are often used in anion-exchange columns. A solution containing ions is required as the mobile phase because electroneutrality must be maintained within the column at all times. As a sample ion is eluted from an ion-exchange site, it must be replaced by another ion.

Fewer detectors can be used successfully with ion-exchange chromatography than can be used with LLC and LSC. Many inorganic ions do not absorb or fluoresce uv-visible radiation, and not all ions are electroactive. Consequently, fractions of eluate from an ion-exchange column are usually collected in containers (e.g., test tubes) either at fixed time intervals or after fixed volumes have flowed from the column, and each fraction is assayed for each of the sample components. That tedious procedure is often made more bearable with the aid of an automatic fraction collector. In some cases continuous monitoring of the effluent from an ion-exchange column is possible with one of the detectors used for LLC or LSC.

Flow-through electrical conductivity detectors are especially well suited for use with ion-exchange chromatography. Because all ions contribute to the electrical conductivity of a solution, electrical conductivity detectors should respond to all ions. Unfortunately, it is not always possible to use conductivity detectors with ion-exchange chromatography because the relatively high concentration of mobile-phase ions in the eluate masks the detector's response to the considerably more dilute sample ions. The techniques used with *ion chromatography* have overcome the problem.

Ion Chromatography

In a variation of ion-exchange chromatography, known as *double-column ion chromatography*, a conductivity detector is used after removal of the mobile-phase ions from the eluate. After separation of the sample ions in an ion-exchange column, the mobile-phase ions are removed in a second ion-exchange column. The eluate then contains only sample ions, which are detected with a conductivity detector. Ion chromatography with chemical suppression (use of the second column) was first proposed as an analytical technique by Small, Stevens, and Bauman.[15]

A block diagram of the apparatus used for ion chromatography is shown in Fig. 25-5. The mobile phase is pumped through an injector, where the sample is added, and into an ion-exchange column known as the *separating column*. The separating column is used to separate the sample component ions. Upon exiting from the separating column, the eluate flows into a second ion-exchange column called a *stripping* (or *suppressor*) *column*. The stripping column is used to remove selectively only the mobile-phase ions from the eluate of the separating column. The remaining sample ions are detected in the conductivity detector.

If the electrolyte in the mobile phase is hydrochloric acid (often used during the separation of cations), an anion-exchange resin in the hydroxide form is used in the stripping column. The chloride from the dilute hydrochloric acid in the mobile phase exchanges with hydroxide on the column and then reacts with the hydronium ions (from the hydrochloric acid) to form water. The net result is the removal of both mobile-phase ions from the solution. Because the stripping column is an anion-exchange column, it does not affect the separated cations in the sample.

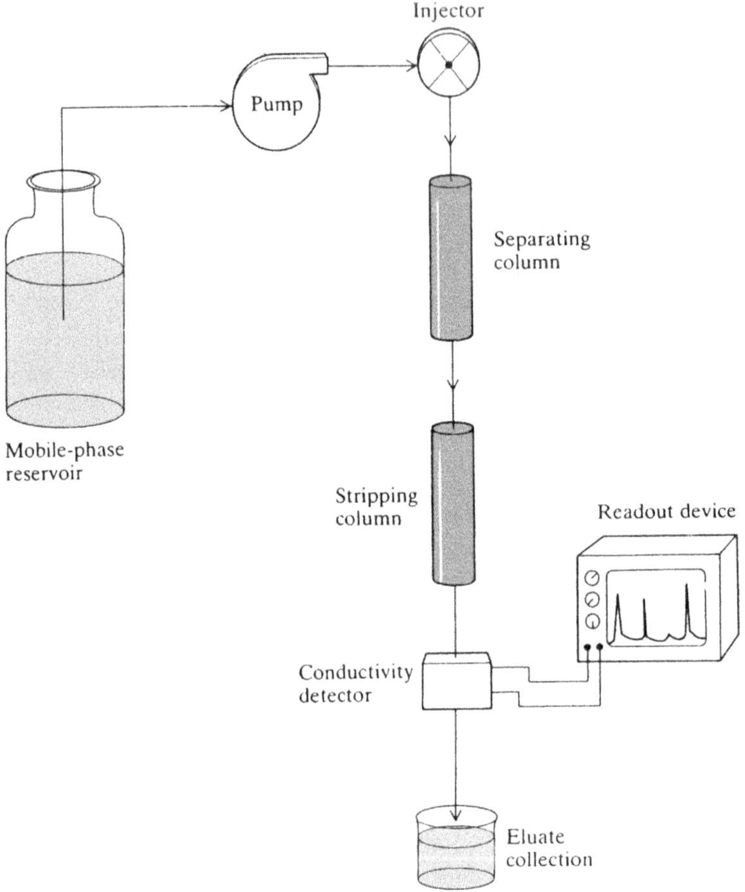

Figure 25-5 Block diagram of the apparatus that is used for one type of ion chromatography.

If anions are to be separated and the mobile phase contains dilute sodium hydroxide, the anion-exchange separating column is followed by a cation-exchange stripping column in the hydrogen form. On the resin of the stripping column, the sodium ions exchange with H^+ which subsequently reacts with hydroxide to form water and effectively remove the mobile-phase ions (Na^+ and OH^-) from the solution. Other chemical reactions also can be utilized in the stripping column to remove mobile-phase ions. As an example, if the mobile phase contains silver nitrate (sometimes used as an electrolyte in the mobile phase for cationic separations), the stripping column can contain an anion-exchange resin in the chloride form. The nitrate exchanges with chloride, which reacts with silver ion in the solution to yield a silver chloride precipitate. The reaction removes the mobile-phase ions from the solution. Unlike separating columns, stripping columns eventually become exhausted and must be replaced or regenerated. After the eluate exits the stripping column, it flows through an electrical conductivity detector which monitors the conductance of the eluate. The conductivity is continuously recorded on a strip-chart recorder or other readout device.

Stripping columns have been designed which continuously refurbish the reactant ion and which consequently eliminate the need to remove the column to regenerate the reactant. Typically the stripping column contains two concentric compartments. The effluent from the separating column flows downward through the inner compartment where the mobile-phase ions are exchanged as previously described. Simultaneously a solution of the ions that were consumed from the resin of the stripping column flows upward through the outer compartment. As ions are used in the inner compartment, they are replaced with ions from the outer compartment that diffuse across the membrane between the compartments. The device is a *hollow-fiber suppressor.* More detailed descriptions of ion chromatography are available in References 16 and 17. A description of a commercially available instrument that uses a hollow-fiber suppressor is given in Reference 18. The use of conventional HPLC instrumentation for ion chromatography is described in Reference 19.

An alternative to the use of ion chromatography with chemical suppression of the background signal is *single-column ion chromatography* which relies upon electronic suppression of the background signal from the conductivity detector. A diagram of the apparatus and a photograph of a commercially available instrument are shown in Fig. 25-6. The apparatus uses a pump to force the mobile phase through the separating column as in double-column ion chromatography; however, no stripping column is used. Effluent from the separating column flows directly into a conductivity detector.

The problem of a large background signal owing to the high concentration of electrolyte in the mobile phase is overcome by use of a relatively low-concentration electrolyte solution with a low-capacity column, and a detector with the capability to suppress (zero-out) large signals. Typically the detector is capable of measuring a conductivity change of 1 part in 30,000. Use of a single column eliminates the relatively large dead volumes often associated with use of a stripping column. It also eliminates the additional expense associated with the stripping column and its regeneration.

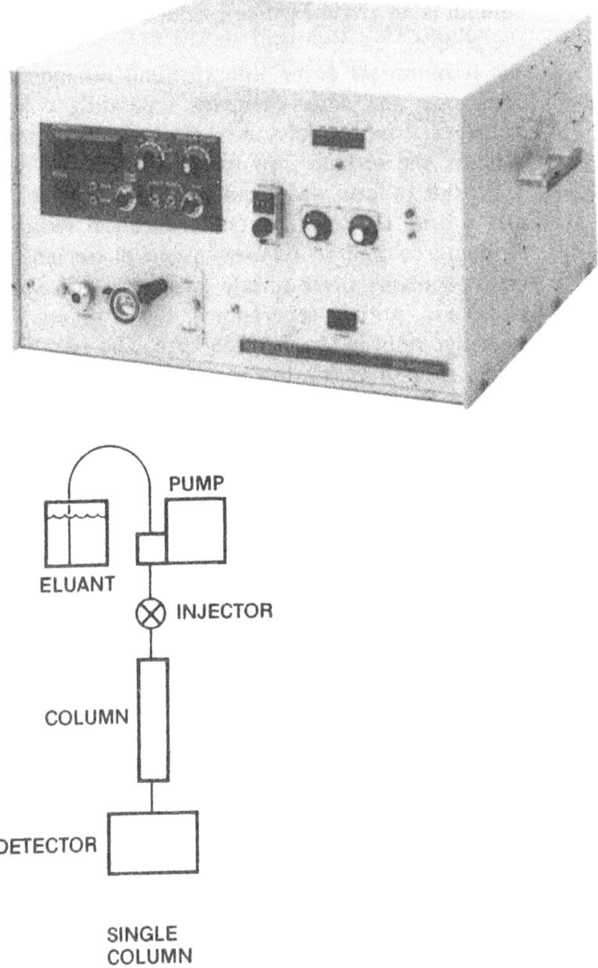

Figure 25-6 Diagram of the components in a single-column ion chromatograph and a photograph of a commercial instrument. *(Courtesy of Wescan Instruments, Inc., Santa Clara, California.)*

Applications of Ion-Exchange Columns

Several important uses of ion-exchange columns exist in addition to the chromatographic separation of ionic mixture components. Water softeners use ion-exchange columns to replace Ca^{2+} and Mg^{2+} with Na^{+}. Sodium ions do not have the undesirable properties of the calcium and magnesium ions that cause water hardness. Ion-exchange columns are sometimes used to remove interfering ions prior to performing an assay.

Ion-exchange columns have been used as a tool to concentrate trace levels of ions. A relatively large volume of a dilute solution of the ion is passed through an ion-exchange column, where the trace ion adheres to the ion-exchange sites. The ion is subsequently displaced from the column with a relatively small volume of a more concentrated solution of an ion which is more strongly attracted to the column (displacement analysis). The technique can be used to concentrate an ion prior to its assay.

Trace metallic ions can sometimes be concentrated with the aid of an anion-exchange column after addition of a negatively charged complexing agent which forms an anionic complex with the metal. The complex is collected on an anion-exchange column and displaced into a much smaller volume. Anionic complexes of cations are sometimes separated chromatographically using anion-exchange columns and elution analysis.

Chelating ion-exchange resins have also been used to concentrate cations. A chelating ion-exchange resin has a chelating agent, such as iminodiacetate, bonded to the polymeric support. The ion is complexed on the resin and then displaced into a relatively small volume with a more strongly complexing cation. Ion-exchange resins have several less popular uses such as isotopic separations, use in equilibrium and kinetic studies, and reagent preparation. Details of those and other uses can be found in the literature cited in the biennial reviews (even numbered years) found in *Analytical Chemistry*.

HIGH-PERFORMANCE LIQUID CHROMATOGRAPHY

High-performance liquid chromatography (HPLC) uses tightly packed columns containing small particles of the stationary phase. Because a greater surface area of the stationary phase is exposed to sample components in an HPLC column than in traditional LSC, LLC, and ion-exchange columns, the efficiency of the HPLC column is relatively high. HPLC stationary phases typically have packed particle diameters between 2 and 40 micrometers. If the diameter of the particle is less than 2 micrometers, the separative technique is ultra high performance liquid chromatography (UHPLC). HPLC is theoretically identical with LSC, LLC and ion-exchange chromatography. All of the descriptive material presented in earlier sections of the chapter apply to HPLC as well as to traditional LC.

HPLC apparatus is more complicated than that used for the other chromatographic techniques. To obtain useful flowrates through a tightly packed HPLC column, a greater pressure differential between the inlet and outlet of the column is required than in the columns used for the other chromatographic methods. That differential is obtained by increasing the pressure on the inlet side of the column with a pump. The outlet of the column is usually maintained at or near atmospheric pressure. A block diagram of the components of a high-performance liquid chromatograph is shown in Fig. 25-7.

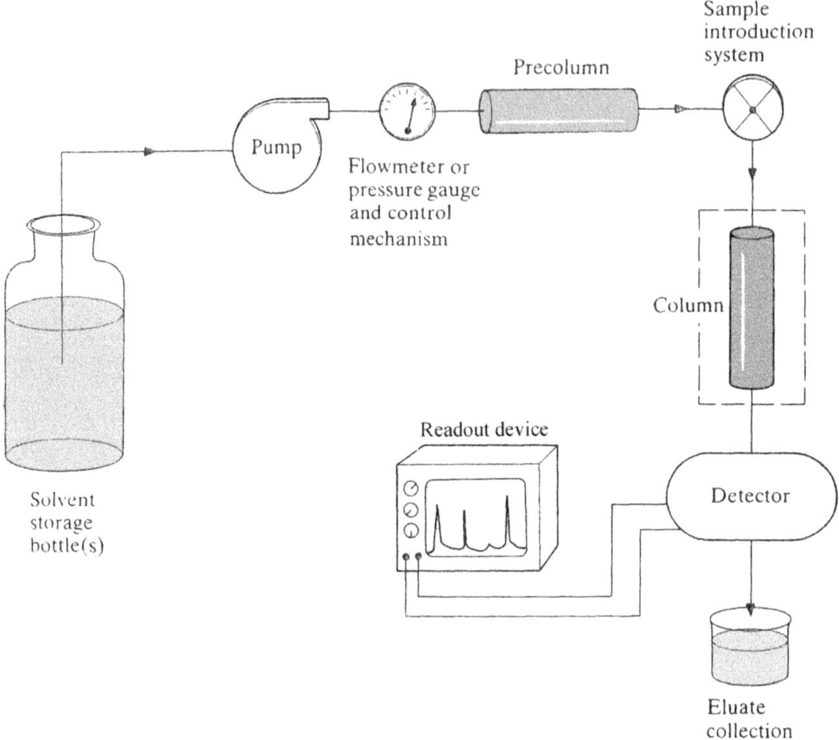

Figure 25-7 Diagram of a high-performance liquid chromatogram. The dashed line indicates an oven that is used in some instruments to control the temperature of the column.

The solvent or mixture of solvents used as the mobile phase must be deaerated prior to use in order to prevent the detector from responding to air bubbles as they exit the column. Deaeration is accomplished either by boiling the solvent(s) prior to use or by applying a partial vacuum to the solvent container. The pumps used in high-performance liquid chromatographs are designed to be nearly pulseless. Consequently, little variation occurs in the pressure or flowrate of the mobile phase on the outlet side of the pump. In most cases HPLC pumps that are used for analysis are adjustable from flowrates of about 0.01 to about 10 cm^3/min. Higher flowrate pumps for separations of large samples on large columns are available. Pumps capable of exerting pressures as high as 40 MPa for HPLC and 100 MPa for UHPLC have been used. Some HPLC pumps are designed to take solvent from a single reservoir, and others can simultaneously pump from more than one reservoir containing different solvents. Because the pumps can continuously vary the ratio of the solvents used as the mobile phase, they can be used for gradient-elution analysis. Generally, gradient-elution analysis pumps are programmable; i.e., the rate at which the composition of the mobile phase changes during development of the chromatogram can be determined by the analyst.

After exiting the pump, the mobile phase is sometimes passed through a precolumn that usually contains the same stationary phase as that in the separating column. The precolumn is used to collect any impurities in the mobile phase that might become irreversibly attached to the column and to saturate the mobile phase with the stationary phase. Although precolumns are not used in all chromatographs and are not essential to the separation, they do provide protection for the separating column and can considerably extend the column's useful life. Because columns are often relatively expensive, it is generally worth while to use a precolumn.

After exiting the precolumn, the mobile phase passes through an injector that is used to introduce the sample mixture to the column. The injector is positioned just before the inlet to the column. In most cases the sample is a solution containing several solutes which are to be separated. The injector usually is a valve which in one position directs the flow of mobile phase directly to the column and in another position passes the mobile phase through a sample loop and then into the column. While the valve is so adjusted that the mobile phase passes directly into the column, the sample is placed in the sample loop with a syringe. Sample loops are carefully calibrated to contain fixed volumes (often between 10 μL and 2 mL). The loop is usually filled to overflowing with sample. If a change in the sample volume is desired, the sample loop can be changed to one containing the proper volume. With some injectors, it is possible to use a calibrated syringe to add any desired sample volume which is less than that of the sample loop. After the sample loop has been loaded, the valve is switched to the inject position and the mobile phase flushes the sample from the sample loop into the column. A diagram of a sample-introduction system is shown in Fig. 25-8.

The sample is swept along with the mobile phase into the column, where separation of the sample components occurs. Because the retention volumes of the sample components are dependent upon the column temperature, the column is sometimes placed in a thermostated oven. For most work the oven is not required.

As the eluate leaves the column, it enters a flow-through detector, where it is continuously monitored. The detectors used for HPLC were described earlier in the appropriate sections devoted to LSC, LLC, or ion-exchange chromatography. Ultraviolet absorption, refractive index, fluorescence, electrochemical and evaporative light scattering detectors are especially popular. The detector is designed to hold a relatively small volume (usually 5 to 10 μL) and is placed as near to the exit side of the column as possible. The electrical signal from the detector is amplified, if necessary, and routed to a recorder or other readout device, which records the developed chromatogram. Effluent from the detector is collected in a suitable container for later use or is discarded.

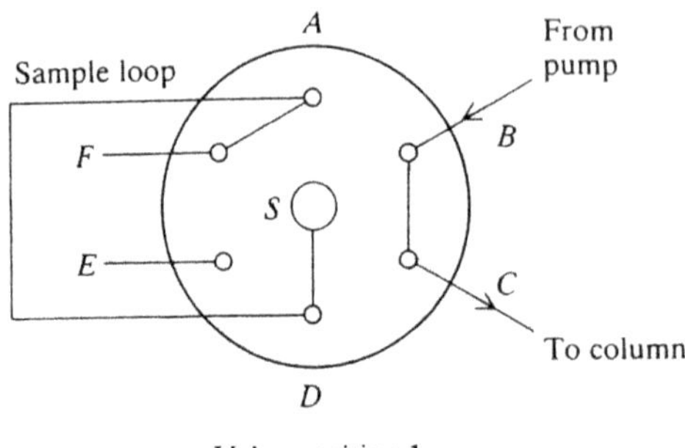

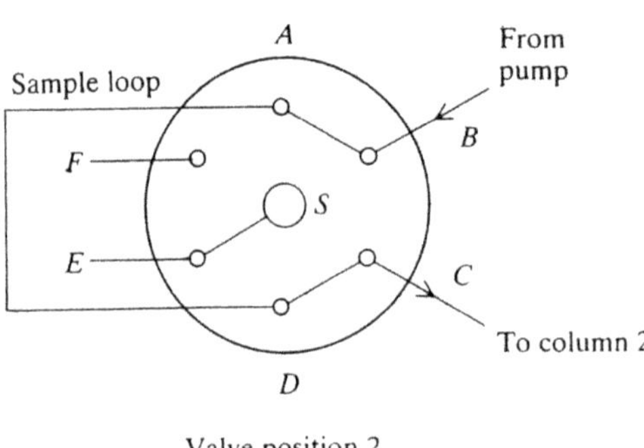

Figure 25-8 Diagram of a sample-introduction system for HPLC. The arrows indicate the direction of mobile-phase flow through the valve. Valve position 1 is used when the sample loop is being filled. Valve position 2 is used to introduce the sample that is in the loop into the mobile-phase stream. Valve position 2 is established by rotating the valve 60° clockwise from valve position 1. The fixed tubes at E and F go to a waste-disposal bottle. The entrance hole at S is used to introduce the sample to the same loop.

HPLC Stationary Phases

HPLC can be performed by using the stationary phases that were described earlier for LSC, LLC, or ion-exchange chromatography. Generally, the stationary phase is packed into a stainless steel column of 10-, 15-, 25-, 30-, 50-, or 100-cm length with an internal diameter which is usually between 1 and 6 mm for analytical columns. *Microbore* columns have an inner diameter of 1 mm or less and typically have a length of 25 cm. Often they are glass-lined. They have the advantages of requiring smaller sample volumes and using less solvent, but are generally less efficient than regular columns and sometimes require instrumental modifications. Larger columns are used for preparative separations of larger quantities of sample components. Both normal and reversed-phase HPLC are used. Among the factors that must be considered when a stationary phase for HPLC is chosen are the nature of the substances to be separated, the required efficiency of the column, and the *loading capacity* of the column. The loading capacity is the maximum sample size which can be passed through the column without altering the sample's distribution coefficient.

Solid adsorbents, such as those used for LSC, are often preferred for the separation of relatively low molecular weight (less than 2000) low-polarity compounds. Of the solid adsorbents used as HPLC stationary phases, silica gel is the most popular. The SiOH groups on silica gel make the gel weakly acidic and account for the attraction of basic compounds to it. Generally silica gel attracts bases in order of their strength; i.e., strong bases are retained on silica gel columns longer than are weak bases (refer to the previous description of the effect of pH).

Alumina is another solid adsorbent which is widely used as a column packing material. Because alumina is basic, it retains acidic compounds. A polyamide stationary phase (nylon-6) is often chosen as the solid adsorbent for the separation of compounds which can form hydrogen bonds to the amides on the stationary phase. Other high-performance LSC stationary phases that can be used generally are polar bonded-phase materials such as diol-silica, cyano-silica, and amino-silica. In those cases a functional group is chemically bonded to a polar support (silica in the indicated cases). Silica gel and alumina are approximately equal in their ability to adsorb polar components. The bonded-phase materials adsorb components to a lesser extent than either silica or alumina. Generally diol-silica is a better adsorbent than amino-silica, which is a better adsorbent than cyano-silica. Adsorption to LSC stationary phases is described in Reference 20.

Silica gel is used as a packing material in the form of pure particles and as a *pellicle* on a solid support. A pellicle is a thin layer or coating on a surface. The pellicles used in HPLC are chemically bonded to the surface of the support material (usually glass beads). Generally, the pellicles used for HPLC have a thickness of about 1 μm on glass beads which have a diameter of about 40 μm. HPLC pellicular materials are larger and more easily packed into the column than are the microparticulate pure adsorbents, but the resulting columns are only 2 to 10 percent as efficient as the columns packed with pure adsorbent. The decreased efficiency of pellicular columns is caused by the decreased surface area of the stationary phases in those columns.

Even with the lower efficiency, most separations that are normally achieved with HPLC can be made by using pellicular columns. Alumina is most often used in the pellicular form, and polyamide can only be used in the pellicular form because it is compressible. Some chromatographers have used various forms of carbon as stationary phases. The use of carbon as a stationary phase for HPLC is presently not widespread. Its use is described in Reference 21. A list of several adsorbents that are used as stationary phases in HPLC is given in Table 25-8.

Table 25-8 Several commercially available adsorbents which are used as stationary phases in HPLC

Adsorbent	**Form †**	**Particle diameter, μm ‡**	**Trade name**
Silica gel	M	4	MicroSil Silica Gel§
	M	5	Partisil 5¶
	M	10	Partisil 10¶
	M	20	Partisil 20¶
	P	40	HS Pellosil ¶
	P	40	HC Pellosil ¶
	M	5, 10, 30	LiChrosorb S1-60 ††
	M	5, 10, 30	LiChrosorb Sl-l00††
	M	5, 10, 15, 20	Polygosil-60‡‡
	M	40	Chromosorb LC-1§§
	M	5, 10, 20	LiChrosphereS1-l00 ††
	M	10	Vydac 101IR¶¶
	M	10	Vydac 101TP¶¶
	M	5, 10	Spherisorb S-W†††
	M	10	μPorasil ‡‡‡
Alumina	M	5, 10, 30	LiChrosorb Alox-T††
	M	5, 10, 20	LiChrosorb Alox 60-D††
	M	40	Chromosorb LC-3§§
	M	5, 10, 20	Spherisorb AY†††
	P	40	HS Pellumina¶
	P	40	HC Pellumina¶
Polyamide	P	40	Pellamidon¶

† M represents porous microparticulate; P represents pellicular.
‡ The diameters are average particle diameters.
§ Manufactured by the Micromeritics Instrument Corp.
¶ Manufactured by Whatman, Inc.
†† Manufactured by E. Merck.
‡‡ Manufactured by Macherey-Nagel.
§§ Manufactured by Johns-Manville.
¶¶ Manufactured by the Separations Group.
††† Manufactured by Phase Separations Limited.
‡‡‡ Manufactured by Water Associates.

If the molecular weights of the sample components are large (drugs, barbiturates, surfactants, etc.), a liquid stationary phase is often used for HPLC. Reversed-phase stationary phases appear to be the most generally useful of the stationary liquid phases. Reversed-phase HPLC can be used for well over half of the separations that are normally made with HPLC, including many of the separations that can be made on silica gel and on ion-exchange resins. Ion-exchange-resin stationary phases are useful whenever ionic species in solution are to be separated.

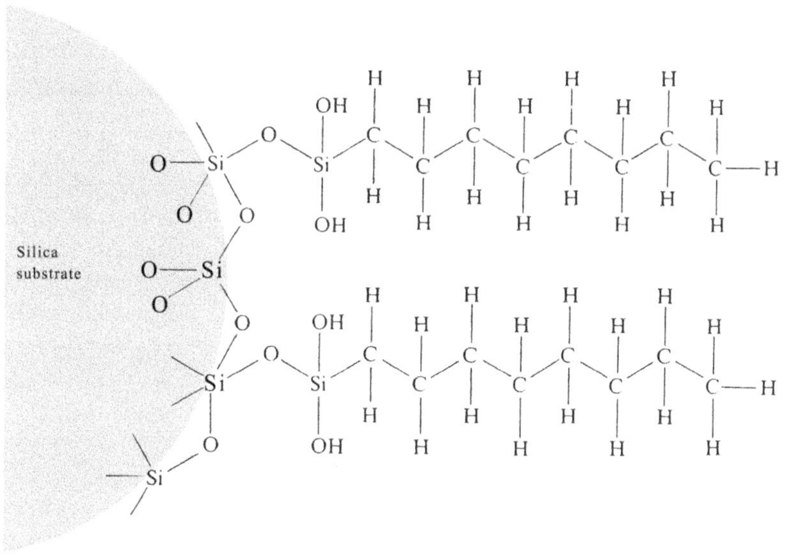

Figure 25-9 Diagram of C_8H_{17} groups bonded to silica. The material is used as a reversed-phase HPLC stationary phase.

Both stationary liquid phases and stationary ion exchangers are normally chemically bonded to an inert support to form either pellicular or coated microparticulate packing material for HPLC. In some cases the bonded polyamide listed in Table 25-8 can be considered to be a polar, stationary liquid phase. Other, stationary polar liquids include bonded amino compounds, bonded peptides, bonded nitriles, and bonded cyano-amino compounds. In most cases the compounds are chemically bonded to porous silica or silica gel. Reversed-phase HPLC stationary phases are usually either bonded octyl groups (C_8H_{17}) or bonded octadecyl groups ($C_{18}H_{37}$). In either case the stationary nonpolar liquid is attached to silica gel (the support) through stable Si—O—SiR bonds (R is C_8H_{17} or $C_{18}H_{37}$). A diagram illustrating the bonding of C_8H_{17} to the silica support is shown in Fig. 25-9.

Both normal and reversed-phase liquids extend outward from the silica or silica gel to which they are attached. If the surface of the silica gel is not completely covered by those groups, separation can occur by partition between the two liquid phases (stationary and mobile) and by adsorption on the uncovered silica gel. The amount of surface coverage of the support material is characterized by the amount of *carbon loading*. The carbon load is the percentage (w/w) of the bonded compound on the stationary phase. As the carbon load increases, the amount of exposed surface area of the supporting material decreases. Often the carbon load is in the range of 3 to 18 percent. A list of several stationary liquid phases which are used for HPLC is given in Table 25-9. A description of the manner in which reversed-phase columns can be selected for a particular analysis is given in References 22 and 23.

Table 25-9 Several commercially available liquids that are used as stationary phases in HPLC

Liquid group	Type	Particle diameter, μm	Trade name
Amine	Normal	10	MicroSil NH_2 †
		10	LiChrosorb NH_2 ‡
		5, 10	Polygosil 60-D-NH_2 §
		5, 10	Nucleosil NH_2§
		0	μ Bondapak NH_2 ¶
		10	Spherisorb S5NH††
Nitrile	Normal	10	MicroSil CN †
		5, 10	Polygosil 60-D-CN§
		10	Vydac 501TP ‡‡
		5	Spherisorb S5CN††
		5, 10	Nucleosil CN§
		10	Chromosorb LC-8§§
		10	μ Bondapak CN¶
Amine and nitrile	Normal	10	Partisil-l0 PAC¶¶
		40	Co :Pell PAC¶, †††
Octyl (C_8H_{17})	Reversed phase	3	Zorbax C8‡‡‡
		4	MicroSil C8†
		5, 10	LiChrosorb RP-8‡
		5, 10	Polygosil-60-D-C8§
		5, 10	Nucleosil C8§
		10	Vydac 2081 R ‡‡
Octadecyl ($C_{18}H_{37}$)	Reversed phase	4	MicroSil C18†
		5, 10	Polygosil-60-D-C18§
		5, 10	LiChrosorb RP-l8‡
		5, 10	Nucleosil C18§
		10	μ Bondapak C18¶
		10	Vydac 201 IR‡‡
		5	Partisil-5 ODS¶¶
		10	Partisil-l0 ODS¶¶
		40	Co:Pell ODS¶¶, †††

† Manufactured by Micromeritics Instrument Corp.
‡ Manufactured by E. Merck.
§ Manufactured by Macherey-Nagel.
¶ Manufactured by Waters Associates.
†† Manufactured by Phase Separations Ltd.
‡‡ Manufactured by the Separations Group.
§§ Manufactured by Johns-Manville.
¶¶ Manufactured by Whatman, Inc.
††† A pellicular on glass beads.
‡‡‡ Manufactured by Du Pont Company.

Many biologically active materials have chiral centers which result in the formation of stereoisomers. Enantiomers are mirror-image stereoisomers. The separation of enantiomers is difficult owing to their similar physical and chemical properties. The separation sometimes can be accomplished with chiral chromatography. The stationary phases which are used for chiral chromatography contain chiral agents which have been attached to solid supports such as silica gel. The bonded chiral agents preferentially discriminate between the enantiomers thereby resulting in their separation. A review of the use and theory of chiral separations is available in Reference 24.

Ion-exchange resins that are chemically bonded to a solid support have been used as stationary phases in HPLC. Generally, the ion exchanger is chemically bonded to silica gel or silica microparticulates. The ion-exchange sites used in HPLC are nearly always strong cation or anion exchangers. A list of several ion-exchange stationary phases which are used for HPLC is given in Table 25-10.

Table 25-10 Several commercially available ion exchangers used as stationary phases in HPLC

Type	Particle diameter, μm	Trade name
Anion exchanger	10	Partisil-l0 SAX †
	10	LiChrosorb AN ‡
	10	Ionex-SB 10§
	10	Vydac 301TP ¶
	5, 8	Vydac 300 VHP¶
	5, 8	Vydac 301 VHP¶
	5, 10	Nucleosil SB§
	40	AS Pellionex SAX †, ††
	10	Mono Q††
Cation exchanger	10	Partisil-10SCX†
	10	LiChrosorb KAT †
	10	Ionex SA10§
	5, 8	Vydac 400 VHP¶
	10	Vydac 401TP¶
	5, 10	Nucleosil SA§
	40	HC Pellionex SCX †, ††
	5	Macronex MC-200§§

† Manufactured by Whatman, Inc.
‡ Manufactured by E. Merck.
§ Manufactured by Macherey-Nagel.
¶ Manufactured by the Separations Group.
†† A pellicular on a glass bead.
‡‡ Available from Pharmacia Fine Chemicals.
§§ Available from Interaction Chemicals, Inc.

The efficiency of the column depends upon the size of the particles used as the stationary phase. As the particle size decreases, the exposed surface area of the stationary phase increases. That causes increased column efficiency, increased retention volumes, and increased retention times. Unfortunately, decreasing the particle size of the packing material also causes increased difficulty in packing the column and an increased pressure differential through the column at a particular flowrate. As an example of the effect that particle size has on column efficiency, the efficiencies of Partisil-20 (20-μm particles), Partisil-l0 (10-μm particles), and Partisil-5 (5-μm particles) are about 10,000, 24,000, and 40,000 plates/m respectively. At a constant column pressure, the flow rate through the column generally is inversely proportional to the square of the particle diameter. Particles with diameters of 20 μm or greater can generally be dry-packed into HPLC columns with little trouble, but smaller particles must be packed as a slurry. Special equipment is required to pack HPLC columns with 10-μm or smaller particles. When the smaller particle sizes are required, many analysts prefer to purchase prepacked columns. A detailed procedure for packing HPLC columns is provided in Reference 25.

HPLC Mobile Phases

The mobile phases used for HPLC can be any of those described earlier (Table 25-3). For solid-adsorbent and liquid stationary phases (both normal and reversed phase), mixed solvents are often used as the mobile phase. Generally, the mobile phase consists of a mixture of a polar solvent, such as an alcohol, and a nonpolar solvent, such as a hydrocarbon. As described earlier, some control over the retention volume of the sample components can be exercised by controlling the polarity of the mobile phase. Gradient-elution HPLC, in which the polarity of the mobile phase is continuously altered by changing the ratio of the solvents in the mobile phase, is probably the most useful way to control retention volumes (or times) of sample components. A detailed description of the use of gradient elution in reversed-phase HPLC can be found in Reference 26. A comprehensive description of the manner in which separations can be optimized is provided in Reference 27.

The mobile phase must be chosen so as not to interfere with the measurement by the detector. For example, if an ultraviolet-absorption detector is used, the solvent cannot absorb ultraviolet radiation. Mixtures of methanol, ethanol, or propanol with heptane, and of chloroform with heptane are popular choices as HPLC mobile phases. Reversed-phase HPLC often uses a mixture of methanol or acetonitrile with water as the mobile phase. A comparison of the methods used to determine void volume in reversed-phase HPLC is given in Reference 28.

ION-PAIR AND ION-SUPPRESSION CHROMATOGRAPHY

Ion-pair chromatography (IPC) is also called paired-ion chromatography. It is a chromatographic method for the separation and analysis of ionic species on liquid chromatographic columns with a liquid stationary phase. Ion-pair chromatography is also useful for separations of mixtures containing both ionic and nonionic species. Generally, a reversed-phase packing material is used as the stationary phase in the column, although normal-phase packing materials also have been used.

Ion-pair chromatography is identical to reversed-phase HPLC except that the mobile phase contains a counterion to the ionic analyte components. The counterion is a cation if the analyte ions are anions and is an anion if the analyte ions are cations. Three alternative explanations are used to explain the separations obtained by using ion-pair chromatography. The *ion-pair model* assumes that counterions (denoted by C) react with the sample ions (denoted by S) to form neutral undissociated species (CS or SC) for anionic and cationic sample ions respectively:

$$C^+ + S^- = CS(\text{mobile phase}) \overset{K}{=} CS(\text{stationary phase}) \qquad (25\text{-}5)$$

$$S^+ + C^- = SC(\text{mobile phase}) \overset{K}{=} SC(\text{stationary phase}) \qquad (25\text{-}6)$$

The neutral species are partitioned between the liquid stationary and liquid mobile phases (K is the distribution coefficient). Separation is based upon the relative values of the distribution coefficients of the different neutral species.

The *ion-exchange model* assumes that the counterions are adsorbed or absorbed into the stationary phase to yield a stationary ion exchanger. The ion-exchange sites are those corresponding to the counterions. The sample ions are separated as on an ion-exchange column. The *ion-interaction model* assumes both adsorptive and electrostatic interactions between the analyte ions and the column. The ion-interaction model appears to offer the best explanation for retention during ion-pair chromatography. A more detailed description of the three models can be found in Reference 29.

If the analyte ions are conjugate acids of weak bases or conjugate bases of weak acids, a form of ion-pair chromatography known as *ion-suppression chromatography* can be used. In ion-suppression chromatography the analyle ion is converted to the electrically neutral acid or base by adjusting the mobile phase to the proper pH. In that case H^+ or OH^- is the counterion. The neutral species are partitioned between the stationary phase and the mobile phase. Because the silica gel support in some reversed-phase HPLC columns begins to dissolve at a pH greater than 7 or 8, it is generally necessary to hold the pH of the mobile phase at values less than 7.

In many cases the counterions used for the separation of organic anions by ion-pair chromatography are quaternary amines such as tetrabutylammonium, cetyl trimethylammonium, and triethylalkyl quaternary amines. The counterions used for the separation of organic cations are normally alkyl sulfonates with carbon chain lengths of five or more, or aryl sulfonates. Examples of sulfonate salts that are used as the IPC reagent include sodium 1-heptane sulfonate and sodium 1-octane sulfonate. Generally the concentration of the counterion in the mobile phase is about 0.005 *M*. The counterion chosen for a particular separation is dependent upon the ions to be separated. A procedure that can be used to choose an appropriate counterion is described in Reference 30.

It is sometimes necessary to adjust the pH of the mobile phase to ensure that the analyte species are in an ionic form so that ion pairs can be formed with the counterions. If the addition of a buffer is required, the buffer's concentration should be kept as low as possible to prevent interference with the counterions. The mobile phase solvents used for IPC are those used for reversed-phase HPLC. Mixtures of methanol and water are most popular, but other solvents such as acetonitrile and water mixtures are also used. Separations at a pH greater than 11 are possible without destroying the silica gel support if the mobile phase consists of primary, secondary, or tertiary alkylamines such as triethylamine in acetonitrile and water mixtures.

PAPER CHROMATOGRAPHY

Paper chromatography (PC) is a form of plane chromatography in which a piece of filter paper or cellulose is used as the stationary phase and a liquid solvent is used as the mobile phase. Of the several ways in which a paper chromatogram can be developed, *ascending development* (Fig. 25-10) is the most popular.

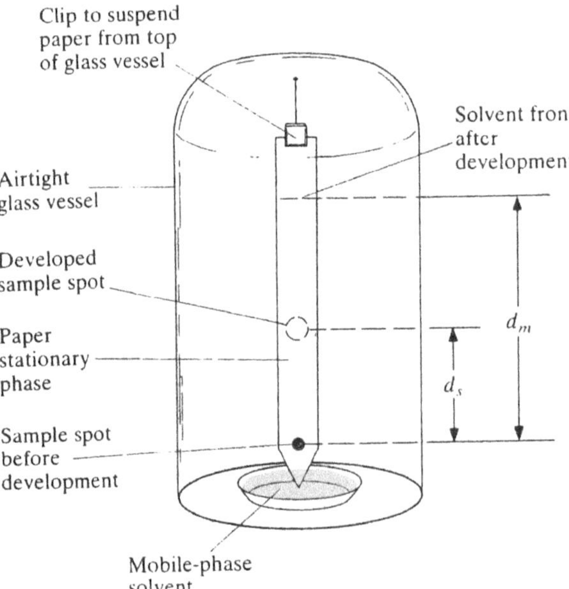

Figure 25-10 Diagram of the apparatus that is used for ascending development of a paper chromatogram. d_m, distance the solvent front moved during development; d_s, distance a sample spot moved during development.

During ascending development, a rectangular strip of paper is used as the stationary phase and the mobile phase flows upward through the paper. Generally, the strip is between 3 and 5 cm wide and about 20 to 30 cm long. A V-shaped notch is often cut in one end of the strip, and an eye dropper, syringe, pipet, Pasteur pipet, or glass tube is used to place a spot of sample solution between the V at the widest point in the notched end of the paper, as shown in Fig. 25-10.

In order to minimize the spot size after development, the sample is placed on the paper in a spot of as small an area as possible. As the chromatogram develops, the sample component diffuses to the regions of lower concentration bordering the spot, thereby increasing the size of the developed spot. After the sample solution has been spotted on the paper and the solvent has been evaporated, the pointed tip of the paper strip is dipped into a vessel containing the mobile phase. The paper strip and mobile-phase reservoir are usually enclosed in an airtight glass vessel. The air within the vessel is kept saturated with vapor from the mobile phase in order to prevent evaporation of the mobile phase from the paper strip. To ensure saturation, pieces of filter paper that are soaked with the mobile phase are often hung inside the glass vessel.

The mobile phase is pulled upward through the paper strip by capillary action. The V in the end of the paper strip directs the mobile phase over the sample spot. As the mobile phase passes over the sample, the sample components are partitioned between the mobile-phase solvent and a stationary liquid phase which is attached to the paper. In normal-phase PC, the stationary phase is water which is hydrogen-bonded to the polar oxygen atoms in the polymeric anhydroglucose units (anhydroglucose is $C_6H_{10}O_5$) of the cellulose. The mobile phase is a solvent which is less polar than water (Table 25-3). In reversed-phase PC the stationary phase is a nonpolar liquid, such as mineral oil, which has been coated on the paper and the mobile phase is a relatively polar solvent such as water or methanol. Separation primarily depends upon the relative distribution coefficients of the sample components between the two liquid phases.

With other types of paper chromatography, a notch is normally not cut in the end of the paper stationary phase which dips into the mobile-phase reservoir. During *descending development* (Fig. 25-11), the mobile phase is added to the top of the strip (rather than the bottom) and descends through the paper. The major advantage of descending development as compared with ascending development is the greater distance the mobile phase can move during the development and the resulting increased separation between the sample components. The upward movement of the solvent during ascending development is limited by the downward pull of gravity on the solvent. In practice, descending development is primarily used to separate the sample components that would not have migrated very far from the original sample spot during ascending development. To enable descending development to be carried out until the mobile phase flows off the bottom edge of the paper, the lower paper edge is often cut into notches. That aids the smooth flow of the mobile phase from the paper.

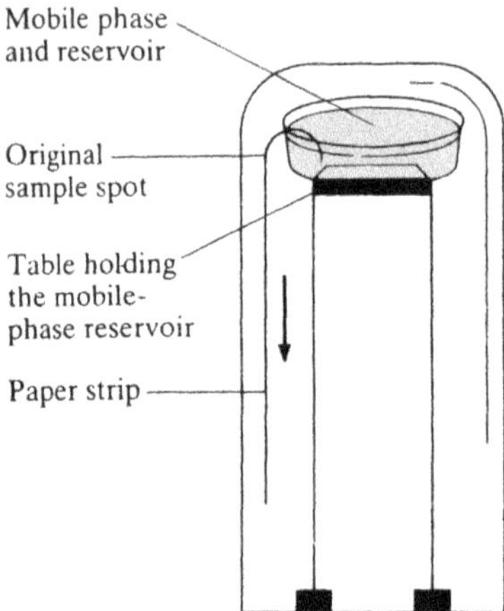

Figure 25-11 Side-view diagram of the apparatus that is used for descending development of a paper chromatogram. The mobile phase travels downward as indicated by the arrow.

PC can also be done by using *radial development*. The sample is spotted in the center of a circular piece of the stationary phase and the mobile phase is slowly added with a pipet, syringe, or eye dropper to the dried spot. The outward motion of the mobile phase from the sample spot causes rings of separated sample components. A variation of the radial development method is the *ring-oven technique*, in which the circular paper is supported at its outer edge on an oven in the shape of a ring (Fig. 25-12). The oven is electrically heated to a temperature that is sufficient to evaporate the mobile phase as it reaches the edge of the filter paper. The evaporation also can be used to concentrate the sample components on the edge of the paper.

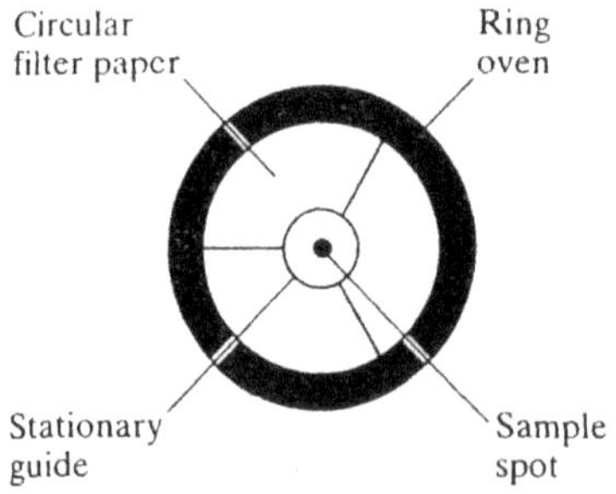

Figure 25-12 Top view of a ring oven. The sample spot and mobile phase are placed on the circular filter paper with the aid of a stationary guide which is elevated above the paper and the ring oven.

When it is impossible to separate all of the sample components with a single solvent or a single mixture of solvents, it is advantageous to use *two-dimensional* development. Generally, a square sheet of cellulose is used as the stationary phase. The sample is spotted on one corner of the stationary phase and developed with one mobile phase as in ascending PC. After development, the paper is dried, rotated 90° so that the edge nearest the developed sample is down, and redeveloped in a second solvent. By using two-dimensional development, it is possible to perform separations that are not possible with either mobile phase in one dimension. The process is illustrated in Fig. 25-13. Two-dimensional development is not limited to use for paper chromatography. It can be adapted to many other chromatographic techniques. A description of the use of two-dimensional developments is provided by Giddings.[31]

PC Spot Location

If the sample components are colored, the developed chromatographic spots can be located by sight. Unfortunately that is rarely the case, and it is usually necessary to use a physical or chemical detective method to find the spots. A number of *visualizing agents* for different classes of chemical compounds are available which can be sprayed on the developed and dried chromatogram or into which the developed chromatogram can be dipped in order to locate the spots. Visualizing agents are chemical compounds that react with the sample components to form a visible reaction product. Several visualizing agents that are used for paper chromatography are listed in Table 25-11.

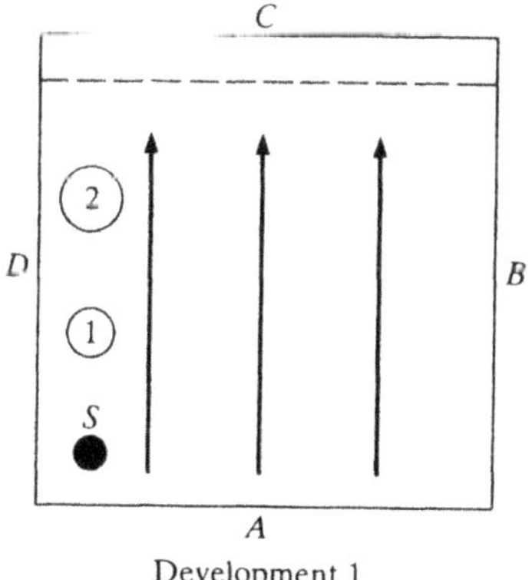

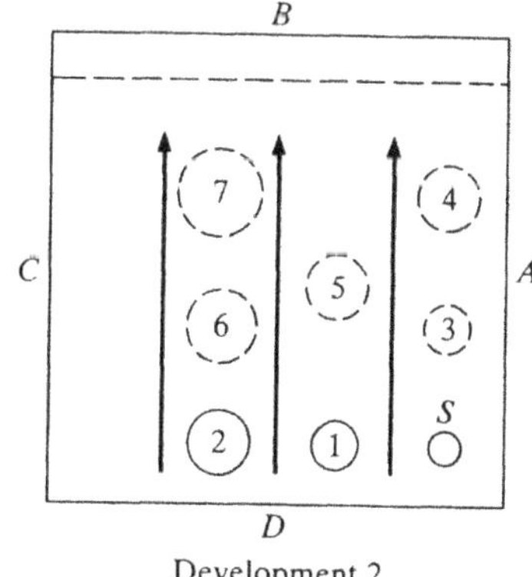

Figure 25-13 Sketches illustrating the operation of two-dimensional plane chromatography. The arrows indicate the direction of mobile-phase flow during development. During development 1 the first mobile phase is used to partially separate the sample *S* into three portions (*S*, 1, and 2). The plane is rotated (note the labeled edges) and a second mobile phase is used to further separate the sample into a total of five components (3 to 7), as indicated by the dashed circles.

Table 25-11 A list of selected categories of compounds that can be separated by using paper chromatography and a visualizing agent that can be used for each category

Except where indicated in a footnote, the reagent is sprayed onto the paper strip

Compound	Visualizing agent	Normal spot color
Alkaloids	Dragendorff's reagent †	Orange, red
Amines	Ninhydrin	Purple
Amino acids	Isatin	Blue
	Ninhydrin-pyridine ‡	Purple, red, brown
Carbohydrates	Aniline-diphenylamine	Blue, brown
Carbonyls	1% KOH or NaOH in EtOH	Red, brown
Fatty acids	Bromophenol blue or bromocresol purple	Indicator color
Guanidines	Sakagechi reagent§	Orange, red
Imidazoles	1% I_2 (w/v) in CCl_4	Brown
Naphthalene derivatives	Diazotized *p*-nitro-aniline¶	Yellow, orange, red, brown, violet
Phenolic acids	Diazotized *p*-nitro-aniline¶	Yellow, orange, red, brown, violet
Phenols	Diazotized sulfanilic acid ††	Yellow, orange, red, brown, violet

† Dissolve 0.8 g of bismuth subnitrate in 50 mL of 25% (v/v) acetic acid. Add 50 mL of 40% (w/v) aqueous potassium iodide, 200 mL of glacial acetic acid, and 2 L of water.

‡ Dissolve 10 mL of a 0.2% (w/v) acetonic solution of ninhydrin in 490 mL of pyridine. The sprayed or dipped paper is heated to 100°C.

§ Add 0.2 mL of a 50% (w/v) aqueous solution of urea to 100 mL of a 0.1% 8-hydroxyquinoline solution in acetone. Dip the developed strip in the mixture and allow to dry. Dip the dried paper strip in a solution prepared by adding 0.3 mL of bromine to 100 mL of 0.5 *M* sodium hydroxide.

¶ Dissolve 1.5 g of *p*-nitroaniline in 45 mL of concentrated hydrochloric acid. Add the solution to 950 mL of water. Add 20 mL of 5% (w/v) aqueous sodium nitrite and 1 L of 10% (w/v) aqueous sodium carbonate. Dip the paper in the solution and allow to dry.

†† Dissolve 9 g of sulfanilic acid in 90 mL of concentrated hydrochloric acid and 900 mL of water. Add 1 L of 5% (w/v) aqueous sodium nitrite and 2 L of 10% aqueous sodium carbonate. After spraying onto the paper, dry the paper in an oven.

Many organic compounds absorb ultraviolet radiation. A spot that contains one of those compounds can be located by holding the developed chromatogram under an ultraviolet lamp. Absorbing species reveal themselves as dark spots. Some other organic compounds can be located by their ability to fluoresce or quench the fluorescence of a fluorescent species which has been coated on the paper either before or after development. In each case the developed paper strip is held under an ultraviolet lamp. Fluorescing compounds appear as light spots, and compounds which quench the fluorescence of a reagent sprayed on the paper appear as dark spots on a light background.

If the sample components are radioactive or can be made so by labeling them with a radioactive tracer, the radioactivity of the paper can be used to locate the sample components. If the sample components are nutrients or poisons for bacteria, the rate of growth or lack of growth in a bacterial culture can be used to locate component spots. Various instrumental methods such as spectrometry, polarography, conductivity, and flameless atomic absorption have also been used as detective techniques. In some cases the best method is to cut the developed paper strip into sections which are perpendicular to the direction of flow of the mobile phase and assay each section individually.

In any case, after the spot has been located it is usually necessary to immediately mark the boundary of each spot *with a pencil.* Ink is usually soluble in the mobile phase. The location of the solvent front and the original sample spot should be marked immediately upon stopping the development. The solvent front is, of course, not visible after the paper dries. The location of the solvent front is needed to measure the retardation factor of each sample component, as will be described in the next section. If it is not possible to keep the original paper strip but a permanent record of the chromatogram is needed, a photocopy of the developed strip can be made.

Analysis by Paper Chromatography

Solute retention on a developed plane chromatogram (paper or thin-layer) is generally easier to measure as the *retardation factor* R_f rather than as the retention volume or retention time. The retardation factor is defined as the ratio of the distance d_s moved by the sample component to the distance d_m moved by the solvent front:

$$R_f = \frac{d_s}{d_m} \qquad (25\text{-}7)$$

The distance moved by the sample component is generally measured from the center of the original sample spot to the center of the developed component spot. Relative retardation factors ($R_f/R_{f,\,\mathrm{ref}}$) or retardation factors of individual sample components can be compared with those of standards as a method of qualitative analysis. In each case the standards must be developed under identical conditions with those used for the sample. Often that is accomplished by developing the analyte and standard chromatograms simultaneously on the same planar stationary phase.

Quantitative analysis by PC is less popular than qualitative analysis. Quantitative analysis can be performed by measuring either the dimensions or a physical property of each developed spot. Usually, quantitative analysis is done by comparing spot dimensions (or areas) with those of standards developed under identical conditions, by measuring the amount of radiation reflected from or transmitted by the spot and comparing it with standards, or by cutting each sample spot from the paper, extracting the analyte with an appropriate solvent, and assaying the resulting solution. Of those methods, the last is the most time-consuming but also the most accurate. Results obtained with the other techniques have a typical relative accuracy of about ± 25 percent.

THIN-LAYER CHROMATOGRAPHY

Thin-layer chromatography (TLC) is a form of plane chromatography in which the stationary phase is a thin layer of a solid on a glass, plastic, or aluminum plate. The plate provides the solid support for the stationary phase. A slurry of the stationary phase is prepared and loaded into a *spreader*, which applies the slurry to the plate in a uniform thin layer (usually between 0.10 and 10 mm thick). After drying, the plate is ready to use. Some stationary phases are activated by heating them prior to use. Generally, TLC plates have dimensions of about 20 by 20 cm, although they can be prepared with any desired dimensions. Some workers use microscope slides for TLC.

TLC plates can be either prepared by the analyst or purchased with the stationary phase attached. Generally the commercially prepared TLC plates use a plastic support. Commercially prepared plates are more convenient but more expensive. Sometimes it is necessary to add a *binder* to the slurry of the stationary phase prior to application to the plate. The purpose of the binder is to secure the stationary phase to the support. Plaster of Paris [$(CaSO_4)_2 \cdot H_2O$] is a common binder, because it forms gypsum ($CaSO_4 \cdot 2H_2O$) upon addition of water.

The sample is spotted or streaked on the TLC plate in the manner used for PC. The chromatogram is usually developed with one or more of the solvents described earlier for LLC or LSC. Although ascending development is most popular, descending, radial, and two-dimensional development are also used. In any case, the plate is enclosed in an airtight glass chamber similar to that used for PC.

The mechanism of separation depends upon the nature of the stationary phase. If an adsorbent is used as the stationary phase, less polar compounds migrate more rapidly along the plate and yield larger retardation factors.

Stationary liquid phases separate sample components according to their relative solubilities in the stationary and mobile phases.

Generally a TLC development requires less time than a PC development because the mobile phase flows more rapidly over a TLC plate than it flows through cellulose. As a consequence, sample spots do not spread as much during TLC development as during PC development, and resolution after TLC development is generally better than that obtained after PC development. Because of those advantages, TLC is a more popular method of analysis than PC. References 32 - 36 provide more details on the use of TLC for analyses.

TLC Stationary Phases

Solid adsorbents such as silica gel and alumina are popular choices for the TLC stationary phase. A stationary liquid phase similar to that used in PC can also be used in TLC. If cellulose is chosen as the stationary phase, the mechanism of separation between the stationary aqueous phase and the mobile phase is identical with the separation mechanism in PC. Other stationary phases which have been used in TLC are charcoal, magnesium oxide, magnesium hydroxide, magnesium silicate, lime [$Ca(OH)_2$], polyamide (e.g., polyacrylonitrile), and various reversed-phase stationary phases such as silicone. In some cases a complexing agent is added to the stationary phase to aid in the separation. Ion-exchange stationary phases such as polyethylene amine also can be used in TLC. In addition to the substance used as the stationary phase, a binder, fluorescent indicator, or other substance is sometimes added to the plate. Plaster of Paris, starch, and carboxymethyl cellulose are often used as binders.

The stationary-phase particle size that is used has an effect upon the efficiency of a TLC plate in a manner analogous to that described earlier for the effect of the stationary-phase particle size on LLC and LSC efficiency. In *high-performance thin-layer chromatography* (HPTLC), the particle size of the stationary phase is about the same as that used in HPLC (5 to 10 μm). The decreased particle size compared with that used in normal TLC (about 40 μm) increases the efficiency of the separation. Silica gel is commonly used as the stationary phase for HPTLC. Reversed-phase HPTLC stationary phases which contain chemically bonded 2-, 8-, and 18-carbon hydrocarbons and ion-exchange HPTLC stationary phases are available. The use of reversed-phase HPTLC is described in Reference 37. Separations which require a distance of 15 cm on silica gel G TLC plates typically require 5 cm or less on HPTLC plates and can be completed in one-third or less of the time required on TLC plates.[38] Typical development times for HPTLC are between 3 and 20 min as compared to between 30 and 200 min for TLC.

The sample volume that is used for an HPTLC development is typically between 0.1 and 0.2 μL as compared to 1 to 5 μL for a TLC development. The diameter of the sample spot before development in HPTLC is normally between

1 and 1.5 mm as compared to between 3 and 6 mm for TLC. A description of HPTLC is available in Reference 39.

If the HPTLC chromatogram is developed radially, the technique is sometimes referred to as *high-performance radial chromatography* (HPRC). In that case the sample is spotted in the center of an HPTLC plate and a pump is used to add mobile phase to the spot at a rate which is usually between 0.1 and 1 μL/s. Some advantage is achieved by supporting the stationary phase on the outside of a glass rod.[40] Two-dimensional separations are possible with TLC and HPTLC.

Extremely small analyte masses (10^{-9} to 10^{-12} g) are used during an HPTLC development. Generally, HPTLC plates are 5 × 5 cm or 10 × 10 cm. Some HPTLC plates have a leading edge of a normal TLC plate connected to an HPTLC plate. With those plates the sample is spotted in the normal-TLC portion and eluted into the HPTLC portion of the plate with mobile phase. The plates make spotting simpler than in HPTLC alone, because the sample can be spotted as in normal TLC. As much as 5 μL can be spotted using the combined TLC-HPTLC plate. HPTLC can be used as a rapid screening technique prior to analysis by HPLC or UHPLC with the same stationary phase. HPTLC is often used to find the mobile phase that is best suited for separation in an HPLC silica gel column. The separations obtained with HPTLC are comparable with those obtained with HPLC.

TLC Zone Detection and Analysis

If the sample component absorbs uv radiation or fluoresces, the developed spot can be located by holding the TLC plate under an ultraviolet lamp. Some spots can be made visible, absorbing, or fluorescent by spraying or dipping the plate in a chemical reagent. An iodine solution is used as a visualizing agent for many organic compounds. Selected categories of compounds that can be separated by TLC and visualizing agents that can be used for each category are listed in Table 25-12. Many of the reagents listed in Table 25-11 that are used in PC can also be used in TLC. A longer list of visualizing agents can be found in Reference 34.

Fluorescamine is sometimes used to make organic amines fluoresce. Radioactive sample components and those which can be made radioactive by labeling them with a radioactive element, e.g., tritium, can be located with radioactivity detectors which are scanned across the TLC plate. Fluorescent quenching by a sample component on TLC plates sprayed with a fluorescing reagent can be used to locate some sample spots. In a few cases it is possible to spray the developed TLC plate with a reagent which forms an electrochemiluminescent compound with the sample component. Application of the proper potential to the sprayed TLC plate results in a visible spot.

Table 25-12 A list containing selected compounds that can be separated by using thin-layer chromatography and a visualizing agent that can be used with each category

Compound	Visualizing agent	Band color
General organic	1% iodine in volatile solvent	Brown
	50% aq H_2SO_4	Brown, black
	50% H_2SO_4, 5% HNO_3	Brown, black
Organic acid	0.3% bromocresol green in H_2O/CH_3OH	Yellow
	0.04% bromocresol purple in H_2O/C_2H_5OH	Yellow
	0.5% bromophenol blue in 0.2% aq citric acid	Yellow, purple
Alcohols	6% cerium(IV) ammonium nitrate in 2 *M* HNO_3	Brown
Aldehydes, ketones	0.4% 2,4-dinitrophenyl-hydrazine in 2 *M* HCl	Yellow, red
Alkaloids	5% cobalt(II) chloride/ 15% ammonium thiocyanate	Blue
	Formic acid vapor	Fluorescent blue under uv
Amines, amides	0.1% Alizarin in EtOH	Violet
	1% potassium persulfate in aq 0.001 *M* $AgNO_3$	Blue, yellow, green, pink, violet
Amino acids, peptides, proteins, enzymes	0.2% 1,3-naphthoquinone-4-sulfonic acid (sodium salt in 5% Na_2CO_3)	Pink, red
Carboxylic acids	0.1% 2,6-dichlorophenol-indophenol in 95% EtOH	Pink
Hydrocarbons	0.05% sodium fluorescein in 50% CH_3OH	Sometimes fluorescent under uv
Phenols	1% *p*-nitrosodimethylaniline in 50% EtOH	Varied
Al, Ba, Ca, Fe, Li, Se, Th, Ti, Zr, Zn, NH_4^+	Saturated alizarin in EtOH	Red, violet
Ag, As, Au, Bi, Cd, Co, Cu, Hg(I), Hg(III), Ni, Pb, Pd, Pt, Sb, Sn, Ti, V	aq Ammonium sulfide	Brown, yellow, orange
Cl^-, Br^-, I^-	0.04% bromocresol purple in 50% EtOH (pH 10)	Purple

Qualitative analysis is done, as in PC, by comparing R_f s or relative R_f s with those of standards that have been developed under identical experimental conditions. Quantitative analysis can be performed in any of several ways. The reflectance, transmittance, fluorescence, or fluorescent quenching of a spot can be measured and compared with that of standards. Both single- and double-beam instruments are used for those measurements. A scanning densitometer is often used to measure the absorbance or fluorescence of a sample component on a TLC or HPTLC plate. A beam of monochromatic radiation is scanned across the plate and the reflected or scattered radiation is monitored with a photomultiplier tube. Generally the accuracy of the results obtained while using those detective methods with TLC is considerably better than the accuracy of the same methods while using PC. Measurements of other physical properties, such as the amount of radioactivity, sometimes can be used for quantitative analysis. Spots can also be removed from the TLC plate and assayed in a convenient manner after extraction of the sample component into a suitable solvent.

SIZE-EXCLUSION CHROMATOGRAPHY

Size-exclusion chromatography (SEC) is a chromatographic technique in which separation is based on the molecular size of the sample components. The stationary phase consists of porous particles that have a fixed pore size. A sample component which is too large to enter the pores is immediately swept from the column with the mobile phase. A sample component which is small enough to enter the pores completely is held on the column longest, and a component which can partially enter the pores has an intermediate retention volume. Component separation is dependent upon the ability of the components to enter the pores of the stationary phase.

Because the molecular size of organic and biochemical compounds generally increases with molecular weight, it is convenient to express molecular sizes in terms of molecular weights. The molecular weight of the smallest particle which is too large to enter partially the pores of a particular stationary phase is the *exclusion limit* of the stationary phase. It ranges with the stationary phase between about 400 and about 50×10^6. The molecular weight of the largest substance that is able to enter the pores completely is the *permeation limit* of the packing material. The stationary phase is chosen such that the particles to be separated fall between the permeation limit and the exclusion limit for the stationary phase.

The equation which best describes size exclusion chromatography is obtained by substituting the capacity factor from Eq. (24-7) ($k' = KV_s/V_m$) into Eq. (24-9) [$V = V_m(1 + k')$] to obtain

$$V = V_m\left(1+\frac{KV_s}{V_m}\right) = V_m + KV_s \tag{25-8}$$

A more thorough description of the principles of SEC and the accompanying mathematical equations can be found in Reference 41.

For species that are too large to enter the pores of the stationary phase, i.e., that are larger than the exclusion limit, the retention volume V equals the mobile-phase volume V_m (also called the *void volume*). The distribution coefficient K in that case is zero; that is, $K = C_s/C_m = 0$. For species that are sufficiently small to enter the pores completely, the concentration of the species inside the stationary phase is identical to that outside the stationary phase, and the distribution coefficient is 1.

The void volume of a column is determined by measuring the retention volume of a species that has a molecular weight greater than the exclusion limit of the stationary phase. The stationary phase volume V_s (also called the total pore volume) is determined for a column with known void volume by measuring the retention volume of a species that has a molecular weight less than the permeation limit of the stationary phase. Because K is a 1 for that species and V and V_m are known, V_s can be calculated by using Eq. (25-8). The distribution coefficient for a species that can partially enter the pores of the stationary phase must be between 0 and 1 and can be obtained from V_s, V_m, and the retention volume of the species by substitution into Eq. (25-8).

Sample problem 25-2 A molecular species with a molecular weight of 80,000 was passed through a column containing a stationary phase that had an exclusion limit of 40,000. The retention volume was 25 mL. A low molecular weight compound that was small enough to enter the pores of the stationary phase completely had a retention volume of 152 mL and a sample component had a retention volume of 102 mL. Calculate the void volume V_m, the total pore volume V_s, and the distribution coefficient of the sample component.

SOLUTION For the compound with molecular weight 80,000, $K = 0$ and $V = V_m = 25$ mL. For the low molecular weight compound, $V = 152$ mL and $K = 1$. Substitution of those values into Eq. (25-8) allows calculation of V_s:

$$V = V_m + KV_s$$
$$152 \text{ mL} = 25 \text{ mL} + (1)\ V_s$$
$$V_s = 127 \text{ mL}$$

The distribution coefficient of the sample component can be obtained by substituting the values of V_m, V_s, and the retention volume into Eq. (25-8) and solving for K:

$$102 \text{ mL} = 25 \text{ mL} + K(127 \text{ mL})$$
$$K = 0.61$$

Stationary phases that are used for size-exclusion chromatography fall into three categories. *Soft gels* consist of porous polymeric materials such as cross-linked dextran (e.g., Sephadex, which is available from Pharmacia Fine Chemicals) or polyacrylamide gels (e.g., Bio-Gel P, which is available from Bio-Rad Laboratories). When a soft gel is used as the stationary phase in size-exclusion chromatography, the technique is *gel filtration chromatography. Semi-rigid gels* are more rigid than soft gels; i.e., they do not expand or contract as easily as the soft gels. Some cross-linked polystyrene beads (polymeric styrenedivinylbenzene) are classified as semirigid gels (e.g., Bio-Beads S from Bio-Rad Laboratories). When a semirigid gel is used as the stationary phase, the technique is *gel permeation chromatography*. Semirigid gels are often used for size-exclusion chromatography with nonaqueous mobile phases.

The final category of stationary phases are the *highly rigid gels* and glasses, which have pore sizes that do not change with solvent or with applied pressure. Typically, those stationary phases are porous glass beads or porous silica spheres which contain various chemically bonded functional groups. The major advantage of the rigid stationary phases is an ability to be used with high column-inlet pressures. The rigid stationary phases can be used for *high-performance size-exclusion chromatography* (HPSEC) (a technique analogous to HPLC in which separation occurs according to molecular weight). A description of several gels that are used for SEC is available in Reference 42.

Soft-gel and semirigid-gel stationary phases are added to the mobile-phase solvent and allowed to swell to maximum size prior to being packed as slurries into solvent-filled columns. Size-exclusion chromatographic detectors are similar to those that are used for other forms of liquid chromatography. In addition to water, solvents that are often used as mobile phases for SEC include tetrahydrofuran, chloroform, and *N,N*-dimethylformamide.

Typical classical SEC columns have a length of 1.2 m and are packed with stationary-phase particles that have a diameter between 40 and 70 μm. A typical sample volume is between 50 and 100 mL and often requires several hours for chromatographic development. HPSEC columns typically have a length between 0.3 and 0.6 m and are packed with particles that have a diameter between 5 and10 μm. A typical sample volume is between 1 and 10 mL and development time is 30 min or less.

Popular detectors that are used for SEC measure the difference in refractive index between the mobile phase and the column effluent, absorption in the uv or ir region, or low-angle laser light scattering (LALLS). In addition, Fourier transform ir detectors and high-speed uv-visible detectors that use diode arrays are used to record the spectrum of a sample component as it is eluted from the column. Various electroanalytical detectors also have been used. Some chromatographers prefer to collect fractions from the column and individually assay each fraction.

SEC is used to separate and assay relatively high molecular weight compounds such as biochemical compounds (peptides, proteins, etc.) and polymers. In most cases compounds with a molecular weight below about 100 are small enough to penetrate the holes in the packing material completely and consequently cannot be separated from each other. Because separation is based upon the size of the sample components, which in turn is often related to the molecular weights of the components, SEC can be used to estimate molecular weights. Generally, the column is "calibrated" by measuring the retention volumes of several standard compounds which have the same general structure as the sample and which have known molecular weights. A working curve is prepared by plotting retention volumes of the compounds as a function of the logarithms of the molecular weights (or the reverse). The working curve is often linear and is used to determine the molecular weight of the sample compound from the retention volume of the compound on the column.

Another chromatographic technique that separates components according to size is *hydrodynamic chromatography.*[43] It is used to separate colloidal particles such as clays, viruses, paints, and polymeric materials such as silver halides, silica, iron(III) oxide, and carbon black. Generally the stationary phase consists of beads of a cation-exchange resin or a copolymer that have diameters between 15 and 20 μm. Because the separated components are too large to fit within the holes of the stationary phase, separation is based upon completely different principles than those which apply to SEC. The mobile phase typically is an aqueous solution of a buffer and surfactant. The ionic strength of the mobile phase must be carefully controlled. As with SEC, the larger diameter particles are eluted first.

AFFINITY CHROMATOGRAPHY

The stationary phase that is used for *affinity chromatography* contains a functional group that specifically bonds to a single chemical or biochemical species or group of species. The functional group is chemically attached to an inert and insoluble support. A substance in the mixture that chemically bonds to the functional group is held on the column while other mixture components exit the column with the mobile phase. The chemically bonded component can be completely and rapidly separated from other components of a mixture. In most cases the bond between the stationary phase and the sample component is that between a ligand (the stationary phase) and the substance with which the ligand complexes (the analyte). After the nonbonded components have been flushed from the column, the mobile phase is altered in such a way as to break the bonds between the sample component and the stationary phase, thereby allowing the retained component to be eluted from the column. The alteration is often accomplished by varying the pH or ionic strength of the mobile phase or by using a gradient elution technique.

The solid support to which the stationary functional group is bonded is usually agarose (a linear polysaccharide composed of galactose and 3,6-anhydrogalactose) or an agarose derivative. The bonded functional group varies with the type of species to be separated. Generally, the functional group is an enzyme that specifically binds to a particular sample component. Either the stationary-phase functional group can be chemically bonded to the solid support by the analyst or the column can be purchased with the functional group already bonded to the support.

ELECTROPHORESIS

Electrophoresis uses an electric potential to separate relatively large biochemical ions according to their charge-to-mass ratios. Although a solution of an electrolyte is required, the separative method does not require a mobile phase. Because the sample ions are not partitioned between a stationary phase and a mobile phase, electrophoresis cannot be classified as a form of chromatography. It is described in the chapter, however, because it more closely resembles liquid chromatography than the topics described in other chapters.

The separation is most often performed on a paper or gel support that is permeated with an electrolyte solution. The solution is usually buffered. The sample is spotted on the gel or paper, and an electric potential from a power supply is applied across the support. Positive ions in the sample mixture are attracted to the negative electrode (cathode) and negative ions are attracted to the positive electrode (anode). The rate of migration of the ions through the electrolyte solution on the support increases as the charge on the ion increases, but it decreases with decreasing ionic mobility in the solution, i.e., with increasing size or mass of the ion. Consequently, the separation is based on the charge-to-mass ratios of the sample components. After the separation has occurred, the paper or gel is removed from the apparatus and the separated components are located.

The power supplies used for electrophoresis can be operated at a constant potential difference, a constant current, or a constant power (potential × current). Most power supplies that are used for electrophoresis can he operated between 0 and 500 V, although some are capable of providing as much as 2000 V. Typically the power supplies provide up to 200 mA of current. Some commercially available power supplies can provide currents as high as 300 or 400 mA. Those in which the power can be controlled are usually capable of operation between 0 and 200 W. The separation proceeds more rapidly with higher applied potential, current, or power. The upper limit at which the power supply can be operated for a particular separation is usually dependent upon the amount of heat generated by the current that flows through the paper or gel. Too much heat can damage the analyte ions.

Paper Electrophoresis

The support that is used during separations by paper electrophoresis is either cellulose acetate or filter paper as in PC. Two popular forms of electrophoresis on paper are used. In the first type, the paper strip is permeated with a buffer-electrolyte solution and supported on a metallic or glass plate. The sample is spotted at the desired position on the strip, and a second metallic or glass plate is placed on top of the paper strip. The plates serve to keep the paper moist with the buffered solution and to dissipate heat away from the paper during the electrophoresis. Because metals conduct heat better than glass, it is better to use metallic plates. The disadvantage of the use of metallic plates is their opacity.

The ends of the paper strips extruding from between the plates are dipped into vessels that contain the buffered solution. Electrodes are placed in the vessels and attached to the power supply as shown in Fig. 25-14. After the sample ions have migrated toward the electrode of opposite charge, but before the ions reach the buffered solution in the vessel containing the electrode, the power supply is turned off and the paper strip is removed and dried. Component spots are located as in paper chromatography.

The second form of paper electrophoresis is *continuous-flow paper electrophoresis*. A solution containing the analyte ions and the buffer electrolyte is continuously added to the top of the sheet of paper (Fig. 25-15). The electrodes are attached to the vertical edges of the paper. As the sample descends through the paper, ions migrate horizontally toward the electrode of opposite charge, and eventually drip from the bottom of the paper after having been horizontally displaced from the position at which the sample was added. Notches which are cut in the bottom edge of the paper act as drip points at which the separated ions can be collected. The paper is sometimes sandwiched between two plates as in the type of paper electrophoresis described previously. Continuous-flow paper electrophoresis is one of the few continuous-flow methods for separating ionic sample components.

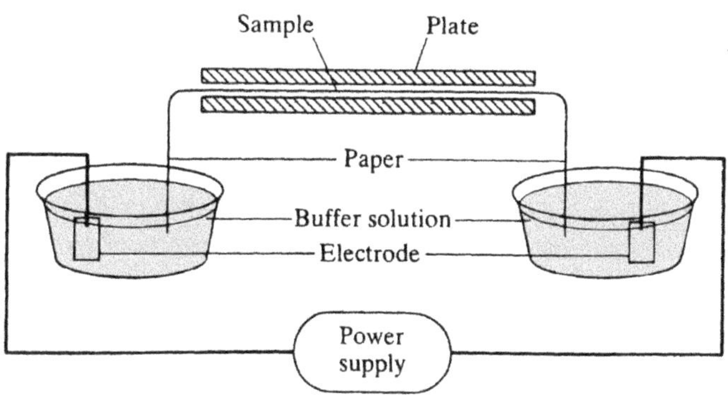

Figure 25-14 Diagram of the apparatus that is used for one form of paper electrophoresis.

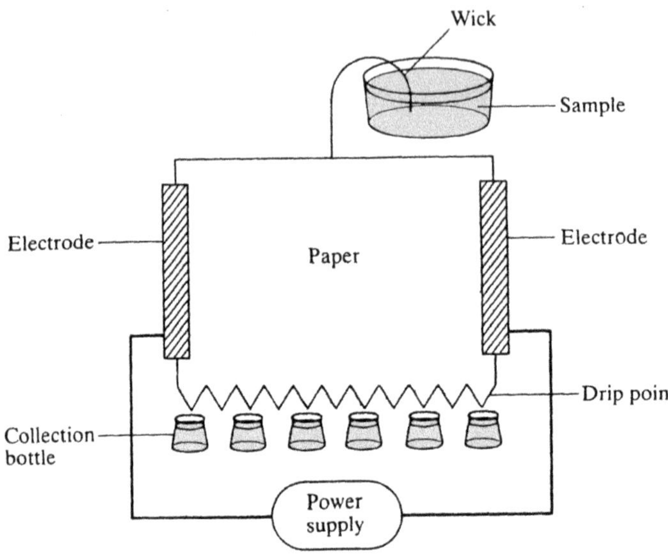

Figure 25-15 Diagram of the apparatus that is used for continuous-flow paper elctrophoresis.

The apparatus that is used for continuous-flow paper electrophoresis can be used for separations by an elution rather than by a continuous-flow method. The sample is spotted at the top of the paper and only the buffered solution is continuously added. If the electrophoresis is stopped before the sample components drip from the bottom of the paper, the technique combines the separating ability of descending paper chromatography (separation vertically) with that of electrophoresis (separation horizontally).

Gel Electrophoresis

Gel electrophoresis is the form of electrophoresis most often used for the separation of biochemical ions. Generally the gel is agar, agarose, or polyacrylamide. The gel is either formed inside thin-walled glass tubes, which are typically 10 cm long, or supported on a glass plate, which usually has dimensions of 8.2 × 8.2 cm. The glass walls in the tubes are relatively thin to allow rapid dissipation of heat from the gel during the separation. If a glass plate is used as the support, the plane of gel is called a *gel slab*. When a gel slab is used, the sample can be spotted in any desired location. When tubes are used, the sample is added with a syringe or micropipet to the end of the gel farthest away from the electrode of charge opposite that of the ions to be separated. Several forms of gel electrophoresis are used which differ in mechanism of the separation.

In regular gel electrophoresis the pores in the gel are large enough for all of the sample components to enter easily. The separation in that case is entirely based on the relative motion of the sample ions toward the oppositely charged electrode. After the ions have been separated but before they reach the edge of the gel, the power supply is turned off and the zones of separated ions are located.

Zone detection is usually accomplished either by adding a dye, such as methylene blue for nucleic acids or Coomassie Brilliant Blue R-250 for proteins, that colors the ions or by adding a *tracking dye*, such as bromocresol green for DNA, which migrates through the gel at the same rate as the sample ions. If the separated ions are to be collected for identification or other study, the portion of the gel containing the component can be cut from the gel and removed. If the gel is held in a tube, the tube is carefully broken, the intact gel removed, and a slice containing the component is withdrawn.

The gel pore size that is used for *gradient gel electrophoresis* decreases with distance through the gel. In that form of electrophoresis the separation is based on the charge-to-mass ratio of the ions, as in regular gel electrophoresis, and upon the ability of the gel to act as a sieve. Smaller ions are capable of traveling farther through the gel than the larger ions, which are prevented from passing through the portion of the gel with a pore size less than their diameter. Polyacrylamide gels are generally used for that form of electrophoresis. The pore size is adjusted by controlling the amount of polyacrylamide in the gel. The method has proved to be especially useful for separation according to the size of protein ions. Many protein ions have about the same charge-mass ratio and consequently are difficult to separate by regular gel electrophoresis.

The final form of gel electrophoresis which will be described uses a gel through which a continuous variation in pH exists. The variation is accomplished by adding a series of ionic species of differing size known as *ampholytes* to the gel and using electrophoresis to cause the ampholytes to migrate into the gel. The ampholytes are usually polyaminopolysulfonic acids which contain both basic (amino) and acidic functional groups and which serve as pH buffers. Because each ampholyte migrates into the gel a different distance and has a different pH associated with it, the pH continuously varies from one end of the gel to the other. Lower pH is found near the anode, and the pH increases as the distance from the cathode decreases.

The protein ions that are separated by using that technique (known *as isoelectric focusing*[44]) migrate into the gel until they reach the pH region at which the protein exists as a neutral species. That point, of course, depends upon the K_a for the individual protein. When the protein becomes a neutral molecule, the migration ceases because the charged electrode has no attraction for a neutral species. The pH at which the sample component becomes neutral is the *isoelectric point pI*. Isoelectric focusing separates sample components according to their isoelectric points. Components can be separated if their *pI*s differ by 0.01 or more. The pH at various points in the gel is determined by measurement with a miniature pH and reference electrode pair. Polyacrylamide gels are often used for isoelectric focusing.

IMPORTANT TERMS

Adsorptive isotherm
Affinity chromatography
Ascending development
Binder
Brockmann activity scale
Carbon load
Chiral chromatography
Continuous-flow paper electrophoresis
Counterion
Descending development
Displacement analysis
Electrophoresis
ELSD
Exclusion limit
Flowrate
Gel electrophoresis
Gel filtration chromatography
Gel permeation chromatography
Gel slab
Gradient gel electrophoresis
Hollow-fiber suppressor
HPLC
HPRC
HPTLC
Hydrodynamic chromatography
Ion chromatography
Ion-exchange chromatography
Ion-exchange resin
Ion-pair chromatography
Ion-suppression chromatography
Isoelectric focusing
Isoelectric point
LCEC
LLC
Loading capacity
LSC
Paper chromatography
Paper electrophoresis
Pellicle
Permeation limit
Plane chromatography
Radial development
Retardation factor
Reversed-phase chromatography
Ring-oven technique
Size-exclusion chromatography
Stripping column
Thin-layer chromatography
Two-dimensional development
UHPLC
Visualizing agent

PROBLEMS

Liquid-Solid Chromatography

25-1 What is the expected order of elution of the following compounds from an alumina column by a liquid mobile phase: *n*-butanol, 1-butyl chloride, *n*-hexanoic acid, *n*-hexane, 2-hexene?

25-2 Predict the order of elution of the following compounds from a calcium oxide column: *n*-butanol, methanol, *n*-hexanol, ethanol.

25-3 List the following compounds in their expected order of elution from an LSC column: benzoic acid, benzene, chlorobenzene, phenol.

Ion-Exchange Chromatography

25-4 Predict the order of attraction of the following anions to an anion-exchange column: Cl^-, I^-, F^-, Br^-.

25-5 Predict the probable order of elution of the following ions from a cation-exchange column: Ca^{2+}, Ba^{2+}, Mg^{2+}, Be^{2+}, Sr^{2+}.

25-6 Arrange the following ions according to their probable order of elution from a cation-exchange column: Na^{+}, K^{+}, Li^{+}, Cs^{+}, Rb^{+}.

Plane Chromatography

25-7 A sample containing two components was separated by paper chromatography. After the chromatogram had developed for 1 h, the solvent front had moved 10.3 cm from the original sample spot; component A had moved 7.5 cm; and component B had moved 2.3 cm. Calculate the retardation factors for components A and B.

25-8 After TLC development, the solvent front had moved 15.0 cm; component 1 had moved 11.6 cm; component 2 had moved 5.6 cm; and component 3 had moved 1.5 cm from the original sample spot. Calculate R_f for each component.

Size-Exclusion Chromatography

25-9 The retention volume of a protein with a molecular weight of about 120,000 on a gel column that had an exclusion limit of 80,000 was 15 mL. The retention volume of naphthalene (MW 128) on the same column was 124 mL. A particular sample component had a retention volume of 87 mL. Assuming naphthalene has a molecular weight less than the permeation limit of the gel, calculate the total pore volume and the distribution coefficient of the sample component.

25-10 For a particular gel-filtration chromatographic column, sucrose is small enough to completely enter the gel pores. The high molecular weight compound, Blue Dextran, was larger than the exclusion limit for the gel. The distribution coefficient of cytochrome c on the column was 0.81. Calculate the retention volume for cytochrome c if the retention volume of sucrose is 195 mL and the retention volume of Blue Dextran is 39 mL.

25-11 Estimate the molecular weight of chymotrypsinogen from the following data obtained with SEC.

Compound	Molecular weight	V, mL
Cytochrome c	13,000	140
Myoglobin	16,900	133
Ovalbumin	46,000	107
Urease	483,000	45
Chymotrypsinogen		125

25-12 The retention volumes of polyethylene oxide (PEO) polymers were monitored with an SEC column using dimethylformamide as the mobile phase. Use the tabulated results to estimate the molecular weight of a PEO polymer that has a retention volume of 13.5 mL.

Molecular weight	*V*, mL
1.1×10^3	17.2
5.1×10^3	16.3
2.0×10^4	15.1
7.7×10^4	14.1
7.0×10^5	12.7

REFERENCES

1. Lehrer, R.: *Amer. Lab.*, **13**(10): 113(1981).
2. Conlon, R. D.: *Instrumn. Res.*, **1**: 90(1085).
3. Borman, S. A.: *Anal. Chem.*, **54**: 327A(1982).
4. Dong, M. W., and J. L. DiCesare: *LC*, **1**:223(1983).
5. Green, R. B.: *Anal. Chem.*, **55**: 20A(1983).
6. Dorn, H. C.: *Anal. Chem.*, **56**: 747A(1984).
7. Roston, D. A., R. E. Shoup, and P. T. Kissinger, *Anal. Chem.*, **54**: 1417A(1982).
8. Bratin, K., C. L. Blank, I. S. Krall, C. E. Lente, and R. E. Shoup, *Amer. Lab.*, **16**(5): 33(1984).
9. Caudill, W. L., A. G. Ewing, S. James, and R. M. Wightman: *Anal. Chem.*, **55**: 1877(1983).
10. Vohra, S. K.: *Amer. Lab.*, **13**(5): 66(1981).
11. Rocklin, R. D.: *LC*, **2**: 588(1984).
12. Majors, R. E.: *LC*, **1**: 488(1983).
13. Ackermann, B. L., M. J. Berna, J. A. Eckstein, L. W. Ott and A. F. Chaudhory in Annual Reviews of Analytical Chemistry, E. S. Yeung and R. N. Zare (eds.), Annual Reviews, Palo Alto, California, **1**: 357(2008).
14. Berg, E. W.: *Physical and Chemical Methods of Separation*, McGraw-Hill, New York, 1963.
15. Small, H., T. S. Stevens, and W. C. Bauman: *Anal. Chem.*, **47**: 180(1975).
16. Small, H.: *Anal. Chem.*, **55**: 235A(1983).
17. Fritz, J. S.: *LC*, **2**: 446(1984).
18. Johnson, E. L.: *Amer. Lab.*, **14**(2): 98(1982).
19. Girard, J. E., and J. A. Glatz: *Amer. Lab.*, **13**(10): 26(1981).
20. Snyder, L. R.: *LC*, **1**: 478(1983).
21. Unger, K. K.: *Anal. Chem.*, **55**: 361A(1983).
22. Snyder, L. R., and P. E. Antle: *LC*, **2**: 840(1984).
23. Snyder, L. R., and P. E. Antle: *LC*, **3**: 98(1985).
24. Ward, T. J. : *Anal. Chem.*, **78**: 3947 (2006).
25. Medina, J. L., and S. Dave: *Amer. Lab.*, **15**(1): 83(1983).
26. Snyder, L. R., M. A. Stadalius, and M. A. Quarry: *Anal. Chem.*, **55**: 1412A(1983).
27. Glajch, J. L., and J.J. Kirkland: *Anal. Chem.*, **55**: 319A(1983).
28. Krstulovic, A. M., H. Collin, and G. Gulochon: *Anal. Chem.*, **54**: 2438(1982).
29. Bidlingmeyer, B. A.: *LC*, **1**: 344(1983).

30. Perry, J. A., L. J. Glunz, T. J. Szczerba, and V. S. Hocson: *Amer. Lab.*, **16**(10): 114(1984).
31. Giddings, J. C.: *Anal. Chem.*, **56**: 1258A(1984).
32. Issaq, H. J., and E. W. Barr: *Anal. Chem.*, **49**: 83A(1977).
33. Lott, P. F., J. R. Dias, and S.C. Slahck: *J. Chrom. Sci.*, **16**: 571(1978).
34. Touchstone, J. C., and M. F. Dobbins: *Practice of Thin Layer Chromatography*, Wiley, New York, 1978.
35. Sherma, J.: *Anal. Chem.*, **76**: 3251 (2004).
36. Sherma, J.: *Anal. Chem.*, **78**: 3841 (2006).
37. Hauck, H. E., and W. Jost: *Amer. Lab.*, **15**(8): 72(1983).
38. Seiler, N., and B. Knodgen: *J. Chrom.*, **131**: 109(1977).
39. Fenimore, D. C., and C. M. Davis: *Anal. Chem.*, **53**: 252A(1981).
40. Ranger, H. O.: *Amer. Lab.*, **13**(11): 146(1981).
41. Barth, H. G.: *LC*, **2**: 24(1984).
42. Kato, Y.: *LC*, **1**: 540(1983).
43. Small, H., and M. A. Langhorst: *Anal. Chem.*, **54**: 892.A(1982),
44. Hoyle, M. C.: *Amer. Lab.*, **11**(12): 32(1979).

BIBLIOGRAPHY

Snyder, L. R., and J. J Kirkland: *Introduction to Modern Liquid Chromatography*, 2d ed., Wiley, New York, 1979.

CHAPTER

TWENTY-SIX

GAS CHROMATOGRAPHY

The mobile phase in gas chromatography GC is the *carrier gas*. The stationary phase can be either a solid adsorbent or a liquid coated or bonded on a solid support or on the walls of the column. If the stationary phase is a solid, the technique is *gas-solid chromatography* (GSC). If the stationary phase is a liquid, the technique is *gas-liquid chromatography* (GLC). Although both techniques are used for analysis, GLC has been more widely used. Separation by using a solid stationary phase is primarily based upon the relative adsorption of the sample components on the solid. Separation by using a liquid stationary phase can be based on either relative solubilities of sample components in the stationary phase or a combination of relative solubilities in the stationary phase and adsorption on the solid support of the stationary phase.

In any case, the sample used in a gas chromatographic separation must either be a gas or be capable of being converted to a gas at the temperature of the column. Gas chromatography has proved to be one of the most widely used methods for the separation and analysis of organic compounds. It can be used to separate many complex mixtures such as that in gasoline (Fig. 26-1). As with the several forms of liquid chromatography, gas chromatography provides both qualitative and quantitative analytical information. Generally, gas chromatography cannot be used for the analysis of inorganic salts, because inorganic salts cannot be converted to gases at the temperatures at which gas chromatographic columns are operated. It has been estimated that up to 20% of existing molecules can be assayed by GC.

RETENTION TIME AND RETENTION VOLUME

Much of the theory of gas chromatography is identical with that of the other forms of chromatography and is described in Chapter 24. The separated

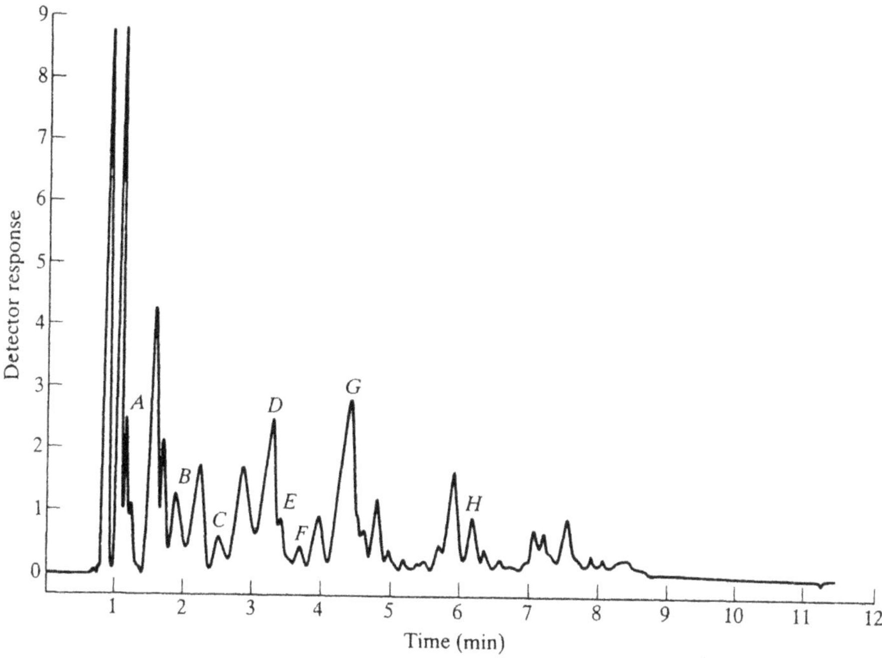

Figure 26-1 A gas chromatogram of a gasoline sample. The labeled peaks correspond to *n*-pentane (*A*), *n*-hexane (*B*), benzene (*C*), isooctane (*D*), *n*-heptane (*E*), methylcyclohexane (*F*), toluene (*G*), and *o*-xylene (*H*).

components of a sample mixture are characterized by individual retention times or retention volumes. Resolution between chromatographic peaks and relative retention in chromatography are defined in Chapter 24 for all forms of chromatography. If gas chromatograms, which are recorded at different carrier gas flowrates, are to be compared, it is advantageous to use retention volumes rather than retention times, because retention volumes are independent of flowrate. Retention times are inversely proportional to flowrate; i.e., as the flowrate of the mobile phase increases, the retention time for a particular component decreases.

All of the equations found in the THEORY section of Chapter 24 apply to GC as well as to LC. Some chromatographers refer to the relative retention as defined by Eq. (24-4) as the retention factor. The efficiency of GC columns is expressed as theoretical plates and the height equivalent of a theoretical plate. The values can be calculated by using Eqs. (24-11), (24-14) and (24-15).

Because gases are compressible, the measured retention volume of a GC sample component is dependent not only upon the column temperature but also upon the inlet and outlet pressures of the column. In some cases it is dependent on the way in which the flowrate through the column is measured. The amount of stationary phase packed within the column also has an effect upon the retention volume of a component. If retention volumes measured by different laboratories are to be compared, it is necessary to correct all measured retention volumes to a specific set of experimental conditions. Gas chromatographic flowrates often are measured at room temperature after the gas has flowed through the detector. For isothermally (constant temperature) operated columns, the measured flowrate F_m can be corrected for the difference between the absolute room temperature T_R (at which the flowrate was measured) and the absolute column temperature T_c by use of

$$F_c = \frac{F_m T_c}{T_R} \tag{26-1}$$

Equation (26-1) can be derived from the definition of flowrate (the volume flowing past a location divided by the time of flow) and Charles' law ($V \propto T$). The *adjusted flowrate* F_c is the flowrate through the column. T_c and T_R are the absolute column and room temperatures respectively.

If a bubble flowmeter is used, the gas generally passes over an aqueous soap solution and must be corrected for the vapor pressure of water P_w at the temperature of the flowmeter. That can be accomplished by adding a second correction factor to Eq. (26-1):

$$F_c = \frac{F_m T_c (P_o - P_w)}{T_R P_o} \tag{26-2}$$

In Eq. (26-2), P_o is the pressure at the outlet of the column (atmospheric pressure) and $P_o - P_w$ is the pressure at the flowmeter owing to the column effluent. The vapor pressure of water at each of several temperatures is listed in Table 26-1. Of course, the correction is not required if a flowmeter is used which does not pass the column effluent through water.

The adjusted flowrate F_c from Eq. (26-1) or Eq. (26-2) can be used to calculate the retention volume V of a particular component with a retention time t by use of

$$V = F_c t \tag{26-3}$$

That retention volume is adjusted to the temperature of the column but corresponds to the volume at the outlet pressure of the column. The average pressure within the column $\overline{P}$ can be shown to be given by[1]

$$\overline{P} = \frac{P_o}{j} \tag{26-4}$$

The *pressure drop correction factor* j is related to the inlet pressure of the column P_i and the outlet pressure of the column P_o by

$$j = \frac{3}{2}\left[\frac{(P_i/P_o)^2 - 1}{(P_i/P_o)^3 - 1}\right] \tag{26-5}$$

Table 26-1 Vapor pressure of water P_w at selected temperatures

Temperature, °C	P_w, torr
15	12.8
16	13.6
17	14.5
18	15.5
19	16.5
20	17.5
21	18.7
22	19.8
23	21.1
24	22.4
25	23.8
26	25.2
27	26.7
28	28.4
29	30.0
30	31.8
35	42.2

The retention volume corrected to the average pressure within the column is the corrected retention volume $V°$:

$$V^o = \frac{VP_o}{\overline{P}} = F_c\, tj \tag{26-6}$$

The *adjusted retention volume* V' is the difference between the retention volume [at the outlet pressure of the column, Eq. (26-3)] of the sample component and the retention volume of the mobile phase:

$$V' = F_c\, t - F_c\, t_m = F_c\,(t - t_m) \tag{26-7}$$

The retention time of the mobile phase t_m in GLC is usually measured as the retention time of air which is injected with the sample. Air is not retarded on GLC columns. The net retention volume V_n is the adjusted retention volume corrected to the average column pressure:

$$V_n = V'j = F_c\,(t - t_m)j \tag{26-8}$$

The net retention volume has been corrected to the flowrate at the column temperature and the average pressure within the column and for the retention time of the mobile phase.

In order to make retention volumes comparable between laboratories, two additional corrections are required. The retention volumes should be corrected for the amount of stationary phase packed into the column and to a constant temperature. The net retention volume can be corrected for those two variables to yield the *specific retention volume* V_g:

$$V_g = V_n\left(\frac{273}{W_L\,T_c}\right) = \frac{273F_c(t - t_m)j}{W_L\,T_c} \tag{26-9}$$

In Eq. (26-9), W_L is the mass of the stationary phase. Dividing by the mass corrects the retention volume to that expected for one gram of stationary phase. The ratio $273/T_c$ adjusts the retention volume to the value expected at a column temperature of 0°C. The specific retention volume is one recommended way to report gas chromatographic data which might be used by analysts outside the laboratory in which the data were generated. In practice, gas chromatographers rarely are required to use Eqs. (26-1) - (26-9) during normal laboratory assays.

Sample problem 26-1 A sample component had a retention time on a GLC column of 5.34 min. The retention time of air injected with the sample was 0.12 min and the flowrate through the column as measured with a bubble flowmeter at the outlet of the column was 29.5 cm^3/min. The temperature of the flowmeter was 27°C and the vapor pressure of water at 27°C is 27 torr (Table 26-1). The mass of the stationary liquid within the column was 0.957 g; the inlet pressure to the column was adjusted to 30.0 psig (pounds per square inch greater than atmospheric pressure); the outlet pressure was 755 torr (atmospheric pressure); and the column temperature was 125°C. Calculate the adjusted flowrate, the pressure drop correction factor, the average pressure within the column, and the specific retention volume of the sample component.

SOLUTION The adjusted flowrate F_c can be calculated from Eq. (26-2):

$$F_c = \frac{F_m\,T_c(P_o - P_w)}{T_R\,P_o} = \frac{29.5(398)(755 - 27)}{(300)(755)}$$

$$= 37.7 \text{ cm}^3\text{/min}$$

The pressure drop correction factor j is calculated with Eq. (26-5). The units of P_i and P_o must be identical when the equation is used so that j becomes a dimensionless term. Because P_i is 30.0 psig and P_o is 755 torr, it is necessary to convert P_o to the units of P_i (or P_i to the units of P_o):

$$30.0 \text{ lb/in}^2 \times \frac{1 \text{ atm}}{14.696 \text{ lb/in}^2} \times 760 \text{ torr/atm} = 1.55 \times 10^3 \text{ torr}$$

That is the inlet pressure *above* atmospheric pressure. The total inlet pressure is

$$P_i = 1.55 \times 10^3 \text{ torr} + 755 \text{ torr} = 2.31 \times 10^3 \text{ torr}$$

The pressure drop correction factor can be obtained by substitution into Eq. (26-5):

$$j = \frac{3}{2}\left[\frac{\left(2.31 \times 10^3 / 755\right)^2 - 1}{\left(2.31 \times 10^3 / 755\right)^3 - 1}\right] = 0.454$$

The average pressure $\overline{P}$ within the column can be calculated with Eq. (26-4):

$$\overline{P} = \frac{P_o}{j} = \frac{755}{0.454} = 1.66 \times 10^3 \text{ torr}$$

The adjusted retention volume V' is obtained by substitution into Eq. (26-7). Air is not retarded on GLC columns.

$$V' = F_c\,(t - t_m) = 37.7 \text{ cm}^3/\text{min } (5.34 \text{ min} - 0.12 \text{ min}) = 197 \text{ cm}^3$$

The net retention volume V_n can be calculated from V' with the aid of Eq. (26-8):

$$V_n = V'j = (197 \text{ cm}^3)(0.454) = 89.4 \text{ cm}^3$$

The specific retention volume V_g is obtained by substitution into Eq. (26-9):

$$V_g = \frac{273V_n}{W_L T_c} = \frac{273(89.4)}{0.957(398)} = 64.1 \text{ cm}^3/\text{g}$$

EFFICIENCY

The efficiency of a gas chromatographic column is expressed in terms of theoretical plates n or the height equivalent of a theoretical plate H as in liquid chromatography. The definitions of those terms for gas chromatography are identical with the definitions used for liquid chromatography. For gas chromatography sometimes it is convenient to use *rate* theory to express column efficiency rather than the plate theory which was described in Chapter 24 and used in Chapter 25 for LC. Rate theory can also be applied to liquid chromatography. According to rate theory, the height equivalent of a theoretical plate for gas chromatography varies with the velocity v of the carrier gas as shown in the *van Deemter equation*:

$$H = A + \frac{B}{v} + Cv \tag{26-10}$$

Theoretically the equation is applicable only to infinitely long columns. In practical terms that means columns containing at least 100 theoretical plates. Actually the equation works best for columns containing a minimum of 1000 plates.

For a specific set of experimental conditions, A, B, and C in the van Deemter equation are constants. The factors leading to A in the van Deemter equation are sometimes included in the C term, thereby eliminating A from Eq. 26-10. In liquid chromatography A is essentially zero. The A term is partially attributable to the different paths that can be taken through a packed column by individual molecules of a sample component. In nonpacked, open tubular, GC columns, A generally is negligibly small if the sample introduction system is efficient. In the absence of diffusion, molecules that enter the column at different points on the cross section of the packing material follow different paths through the column. Because the paths are not necessarily of identical length, molecules that take a shorter path exit the column before molecules that take a longer path.

In the presence of diffusion, a molecule in one path can diffuse to another path. At slow, mobile-phase flowrates, each molecule spends time in every possible path, consequently traveling an average pathlength and eliminating the effect of multiple paths through the column. At higher flowrates, the multiple paths contribute to band broadening and increased H because the molecules are eluted from the column before they have been able to spend time in all of the possible paths.

Another factor that is sometimes considered to contribute to A is related to the injection system of the chromatograph rather than to the column. Evaporation of a liquid sample in the injector does not occur instantaneously. Rather, the evaporation occurs in a manner that causes the sample to be exponentially introduced to the column. That can contribute to exponential tailing that is observed with some chromatographic peaks. As the flowrate of the carrier gas increases, the sample band entering the column becomes broader.

The second term in the van Deemter equation is the B term. B is equal to $2\gamma D_m$, where the *obstruction factor* γ is a function of the way the column is packed and D_m is the diffusion coefficient of the sample molecules in the mobile (gaseous) phase. Because the diffusion coefficient of a substance in the gaseous phase is typically 10^4 to 10^5 times larger than for the same substance in a liquid phase, it is apparent that diffusion in the mobile phase contributes a great deal more to peak broadening in gas chromatography than in liquid chromatography. As the diffusion coefficient of a sample molecule in the gas phase increases, the amount of peak broadening increases and the efficiency of the column decreases. The B term can be minimized, i.e., diffusion can be kept small, by using high values of the carrier gas velocity v. Increasing the carrier gas velocity decreases the time a sample spends in the column and the time available for band broadening.

The last term in the van Deemter equation, the C term, is inversely related to the time required for equilibrium to be established between the gaseous and liquid analyte component phases within the column. As the time required to establish equilibrium increases, the column efficiency decreases and H increases. At equilibrium the concentrations of a sample component in the mobile and stationary phases are those predicted by the distribution coefficient for the substance. Because equilibrium is not established instantaneously and because the gaseous phase is flowing, sample molecules in the gaseous phase are swept ahead of the center of the band while molecules in the stationary phase are delayed.

C in the van Deemter equation is the sum of a component owing to the stationary liquid phase C_L (in GLC) and a component owing to the mobile gaseous phase C_G. C_L is equivalent to $qk'd_f^2/(1+k')^2D_s$, where k' is the capacity factor, d_f is the thickness of the stationary liquid film on the solid support, and D_s is the diffusion coefficient of sample molecules in the stationary liquid phase. The *shape factor* q compensates for the variation of d_f in the stationary phase. It is usually between 0.13 and 0.67 although occasionally values outside of that range can be encountered. As d_f increases, H increases; i.e., column efficiency decreases. Consequently, the liquid film should be kept as thin as possible in order to maximize efficiency. As D_s increases, the efficiency of the column increases. To keep D_s relatively large, it is necessary to choose a liquid phase which has as low a viscosity as possible at the temperature at which the column is operated.

Although an exact expression for C_G in packed columns is unavailable, it is known that it increases with the square of the diameter of the packed particles and the square of the column diameter. Further, it is affected by the geometry of the column and the packing efficiency of the column as well as by other experimental variables. For open tubular columns, in which the stationary liquid is coated on the inner wall of the column, C_G is equal to $\left[\left(1+6k'+11k'^2\right)d_c^2\right]/\left[96(1+k')^2D_m\right]$ where d_c is the diameter of the column and D_m is the diffusion coefficient of the sample component in the gaseous phase. In packed columns small-diameter packing particles again lead to high efficiency. In open tubular columns the efficiency increases as the radius of the tubular column decreases. Consequently capillary columns often are used.

A plot of H as a function of v has the shape shown in Fig. 26-2. The observed H is the sum of the three terms of the van Deemter equation. Because A is usually independent of mobile-phase velocity throughout the normal range used for GC, its contribution to H is constant. At low flowrates the B/v term is large and the Cv term is small, whereas at high flowrates the B/v term is small and the Cv term is large. Calculus can be used with Eq. (26-10) to find the values of H and v at which a gas chromatographic column is operating at optimum efficiency, i.e., at which H is a minimum. Those values are given in the following equations:

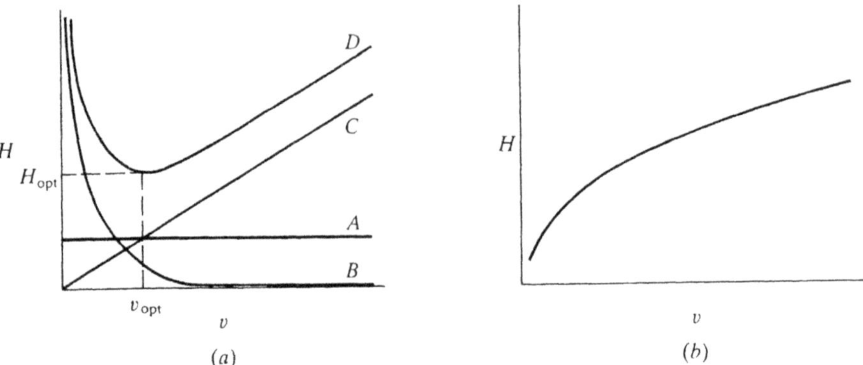

Figure 26-2 A graph of height equivalent of a theoretical plate H as a function of mobile phase velocity v for (*a*) gas chromatography and (*b*) liquid chromatography: curve A, contribution of A to H; curve B, contribution of the B term to H; curve C, contribution of the C term to H; curve D, the sum of A, the B term, and the C term. H_{opt}, the optimum value of H; v_{opt}, the optimum mobile-phase velocity.

$$H_{opt} = A + 2\sqrt{BC} \tag{26-11}$$

$$v_{opt} = \sqrt{\frac{B}{C}} \tag{26-12}$$

and are illustrated in Fig. 26-2. A more thorough description of the factors contributing to the efficiency of a separation in a chromatographic column can be found in Reference 2.

For the purpose of comparison, a plot of H as a function of v for liquid chromatography is shown in Fig. 26-2*b*. The shapes of the plots for gas chromatography and liquid chromatography are considerably different. That is primarily caused by the difference in diffusion coefficients of the sample molecules in the two mobile phases. In gases the diffusion coefficients are 10^4 to 10^5 times larger than in liquids.

APPARATUS

The apparatus used for gas chromatography consists of the several parts shown in the block diagram in Fig. 26-3. The source of the mobile phase is usually a high-pressure gas tank with the appropriate valves and pressure regulators. The injector accepts the sample, vaporizes it if necessary, and introduces the resulting gaseous sample into the inlet side of the column. The separation occurs in the column. The separated components leave the column and flow into the detector, which continuously monitors for their presence. The output from the detector is an electric signal which is displayed by a recorder or other readout device. The flowrate through the column is usually measured at the outlet of either the column or the detector. The injector, column, and detector are electrically heated in most instruments.

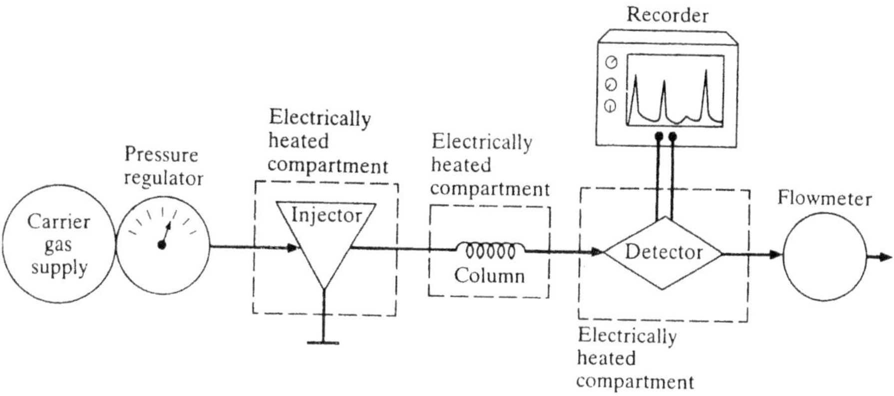

Figure 26-3 Diagram of a gas chromatograph. In many instruments a flow splitter is inserted between the gas supply and the injector, and a second injector and column are included. The effluent from each of the columns passes through a single differential detector. In some instruments a flowmeter is placed between the column and the detector.

Carrier Gases

In most cases the carrier gas (usually helium, hydrogen, nitrogen, or argon) is contained in a compressed-gas cylinder. In some cases another gas is used as the carrier gas. Hydrogen is combustible and should be used with care. The choice of carrier gas is sometimes dependent upon the type of detector used in the gas chromatograph. Helium is the most often used carrier gas.

Injectors

The injector changes the sample to a gas and reproducibly introduces the resulting gaseous sample into the column. The type of injection system used for a particular sample is dependent upon the physical state of the sample. Gaseous samples can be introduced into the column without heating the sample to convert it to a gas. Systems which use no heat and which inject the sample directly into the column are *on-column* inlets. A gaseous analyte can be introduced to the column in any of several ways. For routine gaseous analysis, a valve method is often preferred. The injection system is similar to the valve used to introduce liquid samples into an HPLC column except that the sample loop is considerably larger. In one valve position the carrier gas flows directly into the column. While the valve is in that position the sample loop is flushed and filled with the gaseous sample. Upon switching the valve to the inject position, the carrier gas sweeps through the sample loop and flushes the sample into the column. As an alternative, a gas syringe can be used to inject the sample into an injector designed for liquids. That method usually does not work as well as the valve method, but it is easier and less expensive for the analyst who performs only an occasional assay of a gaseous sample.

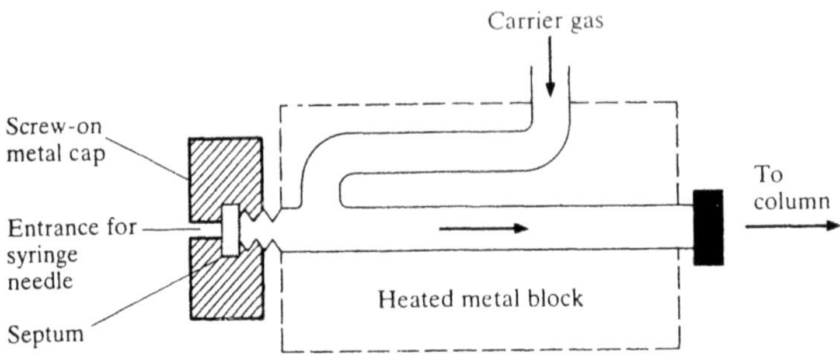

Figure 26-4 Diagram of a liquid-sample injector.

When gaseous samples, such as air samples, are to be assayed for low concentration components, often it is convenient to use *thermal desorption* TD. With this method, the gaseous sample is forced through sorbent tubes prior to entering the column. The tubes are filled with a thermally stable material which reversibly either adsorbs or absorbs the analyte component. Often the tubes are cooled to increase the efficiency of the process. After the analyte component has been concentrated in the tubes, the tubes are mounted in the inlet system where heat is applied to desorb the components as the carrier gas flushes the components into the column. All of the experimental parameters, including flow rate of the gaseous sample through the sorbent tubes, time of flow, initial sorbent tube temperature, desorption tube temperature, etc, are carefuly controlled. TD is one of the most used methods to sample and inject air pollutants into a GC.

The samples most often assayed by gas chromatography are liquids at room temperature. Liquid samples are usually injected with a 1-, 5-, or 10-μL graduated syringe through a septum into a heated flash-evaporation injector. The sample is rapidly vaporized because the injector's temperature is maintained above the boiling point of the highest boiling component in the sample. The carrier gas flushes the vaporized sample from the injector and into the column. Although injectors from different manufacturers differ somewhat in their design, they all function essentially in the same manner. A sketch of the major parts of an injector used for gas chromatography is shown in Fig. 26-4. Typical liquid sample volumes used within gas chromatography vary from about 0.1 to about 5 μL, and the injector temperature is usually held at least 10°C above the column temperature.

The injector shown in Figure 26-4 is a splitless injector, i.e., all of the injected sample enters the column. Split injectors function in the same manner as splitless injectors with the exception that, after the analyte has been vaporized, only a portion of it is flushed into the column. The remainder is removed through one branch of a Y-tube or T-tube which is attached to the outlet from the injector. By controlling the internal diameters of the two branches of the tubing which is attached to each of the alternative paths, the relative flow through each path is controlled. The smaller flow portion is directed into the column. Split-flow injectors are used to decrease the fraction of sample which enters the column. That might be necessary owing to the inability of a narrow bore (capillary) column or the detector to handle relatively large injection volumes. Typical split ratios are between 1:10 and 1:100.

A solid sample is usually more difficult than a liquid sample to convert to a gas. Sometimes a solid can be chemically converted to a volatile derivative. Fatty acids can be converted to volatile methyl esters by reaction with boron trichloride or boron trifluoride in methanolic solutions. The resulting liquid methyl esters of the sample fatty acids can be injected into the chromatograph with a syringe. Other types of chemical reactions which yield volatile reaction products can be used with other types of solid samples.

In some cases pyrolysis equipment is attached to the gas chromatograph and used to vaporize solid samples. The solid sample is placed in the pyrolysis apparatus, where it is heated sufficiently (often to 1000°C) to vaporize the sample or to form volatile sample decomposition products. The vaporized sample is swept into the gas chromatograph with carrier gas.

Pyrolysis gas chromatography has primarily been used to assay polymers and other high molecular weight compounds qualitatively. The chromatogram of the pyrolysis products of a sample is a *pyrogram*. Pyrograms are often complex and are used as "fingerprints" of samples for qualitative analyses. The pyrogram of a sample is compared to those of known materials that were obtained under identical experimental conditions. The process is similar to that used when comparing infrared spectra in the fingerprint region with those of standard substances. Care must be taken when using pyrolysis gas chromatography to follow a standard procedure during the pyrolysis in order to ensure reproducibility. In some cases it is possible to place solid samples directly into the injector, where the samples are volatilized and swept into the column with carrier gas.

A variation of pyrolysis gas chromatography, termed *chromatopyrography* (CPG), is described in Reference 3. The sample is heated in two steps rather than the single step used in traditional pyrolysis gas chromatography. After the sample is placed in the injector port, it is initially heated to 270°C. At that temperature all the volatile components are flash-vaporized and flushed into the chromatographic column with carrier gas. A chromatogram is recorded of the volatile components.

After the chromatogram is recorded, the temperature of the injection port is increased to 1000°C and a pyrogram of the pyrolysis products is recorded. The chromatogram recorded at the 270°C port temperature provides information relating to all of the volatile components in the sample. The port temperature is not sufficiently high, however, to cause decomposition of the polymeric backbone of the sample. At 1000°C the polymer is pyrolyzed and the recorded pyrogram can be used to identify the polymeric composition of the sample. Typical sample sizes vary from about 1 μg to about 1 mg. Although chromatopyrography is normally used to characterize polymers, it can be used to characterize non-polymeric materials.

If multiple samples are assayed by using the same GC procedure, it might be advantageous to utilize an autoinjection system. An automated injection system can relieve the tedium of performing many injections. Several types of autoinjectors are commercially available. The chosen automated injection system depends upon the nature and physical state of the samples.

HEADSPACE GAS CHROMATOGRAPHY

Headspace gas chromatography or *headspace analysis* is used to assay the volatile components of a solid or liquid solution that contains both volatile and nonvolatile components. Because the technique differs from other forms of gas chromatography only in the method of sample injection, it is described at this point in the chapter. Only the gaseous sample components in the *headspace* above the nongaseous portion of the sample are injected into the gas chromatographic column. Because the low-vapor-pressure components do not appreciably exist as gases above the sample, they are not injected into the column. More detailed descriptions of headspace gas chromatography can be found in References 4 and 5.

Sample introduction to the column is accomplished in three steps for liquid solutions. First, a measured volume of the sample is placed in a vial or other appropriate container of predetermined size. A cap that contains a septum is placed on the vial and the vial is placed in a constant-temperature bath. Often the temperature of the bath is adjusted to 80 or 90°C, although other temperatures can be used depending upon the volatility of the sample components. As thermal equilibrium is established, the volatile components of the sample partition themselves between the gaseous and liquid phases. As predicted by Raoult's law, the concentrations of the volatile components in the gaseous phase are proportional to their concentrations in the liquid phase. A partition coefficient can be written for each volatile component that is the ratio of its concentration in the gaseous phase to its concentration in the liquid phase.

After thermal equilibrium has been established, the vial is pressurized with the mobile-phase gas by inserting a metallic tube through the septum in the cap of the sample vial and into the gaseous phase (the headspace) within the vial. The pressure of the gas in the sample vial and the time for pressurization is carefully controlled. After pressurization the tube is withdrawn and a second tube, attached to the inlet of the column, is inserted through the septum into the headspace in the vial. The pressure within the vial causes a portion of the gaseous phase to be injected. The injection period is usually controlled. Finally, the second tube is removed and the chromatogram is recorded. Often an internal standard or the standard-addition technique is used for quantitative analysis of the volatile sample components in order to minimize errors arising from matrix effects.

If the sample is a solid, *multiple headspace extraction* (MHE) can be used to introduce the volatile components of the sample to the gas chromatograph. Essentially the technique consists of repetitively extracting the volatile components from the sample as in headspace chromatography. Because the amount of the volatile components in the sample decreases with each extraction, successive chromatographic peaks decrease in size. Either the extractions are continued until the chromatographic peaks become negligibly small or a series of several (often two to five) extractions are performed and the corresponding chromatograms recorded.

If the extractions are continued until the volatile components in the sample have been exhausted, the sum of the chromatographic peak areas for a particular volatile component is determined and used for quantitative analysis. Theoretically, the area A of each chromatographic peak decreases with the number n of extractions:

$$\ln A = -k(n-1) + \ln A_1 \qquad (26\text{-}13)$$

where A_1 is the area of the peak after the first extraction and k is a constant. If the extractions are not continued until the volatile component is exhausted, a plot of $\ln A$ as a function of n is prepared. The linear plot has a slope of $-k$. The value of k is often obtained from the equation for the best straight line through the data points as determined from a least-squares fit of the data. The value of k is used with

$$\sum A = \frac{A_1}{1-e^{-k}} \qquad (26\text{-}14)$$

to estimate the sum of the areas of all the peaks that would have been obtained if the extractions were continued until the volatile component was exhausted. Descriptions of some of the many applications of headspace gas chromatography can be found in References 6, 7, and 8.

Purge and trap (*PT*) inlet systems combine head-space analysis with the thermal desorption method. An inert gas, which often is helium or nitrogen, is bubbled through an aqueous analyte solution causing the volatile components to be flushed from the solution. The mixture of inert gas and volatile components flows through a tube which contains a sorbent. The sorbent captures and concentrates the volatile components. After sufficient components have been captured, the tube is heated to desorb the analyte components and the components are flushed into the GC column.

Columns

Two categories of columns are used for gas chromatography. *Packed columns* are those which contain particles of the stationary phase packed into a metallic or glass tube. The metallic tubes usually are constructed of stainless steel, but copper and aluminum tubes are also used. A packed column generally has a diameter of between 2 and 10 mm and a length of between 1 and 4 m. The length of the packed column is dependent upon the required number of theoretical plates. To fit into the temperature-controlled oven in the gas chromatograph, the packed column usually must be carefully bent or coiled. The bending or coiling is done in a manner which does not impede the flow of carrier gas through the column. The stationary phase in packed columns is either small particles of a solid adsorbent for GSC or a liquid supported on small particles of a solid for GLC.

Open tubular columns (OTCs) constitute the second category of GC columns. Rather than being filled with particles of packing material, they have the liquid or solid stationary phase coated on the column's inner wall. Typical coating thickness varies from 0.1-10 micrometers. Because the pressure drop through open tubular columns is considerably less than that through packed columns, it is possible to use open tubular columns which are longer than packed columns. Typically, laboratory open tubular columns vary in length between 10 and 100 m. Considerably shorter columns normally are used in portable gas chromatographs which are designed for field use. The inner diameter of open tubular columns often is between 0.05 mm. Most open tubular columns are constructed of glass or fused silica rather than of metal, because metals can catalyze several chemical reactions within the column at the temperatures at which the column is operated. Generally, columns are operated at temperatures about 10 to 25°C above the boiling point of the highest-boiling sample component during isothermal operation. Most separations are better accomplished by using programmed temperature ovens or programmed pressure.

Packed Columns

The function of the solid support in GLC packed columns is to provide an inert surface on which the liquid phase can be coated and to provide an appropriate physical structure for the stationary phase. The solid support must be thermally stable at the temperatures at which the column will be operated, and it should consist of uniformly sized particles. The most common solid supports are derived from diatomaceous earths. Diatomaceous earths are the skeletal remains of the unicellular algae known as diatoms. Diatomaceous earths primarily consist of hydrated silica groups. The large surface area of treated diatomaceous earths is nearly ideal as solid supports for GLC.

Several ways can be used to treat diatomaceous earths which result in good solid supports. If the diatomaceous earth is heated to about 900°C and crushed to particles of uniform size, the result is a pink substance known as Chromosorb P. Chromosorb P possesses polar functional groups which make it a good adsorbent. If Chromosorb P is not coated with a liquid phase, it can be used as the stationary phase in GSC. Because Chromosorb P does not completely lose its adsorbing ability upon being coated with a liquid, it is primarily used as a support for separations of samples containing relatively nonpolar components. Polar sample components can be adsorbed on Chromosorb P as well as dissolved in the stationary liquid coated on the support.

If the diatomaceous earth is heated in the presence of sodium carbonate, a white solid called Chromosorb W is obtained. Chromosorb W suffers the slight disadvantage of being able to be crumbled into small particles if handled roughly. Other treatments of diatomaceous earths result in formation of other solid supports known as Chromosorb A and Chromosorb G. All of the Chromosorb supports are products of the Johns-Manville Corporation. In addition to the products of the diatomaceous earths, several other substances such as powdered Teflon, graphitized carbon black, cadmium chloride, volcanic scoria, Armenian neomberian deposits, activated carbon, many organic polymers, alumina, silica gel, aluminum silicate, magnesium oxide, magnesium carbonate, vermiculite, and glass beads have been used as solid supports. Generally, the diatomaceous earth derivatives are preferred to the other supports. Kieselguhr and Embacel are solid supports which some chromatographers prefer to use. If the solid support is more polar than desired, it can be *deactivated* (made less polar) by coating it with dimethylchlorosilane (DMCS) or by rinsing the support with a solution of a strong acid (usually HCl). Acid washing removes metallic impurities on the support and coats some of the polar sites.

The stationary liquid phase used in GLC packed columns and in some open tubular columns must be a nonvolatile liquid at the temperature at which the column is operated. The sample components must be at least slightly soluble in the liquid phase in order to be separated, and the liquid should, if possible, be a pure compound. Hundreds of liquids exist which meet those requirements and which have been examined as liquid phases for GLC. Generally, liquid phases are classified according to their polarity. Polar sample components are retained longer in stationary polar liquid phases than are nonpolar components, and separation is partially based on the polarity of the sample components.

Generally it is best to choose a stationary phase which has a polarity that is similar to the polarities of the analyte components. The stationary phase and analyte components undergo dipole-dipole interactions. Other factors which can influence retention times on a particular stationary phase include hydrogen bonding and aromatic content of the stationary phase. Retention times of aromatic compounds are greater on stationary phases which have aromatic components.

With nonpolar stationary liquid phases, the components are separated according to their relative solubilities in the liquid and gaseous phases. Generally, as the component increases in molecular weight, its volatility (i.e., its solubility in the gaseous phase) decreases and the time it spends on the column increases. On nonpolar columns, retention times (or volumes) are usually related to the molecular weights or boiling points of the sample components. Generally, the retention time of a component increases with increasing molecular weight and with increasing boiling point.

Polar substances are usually better separated on a polar stationary phase and nonpolar substances on a nonpolar stationary phase. Most GLC separations can be performed with one polar and one nonpolar column; i.e., the components that cannot be separated on the polar column can usually be separated on the nonpolar column. Many chromatographers simultaneously keep polar and nonpolar columns in their dual-column chromatographs. Generally a separation can be accomplished on one of the two columns.

It is obvious that the polarity of the stationary liquid phase is important. Of the several methods which have been used to express the relative polarity of various stationary phases, the method first used by Rohrschneider[9] and later revised by McReynolds[10] appears to be the most successful. In his original paper, Rohrschneider used the retention index I (see the description of the Kovats retention index later in the chapter) for each member of a series of five compounds (benzene, ethanol, methylethyl ketone, nitromethane, and pyridine) on the stationary phase of interest at a column temperature of 100°C. The retention index is related to the retention time of the compound on the particular stationary phase. The change in the retention index ΔI of each reference compound between that observed on a column containing the stationary phase of interest and a column containing squalane is related to the difference in retention times of each reference compound on the two columns. Squalane was chosen as the reference stationary phase because it is essentially nonpolar. It is a 30-carbon paraffin. A value of ΔI is calculated for each of the reference compounds. As the polarity of the stationary phase increases, the retention time of the polar reference compounds increases and ΔI increases. Consequently, higher values of ΔI mean the stationary phase is more polar.

McReynolds used the same approach as Rohrschneider except that 10 reference compounds were chosen and the measurements were made at 120°C. The values of ΔI calculated by McReynolds are referred to as the *McReynolds constants*. Tarjan *et al.*[11] have combined the McReynolds constants of the first five of the McReynolds standards (the five farthest to the left in Table 26-2) to yield an average retention polarity $\overline{P_R}$ which, in a single number, characterizes the polarity of the stationary phase. The average retention polarity is the ratio of the average percentages by which the Is of the first five reference compounds on the particular column exceed the Is of the reference compounds on Squalane to the I on Squalane:

$$\overline{P_R} = \frac{1}{5}\sum\frac{\Delta I}{I_{\text{squalane}}} \times 100 = 20\sum\frac{\Delta I}{I_{\text{squalane}}} \qquad (26\text{-}15)$$

Table 26-2 The McReynolds constants ΔI and average retention polarities $\overline{P}_R$ of several of the more popular GLC stationary phases that are used in packed columns.

Stationary phase	McReynolds constants										$\overline{P}_R$
	Benzene	**Butanol**	**2-Pentanone**	**Nitropropane**	**Pyridine**	**2-Methyl-2-pentanol**	**1-Iodobutane**	**2-Octyne**	**1,4-Dioxane**	***cis*-Hydrindane**	
Squalane	0	0	0	0	0	0	0	0	0	0	0
Apiezon L	32	22	15	32	42	13	35	11	31	33	4.39
Silicone SE-30	15	53	44	64	41	31	3	22	44	−2	6.80
Silicone DC-200	16	57	45	66	43	33	3	23	46	−3	7.11
Silicone DC-550	74	116	117	178	135	81	74	72	128	36	19.25
Silicone QF-1	144	233	355	463	305	203	136	53	280	59	45.99
XE-60 Polyethylene glycol 20M†	322	536	368	572	510	387	282	221	434	148	70.75
Diethylene glycol succinate	492	733	581	833	791	579	418	321	705	237	110.03

† Polyethylene glycol 20M is also known as Carbowax 20M.

Several of the most often used GLC stationary phases, their McReynolds constants, and their retention polarities are listed in Table 26-2. As $\overline{P_R}$ increases, the polarity of the stationary phase increases. Reviews of stationary phases as well as of other aspects of GC can be found in Analytical Chemistry (Reference 12).

Chemically bonded stationary phases similar to those used for HPLC have also been used for gas chromatography. The liquid stationary phase is usually bonded either to small-diameter glass beads as in HPLC, to silica gel, or to the inner wall of open tubular columns. The separation which occurs on columns containing chemically bonded phases is caused by a combination of solvation in the attached liquid phase and adsorption on the solid support. Among the liquid phases which have been chemically bonded and which are commercially available are Carbowax 20M, Carbowax E-4000, cyano diethylene glycol succinate (cyano DEGS), cyano silicone, DEGS, methyl silicone, and nitro DEGS. A description of the procedure used to bond the liquid phase to the solid support and of the applications of bonded phases in gas chromatography can be found in Reference 13. A description of bonded phases in open tubular columns is provided later in the chapter.

The liquid stationary phases that have been described separate components based primarily upon their polarity or their molecular weight. Because both polarity and molecular weight are related to boiling point, the components are often separated according to boiling points. Geometric isomers generally have nearly identical boiling points and consequently are difficult to separate with most stationary phases.

Until recently the preferred stationary phase for separation of isomers was Bentone. It is prepared by treating bentonite with dimethyldioctadecylammonium salts. Bentonite is a clay that has the composition $Al_2O_3(SiO_2)_4 \cdot H_2O$. The treated bentonite is suspended in benzene and mixed with the solid support. As the excess benzene evaporates, the stationary phase coats on the solid support. Apparently separation on Bentone is dependent upon the ability of the components to penetrate into the regions between the layers of the expanded clay. Sample components that can penetrate between the layers are adsorbed on the clay and held on the column longer than components that cannot penetrate the layers. Because penetration between the layers is dependent upon the shape of the component, separation is primarily based upon molecular shape rather than boiling point. It is more common now to use high polarity bonded stationary phases in capillary open tubular columns to separate aromatic isomers.

LIQUID CRYSTALLINE STATIONARY PHASES

Another class of stationary phases that allows separation of geometric isomers is the *liquid crystals*. Some substances exhibit behavior that is intermediate between that expected for a solid and that expected for a liquid. Those substances are liquid crystals. The liquid crystals that are used as gas chromatographic stationary phases are organic compounds that have elongated rigid structures. Often the molecules have polar functional groups on the ends.

The molecules in liquid crystals have more freedom of movement than molecules in a solid, but have less freedom than that normally expected for a liquid. A *lyotropic* liquid crystal is obtained by mixing a polymer or surfactant with a solvent. Lyotropic liquid crystals exhibit the properties of liquid crystals over a relatively large concentration range. A *thermotropic* liquid crystal is obtained by melting an appropriate solid. Thermotropic liquid crystals have been used for gas chromatographic stationary phases. Further heating of a thermotropic liquid crystal causes a change from the liquid crystalline state to the liquid state. The temperature at which the transition occurs is the *clearing point*.

Several types of liquid crystals are possible depending upon the orientation of the molecules within the liquid crystal. In a *nematic* liquid crystal, the molecules are arranged so that the axes corresponding to their lengths are approximately parallel to each other, as illustrated in Fig. 26-5. The solid support upon which the crystals are coated can cause the alignment of molecular axes to be horizontal, vertical, or at some other orientation relative to the surface. The parallel molecules in a nematic liquid are approximately randomly distributed throughout the liquid crystal; i.e., they are not layered. Thermal motion of the molecules in a nematic liquid crystal causes the molecules to deviate from perfect parallelism. The amount of deviation is indicated by a parameter S, which varies from zero for randomly arranged molecules to unity for perfectly parallel molecules. Nematic liquid crystals have values of S between 0.3 and 0.8.

A *cholesteric* liquid crystal consists of a twisted nematic structure. The angle of orientation of the parallel molecules varies with distance through the liquid crystal as illustrated in Fig. 26-5. Otherwise a cholesteric liquid crystal is identical to a nematic liquid crystal.

The final category is the *smectic* liquid crystals. The molecules in a smectic liquid crystal are more parallel than those in nematic liquid crystals (S is greater than 0.8) and the molecules are arranged in layers (Fig. 26-5). Depending upon the angle of orientation of the molecules relative to the layer, eight types of smectic liquid crystals are possible. Each type is given a letter between A and H.

Thousands of organic compounds can be classified as thermotropic liquid crystals within a fixed temperature interval. In order to be useful as a stationary phase for gas chromatography, the liquid crystal should be thermally stable, have a low vapor pressure, and be capable of adhering to a solid support. A list of approximately 100 liquid crystals that have been used as gas chromatographic stationary phases is given in Reference 14. Listings of liquid

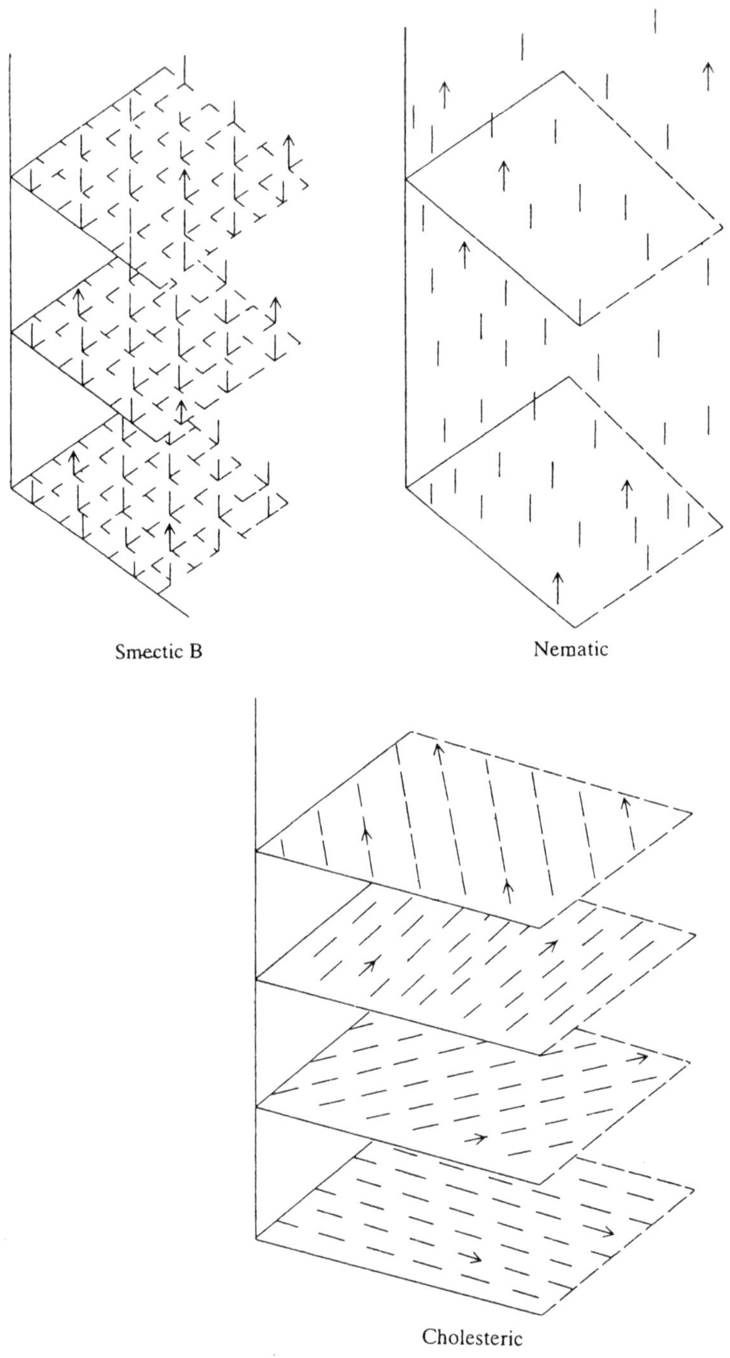

Figure 26-5 Diagrams of the molecular arrangements in some liquid crystals. The alignment within each layer in the cholesteric crystals is not perfect, as shown in the illustration. The arrows indicate the orientation of the molecules.

crystalline stationary phases also can be found in References 15-17. An example of a liquid crystalline stationary phase is *p ,p*-azoxyphenetole,

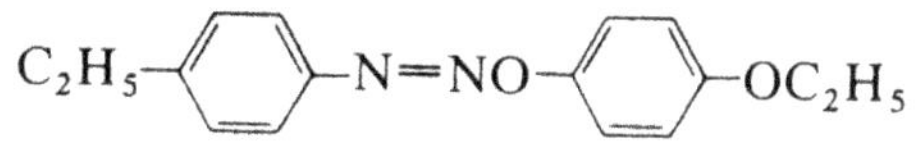

$$C_2H_5-C_6H_4-N{=}NO-C_6H_4-OC_2H_5$$

which has a nematic phase between 138 and 168°C.

Separation on liquid crystalline stationary phases is based upon the geometrical shapes of the sample components, polar interactions between the component and the stationary phase, and dipole-dipole interactions between the component and the stationary phase. Of particular importance are the separations based upon the shapes of the sample molecules. To a first approximation, the separation is based upon the length-width ratio of the component molecules. As the molecule becomes longer and more narrow, it more easily fits between the molecules of the liquid crystalline stationary phase and consequently is retained on the column longer. Similarly, planar molecules are retained longer than nonplanar molecules. Liquid crystalline stationary phases are suited for separating geometric isomers that cannot easily be separated on other stationary phases. As an example the ortho, meta, and para isomers of xylene can be easily separated on liquid crystalline stationary phases. The order of elution is meta, para, ortho. A disadvantage of the use of liquid crystalline stationary phases is the relatively limited temperature range throughout which the phases exhibit liquid crystalline properties. Considerably more detailed descriptions of the use of liquid crystals as gas chromatographic stationary phases can be found in References 14-17.

Open Tubular Columns

Several types of open tubular columns are available. The inner diameter of commercial open tubular capillary columns is between 0.05 and 0.5 mm. The *support-coated open tubular* (SCOT) column has a thin layer of a solid support coated on the inner wall of the column. The stationary liquid phase is coated on the solid support. Although the solid support is often a diatomaceous earth, silica gel, graphite, metal oxides, and silicates also have been used. Because the solid support vastly increases the internal surface area of the column, the ability of the column to handle large samples is increased.

Wall-coated open tubular (WCOT) columns have a thin layer of a stationary liquid phase coated directly on the inner wall of a pretreated glass capillary or fused silica column. If the stationary liquid phase is simply coated on the column wall, it is generally necessary to maintain the thickness of the film at about 1 μm or less in order to prevent *bleeding* of the column. Bleeding occurs when the stationary phase is slowly eluted from the column by the mobile phase. Generally bleeding increases with increasing column temperature. Most commercially available columns have a recommended upper operating temperature. At temperatures above that value, bleeding becomes significant.

Bleeding is prevented in modern WCOT columns by cross-linking the stationary phase within the column and by partially bonding the stationary phase to the column. Short-chain portions of a stationary phase that contains a small proportion of vinyl linkages usually are reacted in the column with a thermally decomposed peroxide. The peroxide provides a free radical that permits cross-linkage. As an alternative, cross-linking can be caused by irradiating[18] the stationary phase with gamma radiation from a source such as ^{60}Co. A fused silica column which has been irradiated with gamma rays is an *irradiated fused silica open tubular* (IFSOT) column.

By using cross-linked stationary phases, it is possible to prepare WCOT columns with film thicknesses up to about 8 μm without significant bleeding. As the thickness of the stationary phase increases, the loading capacity of the column increases but the efficiency of the column decreases and retention times increase. Generally a relatively thick film is used when large samples are required, such as when trace analyses of environmental samples are performed. Normally the film thickness in commercial columns is between 0.1 and 10 μm. A film thickness of 0.25 μm is used for many analytical studies although thicknesses as small as 0.1 μm have been used.

WCOT columns are more efficient than SCOT columns, but their loading capacity is usually less; i.e., they cannot handle as much sample. For a particular stationary phase, the loading capacity of open tubular columns increases with both stationary layer film thickness and column diameter. Other types of open tubular columns include *porous-layer open tubular* (PLOT) columns and *whisker-walled open tubular* (WWOT) columns. A PLOT column has an inner wall coated with a porous layer of the stationary phase and a WWOT column has whiskerlike silica needles coated with the stationary phase on the inner wall of the column.

In *bonded-phase* columns the stationary phase is chemically bonded directly to the walls of the column. Either glass or fused silica columns can be used. Glass is less expensive but contains small amounts of metallic ions that can act as active sites within the column. The metallic sites can increase the retention of polar components. If the column wall is sufficiently thick, the metals can sometimes be removed by leaching, but that process can cause the glass to become brittle. Fused silica columns are used more often than glass columns.

Fused silica is essentially high-purity glass with the composition SiO_2. Fused silica is more expensive than glass but does not contain the metallic impurities present in glass. Consequently, resolution and efficiency are generally higher with fused silica columns. Another advantage of fused silica is its lack of brittleness; i.e., it is easier to handle a fused silica column than a glass column without damaging the column.

Table 26-3 Several stationary phases which are used in open tubular columns

Stationary phase	Composition	Relative polarity
Dimethylpolysiloxane	100%	Nonpolar
Diphenyl-dimethylpolysiloxane	5% phenyl, 95% methyl	Nonpolar
	50% phenyl, 50% methyl	Polar
Arylene-dimethylpolysiloxane	35% arylene, 65% methyl	Polar
Cyanopropylphenyl-dimethylpolysiloxane	50% cyano, 50% methyl	Highly polar
Cyanopropyl-arylene-dimethylpolysiloxane	6% cyano, 94% methyl	Polar
	50% cyano, 50% methyl	Highly polar
Methyltrifluoropropyl-dimethylpolysiloxane	35% fluoro, 65% methyl	Polar
	50% fluoro, 50% methyl	Highly polar
Phenylmethyl-cyanopropylmethyl-	25% phenyl, 25% cyano, 50% methyl	Polar
dimethylpolysiloxane	7% phenyl, 7% cyano, 86% methyl	Slightly polar
Phenylmethyl-vinylmethyl-dimethylpolysiloxane	5% phenyl, 1% vinyl, 94% methyl	Nonpolar
Polyethylene glycol (PEG)	100%	Polar
Base modified PEG	100%	Polar
Acid modified PEG	100%	Polar

The stationary phase in bonded-phase columns is chemically bonded to silicon atoms on the inner surface of the column. Columns that have a bonded stationary phase have a longer lifetime than those that do not have a bonded phase. Most of the stationary phases that have been chemically bonded to the column are similar to those used in packed columns. Several popular bonded stationary phases are listed in Table 26-3. The percentages listed in the table are calculated on a molar basis of the substituents on the Si—O—Si backbones. The chemical structures of some of the bonded stationary phases are shown in Fig. 26-6.

Open tubular columns generally have an inner diameter that is considerably smaller than that of packed columns. Typical internal diameters for WCOT and fused-silica bonded-phase columns vary from 0.05 to 0.53 mm. Because the diameters of open tubular columns are small, they are often referred to as *capillary columns*. Capillary columns generally have a length between 10 and 100 m. Column lengths of 10, 15, 25, 30, 50 and 60 m are popular. Older open tubular columns often had lengths in excess of 100 m. The increased efficiency of modern columns has eliminated the need for longer columns for most assays.

Porous layer open tubular (PLOT) columns are primarily used to separate low molecular weight gases (carbon dioxide, methane, carbon monoxide, air components, argon, oxygen, etc.), saturated hydrocarbons containing 12 carbons or less, and organic solvents. The stationary phases used in PLOT capillary columns consist of small porous particles of a molecular sieve zeolite or other polar adsorbent. The stationary phase usually is attached to the column wall with the aid of a binder. Typical stationary phases are 0.5 nm (pore size) molecular sieves, alumina or alumina which has been deactivated with potassium chloride, sodium sulfate or another substance, polystyrene-divinyl benzene, divinylbenzene-ethylene glycol-dimethylacrylate and activated carbon. Molecular sieve columns primarily are used to separate inorganic and low molecular weight organic gases. PLOT stationary phases separate components based upon each component's ability to be adsorbed on the stationary phase. In addition to adsorption, molecular sieve stationary phases also perform separations based upon the ability of each component to enter the pores in the stationary phases, i.e., smaller components generally have longer retention times.

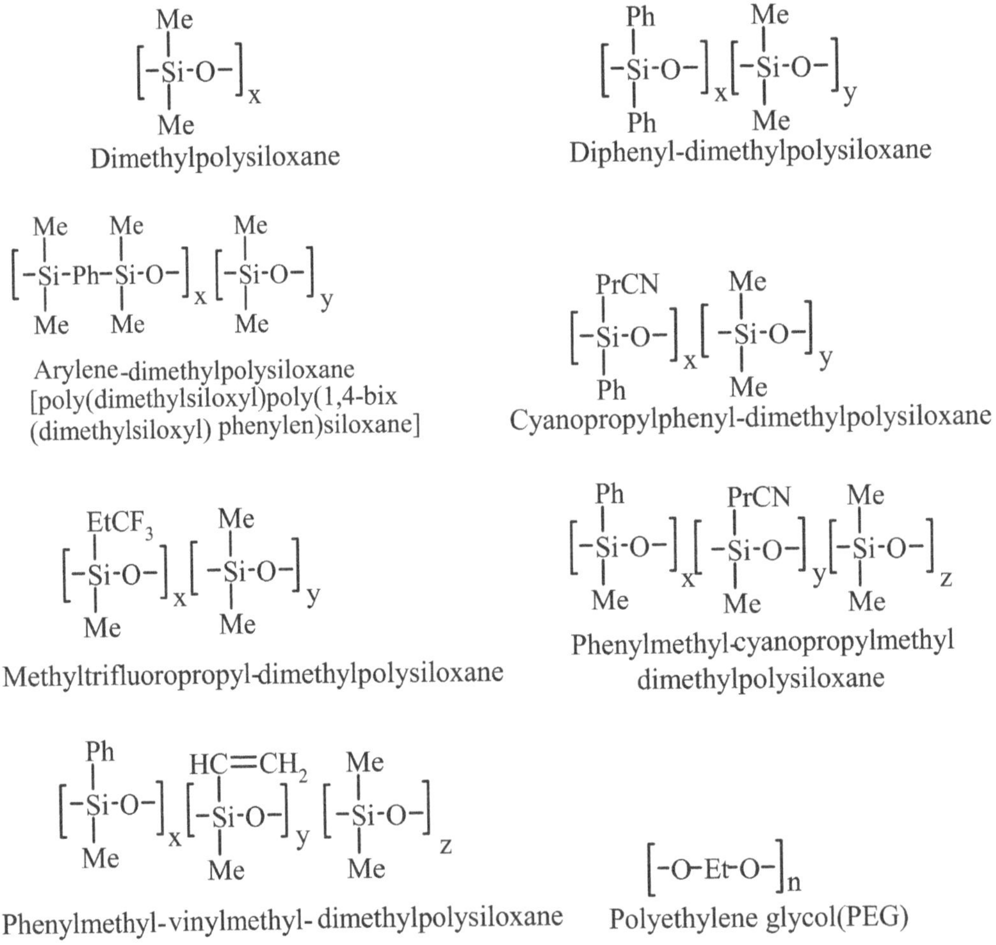

$[-Si(Me)(Me)-O-]_x$

Dimethylpolysiloxane

$[-Si(Ph)(Ph)-O-]_x[-Si(Me)(Me)-O-]_y$

Diphenyl-dimethylpolysiloxane

$[-Si(Me)(Me)-Ph-Si(Me)(Me)-O-]_x[-Si(Me)(Me)-O-]_y$

Arylene-dimethylpolysiloxane
[poly(dimethylsiloxyl)poly(1,4-bix
(dimethylsiloxyl) phenylen)siloxane]

$[-Si(PrCN)(Ph)-O-]_x[-Si(Me)(Me)-O-]_y$

Cyanopropylphenyl-dimethylpolysiloxane

$[-Si(EtCF_3)(Me)-O-]_x[-Si(Me)(Me)-O-]_y$

Methyltrifluoropropyl-dimethylpolysiloxane

$[-Si(Ph)(Me)-O-]_x[-Si(PrCN)(Me)-O-]_y[-Si(Me)(Me)-O-]_z$

Phenylmethyl-cyanopropylmethyl dimethylpolysiloxane

$[-Si(Ph)(Me)-O-]_x[-Si(HC{=}CH_2)(Me)-O-]_y[-Si(Me)(Me)-O-]_z$

Phenylmethyl-vinylmethyl- dimethylpolysiloxane

$[-O-Et-O-]_n$

Polyethylene glycol(PEG)

Figure 26-6 The chemical structures of selected bonded-silicone stationary phases Me, methyl; Et, ethyl; Pr, propyl; Ph, phenyl.

Open-tubular capillary columns have several advantages when compared to packed columns. Because the diameter of the column is small, the number of different paths through the column that a sample component can follow is minimized. That leads to less band broadening (refer to the earlier description of the van Deemter equation). Heat transfer to a column with a smaller diameter is superior to that of the larger packed columns, and the coating on WCOT and bonded-phase columns generally is more uniform than the stationary phase used in packed columns. All of those factors combine to produce better resolution on coated capillary columns than on packed columns.

On-column injections into open tubular columns are sometimes advantageously used. The sample is directly injected into the column rather than into a heated injector. On-column injections are primarily used for compounds that cannot tolerate the high temperature of an injection port or which do not vaporize rapidly and which consequently would produce broad chromatographic bands if placed in a heated injector. A description of the factors to be considered when choosing a fused silica column is found in Reference 19. Capillary columns are used for most assays performed by using GC.

SOLID STATIONARY PHASES

The distribution coefficients of gaseous sample components in packed GSC columns and in PLOT columns containing solid adsorbents are generally considerably greater than the distribution coefficients of the same components in GLC columns. That causes retention times in GSC columns to be considerably longer than those in GLC columns. For that reason, GSC is not used as often as GLC and is normally used only to separate relatively low molecular weight sample components, such as the isomers of butane, acetylene, and ethylene. The solid adsorbents often used in GSC columns include graphitized carbon blacks, charcoals, silica gel and surface-modified silica gel, alumina and surface-modified alumina, Porapak Q, Porapak R, zeolites, ion exchangers such as sulfonated porous polymers, and highly porous synthetic polymers.

DETECTORS

The detectors that are used for gas chromatography are devices which continuously monitor some property of the eluate from the chromatographic column. Some detectors respond to all sample components. Others selectively respond to a limited number of substances. When choosing a detector for use in gas chromatography, the analyst should keep in mind the chemical composition of the samples and the concentrations of the sample components.

Thermal Conductivity Detector

Thermal conductivity is a measure of the ability of a substance to conduct heat. The *thermal conductivity detector* (TCD) is a device that monitors the rate of heat transfer away from an electrically heated wire by the column eluate. The temperature of the wire is related to the rate of heat transfer away from the wire. The temperature of the wire decreases as the rate of heat conduction away from the wire increases. Because the temperature of the wire is directly proportional to the electrical resistance through the wire, it is possible to measure the rate of heat transfer away from the wire by continuously monitoring the resistance of the wire. As the resistance increases, the conduction of heat away from the wire decreases.

The thermal conductivity detector is a differential detector. In the two-filament version shown in Fig. 26-7, pure carrier gas flows through one side of the detector while the column eluate flows through the other. In gas chromatographs which contain two columns, one column is connected to one side of the detector and the other column to the second side. In single-column chromatographs, the carrier gas stream is split into two portions between the gas supply and the injector. Part of the carrier gas passes through the injector and column before entering one side of the detector, and the remaining pure carrier gas flows directly into the other side of the detector. A diagram of a dual-column GC which contains a differential detector is shown in Fig. 26-8.

The relative resistances of the two heated filaments in the TCD are continuously monitored with a Wheatstone bridge (Chapter 2) or a modified version of the bridge. In the four-filament form of the TCD, two of the heated filaments are placed in pure carrier gas and two in the column eluate. Each filament is connected to a separate arm of a Wheatstone bridge. The metallic block of the detector is usually made from stainless steel. It is heated to a constant temperature to prevent changes in its temperature from having an effect on the rate of heat conduction away from the wires. The block temperature is usually adjusted about 10 to 25°C above the column temperature in order to prevent condensation of the sample components in the detector.

The output from the Wheatstone bridge is electronically adjusted such that the recorder displays no deflection while pure carrier gas flows through both sides of the detector. When a sample component passes through one side of the detector and pure carrier gas passes through the other side, the resistance of the wire in the sample side changes depending upon the ability of the sample component to conduct heat away from the filament. The thermal conductivity of the sample component is generally not identical to that of the carrier gas. The change in resistance is monitored and changed to a useful electrical signal (usually a potential). The electrical signal is displayed by the strip-chart recorder or other readout device.

The ability of a gas to conduct heat depends to a great extent upon the average velocity of the gaseous molecules within the detector. As the average velocity of the molecules increases, the gas more rapidly carries heat from the filament. Because light molecules move more rapidly than heavy molecules at a fixed temperature, the thermal conductivity of a gas increases as its molecular weight decreases. In order to maximize the sensitivity of the detector, the carrier gas is chosen so that it has a thermal conductivity which differs as much as possible from that of the sample components. In practice that means a low molecular weight gas is chosen as the carrier gas. Because the lowest molecular weight gas, hydrogen, is highly flammable, most chromatographers prefer to use helium.

The sensitivity of the TCD is affected by the amount of electric current flowing through the filaments. An increase in the filament current leads to a hotter wire, an increase in the temperature differential between the wire and the detector block, and an increase in the sensitivity of the detector. Lower filament currents yield lower sensitivities but longer filament lifetimes. Generally it is advisable to adjust the filament current no higher than is required to achieve the needed sensitivity for a particular assay. Because detector sensitivity varies with filament current, the filament current should not be altered during development of a chromatogram.

The thermal conductivity detector has several advantages in comparison to other GC detectors. The TCD is a universal detector; i.e., it responds to all substances except the carrier gas. The TCD is nondestructive. After a sample component has passed through the detector, it can be collected for later use. The TCD is a rugged detector which can take a good deal of maltreatment and continue to function well. For those reasons, the TCD is a popular GC detector for general use.

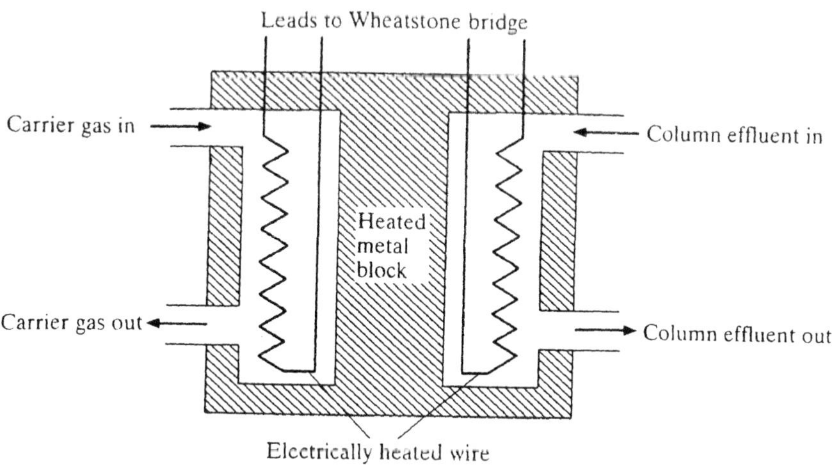

Figure 26-7 Diagram of a two-filament thermal conductivity detector.

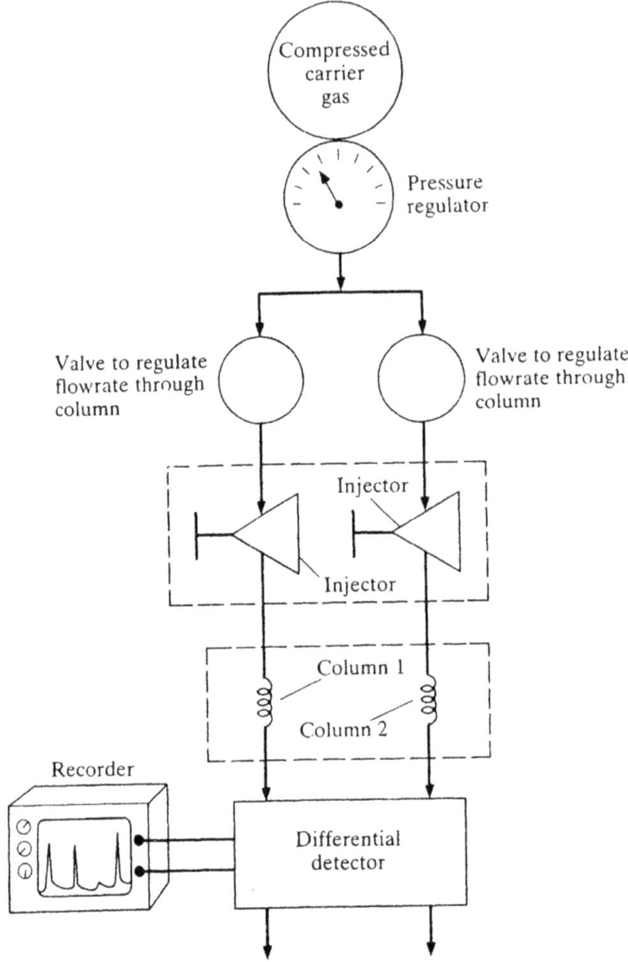

Figure 26-8 Diagram of a dual-column gas chromatograph. The direction of carrier gas flow is indicated by the arrows. The dashed lines around the injectors represent the heated injector block and the dashed lines around the columns indicate a heated oven.

Flame-Ionization Detector

In the *flame-ionization detector* (FID), the eluate coming from the column is combined with hydrogen (the fuel) and air (the oxidant) to form a combustible mixture. The ignited air-hydrogen flame provides sufficient energy to ionize most organic and a few inorganic sample components in the eluate. The relatively low temperature (approximate maximum temperature is 2000 °C) of the air-hydrogen flame is generally sufficient to ionize only carbon compounds and some group 1 metals.

The gaseous cations formed during ionization in the flame are attracted to a negative *collector* electrode and repelled by a positive *repeller* electrode. The repeller electrode is either the metallic burner or an electrode placed near the

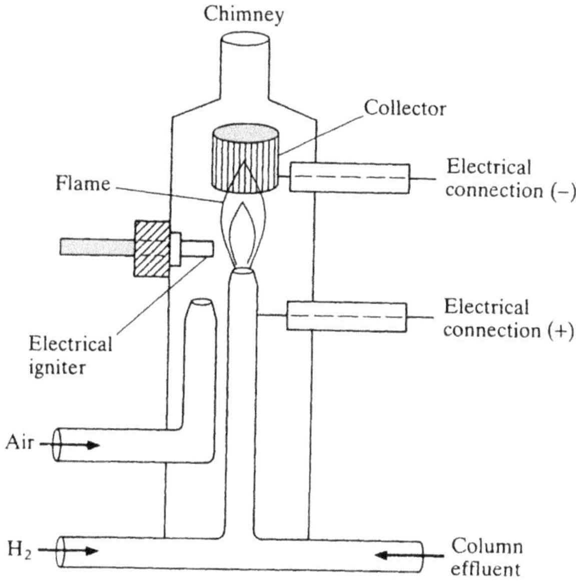

Figure 26-9 Diagram of a flame-ionization detector.

base of the flame. Upon striking the collector electrode, the positive ions cause a current to flow in an external circuit that connects the positive and negative electrodes. The current is amplified and recorded. The current flowing through the circuit is proportional to the number of ions striking the collector, which in turn is proportional to the concentration of ionizable sample components that enter the flame. Because the number of positive ions that are formed in the flame is related to the number of carbon atoms in an organic sample component, the detector's response increases with increasing carbon atoms in the component molecule.

The FID responds only to substances that can be ionized in an air-hydrogen flame. It does not respond to most inorganic compounds including N_2, O_2, CO_2, and water, and its response to organic compounds decreases as the number of oxygen, nitrogen, sulfur, and halide substituents on the compounds increases. The carrier gas used with the FID is not particularly important. Helium and nitrogen are most often used, although other inert gases (neon, argon, krypton, and xenon) also have been used. Although the FID is about a thousand times more sensitive than the TCD, it cannot be used to detect all types of sample components, and it destroys any sample entering the flame. Both single FIDs and dual differential FIDs are used. A diagram of an FID is shown in Fig. 26-9.

Flame Photometric Detector

The *flame photometric detector* (FPD) uses an air-hydrogen flame, as does the FID. Rather than measure the amount of ionization of the sample components in the flame, the FPD uses a photomultiplier (PM) tube to measure the radiation

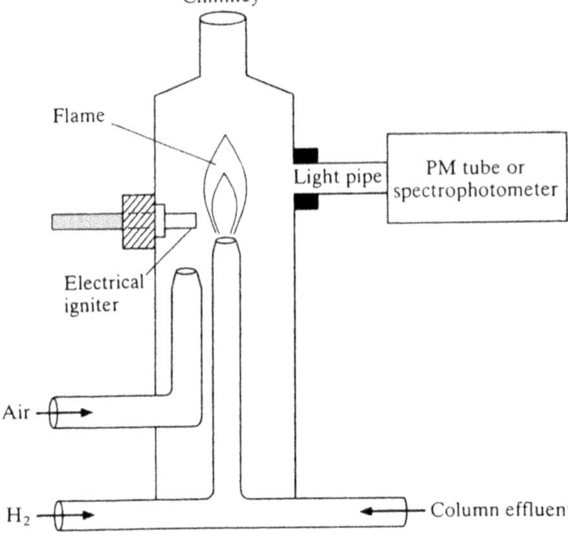

Figure 26-10 Diagram of a flame photometric detector.

emitted in the flame by the sample components. Generally, the FPD is used to detect compounds containing sulfur or phosphorus. It also has been used for organometallic compounds which contain a metallic atom capable of being excited in an air-hydrogen flame, and for compounds containing halogen atoms. A major advantage of the FPD is that it can be made selective for compounds which contain a single element by monitoring only the emitted radiation at the wavelength characteristic of that particular element. That is usually done by placing the appropriate optical filter between the flame and the PM tube. The FPD has become an important detector for the analysis of environmental samples such as those containing pollutant sulfur compounds. A number of practical considerations while using an FPD for routine analyses are described in Reference 20. A diagram of an FPD is shown in Fig. 26-10.

Electron-Capture Detector

In the *electron-capture detector* (ECD), the column effluent flows past a beta source (a radioactive element which emits beta particles). The energy of the collisions between the beta particles and the carrier gas molecules is sufficient to form positive carrier gas ions. The electrons emitted during the ionization are captured by a positive collector electrode, and cause an electric current to flow in an external circuit. The current is amplified and recorded. The beta source is often a ^{63}Ni foil, but tritium and ^{55}Fe are also used. An easily ionized gas, such as methane, is chosen for use with the detector. Usually the carrier gas is a mixture of methane with helium, nitrogen, or argon; or methane is added to the column effluent prior to reaching the beta source.

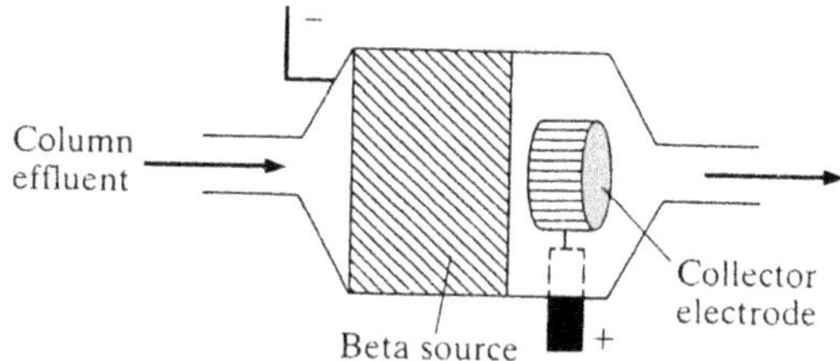

Figure 26-11 Diagram of an electron-capture detector.

When a sample component that has an affinity for electrons enters the detector, some of the electrons emitted by the carrier gas are captured by the component. The resulting decreased current flow in the external circuit is used to indicate the presence of the sample component. The ECD is selective for compounds that capture electrons. Organic compounds which contain electronegative groups, such as nitro groups, phosphorus, oxygen, and halogens, can be monitored with the detector. The ECD also responds to compounds containing lead. The ECD is an excellent detector for pesticides, lead-containing organic compounds, polychlorinated biphenyls (PCBs), and other compounds containing electronegative atoms. The detector does not respond well to saturated hydrocarbons. It is generally more sensitive than the FID or TCD to the compounds to which it does respond. Because the ECD contains a source of radioactivity, a license from the proper regulatory agency is required for its operation in some countries. A diagram of an ECD is shown in Fig. 26-11.

Other Detectors

The *thermionic detector* is a modified version of the FID in which an alkali halide salt tip is placed above or in the flame. The alkali halide increases the sensitivity of the detector for compounds containing nitrogen, phosphorus, sulfur, and halogens. The sensitivity of the detector can be modified by changing the alkali halide. The most popular thermionic detector is the nitrogen phosphorus detector (NPD) which utilizes a glass bead which often contains rubidium. It is highly selective for nitrogen and phosphorus containing compounds. Several types of spectroscopic detectors have been used for gas chromatography. The fluorescent and absorptive detectors appear to be the most popular of them. The column eluate flows through the cell of a spectrophotometer or fluorometer and its absorbance or fluorescence at a fixed wavelength is monitored. In some cases[21] a laser is used as the radiative source for the absorptive or fluorescent measurement.

A Fourier transform infrared (FTIR) spectrophotometer can be used as a GC detector. Generally, capillary columns are used for the separation. FTIR spectrophotometers are useful because they respond sufficiently rapidly to allow an infrared spectrum of a separated sample component to be recorded as the component exits the column. The detector either can be used to continuously monitor the eluate at one or a selected group of wavelengths or an entire spectrum can be repetitively recorded as eluate passes through the detector. A description of the interfacing of an FTIR spectrophotometer to a gas chromatograph and of the instrumental variations and applications for the combined GC/IR technique can be found in Reference 22.

A detector that monitors the absorbance of the column eluate in the far ultraviolet (FUV) region at a wavelength between 120 and 150 nm has been commercially available. In that spectral region nearly all substances absorb, and the detector responds to nearly all substances except noble gases. Other spectroscopic detectors are based on atomic absorption, emission in an inductively coupled plasma, nuclear magnetic resonance, x-ray absorption, and chemiluminescence.

Chemiluminescence detectors are commercially available for nitrogen and sulfur containing compounds. The nitrogen chemiluminescence detector (NCD) utilizes a high temperature burner to convert the nitrogen containing components of the column effluent to nitric oxide (NO). The NO undergoes a chemiluminescent reaction with ozone to yield electromagnetic radiation. The emitted EMR is monitored in the detector with a photomultiplier tube. In a similar manner, the sulfur chemiluminescence detector (SCD) detects radiation emitted from the chemiluminescent reaction of sulfur monoxide (SO) with ozone after formation of SO in a high temperature burner.

A review of the use of microwave-induced electrical discharge detectors can be found in Reference 23. The electrochemical detectors that have been used include conductometric detectors, coulometric detectors, and detectors which use ion-selective electrodes. Generally the column effluent enters the electrochemical detector where it reacts at high temperature to form products which are combined with a solvent. The dissolved reaction products enter the electrochemical cell where the measurement is made. Radiochemical detectors have been used to detect sample components which are naturally radioactive or which have been labeled with radioactive isotopes. In a photoionization detector (PID) some of the components of the column effluent are ionized by the electromagnetic radiation which is emitted from a uv lamp or laser. Diode pumped microchip lasers can be used. Only compounds which have an ionization energy that is less than the energy of the radiation beam are ionized and detected after striking a negatively charged electrode. Ionization sources in addition to EMR sources have been used including high voltage electric discharges, pulsed discharges and thermal ionization sources.

The coupling of mass spectrometry to a gas chromatograph GC/MS is particularly attractive for those laboratories that can afford the instrumentation. Mass spectrometric detection has been used with gas chromatographs that use either packed or open tubular columns. Capillary columns are used most often with GC/MS. The mass spectrometer can be adjusted so that it responds to ions of a fixed mass-charge ratio *m/z* (or a few *m/z*'s), or a complete mass spectrum can be obtained of the column eluate at fairly short time intervals. A chromatogram consists of a plot of the measured or computer-reconstructed total ion current as a function of time. Detectors which function in this manner are mass selective detectors (MSDs). In addition to the chromatogram, a complete mass spectrum of the column eluate at any particular instant is available when the instrument is operated in the scan mode.

The combined chromatographic and mass spectral information can be quite valuable when qualitatively and quantitatively assaying a sample. Normally the chromatographic data (total ion current variation with time or response of another detector) is used for quantitative analysis, and the recorded mass spectrum that corresponds to a particular chromatographic peak is used for qualitative analysis. Modern instruments are computer-controlled and normally compare the recorded mass spectrum to stored libraries of standard spectra for qualitative analysis.

Magnetic sector, quadrupole, Fourier transform, and time-of-flight mass spectrometers have been utilized with GC/MS. Because the time between adjacent peaks that are eluted from modern gas chromatographic columns can be small, it is necessary for the mass spectrometer to correctly record an entire spectrum within a short time span. Scanning magnetic sector spectrometers, quadrupole instruments, FT ion-cyclotron-resonance instruments, and time-of-flight spectrometers[24] have been used. Commercial instruments generally utilize either TOF or quadrupolar mass spectrometers. Use of array MS detectors can potentially increase the rates at which spectra can be obtained. A combined ionic source and mass analysis device, called an *ion-trap detector*, that uses radiofrequency fields to create and store ions has been commercially available at a price significantly lower than other GC/MS instruments. A description of the ion-trap detector can be found in Reference 25.

FLOWMETERS

Several types of flowmeters are used with gas chromatography. The simplest and most inexpensive is the bubble flowmeter shown in Fig. 26-12. The bubble flowmeter is a glass tube calibrated in cubic centimeters. The column eluate enters the flowmeter through a side arm at the bottom of the tube. The lower end of the tube is closed with a rubber bulb containing a soap solution. As the bulb is squeezed, the soap bubbles rise above the level of the side arm attached to the column and the eluate forces the bubbles upward. The time necessary for a bubble to traverse through a fixed volume in the calibrated tube is used to calculate the effluent flowrate (see Sample Problem 26-1).

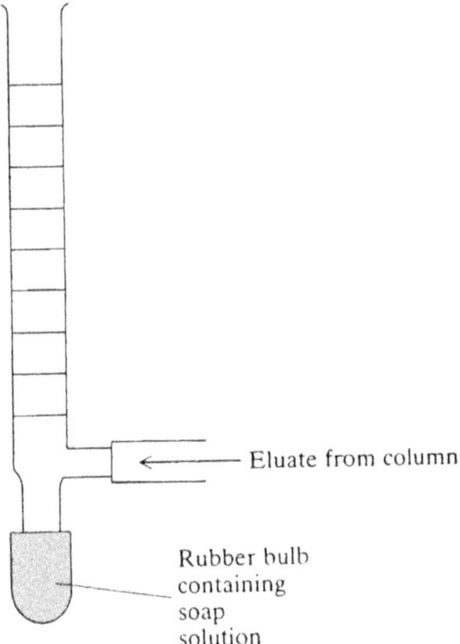

Figure 26-12 Diagram of a bubble flowmeter.

Several more sophisticated devices are available for measuring flowrates in gas chromatography. Such a device is a calibrated tube containing a ball which rises in the tube as gas from the column flows over it from below. The height to which the ball rises is related to the flowrate of the gas. Some flowmeters are placed in the stream of carrier gas between the gas source and the injector rather than after the column. Often a flowmeter of that type contains a valve which permits adjustment of the flowrate. Flowmeters which have digital displays are available.

TEMPERATURE EFFECTS

When operating most gas chromatographs, the analyst has control of the injector, column, and detector temperatures. The injector, in addition to introducing the sample to the column, must vaporize the sample. At injector temperatures below the boiling point of a mixture component, only a portion of the sample is rapidly vaporized and a relatively inefficient separation results. Below the boiling point, the measured number of theoretical plates of the column increases nearly linearly with increasing injector temperature. At injector temperatures above the boiling point, the efficiency of the column is almost constant. Because no advantage accrues from increasing the injector temperature much beyond the sample boiling point, the injector temperature is usually adjusted about 25°C above the boiling point of the highest-boiling sample component. If the boiling points of the sample components are unknown, an injector temperature of 250 °C often is used.

The temperature of the detector is dependent upon the type of detector. The temperature of the TCD block is held only high enough to prevent the sample components from condensing. Because the sensitivity of the detector depends upon the drop in temperature between the heated filaments and the detector block, it is counterproductive to increase the block temperature higher than is absolutely necessary. Generally, the block temperature is adjusted to between 10 and 25°C above the column temperature for columns operated at constant temperature, or about 10 to 25°C above the boiling point of the highest-boiling sample component for columns operated at changing temperature. That ensures that any gaseous component passing through the column will remain in the gaseous state in the detector.

As the column temperature increases, a sample component spends an increasing portion of its time within the column in the mobile phase. A resultant decrease in the retention time and retention volume for the component occurs. Similarly, an increase in pressure across the column causes increased mobile phase flow through the column resulting in decreased retention times. By controlling the column temperature, the analyst can partially control the retention time and retention volume of the sample components. Increasing the column temperature or pressure decreases retention times; decreasing the column temperature or pressure increases retention times. In addition to decreasing retention times, an increase in column temperature or pressure usually decreases the amount of band broadening by decreasing the available time for diffusion in the column. If the sample components are well resolved, the development time can be decreased by increasing the column temperature or pressure. If the sample components are not well resolved, the resolution usually can be increased by decreasing the column temperature or pressure. Increased retention times lengthen the time required to perform the assay and result in broader peaks.

In some cases the first two or three sample components eluted from the column are not well resolved, but the last component is separated from the other components by an undesirably long time. That situation is illustrated in Fig. 26-13, curve *A*. To better resolve the early eluting components, it is necessary to decrease the column temperature. However, an increased column temperature is required to decrease the retention time of the last component. *Temperature or pressure programming* can be used to overcome that problem.

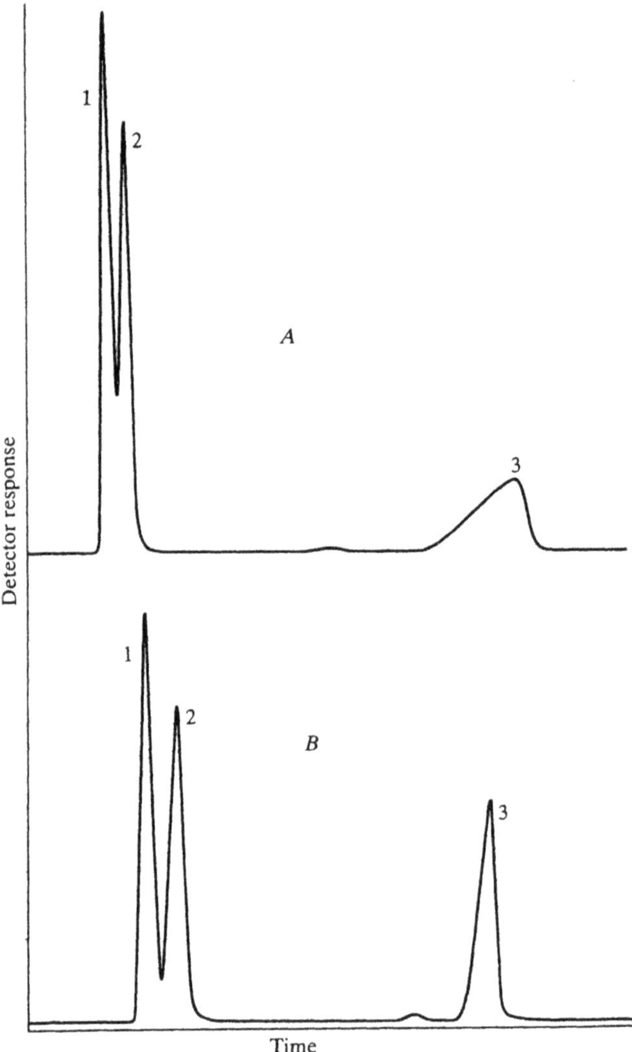

Figure 26-13 Gas chromatograms of a mixture containing equal volumes of methanol (peak 1), ethanol (peak 2), and isoamyl alcohol (peak 3) obtained on a $6\,\text{ft} \times \frac{1}{8}$ in stainless steel column that contained 10% Carbowax 1500 on Chromosorb W. Curve *A*, isothermal chromatogram at a column temperature of 90°C; curve *B*, temperature-programmed chromatogram; initial temperature, 75°C (2 min); final temperature, 150°C (1 min); rate of temperature change is 30°C/min.

Temperature programming is the chromatographic technique by which the column temperature is altered during development of the chromatogram. Gas chromatographs equipped for temperature programming generally permit the analyst to control the initial column temperature, the time at the initial temperature, the final column temperature, the time at the final temperature, and the rate of column temperature increase in going from the initial to the final temperature. Some gas chromatographs permit the temperature increase to be stopped for controlled periods of time within the temperature ramp or even more control. The separation problem illustrated in Fig. 26-13, curve *A*, can be solved by decreasing the initial column temperature for a period of time necessary to resolve the first two peaks, and then increasing the column temperature to a value which shortens the retention time of the component that corresponds to the last peak. The resulting chromatogram is shown in Fig. 26-13, curve *B*. Generally, the analyst should not adjust the column temperature more than about 10°C below the boiling point of the highest-boiling sample component because distorted chromatographic peaks can result, as shown in the last peak in Fig. 26-13, curve *A*. Sometimes that distortion can be avoided by using temperature programming if the final column temperature is sufficiently high and the time at the initial low temperature is not excessively long.

EFFECT OF FLOWRATE

Increasing the carrier gas flowrate through a GC column decreases the retention time of the sample components. Retention volumes are generally independent of flowrate. Flowrates can be used in much the same way that column temperatures or pressures are used to control retention times of components. Increasing the flowrate decreases the retention times and decreasing the flowrate increases retention times. Although the flowrate in most gas chromatographs cannot be automatically altered during development of a chromatogram, some gas chromatographs have been designed to have automatically variable flowrates. Variation is usually accomplished either by increasing the flowrate of carrier gas (often exponentially) by increasing flow through an auxiliary gas line that combines with the main carrier gas line prior to entering the column or by increasing the carrier gas pressure. The first approach is primarily used with capillary (open tubular) columns. If pressure programming is used, the gas chromatographic technique is *chromarheography*. Either technique can be used in a manner analogous to temperature programming.

Generally retention times decrease with an increase in the pressure differential across the column. Conversely, larger retention times can be achieved by decreasing the column pressure. Since normally it is more difficult to design chromatographs which have the capability of programmed pressure than instruments which have the capability of programmed temperature, programmed pressure is used less often than programmed temperature. Some instruments, however, are available commercially which can perform pressure programming. In those instruments it often is possible to utilize temperature programming, pressure programming or a combination of the two types of programming.

QUALITATIVE ANALYSIS

Qualitative analysis using gas chromatography is done by comparing retention times or retention volumes of sample components with retention times or retention volumes of reference compounds developed under identical experimental conditions. That often is done either by measuring the retention time of the sample component and comparing it with the retention times of the separately injected known compounds or by recording the chromatogram of the sample and then recording the chromatogram of the sample to which a known compound has been added. If the spiked sample shows an increase in the area under one of its peaks, that peak is likely due to the added compound. Owing to the fairly large number of compounds that have similar retention times, it is advisable that the analyst use two or more columns packed with stationary phases of different polarity for qualitative analysis of a sample. The chance that different compounds will yield identical retention times on columns of differing polarity is small. Gas chromatography can be combined with other analytical techniques to provide a powerful combined tool for qualitative analysis. It is often used to separate sample components which are then qualitatively assayed by mass spectrometry, infrared spectrophotometry, nuclear magnetic resonance spectrophotometry, or some other analytical method.

If a qualitative analysis is to be performed by comparing sample retention times or volumes with the retention times or volumes of known compounds which were developed on different days, under different experimental conditions, or by a different laboratory, direct comparison of the results obtained with the sample and the results obtained with the standards is generally not possible. Usually relative retention, specific retention volumes, or *Kovats retention indices* are compared.

The Kovats retention index I relates the retention of the sample component to the retention of straight-chain saturated hydrocarbons which elute before and after the sample component. The Kovats retention index is defined in one of the following equivalent ways:

$$I = 100\left[\frac{\log(t - t_m) - \log(t_n - t_m)}{\log(t_{n+1} - t_m) - \log(t_n - t_m)}\right] + 100n$$

$$I = 100\left[\frac{\log(V - V_m) - \log(V_n - V_m)}{\log(V_{n+1} - V_m) - \log(V_n - V_m)}\right] + 100n \qquad (26\text{-}16)$$

$$I = 100\left[\frac{\log \alpha_{s,n}}{\log \alpha_{n+1,n}}\right] + 100n$$

where t is the retention time of the sample component, t_m is the mobile-phase retention time, t_n is the retention time of the saturated hydrocarbon containing n carbons which elutes before the sample component, and t_{n+1}, is the retention time of the saturated hydrocarbon which contains $n + 1$ carbons and which elutes just after the sample component. The corresponding retention volumes are V, V_m, V_n, and V_{n+1}. The relative retention of the sample component with respect to the n-carbon hydrocarbon is $\alpha_{s,n}$, and the relative retention of the $(n + 1)$-carbon hydrocarbon with respect to the n-carbon hydrocarbon is $\alpha_{n+1,n}$. From Eqs. (26-16) it can be seen that the Kovats retention index of an n-carbon straight-chain saturated hydrocarbon is $100n$. In some cases,[26] the straight-chain saturated hydrocarbons used as references in calculating Kovats retention indices are replaced with other types of hydrocarbons.

The Kovats retention index is intended for use with columns operated at constant temperature. At constant temperature the logarithms of the retention volume and retention time increase linearly with increasing carbon number in an homologous series; i.e., a plot of I as a function of log $(t - t_m)$ or log $(V - V_m)$ is linear. That accounts for the use of logarithmic terms in Eqs. (26-16). If the column is temperature-programmed, the linear logarithmic relationship is no longer observed and the Kovats retention index can be replaced by the retention index shown by

$$I = 100\left[\frac{t - t_n}{t_{n+1} - t_n}\right] + 100n = 100\left[\frac{V - V_n}{V_{n+1} - V_n}\right] + 100n \qquad (26\text{-}17)$$

The retention index changes nearly linearly with changes in the bracketed term in Eq. (26-17). Equations (26-16) are used when the column is operated at constant temperature and Eq. (26-17) is used for programmed temperature operation.

Sample problem 26-2 The following data were obtained on a GLC column operated at constant temperature. Calculate the Kovats retention index for the sample component.

Compound	Retention time, min
Air	0.22
n-Hexane	4.13
Sample component	5.20
n-Heptane	6.35

SOLUTION The Kovats retention index can be calculated by direct substitution into Eqs. (26-16), where $t_m = t_{air} = 0.22$ and $n = 6$:

$$I = 100\left[\frac{\log(t - t_m) - \log(t_n - t_m)}{\log(t_{n+1} - t_m) - \log(t_n - t_m)}\right] + 100n$$

$$= 100\left[\frac{\log(5.20 - 0.22) - \log(4.13 - 0.22)}{\log(6.35 - 0.22) - \log(4.13 - 0.22)}\right] + 100(6)$$

$$= 653.8$$

A Kovats retention index between 600 and 700 indicates that the compound elutes between the six-carbon and the seven-carbon hydrocarbon. Likewise, a Kovats retention index of 356 indicates that the sample elutes between the three-carbon and the four-carbon hydrocarbon, etc.

MULTIDIMENSIONAL SEPARATIONS

In planar chromatography it was possible to perform two-dimensional separations by first developing the chromatogram in one mobile phase solvent and then rotating the planar stationary phase containing the developed chromatogram and continuing the development in a second solvent. Similar *multidimensional development* is possible in gas chromatography by using two or more columns which contain different stationary phases. Normally the columns serially are attached to each other. If two columns are utilized, the method is two dimensional development. Two dimensional development in GC differs from two dimensional development in planar chromatography because in GC the stationary phase is changed, whereas in planar chromatography the mobile phase is changed. Multidimensional GC has been used to perform difficult separations which were impossible to perform while using a single column. Since those types of separation are rare, the technique is not used routinely in most laboratories.

Two dimensional chromatography in which back flushing of the first column was used after a single component has been passed to the second column was termed column switching (Reference 27) when it was first developed. The mixture is injected into the first column where partial separation of the components occurs. As a peak of interest leaves the first column, it is automatically injected into the second column through a series of valves. While the remainder of the original mixture is back flushed from the first column with carrier gas, the components under the chosen peak are separated in the second column. Back flushing is the technique in which the direction of mobile phase flow through the column is reversed before all of the components are eluted. The technique in which that portion of the eluate that corresponds to a single peak is injected into a second column is *heart cutting* or simply *cutting*.

The two stationary phases in the two columns are not identical in polarity and are chosen to maximize the efficiencies of the separations. Efficiency is further enhanced by operating each column at the temperature that is best for the particular separation. The combined resolution obtained with two dimensional GC is the product of the individual resolutions obtained on each column ($R = R_1 R_2$). If each component in each of the peaks that is eluted from the first column must be separated, the chromatographic development in the second column is repeated once for each peak eluted from the first column. Back flushing the first column while development takes place in the second column is sometimes necessary to simplify the observed chromatogram by decreasing the number of observed peaks, and to prevent peak overlap of a component with a short retention time on the first column and a long retention time on the second column with the peak of a second component which has a long retention time on the first column and a short retention time on the second column.

QUANTITATIVE ANALYSIS

The *area* under a particular chromatographic peak is directly proportional to the total amount of the component reaching the detector. The amount of the component reaching the detector is equal to the amount of the component injected or is directly proportional to the amount injected when a split injector is used. Consequently, if the sample size is held constant, the area under the component peak is proportional to the concentration of the component in the sample. The proportionality constant is determined with the aid of standards containing a known amount of the analyte component. Either the working-curve method or the standard-addition technique can be used for the purpose. The area under a peak can be measured in any of the ways used while using other methods of analysis. The most popular methods of measuring the peak area are triangulation (approximating the peak shape with a triangle and calculating the area of the triangle), multiplying the peak height h by the peak width at half the peak height $w_{1/2}$ ($A = hw_{1/2}$), use of an electronic integrator, and use of a planimeter. Of those methods, use of an electronic integrator is easiest, provides the most reproducible results and is used most often. If the peaks approximate the shape of triangles with a constant base or if the peaks are so narrow as to have a negligible peak width, the height of each peak is proportional to the peak area and can be used in place of the peak area for quantitative analysis.

It is important to realize that most GC detectors do not respond equally to equal molar amounts or to equal masses of different substances. Consequently, it is necessary to prepare a working curve or to use the standard-addition technique individually for each sample component. Some chromatographers use an internal-standard technique. Typically a constant concentration of an internal standard that is not an analyte component is added to each standard and sample solution. The ratio of the peak height or area of the analyte peak from each solution is measured and divided by that of the internal-standard peak. The resulting ratio is used with the working-curve method or the standard-addition technique. The internal-standard method is particularly useful when the injected volume cannot be accurately measured, when the flowrate cannot be carefully controlled, or when the detector's response is not reproducible.

Supercritical Fluid Chromatography

Supercritical fluid chromatography (SFC) is a form of chromatography that combines some of the advantages of liquid chromatography with the high efficiency associated with gas chromatography. The temperature of the column is maintained at a temperature above the critical temperature of the mobile phase so that the mobile phase cannot exist as a liquid. By use of a relatively high pressure, the mobile phase is sufficiently compressed that it has the good solvating properties of a liquid. Under those conditions the mobile phase exhibits properties that are intermediate between those of a liquid and those of a gas.

In the supercritical fluid the diffusion coefficients of solutes are about 100 times greater than the corresponding diffusion coefficients in liquids. The viscosities of supercritical fluids are approximately the same as those of gases. Consequently, the efficiencies of separations are better than those obtained in liquid chromatography. Compounds that have much higher molecular weights than those which can be separated by using gas chromatography can be separated by using SFC because the solvating ability of supercritical fluids is greater than that of gases. Normally gas chromatographic separations are restricted to compounds with molecular weights less than about 400. No similar restriction applies to separations done by using SFC.

Both packed and open tubular columns have been used in SFC. Recent studies have primarily used capillary, open tubular columns with internal diameters of 100 μm or less and column lengths up to 60 m. The stationary phase must be cross-linked within the column to prevent bleeding. Bonded-phase C8 or C18 HPLC columns are often used. Small sample sizes (typically 0.2 μL) are injected by the valve method to prevent overloading the narrow columns. Pressure programming rather than temperature programming is generally used. As the pressure increases, the mobile-phase molecules are forced closer together and the solvating ability of the mobile phase increases, causing a decrease in retention times.

The temperature of the column often is maintained in an oven identical to those used in gas chromatographs. The remainder of the apparatus is similar to that used for HPLC. The pressure on the column is typically controlled with an HPLC syringe pump. Mobile phases that have been used with SFC include ammonia, benzene, carbon dioxide, isobutane, isopropanol, *n*-butane, *n*-pentane, and nitrous oxide. Carbon dioxide has been used most often. Generally the temperature of the column is maintained about 10 to 15°C above the critical temperature of the mobile phase. As an example, when *n*-pentane (critical temperature is 196°C) is used, the column temperature is typically maintained at 210°C. The pressure on the column is normally between 20 and 40 atm.

Detectors used in supercritical fluid chromatographs include uv absorptive detectors, fluorescent detectors, flame ionization detectors, other spectral detectors, and mass spectrometric detectors. Mass spectrometric detectors are particularly well suited to the technique. Nearly all LC and GC detectors have been used with SFC with the exception of the electrochemical detectors which do not function well in the low polarity of SFC mobile phases. Two dimensional SFC development and combined SFC and GC have been used. More detailed descriptions of the apparatus and the applications of SFC can be found in References 28 - 30.

Other Gas Chromatographic Instruments

Miniaturization of the gas chromatographic components has made it possible to manufacture a miniature gas chromatograph that uses a thermal conductivity detector. The entire chromatograph, with the exception of the tank for the mobile phase, can fit into the palm of a hand. A commercial version of a miniaturized gas chromatograph is available from Microsensor Technology, Inc., of Fremont, California. The device is microprocessor-controlled and can simultaneously operate five columns. The use of small columns and miniaturized detectors in portable GCs have proved valuable for on site analysis of drugs, explosives and environmental contaminants. The upsurge in drug trafficking and terrorism has spurred much of the progress toward development of portable GCs. A review of developments in miniaturization is included in Reference 12.

Work has been done using bonded C18 stationary phases in gas chromatographic, open tubular columns. Use of the bonded phases is analogous to their use in reversed-phase liquid chromatography. The mobile phase used in those columns is usually carbon dioxide to which up to 1 percent of an organic modifier has been added.

IMPORTANT TERMS

Adjusted flowrate
Adjusted retention volume
Average column pressure
Average retention polarity
Back flushing
Bonded-phase column
Bubble flowmeter
Carrier gas
Chromarheography
Chromatopyrography
Column switching
Corrected retention volume
Diatomaceous earths
Electron-capture detector
Far-ultraviolet detector
Flame photometric detector
Flame-ionization detector
GC/MS
Headspace chromatography
Heart cutting
Ion-trap detector
Kovats retention index
Liquid crystals
McReynolds constant
MSD
Multiple headspace extraction
NCD
Net retention volume
NPD
Open tubular column
Packed column
PEG
PID
PLOT column
Pressure drop correction factor
Purge and trap
Retention index
Rohrschneider retention index
SCD
SCOT column
Specific retention volume
Supercritical fluid chromatography
Temperature programming
Thermal conductivity
Thermal conductivity detector
Thermal desorption
Thermionic detector
Van Deemter equation
WCOT column
WWOT column

PROBLEMS

Retention Time and Retention Volume

26-1 The flowrate of the carrier gas exiting a column operated at 125°C was measured with a bubble flowmeter at 23°C and 745 torr and was found to be 43 cm^3/min. Calculate the adjusted flowrate.

26-2 Calculate the adjusted flowrate of the carrier gas under the conditions described in Prob. 26-1 if a flowmeter which did not contain water was used.

26-3 The retention time of a sample component was 6.32 min in a column operated at 140°C. The flowrate through the column was measured with a bubble flowmeter at 27°C and 750 torr and was found to be 35 cm^3/min. Calculate the retention volume of the sample component.

26-4 Calculate the pressure drop correction factor in a column in which the inlet pressure is 30.0 psig and the outlet pressure is atmospheric pressure (752 torr).

26-5 Calculate the average pressure in the column described in Prob. 26-4.

26-6 The retention time of a sample component was 3.79 min in a column operated at 105°C. The flowrate through the column was measured with a bubble flowmeter at 24°C and 765 torr and was found to be 25 cm^3/min. If the inlet pressure of the column was 32.2 psig, calculate the corrected retention volume of the sample component.

26-7 If air injected with the sample described in Prob. 26-6 had a retention time of 0.13 min, calculate the adjusted retention volume V' and the net retention volume V_n of the sample component.

26-8 A sample component had a retention time on a GLC column of 12.35 min. The retention time of air was 0.23 min and the flowrate through the column was found to be 40.0 cm^3/min with a bubble flowmeter at 25°C and 1.00 atm. The inlet pressure to the column was 28.5 psig, the column temperature was 150°C, and the mass of the stationary liquid within the column was 1.12 g. Calculate the specific retention volume of the sample component.

Resolution

26-9 Calculate the resolution between two adjacent peaks if the retention times are 3.65 and 4.10 min and the peak widths are 0.22 and 0.34 min.

Efficiency

26-10 Calculate the number of theoretical plates in a column on which the retention time of a sample component is 7.66 min and the peak width is 47 s.

26-11 Calculate the height equivalent of a theoretical plate if a 3.0-m column were used to obtain the data in Prob 26-10.

26-12 Estimate C_L for a component in a liquid stationary phase that has an average coating thickness of 2.0 μm if the shape factor is 0.45 and the diffusion coefficient of the component in the liquid phase is 1.0×10^{-5} cm^2/s. The retention time of the component on the column is 4.47 min and the retention time of air (not retarded) is 0.13 min.

Headspace Analysis

26-13 The area of a chromatographic peak after the first extraction during headspace analysis was18.3 cm^2. After the second and third extractions the peak areas were 15.6 and 13.3 cm^2 respectively. Estimate the peak area after the fifth extraction.

26-14 Use the data from Prob. 26-13 to determine the sum of the peak areas that would have been obtained if the extractions were continued until the volatile component was exhausted.

Stationary Phases

26-15 Of the stationary phases listed in Table 26-2, which would you choose for the separation of (*a*) pentane, heptane, and decane, and (*b*) ethanol, nitropropane, and pyridine?

Hint: Compare McReynolds constants in (*b*).

26-16 If you wished to use gas chromatography to assay a gaseous mixture of H_2, N_2, O_2, and CH_4, would you use a liquid or a solid stationary phase?

Temperature Effects

26-17 A sample contains 1-butanol (bp 117.5°C), ethanol (bp 78.5°C), and methanol (bp 65.0°C). A polar GLC column was used to separate the sample components and a TCD was used to detect the components. Suggest temperature settings for the injector, column, and detector block.

26-18 A mixture of the four liquids listed below is to be injected into a nonpolar gas-liquid chromatographic column. At what temperature would you adjust the injector, column, and thermal conductivity detector's block? Predict the order of elution from the column.

Component	Boiling point, °C	Molecular weight
C_2H_5OH	78.5	46.1
CH_3OH	65.0	32.0
C_4H_9OH	117.5	74.1
C_3H_7OH	97.1	60.1

Qualitative Analysis

26-19 The following GLC data were obtained. Calculate the relative retention of components B, C, and D with respect to component A.

Sample component	Retention time, min
Air	0.15
A	2.33
B	3.76
C	6.12
D	10.98

26-20 The gas chromatographic data listed below were obtained. Calculate the relative retention of component B with respect to component A.

Sample component	Retention time
Air	24s
A	5.63 min
B	7.53 min

26-21 The following data were obtained by using a chromatographic column operated at constant temperature. Calculate the Kovats retention indices for sample components A, B, and C.

Compound	Retention time, min
Air	0.32
n-Pentane	4.49
A	6.23
n-Hexane	7.08
B	7.78
n-Heptane	11.52
C	16.24
n-Octane	18.11

26-22 From a plot of I as a function of log $(t - t_{m},)$ for the data listed in Prob. 26-21, estimate the retention time of *n*-butane.

Quantitative Analysis

26-23 The following GC data were obtained after injection of 2-μL portions of a sample component and standard solutions of the component. Calculate the concentration of the sample.

Solution concentration, mL/mL	Peak area
0.200	1.43
0.400	2.86
0.600	4.29
0.800	5.73
1.000	7.16
Sample	4.10

26-24 The standard-addition technique was used to assay a sample component by gas chromatography. From the results in the table, determine the concentration of the sample component.

Added concentration, mg/mL	Peak area
0 (sample)	3.72
1.23	7.00
3.47	12.70
4.89	16.30
6.24	20.01
7.15	22.25

REFERENCES

1. James, A. T., and A. J. P. Martin: *Biochem. J.*, **50**: 679(1951).
2. Hawkes, S. J.: *J. Chem. Educ.*, **60**: 393(1983).
3. Hu, J. C. A.: *Anal. Chem.*, **53**: 311A(1981).
4. Drozd, J., and J. Novak: *J. Chromatog.*, **165**: 141(1979).
5. Ettre, L. S., B. Kolb, and S. G. Hurt: *Amer. Lab.*, **15**(10): 76(1983).
6. Jones, E., M. Davis, R. Gibson, B. Todd, and R. Wallen: *Amer. Lab.*, **16**(8): 74(1984).
7. McNally, M. E., and R. L. Grob: *Amer. Lab.*, **17**(1): 20(1985).
8. McNally, M. E., and R. L. Grob: *Amer. Lab.*, **17**(2): 106(1985).
9. Rohrschneider, L.: *J. Chromatog.*, **22**: 6(1966).
10. McReynolds, W. O.: *J. Chromatog. Sci.*, **8**: 685(1970).
11. Tarjan, G., A. Kiss, G. Kocsis, S. Meszaros, and J. M. Takacs: *J. Chromatog.*, **119**: 327(1976).
12. Eiceman, G. A., J. Gardea-Torresdey and F. *Dorma*: *Anal. Chem.*, **78**: 3985 (2006).
13. Driscoll, J. N., and I. S. Krull: *Amer. Lab.*, 15(5): 42(1983).
14. Witkiewicz, Z.: *J. Chromatog.*, **25**: 311(1982).
15. Kelker, H., and E. Von Schivizhoffen, in J. C. Giddings and R. A. Keller (eds.): *Advances in Chromatography*, vol. 6, Marcel Dekker, New York, 1967, pp. 247-297.
16. Janini, G. M., in J. C. Giddings, E. Grushka, J. Cazes, and P. R. Brown (eds.): *Advances in Chromatography*, vol. 17, Marcel Dekker, New York, 1979, pp. 231-277.
17. Baiulescu, G. E., and V. A. Ilie: *Stationary Phases in Gas Chromatography*, Pergamon, Oxford, 1975.
18. Barry, E. F., J. A. Hubball, P. R. DiMauro, and G. E. Chabot: *Amer. Lab.*, **15**(10): 84(1983).
19. DiCesare, J. L., and L. S. Ettre: *Chromatog. Newsl.*, **11**(1): 1(1983).
20. Mowery, Jr., R. A., and L. V. Benningfield, Jr.: *Amer. Lab.*, **15**(5): 107(1983).
21. Green, R. B.: *Anal. Chem.*, **55**: 20A(1983).
22. Griffiths, P. R., J. A. de Haseth, and L. V. Azarraga: *Anal. Chem.*, **55**: 1361A(1983).
23. Risbu, T. H., and Y. Talmi: *CRC Crit. Rev. Anal. Chem.*, **14**: 231(1983).
24. Holland, J. F., C. G. Enke, J. Allison, J. T. Stults, J. D. Pinkston, B. Newcome, and J. T. Watson: *Anal. Chem.*, **55**: 997A(1983).
25. Stafford, G. C., P. E. Kelly, and D. C. Bradford: *Amer. Lab.*, **15**(6): 51(1983).
26. Lee, M. L., D. L. Vassilaros, C. M. White, and M. Novotny: *Anal. Chem.*, **51**: 768(1979).
27. Henry, M. C. and C. R. Yonker: *Anal. Chem.*, **78**: 3909 (2006).
28. Jackson, W. P., B. E. Richter, J. C. Fjeldsted, R. C. Kong, and M. L. Lee, in S. Ahuja (ed.):*Ultrahigh Resolution Chromatography*, American Chemical Society, Washington, 1984, pp. 121-133.
29. Fjeldsted, J. C., and M. L. Lee: *Anal. Chem.*, **54**: 1883(1982).
30. Mueller, F.: *Amer. Lab.*, **15**(10): 94(1983).

CHAPTER

TWENTY-SEVEN

THERMAL ANALYSIS

An analyst who uses a thermal method of chemical analysis monitors a property of the sample as a function of temperature or monitors the temperature change associated with a chemical reaction. While using an analytical method in the first category, the temperature can be increased or decreased linearly, held constant, or programmed as a function of time in much the same manner that is used for gas chromatography. The thermal methods of chemical analysis are divided into categories depending upon the property of the sample that is monitored. Because many properties can be monitored, many types of thermal analysis are possible. The present chapter deals only with the relatively few methods of thermal analysis that are most popular.

Thermogravimetry (TG) most closely resembles classical gravimetric analysis. The mass of the sample is monitored as a function of temperature. *Differential thermal analysis* (DTA) is the method in which the temperature difference between a sample and a nonreactive reference substance is monitored as a function of the controlled temperature of one of the two substances (usually the reference substance). *Differential scanning calorimetry* (DSC) is similar to DTA except that the difference in heat between that which flows into the sample and that which flows into the reference substance is monitored as a function of temperature or of time (if the temperature is not varied). The thermal method in which a mechanical property of a sample is monitored as the sample is subjected to a stress is *dynamic mechanical analysis* (DMA). *Thermomechanical analysis* (TMA) is the technique in which a dimension of the sample is measured as the temperature is altered.

Thermal measurements can be used to locate the endpoints of some titrations. *Thermometric titrations* are those titrations in which the temperature of the titrand is recorded as a function of titrant volume. *Direct-injection enthalpimetry* (DIE) is the technique in which the change in temperature of a sample is monitored after an excess of a chemical reactant is suddenly added to the sample. *Continuous-flow enthalpimetry* (CFE) is the technique in which the temperature change is measured while an excess of a chemical reactant is continuously added to a stream of the sample.

THERMOGRAVIMETRY

While performing a thermogravimetric study, the mass of the sample is measured either as a function of temperature or time. The plot of the mass or the percentage of the initial mass as a function of temperature or time is a *thermal curve*. If the temperature is varied during the study, the mass is plotted as a function of temperature. During isothermal studies or constant-temperature portions of programmed temperature studies, the mass is plotted as a function of time. Any process in which a mass change occurs can be studied with TG. Generally mass changes are caused by loss of a volatile component, such as water, from the sample. TG cannot be used to study processes such as melting that do not involve a mass change. The change in mass can be used for quantitative analysis and the temperature at which the change takes place is useful for qualitative analysis.

The apparatus that simultaneously heats the sample and monitors its mass is a *thermobalance*. A diagram of a thermobalance is shown in Fig. 27-1. Mass measurements are made with a highly sensitive microbalance. Often the balance is an electromechanical device.

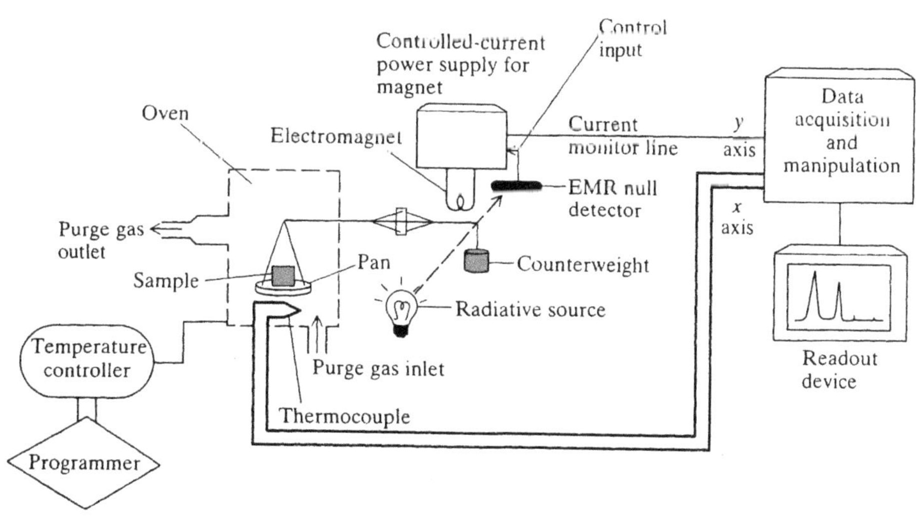

Figure 27-1 A diagram of a thermobalance.

The sample is placed in the *sample boat* on the single pan of the balance. Counterweights are used on the opposite arm of the balance as in a laboratory single-pan balance. The beam of the balance is constructed from quartz. The balance pan typically is constructed from platinum, alumina or ceramic. It is attached to a metallic arm housed between the poles of an electromagnet. Motion of the balance is detected by observing the deflection of a light beam on a photosensitive null detector which is attached to the beam of the balance, or by another method. A deflection of the beam is automatically compensated by a change in current flow through the electromagnet which restores the beam to its original position. The current that flows through the electromagnet is monitored and related to the sample mass. The electrical signal is related to the mass on the balance and is displayed on the readout device. A commercial thermobalance is described in Reference 1. Although samples as large as 1.5 g can be studied with some commercial instruments, samples with masses between 5 and 25 mg are used for most studies. Balance sensitivities of 2 micrograms are typical with an accuracy of 0.02%. Some commercial instruments have a sensitivity of 0.1 microgram with an accuracy of 0.2%.

The sample is enclosed in an electric furnace. The initial temperature, final temperature, and rate of temperature increase between the initial and final temperatures can be controlled on some instruments by dials located on the control panel of the instrument. Computer-controlled instruments are common and offer more versatility. With those instruments the operational parameters are entered on a keyboard. Oven temperature control down to – 170°C can be achieved with liquid nitrogen as the coolant, or down to – 40 or – 70°C, depending upon the instrument, with freon. The rate of temperature increase or decrease often can be adjustable between 0.1 and 200°C/min. Typically the heating rate is adjusted between 5 and 25°C/min. The oven can be operated isothermally or in a programmed mode. Thermal equilibrium is established within 30 s. In some instruments the temperature of the oven can be programmed to stop increasing at each of four temperatures and to hold that temperature for a specified period prior to increasing to the next temperature at the chosen rate of temperature increase. Essentially programming is done as in programmed temperature gas chromatography (Chapter 26). The upper temperature used for TG normally is 1000°C, but in some commercial instruments it is as high as 1500°C.

The temperature of the oven is normally monitored by a thermocouple that is placed near, but not in contact with, the sample. Because the thermocouple is not actually in contact with the sample, some error on the temperature axis of the thermal curve is possible. The accuracy of the temperature measurement varies with the instrument from about 0.5-5°C. The temperature differential between the sample and the thermocouple is dependent upon the rate of temperature increase, the thermal conductivity of the gaseous atmosphere in the oven, and the flowrate of the gas.

At temperatures above 500°C, radiation, rather than convection, can be the major mechanism by which heat is transferred to the sample. Consequently the ability of the sample to absorb thermal radiation at temperatures above 500°C can affect the temperature of the sample. Dark-colored or black samples absorb thermal energy more readily than white samples under otherwise identical conditions and consequently have a higher temperature than white samples.

The thermocouple can be calibrated by mixing the sample with several ferromagnetic substances that have different *Curie points*. The mixture is placed in the balance pan and heated in a magnetic field. A small permanent magnet is placed under or above the pan during the measurements. Because the magnetic field attracts the ferromagnetic substances, the recorded mass is a fixed amount greater or less than the actual mass depending upon the position of the magnet. As the sample is heated, the ferromagnetic substances become more randomly oriented. At the Curie point (a temperature) of each substance, the material becomes nonmagnetic and an apparent sudden change in mass occurs. If the Curie point of each of the several ferromagnetic substances is known, the temperature scale can be calibrated using the locations of the apparent mass changes. Because the transition at the Curie point is reversible, the same reference compounds can be heated and cooled several times. Up to nine Curie-point transitions are used to standardize the temperature scale. Temperature calibrations which are performed in this manner are usually accurate to within 2°C.

In some furnaces it is difficult to use Curie-point transitions for temperature standardization because it is difficult to properly orient the magnet within the oven. In those cases a small rider constructed from a metal that has a known melting point can be hung from the beam of the balance. When the temperature of the oven reaches the melting point of the rider, the metal melts and falls from the beam, causing a sudden change in mass. Another version of the technique[2] uses a platinum rider that is spot welded to a hook constructed from a metal with a known melting point. The rider is hung from the beam directly above the pan of the balance. At the melting point of the hook, the assembly falls into the pan causing a momentary action-reaction deflection on the mass scale. No permanent deflection occurs because the total mass on the beam of the balance is unchanged.

Most studies that utilize a thermal method are performed under a controlled atmosphere. The instrument is normally equipped with a system whose function is to continuously purge the furnace and sample with a gas. Often a pneumatic purging system is used. The gases that are most often used for purging the furnace are nitrogen, helium or argon (when an inert atmosphere is required), or carbon dioxide, air or oxygen (when a reactive atmosphere is desired).

Some instruments are capable of *purge gas switching*. Purge gas switching in modern instruments is performed under computer control. Generally the system is initially purged with nitrogen, helium or argon. After the sample has been heated to the final temperature, the purge gas is switched to carbon dioxide, air or oxygen. Mass decreases owing to loss of volatile sample components are studied under the inert atmosphere and reaction of the sample with carbon dioxide, air or O_2 is studied after the purge gas has been changed. The purge gas is changed by closing the inlet valve to the thermobalance from the first purge gas while simultaneously opening the valve to the second gas. Computer-operated valves can be used for the process.

In older instruments the amplified emf output from the thermocouple was used to drive the x axis of an xy recorder while the mass was recorded on the y axis. Computer-controlled instruments can compare the output from the thermocouple with tabulated values stored in read-only memory and use the corresponding temperatures to prepare the thermal curve. The thermal curve can be displayed on a video display terminal, a printer plotter, or some other computer-controlled readout device. Prior to displaying the thermal curve, the data are often smoothed by *boxcar averaging*. Essentially, each plotted data point is the average of many measured values. Some computers are capable of *multitasking*, i.e., of simultaneously controlling several instruments. A description of modern thermogravimetry can be found in Reference 3. Many modern instruments are attached to an autosampler which permits automated sequential study of multiple samples. Some instruments can be combined with another analytical method, such as gas chromatography (TG/GC), mass spectrometry (TG/MS), FTIR spectrometry (TG/FTIR), or another technique which assays the products of thermal reactions that occur in the thermal analyzer (Reference 4).

Thermogravimetric Analysis

TG can be used for qualitative and quantitative analysis. The *onset temperature* T_{onset} (Fig. 27-2) can sometimes be useful for qualitative analysis. It is the temperature at which a mass loss occurs. It is measured by extrapolating to intersection the linear portions of the thermal curve recorded prior to and during the mass loss. The onset temperature can be compared to values for standards that were measured under identical experimental conditions for qualitative analysis. Often a single sample exhibits several consecutive mass losses and onset temperatures.

Quantitative analysis is done by measuring, from the thermal curve, the mass loss Δm obtained during a process in which a known molecular weight component is vaporized from the sample. The mass loss is obtained from the vertical distance on the thermal curve between the horizontal portions prior to and after the mass loss (Fig. 27-2). For pure samples assignment of mass losses from the thermal curve to particular volatile components can aid in qualitative analysis.

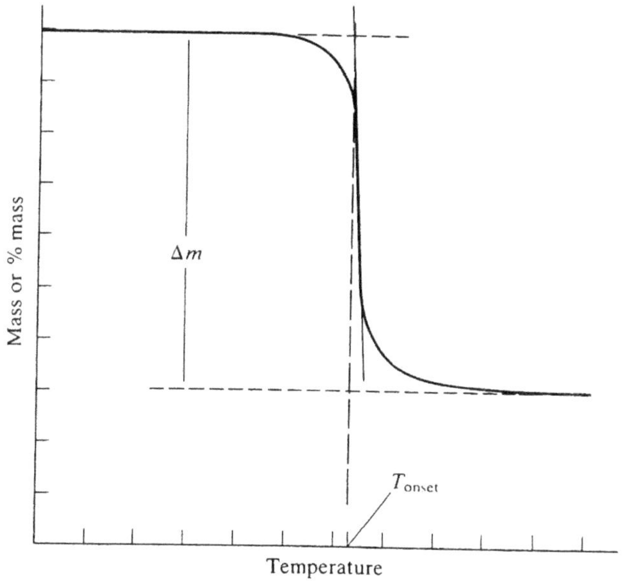

Mass or % mass

Δm

T_{onset}

Temperature

Figure 27-2 A diagram illustrating measurement of the onset temperature T_{onset} from a thermal curve.

Sample problem 27-1 The thermal curve of a 125.70-mg sample that contained a mixture of $CaC_2O_4 \cdot H_2O$ (MW 146.12) and a thermally stable salt had a mass loss Δm of 6.98 mg at an onset temperature of about 140°C corresponding to vaporization of water. Determine the percentage (w/w) of $CaC_2O_4 \cdot H_2O$ in the sample.

SOLUTION The reaction that occurs during the mass loss is

$$CaC_2O_4 \cdot H_2O(s) \rightarrow CaC_2O_4(s) + H_2O(g)$$

From Δm it is possible to calculate the original mass of $CaC_2O_4 \cdot H_2O$ in the sample:

$$6.98 \text{ mg } H_2O \times \frac{1 \text{ mmol } H_2O}{18.015 \text{ mg } H_2O} \times \frac{1 \text{ mmol } CaC_2O_4 \cdot H_2O}{\text{mmol } H_2O}$$

$$\times \frac{146.12 \text{ mg } CaC_2O_4 \cdot H_2O}{\text{mmol } CaC_2O_4 \cdot H_2O}$$

$$= 56.6 \text{ mg } CaC_2O_4 \cdot H_2O$$

The percentage of $CaC_2O_4 \cdot H_2O$ in the sample is

$$\frac{56.6 \text{ mg}}{125.70 \text{ mg}} \times 100 = 45.0\%$$

TG is useful for the quantitative analysis of substances that undergo loss of volatile components at temperatures below 1500°C. Inorganic compounds that lose waters of hydration are examples. Carbonates often undergo transitions to the corresponding oxides and oxalates to the carbonates. Some molecules can lose ammonia. In oxidizing atmospheres (oxygen or air purge gas) oxidation products of the sample can form. As an example, while using oxygen as the purge gas, FeS_2 is converted to $Fe_2(SO_4)_3$, causing a mass gain rather than a mass loss. The $Fe_2(SO_4)_3$ is further oxidized to Fe_2O_3. With nitrogen as the purge gas, FeS_2 is converted to FeS and S. Organic molecules can lose any of several volatile components. Thermal curves are often used to characterize polymeric materials such as polyvinyl chloride and polyethylene. Thermal analysis also is used to study inorganic compounds, explosives, pharmaceuticals and biological materials. A review of applications is available in Reference 5. TG can be used to determine the temperature range within which a precipitated reaction product can be dried during a gravimetric analysis and to study some chemical reactions at elevated temperatures.

Qualitative and quantitative chemical analyses of the stream of purge gas after it has passed through a thermal analyzer (TG, DTA, or DSC) can be used to provide further information related to the thermal process. *Evolved gas detection* (EGD) is the technique in which the amount of a sample component in the purge gas is monitored. A qualitative analysis of the gas is not performed. *Evolved gas analysis* (EGA) differs from EGD in that the identity of the components in the purge gas stream is determined. In either case a nonthermal analytical instrument is interfaced to the thermal analyzer. Mass spectrometers often are interfaced to thermal analyzers. TG/MS and TG/MS/MS (a thermobalance interfaced to tandem mass spectrometers) are both used. Other popular combined techniques are thermogravimetry / flame ionization detection (TG/FID) in which the purge gas stream from the thermobalance is passed through a flame ionization detector similar to that used for gas chromatography (Chapter 26) and TG/FTIR. Some analysts have used thermal analysis in combination with x-ray diffraction in order to examine structural changes that occur during heating (Reference 5).

Derivative Thermogravimetry

Derivative thermogravimetry (DTG) is the technique in which the ratio of the change in mass to the change in heating time of the sample is plotted as a function of the sample temperature. The resulting thermal curve is the first derivative of the TG thermal curve. A typical thermal curve obtained using DTG superimposed on the corresponding thermal curve obtained using TG is shown in Fig. 27-3.

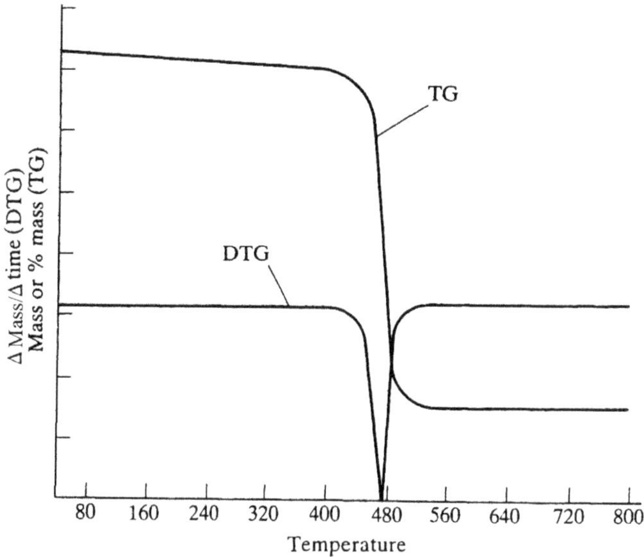

Figure 27-3 A thermal curve recorded by using thermogravimetry (TG) at a heating rate of 20°C/min and the corresponding thermal curve recorded using derivative thermogravimetry (DTG).

Some instruments record both the TG and DTG thermal curves. A derivative thermal curve can be obtained using an electronic circuit, such as an *RC* circuit, or an appropriate operational amplifier circuit (see the appropriate descriptions in Chapters 2 through 4). Alternatively, points on the curve can be mathematically calculated by the computer-controlled data-acquisition system from the recorded TG thermal curve. The points are plotted to yield the curve. The first approach is used for instruments that are not controlled by a computer. The second approach normally is used for microprocessor-controlled instruments. In other ways the apparatus that is used for DTG is identical to that used for TG. Qualitative and quantitative analyses are done by using the temperatures at the peaks and the peak heights respectively.

Differential Thermal Analysis

Differential thermal analysis (DTA) is the analytical technique in which the temperature differential between the sample and a nonreactive reference material is monitored while the two substances are subjected to the identical heating program. The thermal curve is a plot of the temperature difference as a function of the temperature of one of the two substances. Normally the temperature difference is recorded as a function of the temperature of the reference substance.

The sample and reference substance are held in separate containers within the same furnace. Usually the sample cups are constructed from platinum and are fairly close to each other (typically within about 1 cm) in order to ensure identical heating. The temperatures are monitored with thermocouples that are placed in the pedestals that support the sample cups. In some instruments the thermocouples directly contact the substances while in other instruments the thermocouples are physically separated from the substances by narrow pieces of a thermally inert material that often contains platinum or a ceramic.

The reference substance is usually thermally inert throughout the temperature interval used for the study. Alumina is often used. Some analysts use silicon carbide or glass beads as the reference. Glass beads, of course, should not be used throughout the temperature interval in which the glass undergoes a phase change.

The oven and temperature programmer are similar or identical to those used for TG. The upper temperature that is used in DTA is often higher than that used for TG. The major applications of DTA are to high-temperature measurements performed on alloys, ceramics, glasses, and minerals. Those materials typically are examined at temperatures as high as 1600°C. As in TG, a purge gas is used during the studies.

Unlike TG, DTA does not require a change in mass of the sample in order to obtain meaningful information. DTA can be used to study any process in which heat is absorbed or evolved. Among the endothermic processes that can be studied using DTA are melting, boiling, and sublimation. Absorptive processes are often exothermic and can be studied by using DTA. Desorption is normally endothermic.

Thermal curves obtained with DTA show peaks corresponding to processes in which the temperature of the sample is increased or decreased relative to the temperature of the reference substance. Upward deflections usually correspond to exothermic reactions and downward deflections to endothermic reactions. The temperatures at which the peaks are observed in the thermal curve can be used for qualitative analysis by comparison with curves recorded under similar conditions with known materials. Peak areas or peak heights are used for quantitative analyses.

The peak areas obtained while using DTA are directly proportional to both the mass of the sample and the ΔH for the process that corresponds to the peak. The proportionality constant is affected by experimental variables such as the initial temperature of the oven, the heating rate, the particle size, and the purge gas used for the study. For quantitative analysis with DTA or any other thermal method, it is necessary to carefully control all experimental parameters and to maintain identical experimental conditions while recording the thermal curves of the sample and the standards. A good deal of useful information relating to DTA and other thermal methods of analysis can be found in the collection of papers in References 6 and 7.

DIFFERENTIAL SCANNING CALORIMETRY

Differential scanning calorimetry (DSC) is the technique in which the temperatures of a sample and reference substance are controlled at identical values while the amount of heat that is added to one of the two substances in excess of that added to the other substance is monitored. The temperatures of the sample and the reference are identical but can be varied during the study. The thermal curve is a plot of the added heat differential as a function of the temperature if the temperature is changed, or as a function of time for isothermal operation.

The sample and reference substance are placed in separate holders. The two holders are mounted in a massive heat sink that serves to insulate them from each other. The sample and reference holders are constructed from a material, such as a platinum-iridium alloy, that has a low thermal mass.

The sample and reference holders are mounted on individual thermocouples and electric heaters. A difference in temperature between the two holders is sensed by monitoring the two thermocouples and is compensated by applying heat to the lower temperature substance. The difference in electric power that is supplied to the two electric heaters is continuously monitored and converted to a difference in heat added to one of the holders. That difference is plotted as the vertical component of the thermal curve. The axis is calibrated in milliwatts or in millicalories (or another heat unit) per second added to one holder in excess of that added to the other holder. Upward deflections on the thermal curve usually correspond to processes in which additional heat is added to the sample.

As with the other thermal methods, temperature programming in a purge gas atmosphere is normally used. The temperature can be varied from about 15°C to about 1000°C on many instruments. In some instruments the temperature can either be increased or decreased during the study. The addition of a cooling system permits measurements down to – 70°C for systems that use freon as the coolant or down to –170°C for systems that use liquid nitrogen. DSC is more popular than DTA. Because the two techniques are normally used in different temperature ranges, the information obtained from one technique can be used to supplement that obtained from the other technique.

DSC can be used to measure enthalpimetric changes associated with a particular process. It is commonly used to measure specific heats, heats of fusion, and enthalpimetric changes associated with physical and chemical transitions. As with DTA, it can be used for qualitative analysis by comparing peak locations (temperatures) to those of known materials that were recorded under identical conditions. It can also be used for quantitative analysis of a mixture by comparing peak areas to those of standards.

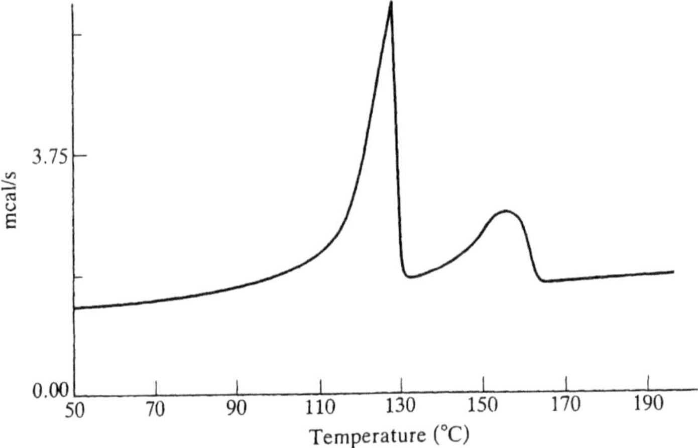

Figure 27-4 A thermal curve of a blend of polyethylene and polypropylene obtained by using differential scanning calorimetry. Upward deflections correspond to endothermic processes. The sample mass was 13 mg and the scan rate was 10°C/min.

In the drug industry DSC is used to determine the purity of some drugs. The thermal curve is recorded and used with a van't Hoff plot to determine purity. The temperature is slowly scanned through the melting region and a plot is prepared of the inverse of the melted fraction of the substance as a function of temperature. The slope of the line is equal to the melting point depression. Enough information is obtained from the curve to solve the van't Hoff equation for the mole fraction of impurity in the sample.

DSC can be used to measure activation energies and rate constants for a particular transition. Often it is used to measure the *glass transition temperature* T_G of a substance. T_G is the temperature at which a transition of a polymer from a glassy state to a liquid state occurs. It corresponds to the temperature at the inflection point in the DSC thermal curve for the transition. All amorphous polymers have a T_G, although those for highly crystalline polymers are sometimes difficult to measure. DSC is used to study and characterize many materials including adhesives, coals, elastomers, fertilizers, foods, glasses, metals, minerals, paints, petroleum products, plastics, polymers, resins, rubbers, semiconductors, steels, textiles, and waxes. A typical DSC thermal curve is shown in Fig. 27-4.

THERMOMECHANICAL AND DYNAMIC MECHANICAL ANALYSIS

Thermomechanical analysis (TMA) is the technique in which a physical dimension of a material, often the length, is monitored as a function of temperature. As with the other thermal methods a thermocouple is used to monitor the temperature of the sample in an oven through which a purge gas flows. Most instruments that are designed to perform TMA use a movable-core *linear variable differential transformer* (LVDT) to transform the dimensional change into an electrical signal. An LVDT is an electric transformer that has a movable core. The change in the output from the LVDT is related to the displacement of its core. The core is in direct physical contact with the sample. As the sample expands or contracts, the core moves within the transformer and the output from the transformer varies. The output from the transformer is related to the dimension (length or width) of the sample that is responsible for movement of the core. TMA can be used to measure expansion coefficients, shrinkage of a polymeric material, glass transition temperatures, and other properties which involve a dimensional change.

Dynamic mechanical analysis (DMA) is the technique in which a mechanical property of a test material is monitored as a function of temperature while the material is subjected to a periodic stress. Studies are performed in a programmable oven under a controlled atmosphere. The sample is clamped between two arms. One of the arms moves relative to the other in a periodic fashion, thereby placing a stress on the sample. The amount and frequency of the motion can be varied. The motion causes the sample to be periodically deformed. The amplitude of the deformation is monitored with an LVDT whose core is attached to the moving arm of the apparatus. The output from the LVDT reflects the frequency and amplitude of motion of the arm.

The system is forced to oscillate at a constant amplitude. The energy that is required to cause the system to oscillate at constant amplitude is related to the damping of the sample. The frequency of the oscillation is related to the modulus of the sample. Modulus is a measure of the amount of stress that must be imposed on an object in order to result in a defined deformation. Either property can be plotted as a function of temperature. The technique is used to measure tensile strength modulus and the damping characteristics of the sample material. A description of a commercial instrument that can be used for DMA as well as for other thermal studies can be found in Reference 8. A review of research papers related to the thermal methods of analysis described up to this point in the chapter can be found in Reference 9.

THERMOMETRIC TITRATIONS

In addition to the uses described earlier in the chapter, it is sometimes possible to locate the endpoint of a titration by measuring the temperature of the titrand as a function of the volume of added titrant. A titration in which the temperature of the titrand is monitored as a function of added titrant volume is a *thermometric titration*. If the chemical reaction that occurs during the titration is exothermic, the temperature of the titrand is expected to rise with the addition of titrant. Similarly, if an endothermic chemical reaction occurs, the temperature falls as the titrant is added. The endpoint corresponds to the titrant volume at which the temperature change, owing to the chemical reaction, stops increasing or decreasing. The reaction vessel must be insulated in order to prevent heat from being absorbed or evolved to the environment. The titration curve is a plot of temperature change ΔT as a function of added titrant volume.

The success of titrations followed by an electroanalytical method, such as potentiometry or amperometry, and of spectrophotometric titrations depends upon the change in Gibbs free energy ΔG for the chemical reaction that occurs during the titration. If ΔG is sufficiently large, the change in the measured property is large and the endpoint is easily located. As an example, from Eq. (22-3) ($\Delta G = -nFE$) it is apparent that the potential change during a potentiometric titration is directly related to ΔG for the titration reaction. As ΔG increases, the change in measured potential becomes greater.

The success of a thermometric titration depends upon the amount of heat ΔH evolved or absorbed during the reaction rather than upon ΔG for the reaction. Consequently, temperature changes in insulated systems can be used to locate the endpoints of some titrations whose endpoints cannot be located by using other methods. An example is the titration of H_3BO_3 with NaOH. The potentiometric endpoint is nearly impossible to locate, whereas the thermometric endpoint is well defined because ΔH for the reaction is relatively large even though ΔG is small.

The titrant in a thermometric titration is generally continuously added with a motor-driven syringe or some other type of controlled-flow buret. The temperature, as monitored during addition of the titrant with a thermistor, is used to prepare the titration curve. It is possible to add the titrant stepwise and to measure the temperature after each addition; however, that procedure requires more time and increases the risk of losing or gaining heat from the surroundings of poorly insulated titration vessels. A thermometric titration curve that utilizes an exothermic chemical reaction is shown in Fig. 27-5, curve *A*, and a curve that utilizes an endothermic initial reaction is shown in Fig. 27-5, curve *B*. The endpoint shown in Fig. 27-5, curve *B*, corresponds to the reaction of one mole of perchloric acid for each mole of sodium malonate to form sodium hydrogen malonate. The temperature rises after the endpoint because the reaction of perchloric acid with sodium hydrogen malonate to form malonic acid is exothermic.

The titration curves shown in Fig. 27-5 are those obtained with an apparatus which continuously adds titrant to the titrand. The output from the thermistor is measured and displayed on the temperature axis of the titration curve. The readout device is a strip-chart recorder that is used at a fixed chart speed or a computer-controlled device.

The recorder is turned on and the temperature of the titrand is monitored prior to addition of any titrant. The portion of the titration curve that corresponds to the change in temperature ΔT with time t prior to start of the titration is the *pretitration* region of the curve. During the titration the rate of addition of the titrant is constant and the time axis on the chart is directly

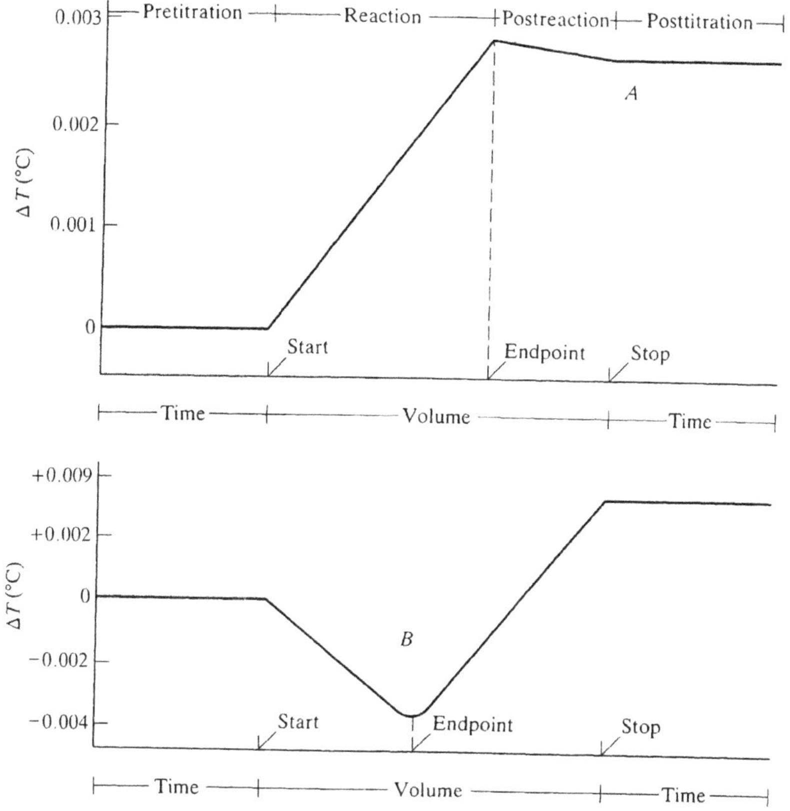

Figure 27-5 Thermometric titration curves: curve *A*, the titration of iron(II) sulfate with potassium dichromate and curve *B*, the titration of sodium malonate ($NaOOCCH_2COONa$) with perchloric acid.

proportional to the added titrant volume. Prior to the endpoint (the *reaction* region of the curve), the temperature of the titrand increases if the chemical reaction is exothermic or decreases if the reaction is endothermic. At the endpoint no further heat is absorbed or emitted by the reaction and the temperature of the titrand ceases to change at the rate observed in the reaction region.

If no further chemical reaction occurs with the addition of more titrant, any observed temperature change in the *postreaction* portion of the titration curve is caused by ΔH changes associated with the addition and dilution of titrant with titrand, changes associated with an increase in the volume of the solution in the titration vessel, and changes associated with heat loss or gain from the surroundings. The endpoint is located by extrapolating the linear portions of the curve in the reaction and postreaction regions to interception. The *posttitration* portion of the curve occurs after titrant delivery stops but while temperature monitoring continues.

The temperature changes that are observed during titrations are dependent upon the heat capacities of the titration vessel and its contents, the enthalpy change for the chemical reaction, and the amount of reactants. Typical temperature changes are in the millidegree region. The addition of titrant to titrand during the titration increases the volume of the solution in the titration vessel and the heat capacity of the contents of the vessel. Consequently, the temperature increase or decrease, resulting from a fixed addition of titrant, near the end of the titration is not as large as that near the start of the titration. To minimize the difference, the concentration of the titrant is usually large relative to the concentration of the titrand. Generally, the titrant concentration is

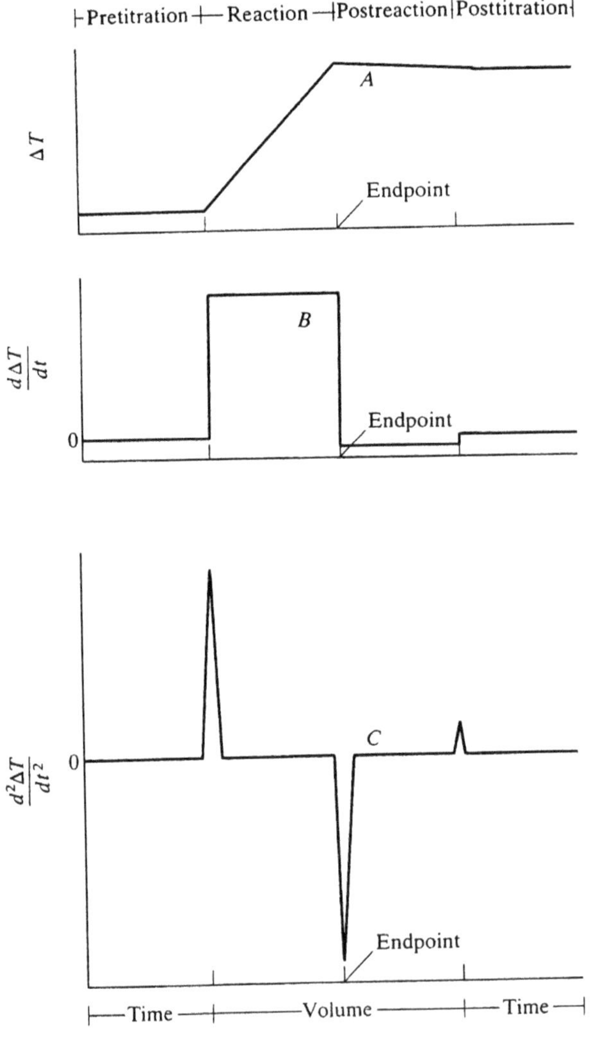

Figure 27-6 Sketches of three forms of thermometric titration curves: curve A, normal; curve B, first derivative; curve C, second derivative. T is temperature and t is time.

sufficiently large that the volume of titrant added at the endpoint is 5 percent or less of the titrand volume. The resulting titration curve does not need to be corrected for dilution.

In some cases analysts prefer to use first- or second-derivative thermometric titration curves. A sketch showing a thermometric titration curve and the corresponding first- and second-derivative curves is shown in Fig. 27-6. Often the second-derivative curve is plotted by the readout device showing all peaks deflecting in the same direction. First- and second-derivative curves are particularly useful when the region around the endpoint of the standard titration curve is highly rounded, thereby making it difficult to locate the endpoint.

Thermometric titrations can be used for quantitative analysis when the ΔH for the titration reaction is sufficiently large to cause a measurable temperature change and when the reaction proceeds essentially to completion. The major use for thermometric endpoint location has been in acid-base titrimetry. Thermometric titrations in nonaqueous solvents and in molten salts have been performed. The results of thermometric titrations have been used to determine the stoichiometry of titration reactions as well as to quantitatively assay samples. A detailed description of thermometric titrations can be found in Reference 10.

ENTHALPIMETRY

Quantitative analysis of a sample can sometimes be performed by measurement of the change in enthalpy that is associated with the addition of an excess of a chemical reactant to the sample. If the chemical reaction is exothermic (or endothermic) and the H for the chemical reaction is known, the change in temperature in the reaction vessel can be used to determine the amount of the limiting chemical reactant in the sample. *Direct injection enthalpimetry* (DIE) is the technique in which a single portion of an excess of one chemical reactant is added to a fixed volume of a sample that contains the other reactant. The experiment is performed in a highly insulated vessel of known heat capacity. The increased temperature in the vessel is used to calculate the amount of heat evolved in the reaction. Because the change in enthalpy for the chemical reaction is known, the concentration of the limiting reactant in the sample can be calculated. DIE has primarily been used for quantitative analyses of reactants in acid-base and complexation reactions. It can be used for analysis of a reactant in any chemical reaction which has a known change in enthalpy. Sometimes DIE is referred to as an *enthalpimetric* titration.

Another enthalpimetric method is continuous-flow enthalpimetry (CFE). CFE is similar to DIE except that an excess of one reactant is continuously added to a flowing stream of the sample. The temperature of the stream is continuously monitored and related to the concentration of the limiting reactant in the flowing sample. CFE is particularly well suited to process control applications in industry (Chapter 28). A description of DIE and CFE can be found in Reference 10.

IMPORTANT TERMS

Boxcar averaging
Continuous-flow enthalpimetry
Curie point
Derivative thermogravimetry
Differential scanning calorimetry
Differential thermal analysis
Direct injection enthalpimetry
Dynamic mechanical analysis
Enthalpimetric titration
Evolved gas analysis
Evolved gas detection
Glass transition temperature
LVDT
Multitasking
Onset temperature
Purge gas
Purge gas switching
TG/FID
TG/FTIR
TG/GC
TG/MS
Thermal curve
Thermobalance
Thermogravimetry
Thermomechanical analysis
Thermometric titration

PROBLEMS

Thermogravimetry

27-1 A thermogravimetric thermal curve was obtained for a sample that contained magnesium sulfate heptahydrate. The mass of the sample was 2.89 mg and the heating rate was 5 K/min. The purge gas was air. A single step was observed in the thermal curve at an onset temperature of 378 K corresponding to formation of magnesium sulfate monohydrate. The mass loss for the step was 0.59 mg. Determine the percentage of $MgSO_4 \cdot 7H_2O$ in the sample.

27-2 The thermal curve corresponding to a sample that contains $Al(OH)(HCOO)_2 \cdot 0.5H_2O$ shows stepwise weight loss at temperatures of 200, 260, and 350°C in either air or nitrogen corresponding to decomposition to Al_2O_3. A 25.00-mg sample had a total weight loss of 22.5 percent of the initial mass of the sample. Determine the percentage of $Al(OH)(HCOO)_2 \cdot 0.5H_2O$ in the sample.

27-3 The palladium complex $Pd(LH_2)Cl_2$, where L is dimethyldithiooxamide, decomposes at an onset temperature of 231°C to yield a polymer containing an average of $n + 2$ palladium atoms for each molecule of the polymer, as shown in the following chemical equation:

$$(n+2)Pd(LH_2)Cl_2 \rightarrow Cl_2PdL(PdL)_nPdLH_2 + 2(n+1)HCl(g)$$

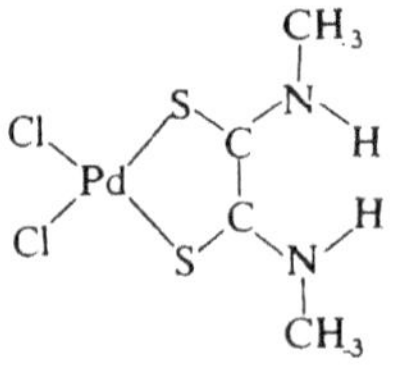

$Pd(LH_2)Cl_2$

The thermogravimetric mass loss from the thermal curve was 19.0 percent while using nitrogen as the purge gas. Determine the average value of n for the polymer.

Thermometric Titrations

27-4 A 50.00-mL portion of a silver nitrate solution was thermometrically titrated with 0.9758 *M* HCl:

$$AgNO_3(aq) + HCl(aq) \rightarrow AgCl(s) + HNO_3(aq)$$

The endpoint of the titration was 1.284 mL. Determine the concentration of silver nitrate in the solution.

27-5 A 25.00-mL portion of a solution containing ethylenediamine was thermometrically titrated with 1.046 *M* hydrochloric acid:

$$H_2NCH_2CH_2NH_2 + HCl \rightarrow H_2NCH_2CH_2NH_3^+ + Cl^-$$

The first endpoint of the titration was 1.048 mL. Determine the concentration of ethylenediamine in the solution.

Ethalpimetry

27-6 A 1.00-mL portion of 1 *M* HCl was added to a 99.0-mL aqueous solution containing pyridine:

$$C_5H_5N + H^+ \rightarrow C_5H_5NH^+, \qquad \Delta H = -4.97 \text{ kcal/mol}$$

The temperature of the reaction vessel increased by 4.55×10^{-3} K. Assuming no heat is absorbed by the vessel, the specific heat of the solution is 1.00 cal/g · K, and the density of the solution is 1.008 g/mL, calculate the concentration of pyridine in the sample.

27-7 DIE was used to assay an aqueous solution containing phenol. A 2.00-mL portion of 1.5 *M* NaOH was added to a 100.0-mL portion of the sample:

$$C_6H_5OH + OH^- \rightarrow C_6H_5O^- + H_2O, \qquad \Delta H = -7.75 \text{ kcal/mol}$$

The temperature increased by 0.0762 K. The specific heat of the final solution was 1.02 cal/g · K and the density of the solution was 1.01 g/mL. Assuming no heat absorption by the vessel, calculate the concentration of phenol in the solution.

REFERENCES

1. Gill, P. S.: *Amer. Lab.*, **16**(1) :39(1984).
2. McGhie, A. R.: *Anal. Chem.*, **55**:987(1983).
3. Earnest, C. M.: *Anal. Chem.*, **56** : 1471A(1984).
4. Menard, K. P., R. Spragg, G. Johnson and C. Sellman: *Amer. Lab.*, 42: 21(2010).
5. Vyazovkin, S.: *Anal. Chem.*, **78**:3875 (2006).
6. Miller, B. (ed.): *Thermal Analysis*, vols. I and II, Wiley, New York, 1982.
7. Sorai, M. (ed.): Comprehensive Handbook of Calorimetry and Thermal Analysis, Wiley, Hoboken, New Jersey, 2004.
8. Gill, P. S., L. C. Thomas, and R. L. Blaine: *Amer. Lab.*, **17**(1):34(1985).
9. Meisel, T., and K. Seybold: *CRC Crit. Rev. Anal. Chem.*, **12**(4):267(1981).
10. Hansen, L. D., R. M. Izatt, and J. J. Christensen, in J. Jordan (ed.): *New Developments in Titrimetry*, Marcel Dekker, New York, 1974, pp. 1-89.

CHAPTER

TWENTY-EIGHT

AUTOMATED ANALYSIS

The use of automation in routine analyses is popular in most medical laboratories, in many industrial laboratories, and in some academic laboratories. In some industrial processes the results from automated analyses are used to automatically control one or more variables in the industrial process. Many reasons exist for using automated rather than manually performed analyses. Many laboratories, especially industrial and medical laboratories, have a work load that is much too large to permit the analyst to spend a significant portion of time on each analysis. The medical technologists in most hospitals and other medical laboratories, as well as many industrial analysts, are unable to do each analysis required of them without using automation. By automating the routine analyses, the analyst is free to devote time to the unusual analysis which requires human flexibility and initiative.

Automation has several advantages in addition to decreasing the workload of the analyst. Because automation decreases the number of analysts required in a busy laboratory and decreases reagent waste, the cost of each assay is usually less in an automated laboratory than in a nonautomated laboratory. An automated analytical system treats each sample identically. No error is introduced owing to a tired or poorly trained analyst. In many cases automation permits laboratory technicians to be used in place of highly trained analysts. The training necessary to perform an automated analysis is usually considerably less than that required to perform the identical nonautomated analysis. That frees the analyst for other duties.

In many cases the analytical results are needed more rapidly than the average analyst can provide them. Automation alleviates the problem because automated analysis is generally more efficient than nonautomated analysis. The time delay in obtaining the result of an analysis is particularly important when a patient's treatment or an industrial plant's operating parameters depend upon the analytical result. In many industrial operations, *process-control* analytical instrumentation is used to monitor a product continuously and to adjust the plant's operating parameters automatically in order to keep the product within desired specifications. The automation of analyses of harmful substances can improve safety records. The major disadvantage of automation is the initial cost of the apparatus. Often the initial cost is considerably greater than that of non-automated apparatus, although the cost for each assay throughout the lifetime of the apparatus is generally less than the cost for each nonautomated assay.

Automated analyses can be classified into two categories. In many cases samples are collected. either automatically or manually, and transported to the laboratory, where the analyses are automatically performed. In the ensuing descriptions such analyses are termed *laboratory analyses*. In other instances the substance is both automatically sampled and automatically transported to the analytical instrument (often through a pipe). In that case the analytical instrument is usually located in a specially designed housing near the sample site rather than in the laboratory. The analytical results are fed through a *control loop* to a device that controls one or more of the variables that affect the sample composition. Those analyses are termed *process-control analyses*. The assembly that performs the analysis and that automatically controls the production apparatus is a *process-control device*. Some analysts define an *automated device* as one which has a control loop and an *automatic device* as one which has none. Because the definitions are not agreed upon, the terms are used interchangeably in the chapter.

Nearly all routine steps in a laboratory procedure can be automated. Among the repetitive steps, in addition to the assay, that have been automated are weighing, mixing, heating, cooling, sampling, diluting, separations, including precipitations and chromatographic separations, solvent extractions and pipeting.

AUTOMATED LABORATORY ANALYSES

Automated laboratory analyzers generally consist of the parts shown in the block diagram in Fig. 28-1. Because one or more of the parts might not be needed for a particular analysis, some automated analyzers do not contain all of the parts shown in the diagram. The sampler is the device into which the collected samples are placed, either manually or automatically, and from which samples are drawn into the remainder of the device. The sampler is often followed by a separator, which removes those sample components that might interfere with the assay. The separated interferences can be discarded or collected for further analysis. Often the separator is a chromatographic column, a precipitator and filter, or a dialysis membrane through which only the assayed substance and noninterfering solution components can pass.

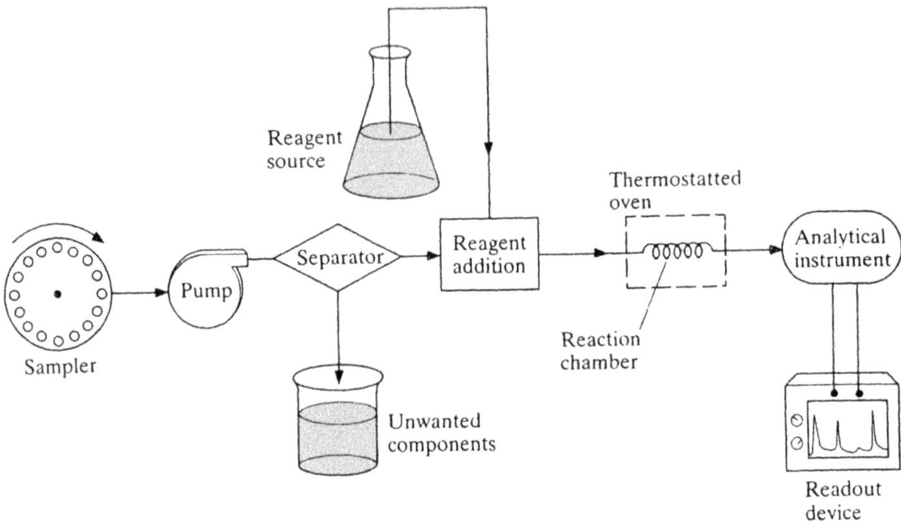

Figure 28-1 Diagram of an automated laboratory analyzer.

In some automated analyses a chemical reagent is added to the sample solution, after removal of interferences, that quantitatively reacts with the sample to yield an easily measurable reaction product. In many cases the reagent reacts with the sample to form a colored product which can be monitored spectrophotometrically. The chemical reagent is stored in a container or reservoir from which predetermined amounts can be withdrawn and added to the sample.

The portion of the apparatus in which the reagent is added is followed by a *reaction chamber* within which the chemical reaction has ample time to occur. In some analyzers the reaction is completed by placing a coil in the flow path of the sample and in others during a time delay prior to insertion of the sample into the analytical instrument. The first approach is often used with continuously flowing streams of samples; the second with systems that carry the sample through the apparatus in individual containers. During the time period in which the chemical reaction occurs, the sample is sometimes held in a controlled-temperature oven or constant-temperature bath, which ensures a constant reaction rate. Generally, the bath or oven is held above room temperature in order to speed completion of the reaction.

After completion of the chemical reaction, the sample passes into the analytical instrument where the concentration of the sample is measured. The result is displayed by the readout device. The analytical instrument can be nearly any of the instruments described in preceding chapters of the text. In many cases the analytical instrument is a spectrophotometer, although a fluorometer, a chromatograph, an electroanalytical device, or some other analytical instrument can be used. The readout device can be a recorder, but in modern analyzers, it is more often a computer which acquires and manipulates data from the analytical instrument. The result is locally displayed, transmitted to a remote computer by using laboratory information management system (LIMS) software, or both transmitted and locally displayed. If LIMS software is used, the remote computer, rather than the local computer, may be used to perform the data manipulations that are required in order to obtain a useful result. Typically the result is reported on a standard form for that particular analysis.

Computerization

The availability of laboratory computers and LIMS software have revolutionized repetitive analysis. Computers are used to perform many of the calculations previously done by the analyst, to accept analytical data from analytical instruments, and to control the individual steps in the analysis. A brief description of computer operation, theory, and programming was included in Chapter 4. In most cases the analyst needs to know little about computers or their programming in order to benefit from their use. Laboratory computers are particularly valuable as aids in analysis because they can accept data directly from an analytical instrument, perform the calculations required to obtain a useful result, and report the result. The entire process is completed much more rapidly and generally more reliably than without a computer. The lines of communication between a laboratory computer, analytical instrumentation, and the analyst are shown in Fig. 28-2.

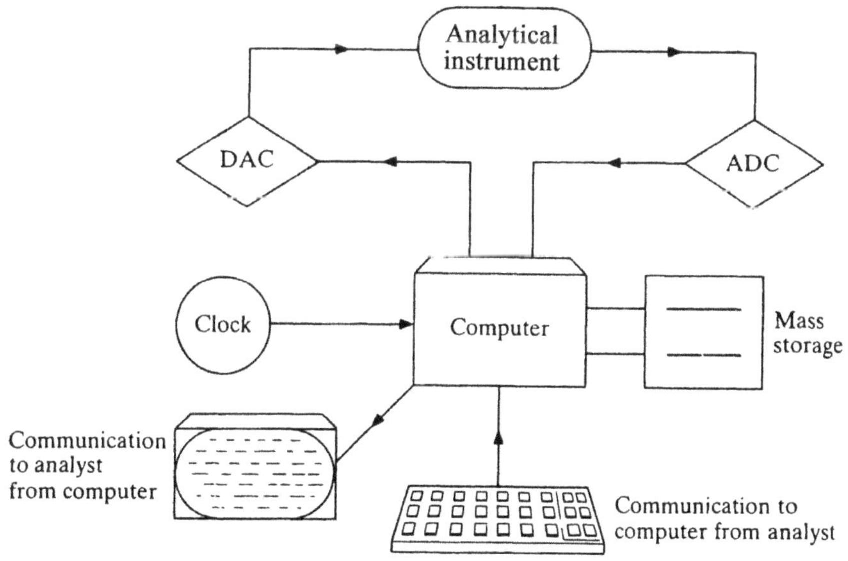

Figure 28-2 Block diagram showing the communication lines between the analyst, the computer, and the analytical apparatus.

All devices that provide information to a computer or that obtain information from a computer are *input/output* (I/O) devices. The I/O device by which the computer and analyst communicate is usually a printer, video display terminal, or keyboard-operated device. The analyst types instructions to the computer on the keyboard and the computer prints the analytical results and other pertinent information. A few computers are voice-activated and audibly communicate to the user.

Because the computer cannot understand the analog signal (Chapter 4) coming from the analytical instrument, that information must be converted to a digital number, prior to entering the computer. The conversion is done with an analog-to-digital converter (abbreviated to A/D converter, or ADC). A description of ADCs was included in Chapter 4. The device accepts the electrical signal (usually a potential) from the analytical instrument and converts it to a binary number which is accepted by the computer. Likewise, instructions from the computer to the analytical instrument normally are converted from a digital binary number to an analog signal. That conversion is done by a digital-to-analog converter (DAC, or D/A converter). The timing of communications between the analytical instrument and the computer is critical. It is accomplished with the aid of a highly accurate clock. The analyst sometimes has control of the rapidity with which the several communications take place. Computer systems which use a clock to control communications with an instrument are said to operate in *real time*.

Laboratory computers are categorized as either microcomputers or minicomputers. Microcomputers can be used as *dedicated computers* or as general-purpose computers. A dedicated computer is one that has a single purpose, e.g., to control the analysis performed by a specific gas chromatograph. A microcomputer is built into many analytical instruments. It is permanently wired to the instrument, usually in such a way that it performs a predefined set of operations such as signal averaging, concentration calculation, and background signal subtraction. In order to make a computer of that type perform any of its programmed functions, it is only necessary to press the appropriate button on a control panel. Microcomputers designed for use with a single instrument are easy to use by analysts and technicians who have no knowledge of computer operations. A computer of the type described here, in which a direct electronic communication between the computer and the apparatus exists, is an *on-line computer*. Dedicated microcomputers are presently being used to control nearly all types of analytical instrumentation including gas chromatographs, gas chromatograph-mass spectrometer combinations, liquid chromatographs, atomic absorption spectrophotometers, voltammetric analyzers, automatic titrators, pH meters, and uv-visible spectrophotometers.

Laboratory microcomputers have been widely used with chromatography. The computer accepts the chromatogram from the chromatograph, measures peak areas and retention times, and sometimes qualitatively and quantitatively identifies each component. The results are usually printed on a strip of paper or are displayed on a video display terminal.

An example of a microcomputer-controlled voltammetric analyzer is the EG and G Princeton Applied Research polarographic analyzer. It can store and recall instructions for up to 12 different analytical methods using Tast polarography, pulse polarography, differential pulse polarography, stripping voltammetry, differential pulse stripping voltammetry, and linear sweep voltammetry. In addition, it can store up to nine complete analytical curves and label report sheets with up to 24 experimental parameters. Another versatile, microprocessor-controlled, electrochemical system is manufactured by Bioanalytical Systems, Inc. Nearly all types of analytical instrumentation are commercially available in versions that are controlled by dedicated microcomputers.

An interesting microcomputer-controlled device is the ChemResearch model 1560 laboratory sample processor manufactured by Instrumentation Specialties Company. The device automates wet chemical procedures such as those which sometimes precede insertion of a sample into an analytical instrument. It can process up to 210 samples through 960 steps that include sample dilutions, reagent additions from as many as 40 different sources, delays for chemical reaction, and sample injection into an analytical instrument.

Minicomputers are larger than microcomputers. Because they are larger, they are capable of handling more complex tasks than those which are handled by microcomputers. Rather than being used as dedicated computers, they are often used to control and to collect data simultaneously from several laboratory instruments. Actually the computers rapidly and sequentially accept the digital signal from each ADC attached to each analytical instrument. The apparatus, including the computer, ADCs, DACs, and computer programs, is a *laboratory management system* (LMS). Laboratory minicomputers are usually designed for general laboratory use. Unlike some dedicated microcomputers, they can be programmed by the analyst to perform several jobs simultaneously. Programming a minicomputer requires, knowledge of a computer language and consequently more training than is required to use a dedicated microcomputer. After the program has been written, the two types of computers are equally easy to use. Most modern personal computers are capable of being used as laboratory minicomputers.

AUTOMATED LABORATORY APPARATUS

Generally, automated laboratory analyzers have some or all of the components diagrammed in Fig. 28-1. The components, design, and operation of the analyzers vary somewhat with the manufacturer. The following six sections are devoted to describing some of the more popular forms of automated analyzers.

The examples used are intended to illustrate different approaches to designing automated equipment. The descriptions are not intended to include examples of all types of automated devices, but rather to illustrate different approaches to the design of automated laboratory equipment.

Continuous-Flow Analyzers

The automated analytical systems manufactured by Technicon Instruments Corporation are some of the most widely used of the automated systems in medical laboratories. They are *continuous-flow systems*; i.e., they are designed to assay individual samples in a flowing stream rather than in discrete containers. Because they are modular in design, individual components of the apparatus can be easily exchanged for other components when the analysis is altered. A block diagram of the parts of the Technicon apparatus is shown in Fig. 28-3.

The sampler is a device into which the samples are placed. It introduces the samples, air, and rinse solution to the pump. Sample introduction is most often accomplished using 40 sample cups on a rotary tray. Other samplers are available for discrete solid samples and for continuously sampling gases and flowing liquids. The discrete liquid sampler first introduces a portion of sample solution into the analytical stream, followed by a block of air, a portion of a wash solution, a second block of air, and then the next sample. In that way each sample is separated from other samples and the wash solution by air pockets.

The substance (sample, wash solution or air) flowing into the analytical stream is usually controlled by a programming cam attached to a motor that operates at constant speed. As the cam rotates, the valve to the desired substance (e.g., the air) is opened while the other valves (to the sample and wash solution) are closed. Further rotation causes the air valve to close while a different valve opens. The ratio of the time that the sample is pumped into the analytical stream to the time that the wash solution is pumped is

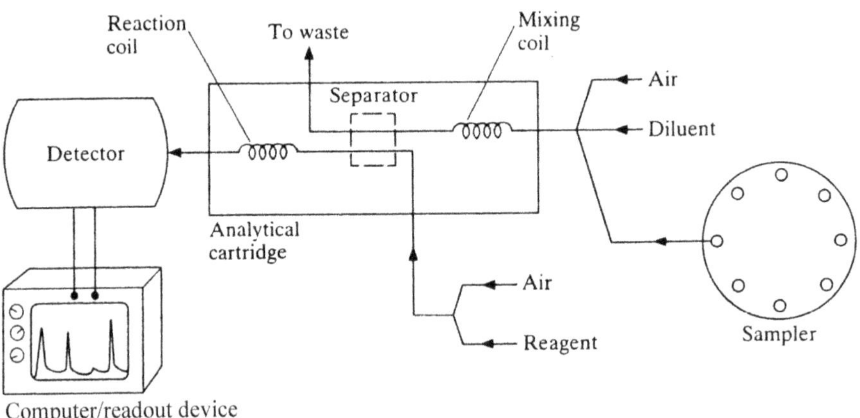

Figure 28-3 Diagram of a Technicon AutoAnalyzer.

controlled by the cam and is normally 2: 1. Ratios as high as 9: 1 are used. In some devices substance flow is controlled by automated valves which are controlled by the computer.

The pump used with the Technicon apparatus is a peristaltic proportioning pump which is capable of pumping fluids through many pieces of flexible tubing simultaneously. A *peristaltic pump* is one in which the liquid is forced through flexible tubing by a bar that constricts the tubing and then travels along the length of the tubing in the desired direction of flow. The liquid is pushed ahead of the constriction. Only the tubing comes in contact with the pumped fluid.

By using tubing of different internal diameters, the flowrate of the fluid can be controlled. As the internal diameter of the tube is increased, the flowrate of fluid through the tube is increased. Consequently, the ratio of the internal diameters of the tubing is related to the proportion of fluid flowing through the tubes; i.e., the pump is a *proportioning pump*. The sample, air, diluent (solvent), and chemical reagents used in the analysis are simultaneously pumped by the peristaltic pump.

After leaving the pump, the sample is mixed with diluent to bring it into the proper concentration range for the particular analysis. The diluted solution enters a portion of the apparatus called the *analytical cartridge*. A different analytical cartridge is used for each type of analysis. The analytical cartridge contains components that are necessary for a specific analysis but that are not required for all analyses. Generally, the first portion of the analytical cartridge contains a helical mixing coil in which convective mixing of the sample and diluent takes place. The diluted sample flows through a separator which is used to remove interferences. In most cases the separator is a dialyzer in which a fixed proportion of the analyte in the sample flows through a dialysis membrane and into a second stream containing a chemical reagent. The separator portion of the apparatus can also contain a phase-separation device.

After passage through the separator, the sample chemically reacts with the reagent with which it is mixed to yield a measurable reaction product. In many cases the reaction product is colored and can be monitored spectrophotometrically. A helical reaction coil permits mixing of the sample with the chemical reagent and delays the measurement for a sufficient period to allow the reaction to occur. The reaction coil is often contained in a heated bath which is adjusted to an optimum temperature for the particular chemical reaction. The reaction product flows from the reaction coil and exits from the analytical cartridge.

After leaving the analytical cartridge, the reaction product enters the detector. The detector measures a property of the solution that can be related to the concentration of the reaction product. The analytical result is displayed by the readout device. The detector can be a photometer, a uv-visible spectrophotometer, a fluorometer, a nephelometer, an infrared spectrophotometer, an ion-selective electrode, an amperometric detector, or some other device. Of the detectors, the photometer and uv-visible spectrophotometer are used more often than the others. The readout device normally is a computerized data acquisition system which acquires the data, performs any required calculations and displays the results.

One form of the AutoAnalyzer is capable of handling up to 150 samples/h. A single sample can be assayed with continuous-flow analyzers for more than one component by use of a multichannel analyzer. Technicon manufactures several sequential multiple analyzers. The Technicon 6/60 performs six sequential assays on each sample at a rate of 60 samples/h. Likewise, the Technicon 12/60 and 18/60 respectively perform 12 and 18 assays on each sample at a rate of 60 samples/h. Although the Technicon analyzers have primarily been used in hospital laboratories, they are also used in many industrial, environmental, agricultural, and research laboratories. They can be designed to handle nearly any type of routine analysis. Another example of a continuous-flow analyzer is the Coulter Industrial Kem-O-Lab automatic analyzer, which can perform up to six assays per sample at a rate of 240 samples/h. A more thorough description of older continuous-flow analyzers is given in Reference 1. Descriptions of modern commercial automated analyzers are available from each of the manufacturers.

Flow Injection Analyzers

The continuous-flow analyzers that were described in the preceding section use blocks of sample separated by plugs of air and a wash solution. The chemical reaction is performed within a block of solution. It is assumed that chemical equilibrium is established during the reaction, or, at the very least, that the reaction has proceeded a constant proportion of the way toward completion. The concentration of the product throughout the block is essentially constant and the detector yields a constant signal while the block of sample passes through it.

Flow injection analysis (FIA) is a continuous-flow technique in which the sample is directly injected into a flowing stream of either a carrier liquid or a solution of a chemical reagent. The sample is not separated from the remainder of the stream by blocks of air and wash solution as in other forms of continuous-flow analysis. A diagram illustrating the major parts of one type of a flow injection analyzer is shown in Fig. 28-4.

The reagent and the carrier are shown in separate containers in Fig. 28-4. In some analyzers the reagent solution serves as the carrier, thereby eliminating the second flow line from the apparatus. Generally, less reagent can be used if the two are separated because in that case the reagent does not have to continuously flow through the apparatus ; it can be added to the stream only when needed to react with a sample. Often peristaltic pumps are used to force the solutions through the analyzer. In some instruments a single proportioning pump is used

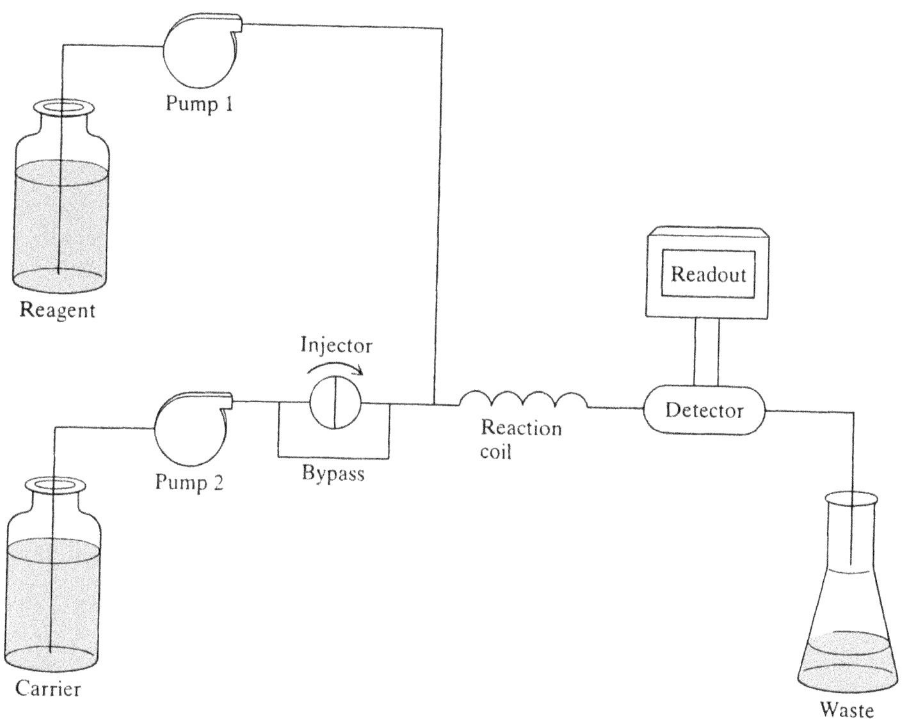

Figure 28-4 A diagram of one form of flow injection analyzer.

in place of the two pumps shown in Fig. 28-4 to simultaneously add reagent and carrier. Precision syringes that are driven by stepper motors can be used in place of pumps.[2]

Several types of injectors are used. In some instruments the injector consists of a rotary valve and a bypass. Often the injector is similar to those used in high-performance liquid chromatography. The injector shown in Fig. 28-4 is the rotary-valve type. While the valve is in the loading position, as shown in the figure, the carrier flows through the bypass around the valve. The resistance to flow in the bypass is greater than that through the open valve because the diameter of the bypass tubing is less than that in the valve. While in the loading position, the valve is filled with sample by using a syringe or other device. The sample size is determined by the volume of tubing in the valve. The tubing must be changed in the same manner that the sample loop is changed in HPLC injectors, in order to change the sample volume. After the valve has been filled, the injection is made by rotating the valve so that the tubing is aligned with the flow line. The carrier flushes the sample out of the valve and into the remainder of the apparatus. Sample volumes in the range between 5 and 100 μL are generally used. Typically the internal diameter of the tubing used in flow injection analyzers is between 0.2 and 1.0 mm. Often the tubing is constructed from polypropylene, some other plastic, or stainless steel. Flowrates between

0.4 and 2.0 mL/min are common. For multiple samples it is possible to automate repetitive injections.

After the sample has been injected into the carrier stream, the chemical reagent is added to the stream (unless it is already present) and the sample reacts with the reagent to form a measurable reaction product. Sometimes the apparatus contains a reaction coil or tube between the injector and the detector that allows extra time for the sample and reagent to disperse into each other. The reaction coil usually has a length between 0.1 and 2.0 m.

The detector responds to the chemical reaction product. Detectors used in flow injection analyzers are identical to those used in classical continuous-flow analyzers. The most popular detectors are visible photometers, spectrophotometers, and ion-selective electrodes (particularly coated-wire electrodes). Other detectors that can be used for FIA include atomic absorption spectrophotometers, conductometers, inductively coupled plasmas, turbidimeters, nephelometers, fluorometers, refractometers, voltammeters, differential potentiometers, and devices that monitor chemiluminescence.

Dispersion of the sample into the reagent stream takes place by longitudinal flow and by radial diffusion. The portion of the sample block in contact with the walls is essentially held in place by the walls while the portion of sample in the center of the tube moves forward with the flowing stream. Diffusion can occur either longitudinally or radially. *Longitudinal diffusion* is along the length of the tube either with or against the flow. *Radial diffusion* occurs perpendicularly to the direction of flow. Although the theory of dispersion during FIA is complex, it is well established. A thorough description of the theory of dispersion during FIA is too lengthy to be included here. An introduction to the theory can be found in References 3 and 4.

It is worth while to mention some of the conclusions from a theoretical study of FIA. The conclusions are expressed as rules in Reference 3. Some of the useful rules are (1) in narrow tubes dispersion can be decreased by decreasing the flowrate; (2) long, narrow reaction coils result in gaussian peak shapes; (3) limited dispersion occurs when the sample is injected into a carrier stream that flows at a minimal practical rate and when the distance between the injector and the detector is small; and (4) dispersion of the sample in a narrow tube increases with pumping rate and with the square root of the distance between the injector and detector or with the square root of the time in the apparatus between injection and detection.

Dispersion is important in some analyses to mix the sample with the reagent. Dispersion broadens the band of sample and decreases the concentration of sample in the center of the band. The width of the band is one of the factors that limits the rate of sample injection. Consequently, dispersion should be kept as small as possible as long as the reaction between the reagent and sample stream can occur sufficiently. Sampling rates as large as 300 samples/h can be used with FIA.

Because the sample disperses as it flows through the apparatus, the detector generally yields a gaussian-shaped peak rather than the constant signal observed during classical continuous-flow analysis. The peak height, the height at a fixed time after injection, the peak area, or some other variable such as the peak width can be measured and related to analyte concentration.

For most analyses working curves are prepared by plotting the detector response to standard solutions as a function of concentration. Often the working curve is stored in the memory of the microprocessor that is used to control the analyzer. FIA has many applications in addition to repetitive routine analyses. A brief description of a few of the applications is described here. More information can be found in References 2 through 6.

Gradient calibration is the technique in which the response of the detector, at some time other than that corresponding to the center of the peak, is used to construct the working curve. In practice the detector response is usually monitored at a fixed time after sample injection. Gradient calibration is particularly useful when the concentrations of the samples and standards cause off-scale detector responses at the centers of the peaks.

If a spectroscopic or voltammetric detector can be rapidly scanned, an entire EMR spectrum or voltammetric scan can be repetitively obtained as the sample passes through the detector. That technique allows the simultaneous determination of more than one sample component. *Stopped-flow* techniques can also be used during FIA. The pump forcing the carrier and sample through the apparatus can be turned off or blocked off while the sample is in the reaction coil or in the detector. If the delay occurs while the sample is in the reaction coil, the sample reacts for a longer period prior to entering the detector. If the flow is stopped while the sample is in the detector, the formation of the reaction product can be monitored as a function of time and used to determine the kinetics of the chemical reaction.

Flow injection titrations are performed by injecting an excess of the chemical reagent (titrant). The width of the peak corresponding to the product of the reaction with the sample is monitored at a fixed proportion of the peak height and is used to calculate the concentration of the sample. The peak width non-linearly increases with concentration of the sample.

The effect of interferences in an analysis can be studied by observing the effect of the interference on a sample peak. The interference and sample are simultaneously injected through two valves that are slightly displaced from each other in the stream. The two substances disperse into each other in the stream. If a negative chemical interference occurs, the area or the height of the sample peak decreases. If a positive chemical interference occurs, the peak area or height increases. If the detector responds to the pure interference, a peak is observed for the sample and a second peak for the interference. Other applications are described in the References.

Discrete-Sample Analyzers

Discrete-sample analyzers are designed to perform analyses of samples in reaction vessels (usually test tubes) in much the same manner that an analyst would manually perform the analyses. Rather than cause the sample to flow through tubing as continuous-flow analyzers do, discrete-sample analyzers carry the sample in discrete containers through several stations in the apparatus. At each station a step in the analysis takes place; e.g, a chemical reagent is added, the solution is stirred, or a measurement is made. Discrete-sample analyzers utilize an assembly-line approach to automation. Many models of discrete-sample automated analyzers are available.

The Automatic Clinical Analyzer (ACA), which has been manufactured by E. I. du Pont de Nemours and Company, is a multichannel analyzer designed for use in hospital laboratories. It is unusual in that premeasured reagents for a specific photometric assay are enclosed by the manufacturer in separate portions of a plastic analytical pack, as shown in Fig. 28-5. A sample in a sample cup is placed in the input tray of the apparatus and an analytical pack for each of the assays that are to be performed on the sample is loaded immediately behind the cup. The top of each analytical pack is labeled by the analyst with pertinent information, such as the identification of the patient, and that information is automatically printed on the report sheet. The top of each analytical pack is binary-coded to tell the computer which analysis is to be performed.

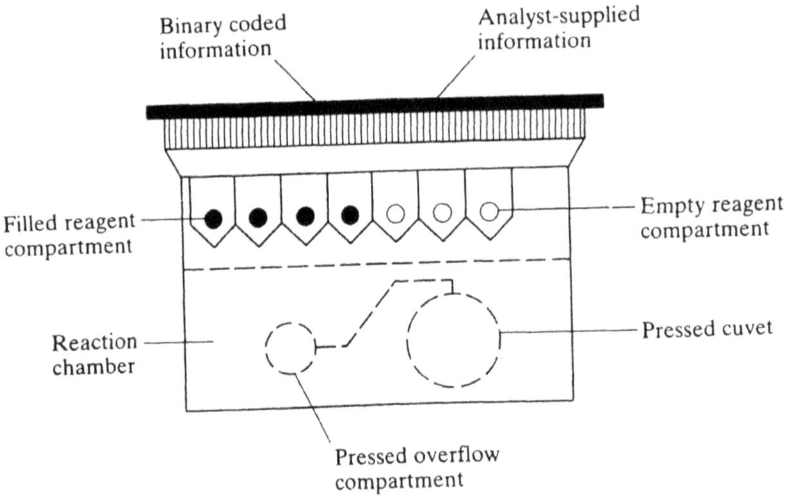

Figure 28-5 Diagram of an analytical pack that is used in the Du Pont Automatic Clinical Analyzer. The dashed lines indicate compartments formed by the combination press and spectrophotometer.

A measured aliquot of sample (from the sample cup) and a diluent are injected into the analytical pack. The two substances drain into the reaction chamber located below the reagent compartments in the pack. The process is repeated for each of the packs following a sample cup. The filled analytical pack is connected to a transport chain which moves the pack through a series of stations within the apparatus. The transport chain is propelled by an electric motor. A specific step in the analysis is performed at each station. At an early station, the pack is heated to 37°C by an electric heater. At the next station the appropriate reagent compartments are broken open as they are squeezed by clamps. The reagents within the compartments flow into the reaction chamber where they mix with the sample. Mixing is facilitated by vibrating the analytical pack.

At succeeding stations the pack is delayed while the chemical reaction occurs; more reagents are added by clamping; and the pack is again vibrated. The entire procedure occurs in a temperature-controlled compartment. Finally, the analytical pack enters a combination press and photometer, where a cuvet is formed in the reaction chamber, as shown by the dashed line in Fig. 28-5; the visible absorbance of the reaction product is then measured. The used pack is automatically removed from the instrument after the photometric measurement is made, and the results are printed on a report sheet.

A chromatographic separation, prior to reagent addition, can be accomplished in the Du Pont ACA by enclosing a small, disposable chromatographic column in the top of the analytical pack. The sample passes through the column on its way to the reaction chamber. The analytical packs used in the ACA contain seven reagent compartments, and therefore as many as seven different reagents or groups of reagents can be added during the assay.

Because each type of assay is accomplished with a specific analytical pack containing the reagents required for that particular assay, and because the computer's operation is directed by the binary code on the top of each analytical pack, the Du Pont ACA can be used to perform a series of assays of different types sequentially. That is not as easily accomplished with most automated analyzers. Another advantage that the Du Pont ACA offers is ease of operation. It does not require the services of a highly trained analyst. To obtain versatility and ease of operation, speed and thrift were sacrificed. The Du Pont ACA generally operates more slowly than other automated analyzers, and the analytical packs are relatively expensive.

A device manufactured by the American Monitor Corporation is another popular automated analyzer. The sampler of the KDA is a rotating sample tray similar to that used in the Technicon system. The tray successively rotates each sample solution into the proper position to start the analysis. An aliquot of the sample automatically is removed from the sample container and placed in a reaction tube. Diluents and chemical reagents are added to respectively bring the sample within the proper concentration range and to form a light-absorbing reaction product. The absorbance of the reaction product is measured with a spectrophotometer. The wavelength at which the spectrophotometer operates, the reagent added, and all other variables of the apparatus are controlled by a dedicated computer. The computer performs the calculations required to obtain the result and displays the results on a computer-controlled readout device.

The American Monitor Corporation's apparatus is not a multichannel device; i.e., it performs a single type of analysis at a time. The type of analysis being performed can be easily changed by proper instruction to the computer. As with the Du Pont instrument, the KDA uses a single type of detector. A spectrophotometer is used as opposed to the ACA's photometer. Analytical results are obtained nearly 15 times more rapidly with the American Monitor apparatus than with the Du Pont ACA.

Centrifugal Force Analyzers

A *centrifugal analyzer* consists of a wheel containing compartments for samples, reagents, and the detector measurement. As the wheel rotates, the solutions in the compartments closer to the center of the wheel are moved by centrifugal force through connecting channels to the outer compartments, which contain other solutions. During the process the sample and reagents are mixed and then forced into the outermost compartment, where a detector measures the concentration of the reaction product.

Because many samples can be assayed both simultaneously and rapidly with a centrifugal analyzer, the device is sometimes referred to as a *parallel-fast analyzer*. Centrifugal analyzers vary in the number of samples that can be simultaneously assayed and in various operating parameters. Perhaps the best known of the centrifugal analyzers is the GeMSAEC analyzer that is manufactured by Electro-Nucleonics, Inc. GeMSAEC is an acronym for the National Institute of *G*eneral *M*edical Sciences and the *A*tomic *E*nergy *C*ommission, who sponsored the work which resulted in development of the analyzer. The Atomic Energy Commission was the precursor of the Nuclear Regulatory Commission in the United States.

A side-view sketch of the rotor used in the GeMSAEC analyzer is shown in Fig. 28-6. Sample solutions are loaded into the compartments located nearest to the center of the wheel. Reagents are loaded in the reagent compartments and the wheel is rotated about its axis. Centrifugal force from the rotation moves the sample over the barrier separating the sample from the reagent, and eventually through a capillary and into a cuvet at the outer edge of the wheel. Mixing is accomplished by the rapid change in rotation rate as the wheel accelerates. Radiation from a monochromator passes vertically through the quartz windows of the cuvets to a photomultiplier tube which measures the transmittance of each sample as it rotates past the spectrophotometric detector. The transmittances of the samples were originally displayed on an oscilloscope, where each sample

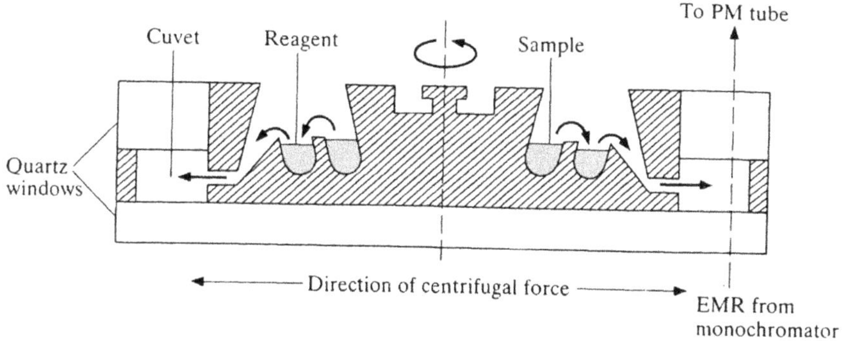

Figure 28-6 Side-view diagram of the rotary wheel that is used in the GeMSAEC automated analyzer. The dashed line indicates the axis of rotation and the curved arrow around the dashed line indicates the direction of rotation. The direction of solution flow is indicated by the arrows connecting the compartments within the analyzer.

appears as a negative peak. The peaks are separated by flat portions. The rotors used in centrifugal analyzers have positions for between 15 and 30 samples. The results are monitored by a computer, which calculates concentrations and averages results after many rotations.

The major advantage that centrifugal analyzers offer is the speed with which they obtain results. The major disadvantage is that each sample and, reagent compartment must be individually loaded prior to the analyses. Perhaps that disadvantage can be overcome by the use of laboratory robots (described later in the chapter) or some other automated device. A centrifugal analyzer can do only a single type of analysis on a loaded series of samples. Different types of analyses are performed by changing the reagents and the wavelength at which the detector is adjusted.

Automatic Titrators

Automatic titrators either add titrant to a titrand solution with a motor-driven syringe or from a buret with an automatically operated valve, or they coulometrically generate the titrant in the titrand. In most cases the progress of the titration is followed potentiometrically. In many cases an ion-selective electrode, such as a pH electrode, is used as the indicator electrode.

Automatic potentiometric titrators are of several types. Some add titrant until a predetermined potential is reached or approached, at which time the titrant addition is stopped. They can be designed to operate from the first or second derivative of the electrode potential as well as from the measured potential. The original titrand concentration is either manually or automatically calculated from the concentration and volume of the added titrant.

Other titrators prepare a titration curve by continuously plotting the potential during the titration as a function of the added titrant volume. In some types of apparatus, the plotting is accomplished by continuously recording the indicator electrode potential, the pH, or the p*I*. The potential can be continuously displayed on a recorder as a function of titrant volume or time, if the titrant is added at a constant rate. Alternatively, a computerized data acquisition system can be utilized to acquire and display the data. With some titrators, first- and second-derivative potentiometric plots are possible.

The titrators of the final category add titrant until a predetermined pH or pI of the titrand solution is reached. When the pH varies from the desired pH because of a chemical reaction in the solution, the titrator again adds titrant to return the pH to the predetermined value. An automatic titrator of that type is a *pH-stat*.

Titrators that are designed to stop titrant addition at the endpoint can operate on any of several principles. In some cases the titration is carried out until a predetermined potential is reached. In other cases the titrator actually locates the endpoint by monitoring the slope (first derivative) of the titration curve. When the slope reaches a maximum, i.e., when the first-derivative titration curve reaches a maximum, the titrant flow ceases. Other titrators use a second-derivative technique to locate the endpoint. In those titrators the titrant flow is stopped when the second derivative of the potential with respect to the volume becomes zero. Many automatic titrators have the ability to sense the approach of an endpoint and to alternately continue and stop titrant flow or decrease the rate of titrant flow as the endpoint is approached. Because the observed titrand potential sometimes lags slightly behind the equilibrium potential, the procedure prevents overshooting the endpoint. Automatic coulometric titrators are generally designed to perform Karl Fischer water titrations by electrolytically generating the iodine in Karl Fischer reagent from iodide.

Most modern titrators are microprocessor-controlled and can perform more than one function. Some titrators can measure either pH or millivolts, and can plot standard and first-derivative titration curves. Often they can be operated as a pH-stat and can sequentially perform multiple sequential titrations of analytes that are in beakers on a rotating tray.

Some titrators are capable of utilizing multiple endpoint location methods such as those that are used in pH, mV, colorimetric, conductometric, and amperometric titrations as well as functioning as a pH-stat. Colorimetric and spectrophotometric titrations are performed with the aid of a probe colorimeter or cell that is inserted into the titrand in the same manner in which an electrode might be inserted. The probe cell can be connected to the colorimeter through an optical fiber cable.

Nearly all of the computer-controlled titrators automatically calculate the concentration of the sample. Many of them can be interfaced to electronic balances. Titrations in which the chemical reaction occurs rapidly can generally be completed in about two minutes.

Robots

A *robot* is a device that can perform a series of programmed physical manipulations. Generally robots are used to move tools, components, or other objects in order to accomplish a work function. Robots are used in chemical analysis for automation. They differ from the automated systems previously described in that they more nearly mimic human movement and in that some robots can be programmed to perform a wide range of functions. Zymark Corporation was the first commercial supplier of laboratory robots.

Laboratory robots consist of a mechanical arm that is placed on or near the laboratory bench. Any of several types of hands can be attached to the end of the arm to allow the device to grasp objects, make dilutions, or perform nearly any other laboratory function. Movements of the robotic arm are computer-controlled. Robots are particularly well suited to performing repetitive manipulations that are boring to humans. They can safely be used in sterile environments in which the presence of a human can be detrimental to the analysis.

Generally robots do not function significantly more rapidly than human technicians and often perform particular steps more slowly than humans. They do have the advantages of not becoming easily distracted and of being able to continuously function for long periods of time. Often the unattended use of a robot permits assays to be performed around the clock rather than only during normal working hours, thereby increasing the total throughput of a laboratory.

Robots can be classified according to the type of motion the robotic arm can perform.[7] The four most common types of robotic arms are the cartesian, cylindrical, revolute, and spherical types. The cartesian-type arm is the simplest and is capable of independent motion along any of the three cartesian (x, y, and z) axes in three-dimensional space. The cylindrical-type robotic arm can access any area in a cylinder around the robot. Of course, the arm cannot access the space occupied by the robot. The cylindrical design is used by the Zymark Corporation for their popular Zymate laboratory robot. The design is illustrated in Fig. 28-7. The arm can move up and down along the axis that defines the height of the cylinder and can rotate around the cylinder at any height. The arm can rotate at any radius within a maximum that is defined by the extended length of the arm and a minimum that is defined by the radius of the body of the robot at the center of the cylinder. Furthermore, the hand of the robot can rotate around the arm.

Revolute-type arms grasp objects from above. They can reach over low-lying obstacles to grasp the desired object. Spherical-type arms are similar to cylindrical-type arms in that they generally approach an object horizontally rather than vertically, as with the revolute type arms. They approach objects from the side rather than from the top. The movement of both the revolute- and

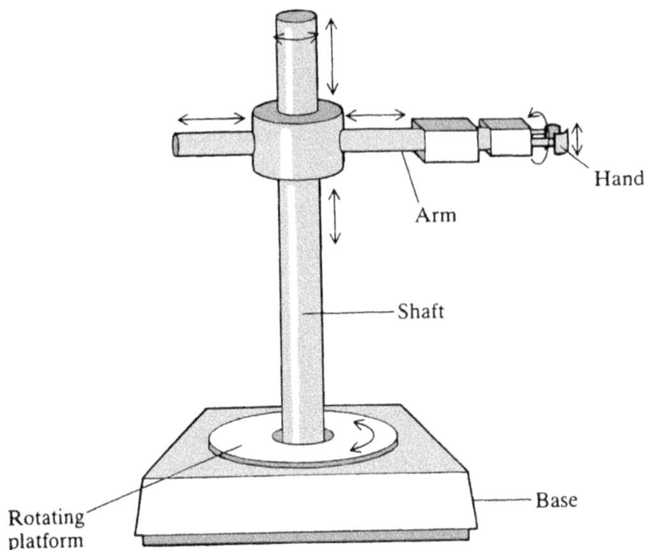

Figure 28-7 A diagram of a cylindrical-type robot. The allowed directions of motion are indicated by the arrows. The hand of the robot can access any location within the cylinder defined by the motion of the hand around the shaft, with the exception of the space occupied by the robot. The motors that drive the robot are not shown.

spherical-type arms sweeps out a hemisphere above the base of the arms rather than a cylinder. They can reach objects directly above the base because the arms can rotate vertically as well as horizontally on the base. A revolute-type arm is used in the F5 robot system that is sold by Thermo Scientific. A diagram of a robot that utilizes a revolute design is shown in Fig. 28-8.

Choice of a particular robotic design is dependent upon the geometrical arrangement of the apparatus and objects that must be accessed by the robot. It is possible to extend the accessible volume of any robot by mounting the robot on a wall, in an inverted position or on a movable platform which can be controlled by a computer.

Motion of a robot can be accomplished in several ways. In industrial applications, where heavy objects must be moved by the robot, hydraulic drive systems are often used. Pneumatic systems are normally used when speed of motion is a primary requirement. Laboratory robots are generally driven by electric servo motors or electric stepping motors. Digital stepping motors are convenient because they can easily be controlled by a computer; however, stepping motors generally cannot be used to move heavy objects. The F5 robot has a maximum payload of 5 kg. AC or DC servo motors can be computer-controlled through a DAC that is attached to the computer. The dc voltage coming from the DAC serves as the drive signal for the motor.

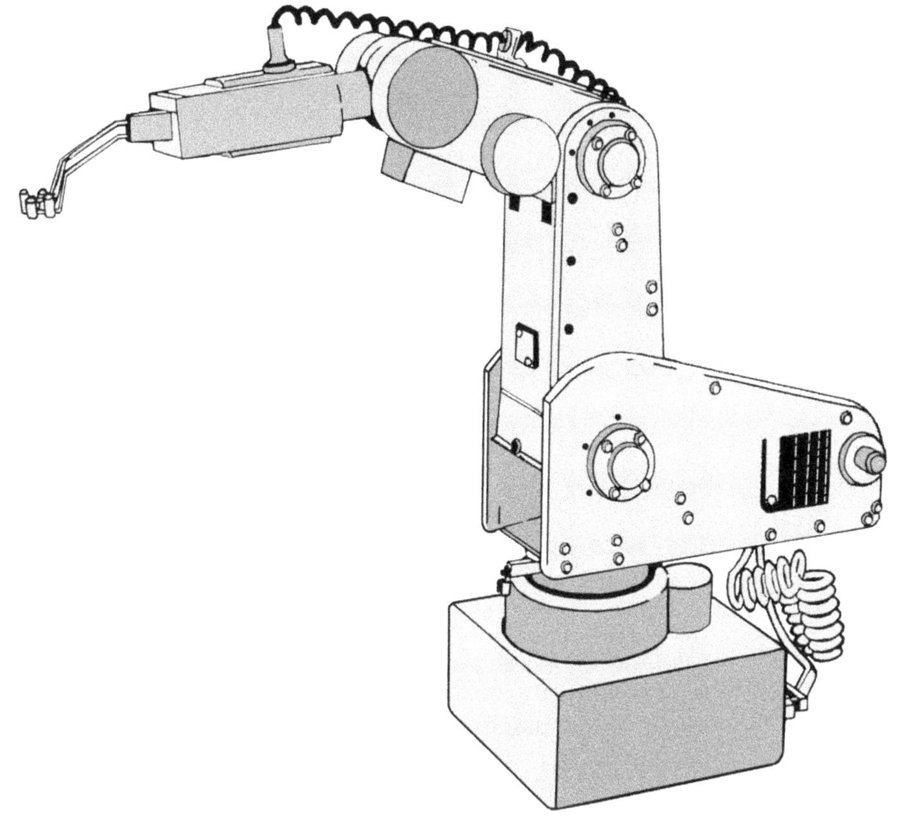

Figure 28-8 A diagram of a revolute-type laboratory robot.

In any case, a separate drive system is generally used for each type of motion that the robot can perform. That allows the robot to simultaneously perform more than one type of motion. For the robot shown in Fig. 28-7, a separate motor would normally be used to rotate the base, to move the arm vertically on the shaft, to move the arm horizontally, to rotate the hand, and to operate the hand mechanism.

A great deal of effort has gone into designing robotic hands that are versatile and that can mimic the many motions that are possible with a human hand. Unfortunately such devices are highly complex and require relatively complex and lengthy computer programs to operate. Consequently, laboratory robots generally have hands that perform a single function, such as grasping or operating a syringe. General-purpose laboratory robots usually have interchangeable hands. After performing one function the robot deposits its hand in a specially designed holder, and mounts another hand that is capable of performing the next analytical function. Humans are not required to remove or to mount the robotic hands. It is regarded as another task that is programmed into the controller of the robot.

Hands that are designed to grasp different types of objects, to filter, to operate a syringe, to operate switches, and to perform other laboratory operations are in use. Special applications sometimes require design of a new hand. In some cases it is advantageous to use multipurpose hands. A multipurpose hand essentially consists of two or more hands simultaneously attached to a single robotic arm. Multipurpose hands are more cumbersome but can save time in exchanging hands. The presence of additional hands on a single arm can sometimes hinder the manipulations of the robot.

The motion of laboratory robots is computer-controlled. The computer controls each of the drive systems in the robot, causing the robot to perform the desired function. Robots are programmed using one of three methods. *Explicit programming* consists of specifying the coordinates (location) to which the hand is to move, the pathway it is to follow, the acceleration, deceleration, velocity of motion, hand orientation during the motion, and any other variables necessary to define a particular robotic function. Explicit programming most nearly resembles classical computer programming. It can be tedious and frustrating and can require a trained programmer.

Lead-through programming and *walk-through programming* are generally easier and less frustrating, particularly to analysts who view robotic programming as a means to an end rather than as an enjoyable task. Lead-through programming consists of using a *training* (or *teaching*) *pendant* or specially defined function keys of a computer keyboard to control the operation of the robot as it performs a function for the first time. A training pendant is a set of robotic controls that are generally mounted in a box about the size of a hand-held calculator. The motions and various functions performed during the analysis are recorded in computer memory and later transferred to a computer information storage device. During later analyses, the recorded motions are read into computer memory and used to control the robot. The robot duplicates its movements during the programming process. Lead-through programming is often used in laboratory robots.

Walk-through programming can only be used for robots that have built-in sensors that provide feedback (see the following section on process control) to the computer. The feedback sensors are activated while the robot is physically moved through the process. The computer records the positions, amount of movement, etc., and stores the information. The stored program is later used to control the robot.

Robotic programming steps are sometimes classified according to *laboratory unit operations* (LUOs). An LUO is a particular function performed during a laboratory analysis. Examples of laboratory unit operations are grinding, separating, weighing, and instrumentally measuring the sample. Work is progressing on the use of robots with *artificial intelligence* which can make decisions concerning future laboratory experiments that are to be performed by the robot.

Without feedback control of the robot, the robotic motions must be highly reproducible and the objects that are manipulated by the robot must be exactly in the proper positions. Robots that operate without feedback are said to operate under *open-loop control*. Robots that operate under open-loop control are often programmed to periodically return to a "home" position that serves as a reference position. If feedback to the computer from some sort of sensors is used, the robot operates under *closed-loop control*. The positioning of objects and the motion of the robot do not need to be as accurate during closed-loop control as during open-loop control.

Although many types of sensors can be used in conjunction with robots, normally the sensors either respond to touch or to sight. Sensors that respond to touch usually consist of microswitches. They are often used in the hand of a robot. When the robot contacts an object, the switch closes. Touch sensors can also be used for other purposes. As an example, a switch in the cell cavity of a spectrophotometer could be used to indicate that the robot had successfully placed a cuvet in the instrument.

Sight sensors generally consist of a beam of light (possibly from an LED or laser) that is reflected from or absorbed by an object in its path. The beam is typically detected by a photodetector that is attached to the apparatus through a fiber optic cable or by a photodiode array. The presence or absence of an object in the radiative beam can be detected. By using photodiode arrays, spatial information, such as width of the object, can be obtained.

Robots have been used to automate many types of chemical analyses. Nearly all forms of repetitive laboratory operations can be performed by robots. It should be emphasized that the use of robotics for automation is not always advantageous. Sometimes an analysis can be more efficiently automated using one of the systems described earlier in the chapter. At other times, it is best to use a robot to prepare samples for and to control a different form of automated analyzer. The major advantages of laboratory robots are their flexibility, their ability to perform tedious tasks that otherwise would be performed by a technician, and their ability to work in hazardous and sterile environments. Numerous experiments[8] have shown that analytical results obtained while using robots are comparable in accuracy and precision to those obtained by using human technicians. For most uses in industrial laboratories, the robot pays for itself in saved labor costs within about one year. Descriptions of many laboratory applications of robots can be found in References 8 through 12.

PROCESS CONTROL

Process-control devices sample a chemical or physical variable in a manufacturing process, measure that variable, compare the measured results with specifications for the particular process, and, if the variable does not meet the desired specifications, adjust a second variable in the process to cause future analytical results to meet specifications. The entire process is automated; i.e., no human intervention is required beyond the initial adjustments. Several reasons exist for using process-control devices. Generally, the devices can respond to changes in a manufacturing process more rapidly than human operators. That keeps the product of the manufacturing process within specifications a greater fraction of the time. In some cases, such as when the product is poisonous or highly flammable, it is dangerous for a human to sample and assay the product. In most cases process-control apparatus can increase efficiency of operation and reduce the amount of human labor.

The measured variable can be either a chemical product of the manufacturing process or some variable which is directly related to the manufactured product. The controlled variable can be a pressure, temperature, flowrate, type of chemical reagent added to the process, or something else. Automatic valves which are operated either electrically or mechanically (as in the pneumatic control systems which are operated by air pressure) can be controlled to adjust pressure, flowrate, and rate of addition of chemical reagents. Generally, the process-control apparatus is housed as near to the process being controlled as possible. That minimizes the time delay between the sampling and the control process.

Automated process-control equipment consists of at least three pieces of equipment arranged to form a control loop, as shown in the block diagram in Fig. 28-9. The sampled variable is measured with an analyzer of some type or with a device that measures a physical property such as temperature. The result of the analysis is compared with predetermined specifications for the measured variable by a *controller*. The analytical result is often transmitted to a recorder, alarm, or similar device located in the control room of the plant. That makes it possible for a human controller in the control room to keep track of the measured variable and correct any failure of the automated controller or automated operator. The *operator* is the device, e.g., the automated valve, which is controlled by a signal from the controller. The operator controls one of the variables in the manufacturing process.

Control loops are divided into two categories. A *feedback control loop* requires less knowledge of the interactions between the reactants and their effects upon product formation in the process. Consequently, it is the more popular of the two types of control loops. The measured variable in feedback control loops is a product of the process, and the controlled variable is one of the reactants entering the process. If the time between the sampling and the corrective action (the *dead time*) is large with respect to the speed of the process, feedback control cannot adequately control the process.

Feedforward control loops monitor one variable at the input of the process while controlling a second input variable. Feedforward control loops can be used only if the relationships between the measured variable, the controlled variable, and the product of the process are known. Feedforward control has the advantage

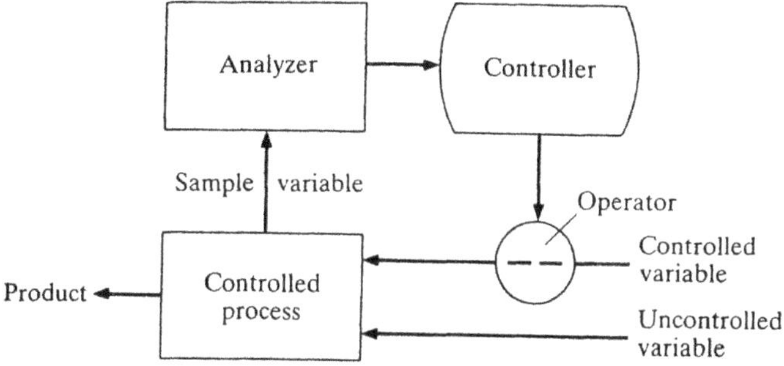

Figure 28-9 Diagram of a control loop that is used in automated process control.

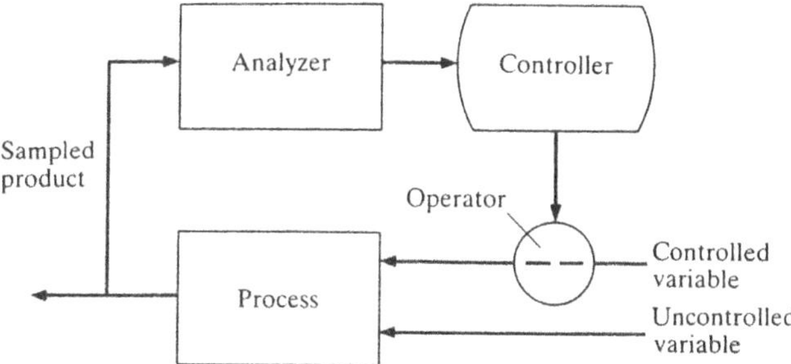

Figure 28-10 Diagram of a feedback control loop.

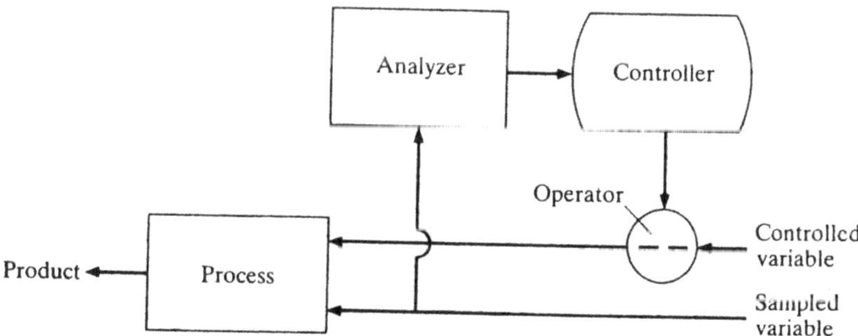

Figure 28-11 Diagram of a feedforward control loop.

of being able to correct an error in the process before the error occurs; i.e., it functions more rapidly than feedback control. The two forms of process-control loops are illustrated by the block diagrams in Figs. 28-10 and 28-11.

The operator in the control loop can perform its function in any of several ways. The simplest control mechanism is *two-position control*, in which the operator is either cmpletely on or completely off. If the operator is an automated valve, the valve could be either completely open or completely shut. If the operator is a heating coil, the coil is completely on or completely off, etc. With *proportional control* the operator is at a position that is directly proportional to the difference (the error) between the actual and the desired analytical results. With proportional control the operator can be completely on, completely off, or at any position between the two extremes. When the error is corrected, the operator returns to its original position. The error must persist in order for the operator to remain in its altered position.

Proportional control can be combined with *integral* and/or *derivative action*. With integral action the response of the operator is related to the time integral of the error, i.e., to the area under a plot of error as a function of time. With derivative action the operator response is related to the rate of change of the error with time, i.e., to the slope (first derivative) of a plot of error as a function of time.

Two-position control leads to fluctuations in the composition of the product and can be used only if the fluctuations can be made sufficiently small to be tolerable. Proportional control is most often used for processes in which changes in the variables in the system are relatively small. Proportional-plus-integral control is generally used when variations within the system can be relatively large. Proportional-plus-derivative control is used when the dead time of the process is relatively large, i.e., when the time required for sampling and analysis is large relative to the rate of change of the process variables. In some cases proportional control is simultaneously combined with integral action and derivative action.

Process-Control Analyzers

Because process-control analyzers must function without constant human supervision, they must be more stable and more rugged than the corresponding laboratory instruments. Sometimes process-control analyzers are categorized as either *nonspecific analyzers* or as *specific analyzers*. Nonspecific analyzers measure a physical property such as temperature, pressure, or thermal conductivity. Such analyzers do not directly provide information about a specific chemical component of the system. Specific analyzers assay one or more individual components of the system. Instruments such as spectrophotometers and chromatographs are specific analyzers. Specific analyzers must be calibrated in much the same manner in which laboratory instruments are calibrated. Usually calibration is done by performing a laboratory assay on the sample assayed with the process-control analyzer and adjusting the process-control analyzer to yield the value obtained in the laboratory. The process is periodically repeated to ensure that the process-control analyzer has not drifted from its calibrated position.

Process-control analyzers can come from nearly any of the categories of instruments described earlier in the text. The more popular analyzers use a spectroscopic, chromatographic, or electroanalytical technique. Among the spectroscopic instruments used for process control are ultraviolet, visible, and infrared absorptive instruments. Turbidimetry, flame emission, fluorescence, chemiluminescence, and refraction have also been used in process control. Refractive-index and infrared-absorptive measurements are used more often in process control than other spectral methods.

Refractive index measurements, differential refractive-index measurements, and critical-angle measurements can be used to measure the concentration of one component in a two-component mixture. The concentration of one component in more complex solutions generally cannot be determined from refractive-index measurements. The refractive index and the critical angle can be measured directly in the process stream. Among other uses, the measurements have been used for process-control assays of sugar in beverages, alcohol in beverages, dissolved solids in beverages, ketchup, jam and jelly, and saturated hydrocarbons in fats and oils.

Although dispersive infrared spectrophotometers are occasionally used, the infrared analyzers that are used for process control are normally nondispersive. As described in Chapter 12, the devices respond to a single substance. Interferences in the measurements that are made by some instruments that use traditional infrared detectors are eliminated by passing radiation from the source through filters that contain a high concentration of the interference. Essentially all of that portion of the infrared spectrum that could be absorbed by the interference is absorbed by the filter, thereby eliminating the effect of the interference on the detector. Infrared absorbance measurements have been used to monitor CO_2, CO, CH_4, $H_2O(g)$, NH_3, ethane, cyclohexane, SO_2, ethylene, and other gaseous components. Liquid solutions also can be monitored by the analyzers. Near-infrared measurements using tungsten-filament lamps as sources are often used to monitor water concentrations. Infrared reflectance can be used to monitor water concentration in some solids, such as in paper. If moisture is present some of the incident radiation is absorbed and less is reflected at the wavelengths characteristic of water.

Gas chromatography probably is the most-used process-control chromatographic technique. Process-control gas chromatographs are generally capable of automatic column switching and backflushing. Backflushing eliminates from the column those long-retention-time components in the sample which are not assayed. The sample is automatically injected into the proper column and partially or completely separated into its components. After the component of interest has reached the detector, the flow of carrier gas through the column is reversed to remove the long-retention-time components from the column and thereby minimize the analytical dead time prior to injection of the next sample. The backflushed components generally flow to the detector in a single block and appear as a single chromatographic peak. Injection into process-control gas chromatographs is usually accomplished with an automatic sampling valve which is designed in a manner similar to that of the sample-introduction valves used for HPLC. The GC detectors most often used with process-control chromatographs are the thermal conductivity detector and the flame-ionization detector. Gas chromatographic process-control analyzers have proved to be especially useful in the control of distillation columns. Process-control gas chromatographs can be operated either with or without temperature programming.

Both liquids and gases have been monitored by process-control gas chromatography. Gaseous sample volumes are normally in the range from 0.5 to 100 mL and liquid sample volumes range from 0.5 to 5 μL. In order to minimize dead time, the gas chromatograph is placed as near to the sampled stream as possible. Generally it is advisable to keep the total sampling and assay time less than 4 or 5 minutes. The controller is often well removed from the instrument, in the laboratory or control room of the industrial plant. Timing sequences during sample injection, development, backflushing, and column switching can be controlled mechanically with rotating cams or servo motors, or electronically. Microcomputers are often used for instrumental control and to perform peak integrations and calculations. Liquid chromatographs have also been used as process-control analyzers.

Of the electroanalytical process-control techniques, potentiometry has been used most often. The pH-meter is probably the most used of the process-control electroanalyzers. Ion-selective electrodes have been used in process control with flow-through cells to monitor H^+, S^{2-}, CN^-, Na^+, Ag^+, water hardness, F^-, and other ions. Amperometry is commonly used to monitor dissolved O_2. Other electroanalytical techniques which have been used in process control include coulometry, conductometry, and automated potentiometric titrimetry. In some process-control applications it has been possible to use radiochemical analyzers, which monitor the radioactivity of the sample.

In some cases modified versions of the automated analyzers that were previously described in the chapter can be used for process control. As an example, the Monitor 650 that is manufactured by Technicon Instruments Corporation is a process-control version of their laboratory continuous-flow analyzer. It uses a visible spectrophotometer to monitor the concentration of the chemical reaction product of the assayed substance.

IMPORTANT TERMS

Analog-to-digital converter
Analytical cartridge
Automated device
Automatic device
Automatic titrator
Centrifugal analyzer
Closed-loop control
Continuous-flow analyzer
Control loop
Controller
Dead time
Dedicated computer
Derivative action
Digital-to-analog converter
Discrete-sample analyzer
Explicit programming
Feedback control
Feedforward control
Flow injection analysis
Flow injection analyzer
Flow injection titration
Gradient calibration
Input/output device
Integral action
Laboratory information management system
Laboratory management system
Laboratory unit operation
Lead-through programming
Longitudinal diffusion
Nonspecific analyzer
On-line computer
Open-loop control
Operator
Parallel-fast analyzer
Peristaltic pump
pH-stat
Process control
Proportional control
Proportioning pump
Radial diffusion
Real time
Robot
Specific analyzer
Stopped-flow techniques
Training pendant
Turnkey operation
Two-position control
Walk-through programming

REFERENCES

1. Snyder, L., J. Levine, R. Stoy, and A. Conetta: *Anal. Chem.*, **48**: 942A(1976).
2. Walser, P. E., and H. A. Bartels: *Amer. Lab.*, **14**(2): 113(1982).
3. Ruzicka, J., and E. H. Hansen: *Anal. Chim. Acta*, **99**: 37(1978).
4. Betteridge, D.: *Anal. Chem.*, **50**: 832A(1978).
5. Ruzicka, J.: *Anal. Chem.*, **55**: 1040A(1983).
6. Ruzicka, J., and E. H. Hansen: *Flow Injection Analysis*, Wiley-Interscience, New York, 1981.
7. Dessy, R. D.: *Anal. Chem.*, **55**: 1100A(1983).
8. Hawk, G. L., and J. R. Strimaitis (eds.): *Advances in Laboratory Automation Robotics 1984*, Zymark Corporation, Hopkinton, Massachusetts, 1984.

9. Hawk, G. L., J. N. Little, and F. H. Zenie: *Amer. Lab.*, **14**(6): 96(1982).
10. Dessy, R.: *Anal. Chem.*, **55**: 1232A(1983).
11. Owen, G. D., and R. A. DePalma: *TrAC*, **4**: 32(1985).
12. Leiper, K. J.: *TrAC*, **4**: 40(1985).

BIBLIOGRAPHY

Foreman, J. K.: *Topics in Automatic Chemical Analysis*, vol. 1, Halsted, New York, 1979.

APPENDIX

A

PROBLEM ANSWERS

CHAPTER 2

2-1 4780 C, 0.0495 F**;** **2-2** 6640 s; **2-3** 0.869 A; **2-4** 0.158 J; **2-5** 33.8 V; **2-6** 0.012 A; **2-7** 3.47×10^{-4} s; **2-8** 5.17×10^{-3} W; **2-9** 2.36×10^{-3} A; **2-10** 0.642 W, 350 Ω; **2-14** 1.75×10^{4} Ω, 17.5 V; **2-15** 5.00 V; **2-16** 115 Ω; **2-17** 7.20 V; **2-18** 2.53 V; **2-19** 74.3 Ω; **2-20** $I = 1.99$ A, $I_1 = 1.00$ A, $I_2 = 0.591$ A, $I_3 = 0.400$ A; **2-21** 2.35 V; **2-22** 0.430 A (25 Ω), 0.213 A (20 Ω), 0.121 A (35 Ω), 0.0944 A (45 Ω); **2-23** 35.1 Ω; **2-24** 0.143 A; **2-25** 3.43 Ω; **2-26** 16, 107, 380, 834 Ω; **2-27** 1.95 V; **2-28** 400 Hz, 11.7 mA (p–p), 4.14 mA (rms); **2-29** 70.7 mV, 141 mV, 0.0167 s; **2-30** 36.8 Ω**;** **2-31** 15.0–1875 Ω; **2-32** 0.00750 C; **2-33** 68.6 mF; **2-34** 0.001, 0.014, 0.069, 0.14, 0.27, 0.57, 0.81, 0.94, 0.99, 1.00, 1.00; **2-35** 72 Hz, 0.0022 s; **2-36** 5.3, 0.27, 16 Hz; **2-37** 1.00, 1.00, 1.00, 0.99, 0.96, 0.82, 0.58, 0.34, 0.14, 0.071, 0.014; **2-38** 53.1, 183 Ω; **2-39** 147 Ω; **2-40** 0.136 A; **2-41** 0.125; **2-42** 220, 54, 110, 160; **2-43** 1.0, 0.99, 0.97, 0.85, 0.62, 0.37, 0.16, 0.08, 0.008; **2-44** 796 Hz, 2.00×10^{-4} s; **2-45** 0.179 Ω; **2-46** 5.29, 63.0 Ω; **2-47** 5.05 Ω; **2-48** 2.97 A; **2-49** 5.81, 0.200, 6.37, 0.549 A; **2-51** 1400 Hz, 0.13 A; **2-52** 0.57 V; **2-53** 270 Hz, 95, 8.4, 8.4 Ω; **2-54** 0.82, 60, 60 A

CHAPTER 3

3-10 1.3 kHz; **3-11** 56 μF; **3-13** 17,300

CHAPTER 4

4-2 645; **4-3** 12.5 Ω; **4-4** 680; **4-5** – 0.547 V; **4-12** 186; **4-13** 907;

4-14 255, 1023, 4095, + 127, + 511, + 2047; **4-15** 10011101;

4-16 1 110 101 001 001; **4-21** ones complements: 010010, 10 001 110,

00 111 000, 11 000 000, 01 010 101, 10 101 011, twos complements: 010 011, 10 001 111, 00 111 001, 11 000 001, 01 010 110, 10 101 100;

4-22 – 13, + 113, – 71, + 63, – 42, + 84 (assuming twos-complement convention is not used);

4-23 111 110, 1 000 101, 110 000, 011 101, 1 000110; **4-24** 25, – 5, – 1, 117, 41

CHAPTER 5

5-1 1.09×10^{15} s–^{1}, 9.17×10^{-16} s, 7.22×10^{-19} J; **5-2** 1.577;

5-3 6.515×10^{14} Hz, 4.316×10^{-19} J; **5-4** 636.8, 434.1 nm;

5-5 3.37×10^{-19} J; **5-6** 2.87×10^{-19} J; **5-7** 2.37×10^{-4} *M*; **5-8** 186;

5-9 77.1, 13.2, 37.2, 60.7, 9.48; **5-10** 0.910, 0.057, 0.349, 0.207, 0.425;

5-11 5 cm; **5-12** 1.68×10^{-4} *M*; **5-13** 3.50×10^{-5} *M*; **5-14** 1.11×10^{-3} *M*;

5-15 2.33×10^{-4} *M*; **5-16** 175 nm; **5-17** 339, 226, 170 nm; **5-18** 51.7°,

51.7°; **5-19** impossible; **5-20** 19.8°; **5-21** 287/*n* nm;

5-22 1.26×10^{3} grooves/mm, 7.92×10^{-4} mm; **5-23** 150, 0 .57 mm/nm,

1.8 nm/mm; **5-24** 0.79 mm/μm; **5.25** 641.6 nm

CHAPTER 6

6-1 5.1035×10^{14} Hz; **6-2** 5.0763×10^{14} Hz; **6-3** 1.96×10^{12} Hz;

6-4 5.0×10^{5} ; **6-5** 1.1×10^{4} ; **6-6** 4.4; **6-7** 5.2; **6-8** 58.6;

6-9 26,400%; **6-10** 3.56 μg/mL; **6-11** 42.3 μg/mL; **6-12** 98.4%

CHAPTER 7

7-1 7.60 μg/mL; **7-2** 9.94×10^{-5} *M*; **7-3** 6.2 ng/mL

CHAPTER 8

8-3 15% increase; **8-4** 25% increase; **8-5** 8.39×10^{-5} *M*

CHAPTER 9

9-1 219 nm; **9-2** 273, 264 nm; **9-3** 249, 237 nm; **9-4** 292 nm; **9-5** 1430;
9-6 0.00253 *M*; **9-7** 4.33×10^{-5} *M*; **9-8** 6.21×10^{-5}*M*;
9-9 7.700×10^{-4} *M*; **9-10** 9.87×10^{-5}, 5.96×10^{-4}*M*;
9-11 83.5% ASA, 6.70% caffeine

CHAPTER 10

10-2 4.41×10^{20} photons, 0.0229 J/s · cm^2; **10-3** 0.00840 J/s · cm^2

CHAPTER 11

11-3 0.59; **11-4** 0.040; **11-5** 2; **11-6** *c*; **11-7** *c*; **11-8** $d > a = b > c > e$;
11-9 no; **11-10** yes; **11-11** 251 mg/L; **11-12** 0.00622%;
11-13 1.55×10^{-5} *M* A, 5.41×10^{-5} *M* B

CHAPTER 12

12-1 1700, 800, 3700 cm^{-1}; **12-2** 3.160, 5.479, 16.0 μm; **12-3** 30; **12-4** 54;
12-5 KCl, KBr, CsBr, AgCl, CsI; **12-6** *n*-heptane; **12-9** 37.1°;
12-10 42.2°; **12-11** 41.3°; **12-12** 47.7°; **12-13** 23.6; **12-14** 34.6;
12-15 2499; **12-16** triethylamine; **12-17** 1-octanol; **12-18** 1-hexene;
12-19 1-iodobutane; **12-20** methyl isobutyl ketone; **12-21** benzyl alcohol;
12-22 wet benzonitrile; **12-23** 0.097 mm; **12-24** 0.024 mm

CHAPTER 13

13-1 0.625 m; **13-2** 829 Hz; **13-3** *a, b, d*; **13-5** PAS in *b, d, e*

CHAPTER 14

14-1 0.345; **14-2** 0.00388 L/mg cm; **14-3** 49 mg/L;
14-4 5.6 mg/L; **14-5** 62.9

CHAPTER 15

15-1 – 0.2°C; **15-2** 0.1, 1°C; **15-3** ±2°C; **15-4** 22.32; **15-5** 16.09, 16.065; **15-6** acetic acid; **15-7** 11.6%; **15-8** 0.69%

CHAPTER 16

16-1 7.093 $\times 10^{-27}$ J; **16-2** 2.038 $\times 10^{-27}$, 2.038 $\times 10^{-27}$ J; **16-3** 1.071 $\times 10^{7}$, 3.076 $\times 10^{6}$, 3.076 $\times 10^{6}$ Hz; **16-4** 0.999 995 75; **16-5** 0.999 990 33; **16-6** 2.7 ppm; **16-7** 1:6:15:20: 15:6:1; **16-8** 1:2: 1, 1: 3 : 3 : 1, 1, ratio of each type is 3: 2: 9; **16-9** nitrobenzene; **16-10** toluene; **16-11** methyl benzoate; **16-12** isopropyl acetate; **16-13** cyclohexane; **16-14** dichloromethane; **16-15** ethylene glycol; **16-16** 3, 3-dimethyl-1-butanone; **16-17** *p*-methyl benzaldehyde (the aldehydic proton is out of the spectral range); **16-18** pyridine; **16-19** 22.7 ± 0.6%; **16-20** 56s^{-1}

CHAPTER 17

17-1 0.34, 0.82, 1.2 T; **17-2** 6.3 $\times 10^{-24}$, 1.5 $\times 10^{-23}$, 2.3 $\times 10^{-23}$ J; **17-3** 0.3368 T; **17-4** 2.150; **17-5** 0.998, 0.995, 0.9930; **17-7** 4, 4, 4, 6, 6; **17-9** 10 peaks, 1: 9 : 36 : 84 : 126 : 126 : 84 : 36 : 9 : 1; **17-10** 12 peaks; **17-11** *a*; **17-12** 3.0 $\times 10^{-5}$ *M*; **17-13** 2.19 $\times 10^{-5}$ *M*

CHAPTER 18

18-1 1.13 $\times 10^{4}$ V; **18-2** 8.16 $\times 10^{3}$ V; **18-3** 0.0496 nm; **18-4** 0.16, 0.23, 0.074 nm ; **18-5** 6.0 : 3.8 : 2.2: 1.0; **18-6** 3.57 $\times 10^{-17}$, 1.94 $\times 10^{18}$, 5.95 $\times 10^{-18}$, 1.47 $\times 10^{19}$ Hz, 840, 155, 50.4, 20.4 pm; **18-7** 631.8 cm^2/g; **18-8** 202.2 cm^2/g; **18-9** 4.67 μm; **18 -10** 2.1 $\times 10^{-3}$ g/cm^2 **18-11** 44.8°; **18-12** 0.87 – 9.8, 0.093 – 1.0, 0.035 – 0.39, 0.024 – 0.27; **18-13** 130; **18-14** 2.0 $\times 10^{5}$; **18-15** 0.673%; **18-16** 1.86%; **18-17** 0.541 nm; **18-18** 0.590 nm

CHAPTER 19

19-2 4.6 $\times 10^{-5}$, **19-3** 7.9 $\times 10^{-5}$ **19-4** 6.9 $\times 10^{2}$ eV; **19-5** 5 eV; **19-6** $BaSO_4$; **19-10** 1.7 $\times 10^{7}$; **19-12** 1123 eV

CHAPTER 20

20-1 $^{6}Li\ (n, t)\ ^{4}He$, $^{9}Be(\gamma, n)^{8}Be$, $^{54}Fe(\alpha, n)^{57}Ni$, $^{37}Ar(e, \nu)^{37}Cl$;
20-2 $2.6 \times 10^{18}\ s^{-1}$; **20-3** 19.2 h; **20-4** $5.02 \times 10^{-3}\ d^{-1}$, 138 d;
20-5 44.72(2.236%), 141.42 (0.70710%); **20-6** 76.5, 2.26 min^{-1}
20-7 21.3, 0.147 s^{-1}; **20-8** 3913, 3927 (a_c); **20-9** 108.4 counts;
20-10 $4.8 s^{-1}$; **20-11** 78 h; **20-12** 0.0117%; **20-13** 0.116*M*;
20-14 4.26×10^{-4} *M*

CHAPTER 21

21-1 629; **21-2** 748.5; **21-3** 1.3×10^{3} **21-4** 860; **21-5** 0.78 T;
21-6 6860 V; **21-7** 2.09 μs; **21-8** 16.8, 20.6, 23.8, 26.6 μs;
21-9 4.8×10^{6} rad/s; **21-10** 26.8 eV; **21-11** $C_9H_6O_2^+$, $C_8H_6O^+$, C_7H_4, $C_7H_5^+$; **21-12** *M* $(^{81}Br)^+$, *M* $(^{79}Br)^+$, $C_6H_{11}CH_2C(CH_3)CH^+$, $^{80}BrCH_2C(CH_3)CH_2^+$, $^{79}BrCH_2C(CH_3)CH_2^+$, $C_6H_{11}^+$, $CH_2C(CH_3)CH_2^+$, $C(CH_3)CH_2^+$; **21-13** $H_3CCH_2CH_2CH_3^+ \rightarrow CH_3CH_2CH_2^+ + CH_3$;
21-14 52.1% *n*-butane, 47.9% *t*-butane

CHAPTER 22

22-1 0.075 *M*; **22-2** 4.20×10^{-3} *M*; **22-3** 0.34 *M*; **22-4** 0.0455 *M*;
22-5 0.91; **22-6** 0.75; **22-7** 9.88×10^{-5} *M*; **22-8** 1.06×10^{-3} *M*;
22-9 9.01×10^{-5} *M*; **22-10** 1.07×10^{-4} *M*; **22-11** 0.03; **22-12** 0.56 V;
22-13 1.8×10^{-4} *M*; **22-14** 0.384 V; **22-15** 5×10^{-13}; **22-16** –0.2114 V;
22-17 0.89 V; **22-18** 0.156 V; **22-19** – 0.114 V; **22-20** 10^{125};
22-21 – 0.793 V; **22-22** 3.4; **22-23** 0.10; **22-24** – 0.1899 V; **22-25** 0.13;
22-26 9.42×10^{-5} *M*; **22-27** 5.73×10^{-4}*M*; **22-28** 0.117 V; **22-29** 1.07 V;
22-30 0.90, 0.89, 0.89, 0.88, 0.88, 0.83, 0.604, 0.576, 0.563, 0.545, 0.527;
22-31 4.80×10^{-2} *M*; **22-32** 0.99 V; **22-33** 15.44%; **22-35** 1.33×10^{-4} *M*;
22-36 1.5×10^{-3} *M*; **22-37** 3.24×10^{-3} *M*; **22-38** 7.15×10^{-4} *M*

CHAPTER 23

23-1 3.50 m*M*; **23-2** 1.34 mg/s; **23-3** $4.7 \times 10^{-6}\ cm^2/s$; **23-4** 2.64 m*M*;
23-5 $6.10 \times 10^{-6}\ cm^2/s$; **23-6** $6.70 \times 10^{-6}\ cm^2/s$; **23-7** $9.49 \times 10^{-6}\ cm^2/s$;
23-8 0.17 m*M*; **23-9** 2, – 0.630 V; **23-10** irreversible; **23-11** 1.90 m*M*;

23-12 20.0 ms; **23-13** 0.160 m*M*; **23-14** 26.4 μA; **23-15** 11.3 μA; **23-16** no; **23-17** 0.15 cm^2; **23-18** 0.146 cm/y **23-19** 5.02×10^{-3} *M*; **23-20** 3.68×10^{-3}*M*; **23-21** 4.40×10^{-5} *M*; **23-22** 1.24×10^{-4}*M*; **23-23** 1.05×10^{-3}*M*; **23-24** 2.25×10^{-3} *M*; **23-25** 8.15 cm^{-1}; **23-26** 3.98 μS; **23-27** 2.52×10^{-3}; **23-28** 1.10×10^{-10}; **23-29** 1.70×10^{-5}; **23-30** 6.00×10^{-3} *M*

CHAPTER 24

24-1 1.55; **24-2** 0.442, 0.529, 0.795, 1.00, 1.60, $r = 0.0865$; **24-3** 11.0, 17.0; **24-4** 26; **24-5** 6.04, 6.93, 14.3 cm^3; **24-6** 76.4 plates, 0.327 cm/plate; **24-7** 1.02×10^3 plates, 0.0245 cm/plate; **24-8** 4.9×10^3 plates, 5.1×10^{-3}cm/plate; **24-9** 0.50 mm/plate; **24-10** 0.10 mm/plate; **24-11** 958 plates; **24-12** 0.92; **24-13** 0.77; **24-14** 2.0, yes; **24-15** 1.2; **24-16** 1.2; **24-17** 2.0; **24-18** 1.5

CHAPTER 25

25-1 hexane, hexene, butyl chloride, butanol, hexanoic acid; **25-2** methanol, ethanol, butanol, hexanol; **25-3** benzene, chlorobenzene, phenol, benzoic acid; **25-4** F^-, Cl^-, Br^-, I^-; **25-5** Be^{2+}, Mg^{2+}, Ca^{2+}, Sr^{2+}, Ba^{2+}; **25-6** Li^+, Na^+, K^+, Rb^+, Cs^+ **25-7** 0.73, 0.22; **25-8** 0.773, 0.37, 0.10; **25-9** 109 mL, 0.66; **25-10** 170 mL; **25-11** 23,000; **25-12** 2.1×10^5

CHAPTER 26

26-1 56 cm^3/min; **26-2** 58 cm^3/min; **26-3** 290 cm^3; **26-4** 0.454; **26-5** 2.18 atm; **26-6** 52 cm^3; **26-7** 110, 48 cm^3; **26-8** 180 cm^3/g; **26-9** 1.6; **26-10** 1.5×10^8 plates; **26-11** 2.0 mm/plate; **26-12** 5.1×10^{-5} s; **26-13** 9.7 cm^2; **26-14** 120 cm^2; **26-15** squalane, silicone QF – 1; **26-16** GSC; **26-17** 143, 128, 138°C; **26-18** 143, 128, 138°C, methanol, ethanol, propanol, butanol; **26-19** 1.66, 2.74, 4.97; **26-20** 1.36; **26-21** 572, 620, 776; **26-22** 2.95 min; **26-23** 0.577 mL/mL; **26-24** 1.42 mg/mL

CHAPTER 27

27-1 47 %; **27-2** 35.0%; **27-3** 4.6; **27-4** 0.02506 *M*; **27-5** 0.04385 *M*; **27-6** 9.33×10^{-4} *M*; **27-7** 0.0103 *M*

APPENDIX

B

ASCII CHARACTERS WITH CORRESPONDING OCTAL, DECIMAL, HEXADECIMAL (HEX.), AND BINARY VALUES

ASCII	Octal	Decimal	Hex.	Binary
NUL	0	0	0	0
SOH	1	1	1	1
STX	2	2	2	10
ETX	3	3	3	11
EOT	4	4	4	100
ENQ	5	5	5	101
ACK	6	6	6	110
BEL	7	7	7	111
BS	10	8	8	1000
HT	11	9	9	1001
LF	12	10	A	1010
VT	13	11	B	1011
FF	14	12	C	1100
CR	15	13	D	1101
SO	16	14	E	1110
SI	17	15	F	1111
DLE	20	16	10	10000
DC1	21	17	11	10001
DC2	22	18	12	10010
DC3	23	19	13	10011
DC4	24	20	14	10100

Table *Contd...*

ASCII	Octal	Decimal	Hex.	Binary
NAK	25	21	15	10101
SYN	26	22	16	10110
ETB	27	23	17	10111
CAN	30	24	18	11000
EM	31	25	19	11001
SUB	32	26	1A	11010
ESC	33	27	1B	11011
FS	34	28	1C	11100
GS	35	29	1D	11101
RS	36	30	1E	11110
US	37	31	1F	11111
SP	40	32	20	100000
!	41	33	21	100001
”	42	34	22	100010
#	43	35	23	100011
$	44	36	24	100100
%	45	37	25	100101
&	46	38	26	100110
’	47	39	27	100111
(	50	40	28	110000
)	51	41	29	101001
*	52	42	2A	101010
+	53	43	2B	101011
,	54	44	2C	101100
–	55	45	2D	101101
”	56	46	2E	101110
/	57	47	2F	101111
0	60	48	30	110000
1	61	49	31	110001
2	62	50	32	110010
3	63	51	33	110011
4	64	52	34	110100
5	65	53	35	110101
6	66	54	36	110110
7	67	55	37	110111
8	70	56	38	111000
9	71	57	39	111001
:	72	58	3A	111010
;	73	59	3B	111011
<	74	60	3C	111100
=	75	61	3D	111101
>	76	62	3E	111110

Table *Contd...*

ASCII	Octal	Decimal	Hex.	Binary
?	77	63	3F	111111
@	100	64	40	1000000
A	101	65	41	1000001
B	102	66	42	1000010
C	103	67	43	1000011
D	104	68	44	1000100
E	105	69	45	1000101
F	106	70	46	1000110
G	107	71	47	1000111
H	110	72	48	1001000
I	111	73	49	1001001
J	112	74	4A	1001010
K	113	74	4B	1001011
L	114	76	4C	1001100
M	115	77	4D	1001101
N	116	78	4E	1001110
O	117	79	4F	1001111
P	120	80	50	1010000
Q	121	81	51	1010001
R	122	82	52	1010010
S	123	83	53	1010011
T	124	84	54	1010100
U	125	85	55	1010101
V	126	86	56	1010110
W	127	87	57	1010111
X	130	88	58	1011000
Y	131	89	59	1011001
Z	132	90	5A	1011010
[	133	91	5B	1011011
\	134	92	5C	1011100
]	35	93	5D	1011101
^	136	94	5E	1011110
—	137	95	5F	1011111
‘	140	96	60	1100000
a	141	97	61	1100001
b	142	98	62	1100010
c	143	99	63	1100011
d	144	100	64	1100100
e	145	101	65	1100101
f	146	102	66	1100110
g	147	103	67	1100111

Table *Contd...*

ASCII	Octal	Decimal	Hex.	Binary
h	150	104	68	1101000
i	151	105	69	1101001
j	152	106	6A	1101010
k	153	107	6B	1101011
l	154	108	6C	1101100
m	155	109	6D	1101101
n	156	110	6E	1101110
o	157	111	6F	1101111
p	160	112	70	1110000
q	161	113	71	1110001
r	162	114	72	1110010
s	163	115	73	1110011
t	164	116	74	1110100
u	165	117	75	1110101
v	166	118	76	1110110
w	167	119	77	1110111
x	170	120	78	1111000
y	171	121	79	1111001
z	172	122	7A	1111010
{	173	123	7B	1111011
¦	174	124	7C	1111100
}	175	125	7D	1111101
~	176	126	7E	1111110
DEL	177	127	7F	1111111

APPENDIX

C

ABBREVIATIONS

Abbreviation	Meaning
A	Absorbance, amplitude, area
a	Absorptivity, polarizability, activity
a_f	Asymmetry factor
AAS	Atomic absorption spectrophotometry
ac	Alternating current
A/D	Analog to digital
ADC	Analog-to-digital converter
AES	Atomic emissive spectrometry, Auger electron spectroscopy
AFS	Atomic fluorescent spectrometry
ALU	Arithmetic-logic unit
ASCII	American standard code for information interchange
ATR	Attenuated total reflectance
B_0	Magnetic flux density
BCD	Binary coded decimal
BL	Bioluminescence
C	Capacitance
CAD	Collision activated decomposition
CARS	Coherent anti-Stokes Raman spectrometry
CFE	Continuous-flow enthalpimetry
CI	Chemical ionization
CIDNP	Chemically induced dynamic nuclear polarization
CL	Chemiluminescence
cls	Complete lineshape (nmr)
CMA	Cylindrical mirror analyzer
cmr	Carbon-13 magnetic resonance
CP	Cross-polarization
CPG	Chromatopyrography
CPU	Central processor unit

Table *Contd...*

Abbreviation	Meaning
CSRS	Coherent Stokes Raman spectrometry
CW	Continuous wave
D	Diffusion coefficient
D/A	Digital to analog
DAC	Digital-to-analog converter
dc	Direct current
DCI	Direct chemical ionization
DIDA	Direct isotopic dilution analysis
DIE	Direct injection enthalpimetry
DKAM	Double known-addition method
DMA	Direct memory access, dynamic mechanical analysis
dme	Dropping mercury electrode
DMM	Digital multimeter
dnmr	Dynamic nuclear magnetic resonance
DSC	Differential scanning calorimetry
DTA	Differential thermal analysis
DVM	Digital voltmeter
E	Steady-state potential
E^{o}	Standard potential
$E^{o\prime}$	Formal potential
e	Instantaneous potential, charge
ECD	Electron capture detector
ECL	Electrochemiluminescence
EGA	Evolved gas analysis
EGD	Evolved gas detection
EI	Electron-bombardment ionization
EIS	Electrochemical impedance spectroscopy
ELDOR	Electron double resonance
ELSD	Evaporative light source detector
EMR	Electromagnetic radiation
ENDOR	Electron-nuclear double resonance
EPROM	Erasable programmable read-only memory
EPXMA	Electron probe x-ray microanalysis
ESCA	Electron spectroscopy for chemical analysis
esr	Electron spin resonance
EXAFS	Extended x-ray absorption fine structure
F	Flowrate, farad, faraday
f	Activity coefficient
FAB	Fast-atom bombardment
FABMS	Fast-atom-bombardment mass spectrometry
FD	Field desorption
FES	Flame emissive spectrometry
FET	Field-effect transistor
FI	Field ionization
FIA	Flow injection analysis
FID	Flame-ionization detector
FPD	Flame photometric detector

Table *Contd...*

Abbreviation	Meaning
FTIR	Fourier transform infrared
G	Conductance
GC	Gas chromatography
GC/MS	Gas chromatography/mass spectrometry
GLC	Gas-liquid chromatography
GM	Geiger-Müller
GSC	Gas-solid chromatography
H	Height equivalent of a theoretical plate
hc	Hollow cathode
HETP	Height equivalent of a theoretical plate
hmde	Hanging mercury drop electrode
HPLC	High-performance liquid chromatography
HPRC	High-performance radial chromatography
HPSEC	High-performance size-exclusion chromatography
HPTLC	High-performance thin-layer chromatography
I	Steady-state current, intensity, ionic strength, Kovats retention
ΔI	McReynolds constant
i	Instantaneous current
i_d	Diffusion current
i_l	Limiting current
i_p	Peak current
i_r	Residual current
icr	Ion cyclotron resonance
ID	Inner diameter
IFSOT	Irradiated fused silica open tubular
IGFET	Insulated gate field-effect transistor
IIDA	Inverse isotopic dilution analysis
IMFP	Inelastic mean free path
INAA	Instrumental neutron activation analysis
I/O	Input/output
IPC	Ion-pair chromatography
ir	Infrared
IRS	Inverse Raman scattering
ISE	Ion-selective electrode
ISFET	Ion-selective field-effect transistor
ISS	Ion-scattering spectroscopy
J	Coupling constant
j	Pressure drop correction factor
JFET	Junction field-effect transistor
k'	Capacity factor
L	Inductance
LALLS	Low-angle laser light scattering
LARIS	Laser-ablative resonant ionization spectroscopy
LC	Liquid chromatography
LC/EC	Liquid chromatography/electrochemistry
LD	Laser desorption, lethal dose
LED	Light emitting diode

Table *Contd...*

Abbreviation	Meaning
LEI	Laser enhanced ionization
LEIS	Localized electrochemical impedance spectroscopy
LIMS	Laboratory information management system
LLC	Liquid-liquid chromatography
LMS	Laboratory management system
LSC	Liquid-solid chromatography
LSI	Large-scale integration
LSIMS	Liquid secondary ion mass spectrometry
LUO	Laboratory unit operation
LVDT	Linear variable differential transformer
m_I	Nuclear magnetic quantum number
m_s	Electron magnetic quantum number
MAS	Magic angle spinning
MHE	Multiple headspace extraction
MOLE	Molecular optical laser examiner
MOSFET	Metal-oxide semiconductor field-effect transistor
MPI	Multiple photon ionization
MS	Mass spectrometry
MSD	Mass selective detector
MSI	Medium-scale integration
mtfe	Mercury thin-film electrode
m/z	Mass-charge ratio
N	Transference number
n	Refractive index, number of theoretical plates
NAZ	Normal analytical zone
NAA	Neutron activation analysis
NCD	Nitrogen chemiluminescence detector
NCI	Negative chemical ionization
nmr	Nuclear magnetic resonance
NOE	Nuclear Overhauser effect
NPD	Nitrogen phosphorus detector
OD	Outer diameter
OTC	Open tubular column
OTE	Optically transparent electrode
Ox	Oxidized chemical species
P	Power
P_R	Retention polarity
PAS	Photoacoustic spectrometry
PC	Paper chromatography
PEG	Polyethylene glycol
PES	Photoelectric spectroscopy
PESIS	Photoelectric spectroscopy of the inner shell
PESOS	Photoelectric spectroscopy of the outer shell
PID	Photoionization detector
PIXE	Particle-induced x-ray emission, proton-induced x-ray emission
PLOT	Porous-layer open tubular
PM	Photomultiplier
PMA	Polymethylacrylate

Table *Contd...*

Abbreviation	Meaning
pmr	Proton magnetic resonance
ppb	Parts per billion
PPI/NICI	Pulsed positive-ion/negative-ion chemical ionization
ppm	Parts per million
PROM	Programmable read-only memory
PT	Purge and trap
PVC	Polyvinyl chloride
PZT	Piezoelectric transducer
R	Resistance, resolution
R_f	Retardation factor
r	Retention ratio
RAM	Random access memory
Red	Reduced chemical species
RIS	Resonant ionization spectroscopy
RNAA	Radiochemical neutron activation analysis
ROM	Read-only memory
RPC	Reversed-phase chromatography
rrde	Rotated ring-disk electrode
RRS	Resonant Raman spectrometry
SA	Specific activity
SAM	Scanning Auger microprobe
SCD	Sulfur chemiluminescence detector
SCOT	Support coated open tubular
SEC	Size-exclusion chromatography
SECM	Scanning electrochemical microscopy
SEM	Scanning electron microscope
SERS	Surface-enhanced Raman scattering
SFC	Supercritical fluid chromatography
SFR	Spin-flip Raman
SIMS	Secondary ion mass spectrometry
smde	Static mercury drop electrode
S/N	Signal-to-noise ratio
SRG	Stimulated Raman gain
SSI	Small-scale integration
STEM	Scanning transmission electron microscope
T	Transmittance, period, absolute temperature (K)
T_G	Glass transition temperature
t_m	Mobile-phase retention time
TCD	Thermal conductivity detector
TD	Thermal desorption
TG	Thermogravimetry
TG/FID	Thermogravimetry/flame ionization detector
TG/FTIR	Thermogravimetry/Fourier transform infrared spectrometry
TG/GC	Thermogravimetry/gas chromatography
TG/MS	Thermogravimetry/mass spectrometry
TGL	Temperature-gradient lamp
TISAB	Total ionic strength adjustment buffer

Table *Contd...*

Abbreviation	Meaning
TLC	Thin-layer chromatography
TMA	Thermomechanical analysis
UHPLC	Ultra high performance liquid chromatography
UPS	Ultraviolet photoelectric spectroscopy
uv	Ultraviolet
V_m	Mobile-phase retention volume
VLSI	Very large-scale integration
WCOT	Wall-coated open tubular
WWOT	Whisker-walled open tubular
X	Reactance
XPS	x-Ray photoelectron spectroscopy
Z	Impedance
α	Relative retention
Δ	Change in or difference between
ε	Molar absorptivity
θ	Cell constant
k	Specific conductance
Λ_m	Molar conductance
λ	Wavelength
μ	Dipole moment, magnetic moment
v	Frequency, kinematic viscosity
ρ	Specific resistance
Σ	Sum of
σ	Standard deviation, cross-sectional area of a nucleus

APPENDIX

D

CONSTANTS

Term	Symbol	Value
Avogadro"s number	N	6.022141×10^{23} mol^{-1}
Bohr magneton	μ_B	9.2732×10^{-28} J/G
		9.2732×10^{-24} J/T
Boltzmann constant	k	1.3806×10^{-23} J/K
Faraday constant	F	96,485 C/mol
Gas law constant	R	8.3143 J/K · mol
		8.3143 V · C/K · mol
		1.9872 cal/K · mol
		0.08205 L · atm/K · mol
g-Factor (free electron)	g_e	2.002319
Nuclear magneton	β	5.0505×10^{-31} J/G
Planck constant	h	6.626068×10^{-34} J · s
Velocity of EMR in a vacuum	c	2.9979×10^{8} m/s

INDEX

A

B

D

E

F

G

H

I

J

K

L

M

N

O

P

Q

R

S

T

U

V

W

X

Z

www.ingramcontent.com/pod-product-compliance
Ingram Content Group UK Ltd.
Pitfield, Milton Keynes, MK11 3LW, UK
UKHW061535190726
13845UKWH00005B/114

9 789385 433207